6
ESSENTIALS
OF FIRE FIGHTING
AND FIRE DEPARTMENT OPERATIONS

Edited By

Frederick M. Stowell, Project Manager/Writer

Lynne Murnane, Senior Editor

BRADY

Brady Publishing, a division of Pearson Education
Upper Saddle River, New Jersey 07458

Published by Fire Protection Publications, Oklahoma State University

IFSTA
INTERNATIONAL FIRE SERVICE TRAINING ASSOCIATION
Validated by the International Fire
Service Training Association

RECYCLABLE

Lower cover photo courtesy of Chris Mickal/Distric Chief, New Orleans (LA) FD Photo Unit.

INTERNATIONAL FIRE SERVICE TRAINING ASSOCIATION

The International Fire Service Training Association (IFSTA) was established in 1934 as a *nonprofit educational association of fire fighting personnel who are dedicated to upgrading fire fighting techniques and safety through training*. To carry out the mission of IFSTA, Fire Protection Publications was established as an entity of Oklahoma State University. Fire Protection Publications' primary function is to publish and distribute training materials as proposed, developed, and validated by IFSTA. As a secondary function, Fire Protection Publications researches, acquires, produces, and markets high-quality learning and teaching aids consistent with IFSTA's mission.

IFSTA holds two meetings each year: the Winter Meeting in January and the Annual Validation Conference in July. During these meetings, committees of technical experts review draft materials and ensure that the professional qualifications of the National Fire Protection Association® standards are met. These Conferences bring together individuals from several related and allied fields, such as:

- Key fire department executives, training officers, and personnel
- Educators from colleges and universities
- Representatives from governmental agencies
- Delegates of firefighter associations and industrial organizations

Committee members are not paid nor are they reimbursed for their expenses by IFSTA or Fire Protection Publications. They participate because of a commitment to the fire service and its future through training. Being on a committee is prestigious in the fire service community, and committee members are acknowledged leaders in their fields. This unique feature provides a close relationship between IFSTA and the fire service community.

IFSTA manuals have been adopted as the official teaching texts of many states and provinces of North America as well as numerous U.S. and Canadian government agencies. Besides the NFPA® requirements, IFSTA manuals are also written to meet the Fire and Emergency Services Higher Education (FESHE) course requirements. A number of the manuals have been translated into other languages to provide training for fire and emergency service personnel in Canada, Mexico, and outside of North America.

If you need additional information concerning the International Fire Service Training Association (IFSTA) or Fire Protection Publications, contact:

Customer Service, Fire Protection Publications, Oklahoma State University
930 North Willis, Stillwater, OK 74078-8045
800-654-4055   Fax: 405-744-8204

For assistance with training materials, to recommend material for inclusion in an IFSTA manual, or to ask questions or comment on manual content, contact:

Editorial Department, Fire Protection Publications, Oklahoma State University
930 North Willis, Stillwater, OK 74078-8045
405-744-4111   Fax: 405-744-4112   E-mail: editors@osufpp.org

Table of Contents

List of Tables

This sixth edition of the IFSTA **Essentials of Fire Fighting** is intended to serve as a primary text for the firefighter candidate or as a reference text for firefighters who are already members of a fire and emergency services organization. This manual addresses the fire fighting objectives found in NFPA® 1001, *Standard for Fire Fighter Professional Qualifications*, 2013 Edition. The job performance objectives in NFPA® 1001 for competencies for minimum emergency medical care and hazardous materials response are addressed in the IFSTA/Brady **Essentials of Fire Fighting and Fire Department Operations** manual.

Acknowledgment and special thanks are extended to the members of the IFSTA validation committee who contributed their time, wisdom, and knowledge to the development of this manual.

IFSTA Essentials of Fire Fighting, 6th Edition Validation Committee

Chair
Russell J. Strickland
Life Member
Singerly Fire Company
Elkton, Maryland

Vice-Chair
Mark S. Pare
Deputy Director
Massachusetts Fire Academy
Stow, Massachusetts

Secretary
Josh M. Stefancic
District Chief
Largo Fire Rescue
Largo, Florida

Committee Members

Raniero L. Angelone
Fire Lieutenant/Paramedic
Indian River County Fire Rescue
Instructor
Indian River State College
Vero Beach, Florida

Claude Beauchamp
Director of Operations
Quebec National Fire Academy
Laval, Quebec, Canada

Paul Boecker III
Captain
Sugar Grove Fire District
Sugar Grove, Illinois

Scott Burrow
Supervisor, Hazardous Materials Training
312th Training Squadron, D.O.D. Fire Academy
Goodfellow Air Force Base, Texas

Mark Butterfield
Fire Science Faculty/Instructor
Hutchinson Community College
Hutchinson, Kansas

Mary Cameli
Assistant Fire Chief
Mesa (AZ) Fire and Medical Department
Mesa, Arizona

Shad Cooper
Fire Instructor
Wyoming Fire Marshal's Office
Green River, Wyoming

Committee Members (continued)

David O. Couvelha
Certification Program Manager
Kansas Fire & Rescue Training Institute
University of Kansas
Lawrence, Kansas

Glenn Davis
Assistant Fire Chief of Training
Helena (MT) Fire Department
Helena, Montana

Earl "Rob" Freese III
Executive Director, Public Safety Training and
  Certification
Department of Public Safety Training and
  Certification
Bucks County Community College
Newtown, Pennsylvania

Russ Grossman
Field Program Coordinator
Iowa Fire Service Training Bureau
Ames, Iowa

Chistopher J. Growley
Curriculum Supervisor
South Carolina Fire Academy
Columbia, South Carolina

Edward E. Hartin
Fire Chief
Central Whidbey Island Fire and Rescue
Coupeville, Washington

Kerby Kerber
Deputy Director
Delco ESTC
Sharon Hill, Pennsylvania

Joshua W. Livermore
Director of Fire Science
Mohave Community College
Bullhead City, Arizona

Jim Lovell
Fire Training Lieutenant
Santa Fe County Fire Department
Santa Fe, New Mexico

Daniel Madrzykowski
Fire Protection Engineer
National Institute of Standards and Technology
  (NIST)
Gaithersburg, MD

Richard L. Merrell
Lieutenant
Fairfax County Fire Rescue Department
Fairfax, Virginia

Edward J. O'Hanlon, Jr
Coordinator Fire/Industrial Training
Montgomery County Fire Academy
Conshohocken, Pennsylvania

Jim Pendergast
Fire Chief
Penhold (Alberta) Fire Department
Penhold, Alberta, Canada

Thomas C. Rullo
Deputy Fire Chief
Mashpee Fire & Rescue Department
Mashpee, Massachusetts

Demond Simmons
Captain
Oakland Fire Department
Oakland, California

Daniel Stout
Captain
Las Vegas (NV) Fire and Rescue
Las Vegas, Nevada

Don Turno
Fire Protection Engineering/Program Manager
Savannah River Nuclear Solutions LLC
Aiken, South Carolina

Karen D. Wilcox Lewis
Captain, Retired
Prince George's County Fire/EMS Department
Largo, Maryland

An important element of the 6th edition of **Essentials of Fire Fighting** is the continuing partnership between IFSTA/ FPP and Brady Publishing, a division of Pearson Education. Brady has a long and outstanding reputation in the fire publishing world and is the *premier* publisher of emergency medical training materials in North America and beyond. Teaming the two most dominant publishers of emergency responder training materials clearly has obvious benefits to the firefighters and agencies that we serve. The resources that Brady brings to this partnership allow IFSTA/FPP to develop additional supplemental on-line and electronic resources for the **Essentials** product line. Their assistance with publishing the expanded **Essentials of Fire Fighting and Fire Department Operations** manual is also a key element of the partnership. Special thanks go out to Brady Publisher Julie Alexander; Editor-in-Chief, Brady Professional & Career Higher Education, Marlene Pratt; Program Manager, Fire Professional & Career Higher Education, Stephen G. Smith; and all the Brady staff who worked to make this partnership happen.

It would not be possible to develop a manual of this type without the assistance and cooperation of numerous individuals, fire departments, and training agencies that rise above and beyond the call of duty to assist us with creating the thousands of photographs needed to illustrate the concepts contained herein.

Much appreciation is given to the following agencies, organizations, and individuals for contributing information, photographs, and technical assistance instrumental in the development of this manual:

Akron Brass Company

Doug Allen

Ames (IA) Fire Department

Ansul Corporation

Jason Arias

Arlington (TX) Fire Department

Danny Atchley

Jocelyn Augustino, FEMA

Tom Aurnhammer

Eddie Avila

Jennifer Ayers

Badger Fire Protection

Robert Bird

Boca Raton (FL) Fire Rescue

Ron Bogardus

Andrea Booher, FEMA

Ted Boothroyd

Bill Branson

Joel G. Brennan

Alan Braun

Bullhead City (AZ) Fire Department

Tim Burkitt, FEMA

California Office of Emergency Services

Tom Clawson

Shad Cooper

Matthew Daly

Edmond (OK) Fire Department

Elkhart Brass Manufacturing Company

Jason Epley

Brent Erb

Bob Esposito

Euless (TX) Fire Department

Florida State Fire College

Nicole Fuge

G&M Body Shop — West, Stillwater, Oklahoma

Dick Giles

Sam Goldwater

Carl Goodson

Trent Hawkins

Win Henderson

Donny Howard

International Association of Fire Fighters (IAFF)

International Association of Fire Chiefs (IAFC)

Iowa Fire Service Training Bureau (IFSTB)

Chiaki Iramina

Steve Irby

Jackson Hole (WY) Fire and EMS

Syeed Janbozorgi

Ron Jeffers

Edwin Jones

Keller (TX) Fire-Rescue

Steven Kerber, UL, Inc.

Bradley A. Lail

A. Lee III

Los Angeles (CA) Fire Department

Michael Mallory

Dan Madrzykowski

Rich Mahaney

Jeff Mangum

Taylor Marr

Pat McAuliffe

McKinney (TX) Fire Department

Mesa (AZ) Fired Department

Chris Mickal

Mine Safety Appliances Company (MSA)

Missouri Fire Rescue Training

Mohave Museum of History and Arts, Kingman,
   Arizona

Rick Montemorra

Ron Moore

Morning Pride Manufacturing

National Cancer Institute

National Fire Fighter Near Miss Reporting System

National Fire Protection Association® (NFPA®)

National Interagency Fire Center

New South Wales (Australia) Fire Brigades

Fire Department, City of New York (NY)

James Nilo

North Richland Hills (TX) Fire Department

Oklahoma Fire Service Training, Oklahoma State
   University

Oklahoma Highway Patrol

OnCue Express, Stillwater, Oklahoma

Owasso (OK) Fire Department

Portland (OR) Fire Department

Richland Hills (TX) Fire Department

Richmond (VA) International Airport

Mike Rieger, FEMA

Liz Roll, FEMA

Tom Ruane

Sand Springs (OK) Fire Department

Mark Schultz

Sonrise Photography, Verdigris, Oklahoma

Southlake (TX) Fire Department

Spinner Publications

William D. Stewart

Stillwater (O K) Fire Department

Stillwater (OK) Medical Center

Stillwater (OK) Police Department

Rhett Strain

Sweetwater County (WY) Fire District #1

Tarrant County (TX) Community College

Task Force Tips

Tulsa (OK) Air National Guard Fire Department

Tulsa (OK) Fire Department

United Plastic Fabricating, Inc., North Andover, MA

United States Government

   Department of Agriculture

   Department of Homeland Security

       Bureau of Alcohol, Tobacco, Firearms, and
          Explosives

       Federal Emergency Management Agency (FEMA)

       National Institute of Standards and Technology
          (NIST)

       U. S. Customs and Border Protection (CBP)

       U.S. Fire Administration (USFA)

   Department of Defense

       U.S. Air Force

       U.S. Coast Guard

       U.S. Marine Corps

       U.S. Navy

   Department of Energy

   Department of Health and Human Services (HHS)

       Centers for Disease Control and Prevention (CDC)

       National Institute for Occupational Safety and
          Health (NIOSH)

   Department of Transportation

   Environmental Protection Agency (EPA)

August Vernon

Brandon Wagoner

Dennis Walus

West Des Moines (IA) Fire Department

Sean Worrell

Wyoming State Fire Marshal's Office

Jake Zlomie

Special thanks go to those individuals who provided assistance in the creation of this manual:

Thomas Beers, Firefighter Cancer Support Network (FCSN)

MSgt Terry J. Dybdahl, USAF (Retired)

Dan Madrzykowski, National Institute of Standards and Technology (NIST)

Vicky Roper, Administrator, Canadian Association of Fire Chiefs

Julie Fortney

Larry Jenkins

Eldon J. Moore

Amanda Roberts

Xzavier Roberts

Special thanks are given to Randy Novak, Dawn Beisner, and Doug Allen of the Iowa Fire Service Training Bureau (IFSTB) for their work on a series of skill sheet photos for this project.

Last, but certainly not least, gratitude is extended to the following members of the Fire Protection Publications staff whose contributions made the final publication of this manual possible.

Essentials of Fire Fighting, 6th Edition, Project Team

Project Manager
Frederick M. Stowell, Senior Editor

Director of Fire Protection Publications
Craig Hannan

FPP Editorial Projects Manager
Clint Clausing

Curriculum Technology Manager/Lead Instructional Developer
Elkie Burnside

Production Coordinator
Ann Moffat

Editorial Staff
Ed Kirtley, IFSTA/Curriculum Projects Coordinator
Cynthia Brakhage, Senior Editor
Jeff Fortney, Senior Editor
Lynne Murnane, Senior Editor
Leslie Miller, Senior Editor
Libby Snyder, Senior Editor
Mike Sturzenbecker, Senior Editor
Chris Hayden, Research Technician

Gabriel Ramirez, Research Technician
David Schaap, Graduate Research Technician
Jake Zlomie, Research Technician
Nancy Trench, FPP Assistant Director
Tara Gladden, Editorial Assistant
Susan F. Walker, Librarian

Illustrators and Layout Designers
Ann Moffat, Production Coordinator
Clint Parker, Senior Graphics Designer
Ben Brock, Senior Graphics Designer
Errick Braggs, Senior Graphics Designer
Missy Hannan, Senior Graphics Designer
Ruth Murdoch, Senior Graphics Designer

Curriculum Developers
Beth Ann Fulgenzi, Instructional Developer
Andrea Haken, Instructional Developer
Lynn Hughes, Instructional Developer
Brad McLelland, Instructional Developer
Melissa Noakes, Instructional Developer
Lori Raborg, Instructional Developer

IFSTA/FPP Photographers
Jeff Fortney
Leslie Miller
Brett Noakes
Clint Parker
Fred Stowell
Mike Sturzenbecker
Mike Wieder

Technical Reviewers, manual
Ron Bowser
Academy Curriculum Manager
Prince William County Department of Fire & Rescue
Nokesville, Virginia

Wes Kitchell
Fire Captain, Retired
Cloverdale, California

Mike McLaughlin
Division Chief
Merced (CA) Fire Department
Sonora, California

Anna Wieringa
Deadwood, South Dakota

Richard Dunn
Columbia Fire Department
Columbia, South Carolina

Technical Reviewers, curriculum
Steve Dunham
Instructor
Fire Service Training, Oklahoma State University
Stillwater, Oklahoma

Steven Martin
Deputy Director
Delaware State Fire School
Dover, Delaware

Technical Reviewers, videos
Scott Burrow
Supervisor, Hazardous Materials Training
312th Training Squadron, D.O.D. Fire Academy
Goodfellow Air Force Base, Texas

Glenn Davis
Assistant Fire Chief of Training
Helena (MT) Fire Department
Helena, Montana

Josh M. Stefancic
District Chief
Largo Fire Rescue
Largo, Florida

Indexer
Nancy Kopper

The IFSTA Executive Board at the time of validation of the **Essentials of Fire Fighting, 6th edition** was as follows:

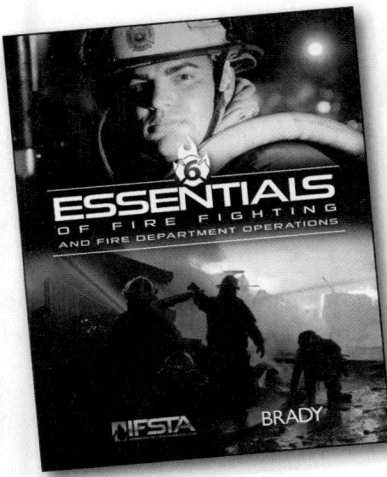

The accompanying student and instructor materials for the 6th Edition of **Essentials of Fire Fighting** are your best source for both teaching and training. Components that were available with the 5th Edition have been updated to provide an even better classroom and training ground experience for students. Components that were previously available only in print form are now available in multiple formats including digital versions on flash drives and apps that can be taken anywhere. Online training is available through ResourceOne, IFSTA/FPP's learning management system. Video skills for all of the skill sheets in this manual are also available for streaming on ResourceOne and for purchase as a DVD package. Online training is also available for the expanded edition of the 6th edition through Brady Books MyFireKit, MyFirefighterLab, Resource Central, and Pearson Online LMS. For more specific information about any of 6th edition components, please call Fire Protection Publications at 800-654-4055 or Brady Books at 800-922-0579. You can also visit our websites: **www.ifsta.org** or **www.bradybooks.com**.

IFSTA Flash Drive Curriculum: Get the most out of face-to-face instruction

- Contains everything needed for a successful course, and the entire package is customizable:

  — **Plan of Instruction** – A planning aid for instructors that identifies the scope and format of the course. It lists resources needed for the course and other useful information for preparing to teach a Firefighter I or Firefighter II course.

  — **Syllabus Course Template** – This template can be customized to meet the needs of individual courses. Instructors can insert the class schedule including dates for tests, quizzes, assignment deadlines, and other information.

  — **Lesson Outlines** – Outlines instructional content and contains review questions and discussion topics at regular intervals to reengage students during presentations.

  — **PowerPoint® Presentations** – Feature a graphical style designed for today's visual learner; slides include images, video clips, and interactive artwork designed to keep students engaged.

  — **"How to Use IFSTA PowerPoints" Video** – This video explains the benefits of using graphically based slides and guides instructors on how to use the features in IFSTA PowerPoints® alongside the Lesson Outlines.

IFSTA Lesson Outlines and PowerPoints® are designed to work together to create a complete classroom experience that maximizes student learning and knowledge retention.

  — **Chapter Tests** – Included both in MS Word format and in a test bank format provided by ExamView™; ExamView™ can create shorter or longer tests that cover multiple chapters, learning objectives, or the entire course. 20 to 75 question test per chapter.

  — **Chapter Quizzes** – Brief review of chapter content. 15 to 20 questions per chapter.

Both tests and quizzes are used for assessment on a chapter-by-chapter basis. Answer keys are provided for both.

  — **Course Workbook Answer Key** – Corresponds with the print or R1 delivered copy of the workbook that instructors can use as homework or precourse assignments.

  — **Skill Evaluation Checklists** – Can be used by instructors to assess student progress on skills training. Includes equipment list and task steps needed to complete each skill.

  — **Learning Activities** – Reinforce key concepts and learning objectives with small group and individual activities.

  — **Clip Art** – Contains all images used in the manual, included for use when customizing your own slides or presentations.

All Essentials Curriculum components contain separate documents for Firefighter I and Firefighter II where appropriate. The Essentials Curriculum covers all 21 chapters of the basic edition and all 24 chapters of the expanded edition.

Course Workbook:
Increase your understanding

- This spiral bound workbook can be used for homework assignments, precourse assignments, or supplemental study.

- The information is covered on a chapter-by-chapter basis and is divided into distinct Firefighter I and Firefighter II sections.

- Various question types including key terms, true/false, fill-in-the-blank, picture identification, matching, multiple choice, short answer, scenario, and crossword puzzles.

Exam Prep: Prepare for Certification or Promotion

- Exam Prep contains over 1,400 multiple-choice questions designed to help students prepare for their certification or promotional exams.

- The information is covered on a chapter-by-chapter basis and is divided into distinct Firefighter I and Firefighter II sections.

- Exam Prep is available as a print, spiral bound book, or as an interactive digital program on USB flashdrive.

Essentials Skill Sheet Handbook:
A training tool designed for the training ground.

- Constructed with a sturdy wire, spiral binding and heavy, resilient paper this handbook contains all 192 skills needed for use in your training environment in one volume.

- The handbook is an excellent resource for departments refreshing their skills or for instructors who need a reference they can use in a training environment when teaching or assessing students.

Skills Video Series:
Hone your skills

- Series includes all 192 skills from the Essentials 6th edition.

- Each video is short and concise and demonstrates one skill.

- You can purchase access to streaming videos on ResourceOne for some or all of the video series.

- Also available for purchase as a DVD series that includes all of the skills in one package.

- An excellent resource for departments who want to quickly review skills with firefighters or for students who are seeking extra practice preparing for skills certification tests.

ResourceOne

ResourceOne: Free Online Training

- Save time and money by using the training method that best suits your needs – face-to-face, online, or hybrid training.
- Administrative control allows you to fully customize the online classroom.
- Upload new documents or customize ours to fit your needs.
- Online gradebook provides an immediate grade history and progress tracking.
- Discussion forum provides easy communication for students and the instructor.
- Secure environment that can be customized for your organization.
- Completely online environment that does not require you or your students to download, license, or install any software.
- Integrated email function so instructors can email students.
- Includes customizable end of course evaluation.

Three levels of access:

- — **Instructor Toolkit** — Available free when you register on ResourceOneContains PowerPoints®, Lesson Outlines, and "How to Use IFSTA PowerPoint® Presentations" video.

- — **Full IFSTA Curriculum Online** — Free to organizations that adopt Essentials orother IFSTA training materials as their manual of choice for their courses; contact us to gain access to the entire, IFSTA flashdrive curriculum in an online environment on ResourceOne at no additional cost.

- — **ResourceOne Classroom** — In addition to the Full IFSTA Curriculum, use IFSTA's course creator to select from premium features on a per student fee basis that allow you to lead a secure, online course for your students.

- Optional, premium content includes:
  - — Skill Videos (Essentials 6th Only)
  - — Course Workbook in a format where students can complete assignments online (key terms, true/false, fill-in-the-blank, picture identification, matching, multiple choice, short answer, scenario, and cross word puzzles)
  - — Exam Prep in a format where students can study online (Hundreds of multiple choice questions not found in the quizzes or tests)

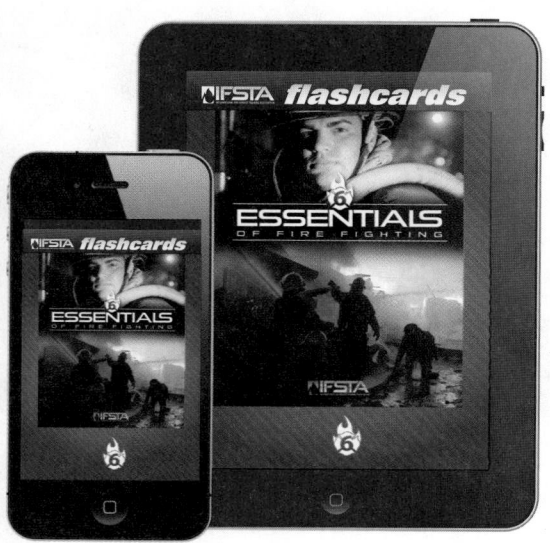

Key Terms Flashcard App: Review Core Content

- This interactive flashcard app allows students to review important vocabulary from the manual and makes studying-on-the-go easier than ever.
- The app allows users to track performance, study multiple chapters simultaneously, and remove or reinsert questions from the flash card deck.
- Flashcard apps are compatible with iPhone, iPod touch, and iPad.

MyBRADYLab™

MyBRADYLab for Essentials of Fire Fighting, 6th Edition:
Digital Student Personalized Study and Remediation

- Optional student resource that offers a personalized, guided study plan to help students identify where they need help and what additional resources are available to help them. Students take a pretest generates a personalized study plan.

- Objectives and activities tied specifically to NFPA® 1001 job performance requirements with specific page references, artwork, and other content from the Essentials manual 6th Edition.

- Includes all of the instructor resources found on Resource Central.

- Acts as a complete online learning management system (LMS) that includes an assignment calendar, gradebook, communication tools and more!

- May be purchased in combination with other supplements as part of a value pack or as a standalone product.

- Designed to help educators help students succeed.

eLearning for Essentials of Firefighting 6e

Pearson Online LMS Supplement:

- Previously known as CourseCompass

- An online, Learning management system (LMS) for instructors or organizations who do not currently have access or hosting capability to administer an LMS of their own.

- Allows instructors to house their classroom management online

- Comes populated with instructor as also available on Resource Central

- Access is free to institutions, but students must purchase an access code when they purchase their Essentials manual

Introduction

Courtesy of District Chief Chris Mickal, New Orleans (LA) Photo Unit.

Welcome to the 6th edition of the **Essentials of Fire Fighting** manual. Since 1977, Fire Protection Publications (FPP) and the International Fire Service Training Association (IFSTA) have published this manual to provide firefighters with the highest quality of training materials possible. In 2008, FPP, IFSTA, and Brady Publishing joined to provide a new and expanded edition of **Essentials**. Both the basic **Essentials of Fire Fighting** and the expanded **Essentials of Fire Fighting and Fire Department Operations** will assist you in meeting the Firefighter I and Firefighter II job performance requirements (JPRs) of the National Fire Protection Association's® (NFPA®) Standard 1001, *Standard for Fire Fighter Professional Qualifications*.

Since its original adoption in 1974, NFPA® 1001 has become the widely accepted standard for firefighters in North America and other parts of the world. The acceptance and recognition of a national standard provides a baseline for professional excellence in the fire service. In order to help firefighters meet this baseline, FPP, IFSTA, and Brady Publishing provide **Essentials of Fire Fighting**, the most accurate and highest quality training materials available. In addition, we are dedicated to developing new editions of this manual to keep pace with updates to the NFPA® 1001 standard.

You should keep in mind that NFPA® 1001 sets forth *minimum* performance requirements. It is perfectly acceptable for any jurisdiction to exceed the specified requirements when applying the standard locally. What is not considered acceptable is for any jurisdiction to reduce the requirements of the standard.

NFPA® 1001 is separated into two levels: Firefighter I and Firefighter II. Some state and local jurisdictions still use a Firefighter III designation in their training and certification programs because NFPA® 1001 had a third level at one time. In most cases these organizations have chosen to rename the NFPA® 1001 Firefighter II level to their own Firefighter III level to meet local needs.

To delineate between Firefighter I and Firefighter II, both the NFPA® 1001 and the IFSTA **Essentials** committees decided that a *Firefighter I* is a person who is minimally trained to function safely and effectively as a member of a fire fighting team under direct supervision. However, a person meeting the requirements of Firefighter I is not considered a "complete" firefighter. A "complete" firefighter, rather, has satisfied the JPRs for both levels I and II. A *Firefighter II* may operate under general supervision and may be expected to lead a group of equally or lesser trained personnel through the performance of a specified task.

The 2013 edition of NFPA® 1001 continues the Job Performance Requirement (JPR) format introduced in the 1997 edition of the standard. *Job Performance Requirements* are based on a job task analysis that identifies what a firefighter actually does on the job or at least should be capable of doing. This JPR format is used in all NFPA® professional qualifications standards.

The 2013 edition of NFPA® 1001 does not reflect a substantial overhaul of the 2008 version of the standard. The new edition does place a slightly increased emphasis on issues such as safe operations at roadway incident scenes, as well as increased firefighter involvement in public fire and life safety functions and community risk reduction.

The information contained in the 6th edition of **Essentials** will help firefighters meet these new requirements of NFPA® 1001. In addition, this edition has separated the Firefighter I and Firefighter II level material into separate sections within specific chapters. This altered format will allow the student to locate and study material required to meet the appropriate level. It will also assist the instructor in presenting material for each level or for a course that combines both levels of instruction.

The list of JPRs covered in each chapter is found at the beginning of each chapter, as are the chapter learning objectives based on the JPRs. More specific directions on where JPRs are covered within each chapter are contained in **Appendix A** and in the curriculum materials for the instructor. NFPA® 1001 does not require that the objectives be mastered in the order in which they appear in the standard. Local agencies may choose the order in which they wish the material to be presented.

This edition of the manual also continues the inclusion of skill sheets as the tool for learning basic skills required by the standard. Separating the written text from the step-by-step procedures makes the manual easier to read and the skills easier to locate. Therefore, skill sheets describing the step-by-step procedures for many of the tasks described in the text are found at the end of that chapter. Note that while the skill sheets contained within the manual do contain written information on all of the important steps in that skill, they do not have a photo or illustration for each step. This was done in order to keep the manual to a reasonable size. The supplemental **Essentials of Fire Fighting Firefighter I and II Skills Handbook** contains all of the skill sheets covered in the expanded edition of the **Essentials** manual, with additional photographs and illustrations to highlight all of the steps in the processes.

Purpose and Scope

The *purpose* of the **Essentials of Fire Fighting**, 6th Edition, is to provide entry-level firefighter candidates with the basic information necessary to meet the job performance requirements (JPRs) of NFPA® 1001, *Standard for Fire Fighter Professional Qualifications*, 2013 edition, for Firefighter I and Firefighter II firefighters. The information is presented in a format that will allow training agencies to develop training programs for each level or a program combining the two levels. Additional information that exceeds the standard requirements based on the experiences of the validation committee and the editors has also been included.

The *scope* of the manual addresses the basic duties assigned to a Firefighter I or Firefighter II firefighter. These duties include fire suppression, rescue and extrication, ventilation, loss control, fire prevention, and public life safety education among others. To fully meet the requirements of NFPA® 1001, the firefighter must also meet some minimal first aid requirements and the requirements for First Responder at the Operational Level of NFPA® 472, *Standard for Competence of Responders to Hazardous Materials/Weapons of Mass Destruction Incidents*. The information needed to meet these requirements is covered in the IFSTA/Brady **Essentials of Fire Fighting and Fire Department Operations** manual.

The first twenty-one chapters in the IFSTA/Brady **Essentials of Fire Fighting and Fire Department Operations** manual are identical to the twenty-one chapters in the basic **Essentials of Fire Fighting** manual. **Essentials of Fire Fighting and Fire Department Operations** contains three additional chapters that cover the first aid and hazardous materials information. All study guide and instructor materials cover the full twenty-four chapters.

The methods shown throughout this manual have been validated by the International Fire Service Training Association (IFSTA) as accepted methods for accomplishing each task. However, they are *not* to be interpreted as the only acceptable methods of accomplishing a given task. Other methods of performing any task may be specified by a local agency. For guidance in seeking additional methods for performing a given task, the student or instructor may consult any of the IFSTA expanded-topic manuals for more in-depth information on a particular topic.

24-Hour Clock

Like many parts of the fire service, IFSTA manuals use a 24-hour clock when referencing time. The 24-hour clock is a convention of timekeeping in which the day runs from midnight to midnight and is divided into 24 hours, numbered from 0 to 23. This method of keeping time is also known as *military time* or *Army time*. In spoken language, the hour mark is followed by the word *hundred*. For example, 9 am is written 0900 and spoken as *o nine hundred*; 2 pm would be written 1400 and spoken as *fourteen hundred*, and 10 pm would be written 2200 and spoken as *twenty-two hundred*, etc. Time between the hours does not use the word *hundred*, however. Those times are written and spoken, for example, like this: 4:30 pm would be written as 1630 and spoken as *sixteen thirty* hours.

Key Information

Various types of information in this manual are given in shaded boxes marked by symbols or icons. See the following definitions:

Firefighter I and II icons — These icons indicate information in a chapter that will meet NFPA® requirements for Firefighter I and Firefighter II. The icons can be found in the chapter table of contents and in the text at the first section to which the icon applies. After an icon appears in a chapter, it will apply to all subsequent material until the appearance of another icon. When only one icon appears in the table of contents and text, the entire chapter is intended to meet the requirements of that level.

Information

Information boxes are used to provide information that needs to be emphasized, highlighted, or separated from the main text for any reason at the editor's discretion. This is a catch-all separation box that can be used for any need-to-know information.

Barrier Tape

In recent years, poly vinyl barrier tape has become popular for this use. It is 2.5 micrometers (mil) thick and comes in 3 inch x 1000 foot (76.2 mm by 304.8 m) rolls. Vinyl barrier tape is a highly visible yellow, with words in black stating "Caution Fire Line Do Not Enter." It is inexpensive, easy to store on the apparatus, and can be abandoned in place at the end of the incident.

Safety Alert

Safety Alert boxes are used to highlight information that is important for safety reasons. Safety-related information that requires an in-depth explanation or is too lengthy for use in a **CAUTION** or **WARNING** box may be emphasized in these boxes. See the following example:

Hazards Determine the PPE

Always wear the correct PPE that is designed and intended to protect you from the specific type of hazard(s) presented by the incident.

Case History

A case history analyzes an event. It can describe its development, action taken, investigation results, and lessons learned. Illustrations can be included.

Courtesy of Los Angeles (CA) Fire Department.

Tactical Ventilation — Planned, systematic, and coordinated removal of heated air, smoke, gases or other airborne contaminants from a structure, replacing them with cooler and/or fresher air to meet the incident priorities of life safety, incident stabilization, and property conservation.

Key Term

A key term is designed to emphasize key concepts, technical terms, or ideas that firefighters need to know. They are listed at the beginning of each chapter, highlighted in bold **red** font, and the definition is placed in the margin for easy reference.

Signal Words

Three key signal words are found in the text: **WARNING**, **CAUTION**, and **NOTE**. Definitions and examples of each are as follows:

- **WARNING** — Warning boxes provide very brief warnings about things or situations that may cause death, destruction, or severe injury. See the following example:

> ///////////////////////
> ## WARNING
> Live-fire training must adhere to the requirements set forth in NFPA® 1403, *Standard on Live Fire Training Evolutions* (Current edition).

- **CAUTION** — Caution boxes provide very brief cautionary statements about things or situations that might cause harm or adverse consequences if not understood or specifically noted. See the following example:

> ///////////////////////
> ## CAUTION
> Fire and emergency responders must be familiar with the physiological, emotional, and technological limitations caused by the use of respiratory protection equipment to prevent injury or death.

- **NOTE** indicates important operational information that helps explain why a particular recommendation is given or describes optional methods for certain procedures. See the following example:

NOTE: This information is based on research performed by the International City/County Managers Association, Inc.

Referenced NFPA® Standards and Codes

One of the basic purposes of IFSTA manuals is to allow fire and emergency services personnel and their departments to meet the requirements set forth by NFPA® codes and standards. These NFPA® documents are referred to throughout this manual. References to information from NFPA® codes are used with permission from National Fire Protection Association®, Quincy, MA 02169. This referenced material is not the complete and official position of the National Fire Protection Association® on the referenced subject, which is represented only by the standard in its entirety.

Chapter Contents

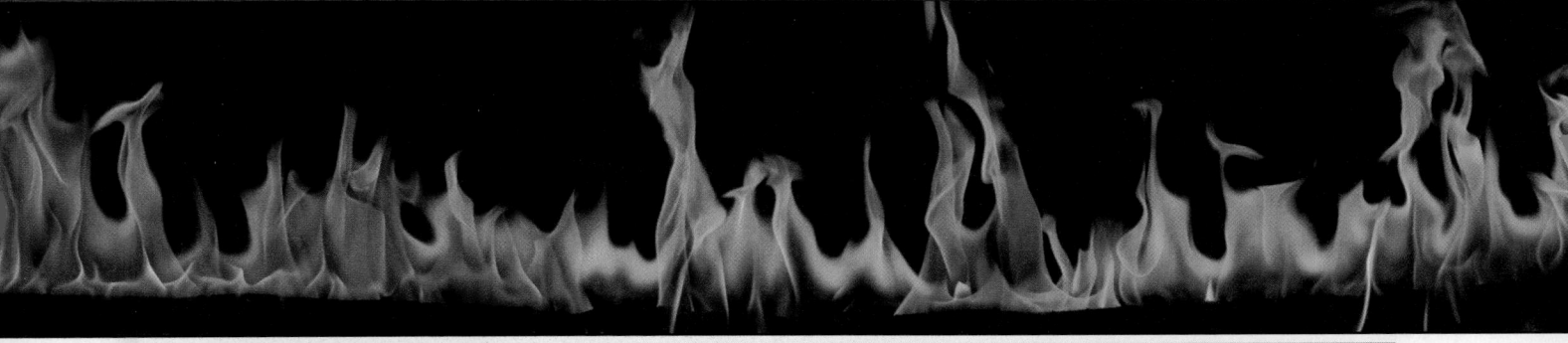

Key Terms

All-Hazard Concept 21
Authority Having Jurisdiction (AHJ) 20
Code 36
Culture 17
External Customers 24
Fireproof 12
Incident Command System (ICS) 28
Internal Customers 24
Interoperability 14
Jargon 18
Line Functions 24
Panic Hardware 12
Policy 35
Procedure 36
Standard Operating Procedure (SOP) 36
Staff/Support Functions 24
Standard 36
Wildland/Urban Interface 26

NFPA® Job Performance Requirements

This chapter provides information that addresses the following job performance requirements of NFPA® 1001, *Standard for Fire Fighter Professional Qualifications* (2013).

Firefighter I
5.1.1
5.1.2

Firefighter II
6.1.1

1. Summarize the history of the fire service.

2. Explain the organizational characteristics, cultural challenges, and cultural strengths that influence the fire service.

3. Describe the mission of the fire service. (5.1.1)

4. Describe the organization of fire departments. (5.1.1)

5. Distinguish among functions of fire companies. (5.1.1)

6. Summarize primary knowledge and skills the firefighter must have to function effectively. (5.1.1, 6.1.1)

7. Distinguish among the primary roles of fire service personnel. (5.1.1, 6.1.1)

8. Describe fire department organizational principles. (5.1.1)

9. Locate information in departmental documents and standard or code materials. (5.1.2)

10. Distinguish between fire department SOPs and rules and regulations. (5.1.1)

11. Explain the ways the fire service may interact with other organizations. (5.1.1)

Chapter 1
Orientation and
Fire Service History

Case History

There are over 34,000 (U.S. 30,500; Canada 3,500) public and private career, combination, and volunteer fire and emergency services organizations in North America. The majority, approximately 29,300 (U.S. 26,000; Canada 3,300), are volunteer organizations. In addition, there are numerous private fire departments that protect industrial complexes as well as military fire departments that serve on bases at home and abroad.

More than 1,260,000 (U.S. 1,150,000; Canada 110,000) men and women serve in these fire and emergency services organizations. They each face the same hazards, risks, and challenges during their service careers. They also share a common bond — the desire to provide professional service to their neighbors and communities. This bond is also shared with everyone who has been a firefighter since the beginning of organized fire fighting in colonial North America. Over the past four centuries, people have joined together to save lives and property from fire and disasters and provide medical assistance to the injured. They have accomplished this mission through emergency response as well as providing fire and life safety education and prevention activities.

Accomplishing these tasks has never been easy. Over the years many firefighters have been injured and far too many have died in the line of duty. Yet they continue to confront hazards, take risks, and sustain their sacrifice. They also continue to take pride in the service they provide and their bond with their fellow firefighters.

You are about to become a member of a respected and essential profession: the fire and emergency service. As a firefighter you will be required to perform strenuous physical labor and be exposed to high levels of mental and emotional stress as well as mortal danger. To carry out your assigned duties, you will have to safely master a variety of skills, make critical decisions quickly, and follow orders.

Your training will include not only physical skills but also knowledge and concepts that you can apply to your work in the fire and emergency services. You will learn the theories of fire behavior, how to analyze a situation and make a risk assessment, and how to keep mentally and physically fit. You will also learn how to gain access to a structure, select proper tools for each task, and apply extinguishing agents to control a fire. These are only a small portion of the topics you will cover in your training.

To begin your training, you will learn about the history and culture of the fire and emergency services in North America, the mission of the service, and the general organization of a fire department, including the positions, roles, and responsibilities of its members. You will learn the duties expected of a firefighter, the department regulations, policies, and procedures that govern your actions, and the national, state/provincial, and local laws and ordinances that apply to you and your department. Finally, you will learn about other agencies that routinely interact with the fire department.

When you complete your basic training, you will have the knowledge, skills, and abilities that are required of an entry-level firefighter. However, this is only the beginning of your professional development as a firefighter. Throughout your career in the fire service, you will continue to learn new skills and techniques that will help you advance through the ranks as far as your abilities and desires will take you.

I Fire Service History and Culture

The history of the fire service is more than a series of events, dates, and names of influential people. These events caused changes in the way the fire service provided services, as well as changes in society. These changes have resulted in the growth of the fire service into the fire and emergency services organization that you are about to become a part of today.

Our history has also helped to create a unique culture. Traditions such as the design of firefighter helmets or the concept of the volunteer fire service are deeply entrenched **(Figure 1.1)**. Other characteristics of our profession are less obvious but just as important, such as the military-style rank structure and the Maltese Cross on uniforms and equipment. The culture is symbolized by images of the heroic firefighter attacking a fire or rescuing a child that reinforce an attitude of selfless sacrifice, bravery, and pride.

Figure 1.1 The symbolism of the heroic firefighter dates to the earliest traditions of the fire service.

Traditions and culture can also be barriers to change, although these barriers are lifting. In North America, the fire service has traditionally been a male profession, but today many departments are led by female chief officers. Departments were also slow to integrate minorities, although today members of all racial and ethnic groups serve with distinction.

Fire Service History

Fire has always been an important element of all cultures. It has been associated with religious and philosophical beliefs. Fire has provided warmth, security, and light, and it has been used to cook food and manufacture tools. But uncontrolled fire is a powerful enemy, capable of destroying both lives and property. Historic fires have sometimes engulfed entire cities, slowing economic growth and the progress of society. This was the case when European colonists arrived in North America.

Colonial North America

In 1608, Jamestown had the first recorded major fire in the New World. It burned down many structures in the settlement and destroyed clothing and provisions. As a result, more than half the settlers died from lack of food and exposure to the severe winter.

In 1631, the city of Boston banned thatched roofs and wooden chimneys to prevent the outbreak of fires. In 1653, the city purchased its first fire engine, before forming the first-ever paid fire company in 1678.

But the first fire organization in North America was formed a few years earlier, in the city of New Amsterdam (later known as New York). In 1647, new governor Peter Stuyvesant appointed the Surveyors of Buildings, a group of men who acted as fire wardens. Although they had no special authority while a fire was in progress, they performed fire prevention work and imposed fines on violators. The New Amsterdam Fire Company was later formed in 1658, with a total of eight members. They patrolled the streets from 2100 hours to sunrise, ready to sound the alarm if a fire was spotted. This company of eight men had access to leather fire buckets, hooks, and small ladders.

Encouraged by the success of Boston's fire company, the citizens of Philadelphia formed a fire society in 1735. Guided by Benjamin Franklin, they formed the Union Volunteer Fire Company the following year. In 1737, New York formed its own volunteer fire department. By this time, new technology was helping to drive the trend toward organized fire protection. Buckets of water were replaced with hand-operated pumps and nozzle-equipped hoses.

Industrial Revolution

In the mid-1800s, hand-operated pumps were replaced by horse-drawn steam pumpers (**Figure 1.2, p. 12**). These remained in service until the 20th Century, when the age of automotive fire apparatus began.

By the end of the 19th Century, the Industrial Revolution had brought further technological advances. Iron structural members had replaced wooden frames, which had limited the height of buildings. Iron was then replaced by steel, which was lighter and stronger. In the cities, steel-frame buildings known as *skyscrapers* could now reach ten to twenty stories.

Figure 1.2 Antique pumpers were innovative in their time and their use continued until automotive apparatus took over the task of fire fighting.

But the use of steel did not eliminate disastrous fires, and in some ways it increased the danger. It allowed large numbers of people to work in tall buildings that had combustible contents and interior finishes, with only limited means of escape. Natural gas and electric lighting also placed ignition sources within the structures, making fire even more likely. Civic leaders realized that they would need to regulate how structures were designed and the types of building materials that should be used.

In 1896, the National Fire Protection Association® (NFPA®) was formed to develop consensus-based codes and standards intended to ensure fire and life safety. The first standard it published was NFPA® 13, which regulated the design and installation of fire protection sprinkler systems in structures. This was followed by the *National Electric Code®*, published in 1897.

Significant Historical Events

Society tends to be reactive rather than proactive when it comes to fire and life safety. In most cases, fire safety-related codes, standards, ordinances, and laws have been enacted only after catastrophes have occurred. The following is a list of fires that resulted in changes to fire and life safety regulations.

- **Iroquois Theater Fire, Chicago (1903)** — A fire in a theater that was designed to be **fireproof** claimed the lives of 602 occupants and injured 250. Combustible scenery, curtains, and interior finish were ignited by an electric spotlight. Most of the dead were suffocated or trampled in the panic to escape. In response, laws were enacted to require **panic hardware** on exit doors, which were also required to swing outward. Other laws placed limitations on the number of theater seats in an aisle, and the amount of combustible material allowed in places of assembly.

- **The Great Fire of 1904, Toronto, Ontario (1904)** — A fire of undetermined origin destroyed 104 buildings in the central business district of Toronto, Ontario. The fire spread rapidly and outstripped the ability of the fire department to control it. Unprotected elevator shafts and stairways contributed to the rapid spread within buildings. The fire lead to major changes in Toronto's building code, including requirements for fire doors and walls, sprinkler systems, larger water mains, additional fire alarm boxes, and removal of overhead obstructions along streets. The size of the fire department was also increased.

- **Triangle Shirtwaist Fire, New York City, New York (1911)** — Fire in a ten-story building used for manufacturing clothing claimed the lives 146 employees, mostly young women who leapt to their death as the fire spread through the combustible contents. The building was thought to have been fireproof. The fire resulted in the NFPA® 101, *Life Safety Code* (originally titled *Building Exits Code*), which established requirements for means of egress (exit).

Fireproof — Obsolete term for resistance to fire; inappropriate because all materials except water will burn. Other terms such as *fire resistive* or *fire resistant* should be used.

Panic Hardware — Hardware mounted on exit doors in public buildings that unlock from the inside and enable doors to be opened when pressure is applied to the release mechanism.

Figure 1.3 The Cocoanut Grove Nightclub fire demonstrated to first responders the necessity of stricter fire laws and wider-spread education. *Courtesy of Spinner Publications.*

- **Cocoanut Grove Nightclub Fire, Boston, Massachusetts (1942)** — A fire in a single-story nightclub killed 492 people. The high death toll was caused by overcrowding, the use of combustible interior finishes and decorations, and the lack of emergency lighting. This tragedy led to stricter fire and life safety requirements for assembly-type occupancies **(Figure 1.3)**.

- **Ringling Brothers and Barnum and Bailey Circus Fire, Hartford, Connecticut (1944)** — Fire in a circus tent claimed the lives of 168 people, many of them children. Paraffin (wax) and gasoline had been used to make the tent waterproof, and exits were inadequate or blocked. The incident resulted in the development of life safety standards to regulate the manufacture and use of tents for public occupancy.

- **Our Lady of the Angels School Fire, Chicago, Illinois (1958)** — This fire killed ninety-two children and three teachers. Unprotected stairwells, a combustible interior finish, and the lack of a fire alarm system were all blamed for the loss of life. The fire brought additional attention to the need for improvements in the design of school buildings, the requirement for fire detection and alarm systems, and the need for enclosed stairwells. New laws also required schools to conduct fire evacuation drills throughout the academic year.

- **MGM Grand Hotel Fire, Las Vegas, Nevada (1980)** — Fire in an unsprinklered casino killed 85 and injured 679 in the 26-story hotel above. Smoke spread into the tower via stairwells, elevator shafts, and ventilation and pipe shafts. Smoke inhalation killed the majority of victims, mainly on the upper floors. Local laws were enacted that required all medium- and high-rise buildings to install sprinkler systems.

- **Station Nightclub Fire, West Warwick, Rhode Island (2003)** — One hundred patrons died and 256 were injured when stage pyrotechnics ignited a fire in this crowded nightclub. The lack of a sprinkler system allowed the fire to spread rapidly through combustible interior finishes. The primary cause of fatalities, though, was that occupants tried to escape through the main entrance rather than using nearby emergency exits. A bottleneck at the main exit caused many to be trampled to death and many others, who were stuck near the door, to die of smoke inhalation. If they had gone to the other exits, they would have been able to escape much more quickly without any deaths from trampling, and fewer (if any) from smoke inhalation. One result of the fire was greater awareness of the need for occupants of any type of assembly occupancy to be aware of and to use the closest exit. The NFPA® also made changes in fire sprinkler and crowd management requirements for nightclubs and other assembly occupancies.

- **Murrah Building Bombing, Oklahoma City, Oklahoma (1995)** — A truck bomb exploded outside this Federal Office Building, killing 168 people in the worst case of domestic terrorism in U.S. history. Society's response to the tragedy has had mixed effects on first responders. Increased emphasis on structural-collapse, confined-space, and search and rescue training and equipment has benefited the fire service. However, the installation of security barriers around government facilities has limited access and increased operational challenges for responders (**Figure 1.4**).

Figure 1.4 The aftermath of the Alfred P. Murrah building bombing resulted in greater preparedness in collapse rescue incidents but new regulations have also created challenges for responders.

- **World Trade Center and Pentagon Terrorist Attacks (2001)** — The events of September 11, 2001, have had a profound effect on the fire and emergency services in the U.S. Most evident was the increase in funding that has provided equipment, apparatus, and training relating to hazardous materials responses and search and rescue. Communication obstacles faced by responders to the tragedy also forced the fire service to improve its ability to communicate with other agencies during major disasters, which is referred to as **interoperability**.

Interoperability — Ability of two or more systems or components to exchange information and use the information that has been exchanged.

These are only a few of the tragic fires, explosions, and incidents that have occurred over the past century in North America. The loss of life and property in these events served as catalysts for dramatic changes to building and fire codes. Today fires are less frequent and less severe due to changes in building design and construction requirements, the addition of fire protection systems, and an increased public awareness of fire safety.

General Trends of Change

The fire service in North America has constantly evolved throughout its history. During the 20th Century, changes took place in the following areas:

Fire prevention and public safety education. The National Commission on Fire Prevention and Control (NCFPC) was authorized by the *Fire Research and Safety Act of 1968*. The purpose of the commission was to determine how to reduce fire loss in America. In 1973, the NCFPC published a report titled *America Burning*, which emphasized the need for public fire and life safety education and prevention programs. Departments began establishing fire and life safety education, fire code inspection and enforcement, and plans review divisions. The efforts of these divisions have helped to reduce fire loss, change personal habits, and improve fire and life safety. In 1987 a second edition, *America Burning Revisited*, described the evolution of fire safety and reemphasized the continuing need for public education. Unfortunately, the statistics on fire loss still indicate that more must be done to create a fire-safe society. Currently, only a small portion of fire department budgets are allocated to fire and life safety education.

Firefighter safety. For many years, firefighter line-of-duty-deaths (LODD) have remained fairly constant, averaging one hundred per year. Total injuries have also stayed at around 80,000 per year, including those that occur while performing both emergency and nonemergency duties. These numbers have remained constant despite the fact that departments have implemented safety, health, and wellness programs **(Figure 1.5)**, changed emergency scene procedures to improve safety, and purchased more advanced protective clothing. As a result, the National Fallen Firefighters

Figure 1.5 Health and wellness programs benefit firefighters by reducing the likelihood and impact of some preventable injuries.

Foundation hosted the Firefighter Life Safety Summit in 2004. This summit resulted in the *Everyone Goes Home®* program and the 16 Firefighter Life Safety Initiatives intended to reduce LODDs by 50 percent by 2014. The point of the program and the initiatives is to educate firefighters in ways that will reduce LODDS and injuries.

Emergency medical services. The need for fire departments to provide Emergency Medical Services has increased over the last few decades. All departments provide some type of First Responder Care and many provide Basic Life Support (BLS) and Advanced Life Support (ALS), as well as transport to higher levels of care. The importance of fire service EMS operations was emphasized in a 2007 report that proposed that the fire service is the best agency to provide EMS. Part of the report states:

The fire service can be configured many ways to deliver prehospital 9-1-1 emergency medical care such as the following general configurations:

- *Fire service-based system using cross-trained/multi role firefighters. Firefighters are all-hazards responders, prepared to handle any situation that may arise at a scene including patient care and transport.*

- *Fire service-based system using employees who are not cross-trained as fire suppression personnel. Single role EMS-trained responders accompanying firefighter first-responders on 9-1-1 emergency medical calls.*

- *Combined system using the fire department for emergency response and a private or "third service" (police, fire, EMS) provider for transportation support. Single role emergency medical technicians and paramedics accompany firefighter first responders to emergency scenes to provide patient transport in a private or third service ambulance.*

While there are pros and cons to the various system approaches, the individual firefighter must always strive to ensure that his or her contribution provides the maximum benefit to the patient and minimum risk to the community.

For the complete white paper and additional information on the importance of EMS to the fire service, see **Appendix B**.

Hazardous materials. In the years following World War II, the manufacture and use of hazardous materials increased exponentially. Incidents during transport became common occurrences, typically involving flammable/combustible liquids, solids, and gases. In response, chemical companies developed new fire suppression agents, and the fire service developed ways to safely respond to leaks and spills. To accomplish this, departments drafted hazardous materials control procedures, trained personnel, and purchased special apparatus, foam nozzles, and personal protective clothing and equipment.

Laws and regulations were also enacted to address the problem. The Superfund Amendments and Reauthorization Act of 1986 (SARA Title III) and OSHA 1910.120 regulate hazardous materials transport, transport vehicle design, and emergency responder training.

Terrorism. As international and domestic terrorism have spread, the fire and emergency services have taken a greater role in preparing for and responding to acts of terrorism. Since the tragic events of September 11, 2001, fire departments have significantly improved their readiness for terrorist attacks and incidents involving weapons of mass destruction (WMDs). Increased government funding has paid for specialized urban rescue equipment, apparatus, and training programs.

Natural disasters (all-hazard mitigation). Over the past fifty years, the fire service has also improved its readiness to respond to a growing variety of natural disasters, which are far more common than terrorist incidents. Firefighters are usually the first to be called and the first to arrive, and in many communities they are the only organization that can address the potential hazards resulting from any type of natural disaster (all-hazard mitigation). An emphasis on disaster preparedness has also resulted in greater coordination and training between emergency response agencies. Additional training, equipment, and established procedures have helped departments prepare for disasters such as floods, earthquakes, and hurricanes.

Professionalization of the fire service. As the fire service has provided a greater variety of services, its leaders have required more advanced education and specialized training. Even mid-level managers, district/battalion chiefs, and company officers often need college degrees. All departments need officers trained in business administration, management, public administration, and emergency management, as well as skills in public speaking and interpersonal skills. To help meet this need, the National Fire Academy offers the Executive Fire Officer Program for chief officers. Departments also offer specialized training to help personnel meet the challenges posed by terrorism and natural disasters. This emphasis on training and education has created a fire service that is more professional than ever before.

Accreditation and Certification

Accreditation and certification are also important aspects of this trend, for both individual responders and their organizations. They are granted through evaluation and testing based on consistently applied, internationally recognized criteria. This ensures that personnel and agencies are better prepared to provide assistance in their own jurisdiction or any other. Certified personnel can also apply for jobs, assist at incidents outside their jurisdiction, and move from agency to agency with greater credibility.

Two important accrediting agencies are the Insurance Services Office (ISO) and the Commission on Fire Accreditation International (CFAI), both part of the Commission for Public Safety Excellence (CPSE). Additional agencies include the International Fire Service Accreditation Congress (IFSAC) and the National Professional Qualifications Board (Pro Board).

Community-based fire protection. Community-based programs are an attempt to better integrate fire departments into the communities they serve. They were developed in Los Angeles during the 1990s to address concerns about the lack of minorities in the ranks. Programs consist of three stages: research, education, and implementation. First, a department gathers information on the demographics of the community it serves. Department personnel then learn about the local community and its specific challenges. Finally, the department design projects specifically tailored to meet the needs of the local population. Many of these projects involve emergency services, such as teaching Cardiopulmonary Resuscitation (CPR) or home fire safety (**Figure 1.6**). Other projects serve a broader goal of breaking racial and socioeconomic barriers and bringing the community together.

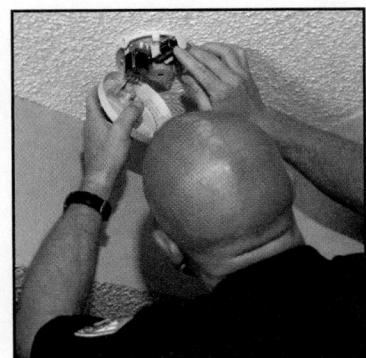

Figure 1.6 Home fire safety is greatly increased by the installation and maintenance of smoke detectors.

Fire Service Culture

The history and traditions of the fire service are the basis for its **culture**. Any type of fire service organization has a set of shared assumptions, beliefs, and values that influence the perceptions and behavior of its members. Your education about this

Culture — The shared assumptions, beliefs, and values of a group or organization.

Jargon — The specialized or technical language of a trade, profession, or similar group.

culture has begun as you learned about the history and symbols of the fire service. This education will continue as you learn the **jargon**, attitudes, and stories that are shared by members of your own department.

Organizational Characteristics

The fire service is sometimes referred to as a paramilitary organization because its organizational characteristics closely resemble those found in the military. These characteristics include:

- A command structure, including the chain of command
- Ranks to define positions within the structure
- Uniforms, badges, and symbols of rank
- An emphasis on teamwork, discipline, and following orders

Like the military, firefighters and their officers are expected to trust the decisions of those in charge, listen with respect, follow orders, and accept risk. This type of professional conduct is expected from every member of the department, and personnel who fail to live up to this standard may be disciplined.

At the same time, firefighters are expected and encouraged to participate in decision-making. Members who follow established policies and the chain of command can make suggestions based on their knowledge and experience or question decisions. You will learn the appropriate time, place, and process for this kind of participation during your training.

Cultural Challenges

Changes in the fire service have brought about personal challenges for firefighters. These challenges include diversity, resistance to change, differences of personal characteristics, and accepting personal responsibility.

Diversity. The membership of the fire service has been changing since the early 1950s. Today's goal is to better reflect the composition of the local community. These shifts have not come easily, but their challenges can be overcome; for example, assigning sleeping arrangements at stations housing men and women has been addressed in a variety of ways **(Figure 1.7)**.

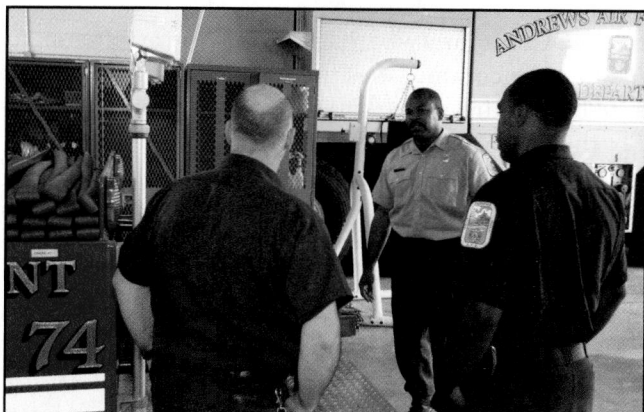

Figure 1.7 Today's fire service actively works to reflect the composition of the local community.

Resistance to change. Many firefighters are resistant to change in all forms, including diversity, revised tactics and regulations, and new procedures for controlling a fire. But as the advantages of change have become clearer, members are slowly accepting the new form the fire service is taking. It is important for new firefighters to remain flexible and begin to embrace change.

Differences of personal characteristics. Our personal characteristics are founded on our world view, that is, how we see ourselves and our culture. We learn our world view from our peers, family, faith, experiences, and education. We tend to associate with people who have similar world views and personal characteristics. As a result, we may resist allowing people that we think are different to join our group. Sometimes these differences are felt across generational lines. One problem this has caused in the fire service is that older members believe that new recruits do not live up to the standards set by more experienced personnel. Peer pressure then forces the new recruits to try harder to prove themselves, which sometimes results in unsafe behavior.

Accepting personal responsibility. It is important that firefighters take personal responsibility for their actions. You must be able to justify your behavior, as you will always be held accountable for your actions. The longer you have served and the higher your rank, the more important this trait becomes. It is important to learn this lesson early in your service, as accountability is the basis for a successful and safe career as a firefighter.

It is your responsibility to make your fire department and the fire service better through your actions and the image you project. You should always treat new personnel as you would want to be treated, thus setting an example of behavior for them to follow.

Cultural Strengths

For many years, the public has viewed the fire service as the most respected of all public servants. This positive public image is based on the cultural strengths exhibited by its members. Any department must demonstrate the following cultural strengths:

- Integrity
- Moral character
- Work ethic
- Pride
- Courage
- Loyalty
- Respect
- Compassion

Integrity. Integrity means following a strict ethical code and doing the right thing simply because it is right, not because it is required. Personal and professional integrity is the basis for all the other personal strengths described here. Another way to think of integrity is that it means doing the right thing even when no one else is watching.

Moral character. Moral character also involves right or just behavior, but with an emphasis on trust. As firefighters, our coworkers, supervisors, and members of the public must be able to trust us to do and say what is right. Our words and actions must reinforce their trust, and any violation of that trust reflects poorly on all firefighters. Once you become a firefighter, you will always be held to a higher standard.

Work ethic. Having a good work ethic means valuing the virtues of hard work and thoroughness. It means doing what needs to be done without being told, doing what you are asked to do without complaint, doing the task completely, and doing it to the best of your ability. It also means being prompt, reliable, and willing to take the initiative. Even tasks that are unpleasant or seem unimportant require your best effort. It is important to remember that throughout your fire service career, you will be judged by the quantity and quality of the work that you do.

Pride. For most firefighters, being a part of the fire service gives them a feeling of self-respect and self-worth. In addition to personal pride, firefighters are proud of their department and of the fire service in general. They demonstrate this by taking pride in their personal appearance, displaying fire service symbols on personal vehicles, wearing nonuniform apparel with the department name or Maltese Cross, and collecting fire service memorabilia.

Courage. Of all the personal characteristics firefighters have, courage is the most obvious. Courage is the ability to confront fear, pain, danger, or uncertainty. Although the image of firefighters racing into a hazardous situation while others are racing out is an overused stereotype, they do place themselves in harm's way to protect others. They confront danger in a controlled and rational way, analyzing the risk, planning the most appropriate response, and relying on their training to do the best job possible.

Loyalty. Firefighters are faithfully loyal to the fire service, their department, and their coworkers. They will risk their own lives to save a trapped or missing firefighter. They will take care of an injured firefighter or the family of a fallen firefighter. They will defend their department and the service when someone attempts to tarnish its image.

Respect. Firefighters have always exhibited an attitude of esteem toward their peers, superiors, and fellow citizens. This attitude, as well as the public service firefighters provide, has always led the public to respect the work of the fire service. In North America, this respect has been heightened since the terrorist attacks of September 11, 2001. The courage shown by the New York Fire Department and the horrific losses they suffered generated sympathy and admiration for the survivors that has carried over to the fire service as a whole.

Compassion. Finally, firefighters care about the citizens they serve, their fellow firefighters, and their families. They show this compassion when they comfort victims while giving medical aid, provide assistance when fire leaves families homeless, and grieve when a department member is killed in the line of duty (**Figure 1.8**).

Fire Service Mission and Organization

The mission of the fire service — the reason the fire service exists — is usually mandated by a law or ordinance enacted by the **authority having jurisdiction (AHJ)**. The AHJ determines what services are needed to protect its citizens and establishes the fire service to meet that need. Different communities require different types of services, so fire departments' missions will vary among different cities, states/provinces, and regions. However, the most basic objectives of the fire service will always include protecting lives, property, and the environment.

The AHJ also establishes the organization of the department. It determines the number of facilities and their locations, types of apparatus, number of personnel, and the overall type of department. The AHJ also establishes a hierarchy or organizational chart, assigning functions and responsibilities to specific jobs and ranks. It also sets requirements for the minimum levels of training and certification necessary to attain those ranks.

Authority Having Jurisdiction (AHJ) — Term used in codes and standards to identify the legal entity, such as a building or fire official, that has the statutory authority to enforce a code and to approve or require equipment; may be a unit of a local, state, or federal government, depending on where the work occurs. In the insurance industry, it may refer to an insurance rating bureau or an insurance company inspection department.

Figure 1.8 Members of the fire service honor fallen firefighters and offer support to their families.

Fire Service Mission

The mission of the fire service is *to save lives and protect property and the environment from fires and other hazardous situations.* As mentioned earlier, this mission has evolved. Today many fire departments take an **all-hazard concept** approach to the services they provide. An all-hazards approach means that the department may provide:

- Fire suppression protection
- Emergency medical services
- Technical rescue services
- Hazardous materials mitigation
- Airport and/or seaport protection
- Emergency management services
- Fire prevention services and public education (engineering, education, and enforcement)
- Community risk reduction
- Fire cause determination

The mission of your local department will depend on the legal mandate that established it, and this mandate will be based on the needs of your community. Your own duty will be to fulfill the stated goals and objectives of your department's mission statement. These statements are part of the department's rules and regulations and are usually posted in every department facility. They are made available to all personnel and should be available to the community you protect.

All-Hazard Concept — Provides a coordinated approach to a wide variety of incidents; all responders use a similar, coordinated approach with a common set of authorities, protections, and resources.

Fire Department Organization

Fire and emergency services organizations are structured to meet the missions they are mandated to fulfill. To understand the organizational structure of your department, you need to know the general organizational structures used, the common types of fire departments, types of staffing, and the separation of departmental duties. You will also need to know the types of fire department units, known as *companies*, and the positions that personnel fill to complete the mission of the department. Finally, you need to know your own duties as a Firefighter I or II.

Organizational Structure

Generally, the organizational structure forms a pyramid/hierarchy with the chief of the department at the tip and the firefighters forming the base. In between, the layers are composed of personnel assigned by rank and duty. As a person moves up the pyramid, the number of positions between the tip and base decreases and their authority and responsibility increase (**Figure 1.9**).

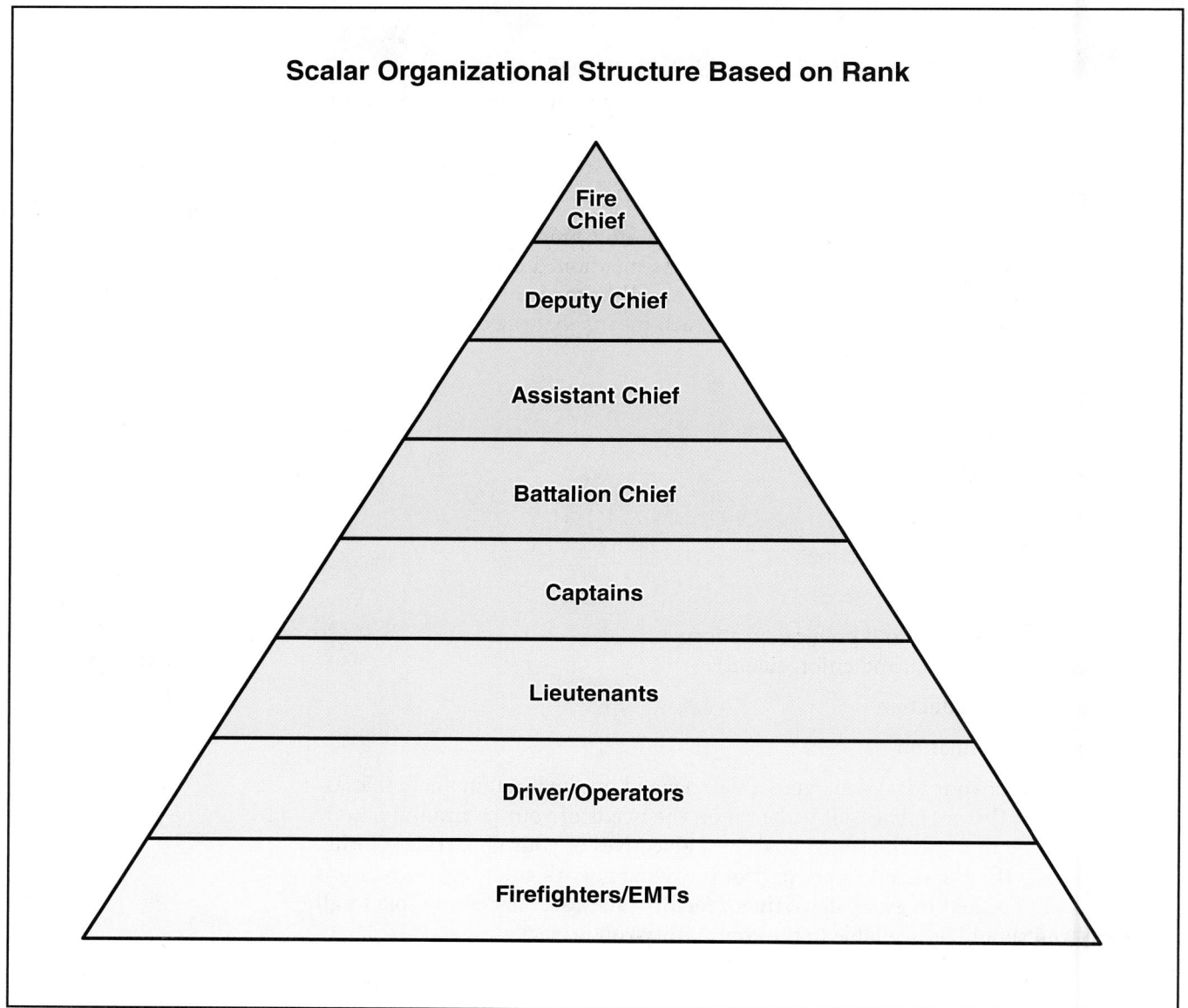

Figure 1.9 A scalar model includes all personnel in an organizational hierarchy.

Types of Fire Departments

Fire and emergency services organizations can be either *public* or *private*. Most organizations in North America are public, meaning they are funded by the community through taxes, fees, grants, fund-raisers, donations, and contracts. The community may be a municipality, county, district, or other area as defined by the AHJ.

Private organizations raise money through contracts, billing for services, and revenue provided by their parent organization. They typically provide services to a single company or facility, and in some cases to a municipality.

Staffing

For any department to be effective, it must have appropriate staffing. The different staffing classifications are based on how often firefighters work and whether or not they are paid. *Career* firefighters work a required schedule and receive pay and benefits for their work. *Volunteer* firefighters do not work from a set schedule but must respond to all incidents to which they are called. They may be required to attend weekly training sessions and station work days. They may be part-time paid, paid-on-call, fully volunteer, or a combination of these. In some jurisdictions, volunteer firefighters may be considered as paid if they receive retirement benefits at the end of their service time.

Career departments. Most large cities and some counties maintain the department's facilities and equipment. They employ full-time, career firefighters and other personnel to provide necessary services. Departments that serve military installations and private industrial sites are also typically career organizations.

Full-time career departments are continually staffed, meaning there is always someone on duty. Emergency responders must live in the fire station while on duty, but administrative staff typically work conventional hours.

Volunteer departments. Volunteer personnel receive minimal or no pay for their work. A volunteer department may be overseen by the local government, or it may be independent and governed by an elected board of directors.

Some volunteer departments are publicly funded. The town or county provides the facility for the fire station, purchases equipment, and pays for its maintenance. However, some volunteer departments rely on other funding sources. They raise money through subscription fees, billing homeowners for the cost of responding to a call, and fund-raising events such as bake sales, pancake breakfasts, or fairs.

In most volunteer departments, the fire station is not continuously staffed. Personnel respond to emergencies from home or their workplace. When summoned, designated personnel go to the fire station and drive the apparatus to the emergency scene, while others report directly to the scene. Most communities in the U.S. and Canada are protected by volunteer organizations (**Figure 1.10, p. 24**).

Part time and paid-on-call firefighters do not live in the fire station. They are summoned to the station or emergency scene by telephone, pager, or community signal. They are paid for responding, usually with an hourly wage or a set fee per response. This approach to compensation may also be used to pay part-time personnel in full-time, career organizations.

Combination departments. Combination departments are staffed by a mixture of career and volunteer firefighters. For example, a primarily volunteer department may have some career, full-time personnel such as the fire chief, EMTs, or driver/operators. At the same time, a career department with mostly full-time career firefighters may also have volunteers who have been trained to provide emergency medical care, scene control, rescue, or fire suppression.

Figure 1.10 Volunteer organizations protect areas that may not have a high enough population or incident rate to fund a full-time department.

Line Functions — Personnel who provide emergency services to external customers (the public).

Staff/Support Functions — Personnel who provide administrative and logistical support to line units (internal customers).

External Customers — Citizens of the service area protected by the organization.

Internal Customers — Employees and membership of the organization.

Separation of Departmental Duties

In some volunteer and most career departments, personnel are divided into two groups: **Line** and **Staff**. *Line personnel* deliver emergency services to **external customers** (the public). *Staff personnel* provide administrative and logistical support for **internal customers** (line personnel), in areas such as finance, maintenance, and training **(Figures 1.11a and b)**.

In some departments, staff officers deliver services to both internal and external customers. For instance, training officers may provide cardiopulmonary resuscitation (CPR) training to the public as well as to firefighters. Both line and staff functions are critical to the successful operation of any organization. In the end, we are all members of the same department and the same profession.

Figure 1.11a Line personnel deliver emergency services to the public, including fire suppression. *Courtesy of Bob Esposito.*

Figure 1.11b Staff personnel deliver services to line personnel, including training.

Fire Companies

The basic unit of fire fighting operations is a *company,* which is commanded by a company officer and typically includes a driver/operator or engineer, firefighter, and/or emergency medical technician. Actual staffing is determined by the AHJ. Multiple companies within a response area are grouped into a *battalion* or *district,* whose day-to-day functions are overseen by the *operations division.*

There are many different types of companies, each of which are organized based on local needs. In large departments, most companies are specialized. That is, they are organized and trained to excel at one particular aspect of the fire service, such as structural fire fighting, rescue, or EMS. In smaller departments, a single company typically performs all of these functions. These are the general types of companies, and their primary duties (**Figures 1.12a-d**):

Figure 1.12a An engine company provides fire suppression, but may also perform ventilation and search and rescue duties.

Figure 1.12c A hazardous materials company may be placarded to indicate the scope of the incidents that may be addressed.

Figure 1.12b A ladder company provides access to upper levels of a structure and performs forcible entry, salvage and overhaul, and utilities control duties. *Courtesy of Ron Moore, McKinney (TX) FD.*

Figure 1.12d An aircraft rescue and fire fighting company provides rescue and fire suppression duties in incidents involving aircraft. *Courtesy of Edwin Jones.*

- **Engine company** — Performs fire suppression duties at structure, vehicle, wildland, and other types of fires, such as providing a water supply and advancing attack hoselines. Additional duties may include search and rescue, extrication, ventilation, and emergency medical care.

- **Truck (Ladder) company** — Performs forcible entry, search and rescue, ventilation, salvage and overhaul, and utilities control, and provides access to upper levels of a structure. May also provide elevated water streams, extrication, and emergency medical care.

- **Rescue squad/company** — Searches for and removes victims from areas of danger or entrapment and may perform technical rescues.

- **Brush company** — Extinguishes ground cover or grass fires and protects structures in areas close to fields and woodlands, referred to as the **wildland/urban interface**.

- **Hazardous materials company** — Mitigates hazardous materials incidents.

- **Emergency medical/ambulance company** — Provides emergency medical care to patients and may transport them to a medical facility.

- **Special rescue company** — Performs technical rescues, including rapid intervention for the rescue of firefighters.

- **Aircraft rescue and fire fighting company** — Performs rescue and fire suppression activities involving aircraft accidents.

Fire Department Personnel

In most jurisdictions, professional qualifications for firefighters are based on the National Fire Protection Association® (NFPA®) Standard 1001, *Standard for Fire Fighter Professional Qualifications*. The NFPA® standard establishes basic criteria and then distinguishes between two levels of competency/professionalism: Firefighter I and Firefighter II. To be considered for employment by a fire department, a candidate must first meet the following criteria:

- Minimum educational requirements set by the AHJ, usually a high school diploma or general educational diploma (GED)

- Age requirement set by the AHJ, which may have a maximum limit based on the local or state/provincial pension law

- Medical requirements set forth in NFPA® 1582, *Standard on Comprehensive Occupational Medical Program for Fire Departments*

- Job-related physical fitness requirements set forth by the AHJ

Candidates must also be able to provide basic medical care, including cardiopulmonary resuscitation (CPR), bleeding control, infection control, and shock management. If they do not already possess these skills, they may also receive this training from the department. Departments that provide EMS may also require that candidates be certified EMS First Responders, EMTs, or paramedics.

In most organizations, personnel are classified as either uniformed or nonuniformed. Uniformed personnel have received basic firefighter training and may perform line or staff functions. Nonuniformed personnel, sometimes referred to as civilians, are not trained as firefighters and do not perform fire fighting or other hazardous activities. The following sections provide an overview of the variety of uniformed and nonuniformed positions that you may hold or come in contact with during your career in the fire service.

Wildland/Urban Interface — Line, area, or zone where an undeveloped wildland area meets a human development area. *Also known as* Urban/Wildland Interface.

Line Functions

Basic firefighters perform line and staff functions and are the foundation of the uniformed part of the fire service. Requirements for this and all other uniformed positions are found in the NFPA® professional qualifications standards. The following is a list of line positions, their primary roles, and the NFPA® standard covering their professional qualifications:

Firefighter. NFPA® 1001, *Standard for Fire Fighter Professional Qualifications,* establishes two levels for firefighter certification, Firefighter I and II. The primary difference between a Firefighter I and a Firefighter II is that a Firefighter I works under the direct supervision of a Firefighter II or company officer. A Firefighter II is trained to coordinate Firefighter I and Firefighter II personnel while working under the supervision of a company officer.

Firefighter I personnel are trained in fire suppression, search and rescue, extrication, ventilation, salvage, overhaul and EMS **(Figure 1.13)**. They must also meet the hazardous materials requirements defined by NFPA® 472, *Standard for Competence of Responders to Hazardous Materials/Weapons of Mass Destruction Incidents.* Specific duties of a Firefighter I as required in NFPA® 1001 include:

Figure 1.13 Overhaul is one of the core skills taught to Firefighter I personnel.

- Locating information in departmental documents and code materials

- Receiving an emergency call and dispatching a response to a reported emergency

- Receiving emergency and nonemergency telephone calls

- Transmitting and receiving messages on the fire department radio system

- Responding to an emergency scene on fire apparatus

- Establishing and operating in work areas at emergency scenes

- Donning and doffing (putting on and taking off) personal protective clothing

- Using self-contained breathing apparatus (SCBA) in emergency situations

- Using ropes and knots to hoist tools and equipment

- Forcing entry into a structure through doors, windows, or walls for search and rescue or fire suppression

- Operating as part of a team at an emergency incident

- Exiting a hazardous area

- Setting up and using ground ladders

- Attacking a vehicle fire

- Extinguishing fires in exterior Class A materials

- Conducting a search and rescue in a structure as part of a team

- Attacking an interior structure fire as part of a team

- Performing horizontal and vertical ventilation on a structure as part of a team——

- Performing overhaul activities at a fire scene
- Performing salvage activities to conserve property
- Connecting a fire department pumper to a water supply
- Using a portable fire extinguisher to extinguish incipient (early) Class A, B, and C fires
- Using lights and portable generators to illuminate an emergency scene
- Turning off building utilities
- Combatting a ground cover fire
- Operating, inspecting, cleaning, and maintaining ladders, ventilation equipment, SCBA, ropes, salvage equipment, and hand tools based on manufacturer's or departmental guidelines
- Cleaning, inspecting, and returning fire hose to service

In addition to coordinating Firefighter I and II, a Firefighter II performs more complex fire fighting tasks, and assumes and transfers command within the **Incident Command System (ICS)**. Duties of a Firefighter II include:

- Completing basic incident reports
- Communicating the need for additional assistance at an incident
- Extinguishing ignitable liquid fires using foam suppression concentrates and nozzles
- Coordinating an interior attack line at a structure fire
- Controlling flammable gas fires
- Protecting evidence and determining fire cause and origin
- Extricating victims trapped in a motor vehicle
- Assisting technical rescue operation teams
- Performing fire safety surveys in a private residence
- Presenting fire safety information to small groups
- Preparing preincident surveys
- Maintaining power plants, power tools, and lighting equipment based on manufacturer's instructions
- Performing annual fire hose service tests

The following are additional line functions in the fire service:

- **Fire apparatus driver/operator** — Personnel trained to drive fire apparatus to and from fires and other emergencies. They must be able to operate pumps and aerial devices, and are responsible for servicing and maintaining the apparatus **(Figure 1.14)**. Driver/operators must meet the minimum requirements of NFPA® 1002, *Standard for Fire Apparatus Driver/Operator Professional Qualifications*.

- **Airport firefighter** — Firefighters specially trained in airport operations and aircraft safety (known as *aircraft rescue and fire fighting* [ARFF]). They must be certified to NFPA® 1003, *Standard for Airport Fire Fighter Professional Qualifications*. Personnel assigned to the operation of aircraft fire fighting vehicles must be certified to NFPA® 1002, *Standard for Fire Apparatus Driver/Operator Professional Qualifications*.

Incident Command System (ICS) — Standardized approach to incident management that facilitates interaction between cooperating agencies; adaptable to incidents of any size or type.

Figure 1.14 The driver/operator of a fire apparatus is responsible for the appropriate operation of the apparatus. *Courtesy of Doug Allen and Ames (IA) FD.*

Figure 1.15 Incidents requiring a rescue component should be assigned to personnel certified to the appropriate specialty area of NFPA® 1006.

- **Hazardous materials technician** — Personnel certified to handle hazardous materials and chemical, biological, radiological, nuclear, or explosive (CBRNE) emergencies. Certification is based on NFPA® 472, *Standard for Competence of Responders to Hazardous Materials/Weapons of Mass Destruction Incidents.*

- **Rescue technician** — Personnel certified to handle technical rescue situations such as high-angle (rope) rescue, trench, structural collapse, confined-space, vehicle and machinery, water, ice rescue, and cave or mine rescues **(Figure 1.15)**. They must meet the requirements of NFPA® 1006, *Standard for Technical Rescuer Professional Qualifications.*

- **Wildland firefighter** — Personnel trained to extinguish fires in outdoor vegetation, including the wildland/urban interface. Certification is based on NFPA® 1051, *Standard for Wildland Fire Fighter Professional Qualifications.*

- **Fire department incident safety officer** — Personnel who monitor operational safety at emergency incidents. This task may be assigned to any qualified fire officer during the incident. The duties are defined in NFPA® 1521, *Standard for Fire Department Safety Officer.*

- **Fire police personnel** — Personnel who assist law enforcement officers with traffic control, crowd control, and scene security at emergency operations. There is not a current NFPA® standard for this position.

- **Fire department officer** — All fire department officers must meet the requirements of NFPA® 1021, *Standard for Fire Officer Professional Qualifications*. Officer types include:

  — Company officers who supervise a fire company in the station and at fires and other emergencies. They may also supervise a group of fire companies within their response area.

  — District/Battalion Chiefs who supervise a group of fire companies and stations. They may also manage units responsible for logistics, fire prevention, training, planning, or other functions.

  — Assistant/Deputy Chiefs who manage a variety of upper level functions, such as emergency operations, administration, fire prevention, or training. These titles and their functions vary between organizations.

  — Fire Marshals who manage the fire prevention, plans review, and investigation divisions. They may hold a chief officer rank and must meet the requirements of NFPA® 1021 and 1037, *Standard for Professional Qualifications for Fire Marshal*.

  — A Fire Chief who is responsible for all operations within the department.

EMS. Departments that provide emergency medical services may employ trained medical responders on engine, truck, or rescue companies, or department operated ambulances. These personnel may be trained to the National EMS Scope of Practice Model. In many cases these personnel also perform the duties of a firefighter. The following list describes the roles they assume when providing medical assistance:

- **Emergency Medical First Responders (EMRs)** — Personnel who provide immediate lifesaving care to critical patients while awaiting additional EMS assistance. They also assist higher level personnel at the scene and during transport.

- **Emergency Medical Technicians (EMTs)** — Personnel who provide basic emergency medical care and transportation for critical and emergent patients. They provide medical treatment using the basic equipment typically found on an ambulance.

- **Advanced Emergency Medical Technicians (AEMTs)** — Personnel who provide basic and limited advanced emergency medical care and transportation for critical and emergent patients. They provide medical treatment using the basic and advanced equipment typically found on an ambulance **(Figure 1.16)**.

- **Paramedics** — Personnel who provide advanced emergency medical care for critical and emergent patients. They provide medical treatment using the basic and advanced equipment typically found on an ambulance.

Fire Prevention. Fire prevention is an important line function that is completely separate from emergency operations. The fire prevention division is in direct contact with the public through building inspections, code enforcement, plans review, and public education. This division is typically headed by a chief officer, who may be called the chief in charge of fire prevention, the fire prevention officer, or the fire marshal. The chief officer may supervise additional officers, who typically serve in the following positions:

- **Fire prevention officers/inspectors** — Uniformed or nonuniformed personnel who perform periodic inspections, enforce building and fire codes, and interact with the public. Inspectors must meet the minimum requirements of NFPA® 1031, *Standard for Professional Qualifications for Fire Inspector and Plan Examiner*, but the standard does not require them to be firefighters or fire officers.

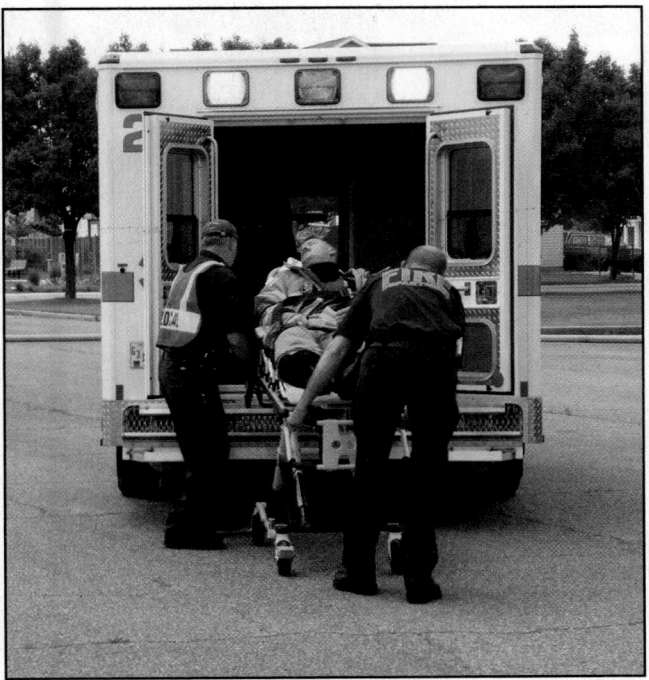

Figure 1.16 Ambulances are equipped with a wide variety of tools and equipment to triage and stabilize patients.

Figure 1.17 Regardless of whether the Plans Examiner is a member of the fire service, he or she must meet the minimum requirements of NFPA® 1031. *Courtesy of Sand Springs (OK) FD.*

- **Plans examiners** — Uniformed or nonuniformed personnel who ensure code compliance by reviewing architectural plans and fire protection system plans, for both new construction and renovations **(Figure 1.17)**. They must meet the minimum requirements of NFPA® 1031, *Standard for Professional Qualifications for Fire Inspector and Plan Examiner.*

- **Fire and arson investigators** — Uniformed personnel who investigate fires and explosions to determine their origin and cause. They must be certified to the minimum requirements of NFPA® 1033, *Standard for Professional Qualifications for Fire Investigator.* In some jurisdictions, they must have the same certification and training as law enforcement officers.

- **Fire and life safety educators** — Uniformed or nonuniformed personnel who inform the public about fire and life safety hazards, fire causes, and precautions, or actions to take during a fire. Personnel should meet the minimum requirements of NFPA® 1035, *Standard for Professional Qualifications for Fire and Life Safety Educator, Public Information Officer, and Juvenile Firesetter Intervention Specialist.*

- **Fire protection engineers/specialists** — Uniformed or nonuniformed personnel who check architectural and fire protection systems plans for proposed buildings to ensure compliance with local fire and life safety codes and ordinances. They may also act as consultants to the fire department administration in the areas of fire department operations and fire prevention.

Figure 1.18 Telecommunicators provide a critical communication link between in-service companies and complete incident reports. *Courtesy of Alan Braun.*

Staff Functions

Staff functions provide critical support to line personnel in order to help an organization carry out its mission. In many departments these functions are performed by nonuniformed civilians. The following list describes some of these staff functions:

- **Fire department health and safety officer (HSO)** — This uniformed or nonuniformed officer oversees the department's occupational safety and health program. These programs are described in NFPA® 1500, *Standard on Fire Department Occupational Safety and Health Program,* and NFPA® 1521, *Standard for Fire Department Safety Officer.* During major incidents, the HSO often acts as the incident safety officer (ISO).

- **Telecommunicators** — These personnel receive emergency and nonemergency phone calls, dispatch units, establish and maintain communications links to in-service companies, and complete incident reports **(Figure 1.18)**. They are trained to the minimum requirements of NFPA® 1061, *Standard for Professional Qualifications for Public Safety Telecommunicator.*

- **Fire alarm maintenance personnel** — These technicians maintain municipal fire alarm systems based on the requirements of NFPA® 72, *National Fire Alarm and Signaling Code®.*

- **Apparatus and equipment maintenance personnel** — These mechanics and technicians maintain fire department apparatus, vehicles, and equipment. NFPA® 1071, *Standard for Emergency Vehicle Technician Professional Qualifications* provides the minimum training required for certification.

- **Information systems personnel** — Personnel who manage electronic databases, such as those used for fire reporting.

- **Clerical staff** — Personnel who provide secretarial, administrative, and record-keeping support.

Large fire departments generally include a training division to provide internal training for entry level personnel, mandated certification training, and periodic refresher courses. Smaller departments may have a chief or company officer who oversees training. State/provincial, regional, and national training academies may also provide services and specialized training courses to small departments. The following positions are normally a part of a training division:

- **Instructors** — Uniformed personnel who train members of the department or other students. They are certified to Firefighter I requirements of NFPA® 1041, *Standard for Fire Service Instructor Professional Qualifications.*

- **Training officer/chief of training** — This position oversees all training activities and supervises personnel assigned to the training division. Training officers are certified to Firefighter II or higher, based on the requirements of NFPA® 1041, *Standard for Fire Service Instructor Professional Qualifications.*

Organizational Principles

All fire and emergency services organizations adhere to the same basic organizational principles. These principles ensure that the department delivers services efficiently and effectively, and completes its assigned mission. To function effectively as a member of your organization, you must operate according to the following organizational principles:

- Chain of command
- Unity of command
- Span of control
- Division of labor
- Discipline

Chain of Command

The chain of command is the formal line of authority, responsibility, and communication within an organization. The chain of command can be shown on an organizational chart with the fire chief or chief executive officer at the top and the firefighters and emergency responders at the bottom. By adhering to these visible authority relationships, organizations also ensure unity of command (**Figure 1.19, p. 34**).

Unity of Command

This principle means that each employee reports directly to just one supervisor. Moving up through the chain of command, all personnel ultimately report to the fire chief.

Span of Control

This principle establishes the maximum number of subordinates or functions that any one supervisor can control, typically three to seven. Five is considered optimum (**Figure 1.20, p. 34**).

Division of Labor

Division of labor is the process of dividing large jobs into smaller jobs in order to make them more manageable, equalize workloads, and increase efficiency. Division of labor is necessary in the fire service for the following reasons:

- To assign responsibility
- To prevent duplication of effort
- To assign specific and clear-cut tasks

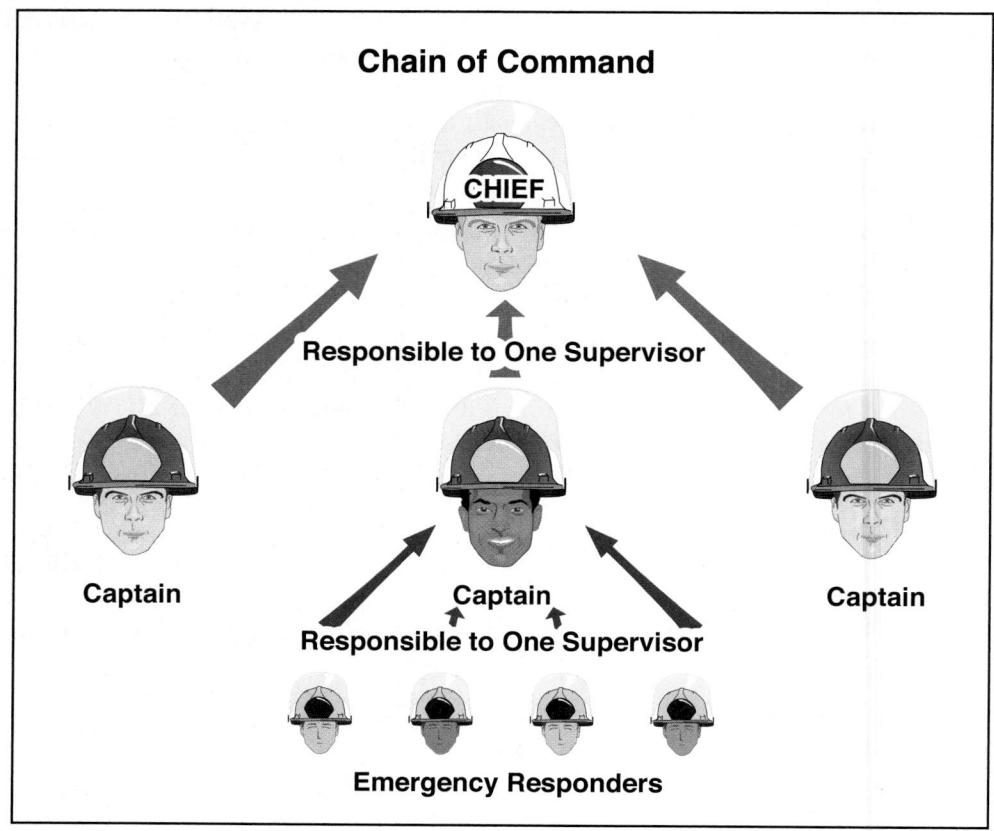

Figure 1.19 The Chain of Command can be portrayed as a simple straight-line relationship between the Chief of the department and the lowest-ranked firefighters, and all personnel between.

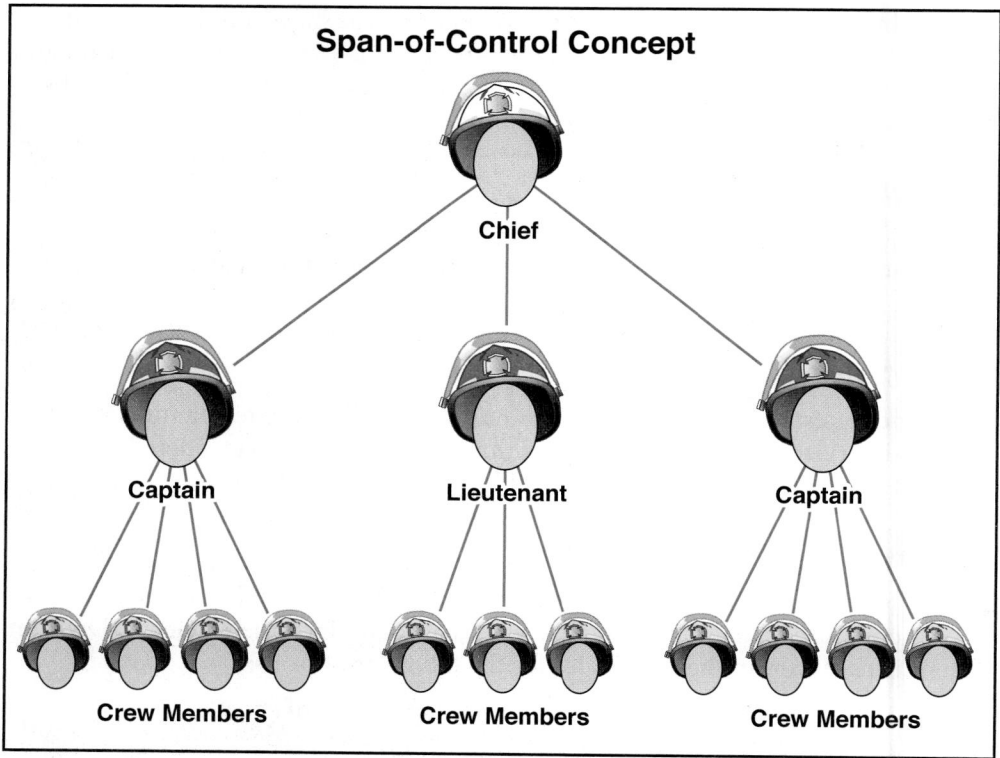

Figure 1.20 Span of control is a tool used to promote organization during the activities required at an incident.

Discipline

Discipline refers to an organization's responsibility to provide leadership, and an individual's responsibility to follow orders. It is administered through rules, regulations, and policies that define acceptable performance and expected outcomes. It can only be properly enforced if rules are clearly written and communicated throughout the organization.

Fire Department Regulations

Written regulations are essential for the successful operation of any fire and emergency services organization. They clarify expectations and delegate authority based on the organization's structure and mission.

Departmental regulations consist of policies and procedures. The department and its employees are also governed by municipal ordinances, state/provincial and federal laws, and the codes and standards mandated by the AHJ.

Additionally, departments that have labor/management agreements with a union or bargaining agent will have a written contract that directly affects their members. A copy of the contract is usually provided to all members and maintained in every workplace.

It is your responsibility to learn and adhere to your department's regulations. You must be able to locate these documents and find the parts that apply to your duties, authority, and responsibility. To ensure that members have access to regulations so that they can fully understand them and be able to comply, organizations do the following:

- Distribute them in written or electronic format.
- Communicate them verbally to all members.
- Post them in a conspicuous place in all facilities.

If you have a question about departmental regulations, it is important to know where to find them. If you do not know, ask your supervisor, who may be able to answer your question directly. If not, he or she can probably tell you where to find the documents so that you can look up the answer for yourself. Regulations are always filed in the administrative office, and each station, facility, or division office typically keeps a copy. New policies or amendments should also be posted on bulletin boards (**Figure 1.21**).

Once you have located the documents, you will have to know where to look for the information you seek. To do so, you must know the difference between policies, procedures, laws, codes, and standards.

Figure 1.21 Policies should be kept current and accessible to those who will be held accountable for them.

Policies

A **policy** is a guide to decision-making within an organization. They are determined by top management, then distributed to lower ranks to be implemented. Policies set boundaries and establish standards of conduct that an organization expects from its members. They address issues such as working hours, emergency response guidelines, and chain of command.

Policies may be created in response to government mandates, such as certification training, or changes to operational needs. Most policies involve written criteria, but some remain unwritten and are known as *organizational norms* or *past practices.* They typically result from an organization's traditional approach to routine tasks and are implied rather than formally stated.

Policy — Guide to decision making in an organization.

Unwritten policies are common when no clear policy exists, or policies are out of date. If either situation occurs, it is important for organizations to create clear written policies. If they do not, there may be dangerous consequences that result in legal liability.

Procedures

Procedures are detailed written plans that list specific steps for approaching a recurring problem or situation. Examples include the steps required to ventilate a roof or handle hazardous materials spills. Most organizations provide personnel with **standard operating procedures (SOPs)** that enable all members to perform specific tasks to the required standard.

Some departments issue standard operating guidelines (SOGs). These are similar to SOPs but may allow firefighters some leeway in particular situations, whereas SOPs are hard and fast rules.

Laws, Statutes, or Ordinances

Policies may be the result of federal mandates, state/provincial regulations, or local government laws and ordinances that directly affect the fire department. For example, equal employment opportunity practices have been mandated by the U.S. Fair Labor Standards Act (FLSA) and the Americans with Disabilities Act (ADA). These practices must be followed by all public career fire departments. The FLSA sets criteria for pay and the ADA affects both hiring practices and building accessibility **(Figure 1.22)**.

U.S. federal laws that pertain to firefighters are included in the *Code of Federal Regulations (CFR)* and include the requirements administered by the Occupational Safety and Health Administration (OSHA). In Canada, the agency responsible for health and safety of federal firefighters is Occupational Health and Safety (OH&S), while each province has its own agency for provincial and local firefighters.

Codes and Standards

A **standard** is a set of principles, protocols, or procedures that is developed by committees through consensus. Standards typically explain how to do something or provide a set of minimum standards to be followed **(Figure 1.23)**. A **code** is a collection of rules and regulations that has been enacted by law in a particular jurisdiction. It may be based on a standard, or it may incorporate an entire standard.

Building codes and standards are enforced by a jurisdiction's building or fire department. Fire and life safety codes regulate construction materials and design, as well as occupant behavior and manufacturing processes. Codes must be up-to-date and written and applied in a consistent manner.

Orders and Directives

Orders and directives are issued from the top of the chain of command and used to implement departmental policies and procedures. They may be written or verbal. They may be based on a policy or procedure, making compliance mandatory. The terms *orders* and *directives* generally mean the same thing, but exact definitions vary between departments.

Procedure — Step-by-step written plan that is closely related to a policy. Procedures help an organization to ensure that it consistently approaches a task in the correct way, in order to accomplish a specific objective and comply with a policy.

Standard Operating Procedure (SOP) — Rule for how personnel should perform routine functions or emergency operations. Procedures are typically written in a handbook, so that all firefighters can become familiar with them.

Standard — A set of principles, protocols, or procedures that explain how to do something or provide a set of minimum standards to be followed. Adhering to a standard is not required by law, although standards may be incorporated in codes, which are legally enforceable.

Code — A collection of rules and regulations that has been enacted by law in a particular jurisdiction. Codes typically address a single subject area; examples include a mechanical, electrical, building, or fire code.

Figure 1.22 The Americans with Disabilities Act sets standards and enforcement for building accessibility accommodations. *Courtesy of Michael Mallory, Tulsa (OK) FD.*

Figure 1.23 Code inspections follow established standards to determine whether a facility is within compliance.

Interacting with Other Organizations

The fire department is one of many organizations that provides services to the public during emergencies. To work efficiently with members of these organizations, you need to know the services they provide to both the public and the fire department. This section will acquaint you with the following:

- Emergency medical services
- Hospitals
- Emergency/disaster management
- Law enforcement
- Utility companies
- Public works department
- Media
- Other agencies

Emergency Medical Services

Emergency medical service (EMS) organizations provide pre hospital patient care and/or transportation to a medical facility. They may be privately operated or an agency of the AHJ. EMS organizations may provide complete medical services, or they may simply supplement the services provided by the fire department. In some jurisdictions, they also provide standby units for fire department training evolutions.

Hospitals

Public and private hospitals can be found in any major urban area and in many smaller communities within North America. Hospitals are regulated by state medical authorities, which determine the services hospitals may provide based on their staffing and facilities. Hospital regulatory agencies include:

- Joint Commission (JC), which establishes standards of care
- State Department of Health Services, which issues operating licenses
- Occupational Safety and Health Administration (OSHA), which establishes levels of accreditation

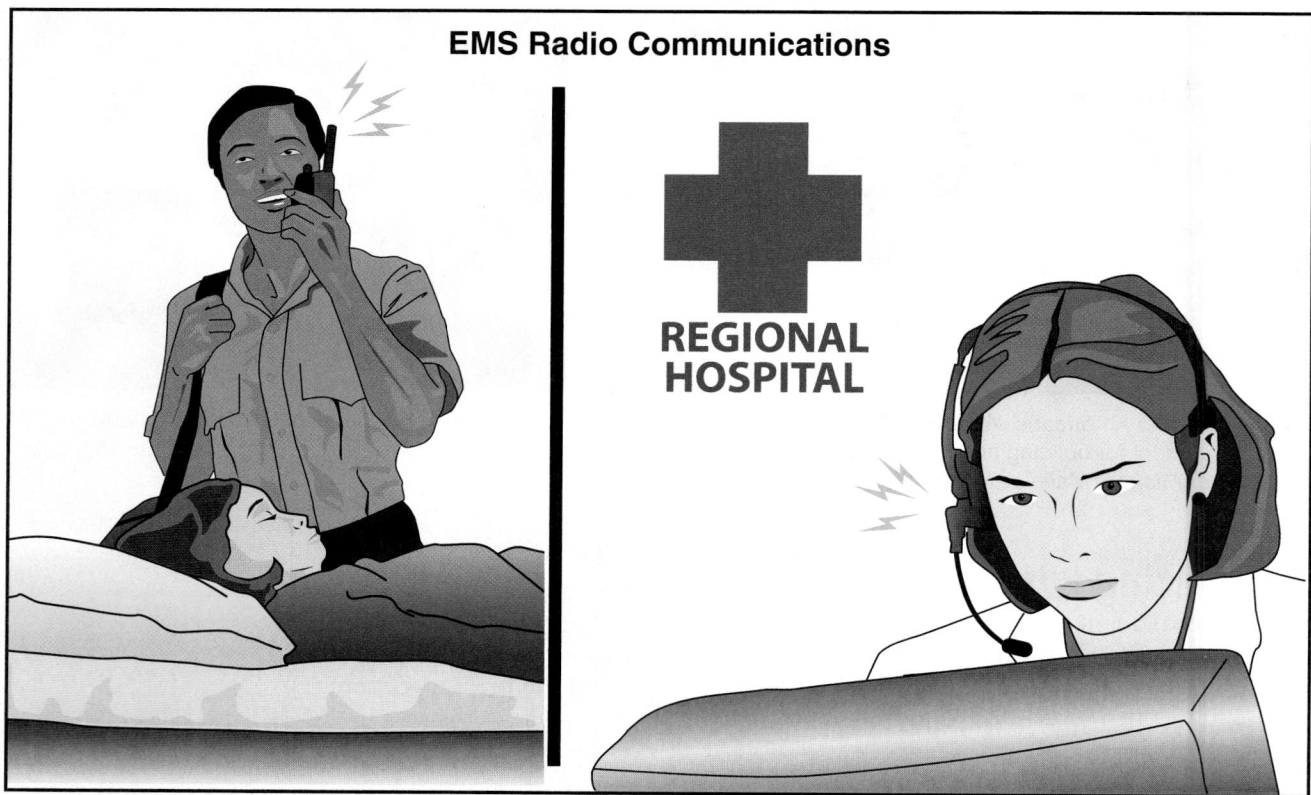

EMS Radio Communications

REGIONAL HOSPITAL

Figure 1.24 Hospitals make many accommodations for emergency medical services including direct information lines.

EMS and fire service personnel typically interact with hospitals and trauma centers that have emergency rooms (ERs), which receive patients via ambulance, helicopter, personal vehicle, or walk-in. These facilities may also provide direct assistance to units at an emergency incident, either by radio communication or by dispatching a medical professional to the scene **(Figure 1.24)**.

Emergency Disaster Management

Emergency management agencies are administered by city, county, and tribal governments. They manage emergency and disaster response by coordinating multiagency activities. In some jurisdictions, the fire department is the designated emergency disaster management authority.

Law Enforcement

Fire departments may interact with law enforcement agencies from all levels of government, such as:

Figure 1.25 Incident support in the form of scene security and traffic control may be provided by all levels of law enforcement agencies.

- Local police department
- County sheriff's office
- State/provincial police or highway patrol
- Federal agencies

These agencies may provide incident scene security, traffic and crowd control, and protection to firefighters They may also handle fire cause determination and investigation, explosives disposal, and intelligence gathering **(Figure 1.25)**.

Utility Companies

Utility companies may be privately owned or a department of local government. Their personnel assist at incidents by shutting off natural gas lines, electricity, or public water mains. They may also train firefighters to perform these tasks. Some water departments also maintain hydrants and perform with hydrant flow tests.

Public Works

Public works departments typically oversee the construction and maintenance of public roads, buildings, and sewers. Public works departments are administered at the municipal, county, and state/provincial level. In some jurisdictions they also oversee the water distribution system. The public works department is a division of the AHJ. It assists the fire department by providing:

- Heavy equipment for confined-space rescue
- Earth-moving equipment
- Flood-control equipment and materials
- Sand for containing spills
- Barriers and signs to divert traffic
- Facility maintenance and repair
- Civil/structural engineers who can determine the structural integrity of a damaged building

Media

The media consists of journalists from print (newspapers and magazines), broadcast (radio and television), and Internet sources. Each can help the fire department in a variety of ways. For example, up-to-date broadcast news can divert traffic from an incident scene or alert the public when large-scale evacuation is necessary. Ongoing coverage also informs the public about fire and life safety information, fire department activities and events, and important changes to department policies and procedures.

When approached by members of the media at an emergency scene, firefighters should never answer questions or offer an opinion unless they have been authorized to do so. Instead, they should direct journalists to an appropriate person, such as the Incident Commander, ranking officer, or Public Information Officer (PIO).

Other Agencies

Many additional agencies assist at emergency or nonemergency incidents, or through administrative channels. They include:

- Public health departments
- Coroner/medical examiner's offices
- National or state/provincial environmental protection agencies
- National or state/provincial forestry departments
- Coast Guard
- National Guard and military reserve units
- District attorney's offices
- Occupational Safety and Health Administration (OSHA)

Figure 1.26 Many agencies may provide support to emergency responders by providing resources for rehabilitation. *Courtesy of Tim Burkitt/FEMA.*

- Federal Aviation Administration (FAA)
- Water resources boards
- Community organizations
- Local Emergency Planning Committee/Office of Emergency Management (LEPC/OEM)
- Red Cross
- Salvation Army **(Figure 1.26)**
- University and federal laboratories

Chapter Summary

You are about to become a member of a profession with a long and proud tradition of service to the community. Whether you are a member of a volunteer, combination, or career fire department, you are carrying on the traditions established by the firefighters who have gone before you.

To perform the duties of a firefighter, you will need to perfect the physical skills described in this manual and taught to you in your training classes. You will also need to know the regulations that govern your particular organization as well as the federal, state/provincial, and local laws that regulate the fire service. You will also need to become knowledgeable in the Incident Command System and the part you play in it.

Finally, you will need to be familiar with the organizations and agencies that interact with the fire department at emergency and nonemergency incidents. This manual is the beginning of your training and will provide you with general knowledge on which to base your future training and service.

Firefighter I

1. How were early fire organizations started?

2. What are some of the areas that have changed significantly in the 20th Century for fire service in North America?

3. How do organizational characteristics, cultural challenges, and cultural strengths influence the fire service?

4. What is the mission of the fire service?

5. What are the three main types of staffing found in the fire service?

6. How are the duties of an engine company different from a rescue squad/company?

7. What is the central difference between line functions and staff functions?

8. What is the primary difference between Firefighter I and Firefighter II duties?

9. How are qualifications for different line positions regulated?

10. What types of staff functions support and supplement line functions?

11. What are the organizational principles of the fire service?

12. What steps can be taken to locate information in department policies?

13. How are policies and procedures different from one another?

14. What other organizations may provide services to the public along with firefighters?

15. What should a firefighter do when approached by members of the media?

Chapter Contents

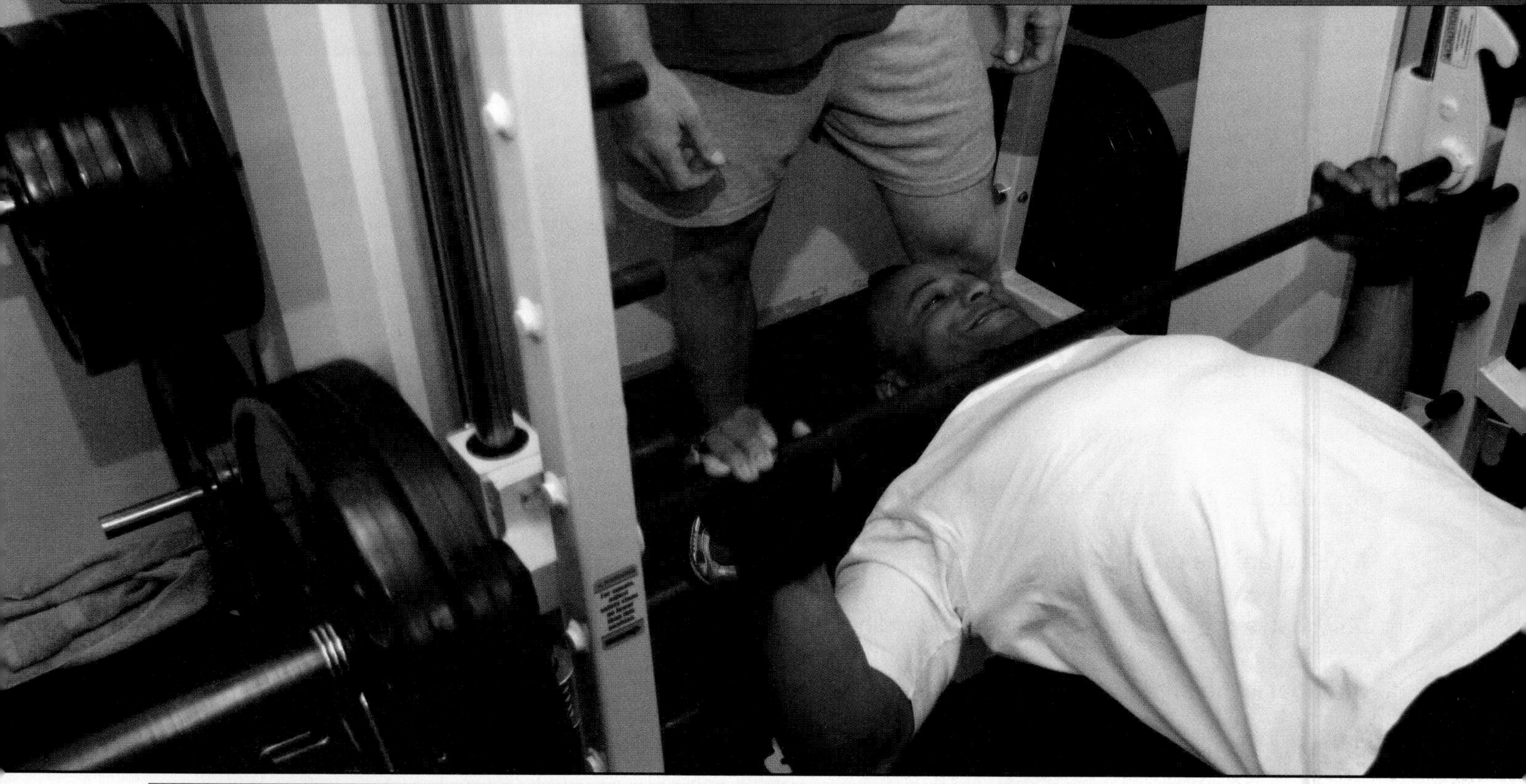

Chapter 2

Key Terms

NFPA® Job Performance Requirements

This chapter provides information that addresses the following job performance requirements of NFPA® 1001, *Standard for Fire Fighter Professional Qualifications* (2013).

Firefighter I
5.1.1
5.3.2
5.3.3
5.3.4
5.3.5

1. List the main types of job-related firefighter fatalities, injuries, and illnesses. (5.1.1)

2. Describe the National Fire Protection Association® standards related to firefighter safety and health. (5.1.1)

3. Identify Occupational Safety and Health Administration (OSHA) regulations and how they relate to firefighters. (5.1.1)

4. Summarize the model that supports the concept of risk management. (5.1.1)

5. Describe fire department safety and health programs. (5.1.1)

6. Summarize firefighter health awareness issues. (5.1.1)

7. Summarize safe vehicle operations. (5.3.2, 5.3.3)

8. Summarize guidelines for riding safely on the apparatus. (5.3.2)

9. Describe ways to help prevent accidents and injuries in fire stations and facilities. (5.1.1)

10. Explain general guidelines for tool and equipment safety. (5.3.4)

11. Describe ways to maintain safety in training. (5.1.1)

12. State the practices a Firefighter I uses for emergency scene preparedness and safety. (5.1.1, 5.3.3)

13. Summarize general guidelines for scene management including highway incidents, crowd control, and cordoning off emergency scenes. (5.1.1, 5.3.3)

14. Explain the importance of personnel accountability. (5.3.5)

15. Respond to an incident, correctly mounting and dismounting an apparatus. (Skill Sheet 2-I-1, 5.3.2, 5.3.3)

16. Wearing appropriate PPE, including reflective vest, demonstrate scene management at roadway incidents using traffic and scene control devices. (Skill Sheet 2-I-2, 5.3.3)

Chapter 2
Firefighter Safety and Health

Case History

On February 20, 2009, a 45-year-old male Oklahoma volunteer firefighter responded to an 800-acre wildland fire. After fighting the fire for approximately 10 hours, he collapsed while assisting with "mop-up" duties. His colleagues administered immediate medical assistance, but he was unresponsive and not breathing. Soon afterwards his pulse stopped, and cardiopulmonary resuscitation (CPR) was ineffective. A shock from an automated external defibrillator (AED) restored his pulse, but he died en route to the hospital.

After performing an autopsy, the medical examiner determined that death had been caused by severe underlying heart disease. National Institute for Occupational Safety and Health (NIOSH) investigators concluded that the physical stress of performing fire extinguishing activities, when coupled with the underlying heart disease, had probably triggered a fatal heart attack.

The NIOSH line-of-duty death (LODD) report recommended that fire departments take the following actions:

- Provide preplacement and annual medical evaluations, consistent with National Fire Protection Association® (NFPA®) 1582, *Standard on Comprehensive Occupational Medical Program for Fire Departments.*

- Incorporate exercise stress tests, following standard medical guidelines, into fire department medical evaluation programs.

- Phase in a comprehensive wellness and fitness program for firefighters, consistent with NFPA® 1583, *Standard on Health-Related Fitness Programs for Fire Department Members.*

- Perform annual physical fitness evaluations consistent with NFPA® 1500, *Standard on Fire Department Occupational Safety and Health Program.*

- Require firefighters to receive medical clearance to wear self-contained breathing apparatus (SCBA) as part of the fire department's medical evaluation program.

Source: NIOSH Firefighter LODD Report F2009-09, February 20, 2009.

As a fire and emergency services responder, it is your duty to protect the lives of the people in your community. To perform that duty, you must first protect yourself through a proactive approach to safety, health, and fitness. You must be aware of the dangers you face not only at emergency responses but also from lifestyle choices. By knowing these dangers, you can implement a strategy of safe behavior and healthy lifestyle choices.

Safety is an important aspect of fire service culture. Your department should have a health and safety plan intended to create a safe working environment. This chapter addresses important elements of these plans and how they affect you. It overviews various types of firefighter fatalities, injuries, and illnesses, and presents strategies for minimizing these dangers. It also discusses the general safety requirements for driving emergency vehicles, working with tools and equipment, and operating at emergency incident scenes.

I Firefighter Fatalities, Injuries, and Illnesses

Knowing that fire fighting is one of the most hazardous professions, many firefighters have traditionally accepted fatalities, injuries, and illnesses as part of their job. This attitude is compounded by the stereotypical image of the firefighter as heroic and fearless in the face of danger. But most fatalities, injuries, and illnesses are preventable, so understanding their causes can help you to avoid them.

The NFPA® and the U.S. Fire Administration (USFA) publish annual statistics on firefighter fatalities and injuries. Over the past decade, the numbers have remained fairly constant, averaging approximately one hundred deaths and eighty thousand job-related injuries (**Table 2.1**). Statistics on job-related illnesses are less reliable because neither organization collects this data, and many illnesses are not reported. In an effort to gain a more accurate picture of firefighter injuries, the National Firefighter Near-Miss Reporting System has been developed. The system takes reports of incidents that could have resulted in serious injuries or fatalities as a means of increasing awareness and generating a safety-based culture in the fire service.

Fatalities

In 2010, the NFPA® recorded seventy-two on-duty firefighter fatalities in the United States, the lowest annual figure since NFPA® began collecting this data in 1977. Of those, forty-four were volunteer firefighters, twenty-five were career, two were employed by state land management agencies, and one was a member of a prison inmate wildland fire fighting crew. The leading cause of death was stress or overexertion (thirty-nine fatalities), a category that includes heart attacks and strokes. Motor vehicle accidents also killed six, all volunteer firefighters (**Figure 2.1**). The remaining causes of death included falls, being struck by an object, being trapped in a structure fire, or gunshot wounds while tending to a medical patient.

Injuries

Firefighter injuries range from relatively minor to those requiring hospitalization and lost duty time. The most recent NFPA® report on firefighter injuries estimated that slightly less than 72,000 firefighters were injured during 2010. Injuries sustained on the fireground accounted for 32,000 while nonfire emergency incidents amounted to almost 14,200. Strains and sprains were the leading types of injuries followed by cuts and bruises (**Figure 2.2, p. 48**). Other injuries occurred during training, during nonemergency incidents, while responding to or returning from incidents, and in various other on-duty activities.

Table 2.1
Firefighter Fatalities and Injuries Recorded Between 2001-2010

The following table gives an overview of the number of on-duty deaths
and injuries that occurred from 2001 to 2010.

Year	Deaths[1]	Fireground Injuries[2]	Total Injuries[2]
2001	105[3]	41,395	82,250
2002	101	37,860	80,800
2003	113	38,045	78,750
2004	119[4]	36,880	75,840
2005	115	41,950	80,100
2006	107	44,210	83,400
2007	118	38,340	80,100
2008	118	36,595	79,700
2009	90	32,205	78,150
2010	72	32,200	72,000

[1] This figure reflects the number of deaths as published in USFA's annual report on firefighter fatalities. All totals are provisional and subject to change as further information about individual fatality incidents are presented to USFA.

[2] This figure reflects the number of injuries as published in NFPA®'s annual report on firefighter injuries.

[3] In 2001, an additional 341 FDNY firefighters, three fire safety directors, two FDNY paramedics, and one volunteer from Jericho (NY) Fire Department died in the line of duty at the World Trade Center on September 11.

[4] The Hometown Heroes Survivors Benefit Act of 2003 has resulted in an approximate 10 percent increase to the total number of firefighter fatalities counted for the annual USFA report on Firefighter Fatalities in the United States beginning with CY2004.

Courtesy of the United States Fire Administration and the Federal Emergency Management Agency.

Figure 2.1 Vehicle accidents involving fire apparatus are partially responsible for fire-fighter fatalities. *Courtesy of Mike Mallory and Tulsa (OK) FD.*

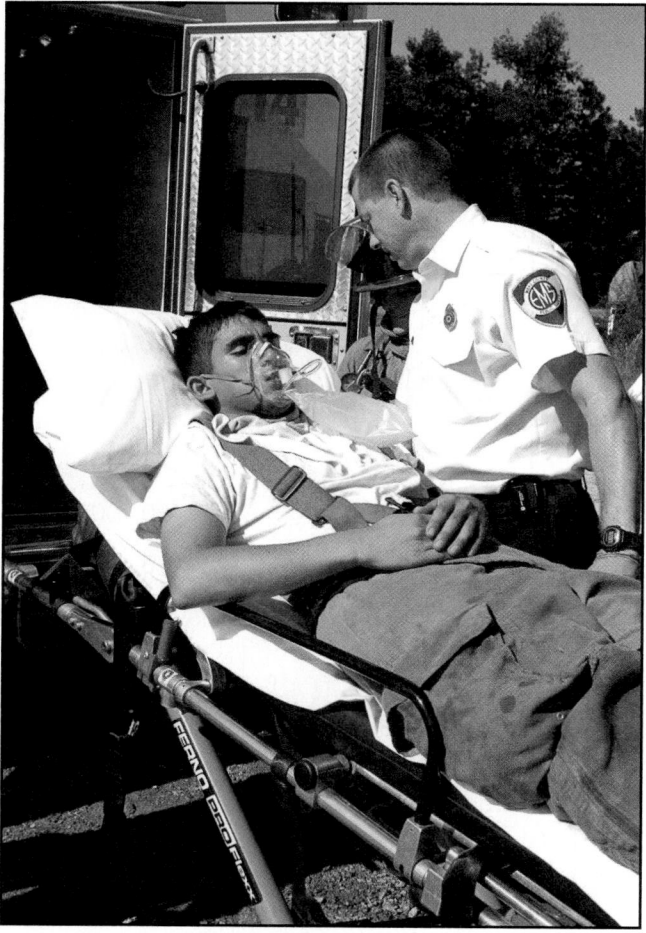

Figure 2.2 Firefighters are susceptible to injuries that range from severe to minor over the course of their employment.

Besides causing physical pain and emotional distress, injuries are also costly to the individual and to the organization. For career departments, injuries result in lost duty time, higher worker's compensation payments, and the need to replace injured personnel. Volunteer departments carrying worker's compensation insurance will also bear additional costs after injuries. But many injured volunteers will receive no compensation for the wages they lose at their full-time jobs.

Many firefighter injuries can be prevented by:

- Providing effective training

- Maintaining company discipline and accountability

- Following established safety-related SOPs

- Using personal protective clothing and equipment

- Maintaining high levels of physical fitness

- Following risk management guidelines

- Using rehabilitation facilities at emergency incidents

Physically fit firefighters are more productive and less likely to suffer strains and sprains, which account for nearly 50 percent of all firefighter injuries. Physical exercise is also an effective way of lowering stress and reducing the risk of cardiac arrest and strokes.

Illnesses

Acute and **chronic** illnesses are another source of lost duty time. Acute illnesses, such as colds and viruses, last only a few days and result in little lost duty time. Chronic illnesses, such as cancer or diabetes, are long lasting and can even be fatal.

Some chronic illnesses are hereditary, but many result from lifestyle choices, working in toxic atmospheres, and exposure to infected patients. Firefighters should protect themselves from the following illnesses:

- Cardiovascular diseases
- Respiratory diseases
- Cancer
- Obesity
- Diabetes
- Drug and alcohol dependence
- Stress-induced hypertension
- Tobacco use/dependence
- Exposure-related diseases, such as AIDS and hepatitis

Cardiovascular Diseases

Cardiovascular diseases are the leading cause of firefighter fatalities, accounting for 45 percent of line-of-duty deaths. Among the work-related causes are:

- Exposure to smoke and chemicals
- Heat stress from fires and high temperatures **(Figure 2.3)**
- Psychological stress
- Long, irregular work hours

Figure 2.3 Heat stress and smoke exposure contribute to cardiovascular disease.

Acute — Sharp or severe; having a rapid onset and short duration.

Chronic — Long-term and reoccurring.

These causes can be **mitigated** by departmental policies, procedures, and equipment. It is your responsibility to adhere to these policies and procedures, wear required personal protective equipment (PPE), and reduce exposure to **hazards**.

Non-job-related causes of cardiovascular disease include obesity, tobacco use, and lack of physical fitness. These can be controlled by eating a healthy diet, discontinuing tobacco use, and exercising regularly.

Respiratory Diseases

Common respiratory diseases include asthma, lung cancer, and chronic obstructive pulmonary disease (COPD). In the general population, respiratory diseases are typically caused by cigarette smoking. But firefighters are also exposed to smoke, gases, and chemicals (such as carbon monoxide and phosgene) that can cause chronic respiratory diseases. The only way to avoid these deadly hazards is to wear proper respiratory protective equipment.

Chemicals Found in Smoke that Contribute to COPD

- Carbon monoxide
- Sulfur dioxide
- Hydrogen chloride
- Phosgene
- Nitrogen oxides
- Aldehydes
- Particulates

Always follow departmental regulations for wearing respiratory protective equipment. During structural fire fighting, for example, never remove your SCBA until your supervisor or the Incident Commander (IC) says it is safe to do so. Always wear protective equipment in situations that are **immediately dangerous to life and health (IDLH)**. Always choose the appropriate level of protection during medical emergencies or when working around paints, thinners, cleaners, dust, or particulates (**Figure 2.4**).

Finally, it is important not to smoke or to expose yourself to secondhand smoke. A healthy lifestyle ensures that you have the lung capacity to perform your duties.

Figure 2.4 Hazards must be accommodated, even while performing routine maintenance tasks.

Cancer

Recent studies indicate that firefighters are at greater risk of contracting four types of cancer: testicular cancer, prostate cancer, non-Hodgkin's lymphoma, and multiple myeloma, a cancer of the bone marrow. The cause of these cancers is exposure to carcinogens that are present in all fires and in the exhaust fumes of apparatus.

To protect your health, you must follow departmental safety policies, wear the correct level of respiratory protection, and thoroughly clean your PPE when it is contaminated by smoke. Helmet liners, SCBA facepieces, and coat collars should always be cleaned to prevent contaminants from coming in contact with your face. For further information see **Appendix C**.

Obesity

Avoid obesity by exercising and eating a healthy diet. A 2010 study found that up to 40 percent of firefighters nationwide are overweight, 6 percent above the national average in the general population. Being overweight or obese puts a person at a higher risk of:

- Type 2 diabetes
- Coronary heart disease
- Stroke
- Hypertension
- Some forms of cancer (breast, colorectal, endometrial, and kidney)

Diabetes

Diabetes is marked by high levels of sugar in the blood. It is divided into two classifications: Type 1, which is controlled by insulin, and Type 2, which can be controlled through exercise, diet, and oral medication. Until recently, NFPA® 1582, *Standard on Comprehensive Occupational Medical Program for Fire Departments*, prohibited persons with diabetes from being hired as firefighters. But since 2007, persons who are under a physician's care and have control over their diabetes may be hired as firefighters and medical responders.

To prevent yourself from becoming diabetic, you must exercise, eat properly, and have an annual medical examination. You must also be aware of the symptoms of diabetes and of your family medical history.

Drug and Alcohol Abuse

Drugs and alcohol impair your ability to function and slow your reaction time. Most departments prohibit firefighters from responding to an emergency if they have consumed alcohol beforehand, or if they are taking prescription drugs that cause impairment.

Many fire departments prohibit all tobacco use by candidates for employment. Some departments also prohibit candidates from using alcohol.

It is up to you to control your use of alcohol when you are on-call or prior to reporting for duty. If you have been drinking alcohol, you must not respond to a call or report to the station. The same is true if you are taking medication that causes impairment.

CAUTION

Drinking alcohol can slow reaction times and impair
judgment. Alcohol and fire fighting do NOT mix.

Stress-Induced Hypertension

Fire fighting is extremely stressful. Adrenaline begins to rush after the initial alarm and sometimes does not stop until the conclusion of the incident. Over time, the physical and emotional stress of responding to emergencies can cause physical damage.

Stressor — Any agent, condition, or experience that causes stress.

Both physical and emotional stress refer to the body's reaction to a **stressor**. Firefighters are exposed to stressors such as witnessing traumatic events and the pressure to react correctly in an emergency. They are also subject to the same stressors as the general population, such as the death of a loved one, or domestic and financial problems.

Stress can cause physical symptoms such headaches, nausea, and weakness in the legs. See **Table 2.2** for a list of physical, emotional, mental, and behavioral signs of stress. The damage stress does to your body accumulates over time and can take years to develop. One serious long-term consequence of stress is hypertension (high blood pressure).

Controlling stress can be difficult. However, if you are in good physical condition, have a positive mental attitude, and relax whenever possible, you can mitigate the effects of stress **(Figure 2.5, p. 54)**. One effective method is to develop a personal stress management program. A sample program is provided in the following textbox.

A Personal Stress Management Program

General concepts:

- Be positive and don't dwell on the negative aspects of a situation.
- Be good to yourself and your body.
- Plan enjoyable activities.
- Take regular breaks from stressful activities.

Physical activity:

- Start a physical activity/fitness program. 20 minutes of aerobic activity three times per week is recommended.
- Exercise with a partner or group.

Nutrition:

- Select healthy foods based on the U.S. Department of Agriculture's MyPlate guide.
- Eat normal-sized portions on a regular schedule.

Social support:

- Make an effort to socialize. Being with friends can help you to feel less stressed.
- Be good to yourself and others.

Relaxation:

- Use relaxation techniques, such as yoga, meditation, or listening to music.
- Listen to your body when it tells you to slow down or take a break.
- Get enough sleep. Good sleep habits are one of the best ways to manage stress.
- Take time for personal interests and hobbies.

Table 2.2
Warning Signs and Symptoms of Stress

Cognitive Symptoms
- Memory problems
- Inability to concentrate
- Poor judgment
- Dwelling on the negative
- Anxious thoughts
- Constant worrying

Physical Symptoms
- Diarrhea or constipation
- Nausea, dizziness
- Loss of sex drive
- Frequent colds
- Pain in the chest, shoulders, neck, or low back
- Stomach/abdominal pain
- Muscle tension, spasms, or nervous tics
- Unexplained rashes or skin irritations
- Sweaty palms
- Sweating when not physically active
- 'Butterflies' in stomach
- Indigestion
- Inability to sleep or excessive sleep
- Shortness of breath
- Holding breath
- Loss of energy

Emotional Symptoms
- Moodiness
- Irritability or short temper
- Agitation, inability to relax
- Feeling overwhelmed
- Feeling lonely and isolated
- Depression or general unhappiness

Behavioral Symptoms
- Eating more when you are not hungry
- Sleeping too much or too little
- Isolating yourself from others
- Procrastinating or neglecting responsibilities
- Using alcohol, cigarettes, or drugs to relax
- Nervous habits (nail biting, pacing)
- Feeling frustrated at having to wait for something
- Feeling restless
- Being easily confused
- Negative self-talk
- Feeling you can't cope
- Difficulty making decisions
- Having emotional outbursts
- Generally feeling upset
- Lack of sense of humor

Figure 2.5 Recreational sports contribute to overall well-being by helping control stress levels.

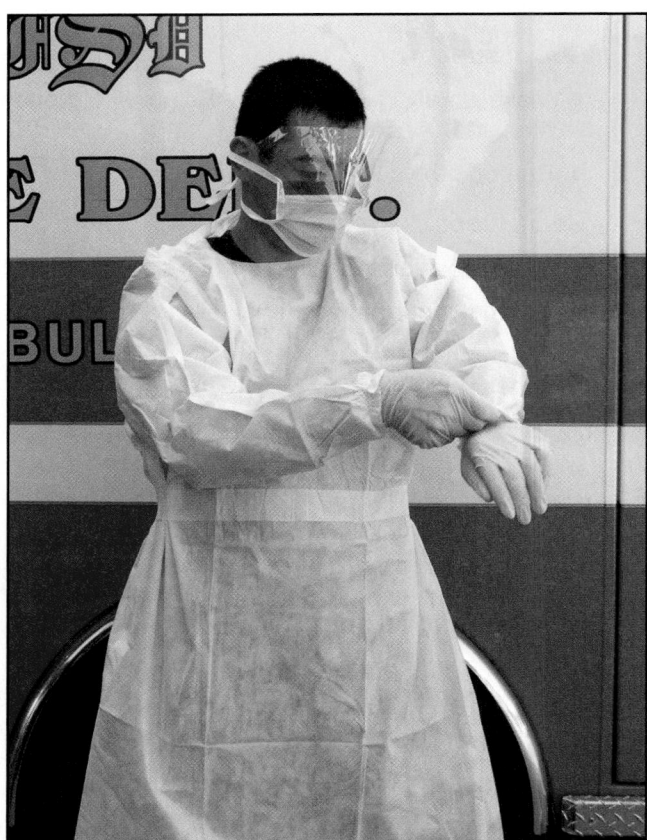

Figure 2.6 Appropriate protective clothing includes body substance isolation equipment to prevent potentially dangerous medical exposures.

Employee Assistance Program (EAP) — Program to help employees and their families with work or personal problems.

If your personal stress management program is not working, you should seek assistance from the **employee assistance program** in your organization. If your organization does not have one, speak to your physician.

Tobacco Use/Dependence

Any type of tobacco use can cause cancer, heart disease, and COPD, and make other chronic diseases last longer. Many departments prohibit smoking within all facilities, require that probationary firefighters do not smoke, and offer smoking cessation programs for members who do smoke.

If you do not smoke, do not start. If you do, take advantage of one of the many programs offered by your employer, the International Association of Fire Fighters (IAFF), or your physician for help in stopping smoking.

Exposure-Related Diseases

Treating medical patients can expose you to AIDS, hepatitis, tuberculosis, and many other diseases. Some firefighters have even contracted antibiotic-resistant strains of bacteria, such as methicillin-resistant *Staphylococcus aureus* (MRSA). To protect yourself, always wear appropriate protective clothing and respiratory protection **(Figure 2.6)**, and use body substance isolation (BSI) methods to treat all patients. Departments may also provide vaccinations against some diseases, such as hepatitis.

Fire Service Safety Standards, Regulations, and Initiatives

Numerous occupational safety standards and regulations address the inherent danger of fire fighting. The most prominent of these come from the NFPA® and the U.S. Occupational Safety and Health Administration (OSHA). Additional safety programs are also available through the National Fallen Firefighters Foundation's Everyone Goes Home® program and the International Association of Fire Chiefs' annual safety stand-down.

NFPA® Standards

Most NFPA® standards establish criteria for building materials, equipment, and fire and life safety systems. However, many relate to firefighter safety and health. Some establish design criteria for protective clothing and equipment, such as:

- NFPA® 1971, *Standard on Protective Ensembles for Structural Fire Fighting and Proximity Fire Fighting*
- NFPA® 1975, *Standard on Station/Work Uniforms for Emergency Services*
- NFPA® 1977, *Standard on Protective Clothing and Equipment for Wildland Fire Fighting*

Others define safe training practices and programs, including:

- NFPA® 1403, *Standard on Live Fire Training Evolutions*
- NFPA® 1404, *Standard for Fire Service Respiratory Protection Training*
- NFPA® 1407, *Standard for Training Fire Service Rapid Intervention Crews*
- NFPA® 1410, *Standard on Training for Initial Emergency Scene Operations*
- NFPA® 1451, *Standard for a Fire Service Vehicle Operations Training Program*

There are also standards that set requirements for the care and maintenance of personal and respiratory protection equipment, such as:

- NFPA® 1851, *Standard on Selection, Care, and Maintenance of Protective Ensembles for Structural Fire Fighting and Proximity Fire Fighting*
- NFPA® 1981, *Standard on Open-Circuit Self-Contained Breathing Apparatus (SCBA) for Emergency Services*

Finally, there are NFPA® standards that establish safety programs that departments must adopt, including:

- NFPA® 1581, *Standard on Fire Department Infection Control Program*
- NFPA® 1582, *Standard on Comprehensive Occupational Medical Program for Fire Departments*
- NFPA® 1584, *Standard on the Rehabilitation Process for Members During Emergency Operations and Training Exercises*

The most comprehensive NFPA® standard dealing with firefighter safety and health is NFPA® 1500, *Standard on Fire Department Occupational Safety and Health Program*. This standard specifies the minimum requirements for a fire department safety and health program and may be applied to any fire department or similar organization, public or private. The basic concept of NFPA® 1500 is to promote safety throughout the fire service.

Because it is a minimum standard, any department or jurisdiction is free to exceed the standard's specified requirements. Many governing bodies in both the U.S. and Canada have chosen to adopt one or more of the NFPA® standards into their laws and ordinances.

Among the topics covered in NFPA® 1500 are the following:

- Safety and health-related policies and procedures
- Training and education
- Fire apparatus, equipment, and driver/operators
- Protective clothing and protective equipment
- Emergency operations
- Facility safety
- Medical and physical requirements
- Member assistance and wellness programs
- Critical incident stress management (CISM) program

Safety and Health-Related Policies and Procedures

The departmental safety and health program must address all anticipated hazards to which the members might be exposed. Besides the obvious hazards associated with fire fighting, this includes hazardous materials releases, communicable diseases, energized electrical equipment, and driving apparatus during emergency responses. The program must also include provisions for dealing with a variety of nonemergency issues, such as alcohol and drug abuse, both on and off duty.

Training and Education

NFPA® 1500 also outlines requirements of a fire department training program. The goal of the program must be to prevent occupational fatalities, injuries, and illnesses. Requirements include initial training for new recruits, methods for becoming proficient in firefighter duties, and a process for evaluating firefighter skills and knowledge. Personnel may not be assigned to fire fighting duties until they have completed the required skills evaluation and training, which must meet the requirements of the NFPA® 1000 series of professional qualification (ProQual) standards. All such training must be thoroughly documented.

Annual proficiency training and evaluations are required to ensure that members maintain their knowledge and skills **(Figure 2.7)**. Training is also required whenever policies, procedures, or guidelines are updated.

Fire Apparatus, Equipment, and Driver/Operators

NFPA® 1500 establishes safety requirements for all uses of fire department apparatus. Design requirements include restraint devices (seat belts) for all personnel driving or riding in the apparatus. Maintenance and inventory records for all equipment must be maintained by the department.

Figure 2.7 Training and evaluation occur regularly to ensure that all personnel are able to perform assigned tasks correctly and efficiently.

The standard also establishes minimum requirements for driver/operators. It requires that they wear a seat belt, obey all traffic signals and regulations, and be thoroughly trained before operating the apparatus in an emergency (**Figure 2.8, p. 58**).

Protective Clothing and Protective Equipment

NFPA® 1500 requires all personnel operating in an IDLH atmosphere or hazardous area to be fully equipped with personal protective equipment, including SCBA. Fire departments should provide all members with at least one full set of personal protective clothing (preferably two) and appropriate protective equipment. Protective clothing must meet the current edition of the NFPA® design standard for the clothing.

Protective equipment refers primarily to self-contained breathing apparatus (SCBA), supplied-air respirators (SARs), and other respiratory protection. It may also refer to body armor for personnel responding to incidents when there is a potential for violence toward responders (**Figure 2.9, p.58**). Personnel who respond to medical emergencies must be provided with protective clothing that protects against blood and airborne pathogens.

Emergency Operations

NFPA® 1500 requires that emergency operations be managed through some form of incident management system (IMS), also known as an incident command system (ICS). The system must include a **risk management plan** and a personnel accountability system.

The standard limits emergency operations to those that can be safely conducted by personnel on scene. Emergency operations must include rapid intervention crews for firefighter rescue, rehabilitation (**rehab**) facilities, and postincident analysis.

Risk Management Plan — Written plan that analyzes the exposure to hazards, implements appropriate risk management techniques, and establishes criteria for monitoring their effectiveness.

Rehabilitation (Rehab) — Allowing firefighters to rest, rehydrate, and recover during an incident.

Figure 2.8 An apparatus driver/operator takes on liability for personnel safety, apparatus functionality, and adherence to applicable laws and regulations.

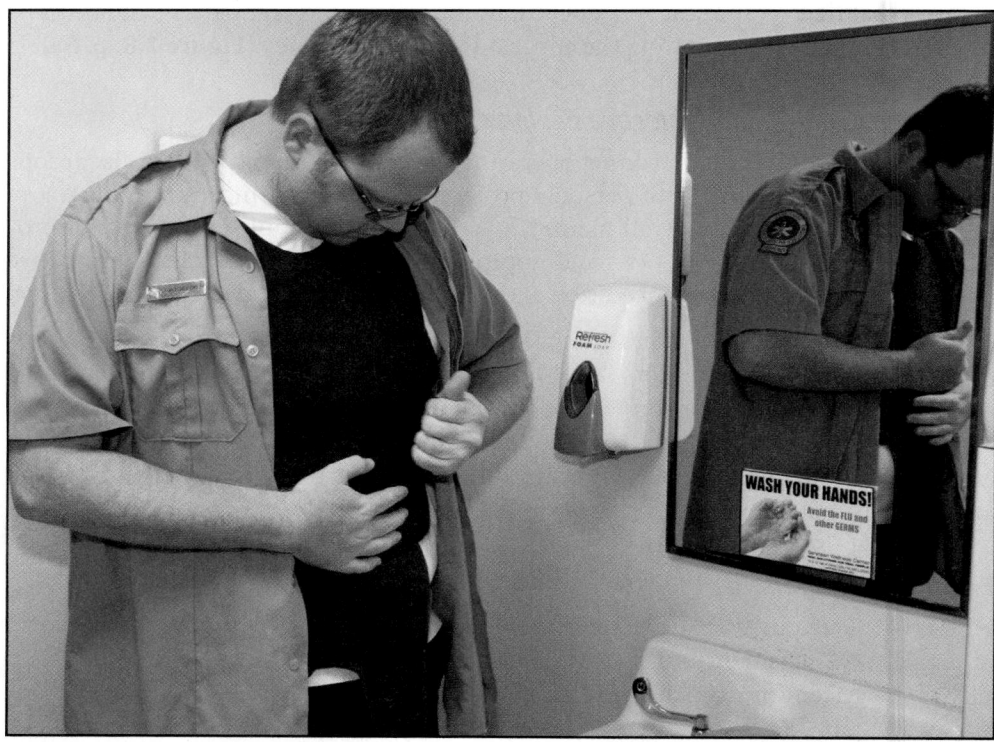

Figure 2.9 Protective equipment may include body armor that protects responders from potentially violent incidents.

Facility Safety

NFPA® 1500 sets minimum design requirements for fire department facilities (fire stations, training centers, administration buildings, and maintenance shops) that meet NFPA® 101, *Life Safety Code*®. Facilities must have a space and means for cleaning, disinfecting, and storing infection control devices. The standard also requires inspection, maintenance, and prompt repairs and prohibits all smoking in the facilities.

Medical and Physical Requirements

NFPA® 1500 requires medical evaluations to ensure that candidates can perform the duties of a firefighter. It also prohibits any firefighter who is under the influence of alcohol or drugs from participating in fire department operations. It requires fire departments to develop physical performance standards for hiring emergency responders, and annual medical exams to verify continued fitness.

The standard also requires the establishment of job-related physical fitness standards and a fitness program that allows members to maintain the required level of fitness **(Figure 2.10)**. A designated physician must be available to advise personnel about health and wellness issues. Departments must document all on-the-job injuries and exposures and operate an infection control program. Medical records for all personnel must be kept in a confidential health database.

Member Assistance and Wellness Programs

Departments must establish an employee assistance program (EAP) to help firefighters and their families with substance abuse, stress, and other personal problems. They must also operate a wellness program to help with health-related issues, such as quitting smoking.

Critical Incident Stress Management

The standard requires that departments establish a critical incident stress management (CISM) program. Its goal is to provide counseling for personnel involved in highly stressful incidents, such as child fatalities, mass casualty incidents, or the death of a fellow firefighter.

Figure 2.10 Physical fitness levels can be measured against a standard.

OSHA Regulations

The Occupational Safety and Health Administration (OSHA) is a division of the U.S. Department of Labor. Its regulations are contained in Title 29, Chapter XVII of the *Code of Federal Regulations (CFR)* and are designed to ensure that workplaces are free from hazards that can cause death or serious injury. They address topics such as hazardous materials, emergency operations, and personal protective equipment.

For firefighters, OSHA regulations apply only to federal and private sector employees. However, twenty-five U.S. states and two territories operate occupational health and safety plans that provide equivalent protection to federal OSHA standards **(Figure 2.11)**. Even in non-OSHA states, many local agencies follow OSHA regulations because they are nationally recognized safety standards.

Canadian firefighters should refer to their provincial safety and health requirements and Canada's National Occupational Safety and Health (CanOSH) regulations. Although most Canadian requirements are based on NFPA® standards, additional Canadian Centre for Occupational Health and Safety (CCOHS) requirements may apply.

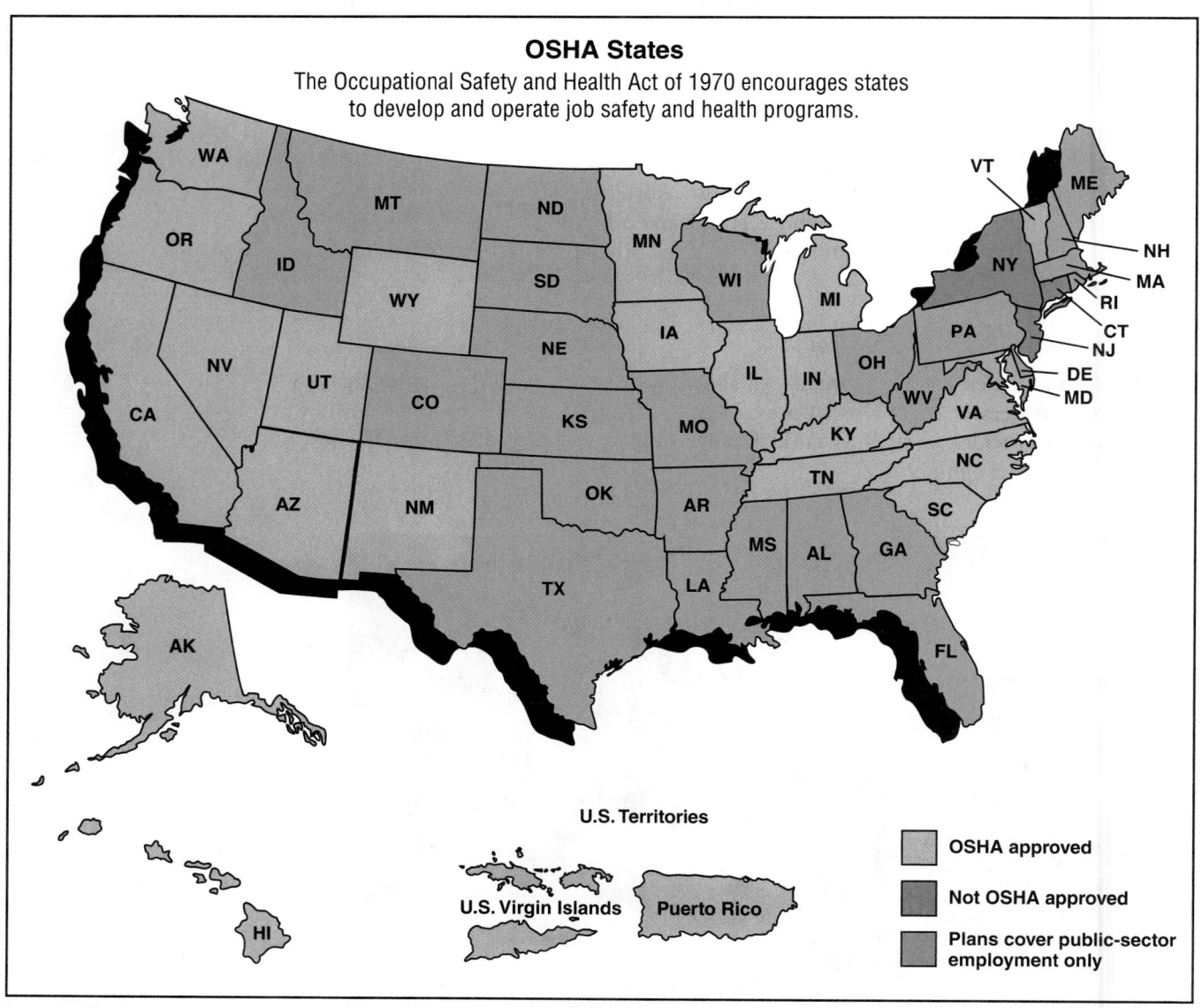

OSHA States

The Occupational Safety and Health Act of 1970 encourages states to develop and operate job safety and health programs.

U.S. Territories

U.S. Virgin Islands   Puerto Rico

OSHA approved

Not OSHA approved

Plans cover public-sector employment only

Figure 2.11 OSHA regulations provide a layer of protection to employees, but they are not equivalent across all states.

All firefighters must comply with the occupational safety and health regulations that apply in their jurisdiction, which become your department's **standard of care**. Understanding these regulations will increase your awareness of workplace safety.

Standard of Care — Level of care that all persons should receive; care that does not meet this standard is considered inadequate.

Everyone Goes Home®

The National Fallen Firefighters Foundation, a nonprofit organization, was created in 1992 by Congress to lead a nationwide effort to honor America's fallen firefighters. One of the ways that the Foundation honors fallen firefighters is by preventing line-of-duty deaths. The Foundation is committed to helping the U.S. Fire Administration meet its stated goal of reducing firefighter fatalities by 25 percent within five years and by 50 percent within ten years. In 2004, the Foundation developed 16 Firefighter Life Safety Initiatives to provide the fire service with a blueprint for making changes **(Figure 2.12, p. 62)**. The Foundation believes that adoption of the life safety initiatives is a vital step in meeting these goals. The 16 Firefighter Life Safety Initiatives, also known as *Everyone Goes Home®*, are:

1. Define and advocate the need for a cultural change relating to safety; incorporating leadership, management, supervision, accountability and personal responsibility.

2. Enhance personal and organizational accountability for health and safety.

3. Focus greater attention on the integration of risk management with incident management at all levels, including strategic, tactical, and planning responsibilities.

4. All firefighters must be empowered to stop unsafe practices.

5. Develop and implement national standards for training, qualifications, and certification (including regular recertification) that are equally applicable to all firefighters based on the duties they are expected to perform.

6. Develop and implement national medical and physical fitness standards that are equally applicable to all firefighters, based on the duties they are expected to perform.

7. Create a national research agenda and data collection system that relates to the initiatives.

8. Utilize available technology wherever it can produce higher levels of health and safety.

9. Thoroughly investigate all firefighter fatalities, injuries, and near misses.

10. Grant programs should support the implementation of safe practices and/or mandate safe practices as an eligibility requirement.

11. National standards for emergency response policies and procedures should be developed and championed.

12. National protocols for response to violent incidents should be developed and championed.

13. Firefighters and their families must have access to counseling and psychological support.

14. Public education must receive more resources and be championed as a critical fire and life safety program.

15. Advocacy must be strengthened for the enforcement of codes and the installation of home fire sprinklers.

16. Safety must be a primary consideration in the design of apparatus and equipment.

Figure 2.12 A culture of safe and responsible best practices will yield positive results for everyone involved.

Figure 2.13 Fire companies are encouraged to schedule routine training days that focus on safety and wellness topics in addition to the more common skills training classes.

IAFC Safety Stand-Down

In response to the continuing level of job-related fatalities, injuries, and illnesses, the International Association of Fire Chiefs (IAFC) initiated a Fire Fighter Safety Stand-Down in 2005. The purpose of the annual day-long stand-down (usually held on each work shift in the third week in June) is to focus on firefighter safety. All non-emergency work ceases and safety training sessions are held. Topics typically include safety-related procedures, policies, and skills (**Figure 2.13**).

Risk Management

NFPA® 1500 requires that fire and emergency service organizations implement a risk management plan. This plan is an established set of decision-making criteria, based on a thorough assessment of benefit and risk.

The Phoenix (AZ) Fire Department (PFD) has implemented an exemplary risk management plan. The plan is intended to help PFD officers make reliable decisions during an emergency response. Its essence is as follows:

- Each emergency response is begun with the assumption that *responders can protect lives and property.*

- Responders will *risk their lives a lot, if necessary, to save savable lives.*

- Responders will *risk their lives a little, and in a calculated manner, to save savable property.*

- Responders will *NOT risk their lives at all to save lives and property that have already been lost.*

Phoenix Model Simplified

- Risk a lot to save a lot.
- Risk a little to save a little.
- Risk nothing to save nothing.

One of the ways that the PFD applies its concept of risk management on a daily basis is through a list of key behaviors that are expected of all personnel. Most behaviors apply to emergency operations, but even nonemergency-related behaviors help to instill a *safety-first* mindset. The full list of these key behaviors is as follows:

- Think.
- Drive defensively.
- Drive slower rather than faster.
- Stop if you can't see at intersections.
- Don't run for a moving rig.
- Always wear your seat belt/safety strap.
- Wear full protective clothing and SCBA.
- Don't *ever* breathe smoke.
- Attack with a sensible level of aggression.
- Always work under sector Command — no freelancing.
- Keep the crew intact.
- Maintain a communications link to Command.
- Always have an escape route (hoseline/lifeline).
- Never go beyond your air supply.
- Use a big enough and long enough hoseline to accomplish the task.
- Evaluate the hazard — know the risk you are taking.
- Follow standard fireground procedures — know and be a part of the plan.
- Vent early and vent often.
- Provide lights for the work area.
- If it's heavy, get help.
- Always watch your fireground position.
- Look and listen for signs of collapse.
- Rehab fatigued companies — assist stressed companies.
- Pay attention at all times.
- Everybody takes care of everybody else.

These behaviors can prevent firefighter casualties. They may also help you survive to enjoy a service retirement instead of a disability retirement.

Fire Department Safety and Health Programs

To comply with NFPA® 1500, fire departments must implement a firefighter safety and health program, whose main goals are to:

- Prevent human suffering, fatalities, injuries, illnesses, and exposures to hazardous atmospheres and contagious diseases.
- Prevent damage to or loss of equipment.
- Reduce accidents and hazardous exposures, and their severity

To be effective, the safety and health program must be promoted and practiced at all levels throughout the organization, from the fire chief to the newest recruit. Anyone who fails to comply with the rules of the safety program sets a negative example for others. It is not enough to talk about and teach safety procedures; they must be practiced and enforced.

Safety and Health Program Components

Any program must include departmental safety and health policies, safety and health training for all personnel, and an accident prevention program. A safety and health committee must be established, to be headed by the health and safety officer (HSO).

Firefighter Health Considerations

In addition to adopting safe behaviors, you must also adopt a healthy lifestyle. In the case history at the beginning of this chapter, the physical stress of emergency operations contributed to the firefighter's line-of-duty death (LODD). But if you exercise and eat a healthy diet, your body will be better able to withstand stress.

In career departments, firefighters are typically required to participate in a physical fitness program. In volunteer departments, firefighters must establish their own regimen to ensure that they are fit for duty.

Figure 2.14 Good nutrition practices have far reaching benefits.

Good nutrition is also important, because eating the right foods in the right amounts will prevent you from becoming overweight **(Figure 2.14)**. As previously discussed, being overweight can lead to cardiovascular disease, diabetes, and musculoskeletal injuries. These conditions affect your overall health and may make you unfit for duty.

To maintain your personal health, follow these general guidelines:

- Stay informed about job-related health issues.
- Wear incident-appropriate personal protective clothing and respiratory protection.
- Clean all PPE at least twice annually and remove heavy contamination after each use.
- Follow recommendations for vaccination against hepatitis B.
- Use precautions to avoid exposure to airborne and bloodborne pathogens.
- Use proper lifting techniques to avoid muscle strains and other related injuries **(Figure 2.15)**.
- Use lifting tools or get help with lifting heavy objects.
- Clean, disinfect, and properly store tools and equipment used in patient care.
- Maintain a regular exercise program.
- Maintain a diet low in cholesterol, fat, and sodium.
- Reduce heart attack and stroke risk by maintaining blood pressure and cholesterol levels within acceptable limits.
- Reduce cancer risk by eliminating the use of all tobacco products.
- Have regular physicals and medical checkups.

For more information on firefighter fitness and health considerations, refer to NFPA® 1500 and the IFSTA **Occupational Safety, Health, and Wellness** manual.

Figure 2.15 The placement of heavy items requires the use of proper lifting techniques to avoid injuries.

Additional Resources

● USFA Emerging Health and Safety Issues in the Volunteer Fire Service

● USFA Health and Wellness Guide for the Volunteer Fire and Emergency Services

● NVFC Heart-Healthy Firefighter Resource Guide

● IAFF Fit to Survive

Employee Assistance and Wellness Programs

An employee assistance program (EAP), sometimes known as a wellness program, provides services to both firefighters and their families. An EAP offers easily accessible, confidential assistance with personal problems that can affect job performance. It provides education, counseling, and referrals to professional services, for concerns such as:

● Alcohol abuse

● Drug abuse

● Personal problems

● Stress

● Depression

● Anxiety

● Marital problems

● Financial problems

A program should also offer counseling and education for firefighters struggling with health-related problems, such as:

- Nutrition
- Hypertension
- Cessation of tobacco use
- Weight control
- Physical conditioning

Apparatus and Vehicle Safety

Motor vehicle accidents, which were the second leading cause of firefighter fatalities in 2010 occur when personnel are responding to and returning from emergency incidents, either in fire department apparatus or in personally owned vehicles (POVs). Accidents may also occur when operating department vehicles for nonemergency tasks.

There are five basic causes for fire apparatus collisions, namely:

- Improper backing of the apparatus
- Excessive speed by the fire apparatus driver/operator
- Lack of driving skill and experience by the fire apparatus driver/operator
- Reckless driving by the public
- Poor apparatus design or maintenance

Always follow posted speed limits and practice proper driving techniques, particularly when backing the apparatus **(Figure 2.16)**. While you cannot control other drivers, you can anticipate what they might do; stay aware of your situation and be prepared to take evasive action. Although you are not responsible for the design of your apparatus, you are responsible for inspecting and reporting maintenance needs in order to keep it in a safe condition.

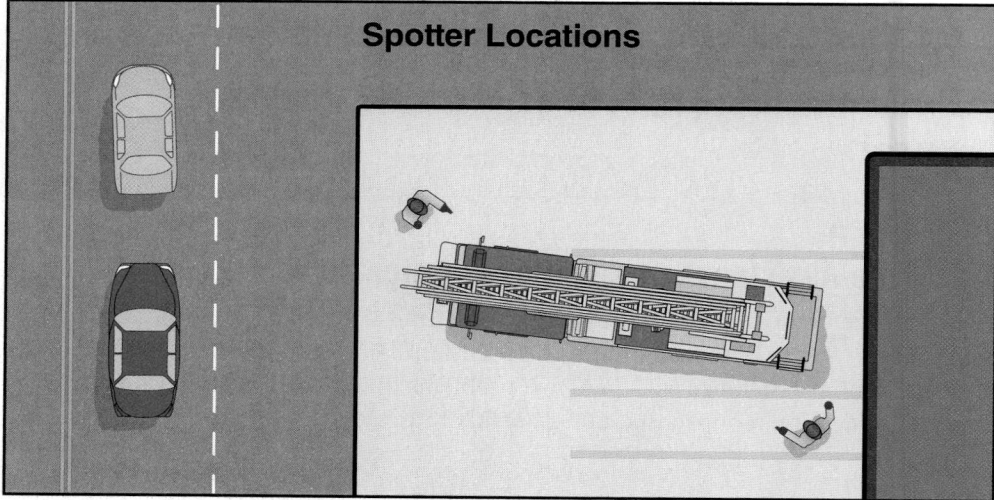

Figure 2.16 Oversized vehicles, such as aerial apparatus, will require assistance from personnel outside of the apparatus to complete maneuvers.

Seat belts must be worn at all times once an apparatus or vehicle is in motion (**Figure 2.17**). If you are the driver/operator, fasten your seat belt. Before moving the vehicle, make sure that all passengers are properly seated with their seat belts fastened.

The use of cell phones is an increasing cause of accidents. Numerous studies indicate that even hands-free cell phone use can be distracting. If you have to make or receive a call when driving, pull to the side of the road or into a parking lot first. But if possible, wait until you reach your destination before making or responding to a call.

Never operate a vehicle for which you are not certified. In most states/provinces, a commercial driver's license (CDL) is not required to operate a fire apparatus, ambulance, or staff vehicle. However, a CDL is always required to operate a fuel truck or passenger bus. Check with your state/provincial motor vehicle department to determine the specific licensing requirements.

Safe Vehicle Operation

Safe vehicle operations begin with **situational awareness**. You must be aware of other drivers and your surroundings, and anticipate potential hazards. You must also be familiar with the state/provincial driving laws and the ordinances of your jurisdiction.

You must also be aware of your own physical, mental, and emotional condition. If you are taking medication or you are ill, your reaction time may be slowed or your vision impaired. In either case, you must not get behind the wheel to operate the vehicle.

Defensive Driving Guidelines

Driving defensively means anticipating potential hazards and developing strategies to deal with them. These skills are especially important when responding to an emergency. Follow these defensive driving guidelines:

- Obey all traffic laws.
- Drive attentively, watching out for other drivers.
- Take extra caution at intersections.
- Be aware of your vehicle's blind spot.
- Stay out of other people's blind spots.
- Keep your eyes moving.
- Anticipate potential hazards by thinking ahead.
- Be prepared to stop or swerve to avoid a collision.
- Communicate your intentions by using turn signals early.
- Always maintain a safe distance from the vehicle in front of you (**Figure 2.18, p. 68**).
- Adapt your speed and vehicle control to existing weather conditions.
- Always wear seat restraints.
- Avoid distractions.
- Stay alert.
- Practice whenever possible to develop your skills.

Figure 2.17 While an apparatus is in motion, seat belts must be worn properly by all people with access to them.

Situational Awareness — Perception of the surrounding environment, and the ability to anticipate future events.

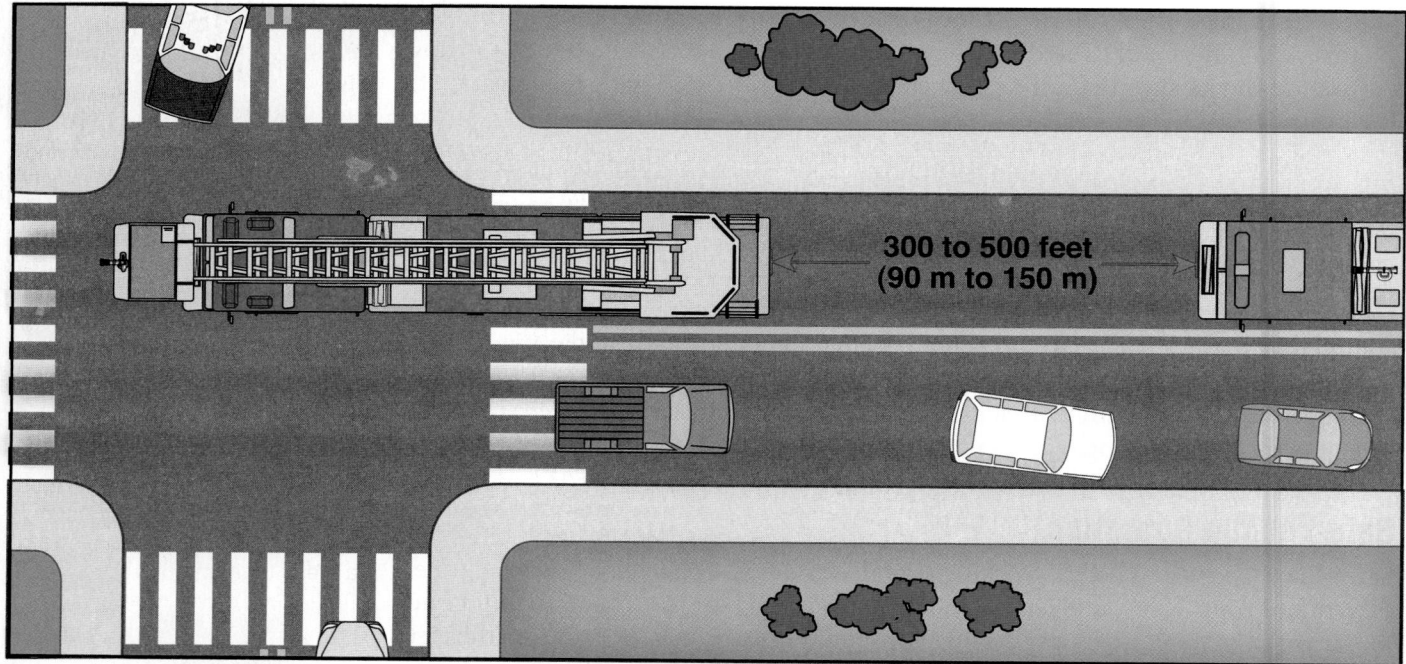

Figure 2.18 Emergency vehicles should maintain a safe traveling distance, particularly when other large vehicles are involved.

Responding in Private Vehicles

State/provincial laws that regulate how volunteers may respond to emergency incidents vary. Some permit volunteers to employ warning lights on their vehicles and respond as though they are emergency apparatus, but others require strict adherence to posted speed limits **(Figure 2.19)**. You must be familiar with your state/provincial laws and adhere to them.

Remember to remain calm when responding. You cannot assist a victim or extinguish a fire if you are involved in a motor vehicle accident. According to the USFA, accidents while responding to/returning from emergencies are the second leading cause of firefighter fatalities. In over 25 percent of these fatal crashes, firefighters were driving a private vehicle.

Fire Apparatus and Department Vehicles

Pumpers, water tenders, and aerial apparatus have unique characteristics. They are larger, heavier, and have a higher center of gravity than most vehicles. This makes them more difficult to handle and means that they require greater stopping distance and may become unbalanced on a curve. In emergency or nonemergency situations, follow these safe driving techniques:

- Slow down. Speed is less important than arriving safely.
- Approach intersections and railroad crossings cautiously.
- Realize that not everyone will see, hear, or react to you.
- Take the safest route to an emergency, avoiding dangerous routes.
- Use extra caution on wet or slick roads.
- Do not grind gears or ride the clutch on standard transmissions.
- Use a low gear when starting from a dead stop.
- Keep all the wheels on the road surface at all times.

Figure 2.19 Volunteer firefighters may be permitted the use of warning lights that will mark their personally owned vehicles as official emergency apparatus. *Courtesy of Mike Mallory and Tulsa (OK) FD.*

Figure 2.20 Bridges may not be reinforced adequately to support the weight of apparatus.

- If the wheels do leave the road surface, slow to 20 mph (32 km/h) or less before bringing them back down.
- Slow down when approaching a curve. Even the posted speed limit may be too fast for a fire apparatus to negotiate the curve safely.
- Take nothing for granted.
- Take routes that avoid school zones and playgrounds.
- Never drive against the flow of traffic unless it is absolutely necessary.

You should be aware of the maximum allowed vehicle weights for bridges in your response area. Some bridges may not be strong enough to support the weight of fire apparatus, especially water tenders and aerial apparatus **(Figure 2.20)**. Bridges and overpasses may also have limited overhead clearance that prevents water tenders and aerial apparatus from passing underneath. You may have to take an alternate route to the incident scene.

Finally, you must know how the motor vehicle laws for civilians will affect their actions toward apparatus on emergency responses. Most state laws require civilians to pull over to the right and stop for emergency vehicles; therefore, you should always pass on the left.

Emergency responses. Before an apparatus starts, the driver/operator and all passengers must be seated, fully belted, and in their protective clothing. Hearing protection must be worn, although wearing fire helmets is only permitted in open cab apparatus.

While en route to the incident scene:

- Make sure that all visible warning devices are on and functioning.
- Leave the station slowly.
- Approach intersections cautiously.
- Come to a complete stop at red lights and stop signs.
- Cautiously move into the opposing lane of traffic at the intersection if all traffic is stopped.

- Use extreme caution at railroad crossings.

- Aim high when steering (watch further down the roadway).

- Always use headlights, even during the daytime.

- Use a variety of audible warning sounds, because other drivers notice a change in sounds.

- Multiple emergency vehicles should travel at least 300 – 500 feet (91 m to 150 m) apart.

- Do not blind oncoming drivers with your lights.

- Turn off wig-wag lights at night.

- Always use your turn signals.

Important principles of braking and stopping include:

- The faster you are moving, the longer it takes to stop.

- *Driver-reaction distance* is the distance the apparatus travels from when the driver/operator realizes the need to stop until the driver/operator's foot touches the brake pedal.

- The *braking distance* is the distance the apparatus travels from when the driver/operator touches the brake pedal until the vehicle is completely stopped.

- *Total stopping distance* is the sum of the driver-reaction distance and the braking distance (**Figure 2.21**).

- Slow down on wet or icy roadways because braking distance is always greater under these conditions.

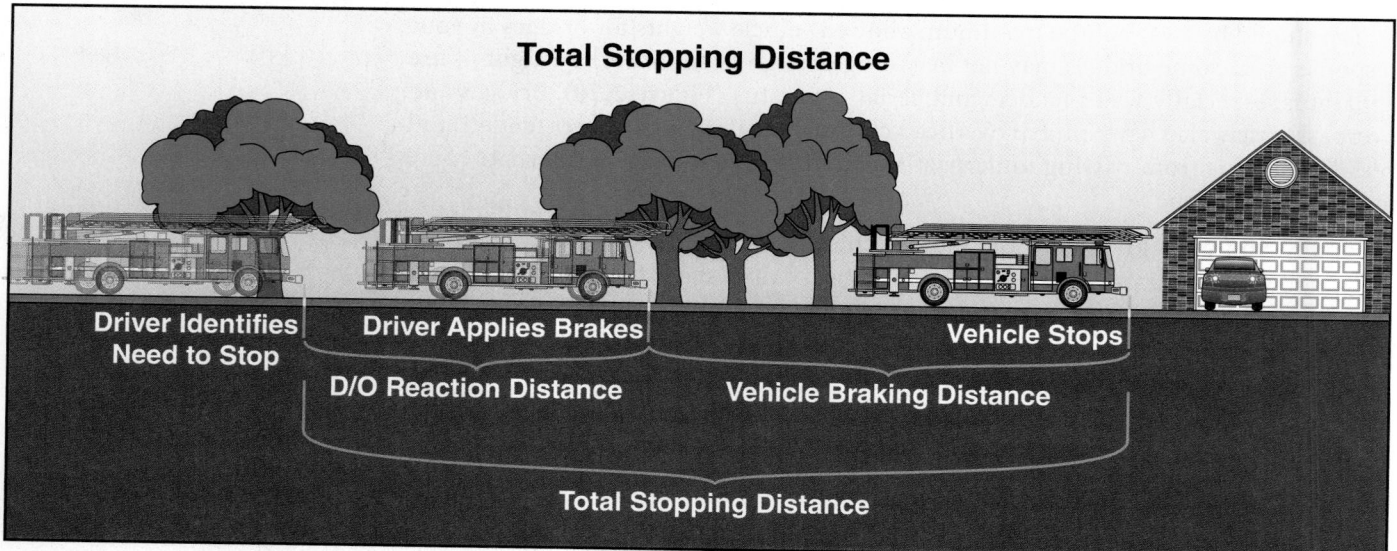

Figure 2.21 The total stopping distance includes the distance the apparatus travels while the driver/operator determines the need to stop the vehicle.

Figure 2.22 Internal baffles allow liquid to travel between compartments in a controlled manner. *Provided courtesy of United Plastics Fabricating, Inc. All rights reserved.*

Types of Water Tenders

In many jurisdictions, firefighters must drive water tenders (mobile water supply) to emergency incidents alone. There are two types of water tenders in operation today: custom-built apparatus and home-built apparatus. Custom-built water tenders have baffled water tanks capable of containing 1,000 gallons (4 000 L) or more of water. Home-built water tenders may be former military fuel tankers that may or may not be properly baffled **(Figure 2.22)**. Internal **baffles** prevent the water from sloshing and shifting the center of gravity of the apparatus.

Baffle — Intermediate partial bulkhead that reduces the surge effect in a partially loaded liquid tank.

If you are operating a water tender, remember that the vehicle is much heavier and more difficult to control. Do not exceed the speed limit for the apparatus or the existing conditions and slow when entering curves or turns. If the wheels leave the pavement, slow down before attempting to pull back on to the roadway.

CAUTION
All fire apparatus handle differently when loaded with water.

Nonemergency operations. Nonemergency operations include returning from incidents and going to the training or maintenance center. Besides the defensive driving skills already mentioned, other considerations include:

- Never allow anyone to ride on the exterior of the apparatus. The exception to this is slow-speed (less than 5 mph [8 km/h]) hose loading operations. The apparatus must be driven forward while doing this.

- Always use spotters when backing the apparatus. Maintain visual contact with the spotters. One spotter may use a portable radio to talk to you **(Figure 2.23)**.

  Spotters should follow these safety guidelines:

- Wear PPE or reflective vests.

- Always remain visible to the driver/operator.

- Never shine a hand light into the mirror.

- Use your department's hand signals for communicating with the driver.

  Driver/operators must also learn to avoid or combat skids:

- Do not drive too fast for conditions.

- Anticipate weight shifts and possible obstacles.

- Pump brakes on old apparatus; let up and pump again if skid begins.

- Newer apparatus equipped with antilock braking systems (ABS) maintain constant pressure on the brake. Be prepared to let up if skidding does occur.

- Do not release the clutch on a standard transmission until the vehicle is under control.

- On slick roads, test the brakes at low speed to see if vehicle is likely to skid.

Safety on the Apparatus

Passengers in an apparatus can be at risk during accidents. Follow your organization's SOPs as well as the requirements listed in NFPA® 1500 for riding in an apparatus. Guidelines for safely riding on, mounting, and dismounting an apparatus include:

- Never stand on or in a moving apparatus. The only exception is while loading hose as the apparatus moves slowly forward (5 mph [8 km/h] or less).

- Always be seated and securely belted in before the apparatus moves.

- Always wear hearing protection or radio headsets.

- Secure all loose tools and equipment.

- Close cab doors securely.

- On unenclosed apparatus, close safety gates or bars securely.

- Use steps and handrails when mounting and dismounting the apparatus. Steps for mounting and dismounting are described in **Skill Sheet 2-I-1**.

Figure 2.23 Apparatus have large blind spots, so the assistance of spotters is necessary during backing maneuvers. *Courtesy of Pat McAuliffe.*

The Seat Belt Pledge

In an effort to reduce responder deaths in motor vehicle accidents, many departments ask their personnel to take the National Fire Service Seat Belt Pledge.

"I pledge to wear my seat belt whenever I am riding in Fire Department vehicles. I further pledge to insure that all my brother and sister firefighters riding with me wear their seat belts. I am making this pledge willingly; to honor Brian Hunton, my brother firefighter, and because wearing seat belts is the right thing to do."

Safety in Fire Stations and Facilities

Unsafe conditions in the fire station endanger both firefighters and visitors, who are the responsibility of the fire department while on department property (**Figure 2.24, p.74**). But most injuries are caused by unsafe behaviors rather than the condition of the facility. The most common facility-related injuries are strains, sprains, broken limbs, abrasions, concussions, and exposure to airborne contaminants. But most injuries can be prevented if you follow appropriate safety procedures.

Figure 2.24 All personnel should take responsibility for the safe working and visiting conditions in a fire station.

Figure 2.25 Good housekeeping practices, complete with warning signs for temporary hazards, serve as preventative measures against injuries.

Back strains, for example, are most commonly caused by improper lifting technique. This type of injury results in the most expensive worker's compensation claims. Improper technique can also cause bruises, fractures, and other types of strains. It can also cause damage if the improperly lifted equipment is dropped. The proper technique involves keeping your back as straight as possible and lifting with your legs. If an object is too heavy or bulky to lift or carry alone, always get help.

Other injuries may result from common accidents such as slips, trips, and falls, which are typically caused by slippery or cluttered surfaces (**Figure 2.25**). One way to prevent such accidents is to stress good housekeeping. Accidents are less likely to occur if floors are clean, aisles are unobstructed, and stairs are well lit.

Cuts and abrasions can result from cooking accidents, cleaning or repairing equipment, and performing facility maintenance. Be careful when using sharp objects, and always wear recommended safety equipment to protect your hands, ears, eyes, and face.

Apparatus exhaust and chemical cleaners are the main sources of airborne contaminants. Avoid exposure by wearing proper respiratory protection, changing the facility's air filters, and using apparatus exhaust capture systems. If you are cleaning contaminated equipment, always wear appropriate protective clothing

In multistory fire stations, firefighters reach the apparatus room quickly by using stairways, slides, and slide poles. Proper cleaning and maintenance will prevent these parts of the station from becoming hazardous. Slide poles are discouraged by the NFPA®, but if you are required to use one, adhere to local safety procedures to prevent skin burns, sprains, or broken ankles.

Firefighters should also avoid rowdy or rough play and other activities that may cause an accident or injury. Thoughtless or careless behavior has no place in a fire department facility. Think before you act.

Figure 2.26 An imperfection in any part of a tool can create a dangerous situation for anyone using the tool or near the area where the tool is being used.

Safety with Tools and Equipment

NFPA® 1500 establishes safety requirements for all uses of tools and equipment. Improperly used or maintained tools and equipment create a serious hazard, whether in the fire station, during training, or at an emergency scene (**Figure 2.26**).

Always use appropriate PPE, such as gloves, vision protection, and hearing protection. Using PPE is fundamental for safe work practices. Although PPE does not take the place of good tool engineering, design, and use, it does provide some protection against hazards.

Firefighters using lawn care equipment around the fire station must follow standard safety procedures for power tools and wear standard protective equipment. This typically consists of gloves, safety boots, eye protection, and hearing protection.

Unpowered Tools

Unpowered tools (also known as *hand tools*) are the most common tool type. Most are designed to be carried and used by one person although some require two people to operate. Examples of unpowered tools include:

- Pike poles
- Sledgehammers
- Flat-head and pick-head axes
- Claw hammers and roofing hammers
- Picks
- Shovels
- Bolt and wire cutters
- Pry bars
- Halligans
- Battering rams

Power Tools

Power tools, such as drills and saws, can cause serious, life-threatening injuries if used improperly. Do not operate a power tool unless you have read the manufacturer's instructions and fully understand them. Only trained, authorized personnel should repair power tools. A repair log must document all repairs and the name of the person performing the repairs (**Figure 2.27**).

Figure 2.27 A repair log is an essential resource for maintaining the good condition of equipment.

Any electrical tool not marked "double insulated" should have a three-prong plug. For firefighter safety, the third prong must connect to an electrical ground while the tool is in use. Never bypass the ground plug or remove the third prong. Without these safety features, an electrical short can cause death or serious injury.

These safety guidelines apply to both hand tools and power tools:

- Wear appropriate PPE.

- Remove loose clothing and keep long hair clear of operating tool heads.

- Remove jewelry, including rings and watches.

- Select the appropriate tool for the job.

- Follow manufacturer's instructions.

- Inspect tools before use to determine their condition.

- Do not use badly worn or broken tools.

- Provide adequate storage space for tools and always return them promptly to storage after use.

- Inspect, clean, and put all tools in a ready state before storing.

- Get the approval of the manufacturer before modifying any tool.

- Use intrinsically safe tools when working in potentially flammable atmospheres, such as around a vehicle's fuel system.

- Do not remove safety shields or compromise built-in safety devices.

Power Saws

Most power saw accidents can be prevented by following these safety rules:
- Choose the right type of saw for the task (**Figure 2.28**).

- Never force a saw beyond its design limitations.

- Wear proper protective equipment, including gloves and vision and hearing protection.

- Remove loose clothing and contain long hair that could become entangled in the saw.

- Have hoselines in place when cutting materials that generate sparks.

- Never use a power saw in a potentially flammable atmosphere.

- Keep bystanders out of the work area.

- Follow manufacturer's procedures for proper saw operation.

- Allow gasoline-powered saws to cool before refueling.

- Keep blades and cutting chains well sharpened.

- Use extreme caution when operating any saw above eye level.

Safety in Training

According to an NFPA® study, one hundred firefighters died in training-related incidents between 1996 and 2005. While it is important to train in conditions that are close to actual emergencies, it is unacceptable for training to result in fatalities or injuries. You must follow the orders of your superiors and training officers, adhere to safety policies and procedures, and be aware of your own physical limitations.

Figure 2.29 Training is most efficient when personnel practice using all of the tools that will be necessary and available at a fire scene.

Figure 2.28 Each type of power saw has unique features that must be matched to the task at hand.

Maintaining Personnel Safety

Maintaining personnel safety during **training evolutions** includes using the appropriate PPE, applying situational awareness, being healthy and in good physical condition, and adhering to all safety regulations.

All personnel participating in training should be fully clothed in the appropriate protective gear. PPE must be used for raising ladders, deploying hose, operating extrication tools, or performing any other activity that simulates actual emergency scene conditions (**Figure 2.29**).

Safety in training is also influenced by your physical condition. Physical discomfort or illness can make you less alert, and therefore more prone to injury. If you have flu symptoms, colds, persistent headaches, or other symptoms indicating severe physical discomfort or illness, you should consult a physician for evaluation before participating.

Rowdy play or other unprofessional conduct is not allowed because it is distracting and can lead to accidents and injuries. You must remain focused on the training exercise at all times.

Training Evolution — Operation of fire and emergency services training covering one or several aspects of fire fighting. *Also known as* Practical Training Evolution.

Figure 2.30a A specially designed burn building provides a controllable environment for smoke and live fire evolutions, and includes the benefit of flexible scheduling for training exercises.

Figure 2.30b An acquired structure provides responders a realistic environment for live fire evolutions.

Live Fire Training

Realistic training is important because how you train determines how you will perform in the field. However, realism sometimes increases the potential for accidents and injuries. Some departments conduct live fire exercises using specially designed burn buildings, gas-fired props, and exterior props at training centers. Others conduct such training in structures acquired for that purpose **(Figures 2.30a and b)**. To ensure that these exercises are safe, they must be conducted according to the requirements of NFPA® 1403, *Standard on Live Fire Training Evolutions*.

Maintain your situational awareness at all times during live fire training. Apply your knowledge of fire behavior, listen to your instructors, and remain with your team or partner. Above all, remain calm and take slow, regular breaths. Breathing rapidly will use up your air supply quickly and reduce the time you have to work in an IDLH atmosphere. Be aware of your physical condition. If you feel that you are getting stressed or overheated, tell your supervisor. Take advantage of the rehabilitation breaks that are provided, and stay hydrated by drinking fluids.

Maintaining and Servicing Equipment

Equipment used for training evolutions often wears out sooner than equipment used less frequently on emergency calls. You must inspect and clean the tools, equipment, and PPE that is assigned to you for training. If an item needs to be repaired or replaced, report it to your supervisor. Clean your PPE thoroughly when it becomes contaminated with smoke or soot. Disinfect your SCBA facepiece, CPR manikins, and medical equipment after each training session. Remember that the way you take care of training equipment will be the way you maintain your equipment when on duty.

Emergency Operations Safety

Protecting your community is your primary mission. You will have to perform tasks during emergency operations that will place you in hazardous, high-risk situations. You will have to accomplish these tasks as safely as possible without injury to yourself. The following sections contain methods to help you safely meet your mission.

National Incident Management System

An Incident Command System (ICS) is an organizational model that applies to any type of emergency. It can be used at a single-unit incident that lasts only a few minutes, or at a complex, week-long incident involving multiple agencies and mutual aid units.

Most ICS models are based on the National Incident Management System–ICS (NIMS-ICS). NIMS-ICS is the standard model for Incident Command in the U.S. It provides a framework that enables agencies from different jurisdictions to work together effectively.

At any incident, you must work within the ICS, under the overall command of the Incident Commander (IC) and direct command of your supervisor. The IC assesses risk, determines an appropriate strategy, and assigns specific tactical objectives. As part of the ICS, you will:

- Report to only one supervisor.
- Operate as part of a team.
- Carry out your orders.
- Remain constantly aware of your situation.
- Advise your supervisor of any changes in the situation.
- Manage your air supply.
- Enter and withdraw from the hazardous area as a team.

All training must be conducted using the ICS model adopted by your department. This will help you become familiar with the structure and functions of the ICS.

Preparedness

Preparedness is an important aspect of firefighter safety. If you are a volunteer firefighter, perform regular inspections to make sure equipment is ready for use **(Figure 2.31)**. If you are a career firefighter, report for duty in uniform, well-rested, and mentally alert. Make sure that your tools, equipment, and PPE are in their proper location and ready for use. Check your SCBA to make sure it is fully functional and the air cylinder is full. For EMS equipment, restock supplies, replace out-of-date materials, and update accountability equipment.

Emergency Scene Safety

Although responding to any emergency involves risk, you can minimize your exposure to risk by following these operational guidelines:

- Follow your supervisor's orders.
- Wear appropriate PPE.
- Work as a team.
- Follow departmental SOPs.
- Maintain communications with team members and Command.
- Do a risk/benefit analysis for every action.

Figure 2.31 Volunteer firefighters are responsible for the good condition of their own equipment.

- Employ safe and effective tactics.

- Never operate alone or without supervision.

- Perform an initial assessment and maintain situational awareness.

Situational awareness is critical at the scene of any incident. Your safety depends on being able to understand what is going on around you and recognize potential threats. All firefighters must train themselves to develop this vital skill.

Structural fire fighting is particularly hazardous. In addition to maintaining situational awareness, you must follow orders, follow departmental SOPs, and apply these basic techniques:

- Scan the outside of the building before entry to locate windows and doors that could be used as escape routes.

- Wear full PPE, including SCBA.

- Manage your air supply.

- Bring appropriate tools and equipment.

- Stay in physical, vocal, or visual contact with other members of your team.

- Maintain radio contact with Command or with others outside the building.

Safety at Roadway Incidents

At motor vehicle fires and accidents, both firefighters and victims are in danger of being struck by traffic. To ensure your safety you must be visible, work within a protected area, and exercise situational awareness.

Your first form of protection is to be visible to the drivers of passing vehicles. U.S. Department of Transportation (DOT) regulations require all personnel at roadway incidents to wear high-visibility vests. The vests must have reflective trim and five-point breakaway fasteners at the shoulders, side, and waist to meet NFPA® safety standards. Reflective trim on structural PPE does not provide enough visibility to meet these standards (**Figure 2.32**). But if you are also performing fire fighting or hazardous materials activities at the scene, you cannot wear the high-visibility vest over your PPE. Don the vest only after you have completed these activities.

Apparatus lights and scene lighting also contribute to visibility. But they can make the situation more hazardous if they are improperly used. For example, during the early stages of an incident, emergency vehicle warning lights help ensure the safety of responders, victims, and drivers approaching the scene. But these lights do not provide effective traffic control and are often confusing to civilian motorists, especially at night. Some departmental SOPs limit their use and require responders to:

- Turn off all forward facing lights, including headlights

- Minimize flashing lights on the vehicle's sides and rear

- Turn off lights that face approaching traffic, to avoid blinding or distracting drivers

- Turn off all headlights, unless they are being used to illuminate the work area or warn motorists that the vehicle is in an unexpected location

Floodlights must also be used with caution. If used improperly, they can easily blind nearby motorists. Always raise them to a height that allows light to be directed toward the scene, but not into the eyes of approaching drivers (**Figure 2.33**).

To establish a protected work area, emergency vehicles form a protective barrier between oncoming traffic and working personnel. Apparatus should be parked at an angle, with the front wheels turned away from the scene so that the vehicle will not

Figure 2.33 Floodlights should be used carefully to not create a traffic hazard while providing lighting to a work scene.

Figure 2.32 The addition of a reflective vest is required at roadway incidents to increase the visibility of first responders.

strike working personnel if it is hit from behind (**Figure 2.34, p. 82**). To protect the pump operator, position the apparatus with the pump panel on the protected side away from traffic.

Additional apparatus can also form a second barrier 150 to 200 feet [45 to 60 meters] behind the first. Any apparatus not used for this purpose must be moved to the shoulder of the roadway. If the accident occurred at an intersection, multiple apparatus are required to protect all sides of the work area.

Signs and traffic cones are also used to detour traffic around the emergency scene as it approaches. Deploy them to close off at least one lane of traffic next to the scene. Request traffic control assistance from local law enforcement or fire police.

General guidelines for maintaining situational awareness at a roadway incident are:

- Look before you move.
- Keep an eye on moving traffic.
- Walk facing oncoming traffic.
- Follow departmental SOPs.

Exiting your vehicle can be particularly hazardous, so always use extreme caution. Whenever possible, both drivers and passengers should mount and dismount on the side of the vehicle that is not exposed to oncoming traffic. If you must dismount on the exposed side, watch for oncoming vehicles before opening your door, and wait for a break in traffic before exiting.

Figure 2.34 Apparatus may park at an angle to protect rescuers and victims at a motor vehicle accident.

Scene Management

The goal of scene management is to create a work area in which emergency responders can perform their duties safely and effectively. This is accomplished by controlling access to the scene, through **crowd control** and the establishment of control zones.

Crowd Control

Law enforcement is usually responsible for controlling bystanders, but in some cases firefighters must perform this task themselves. Left uncontrolled, spectators can wander through the scene and interfere with working personnel. They may also injure themselves, requiring medical attention from responders who should be focused on the original emergency.

Anyone can become highly emotional at an emergency scene, but friends or relatives of a victim can be especially difficult to deal with. Treat them with sensitivity and understanding, and gently but firmly restrain them from getting too close. Allow them within the cordoned area, but not close enough to interfere with responders. A firefighter or other responsible person should also stay with them until the victim is removed from the scene.

Control Zones

Establishing control zones is typically the best way to secure the scene. The area is cordoned off with rope, fireline, or caution tape; these can be tied to any stationary object except a vehicle, because vehicles may have to be moved as the incident progresses **(Figure 2.35)**. During incidents on streets and highways, scene management may also be critical to preventing firefighter injuries or fatalities. Procedures for scene management at roadway incidents are illustrated in **Skill Sheet 2-I-2**.

Control zones are commonly labeled hot, warm, and cold, with the cold zone being the most distant from the incident **(Figure 2.36, p. 84)**. The outer boundary of the cold zone does not have to be any specific distance away from the scene, but it must be far enough to ensure that bystanders are not in danger. Other zone boundaries should take into account how much room personnel need to work and the degree of hazard, as well as wind, weather, and topography.

Figure 2.35 Control zones must be clearly marked and enforced to keep all responders as safe as possible.

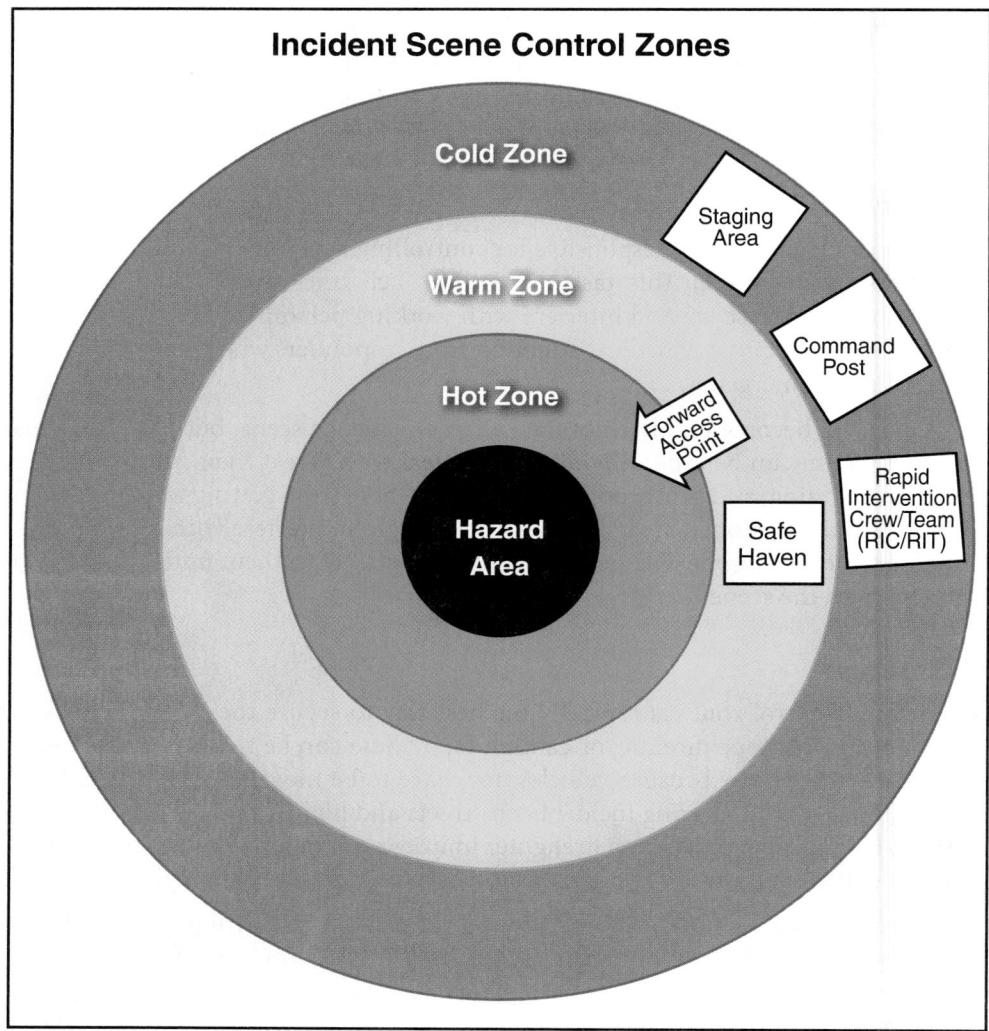

Incident Scene Control Zones

Cold Zone

Warm Zone

Hot Zone

Hazard Area

Staging Area

Command Post

Forward Access Point

Rapid Intervention Crew/Team (RIC/RIT)

Safe Haven

Figure 2.36 Control zones are established in a concentric order with the most dangerous areas in the central area, staging around that, and crowd control beyond the outside perimeter.

Collapse Zone — Area beneath a wall in which the wall is likely to land if it loses structural integrity.

Another type of control zone is the collapse zone, an area that must be kept clear in case a wall or other piece of the structure collapses into it. **Collapse zones** are established under any of the following conditions:

- Prolonged heat or fire have weakened the structure
- A defensive strategy has been adopted
- Interior operations cannot be justified

Most structural collapses involve brick or masonry block, either from structural components or veneer. Church steeples, water tanks, chimneys, and false facades or overhangs extending above the top of the structure must be viewed as potential collapse hazards, even if the structure itself is not. Be aware that structural collapses can occur well after the fire is extinguished. Once a collapse zone has been established, it must be strictly enforced and observed by all personnel.

A collapse zone must be established adjacent to any exposed exterior wall. Apparatus and personnel operating master stream appliances must not be positioned in the collapse zone. The size of the collapse zone depends on other exposures, the type of building construction, and the safest location for apparatus and personnel (**Figure 2.37**). Traditionally, collapse zones must extend a distance of 1½ times the height of the structure. A 30-foot (10 m) tall structure would therefore require a collapse zone at least 45 feet (15 m) from its base. However this rule of thumb may not be practical for taller structures.

Because the collapse zone extends the full length of all of the affected walls, the safest location for defensive operations is at the corner of the building. Master streams and apparatus can be located in the area formed by a 90-degree arc from the wall intersection as long as they are far enough away that flying debris will not strike them.

Personnel Accountability

Accountability systems are designed to track personnel, both in and out of the hot zone. All departments use some type of accountability system at every incident. All personnel should be trained in its use, and it should be part of all training exercises (**Figure 2.38**).

An accountability system can save your life if your SCBA malfunctions, you become lost or trapped, or there is an unexpected change in fire behavior. If the IC does not know the location of personnel involved in the operation, it is impossible to determine whether someone is trapped inside. Firefighters have died because they were not known to be missing until it was too late.

There are a variety of accountability systems available. They include passport systems, SCBA tag systems, and computer-based electronic systems. However, the systems are only effective if they are properly implemented by the IC and used by the personnel at the incident.

Figure 2.37 A collapse zone must be established around any fire damaged structure. *Courtesy of Ron Moore/ McKinney (TX) Fire Department.*

Figure 2.38 Accountability systems provide a mechanism for ensuring the safe return of responders from a fire scene.

Passport System

In a passport system, sometimes called a *tag* system, company officers have a passport listing for every member of their crew. Before entering the hot zone, firefighters give their passports/tags to their supervisor or a designated Accountability Officer (AO). Passports are then attached to a control board or personnel identification chart. Firefighters then collect their passports after leaving the hot zone. This system allows Command to know which companies and which personnel are operating within the hot zone.

SCBA Tag System

A tag attached to individual SCBAs provides more accountability data and greater safety. Before entering the hot zone, firefighters give their tags to an AO who records time of entry and the expected time of exit. Exit time is based on the air pressure in the lowest-reading SCBA in the team. The AO also does a brief check to ensure that all protective equipment is on and functional. Firefighters leaving the hot zone take back their tags so that the AO knows who is still inside the hot zone. On extended operations, relief crews are sent in the hot zone before interior crews run low on air.

Computer-Based Electronic Accountability Systems

Computer-based electronic accountability systems provide an even greater level of safety. These rapidly deployable systems use radio-based tracking or radar-based transmitters attached to PPE. Most systems sound an alarm if a firefighter becomes immobile or calls for assistance. They can also sound a MAYDAY or evacuation alarm and verify that it has been received by the IC or other firefighters.

Figure 2.39 A digital personnel tracking system offers responders another layer of safety in dangerous environments.

Some SCBA manufacturers have developed units with digital accountability features. The SCBA is stored in a riding position on the apparatus. At the beginning of the work shift, personnel "log-in" to their assigned SCBA (**Figure 2.39**). When they arrive at an incident, their position and air supply automatically register on the IC's tracking software within the command unit.

Electronic systems should never fully replace manual accountability systems, such as passport and SCBA tag systems. They should only be used as a supplemental safety measure.

Chapter Summary

Firefighter safety is essential to your mission of protecting the public. Protect yourself during emergencies by relying on your situational awareness, wearing required PPE, following orders, and following your department's safety policies and procedures. Above all, take responsibility for your own safety. You cannot protect the community unless you first protect yourself.

Firefighter I

1. What types of job-related injuries and illnesses can a firefighter expect to encounter?

2. What topics does NFPA® 1500 cover regarding firefighter safety and health?

3. Federal OSHA regulations apply to what specific groups of firefighters?

4. How do the key behaviors of driving defensively, keeping the crew intact, and following standard fireground procedures support the concept of risk management?

5. What are the main goals of a safety and health program?

6. What areas can an Employee Assistance and Wellness program assist with?

7. What key defensive driving skills can help promote safe vehicle operations?

8. What are the guidelines for safely riding on, mounting, and dismounting an apparatus?

9. How can firefighters prevent most back and leg strains related to injuries at fire stations and facilities?

10. What are two general guidelines to follow that can improve tool and equipment safety?

11. What three steps can you take to maintain personnel safety during training?

12. What fundamental rules can help minimize risks at an emergency scene?

13. What protective equipment is available to increase your safety at emergency scenes?

14. What types of control zones may be used to establish scene security?

15. Why is proper use of your personnel accountability system so important?

SKILL SHEETS

2-1-1
Respond to an incident, correctly mounting and dismounting an apparatus.

Step 1: Don appropriate personal protective equipment (PPE).

Step 2: Mount apparatus using handrails and steps per local procedures.

2-I-1
Respond to an incident, correctly mounting and dismounting an apparatus.

SKILL SHEETS

Step 3: Sit in a seat within the cab and fasten safety belt. Follow all local safety regulations.

Step 4: Remain seated with safety belt fastened while vehicle is in motion.

Step 6: Dismount apparatus using handrails and steps per local procedures.

Step 5: When vehicle comes to a complete stop, unfasten safety belt and prepare to dismount.

SKILL SHEETS

Wearing appropriate PPE, including reflective vest, demonstrate scene management at roadway incidents using traffic and scene control devices.

2-1-2

Step 1: Don appropriate personal protective equipment (PPE), including reflective vest.

Step 2: Set up traffic cones and scene control devices appropriate to the assignment following local procedures.

Step 3: Set up established work areas.

2-1-2

Wearing appropriate PPE, including reflective vest, demonstrate scene management at roadway incidents using traffic and scene control devices.

SKILL SHEETS

Step 4: Perform tasks as directed to complete the assignment.

Step 5: Follow local procedures for highway incidents.

Step 6: Remove traffic cones and scene control devices.

Chapter Contents

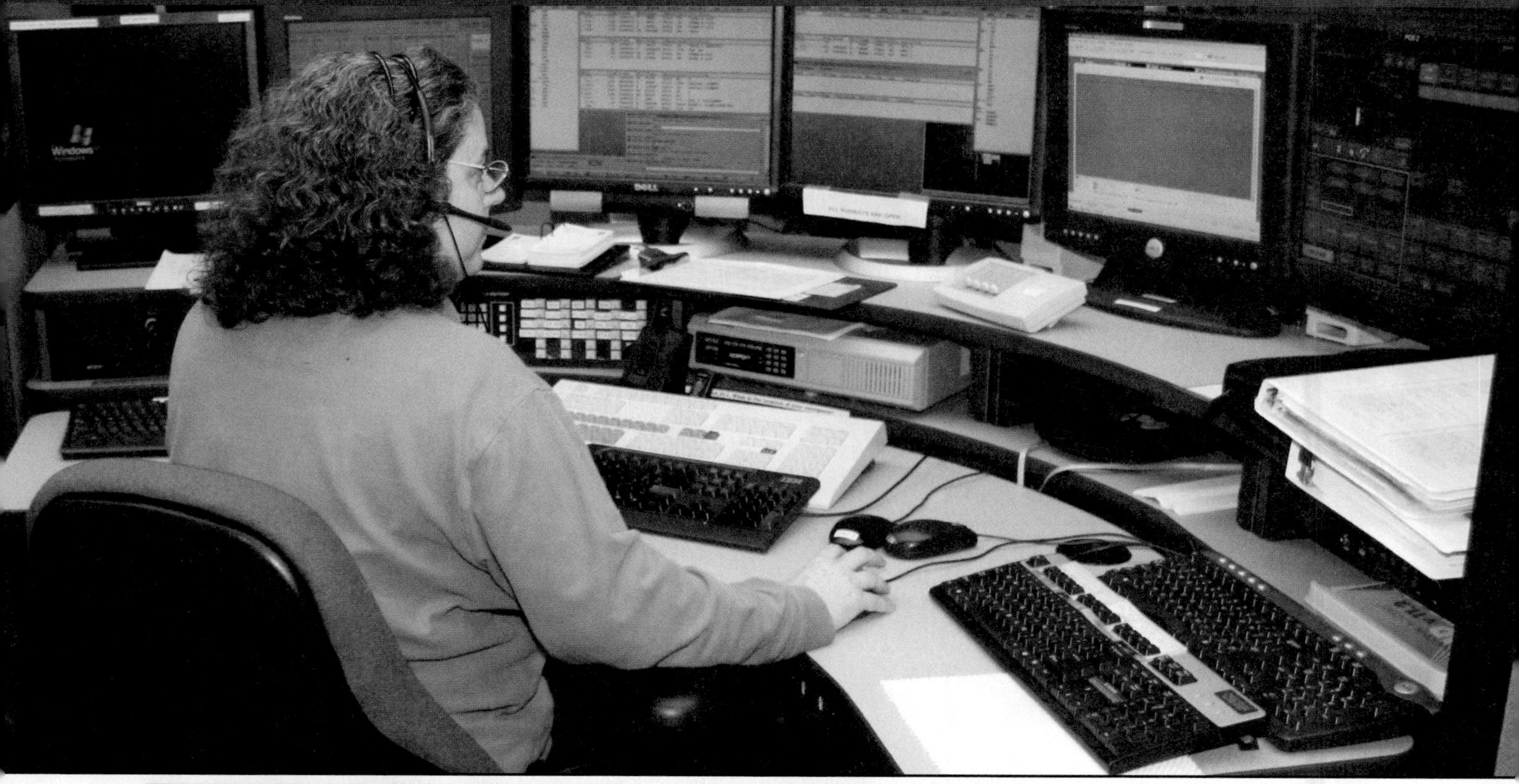

Key Terms

NFPA® Job Performance Requirements

This chapter provides information that addresses the following job performance requirements of NFPA® 1001, *Standard for Fire Fighter Professional Qualifications* (2013).

Firefighter I

5.2.1

5.2.2

5.2.3

Firefighter II

6.2.1

6.2.2

Firefighter I Chapter Objectives

1. Explain the procedures for receiving emergency and nonemergency external communications. (5.2.1, 5.2.2)

2. Describe the information required to dispatch emergency services. (5.2.1, 5.2.2, 5.2.3)

3. Describe the systems used for internal communications. (5.2.1, 5.2.2)

4. Explain radio limitations that may impact internal communications. (5.2.3)

5. Describe radio procedures used for internal communications. (5.2.1, 5.2.2, 5.2.3)

6. Handle emergency and nonemergency calls. (Skill Sheet 3-I-1, 5.2.1, 5.2.2)

7. Use a portable radio for routine and emergency traffic. (Skill Sheet 3-I-2, 5.2.1, 5.2.3)

Firefighter II Chapter Objectives

1. Discuss the aspects that make up on-scene communications. (6.2.2)

2. Explain the information gathered by post incident reports. (6.2.1)

3. Create an incident report. (Skill Sheet 3-II-I, 6.2.1)

Chapter 3
Fire Department Communications

Case History

In July 2007, a fire officer and a driver/operator died during a fire in a 1950s single-story wood–frame house in Contra Costa County, California. The firefighters were members of the first unit to arrive at the incident and were attempting to locate and rescue two elderly residents reported to be in the structure. Upon arrival, the officer established Command, then passed Command to what he thought was the next arriving unit, and advanced an attack hoseline into the front of the structure. After controlling the fire in the front portion of the house, the two firefighters proceeded to make a search of the bedrooms, one of which was later determined to be the source of the fire. Other firefighters searched the kitchen area, locating and removing one of the victims. Rapid fire development caused by the improper use of positive-pressure ventilation and the ignition of fire gases occurred at this point, trapping and killing the two firefighters in the bedroom area. The second victim was later located in the kitchen area.

Investigators of Contra Costa County Fire Protection District (CCCFPD) and the National Institute of Occupational Safety and Health (NIOSH) determined that there were multiple factors that contributed to these line-of-duty deaths (LODDs). However, the reports did determine that poor communications played a major role. The CCCFPD report stated, "Clear and concise radio communications combined with good radio discipline are key elements of safe and efficient emergency operations." Among the radio communications problems, the following were noted:

- The private alarm company that received the initial alarm delayed notifying the fire dispatch center.

- The dispatch center, which was experiencing a high volume of radio traffic, delayed the dispatch of additional units.

- The outgoing officer did not receive an acknowledgment when passing Command. This resulted in a lack of formal Incident Command and a delay in determining that the firefighters were missing when fire conditions rapidly changed.

- Without an established and operating Incident Command System, information was not relayed to units on scene in a timely fashion and strategic decisions were delayed.

Fire department communications can be divided between external and internal communications. External communications are requests from the public for both emergency and nonemergency assistance. These requests may be received at a central point and then relayed to the appropriate fire department unit or individual. Internal communications consist primarily of radio transmissions between units and individuals during emergency operations. Your effectiveness as a firefighter will depend on your knowledge of your local communications system and how it operates. Your personal safety will depend on your ability to operate the radio that is assigned to you during emergency incidents.

☐ External Communications

In the majority of the United States and Canada, the public can dial 9-1-1 for help in an emergency. In some communities and rural areas, individuals must dial a conventional seven- or ten-digit telephone number to report an emergency. In either case, the public connects to a telecommunications center that gathers the necessary information and dispatches the appropriate emergency response.

Nonemergency calls are handled on dedicated conventional or business telephone numbers outside of the 9-1-1 system. Nonemergency calls are generally routed to the division of the fire department that can best handle them, such as the fire prevention division. Sometimes, however, emergency calls may be received on nonemergency telephone lines or directly by individuals walking into the fire station. You must know your local protocol for handling emergency calls received on nonemergency lines or from walk-in reports.

Receiving Emergency Calls

The system for receiving emergency calls varies among communities, states/provinces, and regions regardless of the telephone number that is used. There are two broad categories of telecommunications systems: **(Figure 3.1)**:

- **Emergency Service Specific Telecommunications Center** — Separate telecommunications or dispatch centers that the fire department, emergency medical service, or law enforcement agency operates.

- **Public Safety Answering Point (PSAP)** — Central location that takes all emergency calls and routes the call to the fire, emergency medical, or law enforcement dispatcher.

Regardless of which category of communications center an area uses, certain equipment and procedures should exist to properly manage emergency calls. The sections that follow discuss these procedures and equipment.

Communications Center Equipment

Depending on local requirements and capabilities, telecommunications centers contain a variety of equipment required for the handling of emergency calls. Some of the more common pieces of communication equipment include the following:
- Two-way radio system for communicating with mobile and portable radios at the emergency scene as well as base station (fixed locations) radios in fire stations or other department facilities
- Telecommunications Device for the Deaf (TDD), Teletype (TTY), and Text phone for receiving calls from hearing-impaired individuals **(Figure 3.2, p. 98)**

Public Safety Answering Point (PSAP) — Any location or facility at which 9-1-1 calls are answered either by direct calling, rerouting, or diversion.

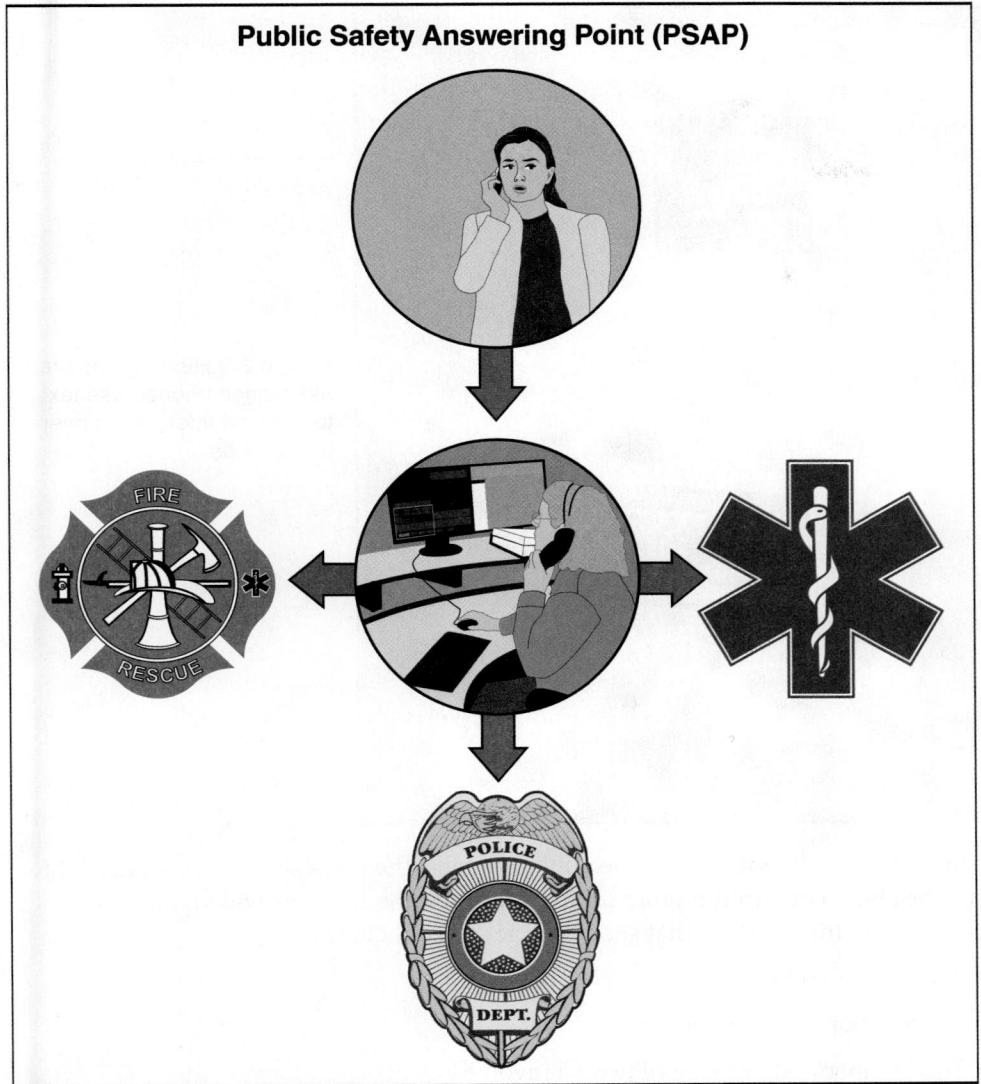

Public Safety Answering Point (PSAP)

Figure 3.1 A Public Safety Answering Point serves as a contact between a person expressing a problem and an organization with a plan of action.

- Tone-generating equipment for dispatching resources
- Telephones for receiving both emergency and nonemergency calls
- Direct-line telephones for communications with fire department facilities, hospitals, utilities, and other response agencies
- Computers for dispatch information and communications
- Recording systems or devices to record telephone calls and radio transmissions
- Alarm-receiving equipment for municipal alarm box systems and private fire alarm reporting systems

Processing Emergency Calls

Emergency calls must be handled quickly to ensure the safety of the community. Minimum requirements for receiving, processing, and dispatching emergency responders are included in NFPA® 1221, *Standard for the Installation, Maintenance, and Use of Emergency Services Communications Systems*. Telecommunicators, also referred to as dispatchers, are trained to obtain the correct information quickly and

Figure 3.2 Hearing-impaired assistance phones use text to transmit information over phone lines.

accurately. In situations where the public contacts the fire station directly, firefighters must be able to obtain the same information that the caller provided. Based on local protocol, the information that should be gathered includes:

- The type of emergency
- The location of the emergency
- The number and location of people involved
- The name and location of the caller
- The caller's callback number
- The cross street, building name, neighborhood, area of city/county, or any nearby landmarks

Over 96 percent of the U.S. has access to Enhanced 9-1-1 (E-9-1-1) systems through landline telephone connections. These systems combine telephone and computer equipment including computer-aided dispatch (CAD) to provide the dispatcher with instant information such as the caller's location and phone number, directions to the location, and other information about the address. When the dispatcher answers the telephone, the computer displays the street address that is associated with the telephone number the call is being made from using an **Automatic Location Identification (ALI)**, which relies on **Global Positioning System (GPS)** data.

Public Alerting Systems

Public alerting systems are those systems that anyone can use to report an emergency. Besides telephones, these systems include two-way radios, wired telegraph circuit boxes, telephone fire alarm boxes, and radio alarm boxes.

Automatic Location Identification (ALI) — Enhanced 9-1-1 feature that displays the address of the party calling 9-1-1 on a screen for use by the public safety telecommunicator. This feature is also used to route calls to the appropriate public safety answering point (PSAP) and can even store information in its database regarding the appropriate emergency services (police, fire, and medical) that respond to that address.

Global Positioning System (GPS) — System for determining position on the earth's surface by calculating the difference in time for the signal from a number of satellites to reach a receiver on the ground.

Radio. On occasion, an emergency may be reported by radio. This type of report is most likely to come from fire department personnel or other government workers who carry radios. The firefighter or dispatcher monitoring the radio gathers the same kind of information that would be taken from a telephone caller.

Wired telegraph circuit box. Historically, many cities installed street corner alarm boxes to allow citizens to report a fire when businesses were closed and there were no public telephones in the neighborhood **(Figure 3.3)**. Pressing a lever on the door of the box transmits a unique telegraph code that identifies location of the activated box. While these boxes are very reliable, they only transmit their locations not the nature of the emergency. In addition, these boxes are notorious for malicious false alarms. Because of the availability of public telephones and cellular phones, the need for these systems has greatly diminished. Given the expense of testing, maintenance, and replacement parts for telegraph systems and the reduced need for them, most cities have eliminated them.

Telephone fire alarm box. Telephone fire alarm or call boxes are equipped with a telephone for direct voice contact with the telecommunications center. Some municipalities use a combination of telegraph and telephone-type circuits. The pull-down hook is used to send the coded signal and a telephone is included to allow the caller to provide additional information to the telecommunications center.

Radio fire alarm box. A radio fire alarm box contains an independent radio transmitter with a battery power supply **(Figure 3.4)**. Some systems include a small solar panel at each box for recharging the unit's battery. Others feature a spring-wound alternator to provide power when the operating handle is pulled.

Figure 3.3 A wired telegraph circuit box served as a safety device for civilians in a neighborhood.

Receiving Nonemergency Calls

Nonemergency calls received at fire department facilities range from inquiries and requests for assistance to personal calls from family members or friends. Each department will have its own procedures for answering nonemergency calls. For this reason, it is important to know and follow your departmental procedures. Remember that you are representing your department; always be professional and courteous when answering the telephone:

Figure 3.4 Radio fire alarm boxes are self-contained units that may be found along highways or streets.

Figure 3.5 A watch room or booth serves as the communications center for a fire station.

- Answer calls promptly.

- Be pleasant and identify the department, station or facility, unit, and yourself. For example, "Good morning, City of Tompkinsburg, Station 16, Firefighter Jones speaking." It is SOP in some departments to add, "How may I help you?"

- Be prepared to record messages accurately by including date, time, name of caller, caller's telephone number, message, and your name.

- Never leave the telephone line open or a caller on hold for an extended period of time.

- Post the message or deliver it promptly to the person to whom it is directed.

- If you cannot answer the caller's question, refer them to someone who can or say, "I will get that information for you and call you back shortly." Then follow up on the request.

- End calls courteously. Disconnect according to local protocol.

On occasion, you may receive a call from someone who is angry or upset. When handling these types of calls, you must remain calm and courteous. Never become confrontational. Be pleasant and take the necessary information. Then refer the caller to the appropriate officer or division that can assist the caller. In many departments, the Public Information Officer (PIO) is the contact person for nonemergency or complaint calls. You should become familiar with the functions and personnel in each division of your department so you can refer callers efficiently. Handling emergency and nonemergency calls is discussed in **Skill Sheet 3-I-1**.

In some jurisdictions fire stations have a watch room or booth that is staffed throughout the work shift **(Figure 3.5)**. This area contains the radio communications equipment for receiving alarms from the telecommunications center, telephones, TDD/TTY devices, and station intercommunications (intercom) equipment. Local

Figure 3.6 An audible alarm may be used to signal a dispatch to an entire station.

protocol usually requires one member of the crew to remain in the watch room at all times. A bed may be located in or near this space. When you are assigned to monitor the watch room, you will have the following responsibilities:

● Listening to all radio communications

● Answering the telephone

● Acknowledging the receipt of alarms

● Notifying crew members of telephone calls and messages

Dispatching Emergency Services

Once an emergency has been reported, the information must be transmitted to the responding units or personnel. The more time that it takes for receiving, processing, and dispatching units, the greater the potential severity of damage and injuries.

Dispatch begins with some form of alert to the stations, apparatus, or individuals **(Figure 3.6)**. Alarm notification may be one or a combination of the following:

● Visual
  — Station lights
● Audible
  — Vocal alarm
  — Station bell or gong
  — Sirens
  — Whistles or air horns
● Electronic
  — Computer terminal screen with alarm or line printer
  — Direct telephone connection with telecommunications center
  — Radio with tone alert
  — Scrolling message boards

Figure 3.7 Pagers continue to be used in the fire service to alert volunteer responders.

Figure 3.8 An alerting device mounted at the top of a tall building or other feature provides as much range as possible when summoning responders and alerting a community of danger.

— Television override

— Radio

— Pagers

— Cellular telephones

— Text-messaging receivers

— Home electronic monitors

— Landline telephones

Volunteer fire departments may use pagers to alert members of an emergency. Each pager or group of pagers can be set to a specific frequency (**Figure 3.7**). Dispatch can send alert codes to these specific frequencies. When the pager receives its codes, it alerts the wearer by tone, light, and/or vibration. The pager will then either relay a voice message or display an alphanumeric message sent to it. Usually when a number of different departments or public safety agencies share the same dispatch frequency, it is desirable to set pagers to the alert setting to avoid hearing unwanted radio traffic.

Sirens, whistles, and air horns are most commonly employed in small communities (**Figure 3.8**). The alerting device is mounted on a water tower, radio tower, or top of a tall building. These devices produce a signal that everyone in the community can hear. Civilians will be aware that emergency traffic may be on the streets; however, some may also be inclined to follow the apparatus and congest the emergency scene.

In addition to the alerting signal, information regarding the emergency must be broadcast to department members. The information is transmitted vocally, by radio, or visually via a printer, text receiving device, or on a computer screen located in the station or apparatus. The broadcast should include information received from the caller and information from the preincident plan developed for the specific address or a similar facility. Information gathered during preincident surveys and fire prevention inspections may be located in a preplan book carried in the apparatus. Basic information to be broadcast generally includes:

- Units assigned
- Type of emergency
- Address or location
- Dispatch time
- Current conditions, such a wind direction/ speed and road closures
- Units substituted in to the normal assignment, if any

When the dispatch transmission is complete, assigned units confirm receipt of the information according to local protocol. This confirmation may include pressing a button on the alarm panel in the station, keying the radio microphone, or vocally acknowledging the dispatch. The telecommunications center will then confirm that all units have been notified and are on the assignment.

Internal Communications

Internal communications include messages transmitted within the fire department and between the department and other agencies during emergencies including:

- Dispatch transmissions
- Transmissions between the telecommunications center and units at the emergency scene
- Transmissions between the Incident Commander, Command staff, and units at the scene
- Transmissions between the Incident Commander, Command staff, and other agencies
- Transmissions between units and between members of units at the scene

Most internal communications are radio communications. Radio communications are essential to safe and efficient emergency scene operations. Fire department radio systems are used to alert units of an emergency, coordinate tactics at the emergency, request additional resources, and monitor the activities of units and individuals. In many departments, all facilities, apparatus, official vehicles, and personnel are assigned radios during emergencies or on a daily basis. Personnel are trained in local radio procedures including periodic radio tests, and both emergency and nonemergency radio operations.

Internal communications require you to have general knowledge of:

- Radio systems and how they work
- Limitations of radio communications
- Fixed, mobile, and portable radios assigned to you

NOTE: Emergency transmissions that are used when a firefighter is in trouble are described in Chapter 9, Building Search and Victim Removal.

Radio Systems

The radio systems used in the fire service can be classified in multiple ways:

- By their location and size: fixed, mobile, or portable
- By the type of signal used: analog or digital
- By the transmission signal: direct, repeated, or trunked

Figure 3.9 Radios located in a fixed location, such as a fire station watch room, are known as base station radios.

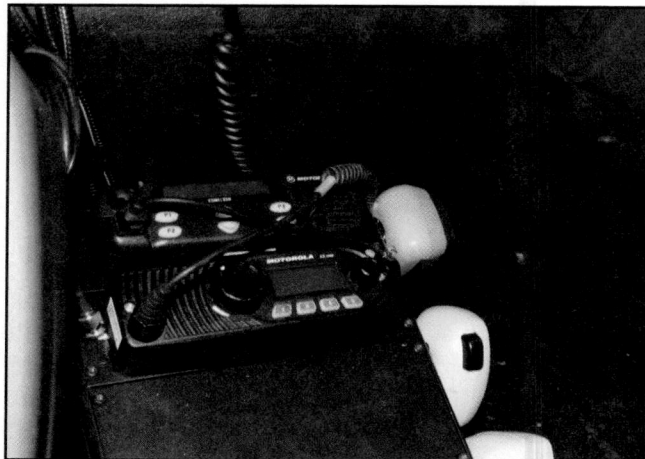

Figure 3.10 Mobile radios may be located in the cab of a vehicle. *Courtesy of James Nilo.*

Location and Size

Base Station Radios — Fixed, nonmobile radio at a central location.

Radios used in fixed locations such as fire stations, telecommunications centers, training centers, or administrative offices are referred to as **base station radios** (**Figure 3.9**). Base stations have stable, powerful transmitters and interference-resistant receivers that provide better performance than mobile and portable radios. Base station equipment includes a receiver, transmitter, antenna, microphone, and speakers. The power source for the base station is the electrical system for the building and is usually connected to an emergency generator in case of power loss. The system may also be connected to the alarm notification system of the fire station.

Mobile radios are mounted in fire apparatus, ambulances, and staff vehicles and are powered by the vehicle's electrical system (**Figure 3.10**). The receiver and transmitter are usually located in the cab within reach of the officer and driver/operator. Headset connections are usually provided for all riding positions. Pumping apparatus have additional connections at the pump panel while aerial devices have connections on the turntable and in the elevating platform. An external antenna is mounted to the vehicle. Mobile radios have better performance than portable radios but are not as powerful as fixed location radios.

Intrinsically Safe Equipment — Equipment designed and approved for use in flammable atmospheres that is incapable of releasing sufficient electrical energy to cause the ignition of a flammable atmospheric mixture.

Less powerful than the fixed and mobile radios, portable radios are handheld devices (**Figure 3.11**). They are designed to withstand heat, moisture, and physical impact. They are powered by rechargeable or replaceable battery packs. Rechargers are usually found in fire stations and apparatus. There is an external antenna attached to the top of the radio that is protected by a rubber jacket. Controls on the radio include knobs for changing channels, adjusting the volume, and a push-to-talk switch for transmitting. These radios also have an orange or red EMERGENCY button that is programmed to transmit a distress signal (**Figure 3.12**). Some models of radios are equipped with a keypad similar to the one found on telephones. Portable radios that are used in hazardous atmospheres must be **intrinsically safe** for that environment (**Figure 3.13**).

Distribution of portable radios depends on local protocol. In some departments only the officer in charge of the unit is assigned a radio. Other departments assign one radio to each member of the company.

Figure 3.11 Portable radios must be durable enough to survive dangerous situations.

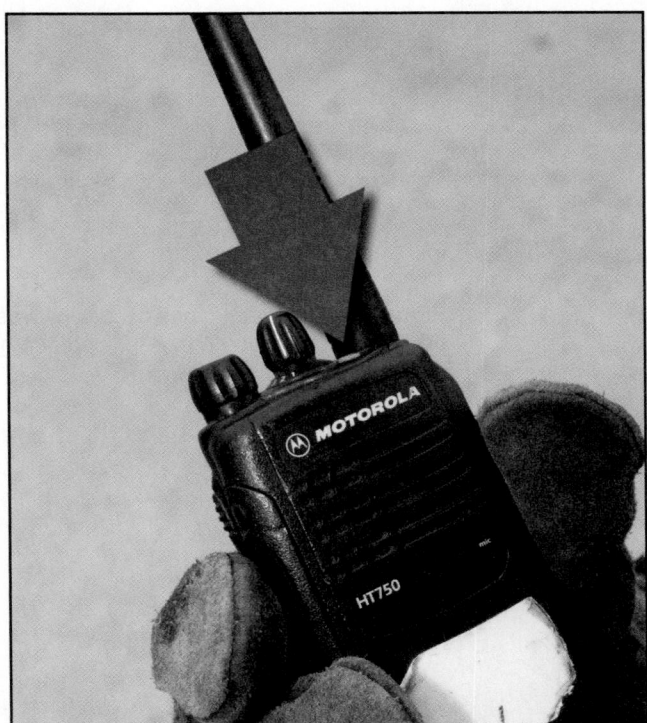

Figure 3.12 A preprogrammed emergency button is included on many styles of portable radios.

Figure 3.13 A portable radio that is intrinsically safe should be labeled by the manufacturer as such.

Signal Type

Radio signals travel between the transmitter and receiver in either analog or digital format. Both formats transmit over either amplitude modulation (AM) or frequency modulation (FM) carrier waves **(Figure 3.14)**. AM waves vary the strength of the signal to reflect the speaker's voice and are sometimes referred to as *medium wave signals.* FM waves allow for the signal to change based upon microphone audio. The signal is filtered to remove any frequencies over the human voice without making changes to the signal, so FM waves do a better job of cancelling naturally occurring noise.

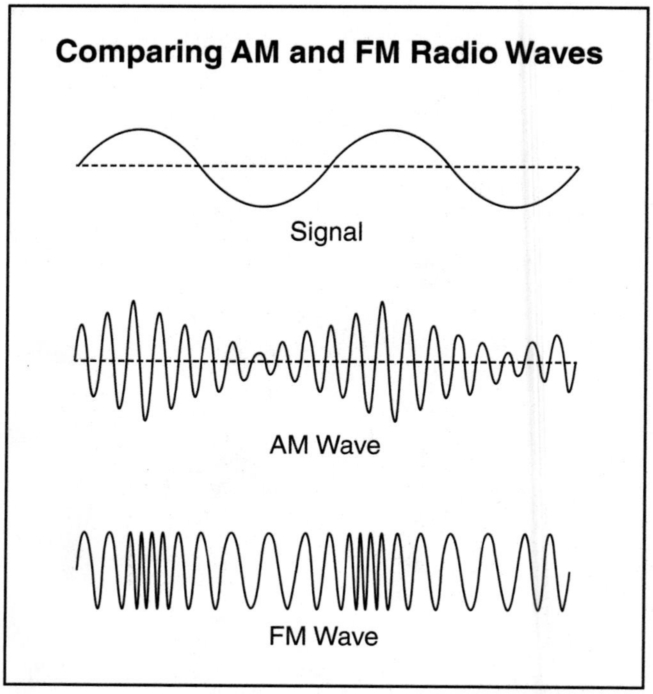

Comparing AM and FM Radio Waves

Signal

AM Wave

FM Wave

Figure 3.14 Different types of radio waves may be used to transmit the frequency of the human voice over long distances.

Digital transmission systems have supplemented and, in some cases, replaced analog transmission systems. Digital radios have improved audio quality and make a better use of their assigned frequency or band. A digital transmitter converts the voice into digital data packets which are then broadcast to the receiver.

Signal Transmission

Direct communication refers to the straight line travel of radio signals between the antenna connected to the transmitter and the antenna connected to the receiver. Because direct communication requires straight line transmission, terrain and buildings may block radio signals. However, direct communication does allow the same radio channel to be used by other groups that are located at a greater distance from the first group. When one radio transmits and another receives, the type of communication is called a simplex system (**Figure 3.15a**). A simplex system is a two-party system operating on one frequency because only one radio can transmit at a time. When the first speaker is finished, the second can press the talk button and respond. Portable radios include this function as a *talk-around mode* which permits units at an incident to communicate directly.

To overcome the problem of barriers to direct communications, a system of repeaters can be used. Repeaters, usually located on high points within the broadcast area, are used to increase the range of the radio system and to send signals over tall barriers. Known as half-duplex communication, the repeater system uses two radio frequencies for communication (**Figure 3.15b**). The transmitting radio transmits on Frequency 1, and that signal is received by the repeater. The repeater then repeats the transmission on Frequency 2, and this signal is received by the receiving radio.

The full-duplex system allows radio communication in both directions simultaneously such as communications on a landline telephone (**Figure 3.15c**). This system does allow for more rapid transmission of information because receivers can respond very quickly to messages. However, communications can become garbled when using this system because multiple microphones are broadcasting information at the same time.

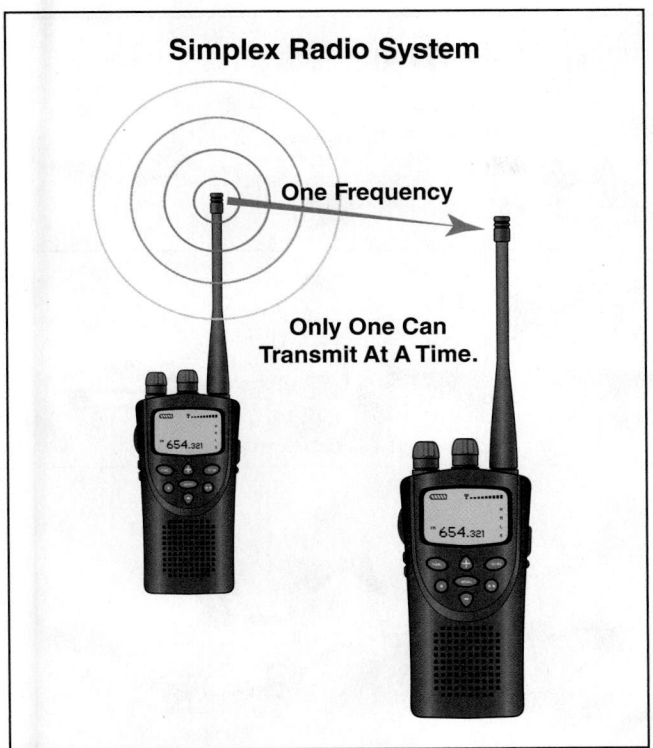

Figure 3.15a A simplex radio system cannot receive and send a signal at the same time.

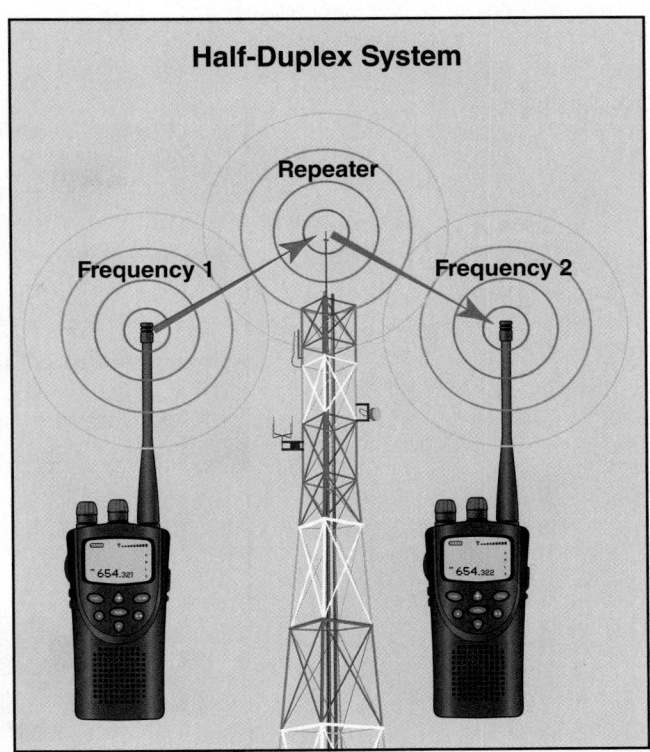

Figure 3.15b A half-duplex radio system uses a series of repeaters to communicate between radios set at different frequencies.

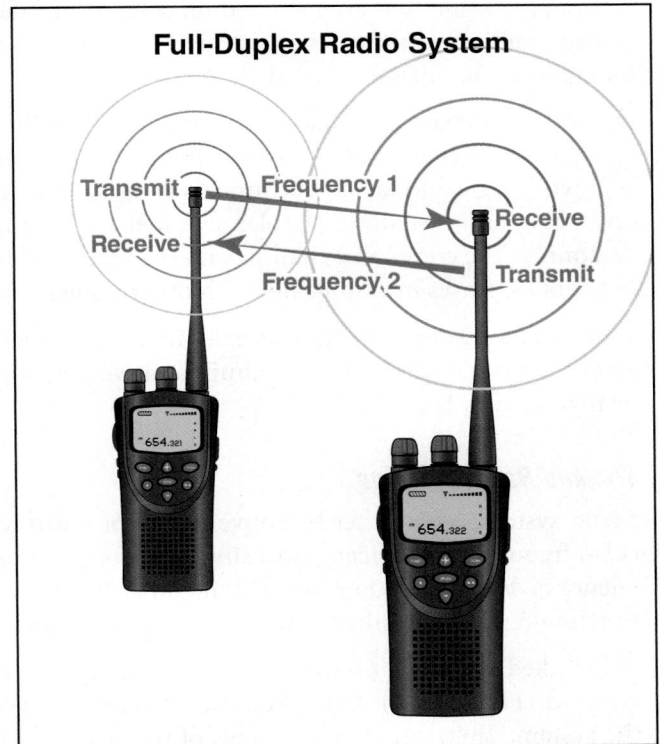

Figure 3.15c A full-duplex radio system allows for real-time communication but may become garbled if too many microphones broadcast at once.

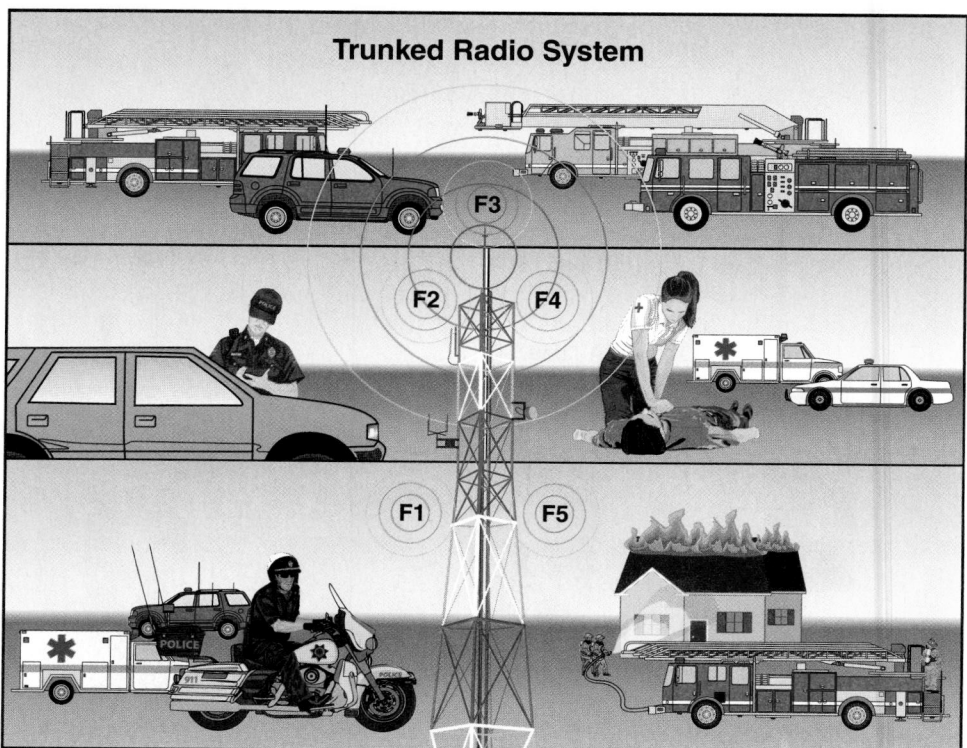

Figure 3.16 A trunked radio system uses repeaters to determine which frequency to use to route calls through the system.

Fireground Channels

Organizing radio operations based on a variety of radio channels is a generally accepted practice in the fire service today. Modern radio systems are designed to operate on multiple channels as needed.

Most fire departments have a channel assigned for dispatching only. When units arrive at the incident, a Command channel is assigned to the Incident Commander (IC) while a second tactical channel is assigned for fireground operations. This second channel is usually a simplex channel for communication between units and personnel. The command channel and the tactical channels may be expanded as the incident increases in complexity with other agencies being added to it.

Nonemergency channels also exist in some departments for use by the training center, code enforcement, and administrative personnel. Use of these channels will be regulated by the AHJ.

Trunked Radio Systems

Radio systems may either be conventional or trunked. In a conventional system, a radio frequency is dedicated to a single function such as operations. Because the frequency is dedicated to only one use, no other function or unit can transmit on it even when no one is transmitting on it. This results in a waste of resources.

Trunked systems improve efficiency by assigning transmissions to available frequencies (**Figure 3.16**). A trunked system depends on repeaters to route calls through the system. Therefore, high volumes of traffic can be handled on multiple channels when large-scale or numerous operations are underway.

The term talkgroup is used to distinguish among physical frequencies or channels in conventional radio systems. In a trunked system, turning on a radio notifies the trunking system controller. The controller assigns a channel to that radio. Having received a channel, the radio is then ready to operate on that channel. Activating the press-to-talk button sends a request to the controller. If the system is available, the controller sends a message that emits a three-beep tone sequence telling the firefighter to continue with the transmission. If the talkgroup is not available, a failure tone sounds, usually described as a *bonk*.

Other features of a trunked radio system include:

- **Multigroup call** — Used to transmit calls to two or more talkgroups; once the connection is established, the talkgroups can communicate with each other.

- **Private call** — Permits one radio to call another much like a telephone call; the conversation is private and will not be heard by others on the channel or talkgroup.

- **Dynamic regrouping** — Emergency alert feature that, if activated, sends a signal to the agency's dispatch center.

It is important for you to understand what your agency provides. Some radios do not have emergency buttons and some dispatch centers do not monitor all radio channels on your portable radio.

Radio Communications

The Federal Communications Commission (FCC) regulates all radio communication in the United States. The FCC issues radio licenses to fire departments that operate radio equipment. Depending on the radio system in a particular locality, one license may cover several departments that operate a joint system. Local department rules should specify who is authorized to transmit on the radio. It is a federal offense to send personal or other unauthorized messages over a designated fire department radio channel. Information on use of a portable radio for routine and emergency traffic can be located in **Skill Sheet 3-I-2**.

NOTE: In Canada, the Canadian Radio-Television and Telecommunications Commission regulates radio communications.

With the adoption and implementation of the National Incident Management System (NIMS), most emergency response agencies in the U.S., including fire departments, have eliminated the use of *ten-codes* and other local terminology historically used for radio transmissions. Ten-codes substituted a number for a specific activity or condition. Most fire departments now use **clear text** without number codes of any kind. Clear text is a standardized set of emergency-specific words and phrases. Canada does not have a national requirement for clear text transmissions; therefore, some departments use ten-codes while others use clear text.

Clear Text — Use of plain language, including certain standard words and phrases, in radio communications transmissions.

Radio Limitations

As the fire service has increased its dependency on portable radios, limitations to their use have become more apparent. There are four main limitations or barriers to all radio transmissions (**Figure 3.17, p. 110**):

- Distance
- Physical barriers
- Interference
- Ambient noise

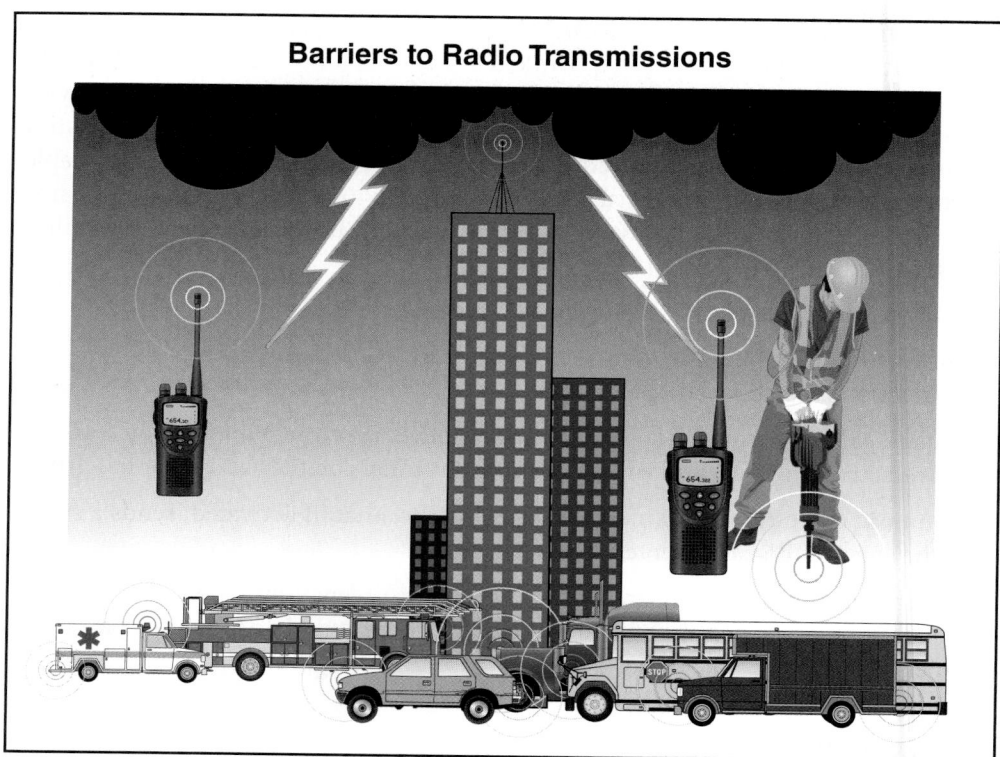

Figure 3.17 Barriers to communication via portable radios are commonly present at an incident scene.

Distance

Radio signals travel in a straight line. The distance the signal will travel depends on the power of the transmitter and receiver and the height of the broadcast and receiving antennas. Some jurisdictions use repeaters to increase the area of coverage. Static and messages that are broken up are an indication that the receiver is near the limit of the transmission range.

Physical Barriers

Any physical barrier between the transmitter and the receiver can block the signal. The signal may be totally blocked, partially blocked, or reflected. While units working within tunnels, basements, or structures may be able to communicate with each other using the talk-around function, they may not be able to talk to the IC or the telecommunications center. At times your own body can act as a physical barrier. You must be aware and consider repositioning yourself to see if your signal can be sent, especially while working within a structure. To overcome physical barriers, you may need to turn your body 90 degrees, lift the portable radio higher, or raise the antenna up straight.

Most people have experienced "deadzones"; namely, the loss of cellular telephone service when traveling in remote areas or inside structures. The same phenomenon occurs with portable radios. It may not be possible to overcome deadzones in rural areas without the addition of repeaters in apparatus or the addition of fixed repeaters. Inside structures, moving to an outside wall, roof, window, or doorway can improve reception. Inspectors or other fire personnel should have performed radio checks during preincident planning surveys to ensure that radios can be used in all areas of the building. In large metal and concrete buildings, radios may not work and runners or other means of communication may be required.

Interference

Interference can originate from the following sources:

- Another powerful radio signal
- Vehicle ignitions
- Electric motors
- High-voltage transmission lines
- Computers
- Equipment that contain microprocessors
- Cellular telephone towers or transmitters
- High-power radio sites such as television and radio stations

Reducing or eliminating interference is primarily the manufacturer's responsibility, although the AHJ and the firefighter bear some limited responsibility. Manufacturers design high-quality transmitters, receivers, and repeater systems to filter out interference. The fire department administration or AHJ should specify and purchase the best quality radio systems that are available within the resources they have. You should be aware of the causes of interference and watch for external causes, such as electric motors, that can be shut off.

Ambient Noise

The emergency scene is filled with ambient noise that can make radio communications difficult. New technology has developed noise-canceling microphones that may help in some instances. Overcoming ambient noise is the responsibility of each person operating a mobile or portable radio at the scene (**Figure 3.18**). The following are some examples of overcoming ambient noise:

Figure 3.18 Sounds created by sirens, engines, strategic commands, and interpersonal communication between responders and bystanders generates layers of ambient noise that make radio communications difficult. *Courtesy of Ron Moore and McKinney (TX) FD.*

- Turn off apparatus audible warning devices when they are no longer needed.

- Move away from noise-emitting equipment when transmitting.

- Follow radio procedures at all times.

- Move to a location that blocks wind noise.

- Use your body or PPE to create a wind barrier when transmitting.

Radio Procedures

Because the radio channel is available to all units and personnel at the scene, the telecommunications center, and in some cases all fire department facilities and the public, it is important to follow local protocol for sending a message. Depending on the local radio system and protocols, frequencies may be monitored. All recorded transmissions become part of the official record on the incident. They may also be used in court or by the news media.

Communication Model

When using radio communications, the communications model begins every time one person opens or *keys* a microphone and begins to speak. The generally accepted communication model consists of six basic elements **(Figure 3.19)**:

- **Sender** — The person who initiates the message using both verbal and nonverbal communication.

- **Message** — The content that the sender is trying to communicate. The message may consist of information intended for multiple human senses (sight, hearing, taste, smell, touch).

- **Receiver** — Individual or individuals to whom the sender is attempting to communicate.

- **Feedback to the sender** — Reaction of the receiver to the message and its tone. If this feedback is verbal, the receiver becomes the sender and relates a new message to the original sender, who becomes the receiver. Receiving feedback allows the original sender to confirm reception of the message and to assess the receiver's level of understanding.

- **Interference** — Anything that may prevent the receiver from completely understanding the message.

Everyone at the emergency scene should follow two basic rules to control communications. First, units or individuals must identify themselves in every transmission as outlined in the local radio protocols. The protocol may be to identify yourself first and then the receiver as in *Engine One to Dispatch* or to identify the person or unit you are calling first as in *Dispatch from Engine One*. Second, the receiver must acknowledge every message by repeating the essence of the message to the sender.

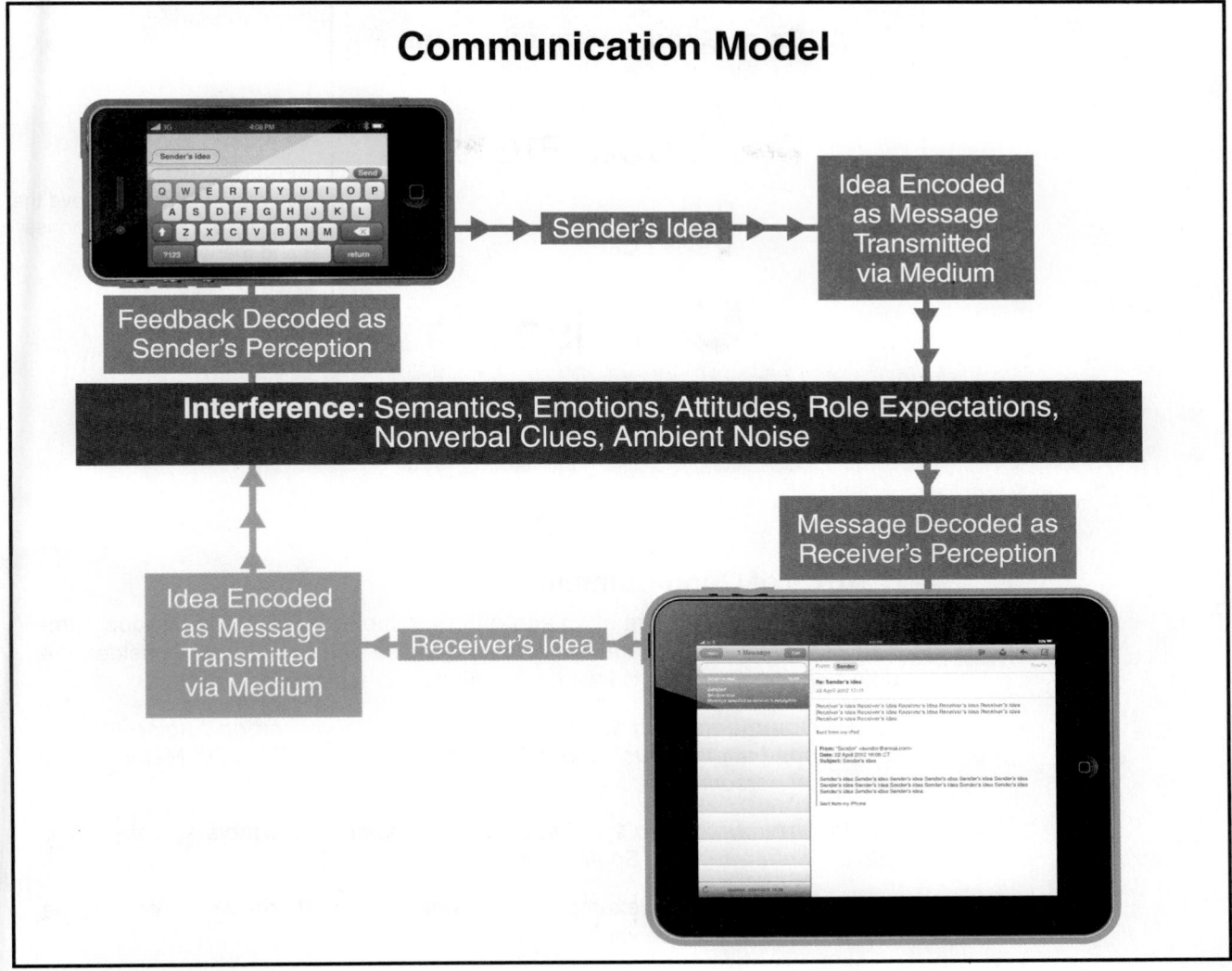

Communication Model

Sender's Idea

Idea Encoded as Message Transmitted via Medium

Feedback Decoded as Sender's Perception

Interference: Semantics, Emotions, Attitudes, Role Expectations, Nonverbal Clues, Ambient Noise

Message Decoded as Receiver's Perception

Idea Encoded as Message Transmitted via Medium

Receiver's Idea

Figure 3.19 Communications between individuals are filtered through several preexisting standards before being interpreted.

Example:

Engine 4: Communications from Engine 4. We are on scene and have a trash fire. Engine 4 can handle. Return all other units.

Communications: Engine 4, I copy you have a trash fire, and you will handle. Other units can be canceled.

Requiring the receiver to acknowledge every message ensures that the message was received and understood. This feedback can also tell the sender if the message was not correctly understood and further clarification is necessary.

Another good practice is to key the microphone and wait for a second or two for the signal to capture an antenna before starting your message. Keying the microphone and immediately speaking often results in a message being cut off at the start and the receiver asking for the message to be resent, thus wasting radio traffic.

ABCs of Good Communications

Accurate

Brief

Concise

ABCs of Good Communications

When transmitting information and orders, adhere to the ABCs of good communications: Be **A**ccurate, **B**rief, and **C**oncise **(Figure 3.20)**. Consider the difference between the following two radio communications:

Radio Communication 1: *"This is Lieutenant Thompson on Engine 57 portable. I need another truck company at this location, 1400 South Memorial, for additional personnel."*

Radio Communication 2: *"Dispatch from Engine 57 portable — assign one truck company to 1400 South Memorial."*

Notice that the second example is accurate, brief, and concise. It also begins with the title of the receiver.

To ensure that the message is heard and understood, you should follow these guidelines:

- Know what you are going to say before you open the microphone.

- Use a moderate rate of speaking focused on clear understanding and do not use pauses or verbal fillers such as *"ah"* or *"um."*

- Use a moderate amount of expression in speech — not a monotone and not over-emphasized — with carefully placed emphasis. Avoid excitement or shouting over the radio and be careful to articulate properly. Attempt to pronounce words correctly.

- Use a vocal quality that is not too strong or weak. Finish every comment, and avoid trailing off near the end of the transmission. Keep the pitch in a midrange — not too high or too low. Avoid using slang or regional expressions, and strive for a good voice quality.

- Do not chew gum or eat food while transmitting a message. Be confident in what you say, and position the microphone appropriately to make the best use of the system.

- Be concise and to the point; do not talk around the issue. Give the information required in a logical and complete manner.

- Do not transmit until the radio frequency is clear.

- Remember that emergency transmissions have priority over any routine transmission.

- Do not use profane or obscene language on the air.

- Hold the radio/microphone 1 to 2 inches (25 mm to 50 mm) from your mouth or self-contained breathing apparatus (SCBA) voice port when transmitting.

- Speak directly into the microphone and not across it.

- Hold the microphone according to the manufacturer's recommendations.

- Repeat the message back to the sender to ensure that they know you received and understood it.

- Place the microphone against your throat if you cannot be understood through your SCBA facepiece. DO NOT REMOVE YOUR FACEPIECE TO TALK INTO THE MIC.

- Practice communicating with your portable radio while wearing your SCBA before you have to use it at an emergency.

WARNING
DO NOT REMOVE YOUR FACEPIECE
TO TALK INTO THE MIC.

Experience at emergency incidents has shown that background noise and personal safety equipment can significantly affect the ability to hear and understand radio transmissions. To improve your ability to hear and be heard, you should follow these radio communications best practices:

- Speak with a loud, controlled, and clear voice.

- Shield the microphone from noise, water, and debris. You can cup your hand over the microphone or use your helmet brim to protect it (**Figure 3.21, p. 116**).

- Locate your microphone or radio as far as possible from your personal alert safety system (PASS) device, low-pressure alarm, or other noise-generating equipment.

- Avoid laying the microphone on the seat of the vehicle because the transmission button may be pressed inadvertently.

- Position the antennae vertically for best transmission results.

- Do not shout. If the receiver did not understand your first transmission, raising your voice will not make it any clearer and will possibly make the message worse, wasting radio traffic.

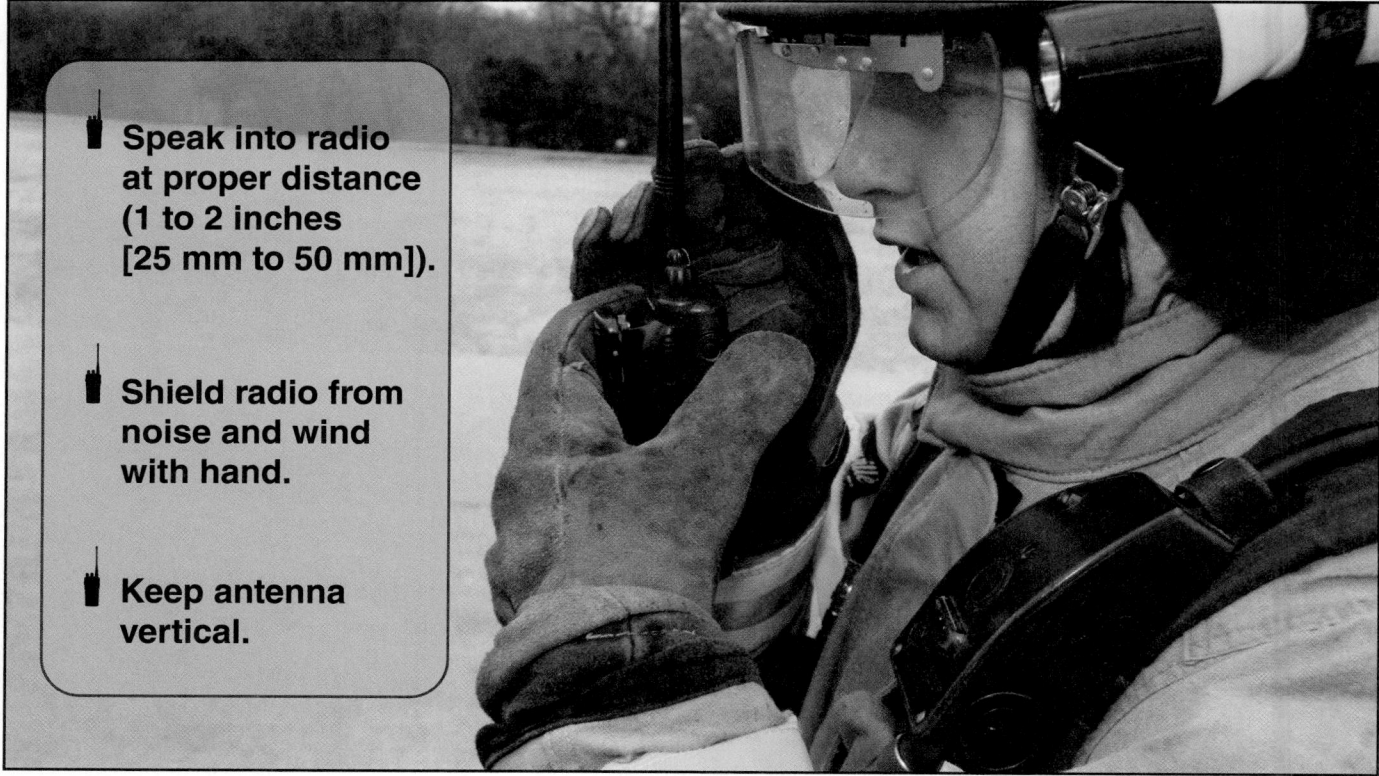

Speak into radio at proper distance (1 to 2 inches [25 mm to 50 mm]).

Shield radio from noise and wind with hand.

Keep antenna vertical.

Figure 3.21 A radio must be held 1-2 inches (25 to 50 mm) away from the face and in an alignment that will minimize signal interference and protect against ambient noise and physical interference.

❚❚ On-Scene Communications

Strategic and tactical decisions, the placement of apparatus, the assignment of personnel, and the need for additional resources will be based on the information contained in the initial and subsequent radio broadcasts. The sections that follow discuss various on-scene communications that affect decision making.

Arrival Report

If you are the first person to arrive on the scene, you will provide an initial radio report. The initial report provides other responding units and the telecommunications center with a description of the conditions as they appear to you. Based on local protocol, this radio report may be referred to as:

- Arrival report
- Report on conditions
- Brief initial report
- On-scene report
- Situation report
- Size-up report
- CAN (condition, actions, needs) report

Regardless of what it is called, the initial report:

- Establishes the initial arrival time
- Informs other responding units of current conditions
- Describes the actions that are being taken
- Describes the actions that need to be taken by other responding units

Upon arrival, you may encounter one of two situations:

1. There is no visual evidence of an emergency, which will require more investigation.
2. Obvious signs of an emergency exist that will require immediate action.

Regardless of the situation, it is the responsibility of the first responding firefighter to report the information. In the case of a potential nonemergency, the firefighter may broadcast a follow-up communication after further investigation before requesting a full response.

Depending on local protocol, the arrival report should contain:

- Unit (or individual) arriving on scene
- Correct address of the incident if different from the dispatch address
- Description of conditions found at the scene
- If a structure fire or collapse, a description of the structure including:
  — Number of floors
  — Type of construction
  — Type of occupancy
- Operational strategy
- Intended initial actions
- Water supply need or availability
- Establishment of Command:
  — Specify who is the Incident Commander (IC)
  — State the name of the incident (may be the address or name of occupancy)
  — State location of the Command Post when applicable
  — State location(s) of staging areas for arriving units
- Calls for additional resources
- Routing instructions to arriving units
- Assignments to arriving units
- Tactical radio channel for use at the incident if required by local protocols

NOTE: In some jurisdictions, the telecommunications center will provide the radio channel to responding units.

Progress Report

Once emergency operations begin, the IC should transmit progress or status reports to provide the telecommunications center with a continuous record of actions at the emergency scene. Progress reports generally contain the following information:

- Transfer of Command
- Change in Command Post location

Figure 3.22 Tactical progress reports should include air consumption.

- Progress (or lack of) toward incident stabilization
- Direction of fire spread
- Exposures by direction, height, occupancy, and distance
- Any problems
- Anticipated actions
- Need for additional resources

The telecommunications center may also be assigned the task of providing periodic time transmissions at designated intervals, such as 15 minutes. These time checks have two important uses. First, they are used to assist in air management to remind interior units to read their air capacity gauges. Second, time checks are used to assist the IC and Command Staff to gauge the effectiveness of the strategies and tactics that are being used to control the situation.

Tactical progress reports also provide an opportunity to check individual air supplies. When giving a progress report, it is good practice to include your air consumption **(Figure 3.22)**. Doing so allows you to keep track of your crew's air management. You and your crew should be outside the structure before anyone's low air alarm activates.

Individual units operating at the incident provide their own tactical progress reports to the IC. These reports update the IC on each team's progress and provide information to the telecommunications center and other units. Like all other radio reports, these must be accurate, brief, and concise. The form these reports take is usually based on local protocol or policy.

Requesting Additional Resources

Changes in the conditions at some emergency incidents may make it necessary to request additional resources. Normally, only the IC may request multiple alarm assignments or order additional resources. Depending on who arrives first, the IC may be a chief officer, company officer, or a firefighter.

You need to know the local procedure for requesting additional resources. You must also be familiar with procedures for requesting multiple or special alarm signals and know what resources each additional **alarm assignment** will include; that is, the number and types of units that respond to each additional alarm.

When multiple alarms are requested for a single incident, maintaining communications with each unit becomes more difficult as radio traffic increases. To reduce the work load on the telecommunications center, a radio-equipped mobile communications vehicle can be used at large incidents (**Figure 3.23, p. 120**). Mobile communications centers monitor the assigned tactical channel, monitor radio channels used by other agencies or services that do not have access to the primary tactical channel, and provide portable radios to personnel from other agencies. The unit also maintains contact with the communications center.

Alarm Assignment —
Predetermined number and type of fire units assigned to respond to an emergency.

Emergency Radio Traffic

Emergency radio traffic includes urgent broadcasts intended to warn personnel at an emergency scene of an impending hazard. The broadcast may originate from the Command Post or from any unit or individual at the scene. Emergency radio traffic begins with a clear statement such as "Dispatch, Engine One, Emergency Traffic!" meant to alert all listeners to pay attention and stay off the air. Dispatch, or the IC, would acknowledge the statement with a vocal response or use of an alert tone on the tactical channel. The unit originating the transmission then states the type of hazard and the action required such as, "Engine One, west wall showing signs of collapse, cease interior operations and evacuate the structure."

Figure 3.23 A mobile communications vehicle serves to reduce the workload of the telecommunications center. *Courtesy of Ron Moore and McKinney (TX) FD.*

Regardless of the term used, evacuation signals are used when the IC decides that interior operations must cease and all firefighters must immediately withdraw from the building or other hazardous area because conditions have deteriorated beyond the point of reasonable safety. All firefighters should be familiar with their department's SOPs for declaring an evacuation signal. Depending on local protocol, the order to cease interior operations may be referred to as *withdraw, abandon,* or *evacuate.* There are several ways this communication may be made. The two most common are to broadcast a radio message on the incident's tactical channel ordering an evacuation, and to sound audible warning devices on the apparatus at the fire scene in a prescribed pattern.

The evacuation order is broadcast several times to make sure everyone hears it. The use of audible warning devices on apparatus, such as sirens and air horns, will work outside small structures, but they may not be heard by everyone working in a large building. Audible warnings using sirens or air horns can also be confused with those being used by units arriving at the scene.

Emergency traffic can also include distress calls from firefighters. In these situations, local protocol may require the use of the MAYDAY signal. When this term is broadcast, all nonessential operations cease and the assigned rapid intervention crew or team (RIC/RIT) is dispatched to assist the distressed firefighters. Additional information on communication procedures used when a firefighter or unit experiences trouble is in Chapter 9, Building Search and Victim Removal.

Personnel Accountability Report (PAR)

A **personnel accountability report (PAR)** is a systematic method of confirming the status of units (company, group, or division, etc.) operating at an incident. When a PAR is requested, every supervisor must verify the status of those under his or her command and report it. In conditions where smoke or darkness impairs visual confirmation, company officers and team leaders/supervisors may have to rely on touch (physical contact) or hearing (voice contact – NOT radio) to verify each member's status. Other supervisors in the Incident Command System chain of command

Personnel Accountability Report (PAR) — Roll call of all units (crews, teams, groups, companies, sectors) assigned to an incident. The supervisor of each unit reports the status of the personnel within the unit at that time, usually by radio. A PAR may be required by standard operating procedures at specific intervals during an incident, or may be requested at any time by the Incident Commander or the Incident Safety Officer.

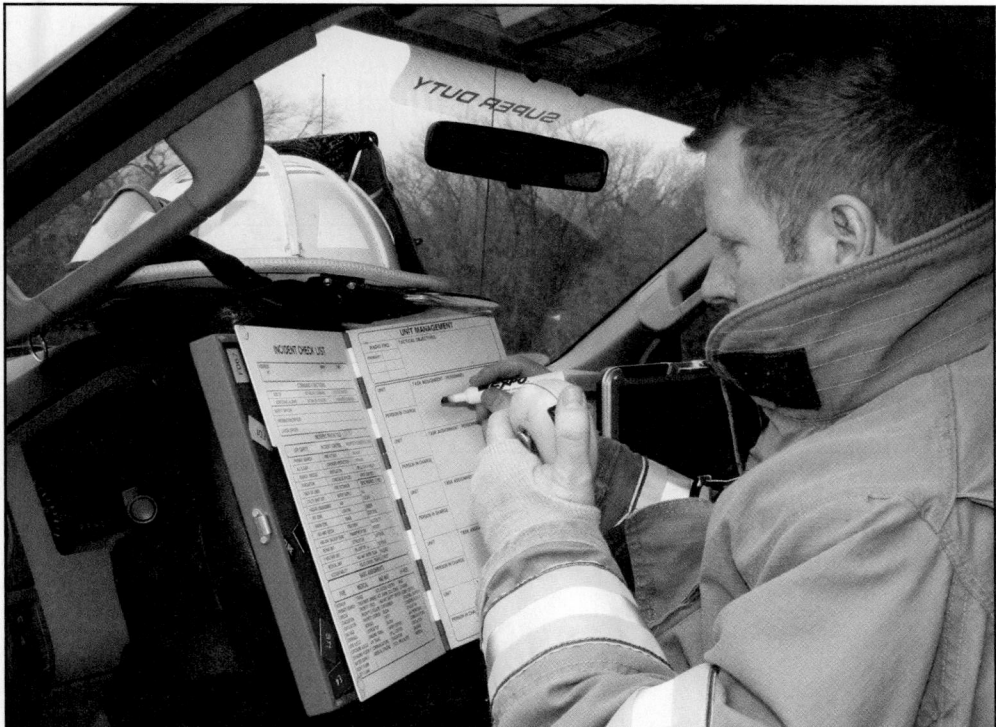

Figure 3.24 Personnel Accountability Reports are used to track personnel through an extended incident.

(Branch Directors, Division Supervisors, etc.) must rely on radio reports from their subordinates **(Figure 3.24)**. One limitation of a PAR check is that it uses a considerable amount of radio time, especially at larger incidents. Such extended radio use can prohibit members from reporting emergency messages as well as a MAYDAY.

The IC can request a PAR at any time, on a regular interval (every 15 minutes), or for any of the following reasons:

- The incident is declared under control.
- There is a change from offensive to defensive strategy.
- There is a sudden catastrophic event such as a flashover, backdraft, or collapse.
- There is an emergency evacuation.
- A firefighter is reported missing or in distress.
- The primary search has been completed.
- The Safety Officer requests it.

Post Incident Reports

Whenever fire department units and personnel respond to an emergency incident, reports are written describing the details of the event. From legal, statistical, and record-keeping standpoints, incident reports are vital to fire department operations. Because reports are available to the public, they must be complete and written in terminology that the general public can understand. Because incident reports are legal documents, there are possible legal consequences if the reports are inaccurate or incomplete.

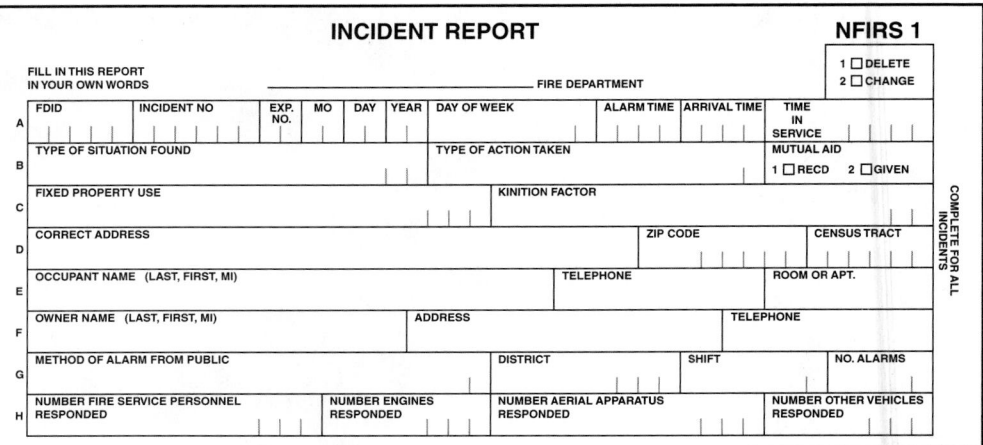

Figure 3.25 An incident report form asks for categories of information that will be an essential part of creating an analysis of the event after the fact.

INCIDENT REPORT   NFIRS 1

Incident reports can be handwritten or entered into a computer. To learn how to create an incident report, refer to **Skill Sheet 3-II-1.** The information gathered on reports is used for:

- Assessing departmental needs
- Budgeting
- Determining trends in types of responses
- Determining trends in firefighter injuries
- Providing information for national fire safety data bases
- Determining trends in fire cause
- Determining requirements for fire and life safety education programs
- Providing information to insurance rating agencies

Incident reports are generally based on a form or checklist that is easy to complete and thorough. The **National Fire Incident Reporting System (NFIRS)**, developed by the United States Fire Administration (USFA), outlines the necessary information needed to complete incident reports. NFIRS is a computer-based system that transfers data from each state to the federal database **(Figure 3.25)**. The system was developed to create data for analyses and to assist in combatting the nation's fire problem. Various types of data can be entered into NFIRS. Each piece of data has a specific code corresponding to the information entered. All fifty states participate in the NFIRS, although not all individual fire departments participate.

Local protocol will list the information that should be included in incident reports. The following is a general list of information that may be required:

- Fire department name, incident number, district name/number, shift number, and number of alarms
- Names and addresses of the occupant(s) and/or owner(s)
- Type of structure, primary use, construction type, and number of stories
- How the emergency was reported (9-1-1, walk-in, radio)
- Type of call (fire, rescue, medical, etc.)
- Action that was taken (investigation, extinguishment, rescue, etc.)
- Property use information (single-family dwelling, commercial occupancy, etc.)
- Number of injuries and/or fatalities

National Fire Incident Reporting System (NFIRS) — One of the main sources of information (data, statistics) about fires in the United States; under NFIRS, local fire departments collect fire incident data and send these to a state coordinator; the state coordinator develops statewide fire incident data and also forwards information to the USFA; begun by FEMA.

Chapter 4
Building Construction

Case History

In December 2010, two career firefighters were crushed to death and nineteen others were injured when the roof of an abandoned and unsecured commercial structure collapsed during fire suppression operations. NIOSH investigators identified a number of contributing factors in this incident that ultimately led to the casualties. One was the condition of the structure and the design deficiency of bowstring truss roof systems constructed prior to 1960. NIOSH recommended that all personnel be trained in the dangers associated with bowstring truss roof systems.

The structure involved in this incident was a one-story commercial structure of Type III ordinary construction originally built in 1926. The structure was approximately 6,700 square feet (622.45 square meters) in size and was constructed with brick masonry walls and separated into two sections. The front section had a flat roof and the back section had a bowstring truss roof. Parapet walls at the front and rear obscured a view of the roof from street level. The unsprinklered structure was located in an area that contained both vacant and abandoned buildings. The structure, formerly used as a commercial laundry, had been abandoned by the owner for more than five years. Both the natural gas and electrical services had been cut off previously.

The flat roof at the front and the bowstring truss roof at the rear both consisted of layers of asphalt rolled roofing covered with tar. The rear portion of the roof was supported by two bowstring trusses parallel to the front street and rear alley. These trusses were supported at both ends by brick pilasters built into the east and west load-bearing walls.

A bowstring truss roof is supported by four load-bearing walls. The ends of the trusses are supported by the side walls. The front and rear walls support roof joists or hip rafters that slope downward from the nearest trusses. A collapsing bowstring truss roof often puts outward pressure on the supporting walls, so that outward wall collapse is likely. While heavy timber roof systems will withstand more degradation by fire than lightweight engineered-wood roof trusses, both types are subject to failure.

Bowstring truss roof systems also suffer from a little-known phenomenon related to inaccuracies in early industry-accepted truss design assumptions. All trusses constructed prior to the late 1960s have a common code deficiency: the bottom chord members have inadequate tensile strength to support roof loads allowed by the current existing codes.

In this incident, the structure had been vacant for several years and had been abandoned by the owner. Citations issued by the city's building department included concerns about deteriorated (rotted) truss members, deterioration of the roof due to water penetration, and penetrated mortar joints along one wall. It is likely that a number of factors contributed to the roof collapse, including the deteriorated condition of the structure, several inches of snow on the roof, the combined weight of firefighters working on the roof, and the effects of ventilation and fire-suppression activities.

Source: NIOSH Report F2010-38

Figure 4.1 The house on the left predates the house on the right by fifty years.

Figure 4.2 The surface area and moisture content of wood components in a structure influence fire conditions.

This case history illustrates why it is important for fire officers and firefighters to know and understand types of building construction and factors that can contribute to structural collapse. Within your jurisdiction, there are structures that may be hundreds of years old located next to ones that were constructed in the past year **(Figure 4.1)**. Each will conform to a different building code intended to provide a certain level of fire safety and structural stability. Each will be constructed from the building materials and architectural design common to the period of construction. You must know the materials used to construct the buildings in your response area, their construction classifications, and the components that make up the structures. You must also know how the building's design and construction affect fire behavior and development inside the structure.

Building Materials

A wide variety of materials are used in the construction of buildings. Some materials, like wood and plastic, will contribute fuel to a fire in the building while other materials, like concrete and masonry, will prevent or limit the spread of fire. All materials react differently when exposed to the heat of a fire. Your knowledge of how these materials react to fire will give you an idea of what to expect from a fire in buildings of a particular type of construction. Common building materials include:

- Wood
- Masonry
- Metals
- Reinforced concrete
- Gypsum
- Lath and plaster
- Glass/fiberglass
- Plastic
- Composite materials

Wood

Wood is the most common building material used in North America and is the main component of a variety of structural assemblies **(Figure 4.2)**. Size and moisture content affect how wood reacts to fire conditions. The smaller the dimensions of the wood, the easier it is to ignite and the faster it will lose structural integrity. Large wooden beams, such as those used in heavy-timber construction, are difficult to ignite and retain their structural integrity even after prolonged exposure to direct flame impingement. Lumber of smaller dimensions needs to be protected by gypsum drywall or other insulation to increase its resistance to heat or fire.

The moisture content of the wood affects the rate at which it burns. Wood with a high moisture content (sometimes referred to as **green wood**) does not ignite as readily nor burn as fast as wood that has been kiln dried or dehydrated by exposure to air over a long period of time. In some cases, wood is pressure treated with fire-retardant chemicals to reduce the speed at which it ignites or burns. However, fire retardants are not always totally effective in reducing fire spread. Pressure treating wood also weakens the wood's load-carrying ability by as much as 25 percent.

Newer construction often contains composite building components and materials that are made of wood fibers, plastics, and other substances joined by glue or resin binders. Such materials include plywood, particleboard, fiberboard, **oriented strand board (OSB)**, and paneling. Some of these products may be highly combustible, can produce significant toxic gases, or can rapidly deteriorate under fire conditions.

Masonry

Masonry includes bricks, stones, and concrete blocks (**Figures 4.3a-c, p. 136**). Brick and stone are generally used to create **veneer walls**, which are decorative covers for wood, metal, and concrete-block load-bearing walls. Masonry is minimally affected by fire and exposure to high temperatures. Bricks rarely show any signs of loss of integrity or serious deterioration. Stones and concrete may lose small portions of their surface when heated, a condition called **spalling**. Concrete blocks may crack, but they usually retain most of their strength and basic structural stability. The mortar between the bricks, blocks, and stone may be degraded by heat and may display signs of weakening.

Metal

Metal building materials commonly include cast iron, steel, aluminum, and other metals. They are used to provide structural support, decorative covering on exterior walls, stairs, door and window frames, ductwork, pipes, and fasteners including nails, screws, and plates. The effect of heat and fire on metal materials will depend on the type of metal and whether or not the metal is exposed or covered.

Iron

Two types of iron can be found in buildings in North America: cast iron and wrought iron (**Figures 4.4a and b, p. 136**). Cast iron was commonly used in the 19th Century for structural support beams and columns, for stairs, balconies, railings, and elevators, and for the facades of buildings. Facades consisted of large exterior wall sections fastened to the masonry on the front of the building. Cast iron stands up well to fire and intense heat, but it may crack or shatter when rapidly cooled with water. During a fire, bolts or other connections that hold cast iron components to the building can fail, causing them to fall. Failure can also result from bolts rusting through or mortar becoming loose around the bolt.

Wrought iron was used in buildings of the early 1800s for nails, straps, tie rods, railings, and balconies. After 1850, it was used for rail and I-beams, channels, and support columns. Today, wrought iron is used for decorations in the construction of gates, fences, and balcony railings. Wrought iron is usually riveted or welded together while cast iron is bolted or screwed.

Green Wood – Wood with high moisture content.

Oriented Strand Board (OSB) — A wooden structural panel formed by gluing and compressing wood strands together under pressure. This material has replaced plywood and planking in the majority of construction applications. Roof decks, walls, and subfloors are all commonly made of OSB.

Masonry — Bricks, blocks, stones, and unreinforced and reinforced concrete products.

Veneer Walls — Walls with a surface layer of attractive material laid over a base of common material

Spalling — Expansion of excess moisture within masonry materials due to exposure to the heat of a fire, resulting in tensile forces within the material, and causing it to break apart. The expansion causes sections of the material's surface to violently disintegrate, resulting in explosive pitting or chipping of the material's surface.

Figure 4.3a Brick may be a misleading indicator of a previous fire situation because it rarely shows signs of heat-related deterioration.

Figure 4.3b Stone may indicate fire exposure through chips and pits in its surface. *Courtesy of Ron Moore and McKinney (TX) FD.*

Figure 4.3c Concrete blocks are often used as a load-bearing member of a structure, and they retain most of their strength and stability in a fire, even when they crack.

Figure 4.4a Cast iron has historically found use in structural facades and in other load-bearing applications.

Figure 4.4b Wrought iron is primarily used as a decorative feature, though it has some historical use as I-beams.

Figure 4.5 Steel is commonly used in large modern buildings. *Courtesy of Ron Moore and McKinney (TX) FD.*

Steel

Steel is the primary material used for structural support in the construction of large modern buildings **(Figure 4.5)**. Steel is also used for stairs, wall studs, window and door frames, and for balconies and railings. It is also used to reinforce concrete floors, roofs, and walls. Steel structural members lengthen (elongate) when heated. A 50-foot (15 m) beam may elongate by as much as 4 inches (100 mm) when heated from room temperature to about 1,000°F (538°C). If the steel is restrained from movement at the ends, it buckles and fails somewhere in the middle. The failure of steel structural members can be anticipated at temperatures near or above 1,000°F (538°C). The temperature at which a specific steel member fails depends on many variables, including the size of the member, the load it is under, the composition of the steel, and the geometry of the member. For example, a **lightweight steel truss** will fail much quicker than a large, heavy I-beam. To reduce the effect of heat on steel structural members, fireproofing materials such as sprayed-on concrete or sprayed-on insulation is used.

From a fire fighting perspective, you must be aware of the type of steel members used in a particular structure. Firefighters also need to determine how long the steel members have been exposed to heat; this gives an indication of when the members might fail. It is important to remember the critical temperature for steel is 1,000°F (538°C); this temperature can easily be reached at ceiling level from the rising heat and smoke. Always consider what effect the heat is having on these structural elements even if you cannot see them.

Elongating steel can actually push out load-bearing walls and cause a collapse. Also, if the walls are able to withstand the elongation, the steel will fail and sag somewhere in the middle, causing collapse of upper floors or the roof. Water can cool steel structural members and stop elongation, which reduces the risk of a structural collapse.

Aluminum

The use of aluminum increased throughout the 20th Century. Initial uses included decorative features such as the tower portion of the Empire State Building in New York City. Other decorative and functional uses include roofing, flashing, gutters, downspouts, window and door frames, and exterior **curtain wall** panels. Aluminum is used in the construction of sun rooms, screened porches, car ports, and awnings found on residential structures. Aluminum studs have replaced wood in many commercial and residential buildings **(Figure 4.6, p. 138)**. Acoustical tile ceilings are supported by aluminum framing and support wires that can create entanglement hazards for firefighters. Aluminum will be affected by heat more rapidly than steel.

Lightweight Steel Truss — Structural support made from a long steel bar that is bent at a 90-degree angle with flat or angular pieces welded to the top and bottom.

Curtain Wall — A nonload-bearing wall, often of glass and steel, fixed to the outside of a building and serving especially as cladding.

Figure 4.6 Commercial and residential structures commonly use aluminum studs. *Courtesy of Ron Moore and McKinney (TX) FD.*

Other Metals

Many other metals may be found in building construction. Tin has been used to produce metal ceiling tiles for over a century and is also used as a roof covering (**Figure 4.7**). Copper is found in wiring, pipes, gutters, and other decorative elements. Finally, lead is still found in pipes, flashing, and as a component of stained glass or leaded glass windows. These metals will fail when exposed to excessive amounts of heat.

Reinforced Concrete

Concrete can be poured in place at the construction site or formed into precast sections and transported to the site (**Figure 4.8**). Reinforced concrete is internally fortified with steel reinforcement bars (**rebar**) or wire mesh. This gives the material the compressive strength (the ability to withstand pressure on the surface) of concrete along with the tensile strength (the ability to withstand being pulled apart or stretched) of steel. While reinforced concrete does perform well under fire conditions, it can lose strength through spalling (**Figure 4.9**). Prolonged heating can cause a failure of the bond between the concrete and the steel reinforcement. Cracks and

Figure 4.7 Many modern structures are covered by metal roofs.

Figure 4.8 Concrete panels may be engineered to nearly any specification necessary for a structure. *Courtesy of Ron Moore and McKinney (TX) FD.*

Figure 4.9 Spalling is a visible indicator of damage to concrete.

Figure 4.10 Gypsum boards have a high water content and are used to shield structural steel elements. *Courtesy of Ron Moore and McKinney (TX) FD.*

spalling in reinforced concrete surfaces are an indication that damage has occurred and that strength may be reduced. In addition, prolonged exposure to chemicals can cause the steel reinforcing bars to corrode and the concrete bond to weaken before any exposure to fire, significantly reducing the time to failure. Therefore, it is very important to know the occupancy history of the structure.

Gypsum

Gypsum, also known as drywall or Sheetrock®, is an inorganic product from which plaster and wallboards are constructed **(Figure 4.10)**. It is unique because it has high water content that absorbs a great deal of heat as the moisture evaporates. The water content gives gypsum excellent heat-resistant and fire-retardant properties. Because it breaks down gradually under fire conditions, gypsum is commonly used to insulate steel and structural members. In areas where the gypsum has failed, the structural members behind it will be subjected to higher temperatures and could fail as a result.

Lath and Plaster

Lath and plaster construction is a process rather than a single material. Horizontal wood strips called lath are nailed to wall studs and covered with a mixture of plaster to form an interior wall finish **(Figure 4.11, p. 140)**. Lath and plaster construction is generally found in buildings constructed prior to the 1950s. Wire mesh was also used to replace the lath for ceilings in some houses. Interior lath and plaster walls can be very difficult to penetrate with axes or hand tools. The walls can also conceal fire within the cavity between surfaces and add fuel to the fire in the form of the studs and lath.

Figure 4.11 Lath and plaster walls may contribute to the fire conditions of a residence.

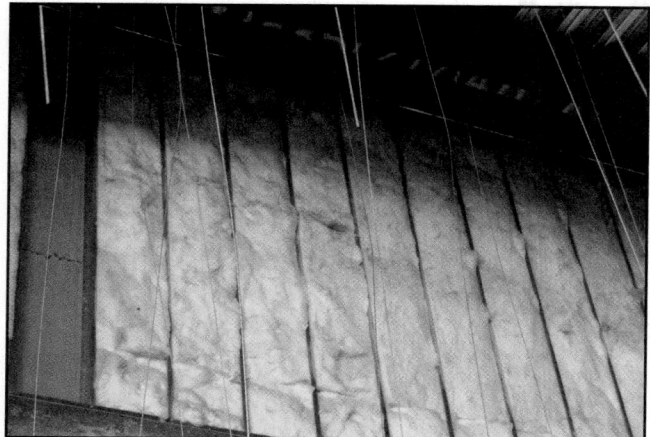

Figure 4.12 Fiberglass insulation may be bound into sheets using combustible adhesives.

Glass/Fiberglass

Glass is not typically used for structural support. It is used in sheet form for doors and windows and in block form for nonload-bearing walls. Wire-reinforced glass may provide some thermal protection as a separation, but conventional glass is not an effective barrier to fire extension. Heated glass may crack and shatter when it is struck by a cold fire stream.

Fiberglass is typically used for insulation purposes and is located between interior/exterior walls and between ceilings and roofs (**Figure 4.12**). The glass component of fiberglass is not a significant fuel, but the materials used to bind the fiberglass may be combustible and can be difficult to extinguish.

Types of Insulation

Besides fiberglass, there are many other types of insulation that you may encounter. Some of them are:

- **Asbestos** — A mineral fiber that was used before 1970 for insulation and as a fire retardant. It was most commonly used for pipe and furnace insulation, around door gaskets on furnaces, and for insulation on steam and hot water pipes. Asbestos is a known carcinogen. Although its use has been banned since 1989, it may still be encountered in older buildings. Since 2000, studies have determined that vermiculite insulation, used in attics and walls, may contain asbestos. Known as *Zonolite®*, this insulation can produce dangerous levels of airborne asbestos when heated.

- **Urea Formaldehyde Foam Insulation (UFFI)** — Originally used in the 1970s for insulating walls, this material caused high levels of formaldehyde emissions when improperly installed. Although it is currently not in use, it may still be found in some older houses.

- **Fiberglass** — A soft wool-like material used as insulation and also for textiles such as drapes. To reduce the loss of heated air, fiberglass was installed inside heating ducts between the 1960s and 1980s.

- **Mineral wool** — The term refers to two different materials: slag wool and rock wool. Accounting for approximately 80 percent of mineral wool produced, slag wool is produced primarily from iron ore blast furnace slag, an industrial waste product. Rock wool is produced from natural rocks. Prior to the 1960s, mineral wool was the most common type of insulation. It is currently becoming more popular as an insulation material.

- **Cellulose** — Most cellulose insulation is approximately 80 percent post-consumer recycled newspaper by weight; the rest is comprised of fire-retardant chemicals and, in some products, acrylic binders. Over time, cellulose loses its ability to be fire retardant. It is also a respiratory irritant when inhaled.

- **Cotton** — A type of insulation made up of cotton and polyester mill scraps, plastic fiber, and borates for fire resistance. It is as effective as fiberglass or cellulose, has fewer documented health risks than fiberglass, and is easy to install. It costs more than fiberglass, however, and is not widely available.

- **Straw** — Straw insulation is used in the exterior walls of **hybrid or natural (green) construction** buildings and houses. The disadvantages are that it can be affected by moisture and may become infected by insects and vermin.

- **Foam** — Foam insulation is applied in rigid boards called extruded and expanded polystyrene (EPS) or blown into wall cavities or voids in spray form. Spray foam can irritate the respiratory system. There are several types of foam insulation including:
  — Polyisocyanurate
  — Polyurethane
  — Polystyrene

There are potential risks to all types of insulation. Some present health risks while others may contribute to fire spread if the fire retardant has deteriorated. Some insulation may emit a toxic or harmful gas when heated. From a personal safety standpoint, it is important to wear the appropriate type of respiratory protection when performing overhaul operations.

Hybrid Construction — Type of building construction that uses renewable, environmentally friendly or recycled materials. *Also known as* Natural or Green Construction.

Plastic

Plastic is used in many forms as a building material. On the exterior of buildings, vinyl siding is used over older siding, foam insulation panels, or other materials. Vinyl siding does not require painting or continued maintenance like wood, and it adds additional insulation when applied properly. Water and sewer pipes are made from varying sizes of plastic pipe and fittings and are used to replace lead pipes (**Figure 4.13**). Other decorative plastic materials include moldings, wall coverings, and mantel pieces. Most plastic will melt and can contribute to the fuel load within a structure.

Composite Materials

The construction industry is using composite materials more frequently than in the past. One reason for this is a shortage of high-quality and large-diameter timber. Another reason is a trend to use recycled materials in place of hardwoods and other materials.

Generally, composite materials are manufactured by combining two or more distinctly different materials. This process results in lightweight materials with high structural strength, resistance to chemical wear, corrosion resistant, and heat resistance. The materials are cost effective and fairly easy to manufacture.

Figure 4.13 Plastic pipes are commonplace in new construction due to their low cost.

Figure 4.14 Particle board outgasses dangerous fumes when heated.

Examples of composite building materials include:

- **Finger-jointed timber** — Small pieces of wood that are joined into longer boards using epoxy resins and glues.

- **Laminated timber** — Also known as plywood or glulam (glue-laminated) wood, these materials are sheets of wood used for roof and floor decking, walls, and stair treads among other uses.

- **Medium density fiberboard (MDF)** — Another type of laminated wood product, MDF is closer in appearance and strength to hardwood. It is used for doors and door-surrounds, decorative moldings, rails, skirtings, and cornices.

- **Particle board** — Made from small particles and flakes generated in the manufacture of lumber, particle board is used for exterior and interior wall panels and furniture **(Figure 4.14)**. Urea formaldehyde is one of the types of glues used to manufacture particle board and can pose a health hazard due to outgassing when heated.

- **Synthetic wood** — This material, produced in sheets and boards, is manufactured from recycled plastic from liquid containers, primarily milk bottles. Synthetic wood is primarily used for exterior rails, stairs, and decks.

Construction Classifications

The type of building construction used in a structure is determined by the architect, structural engineer, or contractor. Locally adopted building codes regulate the type of construction based on the intended use (referred to as occupancy classification), structure size, and the presence or lack of an automatic fire suppression system. The type of building construction is determined by the materials used in the construction and how well the materials resist exposure to fire **(Figure 4.15)**.

Building codes are adopted by the authority having jurisdiction (AHJ) and amended to meet local requirements. Some building codes are locally developed although most are based on nationally accepted model building codes. In the United States, there are currently two major model building codes, the NFPA® 5000, *Building Construction and Safety Code®*, and the International Code Council's (ICC) *International Building Code® (IBC®)* building codes. Canada has one building code that may be adopted by the provincial or local governments.

Not all buildings in your jurisdiction will be regulated by the local building code, however. Factory-built homes, sometimes referred to as mobile or manufactured homes, are generally exempt from local building codes. They are instead regulated by the U.S. federal government through the Department of Housing and Urban Development. It is also possible that federal- or state-owned buildings, including offices, courthouses, university buildings, postal facilities, and other government facilities may be exempt from local building code requirements. Canadian Standards Associations (CSA) and the Standards Council of Canada (SCC) set the design requirements for manufactured housing in Canada.

The following sections describe the general characteristics of each construction type specified in the model building codes for the U.S. and Canada. Although there are minor differences between the model building codes, generally these construction types are common to each. Because local AHJs can amend model codes to meet their needs, you must be familiar with the building codes adopted within your jurisdiction.

Figure 4.15 Building codes regulate many features and processes used in a jurisdiction.

You must also be aware that when existing buildings are renovated, the result may be a structure containing more than one construction method. An example would be an ordinary construction commercial building originally built in the 1940s with a new addition of fire-resistive construction. The result is a structure that appears to be fire resistive but is actually vulnerable to rapid fire development. Renovations may improve the fire and life safety of buildings through the addition of such approved items as fire alarm systems or sprinkler systems. It is also possible that individuals may renovate residential structures to create open floor plans, add living spaces in attics or basements, or change electrical wiring without legally required permits. These modifications create potential hazards for occupants and firefighters.

United States Construction

Both the *IBC®* and the NFPA® classify buildings in five types of construction (Type I through Type V). The types are then further divided into subcategories, depending on the code and construction type. Each construction type is defined by the construction materials and their performance when exposed to fire. Every building is composed of the following building elements:

- Structural frame
- Floor construction
- Roof construction

Figure 4.16 Type I fire-resistive construction is engineered to resist fire for 3-4 hours. *Courtesy of Ron Moore and McKinney (TX) FD.*

Type I (Fire Resistive)

Type I construction (also known as *fire-resistive construction*) provides the highest level of protection from fire development and spread as well as collapse (**Figure 4.16**). All structural members are composed of noncombustible or limited-combustible materials with a high fire-resistive rating. Structural components such as walls, floors, and ceilings must be able to resist fire for a period of 3 to 4 hours depending on the component.

Type I construction can be expected to remain structurally stable during a fire, and it is considered to be the most collapse resistant. Reinforced concrete, precast concrete, and **protected steel** frame construction meet the criteria for Type I construction. Fire walls are used to limit fire spread through structures such as multifamily apartments. However, the protection provided by fire walls can be reduced or eliminated when owners and contractors make unprotected penetrations through these walls for pipes, wires, and ducts. Heat, smoke, and flames can quickly progress through these openings, spreading the fire into compartments that might otherwise have been protected.

Type I structures are often incorrectly referred to as being fireproof. Even though the structure will not burn, the structure may degrade from the effects of fire. Although the use of Type I construction provides structural stability should a fire occur, the addition of combustible materials in the form of contents, furniture, and interior finish can generate sufficient heat over time to compromise the structural integrity of the building.

During a fire, the following conditions may result from Type I construction:

- Compartments can retain heat contributing to the potential for rapid fire development.

- Roofs are extremely difficult to penetrate for the purpose of ventilation due to construction material and design.

- Windows may be nonoperating, causing them to be very difficult to open for ventilation.

Type II (Noncombustible or Limited Combustible)

Buildings classified as Type II construction (also known as *noncombustible* or *limited-combustible construction*) are composed of materials that will not contribute to fire development or spread. Type II construction consists of noncombustible materials that do not meet the stricter requirements of those materials used in the Type I building

Protected Steel — Steel structural members that are covered with either spray-on fire proofing (an insulating barrier) or fully encased in an Underwriters Laboratories Inc. (UL) tested and approved system.

Figure 4.17 Type II construction is engineered to limit the spread of fire, but may not be fire resistive. *Courtesy of Ron Moore and McKinney (TX) FD.*

Figure 4.18 Type III construction has interior features that are made of wood and exterior walls and supports that are noncombustible.

classification (**Figure 4.17**). Steel components used in Type II do not need to be protected for the same lengths of time or have the same **fire-resistance rating** as Type I. Structures with metal framing members, metal cladding, or concrete-block construction of the walls with metal deck roofs supported by unprotected open-web joists are the most common form of this construction type. The fire-resistance rating is generally half that of Type I, or 1 to 2 hours, depending on the component. These buildings are more prone to collapse because they are constructed of lighter-weight materials with lower fire-resistance ratings.

Type II construction is normally used when fire risk is expected to be low, or when fire suppression and detection systems are designed to meet the fuel load of the contents. You must always remember that the term noncombustible does not always reflect the true nature of the structure. Lower fire-resistive ratings are permitted for roof systems and flooring. The fire-resistant metal roof decking may be covered with combustible layers of foam insulation and felt paper covered with asphalt waterproofing. Heated by a fire below the metal deck, these combustible materials can melt and ignite causing a second fire above the roof. Additionally, combustible features, such as balconies and facades, can be included on the exterior of Type II structures.

Type III (Ordinary Construction)

Type III construction (also known as *ordinary construction*) is commonly found in older schools and mercantile, business, and residential structures such as the one discussed in the chapter Case History. This construction type requires that exterior walls and structural members be constructed of noncombustible materials (**Figure 4.18**). Interior walls, columns, beams, floors, and roofs are completely or partially constructed of wood.

Type III construction buildings may contain a number of conditions that can affect their behavior during a fire, including the following:

- Voids exist inside the wooden channels created by roof and truss systems and between wall studs that will allow for the spread of a fire unless **fire stops** are installed in the void (**Figure 4.19, p. 146**).

- Older Type III structures may have undergone renovations that have contributed to greater fire risk due to the creation of large hidden voids above ceilings and below floors that may create multiple concealed voids.

Fire-Resistance Rating — Rating assigned to a material or assembly after standardized testing by an independent testing organization; identifies the amount of time a material or assembly will resist a typical fire, as measured on a standard time-temperature curve.

Fire Stop — Solid materials, such as wood blocks, used to prevent or limit the vertical and horizontal spread of fire and the products of combustion in hollow walls or floors, above false ceilings, in penetrations for plumbing or electrical installations, in penetrations of a fire-rated assembly, or in cocklofts and crawl spaces.

Figure 4.19 Fire stops reduce the amount of open area in void spaces in which fire can travel.

- Structural components may have been removed to change the configuration, or to open up floor space, during renovations. This may reduce the load-carrying capacity of the supporting structural member.

- A change in building use or occupancy may result in additional loads that the building was not designed to carry.

- Prefabricated wood truss systems similar to those used in Type V construction may also be found in new Type III structures. These systems may fail quickly when exposed to fire.

Renovations Can Create Safety Hazards

Renovations to structures can change the interior arrangement, exit paths, and structural integrity of a structure. For their own safety, it is extremely important that firefighters monitor the changes that are continually being made in the buildings within their response areas.

Type IV (Heavy Timber/Mill Construction)

Type IV construction (also known as *heavy timber/mill construction*) is characterized by the use of large-dimensioned lumber (**Figure 4.20**). These dimensions vary depending on the particular building code in use at the time of construction; however, as a general rule these structural members will be greater than 8 inches (203.2 mm) in dimension with a fire-resistance rating of 2 hours. The dimensions of all structural elements, including columns, beams, joists, and girders must adhere to minimum dimension sizing. Any other materials used in construction and not composed of wood must have a fire-resistance rating of at least 1 hour.

Type IV structures are extremely stable and resistant to collapse due to the sheer mass of their structural members. When involved in a fire, the heavy timber structural elements form an insulating effect derived from the timbers' own char that reduces heat penetration to the inside of the beam.

Figure 4.20 Heavy timber construction is very resistant to collapse and combustion.

Figure 4.21 Type V wood frame construction may appear to be Type III because of a veneer, but the actual fire-resistance rating is not comparable.

Exterior walls are constructed of noncombustible materials. Interior building elements such as floors, walls, and roofs are constructed of solid or laminated wood with no concealed spaces. This lack of voids or concealed spaces helps to prevent fire travel.

Modern Type IV construction materials may include small-dimensioned lumber that is glued together to form a laminated structural element (sometimes called glulam elements). These elements are extremely strong and are commonly found in churches, barns, auditoriums, and other large facilities with vaulted or curved ceilings. Glulam beams may fail when exposed to fire because the glue holding the laminates together may be affected by the heat.

Type IV construction buildings may contain a number of conditions that can affect their behavior during a fire:

- The high concentration of wood can contribute to the intensity of a fire once it starts.
- Collapse of masonry walls can be caused by loss of structural integrity of timbers.

Type V (Wood or Stick Frame)

Type V construction is commonly known as *wood frame* or *stick frame* (**Figure 4.21**). The exterior **load-bearing walls** are composed entirely of wood. A veneer of stucco, brick, or stone may be constructed over the wood framing. The veneer offers the appearance of a Type III construction while providing little additional fire protection or structural support to the structure. Perhaps the most common example of this type of construction is a single-family dwelling or residence, although there are many multistory Type V construction apartment buildings in existence.

Load-Bearing Wall — Walls of a building that by design carry at least some part of the structural load of the building in the direction of the ground or base.

Figure 4.22a Aluminum siding is a facade covering that offers long-term protection against weather damage.

Figure 4.22b Wood clapboard may show damage from weather over time.

Figure 4.22c Brick veneer offers superior protection against weather as well as some exposure protection against fire.

Figure 4.22d Shake shingles potentially increase the fire load in a neighborhood.

Stud — An upright post in the framework of a wall for supporting sheets of lath and plaster, wallboard, or similar material.

Type V construction consists of framing materials that include wood 2 x 4 or 2 x 6 inch (50 mm by 100 mm or 50 mm by 152 mm) **studs**. Most structures built in northern climates mandate 6 inch (152 mm) exterior wall cavities for increased insulation. The outside of the framing members is covered with any of the following (**Figures 4.22 a-d**):

- Aluminum siding
- Shake shingles
- Wood clapboards
- Sheet metal
- Cement, stucco, or other masonry products
- Plastic (vinyl) siding

- Planks
- Plywood
- Composite wood (chipboard, particle board, fiberboard)
- Styrofoam (which needs extra bracing) or stucco
- Veneers (brick or stone)
- Asphalt siding (sheets designed to look like brick siding)

Exterior siding is attached by nails, screws, or glue. In the case of stucco, it is spread over a screen lattice that is attached to the framing studs.

Type V construction now includes the use of prefabricated wood truss systems in place of solid floor joists. The truss system creates a large, open void area between the floors of a structure, rather than the closed channel system found with solid wood floor joists. When wood I-beams are used, they are usually constructed of thin plywood or wood composite, glued to 2 x 4 inch (50 mm by 100 mm) size, forming the top and bottom of the truss (**Figure 4.23**). These wood I-beams may have numerous holes cut in them to allow for electric, communication, and utility lines to be extended through them. Under fire conditions these plywood I-beams fail and burn much more rapidly than solid lumber.

Figure 4.23 Wood I-beams have a higher surface area than solid lumber and are a liability in fire conditions.

Manufactured Structures

A manufactured home is a structure that is built in a factory and shipped to the location where it is installed. Manufactured homes take many forms and include mobile homes that have an axle assembly under the frame. Current estimates indicate that manufactured homes make up 25 percent of all housing sales in the U.S. (**Figure 4.24**).

Manufactured homes are not required to conform to the model building codes. They are required to conform to a U. S. Department of Housing and Urban Development (HUD) standard that is similar to Type V construction. The unit's fire resistance will vary depending on age. Manufactured homes built before 1976 have less fire resistance than those of current construction. Lightweight building materials in some manufactured homes are susceptible to early failure in a fire, a major disadvantage. The heat produced by the building's burning contents will cause these materials to ignite or melt rapidly. These contents have the same fuel loads as those found in conventional structures. At the same time, the use of lightweight materials makes forced entry much easier because walls can be quickly breached. Manufactured homes may be anchored directly to a concrete slab or have open crawl spaces beneath them, the latter of which provides an additional source for oxygen during a fire.

Figure 4.24 Manufactured homes are characterized by their modular construction that may be exempt from model building codes.

According to an NFPA® analysis of fires in residential occupancies, there has been a steady decline in fires in manufactured homes since 1980. The analysis compares manufactured homes built before the HUD standard was enacted in 1976 (referred to as *prestandard*) and those constructed after 1976 (*poststandard*). Reductions in fire loss and fatalities can be attributed to construction requirements that include:

- Factory-installed smoke alarms

- Use of flame-retardant materials in interior finishes

- Use of flame-retardant materials around heating and cooking equipment

- Installation of safer heating and cooking equipment

- Installation of gypsum board rather than wood paneling in interior finishes

- Factory-installed fire suppression systems (sprinklers)

Factory-Built Homes

Factory-built homes (also referred to as *manufactured*, *prefabricated*, *modular*, and *industrialized* housing) is a generic term used to describe structures that are partially or completely built in a factory and shipped to the location on which they are to be installed. There are five categories of factory-built homes:

- ***Manufactured homes*** — Of all the types of factory-built homes, manufactured homes are the most common type, almost completely prefabricated prior to delivery, and the least expensive. The HUD code preempts all local building codes and is more stringent than model building codes. Because the HUD code is based on performance standards, it tends to encourage construction innovations. Manufactured homes usually have a permanent steel undercarriage and are delivered on wheels towed by a transport vehicle. They normally range from one-section single-wide homes to three section triple-wide homes.

- ***Modular homes*** — Modular or sectional homes must comply with the same local building codes as site-built homes. Only about six percent of all factory-built structures are modular homes. Modular sections can be stacked vertically and connected horizontally in a variety of ways. The modular section is transported to the site and then attached to a permanent foundation, which may include a full basement.

- ***Panelized homes*** — Panelized homes are assembled on-site from pre-constructed panels made of foam insulation sandwiched between sheets of plywood. The individual panels are normally 8 feet (2.44 m) wide by up to 40 feet (12.19 m) long. The bottom edges of the wall panels are recessed to fit over the foundation sill. Each panel includes wiring chases. Because the panels are self-supporting, framing members are unnecessary.

- ***Pre-cut homes*** — Pre-cut homes come in a variety of styles including pole houses, post-and-beam construction, log homes, A-frames, and geodesic domes. The pre-cut home consists of individual parts that are custom cut and must be assembled on-site.

- ***Hybrid modular homes*** — One of the most recent developments in factory-built homes, the hybrid modular structure includes elements of both the modular design and the panelized design. Modular core units such as bathrooms or mechanical rooms are constructed in the factory, moved to the site and assembled. Preconstructed panels are then added to the modules to complete the structure.

In addition to these types of manufactured homes, prefabricated kit homes have existed for many years. The Sears and Roebuck Company sold over 70,000 kits through its catalog between 1908 and 1940. Today, it is virtually impossible to tell these houses from site-built structures.

In Canada, houses may be constructed from steel shipping containers. Marketed as *Sea Can Houses*, each container is a separate module that is connected to others to construct a complete house.

Canadian Construction

The *National Building Code of Canada (NBCC)* defines three types of building construction:

1. **Combustible construction** — Construction that does not meet the requirements for noncombustible construction.

2. **Noncombustible construction** — Construction in which the degree of fire safety is attained by the use of noncombustible materials for structural members and other building assemblies.

3. **Heavy timber construction** — Combustible construction in which a degree of fire safety is attained by placing limitations on the sizes of wood structural methods and the thickness and composition of wood floors and roofs; also it avoids concealed spaces under floors and roofs.

To enable Canadian code users to understand these definitions, the NBC identifies specific requirements and limitations on materials used for each type of construction within the code. These requirements are listed in table formats that are easy to read and understand based on the occupancy classification and construction type.

In recent years both the harsh winter climate and attempts at increasing energy efficiency have combined to increase the potential for structural collapse in new construction in Quebec, Canada. Residential dwellings may now be built under the design requirements of the *Novoclimat standard* (**Figure 4.25, p. 152**). Basically, it is a standard intended to make new homes more energy efficient and better insulated. The result is an almost airtight structure constructed with smaller dimensional lumber. Air is heated and circulated through a closed system of heating, ventilation, and air conditioning (HVAC) units, thermo-pumps, and other devices. These structures are more likely to fail rapidly under fire conditions because thermal heat and fire gases are trapped within the compartments and structure. The design also makes vertical ventilation more difficult.

Occupancy Classifications

Structures are also classified by their intended use or occupancy. Occupancy classifications are defined by the building code and life safety code adopted by the AHJ. The three primary model building codes in use in North America are NFPA® 5000, *IBC®*, and the *National Building Code of Canada (NBCC)*. **Table 4.1, p. 153** provides a comparison of the occupancy classifications found in these three codes. Structures may be divided into either *single use* occupancies or *separated use* occupancies.

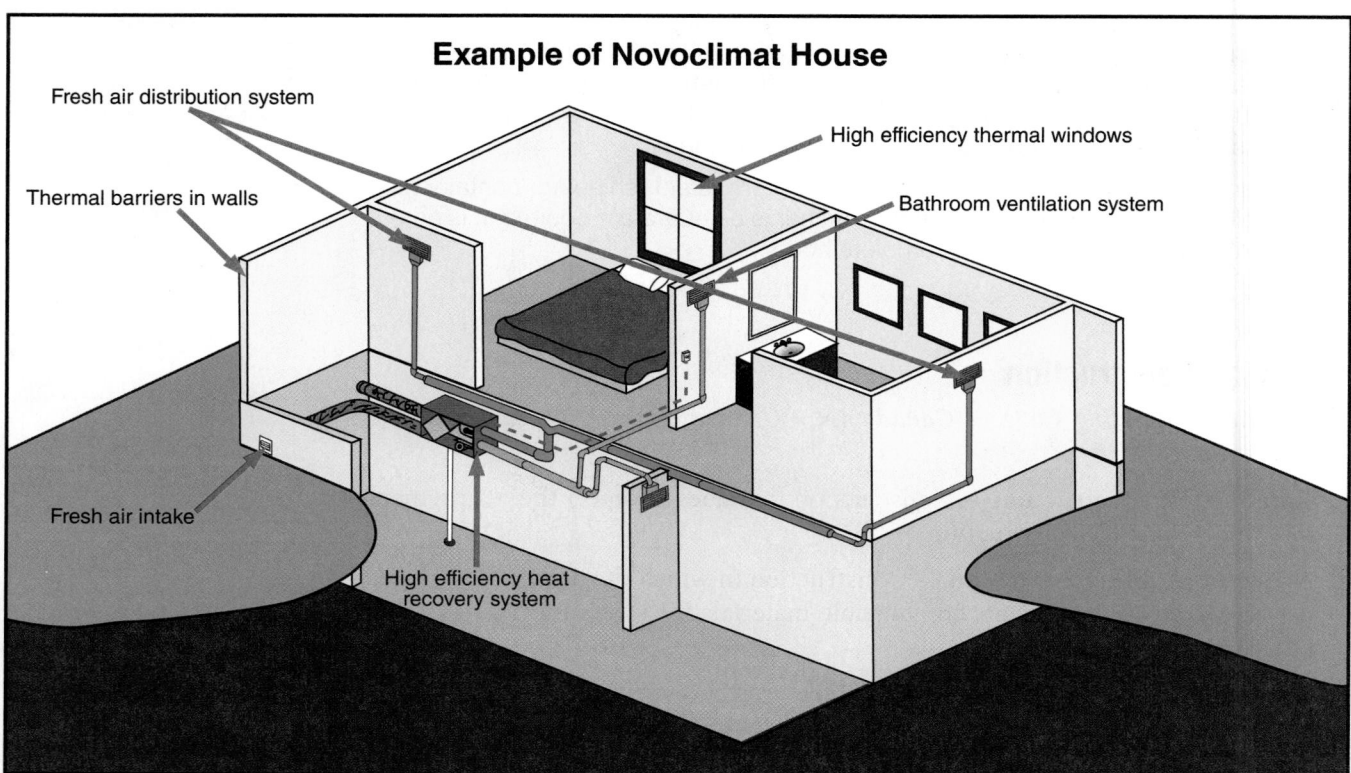

Example of Novoclimat House

Fresh air distribution system

Thermal barriers in walls

Fresh air intake

High efficiency heat recovery system

High efficiency thermal windows

Bathroom ventilation system

Figure 4.25 Residences built under the Canadian Novoclimat standard are well insulated and nearly air-tight.

Single Use

A single-use structure must meet the building code requirements for its intended use. For instance, an office building must meet the requirements found in the *Business Occupancy Classification* while an elementary school must meet the requirements of an *Educational Occupancy* **(Figure 4.26, p. 157)**. Requirements include exit access, emergency lighting, fire protection systems, construction type, and fire separation barriers, among many others.

In many cases, a structure such as an industrial facility will contain multiple types of uses including storage, manufacturing, and offices. The structure is generally classified by its primary function. For example, a furniture manufacturing plant located in a single structure would be classified as a *Manufacturing Occupancy* even though it contains areas for storage of raw materials, finished products, offices, and manufacturing. Fire separation walls and fire protection systems may be required to protect one area from another.

Separated Use

Structures that contain multiple occupancies or use groups must meet the requirements for each individual occupancy classification. That is, in a strip mall, each space is classified by its use and separated from the other units by a fire-rated wall, as required by the building code. The installation of a fire suppression system may also be required.

Within the strip shopping mall, there may be retail outlets (Mercantile Occupancy), offices (Business Occupancy), and restaurants (Assembly Occupancy) **(Figure 4.27, p. 157)**. However, while each occupancy in a separated use structure may have met its own requirements at initial construction, those occupancies may change over time

Table 4.1
Comparative Overview of Occupancy Categories

This table is a general comparative overview of the occupancy categories for three major model code systems. Readers must consult the locally adopted code and amendments for complete information regarding each of these occupancies.

Occupancy	ICC	NFPA®	NBCC
Assembly	**A-1** — Occupancies with fixed seating that are intended for the production and viewing of performing arts or motion picture films. **A-2** — Those that include the serving of food and beverages; occupancies have nonfixed seating. Nonfixed seating is not attached to the structure and can be rearranged as needed. **A-3** — Occupancies used for worship, recreation, or amusement such as churches, art galleries, bowling alleys, amusement arcades, as well as those that are not classified elsewhere in this section. **A-4** — Occupancies used for viewing of indoor sporting events and other activities that have spectator seating. **A-5** — Outdoor viewing areas; these are typically open air venues but may also contain covered canopy areas as well as interior concourses that provide locations for vendors and other commercial kiosks.	**Assembly Occupancy** — An occupancy (1) used for a gathering of 50 or more persons for deliberation, worship, entertainment, eating, drinking, amusement, awaiting transportation, or similar uses; or (2) used as a special amusement building, regardless of occupant load.	**Group A Division 1** — Occupancies intended for the production and viewing of the performing arts. **Group A Division 2** — Occupancies not classified elsewhere in Group A. **Group A Division 3** — Occupancies of the arena type. **Group A Division 4** — Occupancies in which occupants are gathered in open air.
Business	**Business Group B** — Buildings used as offices to deliver service-type or professional transactions, including the storage of records and accounts. Characterized by office configurations to include: desks, conference rooms, cubicles, laboratory benches, computer/data terminals, filing cabinets, and educational occupancies above the 12th grade.	**Business** — Occupancy used for the transaction of business other than mercantile.	**Group D** — Business and personal services occupancies

Table 4.1 (Continued)

Occupancy	ICC	NFPA®	NBCC
Educational	***Educational Group E*** — Buildings providing facilities for six or more persons at one time for educational purposes in grades kindergarten through twelfth grade. Religious educational rooms and auditoriums that are part of a place of worship, which have occupant loads of less than 100 persons, retain a classification of Group A-3.	**Educational Occupancy** — Occupancy used for educational purposes through the twelfth grade by six or more persons for 4 or more hours per day or more than 12 hours per week.	Covered under Group A
Factory Industrial	***Factory Industrial Group F*** — Occupancies used for assembling, disassembling, fabrication, finishing, manufacturing, packaging, repair, or processing operations. - ***Factory Industrial F-1 Moderate Hazard*** (examples include but not limited to: aircraft, furniture, metals, and millwork) - ***Factory Industrial F-2 Low Hazard*** (examples include but not limited to: brick and masonry, foundries, glass products, and gypsum) ***High Hazard Group H*** — Buildings used in manufacturing or storage of materials that constitute a physical or health hazard. - ***High-Hazard Group H-1*** — Detonation hazard - ***High-Hazard Group H-2*** — Deflagration or accelerated burning hazard - ***High-Hazard Group H-3*** — Materials that readily support combustion or pose a physical hazard - ***High-Hazard Group H-4*** — Health hazards - ***High-Hazard Group H-5*** — Hazardous production	**Industrial Occupancy** — Occupancy in which products are manufactured or in which processing, assembling, mixing, packaging, finishing, decorating, or repair operations are conducted.	**Group F Division 1** — High-hazard industrial occupancies **Group F Division 2** — Medium-hazard occupancies **Group F Division 3** — Low-hazard industrial occupancies

Table 4.1 (Continued)

Occupancy	ICC	NFPA®	NBCC
Occupancy Institutional (Care and Detention)	***Institutional Group I*** **Group I-1** — Assisted living facilities holding more than 16 persons on a 24-hour basis. These persons are capable of self-rescue. **Group I-2** — Medical, surgical, psychiatric, or nursing care facilities for more than 5 people who are not capable of self-preservation or need assistance to evacuate. **Group I-3** — Prisons and detention facilities for more than 5 people under restraint. **Group I-4** — Child and adult day care facilities.	**Ambulatory Healthcare** — Building (or portion thereof) used to provide outpatient services or treatment simultaneously to four or more patients that renders the patients incapable of taking action for self-preservation under emergency conditions without the assistance of others. **Healthcare** — An occupancy used for purposes of medical or other treatment, or care of four or more persons where such occupants are mostly incapable of self-preservation due to age, physical or mental disability, or because of security measures not under the occupants' control. **Residential Board and Care** — Building or portion thereof that is used for lodging and boarding of four or more residents, not related by blood or marriage to the owners or operators, for the purpose of providing personal care services. **Detention and Correctional** — An occupancy used to house one or more persons under varied degrees of restraint or security where such occupants are mostly incapable of self-preservation because of security measures not under the occupants' control.	**Group B Division 1** — Care or detention occupancies in which persons are under restraint or are incapable of self-preservation because of security measures not under their control. **Group B Division 2** — Care or detention occupancies in which persons having cognitive or physical limitations require special care or treatment.
Mercantile	***Mercantile Group M*** — Occupancies open to the public that are used to store, display, and sell merchandise with incidental inventory storage.	**Mercantile** — An occupancy used for the display and sale of merchandise.	**Group E** — Mercantile occupancies
Residential	***Residential Group R*** **R-1** — Residential occupancies containing sleeping units where the occupants are primarily transient in nature (boarding houses, hotels, and motels). **R-2** — Residential occupancies containing sleeping units or more than 2 dwelling units where the occupants are primarily permanent in nature (apartments, convents, nontransient hotels, etc…).	**Residential Occupancy** Provides sleeping accommodations for purposes other than healthcare or detention and correctional. **One- and Two-Family Dwelling Unit** — Building that contains not more than two dwelling units with independent cooking and bathroom facilities.	**Group C** — Residential occupancies

Table 4.1 (Concluded)

Occupancy	ICC	NFPA®	NBCC
Residential (continued)	**R-3** — Residential occupancies where the occupants are primarily permanent in nature and not classified as Group R-1, R-2, R-4, or I. **R-4** — Residential occupancies shall include occupancies buildings arranged for occupancy as residential care/assisted living facilities for more than 5 but less than 16 occupants (excluding staff).	**Lodging or Rooming House** — Building (or portion thereof) that does not qualify as a one- or two-family dwelling, that provides sleeping accommodations for a total of sixteen or fewer people on a transient or permanent basis, without personal care services, with or without meals, but without separate cooking facilities for individual occupants. **Hotel** — Building or groups of buildings under the same management in which there are sleeping accommodations for more than sixteen persons and primarily used by transients for lodging with or without meals. **Dormitory** — A building or a space in a building in which group sleeping accommodations are provided for more than sixteen persons who are not members of the same family in one room, or a series of closely associated rooms, under joint occupancy and single management, with or without meals, but without individual cooking facilities. **Apartment Building** — Building (or portion thereof) containing three or more dwelling units with independent cooking and bathroom facilities.	**Group C** — Residential occupancies
Storage	***Storage Group S*** — Structures or portions of structures that are used for storage and are not classified as hazardous occupancies. - ***Moderate-Hazard Storage, Group S-1*** (examples include but not limited to: bags, books, linoleum, and lumber) - ***Low-Hazard Storage, Group S-2*** (examples include but not limited to: asbestos, bagged cement, electric motors, glass, and metal parts)	**Storage Occupancy** — An occupancy used primarily for the storage or sheltering of goods, merchandise, products, vehicles, or animals.	Covered under Group F
Utility/ Miscellaneous	***Utility/Miscellaneous Group U*** These are accessory buildings and other miscellaneous structures that are not classified in any specific occupancy (agricultural facilities such as barns, sheds, and fences over 6 ft [2m]).	—	—

Figure 4.26 Single-use occupancies must meet a specific set of requirements for their intended use.

Figure 4.27 In a strip mall, independent spaces may host a range of occupants without meeting all of the requirements for the most hazardous occupancy.

without proper construction adjustments. For instance, a strip mall that was initially intended to contain only Mercantile Occupancies may have been built with limited fire separations due to the similarity of hazards in the structure. Over time, other types of occupancies, such as a restaurant, may have moved in. The fire separation wall may not have the required rating and the wall may not extend through the roof to form a complete barrier. In addition, unauthorized and non-code-compliant penetrations may have been made in fire walls that will permit the spread of fire and smoke. These conditions emphasize the importance of periodic preincident building surveys by fire companies, fire inspections, building construction plans review, and permit and code enforcement.

Building Components

Regardless of the type of building construction or occupancy, buildings are all composed of the same components. Each of the following has a function and can prevent or contribute to the growth of a fire:

- Foundations
- Floors/ceilings
- Walls
- Roofs
- Stairs
- Doors
- Windows

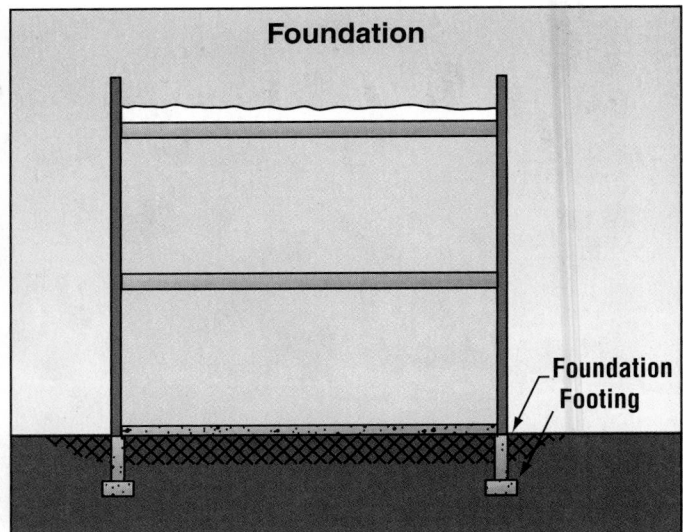

Figure 4.28 The foundation and footings transmit the load of a building to the soil beneath.

Foundation

Foundation Footing

Foundations

A building's foundation is designed to support the weight of the building and all its contents. Foundations may be shallow or deep. A shallow foundation extends a few feet (meters) into the earth around the perimeter of the structure (**Figure 4.28**). The foundation sits on a footing made from poured, reinforced concrete or concrete blocks. The first floor is constructed upon the foundation, taking the form of a solid concrete slab or a stem wall with a wood or metal joist floor that creates a crawl space between the floor and the soil below. Single-story basements can also be constructed on shallow foundations.

Deep foundations are used to support the mass of a large area or tall building. Deep foundations include piers or pilings driven into the soil, drilled shafts, caissons, helical piles, and earth-stabilized columns (**Figure 4.29**). Multiple basement levels may also rest on the piers. Specialized systems may also be used in earthquake-prone regions.

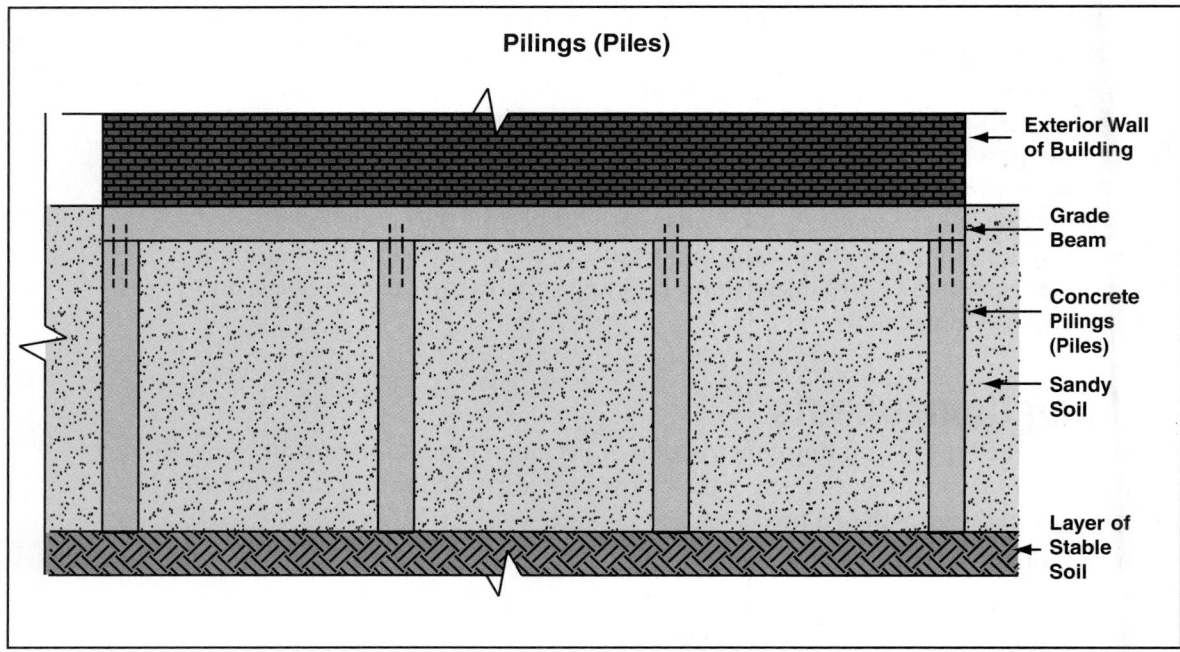

Pilings (Piles)

Exterior Wall of Building

Grade Beam

Concrete Pilings (Piles)

Sandy Soil

Layer of Stable Soil

Figure 4.29 Deep foundations stabilize large or tall structures.

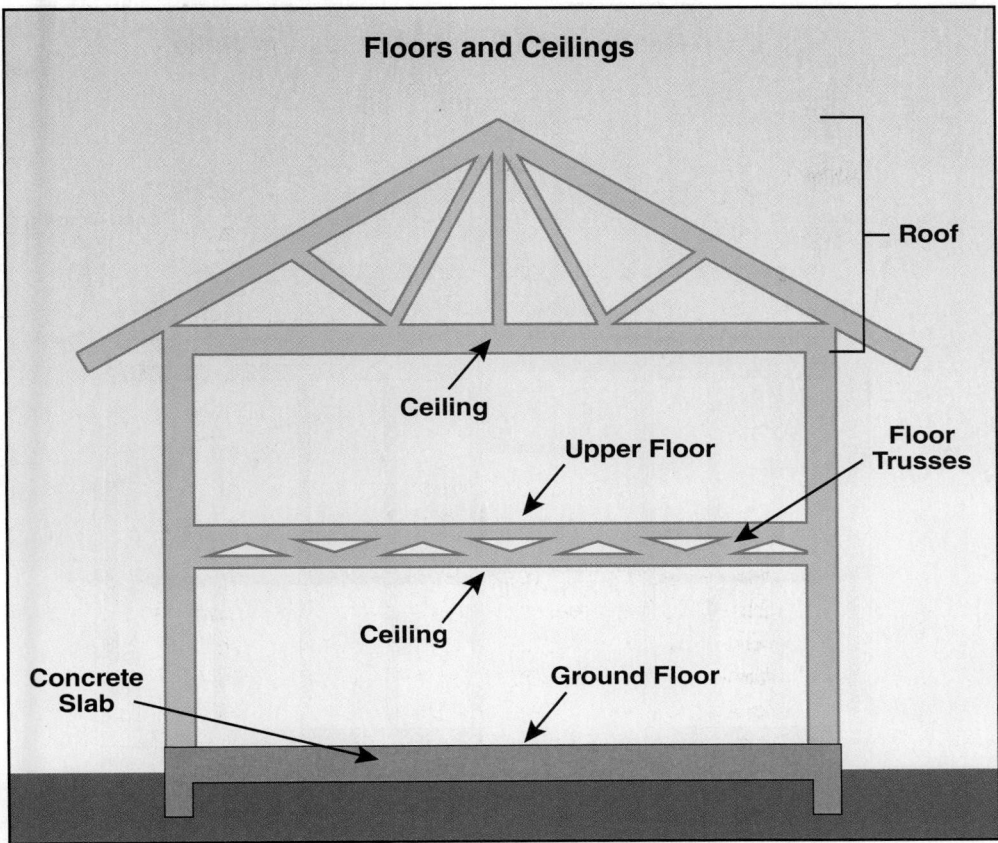

Floors and Ceilings

Roof

Ceiling

Upper Floor

Floor Trusses

Ceiling

Concrete Slab

Ground Floor

Figure 4.30 Floors and ceilings are interrelated in the structure of a building.

Floors/Ceilings

Floors and ceilings form the top and bottom of a compartment with the walls forming the sides. Floors at ground level may consist of a concrete slab or a floor assembly made up of joists and decking over a crawl space or basement **(Figure 4.30)**. Upper floors of a multistory building consist of the joists and decking with the ceiling attached to the bottom. The top level of a building consists of a ceiling, joists or rafters, and the roof above. The space formed by the floor/ceiling or ceiling/roof may contain duct work or open return air plenum, electrical or communications wiring, water or natural gas pipes, or pipes for a fire suppression system. Recessed lighting and speakers for audio systems may also be located above the ceiling. Fiberglass, cellulose, or foam insulation may also be located in the space beneath the roof or under the floors to act as soundproofing.

NOTE: Some states and local jurisdictions have passed laws and ordinances that require buildings other than single-family dwellings to display a warning placard indicating the type of floor or ceiling truss system used in the structure. If your jurisdiction has this requirement, you must be familiar with it and look for the placards when you respond to a structure fire.

Floor and ceiling assemblies may be constructed of a combination of materials. Floors may be poured reinforced concrete; cellular concrete over metal decking; or finished wood, tile, or carpet surface over a wood subfloor attached to metal or wood joists. Ceilings are generally gypsum board, tin tiles, or lath and plaster attached to joists, roof trusses, or beams.

The ceilings in corridors that are designated as exit or egress passageways will have the same fire-resistance rating as the walls in that corridor. The material used to cover the floors is also rated to limit flammability in the corridor. Ratings are indicated in hours.

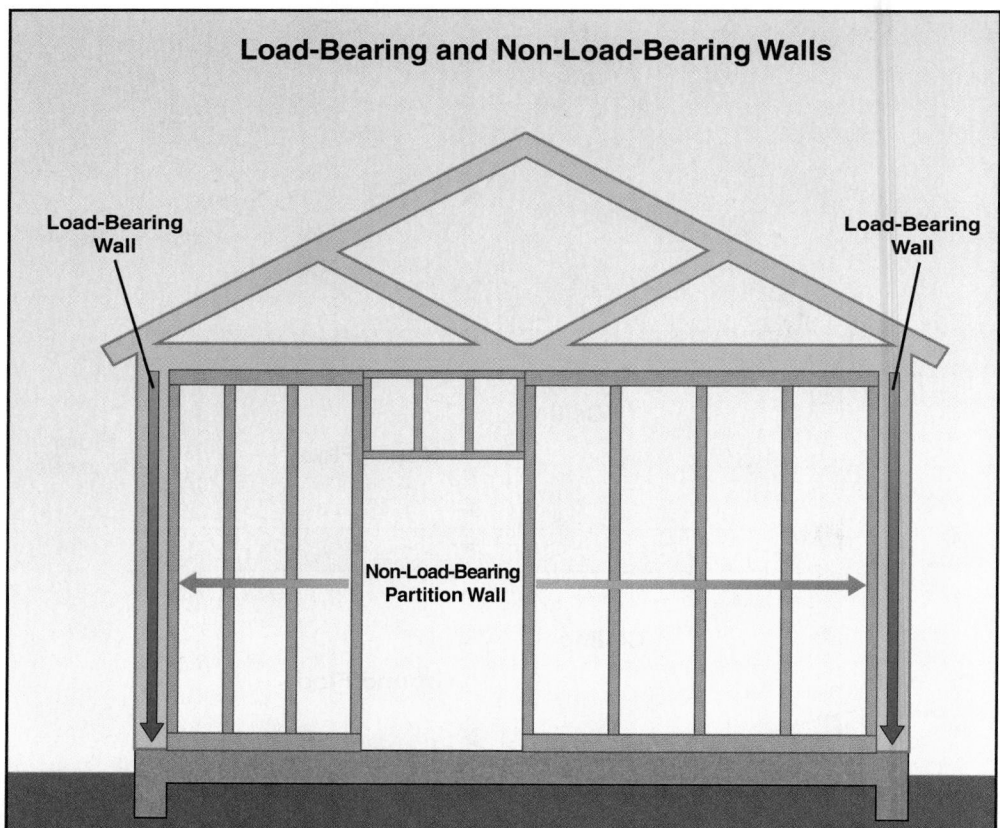

Figure 4.31 Knowing the difference between load-bearing and non-load-bearing walls allows firefighters to make sound decisions during fire evolutions.

Walls

Walls define the perimeter of the building, as well as divide it into compartments or rooms. Exterior walls may be wood or metal siding attached to studs, a single layer of concrete, concrete blocks, or logs. They may also be an assembly of studs with the exterior material on the outside and an interior covering on the inside. Wall assemblies consist of a bottom plate, top plate, vertical studs, and horizontal braces sandwiched between two surfaces made of gypsum or lath and plaster. The cavity formed by the two surfaces may be a dead air space or void or contain some form of insulation.

Interior and exterior walls that support the weight of the structure or structural components are called load-bearing walls. Walls that only support their own weight are **non-load-bearing walls** and may act as a **partition wall** dividing two areas within a structure (**Figure 4.31**). Interior wall assemblies may be rated for a specified fire resistance time depending on the local building code. Rated wall assemblies may be continuous from the floor to the bottom of the next floor. Unrated walls may only extend to the room ceiling creating an open space over multiple rooms.

Because masonry materials do not burn, a variety of masonry walls are used in the construction of **fire walls** (**Figure 4.32**). Fire walls are intended to provide separation that meets the requirements of a specified fire-resistance rating. Fire wall assemblies include the wall structure, doors, windows, and any other protected openings meeting the required protection-rating criteria. Fire walls may be used as **party walls** to separate two adjoining structures or two occupancy classifications within the same structure to prevent the spread of fire from one to the other. Fire wall assemblies can also divide large structures into smaller portions and contain a fire to a particular portion of the structure.

Non-Load-Bearing Wall — Wall, usually interior, that supports only its own weight.

Partition Wall — Interior nonload-bearing wall that separates a space into rooms.

Fire Wall — Fire-rated wall with a specified degree of fire resistance, built of fire-resistive materials and usually extending from the foundation up to and through the roof of a building, that is designed to limit the spread of a fire within a structure or between adjacent structures.

Party Wall — A load-bearing wall shared by two adjacent structures.

Figure 4.32 Masonry walls separate areas that require fire protection rating.

Exterior walls and fire walls are the most difficult to penetrate when attempting to force entry into or to escape an area. Skills required for forcing entry through walls are included in Chapter 11, Forcible Entry. Some interior walls can be penetrated but only to locate hidden fires or to create an escape path.

Roofs

The primary function of the roof is to protect the structure and its contents from the effects of weather. The shape and construction of the roof is intended to provide drainage, support the weight of accumulations of snow, resist the effects of wind, and insulate the interior from external temperature changes. Therefore, the geographic location of the structure can influence the type and construction of the roof. Penetrations or openings in the roof are indications of the general arrangement of the rooms within and may be used to assist in vertical ventilation during a fire.

Roof Types

In general, you will encounter three prevalent types of roof shapes: flat, pitched, and arched. Some buildings have a combination of these roof designs (**Figure 4.33**). Some of the more common styles that compose these types are the gable, hip, gambrel, shed, mansard, and lantern. Less common styles are sawtooth and butterfly roofs. Many residential structures can have a combination of these styles which can contribute to voids underneath and limit the ability of firefighters to access areas under the roof. **Figure 4.34, p. 162** illustrates the various styles of roofs commonly found in North America.

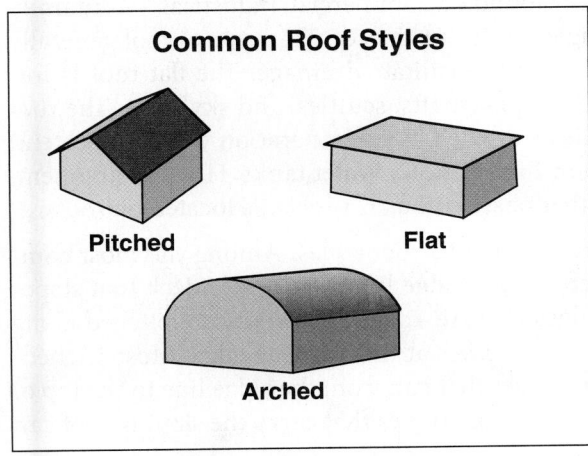

Common Roof Styles

Pitched   Flat   Arched

Figure 4.33 Pitched, flat, and arched roof styles are prevalent in most jurisdictions.

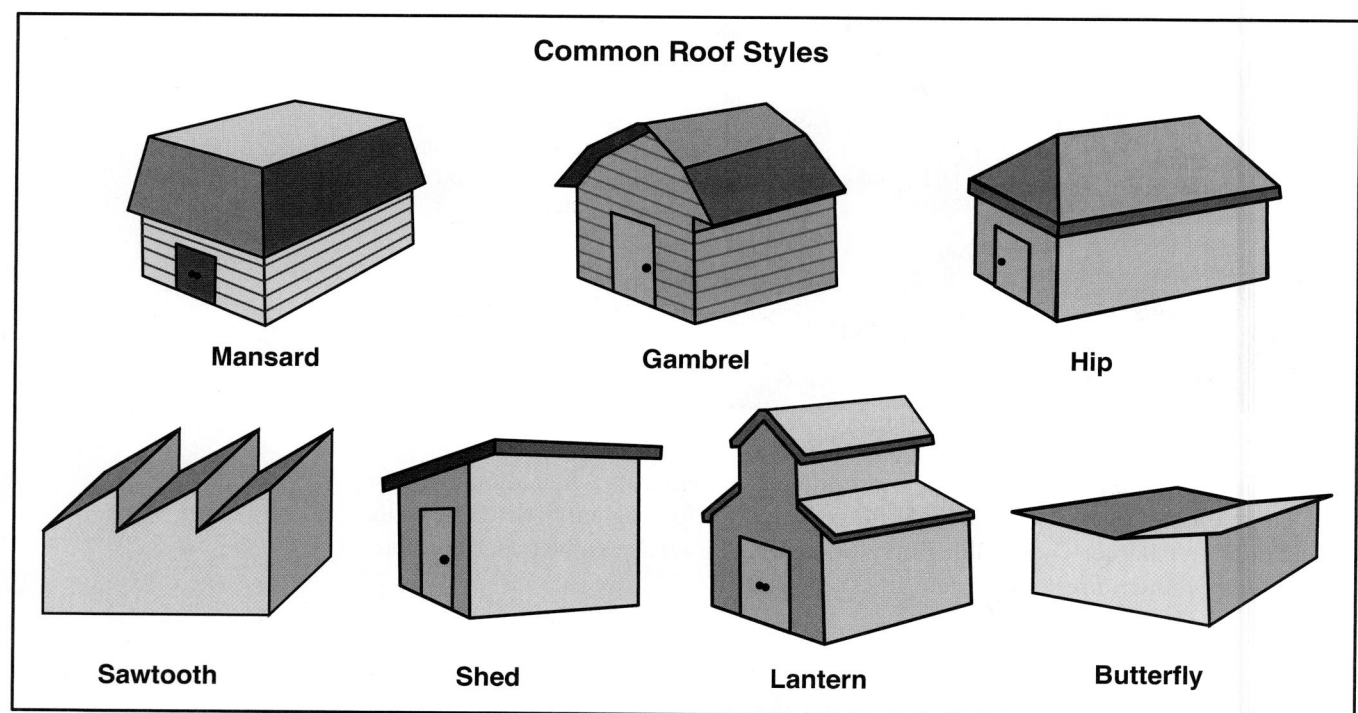

Common Roof Styles

Mansard

Gambrel

Hip

Sawtooth

Shed

Lantern

Butterfly

Figure 4.34 Roof styles that combine pitched, flat, and arched features present unique hazards and benefits to firefighters.

Figure 4.35 A flat roof may be surrounded by a parapet wall to protect and disguise industrial equipment.

Flat roofs. Flat roofs are commonly found on commercial, industrial, multifamily residential structures, and some single-family residences. This type of roof generally has a slight slope toward the outer edge to facilitate drainage. The flat roof is frequently penetrated by chimneys, vent pipes, shafts, scuttles, and skylights. The roof may be surrounded by **parapet** walls or divided by fire separation walls that extend from the foundation to above the roof (**Figure 4.35**). Water tanks, HVAC equipment, antennas, solar panels, signs, and other obstructions may also be located on roofs.

Pitched roofs. There are a number of pitched roof styles. Among the most common are those elevated in the center along a **ridge** line with a roof deck that slopes down to the **eaves** along the roof edges (**Figure 4.36**). Shed roofs are pitched along one edge and the deck slopes down to the eaves at the opposite edge. Most pitched-roof construction involves rafters or trusses that run from the ridge line to the top of the outer wall at the eaves level. The rafters or trusses that carry the sloping roof can

Figure 4.36 Residential pitched roofs often have a center ridge line and eaves.

Figure 4.37 Arched roofs are commonly used in occupancies that require large internal spaces uninterrupted by columns or other supports.

be made of wood or metal. Over these rafters, the roof decking or sheathing material is applied at right angles. Decking, usually in the form of plywood or oriented strand board (OSB), is sometimes applied solidly over the entire roof. In other applications, decking consists of boards or planks set with a small space between them. This is commonly called *skip* sheathing. Pitched roofs usually have a covering of roofing paper applied before the finish surface is laid. The finish may consist of shingles, asphalt roll roofing, metal panels, slate, or tile.

Arched roofs. Arched roofs are ideal for some types of occupancies because they can span large open areas unsupported by columns, pillars, or posts **(Figure 4.37)**. Construction of arched roofs primarily occurred from the late 1800s to the mid-1900s due to the availability of inexpensive lumber. The design of an arched roof depends primarily on the exterior walls to support the weight of the roof. Four types of arched roofs include bowstring, ribbed, diagonal grid (Lamella) and pleated barrel, as illustrated in **Figure 4.38, p. 164**.

Arched Roofs Contribute to Firefighter Casualties

The rapid collapse of arched roofs involved in fire has contributed to many firefighter injuries and fatalities making them one of the most dangerous types of roof construction.

Roof Construction

Roofs are made up of three main components including the roof supporting structure, the roof deck or sheathing, and the roof covering. On some types of structures, such as single-family residences, the roof covering is the most visible component when viewed from the outside. On other types of structures, such as the one described in the Case History, a view of the entire roof may be obstructed by the height of the building, parapet walls, or adjacent buildings. When viewed from inside, the roof supports and deck may be visible or covered by a ceiling assembly which creates a void that may be a few inches (millimeters) or many feet (meters) in depth.

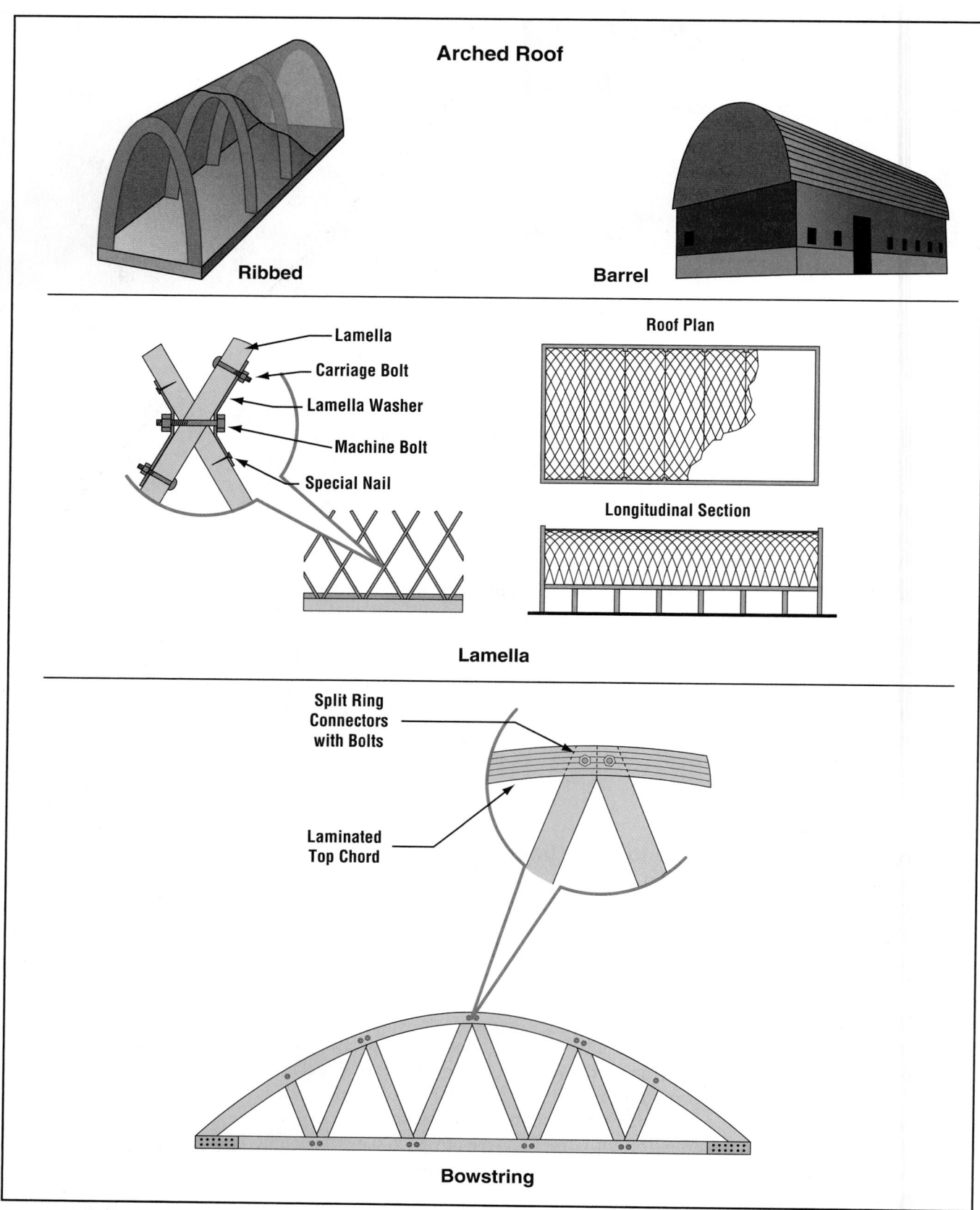

Figure 4.38 Four types of arched roofs are used in structures compliant with specific exterior wall support parameters.

Figure 4.39 Roof beams serve as the framework for the roof decking. *Courtesy of Ron Moore and McKinney (TX) FD.*

Figure 4.40 Metal gussets hold some styles of wood trusses in place.

Roof supports. There are two general types of roof supports used in residential and commercial construction. These are beams and truss assemblies.

Beams are the sections of lumber located directly under the roof decking. On a pitched roof they extend from the ridge line or pole at the peak to each side wall. On flat roofs, the beam extends from wall to wall. The beam may be exposed or concealed behind a ceiling. Beams are generally made of solid timbers, 4 x 4 inch (100 mm by 100 mm) and larger (**Figure 4.39**).

Roof trusses assemblies may be conventional framing constructed on site or pre-manufactured assemblies built in a factory and shipped to the site. Trusses that are constructed on site take longer to construct and consist of top and bottom chords and webbing that extend from the peak to the walls. The ends are connected by a horizontal joist that also has supports between the joist and the **rafters**. Wood trusses are assembled using metal **gusset plates**, also called *gang nails*, that only penetrate about ⅜ inch (9.5 mm) into the wood (**Figure 4.40**). There are a variety of types of trusses including parallel chords, pitched chord, and arched truss, not to be confused with bowstring trusses (**Figure 4.41**). The **parallel chord truss** is generally used to support flat roofs and floor assemblies. They may be constructed from wood or metal. Premanufactured roof truss assemblies are used in the construction of over 60 percent of new homes built in North America.

Rafter — Inclined beam that supports a roof, runs parallel to the slope of the roof, and to which the roof decking is attached.

Gusset Plates — Metal or wooden plates used to connect and strengthen the joints of two or more separate components (such as metal or wooden truss components or roof or floor components) into a load-bearing unit.

Parallel Chord Truss — A truss constructed with the top and bottom chords parallel. These trusses are used as floor joists in multistory buildings and as ceiling joists in buildings with flat roofs.

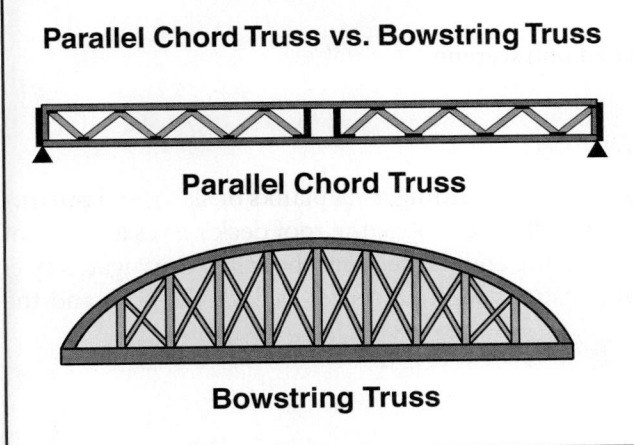

Figure 4.41 Two types of common trusses include the parallel chord truss and the bowstring truss.

The traditional wood-joisted roof uses solid wood **joists** that tend to lose their strength gradually when they are exposed to fire. This loss of strength results in the roofs becoming soft or "spongy" before failure, especially with a wood plank roof deck. Although the softening or sagging of a roof is an obvious indication of structural failure, it should not be considered the only sign of imminent collapse. The plywood or OSB used for roof sheathing can fail quickly and without warning because it is relatively thin, between ½ and ¾ inch (12.7 mm and 19 mm).

In modern practice, box beams and I-beams, also referred to as *wide flange beams*, manufactured from plywood and wood truss joists are often used to support flat roofs and floors. Although these beams provide adequate strength, the thin web portion of plywood I-beams renders them susceptible to early failure in a fire. The relatively slender members used to manufacture a truss are also susceptible to early failure. In addition, the open web design of truss joists permits the rapid spread of fire in directions perpendicular to the truss joist instead of simply along the long dimension of the member.

There has been an increased use of engineered or lightweight construction and trussed support systems in building construction. Engineered construction systems are manufactured from smaller pieces of wood or light gauge steel to form trusses that weigh less than traditional systems made from solid wood or heavy gauge steel. Lightweight construction is becoming more common, and other than Type I and Type IV construction, it can be found in homes, apartments, small commercial buildings, and warehouses. Two of the most common types of engineered construction systems involve the use of lightweight steel or wooden trusses. Lightweight steel trusses are made from long steel chords that are either straight or bent by as much as 90 degrees with either flat or tubular members in the web space. **Lightweight wooden trusses** are constructed of 2 x 3, 2 x 4, or 2 x 6 inch (50 mm by 75 mm, 50 mm by 100 mm, or 50 mm by 150 mm) lumber connected by gusset plates made of wood or metal.

Flat roofs are also supported by open-web steel joists and steel beams. Depending on the fuel load within the structure, unprotected lightweight open-web joists can be expected to fail quickly in a fire.

Roof decks. The roof deck is the portion of roof between the roof supports and the roof covering (**Figure 4.42**). Types of roof decks found in North America include:

- Plywood sheathing
- Oriented Strand Board (OSB)
- Wood tongue and groove
- Corrugated metal
- Sprayable concrete encapsulated polystyrene
- Reinforced concrete
- Double tee preformed concrete

The components of roof decks include sheathing, roof planks or slabs, and **purlins** (**Figure 4.43**). Sometimes, as in concrete deck roofs, the roof deck serves as the roof support. In other cases, the roof covering and the deck are the same. Corrugated steel decking is frequently used in applications where it serves as both the deck and the exterior roof covering.

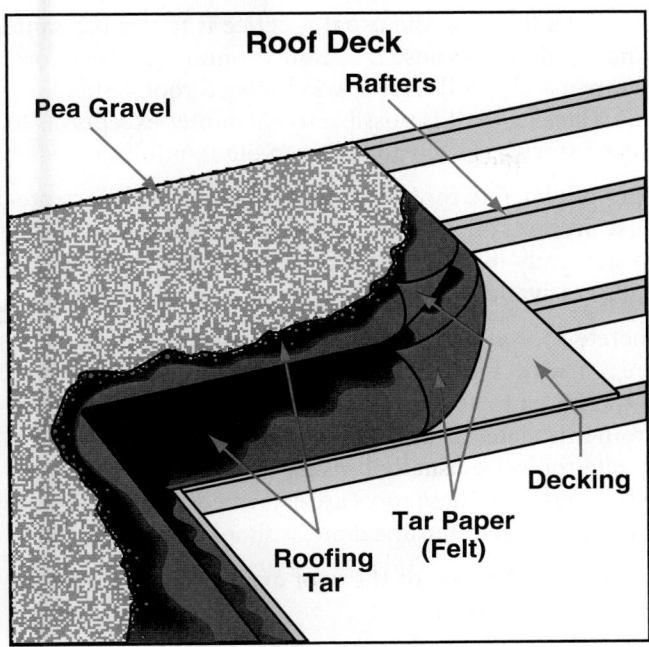

Roof Deck

Pea Gravel

Rafters

Decking

Tar Paper
(Felt)

Roofing
Tar

Figure 4.42 A roof deck is composed of one of several kinds of sheeting that is anchored to the rafters, and then treated or covered to prevent water penetration.

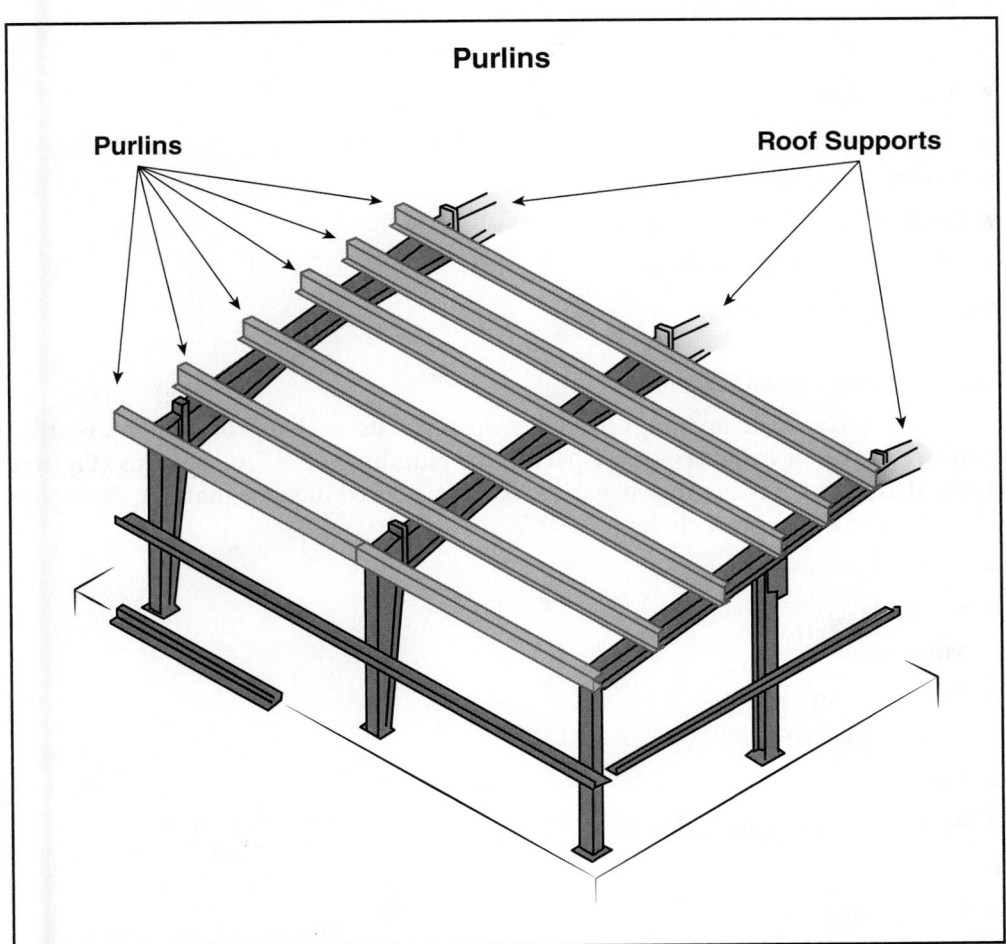

Purlins

Purlins

Roof Supports

Figure 4.43 Purlins cross the roof supports and provide lateral stability for roofing materials.

If a multistory building is to have a flat roof, the usual practice is to use the same structural system for the roof and the floors because it is more economical. Therefore, a building with wood-joisted floors usually will have a wood-joisted roof system and a steel-framed building will have a steel roof. It is possible to encounter exceptions to this general rule, especially where a story has been added to an older building.

There are essentially two types of concrete roofs common in North America: precast and poured-in-place. Because precast concrete roof units can be fabricated off-site and hauled to the construction site ready for use, this type of roof construction is in widespread use. Precast roof slabs are available in many shapes, sizes, and designs.

Another type of precast concrete roof uses lightweight material made of gypsum plaster and Portland cement mixed with aggregates, such as perlite, vermiculite, or sand. This material is sometimes referred to as lightweight concrete. Lightweight precast planks are manufactured from this material, and the slabs are reinforced with steel mesh or rods. Lightweight concrete roofs are usually finished with roofing felt and a mopping of hot tar to make them watertight. These roofs are extremely difficult to penetrate, causing a ventilation problem. When it must be done, it should be a last resort.

Roof coverings. The roof covering is the part of the roof exposed to the weather (**Figures 4.44a-d**). Roof-covering materials include:

- Wooden shingles or shakes (rough cut wood)
- Molded metal or rubber imitation shingles or tiles
- Asphalt shingles
- Asphalt sheets
- Terracotta or concrete tile, slate, synthetic membrane
- Blown-on foam
- Built-up tar and gravel surface
- Metal roof systems or sheets
- Composite materials

Roof Penetrations and Openings

Roof penetrations and openings include a variety of items that provide light, ventilation, access, vapor exhausts, or are part of the plumbing or HVAC systems (**Figures 4.45a-d**). Many roof openings may be locked or secured in some manner. Roof penetrations and openings include:

- Scuttle hatches
- Skylights
- Monitors
- Automatic smoke vents
- Ventilation shafts
- Ventilation fans
- Penthouse or bulkhead doors
- Chimneys
- HVAC exhausts
- Bathroom vent pipes
- Attic vents
- Dormers

Figure 4.44a Wood shingles protect against weather but increase the fire load of a house.

Figure 4.44b Terracotta tiles offer some fire resistance, but are expensive and delicate.

Figure 4.44c Layers of tar and gravel serve to waterproof a roof and prevent the erosion of roofing materials. *Image by R. Bird - Aperture Innovation.*

Figure 4.44d Metal roofs offer some degree of fire resistance.

Figure 4.45a Skylights mark a penetration in a roof that is covered by a different kind of material than the rest of the roofing components.

Figure 4.45b A roof monitor is used to ventilate an attic space.

Figure 4.45c Ventilation fans serve as part of an HVAC system.

Figure 4.45d Chimneys vent smoke and must adhere to applicable rules and guidelines to prevent smoke damage to roofing components.

Penetrations can indicate the location of some types of rooms such as bathrooms or mechanical spaces. Monitors, smoke and attic vents, scuttle hatches, and skylights may be used to gain access to attics and **cocklofts** and may be used to provide an exit point for some types of ventilation. See Chapter 13, Tactical Ventilation, for more information.

Roof Obstructions

Besides the type of roof, its construction, and openings, you must also be aware of obstructions located on the roof. Observing the presence of these items during pre-incident surveys, the initial size-up, and at other times will help in the event the structure needs to be ventilated. Roof obstructions include:

- Green roofs
- Cold roofs
- Photovoltaic roofs
- Rain roofs
- Electrical power lines
- Security
- Structural modifications
- Roof-mounted equipment
- Nonpermitted modifications

Figure 4.46 Green roofs have many benefits including an increase in air quality.

Green roofs. Recent years have seen an increase in designs to protect the environment and conserve energy. One result in the desire for energy conservation has been the development of green roofs or vegetative roof systems. A green roof involves the use of the roof surface of a building for a rooftop garden (**Figure 4.46**). There are several benefits to this use. One is the increased insulating effects between the building interior and the outside. Probably the greatest benefit, however, is in the increase in air quality due to the oxygen-carbon dioxide exchange of growing plants, particularly in urban areas.

A green roof can take several forms. A green roof or rooftop garden can vary from the use of potted plants and flower boxes to a layer of earth with growing plants covering a large area of a roof. Green roofs can be developed on existing roofs and in new construction.

A rooftop garden constitutes a **dead load** on the roof structural system, which must be capable of supporting the load. The layer of earth required for a rooftop garden can vary from a few inches (millimeters) to 1 or 2 feet (0.3 m to 0.6 m). Depending on the depth of the soil, the dead load can vary from 20 pounds per square foot to 150 pounds per square foot (100 kg/sqm to 750 kg/sqm). In new construction, the structural engineer can provide for this load in the structural plans just as is done for snow loads. However, when a garden is planned for an existing roof, the existing structural system must be analyzed to ensure its adequacy.

Under fire conditions the increased load can accelerate structural failure, particularly if the roof is combustible. Green roofs can also interfere with ventilation practices and fire location indicators. Other concerns include:

- The effects of high-velocity winds and uplift wind pressures
- Roof drainage which can add weight to the roof creating a collapse hazard
- The exposure hazard that may be created by dry vegetation on the roof
- The need for a clear space between vegetation and fire walls that penetrate the roof

Cold roofs. A cold roof system is a type of roof installation generally found in cold, snowy climates to prevent ice damming and icicle formation at eaves. A cold roof system is designed to prevent interior heat from escaping into the attic space thus melting the snow on the roof. The installation requires a sheeting layer to be installed above the top chord of the truss or rafter. Above this sheeting layer, a membrane (similar to Tyvek®) is installed to prevent condensation buildup. Above the membrane, two layers of 1 x 4 inch (25 mm by 100 mm) parallel spacers are installed to create a 3-inch (75 mm) void air space, which acts as insulation against the warm air from the home. A second layer of sheeting is installed above the parallel spacers, followed by the roof covering system. The cold roof layering system design can create significant difficulty for firefighters during vertical ventilation operations.

In colder climates, structures that do not have cold roofs are more likely to have ceiling trusses lined with a vapor barrier and covered with at least 12 to 16 inches (300 mm to 400 mm) of insulation. The roof and soffit would then have multiple vents. The membrane over the roof sheathing provides some insulation even though its primary purpose is to protect the plywood sheathing from moisture that may seep in when the snow melts.

Photovoltaic roofs. A photovoltaic (solar energy) system produces clean and reliable energy that can be used in a wide range of applications. Photovoltaic cells in panels can be laid on top of a roof or embedded in the roof **(Figure 4.47)**.

Although solar energy represents a clean source of energy, the electricity generated by the operation of the solar system represents a significant hazard for firefighters. Even if power to the building is shut off, the panels retain a significant amount of electricity. If sunlight is available, the panels will continue to produce power. Research has shown that panels will continue to produce electricity even if they are fire damaged. It is not safe to break photovoltaic cells or skylights that are solar powered. Furthermore, the panels themselves represent a significant tripping and falling hazard.

Figure 4.47 Photovoltaic roofs use solar power to contribute to the energy needs of a building.

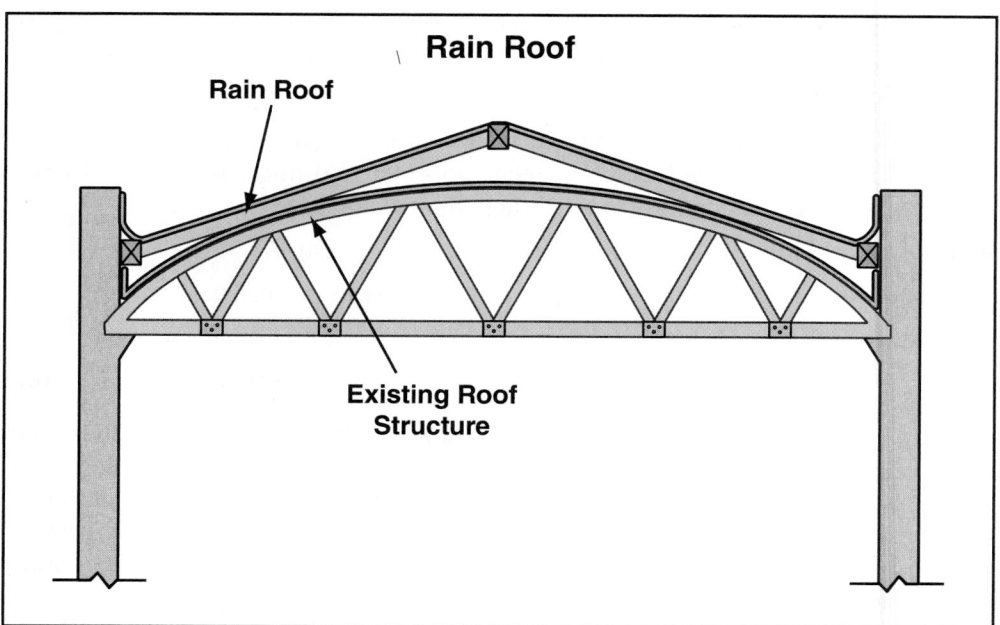

Figure 4.48 Voids are created when a rain roof is installed over an existing roof.

In emergency conditions, electrical shock, inhalation exposure, falls from roofs, and roof collapse always represent serious safety considerations. For these reasons, it is crucial that fire departments conduct thorough preincident planning to identify these structures rather than encountering them during adverse conditions. Solar panels, like other elements on a roof, may not even be visible from the ground on a building with a flat roof.

Rain roofs. **Rain roofs** may be found on commercial buildings, schools, and residential structures **(Figure 4.48)**. Generally, they are pitched roofs placed over older flat roofs for esthetic purposes, to prevent leaks, channel moisture off the roof, and as a more cost-effective alternative to repairing or resurfacing the original roof. Rain roofs may also exist over mobile homes to provide a more permanent appearance. They may be constructed from lightweight metal panels and trusses to form a peak or simply a second flat roof surface made from wood and roofing materials. The void created by the rain roof can conceal a fire and allow it to burn undetected. As the trusses are exposed to fire, they will weaken, increasing the potential for collapse of both the rain roof and the original roof. HVAC units may also be hidden from view under the rain roofs, adding to the collapse potential.

Ventilating a rain roof will not remove smoke from the structure until the original roof is penetrated. However, you must not enter the void area to cut a hole in the original roof because:

- You can become trapped in the void.
- You can be overcome by the heat and smoke trapped in the void.
- You can fall through a weakened original roof.
- You can be caught in an extreme fire condition as the heated gases mix with fresh air.

Rain Roof — A second roof constructed over an existing roof.

Security. In many communities, metal security bars or grilles are mounted over doors and windows to prevent illegal entry. This type of security will slow entry, create emergency exit hazards, and reduce the effectiveness of ventilation tactics. Wired glass may also be encountered in skylights on roof tops. This type of glass is difficult to penetrate and will take time to remove.

Other types of security may also be encountered. For example, in one city, a store owner installed iron plates on the roof of a grocery store to prevent break-ins. During a subsequent fire, the weight of the plates caused the roof to collapse. Installations like this will also prevent any type of timely tactical vertical ventilation. Only thorough preincident surveys have a chance of uncovering barriers like these.

WARNING

Unauthorized security modifications create extreme life safety hazards for firefighters.

Permitted structural modifications. Buildings are constructed to specific requirements developed by trained and certified architects and structural engineers. These requirements, based on locally adopted building codes, are intended to provide occupants with the highest level of fire and life safety possible. Permits are issued by the building official and construction is inspected to ensure compliance with the building code. Over the life of a structure you can expect that modifications will be made to improve and update it. The modifications must meet the local building codes having been inspected and approved by the local building official. The owner/occupant or contractor submits plans that are reviewed and approved by the building official and a permit issued. Modifications may include additions to the structure, removal of non-load-bearing walls or partitions, replacement of fire escapes with enclosed stairways, or sealing windows or doors that are not required for emergency exits. Even permitted modifications can affect fire fighting operations in the structure, making it essential that you are aware of all modifications to structures in your response area.

Nonpermitted modifications. Even though permits and inspections are required by local ordinance, owners or occupants may make unapproved or nonpermitted modifications during construction or renovation that can inhibit effective ventilation and increase the risk of fire extension and structural collapse.

Unapproved and hazardous modifications include:

- Removal of load-bearing interior walls
- Removal of load-bearing pillars or columns
- Removal of roof supports
- Increasing dead load by installing HVAC or mechanical equipment on or under roofs or in attic spaces
- Storing heavy contents on roof support beams or in attic spaces
- Removing or modifying automatic ventilation systems or components
- Removing or modifying code-required fire detection and suppression systems

- Altering attic spaces into living spaces
- Removing fire stops from wall cavities
- Installing or altering interior wall arrangements that affect air flow patterns
- Removing interior doors or sealing exterior openings
- Sealing basement ventilation openings
- Penetrating fire walls
- Installing unapproved rain roofs over flat roofs

The only way that these modifications can be uncovered is through an effective and periodic inspection of all commercial and industrial occupancies in the community. Fire company surveys, while not true inspections, can help to locate and alert code enforcement personnel to changes in structures. Unfortunately, it is impossible to monitor or be aware of modifications that are made to single-family residential structures.

Roof-mounted equipment. Most commercial, industrial, institutional, educational, and some residential structures have equipment mounted on the roofs. This is particularly true of structures with flat roofs. These items add a **live load** to the dead load distributed on the roof, increase collapse hazards, and add to the obstructions that will affect ventilation efforts. Some of these items may also be found on flat roofs underneath rain roofs. Roof-mounted equipment includes (**Figure 4.49**):

- HVAC units
- Water towers
- Telecommunications equipment
  - Telephone towers
  - Radio transmission equipment
  - Television antennas and satellite dishes
- Advertising signs or billboards
- Recreation areas
- Wind generators
- Electrical transformers
- Derricks, hoists, and cranes
- Winches
- Steeples, minarets, spires, and crosses (**Figure 4.50**)
- Electrical lines and weather heads

Many firefighters have been injured and killed because roof-mounted equipment has caused the collapse of a fire-weakened roof. Being aware of fire behavior and how it affects a building as well as the general design and construction of the fire building can help to keep you safe.

Stairs

Stairs provide access to, or egress from, different levels of a structure. Stairs that are a part of the required **means of egress** must provide protection for the occupants as they travel to safety. Stairs meeting these requirements are called *protected* or *enclosed* because they are built to resist the spread of fire and smoke.

Live Load — (1) Items within a building that are movable but are not included as a permanent part of the structure; merchandise, stock, furnishings, occupants, firefighters, and the water used for fire suppression are examples of live loads. (2) Force placed upon a structure by the addition of people, objects, or weather.

Means of Egress — Continuous and unobstructed way of exit travel from any point in a building or structure to a public way, consisting of three separate and distinct parts: exit access, exit, and exit discharge.

Figure 4.49 Roof-mounted equipment increases the dead load.

Figure 4.50 Decorative features such as tall steeples may increase the hazard to emergency responders under fire conditions because of the increased weight on the roof and the increased collapse zone.

Stairs that are not required to be a part of the means of egress system and typically connect no more than two levels are called *access* or *convenience* stairs. Stairs can be classified as either interior or exterior stairs, depending on their location.

The design or layout of a set of stairs may take any of several different forms (**Figure 4.51, p. 176**). Although exterior fire escapes, escalators, and fixed ladders have been used as a means of egress in the past, they are no longer allowed as a required means of egress from normally occupied spaces.

Protected Stairs

Interior protected stairs are critical components of the life safety system of a building. Protected stairs are enclosed with fire-rated construction, usually with either a 1- or 2-hour rating, depending on building height. Also, protected stairs generally serve two stories or more and are part of the required means of egress. They are the primary egress paths from floors above or below ground level and can adversely affect the safety of occupants if they do not maintain a breathable atmosphere.

Exterior Stairs

Exterior stairs may be either open to the air or enclosed. Enclosed exterior stairs must comply with requirements similar to those of interior protected stairs (**Figure 4.52, p. 177**). Open stairs are naturally ventilated but may be partially enclosed from the weather. They typically have at least two adjacent sides open to natural ventilation.

Fire Escapes

Fire escapes are open metal stairs and landings attached to the outside of a building (**Figure 4.53, p. 177**). The lowest flight may consist of a swinging stair section to limit unwanted access. Building codes have not permitted fire escapes in new construction for many decades.

Fire escapes that have been in place for many years may not be able to support the required live load created during emergency evacuations or fire suppression operations. Fire escapes usually are anchored to the building and are not supported at ground level. These anchor points are subject to the freeze-thaw cycle, corrosion from pollution and weather, and temperature changes. The mortar in which the anchors are set may suffer from deterioration or may have originally been inadequate for the expected load.

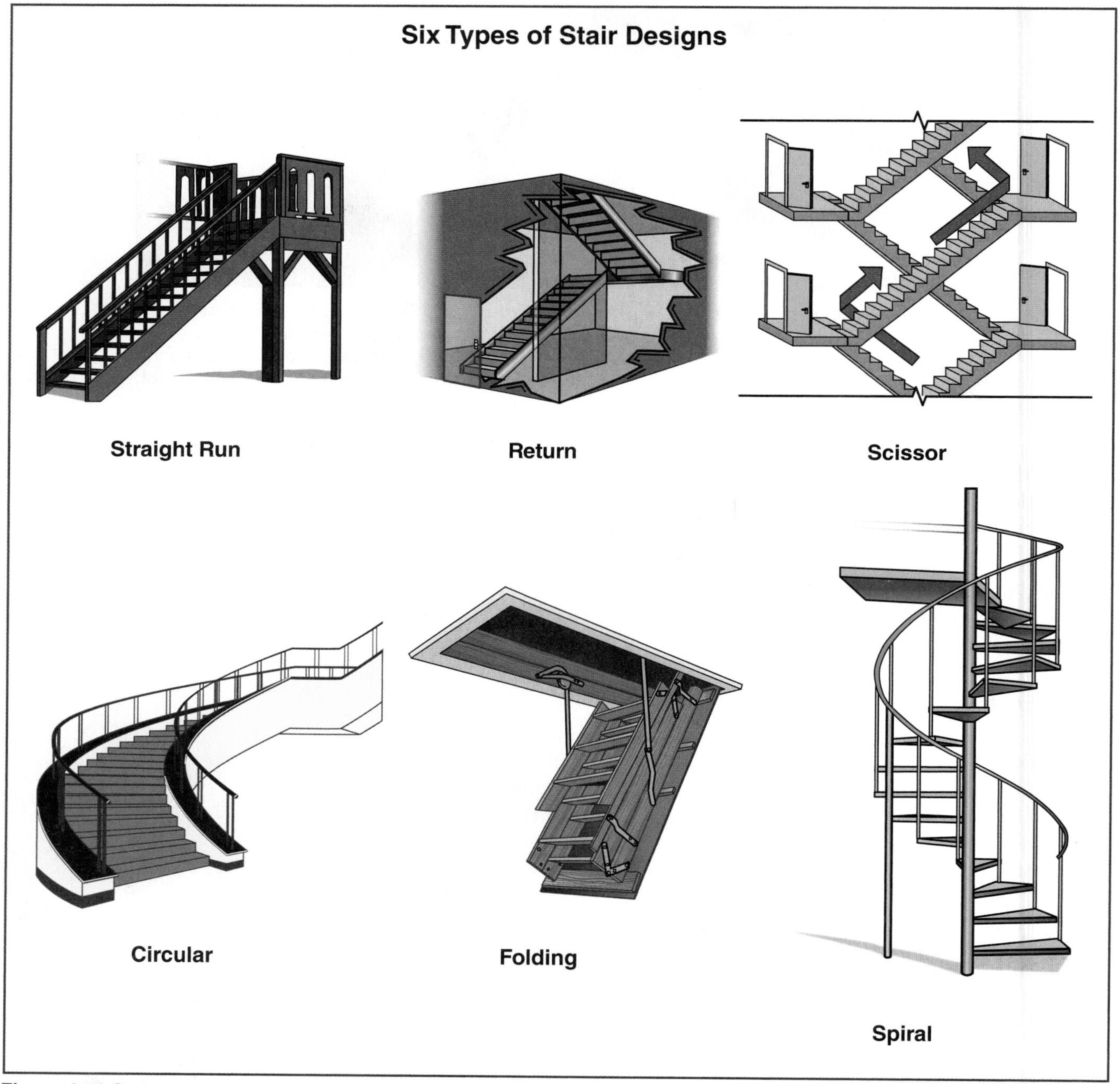

Six Types of Stair Designs

Straight Run

Return

Scissor

Circular

Folding

Spiral

Figure 4.51 Stairs may be designed to meet specific intended functions.

Fire Escapes May Not Support Firefighters

No matter how secure a fire escape appears, it may not be able to hold the weight of firefighters or occupants. The best policy is to use interior stairs if they are available and accessible, followed by fire department ground ladders and aerial devices to access upper floors.

Figure 4.52 Exterior stairs may be open or closed to natural ventilation.

Figure 4.53 Fire escapes are an outdated design of open exterior stairs.

Smokeproof Stair Enclosures

Building codes require a smokeproof stair enclosure under certain circumstances, such as stairs serving a high-rise building. Stair enclosures using either active or passive smoke control may be defined as smokeproof **(Figure 4.54. p. 178)**. A mechanical ventilation system actively keeps a stair enclosure free of smoke, even when a door is open to the fire floor. Activated by automatic fire/smoke detection equipment, a mechanical ventilation system is designed to keep smoke out of the stair enclosure by pressurizing the shaft. The system should be specially designed for the particular installation. A properly designed, installed, and maintained system should allow firefighters to begin suppression operations in one stairwell while occupants use a second one for escape.

Unprotected Stairs

Because unprotected stairs are not enclosed with fire-rated construction, they may serve as a path of spread for fire and smoke. They will not protect anyone using them from exposure to the products of combustion. Building codes typically allow the use of unprotected stairs in buildings when they connect only two adjacent floors above the basement level. These stairs are sometimes referred to as *access* or *convenience* stairs and can be used as part of an exit system in a two-story building **(Figure 4.55, p. 178)**.

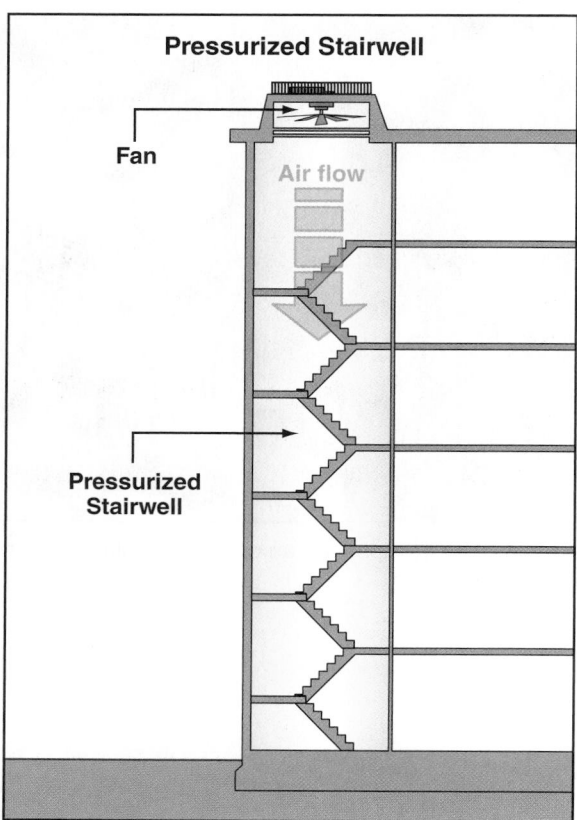

Figure 4.54 A pressurized stairwell incorporates a ventilation system that pushes air into the stairwell, which helps keep smoke from a fire out of the stairwell.

Figure 4.55 Convenience stairs do not provide any fire protection and may be limited in their range.

Doors

Doors vary widely in operation, style, design, and construction. The door manufacturing industry is large enough that it constitutes a specialty within the field of building construction. A single-family dwelling, for example, may have a dozen or more different doors and a large hotel may have hundreds of doors.

Doors may be classified by the way they operate. Generally, the following five types of doors are used in modern building construction:

- Swinging
- Sliding
- Folding
- Vertical
- Revolving

Swinging Doors

A swinging door rotates around a vertical axis by means of hinges secured to the side jambs of the doorway framing (**Figure 4.56**). It may also operate on pivot posts supported at the top and bottom. A swinging door can be either single or double leaf.

Figure 4.56 Swinging doors are common in many occupancies.

Figure 4.57 Sliding doors may have a variety of forms and purposes but all have lateral movement on a track in common.

It may also be single acting, swinging in one direction, or double acting, swinging in two directions. Generally swinging doors are required as exit doors in a means of egress, although other types of doors can be used under very specific conditions.

Sliding Doors

A sliding door is suspended from an overhead track and may use steel or nylon rollers (**Figure 4.57**). Floor guides or tracks are usually provided to prevent the door from swinging laterally. A sliding door can be designed as surface sliding, pocket sliding, or bypass sliding.

A sliding door's main advantage is that it eliminates a door swing that might interfere with the use of interior space. A pocket sliding door, which slides into the wall assembly, is frequently used within residential units because it is out of sight when open. Sliding doors are also used for elevators, power-operated doors in storefront entrances, and fire doors to protect openings that are not a part of the means of egress. Sliding doors are never allowed as a part of a means of egress because they slow the travel of people through the door opening.

Folding Doors

A folding door is hung from an overhead track with rollers or glides similar to those used by a sliding door (**Figure 4.58**). A folding door can be either bifolding or multifolding. Folding doors may be found in residential occupancies, in places of assembly to divide large conference areas into smaller rooms, and as horizontal fire doors. Horizontal fire-door assemblies must meet very specific requirements and be tested and listed for use in a means of egress.

Figure 4.58 Folding doors hang on a track similar to that of a sliding door.

Figure 4.59 Overhead doors are often used in areas with large openings.

Figure 4.60 A revolving door offers an efficient method of maintaining climate control but presents a potential hindrance to firefighters.

Vertical Doors

A door that opens in a vertical plane is known as an overhead door and is often found in industrial occupancies for applications such as loading dock doors, garage doors, freight elevator doors, and fire doors protecting openings that are not part of the required means of egress (**Figure 4.59**). A vertical operating door can be a simple single leaf that is raised in vertical guides along the edge of the doorway, or it can consist of two or more horizontal panels. Vertical rolling doors that consist of interlocking metal slats are commonly used in factories and warehouses.

A door that operates vertically is usually provided with some type of counterbalance mechanism, either actual weights or springs, to help overcome the weight of the door. A vertical door can be raised manually, raised mechanically via chain hoist, or power-operated.

Revolving Doors

A revolving door is constructed with three or four sections or wings that rotate in a circular frame (**Figure 4.60**). A revolving door is designed to minimize the flow of air through a door opening to reduce building heating or cooling losses.

A revolving door prevents the movement of hose or equipment into a building, which can present a problem for firefighters. Furthermore, a crowd of people attempting to flee in an emergency cannot move through a revolving door as quickly as they can through a comparable swinging door. To overcome these restrictions under emergency conditions, the wings of the revolving door are designed to collapse under pressure and provide an unobstructed opening.

Several types of mechanisms hold the wings of revolving doors in place within the door unit. Old models use simple chain keepers or stretcher bars between the wings while new models use spring-loaded, cam-in-groove hardware. Most models employ a collapsing mechanism that allows the wings to open to a *book-fold* position when the wings are pushed in opposite directions.

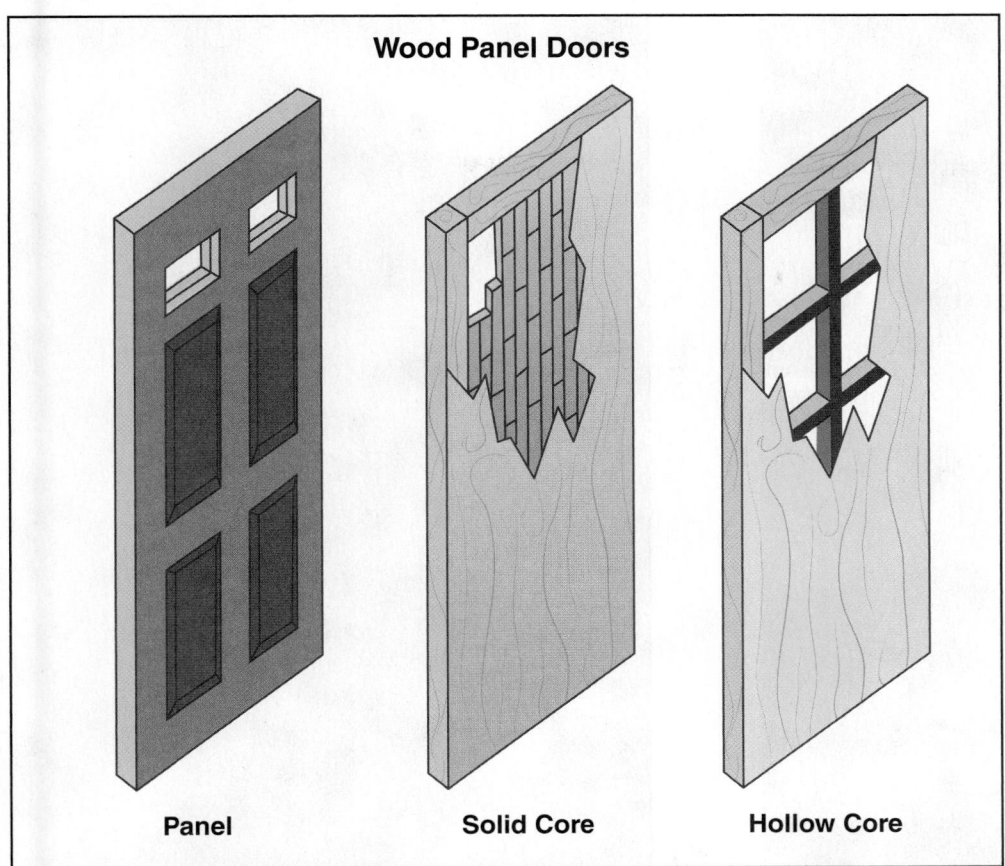

Wood Panel Doors

Panel   Solid Core   Hollow Core

Figure 4.61 Wood panel doors may have a solid or hollow core.

In addition to describing doors by their method of operation, doors can be classified by style and construction material. Door styles are mainly of interest to an architectural designer. However, the construction material of a door influences its effectiveness as a fire barrier and the degree to which it can be forced open during an emergency.

Doors are constructed from wood, metal, and glass. Wood doors may be panel or flush designs and may contain glass components. Aluminum and carbon steel are the metals used most commonly in doors, but stainless steel, bronze, and copper are also used. In addition, doors are sometimes manufactured with a veneer of hardboard, fiberglass, or plastic.

Wood Panel and Flush Doors

A very common type of swinging door is the wood panel door (**Figure 4.61**). A panel door consists of vertical and horizontal members that frame a rectangular area. Thin panels of wood, glass, or louvers are placed within the framed rectangular area.

A flush door (sometimes referred to as a slab door) consists of flat face panels that are the full height and width of the door. The panels are attached to a solid or hollow core. A flush door can be designed with openings to accommodate glass vision panels or ventilation louvers. In the past, flush doors were constructed from one solid piece or slab of wood. Today, a flush door is constructed of wood components finished to present a smooth, unbroken surface on both sides.

Figure 4.62 Glass doors are prevalent in offices and mercantile buildings.

Figure 4.63 Hollow metal doors are common in stairwells and industrial occupancies.

Solid-core doors are formed with an interior core of laminated blocks of wood, particleboard, or a mineral composition. The core is covered with two or three layers of surface material, which is usually plywood. If a wood solid-core door is intended for exterior application where security is a concern, a layer of sheet metal may be attached to the exterior surface to increase resistance to physical attack by an intruder.

A hollow-core door is constructed with spacers between the face panels to provide lateral support. The interior spacers consist of a grid or honeycomb of wood, plastic, or fiberboard. Hollow-core doors are less expensive and lighter than solid-core doors. However, they have minimal thermal or sound-insulating value and usually are used for interior applications.

Solid-core doors are better fire barriers than either panel doors or hollow-core doors. A solid-core door that has not been specifically designed as a fire door will act as a significant barrier to fire if it is closed at the time of the fire.

Glass Doors

Glass doors are used for both exterior and interior applications. They are found in almost all occupancies, but they are most commonly used in office and mercantile buildings. Glass doors can be either framed or frameless (**Figure 4.62**). In a frameless glass door, the door consists of a single sheet of glass to which door hardware such as handles are attached. In a framed door, the glass is placed within and is surrounded by a metal or wood frame with the required door hardware attached to the frame.

Figure 4.64 Fire doors restrict the range of fire involvement.

Figure 4.65 Rated fire doors must be labeled as such.

Building codes require glass doors to be made of tempered glass that resists breakage. In addition, various plastics such as Lexan® or Plexiglas® are often used in framed doors to provide additional security.

Metal Doors

A common type of metal door is a hollow metal door made from steel or aluminum **(Figure 4.63)**. A hollow metal door can be either panel or flush and is normally 1¾ inches (45 mm) thick. A flush door consists of smooth sheet metal face panels 1/20 inch (1 mm) thick. Vertical sheet metal ribs within the door spaced 6 to 8 inches (150 mm to 200 mm) apart separate the face panels of a steel door from one another. A sound-deadening material can be placed between the ribs. An aluminum flush door usually has a core of hardboard and honeycomb-patterned paper.

A metal door can also be constructed of heavy corrugated steel. In this type of door, a steel frame supports one or two corrugated sheets. A door made with two corrugated sheets has an interior core material such as Styrofoam®.

Fire Doors

Fire doors protect openings in fire-rated walls **(Figure 4.64)**. The use of fire doors to block the spread of fire is an established fire-protection technique. Originally found in industrial buildings that date back to the end of the 19th Century, today fire doors are found in all types of occupancies. When properly maintained and operated, fire doors are very effective at limiting the spread of fire and total fire damage. Fire doors differ from ordinary (nonfire doors) in their construction, their hardware, and the extent to which they may be required to close automatically.

To qualify as a rated fire door, the entire assembly, including the door, hardware (hinges, latches, locks, etc.), door seal, and frame, must pass a test by a third-party testing agency. The door assembly is certified as a single unit for a specified time. Fire-resistance classifications, testing, frames and hardware, and construction types are described in the sections that follow. Glass panels in fire doors must achieve the same test results as the rest of the assembly.

Rated fire doors are identified with a label indicating the door type, the hourly rating, and the identifying logo of the testing laboratory **(Figure 4.65)**. Building and fire inspectors use fire door labels to identify fire doors in the field. However, it is not uncommon for the labels to be painted over in the course of building maintenance.

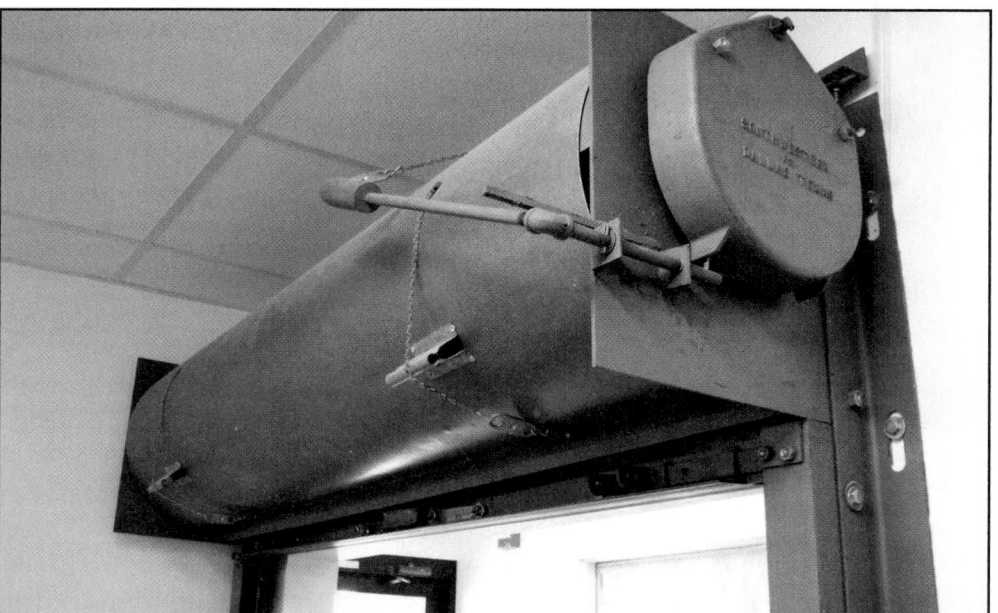

Figure 4.66 Overhead rolling fire doors may be equipped with a fusible link to trigger deployment.

The construction and operation of fire doors depends on the type of occupancy, the amount of space around the door opening, and the required fire-protection rating for the door. Most fire doors will be constructed of metal and may roll, slide, or swing into place when released. Special types of fire-rated fire doors are available for freight and passenger elevators, service counter openings, security (bullet-resisting doors), dumbwaiters, and chute openings.

Rolling steel fire doors. An overhead rolling steel fire door is commonly used to protect an opening in a fire wall in an industrial occupancy or an opening in a wall separating buildings into fire areas **(Figure 4.66)**. An overhead rolling steel fire door may be used on one or both sides of a wall opening. One architectural advantage of an overhead rolling fire door is that it is relatively inconspicuous and does not use wall space next to the opening. This type of door cannot be used on any opening that is required to be part of the means of egress.

This type of door is constructed of interlocking steel slats with other operating components such as releasing devices, governors, counterbalance mechanisms, and wall guides. An overhead rolling door ordinarily closes under the force of gravity when a fusible link melts, but motor-driven doors are available.

Horizontal sliding fire doors. Horizontal sliding fire doors are often found in old industrial buildings, are usually held open by a fusible link, and slide into position along a track either by gravity or by the force of a counterweight **(Figure 4.67)**. Several different materials are used to construct horizontal sliding fire doors. Horizontal sliding doors cannot be used to protect openings in walls that are required parts of a means of egress.

A common type of sliding fire door is a metal-covered, wood-core door. The wood core provides thermal insulation, while the sheet metal covering protects the wood from the fire. Because wood undergoes thermal decomposition when exposed to heat, a vent hole is usually provided in the sheet metal to vent the gases of decomposition. The metals used to cover the wood core include steel, galvanized sheet metal, and terneplate (a metal composed of tin and lead). Fire doors made with terneplate are commonly referred to as *tin-clad* doors; although strictly speaking, the metal used is not pure tin.

Figure 4.67 Horizontal sliding doors are used in some industrial applications to protect openings in walls. *Courtesy of Ron Moore and McKinney (TX) FD.*

Figure 4.68 A fusible link is an effective triggering mechanism for fire containment devices, but it is not the most efficient because it requires certain temperatures to activate. *Courtesy of Ron Moore and McKinney (TX) FD.*

Swinging fire doors. Swinging fire doors are commonly used in stairwell enclosures or corridors that require a fire door. A swinging fire door has the disadvantage of requiring a clear space around the door to ensure closure. However, a swinging door is a good choice when the door is located in a corridor where it must remain open during normal day-to-day operations.

To perform its function, a fire door must be closed when a fire occurs. However, it is normally desirable for fire doors to remain open or to be easily opened to allow for ordinary movement of building occupants. Fire doors can be either automatic or self-closing. An automatic door is normally held open and closes automatically when an operating device is activated. A self-closing door is normally closed and will return to the closed position if it is opened and released.

For a fire door to close, some type of detection device must first sense a fire or the smoke from a fire. The oldest and simplest detection device is a fusible link that melts from the heat of a fire **(Figure 4.68)**. A fusible link has the advantage of being inexpensive, relatively rugged, and very easy to maintain. However, because it depends on heat from a fire, it is slower to operate than devices that react to smoke or the rate-of-temperature rise. A significant amount of smoke may flow through a door opening before a fusible link can release a fire door.

When a smoke detector is used to activate a fire door, the door closes more quickly. It also permits easy testing of the fire door. A smoke detector costs more and requires periodic cleaning. As with all smoke detectors, they must be properly positioned with respect to dead air spaces or ventilation ducts. Some fire doors are designed to close when any component of the fire alarm system is activated.

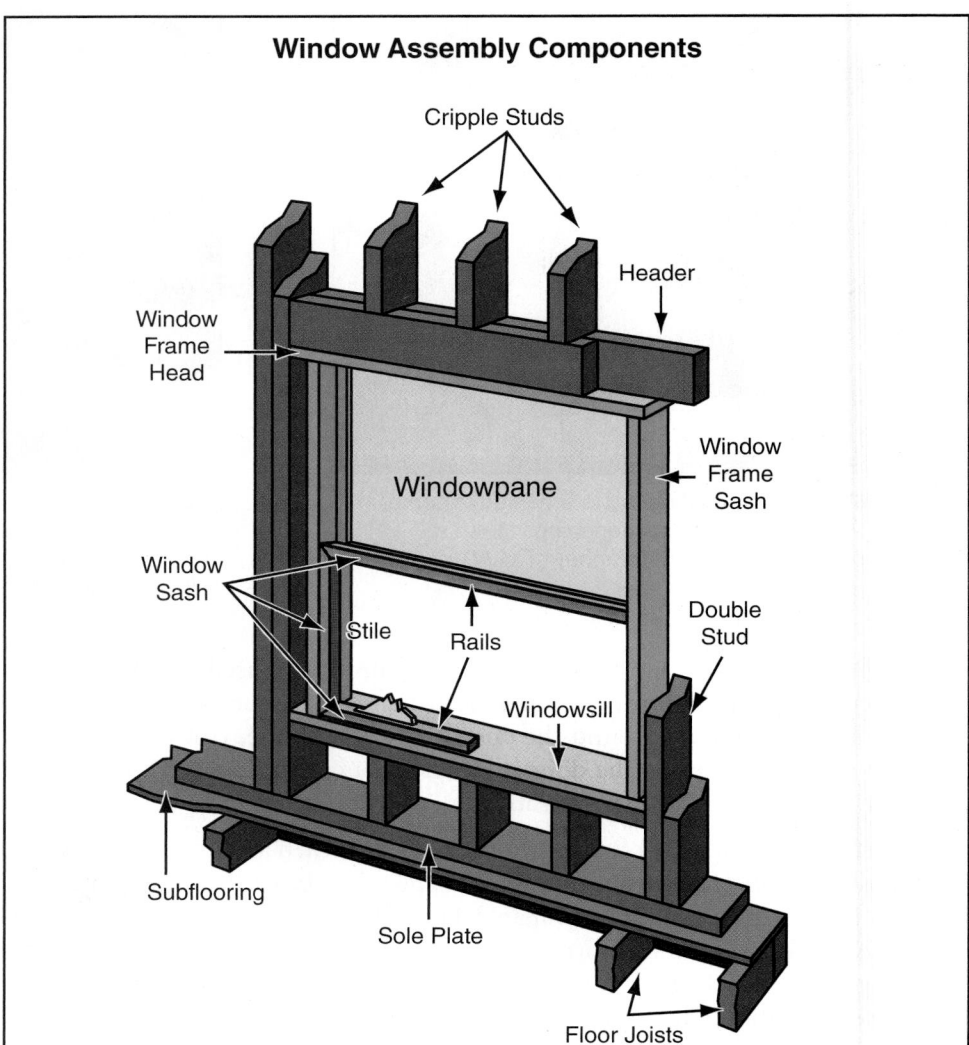

Figure 4.69 A window assembly includes all the hardware necessary to make a complete unit.

Within the figure:

Window Assembly Components

Cripple Studs

Header

Window Frame Head

Windowpane

Window Frame Sash

Window Sash

Stile

Rails

Double Stud

Windowsill

Subflooring

Sole Plate

Floor Joists

Windows

Windows have long been relied upon as a means of light, ventilation, access, and rescue. Factory buildings constructed in the 19th Century were designed with large windows to provide interior light. Many buildings were designed with interior courts or light shafts to provide ventilation.

However, modern buildings often rely on their HVAC systems for ventilation and artificial lighting for illumination. Elimination of windows that can be opened enhances energy efficiency in buildings because it reduces air infiltration around windows. Consequently, some buildings are designed with windows that cannot be opened or without windows altogether, resulting in increased difficulties for tactical ventilation and fire suppression team access.

A window consists of a frame, one or more sashes, and all necessary hardware to make a complete unit (**Figure 4.69**). A window frame includes the members that form the perimeter of a window, and it is fixed to the surrounding wall or other supports. The term *sash* refers to a framed unit that may be included within a window frame, and it may be fixed or movable. The frame is composed of the sill, side jamb, and head jamb. The sill is the lowest horizontal member of the window frame and supports the weight of the hardware and sash.

Figure 4.70 Fixed windows and glass bricks provide light to an occupancy.

All windows contain glass, known as glazing. The glass may be single-, double-, or triple-glazed; that is, there may be one thickness of glass, two thicknesses separated by an inert gas, or three thicknesses separated by voids filled with gas. Some window and door panels may also have retracting shades located in the void.

Windows can be broadly classified into fixed (nonoperable) or movable (operable). Windows that contain both fixed and movable characteristics are generally included in the movable classification.

Fixed Windows

A fixed window consists only of a frame and a glazed stationary sash. A fixed window can be used alone or in combination with movable windows. The large windows found in mercantile occupancies and high-rise office buildings are common examples of fixed windows. Other terms used to describe fixed windows are *display windows*, *picture windows*, and *deadlights* **(Figure 4.70)**. Fixed windows may be found in many applications including over (transom) and around doors, in skylights, in residential applications, and along the fronts of retail shops.

Movable Windows

A movable window is designed in several common configurations **(Figure 4.71, p. 188)**:

- **Double-hung** — Has two sashes that can move past each other in a vertical plane. A double-hung window is commonly used in residential occupancies because it permits circulation through the top and bottom of the window opening. Balancing devices consisting of counterweights, springs, or a spring-loaded coiled tap to hold the movable sashes at the desired position. Windows that use counterweights are found in old buildings.

- **Single-hung** — Has only one sash openable. Balancing devices consisting of counterweights, springs, or a spring-loaded coiled tap to hold the movable sash at the desired position.

- **Casement** — Has a side-hinged sash that is usually installed to swing outward. It may contain one or two operating sashes and can be opened fully by unlatching and pushing or using the mechanical window crank for ventilation.

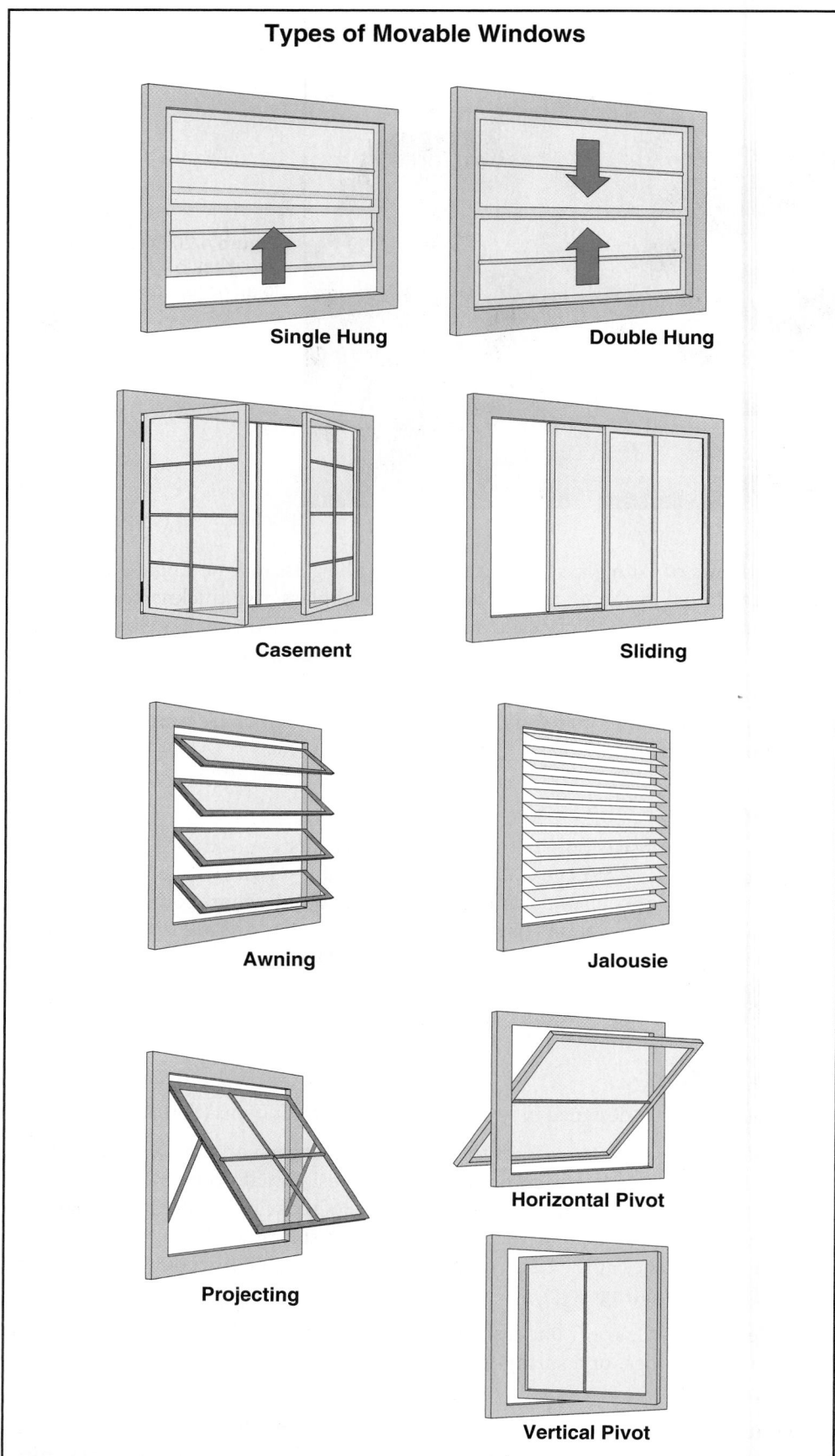

Figure 4.71 Specific configurations of movable windows accommodate the needs of an occupancy.

Types of Movable Windows

Single Hung

Double Hung

Casement

Sliding

Awning

Jalousie

Projecting

Horizontal Pivot

Vertical Pivot

- **Horizontal sliding** — Has two or more sashes of which at least one moves horizontally within the window frame. In a three-sash design, the middle sash is usually fixed; in a two-sash unit, one or both sashes may be movable.

- **Awning** — Has one or more top-hinged, outward-swinging sashes that are opened by unlatching and pushing or using the mechanical window crank. This arrangement permits the window to be open during rain. Hopper windows are similar in design to awning windows except they are hinged at the bottom.

- **Jalousie** — Includes a large number of narrow overlapping glass sections swinging outward (basic concept of the awning window). The individual pieces of glass are about 4 inches (100 mm) in width. The glass sections are supported at their ends by an operating mechanism. Jalousie windows are popular architecturally because the amount of opening can be varied for ventilation without admitting rain.

- **Projecting** — Swings outward at the top or bottom and slides upward or downward in grooves. The projected window usually is operated by a push bar that is notched to hold the window in place.

- **Pivoting** — Has a sash that pivots horizontally or vertically about a central axis. Part of a pivoting window swings inward and part swings outward when it is opened. A window of this design provides the full area of the window opening for ventilation.

Security

Unfortunately, windows may also provide an access point for intruders. Consequently, means are frequently provided to increase the security of windows, especially those windows that are accessible from the ground or adjacent roofs.

A common method for providing window security is to fasten metal bars or screens to the exterior of the window frame or to the building itself. The metal bars may be fastened to the building, embedded in masonry, or mounted on hinges and locked with padlocks or other locking devices. Security windows are available with movable sashes and fixed bars so that the windows can be opened for ventilation while maintaining security of the premises **(Figure 4.72)**.

While preventing unlawful entry is the primary reason for installing security bars or grilles, they also have a negative effect on fire and life safety. The bars or grilles can prevent the escape of trapped occupants or firefighters and can slow the access time for firefighters. When interior operations begin, security bars and grilles must be removed or disabled to ensure firefighter safety in case a rapid egress is required.

Figure 4.72 Security features on windows may accommodate sash movement for ventilation.

Figure 4.73 The process of sizing up a building takes into account several factors that will aid or hinder fireground activities.

ⅠⅠ Hazards Related to Building Construction

To be able to locate and extinguish a building fire, you must know the types of building construction, materials, and components. Knowledge of types of construction will help you to predict fire development, spread, and the effects of fire and fire suppression on the building. You will also be aware of conditions that will lead to structural failure and the indicators of structural instability. You will be able to recognize types of doors and door hardware and select the correct type of forcible entry tool needed to gain access to the building. Knowing the types of roofs, roof supports, and coverings will allow you to judge the safety of a roof prior to accessing it. You will also be able to select the right type of vertical tactical ventilation procedures to use based on the type of roof construction. Finally, knowledge of building construction will help you apply emergency escape techniques if you become lost or trapped inside a building. Being aware of the types of building construction in your jurisdiction will add to your situational awareness, enabling you to stay safe in dangerous situations and allowing you to avoid other hazards.

Sizing Up Existing Construction

The outward appearance of many structures can be deceiving because exterior finishes make buildings appear to be more substantial than they really are. For example, some wood frame structures have a veneer of polystyrene foam shaped to look like stone, masonry, or concrete over which a stucco finish is applied that makes them look like Type I construction buildings. When sizing up a building, look for the following (**Figure 4.73**):

- **Age of the building** — Are there obvious signs of deterioration?
- **Construction materials** — Is it a wood frame, unreinforced masonry, all-metal, or concrete building?
- **Roof type** — Is the roof flat, pitched, or arched? What type of roof covering is visible?
- **Renovations or modifications** — Have additions been made that may create internal hazards?
- **Dead loads** — Are there HVAC units, water tanks, or other heavy objects visible on the roof?
- **Number of stories** — How many stories, above and below ground, are visible? Are there stories below ground on one side of the building that are at ground level on another side?

Figure 4.74 An area with a heavy fuel load has many pieces of combustible materials.

- **Windows** — Can the windows be opened from the inside? Are there security grilles that should be removed from the outside of the building?

Other items are important to observe during size-up, such as the occupancy type, adjacent exposures, the presence or lack of operational fire suppression systems, and fire conditions. Although these are not directly related to the hazards associated with building construction you should be aware of them.

Dangerous Building Conditions

You must be aware of the dangerous conditions created by a fire as well as dangerous conditions that may be created while trying to extinguish a fire. An already serious situation can be made much worse if firefighters fail to recognize the potential of the situation and take the wrong actions.

There are two primary types of dangerous conditions that may be posed by a particular building:

- Conditions that contribute to the spread and intensity of the fire
- Conditions that make the building susceptible to collapse

These two conditions are obviously related — conditions that contribute to the spread and intensity of the fire increase the likelihood of structural collapse. The following sections describe some of these conditions.

Fuel Loading

Fuel load is the maximum heat that can be produced if all the combustible materials, both contents and building materials, in a given area burn **(Figure 4.74)**. Heavy fuel loading is the presence of large amounts of combustible materials in an area of a building. The arrangement of materials in a building directly affects fire development and severity and is considered during preincident surveys when determining the possible duration and intensity of a fire.

Figure 4.75 Furnishings and finishes contribute to the fuel load and smoke generation capability of an occupancy. *Courtesy of Eddie Avila.*

Heavy fuel load is perhaps one of the most critical hazards in commercial, industrial, and storage facilities because the fire can overwhelm the capabilities of a fire suppression system and make it difficult for firefighters to gain access during fire suppression operations. Proper inspection and code enforcement prior to an incident is the most effective defense against these hazards.

Combustible structural components such as wood framing, floors, and ceilings also contribute to the fuel loading in a building. Prolonged exposure to fire may weaken them and increase the chances of collapse.

Furnishings and Finishes

Besides the fuel load, the furnishings and interior finishes can contribute to fire spread and smoke production. Furnishings are the tables, sofas, desks, beds, and other items generally found in all types of occupancies (**Figure 4.75**). The interior finishes include the window, wall, and floor coverings such as drapes, wallpaper, and carpet. The combustibility of furnishings and interior finishes has been identified as a major factor in the loss of many lives in fires.

Roof Coverings

The combustibility of a roof's surface is a basic concern to the fire safety of an entire community. Some of the earliest fire regulations ever imposed in North America related to combustible roof coverings because they were blamed for several conflagrations caused by flaming embers flying from roof to roof.

History has shown that wood shakes in particular, even when treated with fire retardant, can significantly contribute to fire spread. This is a particular problem in wildland/urban interface fires where wood shake roofs have contributed to large fires. Firefighters must use exposure protection tactics to protect combustible roofs on structures adjacent to a fire building or wildland fire.

Large, Open Spaces

Large, open spaces in buildings contribute to the spread of fire throughout. Such spaces may be found in warehouses, churches, large atriums, large-area mercantile buildings, and theaters (**Figure 4.76**). In these facilities, proper vertical tactical ventilation (channeling smoke from a building at its highest point) is essential for slowing the spread of fire. Large spaces may also be concealed from view between roofs and

Figure 4.76 Large open spaces allow fire to travel unchecked.

Figure 4.77 The bowstring truss is distinctive because of its curved top chord.

ceilings and under rain roofs. In these concealed spaces, fire can travel undetected, feeding on combustible exposed wood rafters. When smoke appears through openings in the roof or around the eaves, the exact point of origin may be deceiving.

Engineered and Truss Construction Hazards

In addition to the danger from bowstring construction illustrated in the Case History, firefighters also must be aware of the danger posed by the increased use of engineered or lightweight construction and trussed support systems. Experience has shown that unprotected engineered steel and wooden trusses can fail after 5 to 10 minutes of exposure to fire. These trusses can fail from exposure to heat alone without any flames. For steel trusses, 1,000°F (538°C) is the critical temperature. Unless they are corner-nailed, metal gusset plates in wooden trusses can warp and fail quickly when exposed to heat. Although both steel and wooden trusses may be protected with fire-retardant treatments to enhance their fire resistance, most lack this protection.

Bowstring truss roofs became popular in the 1930s and are easily identified by the roof's arched or curved outline **(Figure 4.77)**. Prior to 1960, the bowstring truss roof design was one of the most common design types for large commercial and industrial structures. The bowstring truss roof was commonly used in facilities such as automobile dealerships and repair facilities, bowling alleys, grocery stores, and industrial complexes wherever large open floor spaces with limited interior supports were needed. The curved top chord members were made either by sawing straight lumber into curved shapes or laminating multiple smaller pieces bent over a jig to the desired shape. Bottom chord members were typically constructed with large, straight lumber members joined with either wood or metal bolted splice plates, located near mid-span, to achieve the required length. The top and bottom chord members were fastened at the truss ends with U-shaped steel heels, or end shoes, bolted to both chord members.

The principles of bowstring truss construction are similar to other types of truss construction. Web members are used to form a series of triangles that transfer tension from the bottom chord and compression from the top chord of the truss onto the load-bearing walls. One difference with the bowstring truss is that the compressional forces within the top chord act to force the load-bearing walls outward as well as downward. Another difference is that the space between trusses is greater than that found between other types of trusses.

Figure 4.78 Construction activities sharply increase the risk of fire in a structure. *Courtesy of Ron Moore and McKinney (TX) FD.*

Figure 4.79 Buildings in the process of demolition may be subject to high risk of rapid fire growth.

Bowstring truss roof systems suffer from a little-known phenomenon related to inaccuracies in early industry-accepted truss design assumptions. One significant design deficiency involves the tensile strength of the bottom chord. Early truss designs assumed wood tensile strength could be defined by bending tests of small straight-grained wood samples free of common wood defects. Prior to the 1960s, large-scale test facilities were uncommon so full-size lumber tests were rarely conducted. During the 1960s, full-size lumber tests revealed that construction grade lumber with natural imperfections such as knots, checks, and irregular grain provides in-service tensile strength significantly less than that predicted by the earlier small-scale, clear-wood tests. Thus, all trusses constructed prior to the late 1960s have a common code deficiency: the bottom chord members may have inadequate tensile strength to support code-prescribed roof loads.

Construction, Renovation, and Demolition Hazards

For a variety of reasons, the risk of fire rises sharply when construction, renovation, or demolition is being performed in a structure. Contributing factors are the additional fuel loads and ignition sources (such as open flames from cutting torches and sparks from grinding or welding operations) brought by building contractors and their associated equipment **(Figure 4.78)**.

Some local fire codes mandate that standpipe systems must remain in operation during the demolition of multistory buildings. Unfortunately, contractors do not always adhere to these requirements. The result is that inoperative standpipes and sprinkler systems have become a contributing factor in fires in buildings under demolition.

Buildings under construction are subject to rapid fire spread when they are partially completed because many of the protective features such as gypsum wallboard and automatic fire suppression systems are not yet in place. Some firefighters think of buildings under construction with exposed wooden framing as the equivalent of a vertical lumberyard. The lack of doors or other barriers that would normally slow fire spread are also contributing factors to rapid fire growth.

Buildings that are being renovated, demolished, or are abandoned are also subject to faster-than-normal fire growth **(Figure 4.79)**. Breached walls, open stairwells, missing doors, and deactivated fire suppression systems are all potential contributors. The potential for a sudden building collapse during fires in these buildings is also a serious consideration. Arson is also a factor at construction or demolition sites because of easy access into the building.

Due to the cost of new construction, renovating old buildings is common in many areas. Hazardous situations may arise during renovation because occupants and their belongings may remain in one part of the building while work continues in another. Fire detection or alarm systems may be taken out of service or damaged during renovation. If good housekeeping is not maintained, accumulations of debris and construction materials can block exits. This may impede occupants trying to escape from the building in an emergency as well as making entry by firefighters more difficult. The contractors or owner/occupants performing the renovations do not always follow local building codes.

Structural Collapse

The structural failure of a building or any portion of it resulting from a fire, snow, wind, water, or damage from other forces is referred to as **structural collapse**. Structural collapse can also be the result of an explosion, earthquake, flood, or other natural occurrence. However, natural or explosion-caused collapses usually occur without warning prior to an emergency response, while fire-caused collapses often occur during emergency operations. The ability to understand how fire can cause building elements to deteriorate resulting in a collapse is extremely important for fire officers and firefighters. Collapse potential should be considered during preincident surveys and throughout the size-up process until the situation is mitigated.

Structural Collapse — Structural failure of a building or any portion of it resulting from a fire, snow, wind, water, or damage from other forces.

Structural Collapse Factors

There are many factors that should be considered when trying to determine the potential for structural collapse. Some of these include renovations, additions, alterations, age of the structure, weather, and loads placed on the structure. Along with these, the following factors should be considered when determining the potential for structural collapse:

- Construction type
- Length of time fire burns
- Stage of the fire
- Contents
- Amount of water used to extinguish the fire

Construction Type

In North America, examples of structural collapse involving high-rise buildings or Type I construction buildings are very limited. Strict building codes have ensured that structural members exposed to fire and high temperatures will remain sound until the fire is extinguished. Structural collapses due to earthquakes generally involve smaller buildings such as the one- to four-story buildings in the Marina Section of San Francisco in 1989. Some of these buildings simply fell over or against another building.

Church steeples, water tanks, chimneys, and false facades that extend above the top of the structure must be viewed as a potential collapse hazard even if the structure is not. Collapses usually involve brick or masonry block and may be structural components or veneer. Structural collapses are not limited to the actual emergency and can occur well after the fire is extinguished. All personnel must ensure the structural stability of the site before entering it.

Until recently there has not been any documented evidence of how long structural members will remain intact when exposed to fire. This lack of documentation has been due to the variety of types of loads or forces that each component may be subjected to during a fire. In 2000 the National Institute of Standards and Technology (NIST) and the United States Fire Administration (USFA) began testing structural components to determine how firefighters could predict potential structural collapse. Although the analysis is still ongoing, some of the results have been released. Tests in both residential and commercial structures indicate that there is very little difference between collapse times for steel bar joist-supported roofs and wood truss-supported roofs. Both are prone to very rapid collapse.

You must remember that collapse of structures using lightweight construction can occur earlier in the incident and may not provide you with the warning indicators listed above. A thorough preincident survey and size-up of the incident scene will provide you with some indication of the presence of lightweight construction.

Hazards of Truss Systems

Firefighters may be injured or killed when fire-damaged roof and floor truss systems collapse, sometimes without warning. Firefighters should take the following steps to minimize the risk of injury or death during structural fire fighting operations involving roof and floor truss systems:

- Know how to identify roof and floor truss construction.

- Immediately report the presence of truss construction and fire involvement to the Incident Commander (IC).

- Use a thermal imager as part of the size-up process to help locate fires in concealed spaces.

- Use extreme caution and follow standard operating procedures (SOPs) when operating on or under truss systems.

- Open ceilings and other concealed spaces immediately whenever a fire is suspected of being in the overhead space. Apply the following tactics:

  — When responding to any structure where fire has been reported, ALWAYS check the ceiling prior to entering the building. If there is a haze or odor of smoke, the first firefighter in the door checks this area. Use a pike pole to lift ceiling tiles or remove ceiling materials.

  — As you advance into the building, check the ceiling at intervals, approximately every 20 feet (6.1 m).

  — When you enter another portion of the building, repeat the process.

  — Use extreme caution because opening concealed spaces can result in backdraft conditions.

  — If smoke or fire is found above the ceiling, remove all ceiling tiles or material until the source is located.

  — Position between the nearest exit and the concealed space to be opened.

  — Be aware of the location of other firefighters in the area.

  — Do not allow the fire to extend over you or get between you and your exit path.

  — When ceilings are higher than you can reach with a pike pole, use an A-frame or attic ladder to gain the required access.

- Understand that fire-resistance ratings may not be truly representative of real-time fire conditions and the performance of truss systems may be affected by fire severity.

Before emergency incidents, fire departments should take the following steps to protect firefighters:

- Conduct preincident planning and inspections to identify structures that contain truss construction.
- Ensure that firefighters are trained to identify roof and floor truss systems and use extreme caution when operating on or under truss systems.
- Develop and implement SOPs to safely combat fires in buildings with truss construction.

At the emergency incident, use the following procedures to protect firefighters:

- Evacuate firefighters performing operations under or above trusses as soon as it is determined that the trusses are exposed to fire, and move to a defensive mode.
- After extinguishing a fire in a building containing truss construction, use defensive overhauling procedures if the truss system has been affected or weakened by the fire.
- Use outside master or large handline fire streams to soak smoldering trusses and prevent rekindles.
- Report any damaged sagging floors or roofs to command.

Source: National Institute for Occupational Safety and Health (NIOSH): *"Preventing Injuries and Deaths of Fire Fighters Due to Truss System Failures,"* NIOSH Publication No. 2005-123.

Length of Time Fire Burns

The longer a fire burns, the greater the temperature of the fire gases in the upper levels of the structure or compartment. In 1918, the American Society for Testing and Materials (ASTM) developed the standard time-temperature curve that is still used to illustrate the rate of temperature increase in a compartment fire. Applying an estimate of the length of time the fire has been burning to the type of construction, you will be able to get a general idea of how hot the ceiling and roof components have gotten.

Stage of Fire

The stage of the fire can easily indicate the quantity of heat that the structure has been exposed to and the potential for structural collapse. Fires in the incipient stage will not have generated sufficient heat or flame to cause unprotected steel or wood frame construction to collapse. However, collapse potential increases in the growth stage as heat increases in the upper levels of the space and flame spreads to and consumes the combustible structural members. In the decay stage, and during postsuppression activities, collapse becomes very likely due to the weakened state of structural members and the buildup of water.

Contents

Another factor that can contribute to structural collapse is the contents within the structure or on the roof. The contents may contribute to collapse in three ways:

- Generating higher temperatures and rapid combustion that will weaken the structure due to a higher fuel load
- Causing collapse more rapidly due to added dead weight
- Increasing stress on structural members due to increased weight from water retention

Contents include such things as stored materials, furniture, and machinery. Like knowledge of the construction type, knowledge of the contents is gained through preincident surveys and inspections.

While contents within a structure are visible during preincident surveys, storage in concealed spaces and attics is not. Attic storage is often heavier than the ceiling joists have been designed to carry. When the joists are weakened by fire, such storage increases the potential for ceiling joists to fail, putting firefighters at risk. Such storage is common in residential dwellings and has been the cause of firefighter fatalities in industrial fires.

Amount of Water Used to Extinguish Fire

Finally, the quantity of water that is used to suppress the fire can have a direct effect on an unstable structure. Every U.S. gallon (SI liter) of water that is used to suppress the fire adds 8.33 pounds (3.96 kilograms) (Imperial gallon=10 pounds) of weight to floors that may already be weakened. The added weight may cause floors to pancake down, or push walls out, resulting in a complete failure of the structure. As an estimate, 250 gpm (1 000 L/min) adds 1 ton (900 kg) of water per minute to the structure.

Besides the factors listed above, indicators of potential or imminent collapse include:

- Roof sagging, pulling away from parapet walls, or feeling spongy (soft) under foot
- Fire involvement of trusses and other engineered structural components
- Floors sagging or feeling spongy (soft) under foot
- Chunks of ceiling tiles or plaster falling from above
- Movement in the roof, walls, or floors
- Noises caused by structural movement
- Little or no water runoff from the interior of the structure
- Cracks appearing in exterior walls with smoke or water appearing through the cracks
- Evidence of existing structural instability such as the presence of tie rods and starst-hat hold walls together
- Loose bricks, blocks, or stones falling from buildings
- Deteriorated mortar between the masonry
- Walls that appear to be leaning
- Structural members that appear to be distorted
- Fires beneath floors that support heavy machinery or other extreme weight loads

- Prolonged fire exposure to the structural members (especially trusses)
- Structural members pulling away from walls
- Excessive weight of building contents

WARNING

Structural collapse can occur with little warning. If indicators start to appear, collapse is imminent and personnel must withdraw from the structure and the collapse zone.

Actions Taken When Collapse Is Imminent

Immediate action must be taken if firefighters suspect that the collapse of a building is imminent or even likely. First, as you and the rest of your crew are exiting the building, inform Command and all others inside the building of the situation. Second, establish and clear the collapse zone as soon as possible. No personnel or apparatus should be allowed to operate in the collapse zone except to cautiously place unstaffed master stream devices and then immediately withdraw once they are in operation. Third, a roll call or personnel accountability report (PAR) should be conducted to ensure the safety of all personnel. Know and heed any evacuation or other emergency signals used by your department.

Determining Collapse Zone

A collapse or safety zone must be established adjacent to any exposed exterior walls of the structure. Apparatus and personnel operating master stream appliances must not be positioned in the collapse zone. Traditionally, collapse zones have been estimated by taking the height of the structure and multiplying it by a factor of 1½ **(Figure 4.80, p. 200)**. For example, a 3-story structure that is 30 feet (9.14 m) tall would require a collapse zone no less than 45 feet (13.72 m) from the base of the structure. As structures increase in height, this becomes impractical by creating a space that limits defensive fire fighting operations. Therefore, you should consider the following guidelines when determining a collapse zone:

- **Type I** construction high-rise buildings are not as likely to collapse, making the primary concern the hazard of flying glass from windows or curtain walls. In Type I construction, it is the contents of the building burning, not the structure itself. Collapse zones must be determined considering the direction and velocity of wind currents that can carry the glass shards. Structural collapse, if it does occur, will be localized and not structure wide.

- **Type II** construction consists of unprotected steel or noncombustible supports, such as I-beams. When exposed to temperatures above 1,000°F (537.78°C), unprotected steel will expand and twist, pushing out walls, and when cooled will slightly constrict. These movements will cause floors and walls to collapse. Any type of construction that includes brick and block walls supporting unprotected steel bar joists and I-beams is involved in a large number of these collapses.

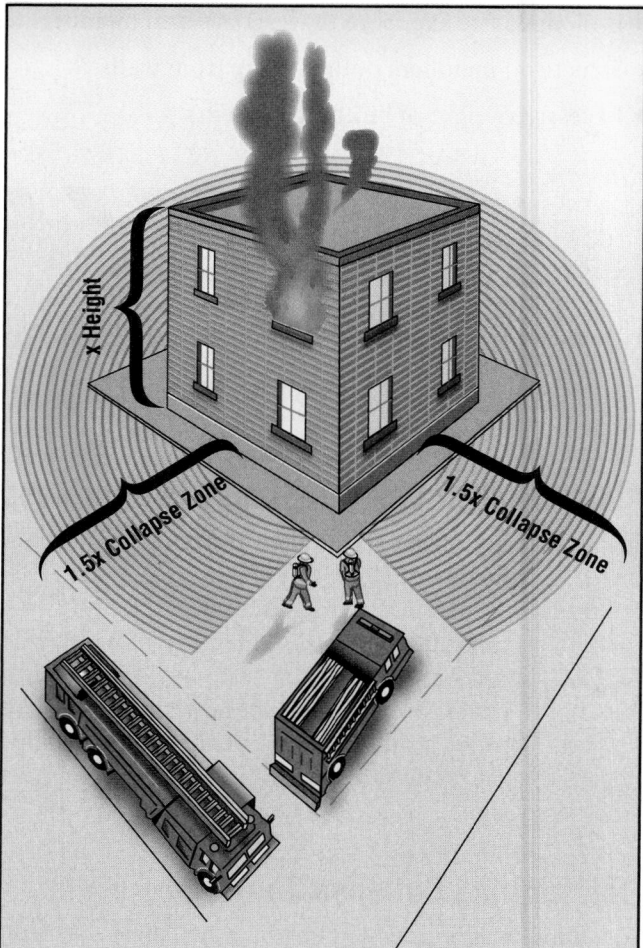

Figure 4.80 The safest position for access to many structures is in the angle formed by the corners.

- **Type III** construction multistory buildings should have a collapse zone of 1½ times the height of the structure. As an example, a building 7 stories (70 feet [21.34 m]) will require a collapse zone of approximately 105 feet (32 m) during defensive operations. In Type III ordinary construction, exterior load-bearing walls are made of concrete, brick, or masonry while interior loads are carried by wood, masonry, or unprotected steel. Masonry construction walls can collapse in one piece or crumble in many parts. The debris can travel a distance and even cause the collapse of other structures or objects.

- **Type IV** heavy-timber or mill construction is one of the least likely to collapse. The weight-bearing capacity of the large-dimension wood members will resist collapse unless they have been affected by a large volume of fire. A collapse zone should be established if the fire is intense or the structure has been weakened by repeated fires over time.

- **Type V** construction collapses are influenced by the style of construction. That is, a multistory platform frame construction structure will generally burn through and collapse inward while a balloon frame construction structure can have full walls fall outward in a single piece. Exterior masonry and veneer walls that are not load-bearing are placed over load-bearing wood walls. Brick veneer attached to the frame can fall straight down (*curtain collapse*) into a pile, or fall straight out as a unit as the ties and supports fail. Although it is rare for a Type V to collapse outward, there is a great danger to firefighters due to interior collapses. Lightweight trusses will fail within 5 minutes when exposed to direct heat.

Balloon and Platform Frame Construction

The terms *balloon frame* and *platform frame construction* are often used in the fire service to describe a construction method used for wood frame residential buildings **(Figure 4.81, p. 202)**. With the advent of lumber mills that could cut 2 x 4 inch (50 mm x 100 mm) lumber and the introduction of cheap mass-produced nails, **balloon frame construction** became common between the late 1800s and the mid-1950s. The speed and ease of building made balloon frame the desired design type for residential construction. Balloon frame construction consists of long continuous studs that run from the sill plate (located on the foundation) to the roof eave line. All intermediate floor structures are attached to the studs. The wall cavities between the studs are open from the ground floor to the attic. A fire in the basement or on the first floor could extend to the attic through this channel.

After World War II, **platform frame construction** became popular. The lack of long wood beams and the speed at which platform buildings could be built replaced balloon frame structures. Floor assemblies create an individual platform that rest on the foundation. Wall assemblies the height of one story are placed on this platform and a second platform rests on top of the wall unit. Each platform creates fire stops at each floor level restricting the spread of fire within the wall cavity. Currently, most houses built have some lightweight construction including the use of trusses, foam sheathing, steel studs, or wood I-beams. The use of steel studs has influenced an increase of balloon framing. Unlike previous balloon frame construction, modern codes require fire-stopping measures that inhibit the spread of fire vertically within walls.

Balloon Frame Construction — A construction method using long continuous studs that run from the sill plate (located on the foundation) to the roof eave line. All intermediate floor structures are attached to the studs. Requires the use of long lumber and generally lacks any type of fire stopping within the wall cavity.

Platform Frame Construction — A construction method in which a floor assembly creates an individual platform that rests on the foundation. Wall assemblies the height of one story are placed on this platform and a second platform rests on top of the wall unit. Each platform creates fire stops at each floor level restricting the spread of fire within the wall cavity.

Collapse zones should be established when:

- There is an indication that the structure has been weakened by prolonged exposure to fire or heat.
- A defensive strategy has been adopted.
- Interior operations cannot be justified.

The size of the collapse zone must take into consideration the following:

- The type of building construction
- Other exposures
- The safest location for apparatus and personnel.

Because the collapse zone extends the full length of all of the affected walls, the safest location for defensive operations is at the corner of the building. Master streams and apparatus can be located in the area formed by a 90-degree arc from the wall intersection as long as they are far enough away that flying debris will not strike them.

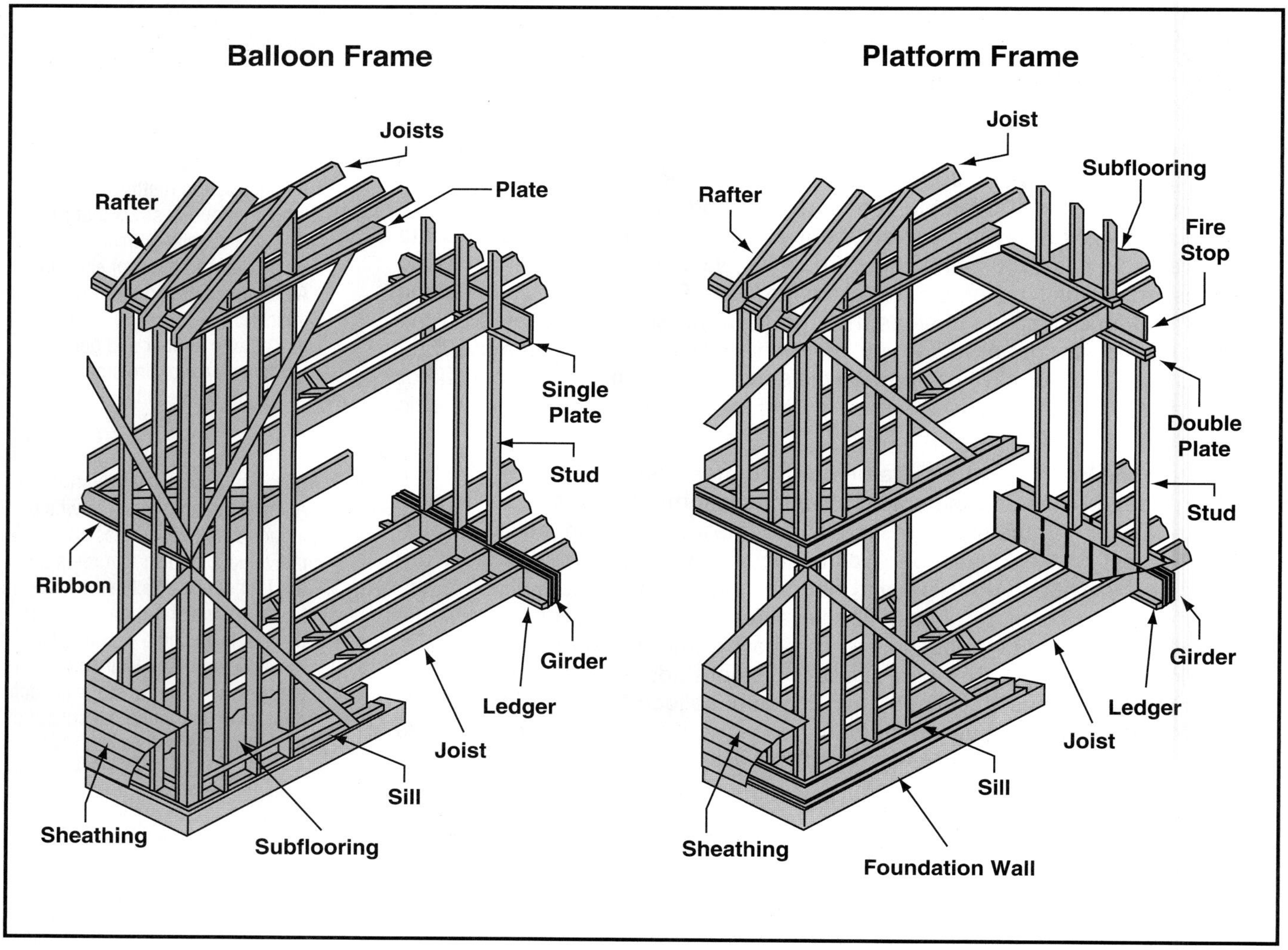

Balloon Frame

- Joists
- Rafter
- Plate
- Single Plate
- Stud
- Ribbon
- Girder
- Ledger
- Joist
- Sill
- Sheathing
- Subflooring

Platform Frame

- Joist
- Subflooring
- Rafter
- Fire Stop
- Double Plate
- Stud
- Girder
- Ledger
- Joist
- Sill
- Sheathing
- Foundation Wall

Figure 4.81 Both balloon frame and platform frame construction are found in many jurisdictions.

Chapter Summary

Firefighters often think of fire as their enemy. When that is the case, quite often the battlefield exists inside buildings. Your safety depends on your ability to maneuver on the battlefield, locate the enemy, and extinguish it. You must know how the building will contribute to, and even control, the spread of fire. You must also understand the effect fire and heat have on structural components and materials, even if you cannot see it occurring, to be able to anticipate the results. Knowing and understanding building construction is as vital as your knowledge and understanding of fire behavior. Further, you must be familiar with the types of construction in your community or response area, be aware of changes to existing structures, and follow trends in building construction, all of which will affect how you fight fire.

Firefighter I

1. What impact can fire have on common building materials?

2. How are different construction classifications affected by fire suppression?

3. What are the main types of occupancy classifications?

4. In what ways can building components impact fire suppression efforts?

Firefighter II

1. What are the main hazards related to building construction during fire suppression?

2. What indicators of building collapse or structural instability may occur during fire suppression?

3. What factors influence structural collapse potential?

4. How is a collapse zone typically measured?

Courtesy of Mike Wieder.

Chapter Contents

Key Terms

NFPA® Job Performance Requirements

This chapter provides information that addresses the following job performance requirements of NFPA® 1001, *Standard for Fire Fighter Professional Qualifications* (2013).

Firefighter I

5.3.10

5.3.11

5.3.12

1. Explain the science of fire as it relates to energy, forms of ignition, and modes of combustion. (5.3.11)

2. Describe the impact of thermal energy on heat, temperature, and heat transfer. (5.3.12)

3. Recognize the physical states of fuel. (5.3.10)

4. Explain the relationship between oxygen content and life safety. (5.3.11)

5. Identify the products of self-sustained chemical reactions. (5.3.11)

6. Explain the factors that affect fire development. (5.3.11)

7. Describe the stages of fire development. (5.3.11)

8. Recognize signs, causes, and effects of rapid fire development. (5.3.11)

9. Describe the methods through which fire fighting operations can influence fire behavior. (5.3.11, 5.3.12)

Chapter 5
Fire Behavior

The traditional mission of the fire service is fighting fires in structures, vehicles, vessels, aircraft, and in the open, including grasslands and forests. In order to provide this service and provide for their own safety, firefighters need to understand the following:

- Fire science
- The combustion process
- Fire behavior and its relationship to various materials and environments
- Classifications of fires and their corresponding extinguishing agents
- Recognition of fire behavior indicators, fire development patterns, and potential for rapid fire development

Combustion — A chemical process of oxidation that occurs at a rate fast enough to produce heat and usually light in the form of either a glow or flame. (Reproduced with permission from NFPA® 921-2011, *Guide for Fire and Explosion Investigations*, Copyright© 2011, National Fire Protection Association®)

Fire — A rapid oxidation process, which is a chemical reaction resulting in the evolution of light and heat in varying intensities. (NFPA® 921)

Heat — A form of energy characterized by vibration of molecules and capable of initiating and supporting chemical changes and changes of state. (NFPA® 921)

Temperature — Measure of a material's ability to transfer heat energy to other objects; the greater the energy, the higher the temperature. Measure of the average kinetic energy of the particles in a sample of matter, expressed in terms of units or degrees designated on a standard scale. *See* Celsius Scale and Fahrenheit Scale.

Fuel — A material that will maintain combustion under specified environmental conditions. (NFPA® 921)

Oxidizer — Any material that readily yields oxygen or other oxidizing gas, or that readily reacts to promote or initiate combustion of combustible materials. (Reproduced with permission from NFPA® 400-2010, *Hazardous Materials Code*, Copyright© 2010, National Fire Protection Association®)

Matter — Anything that occupies space and has mass.

Figure 5.1 The conditions at a fire scene offer indications of fire behavior present and to come. *Courtesy of Bob Esposito.*

I Science of Fire

Firefighters should have a scientific understanding of **combustion, fire, heat,** and **temperature.** Fire can take a variety of forms, but all fires involve a heat-producing chemical reaction between some type of **fuel** and an **oxidizer,** most commonly oxygen in the air. This process is best explained through the study of physical science.

Physical science is the study of **matter** and **energy** and includes chemistry and physics. This theoretical foundation must be translated into a practical knowledge of fire behavior. In order to remain safe, you need to be able to "read" a fire, in other words recognize what is presently happening and predict potential fire behavior (**Figure 5.1**).

The world around you is made up of matter in the form of physical materials that occupy space and have mass. While matter can undergo many types of physical and chemical changes, this chapter will concentrate on those changes related to fire.

A physical change occurs when a substance remains chemically the same but changes in size, shape, or appearance. Examples of physical change are water freezing (liquid to solid) and boiling (liquid to gas).

A chemical reaction occurs when a substance changes from one type of matter into another, such as two or more substances combining to form compounds. **Oxidation** is a chemical reaction involving the combination of an oxidizer, such as oxygen in the air, with other materials. Oxidation can be slow, such as the combination of oxygen with iron to form rust, or rapid, as in combustion of methane (natural gas) (**Figure 5.2**).

Energy

Energy is the capacity to perform work. Work occurs when a force is applied to an object over a distance or when a substance undergoes a chemical, biological, or physical change. In the case of heat, work means increasing the temperature of a substance.

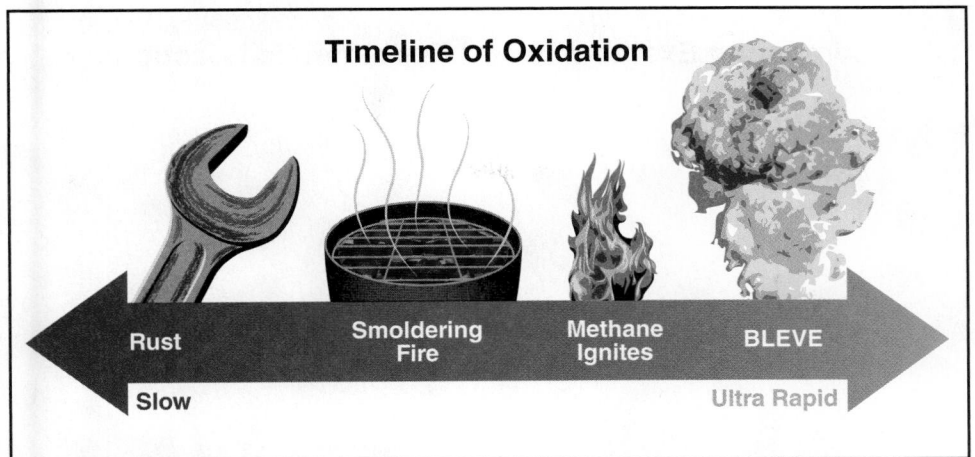

Timeline of Oxidation

Rust | Smoldering Fire | Methane Ignites | BLEVE

Slow | | | Ultra Rapid

Figure 5.2 The timeline of oxidation illustrates the speed difference in types of oxidation between the polar extremes of rust and BLEVE.

Potential | **Kinetic**

Figure 5.3 Potential energy waits in a stored form, as opposed to kinetic energy which actively releases energy. *Courtesy of Dan Madrzykowski, NIST.*

The forms of energy are classified as either potential or kinetic (**Figure 5.3**). **Potential energy** represents the amount of **kinetic energy** that an object can release at some point in the future. Fuels have a certain amount of chemical potential energy before they are ignited. Different fuels can release different amounts of energy over different amounts of time.

Kinetic energy is the energy possessed by a moving object. While a fuel such as wood is not "moving" as you might define it, when heat is introduced, the molecules within the fuel begin to vibrate. As the heat (thermal energy) increases, these molecules vibrate more and more rapidly. The fuel's kinetic energy is the result of these vibrations in the molecules.

There are many types of energy including chemical, thermal, mechanical, electrical, light, nuclear, and sound. All energy can change from one type to another, for example mechanical energy from a machine can convert to thermal energy when friction between moving parts generate heat. In terms of fire behavior, the potential chemical energy of a fuel is converted to thermal energy and released as heat.

Joules (J) — Joules are defined in terms of mechanical energy. It is equal to the energy expended in applying a force of one newton through a distance of one meter. However, it is more useful for firefighters to think about the energy required to increase temperature. 4.2 joules are required to raise the temperature of one gram of water one degree Celsius.

Exothermic Reaction — Chemical reaction that releases thermal energy or heat.

Endothermic Reaction — Chemical reaction that absorbs thermal energy or heat.

Pyrolysis — The chemical decomposition of a solid material by heating. Pyrolysis often precedes combustion.

Vaporization — Physical process that changes a liquid into a gaseous state; the rate of vaporization depends on the substance involved, heat, pressure, and exposed surface area.

Ignition — The process of initiating self-sustained combustion. (NFPA® 921)

Piloted Ignition — Moment when a mixture of fuel and oxygen encounters an external heat (ignition) source with sufficient heat or thermal energy to start the combustion reaction.

Autoignition — Initiation of combustion by heat but without a spark or flame. (NFPA® 921)

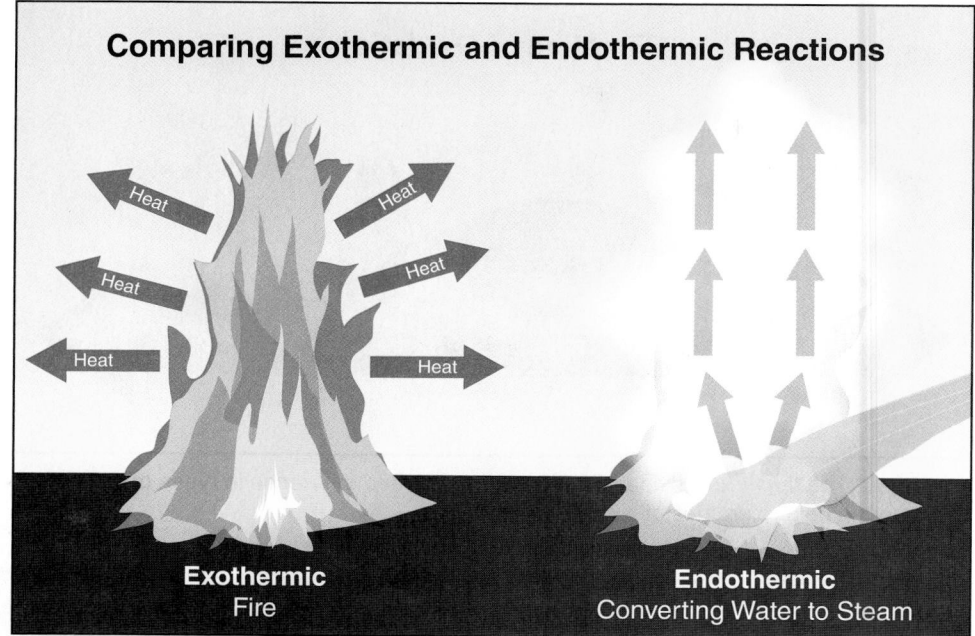

Figure 5.4 Exothermic reactions actively release energy as opposed to endothermic reactions which absorb energy.

The measure for energy is *joules* (J) in the International System of Units (SI). For example, 4.2 joules is the quantity of heat required to change the temperature of one gram of water by one degree Celsius. In the customary system, the unit of measure for heat is the British thermal unit (Btu). A British thermal unit is the amount of heat required to raise the temperature of 1 pound of water 1 degree Fahrenheit. While not used in scientific and engineering texts, the Btu is still frequently used in the fire service. When comparing joules and Btu, 1055 J = 1 Btu.

Chemical and physical changes almost always involve an exchange of energy. A fuel's potential energy is released during combustion and converted to kinetic energy. Reactions that emit energy as they occur are **exothermic reactions**. Fire is an exothermic chemical reaction called combustion that releases energy in the form of heat and sometimes visible light. Reactions that absorb energy as they occur are **endothermic reactions** (**Figure 5.4**). For example, converting water from a liquid to a gas (steam) requires the input of energy and is an endothermic physical reaction. Converting water to steam is an important part of controlling and extinguishing some types of fires.

Forms of Ignition

As fuel is heated, its temperature increases. Transferring sufficient heat causes **pyrolysis** in solid fuels and **vaporization** of liquid fuels, releasing ignitable vapors or gases (**Figure 5.5**). A spark or other external source can provide the energy necessary for **ignition**, or the fuel can be heated until it ignites without a spark or other source. Once ignited, the ignition process continues the production and ignition of fuel vapors or gases so that the combustion reaction is sustained.

There are two forms of ignition: **piloted ignition** and **autoignition** (nonpiloted) (**Figure 5.6**). Piloted ignition is the most common and occurs when a mixture of fuel and oxygen encounter an external heat source with sufficient heat or thermal energy to start the combustion process. Autoignition occurs without any external flame or

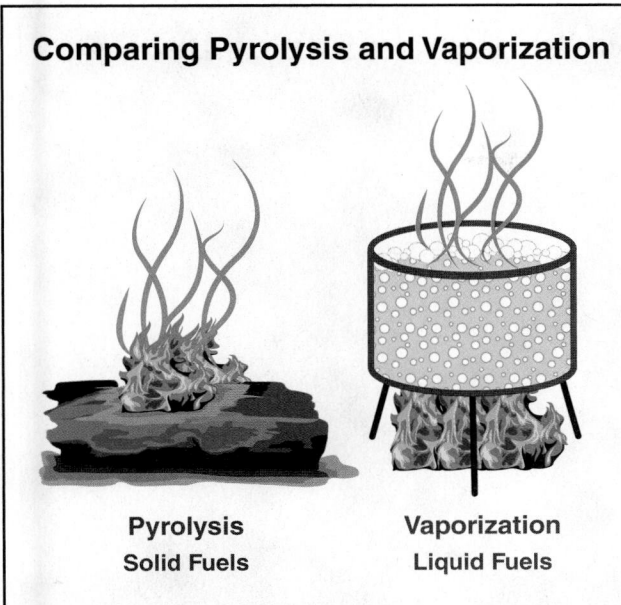

Comparing Pyrolysis and Vaporization

Pyrolysis
Solid Fuels

Vaporization
Liquid Fuels

Figure 5.5 Pyrolysis is the conversion of a solid fuel item into a gas fuel that is capable of supporting combustion, whereas vaporization is the conversion of a liquid to a vapor using the heat energy from combustion.

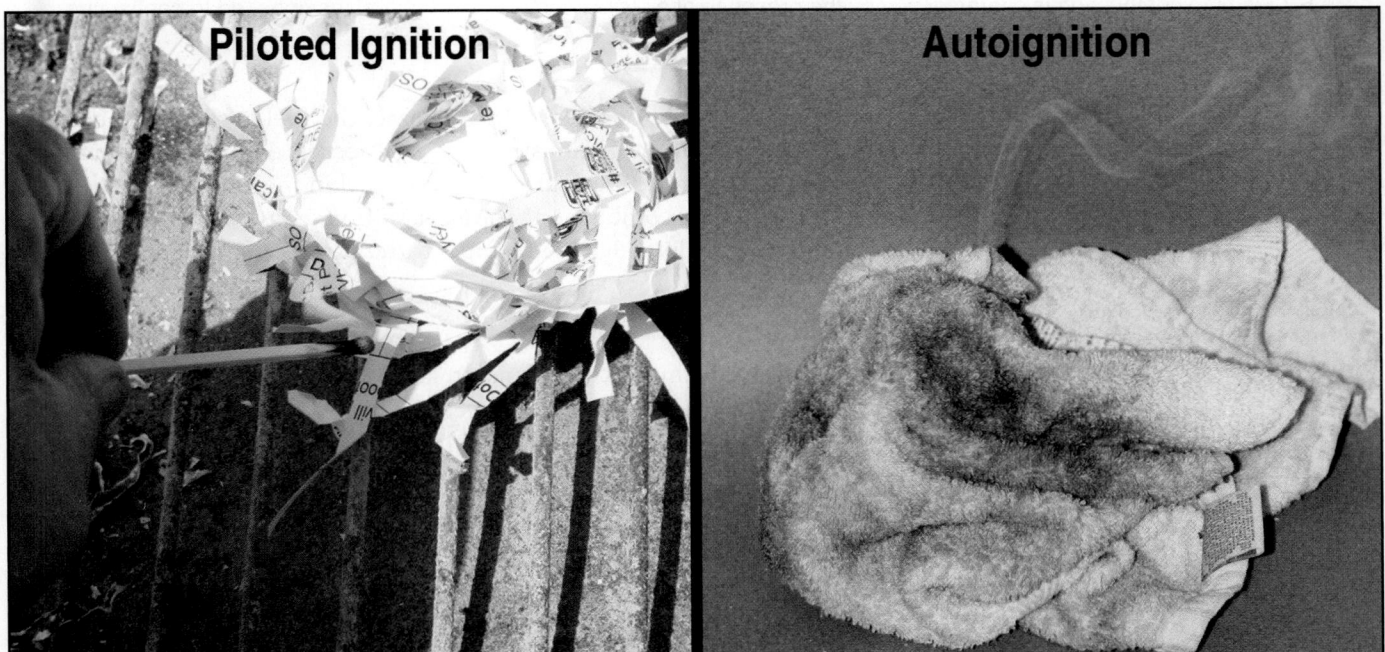

Piloted Ignition

Autoignition

Figure 5.6 Piloted ignition requires the participation of an outside ignition source, whereas autoignition occurs from the confluence of specific conditions.

Flaming Combustion　　　　**Nonflaming Combustion**

Figure 5.7 Flaming combustion is characterized by the presence of a visible flame above the fuel, as opposed to nonflaming combustion which features a lower temperature and smoldering burn.

Autoignition Temperature — The lowest temperature at which a combustible material ignites in air without a spark or flame. (NFPA® 921)

Flame — Visible, luminous body of a burning gas emitting radiant energy including light of various colors given off by burning gases or vapors during the combustion process.

Fire Triangle — A model used to explain the elements/conditions necessary for combustion. The sides of the triangle represent heat, oxygen, and fuel.

Fire Tetrahedron — Model of the four elements/conditions required to have a fire. The four sides of the tetrahedron represent fuel, heat, oxygen, and self-sustaining chemical chain reaction.

Passive Agent — Materials that absorb heat but do not participate actively in the combustion process.

spark to ignite the fuel gases or vapors. In this case, the fuel surface is chemically heated to the point at which the combustion reaction occurs. **Autoignition temperature** (AIT) is the minimum temperature to which a fuel in the air must be heated in order to start self-sustained combustion. The autoignition temperature of a substance is always higher than its piloted ignition temperature.

Modes of Combustion

Fire and combustion are similar conditions. In fact, both words are commonly used to mean the same thing. Combustion, however, is a chemical reaction while fire is a one possible result of combustion. Combustion can occur without fire.

There are two modes of combustion, nonflaming and flaming **(Figure 5.7)**. Nonflaming combustion occurs more slowly at a lower temperature producing a smoldering glow in the material's surface. Flaming combustion is commonly referred to as fire because it produces a visible **flame** above the material's surface.

Fire Models

Two models, the **fire triangle** and **fire tetrahedron**, are used to explain the elements of fire and how fires can be extinguished **(Figure 5.8)**. The oldest and simplest model, the fire triangle, illustrates the three elements necessary for fire to occur: fuel, oxygen, and heat. Remove any one of these elements and the fire will be extinguished.

Research into fire behavior has determined that an uninhibited chemical chain reaction must also be present for a fire to occur. This research resulted in the creation of the fire tetrahedron model to explain fires involving certain types of substances and the types of agents necessary to extinguish them.

Besides fuel, heat, and oxygen, other materials can have a significant effect on both ignition and fire development. **Passive agents** are materials that absorb heat but do not participate actively in the combustion reaction. In building construction, one

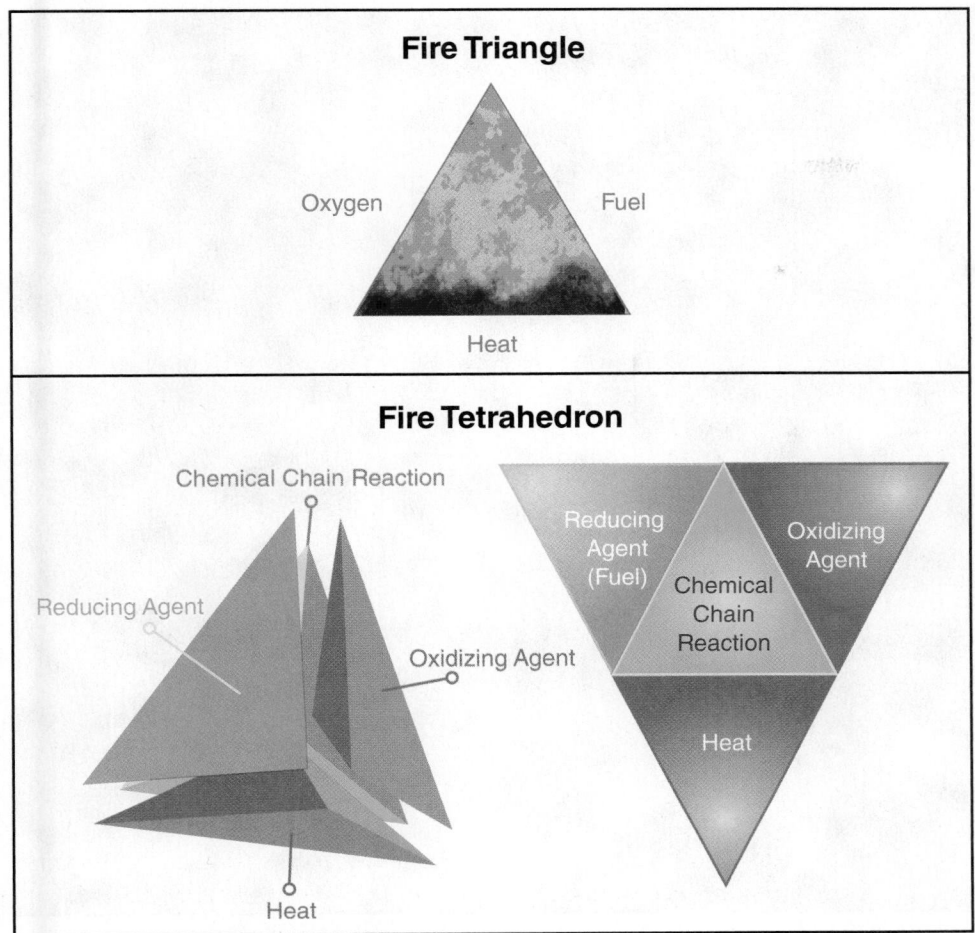

Fire Triangle

Oxygen

Fuel

Heat

Fire Tetrahedron

Chemical Chain Reaction

Reducing Agent

Oxidizing Agent

Heat

Reducing Agent (Fuel)

Oxidizing Agent

Chemical Chain Reaction

Heat

Figure 5.8 The fire triangle illustrates the three components necessary for the existence of fire, and the fire tetrahedron demonstrates the four components required for a self-sustaining fire.

of the most common passive agents is drywall or gypsum board, which contains moisture in the form of hydrates. As the drywall is heated, the moisture is vaporized slowing the increase in the temperature of the gypsum board. Multiple layers of drywall are used to create fire walls that will withstand fire exposure for many hours **(Figure 5.9, p. 214)**.

Outside of structures, the moisture content of vegetation acts as a passive agent in slowing ground cover fires. Relative humidity in the air also plays a part in ground cover fire development. Conversely, vegetation, debris, and dead trees that lack moisture content are easily ignitable and provide fuel creating a fast-moving fire.

Nonflaming Combustion

Nonflaming combustion occurs when burning is localized on or near the fuel's surface where it is in contact with oxygen. Examples of nonflaming or smoldering combustion include burning charcoal or smoldering wood or fabric. The fire triangle is a simple model that can be used to illustrate the elements/conditions required for this mode of combustion.

Figure 5.9 Passive agents, such as drywall paneling, resist fire exposure for many hours. *Courtesy of Donny Howard.*

Flaming Combustion

Flaming combustion occurs when a gaseous fuel mixes with oxygen in the correct ratio and is heated to ignition temperature. Flaming combustion requires liquid or solid fuels to be vaporized or converted to the gas phase through the addition of heat. When heated, both liquid and solid fuels will emit vapors that mix with oxygen producing flames above the material's surface if the gases are ignited. The fire tetrahedron accurately reflects the conditions required for flaming combustion.

Each element of the tetrahedron must be in place for flaming combustion to occur. Removing any element of the tetrahedron interrupts the chemical chain reaction and stops flaming combustion. However, the fire may continue to smolder depending on the characteristics of the fuel.

Products of Combustion

Products of Combustion — Materials produced and released during burning.

As any fuel burns, its chemical composition changes, producing new substances and releasing energy in the form of heat and light. The **products of combustion** are often simply described as heat (energy release, thermal energy) and smoke (new substances) because these are the products that have the most effect on firefighters.

Thermal energy generated during a fire is one product of combustion that heats adjacent fuels and makes them more susceptible to ignition. As they ignite, the fire spreads. Anything can be an adjacent fuel for heat. Persons lacking adequate protection from the heat may suffer burns, damage to their respiratory tract, dehydration, and heat exhaustion.

Table 5.1
Common Products of Combustion and Their Toxic Effects

Carbon Monoxide	Colorless, odorless gas. Inhalation of carbon monoxide causes headache, dizziness, weakness, confusion, nausea, unconsciousness, and death. Exposure to as little as 0.2 percent carbon monoxide can result in unconsciousness within 30 minutes. Inhalation of high concentration can result in immediate collapse and unconsciousness.
Formaldehyde	Colorless gas with a pungent irritating odor that is highly irritating to the nose. 50-100 ppm can cause severe irritation to the respiratory track and serious injury. Exposure to high concentrations can cause injury to the skin. Formaldehyde is a suspected carcinogen.
Hydrogen Cyanide	Colorless, toxic, and flammable liquid below 79°F (26°C) produced by the combustion of nitrogen-bearing substances. It is a chemical asphyxiant that acts to prevent the body from using oxygen. It is commonly encountered in smoke in concentrations lower than carbon monoxide.
Nitrogen Dioxide	Reddish-brown gas or yellowish-brown liquid, which is highly toxic and corrosive.
Particulates	Small particles that can be inhaled and deposited in the mouth, trachea, or the lungs. Exposure to particulates can cause eye irritation, respiratory distress (in addition to health hazards specifically related to the particular substances involved).
Sulfur Dioxide	Colorless gas with a choking or suffocating odor. Sulfur dioxide is toxic and corrosive, and can irritate the eyes and mucous membranes.

Source: *Computer Aided Management of Emergency Operations (CAMEO) and Toxicological Profile for Polycyclic Aromatic Hydrocarbons.*

While the heat from a fire is a danger to anyone directly exposed to it, toxic smoke causes most fire deaths. Smoke is an aerosol comprised of gases, vapor, and solid particulates.

Smoke is the product of incomplete combustion. At a very simple level, for example, complete combustion of methane in air results in the production of heat, light, water vapor, and carbon dioxide. However, in a structure fire, multiple fuels are involved and limited air supply results in incomplete combustion. These factors result in extremely complex chemical reactions producing a wide range of products of combustion including toxic and flammable gases, vapors, and particulates. These products comprise smoke.

Fire gases, such as carbon monoxide, are generally colorless, while vapor and particulates give smoke its varied colors. Most components of smoke are toxic and present a significant threat to human life. The materials that make up smoke vary from fuel to fuel, but generally all smoke is toxic. The toxic effects of smoke inhalation are the result of the interrelated effect of all the toxic products present. **Table 5.1** lists some of the more common products of combustion and their toxic effects.

Carbon monoxide (CO) is a toxic and flammable product of the incomplete combustion of organic (carbon-containing) materials. This gas is probably the most common product of combustion encountered in structure fires. Exposure to it is frequently identified as the cause of death for both civilian and firefighter fatalities. Carbon monoxide acts as a chemical asphyxiant. It binds with hemoglobin in the blood preventing these cells from distributing oxygen to the body.

Hydrogen cyanide (HCN), a toxic and flammable substance produced in the combustion of materials containing nitrogen, is also commonly encountered in smoke, although at lower concentrations than CO. HCN also acts as a chemical asphyxiant

Carbon Monoxide (CO) — Colorless, odorless, dangerous gas (both toxic and flammable) formed by the incomplete combustion of carbon. It combines with hemoglobin more than 200 times faster than oxygen does, thus decreases the blood's ability to carry oxygen.

Hydrogen Cyanide (HCN) — Colorless, toxic, and flammable liquid until it reaches 79°F (26°C). Above that temperature, it becomes a gas with a faint odor similar to bitter almonds; produced by the combustion of nitrogen-bearing substances.

but with a different mechanism of action. HCN acts to prevent the body from using oxygen at the cellular level. HCN is a significant byproduct of the combustion of polyurethane foam, which is commonly used in furniture and bedding.

Carbon dioxide (CO_2) is a product of complete combustion of organic materials. It is not toxic in the same manner as CO or HCN, but it acts as a simple asphyxiant by displacing oxygen. Carbon dioxide also acts as a respiratory stimulant, increasing respiratory rate.

Irritants in smoke are those substances that cause breathing discomfort and inflammation of the eyes, respiratory tract, and skin. Depending on the fuels involved, smoke will contain a wide range of irritating substances. Tests have determined that there are over 20 irritants in smoke, including hydrogen chloride, formaldehyde, and acrolein.

Smoke contains unburned fuel in the form of solid and liquid particulates and gases. Smoke must be treated with the same respect as any other flammable gas because it may burn or explode.

Because the products of combustion are deadly, firefighters must use SCBA when operating in toxic atmospheres. Although the volume and density of smoke is considerably reduced during overhaul, the respiratory hazard is not eliminated. A research study conducted by the Phoenix (AZ) Fire Department determined that hazardous concentrations above published short-term exposure limits are likely to be present during overhaul. In addition, hazardous concentrations may be present in areas outside the structure.

WARNING
Smoke is fuel and is always potentially flammable.
Wear full PPE and SCBA anytime you work in smoke.

Flame is also considered to be a product of combustion. Flame is the visible, luminous body of a burning gas. When a burning gas is mixed with the proper amounts of oxygen, the flame becomes hotter and less luminous. The loss of luminosity is caused by more complete combustion.

Thermal Energy (Heat)

Having a working knowledge of fire behavior requires an understanding of *temperature*, *energy*, and *heat*. These terms are often confused and used interchangeably because the differences between them are not always understood.

Heat is the energy element of both the fire triangle and the fire tetrahedron. Heat is the more commonly used term for **thermal energy**. Heat is kinetic energy transferred from a high-temperature substance to a low-temperature substance. Because heat is kinetic energy, it is always in transit from one location (an open flame) to another (an exposed fuel).

Heat is the thermal kinetic energy needed to release the potential chemical energy in a fuel. As previously mentioned, heat begins to vibrate the molecules in a fuel. As the molecules break down and release vapors, the vapors can ignite and release thermal energy. This new source of thermal energy begins to heat other, uninvolved fuels converting their energy and spreading the fire.

Carbon Dioxide (CO_2) — Colorless, odorless, heavier than air gas that neither supports combustion nor burns; used in portable fire extinguishers as an extinguishing agent to extinguish Class B or C fires by smothering or displacing the oxygen.

Thermal Energy — The kinetic energy associated with the random motions of the molecules of a material or object; often used interchangeably with the terms heat and heat energy. Measured in joules or Btu.

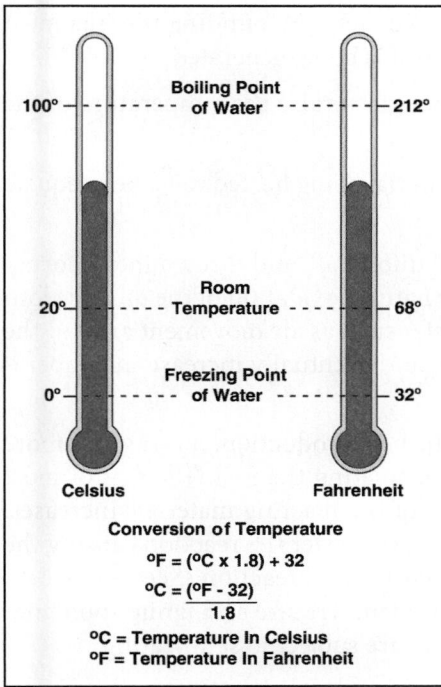

Figure 5.10 Both Celsius and Fahrenheit are commonly used temperature scales, and knowing how to convert between them may save time and effort during an incident response.

Heat and Temperature

Temperature, simply stated, is the measurement of heat. More specifically, temperature is the measurement of the average kinetic energy in the particles of a sample of matter. A block of wood at room temperature has stable molecules and is in no danger of ignition. When the wood is heated, the temperature rises because the molecules have begun to vibrate and move more freely and rapidly.

There are several different scales used to measure temperature. The most common are the Celsius scale, which is used in the metric system, and the Fahrenheit scale, used in the customary system. The freezing and boiling points of water provide a simple way to compare these two scales (**Figure 5.10**).

Sources of Thermal Energy

Chemical, mechanical, electrical, light, nuclear, and sound energy are all sources of thermal energy. They can all transfer heat and cause the temperature of a substance to increase. Chemical, electrical, and mechanical energy are common sources of heat that result in the ignition of a fuel.

Chemical Energy

Chemical energy is the most common source of heat in combustion reactions. When any combustible fuel is in contact with oxygen, the potential for oxidation exists. The oxidation process almost always results in the production of thermal energy.

Self-heating, a form of oxidation, is a chemical reaction that increases the temperature of a material without the addition of external heat. Normally, thermal energy is produced slowly by oxidation and is lost to the surroundings almost as fast as it is generated. The process can be initiated or accelerated by an external heat source such as sunshine. In order for self-heating to progress to **spontaneous ignition**, the material must be heated to its autoignition temperature. For spontaneous ignition to occur, the following factors are required:

Self-Heating — The result of exothermic reactions, occurring spontaneously in some materials under certain conditions, whereby heat is generated at a rate sufficient to raise the temperature of the material (NFPA® 921).

Spontaneous Ignition — Initiation of combustion of a material by an internal chemical or biological reaction that has produced sufficient heat to ignite the material (NFPA® 921).

- The insulation properties of the material immediately surrounding the fuel must be such that the heat cannot dissipate as fast as it is being generated.

- The rate of heat production must be great enough to raise the temperature of the material to its autoignition temperature.

- The available air supply in and around the material being heated must be adequate to support combustion.

For example, rags soaked in linseed oil, rolled into a ball, and thrown into a corner have the potential for spontaneous ignition. The natural oxidation of the oil and cloth will generate heat. If some method of heat transfer such as air movement around the rags does not dissipate the heat, then the cloth could eventually increase in temperature enough to cause ignition.

The rate of the oxidation reaction, and thus the heat production, increases as more heat is generated and trapped by the materials insulating the fuel. The rate of most chemical reactions increases as the temperature of the reacting materials increases. The more heat generated and absorbed by the fuel, the faster the reaction causing the heat generation. When the heat generated by a self-heating reaction exceeds the heat being lost, the material may reach its autoignition temperature and ignite spontaneously. **Table 5.2** lists some common materials that are subject to self-heating.

Table 5.2 Spontaneous Heating Materials and Locations	
Type of Material	**Possible Locations**
Charcoal	Convenience stores Hardware stores Industrial plants Restaurants
Linseed oil-soaked rags	Woodworking shops Lumberyards Furniture repair shops Picture frame shops
Straw and manure	Farms Feed stores Arenas Feedlots

Electrical Energy

Electrical energy can generate temperatures high enough to ignite any combustible materials near the heated area. Electrical heating can occur in several ways, including the following (**Figure 5.11**):

- **Resistance heating** — When electric current flows through a conductor, heat is produced. Some electrical appliances, such as incandescent lamps, ranges, ovens, or portable heaters, are designed to make use of resistance heating. Other electrical equipment is designed to limit resistance heating under normal operating conditions.

- **Overcurrent or overload** — When the current flowing through a conductor exceeds its design limits, it may overheat and present an ignition hazard. Overcurrent or overload is unintended resistance heating.

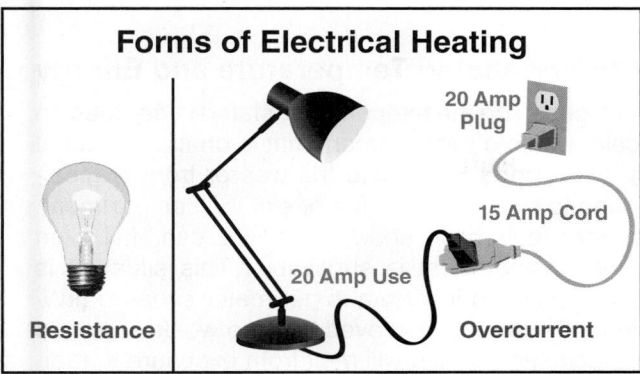

Forms of Electrical Heating

20 Amp Plug

15 Amp Cord

20 Amp Use

Resistance

Overcurrent

Figure 5.11 Forms of electrical heating may serve a domestic purpose or they may be uncontrolled.

- **Arcing** — In general, an arc is a high-temperature luminous electric discharge across a gap or through a medium such as charred insulation. Arcs may be generated when a conductor is separated (such as in an electric motor or switch) or by high voltage, static electricity, or lightning.

- **Sparking** — When an electric arc occurs, luminous (glowing) particles can be formed and spatter away from the point of arcing. In electrical terms, sparking refers to this spatter, while an arc is the luminous electric discharge.

Mechanical Energy

Mechanical energy is generated by friction or compression. The movement of two surfaces against each other creates heat of friction. This movement results in heat and/or sparks being generated. Heat of compression is generated when a gas is compressed. Diesel engines use this principle to ignite fuel vapor without a spark plug. The principle is also the reason that self-contained breathing apparatus (SCBA) cylinders feel warm to the touch after they have been filled.

Heat Transfer

The transfer of heat from one point or object to another is basic to the study of fire behavior. The transfer of heat from the initial fuel package (burning object) to other fuels in and beyond the area of fire origin affects the growth of any fire. Firefighters use their knowledge of heat transfer to estimate the size of a fire before attacking it and to evaluate the effectiveness of an attack. Heat is transferred from warmer objects to cooler objects. Objects at the same temperature cannot transfer heat.

The rate at which heat is transferred is related to the temperature differential of the bodies and the thermal conductivity of the solid material involved. For any given substance, the greater the temperature differences between the bodies, the greater the transfer rate. Transfer of thermal energy is described as **heat flux** (energy transfer over time per unit of surface area) and is typically measured in kilowatts per meter squared (kW/m^2). Heat can be transferred from one body to another by three mechanisms: **conduction**, **convection**, and **radiation**.

Heat Flux — The measure of the rate of heat transfer to a surface, expressed in kilowatts/m^2, kilojoules/m^2 x sec, or Btu/ft^2 x sec. (NFPA® 921)

Conduction — Transfer of heat through or between solids that are in direct contact.

Convection — Heat transfer by circulation within a medium such as a gas or a liquid. (NFPA® 921)

Radiation — Heat transfer by way of electromagnetic energy. (NFPA® 921)

The Importance of Understanding Temperature and Energy

Traditionally, firefighters have focused on the temperature, stated in degrees on the Fahrenheit or Celsius scale, within a compartment that is on fire. Personal protective equipment (PPE) is designed to insulate the wearer from a specified amount of heat long enough to extinguish the fire or exit the compartment. However, recent NIST laboratory tests have shown that PPE can fail even though the temperature in the compartment is survivable. This situation is caused by the radiant energy, expressed in kilowatts per meter squared (kW/m²), from the walls and ceiling. It is currently believed that the weakest part of the PPE system is the SCBA facepiece, which will melt from exposure to radiant energy.

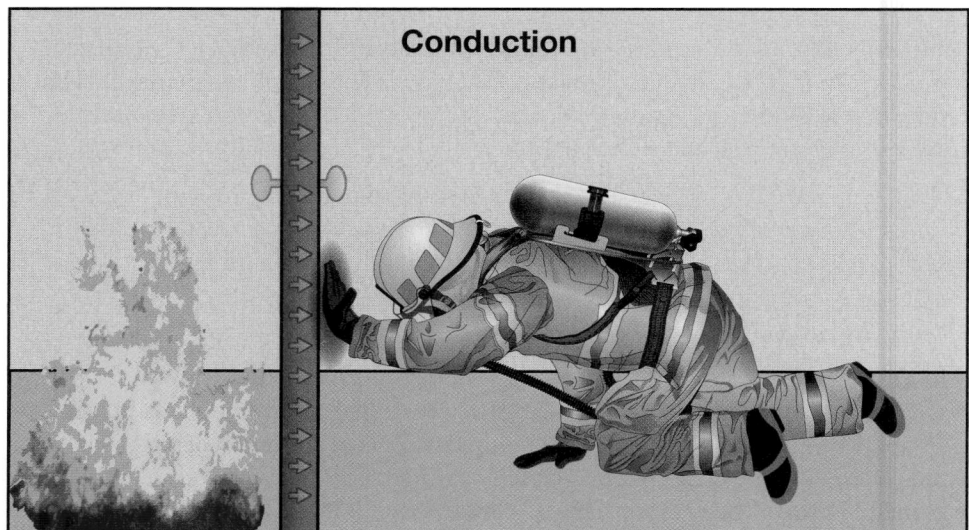

Figure 5.12 Conduction occurs when heat is transferred between solid objects. Heat may also be transered between an object and a person.

Conduction

Conduction is the transfer of heat through and between solids (**Figure 5.12**). Conduction occurs when a material is heated as a result of direct contact with a heat source. Conduction results from increased molecular motion and collisions between the molecules of a substance resulting in the transfer of energy through the substance. The more closely packed the molecules of a substance are, the more readily it will conduct heat. For example, if a metal pipe is heated by a fire on one side of a wall, heat conducted through the pipe can ignite wooden framing components in the wall or nearby combustibles on the other side of the wall.

Heat transfer due to conduction is dependent upon three factors:

- Area being heated
- Temperature difference between the heat source and the material being heated
- Thermal conductivity of the heated material

Table 5.3 shows the thermal conductivity of various common materials at the same ambient temperature (68°F/20°C). As you can see from the table, copper will conduct heat more than seven times more readily than steel. Likewise, steel is nearly forty times as thermally conductive as concrete. Wood is the least able to conduct heat of all these substances.

Table 5.3
Thermal Conductivity of Common Substances

Substance	Temperature	Thermal Conductivity (W/mK)
Copper	68°F (20°C)	386.00
Steel	68°F (20°C)	36.00 – 54.00
Concrete	68°F (20°C)	0.8 – 1.28
Wood (pine)	68°F (20°C)	0.13

Insulating materials slow the conduction of heat from one solid to another. Good insulators are materials that do not conduct heat well. Because of their physical makeup they disrupt the point-to-point transfer of heat or thermal energy. The best commercial insulators used in building construction are those made of fine particles or fibers with void spaces between them filled with a gas such as air. Gases do not conduct heat very well because their molecules are relatively far apart.

Convection

Convection is the transfer of thermal energy by the circulation or movement of a fluid (liquid or gas) **(Figure 5.13)**. In the fire environment, this usually involves transfer of heat through the movement of hot smoke and fire gases. As with all heat transfer, the flow of heat is from the hot fire gases to the cooler structural surfaces, building contents, and air.

Convection may occur in any direction. Generally, the movement will be upward because the smoke and fire gases are heated and **buoyant**. As a fire begins to grow, more air is entrained into the fire and heated, adding to the fire and the rising column containing products of combustion. Convection currents can also move laterally within a structure. Lateral currents are the result of differences in pressure; products of combustion or other gases (air, fuel gases) will move from areas of high pressure into areas of low pressure. Lateral movement can be from the fire area or from the openings on the windward side (where the pressure is higher) of the building to those on the leeward side.

Buoyant — The tendency or capacity to remain afloat in a liquid or rise in air or gas.

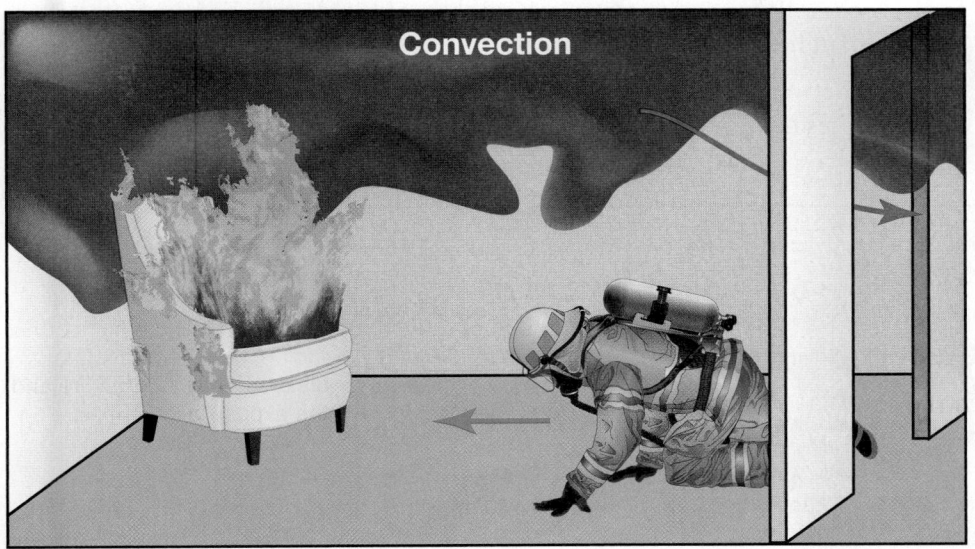

Figure 5.13 Convection is the heating of surfaces and gases through the circulation of fluids.

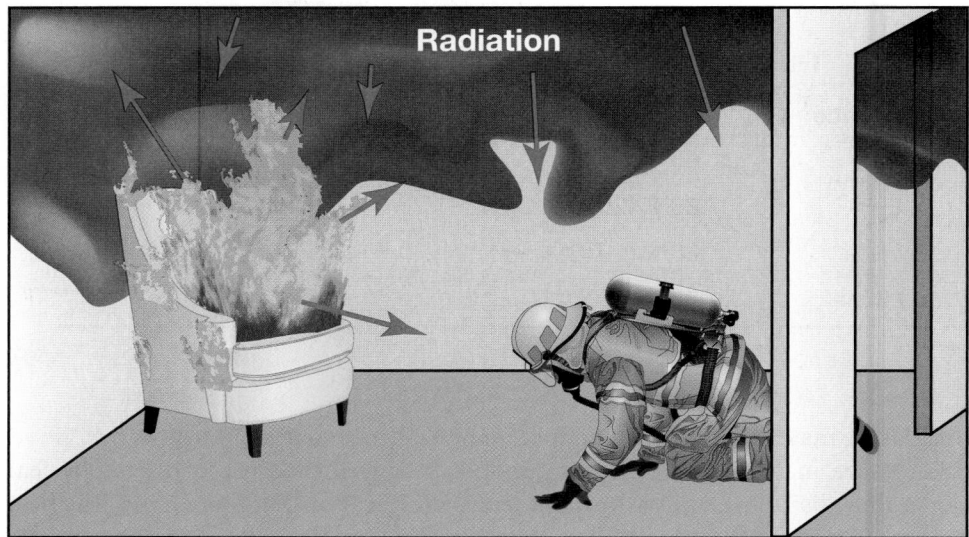

Figure 5.14 Unlike conduction and convection, heat transfer via radiation does not require an additional medium to transmit heat energy.

Firefighters who are working in the **flow path** of the hot smoke and fire gases will feel the increase in temperature as the velocity and/or turbulence increases causing convective heat transfer. This is a similar phenomenon to wind chill, except that in this case, energy is transferred from a hot fluid (gas) to a solid surface (your skin) rather than from a hot surface (your skin) to a cooler fluid (air).

Radiation

Radiation is the transmission of energy as an electromagnetic wave, such as light waves, radio waves, or X-rays, without an intervening medium (**Figure 5.14**). Radiant heat can become the dominant mode of heat transfer when the fire grows in size and can have a significant effect on the ignition of objects located some distance from the fire. Radiant heat transfer is also a significant factor in fire development and spread in compartments.

Wide ranges of factors influence radiant heat transfer, including:

- **Nature of the exposed surfaces** — Dark materials emit and absorb heat more effectively than lighter color materials; smooth or highly-polished surfaces reflect more radiant heat than rough surfaces.

- **Distance between the heat source and the exposed surfaces** — Increasing distance reduces the effect of radiant heat (**Figure 5.15**).

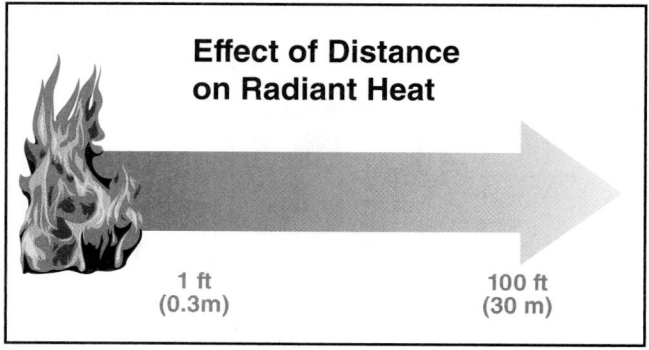

Effect of Distance on Radiant Heat

1 ft
(0.3m)

100 ft
(30 m)

Figure 5.15 The effects of radiant heat decrease as distance between the fire and the exposure increases.

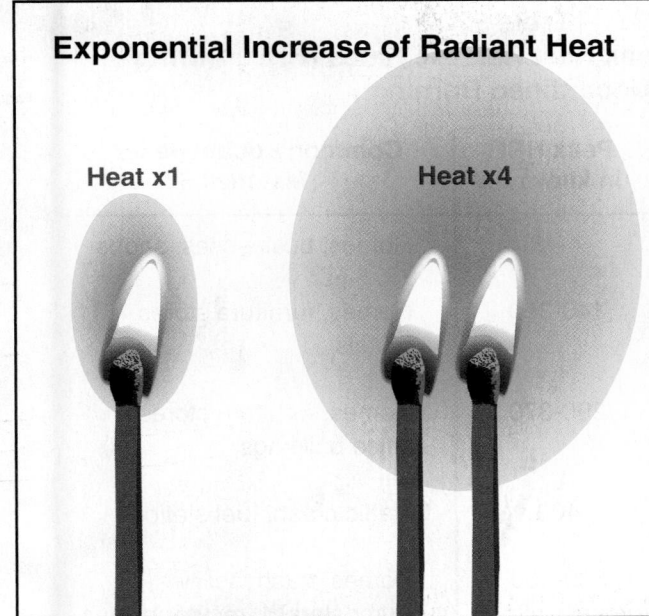

Exponential Increase of Radiant Heat

Heat x1

Heat x4

Figure 5.16 Radiant heat increases exponentially with incremental increases of fuel load.

- **Temperature difference between the heat source and exposed surfaces** — Temperature difference has a major effect on heat transfer through radiation. As the temperature of the heat source increases, the radiant energy increases by a factor to the fourth power (**Figure 5.16**).

Because energy is an electromagnetic wave, it travels in a straight line at the speed of light. The best example of heat transfer by radiation is the heat of the sun. The energy travels at the speed of light from the sun through space (a vacuum) until it collides with and warms the surface of the earth.

Radiation is a common cause of exposure fires (fires ignited in fuel packages or buildings that are remote from the initial fuel package or building of origin). As a fire grows, it radiates more and more energy which is converted to heat when it is absorbed by an object. In large fires, it is possible for the radiated heat to ignite buildings or other fuel packages a considerable distance away. Radiated heat travels through vacuums and air spaces where conduction or convection would normally be disrupted. However, materials that reflect radiated energy will disrupt the transmission of heat. While flames have high temperature resulting in emission of significant radiant energy, hot smoke in the **upper layer** can also radiate significant energy.

Fuel

Fuel is the material or substance that is oxidized or burned in the combustion process. In scientific terms, the fuel in a combustion reaction is known as the **reducing agent**. Fuels may be inorganic or organic. Inorganic fuels, such as hydrogen or magnesium, do not contain carbon. Most common fuels are organic, containing carbon along with other elements. These fuels can be further divided into hydrocarbon-based fuels, such as gasoline, fuel oil, and plastics, and cellulose-based materials, such as wood and paper.

The chemical content of any fuel influences both its **heat of combustion** and **heat release rate (HRR)**. The heat of combustion of a fuel is the total amount of thermal energy released when a specific amount of that fuel is oxidized (burned). In other words, different materials release more or less heat or thermal energy than others depending on their chemical makeup. Heat of combustion is usually expressed in

Upper Layer — Buoyant layer of hot gases and smoke produced by a fire in a compartment.

Reducing Agent — The fuel that is being oxidized or burned during combustion.

Heat of Combustion — Total amount of thermal energy (heat) that could be generated by the combustion (oxidation) reaction if a fuel were completely burned. The heat of combustion is measured in British Thermal Units (Btu) per pound or Megajoules per kilogram.

Heat Release Rate (HRR) — Total amount of heat released per unit time. The HRR is measured in kilowatts (kW) and megawatts (MW) of output.

Table 5.4
Representative Peak Heat Release Rates (HRR) During Unconfined Burning

Fuel Material	Peak HRR in kilowatts	Common Locations for Material
Small wastebasket	4-18	Homes, businesses, shops
Cotton mattress	140-350	Homes, furniture stores, motels
Cotton easy chair	290-370	Homes, furniture stores, office buildings
Small pool of gasoline	400	Traffic crash, fuel stations
Dry Christmas tree	500-650	Homes, trash facilities, Dumpsters™, recycling sites
Polyurethane mattress	810-2630	Homes, furniture stores, motels, dormitories, jails
Polyurethane easy chair	1350-1990	Homes, furniture stores, motels
Polyurethane sofa	3120	Homes, furniture stores, motels, dormitories, office buildings

Adapted from NFPA® 921, 2004 edition

kilojoules/gram (kJ/g). Many plastics, flammable liquids, and flammable gases contain more potential heat or thermal energy than wood (**Table 5.4**). Firefighters can expect to encounter these synthetic materials in modern construction and should be prepared for the higher combustion and heat release rates associated with them.

Power is the rate at which energy is being transferred over time. Another way to describe this is the rate at which energy is converted from one form to another. The standard international (SI) unit for power is the **watt** (W). One watt is one joule per second (J/s).

In terms of fire behavior, power is described as the HRR during combustion. When a fuel is being heated, work is being performed (energy is being transferred). The speed with which this work is occurring, HRR or the rate of heat transfer, is the amount of power being generated. Heat release rate is the energy released per unit of time as a fuel burns and is usually expressed in kilowatts (kW) or megawatts (MW). HRR is dependent on the type, quantity, and orientation of the fuel. HRR is directly related to oxygen consumption because the combustion process requires a continuous supply of oxygen to continue: typically, the more available oxygen, the higher the HRR. Similarly, the HRR decreases as compartments become ventilation-limited. **Figure 5.17** shows an example of the HRR produced by various sizes of fires.

Watt — A unit of measure of power or rate of work equal to one joule per second (J/s).

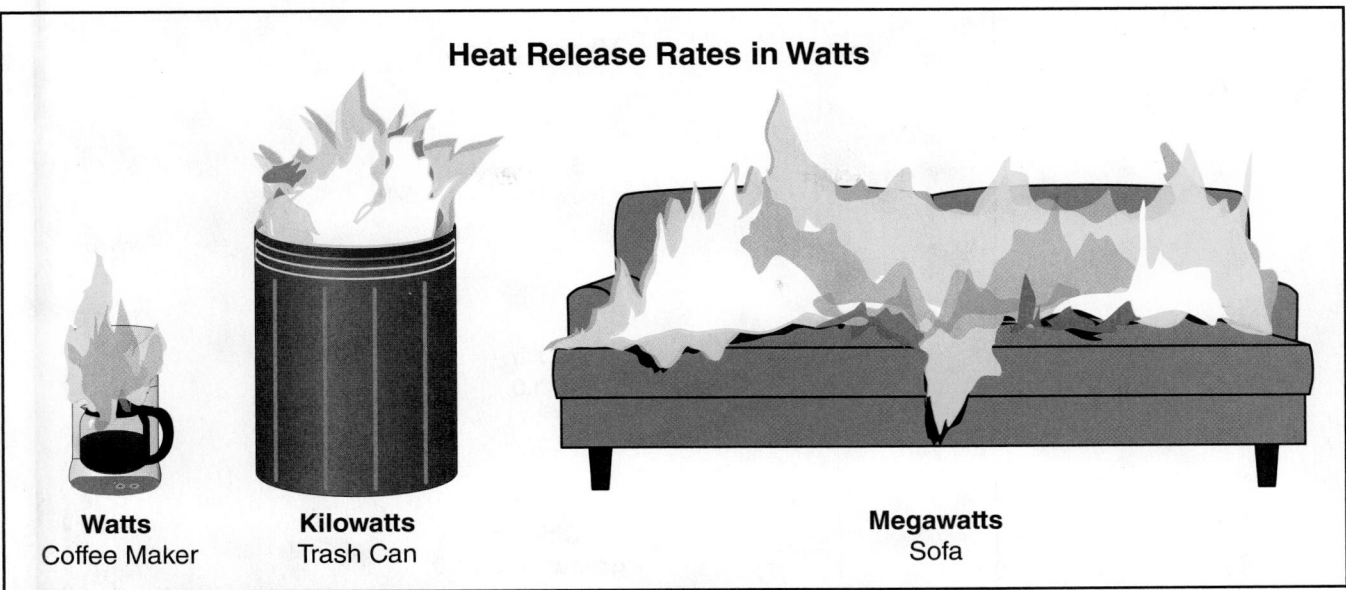

Heat Release Rates in Watts

Watts
Coffee Maker

Kilowatts
Trash Can

Megawatts
Sofa

Figure 5.17 Heat release rates may be measured in watts with a range between watts and megawatts.

Prefixes for Units of Measure: Kilo and Mega

The standard international system of units (SI) specifies a set of prefixes that precede units of measure to indicate a multiple or fraction of that unit. Two common prefixes encountered when discussing energy (Joules) and heat release rate (watts) are kilo and mega. The prefix *kilo* indicates a multiple of 1,000 and *mega* indicates a multiple of 1,000,000 (i.e., a kilowatt is 1000 watts).

A fuel may be found in any of three physical states of matter: gas, solid, or liquid. For flaming combustion to occur, fuels must be in the gaseous state. As previously described, thermal energy is required to change solids and liquids into the gaseous state. *Vapor* is the common term used to describe the gaseous state of a fuel that would normally exist as a liquid or a solid at standard temperature and pressure.

Gaseous Fuel

Gaseous fuels such as methane, hydrogen, and acetylene, can be the most dangerous of all fuel types because they are already in the physical state required for ignition. **Table 5.5** contains characteristics of common flammable gases.

Table 5.5 Characteristics of Common Flammable Gases		
Material	**Vapor Density**	**Ignition Temperature**
Methane (Natural Gas)	0.55	(1004°F) 540°C
Propane (Liquefied Petroleum Gas)	1.52	(842°F) 450°C
Carbon Monoxide	0.96	(1,128°F) 609°C

Source: *Computer Aided Management of Emergency Operations* (CAMEO)

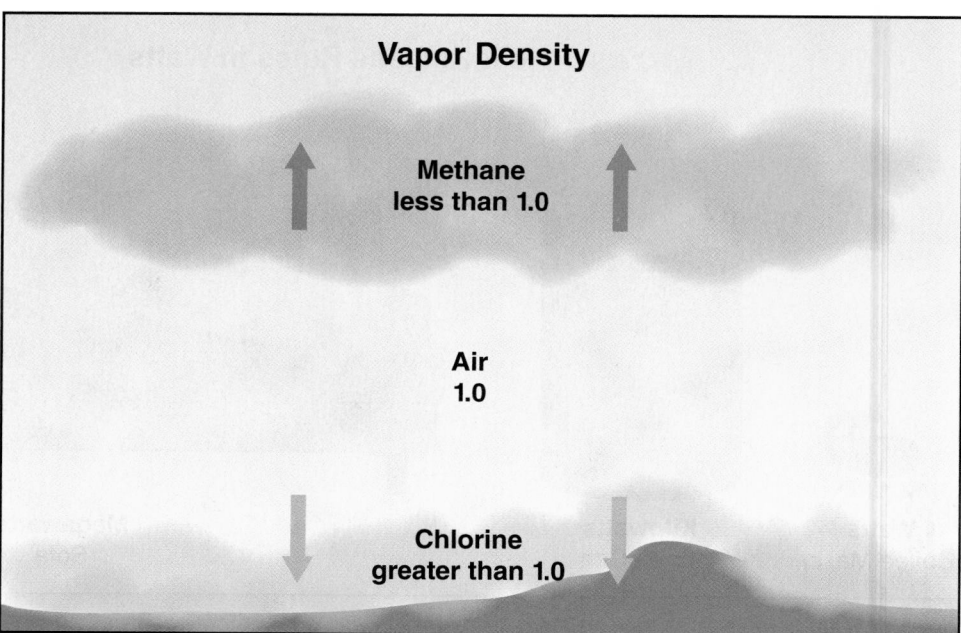

Figure 5.18 The vapor density of a gas provides an indication of where the gas will collect at an incident.

Vapor density describes the density of gases in relation to air. Air has been assigned a vapor density of 1. Gases with a vapor density of less than 1, such as methane, will rise while those having a vapor density of greater than 1, such as propane, will sink **(Figure 5.18)**. These densities presume that the gas and air are at the same temperature (generally specified as 68°F [20°C]). Heated gases expand and become less dense; when cooled they contract and become more dense.

Liquid Fuel

Liquids have mass and volume but no definite shape except for a flat surface or the shape of their container. Unlike gases, liquids will not expand to fill all of a container. When released on the ground, liquids will flow downhill and can pool in low areas. Just as gases are compared to air, the density of liquids is compared with the density of water. **Specific gravity** is the ratio of the mass of a given volume of a liquid compared with the mass of an equal volume of water at the same temperature. Water has been assigned a specific gravity of 1. Liquids with a specific gravity less than 1, such as gasoline and most **flammable liquids**, are lighter than water and will float on its surface. Liquids with a specific gravity greater than 1, such as epichlorohydrin (used in making plastics), are heavier than water causing them to sink **(Figure 5.19)**.

In order to burn, liquids must be vaporized. Vaporization is the transformation of a liquid to vapor or gaseous state. Unlike solids, liquids retain their state of matter partly due to atmospheric pressure. At sea level, the atmosphere exerts a pressure of 14.7 psi (102.9 kPa). In order for vaporization to occur, the escaping vapors must be at a higher pressure than atmospheric pressure. The pressure that vapors escaping from a liquid exert is known as **vapor pressure**. Vapor pressure indicates how easily a substance will evaporate or go into air. Flammable liquids with a high vapor pressure present a special hazard for firefighters.

As a liquid is heated, vapor pressure increases along with the rate of vaporization. For example, a puddle of water eventually evaporates. When the same amount of water is heated on a stove, however, it vaporizes much more rapidly because there

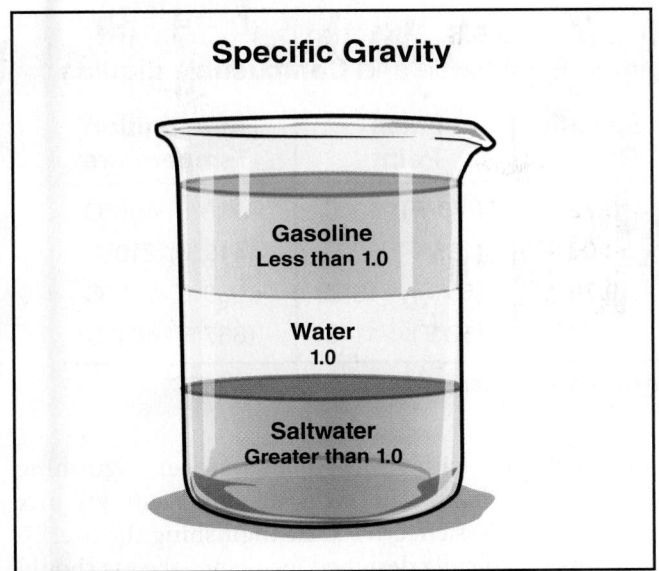

Specific Gravity

Gasoline
Less than 1.0

Water
1.0

Saltwater
Greater than 1.0

Figure 5.19 The specific gravity of a flammable liquid will have an impact on the choice of an extinguishing agent against a fire.

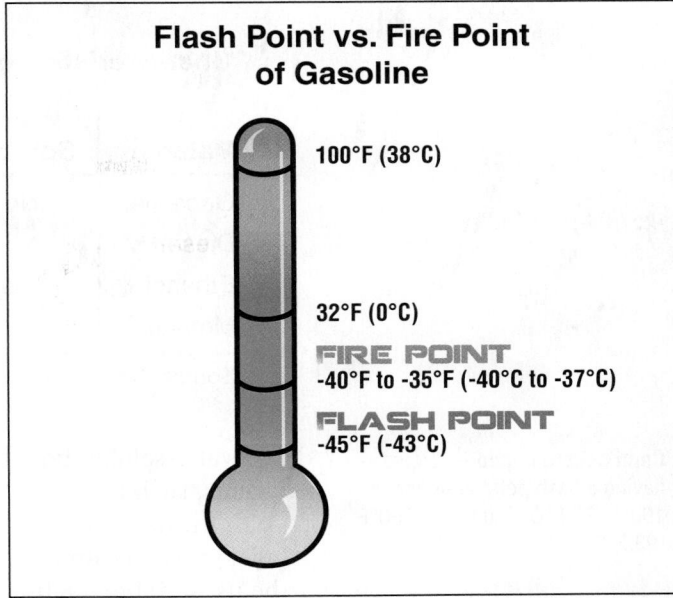

Flash Point vs. Fire Point of Gasoline

100°F (38°C)

32°F (0°C)

FIRE POINT
-40°F to -35°F (-40°C to -37°C)

FLASH POINT
-45°F (-43°C)

Figure 5.20 The flash point allows the creation of a quick fire, whereas the fire point sustains a combustion reaction.

is more thermal energy being applied. The rate of vaporization is determined by the vapor pressure of the substance and the amount of thermal energy applied to it. The volatility or ease with which a liquid gives off vapor influences how easily it can be ignited.

Flash point is the minimum temperature at which a liquid gives off sufficient vapors to ignite, but not sustain combustion. **Fire point** is the temperature at which sufficient vapors are being generated to sustain the combustion reaction (**Figure 5.20**). Flash point is commonly used to indicate the flammability hazard of liquid fuels. Liquid fuels that vaporize sufficiently to burn at temperatures under 100°F (38°C) present a significant flammability hazard.

The extent to which a liquid will give off vapor is also influenced by how much surface area is exposed to the atmosphere. In many open containers, the surface area of liquid exposed to the atmosphere is limited.

A number of other characteristics of liquid fuels are important to firefighters. **Solubility** describes the extent to which a substance (in this case a liquid) will mix with water. Solubility may be expressed in qualitative terms (i.e., slightly, completely) or as a percentage. Materials that are **miscible** in water will mix in any proportion. Liquids such as **hydrocarbon fuels** (i.e., gasoline, diesel, and fuel oil) are lighter than water and do not mix with water. Other liquids (called **polar solvents**) such as alcohols (i.e., methanol, ethanol) will mix readily with water.

Liquids that are less dense (lighter) than water are more difficult to extinguish using water as the sole extinguishing agent. Because the liquid fuel is less dense and will not mix with water, adding water to the liquid fuel may disperse the burning liquid instead of extinguishing it, which could potentially spread the fire to other areas. Liquid fuels that are not water-soluble should be extinguished using the appropriate foam or chemical agent.

Flash Point — Minimum temperature at which a liquid gives off enough vapors to form an ignitable mixture with air near the liquid's surface.

Fire Point — Temperature at which a liquid fuel produces sufficient vapors to support combustion once the fuel is ignited. Fire point must exceed 5 seconds of burning duration during the test. The fire point is usually a few degrees above the flash point.

Solubility — Degree to which a solid, liquid, or gas dissolves in a solvent (usually water).

Miscible — Materials that are capable of being mixed in all proportions.

Hydrocarbon Fuel — Petroleum-based organic compound that contains only hydrogen and carbon.

Polar Solvents — Flammable liquids that have an attraction for water, much like a positive magnetic pole attracts a negative pole; examples include alcohol, ketone, and lacquer.

Table 5.6
Characteristics of Common Flammable and Combustible Liquids

Material	Water Soluble	Specific Gravity	Flash Point	Autoignition Temperature
Gasoline	No	0.72	(-45°F) -43°C	(853°F) 456°C
Diesel	No	>1.00	(125°F) 52°C	(410°F) 210°C
Ethanol	Yes	0.78	(55°F) 13°C	(689°F) 365°C
Methanol	Yes	0.79	(52°F) 11°C	(867°F) 464°C

Source: *Computer Aided Management of Emergency Operations* (CAMEO)

Combustible Liquid — Liquid having a flash point at or above 100°F (37.8°C) and below 200°F (93.3°C).

Water-soluble liquids, however, will mix with some water-based extinguishing agents, such as many types of fire fighting foam. The extinguishing agent will mix with the burning liquid and become much less effective at extinguishing the fire. To avoid this mixture, fire fighting foams specifically designed for polar solvents should be used. **Table 5.6** lists the characteristics of common flammable and **combustible liquids**.

Solid Fuel

Solids have definite size and shape. Solids may also react differently when exposed to heat. Some solids (wax, thermoplastics, and metals) will readily change their state and melt, while others (wood and thermosetting plastics) will not. When solid fuels are heated, fuel gases and vapors are released by pyrolysis, which is the chemical decomposition of a substance through the action of heat. Simply stated, as solid fuels are heated, they begin to decompose and combustible vapors are emitted. If there is enough fuel and heat, the process of pyrolysis generates sufficient quantities of burnable vapors to ignite in the presence of sufficient oxygen or another oxidizer.

The primary fuels commonly found in room or compartment fires are solids such as wood, paper, fabric, or plastic. Pyrolysis must occur to generate the flammable vapors required for combustion. When wood is first heated, water vapor is driven off as the wood dries. As heating continues, the wood begins to pyrolize and decompose into its volatile components and carbon. Pyrolysis of wood begins at temperatures below 400°F (204°C). This is lower than required for ignition of the vapors being given off, which ranges roughly from 1,000°F to 1,300°F (538°C to 704°C). **Table 5.7** outlines the pyrolysis effects within different temperature zones.

The pyrolysis process is similar with synthetic fuels such as plastics and some fabrics. Unlike wood, though, plastics do not generally contain moisture. As a result, pyrolysis occurs in plastics much sooner than in wood because there is no moisture to slow the process.

Unlike liquids or gases, solid fuels have a definite shape and size. This property significantly affects how easily they ignite. The primary consideration is the surface area of the fuel in proportion to the mass, called the surface-to-mass ratio. One of the best examples is that of a large tree:

1. To produce lumber, the tree must be felled and cut into a log. The surface area of this log is very low compared to its mass; therefore, the surface-to-mass ratio is low.

2. The log is then sawn into planks. The result of this process is to reduce the mass of the individual planks compared to the log; the resulting surface area is increased, thus increasing the surface-to-mass ratio.

Table 5.7
Pyrolysis Zone

	Temperature	Chemical Changes
Stage 1	Less than 392°F (200°C)	Moisture is released as the wood begins to dry; combustible and noncombustible materials are released into the atmosphere although there is insufficient heat to ignite them.
Stage 2	392° — 536°F (200°— 280°C)	The majority of the moisture has been released; charring has begun; the primary compound being released is carbon monoxide; ignition has yet to occur.
Stage 3	536° — 932°F (280°— 500°C)	Rapid pyrolysis takes place; combustible compounds are released and ignition can occur; charcoal is formed by the burning process.
Stage 4	Greater than 932°F (500°C)	Free burning exists as the wood material is converted to flammable gases.

Source: Adapted from NFPA® *Fire Protection Handbook*, 19th edition, Volume II, pages 8-35 and 36.

3. The chips and sawdust produced as the planks are sawn into boards have an even higher surface-to-mass ratio.

4. If the boards are milled or sanded, the resulting shavings or sawdust have the highest surface-to-mass ratio of any of the examples.

As this ratio increases, the fuel particles become smaller (more finely divided); for example, shavings or sawdust as opposed to logs. Therefore, their ability to be ignited increases tremendously. As the surface area increases, more of the material is exposed to the heat and generates combustible pyrolysis products more quickly, making the fuel easier to ignite as surface-to-mass ratio increases (**Figure 5.21, p. 230**).

The proximity and orientation of a solid fuel relative to the source of heat also affects the way it burns (**Figure 5.22, p. 230**). For example, if you ignite one corner of a sheet of ⅛-inch (3 mm) plywood paneling that is lying horizontally (flat), the fire will consume the fuel at a relatively slow rate. The same type of paneling in a vertical position (standing on edge) burns much more rapidly because the heated vapors rise and transfer more heat to the paneling.

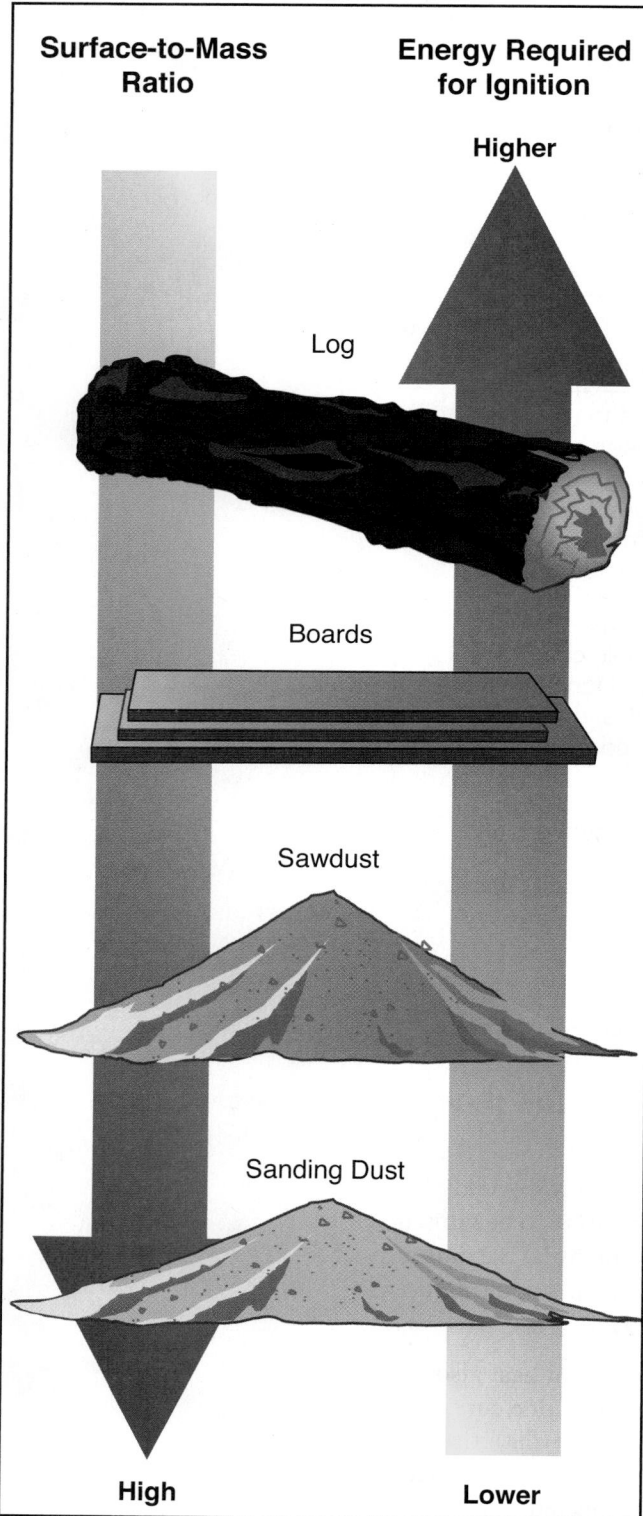

Surface-to-Mass Ratio

Energy Required for Ignition

Higher

Log

Boards

Sawdust

Sanding Dust

High

Lower

Figure 5.21 Fine-grained fuel particles ignite much more readily than large-dimension lumber.

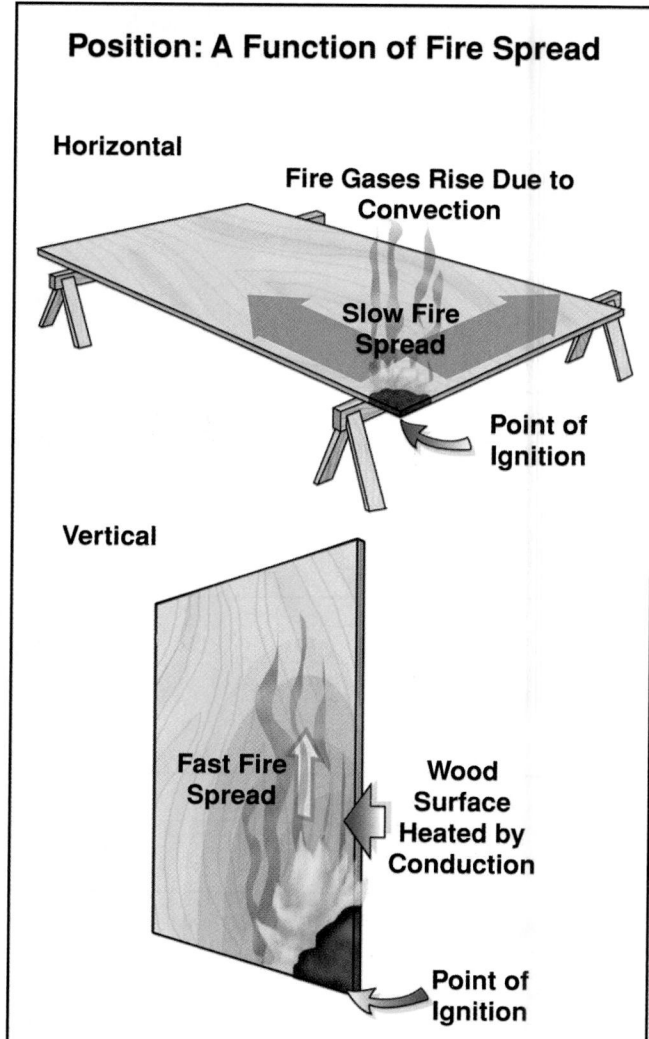

Position: A Function of Fire Spread

Horizontal

Fire Gases Rise Due to Convection

Slow Fire Spread

Point of Ignition

Vertical

Fast Fire Spread

Wood Surface Heated by Conduction

Point of Ignition

Figure 5.22 A board aligned vertically presents much more surface area to pyrolysis from rising gases as compared to a board in a horizontal position.

Table 5.8
Common Oxidizers

Substance	Common Use
Calcium Hypochlorite (granular solid)	Chlorination of water in swimming pools
Chlorine (gas)	Water purification
Ammonium Nitrate (granular solid)	Fertilizer
Hydrogen Peroxide (liquid)	Industrial bleaching (pulp and paper and chemical manufacturing)
Methyl Ethyl Ketone Peroxide	Catalyst in plastics manufacturing

Courtesy of Ed Hartin.

Oxygen

Oxygen in the air is the primary oxidizing agent in most fires. Normally, air consists of about 21 percent oxygen. In addition to oxygen, other oxidizers can react with fuels in much the same way. Oxidizers are not combustible but they will support or enhance combustion. **Table 5.8** lists some common oxidizers.

At normal ambient temperatures (68°F/20°C), materials can ignite and burn at oxygen concentrations as low as 14 percent. When oxygen concentration is limited, the flaming combustion will diminish, causing combustion to continue in the nonflaming mode. Nonflaming combustion can continue at extremely low oxygen concentrations even when the surrounding environment is at a relatively low temperature. However, at high ambient temperatures, flaming combustion may continue at considerably lower oxygen concentrations.

Effects of Oxygen Concentration

Oxygen concentration in the atmosphere has a significant effect on both fire behavior and our ability to survive. Occupational Safety and Health Administration (OSHA) respiratory protection regulation, *29 CFR (Code of Federal Regulations)* 1910.134, defines an atmosphere having less than 19.5 percent oxygen in the air as being oxygen deficient and presenting a hazard to persons not wearing respiratory protection, such as SCBA to provide a supply of fresh air. When the oxygen concentration in the atmosphere exceeds 23.5 percent, this regulation classifies the atmosphere as oxygen enriched and presents an increased fire risk.

When the oxygen concentration is higher than normal, materials exhibit very different burning characteristics. Materials that burn at normal oxygen levels will burn more intensely and may ignite more readily in oxygen-enriched atmospheres. Some petroleum-based materials will autoignite (ignite spontaneously without an external heat source) in oxygen-enriched atmospheres.

Many materials that do not burn at normal oxygen levels will burn readily in oxygen-enriched atmospheres. One such material is Nomex® fire-resistant fabric, which is used in the construction of many types of the protective clothing worn by firefighters. At normal oxygen levels, Nomex® does not burn. When placed in an oxygen-enriched atmosphere of approximately 31 percent oxygen, however, Nomex® ignites and burns vigorously.

Figure 5.23 Oxygen bottles may be found in healthcare facilities and some unexpected locations such as private residences.

Flammable (Explosive) Range — The range between the upper flammable limit and lower flammable limit in which a substance can be ignited.

Upper Flammable (Explosive) Limit (UFL) — Upper limit at which a flammable gas or vapor will ignite; above this limit the gas or vapor is too *rich* to burn (lacks the proper quantity of oxygen). *Also known as* Upper Explosive Limit (UEL).

Lower Flammable (Explosive) Limit (LFL) — Lower limit at which a flammable gas or vapor will ignite and support combustion; below this limit the gas or vapor is too *lean* or *thin* to burn (lacks the proper quantity of fuel). *Also known as* Lower Explosive Limit (LEL).

Fires in oxygen-enriched atmospheres are more difficult to extinguish and present a potential safety hazard to firefighters operating in those atmospheres. These conditions can be found in hospitals and other healthcare facilities, some industrial occupancies, and even private homes where occupants use breathing equipment containing pure oxygen **(Figure 5.23)**.

For combustion to occur after a fuel has been converted into a gaseous state, it must be mixed with air (oxidizer) in the proper ratio. The range of concentrations of the fuel vapor and air (oxidizer) is called the **flammable (explosive) range**. The flammable range of a fuel is reported using the percent by volume of gas or vapor in air for the lower flammable limit (LFL) and for the upper flammable limit (UFL). The **lower flammable (explosive) limit** is the minimum concentration of fuel vapor and air that supports combustion. Concentrations that are below the LFL are said to be *too lean* to burn. The **upper flammable (explosive) limit** is the concentration above which combustion cannot take place **(Figure 5.24)**. Concentrations that are above the UFL are said to be *too rich* to burn. Within the flammable range there is an ideal concentration at which there is exactly the amount of fuel and oxygen required for combustion.

Table 5.9 presents the flammable ranges for some common materials. The flammable limits for combustible gases are presented in chemical handbooks and documents such as the National Fire Protection Association® (NFPA®) *Fire Protection Guide to Hazardous Materials*. The limits are normally reported at ambient temperatures and atmospheric pressures. Variations in temperature and pressure can cause the flammable range to vary considerably.

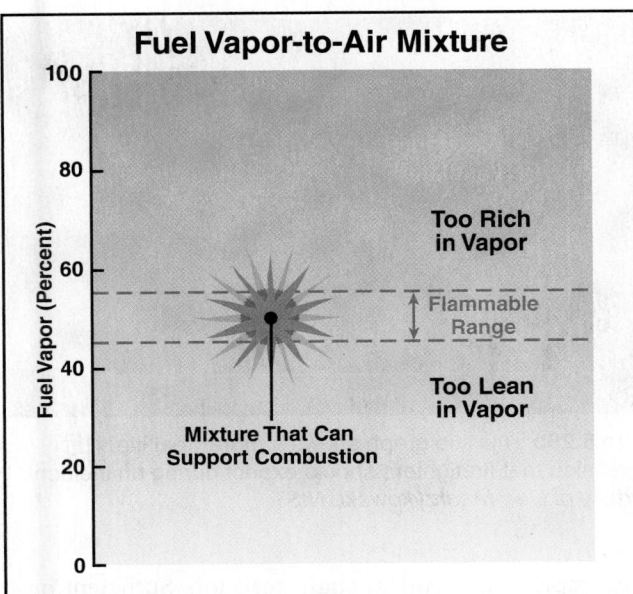

Figure 5.24 The flammable range is a relatively narrow band of conditions between the upper and lower flammable limits.

Fuel Vapor-to-Air Mixture

Too Rich in Vapor

Flammable Range

Too Lean in Vapor

Mixture That Can Support Combustion

Fuel Vapor (Percent)

Table 5.9
Flammable Ranges of Common Flammable Gases and Liquids (Vapor)

Substance	Flammable Range
Methane	5%–15%
Propane	2.1%–9.5%
Carbon Monoxide	12%–75%
Gasoline	1.4%–7.4%
Diesel	1.3%–6%
Ethanol	3.3%–19%
Methanol	6%–35.5%

Source: *Computer Aided Management of Emergency Operations* (CAMEO)

Self-Sustained Chemical Reaction

The self-sustained chemical reaction involved in flaming combustion is complex. Combustion of a simple fuel such as methane and oxygen provides a good example. Complete oxidation of methane results in production of carbon dioxide and water as well as release of energy in the form of heat and light. While this process seems to be quite simple, it is actually quite complex. As flaming combustion occurs, the molecules of methane and oxygen break apart to form **free radicals** (electrically charged, highly reactive parts of molecules). Free radicals combine with oxygen or with the elements that form the fuel material (in the case of methane, carbon and hydrogen) producing intermediate combustion products (new substances), even more free radicals, and increasing the speed of the oxidation reaction.

At various points in the combustion of methane, this process results in production of carbon monoxide and formaldehyde, which are both flammable and toxic. When more chemically complex fuels burn, different types of free radicals and intermediate combustion products are created, many of which are also flammable and toxic.

Free Radicals — Molecular fragments that are highly reactive.

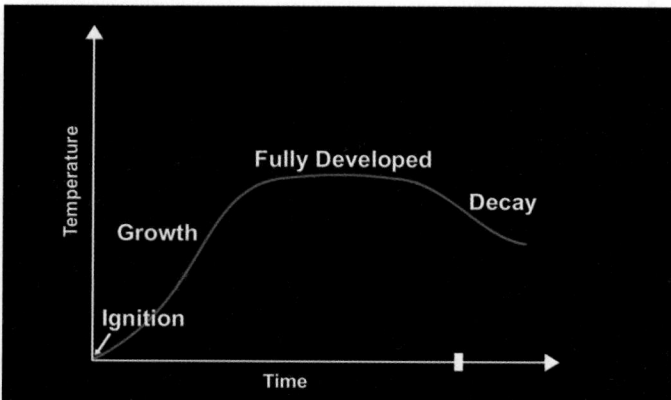

Figure 5.25a This line graph shows the progression of a fire in a lab or controlled environment. *Courtesy of Dan Madrzykowski, NIST.*

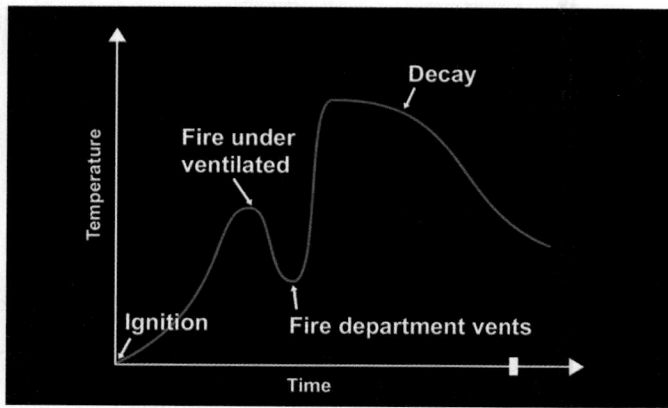

Figure 5.25b This line graph shows a more "real life" progression that firefighters should expect during an incident. *Courtesy of Dan Madrzykowski, NIST.*

Flaming combustion is one example of a chemical chain reaction. Sufficient heat will cause fuel and oxygen to form free radicals and initiate the self-sustained chemical reaction. The fire will continue to burn until the fuel or oxygen is exhausted or an extinguishing agent is applied in sufficient quantity to interfere with the ongoing reaction. **Chemical flame inhibition** occurs when an extinguishing agent, such as dry chemical or Halon-replacement agent, interferes with this chemical reaction, forms a stable product, and terminates the combustion reaction.

Stages of Fire Development

Both unconfined and confined fires will progress through stages or phases. In laboratory simulations, these stages have been shown to be distinct when conditions in the room or compartment remain the same. These stages include incipient, growth, fully developed, and decay. Fires outside the laboratory may not develop through each of the stages in this exact sequence.

The stages of fire development in **Figures 5.25a and b** uses fire behavior in a single compartment to illustrate fire progression. Actual conditions within a building made up of multiple compartments can vary widely. For instance, the compartment of origin may be in the fully developed stage while adjacent compartments may be in the growth stage. In another example, an attic or void space may be in a severely under ventilated decay stage while adjacent compartments are in the growth or fully developed stage. At a fire scene, the stages of fire development are a guide for what *could* occur during the fire but are not a pattern of what *will* occur every time. Firefighters should assess the changing hazards and fire conditions at the incident rather than assume that the fire will follow the same pattern identified in laboratory tests.

Factors That Affect Fire Development

A number of factors influence fire development within a compartment, including:

- Fuel type
- Availability and location of additional fuels
- Compartment volume and ceiling height
- Ventilation

- Thermal properties of the compartment
- Ambient conditions
- Fuel load

Fuel Type

The type of fuel involved in combustion affects the heat release rate (HRR). Fires involving Class B and C fuels will eventually spread to the building contents and structure resulting in a primarily Class A-fueled fire.

NOTE: Classifications of fires are described in detail in Chapter 7, Portable Fire Extinguishers.

In a compartment fire, one of the most fundamental Class A fuel characteristics influencing fire development is surface-to-mass ratio. Combustible materials with high surface-to-mass ratios are much more easily ignited and will burn more quickly than the same substance with less surface area. As described earlier, wood shavings in a lumber mill will ignite and burn more rapidly than the stacked lumber stored there.

Fires involving Class B flammable/combustible liquids will be influenced by the surface area and type of fuel involved. A liquid fuel spill will increase that liquid's surface-to-volume ratio, generating more flammable vapors than the same liquid in an open container **(Figure 5.26)**. The increase of vapor due to the spill will also allow more of the fuel to ignite, resulting in greater heat over a shorter period of time.

Structure fires involving single types of fuels are rare. Modern homes and businesses are largely filled with contents made from petroleum-based materials (plastics, or synthetic fabrics for example). These fuels have a higher heat of combustion and produce higher HRRs than wood alone. Burning synthetic fuels produce products of combustion that contain large quantities of solid and liquid particulates and un-burned gases.

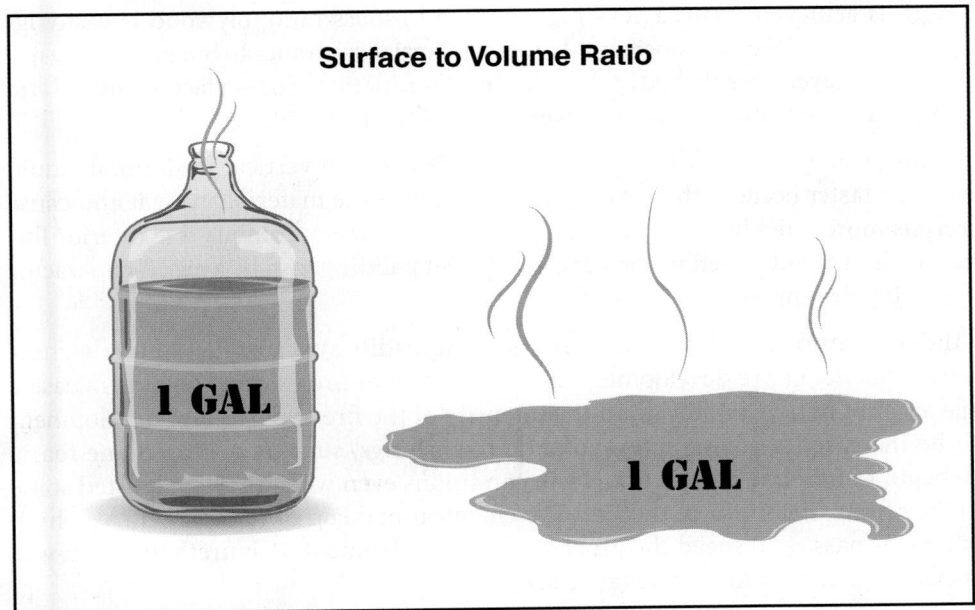

Figure 5.26 Liquid that has spilled onto a flat area will release more vapors than liquid contained in a way that reduces the surface area.

A compartment fire that results from a flammable/combustible gas leak may begin with a rapid ignition of the gas and an explosion. If the fuel source is not controlled, it will continue to burn at the point of release extending to adjacent combustibles. Shutting off the fuel source or controlling the leak will reduce or eliminate the Class B fuel, but the resulting Class A fire will continue to burn until extinguished.

Availability and Location of Additional Fuel

Factors that influence the availability and location of additional fuels include the building configuration, construction materials, contents, and proximity of the initial fire to these exposed fuel sources. Building configuration is the layout of the structure including:

- Number of stories above or below ground
- Compartmentation
- Floor plan
- Openings between floors
- Continuous voids or concealed spaces
- Barriers to fire spread

Each of these elements may contribute to either fire spread or containment. For instance, an open floor plan space may contain furnishings that provide fuel sources on all sides of a point of ignition. Conversely, a compartmentalized configuration may have fire-rated barriers, such as walls, ceilings, and doors, separating fuel sources and limiting fire development to an individual compartment (**Figures 5.27a and b**).

In buildings where the construction materials are flammable, the materials themselves add to the structure's fuel load. For example, in wood-frame buildings, the structure itself is a source of fuel. The orientation of these fuels as well as their surface-to-mass ratio will also influence the rate and intensity of fire spread. Plywood, for instance, consists of several thin layers of wood veneer laminated together while alternating the direction of grains with each layer until the desired thickness and strength is achieved. With a very high surface-to-mass ratio, plywood is easily ignited, even while level and horizontal. As the material continues to burn, the adhesive bond of each layer is weakened and results in delaminating. The surface-to-mass ratio now increases, resulting in rapid consumption of the material.

If this same sheet of plywood were to be oriented in a vertical position, it would burn even faster because the growing fire would heat the material above it and cause pyrolysis more quickly. In addition to structural members, combustible interior finishes, such as wood paneling, carpets, and carpet padding, can be a significant factor influencing fire spread.

The contents of a structure are often the most readily available fuel source, significantly influencing fire development in a compartment fire. When contents release a large amount of heat rapidly, both the intensity of the fire and speed of development will be increased. For example, synthetic furnishings, such as polyurethane foam, will begin to pyrolyze rapidly under fire conditions even when they are located some distance from the origin of the fire. The chemical makeup of the foam and its high surface-to-mass ratio speed the process of fire development. Polyurethane foam will liquefy and continue to burn (**Figure 5.28**).

The proximity and continuity of contents and structural fuels also influence fire development. Fuels located in the upper level of adjacent compartments will pyrolize more quickly from the effect of the hot gas layer. Continuous fuels such as combus-

Open Floor Plan

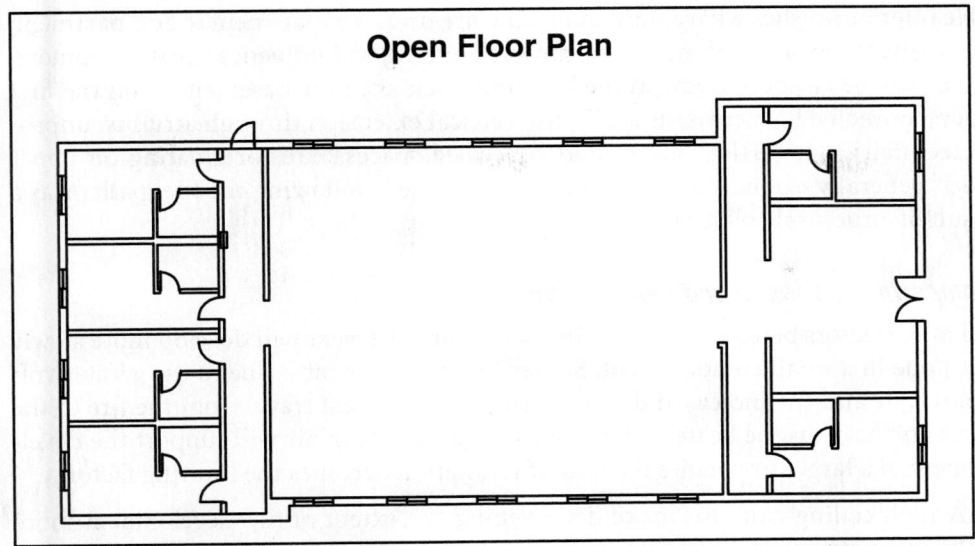

Figure 5.27a An open floor plan features some small compartments and a large open area, often located in the center of the structure.

Compartmentalized Floor Plan

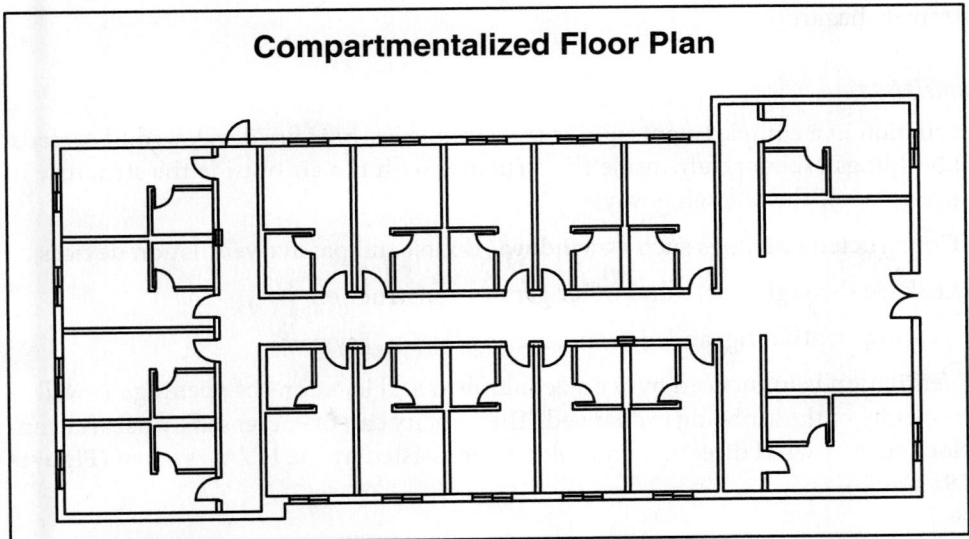

Figure 5.27b A compartmentalized floor plan limits the amount of space and fuel a fire may access.

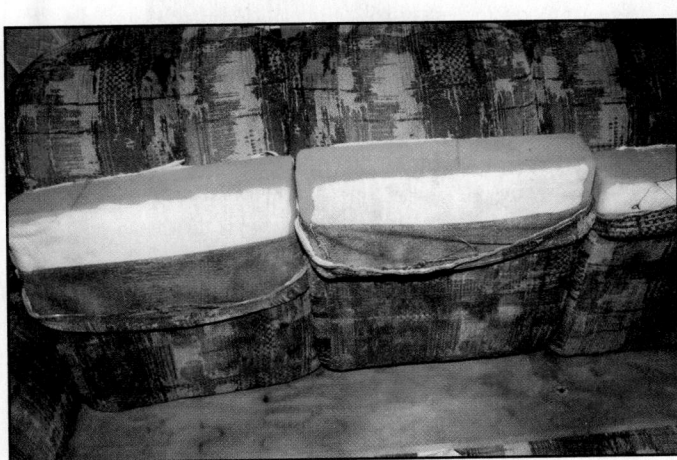

Figure 5.28 Polyurethane foam contained in some furnishings significantly increases the temperature at which a fire will burn in a room.

tible interior finishes will rapidly spread the fire from compartment to compartment. Similarly, the location of the fire within the building will influence fire development. When the fire is located low in the building, such as in the basement or on the first floor, convected heat currents will cause vertical extension through atriums, unprotected stairways, vertical shafts, and concealed spaces. Fires originating on upper levels generally extend downward much more slowly following the fuel path or as a result of structural collapse.

Compartment Volume and Ceiling Height

All other factors being equal, a fire in a large compartment will develop more slowly than one in a small compartment. Slower fire development is due to the greater volume of air and the increased distance radiated heat must travel from the fire to the contents that must be heated. However, a large volume of air will support the development of a larger fire before the lack of ventilation becomes the limiting factor.

A high ceiling can also make determining the extent of fire development more difficult. In structures with high ceilings, a large volume of hot smoke and fire gases can accumulate at the ceiling level, while conditions at floor-level remain relatively unchanged. Firefighters may mistake floor-level conditions for the actual state of fire development. If the large hot gas layer ignites, the situation becomes immediately extremely hazardous.

Ventilation

Ventilation in a compartment significantly influences how fire develops and spreads. All buildings exchange air inside the structure with the air outside the structure in one or more of the following ways:

- Constructed openings such as windows, doors, and passive ventilation devices
- Leakage through cracks and other gaps in construction
- Heating, ventilating, and air conditioning (HVAC) system

Ventilation is influenced by the size, number, and locations of openings as well as the velocity of the air being exchanged. The velocity can be increased by natural conditions such as wind direction and velocity or assisted by the HVAC system (**Figures 5.29a and b**).

Figure 5.29a Fire will be contained when compartment doors are closed, as indicated by the red area of the floor plan. *Courtesy of Dan Madrzykowski, NIST.*

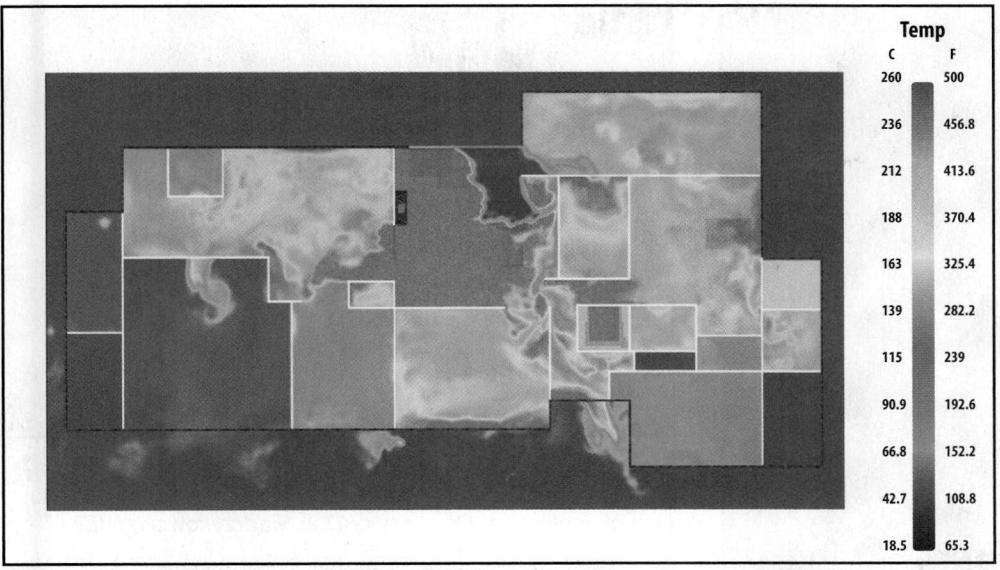

Figure 5.29b When compartment doors and windows are open, heat and flames will spread into other parts of the structure. Notice the increase in temperature in the adjacent compartments, shown in yellow mixed with green. *Courtesy of Dan Madrzykowski, NIST.*

Ventilation vs. Tactical Ventilation

Ventilation is the exchange of air inside a building or compartment with air outside the building or compartment. Tactical ventilation is intentional and conducted by the fire department. It is critical that you understand that:

- Ventilation occurs all the time.
- If smoke is visible outside, ventilation is occurring.
- Any opening made in the building by occupants, the fire, or firefighters changes the ventilation and is likely to influence fire behavior.

Fuel-Controlled — A fire with adequate oxygen in which the heat release rate and growth rate are determined by the characteristics of the fuel, such as quantity and geometry. (NFPA® 921)

Ventilation-Controlled — A fire with limited ventilation in which the heat release rate or growth is limited by the amount of oxygen available to the fire. (NFPA® 921)

Room or compartment fires take two forms: **fuel-controlled** and **ventilation-controlled**. When sufficient oxygen is available, the characteristics and configuration of the fuel control fire development. Under these conditions, the fire is said to be fuel-controlled **(Figure 5.30 p. 240)**. For example, a small fire that is confined to a noncombustible waste basket in a large well-ventilated room will self-extinguish when all the fuel is consumed no matter how much air remains in the room.

When the available air supply begins to limit fire development in a compartment fire, the fire is said to be ventilation-controlled **(Figures 5.31a and b p. 240)**. When a fire develops to the point where it becomes ventilation-controlled, the available air supply will determine the speed and extent of fire development and the direction of fire travel. Fire will have a tendency to grow in the direction of ventilation openings, such as failed windows and doors, because of the introduction of fresh air. The characteristics of the fuel and fuel load in today's typical fires will cause fires to quickly become ventilation-controlled.

Changes in ventilation can alter the ventilation flow path, create rapid fire development, and place firefighters in extreme danger. When the fire becomes ventilation-controlled, the fire's HRR will decrease. If windows or doors fail at this time, the sudden introduction of fresh air creates a rapid increase in the HRR. This rapid increase will also occur when firefighters open a door or window to ventilate the room, enter for fire attack, or to perform search and rescue.

Figure 5.30 In fuel-controlled fires, a lack of fuel will prevent them from spreading.

Figure 5.31a As a fire progress, the available oxygen in a compartment is consumed and the fire begins to cool due to lack of ventilation. *Courtesy of Dan Madrzykowski, NIST.*

Figure 5.31b Access to a new oxygen source causes a fire to grow in the direction of the ventilation opening. *Courtesy of Dan Madrzykowski, NIST.*

Thermal Properties of the Compartment

The thermal properties of the compartment can contribute to rapid fire development. The thermal properties can also make extinguishment more difficult and reignition possible. Thermal properties of a compartment include:

- **Insulation** – Contains heat within the compartment causing a localized increase in the temperature and fire growth

- **Heat reflectivity** – Increases fire spread through the transfer of radiant heat from wall surfaces to adjacent fuel sources

- **Retention** – Maintains temperature by absorbing and releasing large amounts of heat slowly

Ambient Conditions

Ambient Conditions — Common, prevailing, and uncontrolled atmospheric weather conditions. The term may refer to the conditions inside or outside of the structure.

Ambient conditions such as high humidity and cold temperatures can slow the natural movement of smoke. Strong winds place additional pressure on one side of a structure and force both smoke and fire out the opposite side. If a window fails or a door is opened on the windward side of a structure, fire intensity and spread can

Figure 5.32 When a ventilation-controlled fire finds access to a new source of oxygen, its reaction is immediate and violent. *Courtesy of Mike Wieder.*

increase significantly, creating a "blowtorch" effect **(Figure 5.32)**. During fire suppression activities, wind direction and velocity can prevent or assist in ventilation activities.

Cold temperatures can cause smoke to appear white and give a false impression of the interior conditions based upon the color of smoke. Atmospheric air pressure can also cause smoke to remain close to the ground obscuring visibility during size-up.

NOTE: While ambient temperature and humidity outside the structure can have an effect on the ignitability of many types of fuels, these factors are less significant inside a compartment.

Fuel Load

The total quantity of combustible contents of a building, space, or fire area is referred to as the **fuel load** (some documents may use the term *fire load*) **(Figure 5.33, p. 242)**. The fuel load includes all furnishings, merchandise, interior finish, and structural components of the structure. At a scene, you will probably only be able to estimate the fuel load based upon your knowledge and experience. For example, a concrete block structure containing stored steel pipe will have a much smaller fuel load than a wood-frame structure used for storing flammable liquids. Your knowledge of building construction and occupancy types will be essential to determining fuel loads.

Incipient Stage

The **incipient stage** starts with ignition when the three elements of the fire triangle come together and the combustion process begins **(Figure 5.34, p. 242)**. At this point, the fire is small and confined to the material (fuel) first ignited.

Once combustion begins, development of an incipient fire is largely dependent on the characteristics and configuration of the fuel involved (fuel-controlled fire). Air in the compartment provides adequate oxygen to continue fire development:

Fuel Load — The total quantity of combustible contents of a building, space, or fire area, including interior finish and trim, expressed in heat units of the equivalent weight in wood.

Incipient Stage — First stage of the burning process in a compartment in which the substance being oxidized is producing some heat, but the heat has not spread to other substances nearby. During this phase, the oxygen content of the air has not been significantly reduced and the temperature within the compartment is not significantly higher than ambient temperature.

Figure 5.33 The total quantity of combustible materials in one place is referred to as the *fuel load. Courtesy of Ron Moore/ McKinney (TX) Fire Department.*

Figure 5.34 During the incipient stage of a fire, oxygen and fuel are plentiful. *Courtesy of Dan Madrzykowski, NIST.*

Figure 5.35 A fire emits a plume of hot gases and flame that begins the process of pyrolysis on other materials nearby. *Courtesy of Dan Madrzykowski, NIST.*

Plume — The column of hot gases, flames, and smoke rising above a fire; *also called* convection column, thermal updraft, *or* thermal column. (NFPA® 921)

Ceiling Jet — A relatively thin layer of flowing hot gases that develops under a horizontal surface (e.g., ceiling) as a result of plume impingement and the flowing gas being forced to move horizontally. (NFPA® 921)

- During this initial phase of fire development, radiant heat warms the adjacent fuel and continues the process of pyrolysis. A **plume** of hot gases and flame rises from the fire and mixes with the cooler air in the room **(Figure 5.35)**.

- As this plume reaches the ceiling, hot gases begin to spread horizontally across the ceiling in what firefighters have historically called *mushrooming*; however, in scientific or engineering terms it is referred to as a **ceiling jet**.

- Hot gases in contact with the surfaces of the compartment and its contents transfer heat to other materials.

This complex process of heat transfer begins to increase the overall temperature in the room. **Table 5.10** lists the factors that influence the development of fuel-controlled fires.

Table 5.10
Factors Influencing Development of a Fuel-Controlled Fire

Mass and surface area	The greater the surface area for a given mass of fuel, the easier it is for that fuel to be heated to its ignition temperature.
Chemical content	The chemical makeup of the fuel has a significant impact on the heat released during combustion. Many hydrocarbon-based synthetic materials have a heat of combustion that is more than twice that of cellulose materials such as wood.
Fuel load	The total amount of fuel available for combustion influences total potential heat release.
Fuel moisture	While not a factor with all types of fuel, water acts as thermal ballast, slowing the process of heating the fuel to its ignition temperature.
Orientation	Orientation in relation to the fire influences how heat is transferred. For example, a wood wall surface is heated by both convection and radiation, where the floor is more likely to be heated by radiant heat alone.
Continuity	Continuity is the proximity of various fuel elements to one another. The closer (or more continuous) the fuel is, the easier and more rapidly fire will extend. Continuity may be either horizontal (i.e. ceiling surface) or vertical (i.e. wall or rack storage).

Courtesy of Ed Hartin.

In this early stage of fire development, the fire has not yet influenced the environment within the compartment to a significant extent. The temperature, while increasing, is only slightly above ambient and the concentration of products of combustion is low. During the incipient phase, occupants can safely escape from the compartment and the fire could be safely extinguished with a portable extinguisher or small hoseline. It is essential to recognize that the transition from incipient to growth stage can occur quite quickly (in some cases in seconds), depending on the type and configuration of fuel involved.

Growth Stage

As the fire transitions from incipient to growth stage, it begins to influence the environment within the compartment and has grown large enough for the compartment configuration and amount of ventilation to influence it **(Figure 5.36, p. 244)**. The first effect is the amount of air that is entrained into the plume. In a compartment fire, the location of the fuel package in relation to the compartment walls affects the amount of air that is entrained and thus the amount of cooling that takes place. Unconfined fires draw air from all sides and the entrainment of air cools the plume of hot gases, reducing flame length and vertical extension:

- Fuel packages in the middle of the room can entrain air from all sides.
- Fires in fuel packages near walls can only entrain air from three sides.
- Fires in fuel packages in corners can only entrain air from two sides

Therefore, when the fuel package is not in the middle of the room, the combustion zone (area where sufficient air is available to feed the fire) expands vertically and a higher plume results. A higher plume increases the temperatures in the developing

Figure 5.36 During the growth stage of a fire, the fire influences the environment and begins to be influenced by the ventilation conditions of the compartment. *Courtesy of Dan Madrzykowski, NIST.*

Figure 5.37 Firefighters will want to stay below the lowest smoke layer for the best visibility and lowest level of danger available in the compartment.

hot-gas layer at ceiling level and increases the speed of fire development. In addition, heated surfaces around the fire radiate heat back toward the burning fuel which further increases the speed of fire development.

Thermal Layering

The **thermal layering** of gases, sometimes referred to as *heat stratification* and *thermal balance*, is the tendency of gases to form into layers according to temperature (**Figure 5.37**). Generally, the hottest gases tend to be in the upper layer, while the cooler gases form the lower layers. In addition to the effects of heat transfer through radiation and convection described earlier, radiation from the hot gas layer also acts to heat the interior surfaces of the compartment and its contents. Changes in ventilation and flow path can significantly alter the thermal layering.

As the volume and temperature of the hot gas layer increases, so does the pressure. Higher pressure in this layer causes the hot gas layer to spread downward within the compartment and laterally through any openings such as doors or windows. The pressure of the cool gas layer is lower, resulting in inward movement of air from outside the compartment at the bottom as the hot gases exit through the top of the opening. The interface of the hot and cooler gas layers at the opening is commonly referred to as the **neutral plane** because the pressure is neutral where the layers meet (**Figure 5.38**). The neutral plane only exists at openings where hot gases are exiting and cooler air is entering the compartment.

Whenever possible, maintain or raise the level of the hot gas layer above the floor to provide a more tenable environment for firefighters and trapped occupants. Use effective fire control and ventilation tactics to raise the position of the hot gas layer.

Thermal Layering — Outcome of combustion in a confined space in which gases tend to form into layers, according to temperature, with the hottest gases found at the ceiling and the coolest gases at floor level.

Neutral Plane — The level at a compartment opening where the difference in pressure exerted by expansion and buoyancy of hot smoke flowing out of the opening and the inward pressure of cooler, ambient temperature air flowing in through the opening is equal.

Isolated Flames

As the fire moves through the growth stage and becomes ventilation-controlled, isolated flames may be observed moving through the hot gas layer. Combustion of these hot gases indicates that portions of the hot gas layer are within their flammable range and there is sufficient temperature to result in ignition. As these hot gases circulate to the outer edges of the plume, they find sufficient oxygen to ignite. This phenomenon is frequently observed prior to more substantial involvement of flammable products of combustion in the hot gas layer.

Rapid Transition

Rapid transition from the growth stage to the fully developed stage is known as **flashover (Figure 5.39)**. Under laboratory conditions or when a fire is fuel controlled, flashover occurs during the growth stage. However, in an uncontrolled situation, it may be difficult to identify what stage a fire is in, so firefighters should assume that flashover may occur at any time the conditions are right for it to occur. Because flashover can be unpredictable, a detailed description of flashover is addressed later in this chapter under Rapid Fire Development.

Flashover does not occur in every compartment fire. For example, in large-area compartments or where the ceiling is high, flashover may not occur during the transition. Fire development may take an alternative path in a compartment with limited ventilation. The fire may not progress to flashover, but become ventilation-controlled, limiting heat release rate and causing the fire to enter the decay stage while continuing the process of pyrolysis and increasing the fuel content of the smoke.

Most fires that develop beyond the incipient stage become ventilation-controlled. Even when doors and windows are open, there is often insufficient air to allow the fire to continue to develop based on the available fuel. When windows are intact and doors are closed, the fire may move into a ventilation-controlled state even more quickly. While this reduces the heat release rate, fuel will continue to pyrolize, creating extremely fuel-rich smoke.

Hot Smoke
Increased smoke volume causes the neutral plane to descend.

- - - - - - - - - Neutral Plane

Figure 5.38 As a fire grows in intensity, the neutral plane descends because of the influx of hot gases. *Courtesy of Dan Madrzykowski, NIST.*

Flashover — A rapid transition from the growth stage to the fully developed stage.

Figure 5.39 Conditions for a flashover may be difficult to identify at a real fire scene, so firefighters must prepare for the possibility of rapid fire development. *Courtesy of Dan Madrzykowski, NIST.*

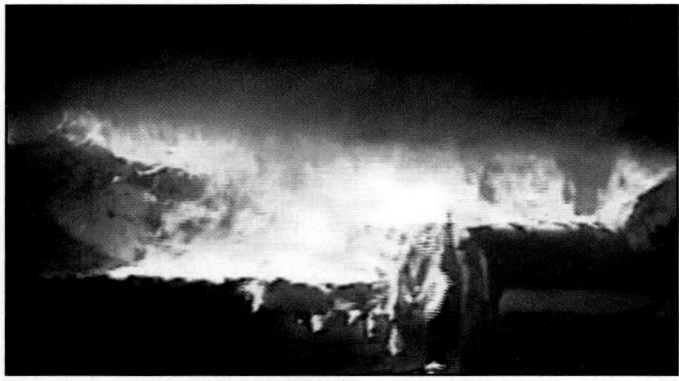

Figure 5.40 All combustible materials actively burn during the fully developed stage. *Courtesy of Dan Madrzykowski, NIST.*

Figure 5.41 During the decay stage, the fire begins to cool and die as fuel and/or ventilation run out of supply. *Courtesy of Dan Madrzykowski, NIST.*

Fully Developed Stage

The fully developed stage occurs when all combustible materials in the compartment are burning **(Figure 5.40)**. During this stage:

- The burning fuels in the compartment are releasing the maximum amount of heat possible for the available fuel and oxygen, producing large volumes of fire gases.

- The fire is ventilation-controlled because the heat release is dependent on the compartment openings. These openings provide oxygen, which supports the ongoing combustion and releases products of combustion. Increases in the available air supply will result in higher heat release.

- Flammable products of combustion are likely to flow from the compartment of origin into adjacent compartments or out through openings to the exterior of the building. Flames will extend out of the compartment openings because there is insufficient oxygen for complete combustion in the compartment.

NOTE: If there are limited or no openings in the compartment, it is unlikely that the fire will reach a fully developed staged due to limited ventilation.

Decay Stage

A compartment fire will decay as the fuel is consumed or if the oxygen concentration falls to the point that flaming combustion is diminished **(Figure 5.41)**. Both of these situations can result in the combustion reaction coming to a stop. However, decay due to reduced oxygen concentration can follow a considerably different path if the ventilation of the compartment changes before combustion ceases and temperature in the compartment lowers.

Consumption of Fuel

As the fire consumes the available fuel in the compartment and the heat release rate begins to decline, it enters the decay stage. If there is adequate ventilation, the fire becomes fuel controlled. The heat release rate will drop, but the temperature in the compartment may remain high for some time. During this stage, the flammable products of combustion can accumulate within the compartment or adjacent spaces.

Limited Ventilation

When a compartment fire enters the decay stage due to a lack of oxygen, the rate of heat release also declines. However, the continuing combustion reaction (based on available fuel and the limited oxygen available to the fire) may maintain an extremely high temperature within the compartment. Temperature decreases but pyrolysis can continue. Under these conditions, a large volume of flammable products of combustion can accumulate within the compartment.

Rapid Fire Development

Over the years, the fire service has coined words and phrases to describe various fire events that result in rapid fire development. Among these events are:

- Flashover
- Backdraft
- Smoke explosion

Rapid fire development has been responsible for numerous firefighter deaths and injuries. Between 1990 and 2000, over 50 firefighters were killed as a result of rapid fire development. To protect yourself and your crew, you must be able to recognize the indicators of rapid fire development, know the conditions created by each of these situations, and determine the best action to take before they occur. In this section, rapid fire development conditions are described along with their indicators.

Flashover

When flashover occurs, the combustible materials in the compartment and the gases produced by pyrolysis ignite almost simultaneously; the result is full-room involvement. Flashover typically occurs during the growth stage of a fire but may occur in the fully developed stage.

Flashover conditions are defined in a variety of ways; however, during flashover, the environment of the room is changing from a two-layer condition (hot on top, cooler on the bottom) to a single well mixed, untenable (even for fully protected firefighters) hot gas condition from floor to ceiling.

The transition period between preflashover fire conditions (growth stage in a fuel-limited case) to postflashover (fully developed stage) can occur rapidly. During flashover, the volume of the fire can increase from approximately ¼ to ½ of the room's upper volume to filling the entire volume of the room and potentially extending out of any openings in the room. When flashover occurs, burning gases push out of openings in the compartment (such as a door leading to another room) at a substantial velocity.

There are four common elements of flashover:

- **Transition in fire development** — Flashover represents a transition from the growth stage to the fully developed stage.

- **Rapidity** — Although it is not an instantaneous event, flashover happens rapidly, often in a matter of seconds, to spread complete fire involvement within the compartment.

- **Compartment** — There must be an enclosed space such as a single room or enclosure.

- **Ignition of all exposed surfaces** — Virtually all combustible surfaces in the enclosed space become ignited.

Two interrelated factors determine whether a fire within a compartment will progress to flashover. First, there must be sufficient fuel and the heat release rate must be sufficient for flashover conditions to develop. For example, ignition of discarded paper in a small metal wastebasket may not have sufficient heat energy to develop flashover conditions in a large room lined with gypsum drywall. On the other hand, ignition of a couch with polyurethane foam cushions placed in the same room is quite likely to result in flashover.

The second factor is ventilation. Regardless of the type, quantity, or configuration of fuel, heat release is dependent on oxygen. A developing fire must have sufficient oxygen to reach flashover, and a sealed room may not provide enough. Heat release is limited by the available air supply. If there is insufficient natural ventilation, the fire may enter the growth stage but not reach the peak heat release rate to transition through flashover to a fully developed fire.

Survival rates for firefighters are extremely low in a flashover. While no exact temperature is associated with flashover, it typically occurs at 1,100°F (593°C) ceiling temperature. At the floor level, heat flux of approximately 20 kW/m² is also typical of flashover conditions. You must be aware of the following flashover indicators to protect yourself:

- **Building indicators** — Flashover can occur in any building; interior configuration, fuel load, thermal properties, and ventilation will determine how rapidly it can occur

- **Smoke indicators** — Rapidly increasing volume, turbulence, darkening color, optical density, and lowering of the hot gas level

- **Air flow indicators** — High velocity and turbulence, bi-directional movement with smoke exiting at top of doorway and fresh air moving in at the bottom, or pulsing air movement

- **Heat indicators** — Rapidly increasing temperature in the compartment, pyrolysis of contents or fuel packages located away from the fire, darkened windows, or hot surfaces

- **Flame indicators** — Isolated flames in the hot gas layers or near the ceiling

One possible and significant indicator of flashover is **rollover**. Rollover describes a condition where the unburned fire gases that have accumulated at the top of a compartment ignite and flames propagate through the hot gas layer or across the ceiling.

Rollover may occur during the growth stage as the hot-gas layer forms at the ceiling of the compartment. Flames may be observed in the layer when the combustible gases reach their ignition temperature. While the flames add to the total heat generated in the compartment, this condition is not flashover. Rollover will generally precede flashover, but it may not always result in flashover. Rollover contributes to flashover conditions because the burning gases at the upper levels of the room generate tremendous amounts of radiant heat which transfers to the fuel packages in the room. The new fuels begin pyrolysis and release the additional gases necessary for flashover.

Backdraft

A ventilation-controlled compartment fire can produce a large volume of flammable smoke and other gases due to incomplete combustion. While the heat release rate from a ventilation-controlled fire is limited, elevated temperatures are usually present within the compartment. An increase in low-level ventilation (such as opening a door or window) prior to upper level ventilation can result in an explosively rapid combus-

Rollover — A condition where the unburned fire gases that have accumulated at the top of a compartment ignite and flames propagate through the hot-gas layer or across the ceiling.

Figure 5.42 Backdraft is a very dangerous stage in a fire during which highly concentrated flammable gases are introduced to oxygen from an entry point below the gases. *Courtesy of Bob Esposito.*

tion of the flammable gases, called a **backdraft** (**Figure 5.42**). Backdraft occurs in the decay stage, in a space containing a high concentration of heated flammable gases that lacks sufficient oxygen for flaming combustion.

When potential backdraft conditions exist in a compartment, the introduction of a new source of oxygen will return the fire to a fully involved state extremely rapidly (often explosively). A backdraft can occur with the creation of a horizontal or vertical opening. All that is required is the mixing of hot, fuel-rich smoke with air. Backdraft conditions can develop within a room, a void space, or an entire building. Anytime a compartment or space contains hot combustion products, potential for backdraft must be considered before creating any openings into the compartment. Backdraft indicators include:

- **Building indicators** — Fire confined to a single compartment or void space, building contents have a high heat release rate.

- **Smoke indicators** — Optically dense smoke, light colored or black becoming dense gray-yellow, although the smoke color alone is not a reliable indicator. Neutral plane rising and lowering similar to a pulsing or breathing movement.

- **Air flow indicators** — High velocity, turbulent smoke discharge, sometimes appearing to pulse or breathe.

- **Heat indicators** — High heat, smoke-stained windows.

- **Flame indicators** — Little or no visible flame.

The effects of a backdraft can vary considerably depending on a number of factors, including:

- Volume of smoke
- Degree of confinement
- Pressure
- Speed with which fuel and air are mixed
- Location where ignition occurs

Backdraft — The explosive burning of heated gases that occurs when oxygen is introduced into a compartment that has a high concentration of flammable gases and a depleted supply of oxygen due to an existing fire.

Do not assume that a backdraft will always occur immediately after an opening is made into the building or involved compartment. If the hot, flammable products of combustion and air mix slowly, a backdraft is unlikely to occur. Mixing of hot flammable products of combustion with air through the action of gravity, air current, pressure differential, and wind effects sometimes takes time so that backdraft may not occur until after air is fully introduced.

You must watch the smoke for indicators of potential rapid fire development including the air current changing direction, neutral plane lifting, or smoke rushing out. To some degree, the violence of a backdraft can be dependent on the extent to which the fuel/air mixture is confined in the compartment: The more confined, the more violent the backdraft will be.

Smoke Explosion

Smoke Explosion — Form of fire gas ignition; the ignition of accumulated flammable products of combustion and air that are within their flammable range.

A **smoke explosion** may occur before or after the decay stage. It occurs as unburned fuel gases come in contact with an ignition source. When smoke travels from the fire it can cool and accumulate in other areas and mix with air. The smoke within its flammable range contacts an ignition source and results in an explosively rapid combustion. Smoke explosions are violent because they involve premixed fuel and oxygen. This is similar to ignition of propane and air within its flammable range. The smoke is generally cool, less than 1,112°F (600°C) and located in void spaces connected to the fire or in uninvolved areas remote to the fire.

Fire Behavior and Fire Fighting Operations

Fire is controlled and extinguished by limiting or interrupting one or more of the essential elements in the combustion process depicted in the fire tetrahedron. Firefighters influence fire behavior in a number of ways:

- Temperature reduction
- Fuel removal
- Oxygen exclusion
- Chemical flame inhibition
- Ventilation and fire behavior

Temperature Reduction

One of the most common methods of fire control and extinguishment is cooling with water. To extinguish a fire by reducing its temperature, enough water must be applied to the burning fuel to absorb the heat being generated by combustion. Water application in sufficient quantities reduces the temperature of a fuel to a point where it does not produce sufficient vapor to burn.

Cooling can extinguish solid fuels and liquid fuels with high flash points. The use of water for cooling is also the most effective method available for the extinguishment of smoldering fires. However, cooling with water cannot sufficiently reduce vapor production to extinguish fires involving low flash point flammable liquids and gases.

In addition to cooling solid and liquid fuels, water can also be used to control burning gases and reduce the temperature of hot products of combustion in the upper layer. This slows the pyrolysis process of combustible materials, reduces radiant heat flux from the upper layer, and reduces the potential for flashover.

Water absorbs significant heat as its temperature is raised, but it has its greatest effect when it is vaporized into steam. When water is converted to steam at 212°F (100°C), it expands approximately 1,700 times. Because of this expansion rate, firefighters should avoid creating too much steam. Excess steam production can reduce visibility, increase the chances for steam burns, and disrupt the thermal balance. Control steam production as follows:

- Use good nozzle technique.
- Apply the appropriate amount of water.
- Apply water using the most effective form (fog, straight, or solid stream based upon existing conditions).

Fuel Removal

Removing the fuel source effectively extinguishes any fire. The simplest method of fuel removal is to allow a fire to burn until all fuel is consumed. While this is not always the most desirable extinguishment method, it is sometimes appropriate. For example, fires involving pesticides or flammable liquid spills may create greater environmental harm if they are extinguished with water, creating substantial runoff and contaminating soil or bodies of water. The best solution may be to allow the fire to burn, minimizing groundwater pollution. A fuel source may also be removed as follows:

- Stopping the flow of a liquid fuel
- Closing valves to stop the emission of gaseous fuels
- Moving solid fuels out of the path of the fire

Oxygen Exclusion

Reducing the oxygen available to the combustion process reduces a fire's growth and may totally extinguish it over time. In its simplest form, this method is used to extinguish stove top fires when a cover is placed on a pan of burning grease. Flooding an area with an inert gas such as carbon dioxide displaces the oxygen and disrupts the combustion process. Oxygen can also be separated from some fuels by blanketing them with foam (**Figure 5.43, p. 252**). Of course, none of these methods work on those rare fuels that are self-oxidizing.

While not generally used for extinguishment in structure fires, limiting the fire's air supply can be a highly effective fire control action. The simplest example of this is when a building occupant closes the door to the fire room before leaving the building. This limits the air supply to the fire and can sometimes prevent flashover.

Chemical Flame Inhibition

Extinguishing agents such as some dry chemicals, halogenated agents (Halons), and Halon-replacement agents, interrupt the combustion reaction and stop flame production. This method of extinguishment is effective on gas and liquid fuels because they must flame to burn. These agents do not easily extinguish nonflaming fires because there is no chemical chain reaction to inhibit. The very high agent concentrations and extended periods necessary to extinguish smoldering fires make these agents impractical in these cases.

Figure 5.43 Foam agents may provide an effective barrier between a fuel and oxygen.

Ventilation and Fire Behavior

In compartment fires, fire control is only one approach to provide a safer environment for firefighters and building occupants. Controlling the movement of smoke and air is also an important strategy.

Unplanned Ventilation

Unplanned ventilation may occur before or after fire suppression operations start **(Figure 5.44)**. Firefighters should expect unplanned ventilation that will significantly influence fire behavior in ventilation-controlled fires.

Unplanned ventilation can result from the wind outside the structure. The wind can increase the pressure inside the structure, drive smoke and flames into unburned portions of the structure and onto advancing firefighters, and upset tactical ventilation efforts. You must be aware of the wind direction and velocity and use it to your advantage to assist in tactical ventilation.

> **WARNING**
> Wind-driven conditions can occur in any type of structure. Wind speeds as low as 10 mph (16 kph) can create wind-driven conditions.

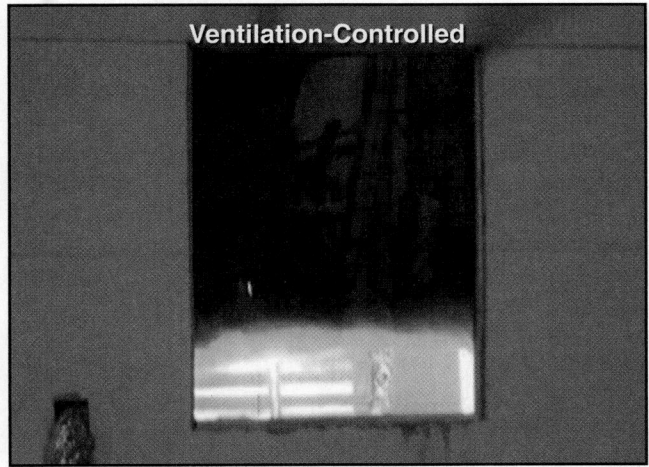
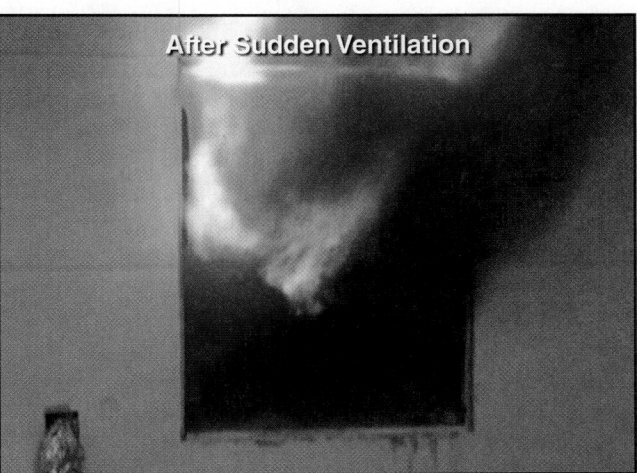

Figure 5.44 Sudden ventilation of a ventilation-controlled fire during conditions including high wind can yield an effect similar to an explosion. *Courtesy of Dan Madrzykowski, NIST.*

Wind is an important factor, but unplanned ventilation is often the result of occupant action, fire effects on the building (such as window glazing), or action other than planned, systematic, and coordinated tactical ventilation. In these cases, situational awareness is essential to ensure your safety and that of other crew members.

Ventilation Strategies

Tactical ventilation is the planned, systematic, and coordinated introduction of air and removal of hot gases and smoke from a building **(Figure 5.45)**. It must be coordinated with fire-suppression operations to prevent unwanted consequences for the hoseline crews. The influences of ventilation on fire behavior are dependent on a variety of factors. Increased ventilation to a ventilation-controlled fire will quickly result in an increase in the heat release rate. Controlling ventilation can be as simple as keeping an exterior door closed until a charged line is in place or as complex as performing vertical ventilation. However, even with coordinated tactical ventilation there will be an increase in the combustion rate when the fire is ventilation controlled.

Tactical Ventilation — Planned, systematic, and coordinated removal of heated air, smoke, gases or other airborne contaminants from a structure, replacing them with cooler and/or fresher air to meet the incident priorities of life safety, incident stabilization, and property conservation.

WARNING
Even coordinated tactical ventilation increases the combustion rate in ventilation-controlled fires.

Figure 5.45 Tactical ventilation uses mechanical and natural ventilation to systematically remove smoke from a building.

Chapter Summary

As a firefighter, you need to understand the combustion process and how fire behaves in different materials and in different environments. You also need to know how fires are classified so that you can select and apply the most appropriate extinguishing agent. Most important, you and your fellow firefighters need to have an understanding of fire behavior that permits you to recognize developing fire conditions and be able to respond safely and effectively to mitigate the hazards presented by the fire environment.

Review Questions

Firefighter I

1. How does the science of fire relate to energy, forms of ignition, and modes of combustion?

2. What impact does thermal energy have on heat, temperature, and heat transfer?

3. What are the physical states that fuel can be found in?

4. How do oxygen content and life safety relate to one another?

5. What products of self-sustained chemical reactions combine to make flammable and toxic substances?

6. What different factors can impact fire development?

7. What are the stages of fire development?

8. What are the signs and causes of a backdraft?

9. How can fire fighting operations impact fire behavior?

Chapter Contents

Key Terms

NFPA® Job Performance Requirements

This chapter provides information that addresses the following job performance requirements of NFPA® 1001, *Standard for Fire Fighter Professional Qualifications* (2013).

Firefighter I

5.1.1	5.3.3
5.1.2	5.5.1
5.3.1	
5.3.2	

1. Describe the purpose of personal protective equipment. (5.1.1, 5.3.3)

2. Describe characteristics of each type of personal protective equipment. (5.3.2, 5.3.3)

3. Summarize guidelines for the care of personal protective clothing. (5.1.1, 5.3.3, 5.5.1)

4. Explain the safety considerations for personal protective equipment. (5.3.1, 5.3.3)

5. Identify respiratory hazards. (5.3.1)

6. Identify types of respiratory protection equipment. (5.3.1)

7. Describe the limitations of respiratory protection equipment. (5.3.1)

8. Explain methods for storing respiratory protection equipment. (5.5.1)

9. Describe general donning and doffing considerations for protective breathing apparatus. (5.3.1, 5.3.2)

10. Summarize general considerations for protective breathing apparatus inspections and care. (5.1.1, 5.5.1)

11. Summarize safety precautions for refilling SCBA cylinders. (5.5.1)

12. Explain procedures for replacing SCBA cylinders. (5.3.1)

13. Explain safety precautions for SCBA use. (5.3.1)

14. Describe nonemergency and emergency exit indicators. (5.3.1)

15. Describe nonemergency exit techniques. (5.3.1)

16. Demonstrate the method for donning structural personal protective clothing for use at an emergency. (Skill Sheet 6-I-1, 5.1.2, 5.3.1, 5.3.2, 5.3.3)

17. With structural personal protective clothing in place, demonstrate the over-the-head method of donning an SCBA. (Skill Sheet 6-I-2, 5.3.1, 5.3.2, 5.3.3)

18. With structural personal protective clothing in place, demonstrate the coat method for donning an SCBA. (Skill Sheet 6-I-3, 5.3.1, 5.3.2, 5.3.3)

19. With structural personal protective clothing in place, demonstrate the method for donning an SCBA while seated. (Skill Sheet 6-I-4, 5.3.1, 5.3.2, 5.3.3)

20. Doff personal protective equipment, including respiratory protection, and prepare for reuse. (Skill Sheet 6-I-5, 5.1.2, 5.3.2, 5.3.3)

21. Demonstrate the steps for inspecting an SCBA. (Skill Sheet 6-I-7, 5.3.2, 5.5.1)

22. Demonstrate the steps for cleaning and sanitizing an SCBA. (Skill Sheet 6-I-7, 5.3.2, 5.5.1)

23. Demonstrate the method for filling an SCBA cylinder from a cascade system, wearing appropriate PPE, including eye and ear protection. (Skill Sheet 6-1-8, 5.3.1)

24. Demonstrate the method for filling an SCBA cylinder from a compressor/purifier system, wearing appropriate PPE, including eye and ear protection. (Skill Sheet 6-I-9, 5.3.1)

25. Demonstrate the one-person method for changing an SCBA cylinder. (Skill Sheet 6-I-7, 5.3.1)

26. Demonstrate the two-person method for changing an SCBA cylinder. (Skill Sheet 6-I-7, 5.3.1)

Chapter 6
Firefighter Personal Protective Equipment

Case History

In 2002, Mark Noble, a nineteen-year veteran of the Olympia (WA) Fire Department, was diagnosed with brain cancer. During his treatment, Mark began to research the connection between firefighters and cancer. He learned that firefighters are exposed to highly toxic substances in virtually every fire and that some toxins can accumulate in the body after repeated exposures. These potentially deadly toxins include asbestos, benzene, polycyclic aromatic hydrocarbons (PAHs), polychlorinated biphenyls (PCBs), carbon monoxide (CO), and other products of combustion that are present during the overhaul phase of a fire.

In his research, Mark found that firefighters are twice as likely as the general public to develop intestinal, liver, testicular, and prostate cancer, non-Hodgkin's lymphoma, and malignant melanoma. They are three times as likely to develop skin, brain, and bladder cancer, as well as leukemia. They are also four times as likely to develop kidney cancer (**Figure 6.1, p.260**).

In 2005, Mark Noble died at the age of 47. The brain cancer that killed him was most likely caused by the toxins he was exposed to in the fire service. Mark loved being a firefighter, but he said that if he had it to do over again, he would wear his SCBA more, and he would be more conscientious about attaching exhaust collection hoses to the apparatus.

Permission to use this information was granted by Mrs. Rebecca Noble and ERGOMETRICS & Applied Personnel Research, Inc., who produced a video interview with Mark during his final months. The video is available at the following Website: www.ergometrics.org.

As a firefighter and emergency responder, you will work in a wide variety of hazardous environments. To protect your life and health, you must wear proper safety equipment at all times. This chapter provides an overview of two critical types of safety equipment: personal protective equipment (PPE) and respiratory protection. You will learn about the different types of this equipment, as well as their functions, proper usage, and maintenance.

Ⅰ Personal Protective Equipment

Personal protective equipment (PPE) is the personal protective clothing (PPC), respiratory protection equipment, and personal alert safety system (PASS) you will wear during emergency responses. It is designed to protect you from hazards and minimize

> **Personal Protective Equipment (PPE)** — General term for the equipment worn by fire and emergency services responders; includes helmets, coats, trousers, boots, eye protection, hearing protection, protective gloves, protective hoods, self-contained breathing apparatus (SCBA), and personal alert safety system (PASS) devices. *Also known as* Bunker Clothes, Full Structural Protective Clothing, Protective Clothing, Turnout Clothing, or Turnout Gear.

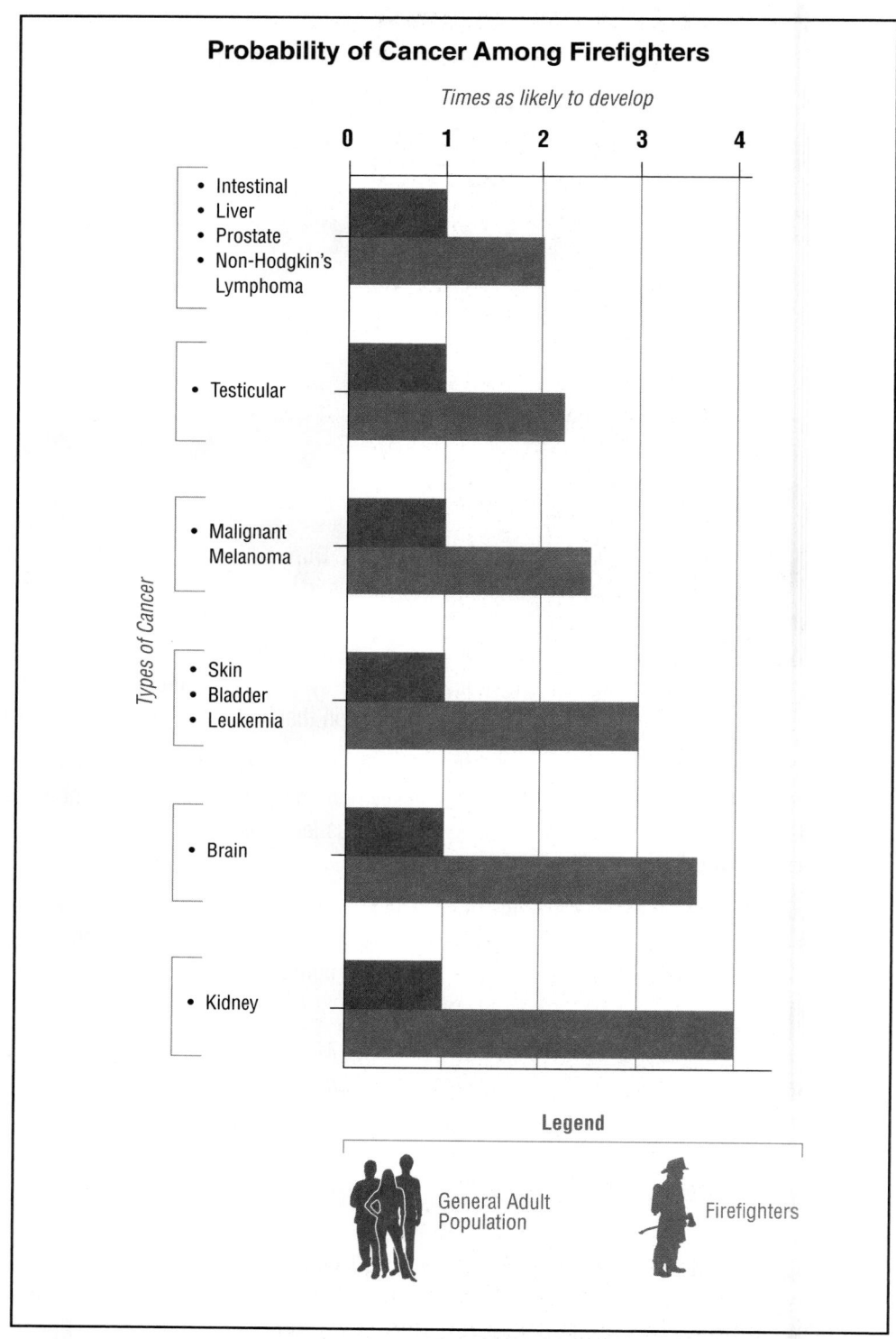

Figure 6.1 Firefighters are exposed to a greater range and concentration of hazards than the general public, and they have a higher likelihood of several types of cancers because of that exposure.

Figure 6.2 Protective equipment and clothing are tailored to the hazards that a first responder will face.

the risk of injury or fatality **(Figure 6.2)**. Your PPC includes helmets, coats, trousers, boots, eye protection, hearing protection, protective gloves, and protective hoods. Its use is mandated by NFPA® 1500, *Standard on Fire Department Occupational Safety and Health Program*, and all equipment must be designed and constructed based on NFPA® standards. Design and construction requirements for SCBA and PASS devices are covered in NFPA® 1981, *Standard on Open-Circuit Self-Contained Breathing Apparatus (SCBA) for Emergency Services*, and NFPA® 1982, *Standard on Personal Alert Safety Systems (PASS),* respectively.

Different emergency operations require different kinds of PPE. For example, some emergency types require full sets of PPE, including respiratory protection, while others require only protective clothing. Types of personal protective clothing include:

- Structural fire fighting protective clothing
- Wildland fire fighting protective clothing
- Roadway operations protective clothing
- Emergency medical protective clothing
- Special protective clothing
- Station and work uniforms

Structural Fire Fighting Protective Clothing

All personal protective clothing designed for **structural** and **proximity** fire fighting must meet the requirements of NFPA® 1971, *Standard on Protective Ensembles for Structural Fire Fighting and Proximity Fire Fighting*. This standard addresses the requirements for helmets, coats, trousers, boots, eye protection, protective gloves, and protective hoods.

Structural Fire Fighting — Activities required for rescue, fire suppression, and property conservation in structures, vehicles, vessels, and similar types of properties.

Proximity Fire Fighting — Activities required for rescue, fire suppression, and property conservation at fires that produce high radiant, conductive, or convective heat; includes aircraft, hazardous materials transport, and storage tank fires.

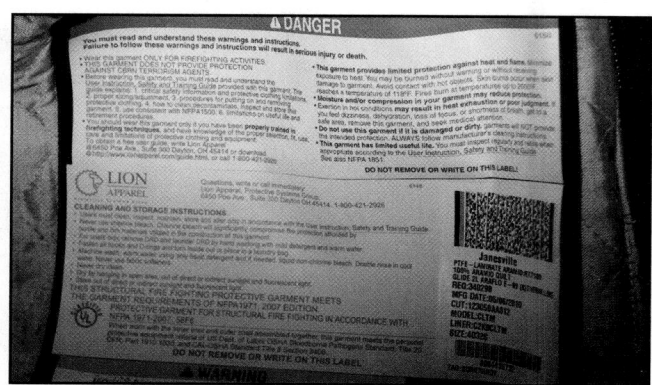

Figure 6.3 NFPA® 1971 regulates the clothing components that comply with the standard.

NFPA® 1971 requires that all components must include a permanent label that shows compliance with the standard **(Figure 6.3)**. This label contains the following statement:

THIS STRUCTURAL FIRE FIGHTING PROTECTIVE (name of component) MEETS THE (name of component) REQUIREMENTS OF NFPA® 1971, (current edition) EDITION.

Labels must also include the following information:

- Manufacturer's name, identification, or designation
- Manufacturer's address
- Country of manufacture
- Manufacturer's identification, lot, or serial number
- Month and year of manufacture
- Model name, number, or design
- Size or size range
- Principal materials of construction
- Footwear size and width (where applicable)
- Cleaning precautions

Personal protective clothing components must be compatible with each other to provide the level of protection intended by the NFPA® standard. Each component is designed to protect you from specific hazards and may not protect you from other types of hazards. For instance, structural personal protective clothing offers no protection against many types of hazardous materials.

Firefighters should never alter their protective clothing. Changing, adding, or removing components may void the manufacturer's warranty, affect your workers' compensation benefits, and endanger your life. Alterations include removing the moisture barrier or liner of coats and trousers, sewing hooks, loops, or clasps to the outer shell, and adding combustible decals to the helmet.

CAUTION

Unauthorized alteration of your PPE may affect the worker's compensation benefits provided to you by your jurisdiction.

Personal Protective Clothing Can Trap Heat and Humidity

Figure 6.4 Protective clothing may create its own hazards including the entrapment of heat.

Structural personal protective clothing is designed to cover all portions of your skin when you are reaching, bending, or moving. It is also designed to prevent heat transfer from the fire to your body. However, one limitation of this design is that it also prevents heat from being transferred *away* from your body. Usually your body cools itself by sweating, but the protective clothing traps body heat and moisture inside the clothing. This may significantly increase your breathing and heart rate, skin temperature, core temperature, and physiological stress (**Figure 6.4**). If environmental conditions allow it, open your protective clothing to permit air flow around your body during authorized breaks. This will lessen heat stress and reduce your heart rate.

Hazards Determine the PPE

Always wear the correct PPE that is designed to protect you from the specific type of hazard(s) presented by the incident.

Figure 6.5a Helmets have developed over time to address specific incident needs.

Figure 6.5b The features of a helmet should be used with the necessary safeguards in place to prevent injuries.

Figure 6.5c Unique styles of helmets should be used according to their specific purposes and the AHJ.

Helmet — Headgear worn by firefighters that provides protection from falling objects, side blows, elevated temperatures, and heated water.

Helmets

One of the primary concerns for firefighters is head protection. **Helmets** are manufactured in a wide variety of designs and styles (**Figures 6.5a-c**). They are designed to provide multiple benefits during structural fire fighting operations, including:

- Preventing heated or scalding water and embers from reaching the ears and neck
- Protecting the head from impact injuries caused by objects or falls
- Providing protection from heat and cold

In addition to providing protection, helmets can also help to identify personnel. Shell color indicates the firefighter's rank, markings indicate the unit, and removable identification labels indicate accountability (**Figures 6.6a-d**). All of these uses are based on a department's standard operating procedures (SOPs).

To properly protect you, your helmet must be worn correctly. Place the helmet on your head, secure the chin strap under your chin and tighten it, and fold the ear flaps down to cover your ears and neck. You must fold the ear flaps down even if you are wearing a protective hood. Some helmets also have a ratchet at the back of the headband to allow you to adjust the fit.

Figure 6.6a Personal identification may be found on the front panel of a helmet.

Figure 6.6b The markings on a helmet often indicate rank and unit.

Figure 6.6c A responder's name or nickname is often marked on the back of his or her helmet.

Figure 6.6d Accountability tags may fit conveniently inside the rear edge of a helmet for easy accessibility before entering an incident scene.

Figure 6.7a SCBA facepieces include eye protection.

Figure 6.7b Helmet mounted faceshields may serve as partial eye protection.

Figure 6.7c Goggles are commonly used in emergency incidents.

Figure 6.7d Safety glasses guard against slow-traveling hazards.

Eye Protection Devices

Eye injuries are some of the most common injuries at emergency incidents, but they are not always reported because they are not always debilitating. Although eye injuries can be serious, they are fairly easy to prevent. Eye protection comes in many forms including SCBA facepieces, helmet-mounted faceshields, goggles, and safety glasses (**Figures 6.7a-d**).

Helmets must be equipped with faceshields or goggles. Faceshields alone do not provide adequate protection from flying particles or splashes and are intended to be used in combination with a primary form of eye protection. NFPA® 1500 requires that goggles or other appropriate primary eye protection be worn when participating in operations where protection from flying particles or chemical splashes is necessary. During fire fighting operations, your primary eye protection is your SCBA facepiece. But in other situations, eye protection is needed when respiratory protection is not required. Some of these situations include:

- Emergency medical responses where exposure to body fluids is possible
- Vehicle extrications
- Wildland and ground cover fires
- Industrial occupancy inspections
- Station maintenance

These situations call for safety glasses or goggles, which protect against approximately 85 percent of all eye hazards. Several styles are available, including some that fit over prescription glasses. Prescription safety glasses are another option, although these must have frames and lenses that meet American National Standards Institute (ANSI) Standard Z87.1, *Occupational and Educational Personal Eye and Face Protection Devices*.

In fire department facilities and maintenance areas, you should be aware of warning signs posted near power equipment requiring the use of eye protection. Always follow your department's safety policies and procedures regarding appropriate eye protection.

Figure 6.8 Protective hoods provide a continuous layer of coverage for the head and neck.

Protective Hood — Hood designed to protect the firefighter's ears, neck, and face from heat and debris; typically made of Nomex®, Kevlar®, or PBI®, and available in long or short styles.

Protective Coat — Coat worn during fire fighting, rescue, and extrication operations.

Protective Hoods

Protective hoods are fabric coverings that protect your ears, neck, and face from exposure to heat, embers, and debris. They cover areas that may not be protected by the SCBA facepiece, helmet, ear flaps, or coat collar. The protective hood's face opening has an elastic edge that fits tightly to the SCBA facepiece, forming a seal. Hoods are typically made of fire-resistant material and are available with long or short skirts **(Figure 6.8)**. The skirts are designed to fit inside the protective coat, forming a continuous layer of protection.

Pull the hood on before the protective coat to help keep the hood's skirt under the coat. To ensure a secure seal between the hood and the SCBA facepiece, secure the facepiece first before pulling up the hood. This way you will not compromise the facepiece-to-face seal.

Protective Coats

NFPA® 1971 requires that all **protective coats** used for structural fire fighting be made of three components: the outer shell, moisture barrier, and thermal barrier **(Figure 6.9)**. These barriers trap insulating air that prevents heat transfer from the fire to your body. They also provide limited protection from direct flame contact, hot water, steam, cold temperatures, and other environmental hazards. Each component is important to your safety and should never be compromised. Removing the liner and wearing only the shell compromises the design of the coat, increases the likelihood of injuries, and voids the manufacturer warranty.

WARNING
All layers of the protective coat must be in place during any fire fighting operation. Failure to wear the entire coat and liner system during a fire may expose you to severe heat resulting in serious injury or death.

Figure 6.9 The three component layers of a protective coat work together to provide resistance against hazards in the environment and are identified as: a) the outer shell, b) the moisture barrier, and c) the thermal barrier.

Protective coats have many design features that provide protection and convenience (**Figures 6.10 a-e, p. 268**). Design features required by NFPA® 1971 include:

- **Retroreflective trim** — Strips of reflective trim on the torso and sleeves make it more visible at night or in low light conditions.

- **Wristlets** — Fabric interface between the end of the sleeve and the palm of the hand that protects the wrist from water, embers, and other debris. Also keeps coat sleeves from riding up when reaching.

- **Collars** — Protects the neck from water, embers, and other debris. The collar must be turned up under the helmet ear flap.

- **Closure system** — Snaps, clips, zippers, or Velcro® fasteners that secure the front of the coat.

- **Drag Rescue Device (DRD)** — Harness and hand loop at the back of the neck that enables a rescuer to grab and drag a downed firefighter.

Coats are typically reinforced in high compression areas, such as the shoulders, and areas prone to wear, such as the elbows. Optional design features such as cargo, radio, or SCBA facepiece pockets are also common, based on local requirements. These optional features must be attached by the manufacturer to meet the NFPA® standard.

Protective Trousers

Protective trousers are constructed from the same fabric, moisture barrier, and thermal layering used in protective coats (**Figure 6.11, p. 269**). High compression areas and areas prone to wear are reinforced, and cargo or patch pockets may be attached for carrying gloves and small tools. Heavy-duty suspenders are used to hold the trousers up. Closure systems are the same as those found on the protective coat.

Protective Trousers — Trousers worn to protect the lower torso and legs during emergency operations. *Also known as* Bunker Pants or Turnout Pants.

Protective Gloves

Protective gloves protect hands and wrists from heat, steam, or cold penetration, and resist cuts, punctures, and liquid absorption. At the same time, gloves must allow enough dexterity and tactile feel for you to perform your job effectively. For instance, gloves must permit you to grasp tools and nozzles or manipulate small objects such as control knobs on portable radios (**Figure 6.12, p. 269**). Properly worn, the gloves cover the wristlet of the protective coat to form a complete seal. Gloves worn for structural fire fighting must be NFPA®-compliant for this type of activity.

Protective Gloves — Protective clothing designed to protect the hands.

Protective Footwear

Fire fighting boots are available in a variety of styles and materials (**Figures 6.13a and b, p. 269**). They protect the foot, ankle, and lower leg from:

- Puncture wounds to the sole caused by nails, broken glass, and other sharp objects

- Crushing wounds to the toes and instep

- Scalding water or contaminated liquids

- Burns from embers and debris

Boots have a steel inner sole and a steel or reinforced toe cap and must be high enough to protect the lower leg. The outer shell may be made of rubber, leather, or other water-resistant material. Thermal, physical, and moisture barriers are required inside the shell. Boot tops fit inside the trouser legs, providing a complete barrier even when you kneel down.

Fire Fighting Boots — Protective footwear meeting the design requirements of NFPA®, OSHA, and CAN/CSA Z195-02 (R2008).

Figure 6.10a Reflective trim catches light and reflects it to enhance the visibility of responders at night or in low-light conditions.

Figure 6.10b Wristlets are built into the ends of coat sleeves to provide a protective interface between sleeves and gloves.

Figure 6.10c Coat collars prevent embers, water, and other debris from getting under the coat.

Figure 6.10d A coat closure system may include more than one mechanism to ensure a complete seal.

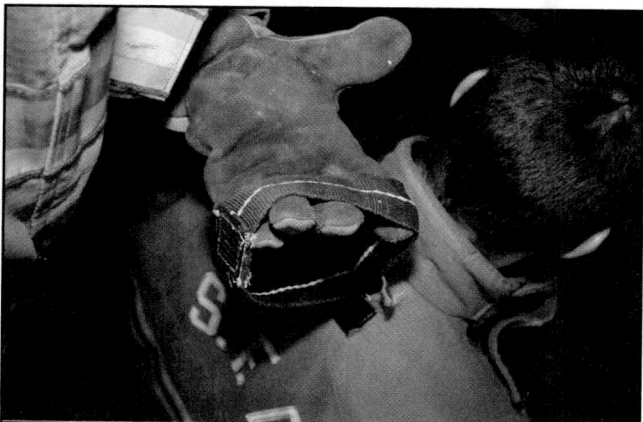

Figure 6.10e A loop on the back of protective clothing serves as a handle to aid in the rescue of a distressed firefighter.

Figure 6.11 Because of their weight, protective trousers are supported in place with suspenders.

Figure 6.12 Protective gloves cover the hands and coat wristlet to provide protection against common hazards.

Figure 6.13a Fire fighting boots protect the foot and ankle from hazards routinely found at an incident scene.

Figure 6.13b Styles of fire fighting boots may vary but their common purpose is to protect a firefighter's feet and lower legs.

Figure 6.14 Ear muffs are one form of hearing protection that can be used during activities with high noise exposure levels.

Hearing Protection Devices

Firefighters are exposed to a variety of loud noises in the fire station, during training, en route to incidents, and at the emergency scene. **Hearing protection** devices guard against temporary and permanent hearing loss. They are not required by NFPA® 1971, but they are required by NFPA® 1500.

To comply with this standard, departments must protect firefighters from the effects of harmful noise. Eliminating or reducing noise is the best solution, but sometimes this is not possible. In these cases, departments must provide hearing protection devices and establish a hearing conservation plan.

Hearing protection is most commonly used when riding on an apparatus where the noise exceeds maximum noise exposure levels (90 decibels in the U.S., 85 decibels in Canada). Intercom/ear protection systems are the most effective for this purpose, because they also allow the crew to communicate with each other or monitor radio communications. Hearing protection is also required during the operation of power tools, generators, and the apparatus pump, and when testing the PASS device (**Figure 6.14**).

However, hearing protection is impractical in some situations, and may even be dangerous. For example, during structural fire fighting, it prevents you from communicating with other firefighters and hearing radio transmissions, changes in fire behavior, or calls from a trapped victim.

Personal Alert Safety Systems (PASS)

Personal alert safety systems (PASS) emit a loud alarm to alert other personnel that a firefighter is in danger. The alarm is activated when a firefighter is motionless for more than 30 seconds, or when a firefighter presses the emergency button. In some models, it may be activated when the temperature exceeds a preset limit. The alarm must be at least 95 decibels (dBA) and must go off continuously for at least one hour.

PASS devices assist rescuers attempting to locate trapped, unconscious, or incapacitated firefighters. They are particularly useful in total darkness, dense smoke, or confined spaces. Some devices are stand-alone units that are manually activated. Other devices are integrated units connected to the SCBA regulator that are automatically activated when the main air supply valve is opened (**Figures 6.15a and b**). SCBA-mounted devices can also be activated manually, without opening the cylinder valve.

PASS devices have at least three settings: *off, alarm,* and *sensing.* They also have a prealarm mode that activates if you are motionless for 30 seconds. This prealarm tone is different from the full alarm tone and is intended to prevent false alarms.

You must learn how to turn the unit from *off* to *sensing* (on) and to manually activate the alarm. You are also responsible for testing, maintaining, and activating your PASS device according to your department's SOPs, manufacturer's instructions, NFPA® 1500, and NFPA® 1982, *Standard on Personal Alert Safety Systems (PASS).*

Wildland Personal Protective Clothing

Personal protective clothing used for structural fire fighting is generally too bulky, heavy, and hot to be practical for wildland fire fighting. NFPA® 1977, *Standard on Protective Clothing and Equipment for Wildland Fire Fighting* contains the specifications for wildland fire fighting personal protective clothing and equipment (**Figure 6.16**). Wildland personal protective clothing and equipment includes:

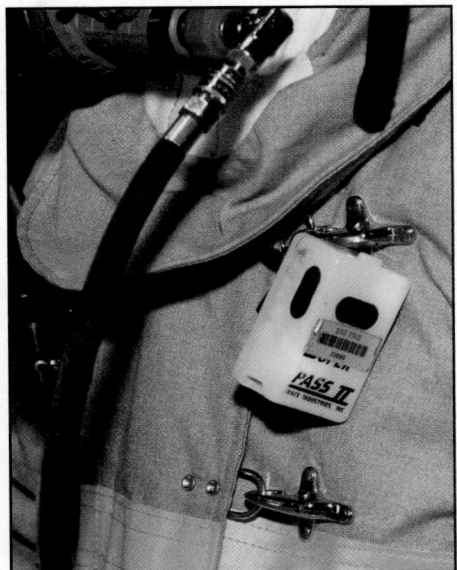

Figure 6.15a A stand-alone PASS device is not integrated with other safety equipment.

Figure 6.15b An integrated PASS unit activates automatically when the air supply valve is opened. *Courtesy of James Nilo.*

Figure 6.16 Wildland personal protective clothing is minimal compared to structural fire fighting clothing.

- **Gloves** — Made of leather or inherently flame-resistant materials. They protect the hand and wrist from sharp or hot objects, temperature extremes, and scalding water.

- **Goggles** — Protect the eyes from ash, embers, dust, and other particulates. Must meet ANSI Z87.1, *Occupational and Educational Eye and Face Protection Devices*.

- **Jackets** — Made of high-strength, flame-resistant fabric, such as Aramid, or treated cotton. May have a thermal liner for use in cold climates. The cuffs close snugly around the wrists, and the front of the jacket must close completely from hem to neck.

- **Trousers** — Made of the same material and design as the jackets. The leg cuffs must close securely around the boot tops.

- **One-piece jumpsuits** — One-piece protective garments are similar in design to the two-piece jacket and trousers ensemble.

- **Long-sleeve shirts** — Protective shirts are worn under the jackets and are of similar design.

- **Helmet** — A lightweight helmet with chin straps that provides impact, penetration, and electrical insulation protection.

- **Face/neck shrouds** — Flame-resistant fabric that attaches to the helmet and protects the face and neck.

- **Footwear** — Typically lace-up safety boots with lug or grip-tread soles. Must be high enough to protect the lower leg. Because the steel toes in ordinary safety boots absorb and retain heat, they are not recommended for wildland fire fighting.

- **Fire shelter** — Fire-resistant aluminized fabric covers that protect the firefighter from convected and radiant heat **(Figure 6.17)**. Its use is required by NFPA® 1500, and its design must meet the United States Department of Agriculture (USDA) Forest Service Specification 5100-606.

- **Load-carrying or load-bearing equipment** — Belt and suspender systems that distribute the weight of the firefighter's equipment including tools, water bottles, and protective fire shelters.

- **Respiratory protection** — Until very recently, the accepted form of respiratory protection has been a cotton bandana or dust mask worn over the nose and mouth. However, studies have shown that these items do not provide sufficient protection. Beginning in 2011, National Institute for Occupational Safety and Health (NIOSH) certified and NFPA® approved air-purifying respirators (APR) and powered air-purifying respirators (PAPR) will be available for wildland fire fighting.

Figure 6.17 A fire shelter reflects the heat of a fire away from the firefighter while in use.

- **Chain saw protection** — Chaps, leggings, or protective trousers made of ballistic nylon fibers that protect the legs **(Figure 6.18)**. According to one source, leg injuries account for a significant number of all chain saw injuries.

Because wildland protective garments will not protect you from extreme heat, you should never wear underclothing made of synthetic materials, such as nylon or polyester, when fighting a wildland fire. These materials melt when heated and can stick to your skin, causing serious burns.

Figure 6.18 Protective chaps may limit some injury while using a chain saw.

> ## WARNING
> Wildland personal protective clothing is not designed, certified, or intended for interior structural fire fighting.

Roadway Operations Clothing

Emergency operations along roadways are extremely dangerous for firefighters and emergency responders. No amount of personal protective clothing can prevent injuries when a person is struck by a rapidly moving vehicle. The best protection is to be visible to other motorists and to work behind a barrier formed by your apparatus, as discussed in Chapter 2.

To increase your visibility to oncoming traffic, you should wear traffic vests with retroreflective trim. Retroreflective trim reflects headlight beams, providing visibility at night or in low light situations **(Figure 6.19)**. Traffic vests are required by federal law at incidents on federally-funded highways. They should be worn over your PPE if possible, or as soon as the situation has stabilized. Vests are not commonly worn during fire suppression or hazardous materials activities but should be worn at all other times.

Figure 6.19 A firefighter wearing a reflective vest is more visible at night than a firefighter wearing standard bunker gear.

Emergency Medical Protective Clothing

When providing medical assistance, emergency responders must protect themselves against exposure to infectious bodily fluids and airborne pathogens. To accomplish this, they must wear emergency medical protective clothing. Most structural and wildland protective clothing does not prevent disease transmission, and fire fighting protective clothing is inappropriate because it can contaminate victims through their open wounds.

Emergency medical protective clothing may be either single- or multiple-use garments (**Figure 6.20**). Single-use garments are disposed of after contact with a patient. Multiple-use garments may be cleaned and reused a specified number of times before disposal. Medical protective clothing must meet NFPA® 1999, *Standard on Protective Clothing for Emergency Medical Operations*. It must also include the following items:

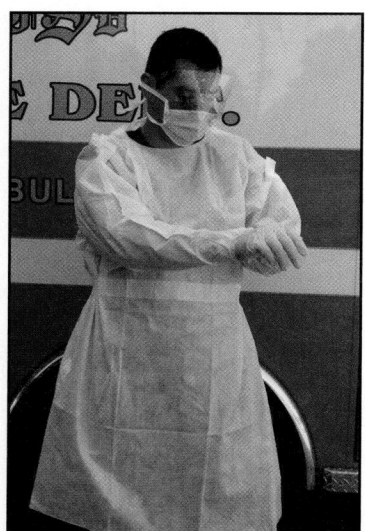

Figure 6.20 Protective clothing includes single-use garments.

- **Utility gloves** — Not used for patient care but do provide a barrier against bodily fluids, disinfectants, and cleaning solutions.

- **Medical examination gloves** — Gloves that are certified for patient care and provide an effective barrier (against infection) up to the wrist.

- **Eye/face protection device** — Faceshield, goggles, safety glasses, or hooded visor that provides limited protection for the eyes and face.

- **Facemask** — A full face device that protects the eyes, face, nose, and mouth.

- **Footwear** — Safety shoes or boots that protect feet and ankles; may be dual certified station/work shoes.

- **Footwear cover** — A single-use item worn over footwear to provide a limited barrier against bodily fluids.

- **Medical garment** — Single- or multiple-use clothing that provides a barrier against bodily fluids; may be sleeves, jackets, trousers, gowns, or coveralls that can be worn over a uniform.

- **Medical helmet** — Head protection that is designed to provide impact, penetration, and electrical insulation protection while working with a patient in a hazardous area. Helmets must meet ANSI design requirements for Type 1 hard hats.

- **Respiratory protection device** — Filter mask that protects the wearer from airborne pathogens.

Special Protective Clothing

Other emergency incident types that require specialized protective clothing include the following:

- **Technical rescue** — Must protect the wearer from physical, thermal, and liquid hazards, as well as infectious diseases. Design criteria are specified in NFPA® 1951, *Standard on Protective Ensembles for Technical Rescue Incidents* (**Figure 6.21 a**). May be dual certified for emergency medical use as defined by NFPA® 1951 and NFPA® 1999. They may also be chemical, biological, radiological, or nuclear (CBRN) certified. While structural personal protective clothing is sometimes worn for technical rescue operations, it is usually too bulky and heavy. Respiratory protection typically consists of air-purifying respirators (APR), SCBAs, or supplied air respirators (SAR).

Figure 6.21c Ice rescue clothing serves functions similar to swiftwater clothing but with more thermal insulation. *Courtesy of Iowa Fire Service Training Bureau.*

Figure 6.21a Technical rescue clothing may include equipment features that are CBRN certified.

Figure 6.21b Swiftwater rescue equipment safeguards against dangers unique to that environment.

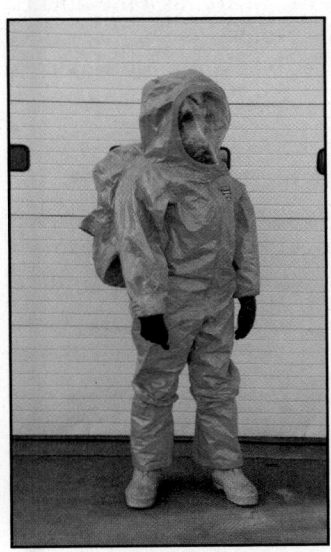

Figure 6.21d Personal protective clothing must be chosen to counteract the specific dangers presented by the substances that will be encountered.

Figure 6.21e Proximity protective clothing reflects radiant heat.

- **Standing/Swift Water rescue** — A full-body wet suit that is buoyant, thermally-insulated, and abrasion/puncture-resistant (**Figure 6.21b**). A rescue helmet is also part of the required ensemble. In addition, a U.S. Coast Guard (USCG) approved personal floatation device (PFD) must be worn in rivers, streams, and lakes, and along shorelines and coastlines.

- **Ice rescue** — Similar to the full-body dry suit that may be used in water rescue, but with more thermal insulation (**Figure 6.21c**). USCG PFDs are mandatory, and some models have watertight hoods, integrated gloves, and attachable boots.

- **Hazardous materials** — Varying types include high-temperature protective clothing and chemical-protective clothing (CPC) that protects against splashes and vapor **(Figure 6.21d, p. 275)**.

- **Chemical, biological, radiological, and nuclear (CBRN)** — CBRN suits protect against these hazard types during accidents and terrorist incidents. They must adhere to NFPA® 1994, *Standard on Protective Ensembles for First Responders To CBRN Terrorism Incidents*. NFPA® 1971 also establishes optional design requirements for structural and proximity ensembles that are certified for CBRN incidents. These ensembles must carry the appropriate CBRN label.

- **Proximity fire fighting protective clothing** — Similar to structural protective clothing but with an aluminized outer shell on the coat, trousers, gloves, and helmet shroud **(Figure 6.21e, p. 275)**. This outer shell is designed to reflect high levels of radiant heat and protect against direct flame contact. Must be resistant to water, impact from sharp objects, and electrical shock. Proximity PPE may be used in some hazardous materials operations.

Station/Work Uniforms

Although station and work uniforms look different between jurisdictions, they are all intended to perform two functions. First, they identify the wearer as a member of the organization. Second, the uniform provides a layer of protection against direct flame contact. For this reason, clothing made of non-fireresistant synthetic materials should not be worn while on duty or under your PPE.

All firefighter station and work uniforms should meet the requirements set forth in NFPA® 1975, *Standard on Station/Work Uniforms for Emergency Services*. The purpose of the standard is to provide minimum requirements for work wear that is functional, will not contribute to firefighter injury, and will not reduce the effectiveness of outer personal protective clothing. Garments addressed in this standard include trousers, shirts, jackets, and coveralls, but not underwear. Underwear made of 100 percent cotton is recommended, because synthetic materials melt when heated and can stick to your skin, causing serious burns.

Clothing certified to meet NFPA® 1975 requirements will have a permanently attached label stating that certification **(Figures 6.22a and b)**. Note that while this clothing is designed to be fire resistant, it is not designed to be worn for fire fighting operations. Structural fire fighting protective clothing must always be worn over these garments when you are engaged in structural fire fighting activities. Wildland protective clothing, depending on design and local protocols, may be worn over station/work uniforms or directly over undergarments. Some station/work uniforms are dual certified as both work uniforms and wildland protective clothing. Dual certified uniforms will always carry the appropriate certification labels.

In many fire departments, safety shoes or boots are part of the station/work uniform. They are required footwear while conducting inspections or doing work around the station. Safety shoes or boots usually have steel toes, puncture-resistant soles, or special inserts. However, footwear used for station duties should not be worn during emergency operations because they might then contaminate living quarters with potentially hazardous substances.

Uniforms can be contaminated/soiled following any emergency response. Therefore, they should not be taken home or washed in personal washing machines or at public laundromats. Contaminated uniforms must be laundered at the fire station or by a contractor.

Figure 6.22a Uniforms worn in a station must conform to NFPA® 1975 because they will serve as a layer of protection under PPE.

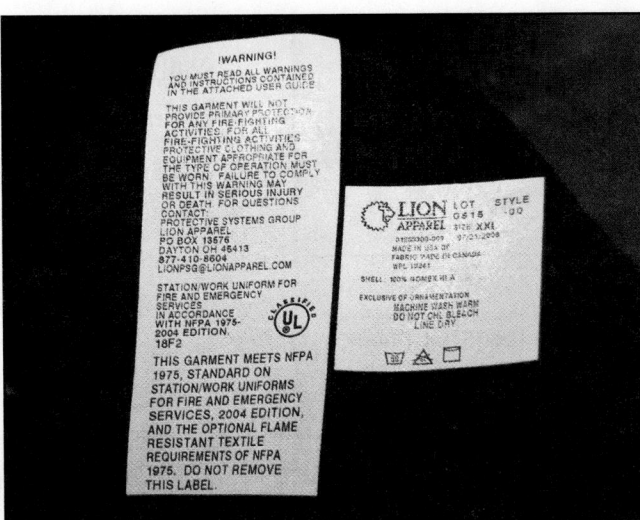

Figure 6.22b Clothing that adheres to NFPA® 1975 should be labeled as such.

Care of Personal Protective Clothing

Your personal protective clothing is your primary barrier protecting you from injury and illness. However, it can also cause injury or illness if it is not properly maintained. Hydrocarbon contamination will reduce the fire resistance of your PPE (**Figure 6.23**). Chemicals, oils, and petroleum products in or on the outer shell can ignite when exposed to fire. Some contaminants can reduce the effectiveness of retroreflective trim, and soot can obscure its visibility. Hydrocarbons, body fluids, and toxins that contaminate PPE can be inhaled, ingested, or absorbed, causing serious and sometimes fatal illness.

You are responsible for the inspection, cleaning, and condition of the PPE assigned to you. Procedures for the care of your PPE are found in your department's SOPs, the manufacturer's instructions, and NFPA® 1851, *Standard on Selection, Care, and Maintenance of Protective Ensembles for Structural Fire Fighting and Proximity Fire Fighting.*

Figure 6.23 Contamination of protective equipment can directly cause health risks and impair the effectiveness of the PPE.

Inspecting

You should inspect your PPE at the beginning of your work shift, after every use, after washing, repair, decontamination, and on a periodic basis, such as weekly or monthly. An annual inspection should be made by a member of your department who is trained in advanced inspection requirements, such as the department's Health and Safety Officer (HSO).

Figure 6.24 A heat- and smoke-damaged faceshield is useless on the fireground and should be replaced before operations.

Conditions that you should look for during a routine inspection include:

- Soiling
- Contamination
- Missing or damaged hardware and closure systems
- Physical damage including rips, tears, and damaged stitching on seams
- Wear due to friction under arms, in the crotch, and at knee and elbow joints
- Thermal damage, including charring, melting, discoloration, and burn holes
- Shrinkage
- Damaged or missing retroreflective trim or reinforcing trim
- Loss of reflectivity of shell on proximity equipment
- Cracks, melting, abrasions, or dents in helmet shell
- Missing or damaged faceshield or hardware (**Figure 6.24**)
- Missing or damaged earflaps or neck shroud
- Loss of watertight integrity in footwear
- Damage to or faulty installation of drag rescue device (DRD)

If your protective clothing requires only routine cleaning that will not cause the item to be removed from service, you should perform that cleaning yourself. If you ever determine that your PPE requires advanced cleaning or decontamination, repairs, or replacement, report this to your supervisor immediately.

CAUTION
Repairs to all PPE must only be performed by persons who are trained and certified to do so.

Cleaning

NFPA® 1851 defines four types of cleaning for personal protective clothing:

- Routine cleaning
- Advanced cleaning
- Specialized cleaning
- Contract cleaning

Types of cleaning are determined by the amount and type of contamination and whether the equipment must be removed from service to perform the cleaning. Many fire departments provide spare sets of PPE to replace units removed from service for cleaning, decontamination, or repairs.

Routine cleaning. Routine cleaning does not require that the clothing be removed from service (**Figure 6.25**). At an incident scene, the process for routine cleaning includes:

- Brushing off loose debris with a broom or soft bristle brush
- Using a gentle spray of water to rinse off debris and soil

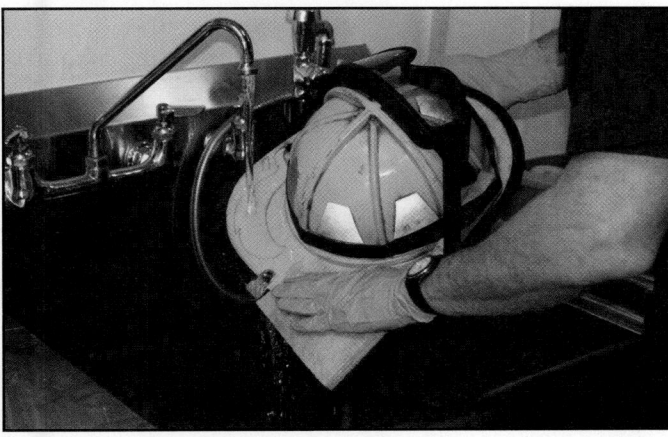

Figure 6.25 Maintain the good order of equipment by keeping it clean.

To remove heavy soil, cleaning should be performed by hand in a utility sink, in the designated cleaning area at the fire station. Whether you are at the scene or in the station, always follow the manufacturer's recommendations and wear appropriate gloves and eye protection.

Advanced cleaning. Personnel trained in the care and cleaning of protective clothing should perform advanced cleaning. A washing machine dedicated to cleaning protective clothing that is designed to handle heavy loads should be used.

> **WARNING**
> Never clean soiled or contaminated protective clothing at home or at a public laundry. This may expose yourself, your family, or others to dangerous contaminants.

Specialized cleaning. Required when clothing is contaminated with hazardous materials or body fluids that cannot be removed by routine or advanced cleaning. May be performed by a trained member of the department or an outside contractor. Clothing that is too contaminated to be cleaned must be removed from service and destroyed.

Contract cleaning. Specialized cleaning performed by the manufacturer, its representative, or a certified vendor. Typically removes accumulated grime or contaminants. Some contractors provide replacement PPE while clothing is being cleaned.

> **WARNING**
> Do not wash contaminated protective clothing in washing machines used for other garments or items. Do not take contaminated protective clothing into the living or sleeping quarters of the fire station or your residence. PPE should not be stored where it can come in contact with vehicle exhausts. PPE that is carried in personal vehicles should be placed in closable garment bags intended for that purpose.

Repairing

Damaged protective clothing must be repaired *immediately*, either by the manufacturer, an approved repair facility, or a trained member of the department. Clothing damaged beyond repair must be removed from service and destroyed. Some damaged clothing may be marked for training only and used in non-live fire training.

Safety Considerations for Personal Protective Equipment

PPE is designed to create a protective barrier between you and your work environment. However, this barrier can also isolate you, preventing you from being aware of important environmental changes and making you overconfident of your own safety.

Keep in mind these important safety considerations that relate to your personal protective equipment:

- Always consider the design and purpose of your protective clothing, and be especially aware of each garment's limitations.

- Moisture in the shell and liner material will conduct heat rapidly, resulting in serious steam burns. Always ensure that the garment is dry before wearing it into a fire.

- PPE insulates you from the heat of a fire. This will protect your life, but it will also delay your awareness of temperature increases.

- Never wear protective clothing that does not fit because it will provide reduced protection. Tight clothing will not close properly, leaving a gap. Loose clothing can hinder mobility and dexterity by bunching up at shoulders, elbows, and knees. It can also snag on debris, create a tripping hazard, absorb contaminants, and reduce thermal protection.

- Make sure that the overlap between coat and trousers is a minimum of 2 inches (50 mm) at the waist when you bend over to a 90° angle (**Figure 6.26**).

- Thermal burns may occur at compression points where the garment layers are pressed together, such as under SCBA shoulder harness, along sleeves in contact with hoselines, and on knees when kneeling on hot debris and embers.

- Radiant heat can rapidly penetrate protective clothing, causing serious burns. If you feel thermal radiant burns developing, withdraw from the area immediately.

- Prolonged exposure to hot environments will cause your body to sweat in order to cool itself. However, the protective clothing liner will retain the moisture produced by sweating, which may cause heat stress or burns. When you feel the symptoms of heat exhaustion, including weakness, dizziness, rapid pulse, or headache, move to a cool, safe area, remove your PPE, and follow established rehabilitation procedures.

- Your PPE is designed to protect you, but it is not designed to protect against extreme fire conditions such as backdraft, flashover, or other extreme fire behavior.

Figure 6.26 A protective coat must overlap the trousers by at least 2 inches (50 mm) no matter which position the firefighter holds during operations.

WARNING
Burns are a function of time and temperature. The longer the exposure and the higher the temperature, the greater the severity of a burn. First degree burns start when skin temperature reaches 118°F (47.8°C). Second degree burns start at 131°F (55°C), and third degree burns start at 152° F (66.7°C).

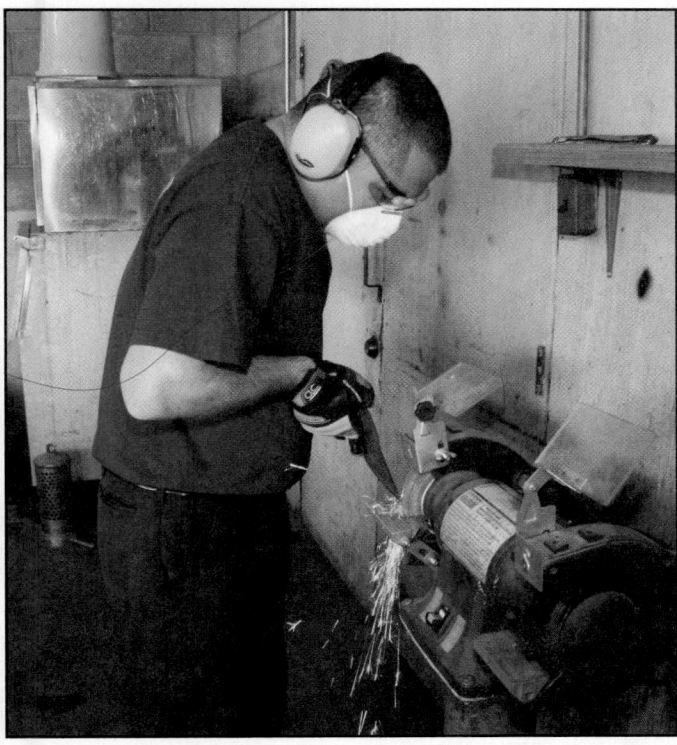

Figure 6.27 Repair work should be assumed to generate particles that must be kept from contact with skin, eyes, and respiration.

Respiratory Protection

The case history at the beginning of this chapter illustrates how inhaling smoke and other products of combustion poses short-term, long-term, and even fatal health hazards. Wearing appropriate respiratory protection is the most effective way to protect your health. Operations requiring respiratory protection include:

- Structural and wildland fires, which produce smoke and other products of combustion

- Medical responses, which may expose you to airborne pathogens

- Confined-space search, rescue, and recovery, which may take place in toxic or low oxygen atmospheres

- Repair work that generates fine particulates such as dust, paint, or metal shavings **(Figure 6.27)**.

Always wear respiratory protection equipment that is appropriate for the type of hazard you are facing. Be sure to use equipment properly so that you are not exposed to **respiratory hazards**.

Respiratory Hazards

Common respiratory hazards include:

- Oxygen deficiency
- Elevated temperatures
- Particulate contaminants
- Gases and vapors
- Airborne pathogens

Respiratory Hazards — Exposure to conditions that create a hazard to the respiratory system, including products of combustion, toxic gases, and superheated or oxygen-deficient atmospheres.

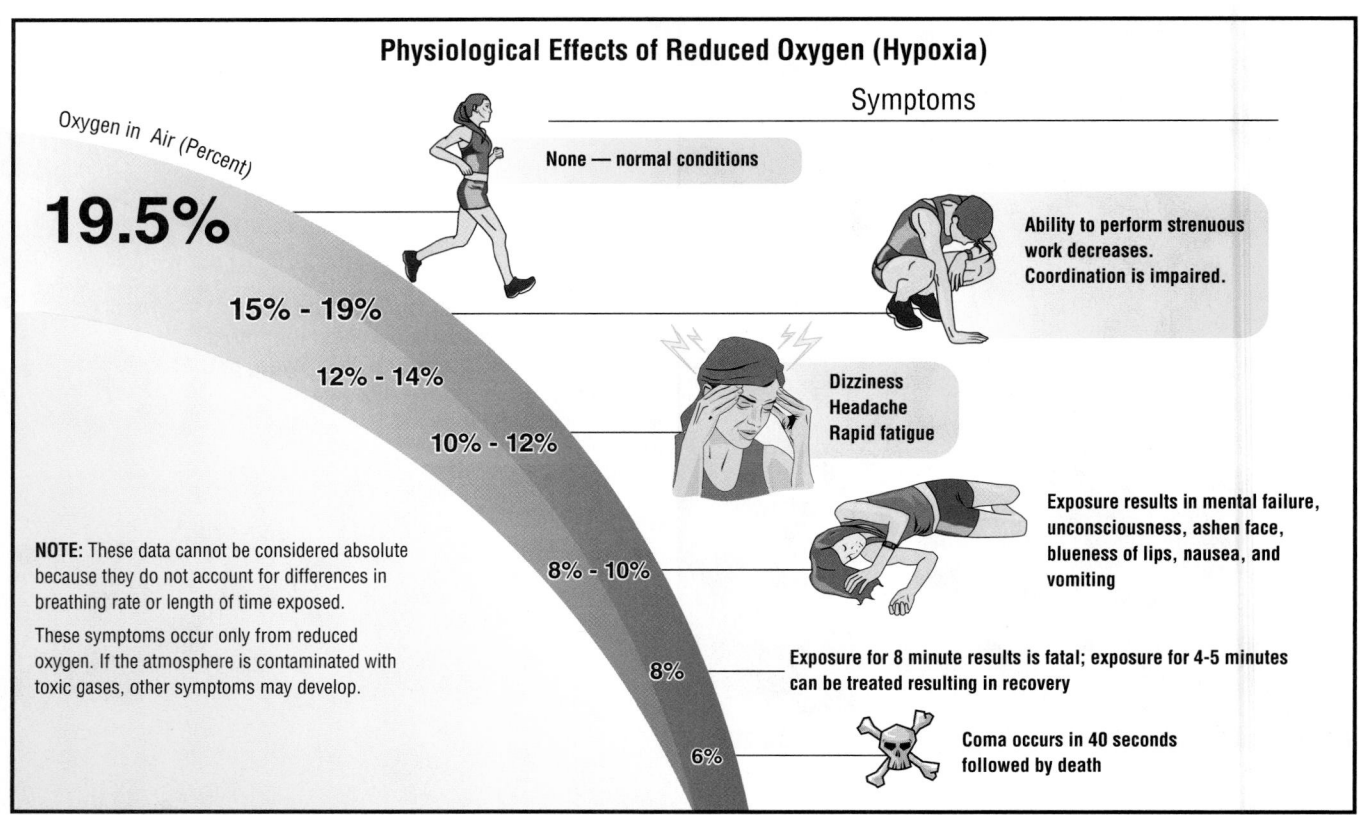

Physiological Effects of Reduced Oxygen (Hypoxia)

Oxygen in Air (Percent)

19.5%

15% - 19%

12% - 14%

10% - 12%

8% - 10%

8%

6%

Symptoms

None — normal conditions

Ability to perform strenuous work decreases. Coordination is impaired.

Dizziness
Headache
Rapid fatigue

Exposure results in mental failure, unconsciousness, ashen face, blueness of lips, nausea, and vomiting

Exposure for 8 minute results is fatal; exposure for 4-5 minutes can be treated resulting in recovery

Coma occurs in 40 seconds followed by death

NOTE: These data cannot be considered absolute because they do not account for differences in breathing rate or length of time exposed.

These symptoms occur only from reduced oxygen. If the atmosphere is contaminated with toxic gases, other symptoms may develop.

Figure 6.28 Oxygen deficiency has increasingly severe effects depending on the level of reduction and the amount of time it is endured.

Respiratory hazards are often found in situations that produce immediate, irreversible, and debilitating effects on a person's health and may result in death. NFPA® 1500 and OSHA classify these situations as immediately dangerous to life and health (IDLH). Prior to entering any structure or area that is or may be IDLH, you must don the correct level of personal protective clothing and respiratory protection.

Oxygen Deficiency

Both NFPA® and OSHA define an **oxygen-deficient atmosphere** as one containing less than 19.5 percent oxygen. When oxygen concentrations are below 18 percent, the human body responds by increasing its respiratory rate. As less oxygen reaches body tissues **hypoxia** occurs. The physiological effects of hypoxia are illustrated in **Figure 6.28**.

Combustion is the most common cause of oxygen-deficient atmospheres. It consumes oxygen and produces toxic gases, which either physically displace oxygen or dilute its concentration. Oxygen-deficient atmospheres also occur in confined spaces such as sewers, chemical storage tanks, grain bins, or underground caverns. They are also found in rooms or compartments where carbon dioxide (CO_2) total-flooding extinguishing systems have discharged.

Some fire departments are equipped with instruments to monitor atmospheres and measure oxygen levels or the presence of toxic gases. Where monitoring is not possible or monitor readings are questionable, SCBA or SAR must always be worn.

Oxygen-Deficient Atmosphere — Atmosphere containing less than the normal 19.5 percent oxygen. At least 16 percent oxygen is needed to produce flames or sustain human life.

Hypoxia — Potentially fatal condition caused by lack of oxygen.

Elevated Temperatures

Exposure to superheated air can damage the respiratory tract. The damage can be much worse when the air is moist. Excessive heat inhaled quickly into the lungs can cause a serious decrease in blood pressure and failure of the circulatory system. Inhaling superheated gases can cause **pulmonary edema** which can cause death from **asphyxiation**. The tissue damage from inhaling hot air is not immediately reversible by introducing fresh, cool air. Prompt medical treatment is required.

Pulmonary Edema — Accumulation of fluids in the lungs.

Asphyxiation — Fatal condition caused by severe oxygen deficiency and an excess of carbon monoxide and/or other gases in the blood.

> ### WARNING
> Inhaling superheated air can cause serious injury or death.

Particulate Contaminants

Particulate contaminants are small particles that may be suspended in the air and are harmful to the respiratory system. Sources of these particulates include:

Particulate — Very small particle of solid material, such as dust, that is suspended in the atmosphere.

- Vehicle exhaust emissions
- Chemical reactions
- Heated metals or metal compounds
- Combustion

According to medical studies, exposure to particulate contaminants causes asthma, lung cancer, cardiovascular disease, and premature death. Smaller particulates are especially dangerous, because particulates bigger than 1 micrometer are filtered by the nasal membranes and do not enter the lungs.

Particulate contaminants may be encountered during the following situations:

- Wildland fires
- Welding and metal cutting operations
- Operation of fire apparatus and small engines **(Figure 6.29)**
- Operations following an explosion or building collapse
- Structural fires, especially during the overhaul phase

Figure 6.29 Diesel exhaust carries small particulates as well as other contaminants that are known to cause health risks.

Figure 6.30a APR units filter the working environment and capture airborne particles.

Figure 6.30b PAPR units include a headpiece, a breathing tube, a battery, and an air blower.

Air-Purifying Respirator — (APR) Respirator that removes contaminants by passing ambient air through a filter, cartridge, or canister; may have a full or partial facepiece.

Powered Air-Purifying Respirator (PAPR) — Motorized respirator that uses a filter to clean surrounding air, then delivers it to the wearer to breathe; typically includes a headpiece, breathing tube, and a blower/battery box that is worn on the belt.

Gas — Compressible substance, with no specific volume, that tends to assume the shape of the container. Molecules move about most rapidly in this state.

Vapor — Gaseous form of a substance that is normally in a solid or liquid state at room temperature and pressure; formed by evaporation from a liquid or sublimation from a solid.

Air-purifying respirators (APRs) and **powered air-purifying respirators (PAPRs)** are generally sufficient to protect you from particulate contaminants **(Figures 6.30a and b)**. Cartridge and canister type APR/PAPRs have half or full facepiece units with replaceable filter elements that capture the particulates. APR/PAPRs are approved for wildland fire fighting but do not protect against toxic gases or heated or oxygen-deficient atmospheres.

Gases and Vapors

Gases and **vapors** may be present at both fire and nonfire incidents. Gases exist at standard temperature and pressure, while vapors result from temperature or pressure changes that affect a solid or liquid. For example, natural gas is found in a gaseous state within the earth, while steam is a vapor created when water is heated.

Gases and vapors can be inhaled, ingested, or absorbed into the body, resulting in illnesses and death. Exposure may cause:

- Cancer
- Cardiovascular disease
- Thyroid damage
- Respiratory problems
- Eye irritation

Fire gases and vapors. Harmful gases and vapors created by combustion include:

- Carbon monoxide
- Carbon dioxide

- Hydrogen cyanide
- Hydrogen chloride
- Hydrogen sulfide
- Nitrous gases
- Phosgene
- Sulfur dioxide
- Ammonia
- Formaldehyde

Of these, carbon monoxide (CO) and hydrogen cyanide (HCN) are responsible for the majority of fire-related fatalities. Carbon monoxide is a colorless, odorless gas that is present in virtually every fire. It is released when an organic material burns in an atmosphere with a limited supply of oxygen.

Carbon monoxide poisoning is a sometimes lethal condition in which carbon monoxide molecules attach to hemoglobin, decreasing the blood's ability to carry oxygen. Carbon monoxide combines with hemoglobin about 200 times more effectively than oxygen does. The carbon monoxide does not act on the body, but excludes oxygen from the blood, leading to hypoxia of the brain and tissues, followed by death if the process is not reversed. **Table 6.1** illustrates the effects of CO on humans.

Hydrogen cyanide is produced by the incomplete combustion of substances containing nitrogen and carbon, such as natural fibers, resins, synthetic polymers, and synthetic rubber. These materials are found in upholstered furniture, bedding, insulation, carpets, and other common building materials. HCN is also released during off-gassing as an object is heated. It may also be found in unexpected places, such as vehicle fires, where new insulation materials give off high amounts of gases and cause fires to last longer.

HCN can be inhaled, ingested, or absorbed into the body, where it then targets the heart and brain. Inhaled HCN enters the bloodstream and prevents the blood cells from using oxygen properly, killing the cell. The effects of HCN depend on the concentration, length, and type of exposure. Large amounts, high concentrations, and lengthy exposures are more likely to cause severe effects, including permanent heart and brain damage or death. HCN is 35 times more toxic than CO. **Table 6.2, p. 286** illustrates the effects of HCN on the human body.

Nonfire gases and vapors. Hazardous materials can produce a potentially hazardous gases and vapors in nonfire emergencies, such as the following:

- Incidents involving industrial, commercial, or warehouse occupancies
- Spills resulting from transportation accidents
- Leaks from storage containers or pipelines

Table 6.1
Toxic Effects of Carbon Monoxide

Carbon Monoxide (CO) (ppm*)	Carbon Monoxide (CO) in Air (Percent)	Symptoms
100	0.01	No symptoms — no damage
200	0.02	Mild headache; few other symptoms
400	0.04	Headache after 1 to 2 hours
800	0.08	Headache after 45 minutes; nausea, collapse, and unconsciousness after 2 hours
1,000	0.10	Dangerous — unconsciousness after 1 hour
1,600	0.16	Headache, dizziness, nausea after 20 minutes
3,200	0.32	Headache, dizziness, nausea after 5 to 10 minutes; unconsciousness after 30 minutes
6,400	0.64	Headache, dizziness, nausea after 1 to 2 minutes; unconsciousness after 10 to 15 minutes
12,800	1.28	Immediate unconsciousness; danger of death in 1 to 3 minutes

*ppm — parts per million

Table 6.2
Lethal Exposure Times and Concentrations of Hydrogen Cyanide (HCN) on Humans

Exposure Time	Concentration in parts per million (ppm)
1 minute	3,404 ppm
6 to 8 minutes	270 ppm
10 minutes	181 ppm
30 minutes	135 ppm
> 30 minutes	20 to 40 ppm

Symptoms Indicating Exposure to Hydrogen Cyanide (HCN)

Initial Concentration	Initial Symptoms	Progressive Symptoms	Final Effect
20 to 40 ppm	Headache, drowsiness, vertigo, weak and rapid pulse, deep and rapid breathing, a bright-red color in the face, nausea, and vomiting	Convulsions, dilated pupils, clammy skin, a weaker and more rapid pulse and slower, slow, or irregular heartbeat, falling body temperature, blue color to lips, face, and extremities, coma, and shallow breathing	Death

At any haz mat incident, always remain at a safe distance (upwind, uphill, upstream) until a risk analysis has been completed. The surrounding atmosphere at these incidents should always be considered dangerous, so SCBAs must be worn until air monitoring demonstrates that the atmosphere is safe.

Hazardous nonfire gases and vapors are always a possibility at transportation incidents and in storage and manufacturing facilities. Common gas and vapor types include:

- **Carbon Dioxide** — Also produced by fire suppression systems

- **Ammonia** — Also produced by air conditioning and cooling systems

- **Sulfur Dioxide** — Also produced by air conditioning and cooling systems

- **Chlorine** — Also found in water treatment facilities, water parks, and swimming pools

- **Pesticides** — Also found in commercial outlets, farms, nurseries, and residences

Toxic gases may also be found in sewers, storm drains, caves, trenches, storage tanks, tank cars, bins, and other confined spaces. Even when toxic gases are not present, the atmosphere in these areas may be oxygen deficient and potentially deadly. Search, rescue, and recovery in these areas require the use of SCBAs and SARs.

Airborne Pathogens

Airborne pathogens are disease-causing microorganisms (viruses, bacteria, or fungi) that are suspended in the air. They may be encountered when assisting victims during medical responses, vehicle extrications, rescue and recovery operations, and terrorist attacks. They cause infection after being inhaled or directly contacted.

Illnesses that can result from exposure to airborne pathogens include:

- Meningitis
- Influenza
- Methicillin-resistant *Staphylococcus aureus* (MRSA)
- Pneumonia
- Tuberculosis (TB)
- Severe acute respiratory syndrome (SARS)
- Measles
- Chickenpox
- Smallpox

Protection against airborne pathogens includes **high-efficiency particulate air (HEPA) filters**, APR/PAPRs, and SCBA/SAR. HEPA filter masks are single-use masks that are certified by NIOSH and designated N95, N99, or N100 **(Figure 6.31)**. Designations indicate the percentage of airborne particles that the masks effectively remove. Note that surgical masks are not approved for use against airborne pathogens. However, they may be used on patients to prevent them from spreading diseases by exhaling, sneezing, or coughing.

Types of Respiratory Protection Equipment

The two categories of respiratory protection equipment are atmosphere-supplying respirators (ASRs) and air-purifying respirators (APRs). ASRs provide breathable air when working in oxygen deficient, toxic, or gas-filled atmospheres, while APRs only filter particulates out of the ambient air. The primary type of respiratory protection that you will use in the fire service is the ASR.

Figure 6.31 A HEPA filter mask removes airborne particles before inhalation.

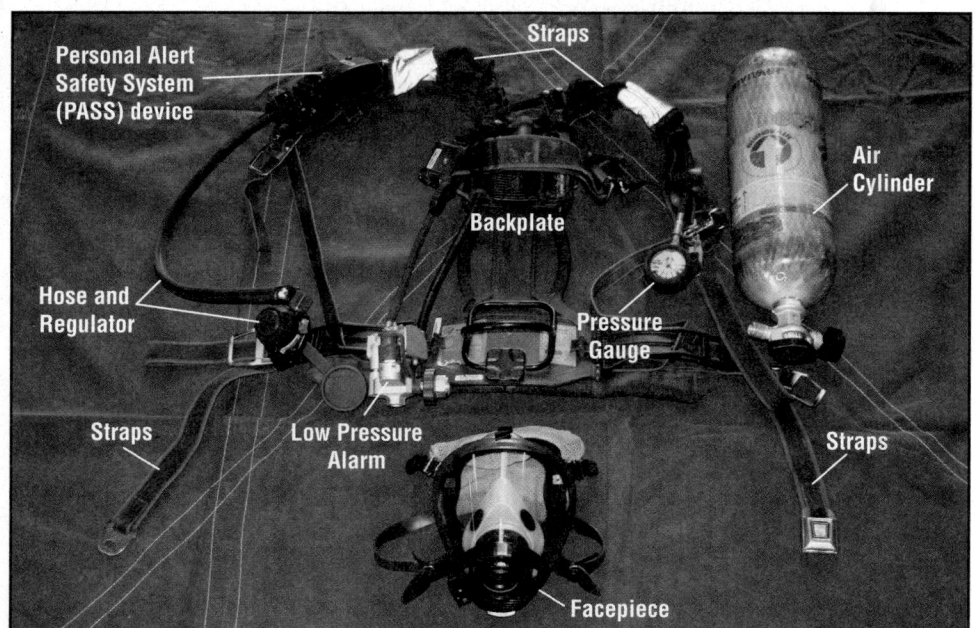

Figure 6.32a An SCBA unit is intended to support a limited-time engagement in high-intensity operations.

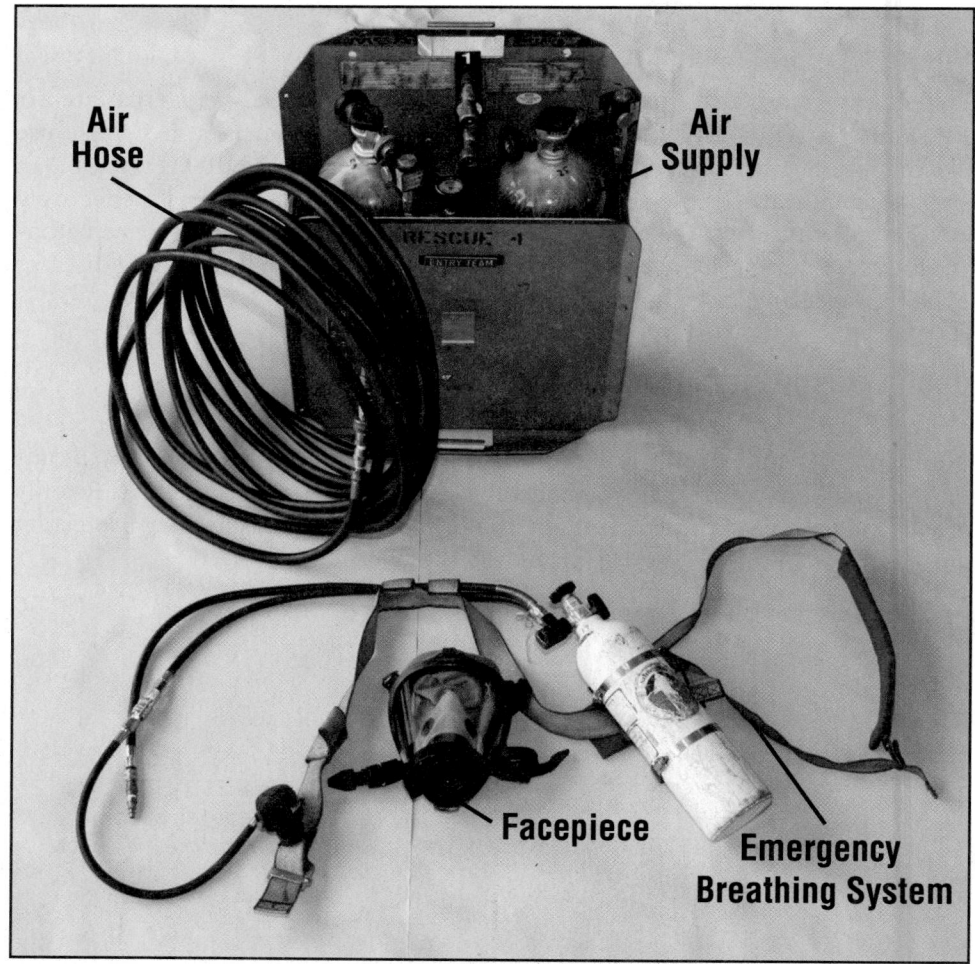

Figure 6.32b A supplied-air respirator system is intended to support longer engagements under conditions that will not damage the equipment.

Atmosphere-Supplying Respirators (ASRs)

Atmosphere-supplying respirators consist of either SCBA or supplied air respirators (SARs) **(Figures 6.32a and b)**. SCBAs carry the air in a cylinder, while SARs are connected to a breathing-air compressor or portable air supply, providing breathable air for a much longer duration.

SARs are used when the firefighter must be in the hazardous area for a long period of time, and there is no danger that fire may damage the air hose. These types of situations include hazardous materials incidents, confined-space rescues, and other technical rescue incidents. They are not used for fire fighting because of the possibility that fire or debris will damage the air-supply hose. They are only used by personnel who are certified in technical rescue functions.

There are two main types of SCBAs: **open-circuit SCBAs**, which use compressed air, and **closed-circuit SCBAs**, which use compressed oxygen **(Figures 6.33a and b)**. In open-circuit SCBA, exhaled air is vented to the outside atmosphere. In closed-circuit SCBA (also known as "rebreather" apparatus), exhaled air stays within the system and is reused. Closed-circuit SCBA are much less common and are mainly used in shipboard operations, extended hazardous materials incidents, some rescue operations, and by industrial fire brigades.

Open-Circuit Self-Contained Breathing Apparatus — SCBA that allows exhaled air to be discharged or vented into the atmosphere.

Closed-Circuit Self-Contained Breathing Apparatus — SCBA that recycles exhaled air; removes carbon dioxide and restores compressed, chemical, or liquid oxygen. Not approved for fire fighting operations. *Also known as* Oxygen-Breathing Apparatus (OBA) or Oxygen-Generating Apparatus.

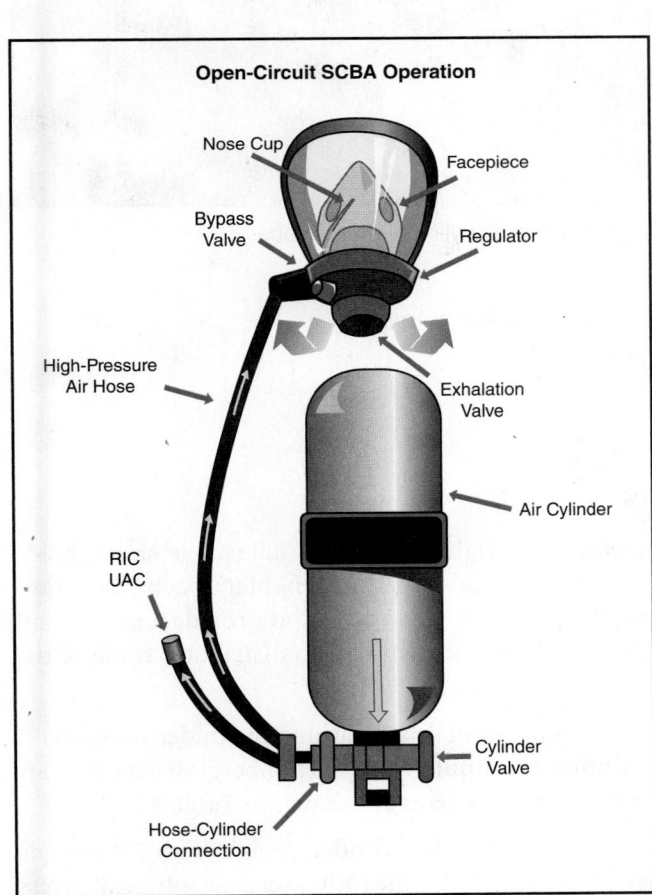

Figure 6.33a An open-circuit SCBA vents exhaled air to the outside atmosphere.

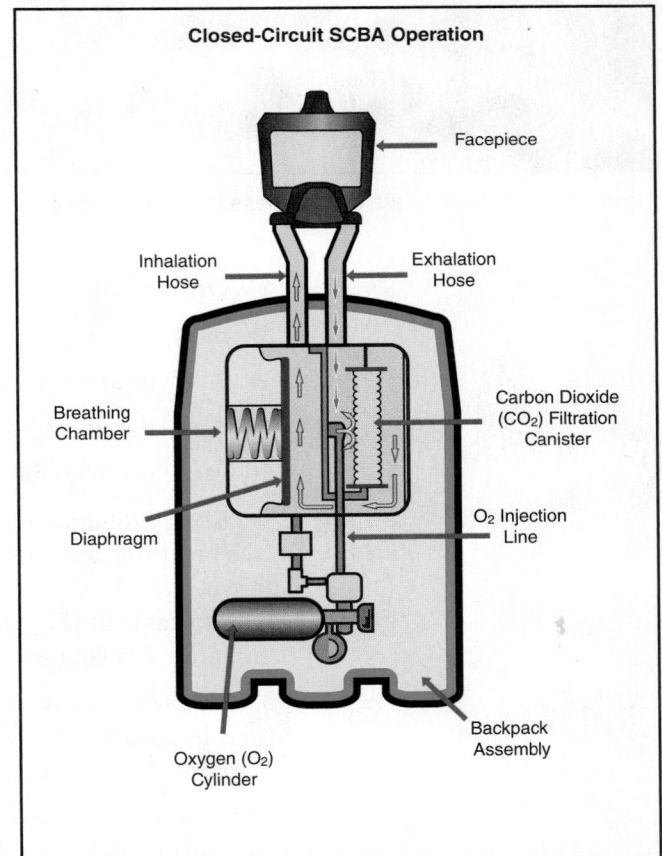

Figure 6.33b A closed-circuit SCBA reuses exhaled air to allow the user to remain in a hazardous environment for a longer duration.

Figure 6.34 The backplate and harness assembly provide stability to the breathing air cylinder while in use.

Open-circuit SCBA consist of four basic components:

- Backplate and harness assembly
- Air cylinder assembly
- Regulator assembly
- Facepiece assembly

Backplate and harness assembly. This rigid frame with adjustable straps holds the breathing air cylinder on the backplate, and onto the firefighter's back. The straps are designed to stabilize the unit, carry part of its weight, and provide a secure and comfortable fit **(Figure 6.34)**. An adjustable waist strap also distributes some of the apparatus' weight to the hips.

Air cylinder assembly. The air cylinder contains breathing air under pressure. It may be constructed of steel, aluminum, aluminum wrapped in fiberglass, or a Kevlar/carbon composite material. Common cylinder sizes are shown in **Table 6.3**.

Depending on size and construction materials, cylinders weigh from 7.9 pounds (3.6 kg) to 18.9 pounds (8.6 kg). This weight significantly increases physical stress during emergency operations.

The cylinder has a control valve, threaded stem and/or quick-connect fitting, and a pressure gauge attached to one end. When in operation, the control valve is opened fully to emit air into the system. The high-pressure hose attaches to the stem and connects the cylinder to the regulator assembly. The pressure gauge displays an estimate of the amount of air in the cylinder in pounds per square inch (psi) (kilopascal [kPa]).

Table 6.3
Breathing Air Cylinder Capacities

Rated Duration	Pressure	Volume
30-minute	2,216 psi (15 290 kPa)	45 ft³ (1 270 L) cylinders
30-minute	4,500 psi (31 000 kPa)	45 ft³ (1 270 L) cylinders
45-minute	3,000 psi (21 000 kPa)	66 ft³ (1 870 L) cylinders
45-minute	4,500 psi (31 000 kPa)	66 ft³ (1 870 L) cylinders
60-minute	4,500 psi (31 000 kPa)	87 ft³ (2 460 L) cylinders

• Rated duration does not indicate the actual amount of time that the cylinder will provide air.

Regulator assembly. Air from the cylinder travels through the high-pressure hose to the regulator assembly. The regulator reduces the high pressure of the cylinder air to slightly above atmospheric pressure and controls air flow to the wearer. When the wearer inhales, a pressure differential is created in the regulator. The apparatus diaphragm moves inward, tilting the admission valve so that low-pressure air can flow into the facepiece. The regulator diaphragm is then held open, which creates the positive pressure. Exhalation moves the diaphragm back to the "closed" position. The regulator may be located on the facepiece, the shoulder harness, or the waist belt harness (**Figure 6.35**).

Depending on the SCBA model, the regulator will have control valves for normal and emergency operations. These are the mainline valve and the bypass valve. On models equipped with both valves, the mainline valve is locked in the open position during normal operations and the bypass valve is closed. On some SCBA, the bypass valve controls a direct air line from the cylinder in the event that the regulator fails. Once the valves are set in their normal operating position, they should not be changed unless the emergency bypass function is needed. The current generation of regulators have only the bypass valve (**Figure 6.36**).

Figure 6.35 The regulator controls the flow of air to meet the respiratory requirements of the user.

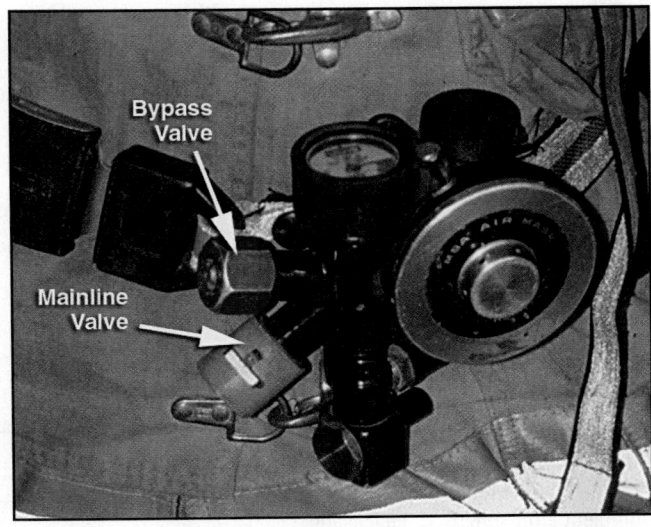

Figure 6.36 The mainline valve and bypass valve on this older belt-mounted regulator are identified by their shape, color, and position.

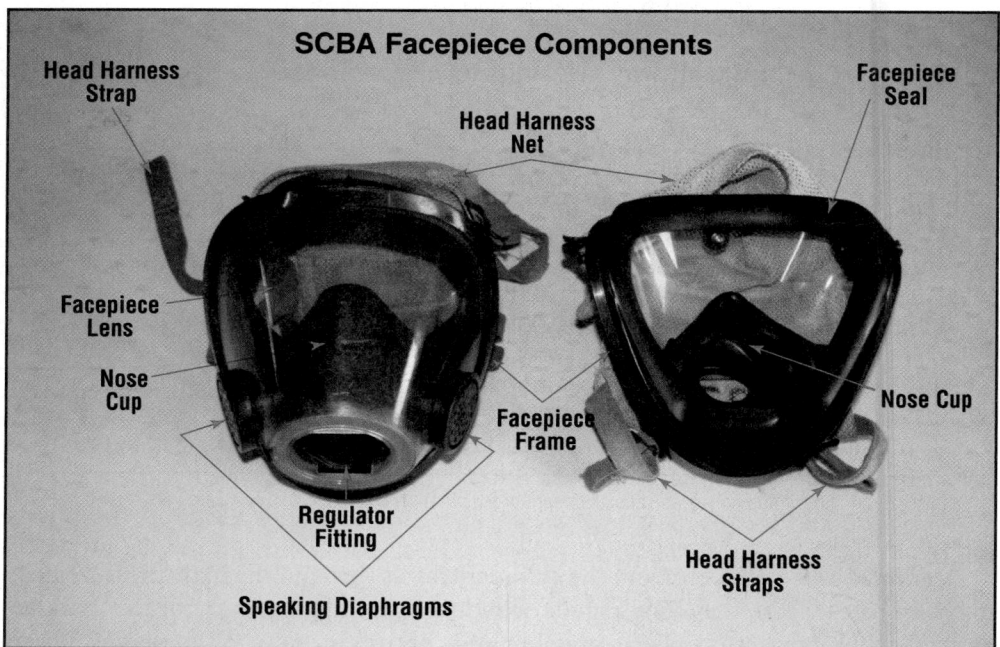

SCBA Facepiece Components

Head Harness Strap

Head Harness Net

Facepiece Seal

Facepiece Lens

Nose Cup

Facepiece Frame

Nose Cup

Regulator Fitting

Head Harness Straps

Speaking Diaphragms

Figure 6.37 A facepiece assembly that fits snugly provides protection against a series of predictable hazards including heat and large airborne particles.

Figure 6.38 The series of lights in a heads-up display provides a constant reminder of the air pressure in a cylinder. *Courtesy of Arlington (TX) Fire Department. Photograph by Jason Arias.*

Qualitative Fit Test (QLFT) — Respirator fit test that measures the wearer's response to a test agent, such as irritant smoke or odorous vapor. If the wearer detects the test agent, such as through smell or taste, the respirator fit is inadequate.

Quantitative Fit Test (QNFT) — Fit test in which instruments measure the amount of a test agent that has leaked into the respirator from the ambient atmosphere. If the leakage measures above a pre-set amount, the respirator fit is inadequate.

Facepiece assembly. The facepiece assembly provides fresh breathing air while protecting the eyes and face from injury. To accomplish these functions, the facepiece must fit tightly to the face. The facepiece assembly consists of (**Figure 6.37**):

- **Facepiece frame and lens** — Made of clear safety plastic and mounted in a flexible rubber facepiece frame. According to NFPA® 1981, all new SCBA facepieces must be equipped with a heads-up display (HUD) (**Figure 6.38**). This feature displays a series of lights on the inside of the facepiece lens indicating the approximate amount of air remaining in the cylinder.

- **Head harness and straps** — This harness, with adjustable straps, net, or some other arrangement, holds the facepiece snugly against the face.

- **Exhalation valve** — Simple, one-way valve that releases exhaled air without admitting any of the contaminated outside atmosphere.

- **Nose cup** — Deflects exhalations away from the lens, reducing fogging or condensation on the lens.

- **Speaking diaphragm** — This mechanical diaphragm permits limited communication by the wearer. It may be replaced by an electronic speaking diaphragm connected to a portable radio.

- **Regulator fitting or hose connection** — Permits the regulator or hose to attach to the facepiece frame.

To ensure that the facepiece has a perfect seal, the wearer must be fit-tested to determine the correct fit (**Figure 6.39**). The test must use the same make, model, style, and size of facepiece that will be worn during emergency operations. OSHA currently accepts two types of tests: **qualitative** and **quantitative**. Both tests provide an adequate assessment of a facepiece's ability to maintain a complete seal to the face.

NFPA® 1500 prohibits beards or facial hair that prevents a complete seal between the facepiece and the wearer's face. Wearing eyeglasses is also prohibited if the side frames pass through the seal area. Eyeglass kits are provided with all full facepiece

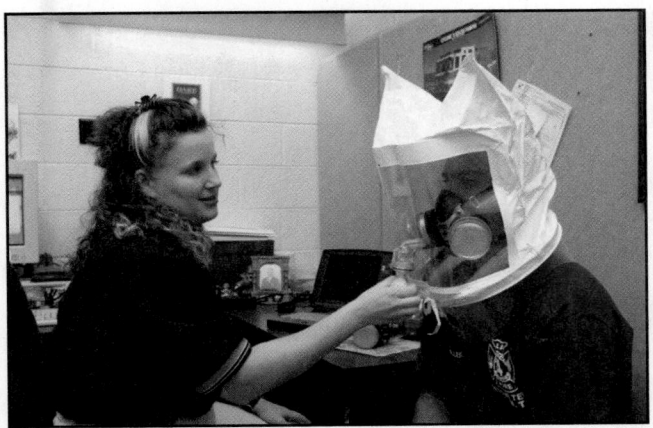

Figure 6.39 The NFPA® requires that each person wearing a respirator must be fit-tested for his or her own equipment.

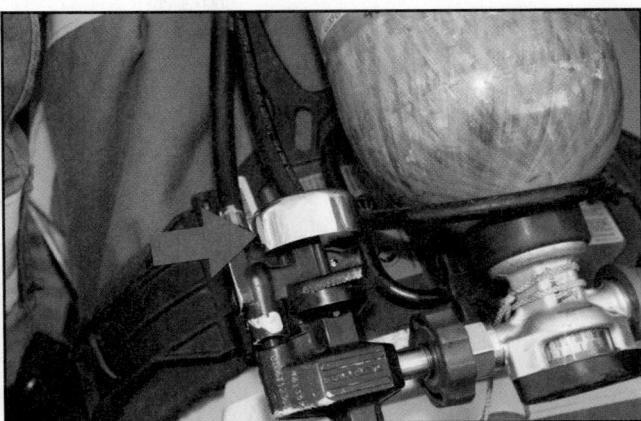

Figure 6.40 An end-of-service-time indicator may take the form of a bell that chimes when the air cylinder is nearly depleted.

masks. Both NFPA® 1500 and *Code of Federal Regulations (CFR)* 1910.134 allow firefighters to wear soft contact lenses while using full facepieces if the firefighter has demonstrated successful long-term (at least 6 months) use of contact lenses without any problems.

Additional components. A remote pressure gauge displays the air pressure within the cylinder. It must be mounted in a visible position. Pressure readings are most accurate at or near the upper range of the gauge's rated working pressures. Low pressure is measured less accurately, so the gauge's readings at this low end of the scale may not match the reading on the regulator gauge. When this occurs, assume that the lowest reading is correct. Also, check to make sure that the equipment is in working order before using it again.

Both NFPA® and NIOSH require that two end-of-service-time indicators (ESTI) or redundant low-pressure alarms be installed on all SCBAs **(Figure 6.40)**. The ESTI's alarm warns the user that the system is reaching the end of its air-supply, typically when it reaches 20-25 percent of the cylinder's capacity. The ESTI has both an audible alarm (like a bell, electronic beep, or high-pitched siren) and a flashing light or physical vibration. The alarm cannot be turned off until the air-cylinder valve is closed and the system is bled of all remaining pressure.

All new SCBA are equipped with a rapid intervention crew universal air coupling (RIC UAC) located within 4 inches (101 mm) of the cylinder outlet. This allows any cylinder that is low on air to be transfilled from another cylinder, regardless of its manufacturer **(Figure 6.41)**. When the cylinders are connected, the air supply equalizes between them. Older SCBA can be retrofitted with the RIC UAC, but this is not required. Thorough training in the use of this feature is required.

Air-Purifying Respirators (APRs)

Air-purifying respirators (APRs) remove contaminants by passing ambient air through a filter, canister, or cartridge. APRs may have full facepieces that provide a complete seal to the face and protect the eyes, nose, and mouth or half facepieces that provide a complete seal to the lower part of the face and protect the nose and mouth **(Figure 6.42, p. 294)**.

> **Code of Federal Regulations (CFR)** — Rules and regulations published by executive agencies of the U.S. federal government. These administrative laws are just as enforceable as statutory laws (known collectively as federal law), which must be passed by Congress.

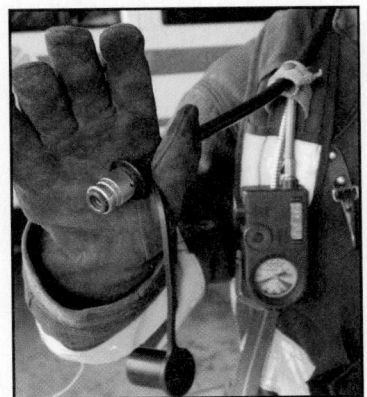

Figure 6.41 The Universal Air Coupling (UAC) allows an RIC/RIT member to connect an air cylinder to the system and transfill the low cylinder. *Courtesy of Kenneth Baum.*

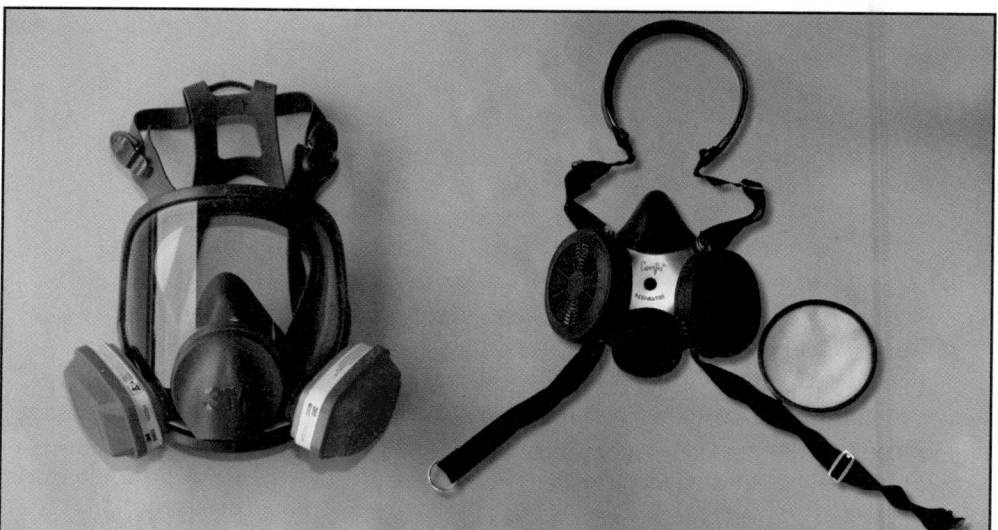

Figure 6.42 Full- and half-face APRs use a filter as a barrier against particulate contamination in the ambient air.

Particulate filters are single-use items that protect the respiratory system from large airborne particulates. They may be used with half or full facepiece masks, and are mounted on one or both sides of the facepiece. When used with a half facepiece mask, eye protection is required.

Particulate filters are regulated by the *Code of Federal Regulations,* specifically Title 42: Public Health, Part 84: Approval of Respiratory Protective Devices. They are divided into nine classes, three levels of filtration (95, 99, and 99.97 percent), and three categories of filter degradation (N, R, and P) that indicate the filter's limitations:

- **N** — Not resistant to oil
- **R** — Resistant to oil
- **P** — Used when oil or non-oil lubricants are present

Particulate filters are used primarily at emergency medical incidents to protect against airborne diseases. They are also appropriate for investigations or inspections involving body recovery; when bird, bat, or rodent excrement is present; at agricultural and industrial accidents; and when working with particulate-producing tools, such as sanders and paint sprayers (**Figure 6.43**).

Limitations of the APR are the limited life of the filters, canisters, and cartridges; the need for constant monitoring of the contaminated atmosphere; and the need for a normal oxygen content of the atmosphere before use. Usage should be restricted to the hazards for which the APR is certified.

APRs should be inspected regularly and cleaned following each use. Filters, canisters, and cartridges should be discarded following use and when they have passed their end of service life date.

NIOSH-certified APRs do not have ESTIs like the ones found on SCBA. Instead, the canisters and cartridges have visual ESTIs only. These indicators show when the air cleanser has become totally saturated and is no longer providing breathable air. Indicators should be checked visually before entering contaminated atmospheres and periodically while wearing the APR. If it appears that the canister or cartridge is reaching its saturation level, exit the area and replace the canister or cartridge.

When an APR is Appropriate

EMS Incidents

Agricultural and Industrial Incidents

Body Recoveries

Around Particulate-Producing Tools

Around Animal Excrement

Figure 6.43 Particulate filters must include an end-of-service-time indicator (ESTI).

Other clues or symptoms that the canister, cartridge, or filter is losing effectiveness are time, taste, smell, and resistance-to-breathing indicators. You should estimate the amount of time that the work in the contaminated area will take and compare it to the manufacturer's estimated life expectancy for the canister, cartridge, or filter. When the estimated time approaches, exit the area. If you can smell or taste the contaminant, the unit is no longer providing the proper level of protection. Finally, filters that have reached their saturation level cause resistance in the breathing process, causing your breathing to become labored. This is an indication that you must leave the work area before removing the facepiece.

Respiratory Protection Limitations

Although they protect you from a variety of hazards, all respiratory protection equipment has limitations. You must be aware of these limitations if you are going to operate safely and effectively in hazardous atmospheres. Limitations may be created by the wearer or by the equipment itself.

Wearer Limitations

The limitations that you have the greatest control over are those created by you, the wearer. These limitations include:

- **Lack of physical condition** — If you are not in good physical condition or if you are overweight, you may deplete your air supply rapidly.

- **Lack of agility** — If you not sufficiently agile, the weight and restriction of the equipment will make you less so, thus making it difficult to accomplish your assigned tasks.

- **Inadequate pulmonary capacity** — You must have sufficient lung capacity to inhale and exhale sufficient air while wearing respiratory protection equipment.

- **Weakened cardiovascular ability** — You must have a strong enough heart to prevent heart attacks, strokes, or other related problems while performing strenuous activity.

Safety Alert

In July 2012, during the production of Essentials 6th edition, the NFPA® issued a safety alert for SCBA facepieces.

See Page 335 for this alert.

- **Psychological limitations** — You must be able to overcome stress, fear, and feelings of claustrophobia while wearing respiratory protection equipment.

- **Unique facial features** — The shape and contour of the face can affect the ability to get a complete facepiece-to-face seal. Weight loss or gain can alter the facepiece seal.

These limitations can be offset through constant training with each type of respiratory protection equipment you use, periodic medical evaluations, and proper fit testing of respiratory protection facepieces. Training will make you more confident and more effective while wearing respiratory protection equipment.

Equipment Limitations

Both ASR and APR/PAPR units have limitations, including:

- **Limited visibility** — The full facepiece can reduce peripheral vision, while facepiece fogging can reduce overall vision.

- **Decreased ability to communicate** — The facepiece can seriously hinder voice communication unless it has built-in voice amplification or a microphone connection.

- **Decreased endurance** — The weight of SCBA units, averaging between 25 and 35 pounds (11 kg and 16 kg), makes you tire more quickly.

- **Decreased mobility** — The increase in weight and the restrictions caused by the harness straps can reduce your mobility.

- **Poor condition of apparatus** — Minor leaks and poor valve and regulator adjustments can result in excess air loss.

- **Low air cylinder pressure** — If the cylinder is not filled to capacity, the amount of working time is reduced proportionately.

You have some control over these limitations through frequent and proper inspections, care, and maintenance. Training with the units that you will use can also help you overcome the weight and mobility factors.

Storing Respiratory Protection Equipment

Methods of storing SCBA vary from department to department, but each department should store equipment so that it can be quickly and easily donned **(Figure 6.44)**. Respiratory equipment should also be protected from contamination, temperature changes, and ultraviolet light, all of which can cause damage. Storage for specific pieces of equipment (such as APR/PAPRs) depends on their size, available storage compartments on the apparatus, and the manufacturer's instructions.

SCBA can be placed on the apparatus in seat, side, and compartment mounts, or stored in carrying cases. If placed in seat mounts, the SCBA should be arranged so that it may be donned without the firefighter having to remove the seat belt.

CAUTION
Do not remove your seat belt in order to don your seat-mounted SCBA.

Figure 6.44 Equipment is secured so it may be donned easily from its storage hook.

Donning and Doffing Protective Breathing Apparatus

Several methods can be used to don an SCBA, depending on how it is stored. The most common include the over-the-head method, the coat method, donning from a seat, and donning from a side/rear external mount or backup mount. Each donning method requires different steps.

Different brands and models also require different steps for securing the SCBA to the wearer. Because the wide variety of SCBA makes it impossible to list these procedures for each manufacturer's model, this section provides only a general description of the four different donning techniques. Make sure to follow the manufacturer's instructions and local SOPs for the particular SCBA assigned to you.

General Donning Considerations

Multiple safety checks must always be made before donning the SCBA. Firefighters with daily shift changes should perform these checks at the beginning of their shifts, then return the SCBA to the apparatus-mounted storage rack or the storage case. Departments that are unable to make daily inspections should perform the following checks immediately prior to donning the SCBA, regardless of how it is stored:

- Check the air cylinder gauge to make sure the cylinder is full. NFPA® 1852, *Standard on Selection, Care, and Maintenance of Open-Circuit Self-Contained Breathing Apparatus (SCBA)*, recommends no less than 90 percent of cylinder capacity **(Figure 6.45, p. 298)**.

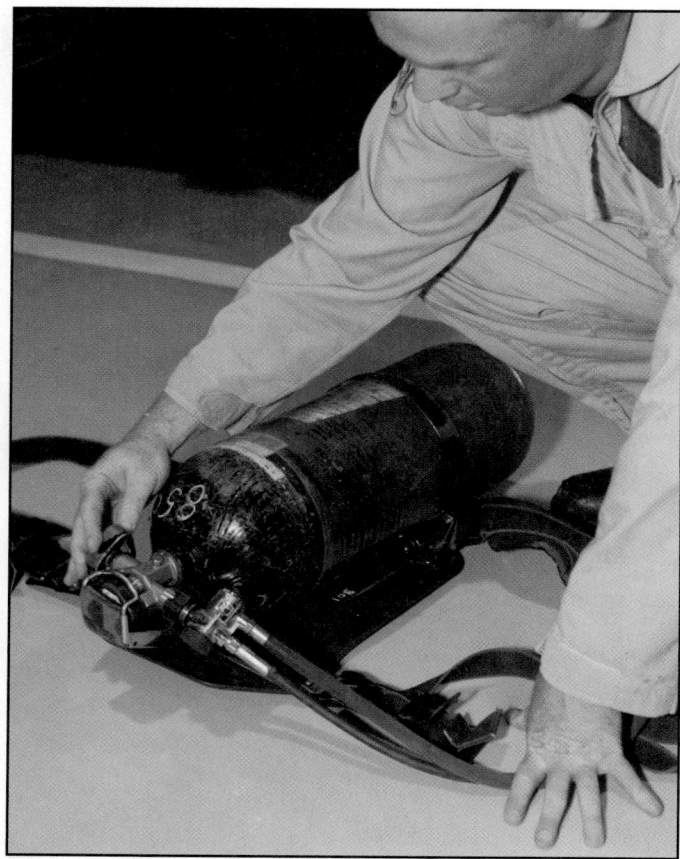

Figure 6.45 Confirm the fill level and proper function of an air cylinder.

Figure 6.46 A remote pressure gauge should be confirmed against the cylinder gauge and read within recommended parameters.

- Check the remote gauge and cylinder gauge to ensure that they read within the manufacturer's recommended limits (**Figure 6.46**).

- Check the harness assembly and facepiece to ensure that all straps are fully extended.

- Operate all valves to ensure that they function properly and are left in the correct position.

- Test the low-pressure alarm.

- Test the PASS device to ensure that it is working.

- Check all battery-powered functions.

Donning an Unmounted SCBA

SCBA stored in cases can be donned using the over-the-head method and the coat method. In both methods, the SCBA must be positioned on the ground in front of the firefighter (either in or out of the case) with all straps extended, ready to don. The steps for donning structural PPE are described in **Skill Sheet 6-I-1**. The steps for donning SCBA using the over-the-head method are described in **Skill Sheet 6-I-2**, and the steps for the coat method are described in **Skill Sheet 6-I-3**.

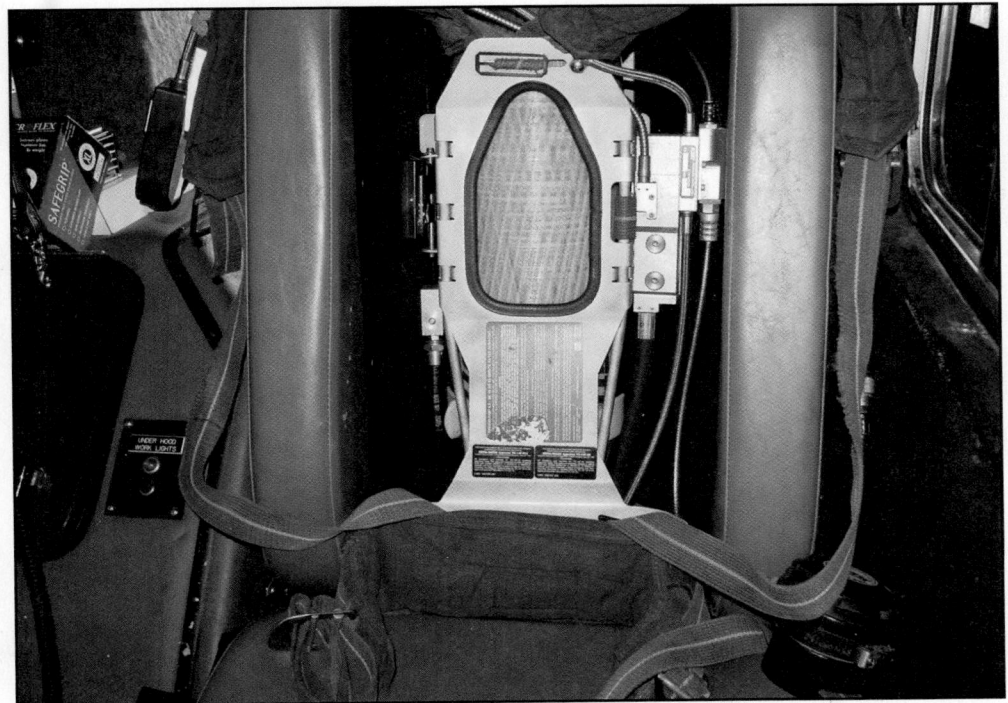

Figure 6.47 Seat-mounted SCBA allow firefighters to don gear while en route. *Courtesy of Kenneth Baum.*

Donning from a Seat Mount

Seat-mounted SCBA permit firefighters to don the unit while seated in the apparatus. But this method should be used only if the firefighter can do so without removing the apparatus seat belt **(Figure 6.47)**.

NFPA® 1901, *Standard for Automotive Fire Apparatus*, requires that the SCBA be held in place by a mechanical latching device. There are a wide variety of mounting brackets available that meet this requirement. The facepiece should be stored in a drawstring or other quick-opening bag, or in a pouch on your protective coat. This will keep it clean and protect it from dust and scratches. (**NOTE:** Do not keep the facepiece connected to the regulator during storage. These parts must be kept separate, in order to check for proper facepiece seal.) The steps for donning SCBA from a seat mount are described in **Skill Sheet 6-I-4.**

CAUTION
Never connect the regulator and breathe cylinder air when seated in the apparatus. This activity will deplete your air supply before you arrive at the incident.

WARNING
Never stand to don SCBA while the apparatus is moving. Standing places both you and other firefighters in danger of serious injury in the event of a fall. NFPA® 1500 requires firefighters to remain seated and belted at all times while the apparatus is in motion.

Figure 6.48 Externally-mounted SCBA units contained in protective covers can be donned quickly. *Courtesy of Ron Bogardus.*

The air cylinder's position in the seat back should match the proper wearing position for the firefighter. The visible seat-mounted SCBA reminds and even encourages personnel to check the equipment more frequently. Because it is exposed, safety checks can also be made more conveniently.

When exiting the apparatus, do so carefully because the extra weight of the SCBA on your back can make slips and falls more likely. Be sure to adjust all straps for a snug and comfortable fit.

Donning from a Side or Rear External Mount

Side- or rear-mounted SCBA are mounted on the exterior of the apparatus (**Figure 6.48**). Although this type of mount does not permit donning en route and requires more time for donning than seat-mounted SCBA, it reduces the chances of slips and falls. It is also faster than donning an SCBA stored in a carrying case, because you do not have to remove the case from the apparatus, place it on the ground, open the case, and pick up the unit. One potential disadvantage of exterior mounting is that the SCBA can be exposed to weather and other physical hazards. However, waterproof covers will minimize the risk of damage.

If SCBA are mounted at the correct height, firefighters can don them with little effort. Having the mount near the running boards or tailboard allows the firefighter to don the equipment while sitting on the running board or tailboard. The donning steps are similar to those for seat-mounted SCBA.

Donning from a Backup Mount

Backup mounts located inside a compartment are protected from the weather and provide the same advantages as side- or rear-mounted equipment. However, some compartment doors may interfere with a firefighter donning SCBA. Other compartments may be located too high on the apparatus, making donning more difficult.

One type of compartment mount has a telescoping frame that extends the equipment outward (**Figure 6.49**). Some of these also telescope upward or downward so that a standing firefighter can don the SCBA more quickly.

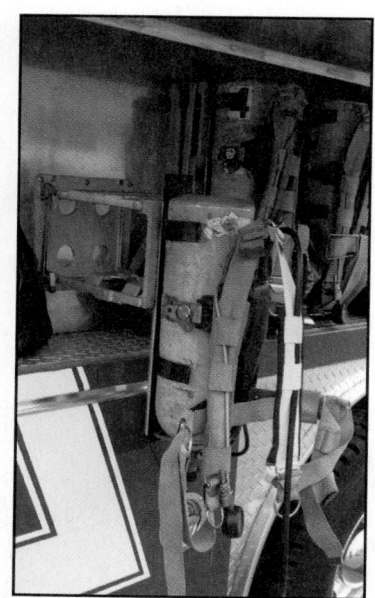

Figure 6.49 SCBA may be mounted on a frame that moves the equipment from an inaccessible storage area to the proper height for easy donning.

The backup mount provides quick access to SCBA (some high-mounted SCBA must be removed from the vehicle and donned using the over-the-head or coat method). The procedure for donning SCBA using the backup method is similar to the method used for seat-mounted SCBA.

Figure 6.50 By pulling facepiece straps simultaneously, a snug fit can be maintained.

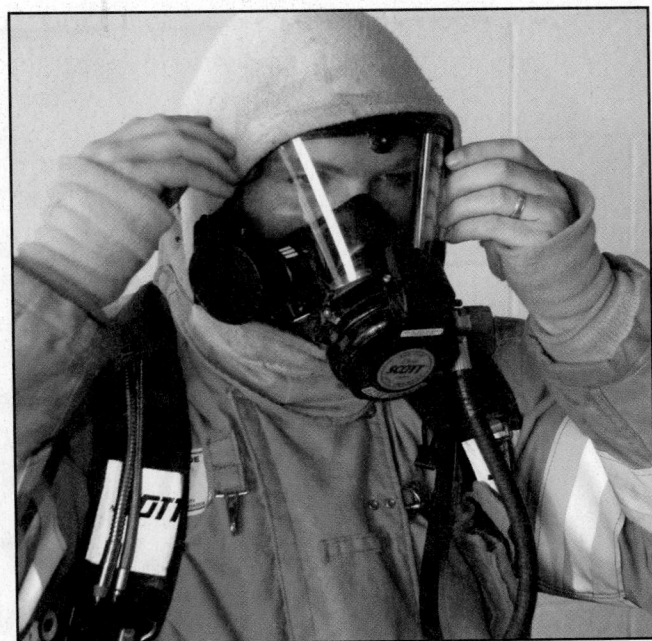

Figure 6.51 The interface between the hood and facepiece should reinforce the unimpaired contact between skin and the facepiece seals.

Donning the Facepiece

Most SCBA facepieces are donned using very similar steps. One important difference is that some facepieces use a rubber harness, while others use a mesh skullcap. Both have adjustable straps, although some models have more straps than others.

Another difference is the location of the regulator. It may be attached to the facepiece or mounted on the waist belt or shoulder harness. The shape and size of facepiece lenses may also differ. Despite these differences, donning procedures for facepieces are essentially the same.

The following are general considerations for donning all SCBA facepieces:

- All straps should be fully extended.

- No hair should come between the skin and the facepiece sealing surface.

- The chin should be centered in the chin cup and the harness centered at the rear of the head.

- Facepiece straps should be tightened by pulling opposing straps evenly and simultaneously to the rear (**Figure 6.50**). Pulling the straps outward, to the sides, may damage them and prevent proper engagement with the adjusting buckles. Tighten the lower straps first, then the temple straps, and finally the top strap if there is one.

- Always check that the facepiece is completely sealed to the face, the exhalation valve is functioning, and all connections are secure. If there is a donning mode switch, check that it is in the proper position.

- The protective hood must be worn over the facepiece harness or straps. All exposed skin must be covered and vision must not be obscured. No portion of the hood should be located between the facepiece and the face (**Figure 6.51**).

- The helmet should be worn with the chin strap secured. Helmets equipped with a ratchet adjustment should be adjusted so that the helmet fits properly (**Figure 6.52**).

Figure 6.52 A ratchet screw on a helmet allows a firefighter to fine tune the fit of the helmet.

Figure 6.53 After removing a facepiece, extending all straps will help make it ready for the next use.

Doffing Protective Breathing Apparatus

Doffing (removal) techniques differ for the different types of SCBA. However, the following actions apply when doffing any brand or model:

- Make sure you are out of the contaminated area and that the SCBA is no longer required.

- Discontinue the flow of air from the regulator to the facepiece.

- Disconnect the regulator from the facepiece.

- Remove the protective hood, or pull it down around your neck.

- Remove the facepiece by loosening the straps and lifting it from your chin.

- Remove the backpack assembly while protecting the regulator.

- Close the cylinder valve.

- Relieve pressure from the regulator according to the manufacturer's instructions.

- Turn off the PASS device.

- Extend all facepiece and harness straps (**Figure 6.53**).

- Check air pressure to determine if the air cylinder needs to be refilled or replaced.

- Clean and disinfect the facepiece.

- Clean the SCBA backplate and harness if necessary.

- Secure the complete unit in its case, seat bracket, or storage bracket.

The steps for doffing personal protective equipment and SCBA and preparing them for use are described in **Skill Sheet 6-I-5**.

Inspection and Maintenance of Protective Breathing Apparatus

The frequency of SCBA inspections is established by NFPA® 1852. Firefighters inspect their SCBA on a schedule determined by their department and based on NFPA®, OSHA, and the manufacturer's requirements. Inspections are typically performed daily, weekly, or whenever firefighters report for duty. However, the period between inspections must not exceed one week. Qualified SCBA repair technicians must inspect the units annually and after any repairs have been completed.

Protective Breathing Apparatus Inspections and Care

Your SCBA requires ongoing inspection and maintenance to protect you properly. You must clean and inspect it after each use, at the beginning of every duty shift, and every week. If repairs are necessary, report this immediately. Extensive repairs and cleaning may require that the unit be taken out of service and replaced by a reserve unit. The steps for inspecting an SCBA are described in **Skill Sheet 6-I-6**.

Inspections

The daily/weekly inspection is divided according to the SCBA components and includes the following:

- **Facepiece** — Perform the following:
  - Inspect the facepiece frame for deterioration, dirt, cracks, tears, and holes
  - Inspect head-harness buckles, straps, and webbing for wear, breaks, or loss of elasticity
  - Inspect the lens for scratches, abrasions, holes, cracks, or heat damage
  - Inspect the HUD for proper operation
  - Inspect the lens for a proper seal with the facepiece frame
  - Inspect the valve seat of the exhalation valve
  - Inspect the springs and covers to ensure cleanliness and ease of operation
  - Inspect the regulator and hose connection points for cleanliness, damage, and proper operation (**Figure 6.54, p. 304**)
  - Inspect the speaking diaphragm for cleanliness and damage

- **Backplate and harness assembly** — Perform the following:
  - Inspect the harness straps and backplate for abrasions, cuts, tears, or heat or chemical-induced damage (**Figure 6.55, p. 304**)
  - Ensure all the buckles, fasteners, and adjustments operate properly
  - Ensure the harness straps are fully extended
  - Inspect the cylinder retention system for proper operation and damage
  - Ensure the cylinder is securely attached to the backplate

- **Breathing air cylinder assembly** — Perform the following:
  - Ensure the cylinder **hydrostatic test** date is current (**Figure 6.56, p. 304**)
  - Inspect the cylinder gauge for cleanliness and damage

Hydrostatic Test — Testing method that uses water under pressure to check the integrity of pressure vessels.

Figure 6.54 A facepiece regulator should be inspected for cleanliness, possible damage, and proper operation. *Courtesy of Kenneth Baum.*

Figure 6.55 Harness straps and backplate may show signs of damage through abrasion or searing. *Courtesy of Kenneth Baum.*

Figure 6.56 The hydrostatic test date indicates when the cylinder had its pressure valves confirmed. *Courtesy of Kenneth Baum.*

— Inspect the cylinder body for cracks, dents, weakened areas, and heat or chemical-induced damage

— Inspect composite cylinders for cuts, gouges, loose fibers, and missing resin material

— Inspect the cylinder valve outlet sealing surface and threads for damage

— Check the valve hand wheel for damage, proper alignment, serviceability, and secure attachment

— Check the burst disc outlet area for debris

— Check the cylinder to ensure that it is full

• **Hoses** — Perform the following:

— Inspect the high- and low-pressure hoses for abrasions, bubbling, cuts, cracks, and heat and chemical-induced damage

— Inspect the hose fittings for cleanliness and damage

— Visually check the high-pressure hose to cylinder "O" ring

— Test the hose connections for tightness

- **Low-pressure alarm** — Perform the following:
  - Inspect the low-pressure alarm and mounting hardware for cleanliness, proper attachment, and damage
  - Test the alarm for proper activation and operation

- **Regulator** — Perform the following:
  - Inspect the regulator controls and pressure relief devices for cleanliness, proper operation, and damage
  - Inspect the housing and components for cleanliness and damage
  - Check the regulator for any unusual sounds during operation, such as whistling, chattering, clicking, or rattling
  - Check the mainline and bypass valve for proper function

- **Pressure indicator gauges** — Perform the following:
  - Inspect the remote pressure indicator gauge for cleanliness and damage
  - Ensure that the pressure readings on the cylinder pressure gauge and remote gauge are within the manufacturer's recommended limits

- **Integrated PASS** — Perform the following:
  - Inspect the PASS device for cleanliness, wear, and damage
  - Ensure all parts are securely attached to the PASS device
  - Test all operating modes for proper operation
  - Test the low battery warning signal for proper operation

Care

You should clean and sanitize your SCBA facepiece immediately after each use to prevent debris from collecting in the exhalation valve and regulator fitting. Dirt or other foreign materials can cause the exhalation valve to malfunction and allow air from the tank to escape, quickly depleting the air supply. Debris can also prevent the regulator from fitting securely to the facepiece. Soot on the facepiece lens can reduce visibility.

Wash the facepiece thoroughly with warm water containing a mild commercial disinfectant and then rinse with clear, warm water (**Figure 6.57**). Take special care to ensure proper operation of the exhalation valve. Dry the facepiece with a lint-free cloth or air dry it. Do not use paper towels to dry the facepiece because they will scratch the facepiece lens. Although facepieces are impact resistant, they scratch easily. Regulators

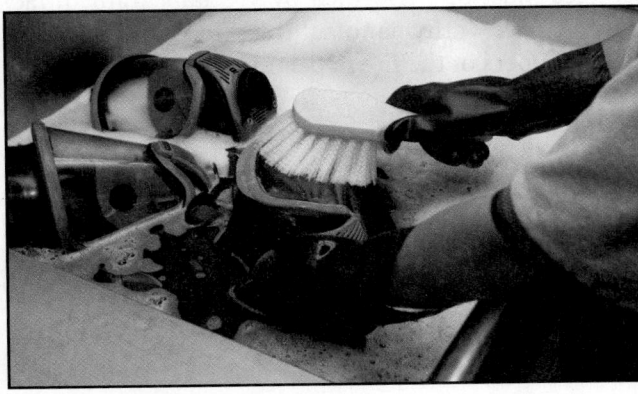

Figure 6.57 Firefighters are responsible for cleaning and maintaining their assigned PPE and reporting repair needs as they occur.

Figure 6.58 Equipment technicians may serve in dual roles, for example as EMTs and maintenance contractors. *Courtesy of Kenneth Baum.*

and low-pressure hoses must not be submerged in water for cleaning. Sanitize the facepiece seal and interior of the facepiece to prevent you from inhaling or coming in contact with contaminants. This is especially important if you share the facepiece with other personnel. As with all maintenance, you should refer to the manufacturer's literature for instructions on proper care.

Facepiece lenses can also fog up internally due to the difference between the inside and outside temperatures. To prevent fogging, some SCBA facepieces are permanently treated with an antifogging chemical. Special antifogging chemicals recommended by the manufacturer can also be applied to the facepiece lens following cleaning.

Many departments issue firefighters with individual facepieces that are not shared with other firefighters. This practice eliminates the risk of spreading germs from one wearer to the next and ensures that the mask is the correct size, providing a complete seal. Even though firefighters have their own assigned facepieces, it is still important that they are cleaned after each use. Keeping facepieces clean and sanitized prevents hydrocarbons from contaminating your skin. When the facepiece is clean and dry it can be stored in a case, a bag, or coat pocket. Regardless of where the facepiece is stored, the straps should be left fully extended to facilitate donning. The procedures for cleaning an SCBA are given in **Skill Sheet 6-I-7.**

Annual Inspection and Maintenance

Annual inspection and maintenance must be performed by specially trained, factory-qualified technicians, in accordance with manufacturer's recommendations. These technicians may be trained members of the fire department or employees of a certified maintenance contractor. **(Figure 6.58)**

SCBA Air Cylinder Hydrostatic Testing

SCBA breathing air cylinders must be stamped or labeled with the date of manufacture and the date of the last hydrostatic test. According to both the U.S. Department of Transportation (49 CFR 180.205) and Transport Canada, the type of material used to construct the cylinder determines the frequency of hydrostatic testing. The testing frequency and life span of cylinders are as follows:

- Steel and aluminum cylinders are tested every five years and have an indefinite service life until they fail a hydrostatic test.

- Hoop-wrapped aluminum cylinders are tested every three years and have a 15-year service life.

- Fully wrapped fiberglass cylinders are tested every three years and have a 15-year service life.

- Fully wrapped Kevlar cylinders are tested every three years and have a 15-year service life.

- Fully wrapped carbon fiber cylinders are tested every five years and have a 15-year service life.

Refilling SCBA Cylinders

Three breathing air sources may be used to refill depleted SCBA air cylinders:

- Stationary fill systems
- Mobile fill systems
- Firefighter Breathing Air Replenishment Systems (FBARS): systems installed in high-rise buildings

Each source must provide Type 1 Grade D quality air to your cylinder, as specified by OSHA and Canadian government requirements. Regardless of the breathing air source, the following safety precautions apply when refilling an SCBA cylinder:

- Check the hydrostatic test date of the cylinder.
- Visually inspect the cylinder for damage.
- Don eye and hearing protection.
- Place the cylinder in a shielded fill station.
- Fill the cylinder slowly to prevent it from overheating.
- Ensure that the cylinder is completely full but not overpressurized.

Skill Sheet 6-I-8 provides a sample procedure for filling an SCBA cylinder from a cascade system. **Skill Sheet 6-I-9** provides a sample procedure for filling an SCBA cylinder from a compressor/purifier.

Filling unshielded cylinders while a firefighter is wearing the SCBA is prohibited. However, a rapid intervention crew or team (RIC/RIT) rescuing a trapped or incapacitated firefighter may be granted an exception to this rule. Even then, the following three criteria must be met before filling a worn SCBA:

- NIOSH-approved RIC Universal Air Connection (UAC) fill options are used.
- A risk assessment has been conducted to limit safety hazards and ensure that necessary equipment is fully operational.
- There is an imminent threat to the safety of the downed firefighter, and immediate action is required to prevent loss of life or serious injury.

Stationary Fill Stations

SCBA breathing air cylinders are filled either from a **cascade system** or directly from a compressor air purification system **(Figures 6.59a and b, p. 308-309)**. Cascade cylinders must meet American Society of Mechanical Engineers (ASME) design requirements. Both systems must be connected to a fill station that holds the SCBA

> **Cascade System** — Three or more large, interconnected air cylinders, from which smaller SCBA cylinders are recharged; the larger cylinders typically have a capacity of 300 cubic feet (8 490 L).

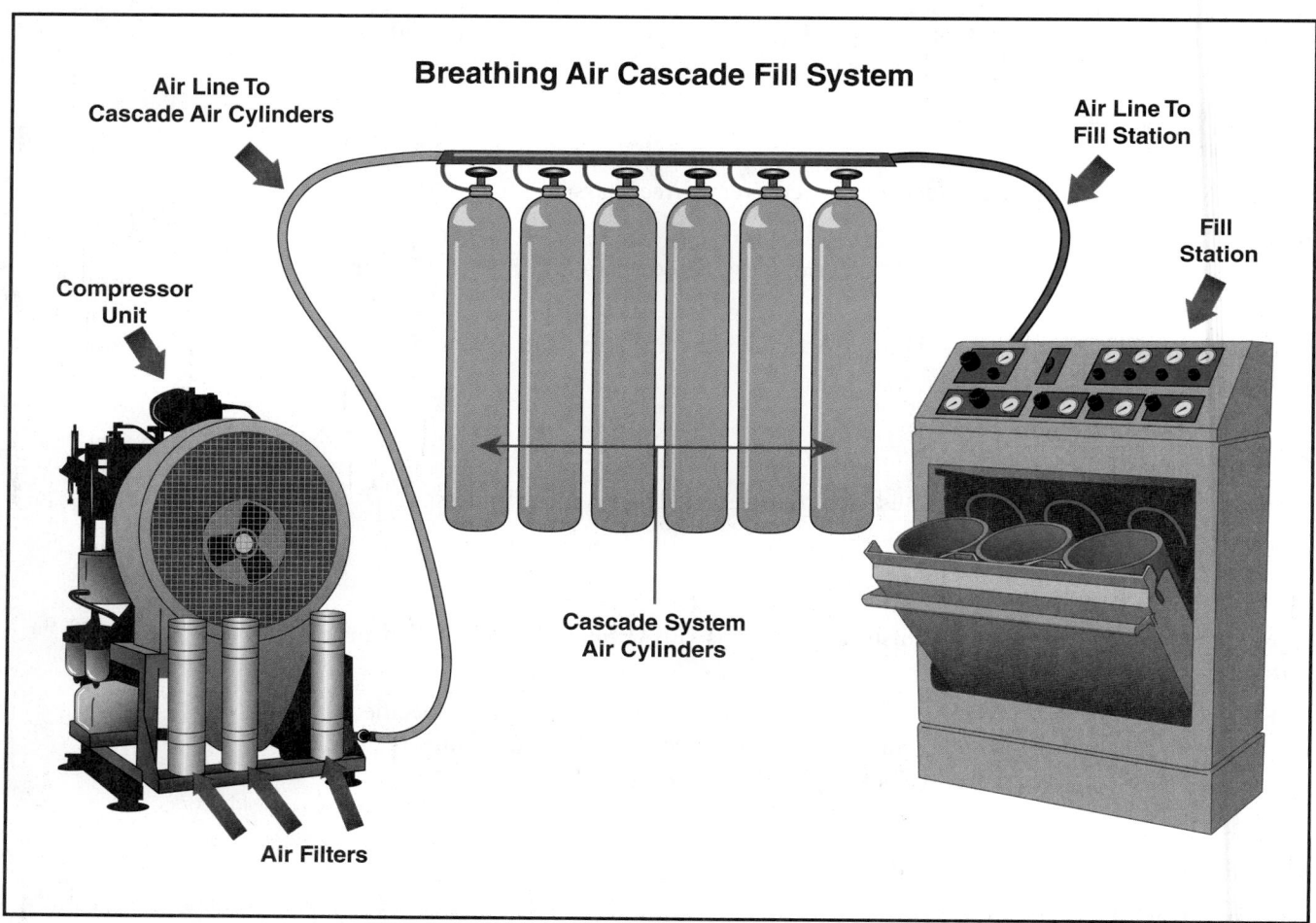

Breathing Air Cascade Fill System

Air Line To
Cascade Air Cylinders

Air Line To
Fill Station

Fill
Station

Compressor
Unit

Cascade System
Air Cylinders

Air Filters

Figure 6.59a A cascade fill system uses three or more large cylinders to refill smaller SCBA cylinders.

cylinders in rupture-proof sleeves during the filling process. If your department fills its own SCBA cylinders, you will be trained how to safely use the system. Some departments do not fill their own cylinders and contract the process out to a qualified breathing air supplier.

Filling procedures should be posted on the fill station and must follow the fill station manufacturer's recommendations to avoid excessive overheating in the cylinder. No matter how the cylinders are filled, the following safety precautions apply:

- Only trained personnel should operate the fill equipment.
- Cylinders must be inspected before filling.
- Hearing and eye protection must be worn during fill operations.
- Cylinders must be placed in the shielded fill station.
- Cylinders must be filled slowly to prevent overheating.
- Cylinders must be filled to their capacity.
- Cylinders must not be overfilled.

The breathing air quality must be regularly tested by a third-party testing facility, and the testing results must be documented. The department's health and safety officer is usually responsible for monitoring the testing and maintaining the documentation.

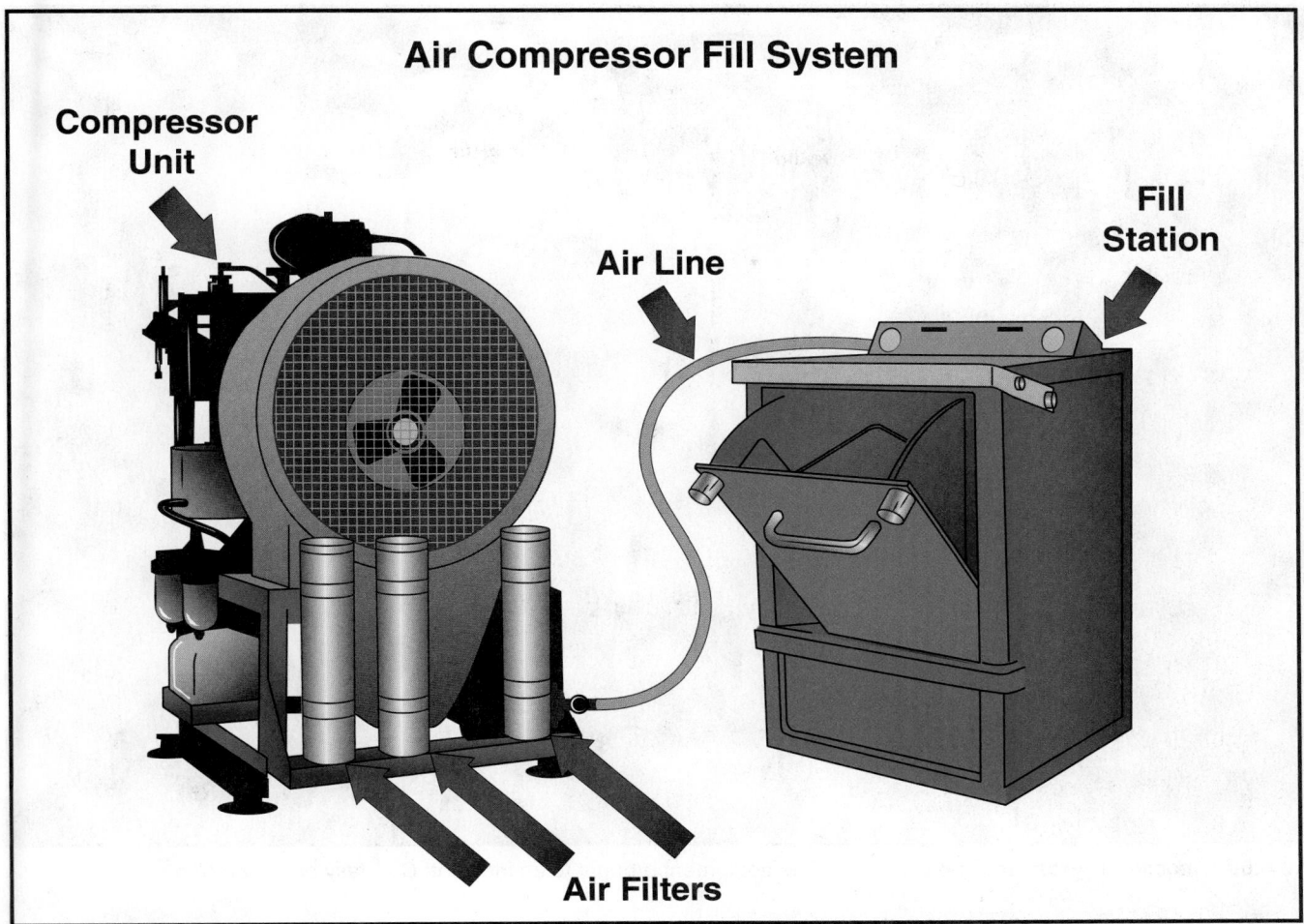

Air Compressor Fill System

Compressor Unit

Air Line

Fill Station

Air Filters

Figure 6.59b A compressor fill system uses an air compressor to generate the pressure necessary to refill cylinders.

Mobile Fill Systems

Mobile breathing air fill systems are designed to refill air cylinders at emergency incidents. They typically consist of a fill station equipped with a breathing-air compressor or cascade fill station and are mounted on a trailer or the apparatus chassis (**Figure 6.60, p. 310**). Cascade cylinders must meet U.S. Department of Transportation (DOT) or Transport Canada specifications.

The operation of the system will be similar to the stationary system previously mentioned. The system may also be designed to support a SAR system or a firefighter breathing air replenishment system (FBARS) installed in a high-rise structure, as described in the next section.

Firefighter Breathing Air Replenishment Systems (FBARS)

In 1988, the Los Angeles City Fire Department responded to a major high-rise fire. Although they eventually brought it under control, to do so they had to carry over 600 replacement SCBA cylinders to the top floors of a 62-story building. Postincident analysis emphasized the fact that this was highly inefficient and a severe strain on personnel. To fight high-rise fires effectively, departments needed a better way to refill SCBA cylinders on upper floors.

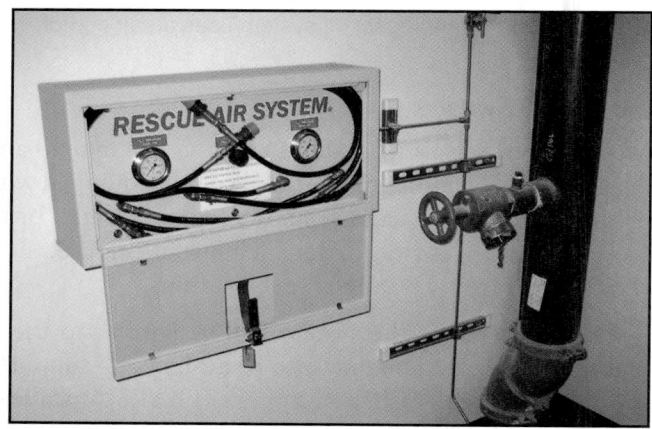

Figure 6.60 A mobile fill system may be included in the equipment brought to an incident. *Courtesy of James Nilo.*

Figure 6.61 An installed breathing air system is required in new construction high-rise buildings. *Courtesy of Brandon Wagoner.*

In response to this problem, many municipalities now require that all newly constructed buildings taller than 75 feet (22.86 m) install Firefighter Breathing Air Replenishment Systems (FBARS) **(Figure 6.61)**. FBARS provide an endless source of breathing air to any floor within the structure from a ground level connection. These systems typically consist of:

- A fire department air connection panel, containing connection fittings, control valves, and gauges, located on the exterior of the structure

- An Emergency Air Storage (EAS) system that provides breathing air if a mobile system is not available to supply the external connection

- Remote air fill panels, containing a certified rupture-proof containment fill station, connection and control valves, and gauges, located in protected stairwells on designated floors

- Interconnected piping certified to carry breathing air under pressure throughout the system

- Low air pressure monitoring switches and alarms, used to maintain a minimum air pressure and warn of pressure loss or system failure

Although not all high-rise buildings have these systems, their installation and use is increasing. In any jurisdiction with tall buildings, firefighters should know which structures have these systems and know how to use them.

Replacing SCBA Cylinders

SCBA backplates are designed for easy removal and replacement of breathing air cylinders. You may be required to replace the breathing air cylinder under the following circumstances:

- During the daily/weekly inspection, if the cylinder contains less than 90 percent of its capacity

- During training exercises

- During long-duration emergency operations

- After any emergency operations

Changing SCBA breathing air cylinders can be either a one- or two-person task. **Skill Sheet 6-I-10** describes the one-person method for changing an air cylinder (**Figure 6.62**). When there are two people, the firefighter with an empty cylinder simply positions the cylinder so that it can be easily changed by the other firefighter. **Skill Sheet 6-I-11** describes the two-person method for changing an air cylinder.

Empty cylinders should be kept separate from full cylinders that have been serviced and are ready for use. Damaged cylinders should be clearly marked and kept separate from both empty and full cylinders.

Figure 6.62 Breathing air cylinders must be replaced at regular intervals and under specific circumstances.

Figure 6.63 Activities that require interaction with a contaminated environment require the use of SCBA.

Using Respiratory Protection Equipment

Properly worn, maintained, and inspected respiratory protection equipment will prevent you from being exposed to airborne hazards. But to use it effectively, firefighters must also be aware of additional safety concerns, such as fatigue, regulating air supply, and proper exit procedures. The following sections describe the safety precautions to be used when wearing SCBA and explain how to properly exit an IDLH atmosphere.

Safety Precautions for SCBA Use

When using SCBA, you should follow these safety precautions:

- Only enter an IDLH atmosphere if you are certified to wear SCBA and have been properly fit tested for the facepiece.

- Closely monitor how you feel while wearing the SCBA. If you become fatigued, notify your supervisor and take a rest before returning to work.

- Remember that your air supply duration can vary, depending on:
  - Air cylinder size and beginning pressure
  - Your physical conditioning
  - The task being performed
  - Your level of training
  - The operational environment
  - Your level of stress
- After entering an IDLH atmosphere, keep your SCBA on and activated until you leave the contaminated area. Improved visibility does not ensure that the area is free of contamination **(Figure 6.63)**. Before you remove your SCBA, the atmosphere must be tested with properly calibrated instruments and found to be safe.
- In any IDLH atmosphere, work in teams of two or more. Team members must remain in physical, voice, or visual contact with each other while in the hazardous area. Radio contact is not sufficient! If available, a thermal imager (TI) can help maintain contact.
- While in the IDLH atmosphere, check your air supply status frequently.
- Exit the IDLH atmosphere before the low air alarm activates to avoid using the reserve air supply.

Exit Indicators and Techniques

You should always be prepared to make a rapid exit or withdrawal. The most common exit procedures are those used at the majority of incidents, typically referred to as nonemergency exit procedures. Less common, but far more important, are emergency exit procedures. These are used in life-threatening situations such as SCBA failures and catastrophic changes during the incident.

Proper exit techniques must be practiced during training and followed during emergency incidents. Nonemergency exit procedures are covered in the following sections, while emergency exit procedures are covered in the firefighter survival section of Chapter 9, Building Search and Victim Removal.

Exit Indicators

There are many circumstances in which firefighters must exit contaminated or hazardous areas. Situations or events that signal the need for exit are called *exit indicators*.

Nonemergency exit indicators occur when:

- The situation is stabilized.
- There is a change in operational strategy.
- It is necessary to replace an air cylinder.
- The Incident Commander (IC) orders a nonemergency withdrawal.
- The assignment is completed.

Emergency exit indicators include:

- Activation of SCBA low-pressure low air alarm
- SCBA failure
- Withdrawal orders issued by the IC or Safety Officer
- Presence of APR/PAPR breakthrough symptoms

- Activation of APR/PAPR end-of-service-life indicators
- Change in concentration of respiratory hazards
- Attaining or exceeding the **permissible exposure limit (PEL)** for a hazardous material while wearing an APR/PAPR
- Changes in environmental conditions, such as temperature, wind direction and speed, and water level and speed, either within or around the site of the incident
- Changes in oxygen level
- Changes in temperature
- Indications of new hazards

At any incident, the IC is responsible for constantly monitoring the environment. When monitoring reveals a potential hazard, such as chemical concentrations that approach the PEL, the IC issues orders to change the required level of respiratory protection or withdraw from the area completely.

One environmental change that firefighters may detect on their own is a change in oxygen level. Oxygen deficiency causes light-headedness, disorientation, loss of coordination, increased breathing rates, and rapid fatigue. If you experience any of these symptoms, report this by radio and evacuate the area immediately. APR canisters or cartridges that are exposed to increased levels of oxygen (above 23 percent) may also have unknown, adverse reactions that require you to evacuate.

Nonemergency Exit Techniques

Nonemergency exit techniques are based on the Incident Command System (ICS) and the accountability requirements of NFPA® 1500. You must be trained in nonemergency exit techniques including the buddy-system, controlled breathing, entry/egress paths, and accountability systems.

Buddy system. In all hazardous atmospheres or situations, firefighters adhere to the buddy system by working in teams of at least two members (**Figure 6.64**). Each team member is responsible for the safety of the other member. At the first sign of any exit indicator (such as orders, low air alarm, or change in conditions), team members

Figure 6.64 All firefighters who will contact a hazardous atmosphere or situation must work in conjunction with a buddy.

must leave as a group or in pairs. Individual members must never be left alone in the IDLH atmosphere. The only time one member may work alone is in a confined space where two members cannot fit. In this instance, the second team member remains outside the area monitoring the **search line**, ready to enter the space if the need for rescue arises.

Search Line — Nonload-bearing rope that is anchored to a safe, exterior location and attached to a firefighter during search operations to act as a safety line.

Controlled breathing. This technique allows for efficient air use in an IDLH atmosphere. Firefighters inhale naturally, through the nose, then forcefully exhale through the mouth, reducing air consumption. It is used primarily with SCBA, but can also be applied to APR/PAPRs, to extend the life of the filter and reduce the intake of toxins after a filter breakthrough.

Practice controlled breathing methods in training until they become second nature. Being conscious of breathing quickly slows breathing down, extending the life of your breathing air cylinder or APR/PAPR. Controlled breathing is an important exit technique because it reduces air consumption during the time required to exit.

Egress paths. When you exit an IDLH area, it is important to use the same path that you used to enter. This path will have familiar landmarks, and it may also be the most direct. This method is especially important in unfamiliar structures because it reduces the possibility that you will become lost or disoriented and allows you to calculate the time it will take to exit the area. Make a habit of this technique by practicing it during training.

However, you must be aware of other means of egress in case your entry route is blocked. One exit technique is to follow hoselines or search lines out of an area. Always apply your situational awareness as you arrive at the scene of an incident and prepare to enter the structure. Look for other possible exit points, note any potential obstructions, and observe the fire conditions visible at the time you enter the IDLH area.

Accountability systems. Accountability systems to keep track of personnel in an IDLH environment are required by NFPA® 1500. Individual departments have different accountability SOPS, but typically firefighters must check in with the accountability officer before entering the IDLH area. All personnel, their locations, and their functions are then noted on a tracking board or some other tracking system. When firefighters leave the IDLH area, they check out with the accountability officer so that they are not counted as missing **(Figure 6.65)**. Many firefighter fatalities can be attributed to improper use of the accountability system.

Figure 6.65 Accountability systems enable officers to confirm that each firefighter has left the IDLH atmosphere.

Chapter Summary

Firefighters are frequently exposed to dangerous environments in which personal protective equipment is necessary to maintain health and safety. Your PPE will protect you from hazards and minimize the risk of injury or fatality, but only if it is properly worn, cleaned, and maintained.

Respiratory protection equipment protects you from toxic gases and vapors, particulates, and disease. But like your PPE, it can only do so if it is properly used, inspected, cleaned, and maintained. In addition to performing all of these tasks, you must know how to choose the type of respiratory equipment that is appropriate for the hazardous atmosphere you face. You must also know how to manage your air supply so that you can exit an IDLH area before your low-pressure air alarm activates.

Review Questions

Firefighter I

1. What is the purpose of personal protective equipment?

2. Why are there differences in the characteristics of structural fire fighting protective clothing and wildland personal protective clothing?

3. What are some basic guidelines for the care of personal protective clothing?

4. What safety considerations do firefighters need to keep in mind when using personal protective equipment?

5. What common respiratory hazards do firefighters face?

6. How do atmosphere-supplying respirators differ from air-purifying respirators?

7. What are some of the limitations of respiratory protection equipment?

8. What should respiratory equipment be protected from during storage?

9. What general considerations need to be taken when donning and doffing protective breathing apparatus?

10. What are the general inspection and care considerations for protective breathing apparatus?

11. What kinds of safety precautions should be taken when refilling SCBA cylinders?

12. What methods can you use to replace an SCBA cylinder?

13. What are the safety precautions taken when using an SCBA?

14. What are common emergency and nonemergency exit indicators a firefighter may encounter during an incident?

15. What are some nonemergency exit techniques firefighters can use?

6-I-1
Demonstrate the method for donning structural
personal protective clothing for use at an emergency.

NOTE: Ensure ensemble is worn according to manufacturer's guidelines or local standard operating procedures.

Step 1: Don pants and boots, which includes suspenders in place.

Step 2: Don hood.

Step 3: Don coat with closure secure and collar up.

Step 4: Don helmet with eye protection on and chin strap in place and fastened.

Step 5: Don structural firefighter gloves.

 SKILL SHEETS

6-1-2

With structural personal protective clothing in place,
demonstrate the over-the-head method of donning an SCBA.

NOTE: The following are general procedures for donning an SCBA. The specific SCBA manufacturer's recommendations for donning and use of the SCBA should always be followed.

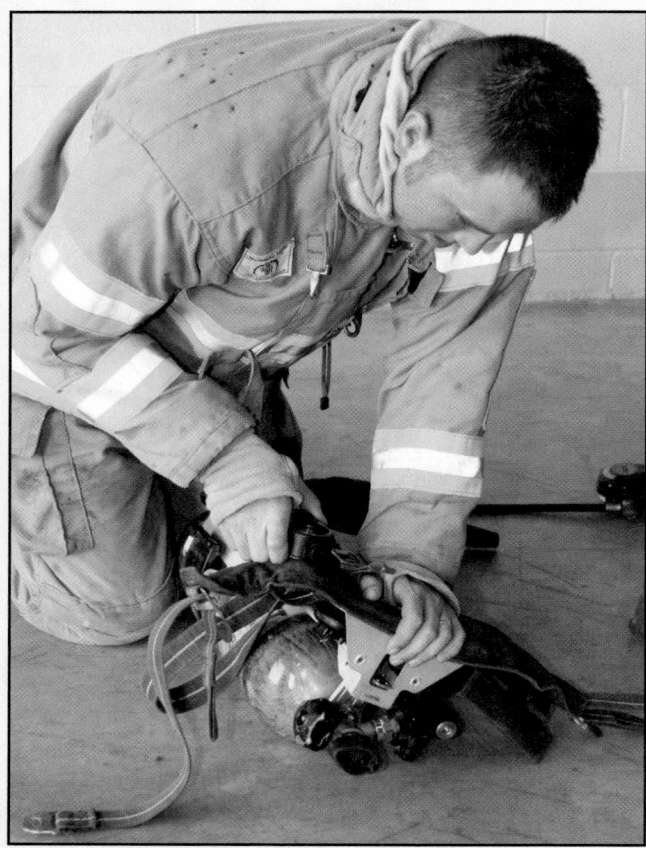

Step 1: Position the SCBA with the valve end of the cylinder away from the body, the cylinder down, and back frame up. All harness straps are fully extended and untangled.

Step 2: Open cylinder valve fully. Listen for the activation of the integrated PASS Alarm if equipped. Listen for the activation of the Low Air Alarm.

Step 3: Check cylinder and regulator pressure gauges. Pressure readings within 100 psi (700 kPa) OR needles on both pressure gauges indicate same pressure.

Step 4: Grab the back frame so that the shoulder straps will be outside of arms. Using proper lifting techniques, raise the SCBA overhead while guiding elbows into the loops formed by the shoulder straps.

Step 5: Release the harness assembly and allow the SCBA to slide down the back.

Step 6: Fasten chest strap, buckle waist strap, and adjust shoulder straps.

Step 7: Don facepiece over the head and securely tighten the straps, pulling the straps straight backwards, not out to the side.

6-1-2

With structural personal protective clothing in place, demonstrate the over-the-head method of donning an SCBA.

SKILL SHEETS

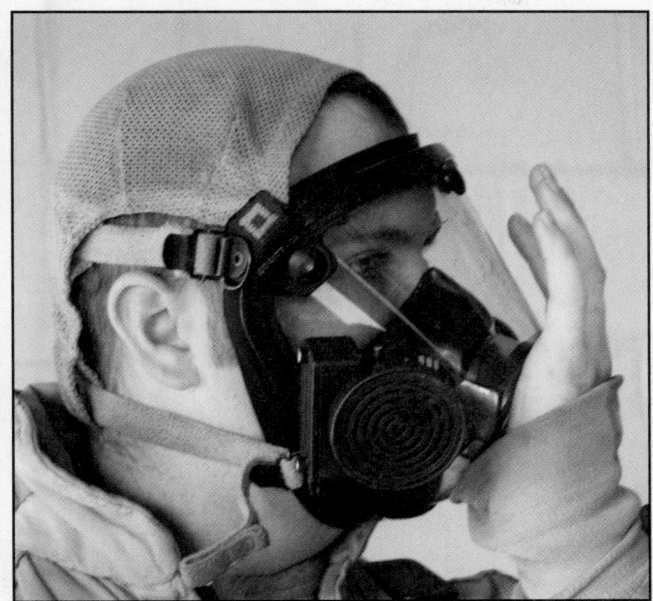

Step 8: After straps are tightened, test the facepiece for a proper seal and operation of the exhalation valve.

NOTE: Not all facepieces are designed for a seal check without the regulator being attached and activated.

Step 9: Don hood, ensure it covers all exposed skin.

Step 10: Connect air supply to facepiece.

Step 11: Activate external PASS device, if not equipped with integrated device.

Step 12: Don helmet, with chin strap secure and adjusted, and gloves.

SKILL SHEETS

6-1-3

With structural personal protective clothing in place, demonstrate the coat method of donning an SCBA.

NOTE: The following are general procedures for donning an SCBA. The specific SCBA manufacturer's recommendations for donning and use of the SCBA should always be followed.

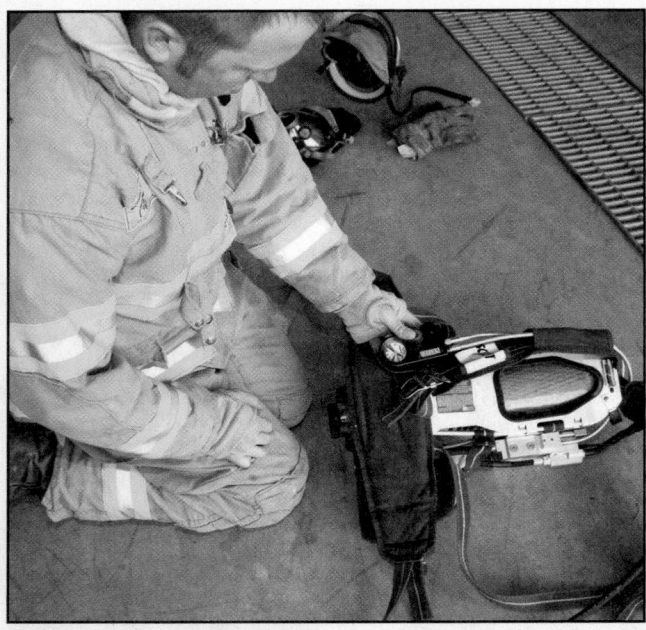

Step 1: Position SCBA with the valve end of the cylinder toward the body, the cylinder down, and back frame up. All harness straps are fully extended and untangled.

Step 2: Open cylinder valve fully. Listen for the activation of the integrated PASS Alarm if equipped. Listen for the activation of the Low Air Alarm.

Step 3: Check cylinder and regulator pressure gauges. Pressure readings within 100 psi (700 kPa) OR needles on both pressure gauges indicate same pressure.

Step 4: Grasp the top of the left shoulder strap on the SCBA with the left hand and raise the SCBA overhead.

Step 5: Guide left elbow through the loop formed by the left shoulder strap and swing SCBA around left shoulder.

Step 6: Guide right arm through the loop formed by the right shoulder strap allowing the SCBA to come to rest in proper position.

6-1-3
With structural personal protective clothing in place, demonstrate the coat method of donning an SCBA.

SKILL SHEETS

Step 7: Fasten chest strap, buckle waist strap, and adjust shoulder straps.

Step 8: Don facepiece over the head and securely tighten the straps, pulling the straps straight backwards, not out to the side.

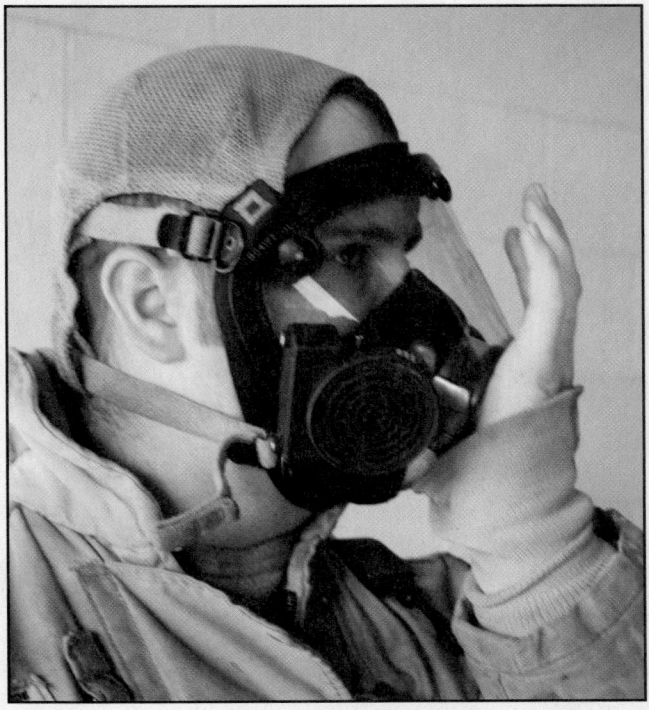

Step 9: After straps are tightened, test the facepiece for a proper seal and operation of the exhalation valve.

NOTE: Not all facepieces are designed for a seal check without the regulator being attached and activated.

Step 10: Don hood, ensure it covers all exposed skin.

Step 11: Connect air supply to facepiece.

Step 12: Activate external PASS device, if not equipped with integrated device.

Step 13: Don helmet, with chin strap secure and adjusted, and gloves.

6-I-4

With structural personal protective clothing in place,
demonstrate the method of donning an SCBA while seated.

NOTE: The following are general procedures for donning an SCBA. The specific SCBA manufacturer's recommendations for donning and use of the SCBA should always be followed.

Step 1: Position body in seat with back firmly against the SCBA and release the SCBA hold-down device.

Step 2: Insert arms through shoulder straps.

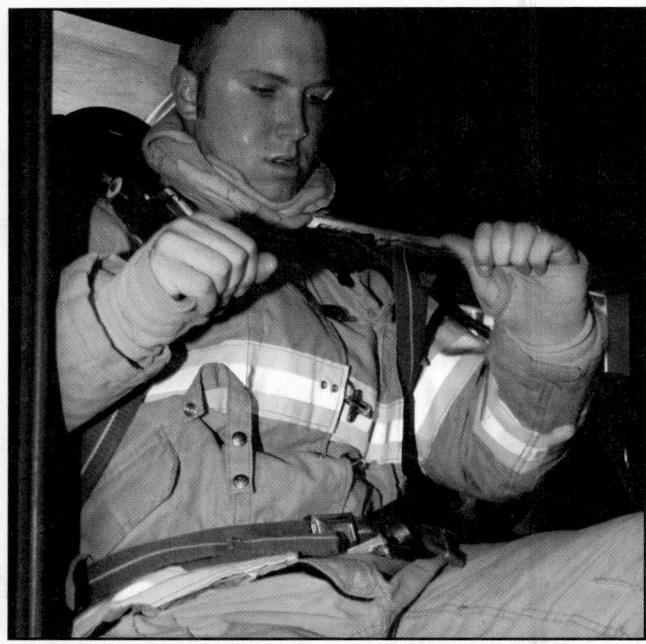

Step 3: Fasten chest strap, buckle waist strap, and adjust shoulder straps.

Step 4: Fasten seat belt before apparatus gets underway.

Step 5: Safely dismount apparatus, using appropriate situational awareness.

Step 6: Open cylinder valve fully.

Step 7: Check cylinder and regulator pressure gauges. Pressure readings within 100 psi (700 kPa) OR needles on both pressure gauges indicate same pressure.

Step 8: Don facepiece over the head and securely tighten the straps, pulling the straps straight backwards, not out to the side.

6-1-4

With structural personal protective clothing in place, demonstrate the method of donning an SCBA while seated.

SKILL SHEETS

Step 9: After straps are tightened, test the facepiece for a proper seal and operation of the exhalation valve.

NOTE: Not all facepieces are designed for a seal check without the regulator being attached and activated.

Step 10: Don hood, ensure it covers all exposed skin.

Step 11: Connect air supply to facepiece.

Step 12: Activate external PASS device, if not equipped with integrated device.

Step 13: Don helmet, with chin strap secure and adjusted, and gloves.

SKILL SHEETS

6-1-5
Doff personal protective equipment, including SCBA and prepare for reuse.

Doff SCBA

Step 5: Return all straps, valves and components back to ready state.

Step 6: Inspect SCBA and facepiece for damage and need for cleaning.

Step 7: Clean equipment as needed and remove damaged equipment from service and report to company officer, if applicable.

Step 8: Place SCBA back in storage area so that it is ready for immediate use.

Step 1: Remove SCBA.

Step 2: Close cylinder valve completely.

Step 3 Bleed air from high-and low-pressure hoses, listen for low air alarm activation.

Step 4: Check air cylinder pressure and replace or refill cylinder if less than 90% of rated capacity.

Doff PPE

Step 1: Remove protective clothing.

Step 2: Inspect PPE for damage and need for cleaning.

Step 3: Clean equipment as needed and remove damaged equipment from service and report to company officer, if applicable.

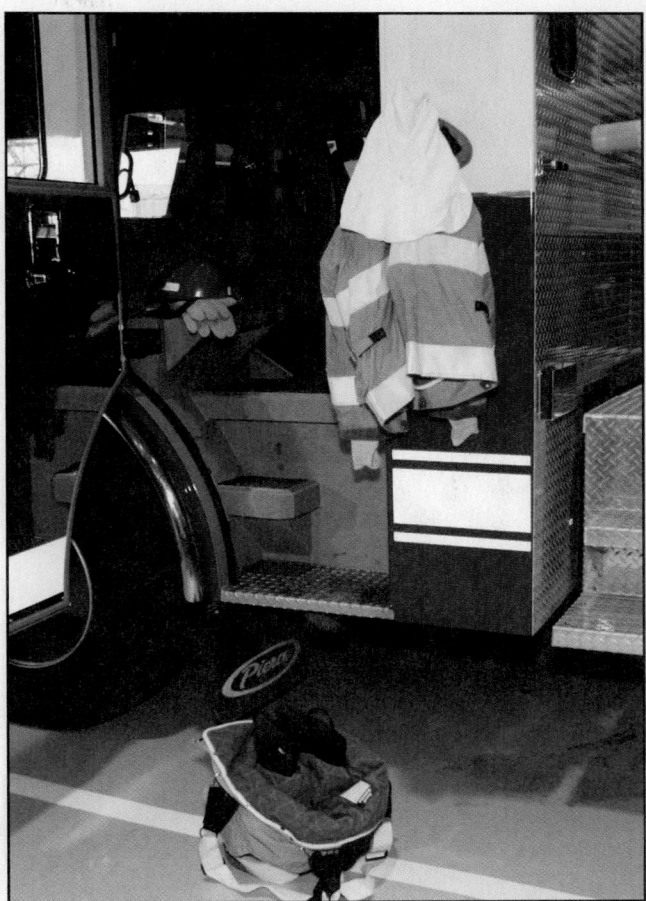

Step 4: Place clothing in a ready state.

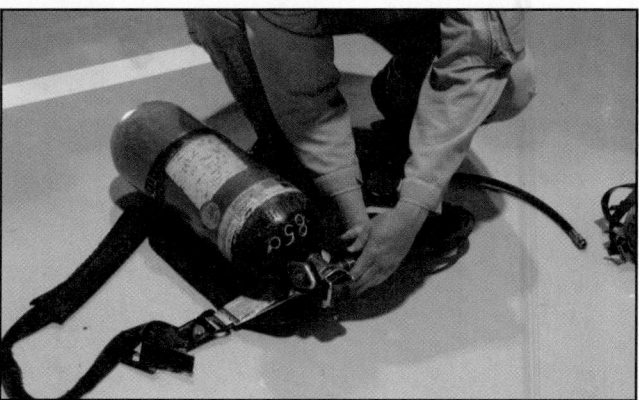

Step 7: Check that gauges and/or indicators (i.e. heads-up display) are providing similar pressure readings. Manufacturers' guidelines determine the acceptable range.

Step 8: Check the function of all modes of PASS device.

Step 1: Identify all components of SCBA are present: harness assembly, cylinder, facepiece, and PASS device.

Step 2: Inspect all components of SCBA for cleanliness and damage.

Step 3: Immediately clean dirty components if found. If damage is found, remove from service and report to company officer.

Step 4: Check that cylinder is full (90 -100 percent of capacity).

Step 5: Open the cylinder valve slowly; verify operation of the low-air alarm and the absence of audible air leaks.

NOTE: On some SCBA, the audible alarm does not sound when the cylinder valve is opened.

Step 6: If air leaks are detected, determine if connections need to be tightened or if valves, donning switch, etc. need to be adjusted. Otherwise SCBA with audible leaks due to malfunctions shall be removed from service, tagged, and reported to the company officer.

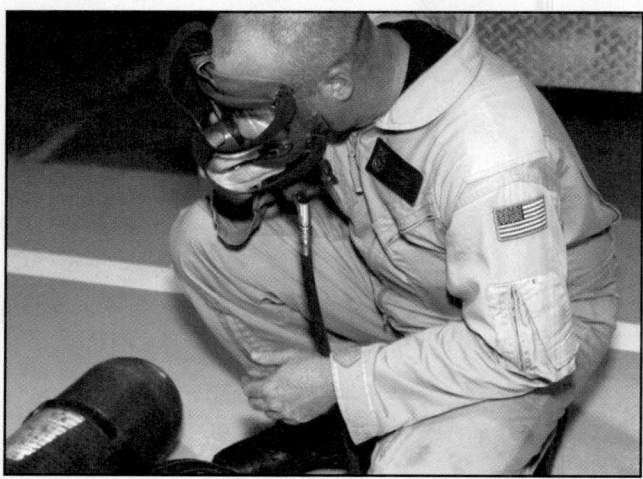

Step 9: Don facepiece over the head and securely tighten the straps, pulling the straps straight backwards, not out to the side.

Step 10: After straps are tightened, test the facepiece for a proper seal and operation of the exhalation valve.

NOTE: Not all facepieces are designed for a seal check without the regulator being attached and activated.

Step 11: Don regulator and check function by taking several normal breaths.

Step 12: Check bypass and/or purge valve, if applicable.

Step 13: Remove facepiece and prepare all components for immediate reuse.

Step 14: Place SCBA components so that they can be accessed quickly for donning in the event of a reported emergency.

6-I-7
Demonstrate the steps for cleaning and sanitizing an SCBA.

SKILL SHEETS

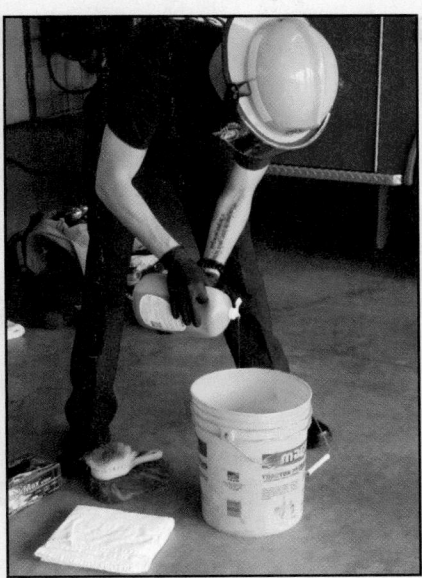

Step 1: Prepare cleaning solution, buckets, etc. according to manufacturer's guidelines and departmental policies.

Step 4: Place all components in a manner and location so that they will dry.

Step 2: Clean all components of SCBA unit according to manufacturer's guidelines and departmental policies.

Step 3: After equipment is clean, inspect for damage. If any damage is noted, report in accordance with local SOPs.

Step 5: Assemble components so they are in a state of readiness.

SKILL SHEETS

6-I-8

Demonstrate the method for filling an SCBA cylinder from a cascade system, wearing appropriate PPE, including eye and ear protection.

NOTE: This skill sheet is only an example. The procedures outlined here may not be applicable to your cascade system. Always check the manufacturer's instructions before attempting to fill any cylinders.

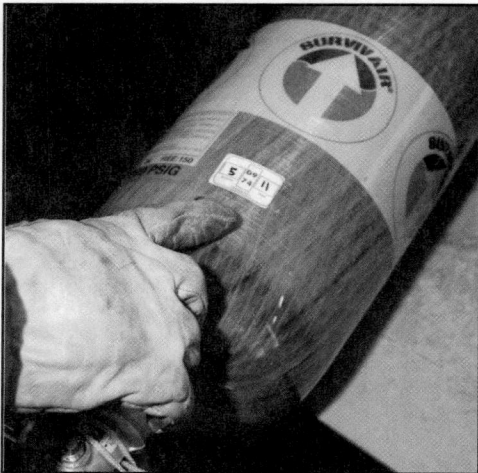

Step 1: Check the hydrostatic test date and recommended fill pressure of the cylinder.

Step 2: Inspect the SCBA cylinder for damage such as deep nicks, cuts, gouges, or discoloration from heat. If the cylinder is damaged or is out of hydrostatic test date, remove it from service and tag it for further inspection and hydrostatic testing.

CAUTION: Never attempt to fill a cylinder that is damaged or that is out of hydrostatic test date.

Step 3: Place the SCBA cylinder in a fragment-proof fill station.

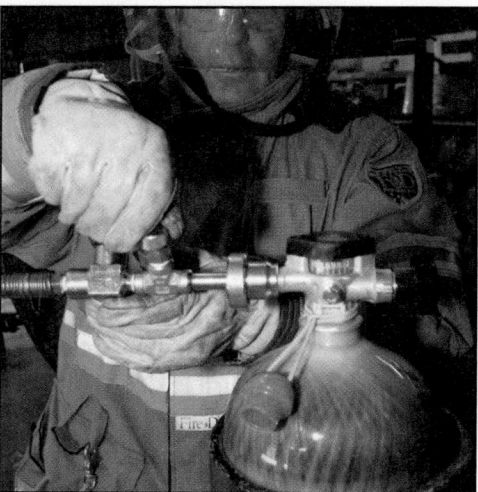

Step 4: Connect the fill hose to the cylinder and close bleed valve on fill hose.

Step 5: Open the SCBA cylinder valve.

Step 6: Open the valve at the fill hose, the valve at the cascade system manifold, or the valves at both locations if the system is so equipped. Check that the regulator setting is appropriate for the cylinder pressure.

NOTE: Some cascade systems may have a valve at the fill hose, at the manifold, or at both places.

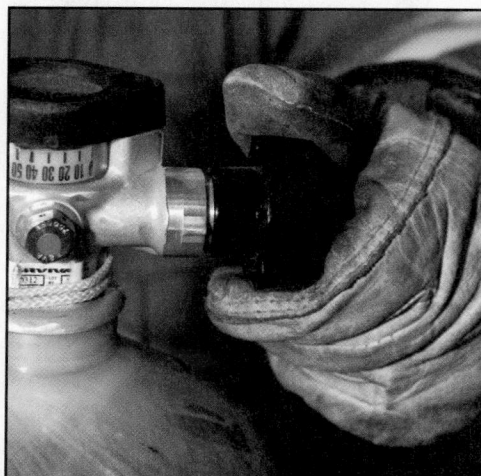

Step 7: Open the valve of the cascade cylinder that has the least pressure but that has more pressure than the SCBA cylinder.

6-I-8

Demonstrate the method for filling an SCBA cylinder from a cascade system, wearing appropriate PPE, including eye and ear protection.

SKILL SHEETS

Step 8: Close the cascade cylinder valve when the pressures of the SCBA and the cascade cylinder equalize.

    a. If the SCBA cylinder is not yet completely full, open the valve on the cascade cylinder with the next highest pressure

    b. Repeat Step 8 until the SCBA cylinder is completely full

Step 9: Close the valve or valves at the cascade system manifold and/or fill line if the system is so equipped.

Step 10: Close the SCBA cylinder valve.

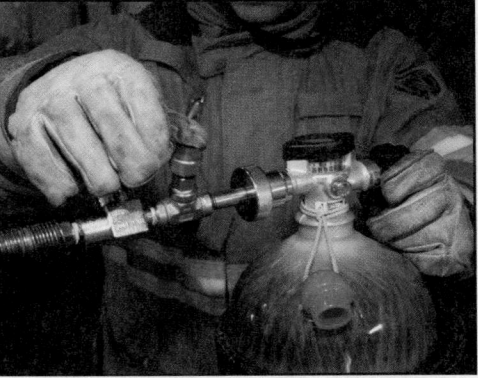

Step 11: Open the hose bleed valve to bleed off excess pressure between the cylinder valve and the valve on the fill hose.

CAUTION: Failure to open the hose bleed valve could result in O-ring damage.

Step 12: Disconnect the fill hose from the SCBA cylinder.

Step 13: Remove the SCBA cylinder from the fill station.

Step 14: Return the cylinder to proper storage.

SKILL SHEETS

6-1-9

Demonstrate the method for filling an SCBA cylinder from a compressor/purifier system, wearing appropriate PPE, including eye and ear protection.

NOTE: This skill sheet is only an example. The procedures outlined here may not be applicable to your compressor/purifier system. Always check the compressor/purifier manufacturer's instructions before attempting to fill any cylinders.

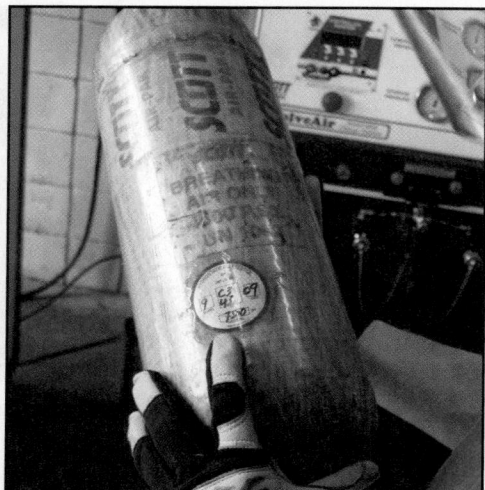

Step 1: Check the hydrostatic test date of the cylinder.

Step 2: Inspect the SCBA cylinder for damage such as deep nicks, cuts, gouges, or discoloration from heat. If the cylinder is damaged or out of hydrostatic test date, remove it from service and tag it for further inspection and hydrostatic testing.

CAUTION: Never attempt to fill a cylinder that is damaged or that is out of hydrostatic test date.

Step 3: Place the SCBA cylinder in a shielded fill station.

Step 4: Connect the fill hose to the cylinder and close bleed valve on fill hose.

Step 5: Open the SCBA cylinder valve.

Step 6: Turn on the compressor/purifier and open the outlet valve.

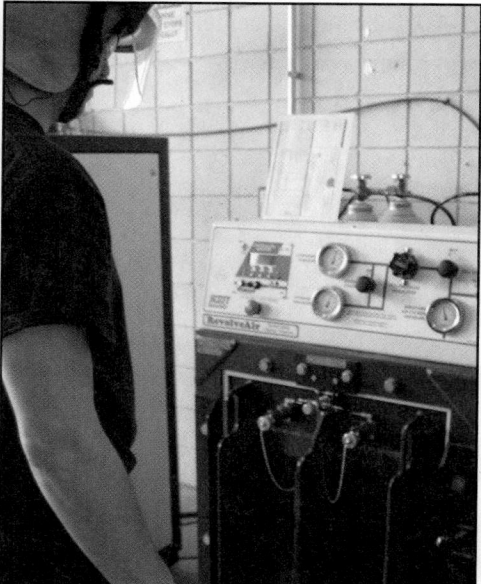

Step 7: Set the cylinder pressure adjustment on the compressor (if applicable) or manifold to the desired full-cylinder pressure. If there is no cylinder pressure adjustment, watch the pressure gauge on the cylinder during filling to determine when it is full.

6-I-9

Demonstrate the method for filling an SCBA cylinder from a compressor/purifier system, wearing appropriate PPE, including eye and ear protection.

SKILL SHEETS

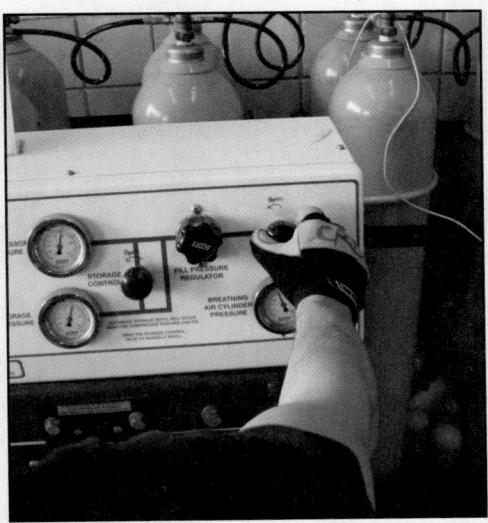

Step 8: Open the manifold valve (if applicable), and again check the fill pressure.

Step 9: Open the fill station valve and begin filling the SCBA cylinder.

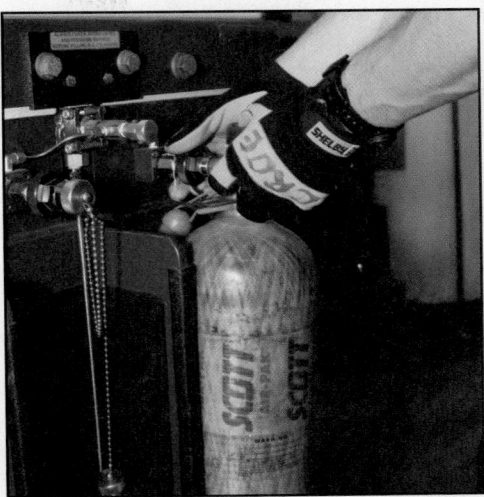

Step 13: Disconnect the fill hose from the SCBA cylinder.

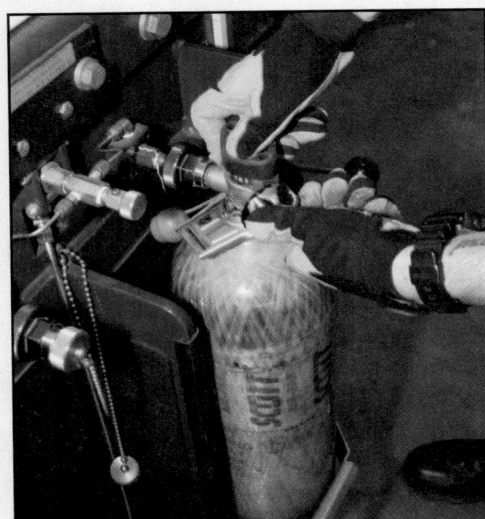

Step 10: Close the fill station valve when the cylinder is full.

Step 11: Close the SCBA cylinder valve.

Step 12: Open the hose bleed valve to bleed off excess pressure between the cylinder valve and valve on the fill station.

CAUTION: Failure to open the hose bleed valve could result in O-ring damage.

Step 14: Remove the SCBA cylinder from the fill station and return the cylinder to proper storage.

SKILL SHEETS

6-I-10
Demonstrate the one-person method for changing an SCBA cylinder.

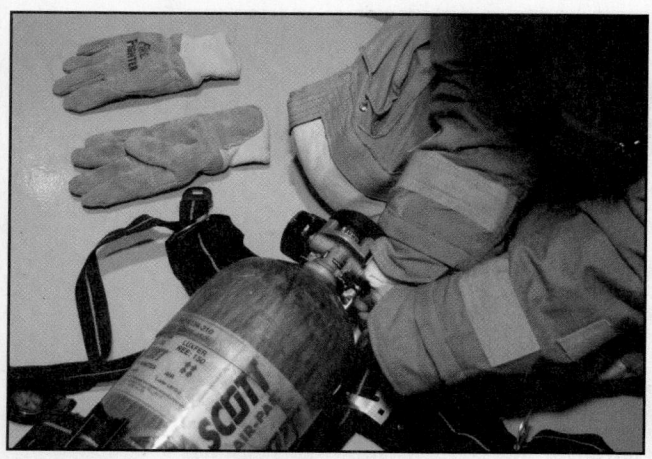

Step 1: Place the SCBA unit on a firm, clean surface.

Step 2: Fully close the cylinder valve.

Step 3: Release air pressure from high- and low-pressure hoses.

Step 4: Disconnect the high-pressure coupling from the cylinder.

Step 5: Remove the empty cylinder from harness assembly.

Step 6: Verify that replacement cylinder is 90-100 percent of rated capacity.

Step 7: Check cylinder valve opening, the high-pressure hose fitting for debris, and the O-ring.

Step 8: Place the new cylinder into the backpack.

Step 9: Connect the high-pressure hose to the cylinder and hand-tighten.

Step 10: Slowly and fully open the cylinder valve and listen for an audible alarm and leaks as the system pressurizes.

NOTE: On some SCBA, the audible alarm does not sound when the cylinder valve is opened. You must know the operation of your own particular unit.

6-I-10
Demonstrate the one-person method for
changing an SCBA cylinder.

SKILL
SHEETS

Step 11: If air leaks are detected, determine if connections need to be tightened or if valves, donning switch, etc. need to be adjusted. Otherwise SCBA with audible leaks due to malfunctions shall be removed from service, tagged, and reported to the officer.

Step 12: Don regulator and take normal breaths.

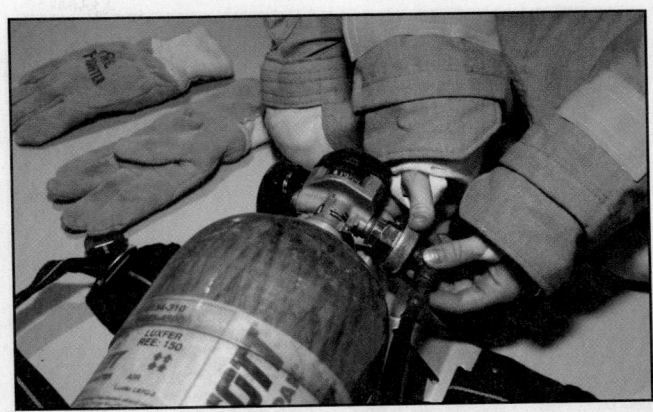

Step 13: Check pressure reading on remote gauge and/or indicators and report reading.

SKILL SHEETS

6-I-11
Demonstrate the two-person method
for changing an SCBA cylinder.

Step 1: Disconnect the regulator from the facepiece or disconnect the low-pressure hose from the regulator.

Step 2: Position the cylinder for easy access by kneeling down or bending over.

Step 3: Fully close the cylinder valve.

Step 4: Release the air pressure from the high- and low-pressure hoses.

Step 5: Disconnect the high-pressure coupling from the cylinder.

Step 6: Remove the empty cylinder from harness assembly.

Step 7: Inspect replacement cylinder and ensure that cylinder is 90-100 percent of rated capacity.

Step 8: Place new cylinder into the harness assembly.

Step 9: Check the cylinder valve opening and the high-pressure hose fitting for debris, clearing any debris by quickly opening and closing cylinder valve, and the O-ring.

Step 10: Connect the high-pressure hose to the cylinder and hand-tighten.

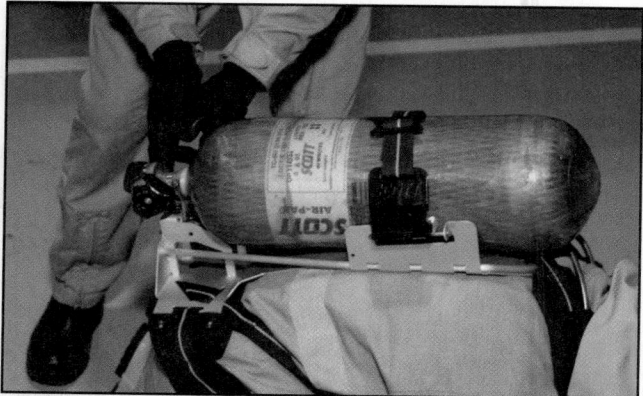

Step 11: Slowly open the cylinder valve fully and listen for an audible alarm and leaks as the system pressurizes.

NOTE: On some SCBA, the audible alarm does not sound when the cylinder valve is opened. You must know the operation of your own particular unit.

Step 12: If air leaks are detected, determine if connections need to be tightened or if valves, donning switch, etc. need to be adjusted. Otherwise SCBA with audible leaks due to malfunctions shall be removed from service, tagged, and reported to the officer.

Step 13: Don regulator and take normal breaths.

Step 14: Check pressure reading on remote gauge and/or indicators and report reading.

NFPA® Safety Alert Issued for SCBA Facepiece Lenses

On July 2, 2012, the National Fire Protection Association® issued the following safety alert:

Safety Alert

Exposure to high temperature environments, which firefighters can encounter during fires they are attempting to extinguish, can result in the thermal degradation or melting of a Self-Contained Breathing Apparatus (SCBA) facepiece lens, resulting in elimination of the protection meant for the user's respiratory system and exposing the user to products of combustion and superheated air.

This alert was based on data gathered by the National Institute for Occupational Safety and Health (NIOSH) while investigating firefighter line of duty deaths between 2002 and 2011. The investigations into three fatalities indicated that firefighters encountered thermal conditions that exceeded the level of protection the facepiece lenses were designed to withstand. At the same time it was determined that the facepiece lens offered the lowest level of thermal protection of any part of the personal protective ensemble. The degradation of the lens resulted in the inhalation of products of combustion and thermal injuries to the firefighter's respiratory system.

The NFPA® will be incorporating new test methods and performance requirements into the 2013 edition of NFPA® 1981, *Standard on Open-Circuit Self-Contained Breathing Apparatus (SCBA) for Emergency Services*. In the meantime, the NFPA® made the following recommendations:

- SCBA facepieces should be inspected before and after each use in accordance with NFPA® 1852, *Selection Care and Maintenance of Open-Circuit Self-Contained Breathing Apparatus*.

- SCBA facepieces that exhibit evidence of exposure to intense heat, such as cracking, crazing, bubbling, discoloration, deformation, or gaps between the lens and frame must be removed from service and repaired or replaced.

- Fire department training programs must contain information on the limitations of respiratory protection, the effects on the facepiece of prolonged or repeated exposures to intense heat, and how to respond to problems that may occur when the facepiece is exposed to intense heat.

- When firefighters and fire officers are evaluating structure fires they must consider the potential for facepiece failure during an interior fire attack. Situational awareness and an understanding of fire behavior are essential to preventing facepiece failure.

- When interior conditions deteriorate, firefighters must be able to recognize the change in conditions and withdraw or seek a safe refuge.

Chapter Contents

Key Terms

NFPA® Job Performance Requirements

This chapter provides information that addresses the following job performance requirements of NFPA® 1001, *Standard for Fire Fighter Professional Qualifications* (2013).

Firefighter I

5.3.16

5.5.1

1. Explain portable fire extinguisher classifications. (5.3.16)

2. Describe types of portable fire extinguishers. (5.3.16)

3. Define the ratings in a portable fire extinguisher rating system. (5.3.16)

4. Explain the considerations taken when selecting and using portable fire extinguishers. (5.3.16)

5. Identify procedures used for the inspection, care, and maintenance of portable fire extinguishers. (5.3.16, 5.5.1)

6. Operate a stored pressure water extinguisher. (Skill Sheet 7-I-1, 5.3.16)

7. Operate a dry chemical (ABC) extinguisher. (Skill Sheet 7-I-2, 5.3.16)

8. Operate a carbon dioxide (CO_2) extinguisher. (Skill Sheet 7-I-3, 5.3.16)

Chapter 7
Portable Fire Extinguishers

Case History

Responding to an early-morning report of a house struck by lightning, the initial company noticed visible smoke several blocks from the incident. Upon arrival the officer reported smoke showing from an upstairs window on the B Side. The officer and firefighter made entry with a water foam extinguisher leaving the driver/operator at the apparatus. The firefighter was wearing full PPE with SCBA while the officer was wearing only a coat and helmet. The team encountered light smoke inside the 1½-story residence and proceeded upstairs. When the team opened a doorway into an unfinished attic, they found the room well-involved in fire. The team attempted to control the fire with the portable water foam extinguisher as fire rolled over their heads. The officer called for an attack hoseline and a supply hoseline from the next-arriving unit. A malfunctioning hydrant caused a delay in establishing a supply line. However, the attack hoseline crew managed to bring the fire under control while operating off the apparatus water supply and began checking for fire extension in the attic.

This incident emphasizes the fact that you must be prepared for the worst possible situation: Always wear complete PPE and respiratory protection when entering the hazard zone. Be aware of the clues that indicate the size of the incident. Remember that portable fire extinguishers are not sufficient to control a structure fire. Follow local standard operating procedures for establishing a water supply during the initial attack.

Source: National Fire Fighter Near-Miss Reporting System.

Portable fire extinguishers are some of the most common fire protection appliances. They can be found in fixed facilities, such as residences, retail stores, and businesses, onboard ships and aircraft, and on vehicles including fire apparatus. Portable fire extinguishers are intended to be used on small fires in the incipient or early growth stage. Primarily intended for occupant use, portable fire extinguishers are also a tool that you will use in a variety of circumstances. You must know about portable fire extinguishers for two reasons: first, so that you can use them and, second, so that you can teach members of your business and residential communities to use them **(Figure 7.1, p. 340)**.

The rapid and accurate use of a portable fire extinguisher can prevent damage and injuries. According to NFPA® 1001, those qualified at the Firefighter I level must know the following about portable fire extinguishers:

- Classifications of types of fires
- Risks associated with each class of fire
- Operating methods of portable fire extinguishers
- Limitations of portable fire extinguishers

Figure 7.1 The proper use of a portable fire extinguisher involves a learned skill set.

Those qualified at the Firefighter I level must be able to:

- Select appropriate extinguisher for size and type of fire
- Safely carry portable fire extinguishers
- Approach fire with portable fire extinguishers
- Operate portable fire extinguishers

This chapter describes the various types of portable fire extinguishers and explains how extinguishers are rated and inspected. Also explained are the steps involved in the selection and use of portable fire extinguishers. For additional information on the rating, placement (location), and use of portable extinguishers, see NFPA® 10, *Standard for Portable Fire Extinguishers*.

I Classifications of Portable Fire Extinguishers

Portable **fire extinguishers** are classified by the type of fire they are designed to effectively extinguish. There are five classes of portable fire extinguishers to match the five classes of fire: Class A, B, C, D, and K. In order to choose the appropriate portable fire extinguisher to use on a given fire, first determine what is burning. Because certain extinguishing agents are only effective on certain classes of fire or fuels, you must be careful in selecting the extinguisher you will use.

Class A

Class A fires involve ordinary combustibles such as textiles, paper, plastics, rubber, and wood. These fuels can be easily extinguished with water, water-based agents such as Class A foam, or **dry chemicals**.

Class B

Class B fires involve flammable and combustible liquids and gases, such as alcohol, gasoline, lubricating oils, and liquefied petroleum gas (LPG). Agents used to extinguish special hazard Class B fires are carbon dioxide (CO_2), dry chemical, and Class B foam.

Class C

Class C fires involve energized electrical equipment. Because water and water-based agents will conduct electrical current, they cannot be used on Class C fires until the electrical energy has been eliminated. Class C extinguishing agents will not conduct electricity making them suitable for electrical fires. Once the power supply has been turned off or disconnected, the fire can be treated as a Class A or B fire.

Class D

Class D fires are those involving combustible metals and alloys such as lithium, magnesium, potassium, and sodium. Some common uses of magnesium are in wheels and transmission components for automobiles and even some metal box springs in beds. These types of fires can be identified by the bright white emissions during the combustion process.

> **CAUTION**
> The use of water or water-based agents on Class D fires will cause the fire to react violently, emit bits of molten metal, and possibly injure firefighters close by.

Class D **dry powder** extinguishers work best on these types of fires; however, do *not* confuse dry powder extinguishers with dry chemical units used on Class A, B, and C fires. It is important to remember that dry chemical agents such as sodium and potassium bicarbonate will react violently with the burning metal if applied to a Class D fire.

> **CAUTION**
> Do not use a dry chemical extinguisher on a Class D fire. The dry chemical often reacts violently with burning metals.

Dry Chemical — Extinguishing system that uses dry chemical as the primary extinguishing agent; often used to protect areas containing volatile flammable liquids.

Dry Powder — Extinguishing agent suitable for use on combustible metal fires.

Wet Chemical System — Extinguishing system that uses a wet-chemical solution as the primary extinguishing agent; usually installed in range hoods and associated ducting where grease may accumulate.

Extinguishing Agent — Any substance used for the purpose of controlling or extinguishing a fire.

Saponification — A phenomenon that occurs when mixtures of alkaline-based chemicals and certain cooking oils come into contact resulting in the formation of a soapy film.

Class K

Class K fires involve combustible cooking oils such as vegetable or animal fats and oils that burn at extremely high temperatures. While most of these fuels are found in commercial and institutional kitchens and industrial cooking facilities, they can also be found in private homes. **Wet chemical systems** and portable fire extinguishers are used to control and extinguish Class K fires.

Types of Portable Fire Extinguishers

Besides classifying portable fire extinguishers by the type of fire they will extinguish, extinguishers are also organized by the type of **extinguishing agent** and the method used to expel the contents. Extinguishing agents use at least one of the following methods to extinguish fire:

- **Smothering** — Excluding oxygen from the burning process
- **Cooling** — Reducing the burning material below its ignition temperature
- **Chain breaking** — Interrupting the chemical chain reaction in the burning process
- **Saponification** — Forming an oxygen-excluding soapy foam surface

Table 7.1 lists the primary and secondary extinguishing methods of various extinguishing agents. Extinguishing agents that work by smothering are ineffective on materials that contain their own oxidizing agent. For example, a fire in magnesium or other combustible metals will flare and intensify if water is applied. **Table 7.2** shows the operational characteristics of different types of portable fire extinguishers described in the following subsections

Table 7.1
Extinguishing Agent Characteristics

Agent	Primary Method	Secondary Method
Water	Cooling	Oxygen depletion
Carbon Dioxide	Oxygen depletion	Cooling
Foam	Oxygen depletion	Vapor suppression
Clean Agent	Chain inhibition	Cooling
Dry Chemical	Chain inhibition	Oxygen depletion
Wet Chemical	Oxygen depletion	Vapor suppression
Dry Powder	Oxygen depletion	Heat transfer cooling

Privately Owned Extinguishers Can Be Dangerous

You should not rely on privately owned portable fire extinguishers found in occupancies. These extinguishers may be obsolete or inoperative as a result of improper maintenance, neglect, or vandalism. You should only use the extinguishers carried on your apparatus.

NOTE: Water-type extinguishers should be protected against freezing if they are going to be exposed to temperatures lower than 40°F (4°C). Freeze protection may be provided by adding antifreeze to the water or by storing in warm areas.

Table 7.2
Operational Characteristics of Portable Fire Extinguishers

Extinguisher	Type	Agent	Fire Class	Size	Stream Reach	Discharge Time
Pump-Tank Water	Hand-carried; backpack	Water	A only	1½-5 gal (6 L to 20 L)	30-40 ft (9.1 m to 12.2 m)	45 sec to 30 min
Stored-Pressure Water	Hand-carried	Water	A only	1¼-2½ gal (5 L to 10 L)	30-40 ft (9.1 m to 12.2 m)	30-60 sec
Aqueous Film Forming Foam (AFFF)	Hand-carried	Water and AFFF	B & C	2½ gal (10 L)	20-25 ft (6.1 m to 7.6 m)	Approximately 50 sec
Halon 1211*	Hand-carried; wheeled	Halon	B & C	Hand-carried: 2½-20 lb (1 kg to 9 kg) Wheeled: to 150 lb (68 kg)	8-18 ft (2.4 m to 5.5 m) 20-35 ft (6.1 m to 10.7 m)	8-18 sec 30-44 sec
Halon 1301	Hand-carried	Halon	B & C	2½ lb (1 kg)	4-6 ft (1.2 m to 1.8 m)	8-10 sec
Carbon Dioxide	Hand-carried	Carbon Dioxide	B & C	2½-20 lb (1 kg to 9 kg)	3-8 ft (1 m to 2.4 m)	8-30 sec
Carbon Dioxide	Wheeled	Carbon Dioxide	B & C	50-100 lb (23 kg to 45 kg)	8-10 ft (2.4 m to 3 m)	26-65 sec
Dry Chemical	Hand-carried stored-pressure; cartridge-operated	Sodium bicarbonate, potassium bicarbonate, ammonium phosphate, potassium chloride	B & C	2½-30 lb (1 kg to 14 kg)	5-20 ft (1.5 m to 6.1 m)	8-25 sec
Multipurpose Dry Chemical	Hand-carried stored-pressure; cartridge-operated	Monoammonium phosphate	A, B, & C	2½-30 lb (1 kg to 14 kg)	5-20 ft (1.5 m to 6.1 m)	8-25 sec
Dry Chemical	Wheeled; ordinary or multipurpose		A, B, & C	75-350 lb (34 kg to 159 kg)	Up to 45 ft (13.7 m)	20 sec to 2 min
Dry Powder	Hand-carried; wheeled	Various, depending on metal fuel (this description for sodium chloride plus flow enhancers)	D only	Hand Carried: to 30 lb (14 kg) Wheeled: 150 lb & 350 lb (68 kg & 159 kg)	4-6 ft (1.2 m to 1.8 m)	28-30 sec
Wet Chemical	Hand-carried	Potassium Acetate	K only	2.5 gal (9.43 L)	8-12 ft (2.4 m to 3.6 m)	75-85 sec

*Rating: Those larger than 9 lb (4 kg) capacity have small Class A Ratings (1-A to 4-A)

All portable fire extinguishers expel their contents using one of the following mechanisms (**Figures 7.2a-c. p. 344**):

- **Manual pump** — The operator physically applies pressure to a pump that increases pressure within the container, forcing the agent out a nozzle at the end of a hose.

- **Stored pressure** — Compressed air or inert gas within the container forces the agent out a nozzle at the end of a hose when the operator presses the handle.

- **Pressure cartridge** — Compressed inert gas is contained in a separate cartridge on the side of the container. When the operator punctures the cartridge, the expellant enters the container forcing the agent out a nozzle on the end of a hose.

Figure 7.2a An extinguisher with a manual pump requires the operator to apply pressure to the pump to increase the pressure in the container.

Figure 7.2b A stored-pressure extinguisher features a prominent pressure gauge.

Figure 7.2c Activating a separate cartridge of compressed inert gas provides the pressure for one type of chemical extinguisher.

Common portable fire extinguishers include:

- Pump-Type Water Extinguishers
- Stored-Pressure Water Extinguishers
- Water-Mist Stored-Pressure Extinguishers
- Wet Chemical Stored-Pressure Extinguishers
- Aqueous Film Forming Foam (AFFF) Extinguishers
- Clean Agent Extinguishers
- Carbon Dioxide (CO_2) Extinguishers
- Dry Chemical Extinguishers
- Dry Powder Extinguishers

Pump-Type Water Extinguishers

Pump-type water extinguishers are intended primarily for use on ground cover fires although they may also be used for small Class A fires. They are designed to be worn on the back with a manually operated trombone-style slide pump **(Figure 7.3)**. The nozzle produces a straight stream, fog, or **water-mist** pattern.

Stored-Pressure Water Extinguishers

Stored-pressure water extinguishers, also called air-pressurized water (APW) extinguishers or pressurized water extinguishers, are useful for all types of small Class A fires **(Figure 7.4)**. They are often used for extinguishing confined hot spots during overhaul operations.

Water-Mist — In the fire service, water mist is associated with a fire extinguisher capable of atomizing water through a special applicator. Water-mist fire extinguishers use distilled water, while backpack pump-type water-mist extinguishers use ordinary water.

Figure 7.3 A pump-operated water extinguisher may be worn as a backpack and produces the same types of stream patterns as hose nozzles.

Figure 7.4 Pressurized extinguishers are useful for small Class A fires.

Water is stored in a tank along with either compressed air or nitrogen. A gauge located on the side of the valve assembly shows when the extinguisher is properly pressurized. When the operating valve is activated, the stored pressure forces water up the siphon tube and out through the hose.

Class A foam concentrate is sometimes added to pump-type or stored-pressure water extinguishers to increase their effectiveness. The addition of Class A foam concentrate serves as a wetting agent that aids in extinguishing deep-seated fires, such as those in upholstered furniture or vehicle seats and wildland fires in densely matted vegetation. Class A foam concentrate in water enhances its effectiveness by reducing the surface tension of water, allowing the water to quickly penetrate the surface.

Water-Mist Stored-Pressure Extinguishers

Although very similar in appearance to standard stored-pressure water extinguishers, water-mist extinguishers use **deionized water** as the agent and nozzles that produce a fine spray instead of a solid stream. Because it is the impurities in water that make it electrically conductive, the deionized water also makes these Class A extinguishers safe to use on energized electrical equipment (Class C). The fine spray also enhances the cooling and soaking characteristics of the water and reduces scattering of the burning materials.

Wet Chemical Stored-Pressure Extinguishers

Similar in appearance to standard stored-pressure water extinguishers, wet chemical fire extinguishers are intended for use on Class K fires involving cooking fats, greases, and vegetable and animal oils in commercial kitchens. Commercial and institutional kitchens may be located in schools, hospitals, cafeterias, restaurants, and catering establishments (**Figure 7.5**). These fire extinguishers contain a special potassium-based, low-pH agent formulated to operate on the principle of saponification in which the agent combines with the oils to create a soapy foam surface over the cooking appliance. Some fire departments carry Class K extinguishers on their emergency response apparatus.

Aqueous Film Forming Foam (AFFF) Extinguishers

Aqueous film forming foam (AFFF) extinguishers are intended for Class B fires (**Figure 7.6**). They are particularly useful in combating fires in or suppressing vapors from small liquid fuel spills.

NOTE: Some manufacturers market AFFF foam concentrates that can be used on Class A fires.

AFFF extinguishers are different from stored-pressure water extinguishers in two ways:

1. The AFFF extinguisher tank contains a specified amount of AFFF concentrate mixed with the water to produce a foam solution.

2. It has an **air-aspirating foam nozzle** that aerates the foam solution, producing a better-quality foam than a standard extinguisher nozzle provides.

The water/AFFF solution is expelled using compressed air or nitrogen stored in the tank with the solution. The resulting finished foam floats on the surface of fuels that are lighter than water. The vapor seal created by the film of water extinguishes the flame and prevents reignition. To prevent the disturbance of the foam blanket when applying the foam, it should not be applied directly onto the fuel; instead, it should be allowed to either gently rain down onto the fuel surface or deflect off a nearby object or surface.

Figure 7.5 Class K extinguishers contain an agent specifically formulated to counteract cooking oil fires. *Courtesy of Ron Moore/ McKinney (TX) FD.*

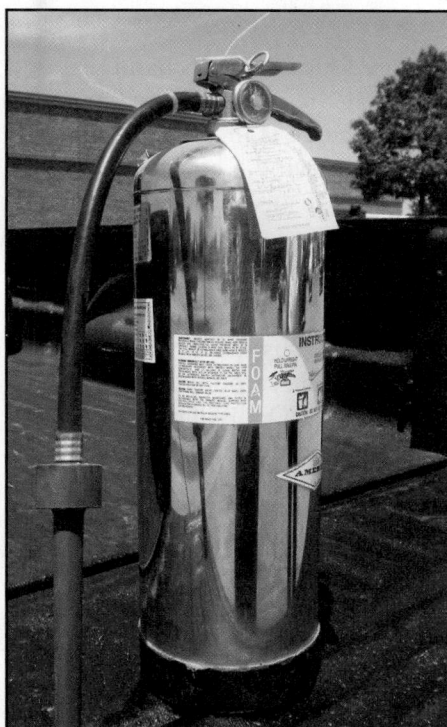

Figure 7.6 Foam extinguishers combat small liquid fuel spills.

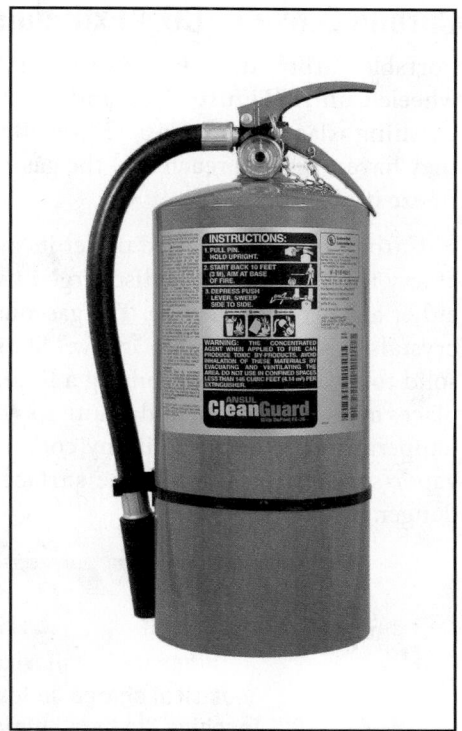

Figure 7.7 A clean agent extinguisher leaves no residue and is effective against Class A-B-C fires. *Courtesy of Ansul Corp.*

AFFF extinguishers are most effective on static pools of flammable liquids. They are not suitable for fires in Class C, Class D, or Class K fuels. They are not suitable for such situations as fuel flowing down from an elevated point and fuel under pressure spraying from a leaking flange. They are most effective on static pools of flammable liquids.

NOTE: AFFF is corrosive and can remove paint from tools and apparatus.

Clean Agent Extinguishers

Clean agents have been developed to replace **halogenated extinguishing agents**. Halon is a generic term for halogenated hydrocarbons and is defined as a chemical compound that contains carbon plus one or more elements from the halogen series (fluorine, chlorine, bromine, or iodine). The two most common Halon extinguishing agents are Halon 1211 (bromochlorodifluoromethane) and Halon 1301 (bromotrifluoromethane). Halons were extremely effective for extinguishing fires in computer rooms, aircraft engines, and areas that contained materials that could easily be damaged by water or dry chemical agents. Unfortunately, halogenated extinguishing agents also have a very damaging effect on the atmosphere's ozone layer. Pressurized with argon gas, Halotron extinguishers are approved by the U.S. Environmental Protection Agency (EPA).

Designed specifically as replacements for Halon 1211, clean agents are discharged as a rapidly evaporating liquid that leaves no residue. Clean agents include FE-36™ hexafluoropropane. hydrochlorofluorocarbon (HCCF), hydrofluorocarbon (HFC), perfluorocarbon (PFC), or fluoroidiocarbon (FIC) **(Figure 7.7)**. These agents effectively cool and smother fires in Class A and Class B fuels, and the agents are nonconductive so they can be used on energized electrical equipment (Class C) fires.

> **Halogenated Extinguishing Agents** — Chemical compounds (halogenated hydrocarbons) that contain carbon plus one or more elements from the halogen series. Halon 1301 and Halon 1211 are most commonly used as extinguishing agents for Class B and Class C fires. *Also known as* Halogenated Hydrocarbons.

Carbon Dioxide (CO_2) Extinguishers

Portable carbon dioxide (CO_2) fire extinguishers are found as both handheld and wheeled units **(Figures 7.8a and b)**. CO_2 extinguishers are most effective in extinguishing Class B and Class C fires. Because their discharge is in the form of a gas, they have a limited reach and the gas can be dispersed by wind. They do not require freeze protection.

Carbon dioxide is stored under its own pressure as a liquefied gas ready for release at any time. The agent is discharged through a plastic or rubber horn on the end of either a short hose or tube. The gaseous discharge is usually accompanied by dry ice crystals or carbon dioxide "snow." Shortly after discharge, this snow changes from a solid to a gas without becoming a liquid. When released, the carbon dioxide gas displaces available oxygen and smothers the fire. Even though CO_2 discharges at subzero temperatures, it has little if any cooling effect on fires. Carbon dioxide produces no vapor-suppressing film on the surface of the fuel; therefore, reignition is always a danger.

Figure 7.8a Liquified CO_2 may be stored in a portable extinguisher.

> **CAUTION**
> When carbon dioxide is discharged, a static electrical charge builds up on the discharge horn. Touching the horn before the charge has dissipated can result in a shock.

Carbon dioxide wheeled units are similar to the handheld units except that they are considerably larger, usually in 50-to 100-pound (23 Kg to 45 Kg) capacities. Wheeled units are most commonly used in airports and industrial facilities. After being wheeled to the fire, the hose (usually less than 15 feet [5 m] long) must be deployed or unwound from the unit before use. The principle of operation is the same as in the smaller handheld units.

Dry Chemical Extinguishers

The terms *dry chemical* and *dry powder* are often incorrectly used interchangeably. Dry chemical agents are for use on Class A-B-C fires and/or Class B-C fires; dry powder agents are used on Class D fires only **(Figure 7.9)**. Dry chemical extinguishers are among the most common portable fire extinguishers in use today. There are two basic types of dry chemical extinguishers:

1. Regular B:C-rated
2. Multipurpose and A:B:C-rated (see Extinguisher Rating System section)

Unless specifically noted in this section, the characteristics and operation of both types are exactly the same. The following are commonly used dry chemicals:

Figure 7.8b Wheeled CO_2 extinguishers are effective against larger Class B-C fires. *Courtesy of Badger Fire Protection.*

- Sodium bicarbonate
- Potassium bicarbonate
- Urea-potassium bicarbonate
- Potassium chloride
- Monoammonium phosphate

During manufacture, dry chemical agents are mixed with small amounts of additives that make the agents moisture-resistant and prevent them from caking (drying or hardening into a mass). This process keeps the agents ready for use even after being stored for long periods, and it makes them free flowing. When refilling an extinguisher, never mix or contaminate dry chemicals with any other type of agent because they may chemically react and cause a dangerous rise in pressure inside the extinguisher.

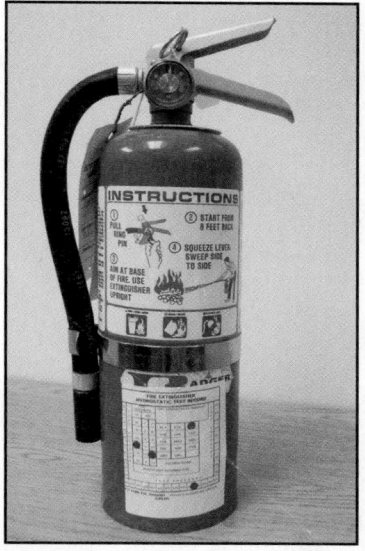

Figure 7.9 Dry chemical agents are effective against Class A-B-C fires.

> ///////
> ### WARNING
> **Never mix or contaminate dry chemicals with any other type of agent.**

The dry chemical agents themselves are nontoxic and generally considered quite safe to use. However, the cloud of chemicals may reduce visibility and create respiratory problems just like any other airborne particulate. Some dry chemicals are compatible with foam, but others are not. Monoammonium phosphate and some sodium bicarbonate agents will cause the foam blanket to deteriorate when applied in conjunction with or after foam to a Class B fire or spill.

On Class A fires, the discharge should be directed at whatever is burning in order to cover it with chemical. When the flames have been knocked down, the agent should be applied intermittently as needed on any smoldering areas. Many dry chemical agents can be mildly **corrosive** to all surfaces.

Corrosive — Capable of causing corrosion by gradually eroding, rusting, or destroying a material.

Handheld Units

There are two basic designs for handheld dry chemical extinguishers: cartridge-operated and stored-pressure. The stored-pressure type is similar in design to the air-pressurized water extinguisher. A constant pressure of about 200 psi (1 400 kPa) is maintained in the agent storage tank. Cartridge-operated extinguishers employ a pressure cartridge connected to the agent tank. The agent tank is not pressurized until a plunger is pushed to release the gas from the cartridge. Both types of extinguishers use either nitrogen or carbon dioxide as the pressurizing gas. Cartridge-operated extinguishers use a carbon dioxide cartridge unless the extinguisher is going to be subjected to freezing temperatures; in such cases, a dry nitrogen cartridge is used.

> ///////
> ### CAUTION
> **When pressurizing a cartridge-type extinguisher, do not place your head or any other part of your body above the top of the extinguisher. If the fill cap was not properly screwed back on, the cap and/or a cloud of agent can be forcibly discharged.**

Wheeled Units

Dry chemical wheeled units are similar to the handheld units but are larger. They are rated for Class A, B, and C fires based on the dry chemical in the unit.

Operating the wheeled dry chemical extinguisher is similar to operating the handheld, cartridge-type dry chemical extinguisher. The extinguishing agent is kept in one tank and the pressurizing gas is stored in a separate cylinder. When the extinguisher is in position at a fire, the hose should first be stretched out completely. Once the agent storage tank and hose are charged, it can make removing the hose more difficult and the powder can sometimes clog in any sharp bends in the hose. The pressurizing gas should be introduced into the agent tank and allowed a few seconds to fully pressurize the tank before the nozzle is opened. Because of the size of the nozzle, the operator should be prepared for a significant nozzle reaction when it is opened. The agent is applied in the same manner as described for the handheld, cartridge-type dry chemical extinguishers.

CAUTION

The top of the extinguisher should be pointed away from the operator and any other nearby personnel when pressurizing the unit.

Figure 7.10 Class D extinguishers use dry powder to isolate the fuel.

Dry Powder Extinguishers

Special dry powder extinguishing agents and application techniques have been developed to control and extinguish fires involving Class D combustible metals. No single extinguishing agent will control or extinguish fires in all combustible metals. Some agents are effective against fires in several metals; others are effective on fires in only one type of metal. Some powdered agents can be applied with portable extinguishers, but others must be applied with either a shovel or a scoop. The appropriate application technique for any given dry powder is described in the manufacturer's technical sales literature. You should be thoroughly familiar with the information that applies to any agent carried on your apparatus.

Class D portable fire extinguishers come in both handheld and wheeled models **(Figure 7.10)**. Whether a particular dry powder is applied with an extinguisher or with a scoop, it must be applied in sufficient depth to completely cover the burning area in order to create a smothering blanket. The agent should be applied gently to avoid breaking any crust that may form over the burning metal. If the crust is broken, the fire may flare and expose more uninvolved material to combustion. Avoid scattering the burning metal. Additional applications may be necessary to cover any hot spots that develop.

CAUTION

Water applied to a combustible metal fire results in a violent reaction that intensifies the combustion and causes bits of molten material to spatter in every direction.

Table 7.3
Portable Fire Extinguisher Ratings

Class	Ratings	Explanations
A	1-A through 40-A	1-A (1¼ gallons [5 L] of water) 2-A (1½ gallons [6 L] of water)
B	1-B through 640-B	Based on the approximate square foot (square meter) area of a flammable liquid fire a non-expert can extinguish
C	No extinguishing capability tests	Tests are to determine nonconductivity
D	No numerical ratings	Tested for reactions, toxicity, and metal burn out time
K	No numerical rating	Tested to ensure effectiveness against 2.25 square feet (0.2 m²) of light cooking oil in a deep fat fryer

If a small amount of burning metal is on a combustible surface, the fire should first be covered with powder. Then, a layer of powder 1 to 2 inches (25 mm to 50 mm) deep should be spread nearby and the burning metal shoveled onto this layer with more powder added as needed. After extinguishment, the material should be left undisturbed until the mass has cooled completely before disposal is attempted.

Portable Fire Extinguisher Rating System

As discussed earlier in this chapter, portable fire extinguishers are classified according to the types of fire for which they are intended. In addition to the classification represented by the letter, Class A and Class B extinguishers are also rated according to performance capability, which is represented by a number. The classification and numerical rating system is based on tests conducted by Underwriters Laboratories Inc. (UL) and Underwriters Laboratories of Canada (ULC). These tests are designed to determine the extinguishing capability for each size and type of extinguisher. **Table 7.3** compares the ratings of each class of portable extinguisher.

Class A Ratings

Class A portable fire extinguishers are rated from 1-A through 40-A. The Class A rating of water extinguishers is primarily based on the amount of extinguishing agent and the duration and range of the discharge used in extinguishing test fires. For a 1-A rating, 1¼ gallons (5 L) of water are required. A 2-A rating requires 2½ gallons (10 L) or twice the 1-A capacity.

Class B Ratings

Portable fire extinguishers suitable for use on Class B fires are classified with numerical ratings ranging from 1-B through 640-B. The rating is based on the approximate square foot (square meter) area of a flammable liquid fire that a nonexpert operator can extinguish using one full extinguisher. The nonexpert operator is expected to extinguish 1 square foot (0.09 m²) for each numerical rating or value of the extinguisher rating.

Class C Ratings

There are not any fire extinguishing capability tests specifically conducted for Class C ratings. Because electricity does not burn, extinguishers for use on Class C fires receive that letter rating because Class C fires are essentially Class A or Class B fires involving energized electrical equipment. The extinguishing agent is tested for electrical nonconductivity. The Class C rating confirms that the extinguishing agent will not conduct electricity. The Class C rating is assigned in addition to the rating for Class A and/or Class B fires.

Class D Ratings

Test fires for establishing Class D ratings vary with the type of combustible metal being tested. The following factors are considered during each test:

- Reactions between the metal and the agent
- Toxicity of the agent
- Toxicity of the fumes produced and the products of combustion
- Time to allow metal to burn completely without fire suppression compared to the time to extinguish the fire using the extinguisher

When an extinguishing agent is determined to be safe and effective for use on a combustible metal, the application instructions are included on the faceplate of the extinguisher, although no numerical rating is given. Class D agents cannot be given a rating for use on other classes of fire.

Class K Ratings

Class K rated extinguishers must be capable of saponifying (converting the fatty acids or fats to a soap or foam) vegetable oil, peanut oil, canola oil, and other oils with little or no fatty acids. Wet chemical agents containing an alkaline mixture, such as potassium acetate, potassium carbonate, or potassium citrate, work by suppressing the vapors and smothering the fire. Any of these agents capable of extinguishing a fire from a deep fryer using these light oils with a surface area of 2.25 square feet (0.2 m²) meet the minimum criteria for Class K rating.

Multiple Markings

Portable fire extinguishers suitable for more than one class of fire are identified by combinations of the letters A, B, and/or C or the symbols for each class. The three most common combinations are Class A-B-C, Class A-B, and Class B-C. All new portable fire extinguishers must be labeled with their appropriate markings. Any extinguisher not properly marked as multipurpose should not be used for any fire other than the type for which it is intended.

The ratings for each separate class of extinguisher are independent and do not affect each other. To better understand the rating system, a common-sized extinguisher, such as the multipurpose extinguisher rated 4-A 20-B:C, should extinguish a Class A fire that is 4 times larger than a 1-A fire, extinguish approximately 20 times as much Class B fire as a 1-B extinguisher, and extinguish a deep-layer flammable liquid fire of 20 square feet (1.8 m²) in area. It must also be nonconductive so it is safe to use on fires involving energized electrical equipment.

Portable fire extinguishers are identified in two ways. One system uses geometric shapes of specific colors with the class letter shown within the shape. The second system, currently recommended in NFPA® 10, uses pictographs to make the selection of the most appropriate fire extinguishers easier. It also shows the types of fires on which extinguishers should *not* be used. **Table 7.4** provides a comparison of the two identification systems.

Table 7.4
Classification of Fire

Class Name	Letter Symbol	Image Symbol	Description
Class A or Ordinary Combustibles	**A** Ordinary Combustibles		Includes fuels such as wood, paper, plastic, rubber, and cloth.
Class B or Flammable and Combustible Liquids and Gases	**B** Flammable Liquids		Includes all hydrocarbon and alcohol based liquids and gases that will support combustion.
Class C or Electrical	**C** Electrical Equipment		This includes all fires involving energized electrical equipment.
Class D or Combustible Metals	**D** Combustible Metals		Examples of combustible metals are: magnesium, potassium, titanium, and zirconium.
Class K or Kitchen	**K** Cooking Oils		Includes unsaturated cooking oils in well-insulated cooking appliances located in commercial kitchens.

Reproduced with permission from Wayne State University, Detroit, MI.

Selecting and Using Portable Fire Extinguishers

In most cases, your supervisor will make the decision on the need to use a portable fire extinguisher and will tell you the correct type to select and use. If, however, you are the only one present and must make the selection on your own, there are a number of factors to consider when deciding which portable fire extinguisher to select for the situation. In addition, there are a number of things to remember when using the extinguisher you selected. This section explains all of these considerations. This information is also important when you are instructing others in the selection and use of extinguishers.

Selecting the Proper Fire Extinguisher

Select extinguishers that minimize the risk to life and property and are effective in extinguishing the fire. To make this selection, consider the following factors:

- Classification of the burning fuel
- Rating of the extinguisher
- Hazards to be protected
- Size and intensity of the fire
- Atmospheric conditions
- Availability of trained personnel
- Ease of handling extinguisher
- Any life hazard or operational concerns

Because of their corrosive particulate residue, do not select dry chemical extinguishers for use in areas where highly sensitive computer equipment is located. The residue left afterward could potentially do more damage to the sensitive electronic equipment than the fire. In these particular areas, clean agent or carbon dioxide extinguishers are better choices.

Using Portable Fire Extinguishers

Portable fire extinguishers come in many types and sizes. While the operating procedures of each type of extinguisher are similar, you should become familiar with the detailed instructions found on the label of the extinguisher. **Skill Sheet 7-I-1** depicts the correct use of a stored-pressure water extinguisher on a Class A fire. **Skill Sheet 7-I-2** depicts the correct use of a dry chemical extinguisher of a Class B fire. **Skill Sheet 7-I-3** depicts the correct use of a CO_2 extinguisher on a Class C fire.

NOTE: Wear full structural or wildland fire fighting personal protective equipment including appropriate respiratory protection when operating any portable fire extinguisher. Even small incipient fires will produce toxic gases that can injure you.

After selecting the appropriate size and type of extinguisher for the situation, make a quick visual inspection. This inspection is necessary to ensure that the extinguisher is charged and operable. This check may protect you from injury caused by a defective or depleted extinguisher. When inspecting an extinguisher immediately before use, check the following **(Figures 7.11a-d)**:

- **External condition** — No apparent damage
- **Hose/nozzle** — In place
- **Weight** — Feels as though it contains agent
- **Pressure gauge (if available)** — In operable range

Figure 7.11a Confirm that the exterior of the extinguisher is undamaged.

Figure 7.11b The nozzle should be in place, secure in its housing, and not blocked by debris.

Figure 7.11c A full extinguisher container is heavier than an empty container.

Figure 7.11d A pressure gauge, if present, should register a reading within the operable range.

If the extinguisher appears to be in working order, you can then use it to extinguish the fire. Approach the fire from the windward side; that is, with the wind at your back.

After performing the visual inspection described earlier, pick up the extinguisher by its handles and carry it to the point of application. Once in position to attack the fire, use the PASS application method (**Figure 7.12**):

P – Pull the pin breaking the thin wire or plastic seal

A – Aim the nozzle at the base of the fire

S – Squeeze the handles together to release the agent

S – Sweep the nozzle back and forth to cover the burning material

Figure 7.12 The PASS application method should be used to place the extinguishing agent on the fire.

Figure 7.13 At an incident or during training, full canisters should stand upright and empty canisters should rest on their sides.

Be sure the extinguishing agent reaches the fire — if it cannot, the agent will be wasted. Smaller extinguishers require a closer approach to the fire than larger units, thus radiant heat or smoke may prevent you from getting close enough for the agent to reach the fire. Adverse winds also can limit the reach of an agent.

Operating an extinguisher close to the fire can sometimes scatter lightweight solid fuels or penetrate the surface of liquid fuels. Apply the agent from a point where it reaches but does not disturb the fuel surface. Releasing the handles will stop the flow of agent.

After the fire is reduced in size, you may move closer to achieve final fire extinguishment. If extinguishment is not achieved after an entire extinguisher has been discharged onto the fire, withdraw and reassess the situation. If the fire is in a solid fuel that has been reduced to the smoldering phase, it may be overhauled using an appropriate tool to pull it apart. A charged hoseline can then be used to soak it to achieve complete extinguishment. If the fire is in a liquid fuel, it may be necessary to either apply the appropriate type of foam through a hoseline or simultaneously attack the fire with more than one portable fire extinguisher. Remember that a portable extinguisher is a first-aid fire fighting appliance and does not take the place of an appropriate-sized hoseline.

If more than one extinguisher is used simultaneously, work in unison with the other firefighters and maintain a constant awareness of each other's actions and positions. Lay empty fire extinguishers on their sides after use. This signals to others that they are empty and reduces the chance of someone taking one and approaching a fire with an empty extinguisher (**Figure 7.13**).

Inspection, Care, and Maintenance of Portable Fire Extinguishers

The inspection, care, and maintenance of fire department portable fire extinguishers are specified by the department's standard operating procedures (SOPs) and based on NFPA® 10 requirements. Extinguishers owned by private companies, organizations, and individuals are regulated by locally adopted codes and standards. Besides being familiar with the SOPs for taking care of extinguishers assigned to your station and apparatus, you should be aware of requirements for extinguishers in buildings on which you do preincident surveys.

Inspection

NFPA® 10 and most fire and life safety codes require that portable fire extinguishers be inspected at least once each year to ensure that they are accessible and operable. Your department SOPs will establish the inspection requirements for your portable fire extinguishers. When inspecting fire extinguishers, there are three factors that determine the value of a fire extinguisher:

- Serviceability
- Accessibility
- Simplicity of operation

Typically, you will inspect the portable fire extinguishers on your apparatus at the beginning of each work period or at least weekly. The following procedures should be part of every fire extinguisher inspection:

- Check to ensure that the extinguisher is in its proper location and that it is accessible.
- Inspect the discharge nozzle or horn for obstructions.
- Check hose for cracks and dirt or grease accumulations.
- Inspect the extinguisher container shell for any physical damage.
- Check to see if the operating instructions on the extinguisher nameplate are legible.
- Check the locking pin and tamper seal to ensure that the extinguisher has not been discharged or tampered with.
- Determine if the extinguisher is full of agent and fully pressurized by checking the pressure gauge, weighing the extinguisher, or inspecting the agent level. If an extinguisher is found to be deficient in weight by 10 percent, it should be removed from service and replaced.
- Check the inspection tag for the date of the previous inspection, maintenance, or recharging (**Figure 7.14**).

If any of the items listed are deficient, remove the extinguisher from service, replace it with an operational extinguisher, and report the need for service in accordance with department SOPs. Only trained personnel should repair or refill portable fire extinguishers.

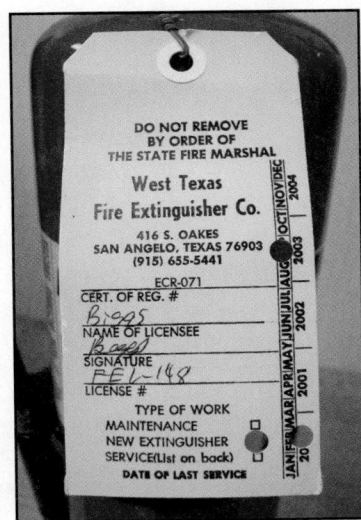

Figure 7.14 Fire extinguishers must be labeled with their maintenance record.

Care

Caring for portable fire extinguishers involves proper handling, storage, and cleaning. The following are guidelines for handling and storage of portable fire extinguishers:

- Never drop or throw a portable fire extinguisher.

- Depending on the size and weight of the extinguisher, carry it diagonally across the body with one hand on the handle and the other on the bottom edge.

- Do not remove the safety pin until you are ready to use the extinguisher.

- Store the extinguisher securely in its apparatus or facility mounting bracket.

- Lay empty extinguishers on their sides to indicate that they are out of service.

- Do not store or stack items in front of wall-mounted extinguishers.

- Shake dry chemical extinguishers monthly to loosen the agent and prevent it from settling.

Clean the extinguisher after each use or periodically. Use warm water and soap to remove dirt, grease, and other foreign material. Avoid using solvents that might damage plastic parts like the gauge face. Remove any corrosion with steel wool or sand paper (**Figure 7.15**).

Always recharge or refill fire extinguishers regardless of the amount of agent used. Recharging must be performed by trained personnel in accordance with the manufacturer's instructions.

Maintenance

Portable fire extinguishers should be removed from service for annual maintenance. This maintenance includes a thorough inspection and disassembly of the unit. Because pressurized portable fire extinguishers operate under internal pressure, they must be hydrostatically tested periodically. NFPA® 10 describes procedures for the hydrostatic testing of extinguisher cylinders as required by both the U.S. Department of Transportation and Transport Canada. The test results must be affixed to the extinguisher shell. The hydrostatic test results on high- and low-pressure cylinders are recorded differently. Maintenance personnel should refer to NFPA® 10 for specific information on extinguisher testing and documentation.

Every six years the dry chemical extinguishing agent should be emptied and the extinguisher refilled. This should be done in a controlled atmosphere to prevent the agent from spreading through the area.

Figure 7.15 An extinguisher must be cleaned periodically to remove dirt and grease.

CAUTION
Never attempt to repair the shell or cylinder of a defective fire extinguisher. Contact the manufacturer for instructions on where to have it repaired or replaced.

Chapter Summary

In the hands of trained personnel, portable fire extinguishers can control or extinguish small incipient or early growth stage fires quickly. Because portable fire extinguishers are effective for ground cover, vehicle, and flammable/combustible liquids fires, you must be familiar with their characteristics and be able to select and use them properly. You should be able to educate the public about extinguishers as well as recognize extinguishers you encounter during preincident building surveys. Finally, you must understand how to inspect, care for, and maintain the portable fire extinguishers assigned to your apparatus and facility.

Review Questions

Firefighter I

1. How are the classifications for portable fire extinguishers divided?

2. What are the differences between wet chemical stored-pressure, aqueous film forming foam (AFFF), and clean agent extinguishers?

3. How do carbon dioxide (CO_2), dry chemical, and dry powder extinguishers differ?

4. How are the ratings used for portable fire extinguishers determined?

5. How should you choose a portable fire extinguisher?

6. When using a portable fire extinguisher, how can you determine the best way to use it?

7. What types of procedures are used to inspect portable fire extinguishers?

8. What are the basic procedures for the care and maintenance of portable fire extinguishers?

Step 1: Size up fire, ensuring that it is safe to fight with an extinguisher.

Step 2: Check that the extinguisher is properly charged.

Step 3: Pull pin at top of extinguisher to break the inspection band.

Step 4: Test to ensure proper operation.

    a. Point nozzle horn in safe direction

    b. Discharge very short test burst

Step 5: Carry extinguisher to within stream reach of fire.

    a. Escape route identified

    b. Upright

    c. Upwind of fire

Step 6: Aim nozzle toward base of fire.

Step 7: Discharge extinguishing agent.

    a. Squeeze handle

    b. Sweep slowly back and forth across entire width of fire

Step 8: Cover entire area with water until fire is completely extinguished.

Step 9: Back away from the fire area.

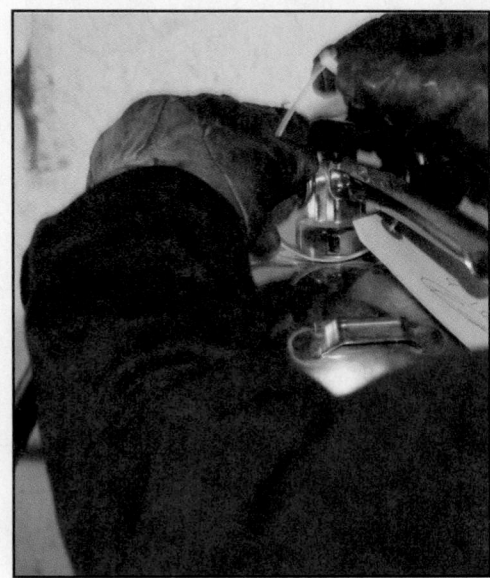

Step 10: Tag extinguisher for recharge and inspection.

Step 1: Size up fire, ensuring that it is safe to fight with an extinguisher.

Step 2: Check that the extinguisher is properly charged.

Step 4: Test to ensure proper operation.

    a. Point nozzle horn in safe direction

    b. Discharge very short test burst

Step 3: Pull pin at top of extinguisher to break the inspection band.

Step 5: Carry extinguisher to within stream reach of fire.

    a. Escape route identified

    b. Upright

    c. Upwind of fire

Step 6: Aim nozzle toward base of fire.

Step 7: Discharge extinguishing agent.

    a. Squeeze handle

    b. Sweep slowly back and forth across entire width of fire

    c. Avoid splashing liquid fuels

Step 8: Cover entire area with dry chemical until fire is completely extinguished.

Step 9: Back away from the fire area.

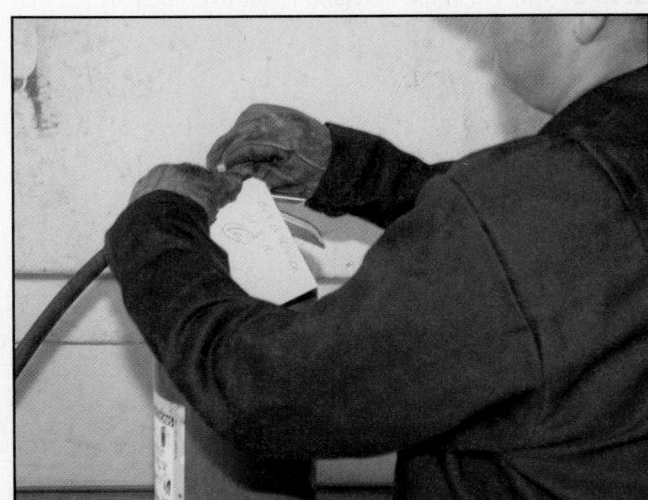

Step 10: Tag extinguisher for recharge and inspection.

Step 1: Size up fire, ensuring that it is safe to fight with an extinguisher.

Step 2: Pull pin at top of extinguisher to break the inspection band.

Step 3: Test to ensure proper operation.

    a. Point nozzle horn in safe direction

    b. Discharge very short test burst

Step 4: Carry extinguisher to within stream reach of fire.

    a. Escape route identified

    b. Upright

    c. Upwind of fire

Step 5: Aim nozzle toward base of fire.

Step 6: Discharge extinguishing agent.

    a. Squeeze handle

    b. Sweep slowly back and forth across entire width of fire

Step 7: Cover entire area with gas cloud until fire is completely extinguished.

Step 8: Back away from the fire area.

Step 9: Tag extinguisher for recharge and inspection.

Chapter Contents

Key Terms

NFPA® Job Performance Requirements

This chapter provides information that addresses the following job performance requirements of NFPA® 1001, *Standard for Fire Fighter Professional Qualifications* (2013).

Firefighter I
5.1.2
5.3.2
5.3.20
5.5.1

1. Compare and contrast the characteristics of life safety rope and utility rope. (5.3.2)

2. Summarize basic guidelines for rope maintenance. (5.5.1)

3. Explain reasons for placing rope out of service. (5.3.20)

4. Describe webbing and webbing construction. (5.3.20)

5. Describe parts of a rope and considerations in tying a knot. (5.1.2, 5.3.20)

6. Describe knot characteristics and knot elements. (5.1.2, 5.3.20)

7. Describe characteristics of knots commonly used in the fire service. (5.1.2, 5.3.20)

8. Select commonly used rope hardware for specific applications. (5.1.2, 5.3.20)

9. Summarize hoisting safety considerations. (5.1.2, 5.3.20)

10. Inspect, clean, and store a rope. (Skill Sheet 8-I-1, 5.5.1)

11. Tie an overhand knot. (Skill Sheet 8-I-2, 5.3.20)

12. Tie a bowline knot. (Skill Sheet 8-I-3, 5.3.20)

13. Tie a clove hitch. (Skill Sheet 8-I-4, 5.3.20)

14. Tie a clove hitch around an object. (Skill Sheet 8-I-5, 5.3.20)

15. Tie a handcuff (rescue) knot. (Skill Sheet 8-I-6, 5.3.20)

16. Tie a figure-eight. (Skill Sheet 8-I-7, 5.3.20)

17. Tie a figure-eight bend. (Skill Sheet 8-I-8, 5.3.20)

18. Tie a figure-eight on a bight. (Skill Sheet 8-I-9, 5.3.20)

19. Tie a figure-eight follow through. (Skill Sheet 8-I-10, 5.3.20)

20. Tie a Becket bend. (Skill Sheet 8-I-11, 5.3.20)

21. Tie a water knot. (Skill Sheet 8-I-12, 5.3.20)

22. Hoist an axe. (Skill Sheet 8-I-13, 5.1.2, 5.3.20)

23. Hoist a pike pole. (Skill Sheet 8-I-14, 5.1.2, 5.3.20)

24. Hoist a roof ladder. (Skill Sheet 8-I-15, 5.1.2, 5.3.20)

25. Hoist a dry hoseline. (Skill Sheet 8-I-16, 5.1.2, 5.3.20)

26. Hoist a charged hoseline. (Skill Sheet 8-I-17, 5.1.2, 5.3.20)

27. Hoist a power saw. (Skill Sheet 8-I-18, 5.1.2, 5.3.20)

Chapter 8
Ropes, Webbing, and Knots

Case History

During a multi company drill at the fire department's training tower, a crew was assigned to rescue a downed firefighter who had fallen through to the floor. To simulate fire conditions, participants had blacked out the facepieces of their SCBA. One rescuer was lowered to the floor below through a 2 x 2 foot (0.61 m by 0.61 m) hole with a life safety rope using the handcuff knot. Before being lowered, the rescuer was attached to a safety line that was being monitored by the officer overseeing the drill. After reaching the downed firefighter, the rescuer attempted to secure several ropes to him so that the firefighters above could lift him through the hole. But with limited visibility, the rescuer struggled to correctly attach the ropes and soon began to run out of air. The rescuer told the crew above that he was short on air and needed to come up. Another rescuer would need to be sent down to complete the rescue. The rescuer stepped in the knots, and the crew above began to raise him back through the floor above. At this point, the rescuer was NOT attached to the safety line, and the person previously tending the safety line was no longer doing so. The transition between the two floors was difficult. The rescuer could not see anyone, and they could not see him due to the blacked out facepieces. He struggled to lift himself out of the hole. No one assisted the rescuer because they could not see him reaching for them. After awhile, the rescuer's legs gave out, and he fell backwards, through the hole, and back down to the floor below. Because his feet were secured in the knots of the rope, the rescuer landed head first. The fall resulted in a broken forearm.

The absence of the safety line during the ascent was a critical factor in this accident. If the safety line had been attached and manned, the rescuer might have been caught before he fell to the floor below. Both the training leader and rescuer should have had greater situational awareness. This is an example of a training exercise that should not have resulted in an accident.

Source: National Fire Fighter Near-Miss Reporting System.

Figure 8.1 Ropes are highly versatile tools used in applications including transferring a tool from the ground to a working floor.

Ropes and webbing are used to perform important fire service tasks. This chapter provides basic knowledge about the proper use of ropes and webbing and shows you how to tie a variety of knots

Rope

Rope is one of the oldest and most common tools used by firefighters **(Figure 8.1)**. You will be trained to use rope in a variety of applications, including:

- Rescues
- Hoisting tools
- Securing unstable objects
- Cordoning off areas
- Gaining access to areas above and below ground level

To use rope effectively, you must know the different types, what they are made from, and how they are constructed. You must also know how to inspect, clean, store, and maintain them, and keep accurate records regarding their use.

Rope Types

Fire service rope is divided into two general classifications: **life safety rope** and **utility rope (Figures 8.2a and b)**. Life safety rope is used to support rescuers and/or victims during emergency incidents or training exercises. It must meet the require-

Figure 8.2a Life safety rope meets strict and specific guidelines and is clearly marked. *Courtesy of Shad Cooper/Wyoming State Fire Marshal's Office.*

Figure 8.2b Utility rope may be stored in a coil in an apparatus compartment. *Courtesy of Shad Cooper/Wyoming State Fire Marshal's Office.*

ments established in NFPA® 1983, *Standard on Life Safety Rope and Equipment for Emergency Services.* Utility rope is used in any situation that does not involve life safety. There is no applicable standard for utility rope, although inspection, cleaning, and maintenance should always follow the manufacturer's recommendations.

In addition to these two primary rope types, NFPA® 1983 also establishes two other classifications: escape rope and water rescue throw line. Both are specialized ropes intended for rescues, whose construction is similar to life safety rope. Escape rope is a single-use rope intended for emergency self-rescue situations. It is designed to carry the weight of only one person, and it must be destroyed after it is used. Water rescue throw line is a floating rope used during water and ice rescues. It can be tied around a rescuer or thrown to a victim (**Figures 8.3a and b, p. 372**).

Escape Rope is a Single-Use Rope

Escape rope must be destroyed after use, even if it is used for training.

Life Safety Rope

NFPA® 1983 specifies that only rope of **block creel construction** using continuous filament virgin fiber for load-bearing elements is suitable for life safety applications. Rope made of any other material or construction must not be used to support firefighters or victims.

NFPA® 1983 requires manufacturers to provide information regarding proper use, inspection and maintenance procedures, and criteria for retiring life safety rope from service. The standard requires that rope must meet the following criteria in order to be reused in life safety situations:

- There are no abrasions or visible damage.
- It has not been exposed to heat or direct flame.
- It has not been subjected to any **impact load**.
- It has not been exposed to the liquids, solids, gases, mists, or vapors of any chemical or other material that can deteriorate rope.
- It has passed an inspection conducted by qualified personnel.

Block Creel Construction — Method of manufacturing rope without any knots or splices; a continuous strand of fiber runs the entire length of the rope's core.

Impact Load — Dynamic and sudden load placed on a rope, typically during a fall.

Figure 8.3a Life safety rope may be used to rappel down the face of a building.

Figure 8.3b A technical rescuer may deploy a throw line as part of rescue evolutions.

A rope log must be kept for every life safety rope. The log must include the product label and manufacturer's instructions, as well as information regarding purchase date, use, maintenance, inspections, and any incident that results in impact loading. Impact loading must be noted because it cannot be detected by inspections, which should be scheduled according to the manufacturer's recommendations.

If a life safety rope has been subjected to an impact load or fails inspection, it must be immediately destroyed; this means it must be altered so that it cannot be mistaken for a life safety rope and unintentionally used again for that purpose. One option is to discard the rope altogether. Another option is to remove the manufacturer's label, cut the rope into smaller lengths, and clearly mark it as utility rope. Life safety rope that has been converted to utility rope is referred to as downgraded.

Utility Rope

Utility rope can be used to hoist equipment, secure unstable objects, or cordon off an area **(Figure 8.4)**. Although the rope industry has standards addressing the properties, care, and use of utility rope, NFPA® does not. Downgraded life safety rope may be used as utility rope if it is still in good condition. Inspect utility rope regularly to see if it is damaged.

Rope Materials

Fire service rope can be constructed from either synthetic or natural fibers. Synthetic fibers are used to construct both life safety and utility ropes while natural fibers are only permitted for utility ropes. The main difference between synthetic and natural fiber ropes is the material used for construction, which affects the use and longevity of each type of rope. Because natural fiber rope loses its strength when it gets wet and rots rapidly, synthetic rope is replacing it in use as utility rope.

Figure 8.4 Utility rope may be used to stabilize a vehicle during extrication evolutions.

Synthetic Fibers

Synthetic fibers include polypropylene, polyester, nylon, Polysteel®, Kevlar™, and Spectra®. **Synthetic fiber rope** has excellent resistance to water, mildew, mold, rotting, shrinkage, and the effects of ultraviolet (UV) light. It also has the following advantages:

- It has a longer life span than natural fiber rope.
- It is very strong yet lightweight.
- It is easy to maintain.

  A disadvantage of synthetic fibers is that they will melt when exposed to heat.

Natural Fibers

Most **natural fiber rope** is made from plant fibers such as manila, sisal, and cotton. Natural fiber rope has the following advantages:

- It is resistant to sunlight.
- It does not melt when exposed to heat.
- It holds a knot firmly.

  However, it has the disadvantage of being prone to mildew and mold. It also deteriorates when exposed to chemicals, and it burns when in contact with embers or open flame.

Rope Construction

The most common types of rope construction are kernmantle, laid, braided, and braid-on-braid **(Figure 8.5, p. 374)**. Kernmantle construction is used for life safety rope while all other types of rope construction are used for utility ropes.

Synthetic Fiber Rope — Rope made from continuous, synthetic fibers running the entire length of the rope; it is strong, easy to maintain, and resists mildew and rotting.

Natural Fiber Rope — Utility rope made of manila, sisal, or cotton; not accepted for life safety applications.

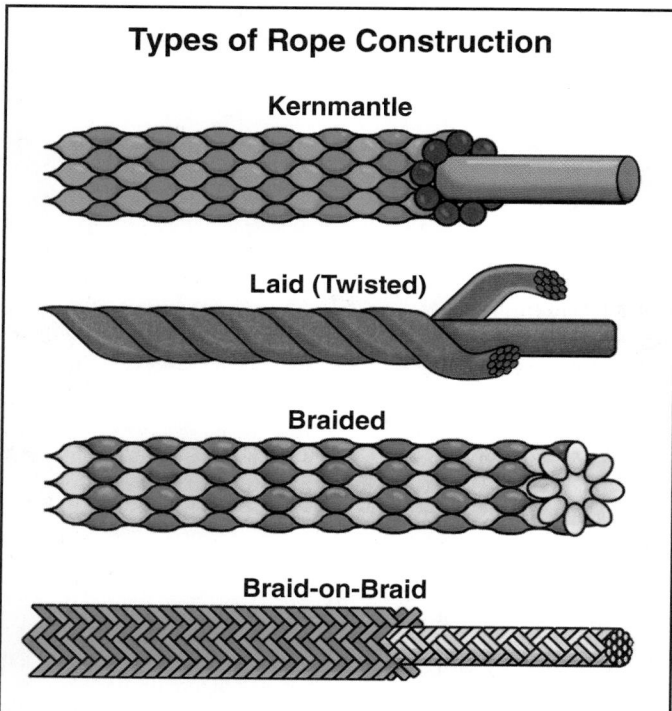

Figure 8.5 Four common types of rope construction may be found in active use.

Types of Rope Construction

Kernmantle

Laid (Twisted)

Braided

Braid-on-Braid

Figure 8.6 Kernmantle rope is defined by its braided sheath covering load-bearing strands.

Kernmantle Rope — Rope that consists of a protective shield (mantle) over the load-bearing core strands (kern).

Kernmantle Rope

Kernmantle is a jacketed synthetic rope composed of a braided covering or sheath (mantle) over a core (kern) of the main load-bearing strands (**Figure 8.6**). The core strands run parallel with the rope's length and works in conjunction with the covering, which increases the rope's stretch resistance and load characteristics. The core is made of high-strength fibers, usually nylon, which accounts for 75 percent of the total strength of the rope. The sheath provides the rest of the rope's overall strength and protects the core from abrasion and contamination.

Dynamic Rope — Rope designed to stretch under load, reducing the shock of impact after a fall.

Kernmantle rope comes in both dynamic (high-stretch) and static (low-stretch) types. **Dynamic rope** is used when long falls are a possibility, such as tower rescue operations. To reduce the shock of impact in falls, dynamic rope is designed to stretch without breaking. Because this elasticity is a disadvantage when trying to raise or lower heavy loads, dynamic rope is not used for rescue or hoisting applications.

Static Rope — Rope designed not to stretch under load.

Static rope is used for most rope-rescue operations. It is designed for low stretch without breaking. According to NFPA® 1983, low-stretch rope must not elongate more than 10 percent when tested under a load equal to 10 percent of its breaking strength. Static rope is used for rescue, rappelling, and hoisting and where falls are not likely to occur or only very short falls are possible.

Laid Rope — Rope constructed by twisting several groups of individual strands together.

Laid (Twisted) Rope

Laid ropes are constructed by twisting fibers together to form strands (**Figure 8.7**), then twisting the strands (typically three) together to make the final rope. Most natural fiber ropes and some synthetic ropes are this type. Laid ropes are used exclusively as utility ropes.

Figure 8.7 Laid ropes may not be used for life safety functions because of their susceptibility to abrasion. *Courtesy of Shad Cooper/Wyoming State Fire Marshal's Office.*

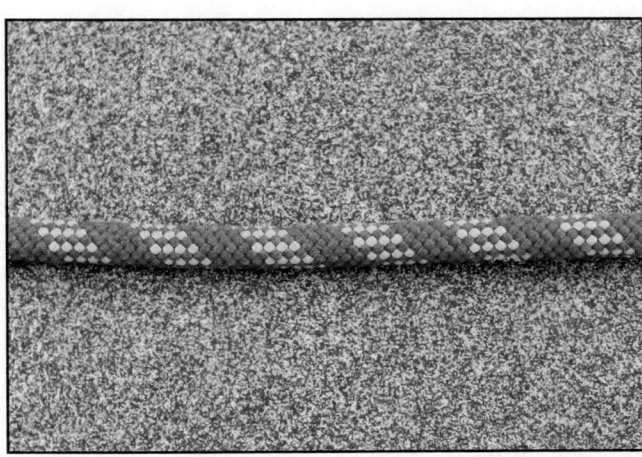

Figure 8.8 Braided ropes may be damaged by abrasion during use. *Courtesy of Shad Cooper/Wyoming State Fire Marshal's Office.*

One disadvantage of laid ropes is that they are susceptible to abrasion and other physical damage. Damage also immediately affects the rope's strength, because such a large proportion of the load-bearing strands are exposed. One advantage of this strand exposure is that it makes laid ropes easy to inspect.

Braided Rope

Braided rope is constructed by uniformly intertwining strands of rope together in a diagonally overlapping pattern (**Figure 8.8**). It is less likely to twist during use than laid rope, but its load-bearing fibers are still vulnerable to direct abrasion and damage. Braided rope is most commonly used as utility rope. Most braided ropes are synthetic, although some use natural fibers.

Braided Rope — Rope constructed by uniformly intertwining strands of rope together (similar to braiding hair).

Braid-On-Braid Rope

Braid-on-braid rope consists of a braided core enclosed in a braided, herringbone-patterned sheath. It is also known as *double-braided rope* and is sometimes confused with kernmantle rope because both are jacketed (**Figure 8.9**).

Braid-on-Braid Rope — Rope that consists of a braided core enclosed in a braided, herringbone patterned sheath.

Figure 8.9 Braid-on-braid rope (the lower rope in this photo) is often used as a utility rope, as opposed to kernmantle rope (the upper rope) which is used in life safety applications.

Figure 8.10 Imperfections and damage on a rope's surface disqualify it from use in its current state.

Figure 8.11 A new rope looks smooth compared to a worn rope's irregularities. *Courtesy of Shad Cooper/Wyoming State Fire Marshal's Office.*

Braid-on-braid rope is very strong. Half of its strength is in the sheath and the other half is in the core. A disadvantage of braid-on-braid rope is that it does not resist abrasion as well as kernmantle rope. Another disadvantage is that the sheath may slide along the inner core of the rope. Braid-on-braid rope is most often used as a utility rope.

Rope Maintenance

All rope must be properly maintained so that it is ready for use when needed. It must be properly inspected, cleaned, stored, and cared for, and a log must be kept that records its use and maintenance history. **Skill Sheet 8-I-1** describes the procedures for inspecting, cleaning, and storing a rope.

Inspecting Rope

All types of rope should be inspected after each use, and unused rope should be inspected at least once a year. Inspections must be documented in the rope log. Ropes are inspected visually and by touch; check for imbedded shards of glass, metal shavings, wood splinters, or other foreign objects that can damage the fibers (**Figure 8.10**). If any of these are found, the rope should be taken out of service and destroyed.

Kernmantle rope. Inspecting kernmantle rope for damage is somewhat difficult because the damage may not be obvious. The inspection can be performed by putting a slight tension on the rope while feeling for lumps, depressions, or soft spots. Soft spots are caused by knots or bends, but they may or may not be signs of permanent damage to the core because core fibers may only be temporarily misaligned. If you feel a soft spot, inspect the outer sheath; if the sheath is damaged, the core is probably damaged as well. However, the core of a kernmantle rope can be damaged without visible evidence on the outer sheath. If there is any doubt about the rope's integrity, it should be removed from service, downgraded to utility status, or destroyed.

In addition to inspecting rope for damage to the core and sheath, inspect the rope for irregularities in shape or weave, foul smells, discoloration from chemical contamination, roughness, abrasions, or fuzziness. Some fuzziness is normal and is not cause for concern, but rope that is excessively fuzzy, either in one spot or overall, should be removed from service. There is no specific guideline for how much fuzziness is too much, so inspectors must rely on their experience and judgment (**Figure 8.11**).

Figure 8.12 Natural fiber rope may indicate wear and damage with visible markers.

Laid rope. Synthetic laid rope should be untwisted so that all sides of each strand can be inspected. In synthetic rope, mildew is not necessarily a problem because the fiber resists rotting and molding. However mildew must always be removed, after which the rope should be cleaned and reinspected. Inspectors should look for the following:

- Soft, crusty, stiff, or brittle spots
- Excessive stretching
- Cuts, nicks, or abrasions
- Chemical damage
- Dirt or grease
- Other obvious flaws

Natural fiber laid rope deteriorates with age. When it reaches the end of its service period, as determined by the manufacturer's recommendations, it must be removed from service. The rope's age can be determined from the rope log. Inspectors should look for the following signs of damage (**Figure 8.12**):

- Ruptured fibers and powdering between strands; these indicate the rope has been overloaded
- Dark red, brown, or black spots between the strands, along with a sour, musty, or acidic odor; these indicate rot and mildew
- Powdering between strands, which indicates internal wear
- Brittle or ruptured fibers, dark red or brown spots, salt incrustation, or swollen areas; these indicate chemical damage
- Rust spots, which occur on ropes used with pulleys or other metal devices
- Accumulations of heavy, greasy materials; these may adversely affect rope strength and reduce holding power

If rotten rope is stored next to new rope, the rot will quickly spread to the new rope. When rot is discovered, both the rotten rope and any surrounding rope must be immediately removed from service, cleaned, and reinspected. Before putting the rope back in storage, be sure to dry and ventilate the storage area.

Figure 8.13 Heat searing damages the integrity of the rope. *Courtesy of Shad Cooper/Wyoming State Fire Marshal's Office.*

Braided rope. Visually inspect braided rope for exterior damage, such as nicks, cuts, heat sears (caused by friction or fire), and excessive or unusual fuzziness **(Figure 8.13)**. Inspect for permanent mushy spots or other deformities by feeling and squeezing the surface of the rope.

Braid-on-braid rope. Visually inspect braid-on-braid rope for heat sears, nicks, and cuts. Feel for lumps, which indicate core damage. If the rope's diameter has shrunk, this may indicate a break in the core. If there is any damage or questionable wear on the sheath, examine it carefully. Check to see if the sheath slides on the core; if it does, cut the end of the rope, pull off the excess material, and then seal the end.

Caring for Rope

Both synthetic and natural fiber ropes can be easily damaged if they are not properly maintained. In addition to the manufacturer's instructions, you should follow these guidelines to ensure that ropes remain in good condition:

- **Avoid abrasion and unnecessary wear** — Rope can be weakened from surface damage caused by chafing or dragging over splintered, rough, or gritty surfaces; constant vibration against apparatus compartment surfaces; and compression when stored in tight spaces.

- **Avoid sharp angles and bends** — Sharp angles, bends, and knots can reduce strength by as much as 50 percent.

- **Protect ends from damage** — Prevent unraveling by properly whipping or taping cut ends.

- **Avoid sustained loads** — Natural fiber ropes have less ability to bear sustained loads than synthetic fiber ropes. If they are subjected to heavy loads for long periods of time, they can break well below the rated load limit. Never exceed the load limit of any rope or subject it to sustained loads for more than two days.

- **Avoid rust** — Keep all synthetic or natural fiber ropes away from rust, which can weaken rope in as little as one to two weeks. If ropes become rust stained, inspect the extent of the stain. If it is halfway through the rope, then rope strength may be reduced by as much as 50 percent, and the rope should be removed from service and destroyed.

- **Prevent contact with chemicals** — Natural fiber rope is extremely vulnerable to chemicals and solvents. Synthetic rope is not entirely resistant to damage from oils, gasoline, paint, and chemicals. Do not let rope contact storage battery solution, washing compounds or solutions, or animal waste. Strong acids, alkalis, and solvents can also damage any rope.

Figure 8.14 Rope may be washed by hand in a sink using a mild detergent. *Courtesy of Shad Cooper/Wyoming State Fire Marshal's Office.*

Figure 8.15 A commercial rope-washing device removes surface debris but cannot address oil and similar contaminants.

- **Reverse ends of the rope periodically** — Uncoil the rope and recoil it with the location of the ends changed. This will ensure even wear along all portions of the rope.

- **Do not walk on rope** — Walking on rope will grind dirt and debris into the strands and will bruise the strands by compressing them.

Cleaning Rope

After use, visually inspect the rope to determine if the rope has been contaminated or soiled. If the rope is contaminated or soiled, use a stiff bristled brush to remove loose surface debris and grime. If additional cleaning is needed, follow the manufacturer's instructions and consult the general cleaning guidelines in the following section.

Synthetic fiber ropes. Synthetic fiber ropes should be washed in lukewarm to warm water, with a mild detergent or fabric softener added to loosen imbedded dirt particles. Bleaches or strong cleansers should not be used. There are three methods for cleaning synthetic rope:

- **Washing by hand** — Place the rope into a utility sink filled with water and detergent, then scrub with a bristle brush (**Figure 8.14**). You can also place the rope in a mesh bag, allow it to soak in the sink, then agitate it by hand to remove grit. When the rope is clean, rinse it thoroughly in clean water to remove detergent.

- **Rope-washing device** — Commercial rope-washing devices consist of a bristle-lined plastic tube that has a garden hose connection on one side (**Figure 8.15**). The rope is manually fed through the device, and multidirectional streams of water clean all sides of the rope at the same time. These devices remove mud and other surface debris, but do not address deeper cleaning problems because they cannot be used with detergent.

Figure 8.16 Clothes washers that do not use a center agitator may be used for rope. *Courtesy of Shad Cooper/Wyoming State Fire Marshal's Office.*

- **Washing machine** — Rope can also be cleaned in front-loading or top-loading clothes washers without center agitators (**Figure 8.16**). Place the rope in a mesh bag to protect the exterior from abrasion, then set the washer on the coolest wash/rinse temperature available and use only a small amount of mild detergent. A washing machine can also be used to rinse a rope that has been cleaned with a high-pressure washer.

Rappel Rope Log — A rope log is used to ensure that rope usage is maintained in accordance with the standards established in the rope grading table. Inspect rope for damage and excessive wear each time it is deployed and again after each use. Immediately retire all suspect ropes.

Unit Rope ID		NSN:		MFR. Lot Number:	
Manufacturer:		Date Manufactured:		Date In Service:	
Color:	Length:	Diameter:	Type Of Rope:		

Date Used	Number of Rappels	Type of Rappels	Rope Grade and Comments	Inspector Initials

Page ----- of -----

Figure 8.17 This example rope log includes fields for all of the pieces of information that must be tracked over the course of the rope's use.

Rope must be dried immediately after it has been washed and rinsed. It can be spread out on a hose drying rack, suspended in a hose tower, or loosely coiled in a hose dryer. Never place it near a heat source or use a heated dryer, because heat can reduce rope's tensile strength. Avoid drying or storing in direct sunlight, as ultraviolet light can also weaken the rope.

Natural fiber ropes. Wipe or gently brush the rope to remove as much of the dirt and grit as possible. Do not use water, as this will damage the rope. Water initially strengthens natural fiber rope, but over time it damages and weakens the fiber. If the rope gets wet during use, dry it thoroughly using the same method described for synthetic rope.

Maintaining a Rope Log

When life safety rope is purchased, it must be permanently identified. Many departments identify new ropes by marking the ends with the unit number and the date it was placed in service. This can be done with a printed label sealed to the rope ends with a liquid compound made for this purpose. A **rope log** must be started and kept throughout the rope's working life (**Figure 8.17**). Dates for all uses, maintenance, and inspections are entered into the log. This information helps determine when the rope should be removed from service. The log should be kept in a waterproof envelope and placed in a pocket that is usually sewn on the side of the rope's storage bag. Rope logs are not required for utility ropes.

Rope Log — Record of all use, maintenance, and inspection throughout a rope's working life; also includes the product label and manufacturer's recommendations.

Rope Storage

Proper rope storage will help ensure that the rope maintains its condition and rated load strength and reaches its life expectancy. Whether stored in a fire department facility or on an apparatus, the rope should be in a clean, dry, unheated area with freely circulating air currents. All rope must be protected from the weather, stored out of direct sunlight, and kept away from chemicals, fumes, and vapors. Ropes should not be stored in the same compartments where gasoline-powered tools or fuel containers are stored. New rope is generally stored on reels until it is placed into service.

The best method for storing rope is to place it into a nylon or canvas storage bag (**Figure 8.18**). The bag makes the rope easy to transport and protects it from abrasion and contamination. The bag can be marked to indicate the type and size of rope and the unit to which it has been assigned. The rope log can also be attached to the bag. An additional advantage of storing synthetic rope in a bag is that the rope can be deployed quickly by holding the end of the rope and throwing or dropping the bag. The weight of the rope inside the bag carries the bag toward the target and the rope unravels from the bag as it travels through the air. The bag may have a drawstring and shoulder straps to make carrying easier.

Figure 8.18 Storage bags with shoulder straps and a drawstring closure are commonly used to contain lengths of rope.

Webbing

Webbing is used in conjunction with rope for both life safety and utility applications. It comes in many forms and may be a simple piece of material, a loop, a ladder belt, or a rescue harness.

Webbing Material and Construction

Webbing is constructed from the same materials used to make synthetic rope, such as nylon. It can be flat or tubular (**Figure 8.19, p. 382**), in either a spiral weave or chain weave design. One-inch (25 mm) webbing is most common in the fire service, although webbing used for other applications may be larger. Any webbing used for life safety applications must be NFPA® compliant.

Webbing — Device used for creating anchors and lashings, or for packaging patients and rescuers; typically constructed from the same material as synthetic rope.

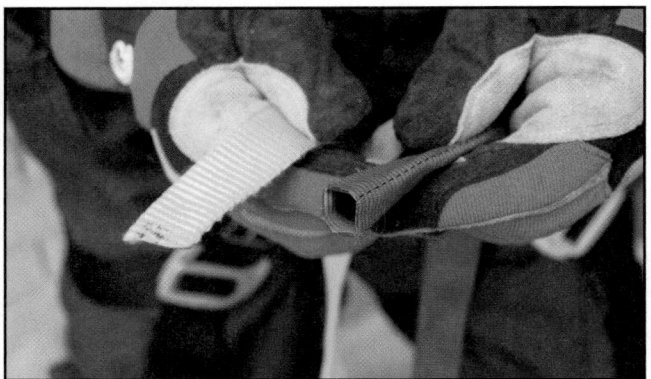

Figure 8.19 Flat webbing and tubular webbing are used almost interchangeably with ropes.

Flat Webbing

Flat webbing is constructed of a single layer of material that resembles an automobile seat belt. Though it is less expensive than tubular webbing, it is stiffer and more difficult to tie into knots. In rescue applications it is mainly used for straps and harnesses.

Tubular Webbing

Because tubular webbing is more supple and easier to tie, it is the most common webbing for rescue applications. There are two types of tubular webbing: edge-stitched and spiral weave. Edge-stitched webbing is formed by folding a piece of flat webbing lengthwise and sewing the edges together. Wear and abrasion can cause edge-stitched webbing to become unstitched. Using a sewing method called *lock-stitching* can prevent this type of damage.

Spiral weave tubular webbing, also known as *shuttle-loom construction*, is preferred for rescue work. Spiral weave webbing is constructed by weaving the tube as a unit, and it is so named because the threads spiral around the tube as it is being woven.

Webbing Uses

Similar to rope, webbing may be unofficially divided into two categories: life safety and utility. Life safety webbing is used:

- To support firefighters during technical rescue operations, either as a rescue harness, or worn as a **ladder belt** while working from a ground ladder or aerial device **(Figures 8.20a and b)**
- To construct technical rescue anchor systems
- To package and secure victims to litters
- To fasten rescue components together

Life safety webbing must be certified to NFPA® 1983 requirements. The standard further describes three classes of rescue harnesses **(Figure 8.21)**:

- **Class I harness** — Also known as a *seat harness*, a Class I harness fastens around the waist and around the thighs or under the buttocks and is intended to be used for emergency escape with a load of up to 300 pounds (1.33 k/N).

- **Class II harness** — A Class II harness fastens in the same manner as Class I harness but is rated for loads up to 600 pounds (2.67 k/N) load. A Class II harness looks exactly like a Class I harness, so the attached label must be used to verify its rating.

- **Class III harness** — Also known as a *full body harness*, a Class III harness fastens around the waist, around the thighs or under the buttocks, and over the shoulders. Like Class II harnesses, Class III harnesses are rated for loads of up to 600 pounds (2.67 k/N).

Ladder Belt — Belt with a hook that secures the firefighter to the ladder.

Figure 8.20a Rescue harnesses are composed of a strategic series of webbing loops with the intention of providing stability during high-angle operations.

Figure 8.20b A ladder belt provides a measure of stability for responders working on an aerial device or platform.

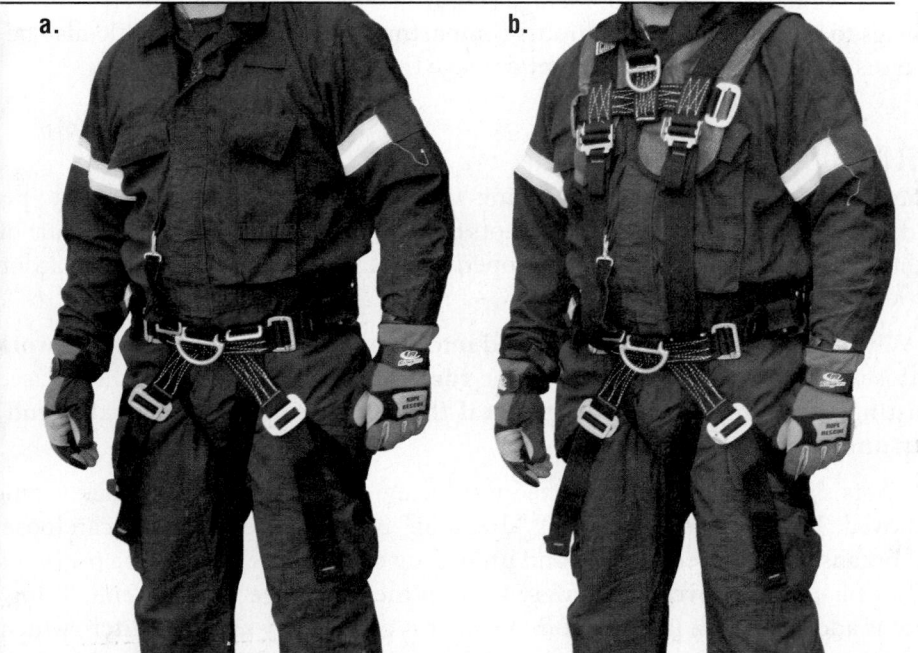

Rescue Harnesses

Uses
To help protect rescuers and victims as they move and/or work in elevated positions during rope rescue operations.

a.

b.

a. Class I (loads up to 300 lb [1.33 k/N]) and Class II (loads up to 600 lb [2.67 k/N])

b. Class III (full body harness)

Figure 8.21 Harnesses made of life safety webbing will be labeled for their rated hauling capacity.

Figure 8.22 Long lengths of webbing may be daisy chained to keep them untangled while in storage.

General purpose utility-type webbing is not regulated by any standard. However the webbing must support the load limit you intend to place on it, plus a safety factor.

Utility webbing is used for a wide variety of applications including:

- Securing hose rolls and bundles
- Raising and lowering tools and equipment
- As part of a search line system
- Securing doors and hatches in an open position
- Carrying hose, SCBA cylinders, and equipment
- Controlling an inward swinging door when it is being forced open
- Pulling an unconscious or incapacitated person out of a hazardous area
- Securing the vehicle roof when it has been folded back during extrication
- Holding a door in place while it is being opened with spreaders during a vehicle extrication

Webbing Care and Maintenance

Care, cleaning, and maintenance of webbing follow the same guidelines used for synthetic rope. Always follow the manufacturer's instructions, especially for life safety harnesses and ladder belts.

Webbing Storage

Many firefighters carry a 20 to 30 foot (6.10 m to 9.14 m) length of utility webbing in their protective coat pocket. This webbing can be quickly tied into a loop, attached to a piece of rope hardware, or wrapped around an object. Long lengths of webbing may be rolled or daisy-chained for storage (**Figure 8.22**). Life safety harnesses are carried in bags to protect them in apparatus compartments. Ladder belts should also be carried inside the apparatus to prevent damage from UV rays and moisture.

Knots

Knots are used with ropes and webbing to join them together, attach them to people and objects, and form loops. Tying knots quickly and correctly is a critical part of fire fighting and rescue operations. Improperly tied knots can be extremely hazardous to both rescuers and victims.

When tying knots, a rope is divided into three parts (**Figure 8.23**). The **working end** is used to tie the knot or hitch. The **running part** is the free end that is used for hoisting or pulling. The **standing part** is the section between the working end and the running part.

Knots should always be tightened until snug, and after tying, all slack should be removed. This process is known as "dressing." But even dressed knots can loosen or fail because of repeated loading and unloading of the rope. One way to prevent such failures is to tie an **overhand safety knot** in the tail of the working end. Tying this knot is addressed in a later section. Another type of safety knot is a **hitch**, which is a temporary knot that can be undone by pulling against the strain that holds it.

Knot — Term used for tying a rope around itself.

Working End — End of the rope used to tie a knot. *Also known as* Bitter End or Loose End.

Running Part — Free end of the rope used for hoisting, pulling, or belaying.

Standing Part — Middle of the rope, between the working end and the running part.

Overhand Safety Knot — Supplemental knot tied to prevent the primary knot from failing; prevents the running end of the rope from slipping back through the primary knot.

Hitch — (1) Temporary knot that falls apart if the object held by the rope is removed. (2) Loop that secures the rope but is not part of a standard rope knot.

Figure 8.23 The divisions of a rope are used to clarify which end should be held in which position for the rope to accomplish the task at hand.

Figure 8.24a A bight is a simple bend in a rope and is the base for many knots.

Figure 8.24b A loop completes a 360 degree bend of the rope.

Figure 8.24c A round turn forms by bending a rope on itself several times.

Elements of a Knot

To be suitable for use in the fire service, a knot must be easy to tie and untie, be secure under load, and reduce the rope's strength as little as possible. A rope's strength is reduced to some degree whenever it is bent; the tighter the bend, the more strength is lost. Some knots create tighter bends than others, reducing the rope's strength to a greater degree.

Bight, loop, and *round turn* are names for the bends that a rope undergoes when tying a knot or hitch (**Figures 8.24a-c**). Knots and hitches are formed by combining these three elements in different ways so that the tight part of the rope bears on the working end to hold it in place. Each of these formations is shown in the following figures:

- The bight is formed by simply bending the rope back on itself while keeping the sides parallel.
- The loop is made crossing the side of a bight over the standing part.
- The round turn consists of further bending one side of a loop.

Figure 8.25 A safety knot pulls the standing end of a rope away from the primary working knot to keep the working knot from failing. *Courtesy of Shad Cooper/ Wyoming State Fire Marshal's Office.*

Figure 8.26 A bowline knot serves many functions in the fire service.

Types of Knots and Hitches

The most common types of knots and hitches used in the fire service include the following:

- Overhand safety
- Bowline
- Half-hitch
- Clove hitch
- Handcuff
- Figure-eight
- Figure-eight bend
- Figure-eight on a bight
- Figure-eight follow through
- Becket bend
- Water knot

Figure 8.27 One or more half-hitches stabilize an object held by another knot or hitch.

Overhand Safety Knots

As an added measure of safety, an overhand safety knot (often just called a safety) can be used when tying any type of knot **(Figure 8.25)**. Although any properly tied knot should hold, it is best to provide the highest level of safety possible. Use of the overhand safety knot eliminates the danger of the running end of the rope slipping back through the knot and causing the knot to fail. **Skill Sheet 8-I-2** describes the procedure for tying the overhand knot.

Bowline

The **bowline** is one of the most important knots in the fire service **(Figure 8.26)**. It is easily tied and untied and is good for forming a single loop that will not constrict. Firefighters should be able to tie a bowline in the open as well as around an object. The method shown in **Skill Sheet 8-I-3** is one method of tying the bowline, although other methods are also effective.

Half-Hitch

The **half-hitch** is particularly useful for stabilizing long objects that are being hoisted. It is always combined with another knot or hitch. For example, one way of hoisting a pike pole is to tie a clove hitch around the pole, then a series of half-hitches along the length of the handle **(Figure 8.27)**.

The half-hitch is formed by making a round turn around the object. The standing part of the rope is passed under the round turn on the side opposite the intended direction of pull. Several half-hitches can be applied in succession if required.

Bowline Knot — Knot used to form a loop; it is easy to tie and untie, and does not constrict.

Half-Hitch — Knot typically used to stabilize long objects that are being hoisted; always used in conjunction with another knot.

Clove Hitch — Knot that consists of two half-hitches; its principal use is to attach a rope to an object such as a pole, post, or hose.

Clove Hitch

The **clove hitch** essentially consists of two half-hitches. Its principal use is to attach a rope to an object such as a pole, post, or hoseline. It is not appropriate for life safety applications because repeated loading and unloading will cause it to fail. The clove hitch may be formed anywhere in the rope, from either end or the middle. If a clove hitch is used for utility applications and will be subjected to repeated loading and unloading, an overhand safety knot should be tied as well.

The two methods of tying this knot are described in **Skill Sheets 8-I-4** and **8-I-5**. When tying it over an object that does not have a free end, the method described in 8-I-4 will not work. Use the method described in 8-I-5 instead.

Handcuff (Rescue) Knot — Knot tied in a bight with two adjustable loops in opposing directions; used during rescues to secure hands or feet, so that a victim can be raised or dragged to safety. *Also known as* Rescue Knot.

Handcuff

The **handcuff (rescue) knot** consists of two adjustable loops formed from a bight **(Figure 8.28)**. During rescues, the handcuff knot is used to secure a victim's hands or feet to raise or drag them to safety. The loops are adjustable until the half-hitches are tightened. See **Skill Sheet 8-I-6** for the steps to tie a handcuff knot.

Figure-Eight

The figure-eight is the foundation knot for an entire family of figure-eight knots **(Figure 8.29)**. It can also be used as a stopper knot so that the rope will not pass through a rescue pulley or the grommet of a rope bag. Refer to **Skill Sheet 8-I-7** for tying procedures.

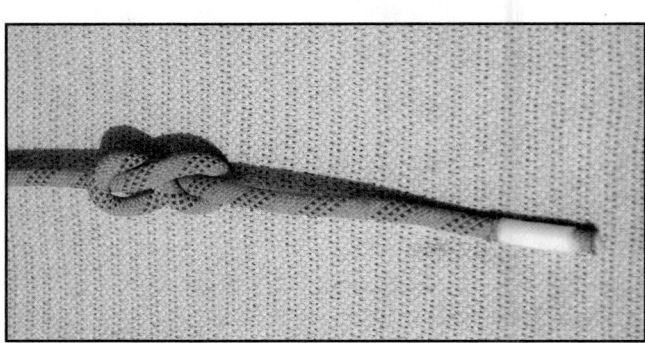

Figure 8.29 A figure-eight knot is considered a basic knot that may be used alone or as the basis for other knots.

Figure 8.28 A handcuff knot is used in victim rescue applications.

Figure-Eight Bend

Also known as the *Flemish Bend*, the figure-eight bend is used primarily on life safety rope to tie ropes of equal diameters together. Refer to **Skill Sheet 8-I-8** for tying procedures.

Figure-Eight on a Bight

The figure-eight on a bight is a good way to tie a closed loop. It is tied by forming a bight in the rope and then tying a simple figure-eight with the bight in the doubled part of the rope. Refer to **Skill Sheet 8-I-9** for tying procedures.

Figure-Eight Follow Through

The figure-eight follow through is used for securing objects. It is basically a figure-eight on a bight that is around the object. Refer to **Skill Sheet 8-I-10** for tying procedures.

Becket Bend

The **Becket bend** is used for joining two ropes of unequal diameters or joining a rope and a chain (**Figure 8.30**). It is unlikely to slip when the rope is wet, which makes it particularly useful in the fire service. However, the Becket bend is not suitable in life safety applications. Refer to **Skill Sheet 8-I-11** for tying procedures.

Water Knot

The water knot is the preferred knot for joining two pieces of webbing, or the ends of the same piece when a loop is needed (**Figure 8.31**). Because the water knot has a tendency to slip, it is important to dress the knot properly and have the webbing as flat as possible when forming the knot. It is also a good idea to allow at least 3 inches (76.2 mm) for the tail. See **Skill Sheet 8-I-12** for tying procedures.

Rope and Webbing Uses

There are five main uses for rope and webbing at an emergency incident:

- Rescue operations
- Hoisting tools and equipment
- As a barrier to indicate the control zone
- As a search line during search operations
- Stabilizing objects

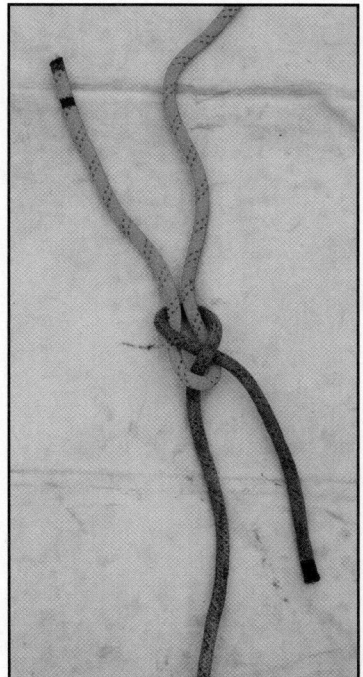

Figure 8.30 A Becket bend may be used to join ropes to each other or to chains.

Becket Bend — Knot used for joining two ropes; particularly well suited for joining ropes of unequal diameters or joining a rope and a chain. *Also known as* Sheet Bend.

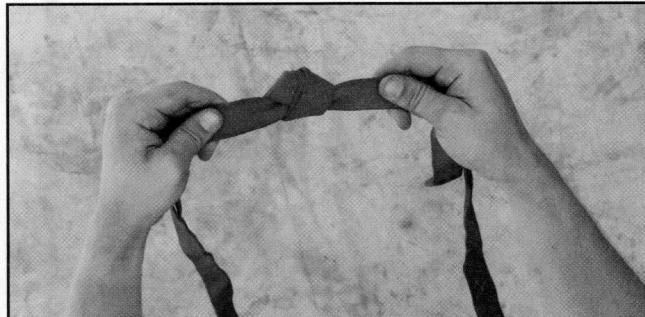

Figure 8.31 The water knot is commonly used for joining two pieces of webbing.

Utility rope is typically used in all of these applications except rescues. Remember that the load carrying ability of the rope or webbing must exceed the weight of the object that will be hoisted or stabilized.

Rescue Uses

Utility rope should never be used during rescue operations. However life safety rope can be used for:

- Rappelling
- Lifting victims and rescuers
- Removing victims from ice and swift water situations

These operations require specialized training beyond the Firefighter I and II levels, but you may be required to assist technical rescue personnel in these operations. You must be able to recognize the life safety ropes, hardware, and equipment used by your department and know how they are used.

Hoisting Tools and Equipment

Rope and webbing are frequently used to raise or lower tools and equipment, with one notable exception. Hoisting pressurized cylinders, such as SCBA cylinders, is prohibited by the Occupational Safety and Health Administration (OSHA) because it is unsafe.

Never Hoist Pressurized Cylinders
Hoisting pressurized cylinders is unsafe and must not be performed.

Tag Line — Non-load-bearing rope attached to a hoisted object to help steer it in a desired direction, prevent it from spinning or snagging on obstructions, or act as a safety line.

To prevent equipment from being dropped, always use proper knots and securing procedures. This prevents damage to the equipment and injury to personnel below. A separate control or **tag line** may be tied to any piece of equipment. The hoisting line may also be tied to the object to serve as a tag line (**Figure 8.32**). Firefighters on the ground use the tag line to prevent the equipment from striking the structure or other objects. When one rope serves as both the tag line and the hoisting line, the knot or hitch-tying methods and methods of hoisting may vary. Keep safety first in mind when selecting the hoisting method.

Rope Hardware

Mechanical Advantage — (1) Advantage created when levers, pulleys, and tools are used to make work easier during rope rescue or while lifting heavy objects. (2) The ratio of the force applied by a simple machine, such as a lever or pulley, to the force applied to the machine by the user.

Carabiners and pulleys are among the most common types of hardware used in hoisting (**Figures 8.33a and b**). A carabiner is a snap link made from aluminum, titanium, or steel, with a sprung or screwed gate that connects ropes to other mechanical gear. A pulley is a simple device used to create a **mechanical advantage** (**Figure 8.34, p. 392**). It consists of a grooved wheel through which a rope can run to change the direction or point of application of a force applied to the rope.

Safety Guidelines

Safety is your primary consideration when hoisting tools or equipment. Always follow these general safety guidelines:

Figure 8.32 A tag line serves to stabilize and control an object while it is being hoisted.

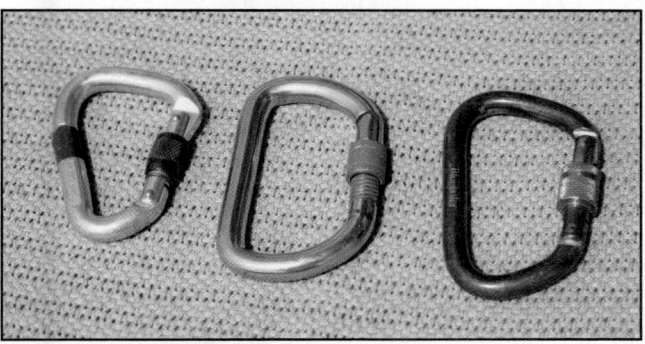

Figure 8.33a Carabiners may be found in many shapes and sizes and should be rated for the activities that they will be expected to aid.

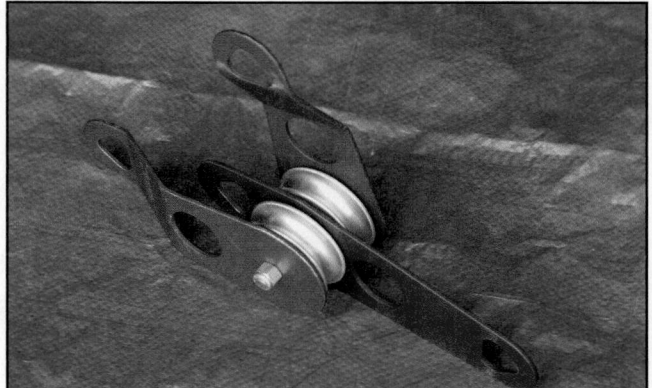

Figure 8.33b A pulley holds a bight and allows the rope to change direction or point of application of a force.

- Before starting a hoisting operation, make sure that you are physically balanced and standing firmly on the ground.
- Use the hand-over-hand method to maintain control of the rope.
- Use an edge roller or padding to protect rope from physical damage when pulling it over sharp edges, such as cornices or parapet walls (**Figure 8.35, p. 392**).
- Use a pulley system for heavy objects.
- Work in teams when working from heights.
- Make sure that all personnel are clear of the hoisting area.
- Avoid hoisting operations near electrical hazards. If this is not possible, use extreme caution.
- Secure the nozzles of any charged hoselines to prevent accidental discharge.
- Use a tag line to help control the hoisted object.
- Avoid hoisting tools and equipment if it is safer to hand-carry them up stairs, a ladder, or an aerial device.

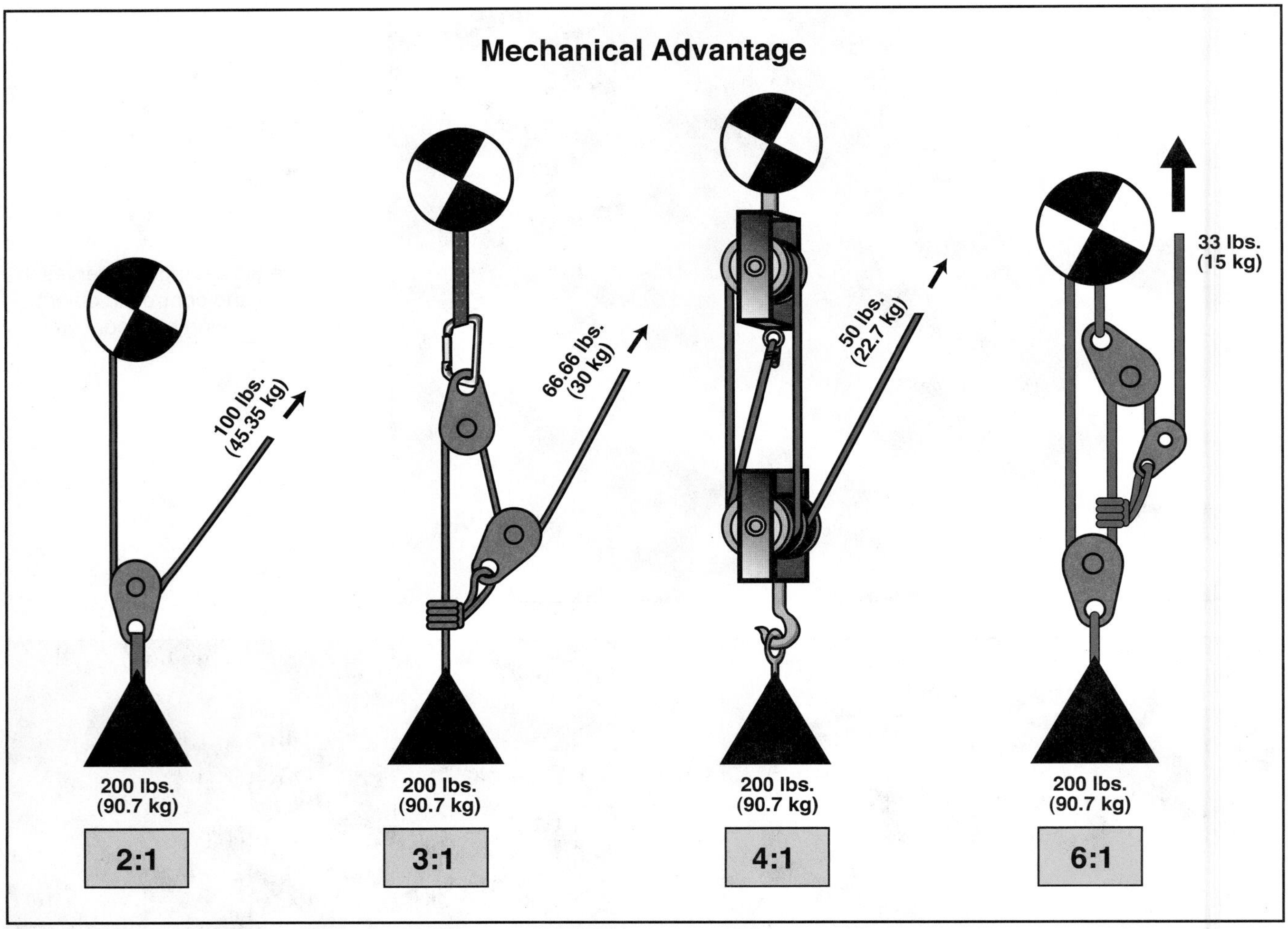
Mechanical Advantage
100 lbs.
(45.35 kg)
200 lbs.
(90.7 kg)
2:1
66.66 lbs.
(30 kg)
200 lbs.
(90.7 kg)
3:1
50 lbs.
(22.7 kg)
200 lbs.
(90.7 kg)
4:1
33 lbs.
(15 kg)
200 lbs.
(90.7 kg)
6:1

Figure 8.34 A pulley may be used by itself or in conjunction with other pulleys to create a mechanical advantage system.

T.F.D.
RESCUE-4
T.F.D.
RESCUE-4
T.F.D.
RESCUE-4
RESCUE-4

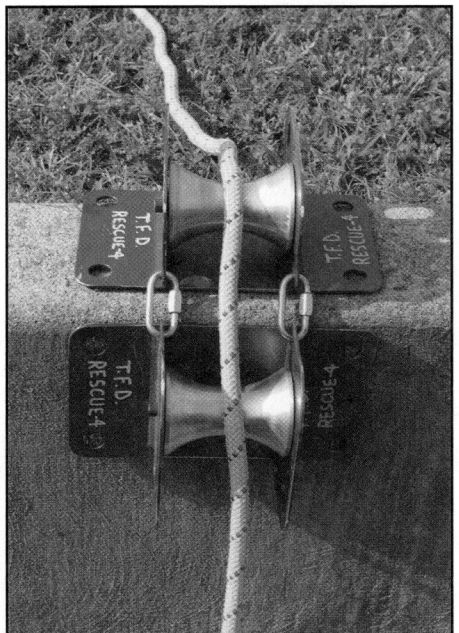

Figure 8.35 Edge rollers are designed to prevent damage to rope as it is pulled over sharp or abrasive edges.

Figure 8.36 A tag line attached to an item being hoisted allows someone on the ground to keep a tool from being damaged or causing damage.

Figure 8.37 A bowline or a figure-eight may be used to lift roof ladders.

Figure 8.38 An uncharged hoseline may be hoisted quickly to upper floors using a length of rope.

Hoisting an Axe

The procedure for attaching and hoisting is the same for either a pick-head axe or a flat-head axe. The hoisting rope can also be used as the tag line. Refer to **Skill Sheet 8-I-13** for more details.

Hoisting a Pike Pole

To raise a pike pole with the head up, tie a clove hitch near the butt end of the handle, followed by a half-hitch in the middle of the handle and another half-hitch around the head **(Figure 8.36)**. Refer to **Skill Sheet 8-I-14** for more details.

Hoisting a Ladder

Tie a bowline or figure-eight on a bight and slip it through two rungs of the ladder, about one-third of the way from the top of the ladder **(Figure 8.37)**. After pulling that loop through, slip it over the top of the ladder. Secure a tag line to the ladder near the foot. Refer to **Skill Sheet 8-I-15** for more details.

Hoisting Hoselines

Hoisting hoselines is often the fastest and safest way to get them to upper floors, but take extra care to avoid damaging the nozzle or coupling **(Figure 8.38)**. Charged hoselines can be hoisted, but it is safer and easier to hoist a dry hoseline. Refer to **Skill Sheet 8-I-16** for more details. If it is necessary to hoist a charged hoseline, **Skill Sheet 8-I-17** explains this procedure.

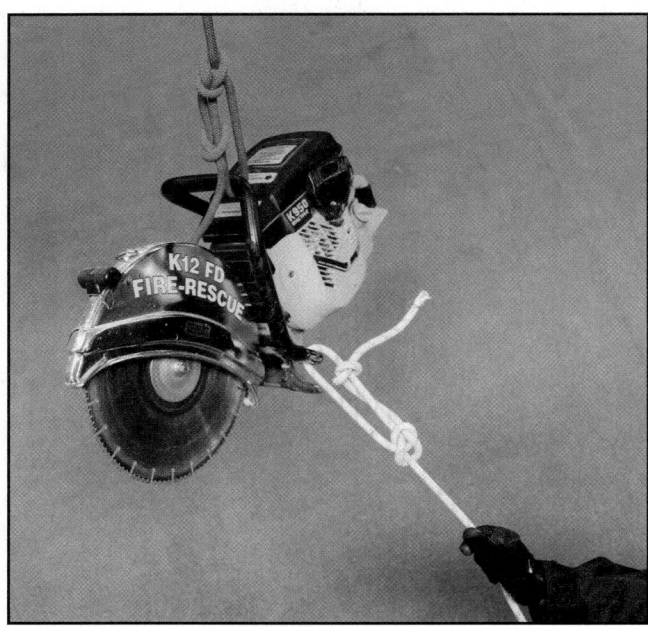

Figure 8.39 Power tools may be safely hoisted using two lengths of rope to haul and stabilize the tool. *Courtesy of Shad Cooper/Wyoming State Fire Marshal's Office.*

Hoisting a Power Saw

To hoist a rotary saw or chain saw, tie a bowline or figure-eight on a bight through the closed handle, then attach a tag line through the handle (**Figure 8.39**). Refer to **Skill Sheet 8-I-18** for more details.

Other Emergency Scene Uses

Besides hoisting and lowering tools and equipment, utility rope and webbing may also be used in other applications. The most common applications are designating control zones, establishing a search lifeline, and stabilizing objects.

Control Zone Perimeter

Utility rope has traditionally been used to establish the perimeter to control access to required control zones (**Figure 8.40**). Clove hitches with half-hitch overhand safety knots are used to tie the rope to trees, sign posts, or other stationary objects.

Barrier Tape

In recent years, polyvinyl barrier tape has become popular for this use. It is 2.5 micrometers (mil) thick and comes in 3 inch x 1,000 foot (76.2 mm by 304.8 m) rolls. Vinyl barrier tape is a highly visible yellow, with words in black stating "Caution Fire Line Do Not Enter." It is inexpensive, easy to store on the apparatus and can be abandoned in place at the end of the incident.

Search Lines

Search lines are used to help search teams working in dark, smoke-filled, or confined spaces. They allow team members to remain in contact with each other and with firefighters at the line's entry point (**Figure 8.41**). They also provide a physical means of finding an exit route. Branch lines are sometimes attached to the main search line,

Figure 8.40 Barrier tape or utility rope may be used to secure a perimeter. *Courtesy of Shad Cooper/Wyoming State Fire Marshal's Office.*

Figure 8.41 Search lines attached to a stable object outside a structure aid RIC/RIT members to find distressed firefighters. *Courtesy of Shad Cooper/ Wyoming State Fire Marshal's Office.*

allowing team members to search larger areas away from the search line while still remaining in contact with the team. For more information about search lines, see Chapter 9.

Object Stabilization

Utility rope and webbing are sometimes used to stabilize an object. For example, they may be used to prevent a vehicle from falling after it has rolled on its side or is suspended over an edge. The rope or webbing is secured to a strong stationary object and then tied to the object to be stabilized. Wire rope or cable from a winch may also be used for this operation. Before any work is performed around the object you must be certain that:

- The rope or webbing and the anchor point are strong enough to hold the weight of the object.

- The knots are tight and safety knots are in place.

- The attachment points at both ends are secure and will not pull free.

- Personnel are clear from the stabilizing line in case it breaks and snaps back.

Chapter Summary

Rope is one of the oldest and most basic tools used by firefighters. It is used for both life safety and utility purposes and may be constructed from synthetic or natural fibers. Synthetic webbing is often used in conjunction with rope.

Firefighters use rope and webbing to hoist tools and equipment, stabilize objects, designate control zones, perform rescues, and escape from life-threatening situations. To use rope and webbing safely and effectively, you must know the various types of ropes and their applications. You must also be capable of tying a variety of knots quickly and correctly. Finally, you must know how to inspect, clean, maintain, and store ropes and webbing so that they are ready for use when needed.

Review Questions

Firefighter I

1. What are the differences in the characteristics of life safety and utility rope?

2. What are the basic guidelines for rope maintenance?

3. Why would a rope need to be placed out of service?

4. What are the two main uses for webbing?

5. What are the three parts of a knot?

6. What are the three elements of a knot that can be combined to create knots and hitches?

7. What are three knots commonly used in your jurisdiction and what are their different uses?

8. What kinds of rope hardware may be encountered when hoisting using rope?

9. What are three safety guidelines that must be used when hoisting tools or equipment?

Inspect Rope

Step 1: While using hands, visually inspect the entire length of the rope for soft, crusty, stiff, or brittle spots; areas of excessive stretching; cuts, nicks, and abrasions; dirt, embedded objects, and other obvious flaws; as well as for cleanliness.

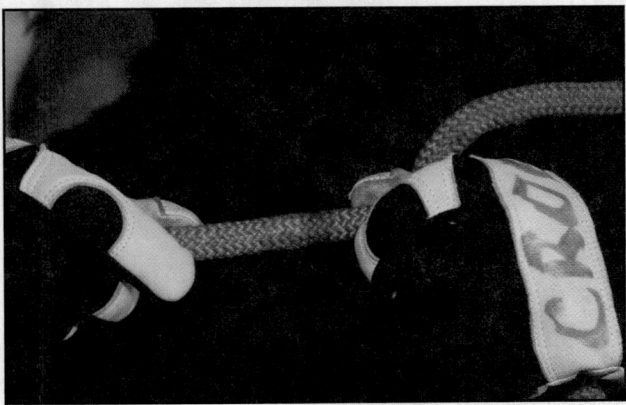

Step 2: Determine if rope has been impact loaded, overloaded, chemically contaminated, or does not meet life-safety reuse requirements.

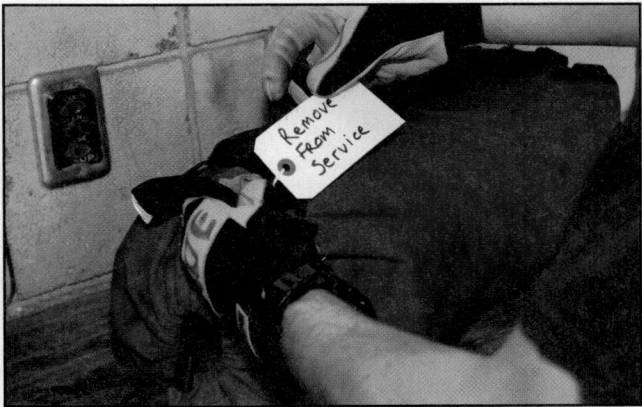

Step 3: Remove any flawed rope from service, disposing of it or labeling it as utility rope per local protocol.

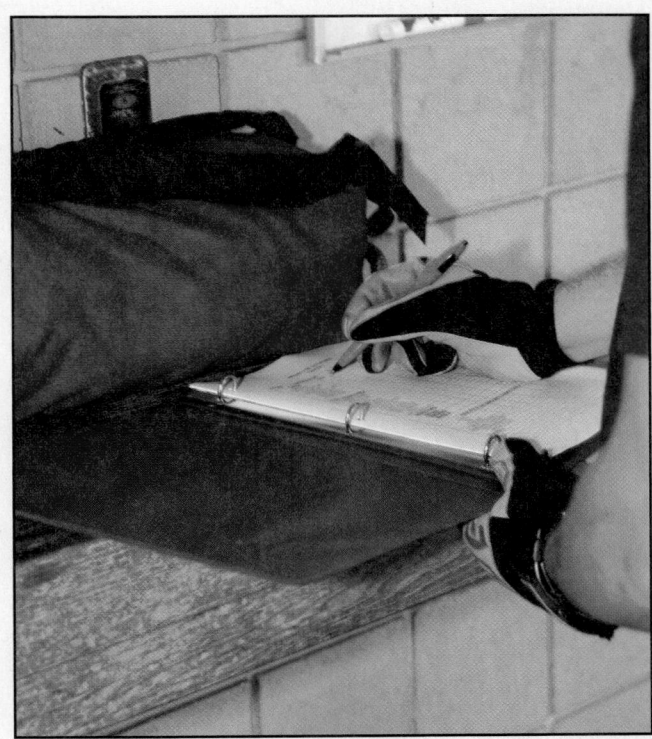

Step 4: Record information in rope logbook.

Clean Rope

Step 1: Clean the rope according to manufacturer's guidelines.

Step 2: Thoroughly rinse the rope.

Step 3: Dry the rope according to manufacturer's recommendations.

Store Rope

Step 1: Store rope per local protocol.

Step 1: Form a loop in the rope.

Step 2: Insert the end of the rope through the loop.

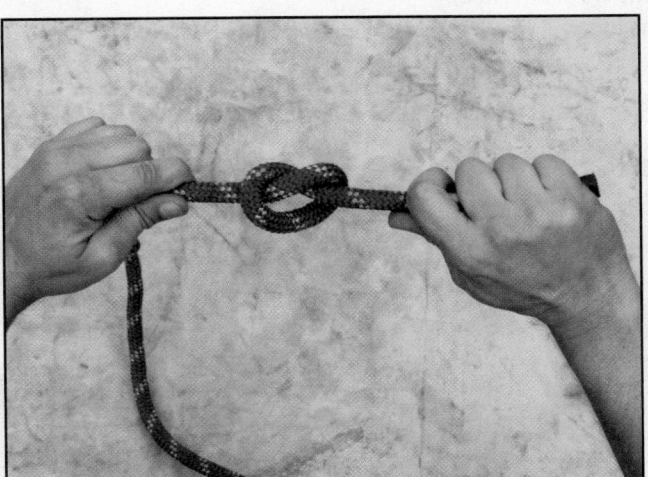

Step 3: Dress the knot by pulling on both ends of the rope at the same time.

Step 1: Select enough rope to form the size of the loop desired.

Step 2: Form an overhand loop in the standing part.

Step 3: Pass the working end upward through the loop.

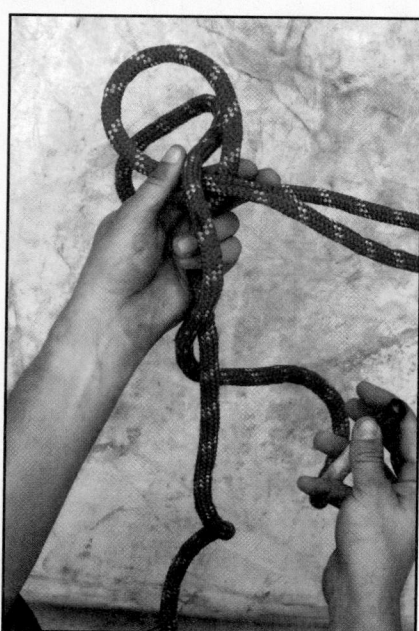

Step 4: Pass the working end over the top of the loop under the standing part.

Step 5: Bring the working end completely around the standing part and down through the loop.

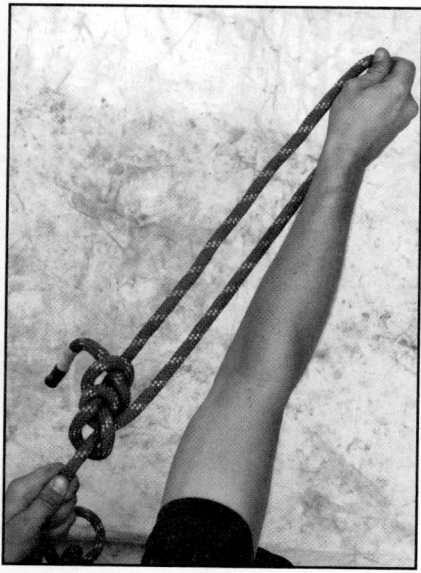

Step 6: Pull the knot snugly into place, forming an inside bowline with the working end on the inside of the loop.

Step 7: Secure the bowline with an overhand safety.

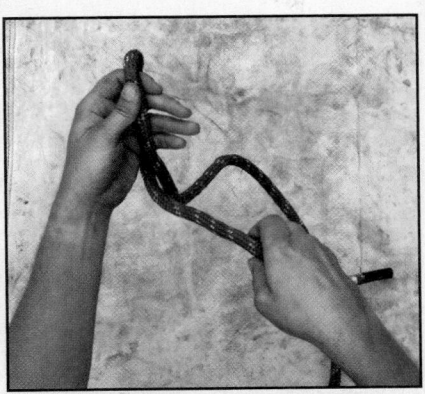

Step 1: Form a loop in your left hand with the working end to the right crossing under the standing part.

Step 2: Form another loop in your right hand (creating a round turn) with the working end crossing under the standing part.

Step 5: Slide the knot over the object.

Step 3: Slide the right-hand loop on top of the left-hand loop.

NOTE: This is the important step in forming the clove hitch.

Step 4: Hold the two loops together at the rope forming the clove hitch.

Step 6: Pull the ends in opposite directions to tighten.

SKILL SHEETS

8-1-5
Tie a clove hitch
around an object.

Step 1: Make one complete loop around the object, crossing the working end over the standing part.

Step 4: Set the hitch by pulling.

Step 2: Complete the round turn about the object just above the first loop as shown.

Step 3: Pass the working end under the upper wrap, just above the cross.

Step 5: Secure with an overhand safety.

Step 1: Form two loops in a length of rope.

Step 2: Simultaneously pull the inner side of each loop through the opposite loop to form two new loops.

Step 3: Place the loops over the victim's hands or feet with the knot between them.

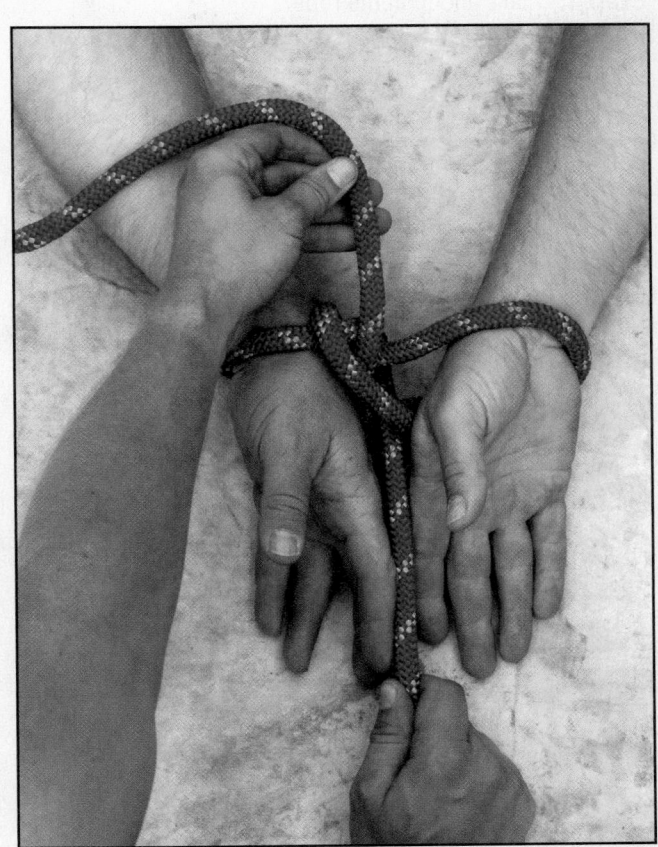

Step 4: Then tighten the knot by pulling the two rope ends away from one another.

Step 1: Make a loop in the rope.

Step 3: Insert the end of the rope back through the loop.

Step 2: Pass the working end completely around the standing part.

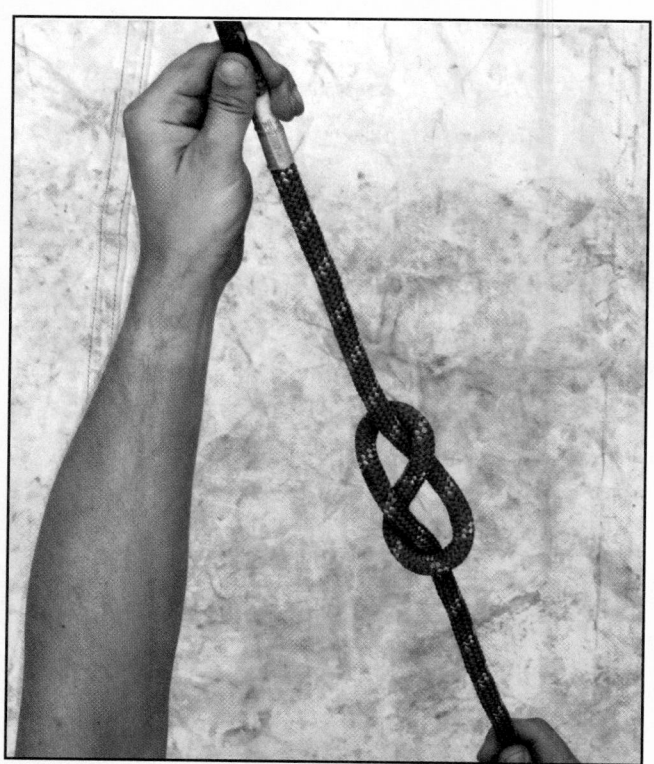

Step 4: Dress the knot by pulling on both the working end and standing part of the rope at the same time.

Step 1: Tie a figure-eight knot on one end of the rope.

Step 3: Use a safety knot, such as the overhand, with this knot.

Step 2: Feed the end of the other rope through the figure-eight knot in reverse. It should follow the exact path of the original knot.

Step 1: Form a bight in the working end of the rope.

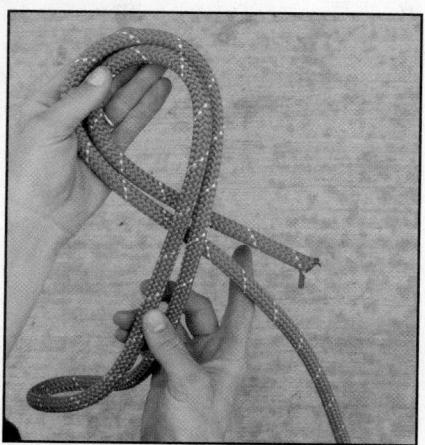

Step 2: Pass it over the standing part to form a loop.

Step 3: Pass the bight under the standing part and then over the loop and down through it; this forms the figure-eight.

Step 4: Extend the bight through the knot to whatever size working loop is needed.

Step 5: Dress the knot.

Step 6: Secure with an overhand safety.

Step 1: Tie a loose figure-eight knot.

Step 4: Exit the rope beside the standing end to complete the knot.

Step 2: Pass the tail end of the rope around the object to be secured.

Step 3: Follow the original figure-eight around the entire knot in reverse.

Step 5: Dress the knot.

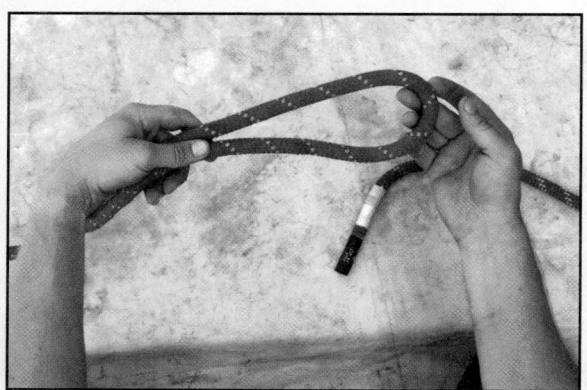

Step 1: Form a bight in one of the ends to be tied (if two ropes of unequal diameter are being tied, the bight always goes in the larger of the two).

Step 2: Pass the end of the second rope through the bight.

Step 3: Bring the loose end around both parts of the bight.

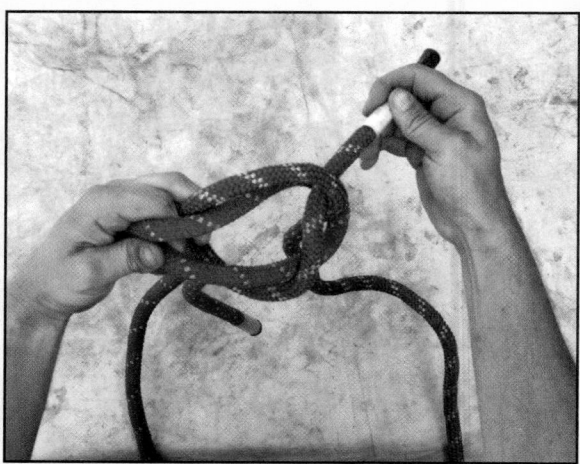

Step 4: Tuck this end under its own standing part and over the bight.

Step 5: Pull the knot snug.

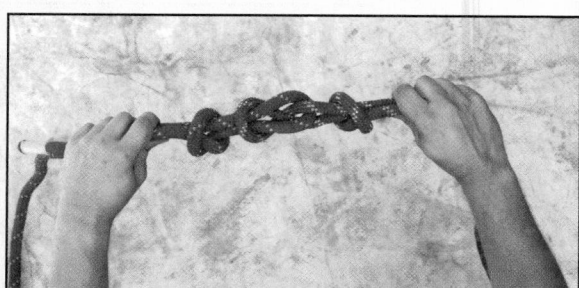

Step 6: Secure with two overhand safety knots.

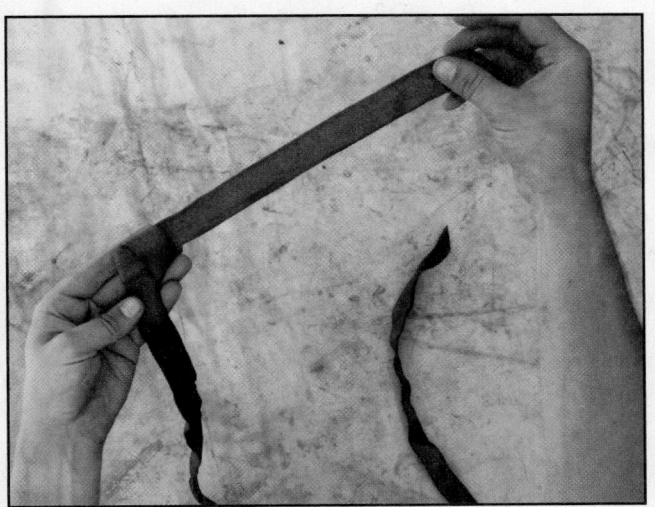

Step 1: Tie an overhand knot loosely in the end of webbing.

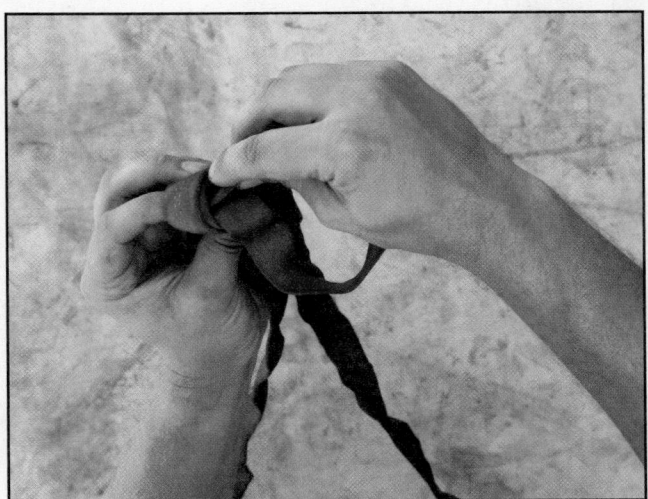

Step 2: Take the opposite end of the webbing and retrace the overhand knot.

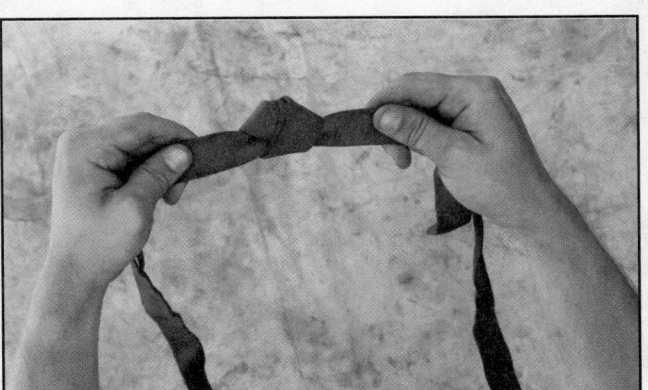

Step 3: Tighten by pulling both sets of ends of the webbing.

Step 4: Dress the water knot so it lays flat and no webbing trace is twisted.

Step 1: Lower an appropriate length of rope from the intended destination of the axe.

Step 2: Tie a clove hitch using the method shown in Skill Sheet 8-I-4.

NOTE: If the rope has a loop in the end, the loop may be used instead of a clove hitch.

Step 3: Slide the clove hitch down the axe handle to the axe head. The excess running end of the rope becomes the guideline.

Step 4: Loop the working end of the rope around the head of the axe and back up the handle.

Step 5: Tie a half-hitch on the handle a few inches (millimeters) above the clove hitch.

Step 6: Tie another half-hitch at the butt end of the handle.

NOTE: Though not required, a safety between the axe and the first half of the hitch will ensure no slipping. Use the running end to create this safety.

Step 7: Hoist the axe.

Step 1: Lower an appropriate length of rope from the intended destination of the pike pole.

Step 2: Secure the rope to the pike pole pointing upward toward the end of the handle using a clove hitch.

Step 3: Leave enough excess running end so that it becomes the guideline.

Step 4: Tie a half-hitch or approved knot around the pike pole in the middle of the handle.

Step 5: Tie a second half-hitch or approved knot around the pike pole under the pike hook.

Step 6: Hoist the pike pole.

Step 1: Lower an appropriate length of rope from the intended destination of the ladder.

Step 2: Make a loop in the end of the rope using a bowline knot.

Step 3: Place the closed loop under the ladder and bring it up between the rung about one-third the distance from the hoisting end.

Step 4: Open the loop and place it over the tip of the ladder.

Step 5: Arrange the standing part under the ladder rungs.

Step 6: Tighten the loop around the beams, pulling the standing part of the rope up behind rungs toward ladder tip.

Step 7: Tie a guideline to the ladder.

Step 8: Hoist the ladder.

Step 1: Lower an appropriate length of rope from the intended destination of the hoseline.

Step 2: Fold the nozzle end of the hoseline back over the rest of the hose so that an overlap of 4 to 5 feet (1.2 m to 1.5 m) is formed.

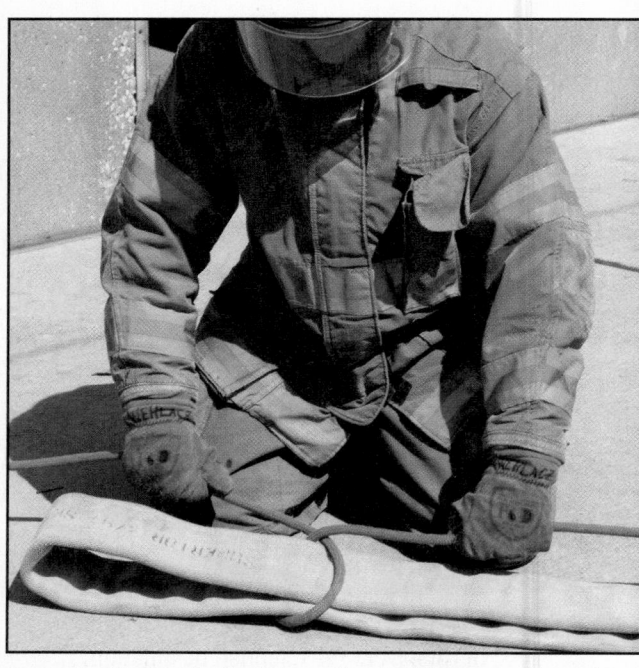

Step 4: Place a half-hitch on the doubled hose about 12 inches (305 mm) from the loop end.

NOTE: With the ties properly placed, the hose will turn on the hose roller so that the coupling and nozzle will be on top as the hose passes over the roller.

Step 3: Tie a clove hitch, with an overhand safety knot, around the tip of the nozzle and the hose it is folded against so that they are lashed together.

Step 5: Hoist hoseline.

Step 1: Lower an appropriate length of rope from the intended destination of the hoseline.

Step 2: Tie a clove hitch, with an overhand safety knot, around the hose about 1 foot (0.3 m) below the coupling and nozzle.

Step 4: Tie a half-hitch around the nozzle to take the strain off the handle.

Step 3: Pass a bight through the nozzle handle and loop it over the nozzle so that the rope holds the nozzle shut while it is being hoisted.

Step 5: Hoist hoseline.

Step 1: Lower an appropriate length of rope from the intended destination of the power saw.

Step 2: Secure the rope to the handle of the power saw using a figure-eight on a bight or a bowline knot. If a bowline is used, it must include an overhand safety knot.

Step 3: Leave enough excess running end so that it becomes the guideline.

Step 4: Hoist the power saw.

Chapter Contents

Key Terms

NFPA® Job Performance Requirements

This chapter provides information that addresses the following job performance requirements of NFPA® 1001, *Standard for Fire Fighter Professional Qualifications* (2013).

Firefighter I

5.2.4

5.3.1

5.3.5

5.3.9

1. Summarize the impact of building construction and floor plans on structural search techniques. (5.3.9)

2. Explain size-up and situational awareness considerations during structural searches. (5.3.9)

3. Summarize safety guidelines for structural search and rescue. (5.3.9)

4. Differentiate between primary and secondary search techniques. (5.3.9)

5. Recognize basic search methods. (5.3.9)

6. Describe victim removal methods. (5.3.5, 5.3.9)

7. Explain firefighter survival methods. (5.3.1)

8. Explain what survival actions firefighters can take when needed. (5.3.9)

9. Describe the actions of a rapid intervention crew or team (RIC/RIT) when locating a downed firefighter. (5.3.9)

10. Demonstrate the procedure for conducting a primary search. (Skill Sheet 9-I-1, 5.3.9)

11. Demonstrate the procedure for conducting a secondary search (Skill Sheet 9-I-2, 5.3.9)

12. Demonstrate the incline drag. (Skill Sheet 9-I-3, 5.3.9)

13. Demonstrate the webbing drag. (Skill Sheet 9-I-4, 5.3.9)

14. Demonstrate the cradle-in-arms lift/carry — One-rescuer method. (Skill Sheet 9-I-5, 5.3.9)

15. Demonstrate the seat lift/carry — Two-rescuer method. (Skill Sheet 9-I-6, 5.3.9)

16. Demonstrate the extremities lift/carry — Two-rescuer method. (Skill Sheet 9-I-7, 5.3.9)

17. Demonstrate the actions required for transmitting a MAYDAY report. (Skill Sheet 9-I-8, 5.2.4, 5.3.5, 5.3.9)

18. Demonstrate the proper procedures for an SCBA air emergency. (Skill Sheet 9-I-9, 5.3.1, 5.3.5, 5.3.9)

19. Demonstrate the actions required for withdrawing from a hostile environment with a hoseline. (Skill Sheet 9-I-10, 5.3.1, 5.3.5, 5.3.9)

20. Demonstrate low profile maneuvers without removing SCBA - Side first technique. (Skill Sheet 9-I-11, 5.3.1, 5.3.5, 5.3.9)

21. Demonstrate low profile maneuvers without removing SCBA - SCBA first technique. (Skill Sheet 9-I-12, 5.3.1, 5.3.5, 5.3.9)

22. Demonstrate the method for breaching an interior wall. (Skill Sheet 9-I-13, 5.3.5, 5.3.9)

23. Demonstrate the steps for disentangling from debris or wires. (Skill Sheet 9-I-14, 5.3.5, 5.3.9)

Chapter 9
Structural Search, Victim Removal, and Firefighter Survival

Case History

During fire fighting operations, a 42-year-old firefighter was killed and another was injured in a partial structural collapse at an apartment complex under renovation. The victim and three crew members were exiting the complex through a breezeway that connected the fire structure to an uninvolved structure when a section of brick veneer from the uninvolved structure collapsed onto the victim and the injured firefighter. The injured firefighter called for help and was freed by other firefighters. Rescue crews did not immediately realize there was another trapped firefighter. A personal accountability report (PAR) was ordered by the Incident Commander (IC) which determined that the victim was missing. A second search located the victim unresponsive and without a pulse beneath a pile of bricks. The victim was extricated, given emergency medical treatment, and transported to a hospital where he was pronounced dead.

The subsequent National Institute for Occupational Safety and Health (NIOSH) investigation determined that the victim was wearing an SCBA with integrated Personal Alert Safety System (PASS) device and a secondary manual PASS device. The integrated PASS was not activated because the SCBA was not in use at the time. The secondary unit had not been manually activated when the wall collapsed. One of the NIOSH recommendations was that personnel be instructed to activate PASS devices prior to entering the hot zone.

This incident illustrates the importance of always activating the PASS device when working in a hazardous area. The device is designed to operate when the wearer is motionless for 30 seconds or longer. Besides smoke inhalation a firefighter can become incapacitated from blunt trauma, as in this case, cardiac arrest, stroke, or fatigue, among other causes. The PASS will not summon assistance if it has not been activated. Only training and practice will ensure that firefighters consciously turn on the device when they enter the hazard area.

Locating and removing victims, generally known as search and rescue, requires firefighters to take calculated risks in hazardous situations. Before you are able to participate in this critical aspect of a firefighter's mission, you must know:

- How to perform a primary and secondary search
- How to use standardized markings to indicate the areas you have searched
- How to work as part of a team
- How to determine the best method for removing victims
- Emergency survival techniques
- General MAYDAY protocol
- How to remove incapacitated firefighters

Ⅰ Structural Searches

The NFPA® reports that in 2009, structural fires killed almost 2,700 civilians and injured 14,800 more. These figures would be far worse if not for the thousands of potential victims firefighters located and removed from burning structures **(Figure 9.1)**. To rescue victims, firefighters must be able to conduct an effective structural search. This requires extensive training and a knowledge of:

- Building construction
- Floor plans and layouts
- Size-up and situational awareness
- Search safety guidelines
- Search procedures
- Thermal imagers
- Search marking systems
- Fire behavior
- Other operations that may influence searches, such as fire control and tactical ventilation

Figure 9.1 Search and rescue activities account for thousands of saved lives and mitigated injuries.

Building Construction

All firefighters must be aware of how building construction affects fire development. This knowledge will help you to predict where fire will spread, how quickly it will develop, and the potential for structural collapse. It will also alert you to safe areas within a structure, design aspects that may prevent you from exiting, and walls or partitions through which you can escape.

Building Floor Plan

To conduct an effective structural search, firefighters must know the layout or floor plan. This knowledge may come from inspections, **preincident surveys**, architectural plans, or personal observation.

Preincident Survey —
Assessment of a facility or location made before an emergency occurs, in order to prepare for an appropriate emergency response.

Many fire departments perform periodic preincident surveys of **target hazards** in their response areas **(Figures 9.2a-d)**. Preincident surveys allow departments to meet the facility's owner/occupants and become familiar with the contents, floor plans, building construction, and manufacturing processes.

Target Hazard — Any facility in which a fire, accident, or natural disaster could cause substantial casualties or significant economic harm, through either property or infrastructure damage.

Figure 9.2a Hospitals shelter a high population of individuals who are unable to care for themselves.

Figure 9.2b Public schools have a high population of people who are dependent on others for resources.

Figure 9.2d High rise buildings may cause major urban damage if compromised.

Figure 9.2c Refineries house highly dangerous chemicals and machinery.

When new buildings are built or existing structures are modified, architectural building plans are submitted to the local authority having jurisdiction (AHJ) for approval. In some jurisdictions, the fire department receives a copy of the plans and is involved in the review and approval process. The floor plan may also be made available **(Figure 9.3)**.

In addition, fire companies should tour buildings under construction or renovation to learn about construction materials and floor plan arrangements. New construction surveys are especially important for residential construction, where the department will not have access for inspections or preincident planning once construction is complete. Another opportunity to observe residential dwellings occurs when the department installs smoke detectors as part of its fire prevention activities.

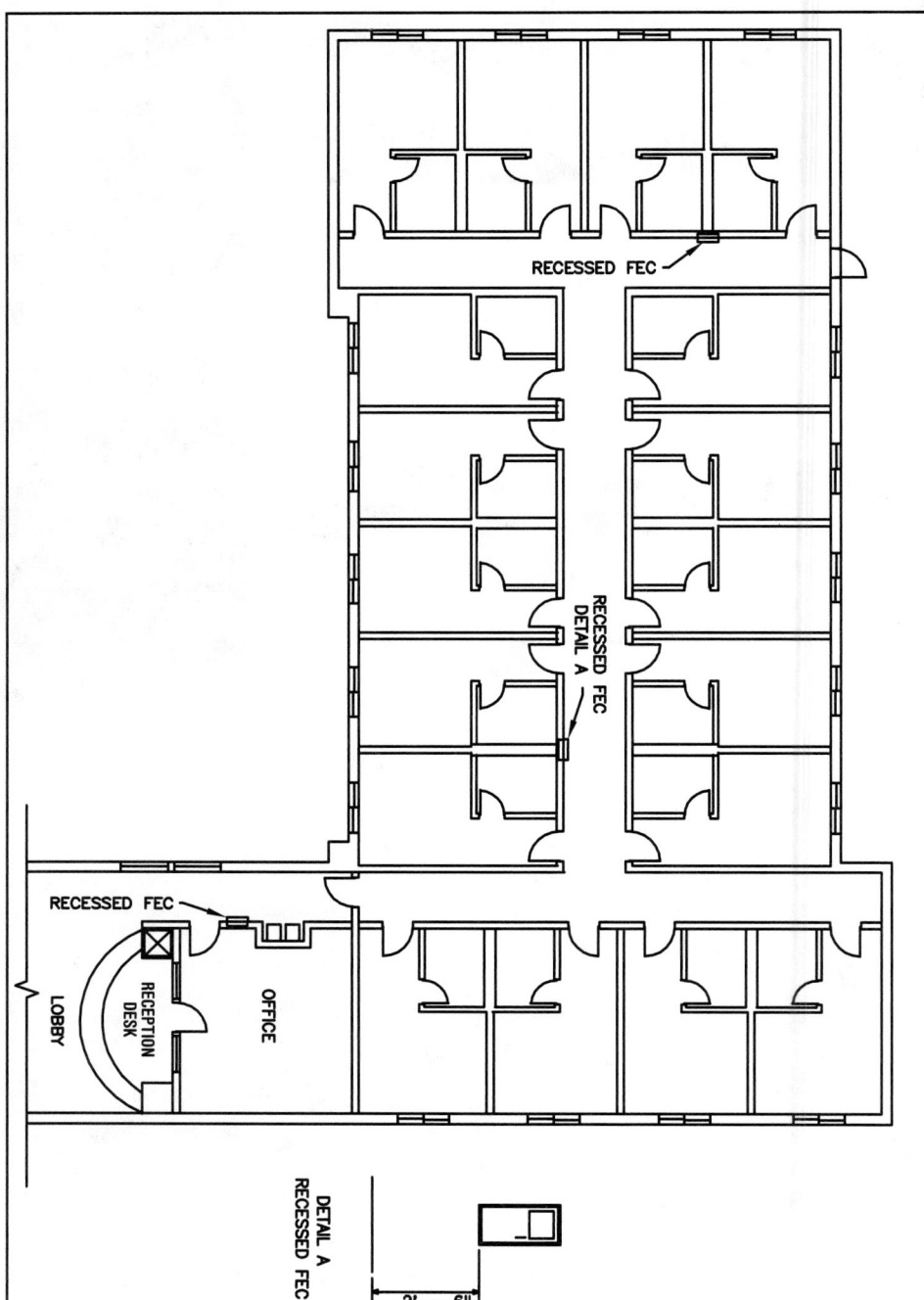

Figure 9.3 A partial floor plan drawing may show locations of fire extinguisher cabinets, labeled in this diagram as FECs.

Figure 9.4 External features of a building give some indication of the use of the spaces inside.

Finally, your personal experience can help you determine the possible floor plan of a structure. Observe the layout of all structures in your response area, including commercial buildings that you visit on a regular basis. Note the similarities between residential structures such as hotels, motels, apartment buildings, and single-family dwellings. After any emergency incident, observe the floor plan and the location of doors, windows, vent pipes, and chimneys **(Figure 9.4)**. These items, visible from the outside, can give you a general idea of the interior floor plan. Remember, though, that interior alterations can drastically change the layout, so be prepared for the unexpected.

You can also gain knowledge by attending local zoning meetings, going to realtors' open houses, and getting to know local building officials. The more you learn, the better prepared you will be when an emergency occurs.

Size-Up and Situational Awareness

Your safety and the safety of your team members depends on your application of size-up and situational awareness. Both of these activities are essential skills that you must learn and practice at every opportunity.

Size-Up

Size-up involves observing the scene of an incident in order to answer the following questions:

- What has happened?
- What is happening?
- What is going to happen?

The first firefighter at the scene must perform the initial size-up and report existing conditions. Throughout the incident, all personnel should also actively monitor conditions so that everyone can stay informed. However, the ultimate responsibility for incident size-up rests with the officer in charge.

Situational Awareness

Begin exercising your situational awareness when you first arrive at the scene. Carefully observe the exterior of the structure for indications of the size and location of the fire, and try to determine whether the building is occupied, based on clues such as vehicles in the driveway or lights visible in windows. Assess the probable struc-

Size-Up — Ongoing evaluation of influential factors at the scene of an incident.

tural integrity of the building, and how long it will take to effectively search the structure. Always identify possible escape routes, such as doors, windows and fire escapes, before entering.

Communication is an important aspect of situational awareness. The fireground is a dynamic environment in which conditions can change rapidly, so tell other members of your company what you observe and pay attention to what they observe. Listen to radio reports from the communication center and other units who may be observing sides of the structure you cannot see. Throughout the incident, keep your team members and supervisors informed of any changes you observe.

After you enter the structure, use your senses to increase your awareness. Listen for sounds that indicate the fire is becoming more intense. Watch for the color of smoke, which indicates the type of fuel and the phase of

Figure 9.5 The handle of a tool may be used to sound the floor to determine its condition.

the fire. Feel walls and doors with the back of your hand to determine whether there is fire on the other side. **Sounding** the floor before you advance will help you determine if it will support your weight **(Figure 9.5)**. Listen for the sounds of structural movement, sagging support members, or obvious structural displacement; these key indicators of structural instability can alert you to impending danger.

Search Safety Guidelines

Search safety guidelines in a structural fire include the following:

- Do not enter a structure in which survivors are not likely to be found **(Figure 9.6)**. If you observe conditions that indicate a lack of survivors, report this to your supervisor.

- If there is a possibility of extreme fire behavior, do not attempt entry until coordinated fire control and ventilation have been implemented.

- Do not **freelance**. Work according to the incident action plan (IAP).

- Maintain radio contact with the Incident Commander (IC).

- Monitor radio traffic for important information or changes in orders.

- Continuously monitor fire conditions that might affect your safety and that of other firefighters.

- Use your department's personnel accountability system.

- Be aware of your entry point and the secondary means of egress from the structure.

Sounding — Striking the surface of a roof or floor to determine its structural integrity or locate underlying support members; the blunt end of a hand tool is used for this purpose.

Freelance — To operate independently of the Incident Commander's command and control.

- Wear full personal protective equipment, including SCBA and personal alert safety system (PASS) device.

- Work in teams of two or more and always remain in physical, visual, or vocal contact.

- When opening or forcing doors, maintain control by placing a strap around the doorknob; this allows you to shut the door quickly if conditions on the other side make it necessary (**Figure 9.7**).

- If you encounter fire in a room, close the door and report the condition.

- Search systematically to increase efficiency and reduce the possibility of becoming disoriented.

Figure 9.6 A fully involved structure does not shelter survivors.

Figure 9.7 A strap around the doorknob allows firefighters to maintain control of a door during searches.

- Where visibility is limited, stay low and move cautiously.

- Continuously monitor the structure's integrity and communicate any changes.

- Mark entry doors into rooms and remember the direction you turned when entering. To exit the building, turn in the opposite direction when leaving the room.

- When visibility is obscured, maintain contact with a wall, hoseline, or search line.

- If possible, have a staffed charged hoseline available when working on the fire floor or the floors immediately above or below. The hoseline can be used for fire suppression, crew protection, and indicating the path out of the structure.

- Coordinate with the IC and ventilation teams before opening windows to relieve heat and smoke.

- Inform your supervisor immediately of any room or rooms that could not be searched.

- Report promptly to your supervisor once the search is complete.

- Keep your supervisor informed of the progress of the fire and the physical condition of the building.

Search Preparations

Before entering any area that is immediately dangerous to life and health (IDLH), prepare yourself and other members of your crew or team (**Figure 9.8**). You must ensure that:

- You know who you report to.

- You have all the necessary tools and equipment, including forcible entry tools, hand light, thermal imagers, and search line.

- Your portable radio is turned on, working properly, and set to the correct fireground channel.

- Your SCBA is turned on, working properly, and contains a full cylinder of air.

- Your PASS device is turned on and working properly.

- You have signed in with the Accountability Officer or Incident Safety Officer.

- You know your assigned duties and the tactical objectives of your crew or team.

- You are aware of the alternate means of egress from the structure.

Search Procedures

Witnesses can be a source of information about occupants who have not been accounted for. Question escaped occupants to obtain information about who might still be inside and their possible location, and the location and extent of the fire. If neighbors are familiar with the occupants' habits and room locations, they can suggest where the occupants are likely to be found. They may also have seen someone near a window before you arrived. Any relevant information from a witness should be relayed to the IC and all incoming units. If possible, all information should also be verified. Never assume that all occupants are out until the building has been searched.

If there are no witnesses, assume that the structure is occupied. Even vacant, boarded-up structures can contain homeless or indigent people (**Figure 9.9**). Many commercial, educational, institutional, and industrial occupancies are occupied 24 hours a day by shift-workers, janitorial staff, or security personnel.

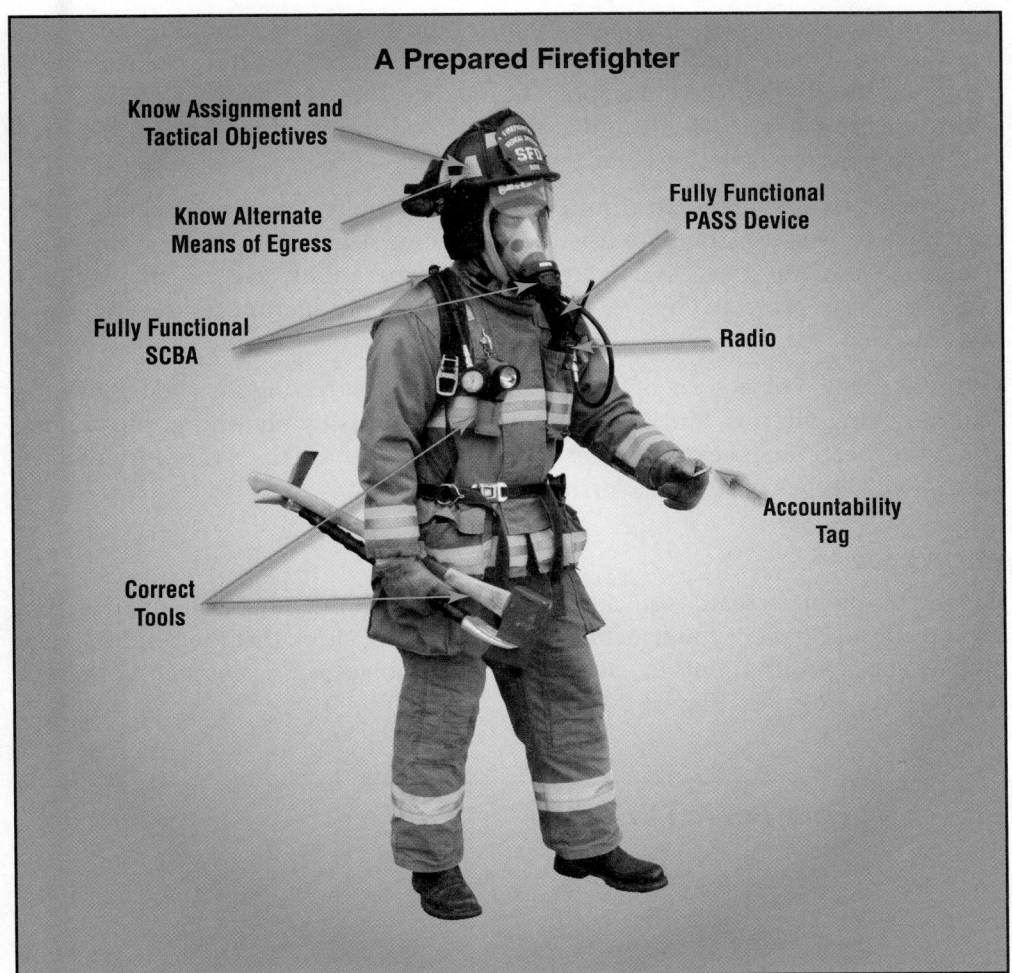

A Prepared Firefighter

Know Assignment and
Tactical Objectives

Know Alternate
Means of Egress

Fully Functional
SCBA

Correct
Tools

Fully Functional
PASS Device

Radio

Accountability
Tag

Figure 9.8 A prepared firefighter
must have the correct gear and
knowledge prepared before
entering an environment that is
immediately dangerous to life
and health.

Figure 9.9 A seemingly vacant
commercial structure may shelter
occupants.

If possible, fire attack and ventilation should be started simultaneously with any interior search. Effective fire control and ventilation create more survivable conditions for firefighters and trapped occupants. They also improve visibility, enabling search crews to quickly find and remove victims. In some cases, fire control may even have to take place before search activities.

However, if resources are limited or if local policy requires, search teams may have to perform the search while advancing a hoseline into the structure. If extinguishment is necessary during search operations, use extreme caution. Firefighters are protected by PPE, but excess steam production can severely burn victims.

There are two objectives of a structural search: **searching for life** by locating and removing victims, and **assessing fire conditions** by obtaining information about the location and extent of the fire. In most structure fires, the search for life requires two types of searches: **primary** and **secondary**.

Primary Search

During the primary search, quickly check the known or likely locations of victims and all affected areas of the structure. While doing so, check that fire conditions are as they appeared from outside, and report any changes you encounter. Search the most critical areas first (**Figure 9.10**), in this order of priority:

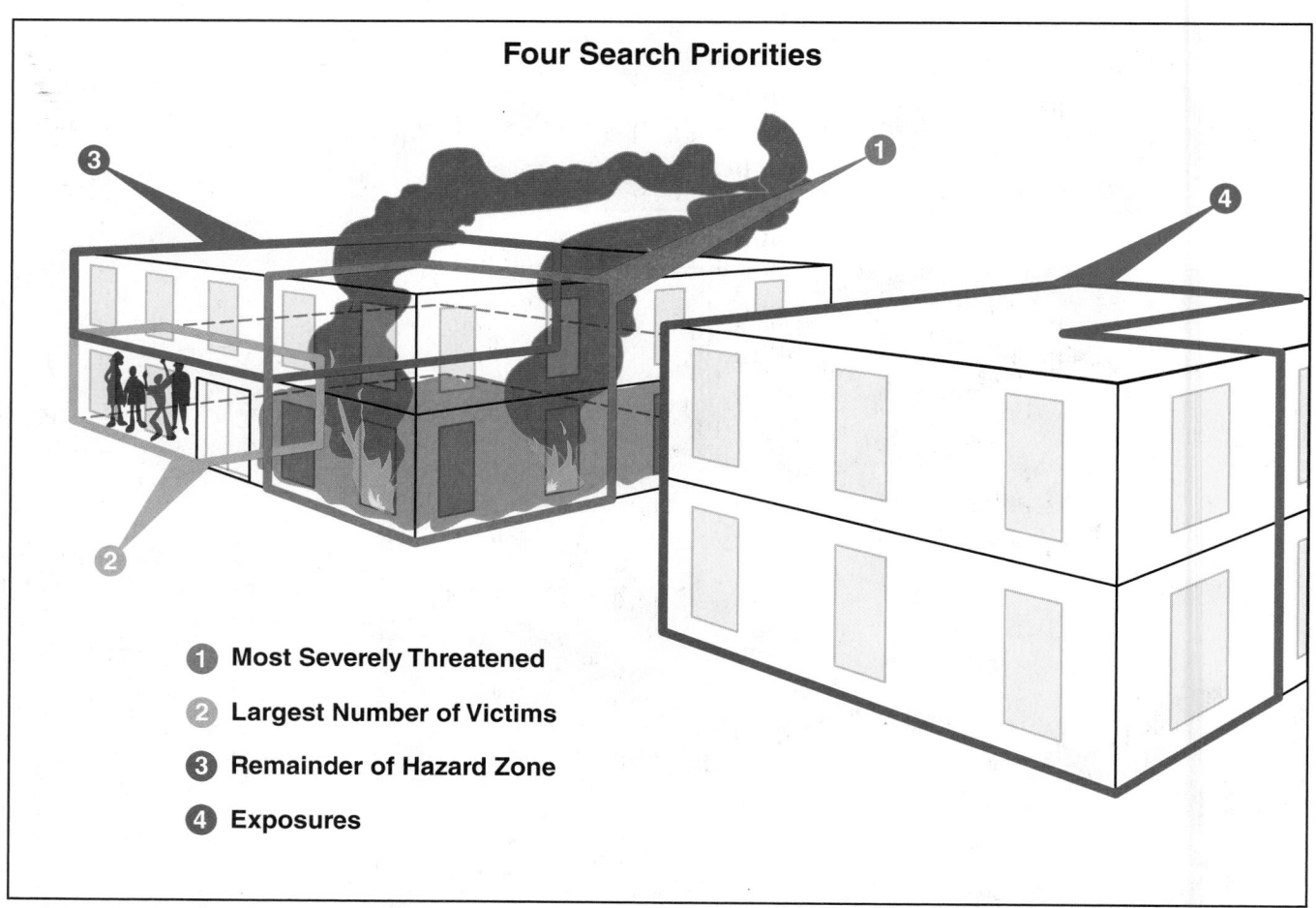

Four Search Priorities

1. **Most Severely Threatened**
2. **Largest Number of Victims**
3. **Remainder of Hazard Zone**
4. **Exposures**

Figure 9.10 The four search area priorities are ranked in order of urgency.

1. **Most severely threatened** — The area closest to the fire on the fire floor and the floor directly above. In multistory structures, the top floor is also considered a severely threatened area due to the accumulation of smoke.

2. **Largest numbers** — Areas that contain the largest possible number of victims.

3. **Remainder of hazard zone** — Areas farthest from the fire on the same level, upper floors, and floors below the fire floor.

4. **Exposures** — Interior and exterior.

During the primary search, always use the buddy system and work in teams of two or more. Rescuers working together can conduct a safer, faster search. When searching in an IDLH atmosphere, maintain physical, visual, or voice contact with other team members. The procedures for conducting a primary search are described in **Skill Sheet 9-I-1**.

Secondary Search

After initial fire suppression and ventilation have been completed, a secondary search is conducted by personnel who did not participate in the primary search (**Figure 9.11**). Using different personnel to conduct the secondary search has the advantage of allowing the search team to use "fresh eyes" and get an unbiased view of the scene. Secondary searches are slower and more thorough than primary searches. However, they must be just as systematic as the primary search to ensure that no rooms or spaces are missed. Structural instability and areas in which the fire is starting to **rekindle** must be reported immediately. The procedures for conducting a secondary search are listed in **Skill Sheet 9-I-2**.

Rekindle — To reignite because of latent heat, sparks, or smoldering embers; rekindling can be prevented by proper overhaul.

Even though the interior of the building may appear to be free of smoke, do not remove your SCBA while conducting the secondary search. Fire gases such as carbon monoxide and hydrogen cyanide may still be present after the fire is extinguished. Air monitoring is the only effective way to determine the presence of toxic gases. Once air monitoring has determined that the atmosphere is safe, your supervisor or safety officer will tell you that you can remove your SCBA.

Search Methods

Each department has its own search procedures, and you will be trained to apply these in a variety of situations. However, most procedures draw on the same general search methods, which can be applied to almost any type of search. Specialized search methods include oriented search, wide area search, and search using thermal imagers.

General Methods

General methods for primary and secondary searches follow a systematic pattern. When you enter a room, turn right or left and follow the walls around until you return to your starting point. As you leave the room, turn in the same direction you did to enter and continue to the next room to be searched (**Figure 9.12, p. 432**). For example, if you turned left when you entered the room, turn left when you leave the room. To remove a victim to safety or to exit the building, turn opposite the direction you turned to enter. Always exit through the same doorway you entered to ensure a complete search.

Figure 9.11 A secondary search is conducted by team members who did not participate in the primary search.

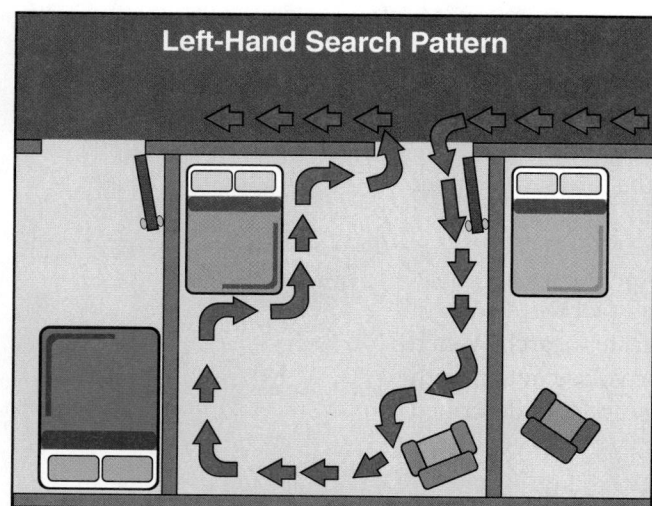

Left-Hand Search Pattern

Figure 9.12 By picking a standard direction to search, a first responder is less likely to become dangerously disoriented in a low-visibility environment.

On the fire floor, start your search as close to the fire as possible and then work back toward the entrance door. This allows your team to reach those in the greatest danger first. People farther from the fire are in less immediate danger, so they can safely wait as your team moves back toward safety. To reach a point nearest the fire, proceed as directly as possible from the entry point. Advancing a hoseline or deploying a search line will provide a way for you to remain oriented so that you can find your way out quickly if fire conditions change.

When rooms extend from a center hallway, both sides should be searched using the oriented-search method discussed. If two teams are available, each one can search opposite sides of the hallway. If there is only one search team, it must search down one side of the hallway and back up the other side (**Figures 9.13 a and b**). Always control the egress passageways so that search teams can escape if conditions change rapidly. This can be accomplished by:

- Wedge doors open to prevent them from shutting behind you or closing on a hoseline.

- Close doors to rooms adjacent to the passageway after they are searched.

- Position hose teams at intervals along the path to cool accumulated gases.

One good search practice is to get low to the floor to perform a quick survey. Thermal layering of heat and the buoyancy of smoke will produce a clear area of vision just above the floor level. Victims, obstacles, or the general layout of a room may be identified more quickly from this perspective.

Depending on conditions, you may be able to walk upright or you may have to crawl on your hands and knees. Walking is preferable if there is minimal smoke and heat. But in heavy smoke or extreme heat, crawling on your hands and knees below the smoke level can increase visibility and reduce the risk of tripping or falling. Although crawling is much slower, an advantage is that it is much cooler near the floor. If you have to use stairs while crawling, proceed head first while ascending and feet first while descending (**Figure 9.14**). Keep your feet and hands as far apart as possible to distribute your weight close to the side of the stairs. That way you can brace yourself if the stairs collapse.

When to Crawl During a Search

If you encounter extreme heat or cannot see your feet through the smoke, you should not be walking upright.

Victims may be found in paths of egress and any areas where they seek shelter from the fire, such as:

- Bathrooms
- Bathtubs
- Shower stalls

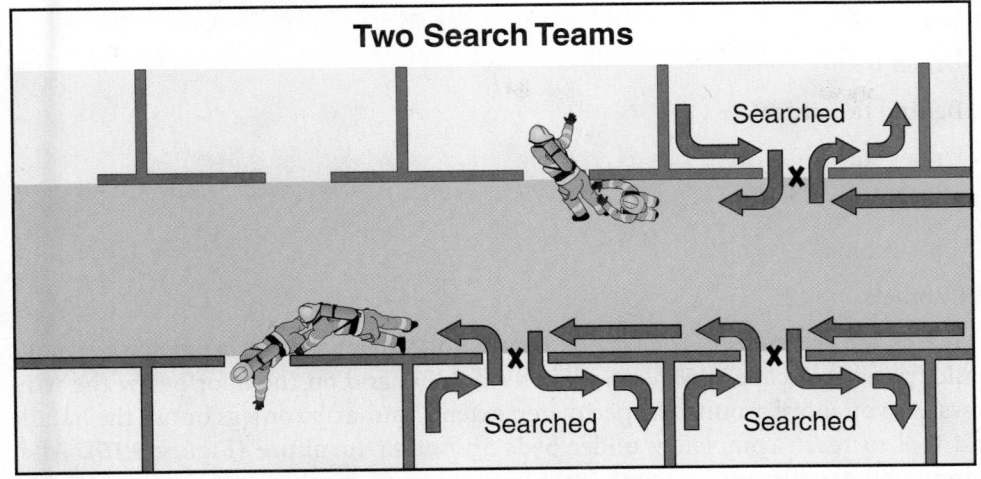

Figure 9.13a When two search teams are available, each team takes one side of a hallway to search.

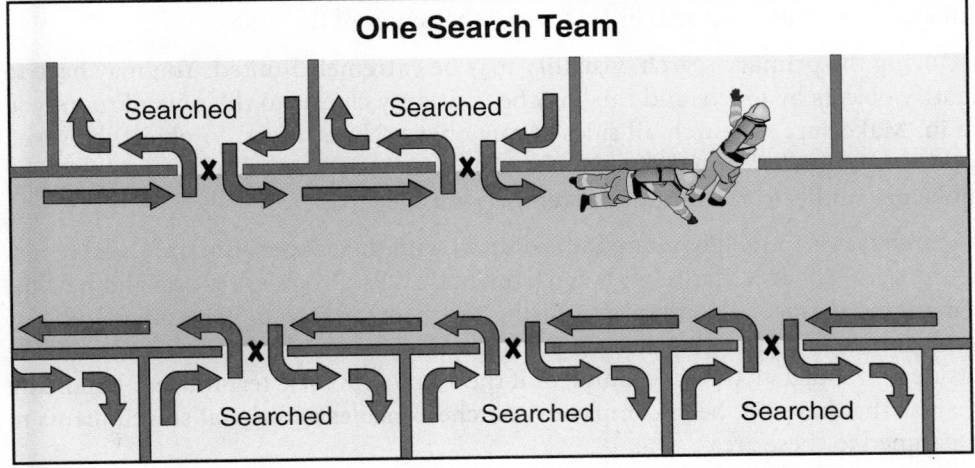

Figure 9.13b When only one search team is available, the team begins on one side of the hallway and works around in a loop.

Figure 9.14 Firefighters descending a staircase during extreme fire conditions should keep their hands and feet close to the edge of the stairs.

- Closets
- Under beds
- Behind furniture
- Under stairs
- Basements
- Attic rooms
- Cabinets

Search the perimeter of each room. Because occupants may be overcome by smoke while trying to escape, always check behind doors and on the floor below the windows. As you move around the perimeter, extend your arms or legs or use the handle of a tool to reach completely under beds and other furniture (**Figure 9.15**). After searching the perimeter, search the middle of the room by placing the tool against the wall and extending your arm or leg toward the center of the room.

During the primary search, visibility may be extremely limited. You may have to identify objects by touch, and this may be your only clue as to the type of room you are in. Make sure to search all sides of any object. Never move an object, however, as this may disorient you. If smoke obscures your vision, report this to the IC. The smoke may indicate that additional ventilation is needed.

Search teams should maintain radio contact with their supervisor or IC and report their progress in accordance with departmental SOPs. Progress reports and new information are especially important during the primary search. For example, the IC should be notified immediately if the fire has spread farther than it appeared from the outside, if trapped victims are found, or if the search has to be terminated. Reporting on areas that have not been completely searched enables additional search teams to be assigned to these areas.

Close the doors to any rooms that are not involved in fire unless the doors are being used for ventilation. This prevents fire from spreading into these rooms. Opening doors and windows can disrupt ventilation efforts and can even spread the fire by drawing it toward the opening.

Figure 9.15 A tool handle extends a searcher's reach under low clearance spaces.

Clear unused hoselines and other equipment from exit pathways. This reduces tripping hazards and generally makes egress less difficult. Be aware of the location of exit pathways in case you have to remove a victim quickly.

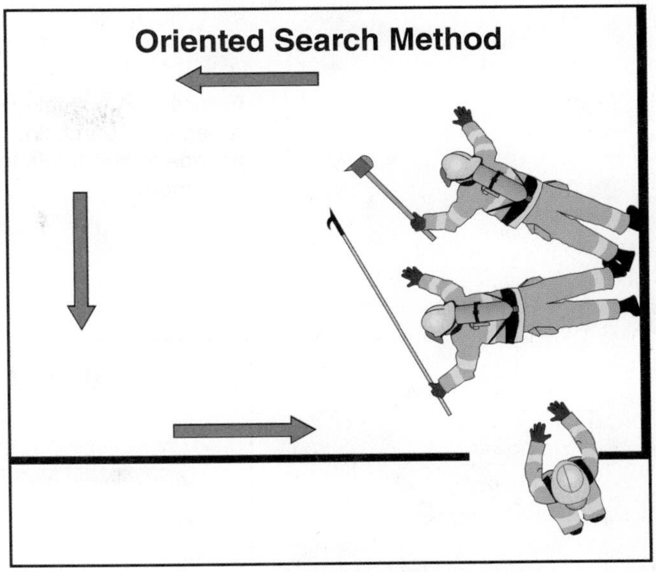

Oriented-Search Method

The oriented-search method is an efficient way for a team to search a room. The team leader remains anchored at the door, wall, or hoseline, while other team members spread out through the room to complete the search **(Figure 9.16)**. All members stay in constant communication with the leader and each other, and update the leader on their progress. They must coordinate their efforts in order to prevent confusion and avoid clustering in one section of the room. After the search is complete, the searchers return to the anchored team leader and proceed to the next room.

Wide-Area-Search Method

A wide-area-search system is sometimes used to conduct a primary search of a large or complex area that is filled with smoke. This system employs a dedicated search line, typically 200 feet (60 m) of ⅜-inch (10 mm) rope with a Kevlar™ sheath to resist heat and abrasion. A minimum of three team members are required, although larger teams can be more effective.

About 10 feet (3 m) outside the entry point to the search area, the end of the search line is tied to a fixed object about 3 feet (1 m) above the floor. A tag indicating the unit or company designation is left at that point **(Figure 9.17)**. An attendant is sometimes stationed at the entry point to maintain communication with the team and monitor their air management.

One team member, usually called the *lead*, picks up the rope bag containing the search line and enters the search area. The lead is accompanied closely (shoulder to shoulder) by another member called the navigator. The navigator directs the lead using a hand light and, if available, a thermal imager (TI) **(Figure 9.18)**. They

Oriented Search Method

Figure 9.16 Oriented search involves one team member maintaining contact with a fixed point and other team members. spreading out in the open area, while all team members remain in constant communication contact.

Figure 9.17 Tags on a search line outside the search area should indicate the unit or company using the line.

Figure 9.18 The lead and the navigator work together to systematically search an area. *Courtesy of Jim Sobota.*

Figure 9.19 A steel ring is tied into search line to provide tether points for searchers.

Figure 9.20 When using a search line to find the exit, the ring will point toward the exit and the knots next to the ring will point toward the fire.

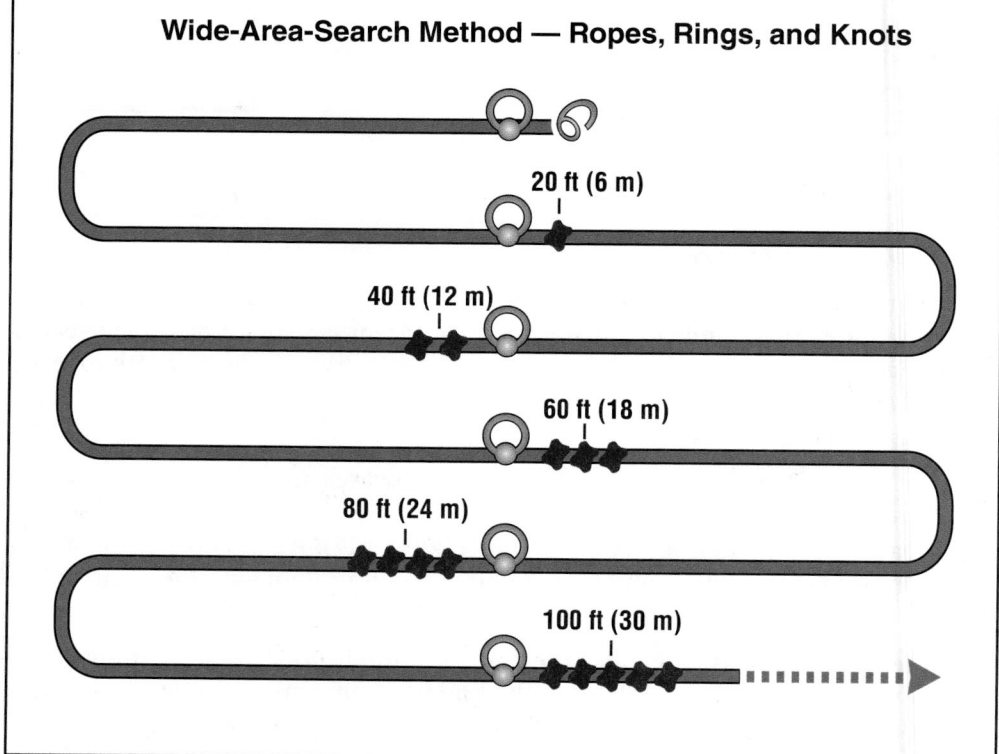

Wide-Area-Search Method — Ropes, Rings, and Knots

20 ft (6 m)

40 ft (12 m)

60 ft (18 m)

80 ft (24 m)

100 ft (30 m)

are followed by one or more radio-equipped searchers. Each searcher carries a tether wrapped around one wrist and a forcible entry tool in the other hand. As the team progresses into the building, the search line pays out behind them and all members maintain contact with the search line.

Every 20 feet (6 m) along its length, a 2-inch (50 mm) steel ring is tied into the search line (**Figure 9.19**). Immediately behind each ring, one or more knots are tied in the search line to indicate distance. After the first ring, one knot indicates 20 feet (6 m) from the beginning of the line. After the second ring, there are two knots indicating 40 feet (12 m) from the beginning of the search line. After the third ring, three knots are tied, and so on (**Figure 9.20**). The knots indicate the distance from the beginning of the search line and they are always *behind* the ring, so they provide a directional indication — knots toward the fire; rings toward the exit.

The rings also provide an anchor point for lateral tethers. The tethers are 20-foot (6 m) lengths of ¼-inch (6 mm) rope with a Kevlar™ sheath. Each tether has a ¾-inch (19 mm) steel ring tied to one end, a knot at the mid-point, and either a nonlocking carabiner or a snap hook on the other end. Each member of the search team carries one tether.

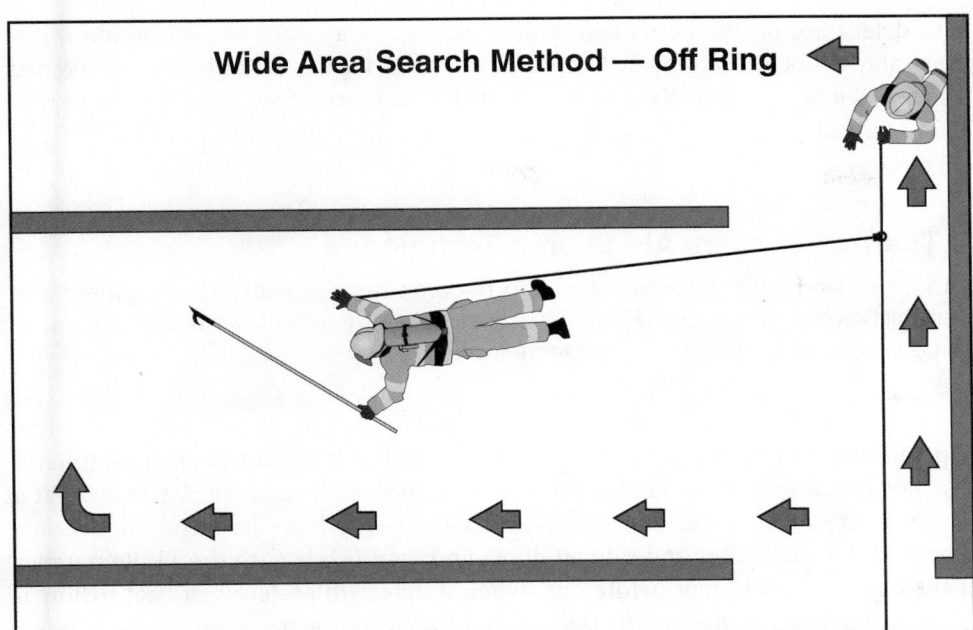

Wide Area Search Method — Off Ring

Figure 9.21 Searching off-ring allows the team member on the search line to move the entire length between the rings while searching.

The tethers enable team members to search areas perpendicular to the search line. They snap their tether onto one of the search line's steel rings, or at any point between the rings, then pay out the tether while moving away from the search line. Reaching the tether's mid-point knot allows the searcher to make a 10-foot (3 m) arc from the attachment point. If nothing is found, the searcher can progress an additional 10 feet (3 m) to the end of the tether **(Figure 9.21)**. There, the searcher can sweep a 20-foot (6 m) arc.If there is still more area to be searched, a second searcher attaches a tether to the ring on the end of the first tether. This enables the second searcher to sweep in a 40-foot (12 m) arc away from the search line. When the searchers return to the search line, they disconnect from the ring and rewind their tethers around their wrists.

Any time team members move off from the search line, they must stay in voice contact with the navigator. The navigator also constantly updates the IC, reporting on fire conditions, what the team has found, and how many knots into the building they have progressed.

Thermal-Imager-Search Method

Thermal imagers (TIs) allow firefighters to see sources of heat through darkness and thick smoke. They are used to locate victims and hidden fires **(Figure 9.22)**. TIs are typically assigned to a chief officer or to specialized units such as heavy rescue companies or rapid intervention teams. In some cases they are assigned to all companies for use in search and overhaul operations.

TIs can also detect heat through barriers, but with important limitations. They cannot detect a person under or behind furniture or on the opposite side of a wall. They also cannot see through water or glass. If a structure is carpeted, TIs may not be

Figure 9.22 A thermal imager shows patterns of temperature in a room where victims may be hidden by debris or dust.

able to detect fire on the floors below. This can create a safety hazard because fire-fighters may think a room is safe to enter when the floor has actually been weakened by the fire below. Another disadvantage is that TIs are very fragile and prone to me-chanical failure.

Thermal Imagers Aid Search Techniques

Practice using your department's TI to become familiar with the equipment's capabilities and limitations. Remember that TIs are an effective search aid, but not a substitute for proper search techniques.

Operate the TI according to manufacturer's instructions and your department's SOPs. Slowly scan the TI around the room close to floor level, then rise to scan at a higher level. If you encounter furniture, use a manual search method to check beneath and behind it. Open closet and cabinet doors and scan inside with the TI. Remember that the camera screen may "white out" when it detects high levels of heat. Allow it time to readjust before proceeding into a room that has shown a high heat level.

Inspect TIs after every use, while following the manufacturer's instructions. Replace batteries as needed, clean them regularly, and make sure they are properly stored. TIs are sensitive electronic tools and must be treated with care. Report any damage or malfunction immediately.

Marking Systems

A consistent room marking system is necessary to ensure a thorough, effective search. Searched rooms can be marked using any of the following:

- Chalk or crayon markers
- Specially designed door markers
- Latch straps over doorknobs

Figure 9.23 A search mark should be placed in a way that will make it visible to subsequent teams, as per local protocols.

Marks should be placed low so that they can be seen under the smoke. They can be placed on the lower third of the door, the lower third of the adjacent wall, or in the landing of adjacent stairs. Marking with latch straps has the added advantage of preventing the door from closing, which might trap the search team. Never mark a room by blocking the door open with furniture, as this can contribute to fire spread. Also, never place the mark inside the room. Subsequent searchers would then have to enter the room to find your marker (**Figure 9.23**).

Some departments have marking systems based on the Federal Emergency Management Agency's Urban Search and Rescue System (FEMA US&R). This sys-tem consists of diagonal marks that would fit into a 2 foot by 2 foot (0.61 m by 0.61 m) square (**Figure 9.24**). A diagonal mark from upper right to lower left indicates that a search is underway. When the search is complete, a second mark is made from upper left to lower right, forming an "X." These marks alert other search teams that they do not have to search that particular room or floor. If a search team fails to report and does not respond to calls for a personnel accountability report (PAR), rescuers can use these marks as a starting point for their search.

Adjacent marks can be used to convey additional information. The search unit is noted to the left of the "X" (for example, *Eng. 2*). Time of completion is noted above the "X" (for example, *09-21-13, 0230 hours*). Hazards are noted to right of the "X" (for example, *weak floor* or *broken glass*). Victims and their condition are noted below the

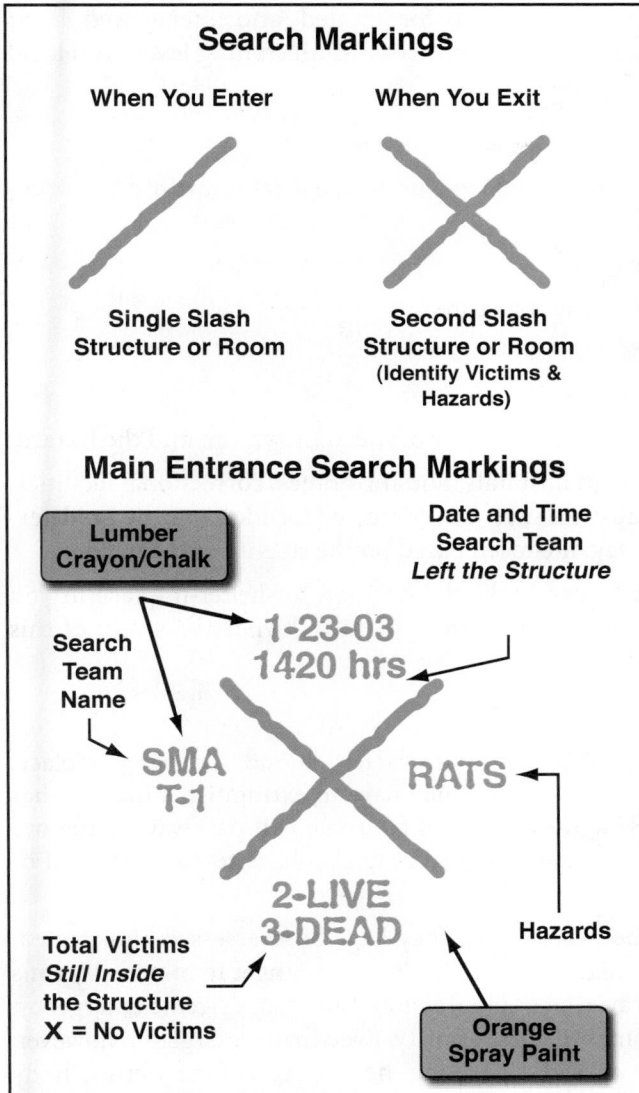

Search Markings

When You Enter

Single Slash
Structure or Room

When You Exit

Second Slash
Structure or Room
(Identify Victims &
Hazards)

Main Entrance Search Markings

Lumber
Crayon/Chalk

Date and Time
Search Team
Left the Structure

1-23-03
1420 hrs

Search
Team
Name

SMA
T-1

RATS

2-LIVE
3-DEAD

Total Victims
Still Inside
the Structure
X = No Victims

Hazards

Orange
Spray Paint

Figure 9.24 The spaces formed by the large "X" are used
to denote (counterclockwise from top) time of search,
name of search team, number of victims, and hazards.

Figure 9.25 Search teams should use the best
available space to mark the findings from their search.
Courtesy of Andrea Booher/FEMA.

"X" (**Figure 9.25**). This can include fatalities still in the room, live victims sheltered in place, or the fact that no victims were located during the primary search.

Victim Removal

Victims located during the search must be separated from the hazard. Depending on conditions, this can be done through self-evacuation, shelter-in-place, or rescue.

Self-Evacuation

Most occupants can evacuate a structure on their own or with minimal assistance. For example, you may have to direct them to an alternate exit or close stairwell doors to maintain the integrity of their exit path (**Figure 9.26**). You may also have to establish a safe haven away from the

Figure 9.26 Some assistance may be
helpful for victims who are otherwise able
to self-evacuate.

structure where occupants can be accounted for, treated, and interviewed. Your actual duties in any self-evacuation setting will depend on staffing levels and local SOPs.

Shelter-in-Place

Sheltering-in-place involves moving victims to a protected location within the structure **(Figure 9.27)**. It is used when:

- The hazard is minor.
- It is safer to keep victims inside the structure.
- Victims are incapacitated and cannot be moved.
- There is limited staffing to assist with evacuation.
- The structure can provide a protective barrier between the victim and the hazard.

Sheltering-in-place is common in hospitals, nursing homes, correctional facilities, high-rises, and high-hazard industrial sites. The protected location may be predetermined during a preincident survey or chosen based on the size-up of the incident.

Only your supervisor or the IC can make the decision to shelter-in-place. If you observe any condition within the structure that might influence the safety of this method, report it immediately.

Rescue

Rescue is required when conditions prevent self-evacuation and sheltering-in-place, or when victims are directly threatened. You may have to extinguish a fire that has cut them off from an exit, provide them with an alternate exit pathway, or remove debris from a victim who has been pinned. You may also have to carry injured or unconscious victims to safety.

Injured victims should not be moved until they have been assessed and treated, unless they (or you) are in immediate danger. The primary danger in moving victims quickly is the possibility of aggravating a spinal injury. But in an extreme emergency, preserving the victim's life becomes the first priority. Even in an emergency, however, never pull a victim sideways — instead, pull along the long axis of the victim's body **(Figure 9.28)**. If the victim is on the floor, pull on the victim's clothing in the neck or shoulder area.

Figure 9.27 Shelter-in-place exists as an option in structures where the inhabitants are not able to leave the confines of the structure and a safe place is defensible inside.

Shelter-in-Place

Improper lifting technique is a common cause of injury (**Figure 9.29**). Keep your back straight and lift with your legs, not your back. One rescuer can safely carry a small child, but two to four may be needed to safely carry an adult. Follow these safety guidelines:

- Lift as a team.
- Focus on keeping your balance.
- Support the head and neck.
- Avoid unnecessary jostling.

Never drag or carry a victim through the hazard zone unless there is no other choice. The various types of carries and drags are described in the following sections.

Incline Drag

The incline drag enables a rescuer to move the victim up or down a stairway or incline. It is especially useful for moving unconscious victims (**Figure 9.30**). This method is described in **Skill Sheet 9-I-3**.

Webbing Drag

In this drag, a rescuer pulls on a section of webbing that has been wrapped around the victim's body. The webbing drag is useful when heat and smoke force you to stay low, or the victim is a downed firefighter wearing an SCBA (**Figure 9.31, p. 442**). This method is described in **Skill Sheet 9-I-4**.

Figure 9.28 Proper moving techniques will minimize strain on a victim's existing injuries and help prevent new ones.

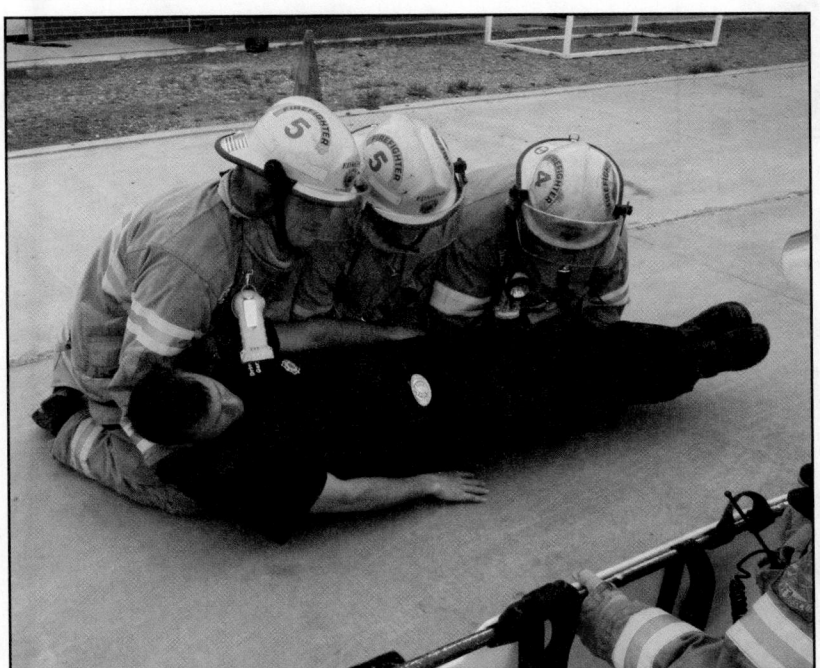

Figure 9.29 Safe lifting techniques minimize danger to victims as well as firefighters.

Figure 9.30 An incline drag may be used with an unconscious victim.

Cradle-in-Arms Lift/Carry

The cradle-in-arms technique is used to carry children or small, conscious adults (**Figure 9.32**). It is not practical for unconscious adults because of the victim's weight and relaxed body, and the difficulty in supporting the head and neck. This method is described in **Skill Sheet 9-I-5**.

Seat Lift/Carry

The seat lift/carry enables two rescuers to carry a conscious or unconscious victim (**Figure 9.33**). This method is described in **Skill Sheet 9-I-6**.

Moving a Victim onto a Litter

Occasionally, rescuers will have the advantage of being able to use a litter to remove a victim. The long backboard litter is common, but other types include the standard ambulance cot, army litter, basket litter, and scoop stretcher (**Figures 9.34a and b**). Similar techniques should be used for moving people onto stretchers and basket litters.

Extremities Lift/Carry

The extremities lift/carry enables two rescuers to move either a conscious or an unconscious victim (**Figure 9.35**). This method is described in **Skill Sheet 9-I-7**.

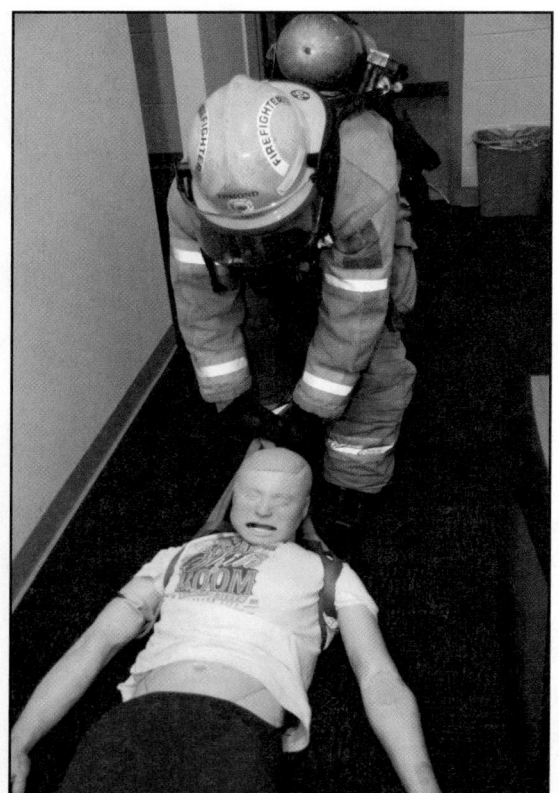

Figure 9.31 A webbing drag allows a rescuer to move a victim while remaining under heat and smoke.

Figure 9.32 Small children may be carried using the cradle-in-arms technique.

Figure 9.33 The seat lift carry allows firefighters to move a victim who is unable to assist in the process.

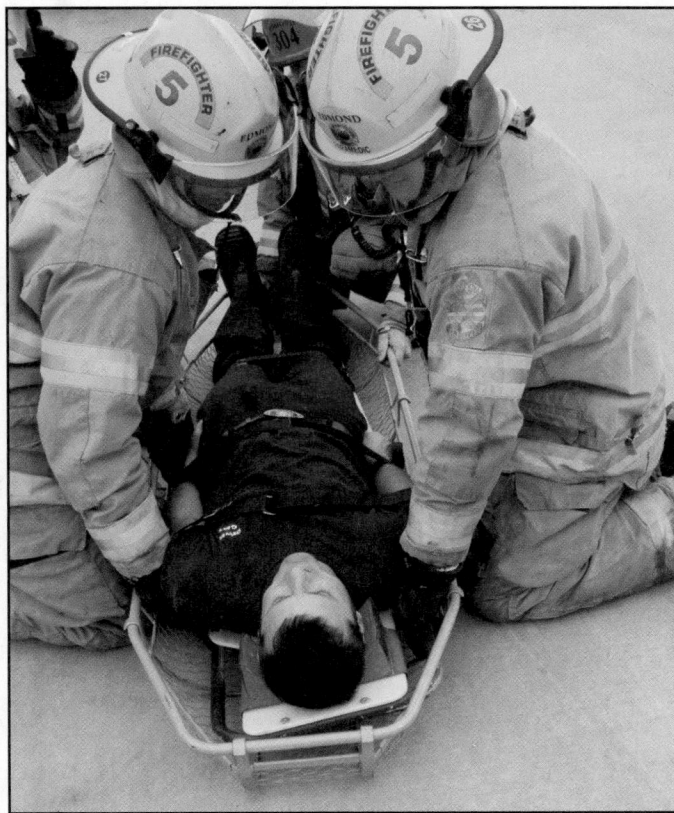

Figure 9.34a Litters enable firefighters to secure a victim to a rigid board or basket that is easier to carry.

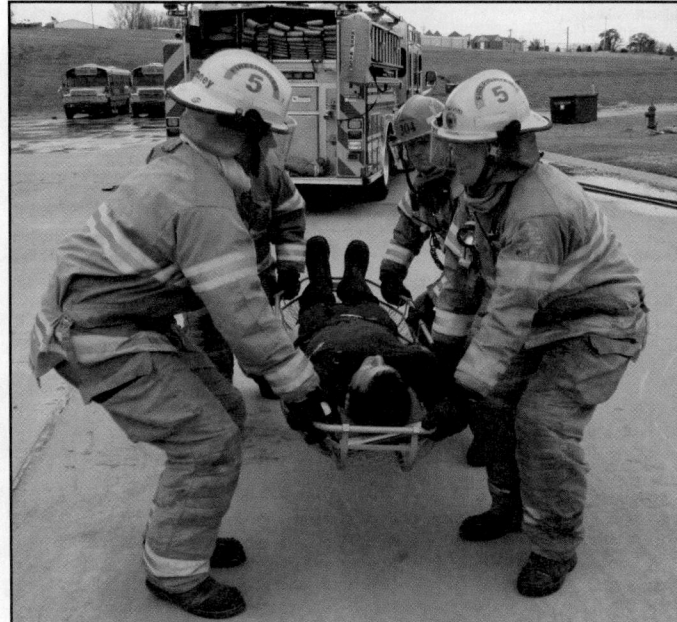

Figure 9.34b Lifting a litter may require several first responders using proper lifting techniques.

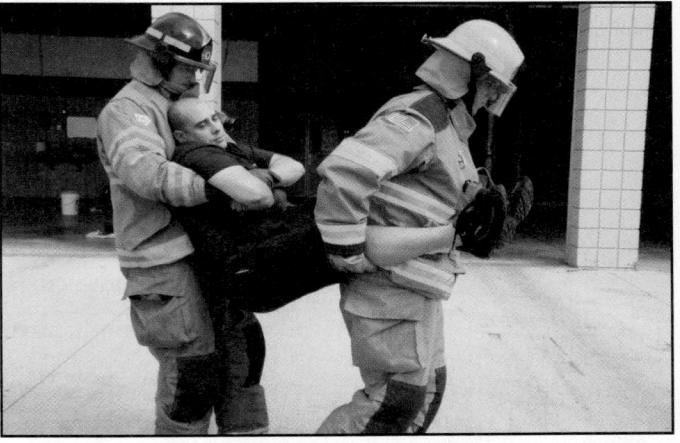

Figure 9.35 Firefighters may use the extremities carry to move an unconscious victim.

Firefighter Survival

Fire fighting is an inherently dangerous profession that requires calculated risk **(Figure 9.36, p. 444)**. To ensure your own survival and that of your fellow firefighters, you must learn to:

- Recognize and avoid potential hazards
- Escape unavoidable hazards
- Rescue lost or trapped firefighters

Prevention-Based Survival

The most important survival technique is avoiding potential hazards. To do so, you must be able to size up a situation before taking action. Guidelines for effective size-up include:

- Read the fire and anticipate the stages of fire development and spread.
- Anticipate the location and extent of the fire.
- Identify the building's construction type and potential for collapse.
- Locate the best means of entry.
- Locate alternate exits.
- Perform a risk/benefit analysis.

Figure 9.36 Fighting fire requires the use of calculated risk and strategy.

Figure 9.37 Environmental conditions profoundly affect fire dynamics. *Courtesy of Matthew Daly.*

- Anticipate how interior conditions may change.

- Determine whether there are enough resources to complete the operation quickly.

- Check your own and your team members' air supply.

- Determine whether you can adhere to the "two-in, two-out" rule, which requires two firefighters available outside the structure to rescue two firefighters working inside.

Other important survival guidelines apply to your actions both before and during interior fire fighting operations. These include:

- Consider both current and projected fire behavior.

- Consider both current and projected structural stability.

- Anticipate how ventilation may affect fire behavior.

- Anticipate how fire dynamics may be affected by environmental conditions, such as wind speed and direction (**Figure 9.37**).

- Always have a plan and a backup plan.

Preparation for Survival

If you cannot avoid a hazardous situation, you must know how to survive it. To do so you must:

- Practice basic fire fighting techniques.

- Practice situational awareness.

- Anticipate the types of survival situations you may face.

- Practice **MAYDAY** and self-rescue techniques.

The basic skills of fire fighting are also essential survival techniques. For example, forcible entry techniques such as forcing windows and cutting debris can also be applied in an escape (**Figure 9.38**). The hoseline used in a fire attack can also protect you from a rapidly progressing fire or help you find your way out of a smoke-filled

MAYDAY — Internationally recognized distress signal.

Figure 9.38 Interior walls may need to be breached to allow firefighters an escape from a hazardous environment.

room. Under normal conditions, air management increases your work time; but if you are lost or trapped, it also increases the amount of time you can survive before escaping or being rescued. The extreme stress of an emergency makes conscious thought difficult, so practice these skills until they become automatic.

Situational awareness will warn you of extreme fire behavior or structural collapse. If you recognize the warning signs, you can withdraw to safety. If you know your own personal physical limitations, you can monitor your stress level, air consumption, and work load and withdraw before you become incapacitated.

To be physically and mentally prepared you must:

- Practice basic skills.
- Train for the hazards you will encounter.
- Practice air management techniques.
- Practice emergency exit techniques.
- Make sure that your PPE matches the hazard and is working properly.
- Know your duties.
- Know your own physical limitations. Watch for signs of fatigue, increased breathing, and increased heart rate.
- Know the limitations of your PPE and air supply system.
- Look out for the members of your crew.
- Listen to your team members.
- Follow orders thoughtfully: if it does not sound right, ask for clarification.

You must be aware of your surroundings:

- Look for signs of key fire behavior indicators, particularly those associated with rapid fire progress, such as flashover, backdraft, and fire gas ignition.
- Listen, feel, and watch for changes in the environment.

Recognition

Recognizing the potential types of MAYDAY situations you may be faced with is the next step in personal survival. Situations that can cause you to have an emergency include:

- **Air emergencies** — Your facepiece is dislodged, you run out of air, or your SCBA malfunctions. It also occurs if your **low-pressure alarm** activates and you are unable to *immediately* exit the hazard area.

- **Lost/disoriented** — You are in extreme darkness, or you lose contact with your partner, your hoseline or search line, or an orientation point such as an entry door.

- **Entanglement** — You are caught on exposed wires, a fallen ceiling grid, or other debris (**Figure 9.39**).

- **Thermal emergencies** — Rapid rise in temperature that exceeds your PPE's level of protection (**Figure 9.40**).

- **Collapse/trapped** — You are unable to exit due to structural collapse. This is most common when the structure has been on fire for an extended period of time; been affected by an earthquake or explosion; or been under construction, renovation, or demolition. It can also occur if the structure has been intentionally damaged to create traps for responders.

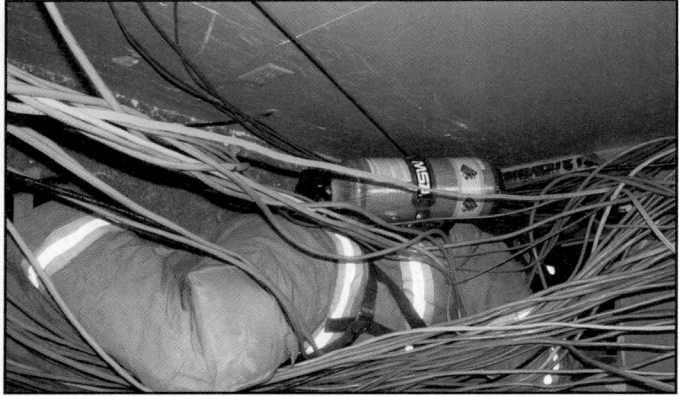

Figure 9.39 Entanglement presents a hazard that will require additional assistance. *Courtesy of Iowa State Fire Training Bureau.*

Figure 9.40 Container fires may foster conditions that will generate a rapid rise in fire temperature. *Courtesy of Rhett Strain/OSU-FST.*

Communication

You increase your chances of surviving if you immediately communicate that you are in danger. First, describe the problem and give your location to your partner, team leader, or supervisor. Remain in place until another firefighter reaches you.

Making a MAYDAY Communication

Managing an emergency while in the hazard zone is a team effort. After (or concurrently with) notification of team members, the member with the emergency takes appropriate corrective action. A member who is not experiencing the problem (generally the supervisor) transmits the MAYDAY message and then assists the member with the problem. If you are separated from your team, this is an emergency in itself and you must be prepared to take corrective action and communicate the MAYDAY yourself.

If you think your life is in immediate danger, transmit your department's MAYDAY signal, activate your PASS device, and communicate your situation to the IC **immediately (Figure 9.41)**. The sooner rescuers know you are in danger, the sooner they can come to your aid. Use the acronym LUNARS to remember what information to provide:

- Location
- Unit
- Name
- Assignment
- Resources needed (such as air or extrication)
- Situation

If possible, communicate your air supply status and any actions you are taking. Describe your location as clearly as possible so that rescuers know where to search for you. Stay in contact with Command and keep them informed if there are changes to your situation. After transmitting the MAYDAY report, activate your PASS device. **Skill Sheet 9-I-8** describes these steps when transmitting a MAYDAY report.

NOTE: An activated PASS device emits a signal that can prevent you from being heard over the radio. Always deactivate your PASS device before transmitting your MAYDAY signal on the radio, then reactivate it after your transmission is finished. Practice this procedure in training.

The word "MAYDAY" is used whenever a firefighter is in immediate danger. When MAYDAY is broadcast, the following actions must immediately be taken:

- All radio traffic ceases and only traffic relating to the MAYDAY is allowed.
- The communication center allocates an available radio channel specifically for the MAYDAY communications.
- Nonessential activities cease and units are directed to assist with searching for the firefighter who has broadcast the MAYDAY.
- The **rapid intervention crew or team (RIC/RIT)** is dispatched to locate the downed firefighter **(Figure 9.42)**.

Rapid Intervention Crew or Team (RIC/RIT) — Two or more firefighters designated to perform firefighter rescue; they are stationed outside the hazard and must be standing by throughout the incident.

Figure 9.41 When calling in a MAYDAY, remember to transmit the information an RIC/RIT will need in order to help you: location, unit, name, assignment, resources needed, and the situation.

Figure 9.42 An RIC/RIT is tasked with standing by at incidents to provide assistance to firefighters in need.

MAYDAY Fireground Response

Although nonessential activities cease, all fire control activities must continue. This is essential to the survival of the firefighter or crew in the MAYDAY situation. Once the MAYDAY situation is resolved, all fireground activities can resume.

Always listen closely when a MAYDAY transmission is made. If you are near the downed firefighter, you may be able to assist with the rescue. Listen closely for your orders after the transmission, and do not freelance. All teams must remain disciplined and focus on their tactical assignments.

If conditions within the hazard zone change rapidly, the IC may give orders for all personnel to exit. You must be familiar with the ways that your department sounds an evacuation signal, such as audible warning devices on the apparatus and radio messages ordering interior crews to immediately exit. Exit orders are typically handled like other emergency traffic and are broadcast several times to ensure that everyone hears them. Warning devices work well in small structures but may not be heard in a large structure, where a radio message may be the only means of notifying personnel. When any evacuation signal is given, all units on scene must give a personnel accountability report (PAR). Exiting crews should proceed to designated safe areas outside the collapse zone.

Air Management

You must be able to regulate your air consumption so that you can exit the IDLH safely. In the past, firefighters stayed in the IDLH until their low-pressure alarm activated. However, this placed them at risk because that amount of air (typically 25 percent of capacity) is insufficient to exit from deep inside a large or complex structure. Even in a small structure, it may not be enough if you encounter an emergency or must wait for assistance.

Point of No Return — Point at which air in the SCBA will last only long enough to exit a hazardous atmosphere.

There are three key principles of air management: always know how much air you have left, know your **point of no return**, and inform the IC if you must exit the structure. To know how much air you have left you must check your gauge regularly, particularly at the following times **(Figure 9.43)**:

- Before entering an IDLH atmosphere
- When moving from one area to another
- After periods of heavy work
- At specific intervals, based on SOPs
- When resting
- Before beginning a new assignment

Figure 9.43 Checking the air gauge regularly will aid in the efficiency of responses.

Your point of no return is based on:

- How much air is required to exit the IDLH
- The lowest cylinder gauge reading of any member of the team
- Your department's SOPs
- Environmental conditions
- Your team's physical and mental condition

Air Management

One air-management system uses a board with columns for the firefighters' names, assignments, cylinder pressure, time of entry, and the estimated time their low-pressure alarm will activate. This time is based on 75 percent of cylinder pressure, in psi (for example, 2216 psi X .75 = 1662). That figure is then divided by 100 to determine the number of minutes the firefighter can remain in the structure (1662/100 = 16.6 minutes). By adding these minutes to the entry time, the air management officer determines each firefighter's exit time (entry at 2230 hrs plus 16.6 minutes = 2247 hrs exit time).

Your point of no return must be based on how much air you consume. Test your consumption rate by simulating emergency conditions during training and measuring how long it takes to consume a full cylinder of air **(Figure 9.44)**.

Individual firefighters can only decide to exit a structure if they become separated from their team, or if there is a catastrophic event such as a structural collapse. Otherwise, the decision to exit is always made by the supervisor or IC. Never leave your team, and never leave a team member alone in the hazard zone. If there are only two team members in the hazard zone, they must leave together.

Effective air management includes knowing how to react in an air emergency. To do so, you must remain calm and follow procedures to determine the cause of the emergency and implement a solution. Practice these procedures until they become second nature. They will help you to conserve air, increase the time you have to escape, and improve the chances that others will be able to locate and assist you. **Skill Sheet 9-I-9** describes the proper procedures for an SCBA air emergency.

Survival Actions

The best survival tactic is to constantly monitor your surroundings, using your situational awareness to help keep you safe. But if a MAYDAY event does occur, you have three possible courses of action: remaining in place, seeking safe haven, and escaping. Whichever you choose, it is vital that you remain calm. Panic is a leading contributor to firefighter death.

Figure 9.44 Tasks may be practiced in training settings to allow rescuers to gauge air usage and time required to complete specific skills.

Remain in Place

Sometimes the only option is to remain in place. Stay calm, breathe slowly to conserve air, and stay low where the temperature is cooler. After completing your MAYDAY report, take these actions to help rescuers locate you:

- Continue to communicate on your radio.
- Activate your PASS device.
- Tap the floor with a tool, or find another way to make noise.
- Shine your flashlight or hand light directly overhead **(Figure 9.45)**.
- If you are unsure of your location, you can momentarily turn off your PASS device and listen for sounds that provide clues to your location, such as traffic, crews working with tools, or the sound of a pumper.

Seeking Safe Haven

Seeking safe haven entails taking actions that either improve your situation or buy you more time to escape. Many times these actions can be performed before you communicate your MAYDAY.

Sheltering techniques include:

- **Staying low to the floor** — Temperatures are cooler and air may be less contaminated.
- **Using your hose stream for protection** — In a thermal emergency, turn your nozzle pattern to a full fog and point it above you while flattening yourself on the floor. This should only be done as a last resort when it is too late to exit. This action can result in steam burns as the water rapidly turns to steam.
- **Closing doors between you and the fire** — This places a barrier between yourself and the fire, allowing you time to find an egress point or breach a wall to escape.
- **Using tools to shore building material** — During a building collapse, you may be able to use your hand tools to shore up building materials. This gives you time to find an egress point or escape the collapse area.
- **Filtering toxic air** — This should be done with your face low to the floor and without removing your regulator. Break the seal of your mask and use your protective hood to filter each breath **(Figure 9.46)**. Be aware that this will not remove toxins such as carbon monoxide (CO) and hydrogen chloride (HCl). It is only a last resort if your SCBA cylinder is completely exhausted.

Escape

Escape is the best option in any of the following circumstances:

- There is an imminent threat of structural collapse.
- There is not a safe place to shelter.
- You have completely exhausted your air supply.
- Extreme fire conditions are about to occur.
- You have been ordered to abandon the structure.

Escape from a hostile environment requires teamwork. Practice emergency procedures as a team so that all members know their individual roles and responsibilities.

Figure 9.45 A portable light helps make a responder more visible in a MAYDAY situation.

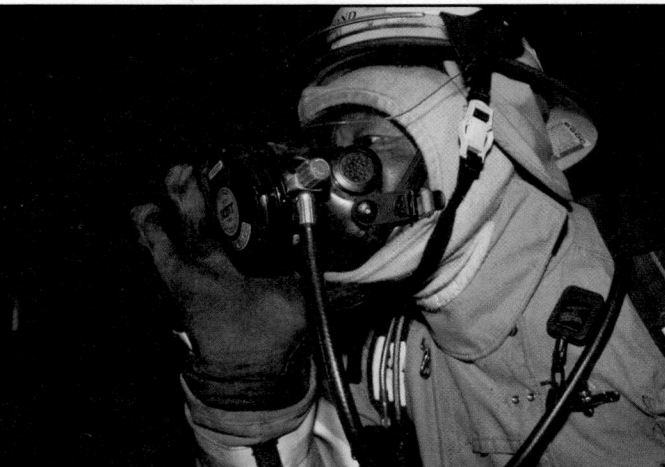

Figure 9.46 A protective hood may be used as a particle-only filter in extreme cases.

If your team is intact, escape as a unit by following your hoseline or search line to your original point of entry. In the event or threat of rapid fire progression, do not leave the nozzle to follow your hoseline out. Withdraw the line and operate the nozzle as needed. **Skill Sheet 9-I-10** describes the procedures for withdrawing from a hostile environment with a hoseline.

Controlling the Fire

You cannot outrun the fire. Use a hoseline to apply water directly on the fire. If the fire is shielded, cool the gases overhead to reduce the potential for the smoke (fuel) to ignite.

If you are separated from your team, follow the hoseline, search line, or wall in the direction from which you came. If you are following a hoseline, feel the first set of couplings you come to. The female coupling is on the nozzle side of the set and the male coupling is on the water source side **(Figure 9.47)**. The male coupling has lugs on its shank while the female coupling has (smaller) lugs on the swivel. Following the hoseline will lead you either to an exit or to the nozzle team. Communicate your progress to your supervisor or to Command.

If you decide to escape, follow these safety guidelines:

* Remain calm.

* If you have a hoseline, maintain control of the nozzle and use it for protection.

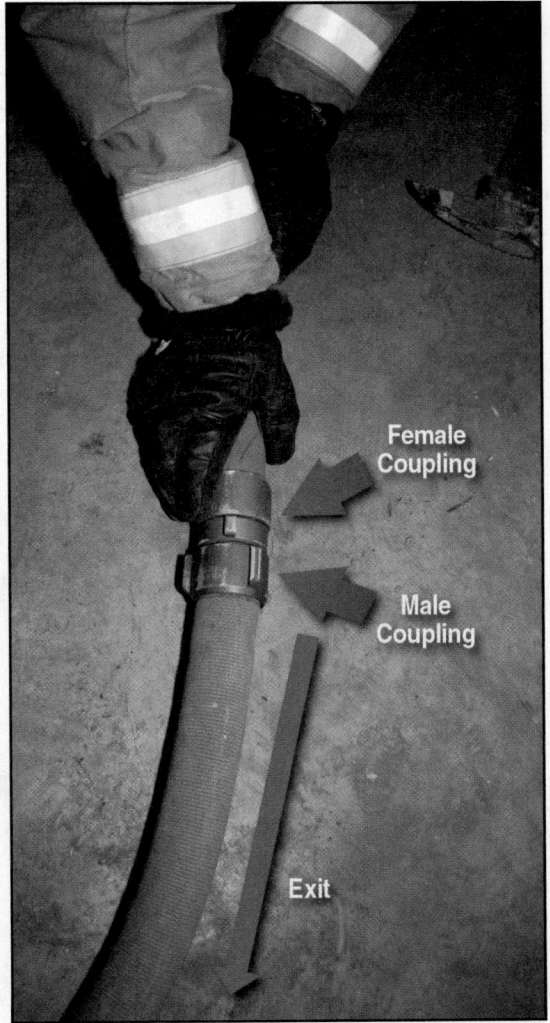

Figure 9.47 Follow the direction the male coupling indicates in order to find the exit and water source.

- Orient yourself to points of reference such as hoselines, search lines, walls, and points of exterior light.

- Stay low.

- Keep in contact with the wall.

- Stay in radio communication.

- Remember which level of the structure you are on.

- Control your breathing.

- Check your air supply frequently and report to Command.

When smoke is dense and low, you may use a duck walk or a low-profile maneuver during your escape. If you can see the floor, you can use a crouched or "duck" walk. This method is slightly faster than crawling but is more dangerous unless you can clearly see the floor in front of you. The low-profile maneuver involves crawling, which is an effective way of moving in areas of low visibility **(Figures 9.48a and b)**. One advantage of crawling is that the area close to the floor is cooler and may have better visibility. It also allows you to feel in front of you with your hand or a tool, warning you of objects or dangerous openings in your path. **Skill Sheet 9-I-11** describes the side technique for low-profile maneuvers without removing SCBA. **Skill Sheet 9-I-12** describes the SCBA-first technique for low-profile maneuvers without removing SCBA.

To search for an exit, locate a wall and crawl along it. With one hand, sweep the floor ahead of you to avoid openings into which you might fall. With your other hand, sweep the wall to find a window. Reach as high as you can without standing. When you locate a window that can be opened, determine whether the window will allow you to exit. Notify Command of your location and ask if opening the window will make interior conditions worse. If there is a door between yourself and the fire, close it before opening the window. If you are on the ground floor, open the window or break it with your forcible entry tool. Make certain the frame is completely clear of glass shards. Climb through the window and lower yourself to the ground feet first.

Figure 9.48a Crawling allows a firefighter to see the condition of the floor and surroundings.

Figure 9.48b Duck walking may be faster than crawling, but it is only safe in conditions where the floor is visible and undamaged.

If you are on an upper story, find out if there is an aerial or ground ladder at or near the window. If there is, climb out and onto the ladder. If there is not, report your location and your need for a means of egress. If you have an escape rope system, secure it to a sturdy interior point, climb through the window, and descend to the ground **(Figure 9.49)**. You should only do this if you have been properly trained in this form of escape.

Breaching an interior wall may give you access to a safe haven or a room from which you can safely exit the building. However, this is a last resort since it requires strength and will use precious air supply. Interior walls are generally made of Sheetrock®, wood, or plaster and lath making them easier to breach than exterior walls. Exterior walls are made of masonry, concrete, metal, or wood and are almost impossible to penetrate.

To breach a wall, remain low and use your forcible entry tool to make an opening **(Figure 9.50)**. Remove enough wall material to make a space large enough to crawl through, then make an opening in the wall on the other side. Use your forcible entry tool to sound the floor on the other side of the wall and to locate any obstructions. You may have to adjust your SCBA to fit through the opening. Loosen your right shoulder strap and waist belt, then shift the SCBA until it is tucked under your left arm. This allows you to exit on your side through an opening with limited clearance.

Figure 9.50 Crawling through stud spaces requires practice under controlled conditions.

Figure 9.49 A properly trained firefighter may use an escape rope system to exit a structure through an upper story window.

Figure 9.51 Backward exits through stud spaces are recommended training evolutions.

You can also exit SCBA-first through walls with standard 16-inch (406.4 mm) stud spaces. Sit back against the open stud space, push your SCBA through followed by one arm and then the next. Push against the wall to pull your body through in a swimming motion (**Figure 9.51**). You should only remove your SCBA as a last resort because this increases the risk of becoming separated from your air supply. **Skill sheet 9-I-13** describes the method for breaching an interior wall.

If you become entangled, immediately broadcast a MAYDAY before attempting to extricate yourself. You may be able to free yourself by using a type of swim stroke to work your way out of the entanglement. It is easier to move back the way you came than to move forward through the entanglement. It is a good practice to carry wire cutters in a pocket that you can access even if your movement is limited. You can use the wire cutters to cut nonelectrical wire to free yourself. **Skill Sheet 9-I-14** describes the procedure for disentangling from debris or wires.

Rapid Intervention

Occupational Safety and Health Administration (OSHA) regulations and NFPA® 1500 require that a rapid intervention crew or team (RIC/RIT) be standing by whenever firefighters are in any hazard zone. RIC/RITs consist of at least two members who are prepared to rescue injured or trapped firefighters. They must be trained in firefighter rescue and equipped with the same PPE as the interior fire fighting crews at the incident. Sometimes more than one RIC/RIT is assigned to stand by, especially if interior crews entered the building at multiple points.

> **CAUTION**
>
> Do not underestimate the time and personnel required to rescue a downed firefighter. Carrying one unconscious firefighter can require four rescuers, and fully removing the firefighter from the hazard zone can require up to twelve rescuers. This process can take as long as 20 minutes to complete.

Mandatory RIC/RIT equipment can be described using the acronym AWARE: Air, Water, A Radio, and Extrication. Teams carry a spare SCBA to provide breathing air for the downed firefighter, a hoseline to create a defensive space, a radio to communicate with Command, and forcible entry tools for extrication. They are also equipped with flashlights or handheld lights, search lines, and a thermal imager (TI) (**Figure 9.52**). Additional equipment may include a litter, power saw, attic ladder, and spare breathing air cylinders.

After any MAYDAY transmission, the RIC/RIT first tries to establish radio contact with the downed firefighter. The IC may also order a brief shutdown of all pumps, generators, fans, and other noise-producing devices so that the firefighter's PASS device can be better heard. If these actions are unsuccessful, the RIC/RIT follows the firefighter's hoseline or search line into the structure, then begins its search from the firefighter's last known location.

The RIC/RIT should stop frequently and briefly remain silent, especially after moving from one room or area to another. This will help them hear the downed firefighter's PASS device sounding. To be completely silent, crew members may even hold their breath for a few seconds when signaled to do so by the search team leader. This may allow the rescuers to hear faint calls for help or the sounds of the downed firefighter's SCBA exhalation valve operating. See the Removing Located Firefighters section near the end of this chapter for more details on RIC/RITs.

Tracking Devices

Digital radio transceivers enable lost or disoriented firefighters to be easily located. These tracking devices were originally developed to locate avalanche victims, but have been adapted for use in the fire service.

The transceivers are about the size of a PASS device and are mounted on the SCBA harness. They operate on 457 kHz and have a range of approximately 100 feet (30 m). They do not interfere with other on-scene radio transmissions (**Figure 9.53**).

Figure 9.52 Basic equipment carried by an RIC/RIT may fit into a duffel bag.

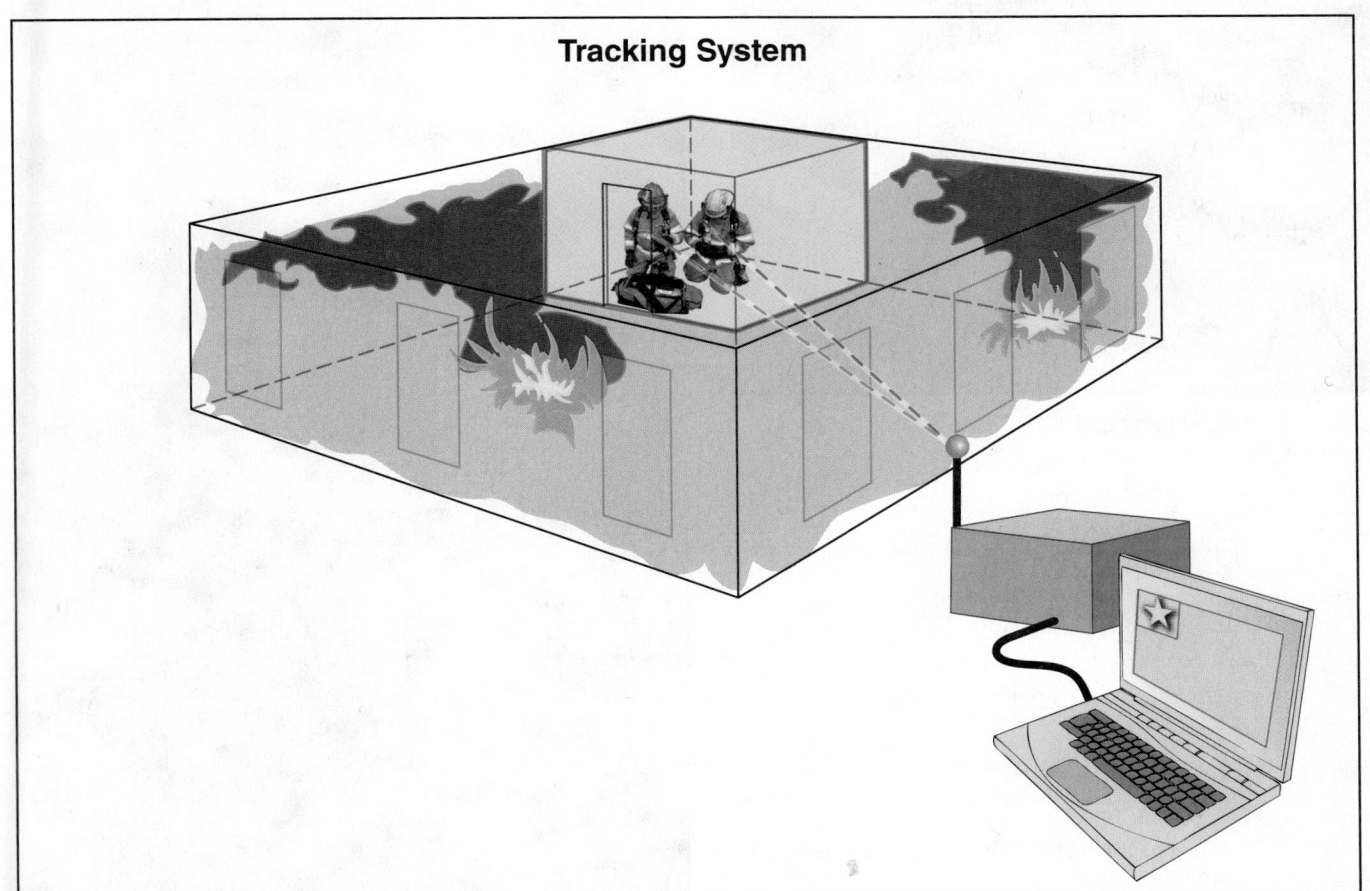

Tracking System

Figure 9.53 A tracking system does not interfere with other on-scene radio transmissions.

Like PASS devices, they are always turned on when entering an IDLH atmosphere. Unlike PASS devices, their low-frequency radio signal is not blocked by walls, floors, or other solid objects. This enables rescuers to easily locate a downed firefighter, even from outside the building. Lost or disoriented firefighters can switch the transceiver from standby mode to a search mode. In this mode the device displays the distance and direction toward another transceiver or to an egress transmitter that has been positioned near an exit.

Removing Located Firefighters

After locating the downed firefighter, the RIC/RIT checks his or her air supply. If necessary, the team may connect a full replacement cylinder. They also deactivate the PASS device and confirm or determine the firefighter's identity (**Figure 9.54**).

Next, the RIC/RIT notifies Command of the firefighter's location and status. If the firefighter is trapped or injured, they will request assistance. While waiting for help to arrive, the team will mitigate any hazards that threaten the downed firefighter's safety. If necessary, they may move the firefighter to a safe haven.

In most cases, exiting the IDLH atmosphere takes priority over stabilizing the firefighter's injuries. If the firefighter is unable to walk, the rescuers should move him or her to safety. Even if the firefighter is uninjured and able to walk, rescuers should escort him or her to a safe location (**Figure 9.55**).

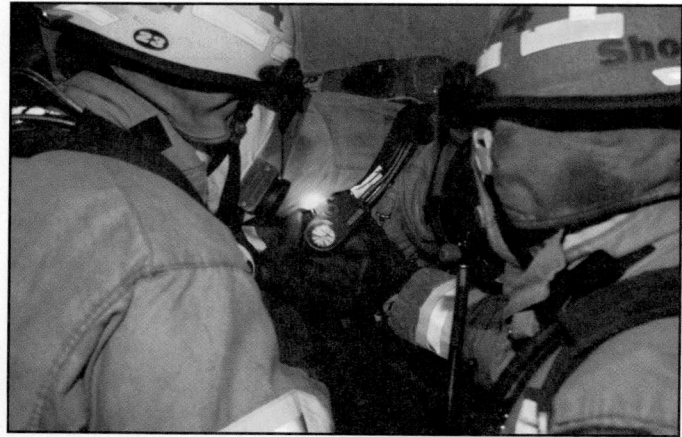

Figure 9.54 The first thing an RIC/RIT checks on a downed firefighter is his or her air supply.

Figure 9.55 When able, an RIC/RIT assists the firefighter to safety.

If the firefighter has a functioning SCBA, the team must move the firefighter carefully so as not to dislodge the mask. If the firefighter does not have a functioning SCBA, the team must either connect the firefighter's facepiece to a functioning SCBA or quickly remove the victim from the hazardous atmosphere.

Buddy Breathing is Not a Recommended Technique

According to both NFPA® and NIOSH, so-called "buddy breathing" techniques are unreliable and more likely to produce two victims instead of one. NIOSH recommends providing respiratory assistance and quickly removing the victim to a clear atmosphere.

WARNING

Never remove your facepiece or compromise the proper operation of your SCBA to share your air supply — not even with another firefighter.

Chapter Summary

Your first priorities at any structural fire are your own survival and that of your fellow firefighters. You must learn to size up a situation, practice situational awareness, manage your air supply, and remove victims to safety. You must know MAYDAY procedures and master self-rescue techniques. Finally, you must be able to locate and rescue downed firefighters as part of a rapid intervention crew or team. You must continually train in all these techniques in order to be prepared for any emergency.

Firefighter I

1. How do building construction and floor plans impact structural search techniques?

2. What information can size-up and situational awareness provide during structural searches?

3. What are five safety guidelines that should be followed during structural search and rescue?

4. What are the main differences between primary and secondary search techniques?

5. What is the general search method used during structural search?

6. When is the appropriate time to use the oriented-search method, wide-area search method, and thermal-imager-search method?

7. What are the main differences in the three types of victim removal methods?

8. What are the three behaviors firefighters must learn and follow to ensure their own survival and that of fellow firefighters?

9. How does a firefighter decide on the best survival action to take if a MAYDAY event does occur?

10. When does a rapid intervention crew or team (RIC/RIT) begin work on an incident scene?

9-1-1
Demonstrate the procedure for conducting a primary search.

SKILL SHEETS

Step 1: Confirm order to conduct primary search with officer or supervisor and establish search pattern to be used.

Step 2: Size up structure to be searched.

Step 3: Search the structure using established search pattern and update IC on progress.

Step 4: Mark rooms that have been searched.

Step 5: Remove any victims and inform IC of victim(s).

Step 6: Exit building when the search is complete.

Step 7: Report completion of primary search to officer or supervisor.

9-1-2
Demonstrate the procedure for conducting a secondary search.

NOTE: A secondary search should be conducted with a different crew than those that did the primary search.

Step 1: Confirm order to conduct secondary search with officer or supervisor and establish search pattern to be used.

Step 2: Size-up structure to be searched.

Step 3: Search the structure using established search pattern, update IC on progress.

Step 4: Mark rooms that have been searched.

Step 5: Remove any victims and inform IC of victim(s).

Step 6: Exit building when search is complete.

Step 7: Report completion of secondary search to officer or supervisor.

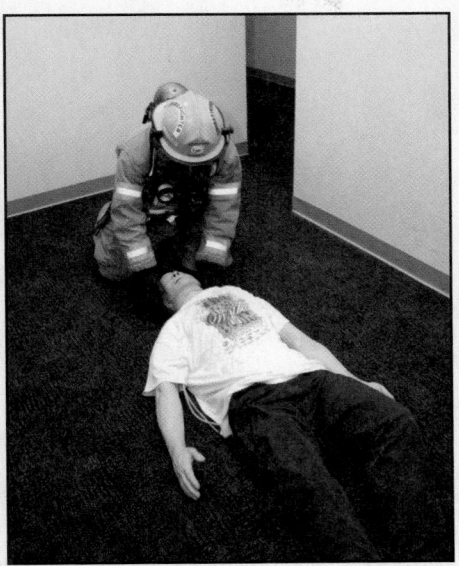

Step 1: Turn the victim (if necessary) so that the victim is supine.

Step 2: Kneel at victim's head.

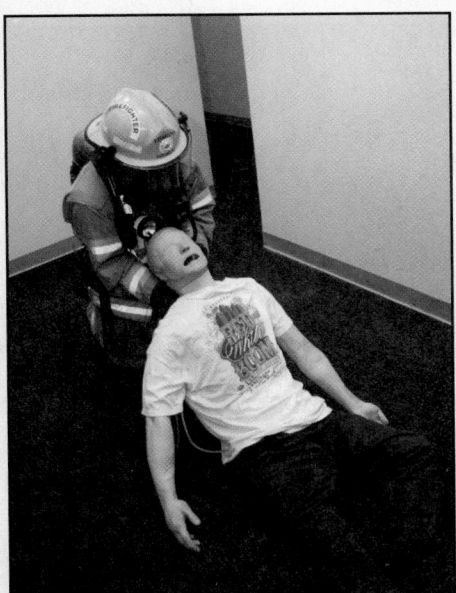

Step 3: Support the victim's head and neck.

NOTE: If head or neck injury is suspected, provide appropriate support for head during movement.

Step 4: Lift the victim's upper body into a sitting position.

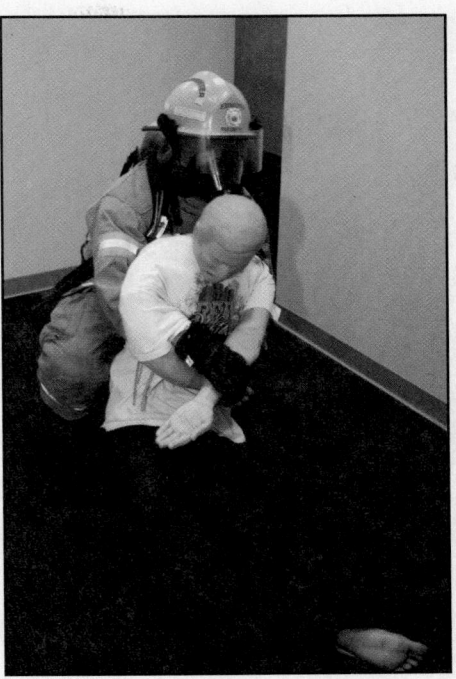

Step 5: Reach under the victim's arms.

Step 6: Grasp the victim's wrists.

Step 7: Stand; the victim can now be eased down a stairway or ramp to safety.

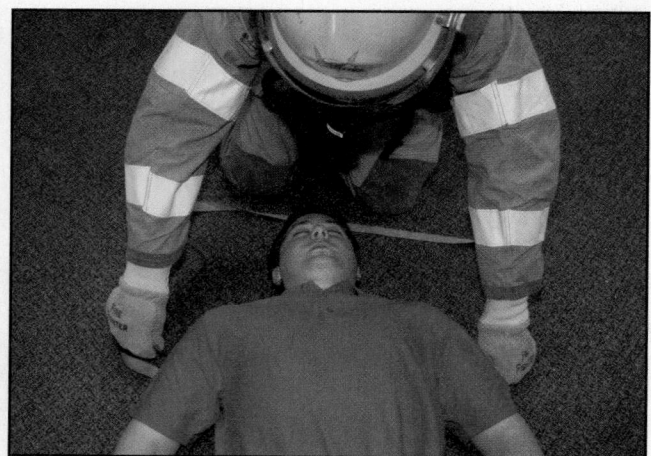

Step 1: Place the victim on his or her back.

Step 2: Slide the large webbing loop under victim's head and chest so that the loop is even with armpits. Position the victim's arms so that they are outside the webbing.

Step 3: Pull the top of the large loop over the victim's head so that it is just past the head.

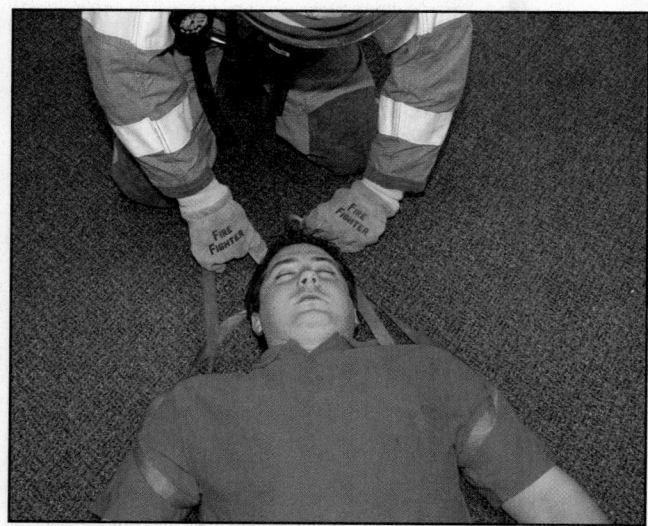

Step 4: Reach down through the large loop and under the victim's back and grab the webbing.

Step 5: Pull the webbing up and through the loop so that each webbing loop is drawn snugly around the victim's shoulders.

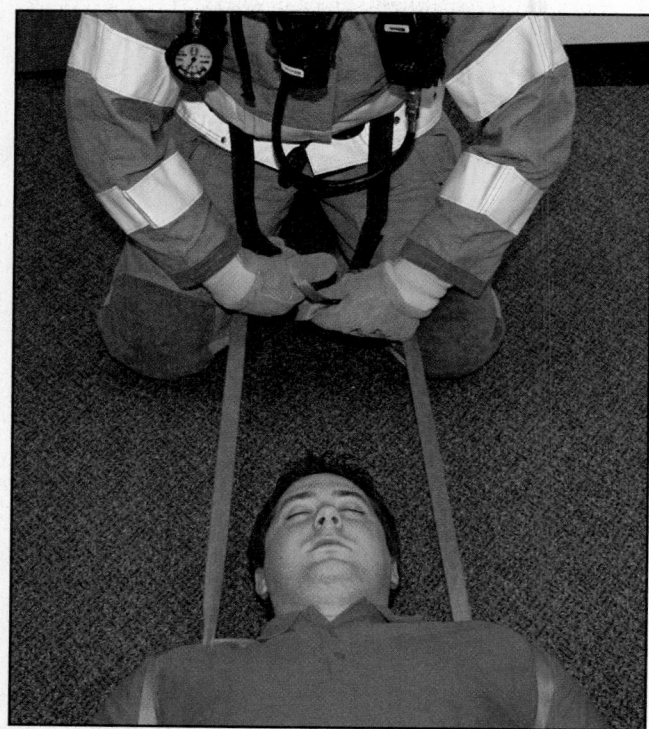

Step 6: Adjust hand placement on the webbing to support the victim's head.

Step 7: Drag the victim to safety by pulling on the webbing loop.

9-1-5
Demonstrate the cradle-in-arms lift/carry - One-rescuer method.

SKILL SHEETS

Step 1: Place one arm under the victim's arms and across the back.

Step 2: Place the other arm under the victim's knees.

Step 3: Lift the victim to about waist height while keeping your back straight.

Step 4: Carry the victim to safety.

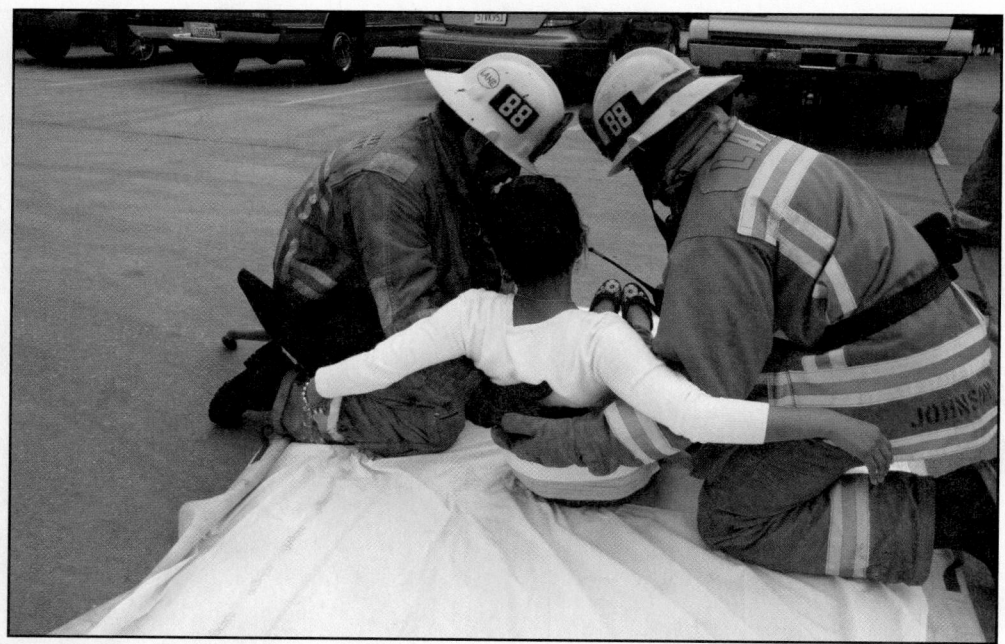

Step 1: Raise the victim to a sitting position.

Step 2: Link arms across the victim's back.

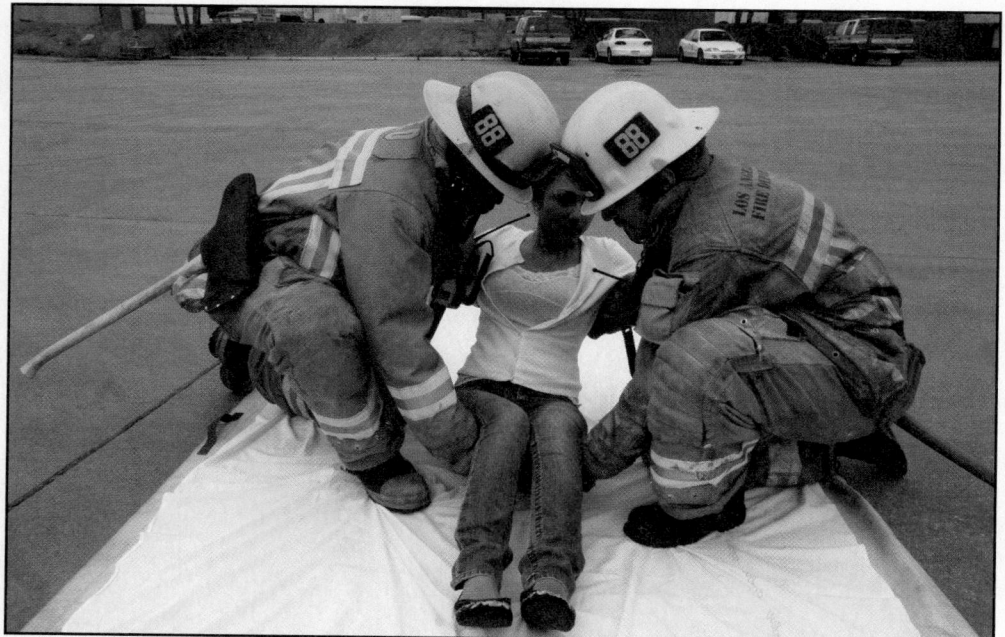

Step 3: Reach under the victim's knees to form a seat.

Step 4: Lift the victim using your legs. Keep your back straight while lifting.

Step 5: Move the victim to safety.

9-1-7
Demonstrate the extremities lift/carry – Two-rescuer method.

SKILL SHEETS

Step 1: Both Rescuers: Turn the victim (if necessary) so that the victim is supine.

NOTE: Keep head and neck stabilized during rolling to prevent spinal injury.

Step 2: Rescuer #1: Kneel at the head of the victim.

Step 3: Rescuer #2: Stand between the victim's knees.

Step 6: Rescuer #1: Push gently on the victim's back.

Step 7: Rescuer #1: Reach under the victim's arms and grasp the victim's wrists as Rescuer #2 releases them. Grasp the victim's left wrist with the right hand and right wrist with the left hand.

Step 8: Rescuer #2: Turn around, kneel down, and slip hands under the victim's knees.

Step 9: Both Rescuers: Stand and move the victim on command from Rescuer #1.

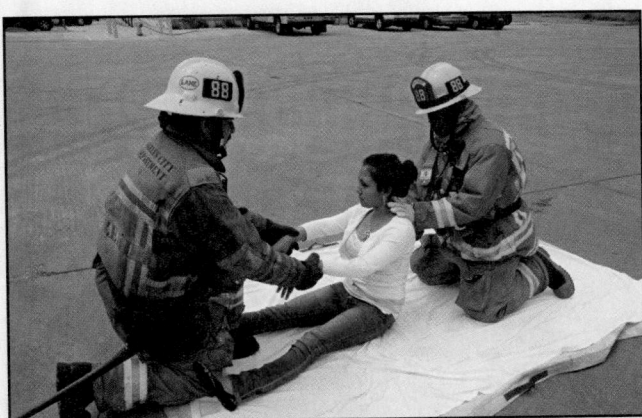

Step 4: Rescuer #2: Grasp the victim's wrists.

Step 5: Rescuer #2: Pull the victim to a sitting position.

Step 1: Determine the need to declare a MAYDAY.

Step 2: Announce *"Mayday Mayday Mayday"* over the emergency communications channel.

Step 3: Provide Command your information per local SOPs.

For example: **LUNARS**

L – Location

U – Unit

N – Name or ID Number

A – Assignment

R – Resources Needed

S – Situation

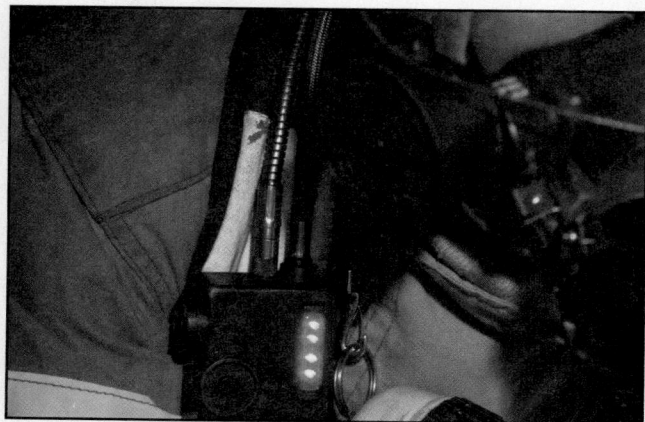

Step 4: Activate PASS device and press radio emergency button, if so equipped.

Step 5: If able, move to a wall and position yourself according to local SOPs.

Step 6: Activate flashlight and point towards the ceiling.

Step 7: Use a tool or other object to make noise; remain calm and conserve air; stay in contact with Command.

9-1-9
Demonstrate the proper procedures to take in an SCBA air emergency.

SKILL SHEETS

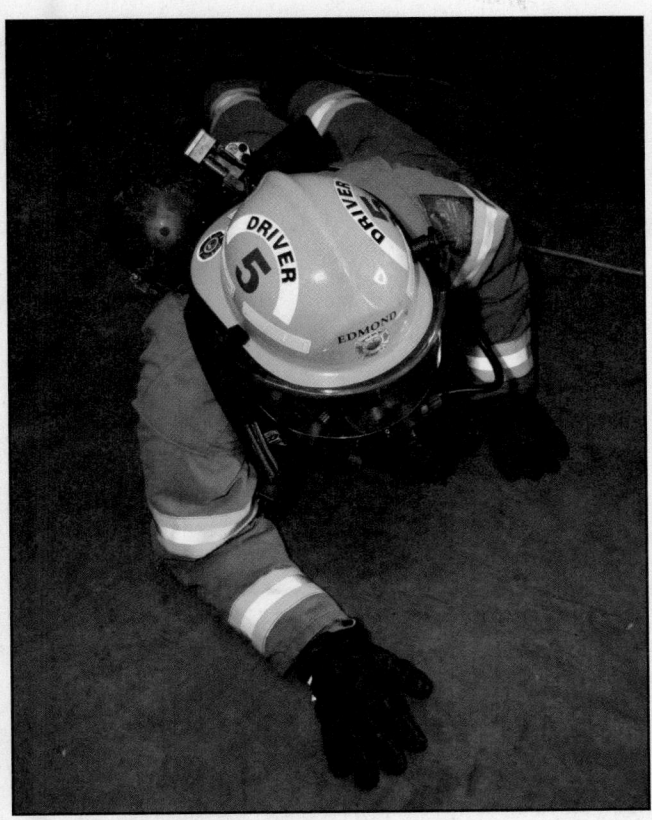

Step 1: Recognize the emergency.

Step 2: Drop face to the ground and ensure cylinder valve is fully open, as well as opening regulator bypass valve as directed by local SOPs.

If bypass resumes air flow:

Step 3: If opening on bypass valve returns air supply, immediately exit the IDLH environment only opening the valve as needed.

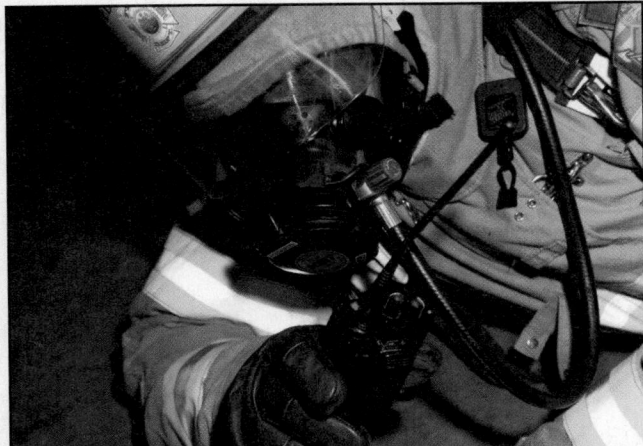

Step 4: Ensure that you communicate your emergency with your crew and command via MAYDAY transmission.

Step 5: Activate PASS device if trapped or disoriented.

SKILL SHEETS

9-1-9

Demonstrate the proper procedures to take in an SCBA air emergency.

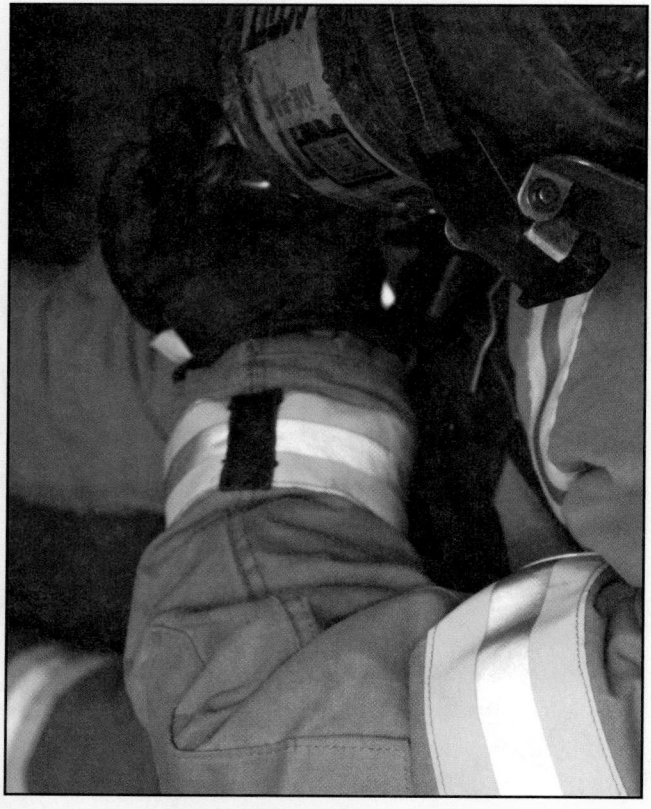

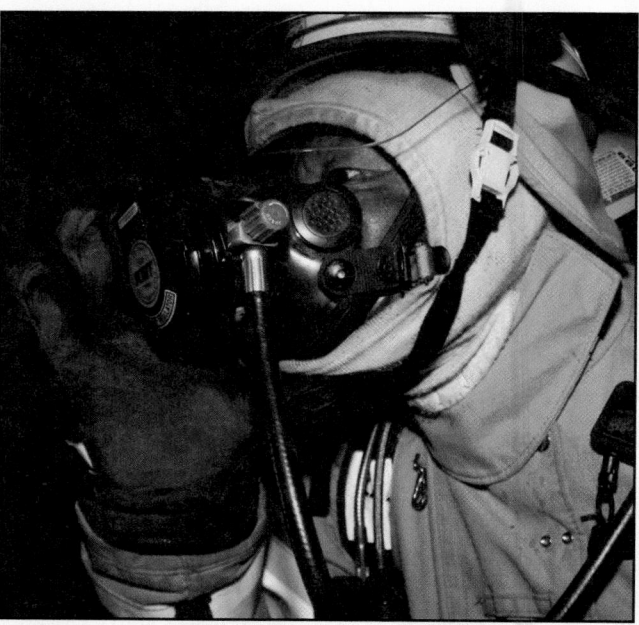

Step 1: Recognize the emergency.

Step 2: Drop face to the ground and ensure cylinder valve is fully open, as well as opening regulator bypass valve as directed by local SOPs.

If bypass results in no return of air flow:

Step 3: If still unable to get air with bypass valve fully open, create gap between chin and mask seal to take breath. Ensure that regulator stays in place and protective hood stays sealed around lower lip of mask.

Step 4: Ensure that you communicate your emergency with your crew and command via MAYDAY transmission.

Step 5: Activate PASS device.

Step 6: Stay low to ground and attempt to escape the IDLH environment. Stop to take breaths as needed, ensuring protective hood remains in place to "filter" air as much as possible.

Step 7: If trapped, stay calm and limit breathing until rescue team arrives.

9-1-9
Demonstrate the proper procedures to take in an SCBA air emergency.

SKILL SHEETS

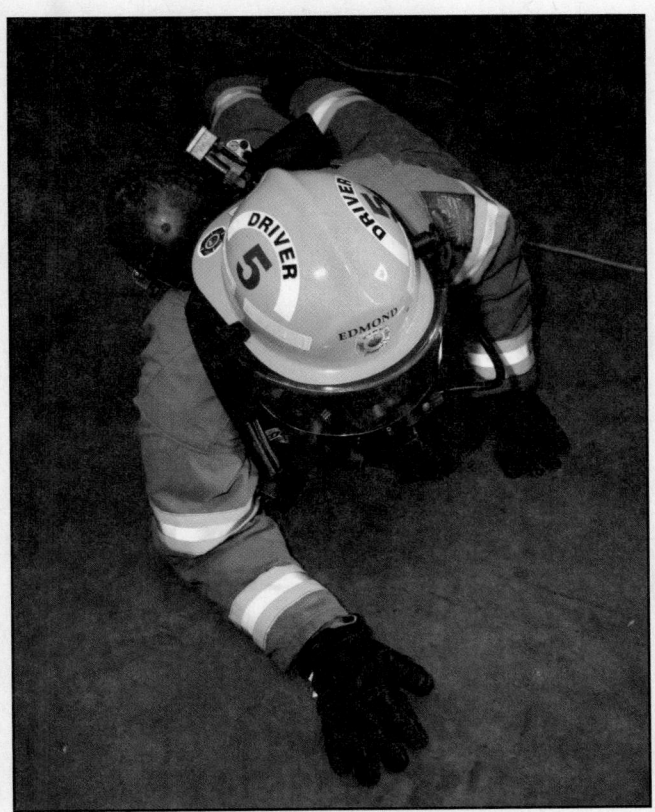

Step 1: Recognize the emergency.

Step 2: Drop face to the ground and ensure cylinder valve is fully open, as well as opening regulator bypass valve as directed by local SOPs.

If air emergency is due to other SCBA malfunction:

Step 3: Stay low to the ground and use emergency breathing techniques as described above.

Step 4: Assess:

    a. Can malfunction be fixed or bypassed?

    b. Can air from cylinder be used if removed from harness?

Step 5: If possible, exit the IDLH area immediately. **OR** If trapped or disoriented, attempt to remove cylinder from pack and utilize air for breathing, conserving after each breath.

Step 6: Communicate via MAYDAY.

Step 7: Maintain calm and activate PASS device.

9-1-10

Demonstrate the actions required for withdrawing from a hostile
environment with a hoseline as a team.

NOTE: These steps are based on the hose team encountering deteriorating conditions beyond their ability to control. If ordered to withdraw by Command, Step 1 would be to acknowledge the order to withdraw and Step 4 would be to advise Command that the crew had exited, current location, and verification of accountability.

Step 1: Recognize deteriorating conditions and advise other members of the hose team.

Step 2: Adjust nozzle to provide maximum flow and operate nozzle to cool the upper layer and prevent extension of flames overhead using long pulses of water fog (if working with a solid stream nozzle, a sweeping pattern across the ceiling may be used).

Step 3: Withdraw the nozzle and hoseline, keeping the nozzle between the hose and the fire as the hoseline is withdrawn. Crew member(s) not on the nozzle should retain forcible entry tools if at all possible. Do not turn back on the fire as you withdraw.

Step 4: After reaching a safe area (e.g., outside the building, floor below), verify accountability for all members of the hose team, determine if anyone is injured.

Step 5: Transmit an emergency traffic message to Command (or supervisor if working for an organizational level other than Command) using the method determined by your local SOPs (for example LUNARS).

9-I-11
Demonstrate low profile maneuvers without removing SCBA - Side technique.

SKILL SHEETS

Step 1: Remain calm and follow steps for MAYDAY transmission.

Step 2: Loosen waist strap and right shoulder strap.

Step 3: Remove right arm from right shoulder strap.

Step 4: Shift SCBA to left side and tuck under left armpit.

Step 5: Ensure that waist strap remains buckled and left arm remains in shoulder strap.

Step 6: Use tool to sound other side of the wall before exiting room.

Step 7: With SCBA tucked tightly under left armpit, lay on right side to create a low profile and attempt escape through restricted opening.

SKILL SHEETS

9-1-12
Demonstrate low profile maneuvers without removing SCBA –
SCBA-first technique

Step 1: Remain calm and follow steps for MAYDAY transmission.

Step 2: Once opening has been made, sound floor on the other side of the opening to ensure safety.

Step 3: Sit with SCBA and back toward the opening.

Step 4: Place one arm and SCBA cylinder into the opening.

Step 5: Using a backstroke technique, swim other arm through the opening.

Step 6: Using both arms and wall board for leverage, pull through the space.

9-1-13
Demonstrate the method for breaching an interior wall.

SKILL SHEETS

Step 1: Remain calm and conserve energy.

Step 2: Communicate emergency to crew and command via MAYDAY transmission.

Step 3: Activate PASS device.

Step 4: If possible, locate hoseline, window, or door to facilitate immediate exit from IDLH environment.

Step 5: Locate nearest interior wall.

Step 6: Use hand tool to create inspection hole in wall board.

Step 7: Use inspection hole to locate stud space.

Step 8: Use hand tool to create hole between stud spaces large enough to fit body through.

Step 9: It may be necessary to remove additional studs, or utilize a "low profile maneuver" to exit through stud space.

Step 10: It may be necessary to remove or work around any wiring, or piping that is running length of the wall.

CAUTION: Use caution if electrical power to building has not been shut down.

Step 11: Using hand tool, sound the floor on the other side of the wall, to ensure soundness of the area.

Step 12: Once the area on the other side has been confirmed safe, use a low profile maneuver, or other technique to exit through the wall.

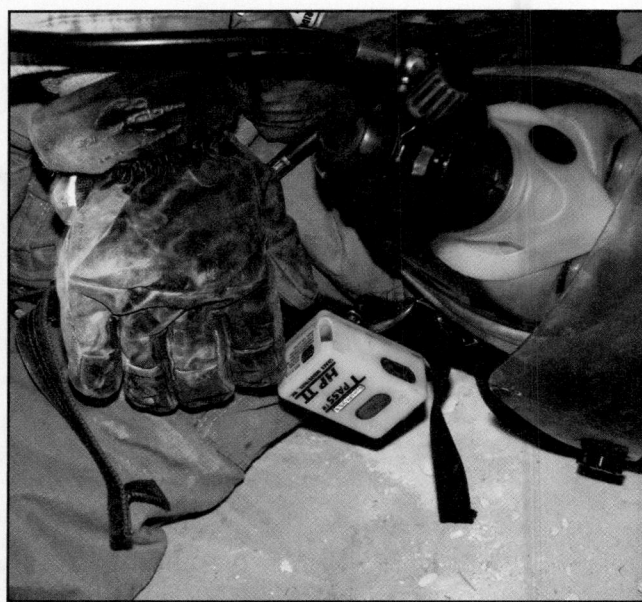

Step 5: Activate PASS device if separated from crew.

Step 1: Recognize the emergency.

Step 2: Communicate to crew and command via MAYDAY transmission.

Step 3: Stop moving and assess situation.

Step 4: Stay calm and control breathing.

9-1-14
Demonstrate the steps for disentangling from debris or wires.

SKILL
SHEETS

Step 6: Decide best steps to take to mitigate situation, conserve energy.

Step 7: If possible, back out of entanglement, or move forward with SCBA pack down and regulator protected with gloved hand.

Step 8: If unable to move backward, or forward, attempt "swim" technique to locate points of entanglement.

Step 9: If possible, utilize a cutting tool to cut out of the entanglement.

Step 10: Consider a partial SCBA removal to assist in locating and removing points of entanglement.

Step 11: If unable to escape, stay calm, communicate with crew, and conserve air until rescue crew arrives.

Chapter Contents

Key Terms

NFPA® Job Performance Requirements

This chapter provides information that addresses the following job performance requirements of NFPA® 1001, *Standard for Fire Fighter Professional Qualifications* (2013).

Firefighter I
5.3.17

Firefighter II
6.4.1
6.4.2
6.5.4

1. Identify types of emergency scene lighting equipment. (5.3.17)

Firefighter II Chapter Objectives

1. Explain considerations for maintenance of electric generators and lighting equipment. (6.4.2, 6.5.4)

2. Describe the types of rescue tools and equipment. (6.4.2, 6.5.4)

3. Explain the uses and limitations of each type of rescue tool. (6.4.1, 6.4.2, 6.5.4)

4. Identify the role of a fire department during vehicle extrication. (6.4.1)

5. Describe safety considerations that must be identified and mitigated during vehicle extrication. (6.4.1)

6. Explain the use of cribbing material during vehicle extrication. (6.4.1)

7. Describe the methods used for gaining access to victims during vehicle extrication. (6.4.1)

8. Explain the role a Firefighter II will play in technical rescue operations. (6.4.2)

9. Describe the various types of technical rescue operations. (6.4.2)

10. Explain the unique hazards associated with each type of technical rescue operation. (6.4.2)

11. Demonstrate the steps for inspecting, servicing, and maintaining a portable generator and lighting equipment. (Skill Sheet 10-II-1, 6.5.4)

12. Prevent horizontal movement of a vehicle using wheel chocks. (Skill Sheet 10-II-2, 6.4.1)

13. Stabilize a vehicle using cribbing. (Skill Sheet 10-II-3, 6.4.1)

14. Stabilize a vehicle using lifting jacks. (Skill Sheet 10-II-4, 6.4.1)

15. Stabilize a vehicle using a system of ropes and webbing. (Skill Sheet 10-II-5, 6.4.1)

16. Stabilize a side-resting vehicle using a buttress tension system. (Skill Sheet 10-II-6, 6.4.1)

17. Remove a windshield in an older model vehicle. (Skill Sheet 10-II-7, 6.4.1)

18. Remove a tempered glass side window. (Skill Sheet 10-II-8, 6.4.1)

19. Remove a roof from an upright vehicle. (Skill Sheet 10-II-9, 6.4.1)

20. Remove a roof from a vehicle on its side. (Skill Sheet 10-II-10, 6.4.1)

21. Displace the dashboard. (Skill Sheet 10-II-11, 6.4.1)

Chapter 10
Scene Lighting, Rescue Tools, Vehicle Extrication, and Technical Rescue

Case History

Fire department units responded to a vehicle accident on an icy highway. The incident required extrication of victims from the vehicle. The highway was closed by law enforcement personnel in the direction of travel in which the vehicle accident occurred. However, there were no physical barriers between the incident scene and the opposing direction of travel. A tour bus approaching from the opposing direction lost control, crossed the median, and entered the immediate accident scene. Fortunately, the bus did not strike any vehicles or personnel.

Emergency responders at the scene felt safe and became complacent because the traffic behind them was blocked. No consideration was given to the traffic coming from the opposing direction. In fact, traffic should have been stopped on both sides of the highway.

Source: National Fire Fighter Near-Miss Reporting System.

Emergency incidents will often require supplemental lighting provided by apparatus mounted lights or portable lighting systems. Firefighter I firefighters must be able to operate this equipment, and Firefighter II firefighters must also be able to maintain it. Rescues from structure fires are just one of many type of rescues that fire departments perform. Some of these rescues, such as confined-space, ice/water, and structural collapse, require specialized training and tools. As a Firefighter I you must be able to use emergency scene lighting equipment under the direction of your supervisor or the Incident Commander (IC). As a Firefighter II, you must be able to use emergency scene lighting equipment and manual and power-operated rescue tools. You must also perform vehicle **extrications** and assist specialized technical rescue personnel. This chapter provides you with the basic knowledge needed to perform these tasks based on the requirements found in NFPA® 1001.

Extrication — Incident in which a trapped victim must be removed from a vehicle or other type of machinery.

⬛ Emergency Scene Lighting Equipment

Emergency scene lighting is required at all incidents that occur at night, in low-light conditions, or inside structures where normal lighting is not available. This equipment includes lights, electrical generators, and auxiliary electrical equipment (**Figure 10.1, p. 480**).

Figure 10.1 Full-scene lighting equipment includes a portable generator, cord, and the lights themselves. *Courtesy of Shad Cooper/Wyoming State Fire Marshal's Office.*

Figure 10.2 Portable electric generators provide electricity to scene lighting and other rescue equipment.

Generator — Portable device for generating auxiliary electrical power; generators are powered by gasoline or diesel engines and typically have 110- and/or 220-volt capacity outlets.

Power Take-Off (PTO) System — Mechanism that allows a vehicle engine to power equipment such as a pump, winch, or portable tool; it is typically attached to the transmission.

Electric Generators

Emergency scene lighting and portable rescue equipment are powered by portable electric **generators**, apparatus-mounted generators, or the apparatus electrical system. Generators are the most common power source used by emergency services personnel **(Figure 10.2)**.

Portable electric generators are powered by small gasoline or diesel engines and have 110- and/or 220-volt capacity outlets. Most are light enough to be carried by two people. They are useful when vehicle-mounted electrical systems are not available.

Vehicle-mounted generators produce more power than portable units **(Figure 10.3)**. They can be powered by gasoline, diesel, propane gas engines, or by hydraulic or **power take-off (PTO) systems**. They typically have 110- and 220-volt outlets and produce more than 50 kW of power. Vehicle-mounted generators with a separate engine have the disadvantage of being noisy, making communication near them difficult. In addition, the exhaust fumes from the generator can contaminate the scene if the vehicle is not positioned downwind.

Hearing Protection Is Required When Using Generators

Electric generators produce high levels of noise. Always wear hearing protection when using this equipment.

Figure 10.4 An apparatus may be equipped with external outlets that are capable of supplying small amounts of electrical power.

Figure 10.3 Vehicle mounted electric generators provide large quantities of power.

Portable lights and equipment may also be powered directly from the apparatus electrical system. If small amounts of power are needed to operate lights and tools, an **inverter** is used to convert the vehicle's 12- or 24-volt direct current (DC) into 110- or 220-volt alternating current (AC) **(Figure 10.4)**. The advantages of this method include fuel efficiency and minimal noise. Disadvantages include constant apparatus exhaust, limited power supply, and limited mobility.

> **Inverter** — Step-up transformer that converts a vehicle's 12- or 24-volt DC current into 110- or 220-volt AC current.

Lighting Equipment

Lighting equipment can be divided into two categories: portable and fixed **(Figures 10.5a and b, p. 482)**. Portable lights are used in building interiors or remote areas of the scene. They range from 300 to 1,000 watts and typically have carrying handles and large bases for stability. Some are mounted on telescoping stands, which allow them to be raised and directed more effectively.

Fixed lights are mounted on a vehicle and wired directly to the vehicle-mounted generator or apparatus electrical system. They are used to provide overall lighting of the emergency scene. These lights are usually mounted on telescoping poles so they can be raised, lowered, or rotated. Some units consist of large banks of lights mounted on hydraulically operated booms. These light banks generally have a capacity of 500 to 1,500 watts per light. The number of fixed light units mounted on an apparatus is limited by the amount of power the vehicle-mounted generator or apparatus electrical system can produce.

> **CAUTION**
> Never connect more lights than the power source can support. Overtaxing the power source results in poor lighting and possible damage to the lights, generator, or electrical system. It may also restrict the operation of other tools using the same power source.

Figure 10.5a Portable lighting can be positioned from the ground.

Figure 10.5b Apparatus-mounted lights provide lighting to an overall emergency scene. *Courtesy of Ron Moore and McKinney (TX) FD.*

Lighting Units Can Cause Injuries

All lighting units produce extreme heat and can cause burns. Be careful when moving lights or turning them off. Bulbs can also explode if they come in contact with water. Never direct lights toward moving traffic because this can blind other drivers.

Auxiliary Electrical Equipment

Auxiliary electrical equipment consists of the following (**Figures 10.6a-d**):

- Electrical cables
- Extension cords
- Receptacles

- Connectors
- Junction boxes
- Adapters
- Ground fault circuit interrupter (GFCI) devices

All auxiliary equipment must be waterproof, **intrinsically safe**, and designed for the amount of electrical current it is intended to carry. Electrical cables and extension cords can only carry a limited amount of electricity. Do not use them with equipment whose power demands exceed the cords' capacity. This creates an electrical hazard and can damage the equipment. Cords may be stored in coils, on portable cord reels, or on apparatus-mounted automatic rewind reels.

Intrinsically Safe — Describes equipment that is approved for use in flammable atmospheres; must be incapable of releasing enough electrical energy to ignite the flammable atmosphere.

Figure 10.6a Electrical cables come in several lengths and purposes. *Courtesy of Shad Cooper/Wyoming State Fire Marshal's Office.*

Figure 10.6b Extension cords may be stored on apparatus-mounted reels. *Courtesy of Ted Boothroyd.*

Figure 10.6c An electrical receptacle must be used responsibly and within the rated limits. *Courtesy of Shad Cooper/Wyoming State Fire Marshal's Office.*

Figure 10.6d A GFI device is used as a safety feature on a generator. *Courtesy of Shad Cooper/Wyoming State Fire Marshal's Office.*

In addition to being waterproof and intrinsically safe, electrical cables and extension cords should have adequate insulation and no exposed wires. Twist-lock receptacles and connectors equipped with grounding wires provide secure, safe connections as long as they are not immersed in water **(Figure 10.7)**.

Junction boxes provide multiple outlets or connections and are supplied through one inlet from the power source **(Figure 10.8)**. All outlets must be equipped with **ground fault circuit interrupter (GFCI)** devices and meet the requirements outlined in NFPA® 70E, *Standard for Electrical Safety in the Workplace®*.

Adapter connections are used to permit different types of plugs and receptacles to be connected. Adapters allow mutual aid departments to operate electrical lights and tools off each other's generators and power sources. Using adapters, fire department lights and tools may also be plugged into standard electrical outlets in structures.

Figure 10.7 Twist-lock fittings prevent a cord from accidentally becoming unplugged during operations.

Figure 10.8 Junction boxes allow electrical equipment of different connection types to be used at some distance from the initial power source.

⚓ Maintenance of Electric Generators and Lighting Equipment

Firefighters must regularly inspect and maintain all portable generators and lighting equipment. Inspections and maintenance must follow manufacturer's instructions and your department's standard operating procedures (SOPs). To meet this obligation, you must:

- Inspect generators, lighting units, and lighting accessories periodically and after each use.

- Review the manufacturer's service manual for specific directions.

- Carefully inspect spark plugs for damage, visible corrosion, carbon accumulation, or cracks in the porcelain **(Figure 10.9)**. Make sure the spark plug wire is tight.

- Replace the spark plug if it is damaged or if the service manual recommends replacement. Ensure proper gap prior to installing. Replace any plug with signs of **arcing**, indicated by the presence of carbon soot around the ground electrode **(Figure 10.10)**.

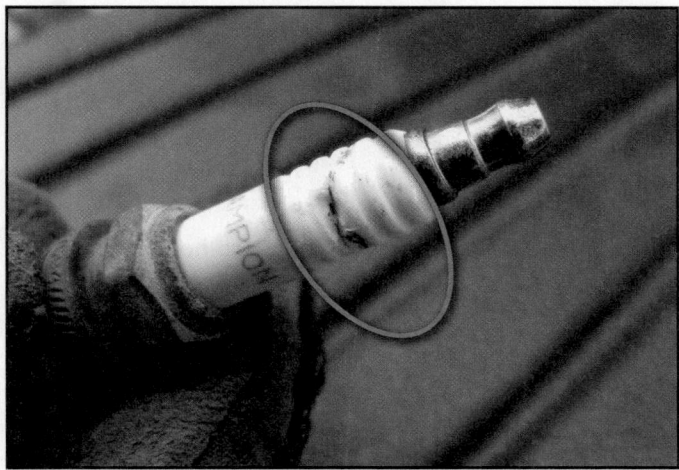

Figure 10.9 Damaged spark plugs create an electrical hazard.

Figure 10.10 Carbon soot that has built up around a ground electrode will prevent a secure connection and lend itself to arcing which creates a hazard.

- Check the generator carburetor and identify any signs of fuel leaks.
- Check fuel level and refill as needed.
- Visually inspect the fuel in the tank to ensure that it is not contaminated. Discard contaminated fuel in an approved manner.
- Check oil level and refill as needed.
- Start the generator and run any tests required in the service manual. If a problem is found with the generator, consult the manual to determine the proper action. Only qualified service personnel or a licensed electrician should perform repair work on the generator.
- Avoid starting a generator while under a load (lighting or other equipment turned on and plugged in). Starting a generator while under a load can damage the electrical system.
- Do not run the generator for a long period of time without a load. This action will overheat and damage the generator.
- Inspect all electrical cords for damaged insulation, exposed wiring, or missing or bent prongs **(Figure 10.11)**.
- Test the operation of the lighting equipment. Connect each light to the generator one light at a time to prevent overloading. Avoid looking directly into the lights when they are powered.
- Replace lightbulbs as necessary. Shut off the power and allow the bulb to cool before replacing. If the bulb must be replaced immediately, wear leather gloves to prevent being burned. Discard faulty bulbs in an approved manner.
- Clean the work area and return all tools and equipment to the proper storage areas.
- Document the maintenance on the appropriate forms or records.

Skill Sheet 10-II-1 describes the steps for inspecting, servicing, and maintaining a portable generator and lighting equipment.

Figure 10.11 This high-voltage wire shows the component pieces that should NOT be exposed during standard operation.

Some types of equipment and some kinds of maintenance are not your responsibility. The driver/operator typically inspects and maintains apparatus electrical systems and apparatus-mounted lights and generators. Detailed maintenance and modification of any lighting equipment must be performed by qualified technicians.

Rescue Tools

Rescue tools can be classified based on two criteria: their power source and use. Their power source can be either manual or power-operated **(Figures 10.12a and b)**. They may be used for:

- **Cutting** — Removing materials or debris to free a victim

- **Stabilizing** — Ensuring that a vehicle or structural member will not move during a rescue

- **Lifting** — Raising a vehicle, vehicle component, or structural member off a victim or raising the victim out of a space

- **Pulling** — Dragging away materials to free a victim

- **Other Activities** — Securing materials in place or breaking up materials to free a victim

This chapter describes power-operated and manual tools used for any purpose during vehicle extrication or technical rescue. Manual forcible entry tools that may also be used for rescue are explained in Chapter 11.

Vision and Hearing Protection Required

Eye and hearing protection must be worn when operating all types of rescue tools.

Power Sources

The power source for power-operated rescue tools may be electric, hydraulic, or pneumatic. Some tools combine both hydraulic and electric power. Criteria for powered rescue tools are established in NFPA® 1936, *Standard on Powered Rescue Tools*. Tools produced by different manufacturers may have very different capabilities, and the same is true of tools that use different power sources. For example, a power saw produced by one manufacturer may be able to perform tasks that a saw produced by another manufacturer cannot. Similarly, a hydraulic extrication tool may be unable to perform tasks handled by a pneumatic extrication tool.

Electric

Rescue saws and other tools can be powered by rechargeable batteries, apparatus electrical systems, and portable or vehicle-mounted generators. The main advantage of this power source is that electricity is readily available at the incident scene. Electric tools are lightweight, and battery powered tools are especially portable because they do not require an electric power cord or cable **(Figure 10.13)**. One disadvantage of battery-powered tools is that they may be less powerful than other tool types.

Always follow manufacturer's recommendations for maintaining rechargeable batteries. Keep them fully charged and dispose of them if they are damaged or unable to hold a charge. Some batteries must be completely discharged periodically to ensure they can maintain a full charge.

Figure 10.12a Manual tools are indispensible for the uses they are able to perform at an incident.

Figure 10.12b Power-operated tools provide greater force and speed during rescue operations.

Figure 10.13 Battery-powered tools are almost as portable as nonpowered tools.

Hydraulic

Most powered rescue tools are powered by hydraulic pumps. Pumps may be operated by hand, an electric motor, or a gasoline engine. They may be portable and carried with the tool, or mounted on the vehicle and connected through a hose reel line (**Figure 10.14**).

Inspect pumps regularly for damage. Make sure that hoses and connections do not leak and that connections are clean and work properly. Leave all maintenance to qualified personnel.

Danger of Hydraulic Injuries

Due to the high pressures produced by power-operated hydraulic pumps, you must be aware of the extreme danger of hydraulic injuries. A hydraulic injury can occur when you come in contact with a pinhole leak of pressurized hydraulic fluid on the hoses. The pressurized hydraulic fluid can penetrate your PPE and your skin and enter your bloodstream. Your skin will swell while your body attempts to fight what it thinks is an infection. To treat this injury, your skin must be opened to remove the fluid and relieve the pressure and swelling.

Pneumatic

Pneumatic tools are powered by pressurized air from compressed air cylinders, such as SCBA cylinders, vehicle-mounted cascade systems, or portable or vehicle-mounted air compressors (**Figure 10.15**). If SCBA cylinders are used to supply the tools, an adequate quantity of cylinders must be on hand to meet the tool's high demand for air.

Follow the manufacturer's recommendations for general care and inspection. All maintenance must be performed by qualified personnel.

Figure 10.14 Hydraulic pumps may be portable for ease of use at an incident scene. *Courtesy of Owasso (OK) Fire Department.*

Figure 10.15 Pneumatic tools connected to SCBA cylinders use a high quantity of compressed air but are very portable.

Powered Rescue Tools

There are four basic types of hydraulic and electric powered rescue tools: spreaders, shears, combination spreader/shears, and extension rams.

Spreaders. Powered hydraulic spreaders were the first powered hydraulic tools used in the fire service. When combined with chains and adapters, they are used for pushing and pulling. Spreaders produce tremendous force at their tips, which may spread as much as 32 inches (800 mm) apart (**Figure 10.16**).

Shears. Hydraulic shears or cutters can cut almost any metal object that fits between the blades. They are also used to cut plastics, wood, and other materials. The blades typically have an opening spread of approximately 7 inches (175 mm) (**Figure 10.17**).

Combination spreader/shears. This tool has removable spreader tips that can be replaced with a set of shears. It is excellent for a small rapid-intervention vehicle or for departments with limited resources. The combination tool is less expensive than individual shears and spreaders, but it cannot cut or spread as forcefully as the individual units (**Figure 10.18**).

Figure 10.16 Powered hydraulic spreaders serve many rescue functions.

Figure 10.17 Hydraulic shears can cut almost anything that will fit between the blades.

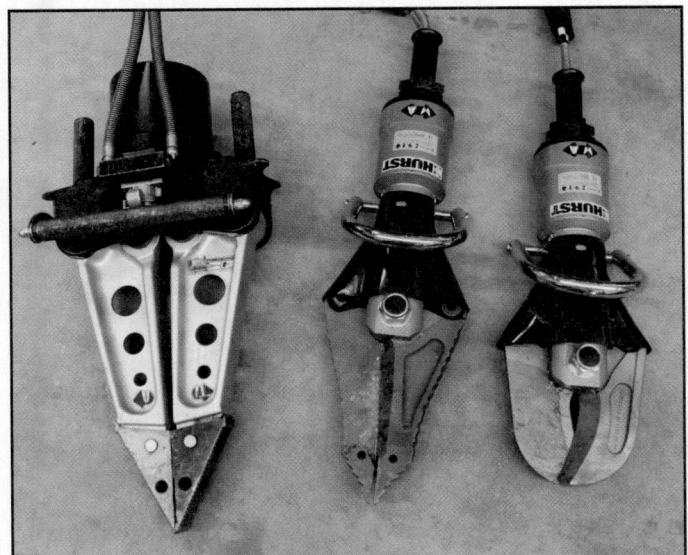

Figure 10.18 Spreaders, combination spreader/shears, and shears are commonly used at an incident scene. *Courtesy of Shad Cooper/Wyoming State Fire Marshal's Office.*

Extension rams. Extension rams are designed primarily for pushing, but they can also be used for pulling. They are typically used when objects must be pushed farther than the maximum opening distance of hydraulic spreaders; for example, when displacing the dashboard of a vehicle, which is referred to as a dash rollover or dash roll up. The largest of these tools can extend from a closed length of 3 feet (1 m) to an extended length of nearly 5 feet (1.5 m). Their opening force (used for pushing) is about twice as powerful as their closing force (used for pulling) **(Figure 10.19)**.

Currently, there is only one electric powered rescue tool system available on the market **(Figure 10.20)**. It is lightweight and compact, but not as powerful as comparable hydraulic tools.

Pneumatic and manual rescue tools are used for lifting, pushing, pulling, hammering, chiseling, and cutting. They are lightweight, inexpensive, and more portable than electric and hydraulic tools. However, they are also slower, less powerful, and more labor-intensive.

Most are vehicle repair tools that have been adapted for use in the fire service **(Figure 10.21)**. Examples include Porta-Power brand tools and come-alongs, which will be discussed later in this chapter. A variety of accessories enable these tools to be used for many different purposes.

Figure 10.19 Hydraulic extension rams are commonly used in extrication evolutions to create and hold an opening.

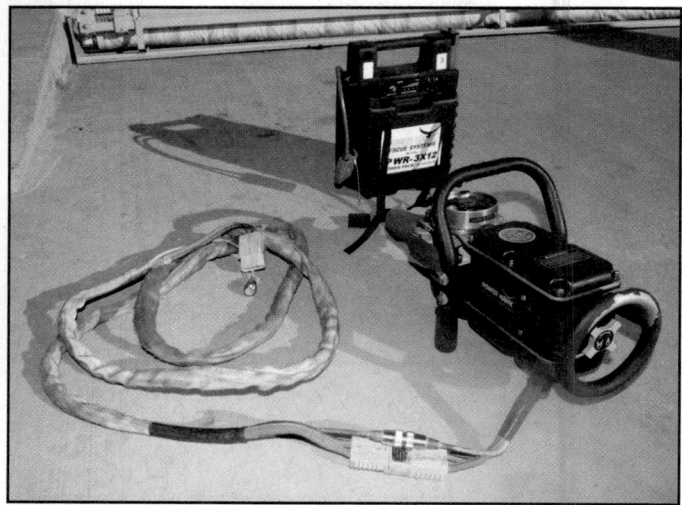

Figure 10.20 The Power Hawk system is the only electric-powered rescue tool system on the market.

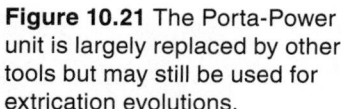 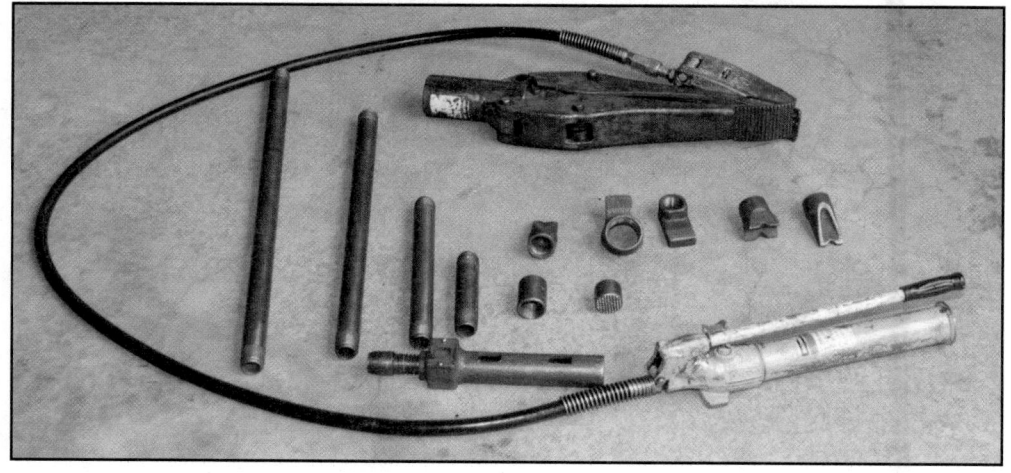

Figure 10.21 The Porta-Power unit is largely replaced by other tools but may still be used for extrication evolutions.

Cutting Tools

Cutting tools are used to cut material away from a trapped victim. Most cutting tools are power saws, which are faster and easier to handle than powered shears. Saws are also more powerful, and there are some exotic metals that only a saw can cut through. Power saws can be gasoline, electric, or battery powered. The most common types are reciprocating, rotary, circular, and whizzer saws.

CAUTION

Always wear full PPE including eye and hearing protection when operating any power saw. Do not force the saw beyond its design limits; you may be injured and/or the saw may be damaged. Follow manufacturer's safety recommendations and departmental SOPs.

WARNING

Never use a power saw in a flammable atmosphere. The saw's motor or sparks from the cutting can ignite flammable gasses or vapors causing an explosion or fire.

Reciprocating Saw

The reciprocating saw is a powerful, versatile, and highly controllable saw. It has a short, straight blade that moves in and out, like a handsaw. It can use a variety of blades for cutting different materials. When equipped with a metal-cutting blade, it is ideal for cutting sheet metal body panels and structural components on vehicles (**Figure 10.22**).

Rotary Saw

Rotary saws used in the fire service are typically gasoline powered, with different blades for cutting wood, metal, or masonry (**Figure 10.23**). Large-toothed blades are used to make quick, rough cuts, while fine-toothed blades are used for precision cutting. Blades with carbide-tipped teeth are superior to standard blades because they are less prone to dulling. Rotary saw blades can spin at up to 6,000 revolutions per minute (rpm).

Figure 10.22 A reciprocating saw can cut straight lines in small spaces.

Figure 10.23 Rotary saws may be equipped with customized blades to accomplish a range of tasks.

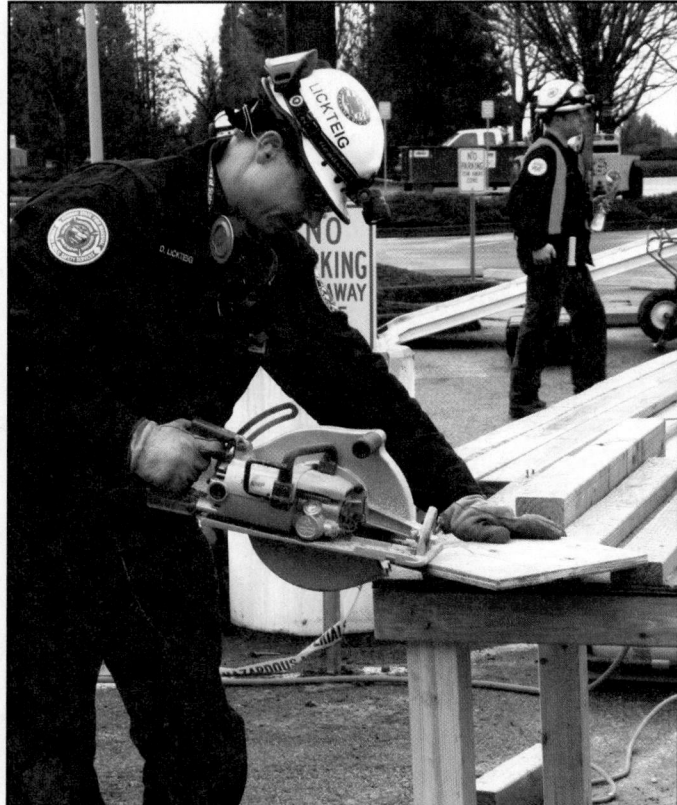

Figure 10.24 Circular saws are commonly available and easy to handle.

Circular Saw

Electric circular saws are versatile, lightweight, and easy to handle (**Figure 10.24**). Small battery-powered circular saws are also available.

Whizzer Saw

The whizzer saw weighs about one-tenth as much as a circular saw (2 lbs. or 0.9 kg). It is quiet, highly portable, and very easy to handle. It is often used for delicate cutting operations, such as removing rings from swollen fingers.

The saw's 3-inch (75 mm) Carborundum® blade cuts through case-hardened locks and up to ¾-inch (20 mm) of steel. It also has a clear Lexan® blade guard to protect both operator and victim. It is driven by compressed air at 90 psi (630 kPa) from an SCBA cylinder with a regulator and will run approximately 3 minutes on a full air cylinder (**Figure 10.25**).

Air Chisel

These pneumatic-powered tools operate at air pressures between 90 and 250 psi (630 and 1 750 kPa). In addition to cutting bits, special bits are also available for operations such as breaking locks or driving in plugs (**Figure 10.26**). Air chisels are used to cut medium- to heavy-gauge sheet metal and remove rivets and bolts. A tool with more air at higher pressures is required to cut heavier-gauge or exotic metals. Follow manufacturer's recommendations for general maintenance and keeping cutting tips sharp.

Figure 10.25 A whizzer saw may run for 3 minutes on an SCBA cylinder.

Stabilizing Tools

Before you begin a rescue or an extrication, you must ensure that the scene is stabilized. The vehicle, object, structural component, or trench wall must not be able to move, as this can injure the victim or a rescuer. **Stabilization** tools include jacks, buttress tension systems, wheel chocks, and cribbing.

Jacks used in stabilization can also be used for lifting. Jacks should always be placed on a flat, level surface, such as a roadway. If a jack must be positioned on a soft surface, place a solid base of cribbing, a flat board, or a steel bearing plate under the jack to prevent it from sinking into the surface (**Figure 10.27**).

Stabilization — Preventing unwanted movement; accomplished by supporting key places between an object and the ground (or other solid anchor points).

Hydraulic Jack

Hydraulic jacks are designed for heavy lifting, but when used in conjunction with cribbing, they can also be used for stabilization (**Figure 10.28, p. 494**). Read the manufacturer's manual to determine the jack's weight capacity.

Nonhydraulic Jack

Nonhydraulic jacks are much less powerful than hydraulic jacks. There are two main types: screw jacks and ratchet-lever jacks.

Screw jack. Screw jacks can be extended or retracted by turning the threaded shaft. Two commonly used types of screw jacks are the bar screw jack and the trench screw jack. Both jacks have a male-threaded stem (similar to a bolt) and a female-threaded component (**Figure 10.29, p. 494**).

Bar screw jacks are typically used to support collapsed structural members. These heavy-duty devices are not normally used for lifting; their primary use is to hold an object in place. The shaft is turned with a long bar that is inserted through a hole in the top of the shaft.

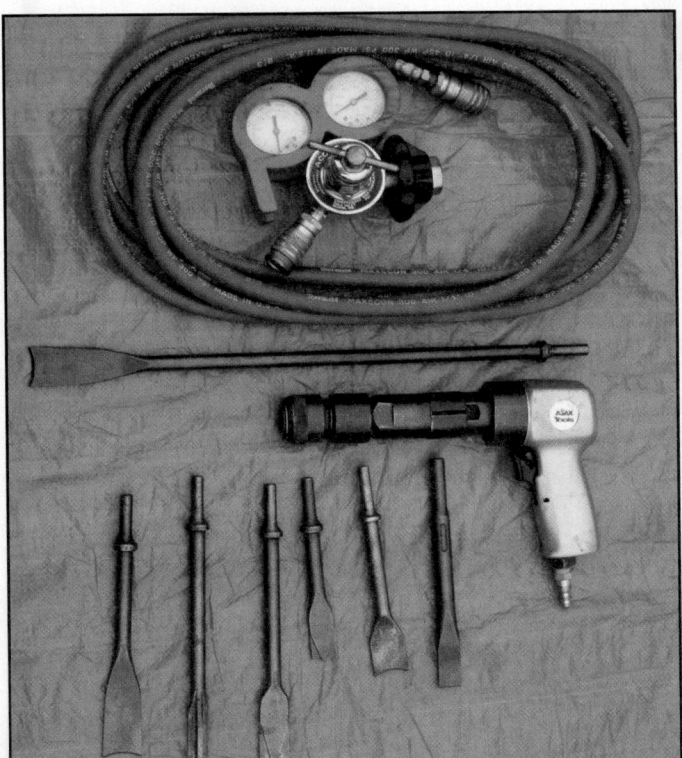

Figure 10.26 Air chisels may be customized with tips for a range of specific uses.

Figure 10.27 A jack requires solid footing for stability during lifting operations.

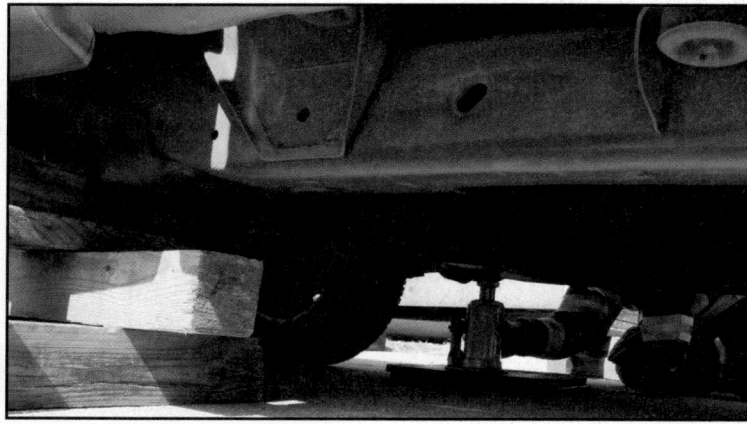

Figure 10.28 Hydraulic jacks are used in extrication functions and should be supplemented with cribbing.

Figure 10.29 Screw jacks can be adjusted to the necessary length by turning the threaded shaft.

Figure 10.30 Ratchet-lever jacks lift items on the carriage in a straight line.

Trench screw jacks often replace wooden cross braces during trench rescue. They are durable, inexpensive, and easy to use. Each jack has two swivel footplates. The first has a stem that is inserted into a section of a 2-inch (50 mm) steel pipe up to 6 feet (2 m) long. The second has a threaded stem that is inserted into the other end of the pipe. An adjusting nut (with handles) on the threaded stem is turned to vary the length of the jack and to tighten it between opposing members in a shoring or stabilizing system.

Ratchet-lever jack. Ratchet-lever jacks are also known as *high-lift jacks*. These medium-duty jacks are used primarily for stabilization and lifting, but they can be modified for use in pushing or pulling operations during vehicle extrication. They are the least stable of all types of jacks. If their load shifts, ratchet-lever jacks can fall over, allowing the load to suddenly drop to its original position. They are also prone to failure under heavy loads and sometimes release if less than 100 pounds (45.36 kilograms) is resting on the tongue.

Ratchet-lever jacks consist of a rigid I-beam with perforations along the side and a jacking carriage that fits around the I-beam (**Figure 10.30**). The geared side of the I-beam has two ratchets; one holds the carriage in position while the other works with a lever to move the carriage up or down.

WARNING

Never work under a load supported only by a jack. If the load shifts or the jack fails, you can be severely injured, or even killed. Loads should always be supported by properly placed cribbing. Live by the saying, "Lift an inch, crib an inch." As the jack lifts the load 1 inch (25 mm), add one 1 inch (25 mm) of cribbing.

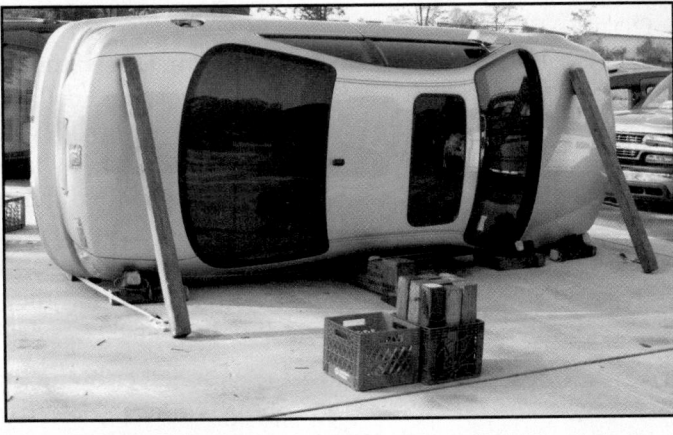

Figure 10.31a A buttress system consists of rods and ties that support the balance of a vehicle while it is in an unstable position.

Buttress Tension System

A buttress tension system is used to stabilize a vehicle that is resting on its side or top. It may consist of a minimum of three 4 x 4 inch (101 by 101 mm) posts wedged between the ground and the vehicle, or it may be a commercial system composed of metal rods and straps. Two posts are placed on the bottom of the vehicle and one is placed on the top. Their exact placement is determined by the condition and weight of the vehicle, the stability of the soil, and the condition of the victims. The purpose of the system is to provide a larger footprint for the vehicle and prevent it from moving or tipping over during extrication (**Figures 10.31a and b**). See the Vehicle Extrication section for more information about this device.

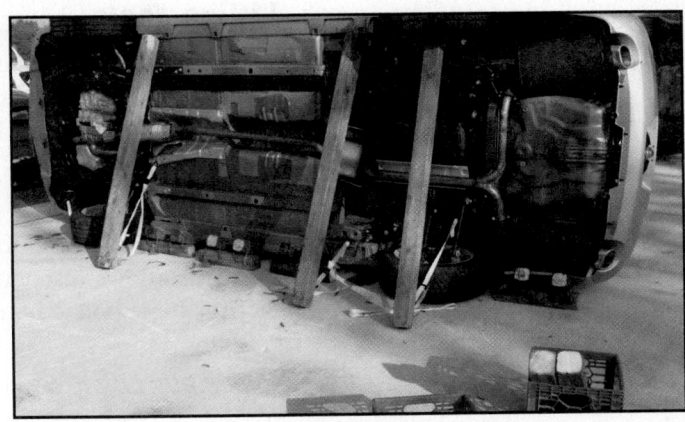

Figure 10.31b The placement of a support system is determined by the condition of the vehicle and its other supports.

Wheel Chocks

Wheel chocks prevent emergency vehicles from moving when they are parked. When placed against the downhill side of the rear tires, they can hold a vehicle in place on a 10 to 15 percent grade. During extrication, they can be also used to stabilize vehicles that have been involved in an accident.

Wheel Chock — Block placed against the outer curve of a tire to prevent the apparatus from rolling; can be wooden, plastic, or metal. *Also known as* Wheel Block.

Wheel chocks are constructed of aluminum, hard rubber, wood, or urethane plastic (**Figure 10.32**). They are designed to resist corrosion from oils, fuels, and solvents. Chocks typically have a pad or traction cleat on the bottom, as well as handles, ropes, or grab holds to make them easier to carry. Chocks carried on fire apparatus must comply with NFPA® 1901, *Standard for Automotive Fire Apparatus;* NFPA® 1906, *Standard for Wildland Fire Apparatus*; and the Society of Automotive Engineers standard SAE-J348.

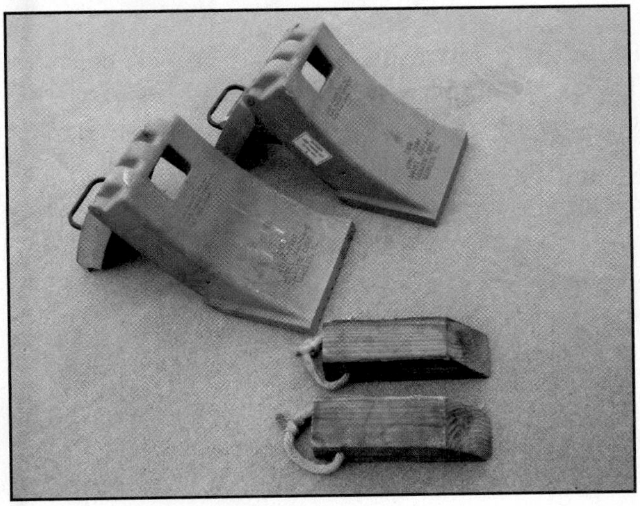

Figure 10.32 Wheel chocks prevent unanticipated movement by an apparatus or other vehicle.

Cribbing — Wooden or plastic blocks used to stabilize a vehicle during vehicle extrication or debris following a structural collapse; typically 4 x 4 (100 mm) inches or larger and between 16 to 26 inches (400 mm by 650 mm) long.

Cribbing Materials

Cribbing is used to stabilize a vehicle during extrication or to stabilize debris following a structural collapse. It consists of wooden or plastic blocks 16 to 36 inches (400 mm to 900 mm) long, which typically measure 2 x 4 inches (50 mm by 100 mm), 4 x 4 inches (100 mm by 100 mm), and 6 x 6 inches (150 mm by 150 mm). Block ends may be painted with different colors indicating different lengths. However, the flat surface of the block is never painted because paint can hide defects and cause the surface to be slippery when wet. Individual blocks may have a rope or webbing handle stapled to the end so that they can be easily carried or safely removed from under objects. Wooden blocks can be locally constructed or commercially purchased.

Wooden cribbing is made from construction grade lumber. However, plastic cribbing is often preferred because it is lighter and lasts many times longer (**Figure 10.33**). It also cannot be contaminated by absorbing fuel, oil, or other substances. One disadvantage of plastic cribbing is that it may slip under wet conditions.

Lifting Tools

Lifting tools are used to lower rescuers, remove an object from a trapped victim, or lift a victim out of a hole or confined space. Commonly used rescue tools include tripods and pneumatic lifting bags.

Tripods

Rescue tripods are used to create an anchor point above a manhole or other opening. This allows rescuers to be safely lowered into confined spaces and rescuers and victims to be hoisted out of them (**Figure 10.34**).

Pneumatic Lifting Bags

Pneumatic lifting bags are air-pressurized devices that give rescuers the ability to lift or displace objects that cannot be lifted with other rescue equipment. There are three basic types of lifting bags: high-pressure, medium-pressure, and low-pressure (**Figure 10.35**).

High-pressure bags. High-pressure bags are constructed from a tough, neoprene rubber exterior reinforced with steel wire or Kevlar™ Aramid fiber. When deflated, the bags lie completely flat and are about 1 inch (25 mm) thick. Their surface area ranges from 6 x 6 inches (150 mm by 150 mm) to 36 x 36 inches (900 mm by 900 mm), and they inflate to a height of 20 inches (500 mm). As it inflates, it loses both stability and lifting power.

Low- and medium-pressure bags. Low- and medium-pressure bags are considerably larger than high-pressure bags, and can inflate to a greater height, up to 6 feet (2 m). They are used to lift and stabilize large vehicles or objects. They are most stable when fully inflated.

Lifting bag safety rules. Operators should follow these safety rules when using pneumatic lifting bags:

- Always plan the lifting operation before you begin. Make sure you have an adequate air supply and sufficient cribbing materials.

- Be familiar with the equipment's operating principles, methods, capabilities, and limitations.

- Follow the manufacturer's recommendations for the specific system used.

- Keep all components in good operating condition.

- Make sure all safety seals are in place.

- Position the bag on or against a solid surface.

- Keep sharp objects away from the bag as it inflates.

- Never inflate a bag without a load.

- Inflate slowly and continually monitor the load for signs of shifting.

- Never work underneath a load supported only by air bags.

- Use enough cribbing to support the load in case of bag failure.

Figure 10.34 Rescue tripods allow a rescuer in a harness to enter and exit a confined space below ground.

Figure 10.33 Wood cribbing has some design limitations that are overcome in plastic cribbing and vice versa.

Figure 10.35 Air bag lifting capacities should be visibly marked on each bag.

Figure 10.36 Air bags and cribbing work together to create a safe void space under a heavy item.

- Use at least three pieces of cribbing per layer, and make sure the top layer is solid. Openings in the center of the cribbing may cause the bag to shift or rupture (**Figure 10.36**).

- Never let the bag contact materials hotter than 220°F (104°C).

- When stacking bags, inflate the bottom bag first and put the smaller bag on top. Never stack more than two bags. Single, multicell bags are more effective.

- Fuel and other petroleum products can weaken the bags and shorten their working life. Never store, use, or place a bag in an area where contact may occur.

CAUTION

Test and inspect bags regularly. If there are any signs of damage or deterioration, remove them from service immediately.

WARNING

If you place anything between the bag and the lifted object, it must be made of pliable material, such as a folded salvage cover. Plywood or other rigid material can be forcefully ejected if the bag distorts under pressure.

Pulling Tools

Pulling tools are used to pull vehicles apart, pull an object away from a trapped victim, or stabilize a vehicle that is resting over an edge. Typical pulling tools include winches, chains, and come-alongs. The powered rescue tools mentioned earlier can also be used to pull apart vehicle components.

Winches

Winches are typically mounted on the front, rear, or side of a vehicle. Compared to other pulling devices, they are stronger, faster to deploy, and have a greater travel or pulling distance. They are typically powered by an electric or hydraulic motor, or a PTO system (**Figures 10.37a-c**). They are used in conjunction with chains and/or cables.

Figure 10.37a Electric winches may be mounted on the front of a vehicle.

Figure 10.37b A hydraulic winch combines a hydraulic hauling system with a vehicle chassis and engine to provide mobility and pulling power.

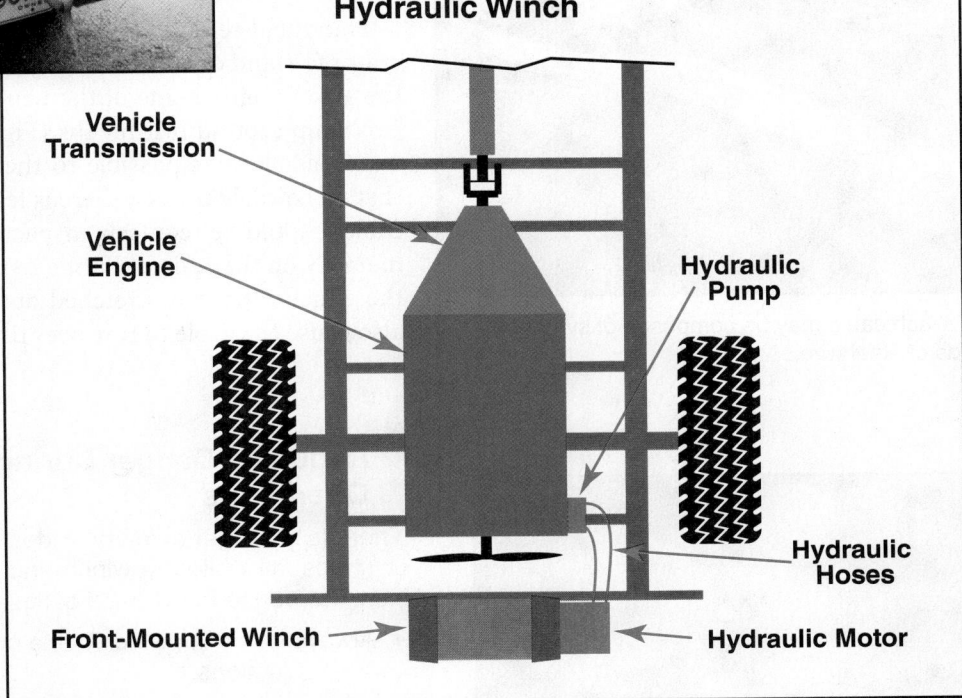

Hydraulic Winch

Vehicle Transmission

Vehicle Engine

Hydraulic Pump

Hydraulic Hoses

Front-Mounted Winch

Hydraulic Motor

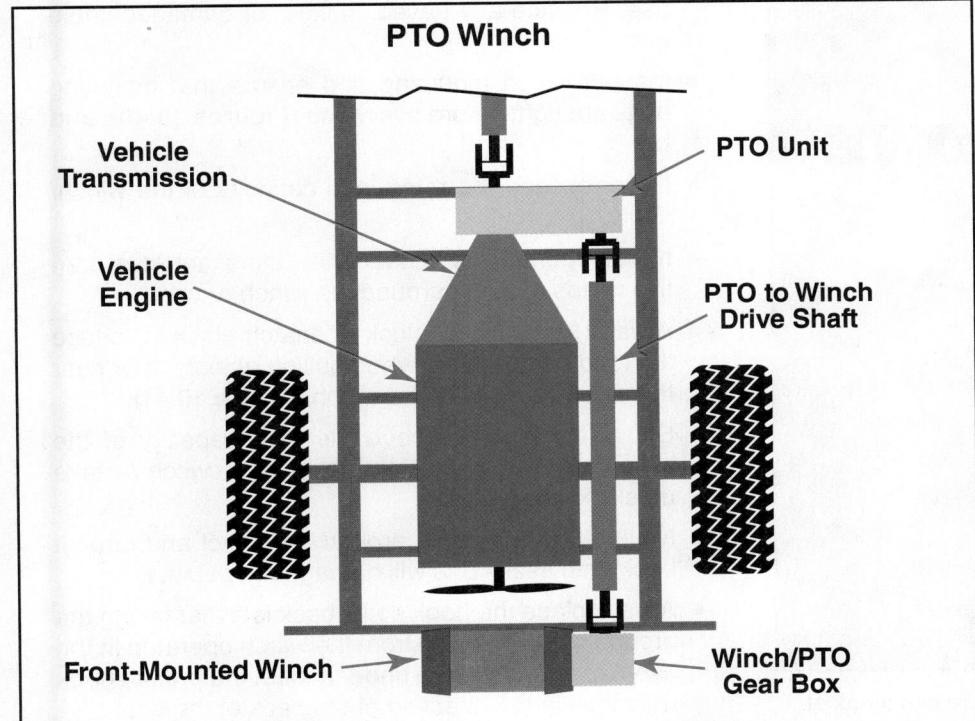

PTO Winch

Vehicle Transmission

Vehicle Engine

PTO Unit

PTO to Winch Drive Shaft

Front-Mounted Winch

Winch/PTO Gear Box

Figure 10.37c A PTO-driven winch rig is bulkier than a hydraulic winch.

Winch cables are made from steel or synthetic fiber (**Figure 10.38**). Steel cable is made from thin strands of steel wire wound together. It is durable and long lasting, but also extremely heavy and rigid, making it difficult to handle. Synthetic fiber cable is lighter and stronger than steel. It floats in water, resists ultraviolet light, and is not affected by temperature variations. If it breaks, it does not recoil or whip like steel cable, making it safer to use.

Handheld remote-control devices allow the winch operator to stand outside the danger zone. The danger zone is the area on either side of the winch cable where the cable can whip around if it breaks (**Figure 10.39**). Position the winch as close as possible to the object being pulled, so that if the cable breaks, there is less cable to recoil. Winch cables should be regularly inspected because they develop memory on the coil, returning to the coiled form it had on the winch after it is stretched out. Vehicle vibrations can also cause the cable to fray over time.

Figure 10.38 Winch cable may be composed of synthetic fibers or strands of steel wire.

DANGER ZONE

Figure 10.39 A winch cable under pressure can break at any point along its length, and if it does so it will whip back toward the winch box.

Reducing Danger During Winch Operations

There is a danger of injury or death if the cable, hooks, or straps fail while the winch and cable are under tension. To reduce the chance of injury, you should:

- Always follow the winch/cable manufacturer's operating instructions.

- Inspect the winch/cable regularly and prior to each use. Replace any frayed, kinked, or damaged cable immediately.

- Inspect winch mounting and ensure that mounting bolts are tight before every use (**Figures 10.40a and b**).

- Never exceed the rated load capacity of the winch/cable.

- Never operate the winch when there are less than five wraps of cable around the winch drum.

- Always use a pulley block or snatch block to reduce the load on the cable when pulling objects at or near the rated capacity of the winch (**Figure 10.41**).

- Shock loads can exceed the load capacity of the winch/cable. Use the remote control switch to take up slack intermittently.

- Never wrap the cable around an object and hook it back onto itself. This will damage the cable.

- Always place the hook so its back is either facing the ground or facing away from the winch operator. In the event of a hook failure under a load, the broken hook will move in the direction of the back of the hook.

Continued

- Keep the duration of winching pulls as short as possible. Do not pull for more than one minute when operating at or near the rated load capacity of the winch/cable.

- Never step over or stand near a cable that is under tension.

- Keep yourself and others a safe distance to the side of the cable under tension. Ensure that no one is standing in front of or behind the winch or the anchor point.

- Use heavy-duty gloves to protect your hands when handling the cable. Never let the cable slide through your hands.

- Drape a blanket, coat, or tarp over the cable approximately 5 to 6 feet (1.5 to 1.8 m) from the hook. This makes the cable more visible to nearby personnel and reduces the recoil force of the cable if it breaks **(Figure 10.42)**.

- Operate the remote control from the cab of the winch-mounted vehicle. Open the vehicle hood for added protection against recoil.

- Stay farther away from the winch than the length of the cable from the winch to the load.

- Never use a winch to pull a living person.

Figure 10.40a Damaged cable cannot be expected to hold the same amount weight as nondamaged cable, and must therefore be retired.

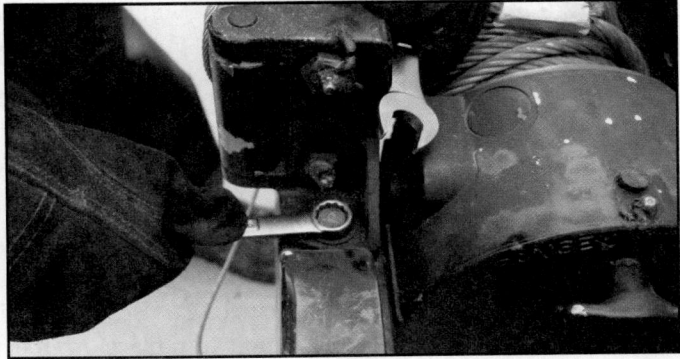

Figure 10.40b Loose winch mounting bolts may require simple tightening before use.

Figure 10.41 A pulley block helps reduce the load on a high tension cable.

Figure 10.42 Draping a flexible weight over a tensioned cable improves the safety of operations.

Come-Alongs

Come-alongs are portable cable winches operated by a manual ratchet lever (**Figure 10.43**). A come-along must be attached to a secure anchor point, after which its cable is attached to the object that must be pulled. The ratchet lever is then used to rewind the cable, pulling the object back toward the anchor point. Come-alongs typically have a load capacity ranging from 1 to 10 tons (0.9 t to 9.1 t).

> **WARNING**
> Use only the operating handles provided by the come-along's manufacturer. These handles are designed to fail before the cable. Never use a prybar or other tool instead.

Chains

Chains are used in conjunction with both winches and come-alongs. The two main chain types are alloy steel chain and proof coil chain, also known as *common* or *hardware chain* (**Figure 10.44**). Only alloy steel chain should be used in rescue operations. It is designed to resist abrasion, corrosion, and effects of hazardous atmospheres.

Tools Used in Other Activities

Power tools can also be used for other activities associated with rescue operations. Pneumatic nailers and impact wrenches can be very handy in a variety of situations (**Figures 10.45a and b**).

Pneumatic Nailers

Pneumatic nailers are used to drive nails or heavy-duty staples into wood. They are especially useful for nailing wedges and other wooden components of cribbing systems into place or securing canvas or vinyl covers over roof openings.

Figure 10.43 A come-along is operated manually to move an object toward a fixed anchor point.

Figure 10.44 Alloy steel chain is the only approved type of chain in a rescue operation.

Figure 10.45a A pneumatic nailer attached to an SCBA cylinder is portable and maneuverable in tight places.

Figure 10.45b An impact wrench can be used to forcibly separate metal panels. *Courtesy of Owasso (OK) Fire Department.*

Impact Tools

Pneumatic impact tools or wrenches have a square drive onto which a socket is attached. The socket can then be applied to a nut or bolt head of the same size to tighten or loosen it quickly. These tools are ideal for disassembling machinery in which a victim is entangled.

Rescue Tool Maintenance

For inspection, care, and maintenance, always follow the manufacturer's recommendations and departmental SOPs. Follow these general guidelines:

- Check all fluid levels.
- Use only the recommended types of lubricants, hydraulic fluids, and fuel grades.
- Keep battery packs fully charged.
- Inspect power tools at the beginning of each work shift and make sure that they start.
- Inspect saw, chisel, and cutter blades regularly. Replace blades that are worn or damaged **(Figure 10.46, p. 504)**.
- Check all electrical components, such as cords and portable receptacles, for cuts or other damage.

- Make sure that all protective guards are functional and in place.

- Make sure that fuel is fresh. A fuel mixture may separate or degrade over time.

- Inspect hydraulic and fuel supply hoses for damage.

- Inspect hydraulic hose couplings (quick disconnect fittings) to ensure that they are clean and functional.

- Make sure that all parts and support items are easily accessible.

Vehicle Extrication

Most rescues involve extricating a trapped victim after a motor vehicle accident (**Figure 10.47**). To perform a safe, effective extrication, you must be able to:

- Perform scene size-up

- Assess the need for extrication

- Stabilize the vehicle

- Secure the vehicle's electrical system

- Recognize different types of passenger safety systems

- Determine the best way to gain access to victims

- Perform the correct type of extrication based on the vehicle type and condition

Scene Size-Up

The first step of an extrication is to size up the incident scene. Size-up begins during dispatch and continues throughout the incident. A careful assessment of the scene is necessary to:

- Prevent injury to rescuers

- Prevent further injuries to victims

- Identify potential hazards

- Clarify required tasks

- Identify needed resources

    As you approach the scene, be observant and try to answer the following questions:

- What are the traffic hazards, and what types of traffic control devices are needed?

- How many and what types of vehicles are involved?

- What type of fuel or power system (hybrid or electric) do the vehicles use?

- Where and how are the vehicles positioned?

- Are the vehicles located on the roadway?

- How many victims are there, and what is their status?

- Is there a fire or potential for a fire?

- Is there a fuel or fluid leak? What control methods need to be implemented?

- Are there any hazardous materials involved?

- Are there any utilities, such as water, gas, electricity or downed power lines, that may have been damaged? If so, do they pose a hazard to victims or rescuers?

- Is there a need for additional resources?

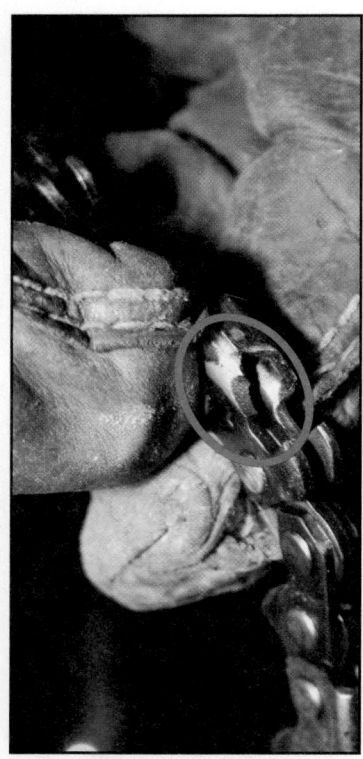

Figure 10.46 A damaged saw chain link creates preventable hazards.

Controlling Hazards

Traffic is the primary hazard at any motor vehicle accident, but additional dangers are created by fire, bloodborne pathogens, sharp objects, and environmental conditions. Reduce your risk of injury by always following proper safety procedures.

Apparatus should be parked so that they form a protective barrier between the scene and ongoing traffic from all directions **(Figure 10.48)**. Signs and cones should be deployed to detour traffic around the scene, and law enforcement officers can provide additional traffic control assistance. You can also protect yourself from traffic hazards by maintaining constant situational awareness. Keep an eye on traffic flow and always stay within the protective barrier provided by the apparatus. Wear your retroreflective vest at all times unless you are directly involved in fire fighting or extrication.

> **WARNING**
> Before you dismount the apparatus, size up the traffic flow and other scene conditions. Remember that drivers are often distracted by emergency vehicles and flashing lights.

Figure 10.47 During extrication operations, contingencies, such as a sudden fire in the engine compartment, should be planned for.

Figure 10.48 Apparatus responding to an incident should stage in a way that protects victims down-lane. *Courtesy of Jennifer Ayers/Sonrise Photography.*

Fire is another potential hazard (**Figure 10.49**). If open flames are present, extinguish them immediately. Isolate spilled fuels and other ignition sources before addressing other operational concerns. Other sources include vehicle batteries, undeployed air bags, downed power lines, and energy-absorbing struts. To isolate these hazards, you must:

- Disconnect vehicle batteries
- Deactivate undeployed air bags
- Cordon off downed power lines or ground level transformers
- Protect shock absorbers and struts from excessive heat and/or physical damage
- Avoid pyrotechnic seat belt **pretensioners**

NOTE: Pyrotechnic seat belt pretensioners are described later in this chapter.

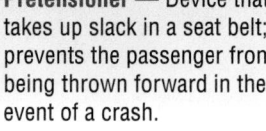

Pretensioner — Device that takes up slack in a seat belt; prevents the passenger from being thrown forward in the event of a crash.

Disabling Air Bags

Although some air bags can be disabled by removing the ignition key, disabling is usually done by disconnecting the battery. However, the air bag may still be dangerous for up to 60 minutes because of the capacitors in the electrical activation system.

Additional hazards may be found in the vehicle's trunk and interior. Flammable adhesives, pressurized solvents, or flammable liquids (such as gasoline) are commonly encountered hazards. Illegal substances used to produce methamphetamine (meth) are also highly flammable. In extreme cases, vehicles have been found to contain entire mobile meth laboratories.

Figure 10.49 Vehicle fires may occur during other emergency response activities. *Courtesy of Bob Esposito.*

Vehicle wheels and tires, either on the vehicle or in the trunk, may also create a hazard. When exposed to fire, alloy wheels made with magnesium (also known as *mag wheels*) can burn with intense heat. Wheels made from pure magnesium are no longer produced, but may still be encountered. High-pressure tires can also create an explosion hazard when punctured or exposed to fire.

Other vehicle components made with magnesium can also cause high heat fires **(Figure 10.50)**. These components include:

- Valve covers
- Steering columns
- Mounting brackets on antilock braking systems
- Transmission casings
- Engine blocks
- Frame supports
- Exterior body components

Bodily fluids resulting from patient injuries can contaminate the interior and exterior of the vehicle. Follow bloodborne pathogen procedures and wear appropriate protective clothing and equipment to avoid exposure. Your boots and the cuff of your protective trousers can become contaminated from standing in blood and other bodily fluids. Be careful to avoid possible contaminants when removing your boots and protective trousers, and follow your department's SOPs for decontaminating this protective equipment.

Broken glass and sharp metal edges can create a hazard during and after the incident. Cover sharp edges with short sections of used fire hose, clear broken pieces of glass away from the door sills and windshield edges, and wear full personal protective clothing while performing vehicle extrication **(Figure 10.51)**. Also, carefully remove any glass or metal shards embedded in your boots following the incident.

Figure 10.50 Magnesium components in older vehicles are now known to generate high heat fires that cannot be safely extinguished with water.

Figure 10.51 Full protective gear is essential while breaking glass windows to prevent injury from flying shards.

Environmental conditions can also create hazards. If roads are icy, be extra careful when dismounting the apparatus. Protect other incoming responders by spreading absorbent on the ice until you are able to use sand or salt.

Vehicle Fuel Types

During the size-up, you must determine the type of fuel system in the vehicles you are approaching. The most common conventional fuels are gasoline and diesel, but alternative fuels include:

- Methanol
- Ethanol
- Reformulated gasoline
- Reformulated diesel (for trucks only)
- Natural gas
- Propane
- Hydrogen
- Electricity (including total electric and hybrid electric vehicles)
- Biofuels
- Coal-derived liquid fuels
- Alcohol blended with other fuels

Both conventional and alternative fuels are easily ignitable. Leaks from the fuel lines and tanks must be controlled and fire protection must be established to prevent the ignition of fuel vapors. Minimal fire protection could be a single firefighter standing by with a portable extinguisher, but it may be necessary to activate foam-generating systems and deploy several fully charged hoselines (**Figure 10.52**). Follow your department's SOPs, but at least one 1½-inch (38 mm) hoseline should typically be charged and ready for use.

By the end of 2010, over sixty vehicle manufacturers were designing or building electrically powered vehicles. Electrically powered vehicles may be classified as either total electric vehicles (EV) or hybrid electric vehicles (HEV) (**Figures 10.53a and b**). Total electric vehicles have a relatively short driving range. They are powered by a bank of batteries, which must be plugged into a charging station to recharge. Hybrid electric vehicles are powered by multiple propulsion systems, such as gasoline-electric hybrids that have internal combustion engines and electric motors.

Vehicle Electrical Systems

The electrical system of conventional fuel vehicles is designed to store and deliver the electricity needed to start the engine and to power and operate the various electrical components such as air conditioning, radios, power windows, and power seats. A typical vehicle electrical system is composed of a battery that stores the electricity; an alternator that produces the electricity; wiring; fuses that protect the electrical system; and lights, fans, and other ancillary equipment (**Figure 10.54, p. 510**).

Most vehicle electrical systems are either 12-or 24-volt systems. Passenger vehicles and light trucks usually have 12-volt systems while larger trucks, recreational vehicles, and military vehicles operate on 24-volt systems.

The danger associated with electric and hybrid vehicles is the high voltage stored within the batteries and running through wiring connected to the vehicle's electric motor. These wires can carry as much as 650 volts of DC. Hybrid electric vehicles

Figure 10.52 A standby team should be available while working on a vehicle that may have a compromised electrical system.

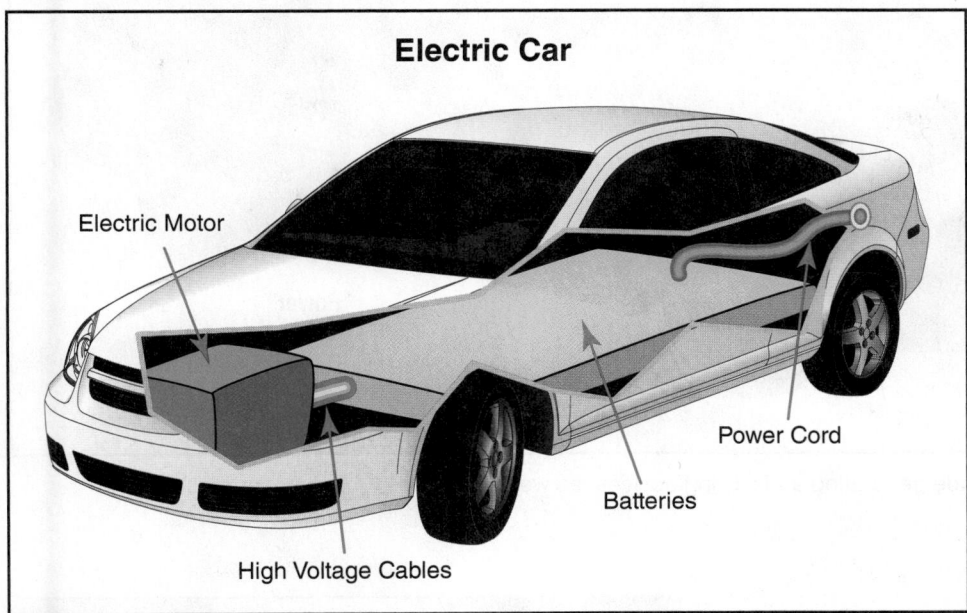

Electric Car

Electric Motor

Power Cord

Batteries

High Voltage Cables

Figure 10.53a An electric vehicle (EV) may have a shorter drive range than a traditional vehicle due to the battery life span.

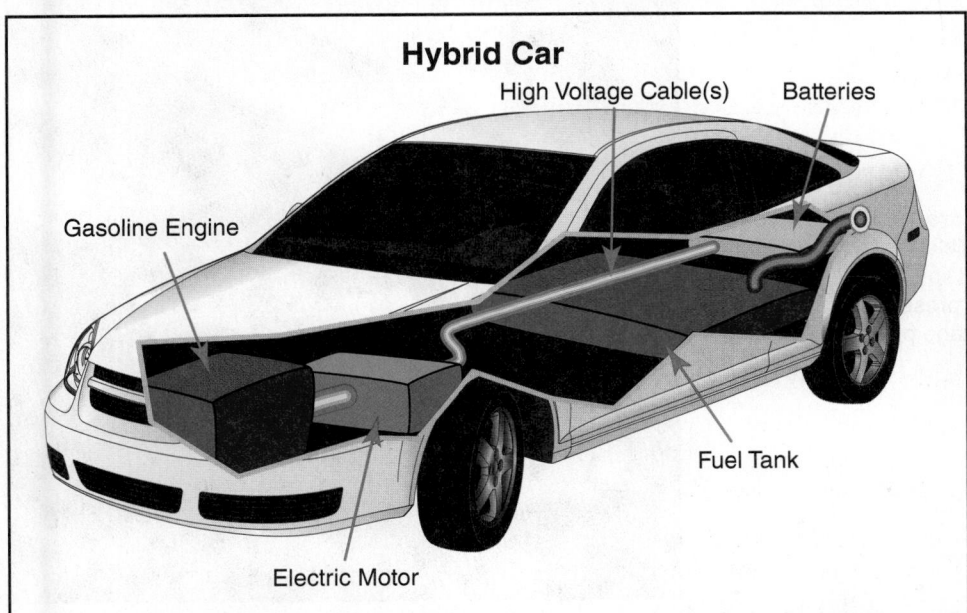

Hybrid Car

High Voltage Cable(s)

Batteries

Gasoline Engine

Fuel Tank

Electric Motor

Figure 10.53b Hybrid vehicles (HEV) contain hazards common to both electric and traditional vehicles.

may be identified by a nameplate or logo. Rescue personnel attempting to isolate the electrical power system and batteries are exposed to the danger of electrical shock. To assist emergency responders in recognizing the high voltage power wires and cables, most hybrid vehicles have wiring that is color-coded orange and covered with orange shielding or orange tape (**Figure 10.55**). The 36-volt system on the Saturn, however, is color-coded blue. General Motors classifies the 36-volt system as intermediate voltage.

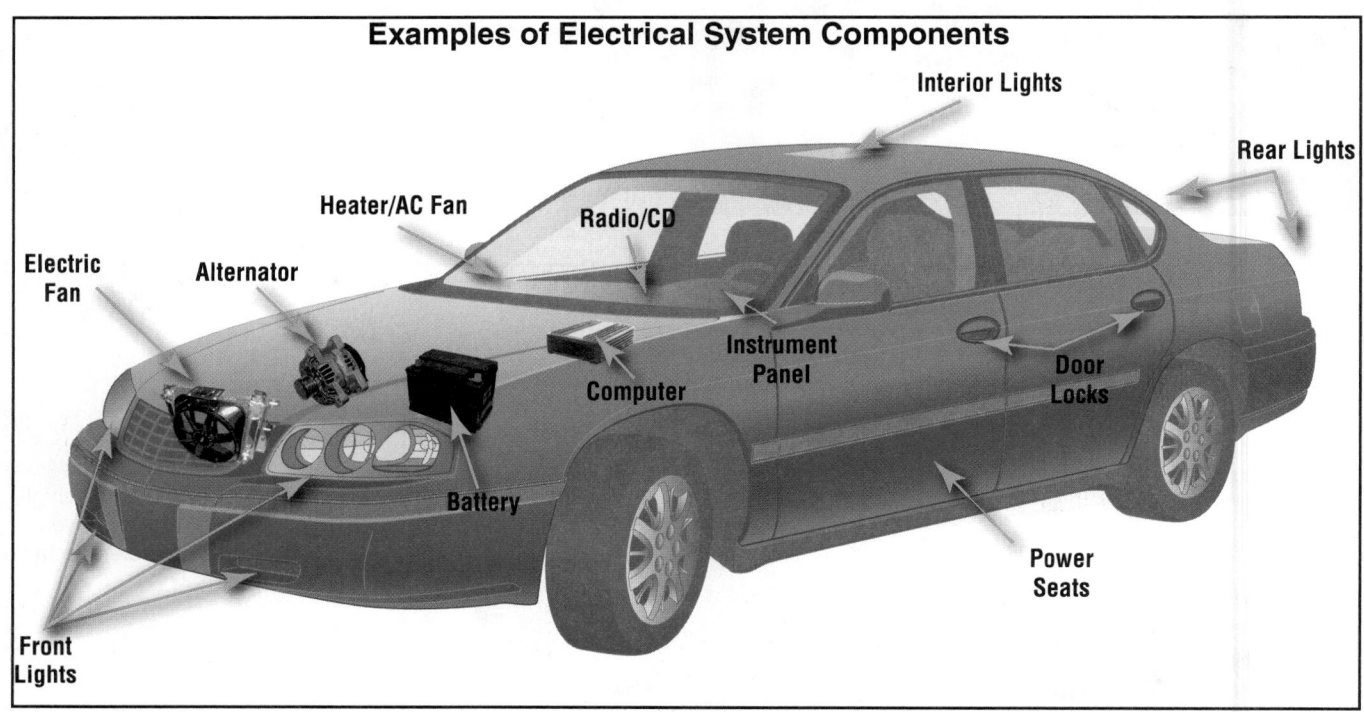

Examples of Electrical System Components

Figure 10.54 Electrical systems include generating and storing devices, as well as devices that use energy during the operation of the vehicle.

Figure 10.55 Most hybrid vehicles use orange wiring and shielding to indicate the presence of high voltage power wires.

Vehicle Voltage Chart

The following are the voltages found in electrical systems of a variety of vehicles:

12 Volt – ALL vehicles

36 Volts – Saturn Vue

42 Volt – Some conventional and hybrid models

72 Volt – Neighborhood Electric Vehicles (NEV)

144 Volt – All Honda hybrids

300 Volt – Toyota first generation Prius hybrids

500 Volt – Toyota second generation Prius

650 Volt – Toyota Highlander SUV, Lexus RX 400h and GS 450h hybrid

Assessing the Need for Extrication

Before beginning an extrication, the entire scene must be thoroughly assessed. In the immediate area around the vehicles, one crew member must:

- Assess the condition and position of the vehicles
- Determine extrication tasks that may be required
- Note any hazardous conditions

Meanwhile, another crew member surveys the entire scene, checking for:

- Other involved vehicles that may not be visible (over an embankment, for example)
- Victims who might have been ejected from the vehicles
- Damage to structures or utilities that may present a hazard
- Other circumstances that require special attention

At night or in low light, a thermal imager (TI) can be helpful in locating victims who have been ejected.

A firefighter who is trained in emergency medical care will:

- Determine the number of victims
- Assess their injuries
- Assess the extent of their entrapment

All of this information helps the Incident Commander (IC) determine the re-sources needed to stabilize the incident and the order in which victims should be extricated. Seriously injured victims must receive higher priority than those with minor injuries. Victims who are not trapped should be removed first to make more working room for rescuers who are trying to remove those who are entrapped. As each assessment is completed, crew members should report the information to the IC. This report should also include the type of accident, such as rollover, head-on, side impact, and so on. The report should also include other identifiers (engine intrud-ing into passenger compartment, unrestrained passengers, deformed steering wheel) that are helpful for hospital emergency room staff when determining likely injuries **(Figure 10.56, p. 512)**.

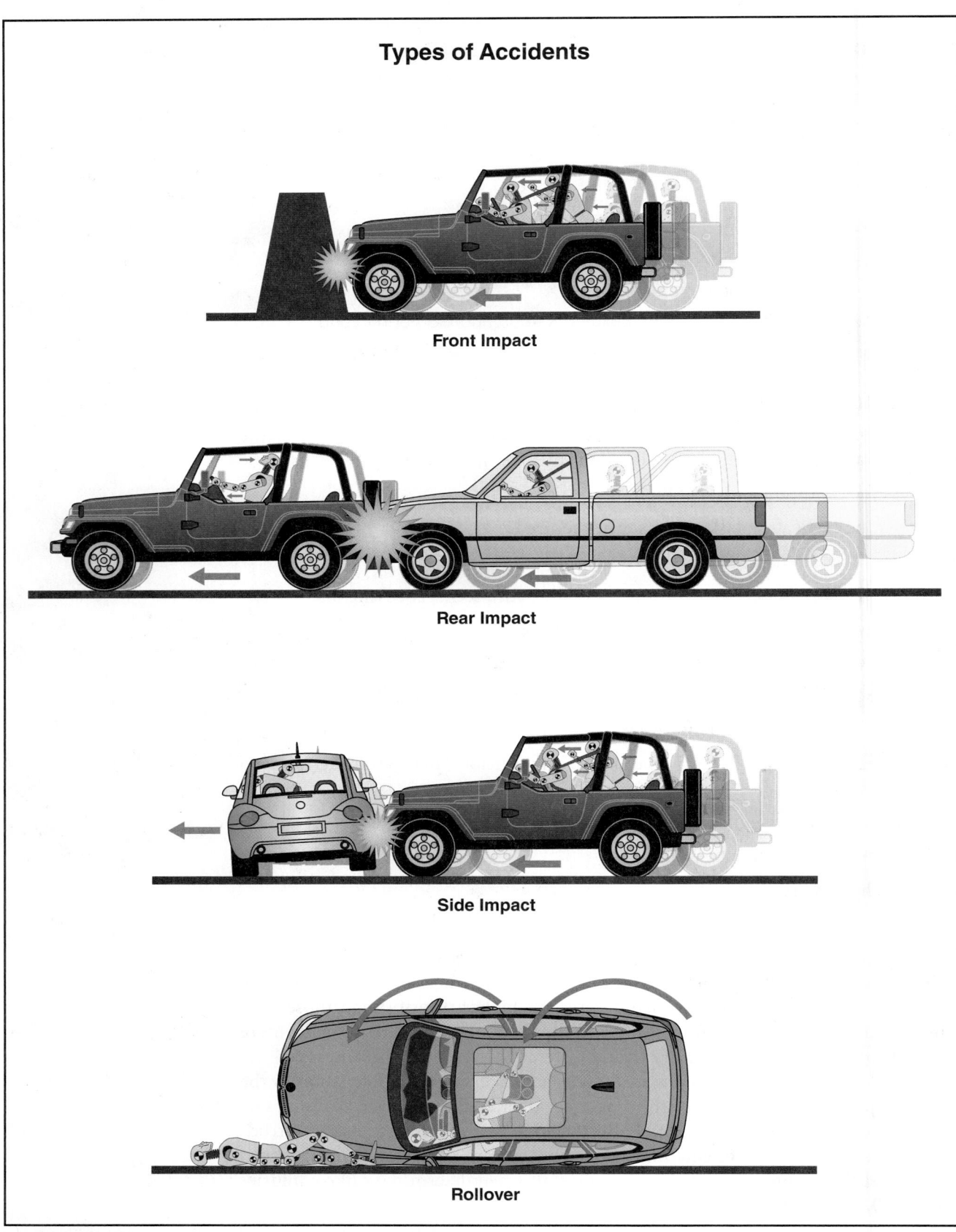

Types of Accidents

Front Impact

Rear Impact

Side Impact

Rollover

Figure 10.56 Knowledge of common types of vehicle impact injuries helps emergency responders to determine the correct response.

Stabilizing the Vehicle

Once the scene has been assessed, the vehicle must be stabilized. Use cribbing, wheel chocks, lifting devices, and buttress tension systems to support key points between the vehicle and the ground (**Figure 10.57**). The goal of stabilization is to prevent the vehicle from moving unexpectedly because movement can cause severe injury to both victims and rescuers.

Never move the vehicle or attempt extrication until the vehicle has been fully stabilized. This is particularly true if the vehicle is on its side, upside down, or in danger of falling over a cliff, bridge, overpass, or embankment (**Figures 10.58a–c**). Never place any part of your body under the vehicle while putting the stabilizing devices in place.

Figure 10.57 Several common resources, including cribbing, wheel chocks, and lifting devices should be used in tandem to support a vehicle before operations begin.

Figure 10.58b A vehicle on its side must be buttressed for stability at the least.

Figure 10.58a A vehicle that has been subject to a front-end collision may end with its wheels on the ground, but it may still require some stabilization to prevent sudden movement during extrication operations. *Courtesy of Bob Esposito.*

Figure 10.58c Unique configurations of vehicles require specialty tools and skills to create a stable environment for rescue activities.

Using Wheel Chocks

Even a vehicle with all wheels on the ground must be stabilized. One simple approach is to deflate the tires. But if deflating the tires is insufficient or impossible, chocking can accomplish the same result. Chocking can be accomplished with standard apparatus wheel chocks, pieces of cribbing, or other similar-sized objects such as bricks or rocks (**Figure 10.59**). **Skill Sheet 10-II-2** describes the steps for preventing horizontal movement of a vehicle using wheel chocks.

General wheel chock guidelines:

- Place chocks in front of and behind tires.
- Place chocks on the downhill side of a vehicle on an incline.
- Place chocks on both sides of the tires if the ground is level or the direction of the grade is undetermined.
- Test and apply the parking brake before placing chocks.
- Center chocks snugly and squarely against the tread of each tire.

If the vehicle's mechanical systems are operable, they can be used to supplement the wheel chocks. If possible, shift automatic transmission to park, or put manual transmission in first gear. Turn the ignition off, and engage the vehicle's parking or emergency brake.

> ////////////////////
> **CAUTION**
> Never rely on the vehicle's mechanical systems as the only means of stabilization. Use them only in conjunction with other methods.

Figure 10.59 Chocking is a non-damaging method of stabilizing and immobilizing a vehicle's wheels. *Courtesy of Jennifer Ayers/Sonrise Photography.*

Using Cribbing Materials

Cribbing is typically used in a box formation (**Figure 10.60**). Cribbing pieces are typically pushed into position with a mallet or with another piece of cribbing. Wedges may be necessary to ensure solid contact between the cribbing and the vehicle. Cribbing can also be placed under the sides of a vehicle to prevent lateral movement. **Skill Sheet 10-II-3** describes the proper method for stabilizing a vehicle using cribbing.

Using Lifting Devices

Lifting jacks and pneumatic lifting bags are used to:

- Raise a vehicle that is on its roof or side
- Support the frame of the vehicle
- Gain access to the interior by lifting an object that is resting on the vehicle

An advantage of using lifting jacks is that they can be adjusted to the required height and may be inserted into a tight space. Disadvantages include the following:

- They are time-consuming to place.
- They may limit access to the vehicle.
- They may shift, allowing the vehicle to move.

Pneumatic lifting bags can also be used to temporarily stabilize the vehicle. To be effective, at least two pneumatic lifting bags are needed. They may be positioned with one on each side of the vehicle, or with one in the front and one in the rear (**Figures 10.61a and b**).

Because jacks and lifting bags can be damaged or jarred loose, solid cribbing must be used to supplement them (**Figure 10.62, p. 516**). **Skill Sheet 10-II-4** describes the proper method for stabilizing a vehicle using lifting jacks.

Figure 10.60 Cribbing may be used in a variety of applications including vehicle stabilization.

Figure 10.61a Pneumatic lifting bags should be used in pairs at opposing sides of an unstable vehicle.

Figure 10.61b Pneumatic bags may be placed on adjacent sides of a vehicle while checking to ensure the stability of the load.

Figure 10.62 Solid cribbing supplements the work of a jack.

Using Struts and Buttress Tension Systems

When vehicles are found upside down, on their side, or on a slope, rescuers should use whatever means are available to stabilize the vehicle. Traditional methods or buttress tension systems can be used to accomplish this.

Traditionally, a combination of cribbing, ropes, webbing, chains, and winch cables can be used to accomplish these types of stabilization tasks. Secure the ropes, webbing, or chains between the frame of the vehicle and a secure object such as a tree, railing, or other substantial object (**Figure 10.63**). You can also use the cable and winch on a fire apparatus or tow truck to secure the vehicle. Be aware that winch cables under tension can be dangerous if they break. Ensure that cables, chains, webbing, ropes, and winches have a safety margin in excess of the weight being secured. Ropes and webbing need to have a sufficient load capacity. **Skill Sheet 10-II-5** describes the method for stabilizing a vehicle using a system of ropes and webbing.

Adjustable struts can be used to stabilize a vehicle that is on its side. Struts typically consist of a square tube attached to a base plate, which spreads the vehicle load. The lower end of the tube houses another tube that telescopes out. A series of holes runs down the sides of both tubes, and a pin can be inserted through the holes to hold the tubes at the desired length. Any space remaining between the top of the tube and the bottom of the vehicle can be taken up with a screw jack in the end of the tube. A 4 x 4 inch (100 by 100 mm) wood post may also be used in place of the adjustable struts to support the vehicle.

Figure 10.63 The choice of an anchor may have implications during the evolution of stabilization tactics.

The strut system has led to the buttress tension system, which is used to stabilize vehicles that are upside down or lying on their sides. Use a minimum of three buttress tension posts. Start by placing two on the least stable side of the vehicle. The longest side of the unit extends at a 50 to 70 degree angle from the ground to a high point on the vehicle. The two tension straps extend from the base of the long leg to points on the vehicle (**Figure 10.64**). **Skill Sheet 10-II-6** describes the method for stabilizing a vehicle using a buttress tension system.

Securing the Electrical System

The vehicle's electrical system must be shut down to eliminate a potential source of ignition. In newer vehicles, cutting power is especially important because electrical power can activate the vehicle's restraint systems, further injuring the victim and creating an additional risk of ignition (see the next section of this chapter for details).

Before shutting off the electrical power, lower power windows and unlock power doors. Move power seats back only if the victim's medical condition has been evaluated and it is safe to do so.

The simplest method of shutting down electrical power is to turn off the ignition and remove the key. This method will work on any type of car, both electric and nonelectric. However, if the ignition is not accessible, there are different procedures for shutting down power in electric and nonelectric vehicles.

Nonelectric Vehicles

Nonelectric vehicles include those powered by conventional fuels (gasoline, diesel) and alternative fuels (such as biofuels, ethanol, or natural gas). If the ignition is not accessible, disconnect or isolate the negative cables to the vehicle's 12-volt battery (**Figure 10.65**). **Cut the negative cable first, then the positive.** Remove approximately 2 inches (50 mm) of each cable.

CAUTION

Shutting off power does not mean that all systems are immediately safe. Restraint systems can work for up to an hour after shutdown.

Figure 10.64 Buttress tension straps and posts stabilize a vehicle that has come to rest on its side.

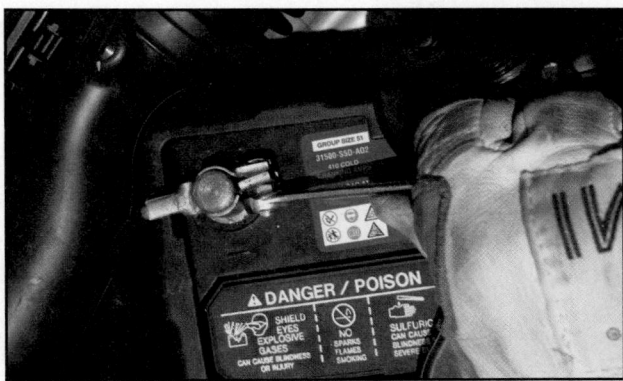

Figure 10.65 Disconnect the battery's negative cable before interacting with the positive cable.

Electric Vehicles

This includes both fully electric and hybrid electric vehicles. General safety guidelines for extrication include:

- Always assume that the vehicle is powered up despite the lack of engine noise.
- Place wheel chocks in front and behind tires to prevent unexpected vehicle acceleration.

- Place the transmission in park, turn off ignition, and remove key to disable the high voltage system.

- Disconnect the 12-volt electrical system in the same way you disconnect the battery in a conventional vehicle. Cutting the 12-volt negative and positive cables will isolate the high-voltage system.

- Because the vehicle can hold a charge from capacitors that can cause it to start, remove the key to a safe distance of 25 feet (7.62 m) from the vehicle.

- Stabilize the vehicle to prevent unexpected air bag deployments.

- Never touch, cut, or open any orange cables or components protected by orange shields. Orange cables and components contain high-voltage charges.

- Never touch, cut, or open any blue cables. Blue cables contain intermediate voltage charges.

- Remain a safe distance from the vehicle if it is on fire.

- Always wear an SCBA during and after fire suppression. A fire in the high-voltage battery pack of an electric car will produce toxic fumes.

- Be aware that metal tools, metal buckles on personal protective equipment, and metal jewelry can cause electrical shock if they come in contact with an energized portion of the vehicle.

- Consider the electrical system unsafe for at least 10 minutes after the ignition has been shut down, and be aware that it may hold a charge for up to 24 hours.

- Contact local auto dealerships for more information about their electric and hybrid electric vehicles.

- Review the Manufacturer Specific Emergency Response Guidelines of common electric vehicles in your response area.

Voltage Can Kill

If you are standing in water or your skin is wet, an electric charge of 20 volts can be deadly. If water is not present, a charge of 60 volts is deadly.

Passenger Safety Systems

Seat belts and more advanced passenger safety systems pose an immediate danger to both victims and rescuers **(Figure 10.66)**. These systems include:

- Supplemental Passenger Restraint Systems (SPRS)

- Side-Impact Protection Systems (SIPS)

- Head Protection Systems (HPS)

- Extendable Roll Over Protection Systems (ROPS)

As discussed in the previous section, shutting down the electrical system can cut the power to these safety devices. But some can be activated in other ways, so you must know how to disable them individually.

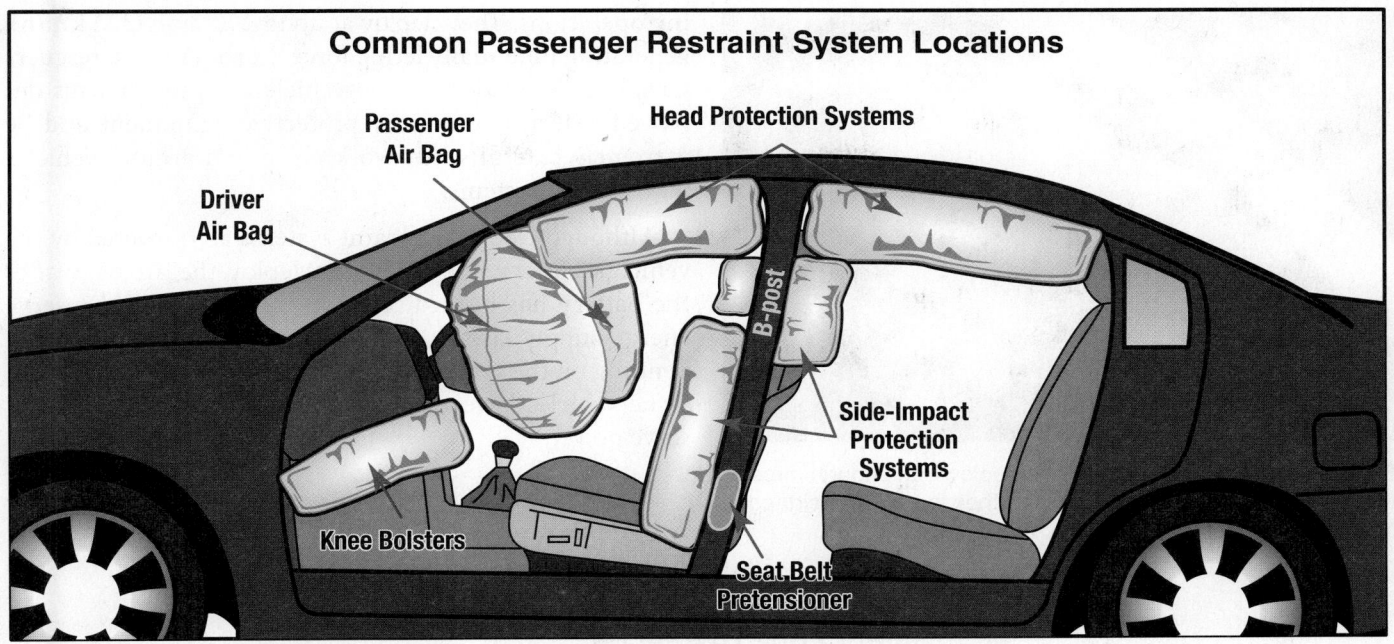

Common Passenger Restraint System Locations

Driver
Air Bag

Passenger
Air Bag

Head Protection Systems

B-post

Side-Impact
Protection
Systems

Knee Bolsters

Seat Belt
Pretensioner

Figure 10.66 Passenger restraint systems may deploy without warning.

Seat Belts

Modern seat belts include devices called pretensioners that lock the belt during a crash to prevent further travel of the belt and the person wearing it. These devices are pyrotechnic, so if they are not disabled they present a significant ignition hazard. Pretensioners are hidden inside the **B-posts** or the center console **(Figure 10.67)**. They can be accessed by removing B-posts' interior trim. Side-impact protection systems (SIPS) (also known as *side curtain air bags*) control modules may also be located in the B-post area. Because the pretensioners and SIPS control modules are hazardous, you should never cut into these units during extrication. Techniques to disable seat belts and seat belt pretensioners include cutting the seat belt webbing, unbuckling, and retracting the seat belts.

B-Post — Post between the front and rear doors on a four-door vehicle, or the door-handle-end post on a two-door car.

WARNING

Never cut a cable or touch a component with yellow/black or orange tapes, insulation, or tags. Doing so may cause electrocution or activate undeployed air bags.

Supplemental Passenger Restraint Systems (SPRS) and Side-Impact Protection Systems (SIPS)

Modern technology has added increased collision protection for vehicle occupants by means of Supplemental Passenger Restraint Systems (SPRS) and Side-Impact Protection Systems (SIPS), commonly called air bags or side curtain air bags **(Figure 10.68, p. 520)**. These air bag systems may be triggered by extrication or fire fight-

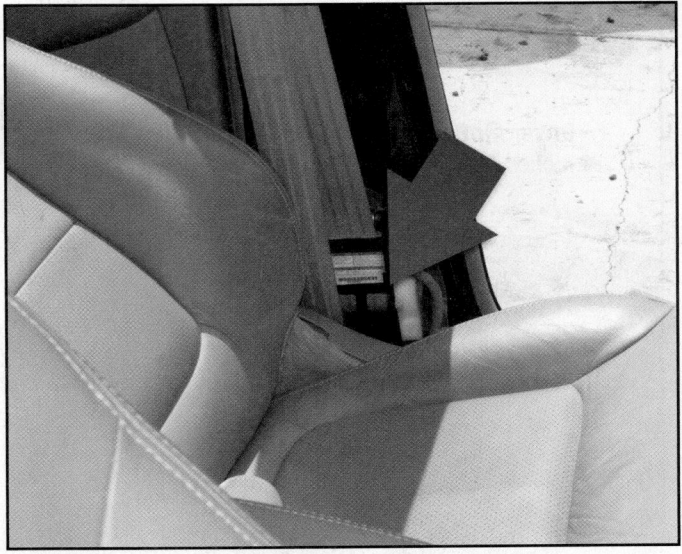

Figure 10.67 Pretensioners are pyrotechnic and should be avoided or protected during extrication evolutions.

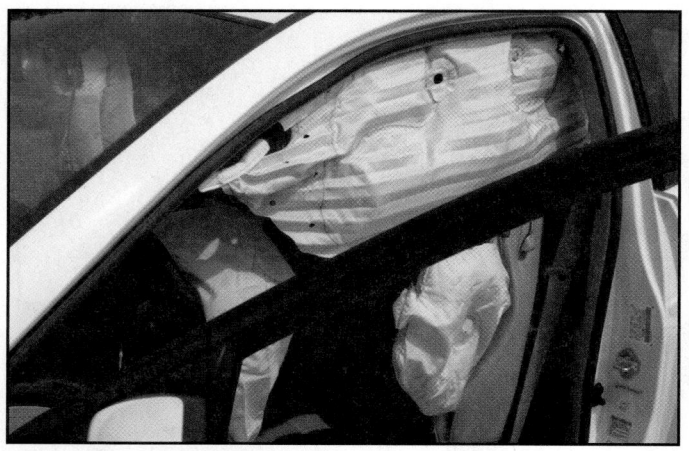

Figure 10.68 Side curtain air bags deploy with enough force to cause injury or worse in situations they were not designed to mitigate.

ing operations. They deploy at up to 200 mph (322 km/h), generating potentially lethal force. In some cases, rescuers have been ejected from the vehicle after its air bags deployed. Always wear your protective equipment and be extremely careful when working in and around vehicles with air bag systems.

Although electric restraint systems are powered by the vehicle's battery, they can still deploy the air bags after the battery has been disconnected. The systems' reserve energy supply lasts anywhere from 2 to 60 minutes, depending on the vehicle's make and model. For this reason, some vehicles have a key-operated switch that drains reserve power.

Safety Zone Distances

Use the 5-10-12-18-20 rule to maintain safe working distances during extrication. Stay at least:

- 5 inches (127 mm) away from side-impact air bags and knee bolsters
- 10 inches (254 mm) away from driver frontal air bags
- 12-18 inches (304 to 457 mm) away from side-impact curtains (which deploy down from the headline)
- 20 inches (508 mm) away from passenger frontal air bags

These distances are general guidelines only. Actual safety zone distances will vary depending on make and model (Figure 10.69).

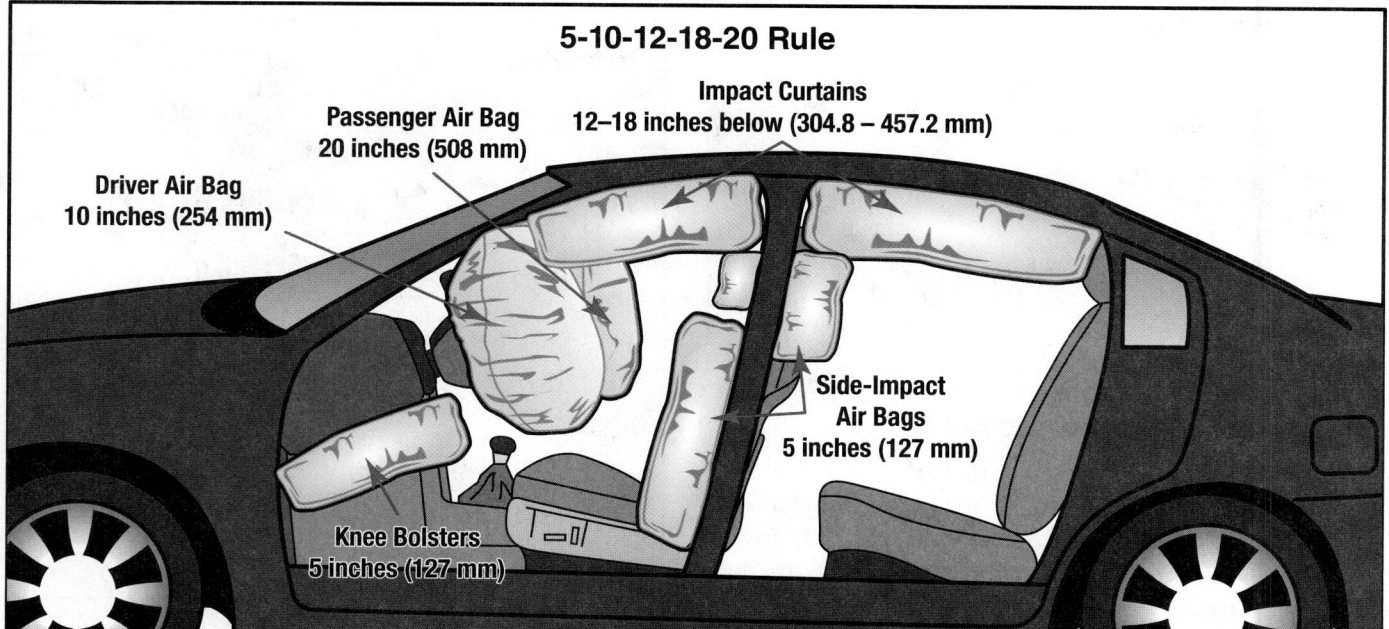

Figure 10.69 In addition to the locations, knowledge of the size of each type of restraint device will help responders access areas strategically.

SIPS restraints are often mechanically operated, so you cannot disable them by disconnecting the battery. If you accidentally apply pressure to the system's mechanical sensors, the air bags may deploy. To prevent accidental deployment, you must cut the connection between the sensors and the control unit, which is usually under the dashboard or in the center console. Disconnection procedures vary depending on the vehicle's make and model.

Head Protection Systems (HPS)

During a side-impact collision, HPS deploy air bags from just above the top of the door frame (**Figure 10.70**). There are two main types, window curtains and inflatable tubes. Window curtains automatically deflate shortly after deployment, but inflatable tubes do not. A knife or sharp tool is typically used to puncture and deflate the tube.

Figure 10.70 Head protection systems deploy automatically in case of side impact.

Rescuers working through a window opening may be directly in the deployment path of HPS air bags. This danger can be mitigated by removing the roof, but be careful not to cut into high-pressure cylinders and other devices used in conjunction with the HPS.

Extendable Roll Over Protection Systems (ROPS)

Extendable ROPS are designed to deploy automatically if the vehicle rolls over. They are found behind the front seat of small sports cars and in the rear window deck of convertibles (**Figure 10.71**). If you are working in a vehicle with an undeployed ROPS, you must stay in a safe working position (**Figure 10.72, p. 522**). Accidental deployment can cause serious injury.

These safety devices can deploy rapidly when the vehicle's inclinometer senses that the vehicle is approaching an angle of 62 degrees laterally or 72 degrees longitudinally, if the vehicle experiences a 3G acceleration force, or becomes airborne

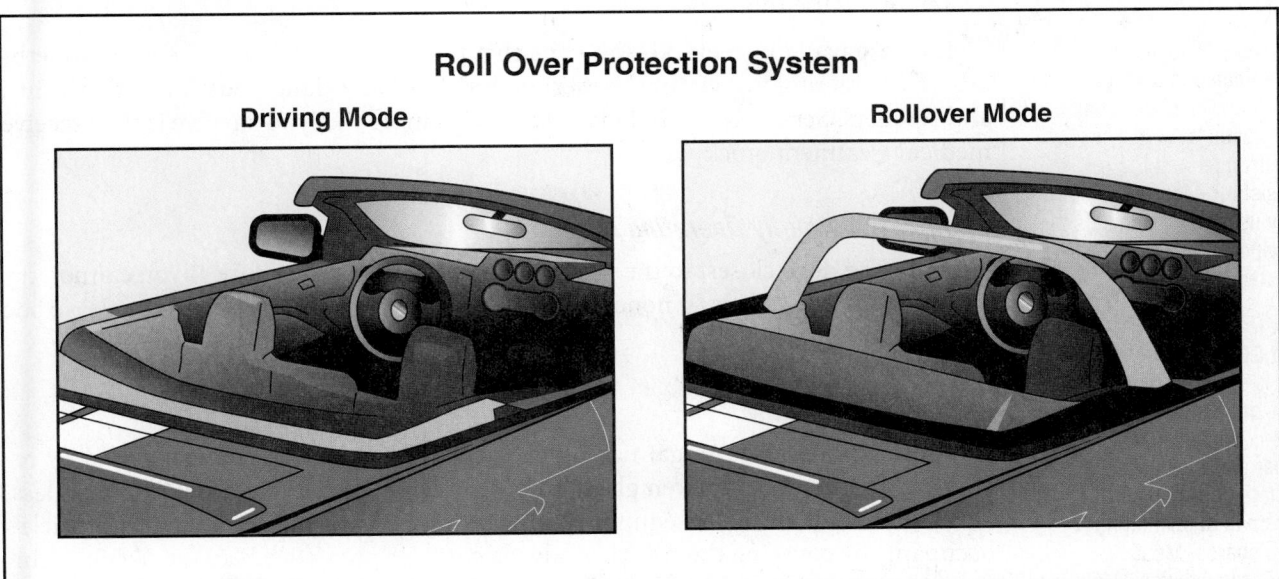

Figure 10.71 An active rollover protection system disengages while the vehicle is in driving mode.

Roll Over Protection System

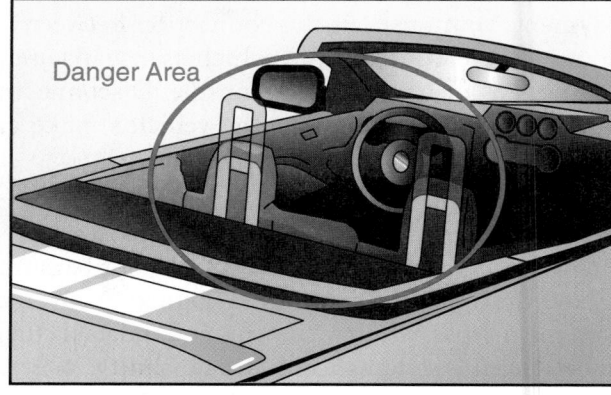

Figure 10.72 Safe working positions are clear of the area where a roll over protection system could engage without notice.

achieving weightlessness for at least 80 milliseconds (such as during a lifting operation) **(Figure 10.73)**. To disable these devices power down the vehicle as soon as possible. In some cases you may have to deploy them intentionally to prevent them from being a hazard during extrication.

Gaining Access to Victims

Once the vehicle is stable and its electrical system is secure, rescuers must gain access to the victims. The four basic methods are **(Figures 10.74a-d)**:

- Opening a normally operating door
- Removing a window
- Prying open a door
- Removing the roof

The best available method is the one that is simplest and fastest. Lengthier methods of access prolong victims' suffering and are more dangerous for both victims and rescuers. Seriously injured victims are also more likely to survive if they receive medical treatment quickly.

Opening a Normally Operating Door

Examine the door closest to the victim and try to open it normally. If you cannot, try the vehicle's other doors. If none of the doors open normally, use a tool to release one of the locks.

Removing a Window

Windows are removed to gain access to victims or reduce the danger posed by remaining fragments of broken glass. To protect yourself against loose or flying glass, always wear full protective equipment, including eye protection. Protect the vehicle's occupants by covering them with a salvage cover or protective blanket. Vehicle windows can be made from **safety glass** or **tempered glass**. Procedures for removing both types are detailed below.

Safety Glass — Two sheets of glass laminated to a sheet of plastic sandwiched between them; the plastic layer makes the glass stronger and more shatter resistant. Most commonly used in windshields and rear windows. *Also known as* Laminated Glass.

Tempered Glass — Treated glass that is stronger than plate glass or a single sheet of laminated glass; safer than regular glass because it crumbles into chunks when broken, instead of splintering into jagged shards. Most commonly used in a vehicle's side and rear windows.

ROPS Deployment

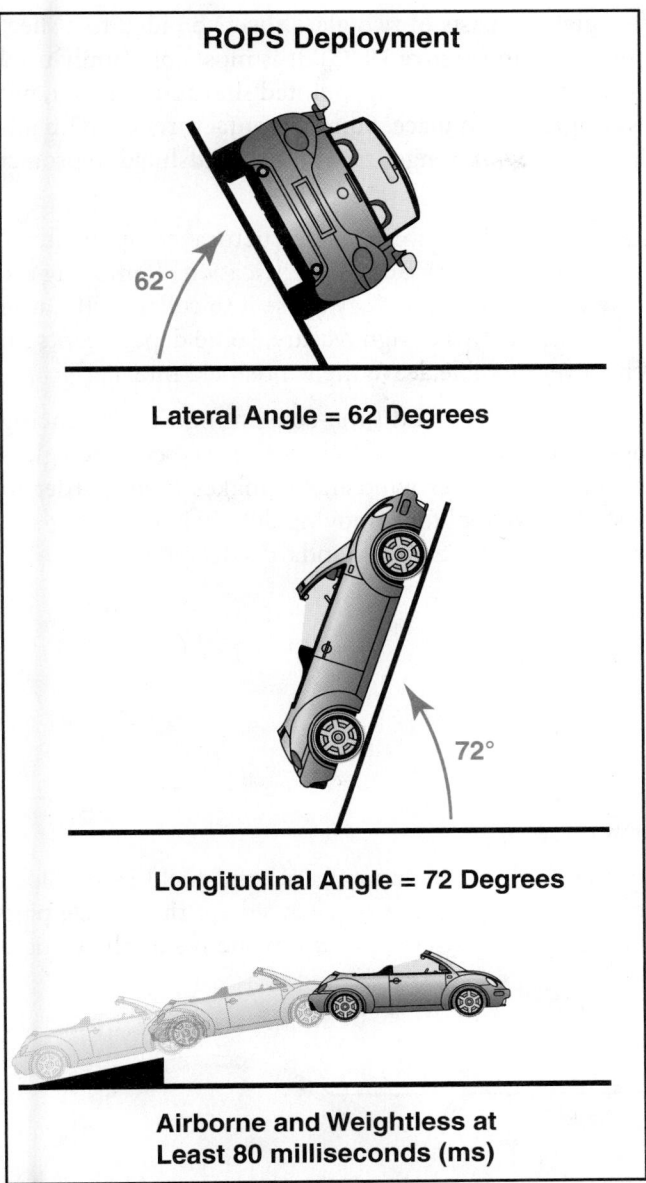

Lateral Angle = 62 Degrees

Longitudinal Angle = 72 Degrees

Airborne and Weightless at
Least 80 milliseconds (ms)

Figure 10.73 Roll over protection systems safeguard against passenger injury by engaging under a specific set of circumstances.

Figure 10.74a As in buildings, the first attempted entry into a vehicle should include checking the functionality of the doors. *Courtesy of Alan Braun/ Missouri Fire Rescue Training.*

Figure 10.74b Removing a window is the second logical step in gaining entry into a vehicle.

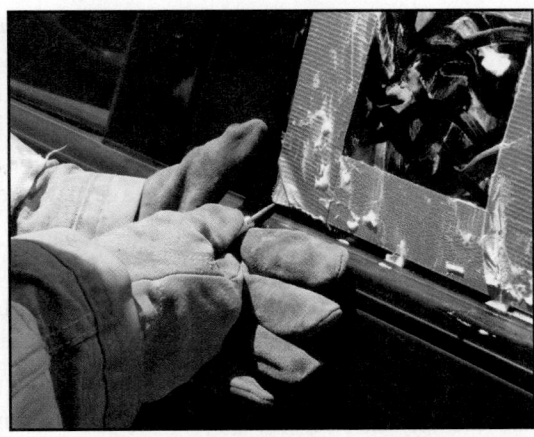

Figure 10.74c If a door doesn't open normally and a window is not feasible, prying open a door may then be attempted.

Figure 10.74d Removing a roof should only be attempted after exhausting the other alternatives.

Removing safety glass. Safety glass consists of two glass sheets bonded to a sheet of clear plastic sandwiched between them **(Figure 10.75)**. It is most commonly used for windshields and rear windows. It breaks into long, pointed shards on impact, but the plastic laminate keeps most fragments in place. Some manufacturers add an additional layer of laminating plastic on the passenger side of the windshield to protect against lacerations.

Because many automobiles now use the windshield as a structural component, removing windshields is no longer standard practice during extrication. Removing the windshield seriously weakens the vehicle body and may cause it to collapse. Because it is difficult to tell which vehicles include this design feature, some departments no longer remove windshields. Whenever possible, leave the windshield intact.

Removing safety glass is more complicated and time-consuming than removing tempered glass. Safety glass does not disintegrate and fall out the way tempered glass does, and the extra layer of laminate on newer windshields makes them harder to chop through. Saws are the most effective tool for removing doubly laminated glass, but any of the following tools may be used to remove standard safety glass:

- Axe
- Air chisel
- Hay hook
- Reciprocating saw
- Coarse blade handsaw
- Windshield cutter or glass saw

Skill sheet 10-II-7 describes the method for removing a windshield in an older model vehicle. This method requires one firefighter on each side of the vehicle plus two additional firefighters to hold a blanket or cover over anyone inside the vehicle **(Figure 10.76)**.

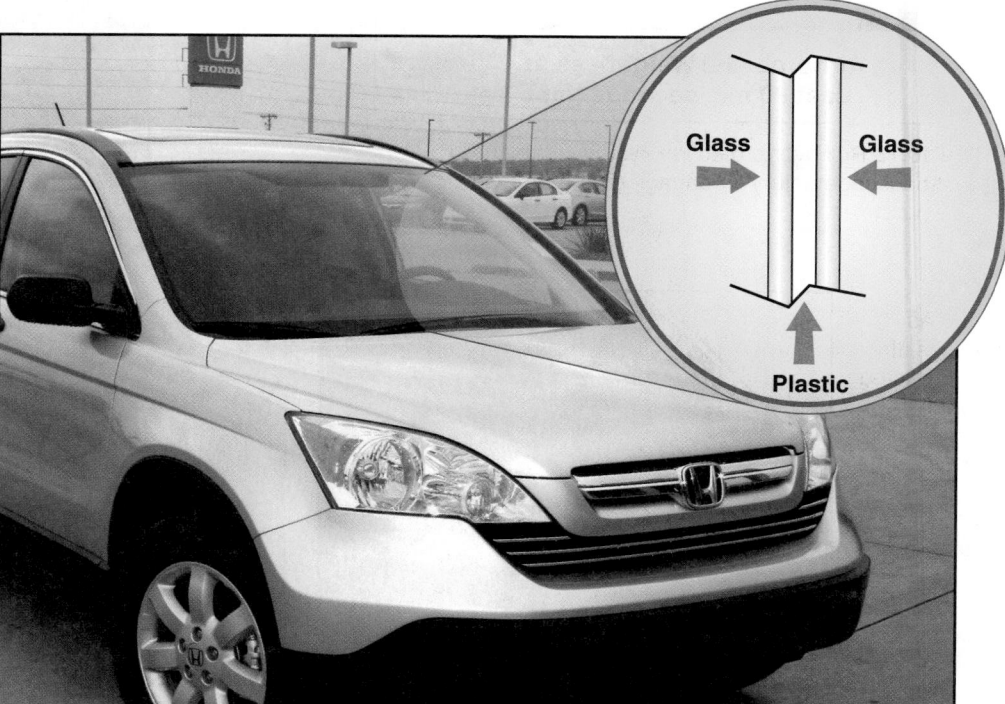

Figure 10.75 Safety glass is a required feature of automotive windshields.

Removing tempered glass. Tempered glass is most commonly used in side and rear windows. On impact, small fractures spread throughout the glass, which separates into many small pieces. This eliminates the long, pointed shards produced by shattered safety glass, but the small pieces can create nuisance lacerations and enter eyes or open wounds.

Removing tempered glass is relatively simple. The lower corner of the window breaks easily when struck with a sharp, pointed object, or when a spring-loaded center punch is pressed into it. Always use your opposite hand to support the hand holding the center punch (**Figure 10.77**). This prevents you from putting your hands through the glass when it breaks and prevents the center punch from hitting a victim who is close to the window. A non-spring-loaded center punch or Phillips screwdriver can also be driven into the window with a hammer or mallet (**Figure 10.78**). If nothing else is available, a Halligan tool or the pick end of a pick-head axe can also be used to break the glass. **Skill sheet 10-II-8** describes the method for removing a tempered glass side window.

> ## CAUTION
> Do not allow the tool to penetrate far enough to injure victims inside the vehicle.

Figure 10.76 Removal of a windshield presents implications that will require the assistance of four firefighters to accomplish safely. *Courtesy of Jennifer Ayers/Sonrise Photography.*

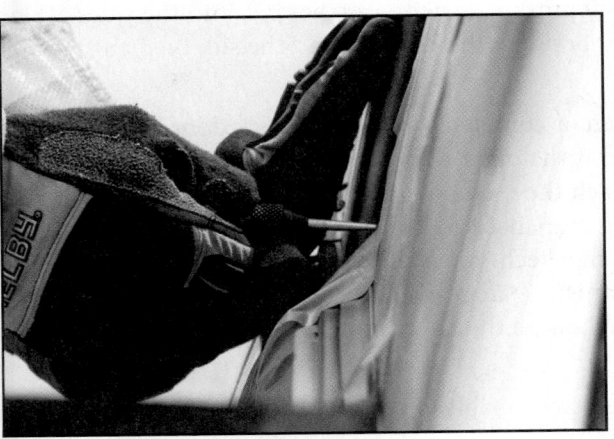

Figure 10.77 Maintaining vehicle stability while using a spring-loaded center punch is essential in safely breaking a tempered glass window.

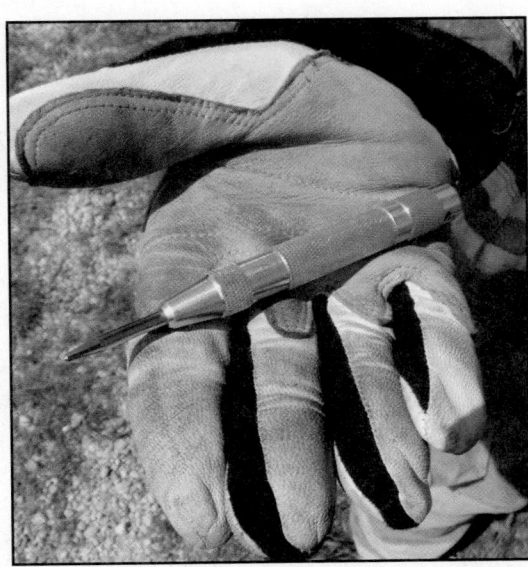

Figure 10.78 A center punch achieves the purpose of knocking out a window by concentrating force on a small point.

Most broken glass will drop straight down onto the floor or seat, so always break the window that is farthest away from the victim. Make sure the victim is covered with a blanket or salvage cover before you break the glass.

Simple materials can also be used to contain the tempered glass during removal. If contact paper is applied to a window, most of the pieces will stick to the paper, allowing the window to be removed as a unit. The same effect is produced by applying duct tape, then spraying the glass surface with an aerosol adhesive. Within seconds this creates a coating layer, binding the glass together and allowing it to be removed in one piece. Rescuers can also use duct tape to form handles that help them carry or control the broken glass (**Figure 10.79**). All of these methods require more time, so use them only if time and patient care are not critical.

Figure 10.79 Duct tape is a simple resource that can be used in some tempered glass removal applications.

Rear windows can be made from tempered or safety glass. If the rear window does not respond to techniques for removing tempered glass, it is probably safety glass.

Defrost Films on Windshields and Rear Windows

Electric vehicles use high-voltage films to defrost windshields and rear windows. These films can cause electric shock.

Prying a Door Open

In some older cars, a lockout tool (also known as a *Slim Jim*) can be used to disengage the door lock. This tool is a strip of metal approximately 24 inches (600 mm) long and ¾ to 1½ inches (20 mm to 40 mm) wide, with notches cut into the sides at one end. Insert the tool into the door between the window and the sill, then engage and release the lock.

A spreader can be used to open or completely remove a stuck door by inserting it into the crack on the hinge side. If the outer door panel is plastic, you may have to remove this panel in order to reach the metal door frame. If the interior molding is also plastic, remove the molding and check for curtain bag initiators or composite metal frames (**Figure 10.80**). Other techniques for opening doors include cutting hinges, breaking the latch mechanism (**Nader pin**), or compromising the door locks (**Figure 10.81**). For more information on these techniques, see the IFSTA **Principles of Vehicle Extrication** manual.

Nader Pin — Bolt on a vehicle's door frame that the door latches onto in order to close.

NOTE: Not all shears or cutters are capable of cutting the Nader pin.

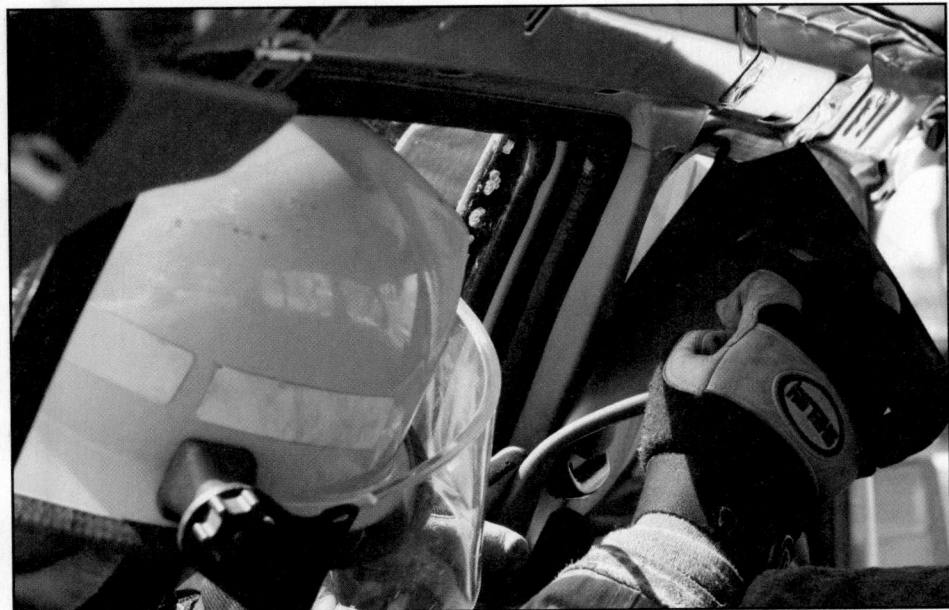

Figure 10.80 Pulling out the interior trim allows a spreader to have unimpeded access to the steel frame of a vehicle door.

Figure 10.81 Component pieces of a door may need to be broken or disengaged to open the door.

Removing the Vehicle Roof

Removing the doors and roof from a **unibody construction** vehicle can seriously compromise its structural integrity **(Figure 10.82, p. 528)**. Always place a step chock or other support under the B-post of these vehicles before removing the roof. Windshields, A-posts, and the forward edge of the roof also contribute to structural integrity. Leave these intact and cut the roof just behind the A-posts. The roof can then be removed by cutting the remaining door posts and lifting the entire roof off as a unit. Cut just below the roof level to avoid cutting into the seat belt pretensioners (commonly found in the B-posts) or side air bag gas cylinders. **Skill Sheet 10-II-9** describes two approaches to removing the roof from an upright vehicle.

> **Unibody Construction** — Method of automobile construction in which the frame and body form one integral unit; used on most modern cars. *Also known as* Bird Cage Construction, Integral Frame Construction, or Unitized Construction.

Cover Sharp Edges with Old Fire Hose

Cut posts leave behind very sharp edges. Use short sections of 3-inch (75 mm) fire hose to cover these hazards.

Door/Roof Posts

Figure 10.82 Door/roof posts are identified by letter starting with the A-post at the front of the passenger compartment.

If the vehicle is on its side, gaining access through the roof is particularly effective. Use an air chisel or reciprocating saw to make a vertical cut in the roof panel, starting about 6 inches (152 mm) from the edge of the windshield and moving down toward the ground. Make a similar vertical cut starting about 6 inches (152 mm) away from the edge of the rear window. Make a horizontal cut to connect the top ends of these vertical cuts, then flap down the roof panel to expose the headliner **(Figures 10.83 a-d)**. Cut the headliner support struts with shears or bolt cutters. If you need to remove the headliner, use a knife to cut the same pattern you made in the roof panel. **Skill Sheet 10-II-10** describes the procedures for flapping a roof when the vehicle is on its side.

Cutting Posts

Case-Hardened Steel — Steel used in vehicle construction whose exterior has been heat treated, making it much harder than the interior metal.

Many modern vehicle components are made from **case-hardened steel** or exotic metals, such as boron, magnesium, titanium, and martensite. These metals are lighter, stronger, and result in greater fuel efficiency, but their strength makes them very difficult to cut through. Hand tools cannot cut through any posts made from exotic metals. Reciprocating and rotary saws may cut through boron metal posts, but the cutting does permanent damage to the saw blade **(Figure 10.84, p. 530)**. In many cases, only specially designed shears can cut through roof support posts made from these materials.

CAUTION
Locate air bag activating mechanisms before cutting door posts.

Roof support posts typically house seat belt pretensioners or side curtain air bags. Make sure you do not activate these safety devices while cutting the posts. Removing the interior trim before cutting is recommended.

Displacing the Dashboard

This extrication technique is common after front-end collisions, in which victims are often pinned under the steering wheel or wedged under the dashboard. To extricate victims in the front seat, you must roll or displace the entire dashboard away from

Figure 10.83a The first cut through a vehicle roof should be vertical and about 6 inches (152 mm) from the windshield.

Figure 10.83b The second cut is parallel to the first cut and about 6 inches (152 mm) from the rear window.

Figure 10.83c The third cut connects the previous two cuts across the top.

Figure 10.83d After the cuts are completed, the roof can be pulled outward.

Figure 10.84 Power saws are the most effective on-site tools against exotic metal posts.

them. The most appropriate method for accomplishing this task depends on available tools, local SOPs, and the vehicle's condition. But the following method is generally accepted in almost all situations.

After removing the door, make a relief cut in the lower part of the A-post. Position an extension ram or hi-lift jack in the doorway, between the base of the B-post and the side of the dashboard. Insert cribbing or other supports under the base of the A-post on unibody vehicles, or between the frame and the body on full-frame vehicles. This prevents the dashboard from returning to its original position after the ram or jack rolls the dashboard forward (**Figure 10.85**). Once the victims have been extricated, these rescue tools can be retracted and removed.

This procedure may be accomplished without removing the windshield or flapping the roof and often by removing only one door. **Skill Sheet 10-II-11** describes this method for displacing (also known as rolling) the dashboard.

In some situations, rolling the dashboard may be unnecessary or impossible. In those cases, it may be necessary to cut the steering wheel post to extricate the driver. If it is accessible, the steering wheel post can be cut to remove the entire assembly. The steering column may also be lifted off the victim by removing the windshield, placing a hook or chain around the column, and using a power lifting tool to raise it straight up (**Figure 10.86**).

Cutting the steering wheel ring can be hazardous to the victim and to responders. There are two primary hazards associated with cutting the steering wheel ring: accidental deployment of the air bag and the resulting spring action of the wheel when cut.

Figure 10.85 Hydraulic rams may be used to roll back a dash after vehicle's structure has been damaged.

Figure 10.86 Firefighters may use lifting tools to pull a steering column away from a victim.

Technical Rescue Incidents

Technical rescuers must meet the requirements of NFPA® 1006, *Standard for Technical Rescuer Professional Qualifications*. As a Firefighter II, you may have to assist technical rescuers in a variety of rescue types **(Figure 10.87)**. Your assistance takes two forms: initial actions that are performed at any rescue scene and tasks related to specific incident types. These specific tasks include:

- Rope rescue
- Structural collapse
- Confined space
- Vehicle extrication
- Water rescue
  — Ice
  — Surface
  — Dive
  — Swiftwater
  — Surf
- Wilderness rescue
- Trench and excavation rescue
- Machinery rescue
  — Industrial
  — Agricultural
  — Elevators and escalators
- Cave, mine, and tunnel rescue

Figure 10.87 First responders may be tasked to assist technical rescuers.

Initial Actions

The first responders to reach the scene must perform the following critical tasks:

- Size up the situation.
- Communicate information.
- Stabilize the situation.
- Stabilize the victim.
- Establish scene security.

If you are a member of a unit, your supervisor will assign you tasks from this list. However, it is still important for you to know each of them. Never attempt to perform any rescue tasks for which you are not qualified or equipped.

Size-Up

Just as you would with any other emergency, you must size up the scene of any rescue. Size-up is an ongoing evaluation of:

- What has happened?
- What is happening?
- What is likely to happen?
- What resources will be needed to resolve the situation?

Figure 10.88 The dispatch report should be verified immediately upon arrival on the scene.

Size-up begins during the initial dispatch. The dispatcher should tell you generally what has happened, including the location, type of rescue, weather conditions, units dispatched, and possibly the number of victims. Some information may be inaccurate because it comes from witnesses or victims, not emergency responders. You may not know that a rescue is required until you reach the scene.

Upon arrival, verify the accuracy of the dispatch report by observing the scene yourself (**Figure 10.88**). Note any hazards that will affect victims or the rescue team. Assess what is likely to occur if no action is taken, and determine the priority of actions that should be taken. For instance, it may be necessary to extinguish a fire before rescue operations can begin. Determine whether additional resources must be dispatched to the scene, such as heavy equipment to move debris, an aerial device to access victims, or a heavy-duty wrecker to move a vehicle.

Communicate the Information

Your next task is to communicate the information you have gathered to the dispatch center and all responding units. The arrival report (sometimes called a situation report) for technical rescue incidents should contain the following:

- Unit arriving on scene
- Correct address of the incident if different from the dispatch report
- Description of conditions found at the scene
- Special considerations
- Intended initial actions

- Water supply if needed
- Establishment of Command
- Any additional resources needed, including medical units

Describe conditions clearly so that other fire crews know what to expect when they reach the scene. Provide details about barriers to access and victims in need of assistance. Include a command statement, which indicates who is in initial command of the incident. State the location of the Command Post and give a description of initial actions. If the operation will be managed over a specific radio frequency, the communications center will tell responding units to switch to that frequency during dispatch.

When a superior officer arrives, you officially transfer Command to that officer. Provide a description of what you have observed and your initial actions.

Stabilize the Situation

Stabilizing the situation means preventing it from getting worse. This involves activities such as blocking traffic, shutting off utilities, or suppressing fires that can affect rescue operations (**Figure 10.89**). Incidents involving machinery may require that the equipment remain on until an expert determines that it is safe to shut it down.

Stabilize the Victim

After the scene has stabilized, the next priority is to provide basic patient care to any accessible victims (**Figure 10.90**). However, you must not put yourself at risk in an attempt to access a victim that will require the skills of a qualified technical rescue firefighter.

Figure 10.89 Stabilization tactics include lockout and tagout of utilities that may be involved in or potentially contribute to the hazard of the incident.

Figure 10.90 Basic care to victims must wait until stabilization procedures are completed. *Courtesy of Alan Braun/Missouri Fire Rescue Training.*

Establish Scene Security

The Incident Command System requires that any scene have a well-maintained perimeter or barrier around the incident. This is intended to:

- Provide a controlled work space
- Protect bystanders from hazards at the incident
- Ensure the use of the personnel accountability system
- Ensure that victims are accounted for
- Protect evidence in the event of a suspicious incident
- Prevent further collapse of a structure or trench due to vehicle vibrations

The Incident Commander determines the location of the outer perimeter, which you may be assigned to mark. Stretch utility rope or barrier tape between any available objects, such as signs, trees, utility poles, or parking meters. Leave a controlled opening near the Command Post. The Accountability Officer will monitor this entry point. Another opening may be necessary to provide access for ambulances and other emergency equipment.

An incident scene is typically divided into three control zones labeled hot, warm, and cold **(Figure 10.91)**. Their size, shape, and distance from the hazard depends on weather conditions, general topography, the amount of room needed by working personnel, and the nature of the hazard.

The hot zone is the most critical area of the scene and includes site of the actual emergency. To limit crowds and confusion, only personnel who are directly involved in resolving the emergency are allowed within the hot zone. Personnel in the hot zone must:

- Sign in to the personnel accountability system
- Wear personal protective equipment (PPE) designed for the specific hazard
- Be trained to manage the situation

The warm zone is immediately outside the hot zone. Access is limited to personnel who are directly supporting the work being performed in the hot zone. All personnel must be in full PPE and ready to enter the hot zone. At hazardous materials incidents, a decontamination station is usually assembled in this zone.

The cold zone is the furthest away from the incident. Access is limited to working personnel, and the outer boundary forms the crowd-control line for the general public.

Incident Specific Tasks

Each type of technical rescue incident has its own requirements for equipment, procedures, and personal protective clothing. To assist the technical rescue team when they arrive, you must be able to recognize, locate, and sometimes operate their equipment. You must also be able to recognize the hazards associated with each type of incident and methods for mitigating it.

Rope Rescue Operations

Rope rescue operations involve the use of life safety rope, harnesses, tripods, and accessories to access and remove victims. You may be assigned to retrieve this equipment. In general, rope rescue is divided into high angle urban/structural and

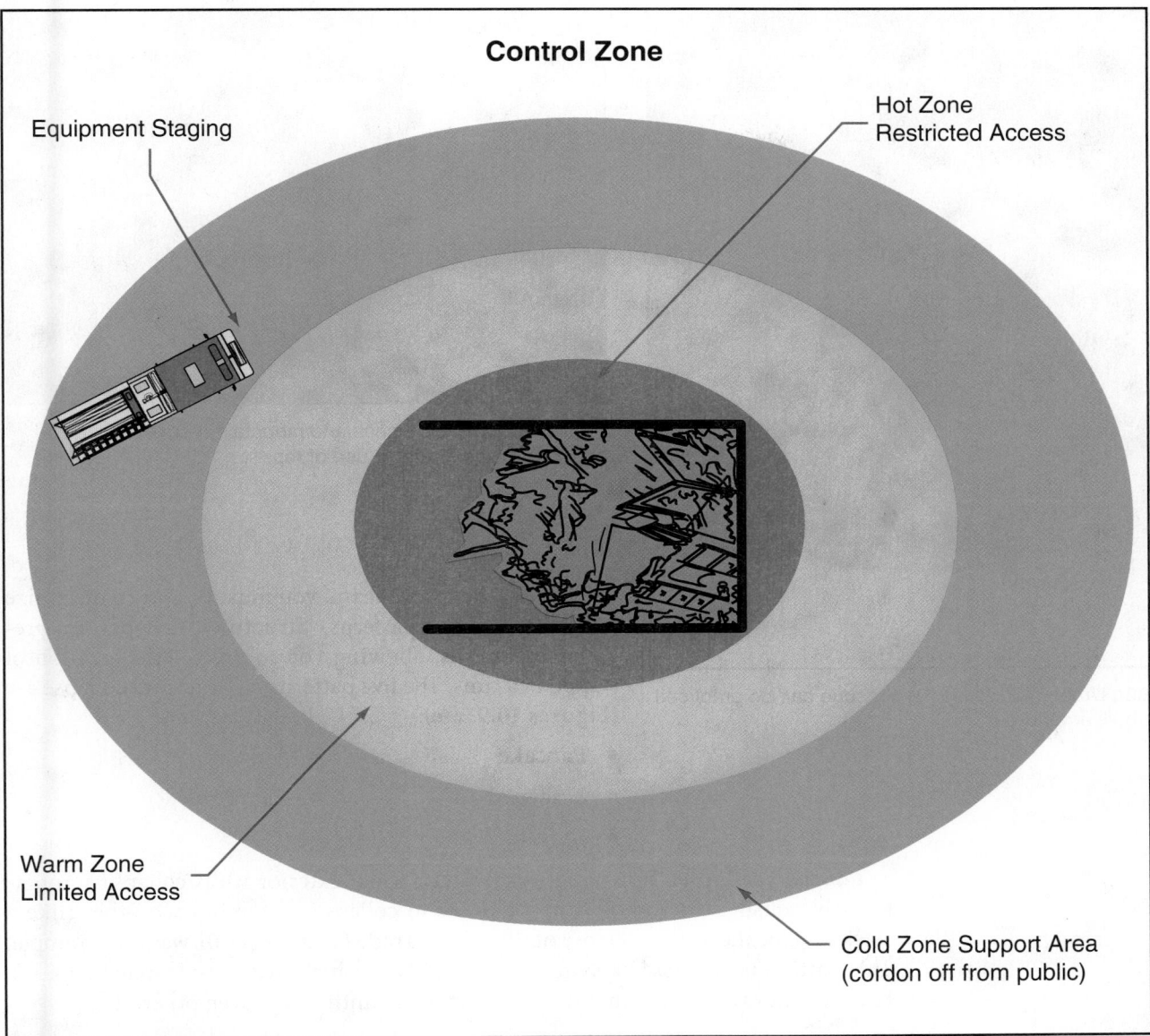

Control Zone

Equipment Staging

Hot Zone
Restricted Access

Warm Zone
Limited Access

Cold Zone Support Area
(cordon off from public)

Figure 10.91 Control zones divide an incident scene into hazard and activity level areas.

wilderness/mountain rescue (**Figures 10.92a and b, p. 536**). An example of a high-angle rescue would be rescuing injured workers from a scaffolding on the side of a structure. Wilderness/mountain rescue may involve removing a victim from the base of a cliff or steep slope. The main hazard posed to rescuers is that they are working without a safety net in the same environment as the victims.

Structural Collapse Rescue Operations

Structural collapse may be caused by fire, extreme weather, earthquakes, explosions, or the deteriorated condition of an aging structure. The first priority at the scene is to help get untrapped victims to a safe area. The next priority is to extricate victims who are lightly trapped by collapse debris. After these victims are taken care of, technical rescue teams attempt to rescue victims who are trapped deep within the rubble.

Figure 10.92a Urban high angle rope rescue can be practiced on purpose built structures.

Figure 10.92b Wilderness and mountain rescue is made possible by the strategic use of ropes.

To assist the rescue team, you must be able to recognize different collapse patterns. Structures collapse in predictable patterns, allowing you to predict the location of trapped victims. The five patterns of structural collapse are **(Figures 10.93a-e)**:

- Pancake
- A-frame
- V-shaped
- Cantilever
- Lean-to

Pancake collapse. This pattern can occur when exterior walls collapse simultaneously, causing the roof and upper floors to collapse on top of each other (like a stack of pancakes). The collapse of the World Trade Center in 2001 was one example. This pattern is the least likely to contain voids in which live victims may be found. However, always assume that there are survivors until it is proven otherwise.

Figure 10.93a A pancake collapse is defined by near-parallel floor structures with debris between that may hold void spaces.

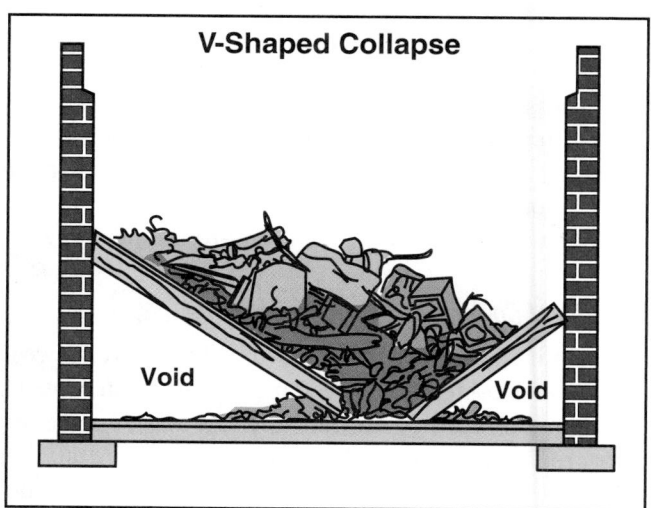

Figure 10.93b A V-shaped collapse creates void spaces on either side of a cave-in at the center of the structure.

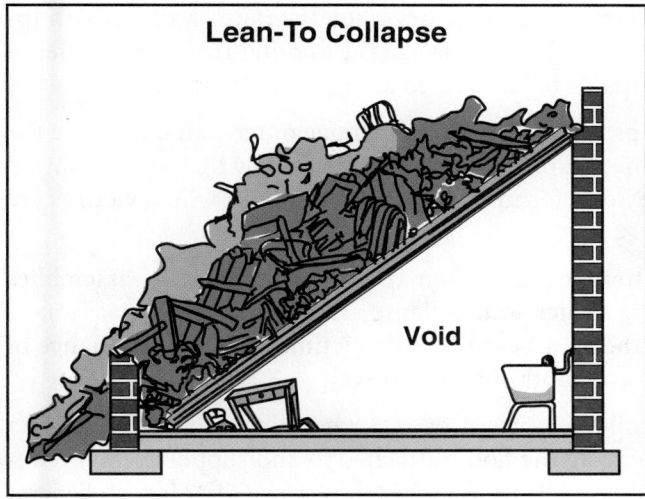

Lean-To Collapse

Void

Figure 10.93c A lean-to collapse is characterized by sections of the wall holding their original position while other sections of the wall fall in toward the definitive event.

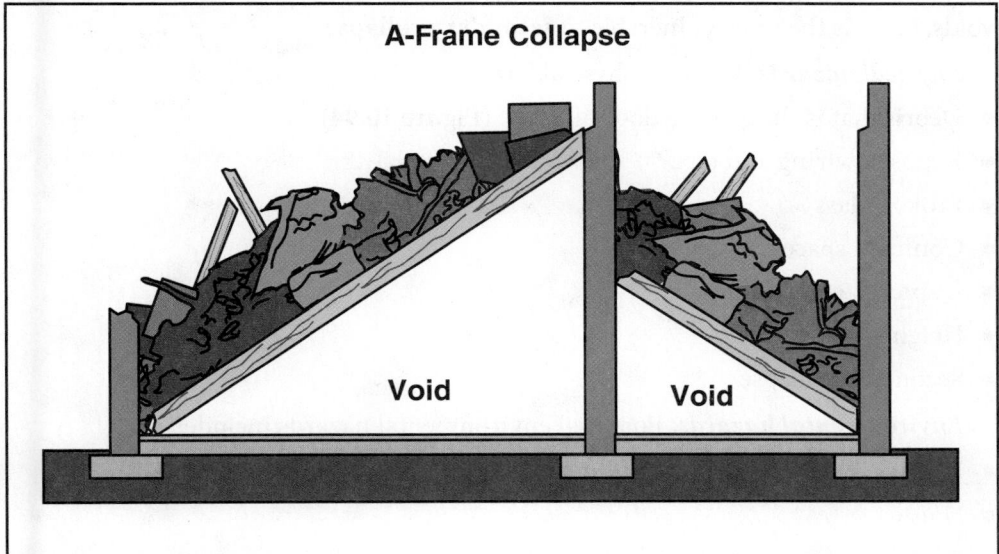

A-Frame Collapse

Void   Void

Figure 10.93d An A-frame collapse is characterized by fallen walls near the outside of the structure and intact void spaces along an interior structural wall.

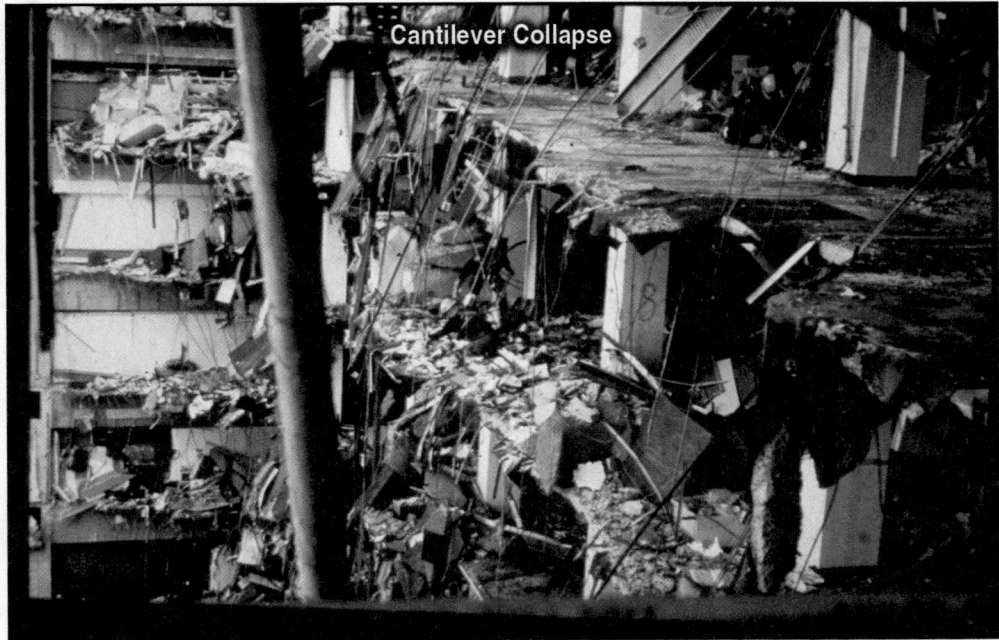

Cantilever Collapse

Figure 10.93e The Murrah Building in Oklahoma City demonstrated a cantilever collapse.

V-shaped collapse. This collapse pattern occurs when the outer walls remain intact and the upper floors and/or roof structure fail in the middle. This pattern offers a good chance of habitable void spaces being created along the outer walls.

Lean-to collapse. This collapse pattern occurs when one outer wall fails while the opposite wall remains intact. The side of the floor or roof assembly that was supported by the failed wall drops to the floor, forming a triangular void in which victims are likely to survive.

A-frame collapse. This pattern occurs when the floor and/or roof assemblies on both sides of a load-bearing center wall collapse. This creates a pair of lean-to collapses on opposite sides of the load-bearing wall. Victims have a good chance of surviving within the void spaces on both sides of the wall.

Cantilever collapse. This collapse pattern occurs when one or more walls of a multistory building collapse, leaving the floors attached to and supported by the remaining walls. One example of this pattern was the Oklahoma City Murrah Federal Building following the 1995 explosion. This pattern offers a good chance of habitable voids, but it is the most vulnerable to **secondary collapse**.

Physical hazards. Primary physical hazards include:

- Debris that is sharp, jagged, or unstable **(Figure 10.94)**
- Exposed wiring and rebar
- Broken glass
- Confined spaces
- Unprotected openings
- Heights
- Secondary collapse

Environmental hazards. Potential environmental hazards include:

- Fire
- Noise
- Darkness
- Temperature extremes
- Adverse weather conditions
- Damaged and leaking utilities
- Atmospheric contamination
- Hazardous materials contamination

Confined-Space Rescue Operations

The Occupational Safety and Health Administration (OSHA) defines a *confined space* as one that:

- Is large enough to enter
- Has limited means of entry and exit
- Is not designed for continuous occupancy

Confined spaces in which firefighters must perform rescues include **(Figure 10.95)**:

- Tanks/vessels
- Silos/grain elevators

Figure 10.94 Debris from a demolition event presents several physical hazards.

- Storage bins/hoppers
- Utility vaults/pits
- Aqueducts/sewers
- Cisterns/wells
- Coffer dams
- Storage tanks

Firefighters without confined-space rescue training are limited to performing non entry rescues and support functions outside of the space. Trained personnel can perform rescues inside the space, but are limited to operations within the scope of their specific qualifications.

Atmospheric hazards in confined spaces include:

- Oxygen deficiency due to inadequate ventilation
- Flammable gases and vapors
- Toxic gases
- Extreme temperatures
- Explosive dusts

Physical hazards include:

- Limited means of entry and egress
- Tight constricted spaces
- Cave-ins or unstable support members
- Standing water or other liquids
- Utility hazards, such as gas, sewage, and electricity

Valuable sources of information include preincident plans and knowledgeable people at the scene. Preincident plans should describe lighting, ventilation, and communication at the scene. They should also have details relevant to protecting victims and rescuers and controlling utilities and other hazards. Plant or building supervisors may be able to tell you the number of victims, their probable location, and potential hazards.

Figure 10.95 Confined spaces contain a range of challenges that can include dangers typical of other types of technical rescues.

Figure 10.96 When the supplied air respirator system is functional and ready for operations, the firefighter monitoring the system should signal the team.

Rescuers may not be able to wear SCBA because of space limitations. Supplied air respirators (SAR) are typically used, especially in extended rescue operations. Hoses up to 300 feet (91 m) long connect rescuers' facepieces to an air cylinder or breathing-air compressor outside the entrance. You may be assigned the task of setting up and monitoring the SAR system for rescue personnel **(Figure 10.96)**.

All rescuers have a search line attached to their rescue harness. This line must be constantly monitored at the entrance, and a communication system between inside and outside personnel must be prearranged. Portable radios may not work within the space, so hard-wired telephones are usually preferred.

One basic method of signaling with the search line is the *O-A-T-H* method. One tug on the search line represents the letter *O* which stands for *OK*. Two tugs represent the letter *A,* which stands for *Advance*. Three tugs represent the letter *T,* which stands for *Take-up* (eliminate slack). Four tugs represent the letter *H,* which stands for *Help*. This system must be practiced often to be effective.

Electrical equipment such as flashlights, portable fans, portable lights, and radios must be intrinsically safe for use in flammable atmospheres. A rapid intervention crew or team (RIC/RIT) qualified for confined-space operations must be standing by while rescuers are working inside the confined space.

Vehicle Rescue Operations

Depending on the local policy and procedures, vehicle extrication may be assigned to a technical rescue team or to an engine or truck company trained and equipped to perform this type of task. If you are not a member of a unit that is trained to perform vehicle extrication, you will assist the rescue team members in setting up their equipment, providing care to the victims, standing by with a charged hoseline, and providing a security barrier around the incident. Remember that you must wear the correct PPE when working near the damaged vehicle and you must wear an approved retroreflective vest if you are not engaged in fire suppression or extrication.

Water Rescue Operations

Water rescue operations include the following conditions:

- Ice
- Surface
- Dive
- Swiftwater
- Surf

Locations where water rescue may be required include:

- Swimming pools
- Ponds
- Lakes
- Rivers
- Streams
- Coastal shorelines
- Swamps
- Drainage canals

- Low-head or low-water dams
- Water treatment facilities

The first task during size-up is to determine whether the incident requires rescue or recovery operations. Rescues are incidents where the victim may be saved, typically if they are stranded or floundering. Recoveries are incidents where the victim has been submerged for a long period of time and is likely dead. The operation's main goal is to recover the body.

Personal flotation devices (PFDs) are mandatory for all personnel entering the water, working within 10 feet (3 m) of the water's edge, or riding in waterborne craft. PFDs must be approved by the U. S. Coast Guard or Transport Canada. Firefighters assisting the rescue team may wear structural PPE for warmth, but they should not approach the water's edge (**Figure 10.97**). If you fall into the water, structural PPE can quickly become water-logged and pull you underwater.

WARNING
Only qualified personnel should attempt a water rescue.

Water rescue hazards that you should describe in the situation report include:

- Undercurrents
- Unstable or slippery soil at water's edge
- Debris
- Sink holes
- Quicksand
- Sharp rocks
- Extreme temperatures
- Chemical or biological contamination
- Poisonous or dangerous reptiles

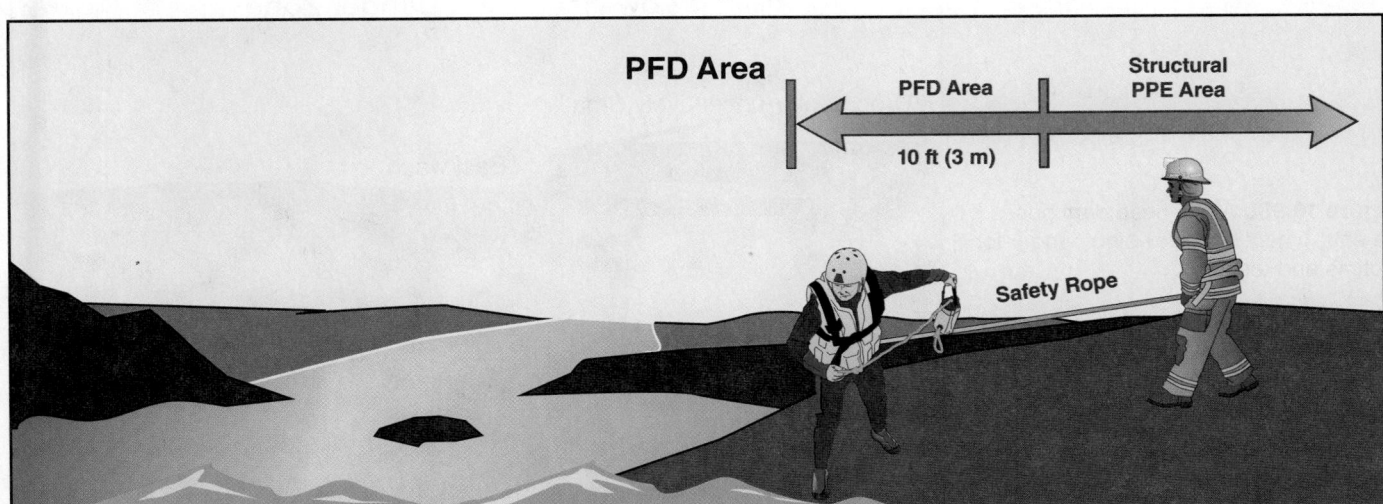

Figure 10.97 Water operations must be approached with the correct gear to prevent injuries to rescuers and victims.

Additional hazards are specific to ice rescues, such as thin and unpredictable ice. Just because ice appears thick does not mean that it is strong; the fact that there is a victim in the water is proof that the ice is weak. Victims are also unlikely to be able to help with their own rescue. Their hands are extremely cold and possibly frozen, making it difficult to grasp a rope or other aid. Their heavy, wet clothing also makes it difficult to keep their heads above water.

The victim will almost certainly be suffering from **hypothermia**, so it is critical to have an advanced life support unit on scene to start immediate patient care. Immersion in ice water causes the body's core temperature to drop dramatically. Rescuers must remove victims quickly to increase their chance of survival.

Low-head or low-water dams are extremely dangerous for both victims and rescuers. These dams create a pool of standing water in a river or stream. Water recirculates as it passes over the face of the dam, creating powerful undercurrents (**Figures 10.98 a and b**). Debris trapped against the upstream face of the dam poses an additional hazard.

Hypothermia — Abnormally low body temperature.

Low-Head Dam — Wall-like concrete structure across a river or stream that is designed to back up water; allows water to flow over the crest and drop into a lower level. *Also known as* Low-Water Dam.

Hazards Created by a Low-Head Dam

Low-head dams are extremely hazardous work areas. Exercise extreme caution when performing any operation nearby.

Figure 10.98a A low-head dam creates a dangerous drowning hazard for anyone who gets caught in the undercurrent. *Courtesy of Jake Zlomie.*

Figure 10.98b A low-head dam poses an entrapment and drowning danger for victims and rescuers.

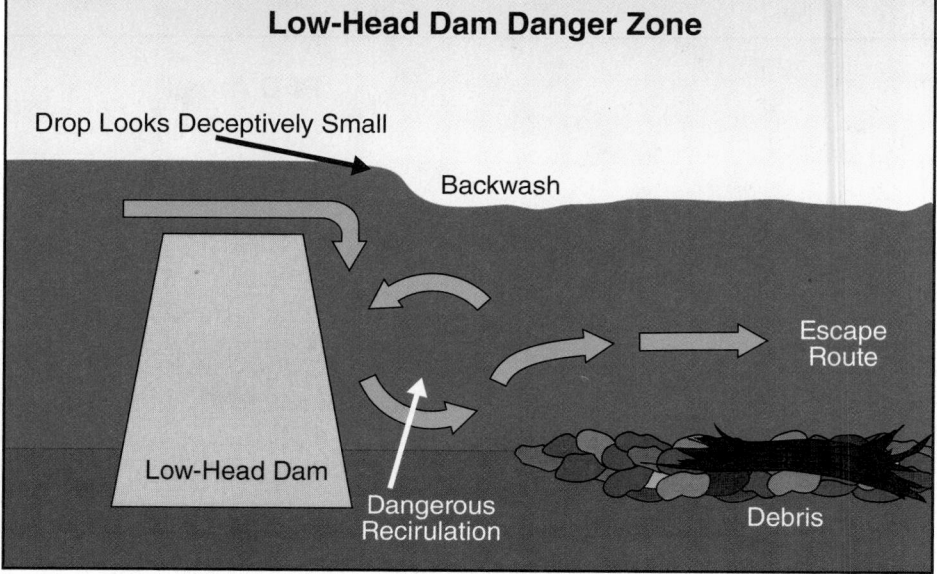

Low-Head Dam Danger Zone

Drop Looks Deceptively Small

Backwash

Low-Head Dam

Dangerous Recirulation

Debris

Escape Route

Wilderness Rescue Operations

Wilderness rescues take place in rugged, inaccessible terrain. Technical rescuers typically use ropes and harnesses to descend to the victim, stabilize any injuries, and remove the victim in a basket or stretcher. Assisting during wilderness rescue includes such tasks as establishing rehab facilities and carrying tools and equipment to a point near the victim.

Hazards include climate extremes and the possibility of a long-term search before the victim is located. Heat stress and dehydration are especially common, so wearing structural PPE is not recommended in hot climates. Structural boots do not protect feet and ankles from the hazard posed by the rough terrain. Large quantities of water should be available, and rehab facilities are mandatory.

Trench Rescue Operations

Trench rescue teams must be skilled in shoring and stabilizing trench walls (**Figure 10.99**). When assisting at a trench rescue, you will most likely be assigned to monitor for hazardous atmospheres or create a safe zone around the trench. Vibrations can cause secondary cave-ins, so all bystanders, nonessential personnel, apparatus, and heavy equipment must be kept well back from the trench lip.

Follow these safety guidelines when assisting with trench rescues:

- Do not enter the trench.
- Cordon off an area 100 feet (30 m) in each direction from the trench.
- Eliminate sources of vibration within 500 feet (150 m) of the trench, such as apparatus and heavy equipment.
- Place exit ladders no more than 50 feet (15.24 m) apart, with the initial ladder near the location of the victim.
- Ladders should extend at least 3 feet (1 m) above the top of the trench (**Figure 10.100**).
- Secure any exposed utilities.
- Be careful when handling tools. Dropped or mishandled tools can injure both rescuers and victims.

Figure 10.99 Trench rescue teams are specially trained to stabilize and shore trench walls.

Figure 10.100 Safe ladder placement contributes to an effective trench rescue operation.

- Be aware of additional hazards, such as underground wiring, water lines, or toxic or flammable gases.
- If the trench is contaminated or oxygen-deficient, set up ventilation fans to allow rescuers to continue working.

Machinery Rescue Operations

Machinery rescues involve a victim who has been caught between parts of a machine. The types of injuries that result make these incidents extremely stressful. Machinery rescues may occur at:

- Machine shops
- Manufacturing facilities
- Agricultural sites **(Figure 10.101)**
- Lumber mills
- Scrap metal recycling facilities
- Construction or demolition sites
- Shipyards
- Railroad yards
- Transportation terminals

    When sizing up the situation, note the following:

- Victim's medical condition and degree of entrapment
- Type of machinery
- Number of rescue personnel needed
- Extrication equipment needed
- Presence of fire or hazardous material
- Scene safety issues
- Precautions necessary before securing power to the machinery

Figure 10.101 Make sure ignition switches are clearly labeled while rescue evolutions are underway near agricultural equipment.

Stabilize the machine with cribbing, chains, or heavy-duty nylon webbing, then shut off the machine's power. Use a **lockout/tagout device** to secure the power switches.

Rescuers may need outside expertise to resolve the situation. Plant personnel who use the machinery are usually good sources of information, but an off-site expert (such as the machinery manufacturer) may be required. Off-site experts should be identified in the department's preincident plan.

> **Lockout/Tagout Device** — Device used to secure a machine's power switches in order to prevent accidental re-start of the machine.

Elevator and Escalator Rescue Operations

In some jurisdictions, these rescues are not limited to technical rescue personnel. Firefighters can often extricate victims quickly and safely if they are equipped with ladders, manual and power tools, and forcible entry tools. But more hazardous types of elevator rescues require certified technical rescuers.

Elevator rescue. Elevators are classified based on their primary use. Passenger elevators typically are small, accelerate quickly, and have automatic controls. Freight elevators are slower, with large access doors and either manual or automatic controls. They can carry up to 3 tons (3.05 t) and can be as large as 12 x 14 feet (3.66 by 4.27 m). Construction elevators are similar to freight elevators, but they are temporary installations used during building construction.

Elevator operating systems use either cables or hydraulics. Cable systems are used in structures of any size. They consist of:

- A fully enclosed elevator shaft
- An elevator car
- Cables attached to the car
- Counterweights
- Vertical tracks
- Emergency safety brakes
- An equipment room for the electrical equipment

One variation of the cable elevator is known as an observation elevator. It is mounted on the wall of an atrium, or on the exterior wall of a structure, to provide passengers with a scenic view (**Figure 10.102**).

Figure 10.102 An observation elevator offers passengers a scenic view while traveling to their destination.

Hydraulic elevators have ram pistons beneath the car that raise and lower it, with hydraulic oil providing the necessary hydraulic pressure. This operating system is used in structures less than six stories high. The mechanical room containing the power unit is usually located on a lower level of the building, within 100 feet (30 m) of the elevator shaft. Some hydraulic systems also have the cables and counterweights used in cable systems.

Elevators may become inoperative because of:

- Physical damage to the elevator shaft caused by earthquakes, explosions, or other events
- Loss of electrical or hydraulic power
- Overheated circuits, switches, or relays
- Jammed or broken cables
- Activation of the emergency stop button
- Short circuit in electrical system
- Elevator car or shaft door left ajar

During size-up, ask building representatives and witnesses for detailed information about the situation. Establish communications with the elevator's occupants, either by telephone, calling out to them from the nearest floor, or establishing an intercom link between the elevator's car and control room. They may be under tremendous stress, so reassure the occupants that you are working to extricate them.

Locate a building maintenance person or call the elevator service company for technical assistance. Next, a firefighter equipped with a portable radio should be sent to the elevator equipment room while the rest of the crew remains at the elevator door on the floor nearest the inoperable elevator car.

If the occupants require immediate medical attention, begin extrication. Avoid forcible entry methods that can damage the elevator and endanger passengers and firefighters. Request assistance from the technical rescue team if necessary.

If the occupants' medical condition is stable, ask them to check the status of the *Emergency Stop* Button. If the elevator is stalled due to a malfunction, such as an overheated relay or loss of power, the *Emergency Stop* Button must be activated before power can be restored. Tell the occupants to press the *Door Open* Button, which may cause the elevator to resume operation.

A member of the building's maintenance staff should check the electrical circuits to see if the elevator has power. If a staff member is not available, the firefighter who was sent to the equipment room can perform this task. Circuits that have tripped due to overheating can occasionally be reset. If the elevator has power, turn the power off for at least thirty seconds, then turn it back on again. This may reset the relays and reactivate the elevator. Once power has been restored, instruct the occupants to press the *Door Open* Button.

Elevators equipped with a recall system can be returned to the ground floor by inserting a key into the control panel on the ground floor. If the doors do not open when the elevator arrives at the ground floor, ask the occupants to press the *Door Open* Button.

Escalator rescue. Also called *moving stairways,* escalators are chain-driven mechanical stairways that move continuously in one direction. The steps are linked together and each step rides a track. The flexible rubber handrails move at the same rate as the stairs. Each escalator in a structure is an individual installation with separate machinery and controls. The drive unit is usually located under the upper landing and is covered by a landing plate. One variation is the moving walkway/sidewalk, which is similar to a conveyor belt and transports people through large structures, such as airport terminals and pedestrian tunnels (**Figure 10.103**).

Many escalators have manual stop switches located on a nearby wall or on the handrail support at the top and bottom of the unit (**Figure 10.104**). Activating the switch slowly stops the stairs and sets an emergency brake. A key-operated switch located in a covered compartment at the bottom of the escalator is used to restart the escalator after the stop button is activated. The stairs should be stopped during rescue operations. As with the elevator, an escalator mechanic should be requested to restart the escalator after the victim is removed. Extrication operations involving moving walkways or sidewalks are similar to escalators.

The majority of escalator rescues result from victims' fingers and toes becoming caught between the step treads and guard plates, or sandals becoming wedged between the treads. To extricate a victim, remove all other passengers from the escalator and use hand pressure to move the treads backward, which allows trapped fingers or toes to be easily freed. Some types of escalators also have a hand crank or wheel that can be operated to move the treads backward. The crank is located under the covered landing plate. After extrication, the escalator is placed out of service until a service technician can work on it.

Figure 10.103 Escalators present unique rescue situations.

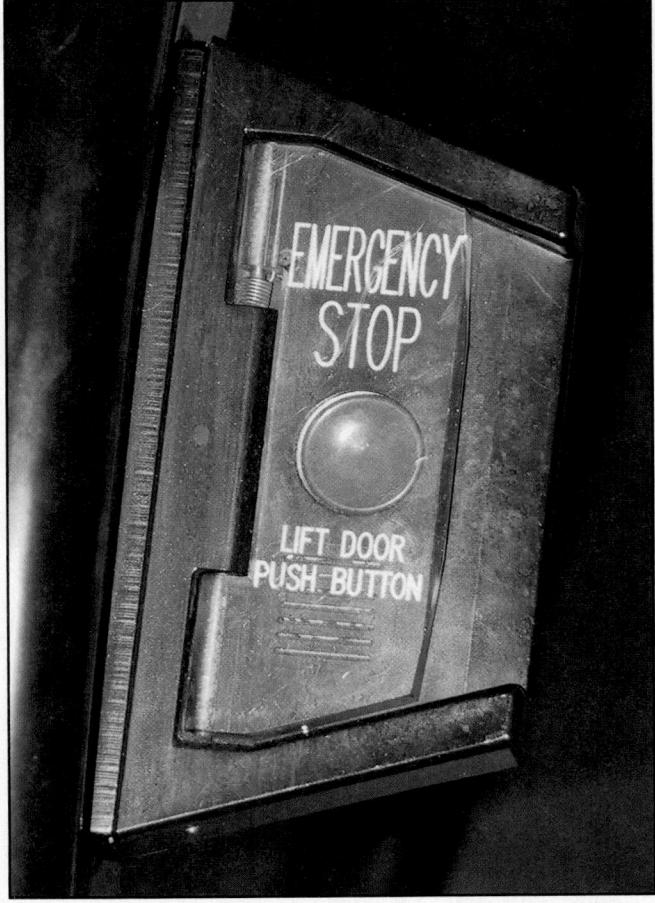

Figure 10.104 An emergency stop button allows passengers to have some control over a malfunctioning escalator.

Cave, Mine, and Tunnel Rescue Operations

Rescues in caves, mines, and tunnels require specialized rescue training and equipment similar to that required for confined-space operations. As a Firefighter II, you should be prepared to assist by monitoring communication channels and search lines, operating SAR systems, and assisting victims once they have been removed from the hot zone. Hazards of these sites include:

- **Caves** — Toxic gases, oxygen deficiency, tight spaces, sharp rocks, potential for cave-ins, lack of available light, standing or swift running water

- **Mines** — Toxic gases, oxygen deficiency, explosive atmospheres, cave-ins, abandoned tools and equipment, lack of available light, standing water

- **Tunnels** — Toxic gases, oxygen deficiency, smoke, fire, tangled debris, electrified rails or wires, lack of available light, standing water or other fluids, biological waste in sewers

Electrical Rescue Operations

In some jurisdictions, incidents involving electricity are not considered technical rescue situations. However, rescues involving energized power lines or equipment are some of the most common and dangerous situations to which firefighters are called. Improper actions can injure or kill you instantly. Whenever you respond to any situation involving electricity, you should **always** do the following:

- Assume that electrical lines or equipment are energized. A power line in contact with a telephone line or a wire fence (out of sight of rescuers) can energize the entire length of the line or fence.

- Establish scene security and deny unauthorized entry.

- Call for the electric company to respond.

- Stand by until the electric company arrives.

- Allow only electric company personnel to cut electrical wires.

Figure 10.105 It may be difficult to estimate the radius of the ground gradient without the correct tools for the job, so it is important to maintain a safe distance.

Electrical wires on the ground can be dangerous without even being touched. Downed lines can energize wire fences or other metal objects with which they come in contact. When an energized electrical wire comes in contact with the ground, current flows outward in all directions from the point of contact. As the current flows away from the point of contact, the voltage drops progressively. This energized area is called the **ground gradient**. Depending on voltage and other variables, such as ground moisture, it can extend for several yards (meters) from the point of contact **(Figure 10.105)**. To avoid this hazard, estimate the distance between two nearby power poles, and stay that distance away from the downed line until you are sure the power has been shut off.

One type of electrical rescue involves a vehicle striking a power pole, severing a high-voltage line that falls onto the vehicle. The first priority is to contact the electric company to shut off the power. Once that is done, the incident becomes a vehicle extrication incident.

Electrified subway or train tracks also pose a hazard to firefighters. Passengers may fall onto or between the rails. Firefighters may also have to enter tunnels to assist passengers from stalled trains or subway cars. Never work around these tracks unless it is proven that the power has been shut off.

Ground Gradient — Electrical field that radiates outward from where the current enters the ground; its intensity dissipates rapidly as distance increases from the point of entry.

///////////////////////

WARNING
When you approach a downed power line, a tingling in your feet indicates that the ground beneath you is electrified. If you feel this, keep your feet together and hop away from the line.

Chapter Summary

Your duties as a firefighter include using and maintaining rescue tools and lighting equipment. You must also know how to work as part of a team while performing vehicle extrications and assisting technical rescuers. This chapter provides only basic information; you must practice these duties to make sure that you can accomplish these tasks quickly and effectively.

Firefighter I

1. What types of emergency scene lighting equipment can be used for rescue incidents?

Firefighter II

1. What types of tasks may be required to maintain electric generators and lighting equipment?

2. What are the main uses of rescue tools?

3. What are some common limitations of rescue tools?

4. What role does the fire department play in size-up and assessing the need for extrication after a motor vehicle accident?

5. What safety considerations must be identified and mitigated during vehicle extrication operations?

6. How can cribbing material be used during vehicle extrication?

7. What methods can be used to gain access to victims during vehicle extrication?

8. What role does a Firefighter II play during technical rescue operations?

9. What are the different rescue practices and goals used during technical rescue?

10. How do the hazards for structural collapse rescue operations compare to hazards of rope rescue operations?

10-II-1

Demonstrate the steps for inspecting, servicing, and
maintaining a portable generator and lighting equipment.

SKILL SHEETS

Step 1: Review the manufacturer's service manual for specific directions.

Step 2: Carefully inspect spark plugs for damage, visible corrosion, carbon accumulation or cracks in the porcelain, and ensure the spark plug wire is tight. Replace spark plug if it is damaged or if service manual recommends.

Step 3: Check carburetor and identify signs of fuel leaks.

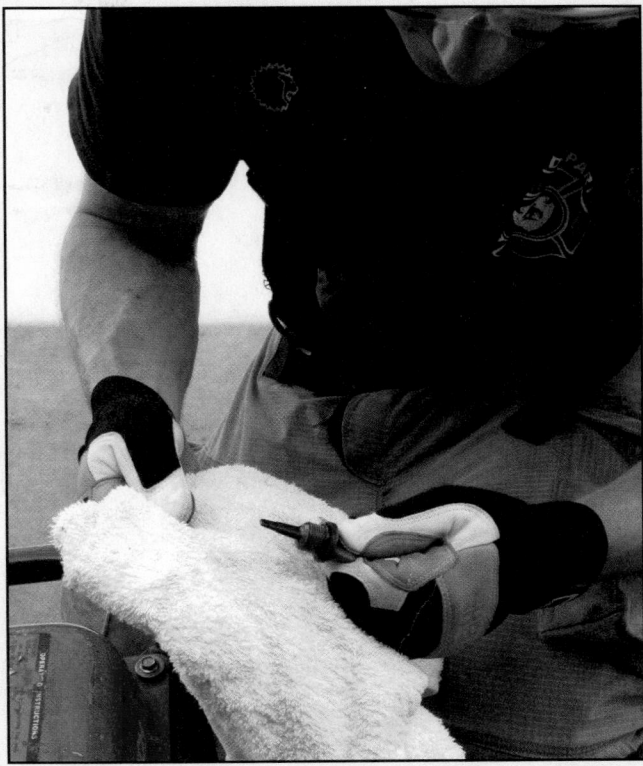

Step 5: Check oil level and refill as needed.

Step 4: Check fuel level and refill as needed.

SKILL SHEETS

10-II-1
Demonstrate the steps for inspecting, servicing, and maintaining a portable generator and lighting equipment.

Step 6: Start generator and run tests as required by service manual.

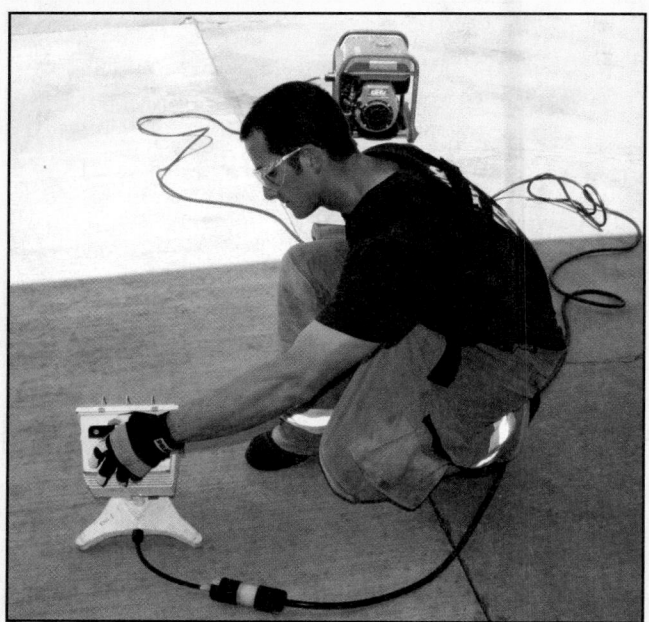

Step 7: Inspect all electrical cords for damaged insulation, exposed wiring, or missing or bent prongs.

Step 8: Test operation of lighting equipment by connecting each light to the generator one light at a time. Replace lightbulbs as necessary, discard faulty bulbs in an approved manner.

Step 9: Clean work area and return all tools and equipment to the proper storage areas.

Step 10: Document maintenance on the appropriate forms or records.

10-11-2
Prevent horizontal movement of a vehicle using wheel chocks.

SKILL SHEETS

Step 1: Determine vehicle's orientation and need for stabilization.

Step 2: Determine vehicle's construction, condition, and integrity.

Step 3: Place chocks in front of and behind tires.

Step 4: Place chocks on the downhill side of a vehicle on an incline.

Step 5: Place chocks on both sides of the tires if the ground is level or the direction of the grade is undetermined.

Step 6: Test and apply the parking brake before placing chocks.

Step 7: Center chocks snugly and squarely against the tread of each tire.

Step 1: Determine vehicle's orientation and need for stabilization.

Step 2: Determine vehicle's construction, condition, and integrity.

Step 3: Determine whether to use a four-point or six-point support.

Step 4: Identify support locations on the vehicle.

Step 5: Determine whether the ground under these support locations will support the vehicle's/equipment's weight.

Step 6: Position sufficient cribbing material at each support location.

Step 7: Construct a crib base appropriate for conditions.

Step 8: Add the next layer of cribbing allowing the ends of the cribbing pieces to extend 3 or 4 inches (75 mm to 100 mm) beyond the individual pieces of the base.

Step 9: Add additional layers as needed, overlapping the crib corners as described above.

Step 10: Use wedges and shims to provide the maximum amount of contact between the crib and the vehicle.

Step 11: Repeat the process until at least four cribs are supporting the vehicle.

Step 12: Evaluate and maintain the integrity of the cribbing.

Step 1: Determine vehicle's orientation and need for stabilization.

Step 2: Determine construction, condition, and integrity of the vehicle.

Step 3: Determine whether to use a four-point or six-point support.

Step 4: Identify support locations on the vehicle.

Step 5: Determine whether the ground under these support locations will support the weight of the vehicle and equipment.

Step 6: Ensure that the opposite side or end of the object to be lifted is resting on cribbing.

Step 7: Select the lifting device to be used.

Step 8: Position the jack so that it is directly beneath a solid portion of the vehicle frame, yet can be operated without rescuers needing to lie beneath the vehicle.

Step 9: As the vehicle starts to lift, construct at least one box crib or insert at least one step chock in the area of the lifting.

Step 10: Once the jack has reached its maximum travel distance and sufficient cribbing is in place, lower the jack until the vehicle is resting firmly on the cribbing.

Step 11: Retract the jack and add additional cribbing beneath it to raise the vehicle further, if necessary.

Step 12: Evaluate and maintain the integrity of the cribbing.

10-11-5
Stabilize a vehicle using a system of ropes and webbing.

SKILL SHEETS

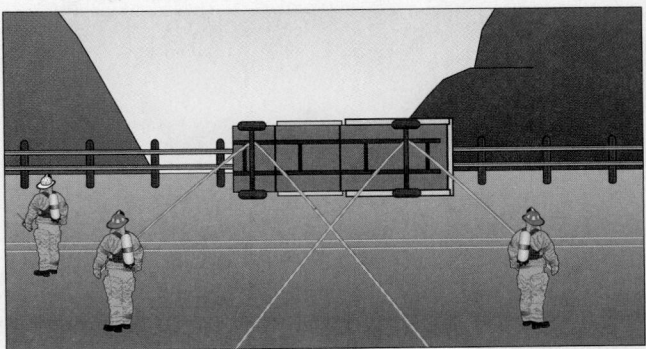

Step 1: Determine vehicle's orientation and need for stabilization.

Step 2: Determine vehicle's construction, condition, and integrity.

Step 3: Determine whether to use a four-point or six-point support.

Step 4: Identify support locations on the vehicle.

Step 5: Ensure that equipment is rated for the anticipated load plus a safety factor.

Step 6: Attach webbing, ropes, or chains to anchor points on the vehicle.

Step 7: Secure the webbing, ropes, or chains to anchor points.

Step 8: Remove slack from the webbing, ropes, or chains.

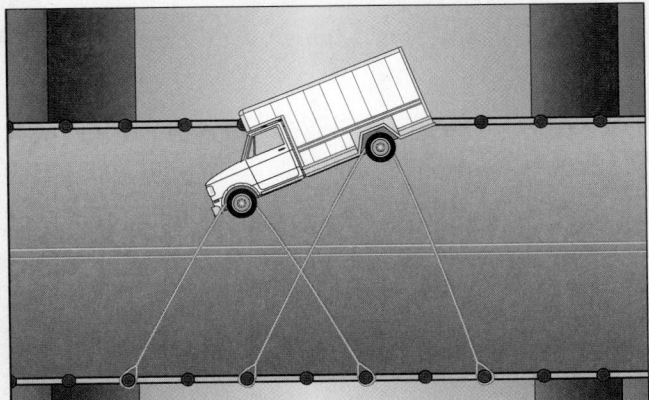

Step 9: Evaluate and maintain the tension of the stabilization equipment used.

SKILL SHEETS

10-11-6
Stabilize a side-resting vehicle using a buttress tension system.

Step 1: Determine vehicle's orientation and need for stabilization.

Step 2: Determine construction, condition, and integrity of the vehicle.

Step 3: Determine whether to use a four-point or six-point support.

Step 4: Identify support locations on the vehicle.

Step 5: Determine whether the ground under these support locations will support the weight of the vehicle and equipment.

Step 6: Manually stabilize and/or place wedges to control vehicle while setting up buttress system.

10-II-6
Stabilize a side-resting vehicle using a buttress tension system.

SKILL SHEETS

Check that straps and tip engagements are tight. Adjust if necessary.

Step 7: Based on situation, location of patient, type and condition of vehicle, and any obstructions, determine which side of vehicle to place the single jack stand, and which side to place the two adjustable stands if using a 3-point setup.

Step 8: With a minimum of two people, adjust and set stands while maintaining situational awareness.

NOTE: Monitor equipment throughout the operation and make adjustments as needed.

Step 9: Set stand(s) on least stable side of vehicle first, then work on opposite side.

Step 10: Engage vehicle with tips as high as possible. Attach base strapping as low as possible. Stands should lean at an angle between 50 to 70 degrees.

Step 11: Once all stands are placed, tighten system up.

Removing the Windshield Seal

Step 1: Cover patients with a blanket, tarp, or fire resistant material to protect them from glass fragments.

Step 2: Identify the method to be used to remove the windshield based upon windshield type, windshield condition, and equipment available.

Step 3: Place the blade of a commercial windshield removal tool under the windshield seal.

Step 4: Hold and stabilize the seal removal tool with one hand, place the other hand on the attached cable and handle and begin to pull toward oneself, ensuring that the blade of the tool remains against the windshield and under the seal at all times.

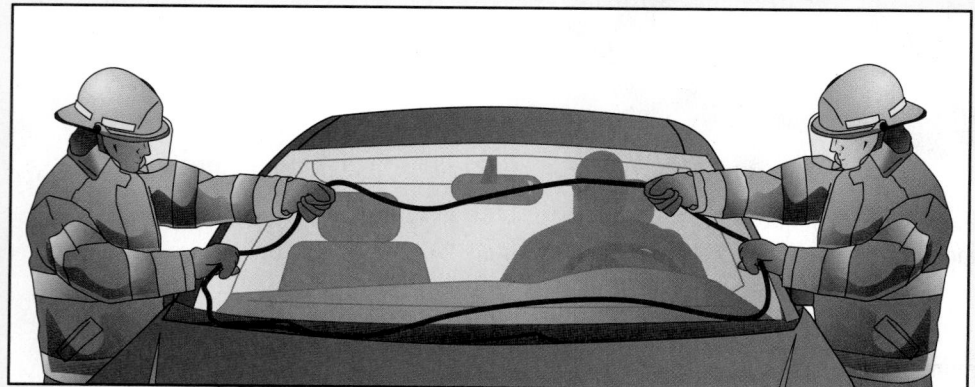

Step 5: Continue until the entire seal has been cut. Upon completion, remove the outer portion of the seal from the windshield.

Step 6: Push the windshield outward from the interior of the vehicle. An alternative option of the removal is to place duct tape handles or suction cups onto the outer portion of the windshield and remove.

Step 7: Upon removal of the windshield, position it away from the rescue scene to ensure safety of personnel.

Cutting the Windshield

Step 1: Cover patients with a blanket, tarp, or fire resistant material to protect them from glass fragments.

Step 2: Identify the method to be used to remove the windshield based upon windshield type, windshield condition, and equipment available.

Step 3: Saw operator cuts two slits in the glass to be removed using reciprocating saw, handsaw, air chisel, or other tool.

NOTE: Ensure appropriate respiratory protection for responders and victims is used.

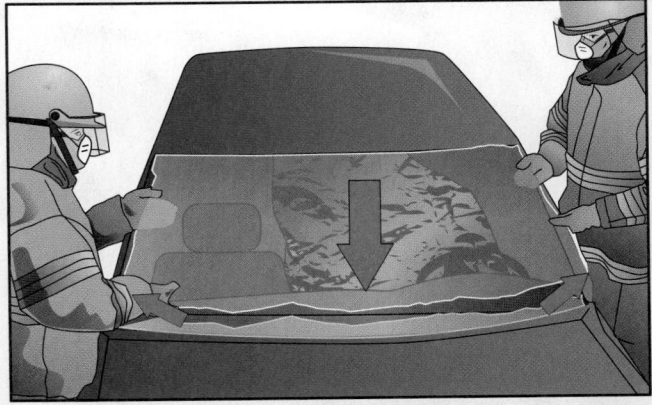

Step 5: Saw operator and other glass-removal team members position themselves on opposite sides of the window.

Step 6: Each team member grasps the glass near bottom cut.

Step 7: Raise the glass moving bottom outward, using care not to break the glass.

Step 8: Remove the glass, pulling down to dislodge from frame and folding back over roof.

Step 4: Operator then cuts the lower portion of the window connecting each side cut near the bottom of window.

Step 9: Place the glass out of the way of operations per local protocol.

Step 1: Select the tool that will be used to break the glass.

Step 2: Ensure patients are protected from glass fragments.

Step 3: Place a center punch or other tool in the lower corner of the window.

Step 4: Brace the hand holding the center punch with the opposite hand to prevent the rescuer from pushing the hand with the punch through the broken glass.

Step 5: Break the window with the punch or other tool.

Step 6: Clear the remaining glass from the window opening.

Removing Glass Method

Step 1: Cut the first post at the furthest point from the patient.

Step 2: Remove glass.

Step 3: Cut the B- and C-posts without cutting into seat belt pretensioners located in the B-posts and any side air bag inflation cylinders that might be located in the C-posts. Assign personnel to support the roof while the posts are being cut so the roof will not fall into the passenger compartment.

Step 4: Cut post closest to the patient last.

Step 5: Remove the roof.

Cutting Across Roof Method

Step 1: Peel back the plastic interior finish and peek inside looking for potential hazards, such as airbags, retractors before cutting.

Step 2: Cut the roof supports/door jams just behind the windshield frame.

Step 3: Continue the cut across the front of the roof behind the windshield frame.

Step 4: Remove the rear window.

Step 5: Cut the B- and C-posts without cutting into seat belt pretensioners located in the B-posts and any side air bag inflation cylinders that might be located in the C-posts. Assign personnel to support the roof while the posts are being cut so the roof will not fall into the passenger compartment.

Step 6: Once all the posts have been cut, lift the roof clear and set it aside.

Flapping the Roof Method

Step 1: Peel back the plastic interior finish and peek inside looking for potential hazards, such as air bags and retractors, before cutting.

Step 2: Cut seat belts and appropriate posts.

Step 4: Flapped roofs should be secured with ropes, chains, straps, or other appropriate material.

Step 3: Use a pike pole or other long object to push the sheet metal down at the bending point and to push the roof up at the front.

Step 3: Push or pull the roof down to provide access to the passenger compartment. Pad any rough edges.

Step 4: If desired, the roof can be removed entirely by cutting the remaining posts. Again, after the cuts are complete, cover any rough edges.

Step 1: Peel back the plastic interior finish and peek inside looking for potential hazards, such as air bags and retractors before cutting.

CAUTION: Ensure that the vehicle is stabilized appropriately and monitor stability throughout operation.

Step 2: Cut the roof posts that are easily accessible and lay the roof down in a manner similar to that used on upright vehicles.

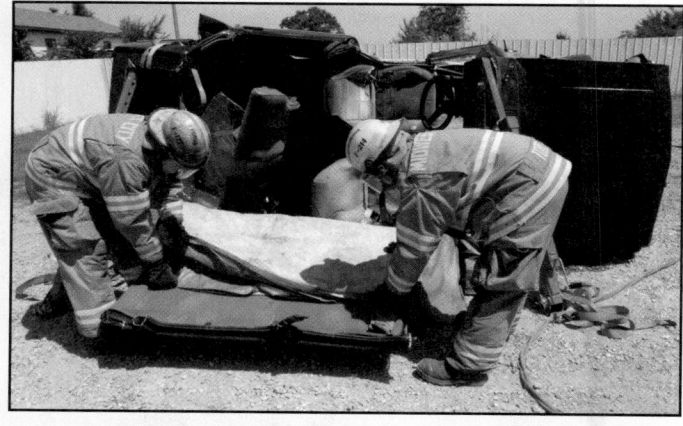

Step 5: Cut the B- and C-posts without cutting into seat belt pretensioners located in the B-posts and any side air bag inflation cylinders that might be located in the C-posts. Assign personnel to support the roof while the posts are being cut so the roof will not fall into the passenger compartment.

Step 6: Once all the posts have been cut, lift the roof clear and set it aside.

NOTE: This skill is also known as *rolling the dashboard*.

Step 1: Remove the front door.

Step 2: Make relief cuts behind the strut mounts to eliminate movement of the front end of the vehicle during this operation.

Step 3: Peel back the plastic interior finish and peek inside looking for potential hazards such as air bags and retractors before cutting.

Step 4: Cut the upper portion of the A-post.

Step 5: Cut the bottom portion of the A-post.

Step 6: Position jacking or ram device between base of the B-post and to an area just above the top hinge on the A-post.

Step 7: Operate the jacking or ram device to move the dashboard.

Chapter Contents

Key Terms

NFPA® Job Performance Requirements

This chapter provides information that addresses the following job performance requirements of NFPA® 1001, *Standard for Fire Fighter Professional Qualifications* (2013).

Firefighter I

5.3.4

5.3.14

5.5.1

1. Explain the basic principles of forcible entry. (5.3.4)

2. Describe the basic construction of locksets. (5.3.4)

3. Describe considerations a firefighter must take when using forcible entry tools. (5.3.4)

4. Indicate steps needed to care for and maintain forcible entry tools. (5.5.1)

5. Explain the ways to force entry through various types of doors. (5.3.4)

6. Identify considerations that need to be taken when forcing entry through locks, padlocks, overhead doors, and fire doors. (5.3.4)

7. Describe forcible entry methods used for windows. (5.3.4)

8. Explain considerations firefighters must take when forcing entry through miscellaneous types of windows and covers. (5.3.4)

9. Describe forcible entry methods for breaching walls. (5.3.4)

10. Explain forcible entry methods for breaching floors. (5.3.4)

11. Indicate methods for forcing fences and gates. (5.3.4)

12. Clean, inspect, and maintain hand tools and equipment. (Skill Sheet 11-I-1, 5.5.1)

13. Clean, inspect, and maintain power tools and equipment. (Skill Sheet 11-I-2, 5.5.1)

14. Force entry through an inward-swinging door – Two-firefighter method. (Skill Sheet 11-I-3, 5.3.4, 5.3.14)

15. Force entry through an inward-swinging door – Cutting the lock out of the door method. (Skill Sheet 11-I-4, 5.3.4, 5.3.14)

16. Force entry through an outward-swinging door – Removing hinge-pins method. (Skill Sheet 11-I-5, 5.3.4, 5.3.14)

17. Force entry though an outward-swinging door – Wedge-end method. (Skill Sheet 11-I-6, 5.3.4, 5.3.14)

18. Force entry using the through-the-lock method. (Skill Sheet 11-I-7, 5.3.4, 5.3.14)

19. Force entry using the through-the-lock method using the K-tool. (Skill Sheet 11-I-8, 5.3.4, 5.3.14)

20. Force entry using the through-the-lock method using the A-tool. (Skill Sheet 11-I-9, 5.3.4, 5.3.14)

21. Force entry through padlocks. (Skill Sheet 11-I-10, 5.3.4, 5.3.14)

22. Use a bam-bam tool. (Skill Sheet 11-I-11, 5.3.4, 5.3.14)

23. Cut a padlock with a rotary saw. (Skill Sheet 11-I-12, 5.3.4, 5.3.14)

24. Force entry through a window (glass pane). (Skill Sheet 11-I-13, 5.3.4, 5.3.14)

25. Force entry through a double-hung window. (Skill Sheet 11-I-14, 5.3.4, 5.3.14)

26. Force a Lexan® window using a rotary saw. (Skill Sheet 11-I-15, 5.3.4, 5.3.14)

27. Force entry through a wood-framed wall (Type V construction) with hand tools. (Skill Sheet 11-I-16, 5.3.4, 5.3.14)

28. Force entry through a wood wall (Type V construction) with a rotary saw or chain saw. (Skill Sheet 11-I-17, 5.3.4, 5.3.14)

29. Breach a wall using a battering ram. (Skill Sheet 11-I-18, 5.3.4, 5.3.14)

30. Force entry through a masonry wall with hand tools. (Sheet 11-I-19, 5.3.4, 5.3.14, 5.3.14)

31. Force entry through a metal wall with power tools. (Skill Sheet 11-I-20, 5.3.4, 5.3.14)

32. Breach a hardwood floor. (Skill Sheet 11-I-21, 5.3.4, 5.3.14)

33. Bridge a fence with a ladder. (Skill Sheet 11-I-22, 5.3.4, 5.3.14)

Chapter 11
Forcible Entry

Gaining access to a structure, facility, or site may require forcing locks, doors, windows, or other barriers. The more quickly firefighters can gain entry, the faster they can control the emergency. To efficiently and effectively gain entrance, you must use your knowledge of building construction, fire behavior, and property conservation. However, you must always size up the situation to determine the correct location to gain entry and the proper technique to use. This chapter identifies the many tools that can be used for structural forcible entry operations. Their proper use, care, and maintenance are crucial to the success of any forcible entry operation. Also described are the characteristics of various types of barriers that may have to be overcome to gain access to buildings during fires or other emergencies.

Basic Principles of Forcible Entry

In the fire service, **forcible entry** commonly refers to the techniques used to gain access into a compartment, structure, facility, or site when the normal means of entry is locked or blocked. When properly applied, forcible entry techniques do minimal damage to the structure or structural components and provide quick access to the

> **Forcible Entry** — Techniques used by fire personnel to gain entry into buildings, vehicles, aircraft, or other areas of confinement when normal means of entry are locked or blocked.

Figure 11.1 Forcible entry refers to the techniques used to gain access to an enclosed area when no other access is available.

emergency **(Figure 11.1)**. Firefighters should determine if alternative means of access are practical before forcing entry into a building. Forcible entry should not be used when normal means of access are available. Forcible entry techniques can also be applied to vehicles, railway passenger cars, aircraft, or ships when necessary. Effective size-up and situational awareness are important to all types of forcible entry.

Where to force entry will be determined by your supervisor or the Incident Commander (IC) based upon the following factors:

- Tactics that must be fulfilled
- Location of the fire or hazard
- Stage of the fire
- Effect on ventilation
- Amount of effort required to force entry

For instance, it may be easier to force a basement window to apply water on a fire than to force entry through a door and advance a hoseline down an interior stairway. Also, the location of an opening can drastically affect fire behavior by adding fresh air to a ventilation-controlled fire. Never force entry without orders to do so.

At every door or window, remember to first "**Try before you pry.**" The door or window may be unlocked and can be opened in a normal manner. Second, especially on commercial and industrial occupancies, look for a lockbox near the main entrance **(Figures 11.2a and b)**. Using a door key or numeric keypad combination found inside a lockbox avoids unnecessary damage to the property and may allow quicker entry than having to force a door or window. In some departments, lockbox information is maintained in the building preincident plans and/or computer-aided dispatch (CAD) data to speed access during incident operations.

Figure 11.2a At every point of entry that may be unlocked, it is important to check to see if nonforcible entry may be possible.

Figure 11.2b Some structures are equipped with fire department lockboxes.

General considerations for forcible entry are as follows:

- **Doors and locks** — Construction, direction of opening (inward or outward), type of frame, type of lock, and mounting of the lock

- **Proper tools** — Knowing the correct tools needed for a particular job; adjusting entry activity based upon available tools; using tools for their intended purposes to make entry

- **Security barriers** — Could include bars, grilles, Lexan® windows, and others; require specialized training, tools, and knowledge to force; block escape routes for both firefighters and occupants; may necessitate making multiple openings for entry **(Figure 11.3)**.

Figure 11.3 Security barriers present an obstacle to firefighters and may require multiple openings to bypass. *Courtesy of Los Angeles Fire Department - ISTS.*

Locksets

Lockset is the term used to describe all types of door latches, locks, and locking devices. The purpose of locksets is to secure doors and prevent unauthorized entry into a room or structure. To perform an adequate forcible entry size-up, you must have an understanding of the types of locks and locking devices that you may encounter during a fire or other emergency.

Door Locks/Latches

Locks and latches are part of the hardware normally found on all exterior doors and many interior doors. Door latches keep the door closed and consist of a handle on both sides of the door and a spring-loaded bar that extends into a receiver in the door frame. It may or may not have a lock as part of the assembly. Locks can be divided into three basic types: mortise lock, cylindrical lock, and rim lock **(Figure 11.4, p. 576)**. Nontraditional types of door locks that provide a higher level of security include multiple bolts, keyless, and electromagnetic locks.

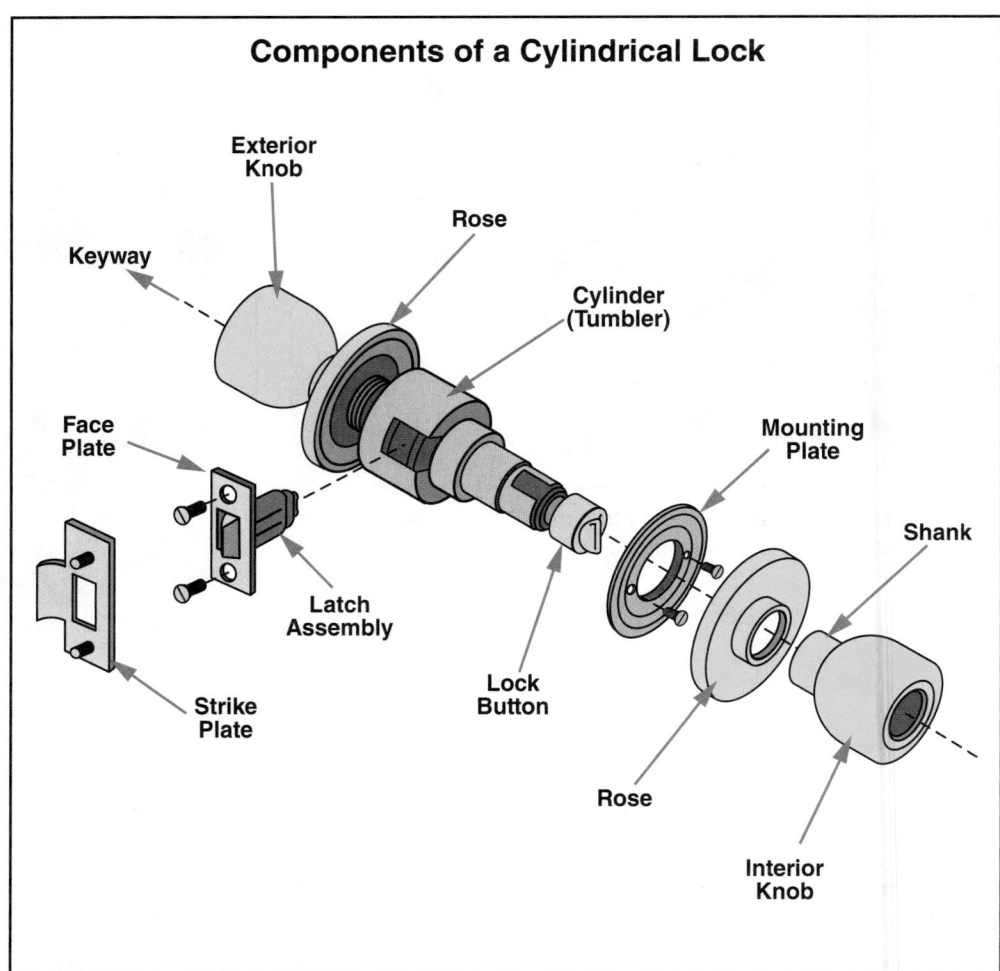

Components of a Cylindrical Lock

Exterior
Knob

Rose

Keyway

Cylinder
(Tumbler)

Face
Plate

Mounting
Plate

Shank

Latch
Assembly

Strike
Plate

Lock
Button

Rose

Interior
Knob

Figure 11.4 A cylindrical lock holds and latches a door in its frame.

Mortise Latch and Lock

The mortise latch and lock assembly is mounted into a cavity in the door edge. Older mortise assemblies have only the latch to hold the door closed, while newer units consist of both a latch and a key-operated dead bolt (**Figure 11.5**). When the mechanism is in the locked position, the bolt protrudes from the lock into a receiver that is mortised into the jamb. The latch may be operated with a doorknob or lever. Mortise locks are used on exterior wood and metal doors and can be found on private residences, commercial buildings, and industrial buildings.

Cylindrical Lock

Cylindrical locks are the most common type of lockset found in residential applications. Their installation involves boring two holes at right angles to one another: one through the face of the door to accommodate the main locking mechanism and the other in the edge of the door to receive the latch or bolt mechanism. There are two types of cylindrical locks: the key-in-knob lock and the tubular dead bolt lock.

The key-in-knob lock has a keyway in the outside doorknob; the inside knob may contain either a keyway or a button (**Figure 11.6**). The button may be a push button or a push and turn button. Key-in-knob locks are equipped with a latch mechanism that is locked and unlocked by both the key and, if present, by the knob button. In the unlocked position, a turn of either knob retracts the spring-loaded beveled latch

Figure 11.5 A mortise latch and lock assembly may consist of just a latch or a latch and a dead bolt.

Lock Bolt

Latch Bolt

Figure 11.6 A key-in-knob lock has a key way **(a)** on the outside doorknob and either another key way or a button **(b)** on the inside knob.

Figure 11.7 A tubular dead bolt lock uses a thumb turn knob on the inside and a key way on the outside of the door.

bolt, which is usually no longer than ¾-inch (19 mm). Because of the relatively short length of the latch, key-in-knob locks are some of the easiest to pry open. If the door and frame are pried far enough apart, the latch clears the strike and allows the door to swing open.

The tubular dead bolt lock is mounted above the doorknob and may have a single cylinder or double action cylinder **(Figure 11.7)**. The single cylinder dead bolt has a keyway on the outside of the door and a thumb turn knob on the inside. The double cylinder lock has a keyway on both sides of the door.

NOTE: The easiest way to breach modern dead bolts in a wood residential door is to force the door itself.

Figure 11.8a A night latch automatically locks the door when the door is closed.

Figure 11.8b A dead bolt uses a manually operated rectangular bolt.

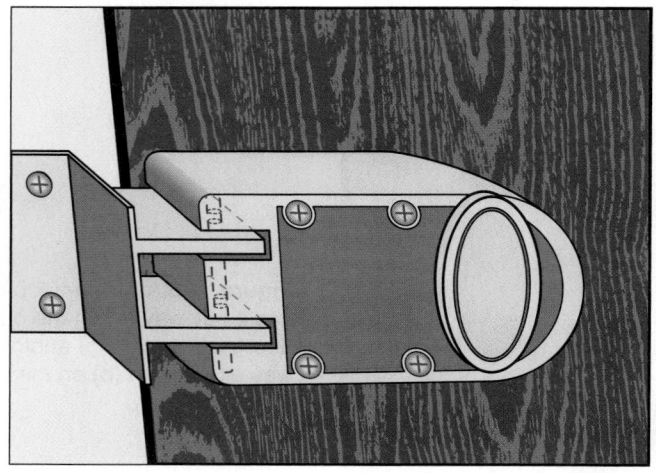

Figure 11.8c A vertical dead bolt is impossible to open by spreading the door from the doorjamb.

Rim Lock

A rim lock is mounted on the interior door surface and is used as a supplemental lock for doors that may or may not have other types of locks. Rim locks operate by turning a thumb turn knob on the inside of the door. This lock is found in all types of occupancies, including houses, apartments, and some commercial buildings. Although some rim locks have a keyway in a cylinder on the exterior of the door, not all do, making it difficult to recognize their presence. There are a variety of rim locks currently available including (**Figures 11.8a-c**):

- **Night latch** — This rim lock has a spring-loaded bolt with a beveled edge facing the door frame. This feature allows the door to lock when it is closed.

- **Dead bolt** — This rim lock has a rectangular bolt that must be manually retracted before the door can be closed and the bolt engaged with the receiver. If the bolt is extended, the door cannot be closed.

- **Vertical dead bolt** — This rim lock has a bolt that slides vertically into the receiver and does not cross the door opening. This feature makes it impossible to open by spreading the door from the doorjamb.

Figure 11.9a Multiple bolt locks provide higher security by using two or more bolts in an edge of the door.

Figure 11.9b Electronic keyless locks provide higher security by maintaining continuous security and controlled access.

Figure 11.9c Electromagnetic locks hold a door securely as long as an electric current passes through the electromagnet and the armature plate.

Higher Security Locks

Higher security locks include (**Figures 11.9a-c**):

- **Multiple bolt locks** — The multiple bolt or multi lock is a dead bolt lock that, when engaged, projects bolts 1 inch (25 mm) into two or more points on one edge of the door. Some versions extend hardened steel bolts into all four edges of the door frame. The lock may have a thumb turn knob or keyway on the inside of the door as well as a keyway on the exterior. A surface-mounted version may also be encountered.

- **Electronic keyless locks** — Found on both exterior and interior doors, keyless or digital lock may have a key pad, card reader, or even a fingerprint-activated screen. The locks are generally battery powered. Some key pads may also have a keyway. They are used for areas that require continuous security and controlled access.

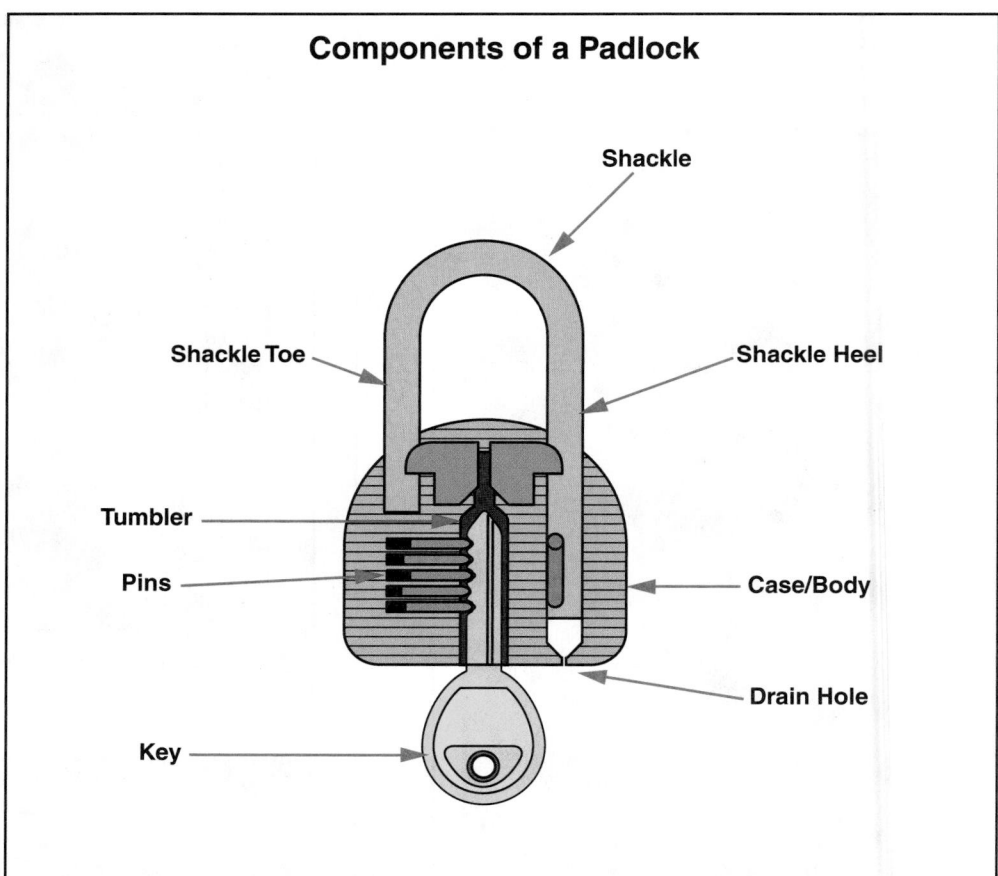

Components of a Padlock

Shackle

Shackle Toe

Shackle Heel

Tumbler

Pins

Case/Body

Drain Hole

Key

Figure 11.10 A padlock is a portable or detachable locking device that may be operated either with a key or a combination.

- **Electromagnetic** — The electromagnetic or magnetic lock consists of an electromagnet attached to the door frame and an armature plate mounted on the door. The door is held shut by an electric current passing through the electromagnet and the armature plate. Shutting off the power will release the door.

Locking Devices

Locking devices may be supplemental to the door lock or used in place of it. Padlocks are the best example of locking devices. Other devices, such as door chains or drop bars, impede entry but are not locks in the traditional sense.

Padlocks are portable or detachable locking devices. There are two basic types of padlocks: standard and heavy-duty. Standard padlocks have shackles of ¼ inch (6 mm) or less in diameter and are not case-hardened steel. Heavy-duty padlocks have case-hardened steel shackles more than ¼ inch (6 mm) in diameter. Many heavy-duty padlocks have what is called "toe and heel locking" where both ends of the shackle are locked when depressed into the lock mechanism. These shackles will not pivot if one side of the shackle is cut. Both sides of the shackle must be cut in order to remove the lock. Padlocks may be key or combination operated (**Figure 11.10**).

Other locking devices include the:

- **Drop bar** — Brackets are bolted or welded to the door and a wood or metal bar rests in the brackets and extends across the door frame (**Figure 11.11**).
- **Door chain** — The door chain is the classic supplemental locking device for residential doors. The chain permits the door to be opened wide enough to see and speak to a visitor but still restrict access.

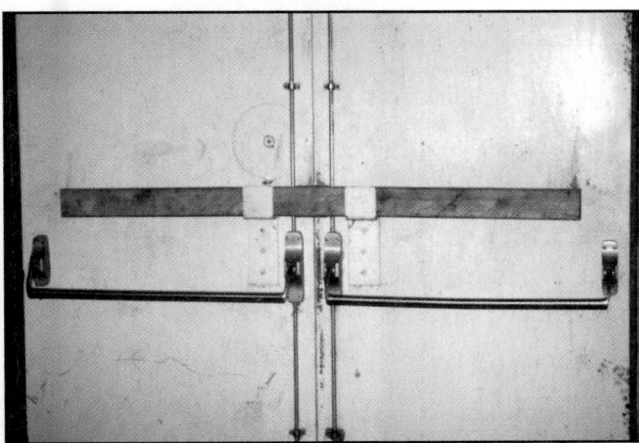

Figure 11.11 Although a violation of fire codes in most areas, drop bars may be found in many commercial occupancies. The one depicted here is mounted on an egress door that also has panic hardware installed on it.

- **Door limiter** — Similar to the supplemental security locks found in hotel rooms, the door limiter consists of a frame-mounted plate with a shaft and knob and a hinged u-shaped shackle that mounts on the door. Like the door chain, the door limiter restricts the opening of the door.

- **Surface bolt** — Surface slide bolts are manually operated supplemental locking devices that can be mounted on most doors and some windows.

- **Internal-mounted bolt** — Flush bolts are installed in the edge of one side of a set of double doors. This permits one side to remain locked while the other door is used for entry and exit. When desired, the bolts can be retracted and both doors opened.

With the exception of the drop bar and the internally mounted bolts, the remainder of the locking devices may be easy to force depending on the tool that is used. It will also be difficult to tell if any of these types of devices are mounted and in use before attempting to force entry.

Forcible Entry Tools

Firefighters must know the characteristics, capabilities, and limitations of the tools available to them before attempting forcible entry. Selection of the proper tool may make the difference in overcoming a barrier. This section describes the various categories of tools used for forcible entry operations. Also included is information on the proper use, care, and maintenance of tools. When using any forcible entry tools, always wear appropriate personal protective equipment (PPE), especially hand, eye, and hearing protection.

CAUTION
Always wear appropriate PPE when using forcible entry tools.

Forcible entry tools can be divided into four basic categories:

- Cutting tools
- Prying tools
- Pushing/pulling tools
- Striking tools

Figure 11.12 A pick-head axe can be used for cutting, prying, and digging.

Figure 11.13 The flat end of a flat-head axe is used in conjunction with other tools, and the blade end may be used on its own.

Cutting Tools

Cutting tools can be powered manually or with another power source. These tools are often specific to the types of materials they can cut and how fast they can cut them. No single cutting tool will safely and efficiently cut all materials. Using a cutting tool on materials for which it was not designed can damage the tool and endanger the operator.

Axes

Axes are the most common types of cutting tools that firefighters use. There are two basic types of axes in common use: the pick-head axe and the flat-head axe. Smaller axes and hatchets are also available for use in salvage and overhaul operations, but these tools are usually too lightweight and inefficient for effective use in forcible entry operations.

Pick-head axe. Pick-head axes are available with either a 6-pound or an 8-pound head (2.7 kg or 3.6 kg) (**Figure 11.12**). The pick-head axe is a versatile forcible entry tool that can be used for cutting, prying, and digging. The head is made of hardened steel; the handles are made of either wood or fiberglass and vary in size. The pick-head axe is very effective for chopping through wooden structural components, shingles and other roof coverings, aluminum siding, and other natural and lightweight materials. The pick end can be used to penetrate materials that the blade of the axe cannot cut easily. The side of the pick-head axe blade can also be used as a striking tool to break windows or as a prying tool to force some doors. Because of its variety of uses, pick-head axes are most often used in structural fire fighting operations.

Flat-head axe. The flat-head axe is the same as the pick-head axe in size, design, and construction, except that a flat striking face replaces the pick end (**Figure 11.13**). The blade of the flat-head axe can be used for all the same purposes as that of a pick-

Figure 11.14 Although they cannot cut case-hardened metals, bolt cutters can shear through thin metal components of locks, doors, and other security features on or around a building.

head axe. Unlike the pick-head axe, the flat-head axe can be used with other tools to force entry. The striking face of flat-head axe is used to strike the other tool, forcing the bit end into a doorjamb or windowsill. Because of its versatility, the flat-head axe is used in both structural and ground cover fire fighting operations.

Metal Cutting Devices

Metal cutting devices are used to cut through heavy-duty locks, metal-clad doors, window security bars and grilles, and similar items. The devices include bolt cutters, cutting torches, and manual or powered rebar cutters.

Bolt cutters. These metal cutting devices are used to cut bolts, iron bars, pins, cables, hasps, chains, and some padlock shackles **(Figure 11.14)**. Bolt cutters are less practical as entry tools than they once were. Modern high-security chains, hasps, and padlock shackles cannot be cut with manual bolt cutters. These materials shatter the bolt cutter's blades or cause the handles to fail under the tremendous pressure that the operator must exert. Other locks are designed to prevent bolt cutters from being inserted into the shackle.

Bolt cutters should not be used to cut case-hardened materials found in locks and other security devices. Face shields and eye protection must always be worn when using bolt cutters to prevent fragments of the cut material from striking the operator's face. Bolt cutters must not be used to cut any energized cables unless the cutters are insulated and designed for that task.

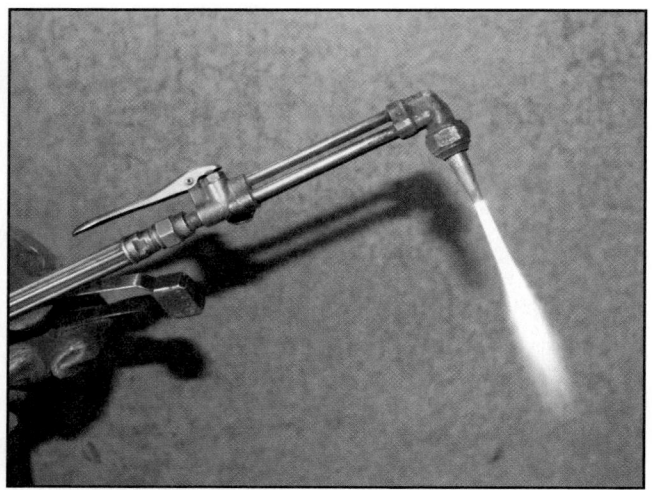

Figure 11.15 Cutting torches offer quick access through security features that may otherwise be impenetrable.

Figure 11.16 Powered and manual rebar cutters are used to create an opening in door or window security bars.

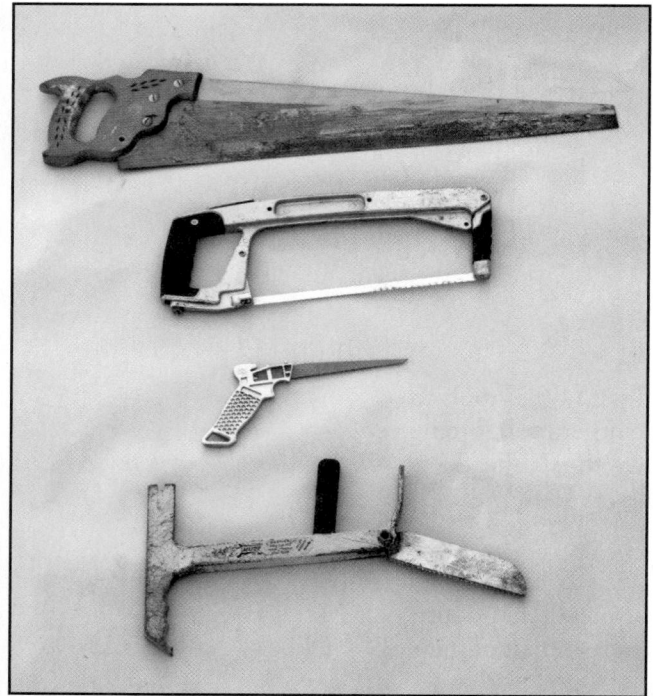

Figure 11.17 Some handsaws retain their usefulness on the fireground, especially when they are matched to a specific job and work space is limited.

Cutting torches. A cutting torch may be needed to cut security bars, grilles, or gates that cannot be easily cut using bolt cutters, rebar cutters, or rotary saws (**Figure 11.15**). Firefighters commonly use oxyacetylene cutting torches, oxygasoline cutting torches, burning bars, and plasma cutters. Training based on the manufacturer's recommendations specific to each cutting and burning device is necessary for safe and efficient operation. A charged hoseline must be in place during the cutting operation to cool the metal and control any sparks that are generated.

Rebar cutters. These hydraulic cutting tools are available in both powered and manual versions (**Figure 11.16**). The manual version requires more energy to use, but can be used in areas beyond the reach of the hydraulic supply hose on powered units. Firefighters can use rebar cutters to cut steel reinforced bars (rebar) in concrete walls or to cut door or window security bars.

Handsaws

Power saws have largely replaced handsaws in the fire service; however, they may be useful when power saws are not available or the work space where a cut must be made is limited. Hacksaws, drywall saws, and keyhole saws are the most common handsaws still in use (**Figure 11.17**). To be proficient in handsaw use, firefighters should know which saw is best suited to the current job and should use good handsaw technique.

Power Saws

Several types of power saws are commonly used in the fire service including the circular saw, rotary saw, reciprocating saw, and chain saw (**Figures 11.18a-d**). Power sources may be a self-contained battery pack, gasoline engine, or electricity from a generator or electrical outlet. You should be familiar with the type of saws used by your department. Always follow both the manufacturer's recommendations and departmental SOPs when operating power saws.

Figure 11.18a Circular saws use a round cutting wheel to make straight line cuts.

Figure 11.18b Rotary saws are usually gasoline powered and require the application of water to prevent sparks from causing problems or the blade from overheating.

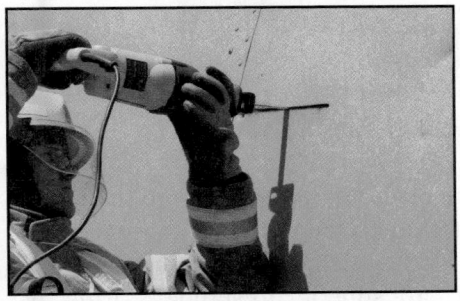

Figure 11.18c Reciprocating saws can make small cuts in confined spaces and are ideal for cutting vehicle body panels.

Figure 11.18d Chain saws serve many purposes and can cut many types of materials.

Always use eye, hearing, and hand protection when operating any power saw. Forcing the saw (or any tool) beyond its design limits can result in property damage and/or injury to the operator. Never use a power saw in a flammable atmosphere. The saw's motor or sparks from the cutting operation can ignite a fire or cause an explosion.

CAUTION
Wear eye, hearing, and hand protection when operating any power saw.

CAUTION
Never force a power saw beyond its design limits. Follow the manufacturer's recommendations.

CAUTION
Do not use a power saw in a flammable atmosphere.

Circular saw. Electric circular saws have many applications in fire fighting, rescue, and overhaul operations. These saws are especially useful in situations where electrical power is readily available and heavier and bulkier power saws are too difficult to handle. Small battery-powered units are also available.

Rotary saw. Rotary saws used in the fire service are usually gasoline powered with changeable blades available for cutting wood, metal, and masonry. When using a rotary saw to cut metal, have a charged hoseline or portable fire extinguisher nearby because of the sparks produced in the cutting operation.

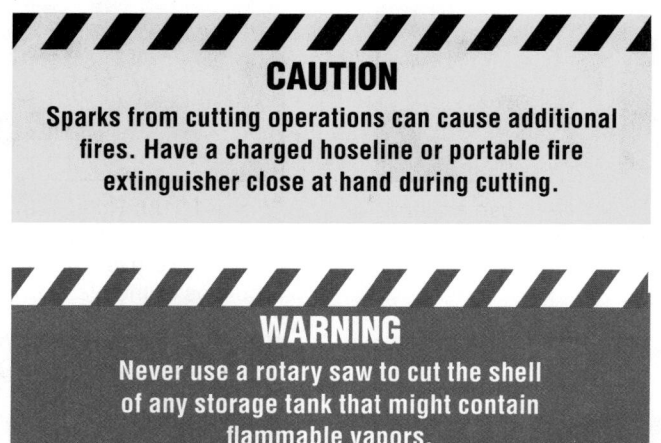

CAUTION

Sparks from cutting operations can cause additional fires. Have a charged hoseline or portable fire extinguisher close at hand during cutting.

WARNING

Never use a rotary saw to cut the shell of any storage tank that might contain flammable vapors.

Depending upon the type of rotary saw, the blades may spin at more than 6,000 rpm. Blades range from large-toothed blades for quick rough cuts to those with fine teeth for a more precise cut. Some blades are made specifically for cutting metal or concrete. Saw blades with carbide-tipped teeth are superior to standard blades because they are less prone to dulling after heavy use.

CAUTION

The blade guards on some rotary saws are not designed for use with carbide-tipped blades. Be sure that the saw is designed for the blades used.

Reciprocating saw. The reciprocating saw is very powerful, versatile, and easy to control. This saw has a short, straight blade that moves in and out with an action similar to that of a handsaw. It can use a variety of blades for cutting different materials. When equipped with a metal-cutting blade, this saw is ideal for cutting sheet metal body panels and structural components on vehicles as well as metal doors and wall panels in structures.

Chain saw. Chain saws are commonly used for forcible entry, ventilation, rescue, and overhaul operations. They can be powered by gasoline engines, electricity, compressed air, or hydraulic power. Cutting chains come in a variety of types for use in cutting wood, concrete, stone, or brick building materials. Chain saws are useful during natural disasters such as tornadoes and ice storms when trees and limbs must be cleared from streets and access routes, and structural collapse debris must be cut to free trapped victims. Saws should be equipped with kickback protection and chain brakes for safety as well as carbide-tipped chains and depth gauges for better saw control.

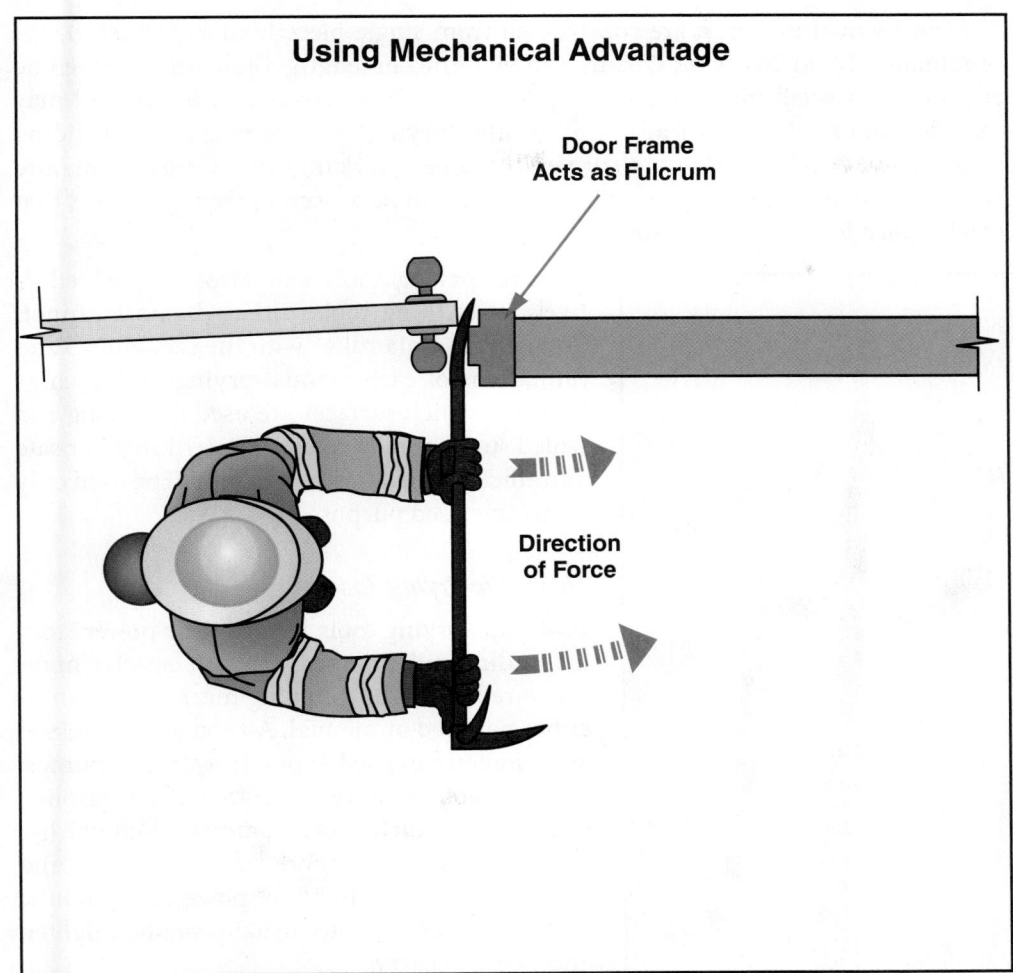

Using Mechanical Advantage

Door Frame
Acts as Fulcrum

Direction
of Force

Figure 11.19 Prying tools provide a mechanical advantage while forcing an opening.

Prying Tools

Prying tools are useful for opening doors, windows, and locks and moving heavy objects. Pry bars and other manually operated prying tools use the principle of the **lever** and **fulcrum** to provide mechanical advantage **(Figure 11.19)**. Force applied to the tool's handle is multiplied at the working end based upon the distance between the fulcrum and the working end. The longer the handle, the greater the force produced at the working end.

Manual Prying Tools

The most common manual prying tools used in the fire service are as follows **(Figure 11.20, p. 588)**:

- Crowbar
- Halligan tool
- Pry (pinch) bar
- Hux bar
- Claw tool
- Kelly tool
- Pry axe
- Flat bar (nail puller)
- Rambar

Lever — Device consisting of a bar turning about a fixed point (fulcrum), using power or force applied at a second point to lift or sustain an object at a third point.

Fulcrum — Support or point of support on which a lever turns in raising or moving something.

Halligan Tool — Prying tool with a claw at one end and a spike or point at a right angle to a wedge at the other end.

Adz — A wedge-shaped blade attached at right angles to the handle of the tool.

Most manual pry tools are constructed from single-piece high-carbon steel, approximately 30 to 36 inches (762 mm to 900 mm) in length. They usually have one end that is beveled into a single wedge or fork. The opposite end of the tool may include a hook, pike tip, or **adz**. Unlike other prying tools, rambars have a sliding weight on the shaft that is used to drive the wedge or fork into an opening. Miniature versions of manual pry tools are also available and have accompanying sheaths that can be worn for carrying the tool.

Some prying tools can also be used effectively as striking tools, although most cannot. You need to be familiar with the capabilities and limitations of each manual prying tool, such as knowing which surfaces are used for prying and which surfaces may be used for striking. For safe and efficient operation, a tool should be used only for its intended purpose.

Hydraulic Prying Tools

Hydraulic prying tools receive their power from hydraulic fluid pumped through special high-pressure hoses. The pumping mechanism can be either powered or manual. Although compressed air is sometimes used to power hydraulic pumps, electric motors or two- or four-cycle gasoline engines are much more common. Manual hydraulic tools require more labor to operate and operate more slowly than powered hydraulic tools; however, they are usually smaller, lighter, and easier to carry.

Hydraulic prying tools that one person can operate have proven to be very effective for forcible entry and extrication operations. These tools can be used in a variety of different operations involving prying, pushing, or pulling. The rescue tools and hydraulic door opener are examples of hydraulic prying tools.

Rescue tools. Hydraulic spreaders and hydraulic rams usually associated with vehicle extrication have some uses in forcible entry (**Figure 11.21**). Hydraulic spreaders can exert force either to spread something apart or to pull heavy objects. Depending on the manufacturer, the tips on these tools can spread as much as 32 inches (800 mm).

Hydraulic rams also have spreading capabilities ranging from 36 inches (900 mm) to an extended length of nearly 63 inches (1 600 mm). Rams can also be used for pushing and pulling. In forcible entry, hydraulic rams can be placed inside a door frame and used to spread the frame far enough apart for the door to swing open.

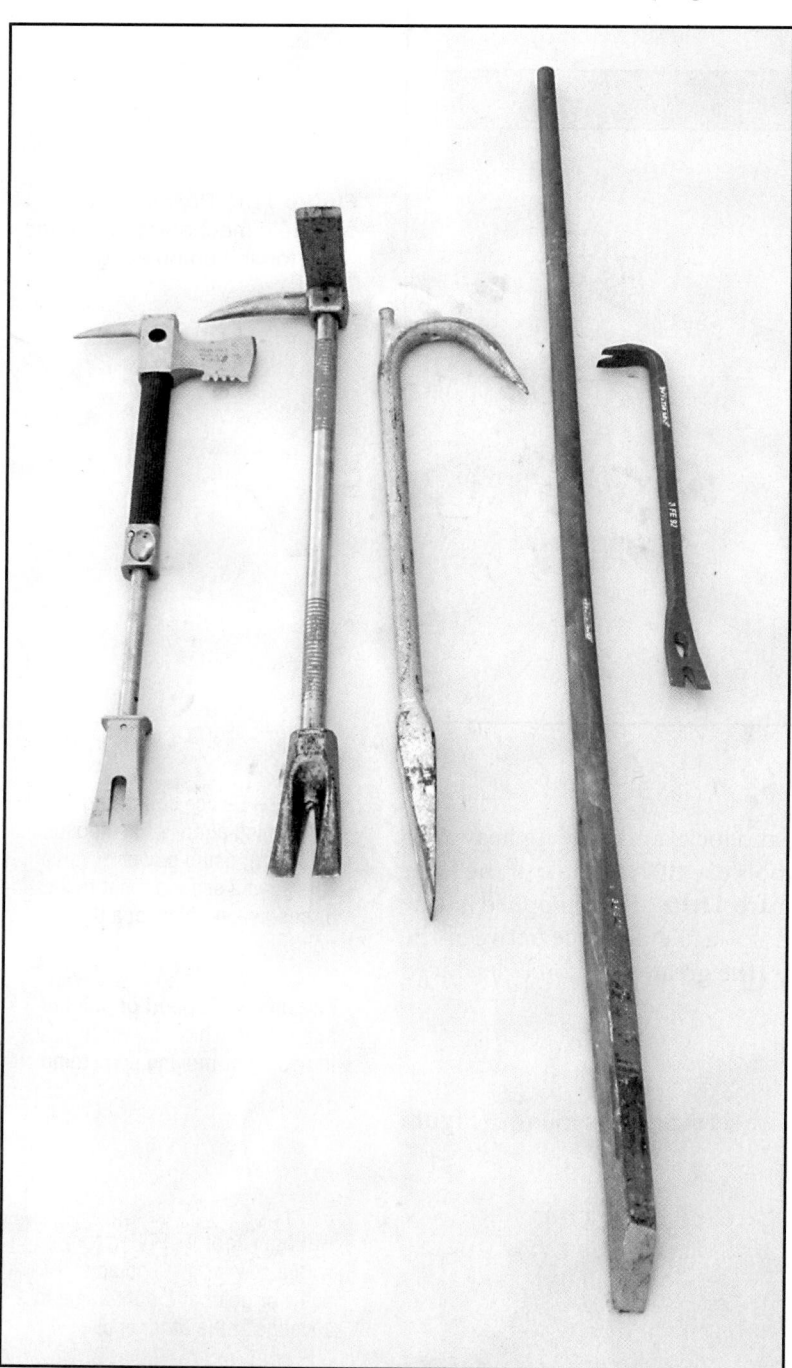

Figure 11.20 Nonpowered manual prying tools are still commonly used in the fire service to break locks, remove hinges, and open doors.

Figure 11.21 Hydraulic spreaders and rams may be used during forcible entry evolutions.

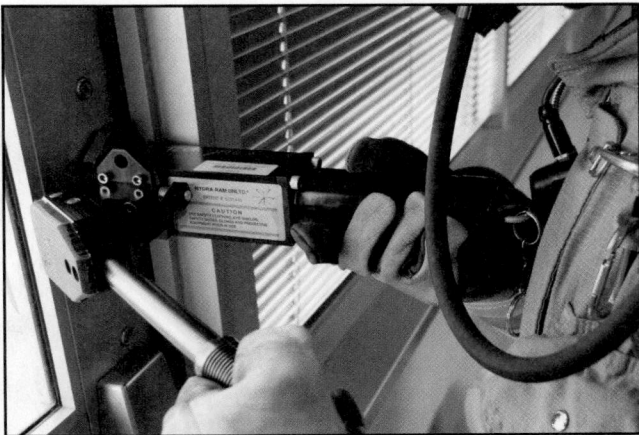
Figure 11.22 Hydraulic door openers are manually-operated spreaders with very fine teeth that may be slipped between a door and its frame.

Hydraulic door opener. This manually-operated spreader is relatively lightweight consisting of a hand pump and spreader device **(Figure 11.22)**. The spreader has intermeshed teeth that can be easily slipped into a narrow opening such as between a door and door frame. A few pumps of the handle cause the jaws of the spreader device to open, exerting pressure on the object to be moved. The pressure usually causes the locking mechanism or door to fail. These are extremely valuable tools when more than one door must be forced, such as in apartments or hotels.

Pushing/Pulling Tools

Pushing and pulling tools have limited use in forcible entry, but in certain instances, such as breaking glass and opening walls or ceilings, they are the tools of choice. This category of tools includes the following:

- Pike pole
- Clemens hook
- Plaster hook
- Drywall hook
- San Francisco hook
- Multipurpose hook
- Roofman's hook
- Rubbish hook

Pike poles and hooks, which are available in various lengths, give firefighters a reach advantage when performing certain tasks **(Figure 11.23, p. 590)**. The plaster hook has two knifelike wings that depress as the head is driven through a ceiling or other obstruction and reopen or spread outward under the pressure of self-contained springs.

With the exception of the roofman's hook, which is all metal, pike poles and hooks should only be used for pushing or pulling, never prying. If a lever is needed, select the appropriate prying tool. Handles of pike poles and hooks are made of wood or fiberglass and may break if used as a lever.

Pike Pole — Sharp prong and hook of steel, on a wood, metal, fiberglass, or plastic handle of varying length, used for pulling, dragging, and probing

Figure 11.23 Pike poles extend the reach of a firefighter and serve pulling and pushing functions.

Striking Tools

A striking tool is a very basic hand tool consisting of a weighted head attached to a handle. Some examples of common striking tools are as follows:

- Sledgehammer (8, 10, and 16 pounds [3.6, 4.5, and 7.3 kg])
- Maul
- Battering ram
- Pick
- Flat-head axe
- Mallet
- Hammer
- Punch
- Chisel

In some cases, a striking tool is the only tool required. In many other situations, the striking tool is used with another tool to gain entry. Striking tools can crush fingers, toes, and other body parts when dropped or used improperly. Poorly maintained striking surfaces may cause metal chips or splinters to fly into the air. Therefore, proper eye protection (safety glasses or goggles in addition to the helmet faceshield) must be worn when using striking tools **(Figure 11.24)**.

The **battering ram** is used to make openings in walls and force doors **(Figure 11.25)**. The battering ram weighs 30 to 40 pounds (13.6 to 18.1 kg) and is made of steel with installed handles and hand guards. One end is forked for breaking ordinary brick and concrete blocks, and the other end is rounded and smooth for battering doors and other types of walls. One to four firefighters holding the attached handles can swing the ram back and forth into a wall or door to create a breach.

Battering Ram — Solid steel bar with handles and guards, a fork on one end, and a blunt end on the other, used to break down doors or create holes in walls. The tool weighs 30 to 40 pounds (13.6 to 18.1 kg) and can be operated by one or more firefighters.

Figure 11.24 Eye protection is essential during striking operations to guard against shards from the item being hit and from possible damage to the striking surface.

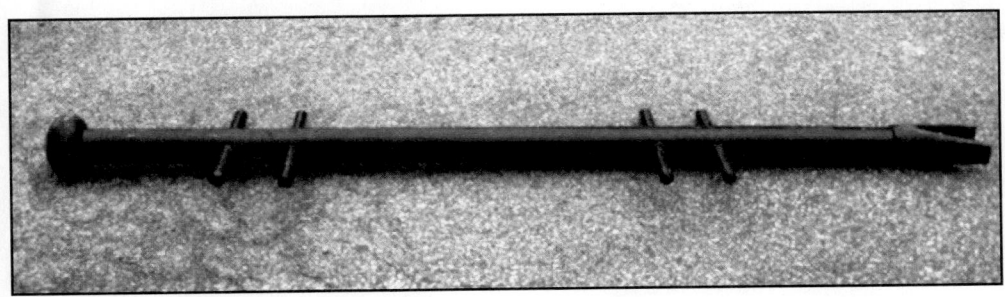

Figure 11.25 Battering rams have a forked end for breaking masonry and a rounded end for breaking doors and other types of walls.

Tools Used in Combination

There is no single forcible entry tool that provides the firefighter with the needed force or leverage to handle all forcible entry situations. In some cases, firefighters must combine two or more tools to accomplish a task. The types of tool combinations carried vary, depending on building construction, security concerns, tool availability, and other factors within a fire department and the area served. Two common tools used in combination for forcible entry are the flat-head axe and the Halligan tool, a combination of tools known as "irons" **(Figure 11.26)**.

The most important factor to consider is selecting the proper tools to do the job. Using tools in situations for which they are not designed can be extremely dangerous. Preincident surveys will help determine which tools will be required to force entry into a particular building or through a particular door, window, or wall in that building.

Tool Safety

Improper use of power and hand tools can result in strains, sprains, fractures, abrasions, and lacerations. To prevent these injuries you must:

- Wear full PPE including hand and eye protection.
- Use only undamaged tools.
- Select the right tool for the type of opening to be made.

Figure 11.26 A set of irons, composed of a flat-head axe and a Halligan tool, function as a unit and are nested together during transport.

- Use tools for their intended purpose only.

- Position yourself so that your weight is balanced on both feet.

- Ensure that you have room to operate the tool properly.

- Be aware that there will be a sudden release of energy when the door, window, or wall is opened.

- Ensure that other personnel are out of the immediate area.

- Be aware of the environment to prevent possible gas or vapor ignitions.

You must become familiar with all the tools you will use, which includes reading and following all the manufacturer's guidelines as well as your department SOPs on tool safety. When tools are not in use, they should be kept in properly designated places on the apparatus. Check the location of tools carried on the apparatus and make sure they are secured in their brackets. Repair or replace damaged tools immediately.

As with other tools, using prying tools incorrectly creates a safety hazard. If a job cannot be completed with a particular tool, do not strike the handle of the tool; use a larger tool. Also, do not use a prying tool as a striking tool unless it has been designed for that purpose.

Unacceptable Practice: Cheater Bars

A cheater bar is a piece of pipe slipped over the handle of a prying tool to lengthen it in order to provide more leverage. Use of a cheater exerts forces on the tool that are greater than the tool was designed to withstand. The additional force can cause the tool to slip, break, or shatter causing serious injury to a firefighter and/or damage to the tool. If a job requires a larger prying tool, use a larger tool.

Rotary saws, power saws, and chain saws must be used with extreme care to prevent injury as follows:

- Match the saw or saw blades to the task and material to be cut.

- Never force a saw beyond its design limitations.

- Always wear full PPE, including gloves, hearing protection, and eye protection.

- Fully inspect the saw before and after use.

- Do not use any power saw when working in a flammable atmosphere or near flammable liquids.

- Maintain situational awareness.

- Keep unprotected and nonessential people out of the work area.

- Follow the manufacturer's guidelines for proper saw operation.

- Keep blades and chains sharpened. A dull saw is more likely to cause an accident than a sharp one (**Figure 11.27**).

- Be aware of hidden hazards such as electrical wires, gas lines, and water lines.

- Remember that the rotating blade on a rotary saw continues to spin after the throttle has been released.

- Use only blades that are manufacturer approved for your saw; blades from different manufacturers may not be interchangeable.

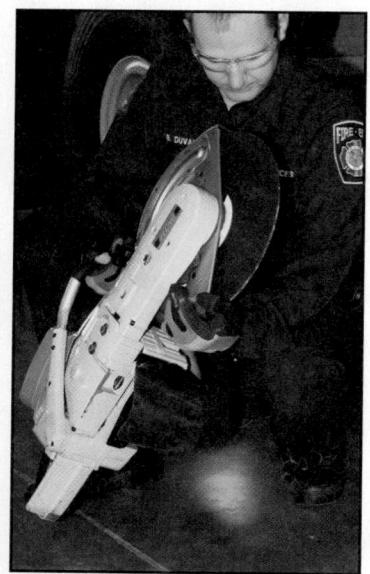

Figure 11.27 Maintenance and repair of blades is vital to ensuring that the tool will work when it is needed.

- Account for the twisting (gyroscopic or torsion effect) caused by the spinning blade of a rotary saw when making cuts in order to maintain control of the saw.
- Start all cuts at full revolutions per minute (rpm) to prevent the blades from binding into the material.
- Store blades in a clean, dry environment.
- Store composite blades in compartments where gasoline fumes will not accumulate because hydrocarbons can attack the bonding material in these blades causing them to deteriorate and violently shatter during use.

Carrying Tools

You must carry forcible entry tools in the safest manner possible. Always take care to protect yourself, other firefighters, and bystanders. Always lift with your legs and not your back when lifting heavy tools or other objects. Get help when transporting heavy tools. Some recommended safety practices for carrying specific tools are as follows:

- **Axes** — If not in a scabbard, carry the axe with the blade away from the body. With pick-head axes, grasp the pick with a hand to cover it **(Figure 11.28)**. Never carry an axe on the shoulder.
- **Prying tools** — Carry these tools with any pointed or sharp edges away from the body. This can be difficult when carrying tools with multiple cutting or prying surfaces such as a bit on one end and an adz on the other.
- **Combinations of tools** — Strap tool combinations together. Halligan tools and flat-head axes can be nested together and strapped.
- **Pike poles and hooks** — Carry these tools with the tool head down, close to the ground, and ahead of the body when outside a structure. When entering a building, carefully reposition the tool and carry it with the head upright close to the body to facilitate prompt use. These tools are especially dangerous because they are somewhat unwieldy and can severely injure anyone accidentally jabbed with the working end of the tool.
- **Striking tools** — Keep the heads of these tools close to the ground. Maintain a firm grip. Mauls and sledgehammers are heavy and may slip.
- **Power tools** — Never carry a power tool that is operating more than 10 feet (3 m); running power tools are potentially lethal weapons. Transport the tool to the area where the work will be performed and start it there. Carry the saw with the blade forward and toward the ground **(Figure 11.29, p. 594)**. To prevent fuel from leaking, ensure that the gas cap is tight and the gasket is in place.

Care and Maintenance of Forcible Entry Tools

Forcible entry tools will function as designed if they are properly maintained and kept in the best condition. Tool failure at an emergency incident can result in potentially harmful delays as well as severe injury or death. Always read the manufacturer's recommended maintenance guidelines for all tools, especially power tools. Follow your department's procedures to report any tools or equipment that need repairs. Tools that are damaged or excessively worn should be removed from service, tagged, and sent to the proper authority for repair or replacement.

Figure 11.28 A pick-head axe should be guarded along both sharp edges.

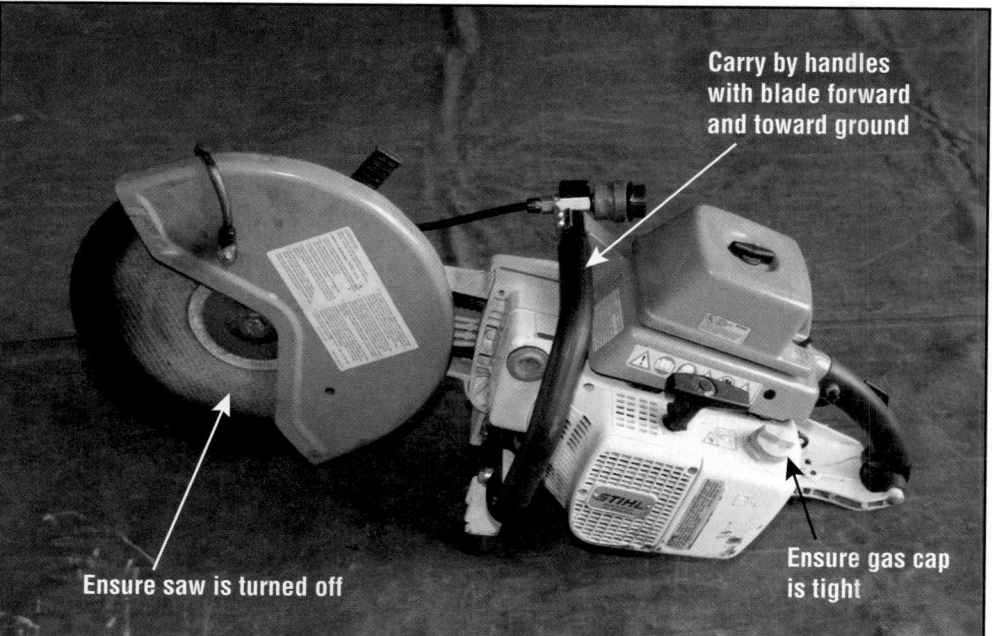

Figure 11.29 One firefighter can easily carry a power saw by the handles after first ensuring that the gas cap is tightly in place and the engine is turned off.

Carry by handles with blade forward and toward ground

Ensure saw is turned off

Ensure gas cap is tight

The sections that follow describe some basic maintenance procedures for various forcible entry tools. Refer to **Skill Sheet 11-I-1** for procedures for cleaning, inspecting, and maintaining hand tools; refer to **Skill Sheet 11-I-2** for procedures for cleaning, inspecting, and maintaining power tools.

Wooden Handles

Care and maintenance of wooden handles includes:

- Inspecting the handle for cracks, blisters, or splinters (**Figure 11.30**).
- Sanding the handle if necessary to eliminate splinters.
- Washing the handle with mild detergent, rinsing, and wiping dry. Do not soak the handle in water because it will cause the wood to swell.
- Applying a coat of boiled linseed oil to the handle to preserve it and prevent roughness and warping. Do not paint or varnish the handle.
- Checking the tightness of the tool head.
- Limiting the amount of surface area used for tool marking. Unit designations can be applied on strips of tape or self-adhesive bar codes on the handle.

Figure 11.30 Care and maintenance of wooden handles includes inspecting the entire surface for abnormalities and damage.

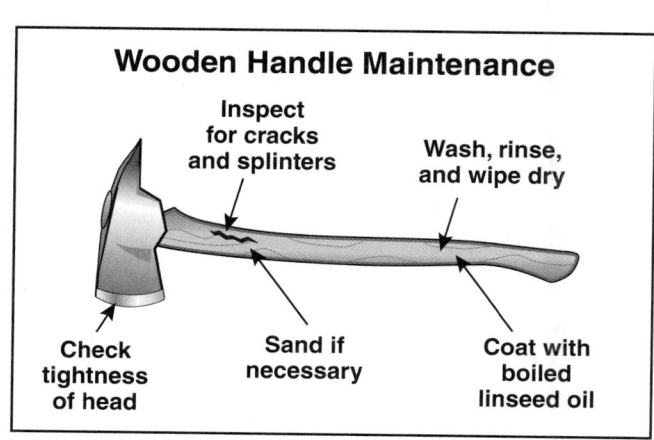

Wooden Handle Maintenance

Inspect for cracks and splinters

Wash, rinse, and wipe dry

Check tightness of head

Sand if necessary

Coat with boiled linseed oil

Fiberglass Handles

Fiberglass handles are easier to maintain than wood handles. Care includes:

- Washing the handle with mild detergent, rinsing, and wiping dry
- Checking for damage or cracks
- Checking the tightness of the tool head

Cutting Edges

Cutting edges on axes require the following care and maintenance:

- Inspecting the cutting edge for chips, cracks, or spurs.
- Replacing axe head when required.
- Filing the cutting edges by hand; grinding weakens the tool.
- Sharpening blade as specified in departmental SOPs. Some axe blades are intentionally left only semi-sharp to make them less prone to chipping.

How well an axe head is maintained directly affects how well it will perform. If the blade is extremely sharp and ground too thin, pieces of the blade may break when cutting gravel roofs or striking nails and/or screws in doors, walls, roof decking, or flooring. If the blade is too thick, regardless of its sharpness, it is difficult to drive the axe head through ordinary objects.

NOTE: Paint should never be applied to the cutting surface of an axe head; this may cause the cutting surface to stick and bind.

Plated Surfaces

Plated surfaces are protected by chromium or another metal applied by an electroplating process. Either wipe plated surfaces clean or wash them using mild detergent and water, rinse, and wipe dry. Inspect these surfaces for damage.

Unprotected Metal Surfaces

Unprotected metal surfaces are the blades, wedges, pikes, handles, and other tool components that have not been electroplated to protect them from rust or corrosion. Instructions for care are as follows:

- Remove dirt and rust with an emery cloth or steel wool.
- Remove burrs from the cutting edge and body with a metal file.
- Do not make the blade edge too sharp; this may cause the blade to chip or break.
- Do not use a mechanical grinder to sharpen the blade edge because it may cause a loss of temper through overheating.
- Oil the metal surface lightly **(Figure 11.31, p. 596)**. Light machine oil works best. Avoid using any metal protectant that contains 1-1-1-trichloroethane. This chemical may damage and weaken the handle.
- Do not apply oil to the striking surface of tools.
- Do not paint metal surfaces — paint hides defects.
- Inspect the metal for chips, cracks, burrs, or sharp edges, and file them off when found.

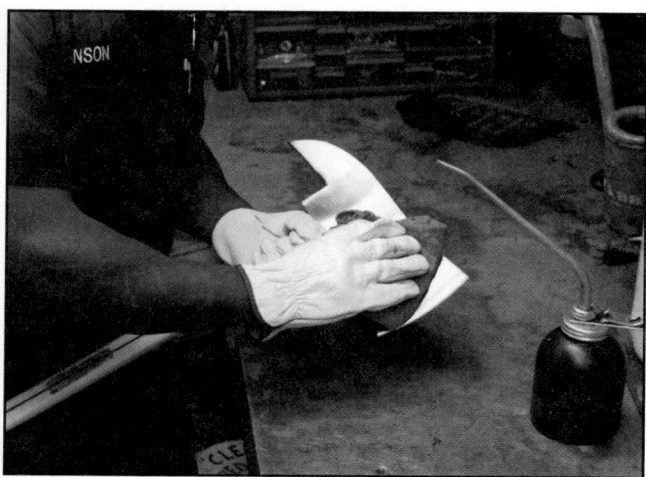

Figure 11.31 A light coat of oil prevents rust and corrosion on metal surfaces.

Figure 11.32 Electrical cord is intrinsically safe only if all of its sheathing and insulation are intact.

Power Equipment

Each power tool will have its own set of instructions for care and maintenance that should be followed. Even minor damage like a loose-fitting filler cap can create potential hazards. General care and maintenance suggestions include:

- Reading and following the manufacturer's instructions.

- Ensuring that rechargeable battery packs are fully charged and ready for immediate use.

- Inspecting power tools periodically and ensuring they will start manually.

- Checking blades for damage or wear.

- Replacing blades that are damaged or worn.

- Checking all electrical components (cords, etc.) for cuts or other damage (**Figure 11.32**).

- Ensuring that the grounding prong has not been removed from three-prong plugs.

- Ensuring that all guards are functional and in place.

- Ensuring that fuel, engine oil, and hydraulic fluid are fresh and at the proper level.

- Checking the condition of all hydraulic hoses and connections.

Forcing Entry Through Doors

Forcing entry through doors is the most conventional method used in the fire service. Once a firefighter has sized up a door, forcible entry can be performed if necessary. Determining the type of forcible entry to use depends on knowledge of door construction, which was presented in Chapter 4, Building Construction, and locks described in this chapter.

Remove Hinge Pins

Force the Door

Force the Lock

Pry Door from Jamb

Figure 11.33 After determining that a door must be forced, use the clues indicated by the door's features to choose the best point of forcible entry.

You should consider the amount of damage that will result from forcible entry. The decision-making process should begin with the minimum of damage and proceed to the maximum amount. Start by trying the door to see if it can be opened in the normal fashion. If that does not work, look for a lockbox that may contain a key or other means of opening the door. You should also look for a door window or sidelight panel that might provide access to the lock on the interior by breaking the glass and activating the interior lock.

If it is obvious that the door will need to be forced, determine if it is quicker to force the lock, remove hinge pins, force the door, or pry the door from the jamb (**Figure 11.33**). The primary factors determining the amount of damage that can be justified are the severity of the emergency and the speed with which entry must be gained.

Rapid-entry lockbox systems provide a means to open locked doors without forcing entry. In areas where these systems are installed, all necessary keys or numeric keypad combinations to unlock the building, storage areas, gates, and elevators are kept in a lockbox mounted at a high-visibility location on the building's exterior. Unauthorized duplication of the master key is prevented because the special key blanks are not available to locksmiths and cannot be duplicated with conventional equipment. Only the fire department possesses a master key that opens all boxes in its jurisdiction.

Maintaining Door Control

During forcible entry, maintaining control of the door is critical. Opening a door changes the ventilation profile and may have an adverse effect on fire behavior. If the door will be damaged to the point that it cannot be closed, steps should be taken to deal with subsequent changes in fire behavior. There may be times you want to block a door open to prevent it from closing against a hoseline or blocking a means of egress.

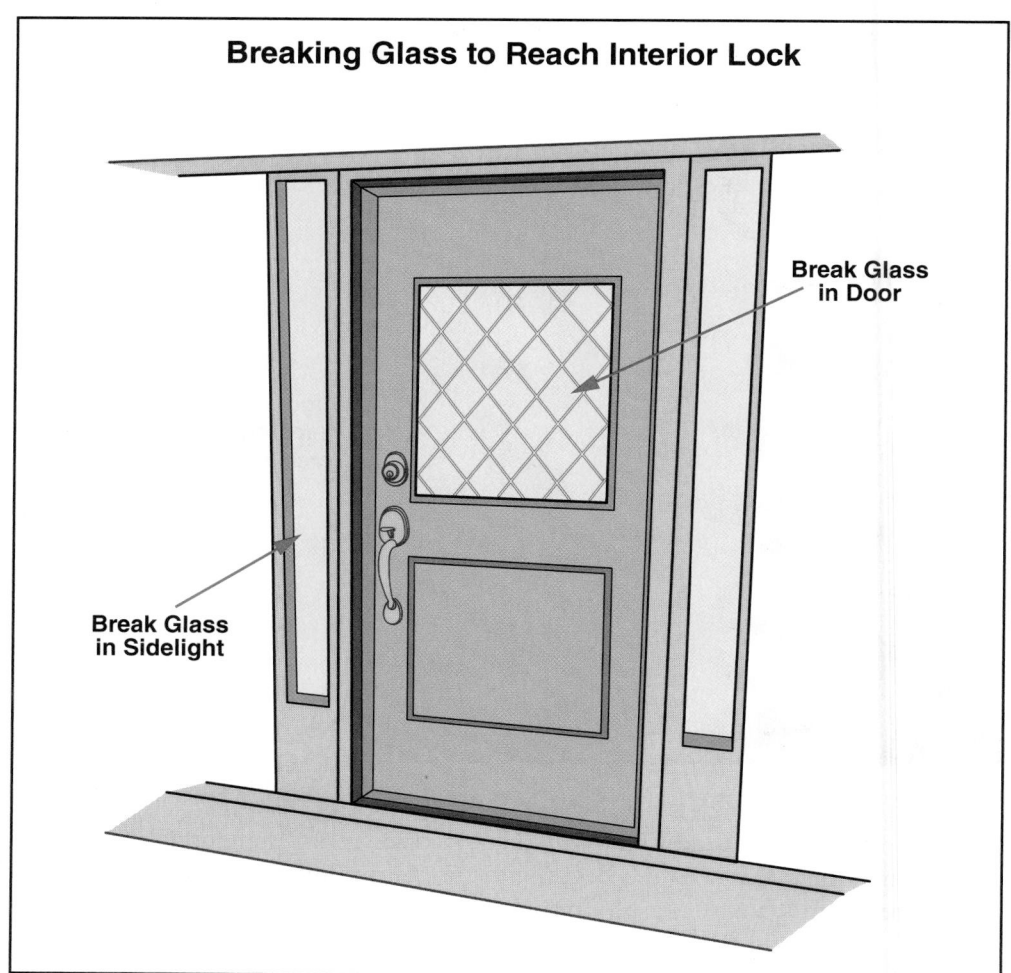

Figure 11.34 Breaking a window next to a door may allow the door to be opened from the inside, preventing damage to the door.

Breaking Door Glass

One of the fastest and least destructive techniques for forcing locked doors is to break the glass in the door or the sidelight next to it **(Figure 11.34)**. Once the glass is broken, the firefighter can reach inside and unlock the door. In some situations, breaking the glass (or what appears to be glass) may be more difficult and costly. For example, tempered glass is very expensive, and Plexiglas™ and Lexan® may resist being broken with conventional hand tools.

Ordinary window glass will shatter into sharp fragments when broken. Firefighters should wear full protective equipment, especially hand and eye protection, to prevent injuries from shattered glass. If firefighters are breaking the glass to gain access into a burning building, they should wear SCBA and have a charged hoseline in place, ready to attack the fire. The techniques used for breaking both door glass and window glass are similar.

Forcing Swinging Doors

The most common type of door is one that swings at least 90 degrees to open and close. Most swinging doors have hinges mounted on one side that permit them to swing in one or both directions; others swing on pivot pins at the top and bottom

Figure 11.35 A single firefighter can open a standard swinging door using a rambar.

of the door. Swinging doors can be either inward- or outward-swinging, or both. Double-acting swinging doors are capable of swinging 180 degrees. An easy way to recognize which way a door swings is to look for the hinges. If you can see the hinges of the door, it swings toward you. If you cannot see the hinges, the door swings away from you. Forcing entry through all types of swinging doors involves basic skills, but requires practice to master.

Inward-Swinging Doors

Forcible entry of single inward-swinging doors requires either one or two skilled firefighters, depending upon the tool or tools used. For example, a single firefighter using a rambar can open most standard swinging doors; however, if a Halligan tool and flat-head axe are used, two firefighters are required **(Figure 11.35)**. **Skill Sheet 11-I-3** describes one technique for forcing a single inward-swinging door.

If the swinging door is metal or metal-clad in a metal frame set in a concrete or masonry wall, other forcible entry techniques may be needed. In some cases, a hydraulic door opener (often called a **rabbit tool**) can be used to force the door open. In other cases, doors resist being pried open so it is necessary to cut around the lock in either of two ways. First, you can use a rotary saw with a metal cutting blade to make two intersecting cuts that isolate the locking mechanism and allow the door to swing. A second method using three intersecting cuts can also produce the same result. **Skill Sheet 11-I-4** describes one technique for cutting the lock out of a single inward-swinging metal door.

Rabbit Tool — Hydraulic spreading tool that is specially designed to open doors that swing inward.

Figure 11.36 Outward-swinging doors may have their hinges removed using a rotary saw or cutting torch.

Figure 11.37 A saw blade can be used to cut a dead bolt through the space between doors.

Outward-Swinging Doors

Sometimes called flush fitting doors, outward-swinging doors present a different set of problems for firefighters. Because the hinges on outward-swinging doors are mounted on the outside, it is often possible to use a nail set and hammer to drive the hinge pins out of the hinges and simply remove the door. If the bottoms of the hinges are solid so that the pins cannot be driven out, it may be possible to break the hinges off with a rambar or Halligan. It may be possible to cut the hinges off with a rotary saw or a cutting torch (**Figure 11.36**).

You can also insert the blade of a rambar or Halligan into the space between the door and the doorjamb and pry that space open wide enough to allow the lock bolt to slip from its keeper. **Skill Sheet 11-I-5** describes one procedure for removing the hinge pins from an outward-swinging door. **Skill Sheet 11-I-6** describes one procedure for forcing the door and doorjamb apart on an outward-swinging door using a rambar or Halligan tool.

Double-Swinging Doors

Double-swinging doors can present a problem depending on how they are secured. If they are secured only by a mortise lock, the doors can be pried apart far enough to let the bolt slip past the receiver. By inserting the blade of a rambar or wedge end of a Halligan-type tool between the doors, they can often be pried apart far enough to allow the bolt to clear the receiver. The blade of a rotary saw equipped with the proper blade can be inserted into the space between the doors to cut the dead bolt (**Figure 11.37**). Some double doors have a security molding or weather strip over the space between the doors. This molding must be removed or a section cut away to allow the blade of the forcible entry tool to be inserted.

Doors with Drop Bars

If a single- or double-swinging door is locked with a drop bar, try one of the following methods to force entry:

- Use a rambar or Halligan tool to spread the space between the double doors. Then insert the blade of a handsaw or other narrow tool into the opening and lift the bar up and out of the stirrups.

- Use a rotary saw to cut the exposed bolt heads that are holding the stirrups on the outside of the door. This will allow the drop-bar to fall away and the door to be opened.

- Insert the blade of a rotary saw into the space between the halves of double doors and cut the security bar.

Tempered Plate Glass Doors

Tempered Plate Glass — Type of glass specially treated to become harder and more break-resistant than plate glass or a single sheet of laminated glass.

In commercial, light industrial, and institutional occupancies, firefighters may be faced with metal-frame doors with **tempered plate glass** panels. These doors are heavy and very expensive. Tempered glass mounted in a metal door frame can be very difficult to break. Unlike regular plate glass, tempered glass resists heat; when broken, it shatters into thousands of tiny cube-like pieces.

Figure 11.38 Using a salvage cover to shield responders from breaking glass is one method of breaching a tempered plate glass door.

If it becomes necessary to break through a tempered plate glass door, strike the glass at a bottom corner with the pick end of a pick-head axe. The firefighter should wear complete PPE including a helmet-mounted faceshield or goggles to protect against eye injury. Some departments place a shield made from a salvage cover as close to the glass as possible, and the blow is struck through the cover **(Figure 11.38)**. Any remaining glass can then be scraped from the frame. Tempered plate glass doors should be broken only as a last resort. Firefighters can also use the through-the-lock method to open tempered plate glass doors as well as other doors.

Forcing Sliding Doors

Sliding doors consist of two glass panels mounted in wood, aluminum, or vinyl-clad material. One panel is stationary while the other slides in a track on small guide wheels or rollers. Sliding doors, sometimes called patio doors, are generally found in single-family residential structures and apartment buildings. Locking devices include latches on the inside of the door and security bars placed in the track. Forcible entry techniques include breaking the glass with an axe or lifting the sliding panel up and out of its track. Attempting to spread the door from the frame will result in the door shattering uncontrollably.

Figure 11.39 A rotary saw is uniquely adept at cutting sheet metal applications including overhead doors.

Figure 11.40 The through-the-lock method of forcible entry may be used after trying conventional methods.

A second type of sliding door is the interior pocket door. It may consist of one -or two-door panels that slide into the adjacent wall. Pocket doors can be forced using the same techniques used to force a swinging door.

Forcing Security Doors and Gates

Security doors and gates will be encountered in a number of forms including:

- Rollup doors, both manual and power operated
- Doors with open steel bars
- Doors that consist of multiple slats that can be closed to form a solid panel

Any type of security door or gate will delay entry and will require planning to determine the most efficient form of entry. Some doors may have padlocks on the outside while others may have locks on the inside. If the lock is inside, there is usually a second means of entry into the structure. Forcible entry techniques include cutting the padlock off, using a rotary saw to make an opening near the lock, or cutting out a section of the panel as in an overhead door **(Figure 11.39)**. The best approach is to know the specific types of security doors and gates in your response area and practice the correct procedure for each.

Through-the-Lock Forcible Entry

The through-the-lock method is preferred for many commercial doors, residential security locks, padlocks, and high-security doors. This technique is very effective and does minimal damage to the door when performed correctly.

Through-the-lock forcible entry requires a good size-up of both the door and the lock mechanism. If the door does not open with conventional forcible entry methods, the through-the-lock entry method can be used (**Figure 11.40**).

On some door locks, the lock cylinder can actually be unscrewed from the door. Storefront doors often have locks that can be unscrewed because it makes it easier for locksmiths to rekey the locks when occupancy changes. If the lock is not protected by a collar or shield, use the procedure described in **Skill Sheet 11-I-7**.

Removing the lock cylinder is only half the job. A key tool must then be inserted to open the lock in the same fashion as inserting a key. The key tool is usually flat steel with a bend on the cam end and a flat screwdriver shaped blade on the other.

Along with standard forcible entry striking and prying tools, special tools may also be needed for this forcible entry technique. Some examples of these special tools are the K-tool, A-tool, J-tool, and shove knife.

K-Tool

The K-tool is useful in pulling all types of lock cylinders (**Figure 11.41**). Used with a Halligan-type tool or other prying tool, the K-tool is forced behind the ring and face of the cylinder until the wedging blades bite into the cylinder (**Figure 11.42**). A metal loop on the front of the K-tool provides a slot in which to insert one end of the prying tool. A firefighter then strikes the top of prying tool with a flat-head axe or other striking tool to set the K-tool. Once set, the prying tool is used to pull the K-tool and the lock cylinder from the door.

When a lock cylinder is located close to the threshold or jamb, the narrow blade side of the K-tool will usually still fit behind the ring. Some sliding glass doors have limited clearance, but only a ½-inch (13 mm) clearance is needed. Once the cylinder is removed, a key tool can be inserted into the hole to move the locking bolt to the open position. **Skill Sheet 11-I-8** describes the method for using the K-tool.

Figure 11.41 The K-tool is designed to pull all types of lock cylinders.

Figure 11.42 A K-tool slides behind the ring and face of a lock cylinder and is then used with a Halligan tool and flat-head axe to remove the lock.

Figure 11.43 An A-tool is a prying tool with cutting edges and may cause more damage to a door than a K-tool, but it is useful against a specific set of protected locks.

A-Tool

In many cases, the A-tool can rapidly accomplish the same job as the K-tool, but it can cause slightly more damage to the door. The A-tool was developed to force entry on locks manufactured with collars or protective cone-shaped covers over them to prevent the lock cylinder from being unscrewed. The A-tool is a prying tool with a sharp notch with cutting edges machined into it **(Figure 11.43)**. The notch resembles the letter *A*.

The A-tool is designed to cut behind the protective collar of a lock cylinder and maintain a hold so that the lock cylinder can be pried out. The curved head and long handle are then used to provide the leverage for pulling the cylinder. The chisel head on the other end of the tool is used when necessary to gouge out the wood around the cylinder for a better bite of the working head. When pulling protected dead bolt lock cylinders and collared or tubular locks, use the A-tool and the procedure described in **Skill Sheet 11-I-9**.

J-Tool

The J-tool is a device made of rigid, heavy gauge wire designed to fit through the space between double-swinging doors equipped with panic hardware. The J-tool is inserted between the doors far enough to allow the tool to be rotated 90 degrees in either direction. A firefighter can then pull the tool until it makes contact with the panic hardware. The firefighter then makes another sharp pull, and the tool should operate the panic hardware and allow the door to open **(Figure 11.44)**.

Shove Knife

This flat steel tool, resembling a wide-bladed putty knife with a notch cut in one edge of the blade, can provide firefighters rapid access to outward swinging latch-type doors. When used properly, the blade of the tool depresses the latch, which allows the door to open.

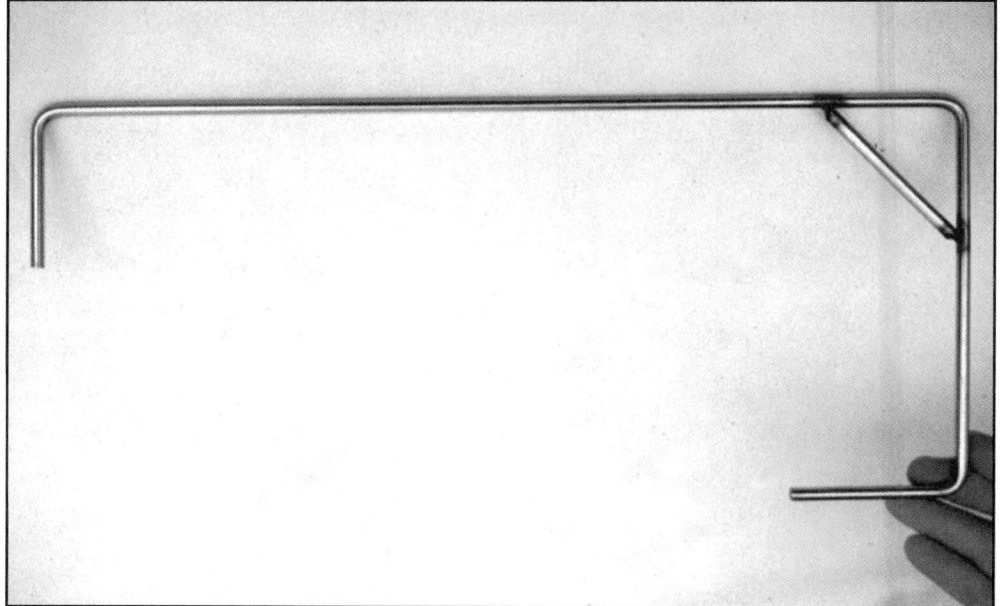

Figure 11.44 A J-tool fits between double swinging doors equipped with panic hardware and is used to activate the door mechanism from the outside.

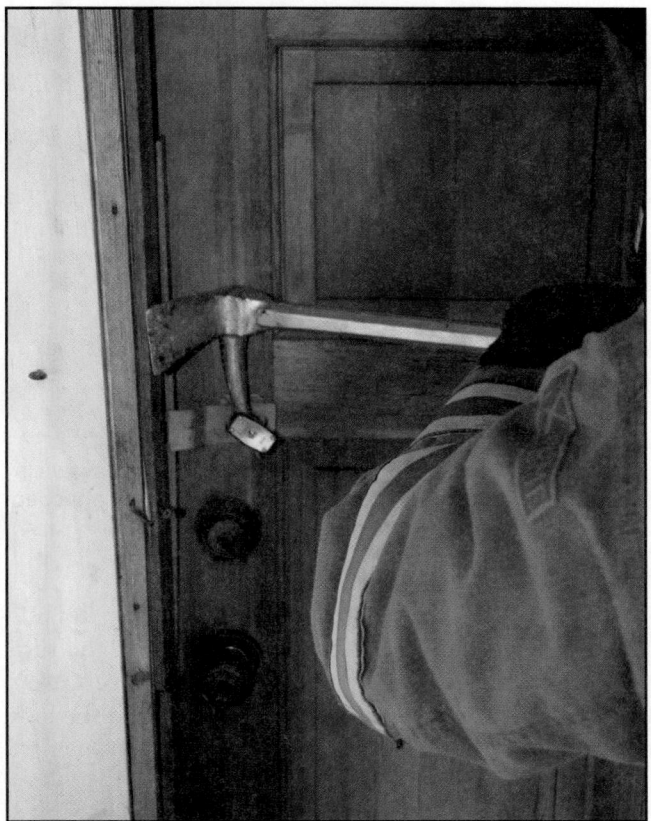

Figure 11.45 Hand tools may be used to force a padlock.

Figure 11.46 A duck-bill inserted into a padlock may break the shackles after sharp force is administered with a maul or flat-head axe.

Forcing Padlocks

Firefighters must be capable of forcing either the padlock itself or the device to which it is fastened. Conventional forcible entry tools can be used to break a padlock or detach the hasp to gain access (**Figure 11.45**). Additional tools are available to make forcible entry through padlocks easier including the following:

- Duck-billed lock breaker
- Hammerheaded pick
- Locking pliers and chain
- Hockey puck lock breaker
- Bam-bam tool

Skill Sheet 11-I-10 describes the process for opening a small lock with a non-case-hardened shackle of ¼ inch (6 mm) or less. If these techniques fail to break the padlock, it is often easier to break the hasp or detach it from the door frame.

If the shackle of the padlock exceeds ¼ inch (6 mm) and the lock, including the body, is case-hardened, the firefighter faces a more difficult forcible entry task. You may need to use either the duck-billed lock breaker or the bam-bam tool as follows:

- **Duck-billed lock breaker** — Wedge-shaped tool that will widen and break the shackles of padlocks, much like using the hook of a Halligan-type tool. Firefighters can insert this tool into the lock shackle and strike the tool with a maul or flat-head axe until the padlock shackles break (**Figure 11.46**).

Figure 11.47 A bam-bam tool screws into a padlock and then pulls the lock tumbler out of the padlock body.

Figure 11.48 Padlocks may be cut using a rotary saw while another firefighter stabilizes the lock using a chain attached with a set of locking pliers.

- **Bam-bam tool** — Uses a case-hardened screw that is screwed into the keyway of the padlock. Once the screw is firmly set, a few firm, quick pulls on the sliding hammer will pull the lock tumbler out of the padlock body (**Figure 11.47**). Firefighters can then insert the flat end of a key tool or a screwdriver into the lock to trip the lock mechanism. **Skill Sheet 11-I-11** describes the process for using a bam-bam tool.

NOTE: This method will not work on Master Locks, American Locks, and other high-security locks. These locks have a case-hardened retaining ring in the lock body that prevents the lock cylinder from being removed.

Using a rotary saw with a metal-cutting blade or a cutting torch may be the quickest method for removing some padlocks. High-security padlocks are designed with heel and toe shackles. Heel and toe shackles will not pivot if only one side of the shackle is cut. Cutting padlocks with a power saw or torch can be somewhat dangerous. One firefighter should stabilize the lock with a set of locking pliers and chain and pull the lock straight away from the hasp. A second firefighter cuts both sides of the padlock shackle with the saw or torch (**Figure 11.48**). **Skill Sheet 11-I-12** describes the process for cutting a padlock with a rotary saw.

Forcing Overhead Doors

When overhead doors must be forced open, it is best to use a rotary saw to cut a square or rectangular opening about 6 feet (1.8 m) high and nearly the full width of the door (**Figures 11.49a-c**). Once firefighters have access to the interior, they should then use the lift mechanism to open the door fully. Cribbing or shoring blocks can be used to prevent overhead doors from closing unintentionally. A pair of vice grips (locking pliers) can be attached to the overhead door rail above head height to keep the door from closing.

> **WARNING**
> All overhead doors should be blocked in the up or open position to prevent injury to firefighters if the built-in control device fails.

Figure 11.49a An overhead door may be opened using a rotary saw to cut an opening.

Figure 11.49b After entering through the opening, a team of firefighters should use the lift mechanism to open the door fully.

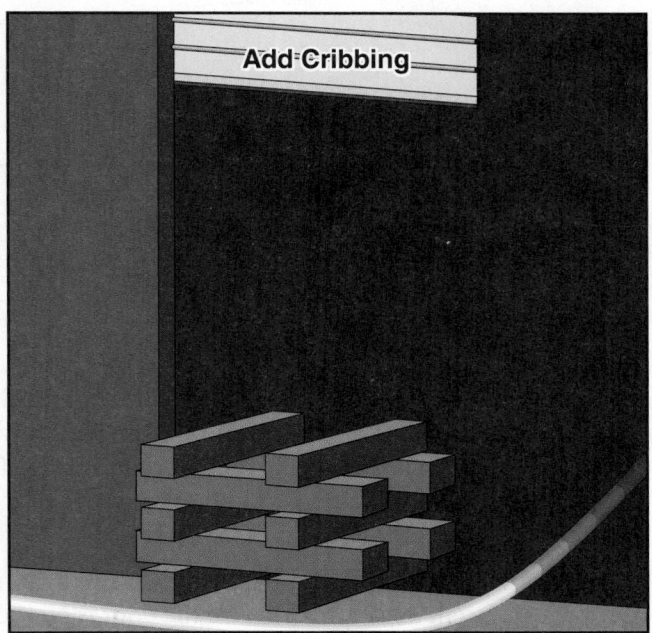

Figure 11.49c A damaged overhead door must be secured in place using cribbing or a pair of vice grips.

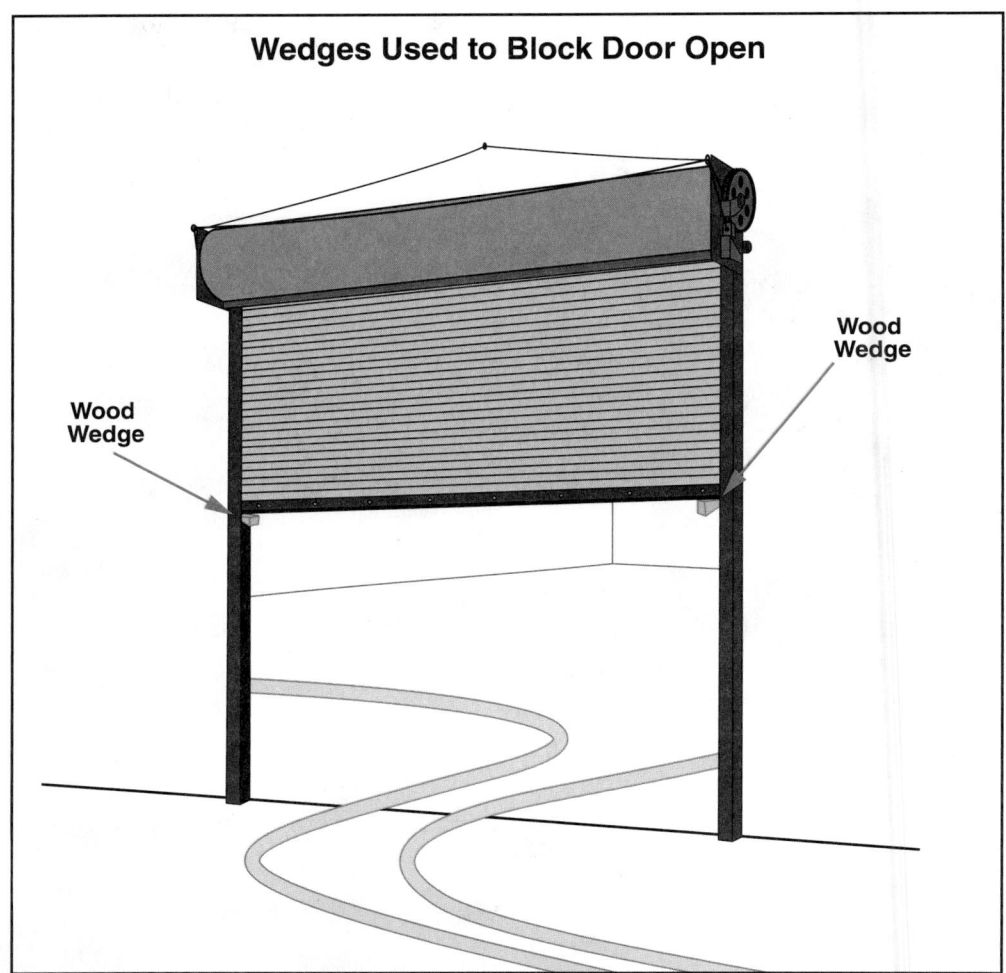

Wedges Used to Block Door Open

Wood
Wedge

Wood
Wedge

Figure 11.50 Wedges keep a fire door from closing so fire hoses may be extended through the opening.

Forcing Fire Doors

Fire doors are movable assemblies designed to cover doorway openings in rated separation walls in the event of a fire in one part of a building. Types of standard fire doors include horizontal and vertical sliding, single and double swinging, and overhead rolling. Fire doors are normally found on the inside of structures, separating one area from another, enclosing a hazardous process or storage area, or protecting a means of egress (exit path such as a protected stairway).

Generally, exterior fire doors are only found where a structure must be protected from an adjacent exposure. Because the door is on the outside of the structure, it will probably be locked. Forcible entry should be similar to any other overhead or sliding door.

Interior fire doors will probably have been manually or automatically activated when the fire was detected. Because these doors only operate when there is a fire, they will not lock in place when closed. A precautionary measure that should be taken when passing through an opening protected by a fire door is to block the door open to prevent it from closing and blocking a means of egress (**Figure 11.50**). Fire doors have also been known to close behind fire attack crews and cut off the water supply in their hoselines.

Figure 11.51 Wire glass is reinforced and must be chopped from the frame.

Forcing Entry Through Windows

Windows are sometimes easier to force than doors even though they are not the best entry point into a burning building. Entry can be made through a window to open a locked door from the inside (**Figure 11.51**). As with doors, size-up of windows is critical to a successful forced entry. Forcing open the wrong window may also disrupt ventilation efforts, intensify fire growth, and draw fire to uninvolved sections of the building.

Breaking Window Glass

Breaking the glass is the most common technique, but it creates certain hazards and obstacles:

- Slows entry into the structure while firefighters clear the glass shards from the frame.

- Creates flying glass shards that may travel great distances from windows on upper floors. Covers floors in glass shards which can make footing treacherous for firefighters advancing charged hoselines.

- Could shower glass on victims inside the structure causing additional injury.

NOTE: When using a pike pole to break a window, position yourself upwind and higher than the window so that falling glass will not slide down the handle toward you.

When windows are broken for entry they can contribute to the spread of fire. To limit the effect of wind entering through openings, wet canvas tarps or fire retardant tarps can be used to cover these openings.

Wire glass is more difficult to break and remove than ordinary window glass because the wire prevents the glass from shattering and falling out of the frame. Use a sharp tool, such as the pick of an axe, to chop wire glass out of its frame

Because windows containing two and three layers of glass are expensive, firefighters must decide if the benefits of breaking the window outweigh the expense of replacing the windows. Multi pane windows are also time-consuming to remove because the glass is held in place by a rubber cement that makes shard removal difficult. General techniques for forcing entry through windows with glass panes are described in **Skill Sheet 11-I-13**.

NOTE: Chapter 4, Building Construction, provides information on the various window types, functions, and materials that firefighters may encounter.

Figure 11.52 Glass block windows are resilient and difficult to force.

Figure 11.53 Double-hung windows have two sashes secured in the middle with thumb-operated locking devices.

Forcing Fixed Windows

A fixed window may be found in single-family residences, mercantile occupancies, and office buildings. Fixed windows may consist of a relatively large solid glass pane, multiple panels, or individual glass blocks formed into a wall. These types of windows may be broken in an emergency as a last resort. Because fixed single-pane windows cover large expanses, breaking them will allow a great deal of air into or out of the building and may seriously affect ventilation efforts. In addition, the time it takes to safely open a fixed window may make it ineffective when other forms of access are available.

Forcing glass block windows or walls should also be performed only as a last resort. These walls are 2 to 4 inches (50.8 mm to 101.6 mm) thick and may be individual blocks held together with mortar or vinyl strips or manufactured panels of blocks up to 47 inches (1 193.8 mm) square **(Figure 11.52)**. These windows may require a sledge hammer or battering ram to break.

Forcing Double-Hung Windows

The double-hung (checkrail) windows are found in residential structures, small office buildings, manufactured houses, and older educational buildings. Manufactured in wood, metal, or vinyl, these windows are made up of two sashes. The top and bottom sashes are fitted into the window frame and operate by sliding up or down. Double-hung windows may contain ordinary glass (single-, double-, or triple-pane), wire glass, Plexiglas™ acrylic plastic, or Lexan® plastic.

In most cases, one or two thumb-operated locking devices located where the horizontal frame members of the top and bottom sashes meet are used to secure double-hung windows (**Figure 11.53**). Surface-mounted window bolts may also be used to fasten the windows more securely.

Forcible entry techniques for double-hung windows depend on how the window is locked and the material of which the sash frames are made. The general technique for forcing a double-hung wood window is described in **Skill Sheet 11-I-14**.

Metal-frame windows are more difficult to pry. The lock mechanism will not pull out of the sash and may jam, creating additional problems. Use the same technique given for a wood-frame window, but if the lock does not yield with a minimal amount of pressure, it may be quicker to break the glass and open the lock manually.

Forcing Single-Hung Windows

Single-hung windows are identical to double-hung windows with the exception that only the bottom panel moves. Locks and locking devices are the same as those found on the double-hung windows. Forcible entry procedures are also the same as those used on the double-hung window.

Forcing Casement Windows

Casement windows are hinged windows with wooden or metal frames. This type of window is sometimes called a crank out window, because the window is opened with a small hand crank. Casement windows consist of one or two sashes mounted on side hinges that swing outward, away from the structure, when the window crank assembly is operated.

Locking devices vary for the casement window from simple thumb-operated devices to latch-type mechanisms. Even if the locking mechanisms are open, the casement window can only be opened using the crank mechanism on the inside (**Figure 11.54**). Single-casement windows can have one or more locking devices and a single crank while double-casement windows can have at least four locking devices as well as two crank devices.

To force open a single-casement window, break the lowest pane of glass and clear shards from the frame. Cut open the window screen behind in the same area. Reach in and unlock the locking mechanism and then operate the crank to open the window frame. Finally remove the screen completely. If the window has one full pane of glass, break it, clear the shards from the frame, and remove the screen. It will be unnecessary to crank it open once the glass is removed.

NOTE: Casement windows should not be confused with awning or jalousie windows (see Awning and Jalousie Windows section), some of which are operated with a hand crank.

Figure 11.54 Casement windows can only be opened using the window crank.

Forcing Horizontal Sliding Windows

Similar to the sliding door, the horizontal window is made with a fixed panel and a sliding panel. The technique for forcing entry through a horizontal sliding window is the same as that used with the sliding door.

Figure 11.55 Jalousie windows consist of a series of narrow glass sections that span the width of the window opening.

Forcing Awning Windows

Awning windows consist of large sections of glass about 1 foot (300 mm) high and as long as the window width. They are constructed with a metal or wood frame around the glass panels, which are usually double-strength glass. Awning windows are hinged along the top rail, and the bottom rail swings out by unlatching and pushing or using the mechanical window crank. The hopper window is similar to the awning except that it hinges at the bottom and opens at the top. Hopper windows are normally used for interior ventilation above a door or window. Forcible entry through awning windows requires either breaking the glass or prying the window up from the frame.

Forcing Jalousie Windows

Jalousie (louvered) windows consist of small sections about 4 inches (100 mm) high and the width of the window (**Figure 11.55**). The individual glass panes are held in the movable frame only at the ends. The operating crank and gear housing are located at the bottom of the window.

Entry through these windows requires the removal of several panes. Because most awning and jalousie windows are relatively small, they offer very restricted access even when all of the glass is removed. As an alternative, if entry must be made through a jalousie window, it may be faster and more efficient to cut through the wall around the entire window assembly and remove it.

Forcing Projecting Windows

Projecting windows are found in factories, warehouses, and other commercial and industrial buildings. These windows often have metal sashes with wire glass and function by pivoting on hinges at the upper corners of the panel. As the window pivots out, it slides down a track allowing an opening at the bottom and top of the window. Forcing entry through projecting windows may be limited to breaking the glass or cutting the window panel out of the frame.

Forcing Pivoting Windows

Similar to the projecting window, the pivoting window has the hinge pins in the middle of the window permitting an equal opening at top and bottom. Latches will be located at the bottom of the window. Forcing entry through these windows follows the same procedure as the projecting window.

Figure 11.56 Hurricane shutters have been adapted to houses in nonhurricane climates as a theft prevention measure. *Courtesy of the Florida State Fire College.*

Forcing Miscellaneous Types of Windows and Covers

The following windows and window openings feature various barriers installed in order to make them either more secure or more structurally sound:

- Hurricane windows and shutters
- High-security windows
- Housing and Urban Development (HUD) windows
- Windows with vacant property systems (VPS)
- Windows with security bars and grilles

Hurricane Windows and Shutters

Windows designed to resist hurricane-force winds use laminated glass combined with an advanced polymer. An ionoplast layer is sandwiched between two layers of glass resulting in a laminated glass that is 100 times as rigid and five times as tear resistant as commonly used high-impact glass. Identifying these windows during preincident planning will help in the selection of the most effective tools and techniques that will be needed when forced entry is required. An axe or the adz end of a Halligan tool can be used to break a hurricane window, similar to breaking the laminated glass in a vehicle windshield. The aluminum window frame may also be cut to remove the window. Both approaches are labor intensive and time consuming.

Hurricane shutters are exterior coverings mounted over windows and patio doors to protect the structure from hurricane-force winds. They may be permanently mounted, such as accordion or roll-up styles, or panels that are installed when a hurricane warning is issued. However, they are no longer found exclusively in hurricane-prone areas; increasingly they are being installed to prevent burglars from gaining entry to a property **(Figure 11.56)**. In many instances these shutters are relatively inconspicuous and blend into the architecture of the house. Remove hurricane shutters using a rotary saw with an aluminum oxide blade. Alternatively, break the lag bolts holding the rail with the adz of a Halligan tool.

High-Security Windows

Similar to hurricane windows, high-security windows have break-resistant plastic panes instead of glass. To be most effective when forcing entry through high-security windows, firefighters must have identified these barriers during preincident planning.

Figure 11.57 Beginning a cut with the saw at full speed will allow the firefighter greater control of the cut and the tool.

Lexan® is one of the plastics used in this application. Lexan® is 250 times stronger than safety glass, 30 times stronger than acrylic, and classified as self-extinguishing. Lexan® is virtually impossible to break with conventional forcible entry hand tools. Polycarbonate windows can be identified several ways. Tapping them gently with a tool produces a dull plastic sound different from glass. Lexan® windows will scratch much easier than glass. Large polycarbonate windows have a wavy surface and some obvious distortion along the sides.

Firefighters can use either a rotary saw or a striking tool to force entry through a Lexan® window. Techniques for forcing entry through Lexan® windows using a rotary saw are given in **Skill Sheet 11-I-15**. Use the following guidelines to force Lexan® using a rotary saw:

- Use a carbide-tipped, medium-toothed blade (approximately 40 teeth). Large-toothed blades will skid off the surface, and smaller toothed blades will melt the Lexan® and cause the blade to bind. If a chain saw is used, it must be equipped with a carbide-tipped cutting chain.

- Start all cuts at full rpm to avoid bounce and vibration of the saw (**Figure 11.57**).

- Wear full PPE including goggles or other eye protection to reduce injury from chips and shards.

- Cut as rapidly as possible without forcing the saw.

- Make the horizontal cuts first, then the vertical cuts.

Using striking or impact tools, such as sledgehammers, axes, rambars, or Halligans, can be effective if the entire pane can be removed through the frame. In many installations, however, the Lexan® pane is bolted or riveted to the frame to prevent punch-through.

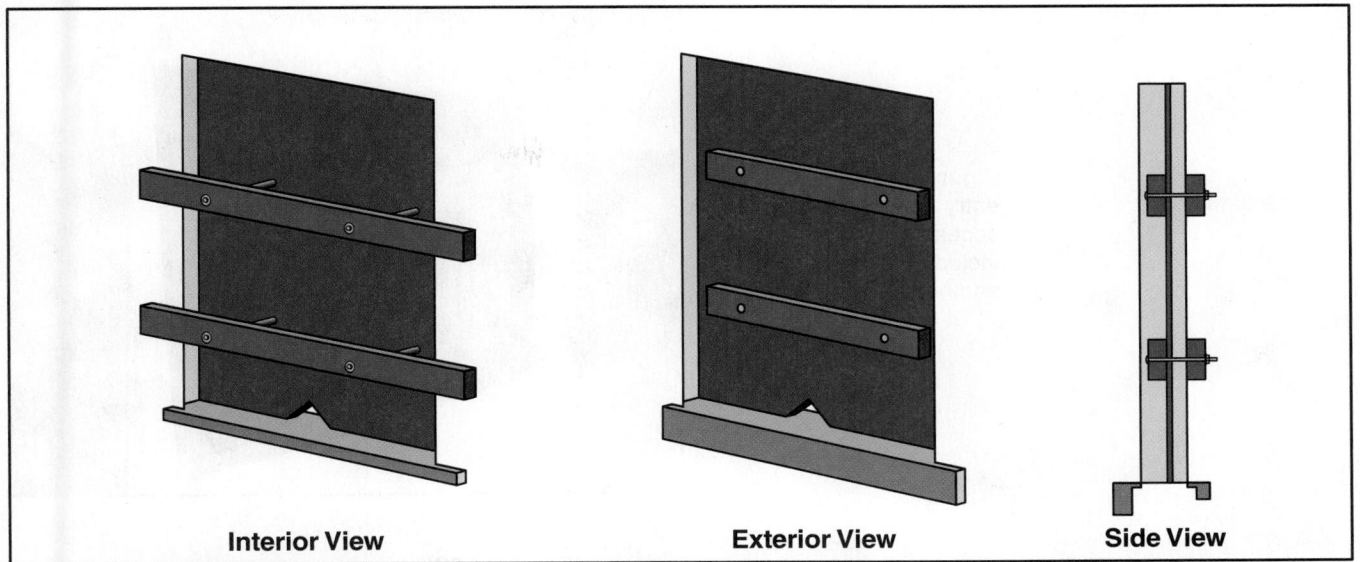

Interior View   **Exterior View**   **Side View**

Figure 11.58 HUD installs plywood over the interior of the window frames using carriage bolts to protect vacant buildings. These installations are often called "strongbacks".

Housing and Urban Development (HUD) Windows

In the U.S., the Department of Housing and Urban Development (HUD) installs plywood over window openings in its vacant buildings to prevent vandalism or break-ins. One sheet of plywood is secured with 2 x 4 inch (50 mm by 100 mm) boards on the inside of the frame, which are connected to similar boards on the outside. The two boards, called "strongbacks", are connected with long carriage bolts (**Figure 11.58**).

Forcible entry can be accomplished in one of two ways:

1. Use a rotary saw to cut off the heads of the carriage bolts, then use a pickhead or punch to push the bolt through the plywood, knocking the board on the inside loose.

2. Use an axe or Halligan to split the wood away from the bolt head and then push the bolt through the plywood.

Vacant Protection Systems (VPS)

Vacant protection systems (VPS) are used by banks, mortgage companies, and building owners to prevent unauthorized entry and vandalism of vacant buildings. Similar to the HUD windows, VPS are metal grates that are secured to the exterior of window openings. Cables are used to pull the grate to the structure and metal strongbacks on the interior. The grates can be removed by cutting mounting tabs or bolt heads off the frame with a rotary saw or the adz of a Halligan tool.

Barred or Screened Windows and Openings

Security bars and grilles are usually mounted to windows on the ground floor or basement level of single-family residential occupancies. They are mounted to the window frame and can be removed quickly using a variety of methods. Security bars and grilles are intended to prevent unauthorized entry but they also create an unintended hazard for both occupants and firefighters. Not only do the bars and grilles impede firefighter access, they also prevent occupants or firefighters who are trapped inside from escaping. If there are sufficient resources available, all security bars should be removed from the building to allow for emergency egress when crews are operating inside the building.

Figure 11.59 Forcible entry methods for gaining access past security grilles include the use of powered and nonpowered tools.

Forcible entry methods include (**Figure 11.59**):

- Removing the mounting bolt heads with the adz of the Halligan tool
- Cutting the bolt heads with a rotary saw with an aluminum oxide blade
- Using the pick end of the Halligan to chip away the masonry around the bar; strike the end of the Halligan with a sledgehammer or maul
- Cutting the bars or grille frame using a rebar cutter

Security grilles or screens may be permanently fixed, hinged at the top or side, or fitted into brackets and locked securely. Those that are hinged can be opened easily if the lock is accessible and can be cut from the frame. If not, the screen fabric can be cut using a rotary saw. Once the screen is removed, the window must be forced as described earlier.

Breaching Walls

Breaching — The act of creating a hole in a wall or floor to gain access to a structure or portion of a structure.

In situations where doors and windows are inaccessible or heavily secured, it may be faster and more efficient to gain access through the structure's wall. Creating a hole in a wall is known as **breaching**. Breaching requires a thorough knowledge of building construction, accurate size-up of the situation, and determination that breaching a particular wall is safe and will accomplish the purpose. Exterior walls will be more difficult to breach than interior walls as described in the following sections.

Exterior Walls

Firefighters must consider possible collapse and safety hazards before breaching exterior walls. Breaching load-bearing exterior walls in fire weakened structures could cause a partial or total collapse. Exterior and interior walls also conceal electrical wires, water pipes, gas pipes, and other components of the building utilities. Firefighters outside the structure will be unable to determine if these concealed components exist in an exterior wall. Exterior walls may be wood frame, brick or concrete block, poured concrete, or metal construction.

Wood Frame Walls

Exterior wood frame walls consist of vertical 2 x 4 or 2 x 6 inch (50 mm by 100 mm or 50 mm by 150 mm) studs covered on the inside with gypsum sheets or lathe-and-plaster while the outside is covered with wood, composite boards, or other materials.

Figure 11.60 Wood framed walls may be opened with an axe or sledgehammer before prying sections large enough for entry with a Halligan or crowbar.

Figure 11.61 The use of a battering ram has limitations and should be restricted to making small holes.

Studs are placed 16, 20, and sometimes 24 inches (400 mm, 500 mm, or 600 mm) apart. The spaces may be void or contain some form of insulation material. Wood siding may be hardwood boards, shake shingles, or panels made of plywood or composite materials.

Firefighters can cut wood framed walls with an axe or shatter them with sledgehammers before prying the wall open using a crow bar or Halligan (**Figure 11.60**). When the exterior wall is opened, the interior wall is then penetrated. **Skill Sheet 11-I-16** illustrates a suggested method for opening a wood framed wall with hand tools. A rotary saw or chain saw is a faster and more efficient method for opening a wood wall and is shown in **Skill Sheet 11-I-17.**

Brick or Concrete Block Walls

The traditional approach to breaching a masonry wall has been to use a battering ram (**Figure 11.61**). Breaching with a battering ram can be slow and labor-intensive. Therefore, it is best suited for opening a small hole in a wall through which water can be applied to a fire on the other side of the wall. While it is possible to use a battering ram to create an opening large enough for firefighters to pass through, it is impractical. **Skill Sheet 11-I-18** illustrates the method for using a battering ram.

Power tools such as rotary saws with masonry blades or pneumatic or electric jackhammers are best for breaching brick and concrete block walls. They are faster and usually require only one person to operate. If these tools are not available and a wall needs to be breached to allow water to be applied to a fire on the other side of the wall,

Figure 11.62 A penetrating nozzle is designed to be driven through a masonry wall.

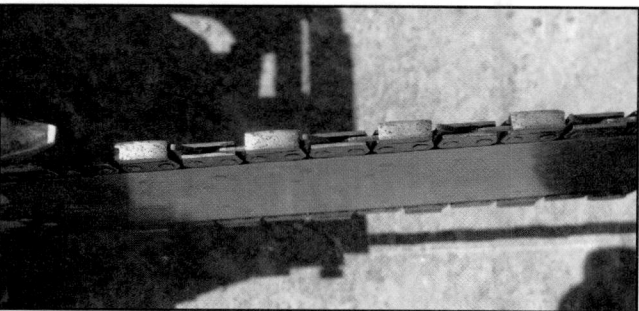

Figure 11.63 Diamond-tipped chains allow chain saws to quickly breach concrete.

a penetrating (drive-in) nozzle can be driven though the wall or an opening can be made large enough to accept a conventional nozzle (**Figure 11.62**). **Skill Sheet 11-I-19** describes the method for breaching a concrete or masonry wall using hand tools.

Concrete Walls

Breaching poured concrete walls is even slower and more labor-intensive than breaching brick or concrete block walls. In addition to the density of the material, the wall is often reinforced with steel rebar. Therefore, breaching concrete walls should be done only when absolutely necessary and no other alternative is available. A number of power tools can be used to breach concrete walls, but one of the fastest and most efficient is a chain saw equipped with a diamond-tipped chain (**Figure 11.63**). If a chain saw is not available, a pneumatic jackhammer can be used.

Metal Walls

Metal walls are common in commercial and industrial occupancies in both rural and urban settings. These walls are usually constructed of overlapping light-gauge sheet metal panels fastened to metal or wooden studs. Nails, rivets, bolts, screws, or other fasteners may be used to attach the panels (**Figure 11.64**). Conventional forcible entry tools, such as an axe, rotary saw, or air-chisel, cut these thin metal panels with relative ease.

Figure 11.64 A rotary saw may be used to cut through thin metal panels.

Make sure no building utilities are located in the area selected for cutting. A charged hoseline or an appropriate fire extinguisher should be available when cutting metal with a rotary saw because of the sparks produced. Cut a square or rectangular opening that is large enough for firefighters to pass through easily. The opening should be at least 6 feet (1.8 m) tall and as wide as needed. If the wall must be breached to allow water to be applied to a fire on the other side of the wall, a penetrating nozzle can be driven through the metal siding. **Skill Sheet 11-I-20** describes the method for breaching a metal wall with a rotary saw.

Interior Walls

Depending on the type of structure and its design, interior walls may be load-bearing or nonload-bearing. Construction materials may be masonry, poured concrete, glass block, lathe-and-plaster, or Sheetrock®. The walls may contain electrical wires, water or gas pipes, or heating and cooling ducts. Your supervisor must determine what effect breaching the wall will have on the structural integrity prior to making the opening. The methods for breaching masonry, concrete, and glass block walls were described previously.

Plaster or Gypsum Partition Walls

Both load-bearing and nonload-bearing interior walls are designed to limit fire spread. This fire resistance is provided by covering the wall with a variety of materials, including gypsum wallboard or lath-and-plaster over wooden or metal studs and framing. Both lath-and-plaster and gypsum wallboard are often relatively easy to penetrate with forcible entry striking hand tools or rotary and chain saws.

Reinforced Gypsum Walls

In some newer buildings, the interior walls in public access areas such as hallways, lobbies, and restrooms are covered with gypsum wallboard that is reinforced with Lexan®. Like other wallboard, reinforced wallboard is attached to the wall frame using drywall nails or screws. Reinforced wallboard looks identical to other wallboard because the Lexan® reinforcement is installed on the back of the wallboard. This wallboard is designed to resist breaching using conventional forcible entry hand tools. If firefighters are to breach this material in a timely fashion, they must know that they are not dealing with ordinary wallboard and must bring power saws with them when they enter the building. The only way they will know that the wallboard is reinforced is if the preincident planning survey identifies the wallboard material.

Breaching Floors

Breaching a floor assembly may be necessary to ventilate an area, apply water to a fire, or rescue occupants trapped by a structural collapse. The breaching tools and methods used will depend on the type of floor construction.

There are almost as many kinds of floors and floor coverings as there are buildings, but subfloor construction is limited to either wood or concrete. Either of these two may be finished with a variety of covering materials. Concrete slab floors are common in residential, commercial, and industrial occupancies. Even upper floors of buildings may be finished with lightweight concrete. The upper floors of multistory residences are usually wooden subfloors over wooden joists or I-beams. It is not uncommon for a floor to be classified according to its covering instead of the material from which it is constructed.

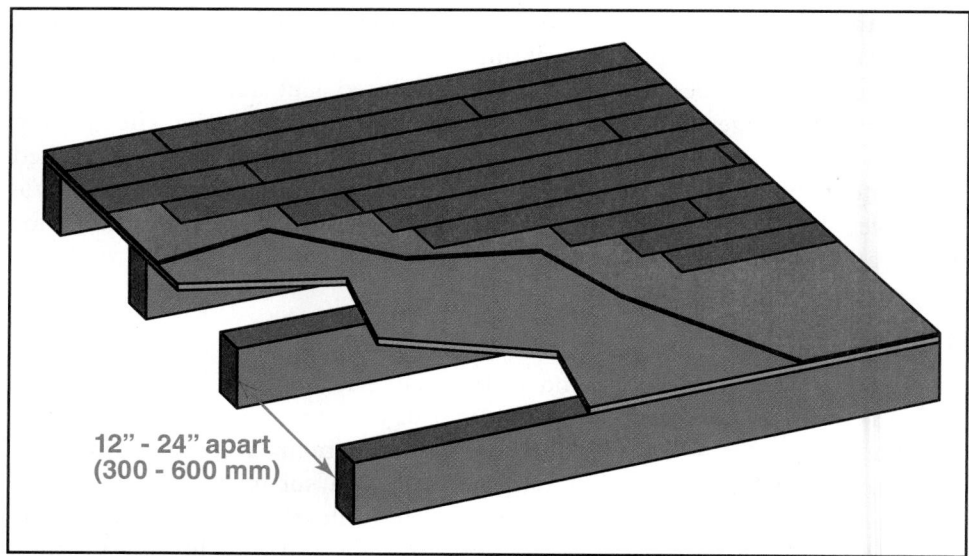

Figure 11.65 A floor should be cleared before cutting operations begin.

The feasibility of opening a floor during a fire fighting operation depends upon how it was constructed and from what material. A wood floor does not in itself ensure that it can be penetrated easily. Many wood floors are installed over a concrete slab. The type of floor construction can and should be determined during preincident planning surveys.

Wooden Floors

Wooden floor joists can be spaced from 12 to 24 inches (300 mm to 600 mm) apart depending upon the distance spanned and the dimensions of the lumber in the joists. Wooden I-beams are generally spaced 24 inches (600 mm) apart regardless of the span. The floor joists are covered by a subfloor consisting of either tongue-and-groove planks or sheets of plywood attached to the joists. Some wooden plank subfloors are laid diagonally to the joists, and the finished floor perpendicularly to the joists. Plywood subflooring is generally laid perpendicularly to the joists. The finish flooring, which may be sheet vinyl, ceramic tile, hardwood, or carpeting, is installed last.

Before a floor is cut, carpets and rugs should be removed or rolled to one side **(Figure 11.65)**. Rotary or circular saws make relatively clean cuts in wooden floors while chain saws make faster, rougher cuts. Use an axe if power saws are not available. If it is necessary to apply water through the floor assembly, a piercing nozzle can be used effectively. If the intention is to use cellar nozzles, Bresnan distributors, or to ventilate the space, a larger opening should be cut. The procedure for opening a wooden floor using an axe is shown in **Skill Sheet 11-I-21**.

Concrete Floors

All concrete floors are reinforced to some degree depending upon where the floor is located and the loads it is designed to support. If a concrete floor must be opened, a number of tools can be used. While it is possible to open a concrete floor using sledgehammers and other hand tools, it is too slow and too labor-intensive to be practical. Concrete-cutting blades are available for most portable power saws, but the most efficient tool may be a pneumatic or hydraulic jackhammer. To open a concrete floor

Figure 11.66 A series of drill holes in concrete may weaken the structure enough to allow a penetrating nozzle into the underlying space.

rapidly, use a jackhammer to cut a line of holes in the floor **(Figure 11.66)**. This method is sometimes referred to as "stitch drilling." A stitch drill can also be used to create an opening for a penetrating nozzle.

Forcing Entry Through Fences and Gates

Residential and commercial property owners often install security fences and gates to prevent unauthorized access. Preincident surveys should provide information on the type of security used and the location of lockboxes.

Forcing Fences

Fences can be made of wood, plastic, masonry, barbed wire, chain-link wire fabric, or ornamental metal. Some are topped with barbed wire or razor ribbon. Fences may also be used to contain livestock, pets, or guard dogs for additional security. As always, adequate size-up and using the most efficient tools and techniques are important when forcing entry through fences and gates.

Remember that some fences in rural areas may be electrified. Take care when cutting these fences to prevent electrical shocks. If possible, de-energize the fence before cutting; if not, take insulation precautions to avoid electric shock or find other means of entry.

Fencing material that is stretched tight can recoil when it is cut, inflicting injuries to firefighters. Stand beside the fence post and cut the wire where it joins the post. When it is cut, it will recoil in the direction of the next post.

Wire fences should be cut near posts to facilitate repair after the incident, provide adequate space for fire apparatus access, and reduce the danger of injury from the recoil of wires when they are cut.

Various fences can be forced in several ways:

- Cut barbed wire fences with bolt cutters.
- Cut chain-link fences with a rotary saw; bolt cutters may be used as well but are slower **(Figure 11.67, p. 622)**. Alternatively, cut the wire bands holding the fence fabric to the posts and lay the fabric on the ground.

Figure 11.67 Nonenergized chain-link fences may be cut with a rotary saw or bolt cutters.

In the case of masonry and ornamental metal fences, it may be easier and faster to go over the fence than through it. A-frame ladders can be used to bridge these fences. The procedure for bridging a fence with a ladder is shown in **Skill Sheet 11-I-22**.

Forcing Gates

Where there are fences used for security there will be access points with locked or controlled gates. Security gates are used for residential housing complexes, industrial sites, construction sites, and agricultural sites. At sites where access is needed continuously, such as industrial sites, the gates may be staffed during hours of operation. In other situations, the gates may be locked with chains and padlocks or other locking devices.

In residential complexes, gates are usually controlled by electronic locks that are activated by a remote opener (similar to a garage door opener), barcode reader, or a keypad. There may also be a lockbox in the vicinity of the gate that contains an opener or keypad code **(Figure 11.68)**. If necessary, and only if allowed by departmental SOPs, entry may be forced by prying the gate open or by using the apparatus bumper to force the gate. Fence gates are often secured with padlocks or chains and can be accessed using the techniques described previously. Gates used to secure patios, swimming pools, or backyards will usually have internal key-operated dead bolt locks. These gates are

Figure 11.68 Gated communities may be protected by an electronic lock with a standard lockbox in the vicinity for the fire department access point.

best accessed using through-the-lock and rim lock techniques. Prying or cutting the gate should be the last method chosen because of the amount of damage that will be done to the gate and fence.

NOTE: Some gates on commercial sites may be provided with special padlocks and/or electronic key switches that are operated with the same key used to access lockboxes.

Chapter Summary

Forcible entry techniques are used by firefighters to gain access into a structure or area when normal means of entry are locked or blocked. When properly applied, forcible entry efforts do minimal damage to the structure or structural components and provide quick access for firefighters. Forcible entry should not be used when a normal means of access is readily available. Firefighters may use forcible entry tools and techniques to breach walls and floors to advance hoselines, apply extinguishing agents, access trapped victims, or ventilate an area. The same tools and methods may be required to exit a structure or compartment in an emergency.

Review Questions

Firefighter I

1. What are the basic principles of forcible entry?

2. What types of locksets may firefighters encounter during forcible entry operations?

3. How can a firefighter know when it is appropriate to use cutting tools and pushing/pulling tools during forcible entry operations?

4. What are some basic tool safety tips firefighters should follow during forcible entry operations?

5. Who cares for and performs maintenance on forcible entry tools?

6. How do the considerations that must be taken when forcing entry through swinging and sliding doors compare?

7. What precautionary methods can be used when forcing entry through overhead or fire doors?

8. How does the process for forcing entry through fixed windows compare to forcing entry through awning windows?

9. What dangers may be present when forcing entry through miscellaneous types of windows and covers?

10. How do forcible entry operations for exterior walls compare to those for interior walls?

11. What does the feasibility of opening a floor during a fire fighting operation depend on?

12. What techniques can be used to force entry through fences or gates?

Tool Cleaning

Step 1: Wash tools with mild detergent or per manufacturer's guidelines. Rinse and wipe dry.

NOTE: Do not soak wooden handles in water because it will cause the wood to swell.

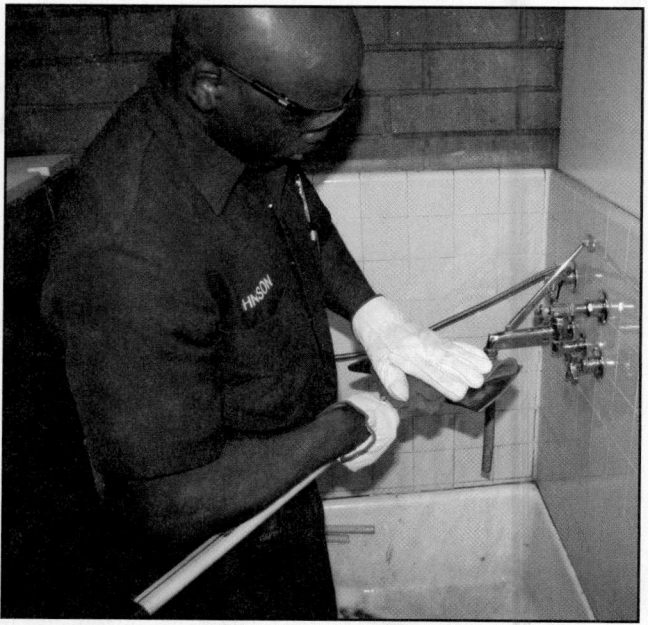

11-I-1
Clean, inspect, and maintain hand tools and equipment.

SKILL SHEETS

Tool Inspection

Step 2: Inspect tools for damage.

Step 3: Inspect parts for tightness and function.

Step 4: Inspect working surface for damage or wear.

Step 5: Inspect tool handles for cracks, splinters, or other damage.

Step 6: Inspect tool head for tightness.

Step 7: Inspect working surface for dullness, damage, chips, cracks, or metal fatigue.

Tool Maintenance

Step 8: Maintain wooden handles.

    a. Repair loose tool heads

    b. Sand the handle to eliminate splinters

    c. Apply a coat of boiled linseed oil to the handle to preserve it and prevent roughness and warping

    d. Do not paint or varnish the handle

Step 9: Maintain cutting edges.

    a. File the cutting edges by hand

    b. Sharpen blade as specified in departmental SOPs

    c. Replace cutting head if required

Step 10: Maintain unprotected metal surfaces.

    a. Keep free of rust

    b. File chips, cracks, or sharp edges

    c. Oil the metal surface lightly, using light machine oil

Tool Cleaning

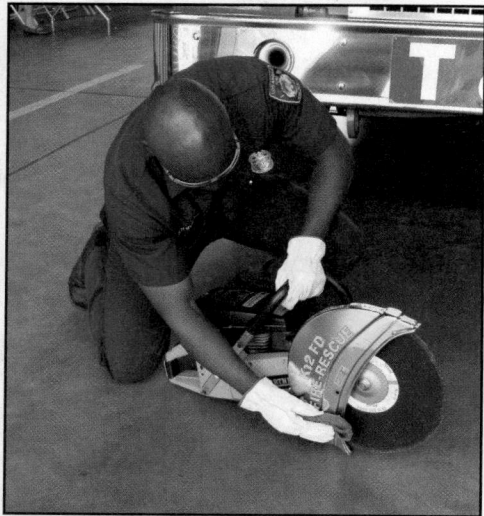

Step 1: Clean tools according to manufacturer's guidelines.

Tool Inspection

Step 2: Inspect tools for damage.

Step 3: Inspect parts for tightness and function.

    a. Ensure that all guards are functional and in place

    b. Check all electrical components for cuts or other damage

Step 4: Inspect working surface for damage or wear.

Step 5: Change a cutting blade on a power tool.

    a. Check blades for damage or wear

    b. Replace blades that are damaged or worn

Step 6: Check fuel level in all power tools and fill as necessary.

    a. Use correct fuel type

    b. Ensure that fuel is fresh

Step 7: Check oil level in all tools and fill as necessary.

Step 8: Start all power tools and keep them running.

    a. Ensure power tools will start manually

    b. Ensure battery packs are fully charged

Step 9: Tag a tool that is out of service.

    a. Place appropriate notification on the tool

    b. Communicate the situation with officer

11-I-3
Force entry through an inward-swinging door — Two-firefighter method.

SKILL SHEETS

Step 1: Firefighter #1: Place the fork of a Halligan bar just above or below the lock with the bevel side of the fork against the door.

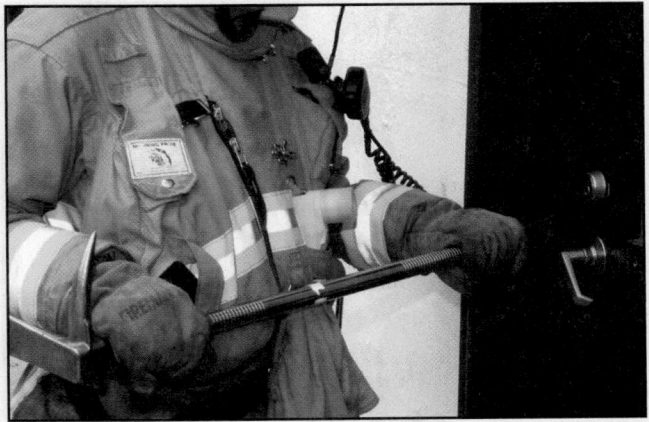

Step 4: Firefighter #2: Drive the forked end of the tool past the interior doorjamb.

Step 5: Firefighter #1: Move the bar slowly perpendicular to the door being forced to prevent the fork from penetrating the interior doorjamb.

Step 6: Firefighter #1: Make sure the fork has penetrated between the door and the doorjamb.

Step 2: Firefighter #1: Angle the tool slightly up or down.

Step 3: Firefighter #2: Strike the tool with the back side of a flat-head axe.

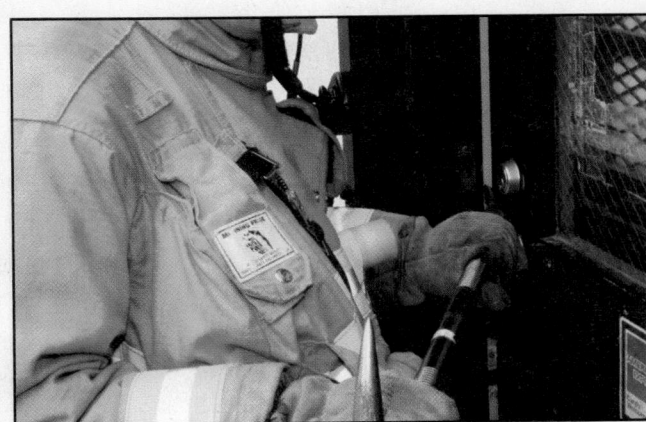

Step 7: Firefighter #1: Exert pressure on the tool toward the door, forcing it open.

SKILL SHEETS

11-I-4

Force entry through an inward-swinging door –
Cutting the lock out of the door method.

Step 1: With a rotary saw, make a horizontal cut through the door above the lock.

Step 3: Make a final vertical cut connecting the two horizontal cuts on the side of the lock away from the door edge.

Step 2: Make a second horizontal cut below the lock.

Step 4: With a flat-head axe or other striking tool, knock the lock through the door.

11-I-5
Force entry through an outward-swinging door –
Removing hinge-pins method.

SKILL SHEETS

Step 1: Start with the top hinge so that heat and smoke will be released at the top of the door.

Step 3: If necessary, twist the Halligan from side to side to loosen the hinge mounting screws.

Step 2: Place the fork end of a Halligan tool between the hinge and the door and pry up or down.

Step 4: Pull the hinge clear of the door.

SKILL SHEETS

11-I-6

Force entry through an outward-swinging door –
Wedge-end method.

Step 1: Firefighter #1: Place the wedge end of the Halligan bar just above or below the lock.

  a. If there are two locks, place the wedge between the locks

Step 3: Firefighter #1: Pry down and out with the fork end of the tool.

  a. Make sure the wedge is sufficiently driven into the space

Step 2: Firefighter #2: Strike the tool using a flat-head axe on the surface behind the wedge, driving the wedge into the space between the door and the jamb.

Step 1: Size up the door and lock.

Step 2: Place a set of locking pliers firmly on the lock cylinder.

Step 5: Insert an appropriate key tool into the lock through the cylinder hole.

Step 6: Manipulate the tool to release the latching mechanism.

Step 7: Open the door.

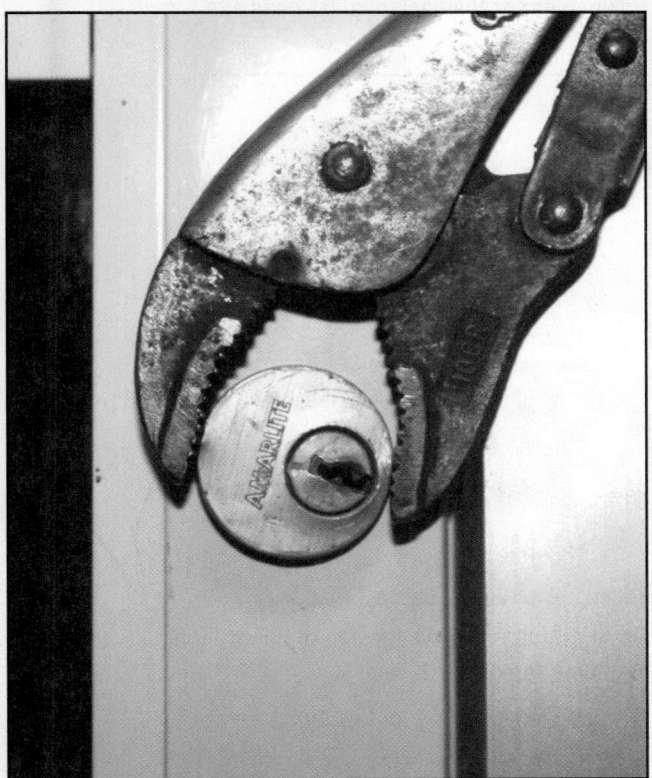

Step 3: Turn the lock cylinder counterclockwise to unscrew it from the door and remove it.

Step 4: Look inside the lock and identify the type of mechanism.

SKILL SHEETS

11-I-8

Force entry using the through-the-lock method using the K-tool.

Step 1: Size up the door and lock.

    a. Make sure the lock is not protected by a collar or shield

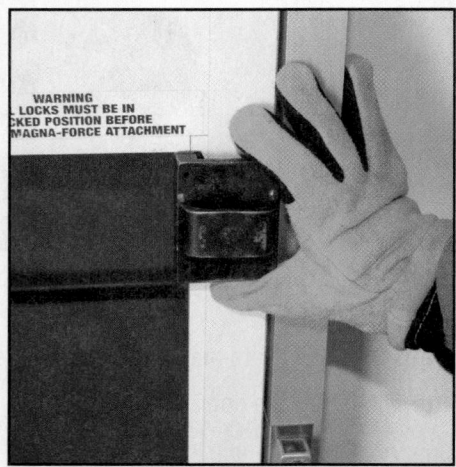

Step 2: Slide the K-tool down over the lock cylinder face.

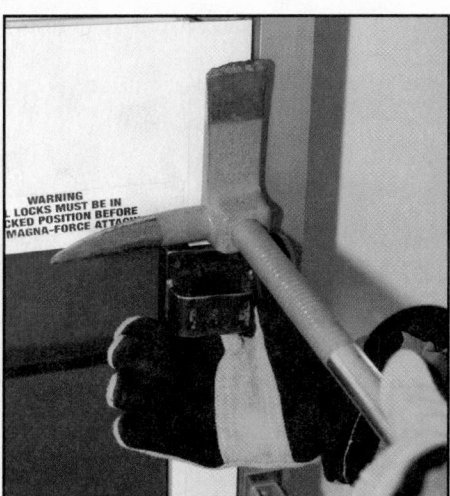

Step 3: Tap the K-tool down with a Halligan bar or the back of a flat-head axe.

Step 4: Insert the wedge end of the pry tool into the strap on the K-tool.

Step 5: Drive the K-tool further into the cylinder.

    a. Make sure the K-tool has an adequate bite into the lock cylinder

Step 6: Pry up on the tool handle to pull the cylinder.

Step 7: Insert a key tool through the cylinder hole to release the latching mechanism.

Step 8: Open the door.

11-I-9

Force entry using the through-the-lock method using the A-tool.

SKILL SHEETS

Step 1: Size up the door and lock.

Step 2: Insert the V-notch of the A-tool between the lock cylinder and the door frame.

    a. The A-tool should be at a slight angle to the lock

Step 3: Tap the A-tool firmly in place behind the lock cylinder.

    a. It may be necessary to drive the A-tool into the frame of the door in order to get behind a tight lock

Step 4: Pry up on the tool and remove the lock cylinder.

Step 5: Insert the key tool into the lock through the cylinder hole.

Step 6: Manipulate the tool to release the latching mechanism.

Step 7: Open the door.

Method One – Hook End

Step 1: Firefighter #1: Insert the hook of a Halligan bar into the shackle of the lock and pull the lock out away from the staple.

Step 2: Firefighter #2: Strike the Halligan bar sharply with a flat-head axe to drive the hook through the lock shackle and break it.

Method Two– Fork End

Step 1: Place the fork of the Halligan bar over the padlock shackle.

Step 2: Twist the lock until the shackle or the hasp breaks.

Method Three– Bolt Cutters

Step 1: Cut the shackle of the padlock, the chain, or the staple with bolt cutters.

    a. Do not attempt to cut case-hardened lock shackles with bolt cutters

Step 1: Using the bam-bam tool with a hardened self-tapping screw attached, insert the screws into the keyway of the padlock and screw in.

       a. Do not overtighten the screw

Step 2: Give the bam-bam tool a sharp rearward blow with the slide.

Step 3: Re-tighten the screw.

Step 4: Give the tool another sharp rearward blow.

Step 5: Repeat until the keyway is removed.

Step 6: Insert the key tool or a screwdriver into the opening and turn until the lock opens.

Step 1: Position the lock against the door or frame exposing both parts of the shackle.

NOTE: Do not attempt to hold or have someone else hold the lock.

Step 2: With a rotary saw, cut both shackles at the same time.

Step 3: Remove the shackle from the door hasp.

Step 1: Size up the situation.

   a. Try window first

   b. Evaluate window construction and locking method

   b. (Single-paned window) At top of pane

   c. To avoid losing control of the tool, do not use excessive force

Step 2: Break the window glass.

   a. (Multiple-paned window) Lowest pane of glass

   d. Keep hands and the tool handle above the point of impact

   e. Use the tool to clean all the broken glass out of the frame once the glass has been broken

Step 1: Size up the situation.

    a. Try window first

    b. Evaluate window construction and locking method

Step 2: Insert the blade of an axe or other prying tool under the center of the bottom sash in line with the lock mechanism.

Step 3: Pry upward on the tool handle to force the lock.

Step 4: Push the lower sash upward to open the window.

Step 1: Size up the situation.

    a. Try window first

    b. Evaluate window construction and locking method

Nonopening Type Method

    a. Cut around edges of entire pane

    b. Remove pane

Step 2: Check the saw blade.

    a. Carbide-tipped

    b. Medium-toothed

    c. Sharp

    d. Undamaged

Step 6: Turn off the saw.

Step 7: Open the window.

    a. Reach through hole or opening

    b. Unlock if applicable

    c. Raise sash and prop open if applicable

Step 3: Ensure that the blade and hand guards are in place.

Step 4: Start and operate the rotary saw per manufacturer's instructions.

Opening Type Method

Step 5: Saw a hole in the window.

    a. Cut triangular shape (three cuts) large enough to reach through

    b. Cut near window lock if applicable

SKILL SHEETS

11-I-16
Force entry through a wood-framed wall
(Type V construction) with hand tools.

Step 1: Confirm order with officer to force entry through wall.

Step 2 Size up the situation.

  a. No other existing entry points available

  b. Wall construction evaluated

  c. Locations of utilities considered

Step 3: Confirm with Command that utilities are off.

Step 4: Remove siding if necessary and locate stud.

Step 5: Cut an inspection hole (small triangle).

Step 6: Make cut utilizing inspection hole.

Step 7: Increase size of hole to allow the passage of fire-fighter (stud may be removed, if necessary).

Step 8: Utilizing the inspection hole, remove wall and insulation material with hand tool and place out of traffic area.

11-I-16
Force entry through a wood-framed wall
(Type V construction) with hand tools.

SKILL SHEETS

Step 9: Using hand tool, push inward and remove interior wall covering.

SKILL SHEETS

11-I-17

Force entry through a wood wall (Type V construction)
with a rotary saw or chain saw.

Step 1: Confirm order with officer to force entry through wall.

Step 2: Wearing full PPE, including eye protection, follow the saw manufacturer's operating instructions.

Step 3: Place the saw blade against the wall at about shoulder height.

Step 4: Begin a cut diagonally to the left ending about a foot (meter) off the ground.

11-I-17
Force entry through a wood wall (Type V construction) with a rotary saw or chain saw.

🔥 **SKILL SHEETS**

Step 5: Repeat with a cut diagonally to the right and the same length.

Step 6: Make a horizontal cut connecting the two diagonal cuts.

Step 7: Using a sledgehammer, flat-head axe, or battering ram, knock the material out from between the cuts.

Step 8: Remove interior wall by pushing inward.

Step 1: Confirm order with officer to force entry through wall.

Step 2: While wearing full PPE, including eye protection, grasp the handles of the battering ram.

Step 3: Place the blunt end against the wall to be breached.

Step 4: Swing the ram back and forth until the wall surface crumbles.

Step 5: When the wall surface crumbles, turn the ram around and strike the wall with the forked end until a hole is made through the wall.

Step 6: Continue striking until the desired size of hole is made.

11-I-19
Force entry through a masonry wall with hand tools.

SKILL SHEETS

Step 1: Confirm order with officer to force entry through wall.

Step 2: Size up the situation.

  a. No other existing entry points available

  b. Wall construction evaluated

  c. Locations of utilities considered

Step 3: Confirm with Command that utilities are off.

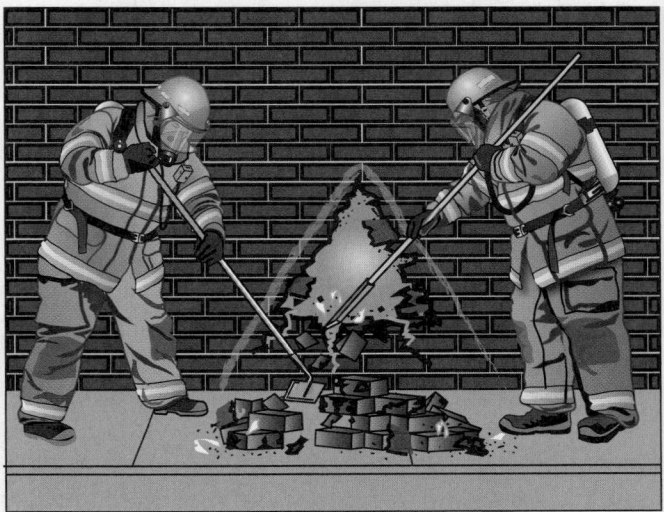

Step 6: Using prying and/or striking tools, begin to remove the highest block first while moving downward and side to side.

Step 4: Determine one block to strike with tool and strike block until it is fractured.

Step 5: Systematically strike and fracture individual block in a triangle pattern until desired hole size is reached.

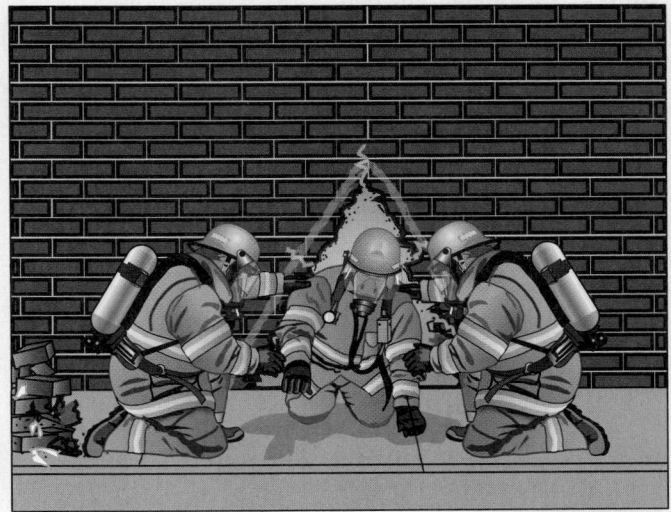

Step 7: Remove wall material and insulation and place out of traffic area.

Step 8: Using tool, push inward and remove interior wall covering, if needed.

SKILL SHEETS

11-1-20
Force entry through a metal wall with power tools.

Step 1: Confirm order with officer to force entry through wall.

Step 2: Size up the situation.

    a. No other existing entry points available

    b. Wall construction evaluated

    c. Locations of utilities considered

Step 3: Confirm with Command that utilities are off.

Step 4: Cut an inspection hole (small triangle).

Step 7: Remove wall material and insulation and place out of traffic area.

Step 8: Using tool, push inward and remove interior wall covering if needed.

Step 5: Locate studs (wall screws indicate) and cut hole near stud.

Step 6: Cutting a triangle, increase size of hole to allow the passage of firefighter.

Step 1: Determine the approximate location and size of hole.

Step 2: Sound the floor to determine the location and direction of the joists.

Step 3: Cut one side of the finished floor using angled cuts.

Step 4: Cut the other side of the finished floor in like manner.

Step 5: Remove the flooring between the cuts with the pick of the axe or other tool.

Step 6: Cut the subfloor using the same technique and angle cuts.

Step 1: Place a roof or straight ladder against the fence at the desired location and correct climbing angle.

Step 2: Carry a roof ladder up the first ladder and place it on the opposite side of the fence next to the first ladder and at the correct climbing angle.

Step 3: Both ladders should extend above the top of the fence by two rungs.

Step 4: Secure the two ladders to the top of the fence with a rope hose tool.

Step 5: Step from the first ladder to the second ladder and climb down.

Courtesy of Bob Esposito

Chapter Contents

Key Terms

NFPA® Job Performance Requirements

This chapter provides information that addresses the following job performance requirements of NFPA® 1001, *Standard for Fire Fighter Professional Qualifications* (2013).

Firefighter I

5.3.6

5.3.9

5.3.11

5.3.12

1. Describe different construction types of ground ladders. (5.3.6)

2. Identify the parts of a ladder including markings and labels. (5.3.6)

3. Recognize the types of ladders used in the fire service. (5.3.6)

4. Explain the considerations addressed by ladder inspection, cleaning, and maintenance. (5.5.1)

5. Describe safety guidelines used when handling ladders. (5.3.6)

6. Explain considerations taken when selecting, lifting, and lowering a ladder. (5.3.6)

7. Describe various methods for ladder carries. (5.3.6)

8. Identify basic considerations and requirements for ground ladder placement. (5.3.6)

9. Describe various methods for ladder raises. (5.3.6)

10. Compare procedures for moving ground ladders. (5.3.6)

11. Explain the methods used to secure ladders. (5.3.6)

12. Describe ladder climbing considerations. (5.3.6)

13. Indicate what methods can be used to work from a ladder. (5.3.9)

14. Explain methods used for assisting a victim down a ladder. (5.3.9)

15. Clean, inspect, and maintain a ladder. (Skill Sheet 12-I-1, 5.5.1)

16. Carry a ladder – One-firefighter low-shoulder method. (Skill Sheet 12-I-2, 5.3.6, 5.3.11, 5.3.12)

17. Carry a ladder – Two-firefighter low-shoulder method. (Skill Sheet 12-I-3, 5.3.6, 5.3.11, 5.3.12)

18. Carry a ladder – Three-firefighter flat-shoulder method. (Skill Sheet 12-I-4, 5.3.6, 5.3.11, 5.3.12)

19. Carry a ladder – Three-firefighter flat-arm's length method. (Skill Sheet 12-I-5, 5.3.6, 5.3.11, 5.3.12)

20. Carry a ladder – Two-firefighter arm's length on edge method. (Skill Sheet 12-I-6, 5.3.6, 5.3.11, 5.3.12)

21. Tie the halyard. (Skill Sheet 12-I-7, 5.3.11, 5.3.12)

22. Raise a ladder – One-firefighter method. (Skill Sheet 12-I-8, 5.3.6, 5.3.11, 5.3.12)

23. Raise a ladder – Two-firefighter flat raise. (Skill Sheet 12-I-9, 5.3.6, 5.3.11, 5.3.12)

24. Raise a ladder – Two-firefighter beam raise. (Skill Sheet 12-I-10, 5.3.6, 5.3.11, 5.3.12)

25. Raise a ladder – Three- or four-firefighter flat raise. (Skill Sheet 12-I-11, 5.3.6, 5.3.11, 5.3.12)

26. Deploy a roof ladder – One-firefighter method. (Skill Sheet 12-I-12, 5.3.11, 5.3.12)

27. Pivot a ladder – Two-firefighter method. (Skill Sheet 12-I-13, 5.3.11, 5.3.12)

28. Shift a ladder – One-firefighter method. (Skill Sheet 12-I-14, 5.3.11, 5.3.12)

29. Shift a ladder – Two-firefighter method. (Skill Sheet 12-I-15, 5.3.11, 5.3.12)

30. Heel a ground ladder. (Skill Sheet 12-I-16, 5.3.11, 5.3.12)

31. Leg lock on a ground ladder. (Skill Sheet 12-I-17, 5.3.11, 5.3.12)

32. Assist a conscious victim down a ground ladder. (Skill Sheet 12-I-18, 5.3.9)

33. Assist an unconscious victim down a ground ladder. (Skill Sheet 12-I-19, 5.3.9)

Chapter 12
Ground Ladders

Fire service ground ladders are an important part of the equipment that you will use at emergency incidents. Ground ladders are carried on all pumpers and aerial ladder apparatus as well as other types of vehicles. Depending on the ladder length, ground ladders require from one to six firefighters to carry and raise into position. While primarily used to access upper stories and roofs of buildings, ground ladders can also be used to reach areas that are below ground level, such as storm drains, confined-space trenches, and pits (**Figure 12.1, p. 654**).

To qualify as a Firefighter I, you must know the following about ground ladders:

- Parts of a ladder
- Hazards associated with setting up ground ladders
- Foundations considered stable for ladder placement
- Different ladder angles for various tasks

Figure 12.1 Ladders are primarily used to access high areas. *Courtesy of Ron Jeffers.*

- Safety limits to the degree of angulation
- Structural components considered reliable for top placement

You must also be capable of performing the following tasks alone or as a member of a team:

- Carrying ground ladders
- Raising ground ladders
- Extending ground ladders and locking the fly
- Determining that a structural component (wall or roof) is capable of supporting a ladder
- Judging extension ladder height requirements
- Placing a ladder to avoid obvious hazards

Ground Ladder Construction

Fire service ladders have similar construction, shape, and design as ladders manufactured for private industry or general use, but they are capable of supporting heavier loads. Their use under adverse conditions requires that fire service ground ladders provide a greater margin of safety than expected from other ladders. Design, construction, and testing specifications for fire service ground ladders are contained in NFPA® 1931, *Standard for Manufacturer's Design of Fire Department Ground Ladders.* NFPA® 1931 requires that folding ladders be designed to support a maximum of

300 lbs (136 kg). Single, roof, combination, and extension ladders must support a maximum of 750 lbs (340 kg). Requirements for use, care, maintenance, and service testing are contained in NFPA® 1932, *Standard on Use, Maintenance, and Service Testing of In-Service Fire Department Ground Ladders*.

Parts of a Ladder

The following terms are used to describe the various parts of fire service ladders **(Figures 12.2 and 12.3, p. 656)**:

- **Beam** — Main structural member of a ladder supporting the rungs or rung blocks
- **Bed section (also called the *base section* or *main section*)** — Lowest and widest section of an extension ladder; while the ladder is being raised or lowered, this section always maintains contact with the ground or other supporting surface

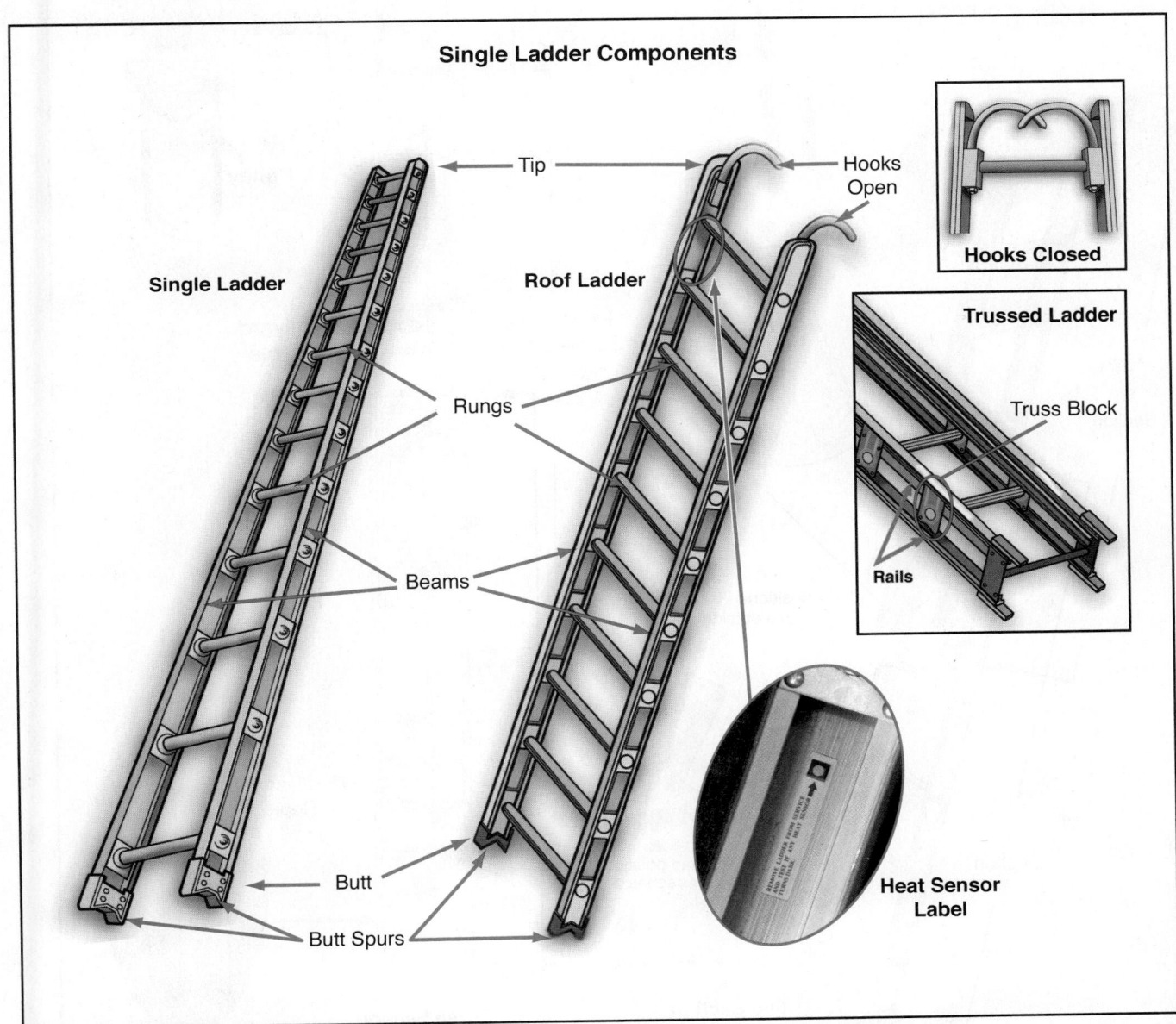

Figure 12.2 Single and roof fire ladders feature several components in common including construction.

Ladder Components

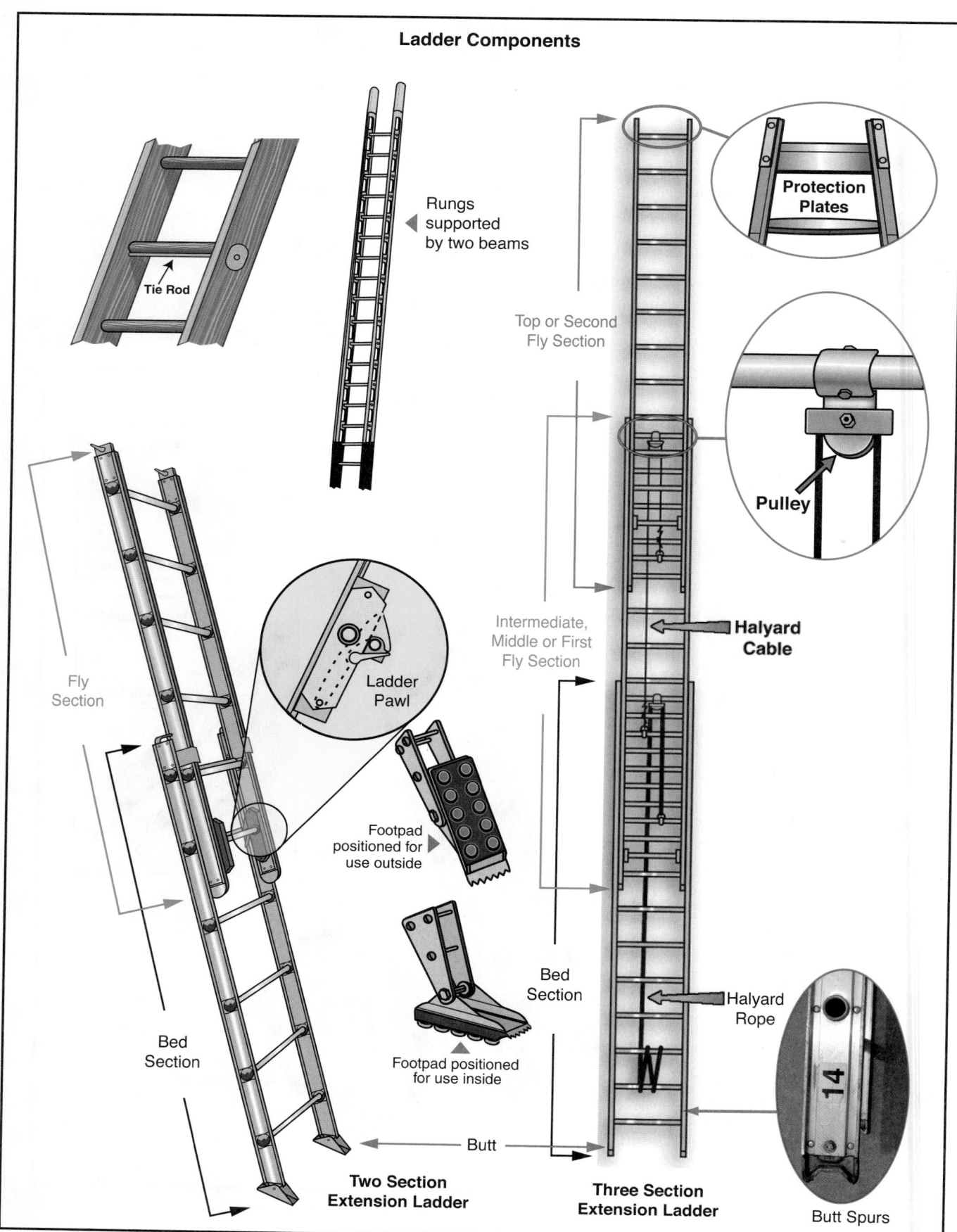

Figure 12.3 Extension ladders share some components with single ladders, and have other features added.

- **Butt (also called *heel* or *base*)** — Bottom end of the ladder; the end that is placed on the ground or other supporting surface when the ladder is positioned

- **Butt spurs** — Metal plates, spikes, or cleats attached to the butt end of ground ladder beams to prevent slippage

- **Fly Section** — Upper section(s) of extension or some combination ladders; the section that moves

- **Footpads (also called *shoes*)** — Swivel plates attached to the butt of the ladder; usually have rubber or neoprene bottom surfaces

- **Guides** — Wood or metal strips, sometimes in the form of slots or channels, on an extension ladder that guide the fly section while being raised

- **Halyard** — Rope or cable used for hoisting and lowering the fly sections of an extension ladder; also called fly rope

- **Heat-sensor label** — Label affixed to the inside of each beam of each ladder section; a color change indicates that the ladder has been exposed to a sufficient degree of heat and should be tested before further use

- **Hooks** — Curved metal devices installed near the top end of roof ladders to secure the ladder to the highest point on a peaked roof of a building

- **Pawls (also called *dogs* or *ladder locks*)** — Devices attached to the inside of the beams on fly sections used to hold the fly section in place after it has been extended

- **Protection plates** — Strips of metal attached to ladders at chafing points, such as the tip, or at areas where it comes in contact with the apparatus mounting brackets

- **Pulley** — Small, grooved wheel through which the halyard is drawn on an extension ladder

- **Rails** — The two lengthwise members of a trussed ladder beam that are separated by truss or separation blocks

- **Rungs** — Cross members that provide the foothold for climbing; the rungs extend from one beam to the other

- **Stops** — Wooden or metal pieces that prevent the fly section from being extended too far

- **Tie rods** — Metal rods located beneath rungs extending from one beam to the other of a wood ladder

- **Tip (top)** — Extreme top of a ladder

- **Truss block** — Spacers set between the rails of a trussed ladder; sometimes used to support rungs

Construction Materials

Fire service ground ladders are constructed of metal, wood, or fiberglass. All construction material must meet the design and testing specifications of NFPA® 1931. Each material has certain advantages and disadvantages. The weight of any ladder will vary depending on the construction material and the length. Heavier ladders require more personnel to safely carry and raise them.

Metal

Metal ladders constructed of heat-treated aluminum are the most common type of ladder currently in fire service use. However, heat-treated does not mean that the ladder is heat resistant. Exposure to radiant heat can cause the ladder to warp and fail. Metal ladders have certain advantages:

- Least expensive construction material
- Easy to repair
- Available in a wide range of sizes
- Available in many styles and types

  They also have certain disadvantages:

- Conduct heat, cold, and electricity very well
- Can fail suddenly when exposed to heat in excess of 200°F (93.33°C)
- Accumulate ice on the rungs in cold weather creating a slipping hazard

Wood

Wood is the oldest construction material used for fire service ground ladders. Douglas Fir is used for the beams, Hickory is used for the rungs, and Red Oak is used for the pulley blocks, slide glides, and other miscellaneous parts. Metal parts are plated for protection while wood parts are coated with marine grade spar varnish. Wooden ladders have some advantages:

- Less likely to conduct electricity
- Do not conduct heat
- Retain strength when exposed to heat
- Have better resistance to flexing and bouncing when being climbed
- Are very durable

  Wooden ladders also have some disadvantages:

- Have the highest cost of all ladders
- May require refinishing of damaged finish, depending on frequency of use
- Can be very heavy

Fiberglass

Fiberglass is the newest and least common material used for fire service ground ladders (**Figure 12.4**). They cost less than wood ladders and more than aluminum ladders. Fiberglass ground ladders have some advantages:

- Will not conduct heat, cold, or electricity
- Have very strong and rigid rails

  Fiberglass ground ladders also have certain, common disadvantages:

- Can suddenly crack and fail when overloaded
- Can burn when exposed to flame

  NOTE: The manufacturer's product manual will contain a list of disadvantages and cautions for the fiberglass ladders that they produce.

Figure 12.4 Fiberglass ladders have some advantages over wood and aluminum ladders.

Ladder Markings and Labels

All fire service ground ladders are required to have markings and warning labels. These markings and labels are factory applied although locally required markings may also be applied.

The following markings are commonly found on fire service ladders:

- NFPA® 1931 requires that the designated ladder length be marked on each beam within 12 inches (305 mm) of the butt plate.

- NFPA® 1931 also requires a manufacturer's name plate with the month and year of manufacture.

- The authority having jurisdiction (AHJ) may require that the apparatus designation or a locally assigned inventory number be stenciled on the beam.

- The tip of the ladder may be painted white or have a strip of reflective tape attached to make the top of the ladder visible in smoky or dark conditions.

- The butt of the ladder is sometimes painted black and the balance point may be indicated with a stripe.

NOTE: Painting wood ladders is not recommended because it can conceal damage to the wood.

All ladders meeting NFPA® 1931 are required to have a certification label affixed to the ladder by the manufacturer indicating that the ladder meets the standard. A variety of warning labels are also required on all types of ladders, including:

- An electrical hazard warning label (**Figure 12.5**)

- A ladder positioning label indicating the climbing angle and the side of the ladder that must be away from the building (**Figure 12.6, p. 660**)

- **Heat sensor labels** are required on metal and fiberglass ladders, placed on the inside of each beam, below the second rung from the tip of each section. The heat sensor is preset to 300°F (149°C) and must have the expiration date indicated on it (**Figure 12.7, p. 660**).

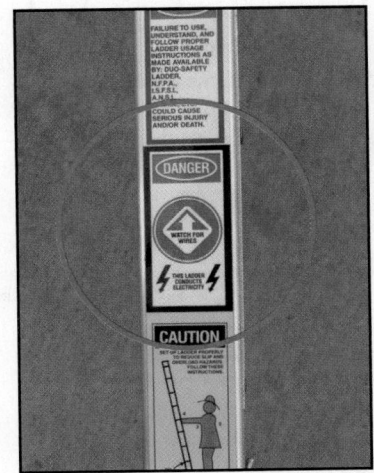

Figure 12.5 Firefighters must be aware that metal ladders conduct electricity.

Heat Sensor Label — Label affixed to the ladder beam near the tip to provide a warning that the ladder has been subjected to excessive heat.

Figure 12.6 A ladder positioning label indicates the proper way to set up a ladder.

Single Ladder — One-section nonadjustable ladder. Also known as Wall or Straight Ladder.

Ladder Types

The fire service typically uses five types of ground ladders:

- Single ladders
- Roof ladders
- Folding ladders
- Extension ladders
- Combination ladders

Commercial ladders, such as step-ladders, may also be carried depending on the local requirements. Commercial ladders may not meet NFPA® requirements for fire service emergency operations.

Single Ladders

Single ladders, sometimes called *wall ladders* or *straight ladders*, consist of one section of nonadjustable or fixed length **(Figure 12.8)**. The overall length of the beams is used to define the length of single ladders, such as a 20-foot (6 m) straight ladder. They are often used for quick access to windows and roofs on one- and two-story buildings. Some single ladders are of the trussed type, a design intended to maximize their strength while reducing weight. Lengths of single ladders vary from 6 to 32 feet (2 m to 10 m) with the more common lengths ranging from 12 to 24 feet (4 m to 8 m).

Figure 12.7 A heat sensor on an aluminum ladder beam indicates if the ladder was exposed to temperatures that could cause failure.

Figure 12.8 A single ladder has only one section of length.

Roof Ladders

Roof ladders are single ladders equipped with folding hooks that provide a means of anchoring the ladder over the ridge of a pitched roof or some other roof part. In position, roof ladders generally lie flat on the roof surface so that a firefighter can stand on the ladder while working. The ladder distributes the firefighter's weight and helps prevent slipping. Roof ladders may also be used as wall or straight ladders. Their lengths range from 12 to 24 feet (4 m to 8 m) **(Figure 12.9)**.

Folding Ladders

Folding ladders are single ladders that are often used for interior attic access. They have hinged rungs allowing them to be folded so that one beam rests against the other. When folded they can be carried in narrow passageways and used in attic scuttle holes and small rooms or closets **(Figure 12.10)**. Folding ladders are commonly found in lengths from 8 to 16 feet (2.5 m to 5 m) with the most common being 10 feet (3 m). NFPA® 1931 requires folding ladders to have footpads attached to the butt to prevent slipping on floor surfaces. You should always wear gloves when closing a folding ladder to protect your hands and fingers from being pinched between the moving metal parts.

Extension Ladders

An **extension ladder** consists of a base or bed section and one or more fly sections that travel in guides or brackets to permit length adjustment. The full length to which it can be extended indicates its size. Unlike single ladders, extension ladders can be adjusted incrementally to the specific length needed to access windows and roofs. Extension ladders generally range in length from 12 to 39 feet (4 m to 11.5 m) **(Figure 12.11, p. 662)**.

Combination Ladders

Combination ladders are designed so that they can be used as a self-supporting stepladder (A-frame) and as a single or extension ladder. Lengths range from 8 to 14 feet (2.5 m to 4.3 m) with the most common length being 10 feet (3 m). The ladder must be equipped with positive locking devices to hold the ladder in the open position **(Figure 12.12, p. 662)**.

Apparatus-Mounted Ground Ladders

NFPA® 1901, *Standard for Automotive Fire Apparatus*, sets the minimum lengths and types of ladders to be carried on pumper apparatus as follows:

- One single (roof) ladder equipped with roof hooks
- One extension ladder
- One folding ladder

The standard does not specify the minimum lengths of these ladders **(Figure 12.13 a, p. 662)**. It does recommend that engines carry a 35-foot (11 m) extension ladder in areas where no ladder trucks are in service.

Figure 12.9 An aluminum roof ladder has stowable hooks that prevent slipping on sloped surfaces or can be folded away so the ladder may be used for ground functions.

Roof Ladder — Straight ladder with folding hooks at the top end; the hooks anchor the ladder over the roof ridge.

Folding Ladder — Single-section, collapsible ladder that is easy to maneuver in restricted places such as access openings for attics and lofts.

Extension Ladder — Variable-length ladder of two or more sections that can be extended to a desired height.

Combination Ladder — Ladder that can be used as a single, extension, or A-frame ladder.

Figure 12.10 Folding ladders may be more easily carried through narrow passageways and rooms.

Figure 12.11 An extension ladder is a common piece of equipment carried to an incident.

Figure 12.12 A combination ladder with a positive locking device can be used as an A-frame ladder and as a single or extension ladder.

Figure 12.13a Abrasion points develop at any point of contact between items on an apparatus.

Figure 12.13b An aerial apparatus must carry a minimum of 115 feet (35 m) of ground ladders.

To comply with NFPA® 1901, aerial apparatus must carry a minimum of 115 feet (35 m) of ground ladders (**Figure 12.13b**). The following ladders may be loaded to meet the length requirement:

- One folding ladder
- Two single (roof) ladders equipped with hooks
- Two extension ladders

Quint fire apparatus — aerial devices mounted on a pumper apparatus — are required to carry a minimum of 85 feet (26 m) of ground ladders. The types of ladders are the same as required for a pumper apparatus.

Ladder Inspection, Cleaning, and Maintenance

Fire service ladders must be able to withstand considerable abuse including sudden overloading, exposure to weather and temperature extremes, and being struck by falling objects. Periodic inspections, service tests, cleaning, and maintenance are critical to ensure the safe operation of ground ladders.

Inspecting and Service Testing Ladders

NFPA® 1932 requires ground ladders to be inspected after each use and on a monthly basis. When inspecting ground ladders, some of the elements that should be checked on all types of ladders include the following:

- Heat sensor labels on metal and fiberglass ladders for a color change indicating heat exposure

  NOTE: Replace heat sensors when their expiration date is reached.

- On ladders without heat sensor labels, heavy carbon (soot) deposits or blistered paint on the ladder tips indicating heat exposure
- Fiberglass ladders for discoloration that could indicate heat exposure
- Rungs for damage or wear
- Rungs for tightness (**Figure 12.14**)
- Bolts and rivets for tightness

  NOTE: Bolts on wooden ladders should not be so tight that they crush the wood.

- Welds for any cracks or apparent defects
- Beams and rungs for cracks, splintering, breaks, gouges, checks, wavy patterns, or deformation
- Any points of contact with apparatus or other ladders where vibration may cause worn areas (**Figure 12.15, p. 664**)

Figure 12.14 The inspection of ground ladders includes checking that the rungs do not move.

⚠️ **WARNING**

Any ladder that has been subjected to direct flame contact, has been exposed to high heat, or has a heat sensor label that has changed color is unsafe for use and should be removed from service for testing.

Figure 12.15 Damage by vibration may occur anywhere a ladder comes into contact with apparatus or other ladders.

In addition to these general inspections, there are some other items that need to be checked, depending on the specific type of ladder being inspected. The following sections highlight some of these items.

Wooden Ladders/Ladders with Wooden Components

Look for the following when inspecting wooden ladders or ladders with wooden components:

- Areas where the finish has been chafed or scraped
- Darkening (blistering or blackening) of the varnish (indicating exposure to heat)
- Dark streaks in the wood (indicating deterioration of the wood)
- Marred, worn, cracked, or splintered parts
- Shoes rounded or smooth
- Water damage

CAUTION
Any indication of deterioration of the wood is reason for the ladder to be removed from service until it can be service tested.

Roof Ladders

When inspecting roof ladders, make sure that the roof hook assemblies operate with relative ease. In addition, the hook assemblies should not show signs of rust; the hooks should not be deformed; and the parts should be firmly attached with no sign of looseness.

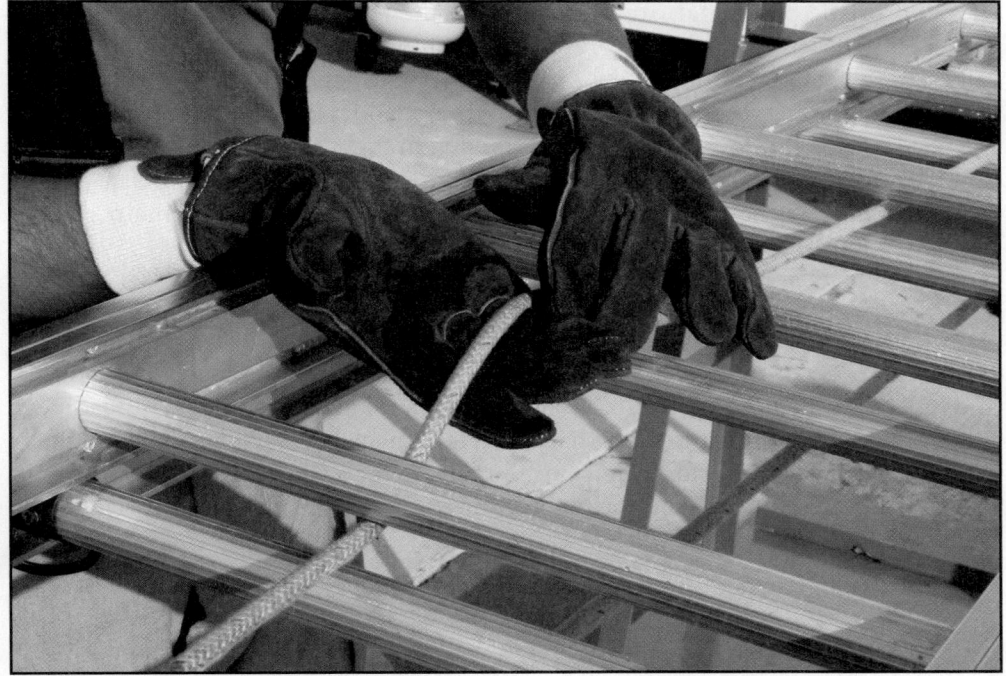

Figure 12.16 All components of an extension ladder must be checked carefully, including the condition of the halyard cable.

Extension Ladders

When inspecting extension ladders, check the following (**Figure 12.16**):

- **Pawl assemblies** — The hook and finger should move in and out freely.

- **Halyard** — If damage or wear is found, the halyard should be replaced.

- **Halyard cable** — Check to see that it is taut when the ladder is in the **bedded position**. This check ensures proper synchronization of the upper sections during operation.

- **Pulleys** — Make sure they turn freely.

- **Ladder guides** — Check their condition and that the fly sections move freely.

If any discrepancies are found, remove the ladder from service until it can be repaired and tested. Ladders that cannot be safely repaired must be destroyed or scrapped for parts. **Skill Sheet 12-I-1** contains general procedures for cleaning, inspecting, and maintaining a ladder.

Because fire service ground ladders are subject to harsh conditions and physical abuse, they must be service tested to ensure that they continue to be fit for use. NFPA® 1932 serves as the guideline for ground ladder service testing. All ground ladders should be service tested before being placed in service, annually while in service, and after any use that exposes them to high heat or rough treatment. The standard has specified tests that the fire department or an approved testing agency should conduct. NFPA® 1932 further recommends that caution be used when performing service tests on ground ladders to prevent damage to the ladder or injury to personnel.

Cleaning Ladders

Regular and proper cleaning of ladders is more than a matter of appearance. Accumulated dirt or debris from a fire may collect and harden to the point where ladder sections cannot function as designed. Therefore, it is recommended that ladders be cleaned after every use.

> **Bedded Position** — Extension ladder with the fly section(s) fully retracted.

A soft bristle brush and running water are the most effective tools for cleaning ladders. Remove tar, oil, or greasy residues with mild soap and water or environmentally safe solvents according to departmental standard operating procedures (SOPs) and manufacturer's recommendations. Anytime a ladder is wet, whether after cleaning or use, wipe it dry. As they clean the ladder, firefighters should look for damage or wear. Any defects should be reported according to departmental SOPs. Where recommended by the manufacturer, occasional lubrication will maintain smooth operation of the ladder.

Maintaining Ladders

It is important to understand the difference between ground ladder maintenance and repair. Maintenance means keeping ladders in a state of usefulness or readiness. Repair means to restore or replace that which is damaged or worn out. All firefighters should be capable of performing routine maintenance on ground ladders according to departmental SOPs and the manufacturer's recommendations. Any ladders in need of repair require the service of a trained ladder repair technician.

General maintenance requirements for ground ladders include the following:

- Keep ground ladders free of moisture.
- Do not store or rest ladders in a position where they are subjected to vehicle exhaust or engine heat.
- Do not store ladders in an area where they are exposed to the weather.
- Do not paint ladders except for the top and bottom 18 inches (457 mm) of the beams for purposes of identification or visibility.

Handling Ladders

Ground ladders can be awkward or difficult to lift, carry, and place because of their weight and length. You must practice safe techniques to prevent personal injury and follow your department's ground ladder SOPs. You must also select the correct ladder for the task you have been assigned to perform.

Ladder Safety

Guidelines for safely carrying, raising, lowering, and working on ladders include the following:

- Develop and maintain adequate upper body strength.
- Wear a full body harness with safety line when training on ladders.
- Operate ladders according to departmental training and procedures.
- Wear full personal protective equipment, including gloves and helmet, when handling and working with ladders.
- Choose the correct ladder for the assigned task.
- Use leg muscles, not back or arm muscles, when lifting ladders below the waist.
- Use an adequate number of firefighters for each carry and raise.
- Do not raise any ladders to within 10 feet (3 m) of electrical wires.

- Secure the tip and anchor the foot of the ladder when in use during training or emergency incidents.

- Grasp extension ladder beams when extending or retracting to prevent fingers from being pinched or caught between sections.

- Check the ladder placement for the proper angle.

- Ensure that the hooks of the pawls are seated over the rungs **(Figure 12.17)**.

- Ensure that the ladder is stable before climbing (both butts in contact, with the ground/roof ladder hooks firmly set).

- Use caution when moving ladders sideways.

- Climb smoothly and rhythmically.

- Never overload the ladder (one firefighter every 10 feet [3 m] or one per section).

- Use a leg lock or ladder belt when working from a ground ladder.

- Relocate a positioned ladder only when ordered to do so.

- Use ladders for their intended purposes only.

- Inspect ladders for damage and wear after each use.

- Secure the foot of unattended ladders to a stationary object using ropes.

Figure 12.17 Pawls hold the fly section in place when securely seated and latched.

It is important that ladders be raised safely and smoothly to avoid injury to firefighters or damage to the ladder. Because speed is often required, movements should be smooth and controlled. Because more than one firefighter may be needed to raise ladders safely and efficiently, teamwork is also important. Individual and team proficiency in handling ladders is developed and maintained through training.

Ladder Selection

The Incident Commander (IC) or supervisor at an incident will usually tell you which ladder to use and/or where to place the ladder. However, in the absence of those orders, you must be able to select an appropriate ladder and a safe location for its placement on your own.

The following factors should be considered when deciding where to place a ladder:

- Needs of the situation
- Ladders available
- Assigned task

- Location of overhead obstructions
- Structural features such as type of roof, wall height, and presence of overhangs
- Wind direction and velocity
- Topography of the area

It is important to remember that when personnel are working on a roof or upper stories, there must be two means of escape requiring at least two ladders at remote locations from each other. Escape routes may be provided by ground ladders or aerial devices.

Selecting a ladder to reach a specific point requires the ability to judge distance. The base of the ladder should be placed away from the building approximately one-quarter of the vertical distance from the ground to the point of contact with the wall. This will provide the optimum climbing angle of approximately 75 degrees. Depending upon the height of the foundation and other factors, a residential story averages about 10 feet (3 m), and the distance from the floor to a windowsill averages about 3 feet (1 m). A commercial story averages 12 feet (4 m) from floor to floor, with a 4-foot (1.2 m) distance from the floor to windowsill. **Table 12.1** is a general guide that can be used in selecting ladders for specific locations.

Table 12.1
Ladder Selection Guide

Working Location of Ladder	Ladder Length
First-story roof	16 to 20 feet (4.9 m to 6.0 m)
Second-story window	20 to 28 feet (6.0 m to 8.5 m)
Second-story roof	28 to 35 feet (8.5 m to 10.7 m)
Third-story window or roof	40 to 50 feet (12.2 m to 15.2 m)
Fourth-story roof	over 50 feet (15.2 m)

Guidelines for ladder length include the following:

- Extend the ladder (a minimum of three to five rungs) beyond the roof edge to provide both a footing and a handhold for anyone stepping on or off the ladder **(Figure 12.18a)**.
- Place the tip of the ladder about even with the top of the window and to the windward (upwind) side of it to gain access to a narrow window or for opening the window for ventilation **(Figure 12.18b)**.
- Place the tip of the ladder just below the windowsill when rescue from a window opening is to be performed **(Figure 12.18c)**.

NOTE: Building walls or parapets that extend more than 6 feet (2 m) above the roof may require the use of an additional ladder to reach the roof deck. A roof or straight ladder can be placed on the roof side of the parapet and next to the extension ladder to assist firefighters to and from the roof.

The next step is to determine how far various ladders will reach based on their designated length. Remember that the designated length is a measurement of the total length of a single section ladder and the maximum extended length of an extension ladder. This is NOT THE LADDER'S REACH, because ladders are set at angles of approximately 75 degrees for climbing. Therefore, the reach will be LESS than the

Figure 12.18b A window to the windward side of a building may be accessed by a ladder placed next to the opening and even with the top of the opening.

Figure 12.18a Extend a ladder three to five rungs above the roofline to provide a handhold and footing for someone stepping on or off the ladder.

Figure 12.18c During operations involving rescue from a windowsill, the proper placement of a ladder has the tip just below the windowsill.

designated length. Single, roof, and folding ladders meeting NFPA® 1931 are required to have a measured length equal to the designated length. In the case of extension ladders, however, the maximum extended length may be as much as 6 inches (150 mm) LESS than the designated length.

Table 12.2 provides information on the maximum working heights of various ground ladders when placed at the proper climbing angle. The following measurements should be noted when considering the information contained in this table:

- For lengths of 35 feet (11 m) or less, reach is approximately 1 foot (300 mm) less than the designated length.

- For lengths over 35 feet (11 m), reach is approximately 2 feet (600 mm) less than the designated length.

Lifting and Lowering Methods

To prevent personal injuries, you must use proper lifting and lowering techniques when handling ground ladders as follows:

- Use the correct number of firefighters for the length and type of ladder to be lifted.

- Bend your knees, keeping your back as straight as possible, and lift with your legs, NOT WITH YOUR BACK OR ARMS (**Figure 12.19, p. 670**).

- Lift on the command of a firefighter who can see the other members of the team (**Figure 12.20, p. 670**).

Table 12.2 Maximum Working Heights for Ladders Set at Proper Climbing Angle	
Designated Length of Ladder	**Maximum Reach**
10 foot (3.0 m)	9 feet (2.7 m)
14 foot (4.3 m)	13 feet (4.0 m)
16 foot (4.9 m)	15 feet (4.6 m)
20 foot (6.1 m)	19 feet (5.8 m)
24 foot (7.3 m)	23 feet (7.0 m)
28 foot (8.5 m)	27 feet (8.2 m)
35 foot (10.7 m)	34 feet (10.4 m)
40 foot (12.2 m)	38 feet (11.6 m)
45 foot (13.7 m)	43 feet (13.1 m)
50 foot (15.2 m)	48 feet (14.6 m)

Figure 12.19 Proper lifting techniques must be used when lifting a ladder.

Firefighter giving commands

Figure 12.20 The firefighter giving commands during ladder operations must be able to see the other members of the team.

- Make it known immediately if you are not ready to lift a ladder when working with a team; lifting should occur in unison.
- Reverse the procedure for lifting when it is necessary to place a ladder on the ground before raising it as follows:
  — Lower the ladder using your leg muscles.
  — Keep your body perpendicular to the ladder and your feet parallel to the ladder so that when the ladder is placed it does not rest on your toes.

Ladder Carries

Ground ladders must be safely and quickly carried from the apparatus to the point where they are to be used. First, the ladder must be properly removed from the apparatus. On pumper apparatus, one or two firefighters should be able to remove the ladder. On aerial apparatus, three or four firefighters may be required. Because there are many different types of apparatus and means of mounting ladders, all carries in this section are demonstrated from the ground. In most cases, the ladders are carried butt end forward.

Removing Ladders from the Apparatus

Ground ladders are mounted on pumpers, aerial ladder apparatus, quints, and specialized apparatus. Ground ladders carried on pumper apparatus may be mounted in the following ways:

- Vertically, in racks on the right side of the apparatus
- Vertically, in a compartment between the hose bed and the right side of the body, accessed from the rear

- Horizontally, in a compartment under the right side of the hose bed, accessed from the rear of the apparatus
- In a mechanically operated rack that lowers the ladder from the top of the hose bed to the right hand side

On aerial and quint apparatus, the ladders may be mounted vertically on the left or right side of the apparatus bed or horizontally in racks within the bed, which are accessed from the rear of the apparatus. Specialized apparatus such as mobile water supply apparatus and aircraft rescue and fire fighting apparatus generally carry ladders vertically on the outside of the apparatus body.

To assist you in using the ground ladders mounted on your apparatus, you must know:

- The types, length, and location of ladders carried on your apparatus
- How the ladders are stored (racked) either with the butt toward the front or the rear of the apparatus
- How ladders are nested together
- How one nested ladder can be removed leaving the other securely in place
- The order in which nested ladders are racked
- Whether the extension ladder's fly is located on the inside or the outside when the ladder is racked on the side of the apparatus
- The method used to secure ladders in place
- The location at which mounting brackets extend through vertically mounted ladders (Many departments find it a good practice to mark ladders to indicate which rungs go in or near the brackets as shown in **Figure 12.21**.)

The procedures for removing ground ladders from the apparatus when the ladders are mounted on the side or top differ from those used when they are mounted in a flat position. To remove vertically mounted ladders, first unlatch the securing devices and lift the ladder off the bracket and into the correct carrying position **(Figure 12.22)**. To remove ladders that are stored internally in compartments, first open the compartment access panel (if there is one) then slide the ladder out to the proper carrying point. When multiple firefighters are required to carry the ladder, they stand on either side of the horizontally racked ladder and take their assigned location as the ladder is pulled out. Once the ladder has been removed from the apparatus, there are numerous ways it can be carried.

Figure 12.21 Ladders may be marked with an indicator to aid the mounting of ladders on an apparatus.

Figure 12.22 Vertically mounted ladders are latched in place with securing devices.

One-Firefighter Low-Shoulder Carry

Using a low-shoulder carry, one firefighter may be able to safely carry some single or roof ladders. One firefighter can also safely carry a 24-foot (7 m) extension ladder although two firefighters are preferred. To perform a low-shoulder carry, rest the ladder's upper beam on your shoulder with your arm between two rungs near the midpoint of the ladder. The butt of the ladder is carried forward. Carry the forward end of the ladder slightly lowered to provide better balance and allow you to see the way ahead. In this position, if the ladder should strike someone, the butt spurs will make contact with their body instead of their head. Do not open the hooks on a roof ladder until you are ready to ascend to the roof. **Skill Sheet 12-I-2** shows the steps for performing the one-firefighter low-shoulder carry.

CAUTION

Carrying the forward end of a ladder at eye level impedes the carrier's balance and visibility and increases the risk of the butt spurs striking someone else in the head.

Two-Firefighter Low-Shoulder Carry

Although the two-firefighter low-shoulder carry may be used with single or roof ladders, it is most commonly used for 24-, 28- and 35-foot (7 m, 9 m, and 11 m) extension ladders. The two-firefighter low-shoulder carry gives firefighters excellent control of the ladder **(Figure 12.23)**. The forward firefighter should place the free hand over the upper butt spur. This is done to prevent injury in case there is a collision with someone while the ladder is being carried. **Skill Sheet 12-I-3** describes the two-firefighter low-shoulder carry.

Figure 12.23 Firefighters have excellent control of the ladder when carrying from the low-shoulder position.

Figure 12.24 The flat-shoulder carry is used primarily on extension ladders up to 35 feet (11 m).

Figure 12.25 The four-firefighter flat arm's length carry enables a long extension ladder to be transported from the apparatus to the fire scene.

Three-Firefighter Flat-Shoulder Carry

The three-firefighter flat-shoulder carry is typically used on extension ladders up to 35 feet (11 m). This method has two firefighters, one at each end on one side of the ladder, and one firefighter on the other side in the middle **(Figure 12.24)**. **Skill Sheet 12-I-4** shows the procedure for carrying the ladder using the three-firefighter flat-shoulder carry.

Three-Firefighter Flat Arm's Length Carry

The three-firefighter flat arm's length carry begins with the extension ladder on the ground, fly section up. The firefighters are positioned with one at each end on one side of the ladder and one firefighter on the other side in the middle. Facing the butt while kneeling, the firefighters grasp the beam and stand holding the ladder at arm's length. Four firefighters can also be used to perform this carry using the positions described for the four-firefighter flat-shoulder carry **(Figure 12.25)**. **Skill Sheet 12-I-5** shows the procedure for carrying the ladder using the three-firefighter flat arm's length carry.

Four-Firefighter Flat-Shoulder Carry

The same flat-shoulder method used by three firefighters for carrying ladders is used by four firefighters except that the firefighters change position to accommodate the fourth firefighter. When four firefighters use the flat-shoulder carry, two are positioned at each end of the ladder, opposite each other **(Figure 12.26)**.

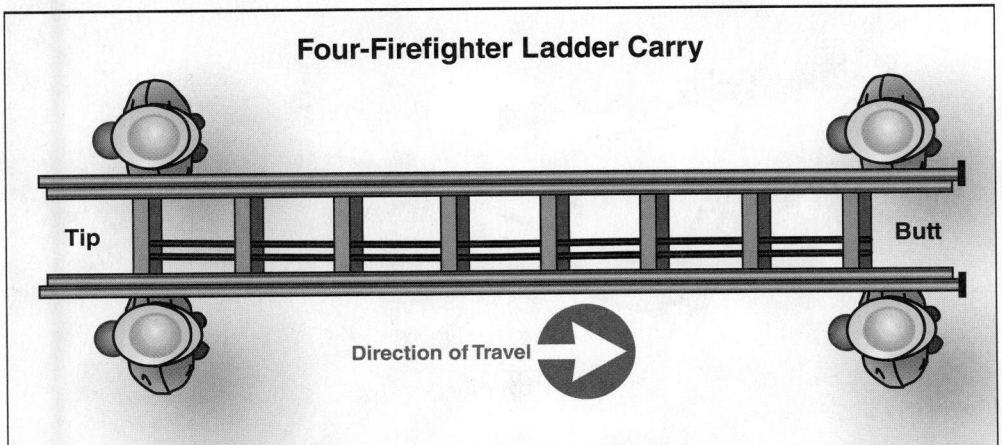

Four-Firefighter Ladder Carry

Tip

Butt

Direction of Travel

Figure 12.26 The flat-shoulder carry uses the same techniques for teams of three or four firefighters, but with a slight change in configuration between the teams.

Figure 12.27 The arm's length on-edge carry is performed with lightweight ladders.

Two-Firefighter Arm's Length On-Edge Carry

The two-firefighter arm's length on-edge carry is best performed with lightweight ladders (**Figure 12.27**). This carry is based on the fact that the firefighters are positioned on the bed section (widest) side of the ladder when it is in the vertical position. **Skill Sheet 12-I-6** shows the procedure for carrying the ladder using the two-firefighter arm's length on-edge carry.

Procedures for Carrying Roof Ladders

The procedures previously described are for carrying ladders butt forward. Because roof ladders are deployed so that the hooks can attach to the roof peak, they may be carried either butt first or tip first. The low-shoulder method is used to carry the roof ladder with the hooks closed to the extension ladder that will be climbed to access the roof (**Figure 12.28**). Both the butt-first and tip-first methods can be used by one or two firefighters.

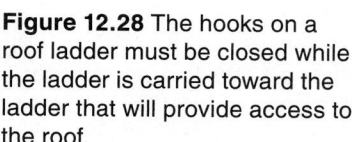

Figure 12.28 The hooks on a roof ladder must be closed while the ladder is carried toward the ladder that will provide access to the roof.

Figure 12.29 Open the hooks when the ladder has reached the staging area near where it will be used.

If the roof ladder is carried butt first from the apparatus, heel it at the base of the extension ladder. Set the ladder down, walk back to the tip and open the hooks **(Figure 12.29)**. Then raise the ladder and rest it on the extension ladder beam, climb the extension ladder, and shoulder the roof ladder at three to four rungs from the tip (with hooks turned out). Carry the ladder the rest of way and deploy it on the roof pitch.

If the roof ladder is carried tip first, carry it to the base of the extension ladder, remove it from the shoulder, placing the butt on the ground, and walk the hands to the tip without laying the ladder down. Open the hooks outward and then return to the center point and shoulder the ladder. Proceed up the extension ladder to the top and deploy the roof ladder.

Placement of Ground Ladders

Proper placement of ground ladders helps to ensure the safety and efficiency of fire-ground operations. The sections that follow contain some of the basic considerations and requirements for ground ladder placement.

Responsibility for Placement

Normally, an officer designates the general location where the ladder is to be placed and the task to be performed, but personnel carrying the ladder frequently decide on the exact spot where the butt is to be placed. Usually, the firefighter nearest the butt is the logical person to make this decision. When there are two firefighters at the butt, the one on the right side is usually the one responsible for placement. Because this guideline may vary from one department to another, firefighters must always follow their department SOPs.

Factors Affecting Ground Ladder Placement

There are two objectives to be met when placing ground ladders:

- Positioning the ladder properly for its intended use.
- Placing the butt the proper distance from the building for safe and easy climbing.

Numerous factors dictate the exact place to position the ladder. If a ladder is to be used for positioning a firefighter to break a window for ventilation, place it alongside the window to the windward (upwind) side. The tip should be about even with the top of the window. The same position can be used when firefighters need to climb in or out of narrow windows or direct hose streams into them.

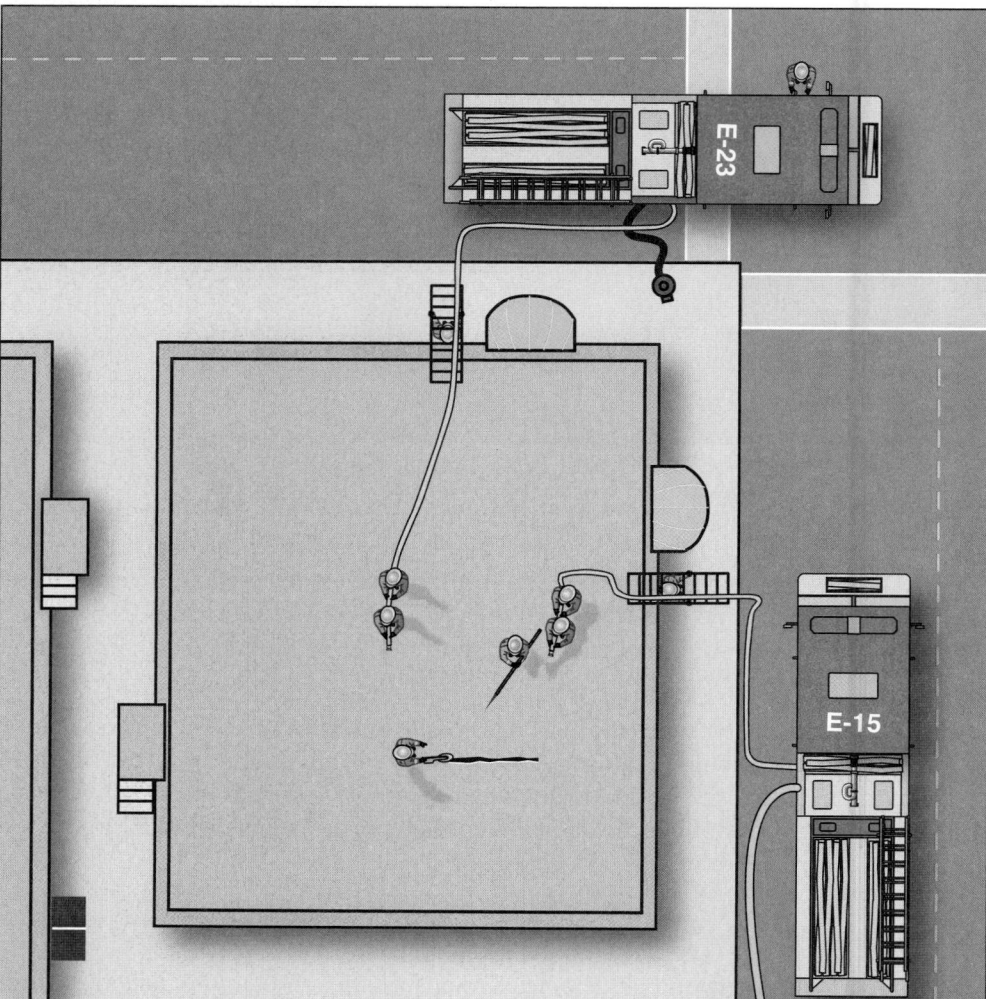

Figure 12.30 Ladders must be placed at two points on different sides of the building.

If a ladder is to be used for entry or rescue from a window, the ladder tip is placed slightly below the sill. If the sill projects out from the wall, the tip of the ladder can be wedged under the sill for additional stability. If the window opening is wide enough to permit the ladder tip to project into it and still allow room beside it to facilitate entry and rescue, place the ladder so that two or three rungs extend above the sill.

Other ladder placement guidelines include the following:

- Place ladders at least two points on different sides of the building (**Figure 12.30**).

- Avoid placing ladders over openings such as windows and doors where they might be exposed to heat or direct flame contact.

- Take advantage of strong points in building construction (such as the corners) when placing ladders.

- Raise the ladder directly in front of the window when it is to be used as a support for a smoke ejector removing cold smoke after a fire has been extinguished. Place the ladder tip on the wall above the window opening.

- Avoid placing ladders where they may come into contact with overhead obstructions such as wires, tree limbs, or signs.

- Avoid placing ladders on uneven terrain or on soft spots.

- Avoid placing ladders in front of doors or other paths of travel that firefighters or evacuees will need to use. Instead, place the ladder to the side of the opening.

- Avoid placing ladders on top of sidewalk elevator trapdoors or sidewalk deadlights. These areas may give way under the added weight of firefighters, their equipment, and the ladder.

- Do not place ladders against unstable walls or surfaces.

The distance of the butt from the building establishes the angle formed by the ladder and the ground. If the butt is placed too close to the building, its stability is reduced because the weight of the person climbing tends to cause the tip to pull away from the building. With the exception of certain rescue situations, when the ladder has been raised into place, the desired angle of inclination is approximately 75 degrees (**Figure 12.31**). A 75-degree angle provides the following benefits:

- Good stability

- Less stress placed on the ladder

- Optimum climbing angle

- Easiest climbing position: the climber can stand perpendicular to the ground, at arm's length from the rungs

If the butt of the ladder is placed too far away from the building, the load-carrying capacity of the ladder is reduced and it has more of a tendency to slip. If placement at such angles becomes necessary, either tie in or heel (steady) the bottom of the ladder at all times. (See Securing the Ladder section for tying in and heeling instructions.)

An easy way to determine the proper distance between the butt of the ladder and the building is to divide the working length (length actually used) of the ladder by 4. For example, if 20 feet (6 m) of a 28-foot (8.5 m) ladder is needed to reach a window, the butt end should be placed 5 feet (1.5 m) from the building (20 feet divided by 4 [6 m divided by 4]) (**Figure 12.32**). Note that only the length of ladder used to reach the window, not the ladder's overall length, is used in this calculation. Exact measurements are unnecessary on the fire scene.

Experienced firefighters develop the ability to visually judge the proper positioning for the ladder. When the ladder is at the proper angle, a firefighter standing straight up on the bottom rung should be able reach straight ahead and grasp a rung directly in front of him or her. Ladders are also equipped with an inclination marking on the outside of the beam which aligns perfectly vertical and horizontal when the ladder is properly set (**Figure 12.33, p. 678**).

Figure 12.31 Ladder placement with an angle of 75 degrees creates an easy climbing angle with good stability and other safety features.

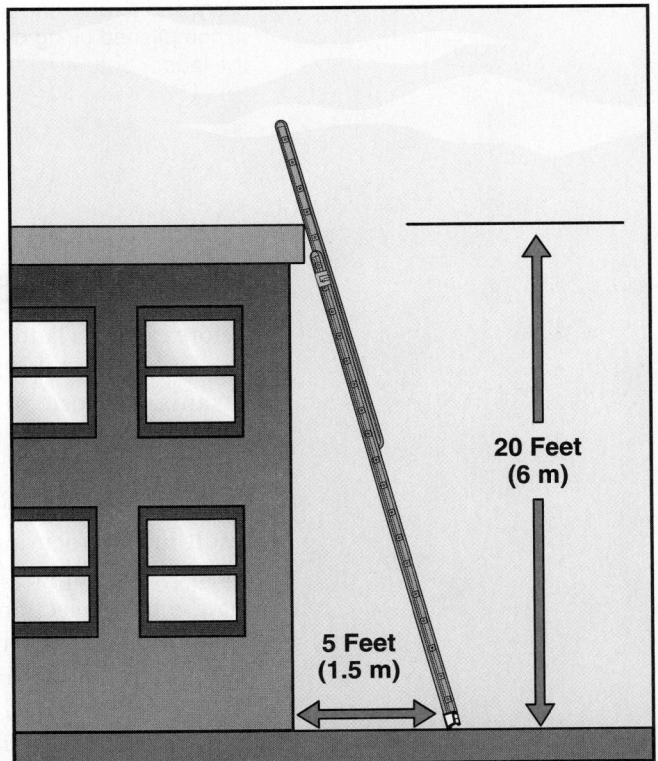

Figure 12.32 A quick calculation to determine the distance between the heel of the ladder and the base of the building is to divide the height of the target by 4.

Figure 12.33 The proper angle of a ladder does not change by incident and may be easily tested by standing on the bottom rung and reaching for a rung at arm level.

Ladder Raises

A properly positioned ladder becomes a means by which rescue and fire fighting operations can be performed. To be most effective, teamwork, smoothness, and rhythm are necessary when raising and lowering fire department ladders. There are numerous ways to safely raise ground ladders. These methods vary depending on the type and size of the ladder, number of personnel available to perform the raise, and weather and topography considerations. The raises described here represent some of the more commonly used methods. In addition, the sections that follow begin with general procedures to follow before raising a ladder.

NOTE: The following sections only contain step-by-step information for raising ladders. In every case, the procedure for lowering the ladder is to reverse the listed steps in the order given.

Transition from Carry to Raise

The methods and precautions for raising single-section and extension ladders are much the same. It is not necessary to place the ladder flat on the ground prior to raising it; only the butt needs to be placed on the ground (**Figure 12.34**). The transition from the carrying position to the raise can and should be done in one smooth and continuous motion.

Figure 12.34 The transition between a carry and a raise can be accomplished using only the ladder butt.

Considerations Before Raising

Before raising a ladder, there are a number of things you need to consider and precautions you must take. Some of the more important ones are the presence of electrical hazards, the position of the extension ladder fly section, and tying the halyard.

Electrical Hazards

Ladders or the people climbing them coming in contact with live electrical wires can result in electrocution leading to death or severe injury. To avoid electrical contact hazards, look up to check for overhead electrical wires or equipment BEFORE making the final selection on where to place a ladder or what method to use for raising it. Look up AGAIN before raising the ladder. The Occupational Safety and Health Administration (OSHA) requires that all ladders must be kept a distance of at least 10 feet (3 m) from all energized electrical lines or equipment. This distance must be maintained while raising the ladder, using the ladder, and lowering the ladder. In some cases, a ladder may come to rest a safe distance from electrical equipment but come too close to the equipment during the actual raise. In these cases, an alternate method for raising the ladder, such as raising parallel to the structure as opposed to perpendicular, may be required (**Figure 12.35**).

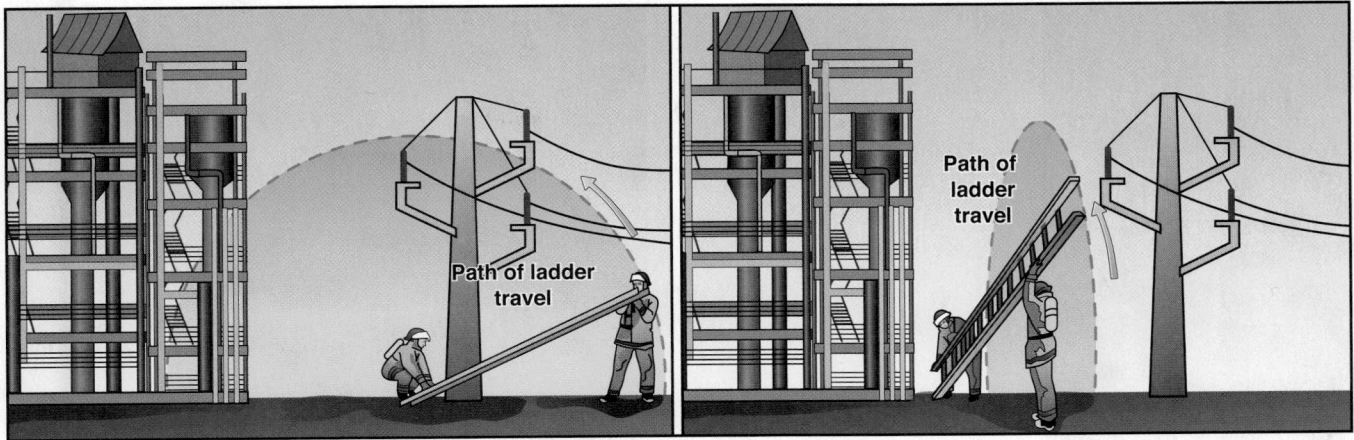

Figure 12.35 Ladder raises should be adapted to the clearances available.

WARNING

All ladders will conduct electricity, especially when wet, regardless of their construction material.

Position of the Fly Section on Extension Ladders

Each ladder manufacturer specifies whether the ladder should be placed with the fly in, toward the structure, or out, away from the structure. This recommendation is based on the ladder design, construction materials, and the fly position at which the manufacturer's tests show it to be strongest. Failure to follow this recommendation could void the ladder's warranty if a failure or damage occurs.

In general, all modern metal and fiberglass ladders are designed to be used with the FLY OUT. Wooden ladders that are designed with the rungs mounted in the top truss rail are intended to be used with the FLY IN (**Figures 12.36a and b, p. 680**). Again, consult departmental SOPs or the manufacturer of the ladder to determine the correct fly position.

Some departments have ladders that are intended to be used with the fly out but prefer that the firefighter extending the halyard be on the outside of the ladder. In this case, firefighters will need to pivot or roll the ladder 180 degrees after it has been extended.

Tying the Halyard

Once an extension ladder is resting against a structure and before it is climbed, the excess halyard should be tied to the ladder with a clove hitch and an overhand safety to prevent the fly from slipping and to prevent anyone from tripping over the rope (**Figure 12.37, p. 680**). In rescue situations where speed is critical, it is not always necessary to wrap the excess halyard before tying off, but it should be placed out of the way. The same tie can be used for either a closed- or open-ended halyard. **Skill Sheet 12-I-7** describes the procedure for tying the halyard.

One-Firefighter Raise

One firefighter may safely raise single ladders and small extension ladders. The following procedures should be used to perform these raises.

Figure 12.36a Extension ladders should be used in accordance with the AHJ regarding the positioning of the fly section.

Figure 12.37 An extension ladder's halyard should be secured to the ladder to prevent the fly from slipping and to secure excess rope.

Figure 12.36b Wooden ladders are designed to be used with the fly section on the inside of the angle formed by the ladder and the structure wall.

One-Firefighter Single Ladder Raise

Single and roof ladders are generally light enough that one firefighter can place the butt end at the point where it will be located for climbing without steadying it against the building or another object before raising. The steps described in **Skill Sheet 12-I-8** should be used to perform the one-firefighter raise for both single and extension ladders.

One-Firefighter Extension Ladder Raise

One method of raising extension ladders with one firefighter is from the low-shoulder carry; however, a different procedure for placing the ladder butt is used. When using the one-firefighter raise from the low-shoulder carry, a building is used to heel the ladder to prevent the ladder butt from slipping while the ladder is brought to the vertical position.

Two-Firefighter Raises

Space permitting, it makes little difference if a ladder is raised parallel with or perpendicular to a building (**Figures 12.38a and b**). If raised parallel with the building, the ladder must be pivoted after it is in the vertical position. Whenever two or more firefighters are involved in raising a ladder, the firefighter at the butt end, called the *heeler*, is responsible for placing it at the desired distance from the building and determining whether the ladder will be raised parallel with or perpendicular to the building. The heeler also gives commands during the operation. There are two basic

Figure 12.38a An extension ladder beam raise may be pivoted parallel to the structure that the ladder will rest against.

Figure 12.38b A flat raise requires more clearance space than a parallel raise.

ways for two firefighters to raise a ladder: the flat raise and the beam raise. **Skill Sheet 12-I-9** describes the procedure for the two-firefighter flat raise. **Skill Sheet 12-I-10** shows the procedure for the two-firefighter beam raise.

Three-Firefighter Flat Raise

As the length of the ladder increases, the weight also increases requiring more personnel for raising the larger extension ladders. Typically, at least three firefighters should be used to raise ladders of 35 feet (11 m) or longer **(Figure 12.39)**. **Skill Sheet 12-I-11** describes the procedure for flat-raising ladders with three firefighters.

To raise a ladder using the beam method with three firefighters, follow the same procedures for the two-firefighter flat raise. The only difference is that the third firefighter is positioned along the beam **(Figure 12.40)**. Once the ladder has been raised to a vertical position, follow the procedures described for the flat raise.

Figure 12.39 The process of raising a ladder 35 feet (11 m) or longer requires three or more personnel.

Figure 12.40 A three-firefighter beam raise uses the same procedures as the two-firefighter method except that the third firefighter is positioned along the beam.

Four-Firefighter Flat Raise

When available, four firefighters can be used to better handle the larger and heavier extension ladders. A flat raise is normally used, and the procedures for raising the ladder are similar to the three-firefighter raise except for the placement of personnel. A firefighter at the butt is responsible for placing the butt at the desired distance from the building and determining whether the ladder will be raised parallel with or perpendicular to the building.

Deploying a Roof Ladder

There are a number of ways to deploy a roof ladder on a pitched roof. Once the roof ladder has been carried up the extension ladder to the roof, the firefighter locks in with one leg or connects a ladder belt to an appropriate rung. Remove the ladder from the shoulder and slide it on the beam, hooks out, up to the peak. When the hooks are over the peak, turn the ladder onto both beams, hooks over the peak and pull down to ensure the hooks have engaged the roof. An alternate method allows the roof ladder to be deployed by sliding it on both beams until the hooks engage the peak. **Skill Sheet 12-I-12** shows the procedure for one firefighter to deploy a roof ladder in position. Two firefighters can deploy the roof ladder as illustrated in **Figure 12.41**.

Procedures for Moving Ground Ladders

In some cases, the basic ladder-raising procedures are not sufficient to get the ladder into its final position for use. In these situations it will be necessary to move the ladder slightly after it has been extended.

Figure 12.41 Two firefighters may deploy a roof ladder from an extension ladder.

Figure 12.42 A two-firefighter pivot is used to rotate a ladder that is safe to be raised by two firefighters.

Pivoting Ladders with Two Firefighters

Occasionally, an extension ladder is raised with the fly in the incorrect position for deployment. When this happens, it is necessary to pivot the ladder. Any ladder flat-raised parallel to the building also requires pivoting to align it with the wall upon which it will rest. Pivot the ladder on the beam closest to the structure. Whenever possible, pivot the ladder before it is extended.

The two-firefighter pivot may be used on any ground ladder that two firefighters can raise (**Figure 12.42**). The procedure described in **Skill Sheet 12-I-13** is for a ladder that must be turned 180 degrees to get the fly section in the proper position. The same procedure is used for positioning a ladder that was flat-raised parallel to the building. In this case, the beam nearest the building is used to pivot the ladder 90 degrees.

Shifting Raised Ground Ladders

Occasionally, circumstances require that ground ladders be moved while vertical. Because they are hard to control, shifting a ladder that is in a vertical position should be limited to short distances such as aligning ladders perpendicular to a building or to an adjacent window.

One firefighter can safely shift a single ladder that is 20 feet (6 m) long or less. The procedure for the one-firefighter shift is described in **Skill Sheet 12-I-14**. Because of their weight, extension ladders require two firefighters for the shifting maneuver described in **Skill Sheet 12-I-15**. Another way to shift a ladder a short distance from side to side is to place the ladder against the building, slide the top of the ladder sideways, and then pick up the butt and move it into position.

Securing the Ladder

For safety, ground ladders must be secured whenever firefighters are climbing or working from them. There are two methods for securing a ladder: heeling and tying in. The process of securing a ground ladder may include any or all of the following:

- Lock the extension ladder locks in place before the ladder is placed against the structure.

- Tie the halyard with a clove hitch and an overhand safety (extension ladder only).

- Prevent movement of the ladder away from the building by heeling and/or securing it with a rope to a nearby firm object.

Heeling

One way of preventing movement of a ladder is to properly heel it. There are several methods for properly heeling, also known as *footing*, a ladder.

One method is for a firefighter to stand beneath the ladder with feet about shoulder-width apart (or one foot slightly ahead of the other). The firefighter then grasps the ladder beams at about eye level and pulls backward to press the ladder against the building (**Figure 12.43a**). When using this method, the firefighter heeling the ladder must wear complete PPE and not look up when there is someone climbing the ladder. The firefighter must grasp the beams and not the rungs and be alert for falling objects or debris.

Another method of heeling a ladder is for a firefighter to stand on the outside of the ladder and chock the butt end with one foot. With this method, either the firefighter's toes are placed against the butt spur or one foot is placed on the bottom rung (**Figure 12.43b**). The firefighter grasps the beams and presses the ladder against the

Figure 12.43a A firefighter heeling a ladder may stand on the inside of the ladder and lean backward.

Figure 12.43b A firefighter heeling a ladder may stand on the outside of the ladder and use his or her foot to chock the butt end.

building. The firefighter heeling the ladder must stay alert for firefighters descending the ladder. **Skill Sheet 12-I-16** illustrates the two methods for heeling a ground ladder. Full PPE with helmet faceshield deployed must be worn when heeling the ladder.

Tying In

Whenever possible, a ladder should be tied securely to a fixed object **(Figure 12.44)**. Tying in a ladder is simple, can be done quickly and is strongly recommended to prevent the ladder from slipping or pulling away from the building. Tying in also frees personnel who would otherwise be holding the ladder in place. A **rope hose tool** or safety strap can be used between the ladder and a fixed object.

Rope Hose Tool — Piece of rope spliced to form a loop through the eye of a metal hook; used to secure hose to ladders or other objects.

Climbing Ladders

Ladder climbing should be done smoothly and rhythmically. You should ascend the ladder so that there is the least possible amount of bounce and sway. This smoothness is accomplished if your knee is bent to ease the weight on each rung. Balance on the ladder will come naturally if the ladder is properly spaced away from the building to create an optimum climbing angle that puts your body perpendicular to the ground (usually a 75-degree angle).

The climb starts after the climbing angle has been checked and the ladder is properly secured. Your eyes should be focused forward, with an occasional glance at the tip of the ladder. Keep your arms straight (horizontal) during the climb; this keeps your body away from the ladder and permits free knee movement **(Figure 12.45)**. When no equipment is being carried, place your hands on the rungs. Grasp the rungs with palms down and your thumbs beneath the rungs. Grasp alternating rungs while

Figure 12.44 Securing a ladder to a fixed object prevents the ladder from slipping or pulling away from its position.

Figure 12.45 Maintaining distance from the ladder while climbing permits free movement.

Figure 12.46 When carrying a tool up or down a ladder, the tool hand may be slid along the beam.

Figure 12.47 A leg lock counts as a safety device while working with both hands on a ladder.

climbing. Coordinate hand and foot movement so that the right hand and left foot are in contact with the ladder as you move the opposite hand and foot to the next rungs. Place feet near the beams with the halyard tied in the center of the rung.

If your feet should slip, your arms and hands are in a position to stop the fall. Climb using your leg muscles and not your arm muscles. Your arms and hands should not reach above your head while climbing because that will bring your body too close to the ladder.

Practice climbing slowly to develop form rather than speed. Speed develops with repetition after the proper technique is mastered. Too much speed results in lack of body control, and quick movements cause the ladder to bounce and sway.

You may be required to carry equipment up and down a ladder during an emergency operation. This disrupts your natural climbing motion either because of the added weight or the need to use one hand to hold the tool. If a tool is carried in one hand, it may be desirable to slide the free hand under the beam while making the climb **(Figure 12.46)**. This method permits constant hand contact with the ladder. Whenever possible, a utility rope should be used to hoist tools and equipment rather than carrying them up a ladder.

Working from a Ladder

Firefighters must sometimes work with both hands while standing on a ground ladder. Either a ladder belt or a leg lock can be used to safely secure the firefighter to a ladder while performing work **(Figure 12.47)**. If you choose to apply a leg lock on a ground ladder, the procedure in **Skill Sheet 12-I-17** should be used.

Figure 12.48 A ladder belt secures a firefighter to a ladder so he or she may work with both hands.

If a ladder belt is used, it must be strapped tightly around the waist (**Figure 12.48**). The hook may be moved to one side, out of the way, while you are climbing the ladder. After reaching the desired height, slide the hook to the center of your body and attach it to a rung. According to NFPA® 1983, *Standard on Life Safety Rope and Equipment for Emergency Services*, a ladder belt is rated only as a positioning device for use on a ladder and does not meet the requirements of life safety harnesses.

Assisting a Victim Down a Ladder

When a ground ladder is intended to be used for rescue through a window, the ladder tip is raised to just below the sill so that it is easier for a conscious victim to climb onto the ladder or for firefighters to lift an unconscious victim onto the ladder. The ladder is heeled and all other loads and activity removed from it during rescue operations. Because even healthy, conscious occupants are probably unaccustomed to climbing down a ladder, they must be protected from slipping and perhaps falling. To bring victims down a ground ladder, at least four firefighters are needed: two inside the building, one or two on the ladder, and one heeling the ladder. The following methods can be performed for assisting a victim down a ladder when it is set at the normal climbing angle; however, these methods work better if the ladder is set at a slightly steeper angle.

The method chosen for assisting a victim down a ladder first depends upon whether the victim is conscious or unconscious. Conscious victims are usually the easiest to lower. Conscious victims can be lowered feet first (facing the building) onto a ladder (**Figure 12.49, p. 688**). The method for assisting a conscious victim down a ladder is shown in **Skill Sheet 10-I-18**.

An unconscious victim can be held on a ladder in the same way as a conscious victim except that the victim's body rests on the rescuer's supporting knee. The victim's feet must be placed outside the rails to prevent entanglement (**Figure 12.50, p. 688**). The rescuer grasps the rungs to provide a secure hold on the ladder and help to protect the victim's head from hitting the ladder.

An alternative lowering position is to use the same hold but have the victim turned to face the rescuer. This position reduces the chances of the victim's limbs catching between the rungs. The unconscious victim facing the rescuer is supported at the crotch by one of the rescuer's arms and at the chest by the other arm. The rescuer may be aided by another firefighter. The methods for assisting an unconscious victim down a ladder are shown in **Skill Sheet 12-I-19**.

Figure 12.49 A conscious victim is assisted down a ladder with his or her face toward the ladder.

Figure 12.50 An unconscious victim is assisted down a ladder facing away from the ladder.

Assisting an Unconscious Victim

An unconscious victim who regains consciousness while being rescued down a ladder may grab the ladder or the rescuer. This sudden reaction by the victim may cause the firefighter to lose his/her grip or footing and increase the risk of falling off the ladder, especially if the victim is facing the firefighter. It is critical that any firefighter assisting an unconscious victim down a ladder observe the victim and be prepared to respond if the victim regains consciousness.

A victim's size also plays a factor in lowering him or her down a ladder. Large victims require more personnel to move and may require more equipment as well. Removing extraordinarily heavy victims requires two rescuers. Two ground ladders are placed side by side. One rescuer supports the victim's waist and legs. A second rescuer on the other ladder supports the victim's head and upper torso. Small children who must be brought down a ladder can be cradled across the rescuer's arms (**Figure 12.51**).

Figure 12.51 A small child may be cradled in a rescuer's arms on the way down a ladder.

Chapter Summary

Ground ladders are an essential tool used to access levels above and below ground. You must know the types of ground ladders your department uses, the parts and construction materials, and how to care for and maintain them. You must be able to select, carry, and place them to effectively gain access and perform your assigned tasks. You must know the correct methods for safely climbing, working from, and assisting victims down ground ladders.

Review Questions

Firefighter I

1. What are the basic parts of fire service ladders?

2. How does a fire service ladder constructed of metal differ from one constructed of fiberglass?

3. What types of markings and labels do fire service ladders have?

4. How do the five types of ladders used in the fire service compare with one another?

5. What types of information do general ladder inspections look for?

6. What are the general maintenance requirements for ground ladders?

7. What are the guidelines for safely carrying, raising, lowering, and working on ladders?

8. What factors must be considered when selecting a ladder placement location?

9. What techniques should be used to prevent personal injuries when lifting and lowering ladders?

10. What information must a firefighter know in order to use ground ladders?

11. What are the two objectives that must be met when placing ground ladders?

12. How can a firefighter determine the proper distance between the heel of the ladder and the building?

13. What considerations must be addressed before raising a ladder?

14. What are two methods of safely moving a ground ladder after it has been raised?

15. How do the two methods used for securing ladders compare with one another?

16. How can a firefighter climb a ladder so that there is the least possible amount of bounce and sway?

17. What methods can be used to secure a firefighter to a ladder when performing work?

18. How many firefighters are needed to bring a victim down a ladder?

Clean

Step 1: Place the ladder flat on the sawhorses, lifting and carrying appropriately.

Step 3: Rinse the ladder thoroughly with clean water.

Step 2: Clean all parts of the ladder with scrub brush and cleaning solution, removing greasy residues with approved cleaners.

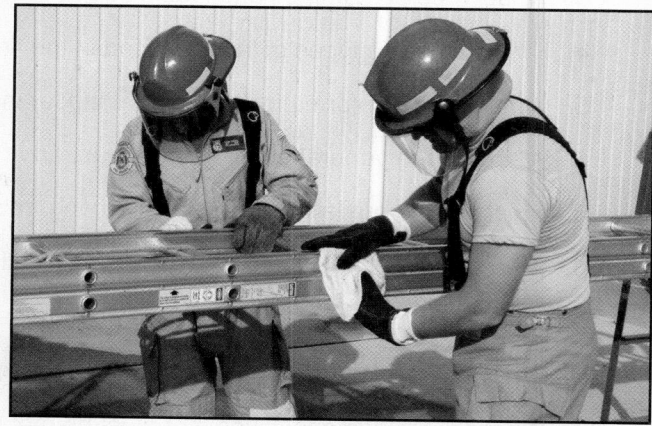

Step 4: Dry the ladder thoroughly with clean, dry cloths.

Inspect

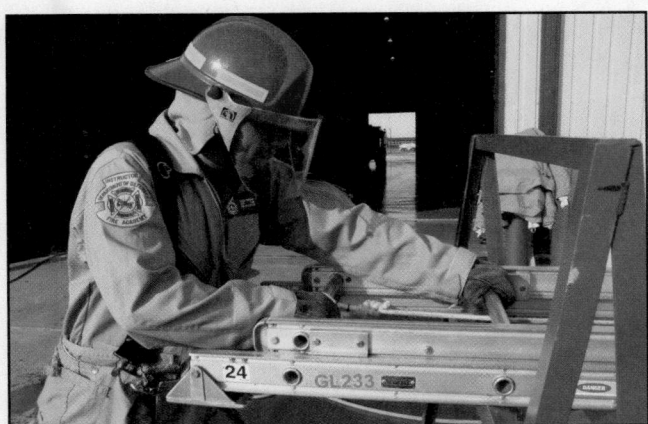

Step 5: Inspect each part of the ladder, noting any:

    a. Looseness

    b. Cracks

    c. Dents

    d. Unusual wear

    e. Bent rungs or beams

    f. Heat damage, deformities or change in sensor label

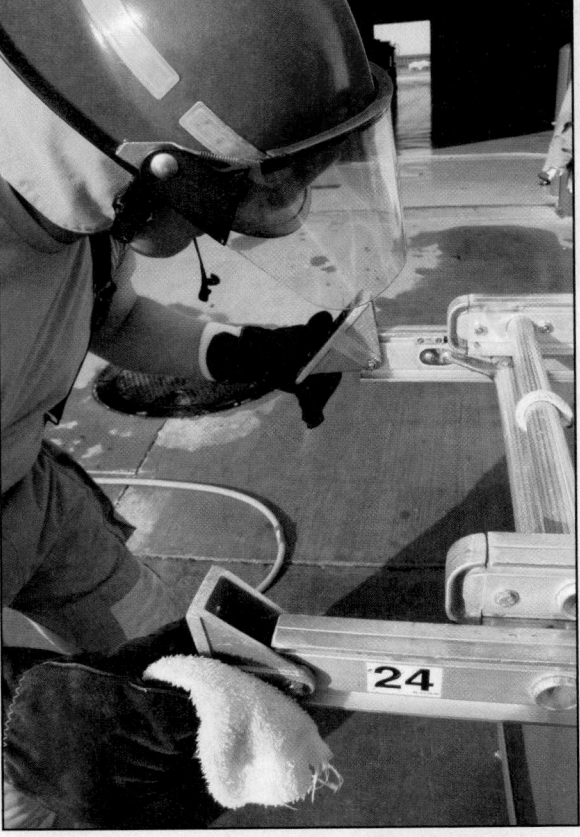

Step 8: Inspect all movable parts (extension, roof, and pole ladders).

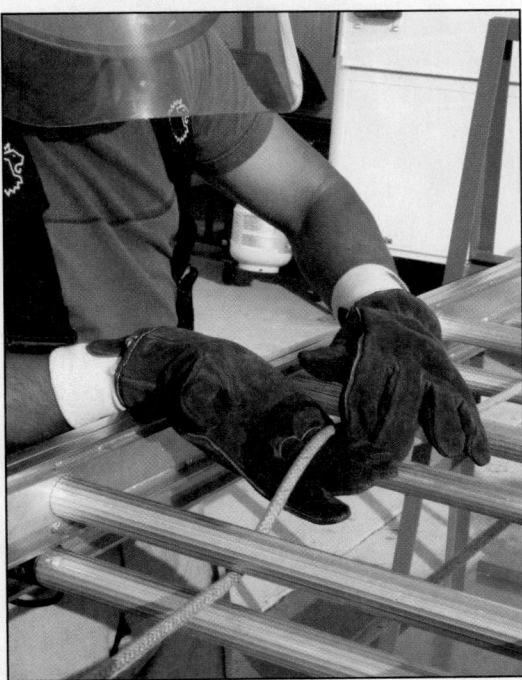

Step 6: Circle any defects found with chalk or grease pen.

Step 7: Inspect the ladder halyard (extension ladders) for:

    a. Fraying or kinking

    b. Snugness of cable when in bedded position

Maintain

Step 9: Lubricate parts as needed and per manufacturer's guidelines.

Step 10: Replace halyard if necessary.

Step 11: Tag and remove from service for any conditions that cannot be corrected with cleaning, inspection, and simple maintenance. Notify officer.

Step 12: Record cleaning, inspection, and maintenance performed.

NOTE: Firefighters must be wearing appropriate PPE before performing this skill.

Step 1: Position yourself at lifting point near the center of the ladder.

Step 2: Kneel beside the ladder.
- **a.** At lifting point
- **b.** Facing ladder tip
- **c.** On knee closest to ladder

Step 3: Grasp the ladder rung opposite your knee.
- **a.** With hand closest to ladder
- **b.** Palm forward

Step 4: Stand the ladder on edge.
- **a.** Pivot on nearer beam, raising farther beam

Step 5: Stand up.
- **a.** Use leg muscles, keeping back straight and vertical

Step 6: Reposition yourself for carrying.
- **a.** As ladder is brought up, pivot toward butt end of ladder
- **b.** Insert other arm through rungs

Step 7: Position ladder for carrying.
- **a.** Upper beam resting on shoulder
- **b.** Butt lowered slightly
- **c.** Steadied with both hands

Step 8: Lower the ladder to the ground.
- **a.** Reverse lifting procedure
- **b.** Body and toes parallel to ladder

12-1-3
Carry a ladder – Two-firefighter low-shoulder method.

SKILL SHEETS

NOTE: Firefighters must be wearing appropriate PPE while performing this skill. Firefighter #1 is located near the ladder butt end and is in command of the lifting operation. Firefighter #2 is located near the ladder tip and is in command of the carrying operation. This skill begins with the ladder lying flat on the ground.

Step 1: Firefighters #1 and #2: Kneel on the same side of the ladder, one at either end, facing the tip.

   a. The knee closest to the ladder is the one touching the ground

Step 2: Firefighter #1: Give the command "Prepare to beam."

Step 3: Both Firefighters: Grasp the ladder beam away from your body.

Step 4: Firefighter #1: Give the command to "Beam."

Step 5: Both firefighters: Pull the ladder into position against them, resting the ladder on its beam.

Step 6: Firefighter #1: Give the command "Prepare to shoulder the ladder."

Step 7: Firefighter #1: Give the command "Shoulder the ladder."

Step 8: Both Firefighters: Stand erect, lifting smoothly and continuously.

Step 9: Both Firefighters: Pivot to face toward the butt end, extending free arm between two rungs to place beam onto shoulders at the same time.

Step 10: Firefighter #2: Now in command of advancing the ladder, gives the command to "Advance."

NOTE: Firefighters must be wearing appropriate PPE while performing this skill. Firefighter #1 is located near the ladder butt end and is in command of the lifting operation. Firefighter #2 is located near the ladder tip and is in command when advancing the ladder. Firefighter #3 is located on the opposite side and at a midpoint of the ladder. This skill begins with the ladder lying flat on the ground.

Step 1: Firefighters #1 and #2: Kneel on the same side of the ladder, one at either end, facing the tip.

> **a.** The knee closest to the ladder is the one touching the ground

Step 2: Firefighter #3: Kneel on the opposite side at midpoint, also facing the ladder tip.

> **a.** The knee closest to the ladder is the one touching the ground

Step 3: All Firefighters: Grasp a rung with the near hand, palm rearward.

Step 4: Firefighter #1: Give the command "Prepare to shoulder the ladder."

Step 5: Firefighter #1: Give the execution command "Shoulder the ladder."

Step 6: All Firefighters: Stand erect, lifting smoothly and continuously.

Step 7: All Firefighters: Pivot to face toward the butt end when the ladder is about chest high.

Step 8: All Firefighters: Extend arm through the ladder to place the beam onto shoulders.

Step 9: Firefighter #2: Now in command of advancing the ladder, gives the command to advance.

12-I-5
Carry a ladder — Three-firefighter flat-arm's length method.

NOTE: Firefighters must be wearing appropriate PPE while performing this skill. Firefighter #1 is located near the butt end of the ladder. Firefighter #2 is located on the same side and near the tip end of the ladder and is in command of the operation. Firefighter #3 is located on the opposite side and at a midpoint of the ladder. This skill begins with the ladder lying flat on the ground.

Step 1: Firefighters #1 and #2: Kneel on one side of the ladder, one at either end, facing the butt end.

> **a.** The knee closest to the ladder is the one touching the ground

Step 2: Firefighter #3: Kneel on the opposite side at midpoint, also facing the butt end.

> **a.** The knee closest to the ladder is the one touching the ground

Step 3: All three Firefighters: Grasp a rung with the near hand, palm rearward.

Step 4: Firefighter #2: Give the command "Prepare to lift the ladder."

Step 5: Firefighter #2: Give the command "Lift the ladder."

Step 6: All three firefighters: Lift the ladder smoothly and continuously to arm's length using leg muscles to stand erect.

Step 7: Firefighter #2: Now in command of advancing the ladder, gives the command to advance.

SKILL SHEETS

12-I-6
Carry a ladder – Two-firefighter arm's length on edge method.

NOTE: Firefighters must be wearing appropriate PPE while performing this skill. Firefighter #1 is located near the butt end of the ladder. Firefighter #2 is located near the tip end of the ladder and is in command of the operation.

Step 1: Both Firefighters: Kneel on the same side of the ladder, facing the ladder butt end.

    a. The knee closest to the ladder is the one touching the ground

Step 2: Firefighter #2: Give the command "Prepare to beam."

Step 3: Both Firefighters: Grasp the ladder beam away from them.

Step 7: Both Firefighters: Grasp the top beam of the ladder with the hand closest to it.

Step 8: Firefighter #2: Give the command "Lift the ladder."

Step 9: Both Firefighters: Lift the ladder smoothly and continuously to arm's length using leg muscles to stand erect.

Step 4: Firefighter #2: Give the command to "Beam."

Step 5: Both Firefighters: Pull the ladder into position against them, resting the ladder on its beam.

Step 6: Once ladder is in position, Firefighter #2, with clear view of Firefighter #1, gives the command "Prepare to lift the ladder."

Step 10: Firefighter #2: Now in command of advancing the ladder, gives the command to "Advance."

NOTE: Firefighters must be wearing appropriate PPE before performing this skill.

Step 1: Wrap the excess halyard around two convenient rungs.

Step 2: Pull the halyard taut.

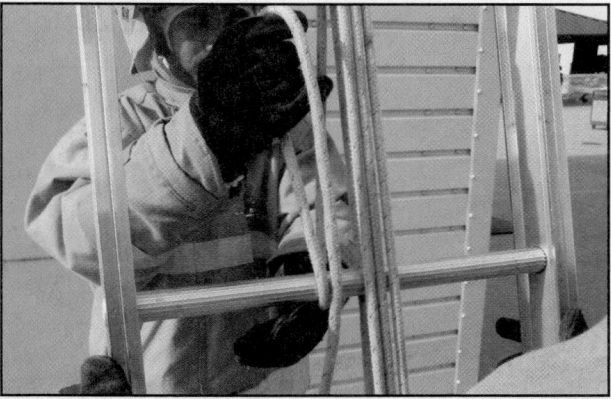

Step 5: Push the halyard underneath and back over the top of the rung.

Step 6: Grasp the halyard with the thumb and fingers.

Step 7: Pull it through the loop, making a clove hitch.

Step 3: Hold the halyard between the thumb and forefinger with the palm down.

Step 4: Turn the hand palm up.

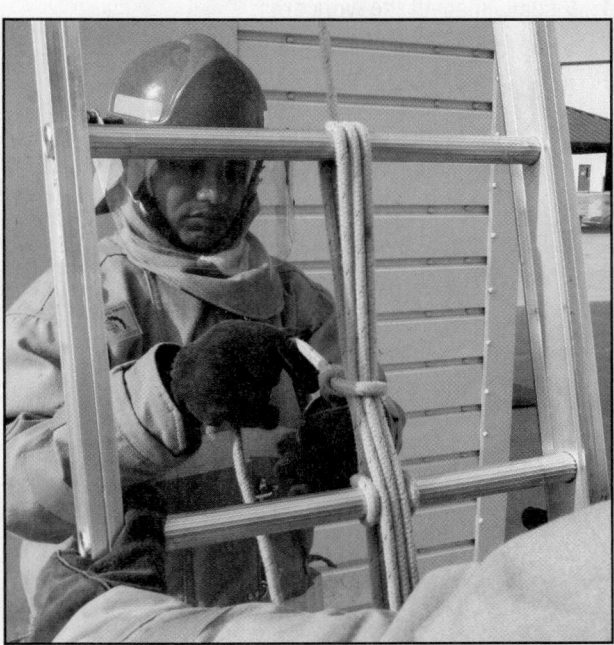

Step 8: Finish the tie by making a half-hitch or overhand safety on top of the clove hitch.

NOTE: Firefighters must be wearing appropriate PPE before performing this skill.

Single Ladder – Beam Method

Step 1: Visually inspect the work area:

    a. Terrain for solid, level footing

    b. Overhead for electrical wires and obstructions

Step 2: Lower the ladder butt to the ground.

    a. Rotate the ladder on the spur until beams are parallel to the building

    b. Raise the ladder up against the wall

    c. Grasp the rungs and pull the butt end away from the wall until it is at a 75 degree climbing angle

    d. Finish positioning the ladder by adjusting the butt to where it is needed

Step 3: Position yourself to raise the ladder.

    a. Grasp rung in front of your shoulder with free hand

    b. Remove other arm from between the rungs

    c. Step beneath ladder and grasp convenient rung with free hand

Step 4: Bring the ladder upright until it rests against the building.

    a. Advance toward the butt end

Step 5: Carefully move the ladder butt end out from the building to the desired climbing angle.

    a. Push against an upper rung

    b. Pull a lower rung

Step 6: Lower the ladder, reversing the raising procedure.

Single Ladder - Flat Method

Step 1: Visually inspect the work area:

    a. Terrain for solid, level footing

    b. Overhead for electrical wires and obstructions

Step 2: Position both butt spurs against the wall.

Step 3: Position yourself to raise the ladder.

    a. Grasp rung in front of your shoulder with free hand

    b. Remove other arm from between the rungs

    c. Step beneath ladder and grasp convenient rung with free hand

Step 4: Bring the ladder upright until it rests against the building.

    a. Advance toward the butt end

Step 5: Carefully move the ladder butt out from the building to the desired climbing angle.

        a. Push against an upper rung

        b. Pull a lower rung

Step 6: Lower the ladder, reversing the raising procedure.

Extension Ladder – Beam Method

Step 1: Visually inspect the work area:

    a. Terrain for solid, level footing

    b. Overhead for electrical wires and obstructions

Step 2: Lower the ladder butt to the ground.

Step 3: Position yourself to raise the ladder.

Step 4: Bring the ladder upright until it rests against the building.

    a. Advance hand-under-hand

    b. Toward the butt

Step 5: Pull the ladder away from the building until in vertical position.

    a. Grasp a convenient rung with both hands

    b. Heel ladder

Step 6: Balance ladder in a vertical position.

    a. One foot at butt of one beam

    b. Ladder steadied with instep, knee, and leg

Step 7: Extend the fly section.

    a. To desired elevation

    b. Use hand-under-hand motion on halyard

    c. Pull halyard straight down

    d. Maintain ladder balance

Step 8: Engage the ladder locks at the desired elevation.

Step 9: Pivot the ladder if necessary until the fly faces out.

Step 10: Lower the ladder against the building.

    a. Grasp beams

    b. One foot against a butt spur or on bottom rung

    c. Gently

Step 11: Tie off the halyard.

    a. Wrap around two convenient rungs

    b. Tie clove hitch

    c. Tie half-hitch or overhand safety on top of clove hitch

Step 12: Pull the ladder butt out from the building.

    a. Push against upper rung

    b. Pull lower rung

    c. Until at proper angle for climbing

Step 13: Secure the ladder for climbing.

Step 14: Lower the ladder, reversing the raising procedure.

Extension Ladder - Flat Method

Step 1: Visually inspect the work area.

    a. Terrain for solid, level footing

    b. Overhead for electrical wires and obstructions

Step 2: Butt both spurs against the wall.

Step 3: Position yourself to raise the ladder.

    a. Grasp rung in front of your shoulder with free hand

    b. Remove other arm from between the rungs

    c. Step beneath ladder and grasp convenient rung with free hand

Step 4: Bring the ladder upright until it rests against the building.

    a. Advance hand-over-hand

    b. Toward the butt

Step 5: Pull the ladder away from the building until in vertical position.

    a. Grasp a convenient rung with both hands

    b. Heel ladder

Extension Ladder - Flat Method

Step 6: Balance ladder in a vertical position.

    a. One foot at butt of one beam

    b. Ladder steadied with instep, knee, and leg

Step 7: Extend the fly section.

    a. To desired elevation

    b. Use hand-over-hand motion on halyard

    c. Pull halyard straight down

    d. Maintain ladder balance

Step 8: Engage the ladder locks at the desired elevation.

Step 9: Pivot the ladder if necessary until the fly faces out.

Step 10: Lower the ladder against the building.

    a. Grasp beams

    b. One foot against a butt spur or on bottom rung

    c. Gently

Step 11: Pull the ladder butt out from the building.

    a. Push against upper rung

    b. Pull lower rung

    c. Until at proper angle for climbing

Step 12: Tie off the halyard.

    a. Wrap around two convenient rungs

    b. Tie clove hitch

    c. Tie half-hitch or overhand safety on top of clove hitch

Step 13: Secure the ladder for climbing.

Step 14. Lower the ladder, reversing the raising procedure.

NOTE: Firefighters must be wearing appropriate PPE while performing this skill. Firefighter #1 is located near the butt end of the ladder. Firefighter #2 is located near the tip of the ladder and is in command of this operation.

Step 1: Both Firefighters: Place the ladder flat on the ground with the butt end toward the structure and approximately ¼ the usable height from the building.

Step 2: Firefighter#2: Check for overhead obstructions and wires.

Step 3: Firefighter#2: Lift the tip of the ladder stepping under the beams and grasp the top rung.

Step 4: Firefighter #1: Heel the ladder by standing on the bottom rung or by placing the toes or insteps on the beam.

Step 5: Firefighter #1: Crouch down to grasp a convenient rung or the beams with both hands.

Step 6: Firefighter #1: Lean back.

Step 7: Firefighter #2: Advance hand-over-hand down the rungs toward the butt end until the ladder is in a vertical position.

Step 8: Firefighter #1: Grasp successively higher rungs or higher on the beams as the ladder comes to a vertical position until standing upright.

Step 9: Both Firefighters: Stand on opposite sides of the ladder.

Step 10: Both Firefighters: Heel the ladder by placing toes against the same beam.

Step 14: Both Firefighters: Lower the ladder gently into position against the structure.

Step 15: Firefighter #2: Place both feet against the butt spurs or on the bottom rung, grasp the rung or beams, and check climbing angle.

Step 16: Firefighter #1: Tie the halyard.

Step 11: Firefighter #2: Grasp the beams, ensuring fingers and hands are on the outside of the beam.

Step 12: Firefighter #1: Untie and grasp the halyard.

Step 13: Firefighter #1: Extend the fly section with a hand-over-hand motion until the tip reaches the desired elevation and engages the ladder locks.

**SKILL
SHEETS**

NOTE: Firefighters must be wearing appropriate PPE while performing this skill. Firefighter #1 is located near the butt end of the ladder. Firefighter #2 is located near the tip of the ladder and is in command of this operation.

Step 1: Firefighter #1: Place the ladder beam on the ground approximately ¼ the usable height from the building.

Step 2: Firefighter #1: Check for overhead obstructions and wires.

Step 3: Firefighter #2: Rest the ladder beam on one shoulder.

Step 4: Firefighter #1: Place the foot closest to the lower beam on the lower beam at the butt end.

Step 5: Firefighter #1: Grasp the upper beam with hands apart and the other foot extended back to act as a counterbalance.

Step 6: Firefighter #2: Advance hand-over-hand down the beam toward the butt end until the ladder is in a vertical position.

Step 7: Both Firefighters: Pivot the ladder to properly position the fly section.

Step 8: Firefighter #1: Untie and grasp the halyard.

Step 9: Firefighter #1: Extend the fly section with a hand-over-hand motion until the tip reaches the desired elevation. Engage the ladder locks.

Step 10: Both Firefighters: Lower the ladder gently into position against the structure.

Step 11: Firefighter #2: Place both feet against the butt spurs or on the bottom rung, grasp the rung or beams, and check climbing angle.

Step 12: Firefighter #1: Tie the halyard.

12-I-11
Raise a ladder – Three- or four-firefighter flat raise.

NOTE: Firefighters must be wearing appropriate PPE while performing this skill. Firefighter #1 is located near the butt end of the ladder, Firefighter #2 is located at the ladder tip, and Firefighter#3 at the ladder mid-point.

Step1: Firefighter #1: Place the ladder butt end on the ground approximately ¼ the usable height from the building.

Step 2: Firefighter #3: Moves to the ladder tip and at the same time rests the ladder flat on shoulders. Firefighter #2: Rest the ladder flat on shoulders as well.

Step 3: Firefighter#1: Check for overhead obstructions and wires.

Step 4: Firefighter #1: Heel the ladder by standing on the bottom rung or by placing the toes or insteps on the beam.

SKILL SHEETS

12-1-11
Raise a ladder – Three- or four-firefighter flat raise.

Step 5: Firefighter #1: Crouch down to grasp a convenient rung with both hands.

Step 6: Firefighter #1: Lean back.

Step 7: Firefighters #2 and #3: Advance in unison, with outside hands on the beams and inside hands on the rungs, until the ladder is in a vertical position.

Step 8: Firefighter #1: Step off bottom rung and stand erect.

Step 9: Firefighters #2 and #3: Each place the inside of a foot against the butt spur.

Step 10: Firefighter #1: Untie and grasp the halyard.

Step 11: Firefighter #1: Place the toe of one foot on the butt spur.

Step 12: Firefighter #1: Extend the fly section by pulling the halyard with a hand-over-hand motion until the tip reaches the desired elevation and engages the ladder locks.

Step 13: All Firefighters: Lower the ladder gently onto the building.

    a. Grasp the beam or a convenient rung

    b. Steady the ladder from the inside position

Step 14: Firefighter #2: Place both feet against the butt spurs or on the bottom rung, grasp the rung or beams, and check climbing angle.

Step 15: Firefighter #1: Tie the halyard.

NOTE: Firefighters must be wearing appropriate PPE before performing this skill.

Step 1: Set the roof ladder down.

Step 2: Open the hooks.

Step 5: Climb the main ladder until your shoulder is about two rungs above the midpoint of the roof ladder.

Step 6: Reach through the rungs of the roof ladder.

Step 7: Hoist the ladder onto the shoulder.

Step 3: Face the hooks outward.

Step 4: Tilt the roof ladder up so that it rests against the other ladder.

Step 8: Climb to the top of the ladder.

Step 9: Lock into the ladder using a leg lock or life safety harness.

Step 10: Take the roof ladder off the shoulder.

Step 11: Use a hand-over-hand method to push the roof ladder onto the roof.

Step 12: Push the roof ladder up the roof until the hooks go over the edge of the peak and catch solidly.

NOTE: Firefighters must be wearing the appropriate PPE while performing this skill. Firefighter #1 is located on the side opposite the structure and is in command of this operation.

Step 1: Both Firefighters: Visually check the terrain and the area overhead.

Step 2: Both Firefighters: Stand on opposite sides of the ladder.

Step 3: Both Firefighters: Grasp the ladder beams with both hands.

Step 4: Firefighter #1: Place a foot against the side of the beam on which the ladder will pivot.

Step 5: Both Firefighters: Tilt the ladder onto the pivot beam.

Step 6: Both Firefighters: Pivot the ladder 90 degrees. Simultaneously adjust positions as necessary.

    a. Repeat the process until the ladder is turned a full 180 degrees and the fly is in the proper position

Step 7: Both Firefighters: Lower the ladder gently into position against the structure.

NOTE: Firefighters must be wearing appropriate PPE before performing this skill.

Step 1: Visually check the terrain and the area overhead.

Step 2: Face the ladder.

Step 3: Heel the ladder.

Step 4: Grasp the beams.

Step 5: Bring the ladder outward to vertical.

Step 6: Shift grip on the ladder, one hand at a time, so that one hand grasps as low a rung as convenient, palm upward.

Step 7: Grasp a rung as high as convenient with the other hand, palm downward.

Step 8: Turn slightly in the direction of travel.

Step 9: Lift the ladder and proceed forward a short distance.

Step 10: Watch the tip as it is being moved.

Step 11: Set the ladder down at the new position.

Step 12: Switch grip back to the beams.

Step 13: Heel the ladder.

Step 14: Lower the ladder into position.

NOTE: Firefighters must be wearing appropriate PPE while performing this skill.

Step 1: Both Firefighters: Visually check the terrain and the area overhead.

Step 2: Both Firefighters: Position on opposite sides of the ladder.

    a. If the ladder is not vertical, it is brought to vertical; if extended, retract the fly section

Step 3: Both Firefighters: Position hands.

    a. One hand grasps as low a rung as convenient, palm upward

    b. Other hand grasps a rung as high as convenient, palm downward

    c. Side grasped low by one firefighter is grasped high by the other

Step 4: Both Firefighters: Lift the ladder just clear of the ground.

Step 5: Both Firefighters: Watch the tip while shifting the ladder to the new position.

Step 6: Both Firefighters: When the new position is reached, set the butt on the ground.

Step 7: Both Firefighters: Re-extend the ladder (if necessary).

Step 8: Both Firefighters: Lower the ladder gently into position against the structure.

NOTE: Firefighters must wear appropriate PPE while performing this skill.

Under the Ladder Method

Step 1: Once ladder is raised, stand beneath it about shoulder-width apart (or one foot slightly ahead of the other).

    a. Wear complete PPE

    b. Do not look up when someone is climbing the ladder

Step 2: Grasp the ladder beams (not the rungs) at about eye level and pull backward to press the ladder against the building.

Step 3: Remain alert for falling objects or debris while others are climbing on the ladder.

In Front of Ladder Method

Step 1: Stand on the outside of the ladder and chock the butt end with one foot.

    a. Wear complete PPE

    b. Either place toes against butt spur or place foot on bottom rung

Step 2: Grasp the beams and press ladder against building.

Step 3: Remain alert for falling objects or debris while others are climbing on the ladder.

NOTE: Firefighters must be wearing appropriate PPE before performing this skill.

Step 1: Climb to the desired height.

Step 2: Advance one rung higher.

Step 3: Slide the leg on the opposite side from the working side over and behind the rung to be locked in to.

Step 4: Hook foot either on the rung or on the beam.

Step 5: Rest on thigh.

Step 6: Step down with the opposite leg.

NOTE: Firefighters must be wearing appropriate PPE before performing this skill.

Step 1: Rescuer and Heeler: Position the ladder.

    a. Tip at the sill of the rescue window

    b. Correct climbing angle

Step 2: Rescuer and Heeler: Secure the ladder.

    a. With rope hose tool

    b. Top and bottom if possible

Step 3: Heeler: Heel the ladder.

Step 4: Rescuer: Climb the ladder.

    a. Until in a position below window for receiving victim

    b. Both feet on one rung

Step 5: Firefighters in building: Lower the victim from the window to the rescuer on the ladder.

    a. Feet first

    b. Facing building

Step 6: Rescuer: Position the victim for carrying.

    a. Forearms under victim's armpits

    b. Hands on ladder rungs in front of victim

Step 7: Rescuer: Descend the ladder.

    a. One rung at a time

    b. Supporting and reassuring victim

NOTE: Firefighters must be wearing appropriate PPE before performing this skill.

CAUTION: No live victims should be used when performing this skill.

Step 1: Rescuer and Heeler: Position the ladder.

    a. Tip at the sill of the rescue window

    b. Correct climbing angle

Step 2: Rescuer and Heeler: Secure the ladder.

    a. With rope hose tool

    b. Top and bottom if possible

Step 3: Heeler: Heel the ladder.

Step 4: Rescuer: Climb the ladder.

    a. Until in a position below window for receiving victim

    b. Raise one foot to next rung

Step 5: Firefighters in building: Lower the victim from the window to the rescuer on the ladder.

    a. Feet first

    b. Facing rescuer

Step 6: Rescuer: Position the victim for carrying.

    a Victim resting on knee

    b. Victim's feet outside beams

    c. Forearms under victim's armpits

    d. Hands grasping ladder beams

    e. Resting victim's chin on victim's chest

Step 7: Rescuer: Descend the ladder.

    a. One rung at a time

    b. Supporting victim on knee

Tactical Ventilation

Chapter Contents

Key Terms

NFPA® Job Performance Requirements

This chapter provides information that addresses the following job performance requirements of NFPA® 1001, *Standard for Fire Fighter Professional Qualifications* (2013).

Firefighter I
5.3.11
5.3.12

1. Describe reasons for tactical ventilation. (5.3.11)

2. Identify considerations that affect the decision to ventilate. (5.3.11, 5.3.12)

3. Explain the critical fire behavior indicators present during tactical ventilation. (5.3.11)

4. Define horizontal and vertical ventilation. (5.3.11)

5. Explain the means for achieving horizontal and vertical ventilation. (5.3.11, 5.3.12)

6. Describe the types of horizontal ventilation. (5.3.11, 5.3.12)

7. Describe the types of vertical ventilation. (5.3.11, 5.3.12)

8. Recognize other types of ventilation situations. (5.3.11)

9. Explain the effects of building systems on tactical ventilation. (5.3.11, 5.3.12)

10. Ventilate using mechanical negative pressure in a window. (Skill Sheet 13-I-1, 5.3.11, 5.3.12)

11. Ventilate using mechanical negative pressure in a doorway. (Skill Sheet 13-I-2, 5.3.11, 5.3.12)

12. Ventilate using mechanical positive pressure. (Skill Sheet 13-I-3, 5.3.11, 5.3.12)

13. Perform horizontal hydraulic ventilation. (Skill Sheet 13-I-4, 5.3.11, 5.3.12)

14. Perform the procedure for sounding a roof. (Skill Sheet 13-I-5, 5.3.12)

15. Ventilate using a power saw to cut an opening. (Skill Sheet 13-I-6, 5.3.12)

16. Ventilate using an axe to cut an opening. (Skill Sheet 13-I-7, 5.3.12)

17. Demonstrate the procedure for opening a flat roof. (Skill Sheet 13-I-8, 5.3.12)

18. Perform the steps for opening pitched roofs. (Skill Sheet 13-I-9, 5.3.12)

19. Demonstrate the procedure for making a trench cut using a rotary saw. (Skill Sheet 13-I-10, 5.3.12)

Chapter 13
Tactical
Ventilation

Case History

On March 30, 2010, a fire in a one-story, single family dwelling resulted in the death of a career firefighter and an occupant as well as the injury of a second firefighter. When emergency units arrived, heavy fire conditions were visible at the rear of the house and moderate smoke conditions existed in the uninvolved interior areas. A search and rescue crew entered the house to search for a civilian who was reported trapped at the rear of the house. Three other firefighters advanced a charged 2½-inch (64 mm) hoseline into the house. Thick, black, rolling smoke banked down to knee level as the hoseline was advanced 12 feet (3.7 m) into the kitchen area.

Additional units were assigned to the roof to perform vertical ventilation and to ground-level windows on the D and B sides to perform horizontal ventilation. When the windows were opened the fire intensified, spreading into the area between the attack hoseline crew and the A-side door. When this occurred, the search and rescue crew saw fire rolling across the ceiling within the smoke. They immediately yelled to the hoseline crew to "get out." The search and rescue crew were able to exit the structure safely and then returned to rescue the injured firefighter and then the victim. The victim received medical care at the scene and was transported to a local hospital where he was pronounced dead.

The subsequent NIOSH investigation into the fatality determined that a contributing factor was the lack of coordination between the ventilation and fire suppression crews. During the incident, uncoordinated ventilation occurred while the hoseline and search and rescue crews were inside the house. The victim and the other firefighters within the small 950 square foot (88 square meters) house were caught between the fire and the ventilation source. One firefighter reported heavy, turbulent, black smoke pushing from a window on the B-side after it was broken. Shortly after, the house sustained an apparent ventilation-induced flashover.

Source: CDC NIOSH LODD Report F-2010-10.

Fires contained within a room or structure consume the available fuel and oxygen as they burn. As the fuel and the oxygen burn, they undergo a chemical reaction that produces heated toxic gases, unburned hydrocarbons, and carbon particles (soot). As the oxygen inside the room or structure is consumed, the fire burns less efficiently and the amount of unburned fuel in the combustion products increases. When fresh air (oxygen) is introduced into the room, the heated gases quickly reignite causing rapid fire development, or backdraft, to occur. Firefighters and occupants cannot survive in a condition of rapid fire development.

To prevent rapid fire development from occurring, you must know how and when to properly release the heated toxic gases using safe and efficient forms of ventilation. Ventilation is the traditional term that refers to the removal of heated air, smoke and fire gases from a structure and replacing them with cooler air. The cooler air allows firefighters to enter the structure to perform search and rescue and fire suppression operations. Proper ventilation decreases the rate of fire spread and increases visibility so firefighters can quickly locate trapped occupants and advance to the seat of the fire. Ventilation also channels heat and toxic gases away from firefighters and trapped occupants.

The concepts presented in this chapter are based on the information found in Chapter 5, Fire Behavior, including compartment fire behavior, stages of fire development, and the burning regime, and Chapter 4, Building Construction, in particular roof design and construction. This chapter describes:

- Reasons for ventilation

- Considerations used to determine if, when, and how to ventilate

- Basics of horizontal and vertical ventilation operations

- Effects of environmental systems in fire situations

I Reasons for Tactical Ventilation

While the term *ventilation* is the traditional word used in the fire service, the activity is more accurately referred to as tactical ventilation. Tactical ventilation is the planned, systematic, and coordinated removal of heated air, smoke, gases or other airborne contaminants from a structure, replacing them with cooler and/or fresher air to meet the incident priorities of life safety, incident stabilization, and property conservation. Tactical ventilation should only be performed when the fire attack hoselines and teams are in place and ready to advance toward the fire (**Figure 13.1**). Successful tactical ventilation depends on:

- Careful planning

- Knowledge of building construction

Figure 13.1 Tactical ventilation must wait until hoselines are in place and the team is ready.

- Knowledge of fire behavior

- Systematic application of procedures for removing the contaminants

- Coordination with other fireground activities

The general reasons for performing tactical ventilation include:

- Reducing interior heat levels
- Decreasing rate of fire spread
- Reducing potential extreme fire behavior
- Improving interior visibility
- Improving firefighter efficiency
- Improving victim survival potential
- Reducing smoke damage and property damage

Correctly implemented, tactical ventilation helps achieve the three incident priorities of life safety, incident stabilization, and property conservation. However, if ventilation is improperly applied, the results can be traumatic to occupants, firefighters, and to the structure itself.

The Effect of Fresh Air on an Oxygen-Deprived Environment

In an oxygen-limited environment, the introduction of fresh air without a coordinated fire attack will result in rapid fire development, the production of more heat, and an increased threat to life safety for occupants and firefighters.

Life Safety

As the highest incident priority, life safety applies to both the occupants who may be trapped in the structure and the firefighters who must enter it to locate and rescue them. Tactical ventilation improves the life safety of firefighters and occupants by:

- Increasing oxygen concentration
- Reducing the concentration of toxic products of combustion
- Reducing temperature
- Increasing visibility to aid in fire fighting operations and primary search operations
- Creating smoke-free paths of egress

Incident Stabilization

Because tactical ventilation reduces interior temperature, potential for extreme fire behavior, and fire spread, it can be effectively combined with fire attack to stabilize an incident. Incident stabilization means controlling and extinguishing the fire and is accomplished in stages:

- Locating the fire
- Confining the fire to the room, area, or structure of origin
- Extinguishing the fire

Property Conservation

When smoke, gases, and heat are removed from a burning structure, the fire can be confined to a specific area. Then, if there are sufficient personnel on scene, effective salvage operations can begin outside the immediate area of the fire even while fire control operations are being conducted. Because tactical ventilation increases the speed with which you can extinguish interior fires, ventilation also reduces the fire damage in a structure. In addition, less water will be needed and, consequently, there will be less water damage to the structure and its contents.

Considerations Affecting the Decision to Ventilate

The decision to ventilate a structure is based on a number of factors including (**Figure 13.2**):

- Risks to occupants and firefighters
- Building construction
- Fire behavior indicators
- Location and extent of the fire
- Type of ventilation
- Location for ventilation
- Weather conditions
- Exposures
- Staffing and available resources

NOTE: The Incident Commander (IC) will make the decision to ventilate the structure.

Figure 13.2 The decision to ventilate a structure must take into account several factors that will each play a significant part in the success of the tactic. *Courtesy of Mike Wieder.*

Risks to Occupants and Firefighters

Timely and effective tactical ventilation will assist search and rescue operations. Life hazards in a structure fire are generally lower if the occupants are awake. On the other hand, if the occupants were asleep when the fire developed and are still in the building, a number of possibilities must be considered:

- Occupants may have been overcome by smoke and fire gases — some may still be alive; others may have perished

- Occupants might have become lost in the structure

- Occupants may be alive and taking refuge in their rooms because the doors were closed

In addition to the hazards that endanger occupants, there are potential hazards to firefighters. The hazards that can be expected from the accumulation of smoke and fire gases in a structure include:

- Visual impairment caused by dense smoke

- Lack of oxygen

- Presence of toxic gases

- Presence of flammable gases

- Possibility of rapid fire development

Building Construction

Over the past 50 years, building construction in North America has changed drastically. In single-family residential structures, the footprint of houses has increased over 150 percent between 1973 and 2008. At the same time lot sizes have shrunk approximately 25 percent, reducing firefighter access and increasing potential exposure risks.

Residential interior layouts and construction materials have also changed. **Figure 13.3** illustrates the differences between older and newer construction. Older structures were composed of smaller compartments, windows that could be opened for ventilation, and empty wall cavities that depended on air pockets to provide insulation.

Figure 13.3 Residential interior layout has changed drastically to incorporate more open spaces.

Modern single-family structures feature open floor plans, high ceilings and atriums, lightweight manufactured structural components, sealed windows, and wall cavities that are filled with synthetic insulation. Construction materials and interior finish consisting of synthetic materials and light composite wood components add to the fuel load of the structure and contribute to the creation of toxic gases during a fire. Because of energy-efficient designs, the structures also tend to contain fires for a longer period of time creating fuel-rich environments. These problems are magnified in large-area residential structures.

Commercial, institutional, educational, and multifamily residential structures also rely on energy conservation measures that increase the intensity of a fire and make the use of tactical ventilation difficult. Open plan commercial structures, such as "Big Box" stores, have high fuel loads in the contents and no physical barriers to prevent the spread of fire and smoke in the space **(Figure 13.4)**.

In addition, the use of plastics and other synthetic materials has dramatically increased the fuel load in all types of occupancies. These synthetic materials produce large quantities of toxic and combustible gases. The heat generated by fires in these types of fuels can escalate rapidly and reach extremely high temperatures.

Knowledge of the building involved is a great asset when decisions concerning tactical ventilation are made. This information can be obtained from the preincident plan, inspection reports, or observation of similar types of structures. Building characteristics to be considered include the following:

- Occupancy classification
- Construction type
- Floor area and compartmentation
- Ceiling height
- Number of stories above and below ground level
- Number and size of exterior windows, doors, and other wall openings

Figure 13.4 Stores that feature tall shelves loaded with commercial goods do not often include physical barriers between sections.

- Number and location of staircases, elevator shafts, dumbwaiters, ducts, and roof openings
- External exposures
- Extent to which a building is connected to adjoining structures
- Type and design of roof construction
- Type and location of fire protection systems
- Contents
- Heating, ventilation, and air conditioning (HVAC) System

Fire Behavior Indicators

As discussed in Chapter 5, Fire Behavior, responsibility for ongoing size-up and risk assessment is not limited to fire officers. Reading the fire and understanding the effect of changes in ventilation is essential to effective tactical operations and firefighter safety. This section reviews critical fire behavior indicators and their relationship to tactical ventilation in the following four major categories:

- Smoke
- Air flow
- Heat
- Flame

Smoke

The following observations about smoke, taken together, can help firefighters obtain a clear picture of the actual, interior fire conditions:

- Volume of smoke discharge (**Figure 13.5**)
- Location of smoke discharge
- Smoke color, density, and pressure
- Movement of smoke

Figure 13.5 Smoke is often a definitive indicator of internal fire conditions. *Courtesy of Bob Esposito.*

Air Flow

Air flow is the movement of air toward burning fuel and movement of smoke out of the compartment. Air flow indicators include velocity, turbulence, direction, and movement of the neutral plane. Air flow is caused by the following:

- Pressure differentials inside and outside the compartment
- Differences in density between the hot smoke and cooler air (**Figure 13.6, p. 738**).

Air flow follows a flow path. There is an inlet vent where the air enters, a flow path (typically through the fire area), and an outlet or exhaust vent. In the case of a single open doorway to a structure fire, the air inlet is the lower portion of the doorway (low pressure below the neutral plane) and the smoke exhaust is the upper portion of the doorway (high-pressure area above the neutral plane). In other cases, the entire open doorway may be the air inlet and the roof vent might serve as the smoke/hot gas exhaust vent. The flow path is the connection between the inlet and the outlet.

Air Flow — The movement of air toward burning fuel and the movement of smoke out of the compartment or structure.

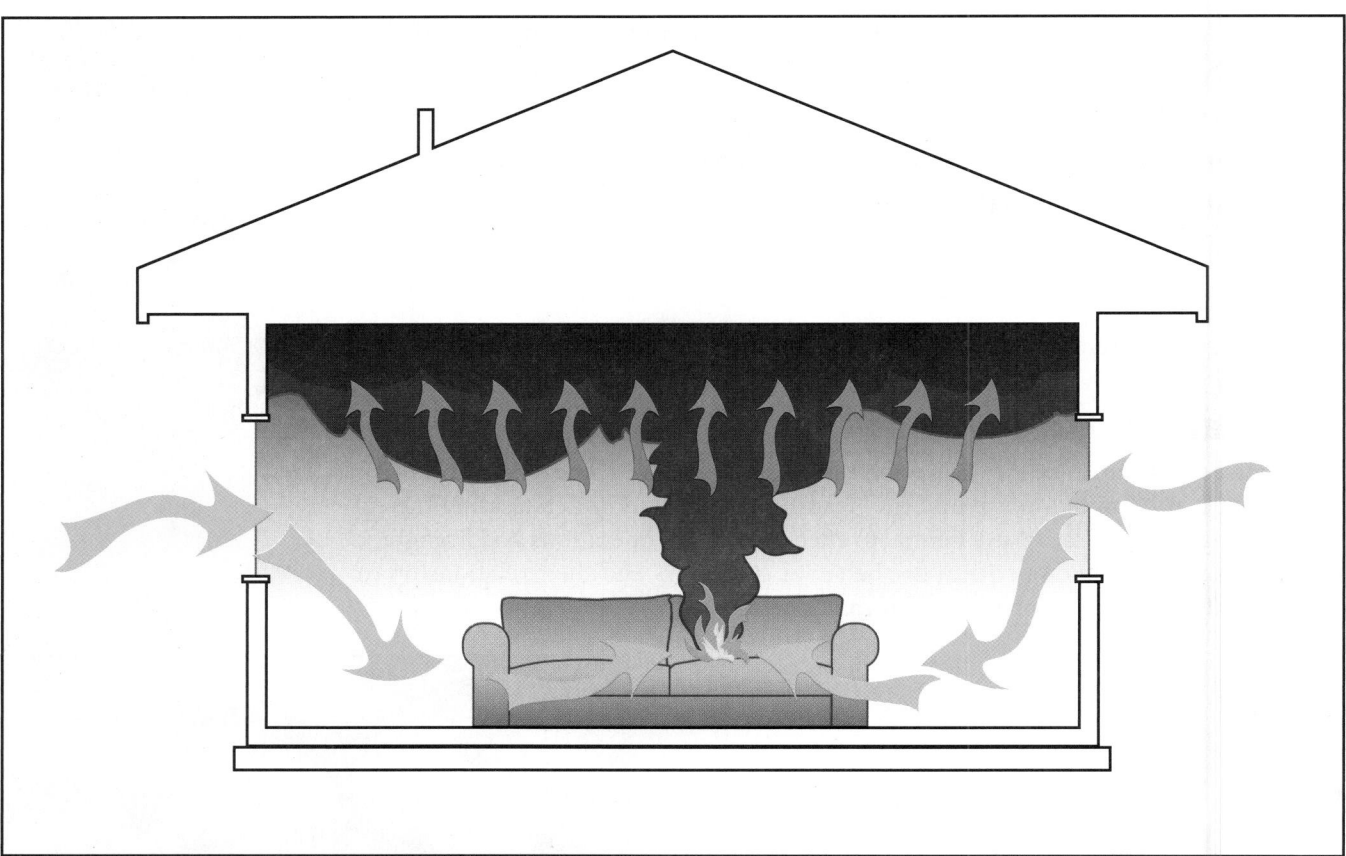

Figure 13.6 Cool air is drawn in toward a fire when the smoke generated by the fire pushes heated air toward the ceiling of a compartment.

Heat

Visual indicators of heat include blistering paint, bubbling roofing tar, and crazed glass. Scanning buildings with a thermal imager or infrared sensor can provide data on internal temperature differences (**Figure 13.7**). You may also determine the presence of increased temperatures through touch or feel on your skin even at a distance.

Flame

Visible flames may provide an indication of the size and location of the fire, for example, fire showing from one window vs. fire showing from all windows on the fire floor. The size and extent of the fire may also be indicated by the effect (or lack of effect) of fire streams on flaming combustion. Flames visible from outside the structure can allow firefighters to assess flame indicators along with ventilation and air flow (**Figure 13.8**).

CAUTION

Do not rely solely on the presence or location of flames to assess an incident.

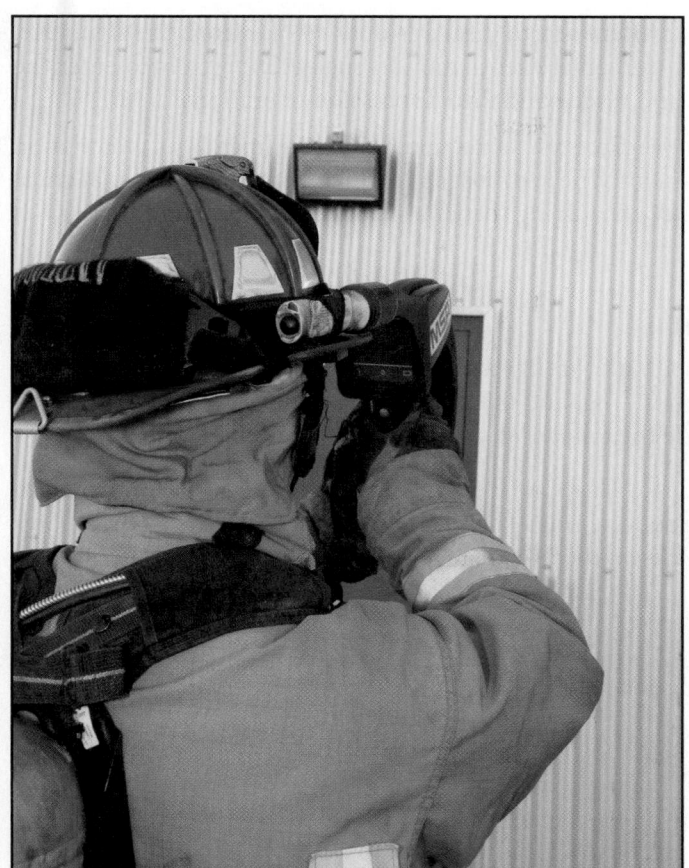

Figure 13.7 A thermal imager may provide valuable information on the current conditions inside a structure.

Figure 13.8 Visible flames may indicate the size and location of the fire. *Courtesy of Mike Wieder.*

Location and Extent of the Fire

A fire may have traveled some distance throughout a structure before firefighters arrive. Therefore, first-arriving units must quickly determine the size and extent of the fire as well as its location.

Tests and experience indicate that creating tactical ventilation openings in an uncoordinated manner can spread the fire to uninvolved areas of the building and cut off escape routes for building occupants. The severity and extent of the fire depend upon a number of factors, including the type of fuel and the amount of time it has been burning, activation of fire detection and suppression systems, and the degree of confinement. The stage of the fire and whether it is fuel or ventilation controlled are a primary consideration in determining tactical ventilation procedures.

Type of Ventilation

To be safe and effective, tactical ventilation operations must be coordinated with other tactical operations including fire suppression and search and rescue. Before orders are given to ventilate a structure, the IC must consider the effects that ventilation will have on the fire's behavior. Fire attack crews with charged hoselines, search and rescue teams, and exposure protection must be in place before tactical ventilation begins. The IC first determines if ventilation is necessary and when, where, and in what form it should be initiated. Conditions present upon arrival will influence ventilation decisions. Some incidents may simply require locating and extinguishing the fire and then ventilating afterward to clear residual smoke from the structure. Other incidents will require immediate ventilation to enable firefighters to enter the building to conduct search and rescue and fire suppression operations.

The type of ventilation (vertical vs. horizontal) and the means of ventilation (natural vs. mechanical) used must be the most appropriate for the situation. Tactical ventilation must be capable of exhausting the volume of heat, smoke, and toxic gases produced by the fire.

Location for Ventilation

Before selecting a place to ventilate, firefighters should gather as much information as possible about the fire, the building, and the occupancy. Factors that have a bearing on where to ventilate include the following:

- Location of occupants
- Availability of existing roof openings, such as skylights, ventilator shafts, monitors, and hatches, which access the fire area
- Location of the fire
- Desired air flow path
- Type of building construction
- Wind direction
- Extent of progress of the fire
- Condition of the building and its contents
- Indications of potential structural collapse
- Effect that ventilation will have on the fire
- Effect that ventilation will have on exposures
- State of readiness of fire attack crews
- Ability to protect exposures prior to ventilating the structure
- Protecting means of egress and access

Weather Conditions

Any opening in a building, whether part of the building design or caused by the fire, allows the surrounding atmosphere to affect what is happening inside the building. Although weather conditions such as wind, temperature, atmospheric pressure, precipitation, and relative humidity can affect tactical ventilation, the most important weather-related influence on ventilation is wind.

Wind conditions must always be considered when determining the proper means and location of tactical ventilation in all types of structures. Wind can blow the fire toward an external exposure, supply oxygen to the fire, or blow the fire into uninvolved areas of the structure (**Figure 13.9**). The means of tactical ventilation selected should work with the prevailing wind and not against it.

CAUTION

A strong wind can overpower the natural convective effect of a fire and drive the smoke and hot gases back into the building.

Figure 13.9 Wind conditions have a profound and significant effect on the behavior of fire. *Courtesy of the Los Angeles Fire Department – ISTS.*

Exposures

When beginning tactical ventilation operations, firefighters must consider both internal and external exposures **(Figure 13.10)**. Internal exposures include the building occupants, contents, and any uninvolved rooms or portions of the building. When ventilation does not release heat and smoke directly above the fire, some routing of the smoke is necessary. The routes the smoke and heated fire gases would naturally travel to exit the building may be the same corridors and passageways that occupants need for evacuation and firefighters need for working.

Ventilation that causes heat, smoke, and sometimes fire to be discharged through wall openings below the highest point of the building creates the danger that the rising gases will ignite portions of the building above the exhaust point. Heat and fire gases may be drawn into open windows or attic vents, and they may also ignite the eaves of the building or adjacent structures.

External exposures such as structures located adjacent to the fire building can be affected by radiation and/or direct flame contact. Window-mounted air conditioning units or HVAC intake vents may draw smoke into adjacent buildings as well. Nearby structures and vegetation can be ignited if hot fire brands or embers are carried aloft by convection. Fire may be drawn into exterior windows or openings of the adjacent exposures.

Figure 13.10 Ventilation activities may cause the fire to extend to other portions of the structure.

Staffing and Available Resources

All tactical ventilation operations require personnel and resources. Staffing requirements range from two firefighters to multiple companies. In a small structure, ventilation may only require two firefighters to open doors and windows to allow the fresh air to enter and the smoke to exit. Additional personnel and companies are required when it is necessary to make a roof opening or activate fans and smoke ejectors.

Resources needed for tactical ventilation include forcible entry tools, power saws, fans or **blowers**, **smoke ejectors**, flexible ducts, stacking and hanging devices, and other support systems, electrical power cords, and generators **(Figure 13.11)**. As the amount and size of ventilation equipment increases, the space on apparatus to store and transport it will increase.

Figure 13.11 Tactical ventilation resources include smoke ejectors and fans.

Types of Tactical Ventilation

There are generally two types of tactical ventilation used for structure fires: horizontal and vertical. Opening doors or windows is one example of **horizontal ventilation** **(Figure 13.12a)**.

Cutting a hole in the roof above the fire or opening existing roof access doors, scuttles, or skylights are all examples of **vertical ventilation (Figure 13.12b)**. For vertical ventilation to be effective, a horizontal inlet opening at or below the level of the fire is needed to provide a flow path for fresh air to enter the structure.

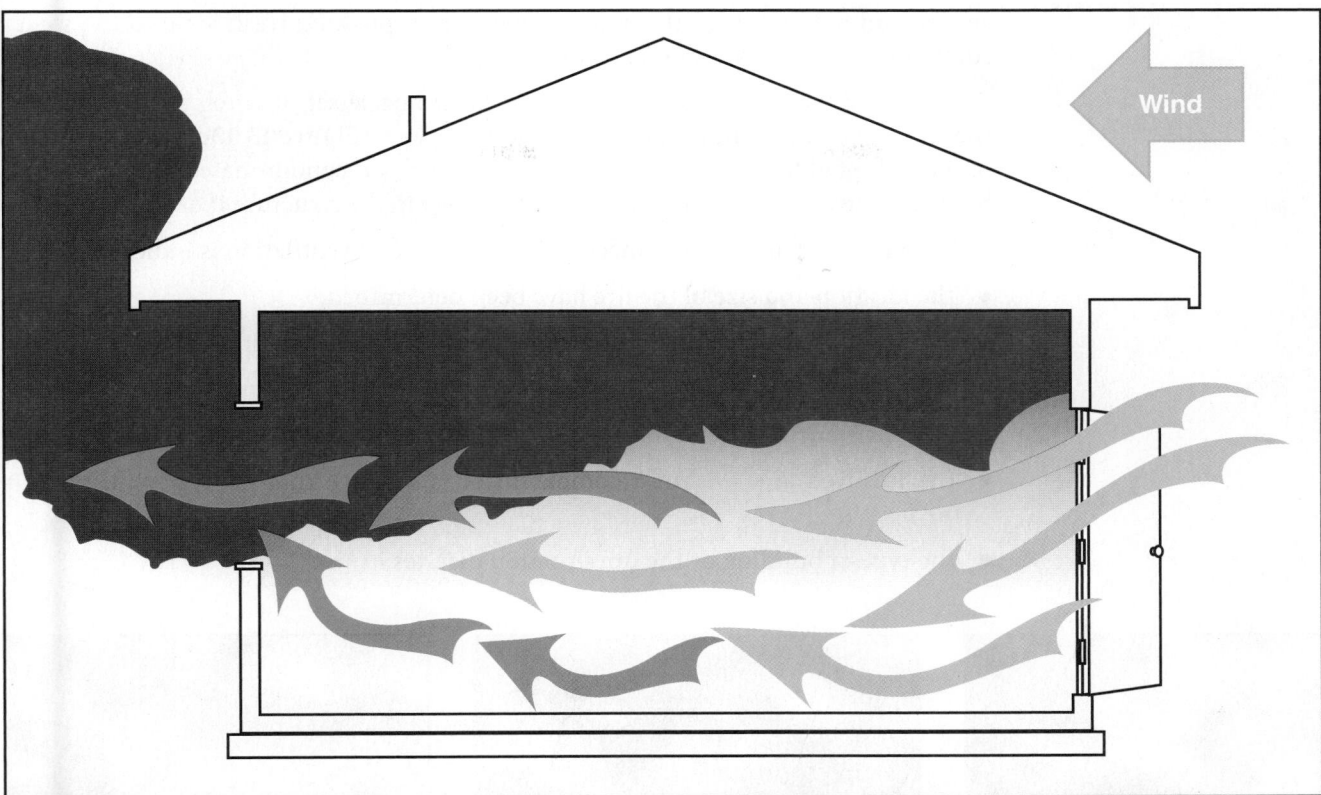

Figure 13.12a Horizontal ventilation pushes products of combustion out of a compartment through an opening at the same level as the incoming fresh air.

Figure 13.12b Vertical ventilation removes smoke from the structure through an opening at a high point of the structure.

Natural Ventilation — Techniques that use the wind, convection currents, and other natural phenomena to ventilate a structure without the use of fans, blowers, and smoke ejectors.

Mechanical Ventilation — Any means other than natural ventilation. This type of ventilation may involve the use of blowers and smoke ejectors.

Hydraulic Ventilation — Ventilation accomplished by using a spray stream to draw the smoke from a compartment through an exterior opening.

The means used for horizontal and vertical ventilation are **natural, mechanical, and hydraulic ventilation**. Natural horizontal ventilation involves opening doors and windows to allow natural air currents and pressure differences to move smoke and heat out of the building. Natural vertical ventilation uses the buoyancy of heated smoke and gases to draw them out of the structure through the roof openings while entraining (pulling or drawing) fresh air into the structure **(Figure 13.13a, p. 744)**.

Mechanical ventilation uses fans, blowers, and smoke ejectors **(Figure 13.13b, p. 744)**. Although mechanical methods can be applied to both horizontal and vertical ventilation, it is most often used for horizontal. The means may involve pulling the smoke and fire gases out through an opening or pushing fresh air into the structure and displacing the smoke and fire gases.

Hydraulic ventilation involves using a spray nozzle set on a fog pattern to draw the smoke out an opening such as a window or door (**Figure 13.13c**). This technique requires firefighters to operate the nozzle within the contaminated atmosphere. It has the disadvantage of increasing water damage to the structure if done improperly.

In general, the need to use mechanical or hydraulic ventilation is indicated when:

- The location and size of the fire have been determined.
- The layout of the building is not conducive to natural ventilation.
- Natural ventilation slows, becomes ineffective, and needs support.
- The fire is burning below ground in the structure.
- The involved area within a compartment is so large that natural ventilation is inefficient.
- The type of building or the fire situation dictates its use.

Figure 13.13a Natural ventilation works with the buoyancy of heated smoke to clear a structure.

Figure 13.13b Mechanical positive pressure ventilation pushes fresh air along a horizontal plane to eject smoke and fire gases.

Figure 13.13c Hydraulic ventilation uses a fog pattern spray to entrain the smoke out of a room.

Horizontal Ventilation

Structures that lend themselves to the application of horizontal ventilation include the following:

- Buildings in which the fire has not involved the attic or cockloft area
- Involved floors of multistoried structures below the top floor, or the top floor if the attic is uninvolved
- Buildings so weakened by the fire that vertical ventilation is unsafe

- Buildings with daylight basements
- Buildings in which vertical ventilation is ineffective

Natural Horizontal Ventilation

When conditions are appropriate, natural horizontal ventilation operations should work with existing atmospheric conditions, taking advantage of natural air flow. Natural ventilation requires no additional personnel or equipment to set up and maintain. When directed by the IC, windows and doors on the **leeward side** of the structure should be opened first to create an exit point. Openings on the **windward side** of the structure are then opened to permit fresh air to enter forcing the smoke toward the **exhaust openings** (**Figure 13.14**). If only a single opening is made, such as opening a door, this vent will serve as both the inlet for air and the exit for smoke.

Mechanical Horizontal Ventilation

When the natural flow of air currents and the currents created by the fire is insufficient to remove smoke, heat, and fire gases, mechanical ventilation is necessary. Mechanical ventilation is accomplished through negative pressure or positive pressure.

Negative-pressure ventilation. **Negative-pressure ventilation (NPV)** is the oldest type of mechanical ventilation. Smoke ejectors are used to expel (pull) smoke from a structure by developing artificial air flow or enhancing natural ventilation. Smoke and fire gases are drawn out of the structure and fresh air is drawn into the structure by the negative pressure the fans have created. Fans can be placed in windows, doors, or roof vent openings to exhaust the smoke, heat, and gases from inside the building to the exterior (**Figure 13.15**).

The fan should be positioned in openings on the leeward side to exhaust in the same direction as the prevailing wind. This technique creates a lower pressure at the inlet allowing fresh air to replace the air being expelled from the building. The procedure for ventilating using mechanical negative pressure in a window is given in **Skill Sheet 13-I-1**. The procedure for ventilating using mechanical negative pressure in a doorway is given in **Skill Sheet 13-I-2**.

Leeward Side — Protected side; the direction opposite from which the wind is blowing.

Windward Side — The side or direction from which the wind is blowing.

Exhaust Opening — Intended and controlled exhaust locations that are created or improved at or near the fire to allow products of combustion to escape the building.

Negative-Pressure Ventilation (NPV) — Technique using smoke ejectors to develop artificial air flow and to pull smoke out of a structure. Smoke ejectors are placed in windows, doors, or roof vent holes to pull the smoke, heat, and gases from inside the building and eject them to the exterior.

Figure 13.14 A plume of smoke may help firefighters identify the direction the wind is blowing.

Figure 13.15 Mechanical negative-pressure ventilation pulls smoke and fire gases out of a room.

Recirculation — Movement of smoke being blown out of a ventilation opening only to be drawn back inside by the negative pressure created by the ejector because the open area around the ejector has not been sealed.

The open areas around a smoke ejector must be properly sealed to prevent air from recirculating back into the building. Atmospheric pressure pushes air back through the open spaces in the doorway or window and pulls the smoke back into the room. This **recirculation** reduces ventilation efficiency. To prevent recirculation, cover the open area around the fan with a salvage cover or other material (**Figure 13.16**).

The flow of smoke and other gases to the exhaust opening should be kept as straight as possible. Every corner causes turbulence and decreases ventilation efficiency. Because smoke and gases accumulate near the ceiling, the smoke ejector should also be located near the top of the opening. Avoid opening windows or doors near the smoke ejector because this action can greatly reduce ventilation efficiency. Remove all obstacles including window screens that may reduce the air flow. Do not allow the intake side of the smoke ejector to become obstructed by debris, curtains, drapes, or anything else that can decrease the amount of intake air.

When ventilating potentially flammable atmospheres, only smoke ejectors equipped with intrinsically safe motors and power cable connections should be used. Smoke ejectors should be turned off when they are moved, and they should be carried by the handles provided for that purpose. Before starting a smoke ejector, be sure that no one is near the blades and that clothing, curtains, or draperies are not in a position to be drawn into the fan blades. The air discharged from the fan should be avoided because of debris that may be picked up and blown by the smoke ejector.

Positive-pressure ventilation. **Positive-pressure ventilation (PPV)** is a mechanical ventilation technique that uses a high-volume fan to create a slightly higher pressure inside a building than that outside. When the pressure is higher inside the building, the smoke inside the building is forced through openings to the lower-pressure area outside. Positive-pressure ventilation requires good fireground discipline, coordination, and tactics.

Positive-Pressure Ventilation (PPV) — Method of ventilating a room or structure by mechanically blowing fresh air through an inlet opening into the space in sufficient volume to create a slight positive pressure within and thereby forcing the contaminated atmosphere out the exit opening.

Uses of Positive-Pressure Techniques

In the initial stage of fire suppression, positive pressure can be used to mechanically exhaust heat and smoke, cool the environment, and improve visibility before firefighters enter the building to begin suppression and/or search and rescue. After the fire has been extinguished, positive-pressure ventilation is used to mechanically exhaust products of combustion and clear the structure of all remaining smoke. It can also be used to prevent smoke and heat flow into an area such as pressurizing stairwells.

Figure 13.16 The use of a salvage cover will prevent recirculation of contaminants during ventilation efforts.

The location where the fan or blower is set up is called the *entry (inlet) opening* or *point*. Once that location is selected, an exhaust opening or point must be created. The size of the exhaust opening varies with the size of the entry opening and the capacity of the blower used (**Figure 13.17**). The exhaust opening may be a window or doorway. Once an exhaust opening has been created, a blower is placed outside the entry opening. For a normal, single, 3-foot wide (0.9 m) door, the distance between the door and the fan should be 4 to 6 feet (1.2 to 1.8 m) (**Figure 13.18**). This distance may need to be altered for larger door openings. It is important that the cone of air from the blower completely covers the doorway opening. To maintain the positive pressure inside, it is important to control the location, number, and size of exterior openings.

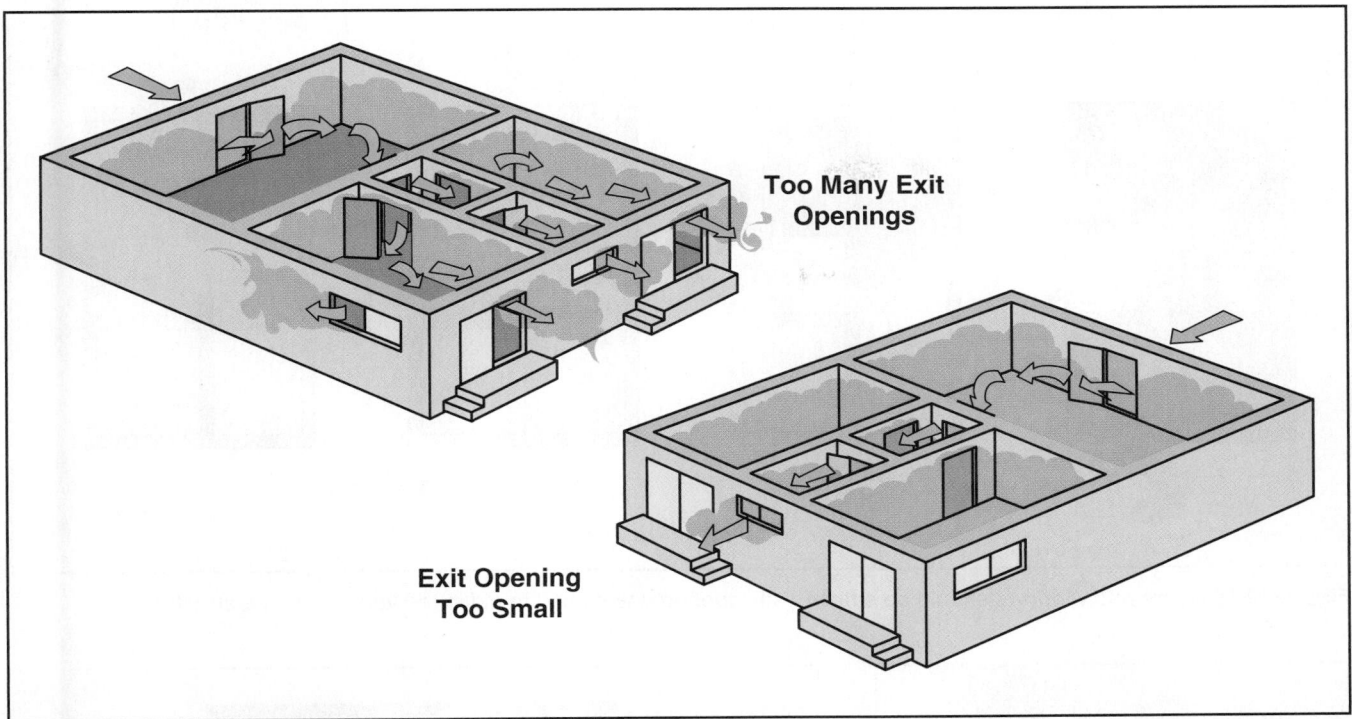

Figure 13.17 Horizontal ventilation is impaired if the area of the opening is too small or too large for the quantity of smoke that is being ejected.

Too Many Exit Openings

Exit Opening Too Small

During post fire suppression activities, PPV can be used to ventilate interior compartments by systematically opening and closing interior doors and exterior windows. This process accelerates the removal of heat and smoke from the building.

When using PPV to ventilate a multistory building, it is best to apply positive pressure at the lowest point. Positive pressure is applied to the building at ground level through the use of one or more blowers. The positive pressure is then directed throughout the building by opening and closing doors until the building is totally evacuated of smoke (**Figure 13.19, p. 748**). If a single fan cannot provide enough pressure and air flow, additional PPV fans can be set up on upper floors and at the entry point to increase the air flow. Smoke can then be systematically removed one floor at a time (starting with the floor most heavily charged with smoke) by selectively opening exit points. Either cross-ventilating floors or directing smoke up a stairwell and out the stairwell rooftop opening are methods for removing smoke one floor at a time (**Figures 13.20a and b, p. 748**). Blowers larger than those typically carried on an engine or ladder truck are also available for use in multistory and large-volume buildings. The procedure for ventilating using mechanical positive pressure is given in **Skill Sheet 13-I-3**.

The main problem in using PPV in aboveground operations is coordinating the opening and closing of the doors in the stairwell being used to ventilate the building. To control

4 – 6 ft
1.2 m – 1.8 m

Figure 13.18 Proper placement of the PPV fan will cover the entire doorway with the cone of air.

Figure 13.19 Doors and windows should be strategically opened and closed to aid in the ventilation of a structure.

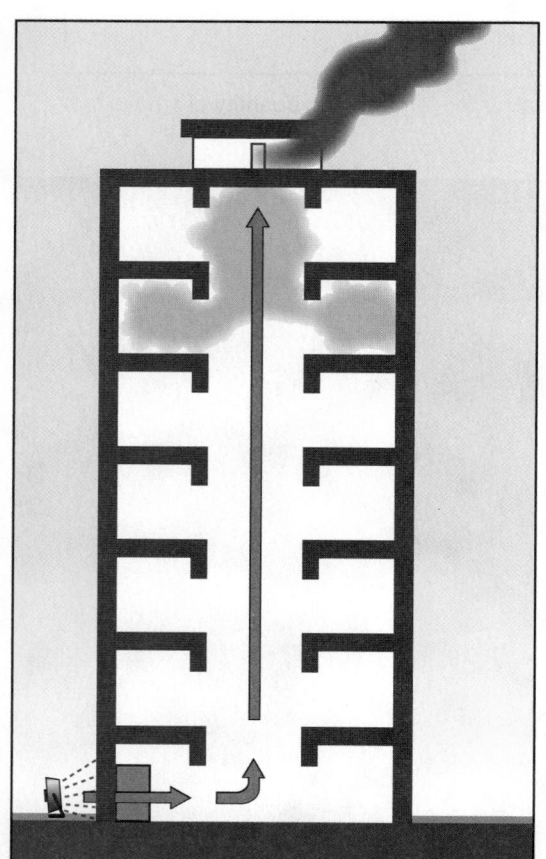

Figure 13.20a Blowers may be used to direct smoke up a stairwell and out a rooftop opening.

Figure 13.20b Cross-ventilating using positive pressure air handling effectively moves smoke to an approved opening.

openings or pressure leaks, one person is placed in charge of the pressurizing process. It is helpful to use portable radios and to have firefighters patrol the stairwell and hallways.

To ensure an effective PPV operation, take the following actions:

- Ensure that your exhaust opening is sufficient to handle the air flow.
- Monitor the operation of the PPV fan.
- Maintain communications between the IC, the interior attack crews, and the PPV operator.
- Take advantage of existing wind conditions.
- Make certain that the cone of air from the fan covers the entire entry opening.
- Reduce the volume of the area being pressurized to speed up the process by selectively opening and closing interior doors.
- Avoid creating unintended horizontal openings.

> **WARNING**
> Improperly applied PPV can change the interior conditions and injure personnel working inside the structure.

The advantages of PPV compared to NPV include the following:

- Firefighters can set up PPV blowers without entering the smoke-filled environment.
- PPV is equally effective with either horizontal ventilation or vertical ventilation because it supplements natural air currents.
- Removal of smoke and heat from a structure is more efficient.
- The velocity of air currents within a building is minimal and creates little, if any, effects that disturb the building contents or smoldering debris. Yet, the total exchange of air within the building is faster than using NPV alone.
- Fans powered by internal combustion engines operate more efficiently in clean air.
- The cleaning and maintenance of fans used for PPV is significantly less than those needed for NPV fans.
- PPV is effective in all types of structures, particularly in large, high-ceiling areas where NPV is ineffective.
- Heat and smoke may be directed away from unburned areas or egress paths.
- Exposed buildings or adjacent compartments can be pressurized to reduce fire spread into them.

The disadvantages of PPV are as follows:

- The structure must be intact.
- Interior carbon monoxide levels may be increased if the exhaust from fans powered by internal combustion engines is allowed to enter.
- Hidden fires may be accelerated and spread throughout the building.

Hydraulic ventilation. Hydraulic ventilation may be used in situations where other types of forced ventilation are unavailable. Hydraulic ventilation is used to clear a room or building of smoke, heat, steam, and gases after a fire has been controlled.

Hydraulic ventilation uses a spray stream from a fog nozzle to entrain smoke and gases and carry them out of the structure through a door or window. To perform hydraulic ventilation, a fog nozzle is set on a wide fog pattern that will cover 85 to 90 percent of the window or door opening through which the smoke will be drawn or pulled. The nozzle tip should be at least 2 feet (0.6 m) back from the opening (**Figure 13.21**). The larger the opening, the faster ventilation will occur. In fact, a master stream device can be set up in an open commercial or industrial doorway such as those on loading docks. The procedure for horizontal hydraulic ventilation is shown in **Skill Sheet 13-I-4**.

There are disadvantages to the use of hydraulic ventilation. These disadvantages include the following:

- There may be an increase in the amount of water damage within the structure, if done incorrectly.

- There will be a drain on the available water supply. This is particularly crucial in rural fire fighting operations where water shuttles are being used.

- In freezing temperatures, there will be an increase in the amount of ice on the ground surrounding the building.

- The firefighters operating the nozzle must remain in the heated, contaminated atmosphere throughout the operation.

- The operation may have to be interrupted when the nozzle team has to leave the area to replenish its air supply.

Precautions Against Upsetting Horizontal Ventilation

You must take care not to upset the effects of horizontal ventilation. For instance, opening a door or window on the windward side of a burning building before creating a ventilation exhaust opening on the leeward side may pressurize the building,

Figure 13.21 Hydraulic ventilation uses the fog nozzle water stream to entrain or carry the smoke out of the building.

2 feet (0.6 m)

intensify the fire, and cause the fire to spread to uninvolved areas. You should also take advantage of air currents established by horizontal ventilation. If an obstruction blocks the established currents in the entry opening, the positive effects of horizontal ventilation may be reduced or eliminated.

Advantages of Mechanical Ventilation

Some of the advantages of using mechanical ventilation in fires and other situations include:

- Supplements and enhances natural ventilation
- Ensures more control of air flow
- Speeds the removal of contaminants
- Reduces smoke damage
- Promotes good public relations

Even in the absence of fire, contaminated atmospheres must be quickly and thoroughly cleared from a building or other confined space. For example, buildings filled with a flammable or toxic gas must be ventilated quickly but safely. Confined spaces containing low oxygen levels can benefit from the introduction of fresh air as the contaminated atmosphere is removed (**Figure 13.22**). In these and many other situations, mechanical ventilation is the best technique to use.

Figure 13.22 Mechanical ventilation may use flexible ducting to reach a contaminated atmosphere.

Disadvantages of Mechanical Ventilation

If mechanical ventilation is misapplied or not properly controlled, it can cause a great deal of harm. Some of the disadvantages of improper mechanical ventilation include the following:

- Can cause a fire to intensify and spread
- Depends upon a power source
- Requires special equipment
- Requires additional resources and personnel

Vertical Ventilation

Vertical ventilation occurs after the IC has:

- Determined the need for ventilation
- Determined that vertical ventilation can be done safely and effectively
- Considered the age and type of construction involved
- Considered the location, duration, and extent of the fire
- Observed safety precautions
- Identified escape routes
- Selected the place to ventilate
- Moved personnel and tools to the roof

Vertical ventilation presents the following increased risks:

- Placing personnel above ground level
- Working on both peaked and flat surfaces
- Working above the fire
- Working on roofs that may have been weakened because of age or fire damage

The IC must assess these risks, implement safety precautions, and determine if the vertical ventilation must be offensive or defensive. Offensive vertical ventilation is intended to aid in reaching and extinguishing the fire while defensive is meant to stop the spread of fire and contain it in one area of the structure.

Safety Precautions

Some of the safety precautions that should be observed include the following:

- Check the wind direction and velocity to determine the effect on exposures.
- Work with the wind at your back or side when cutting a roof opening to protect yourself from heat, smoke, and embers (**Figure 13.23**).
- Note the existence of obstructions or excessive weight on the roof that may contribute to roof collapse.
- Provide a secondary means of escape for crews on the roof (**Figure 13.24**).
- Ensure that main structural supports are not cut while creating a ventilation opening.
- Guard the ventilation opening to prevent personnel from falling into it.
- Evacuate the roof promptly when ventilation work is complete or when ordered to leave.
- Use lifelines, roof ladders, or other means to prevent personnel from sliding and falling off the roof (**Figure 13.25**).
- Make sure that a roof ladder (if used) is firmly secured over the peak of the roof before working from it.
- Exercise caution when working around electric wires, solar panels, and guy wires.
- Ensure that all personnel on the roof are wearing full PPE, including SCBA, and that they are breathing SCBA air.
- Keep firefighters out of the range of those who are swinging axes and operating power saws.
- Remain aware of overhead obstructions within the range of your swing.
- Start power tools on the ground to ensure operation but make sure they are shut off before hoisting or carrying them to the roof.
- When using a power saw, make sure that the angle of the cut is away from the body.
- Extend ground ladders at least three to five rungs above the edge of the roof or top of the parapet wall and secure the ladder to the wall or roof.
- When operating from aerial ladder platforms, the floor of the platform should be even with or slightly above roof level.
- Check the roof for structural integrity before stepping onto it and continue sounding it throughout the operation (**Figure 13.26**).

Wind Direction →

Figure 13.23 Working with the wind from the back protects firefighters from heat, smoke, and embers.

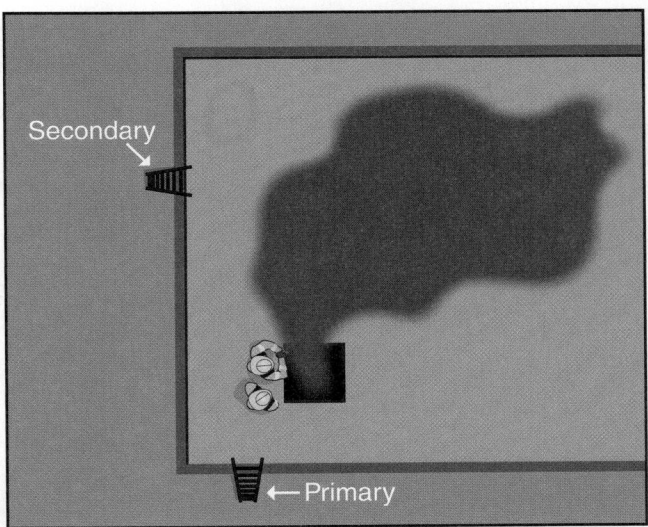

Secondary

Primary

Figure 13.24 Rooftop operations are hazardous and should be accommodated with a secondary means of egress.

Figure 13.25 Roof activities may use ladders as a piece of safety equipment.

Figure 13.26 A roof's ability to hold the weight of a firefighter should never be assumed before those limits are tested.

- Both before and after ventilating, walk on load-bearing walls and strongest points of roof structure whenever possible.
- When the roof has been opened, penetrate the ceiling below to enhance ventilation.

CAUTION

Roof ladders are only meant to prevent slipping and are not intended to be used on fire-weakened roofs.

WARNING

Never direct a fire stream into a vertical exhaust opening when interior attack crews are inside the structure because it will force smoke, heat, and steam down on them.

Before stepping off a ladder, parapet wall, or other place of safety onto the roof of a burning building — especially if the roof surface is obscured by smoke or darkness — firefighters should sound the roof (if possible) by striking the roof surface with the blunt end of a pike pole, rubbish hook, or axe. When struck by a tool, some roofs will feel solid over structural supports and the tool will tend to bounce off the surface. Between the supports, the roof may feel softer and less rigid. The roof may also *sound* solid when struck over a rafter or joist and produce a hollow sound when struck between the supports. By practicing on structurally sound roofs, firefighters can learn to recognize the difference in the feel and the sound of supported and unsupported areas of a roof. The procedure for sounding a roof is given in **Skill Sheet 13-I-5**.

Remember, however, that roofs with several layers of composition shingles or other roof coverings may not respond to sounding. They may sound quite solid when struck with a tool even though the roof supports may have been severely damaged by the fire. Also, roofs covered with tile or slate cannot be sounded; the tiles/slates must be removed to reveal the underlying structure.

Use preincident planning information to identify buildings that have roofs supported by lightweight or engineered trusses. These roofs may fail early in a fire and are extremely dangerous to work on or under.

Be aware of the following warning signs of a possible unsafe roof condition:
- Melting asphalt
- "Spongy" roof (a normally solid roof that springs back when walked upon)
- Smoke coming from the roof
- Fire coming from the roof

NOTE: Some roofs are spongy with no fire involvement. Know the roofs in your response area.

Rotary saws, carbide-tipped chain saws, or a chain saw with adapted features are excellent for roof-cutting operations because they are faster and less damaging than axes or other manual cutting tools (**Figure 13.27**). The saw operator must have good footing and maintain control of the saw at all times. Working on a pitched roof from a roof ladder, a rubbish hook, or Halligan can be used to provide a secure foothold for the saw operator. In most cases, it is safest to turn off the saw when it is being transported to or from the point of operation — especially when moving up or down a ladder. The procedure for ventilating using a rotary saw to cut an opening is given in **Skill Sheet 13-I-6**. The procedure for ventilating using an axe to cut an opening is given in **Skill Sheet 13-I-7**.

The roof ventilation team should be in constant communication with their supervisor or the IC. Responsibilities of the roof ventilation team leader include:

- Ensuring that the roof is safe (sounding, visual observation)
- Ensuring that only the required openings are made
- Directing efforts to minimize secondary damage (damage caused by fire fighting operations)
- Coordinating the team's efforts with those of firefighters inside the building
- Ensuring the safety of all personnel who are assisting with ventilation operations
- Ensuring that there are two means of egress from the roof
- Ensuring an adequate exhaust opening size
- Ensuring that the team leaves the roof as soon as their assignment is completed

CAUTION

Work in groups of at least two, but with no more personnel than absolutely necessary to perform the assigned task.

Figure 13.27 Power saws are able to penetrate roofing quickly and with more precision than hand-powered tools. *Courtesy of Matt Daly.*

Kerf Cut — A single cut the width of the saw blade made in a roof to check for fire extension.

Before cutting any type of ventilation hole, you should cut an inspection hole in the roof. Inspection holes help to determine the location of a fire and the direction of travel of a fire that is located in an attic or cockloft. Inspection holes are used in both offensive and defensive ventilation operations. There are two primary types of inspections holes: the **kerf cut** and the triangle or "A" cut.

The kerf cut is the easiest and fastest inspection hole to cut. Make a single cut in the roof surface using a rotary saw, chain saw, or axe (**Figure 13.28**). The resulting hole should be the width of the saw or axe blade. Although your vision is limited by the size of the hole, you will have an indication of conditions below the roof. A major disadvantage of the kerf cut is that the heat from the fire can cause the tar or membrane material to melt and seal up the cut.

The triangle cut provides the most reliable information of conditions beneath the roof. Using a rotary or chain saw, the cut can be created from a single kerf cut if conditions indicate the need for it. The triangle cut consists of three overlapping cuts that form a triangle or letter "A." First, make diagonal cut from upper left to lower right. Next, make a diagonal cut from the top of the first cut to the lower left. Finally, connect the two lower ends of the previous cuts. The center of the triangle should fall into the space below (**Figure 13.29**). If it is necessary to push the center to free it, always use a tool and never your hand. Remember that heated smoke, gases, and sometimes fire will exit the inspection hole.

Offensive Ventilation Methods

Offensive ventilation involves making an opening over the seat of the fire at or near the highest point of the roof. The type of exhaust opening and the method for making it will depend on the type of roof in which the opening is being made. When cutting an exhaust opening in any type of roof, there are two critical points to bear in mind:

- A square or rectangular opening is easier to cut and easier to repair after the fire.
- One large opening, at least 4 x 8 feet (1.2 m by 2.4 m), is much better than several small ones.

Flat roof. Square or rectangular openings are the most common type of opening made in a flat roof. These openings can be made between the roof trusses or with the truss in the middle of the opening. When the truss is in the middle of the opening, a

Figure 13.28 A kerf cut allows firefighters to assess conditions below without changing the ventilation profile significantly.

Figure 13.29 Triangle cuts present a reliable view of the conditions below a roof.

louver cut is used **(Figure 13.30). Skill Sheet 13-I-8** illustrates the steps for making a louver cut in a flat roof. The steps are as follows:

- Make an initial opening no smaller than 4 x 8 feet (1.2 m by 2.4 m). The exhaust opening may need to be enlarged.

- Identify the location of the rafters.

- Make the short cuts across the top and bottom of the rafter and the long cuts parallel to either side of the rafter.

- Strike the near side to break it loose from the rafters and pull the far side toward you with a roof hook.

Pitched roof, shingle-covered. On shingle-covered pitched roofs, cut a few inches (mm) below the peak on the leeward side. Always cut exhaust openings at or very near the highest point on the roof when possible. Work from a roof ladder with the hooks attached to the ridge line. On extremely steep roofs, it may be necessary to work from an aerial platform. **Skill Sheet 13-I-9** depicts the steps for opening pitched roofs.

Pitched roof, slate or tile-covered. Slate and tile roof covering may be attached to solid sheathing or to *battens* (strips of wood attached to rafters) that have spaces between them. Slate and tile roofs are opened by removing the individual pieces or using a large sledgehammer to smash the slate or tile pieces. If the sub roof is solid, then a ventilation hole is cut in the standard manner for pitched roofs. A hole may not need to be cut in the battens if there is enough space between them for ventilation to take place.

Arched roofs. Procedures for cutting exhaust openings in arched roofs are the same as for flat or pitched roofs except that there is no ridge over which to hook roof ladders. The curvature of the roof prevents roof ladders from lying flat. As soon as you are on the roof, make a kerf cut to locate the arches, observe the truss space, and determine fire involvement before proceeding further. Walk only on the trusses and other strong points when possible.

Figure 13.30 A louvre cut in a roof creates a shield for responders while also opening vertical ventilation.

Figure 13.31 With the use of an aerial ladder, thin metal roofs may be safely opened using a powered saw.

Metal roofs. Thin metal roofs can be sliced open with an axe, carbide tip chain saw, or rotary saw and peeled back **(Figure 13.31)**. Metal cutting tools or power saws with metal cutting blades can be used to open thick metal roofs. On industrial buildings with thick metal roofs, it may be easier and faster to open skylights, monitors, or scuttle hatches. Older buildings may have roofs that are made of large pieces of sheet metal laid over skip sheathing. These can be opened by cutting with a power saw, axe, or a large sheet-metal cutter.

Defensive Ventilation Methods

Trench Ventilation — Defensive tactic that involves cutting an exhaust opening in the roof of a burning building, extending from one outside wall to the other, to create an opening at which a spreading fire may be cut off. *Also known* as strip ventilation.

The trench cut (also referred to as **trench** or **strip ventilation**) is strictly a defensive operation and should NOT be confused with or used as offensive vertical ventilation. Offensive vertical ventilation techniques are used primarily to remove heat, smoke, and gases from the structure and are best done directly above the fire. A trench cut is used to create a fire break that stops the spread of fire in common attic structures or large structures. Though this tactic can be time consuming or physically taxing on personnel, it does work well in large buildings, especially if they have a common cockloft or attic.

When the IC has determined that the main body of the fire is too great to extinguish, he or she will decide to take a defensive stance and abandon efforts to save the currently burning part of the building. At this time, he or she may give the order to make a trench cut. This opening must be created at least 30 feet (9.1 m) ahead of the advancing fire and only after the offensive vertical ventilation opening has been made. Making the offensive opening first allows the heat and smoke to escape and the trench to be completed before the fire front reaches that point. Otherwise, the fire will quickly be drawn to, and burn past, the unfinished trench and continue to spread throughout the building while endangering firefighters on the roof. The trench is created by making two parallel cuts that extend from one exterior wall to the opposite exterior wall, then removing the roof material between the cuts and pushing down the ceiling material below **(Figure 13.32)**. The distance between the cuts should be large enough to prevent fire from burning past the opening, and small enough that it does not compromise the integrity of the roof. A trench cut about 3 to 4 feet (0.9 to 1.2 m) wide should be sufficient. **Skill Sheet 13-I-10** illustrates the steps required to make a trench cut in a flat roof.

If a trench cut is created improperly, it will place firefighters in a very dangerous position, working ahead of the fire. In addition, it may cause the fire to spread more rapidly and potentially destroy the entire structure. To assure your success and safety, it is important that all members involved in the operation:

- Plan ahead.
- Establish communications between the roof ventilation team and the IC.
- Maintain good communication.
- Are aware of the dangers.
- Have a clear understanding of the objective.
- Have a charged hoseline present.
- Wear full PPE and SCBA.
- Have two means of escape from the roof that are remote from each other and do not include crossing over the cut.
- Assign a roof safety officer to observe conditions.
- Cut small inspection holes a few feet from the trench on both the fire side and safe working side. These holes will serve as indicators as to when the fire approaches and also show if the fire has extended to the unburned side.

Figure 13.32 A trench cut is used as a fire break in narrow attic spaces.

Precautions Against Upsetting Established Vertical Ventilation

Ventilation problems can be avoided by well-trained firefighters conducting a well-coordinated attack. Some common factors that can reduce the effectiveness of vertical ventilation are:

- Improper use of mechanical ventilation
- Indiscriminant window breaking
- Fire streams directed into ventilation openings
- Explosions
- Burn-through of the roof, floor, or wall
- Additional openings between the attack team and the upper opening
- Improper location of the vertical ventilation opening

Vertical ventilation is not the solution to all ventilation problems because there are instances where its application is impractical or impossible. In these cases, other strategies, such as the use of strictly horizontal ventilation, must be employed.

WARNING
Do not direct a fire stream into a ventilation opening during offensive interior operations.

Other Types of Ventilation Situations

Most ventilation operations will involve residential occupancies requiring horizontal or vertical ventilation tactics. However, you may be faced with incidents involving basements, windowless buildings, or high-rise buildings that will require a variation on these tactics.

Basement Fires

Basement fires can be among the most challenging situations firefighters will face. Unless the basement has vents installed, heat and smoke will quickly spread upward into the building. Without effective ventilation, access into the basement is difficult because firefighters have to descend through the intense rising heat and smoke to get to the seat of the fire. Access to the basement may be through interior or exterior stairs, exterior windows, or hoistways. Many outside entrances to basements may be blocked or secured by iron gratings, steel shutters, wooden doors, or combinations of these for protection against weather and for security. All of these features may impede attempts at ventilation.

Basement ventilation can be accomplished in several ways. If the basement has ground-level windows or even below ground-level windows in wells, horizontal ventilation can be employed effectively (**Figure 13.33**). If these windows are not available, interior vertical ventilation must be performed. Natural paths from the basement, such as stairwells and hoistway shafts, can be used to evacuate heat and smoke if there is a way to expel the heat and smoke to the atmosphere without placing other portions of the building in danger (**Figure 13.34**). As a last resort, an opening may be cut in the floor near a ground-level door or window, and the heat and smoke can be forced from the opening through the exterior opening using fans (**Figure 13.35**).

Fires in Windowless Buildings

Windowless buildings complicate fire fighting and ventilation operations. Creating the openings needed to ventilate a windowless building may delay the operation for a considerable time, allowing the fire to increase in intensity, consume air and fuel, and spread within the structure.

Figure 13.33 Ventilation may be directed from an upper level toward a below ground fire to horizontally vent the space.

Figure 13.34 Natural ventilation paths may be utilized to mechanically direct smoke through parts of the building that will not spread the hazard.

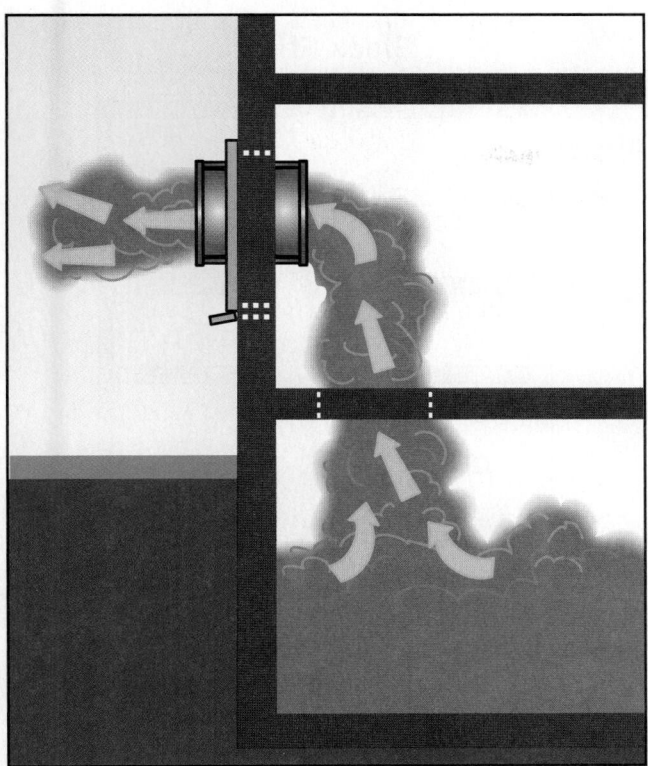

Figure 13.35 If necessary, a basement fire can be vented through a hole cut in the floor above.

Ventilating this type of building can be difficult, and the problems involved vary depending on the size, occupancy, configuration, and type of construction materials used. Windowless buildings usually require mechanical ventilation for the removal of smoke. Most buildings of this type are automatically cooled and heated through ducts. HVAC equipment can sometimes effectively clear the area of smoke by itself; however, unless specifically designed for this purpose, these systems are more likely to cause the spread of heat and fire. In windowless and high-rise structures these systems need to be brought under fire department control prior to fire operations. If they are designed to contain products of combustion and are operating properly, they should be allowed to work; if not, the IC should be notified and steps taken to control the system manually.

High-Rise Fires

High-rises buildings may contain hospitals, hotels, apartments, or offices. Because there are more occupants in high-rise buildings than in other occupancies, life safety considerations are an even higher priority. Tactical ventilation in a high-rise building must be carefully coordinated to ensure the safest and most effective use of personnel, equipment, and extinguishing agents. The personnel required for search and rescue and fire fighting operations in high-rise buildings is often four to six times as great as required for a fire in a typical low-rise building.

Fire, smoke, and toxic gases can spread rapidly through pipe shafts, stairways, elevator shafts, unprotected ducts, and other vertical and horizontal openings in high-rises. These openings contribute to a **stack effect** (natural movement of heat and smoke throughout a building), creating an upward draft and interfering with evacuation and ventilation (**Figure 13.36, p. 762**).

Heated smoke and fire gases travel upward until they reach the top of the building or they are cooled to the temperature of the surrounding air. When this equalization of temperature occurs, the smoke and fire gases stop rising, spread horizontally, and

Stack Effect — Phenomenon of a strong air draft moving from ground level to the roof level of a building. Affected by building height, configuration, and temperature differences between inside and outside air.

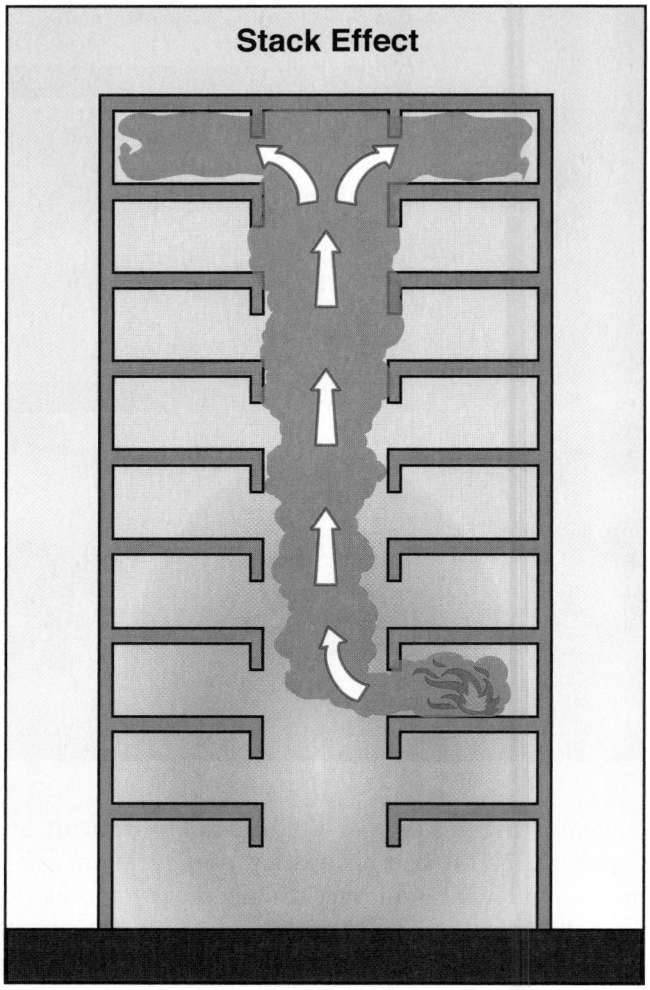

Figure 13.36 If temperatures are hotter inside the structure than the ambient outside temperature, air flow will contribute to the stack effect by drawing air at the bottom of the structure.

Horizontal Smoke Spread — Tendency of heat, smoke, and other products of combustion to rise until they encounter a horizontal obstruction. At this point they will spread laterally (ceiling jet) until they encounter vertical obstructions and begin to bank downward (hot gas layer development).

stratify (form layers). In some cases, such as high-rise buildings, these layers of smoke and fire gases will collect on floors below the top floor. Additional heat and smoke will eventually force these layers to expand and move upward to the top floor of the building. **Horizontal smoke spread** and hot gas layer development can also occur when a vertical exhaust opening is not large enough to exhaust the smoke and gases. Tactics involving horizontal or vertical ventilation assisted by mechanical means must be developed to cope with the ventilation and life hazard problems inherent in stratified smoke. In many instances, ventilation must be accomplished horizontally with the use of mechanical ventilation devices and the building's HVAC systems.

Tactical vertical ventilation in high-rise buildings must be considered during pre-incident planning. In many buildings, only one stairwell penetrates the roof. This stairwell can be used much like a chimney to ventilate smoke, heat, and fire gases from various floors while another stairwell is used as the escape route for building occupants. However, during a fire, the doors on uninvolved floors must be controlled so occupants do not accidentally enter the ventilation stairwell as they are evacuating. Before the doors on the fire floors are opened and the stairwell is ventilated, the door leading to the roof must be blocked open or removed from its hinges. Preventing the door at the top of the shaft from closing ensures that it cannot compromise established ventilation operations. Remember that when ventilating the top of a stairwell, you will be drawing the smoke and heat to you or anyone else in the stairwell between the fire floor and the roof. When an enclosed secondary stairwell is used for evacuating occupants, PPV fans should be located at the bottom floor to pressurize the stairwell and keep smoke from entering it.

In some high-rise structures, ventilation fans are built into the top of the stairwell to assist in ventilation. When activated, these fans draw smoke from the fire floor into the stairwell and out the top. This technique may make it difficult for the fire suppression team to make entry onto the fire floor from this stairwell. The safest and most effective technique may be to pressurize the stairwells with PPV fans to confine the smoke on the floors. Firefighters can advance to the fire floor in a safe atmosphere.

NOTE: Under some conditions, elevator shafts that penetrate the roof may be used for ventilation.

> **WARNING**
> Do not use stairwells or elevator shafts simultaneously for both evacuation and ventilation.

Effects of Building Systems on Tactical Ventilation

Many modern buildings have built-in HVAC systems that can contribute to the spread of smoke, fire, and toxic gases throughout a structure. These systems are usually controlled from a panel in a maintenance and operations center located in the basement or at ground level in the building. There is often a diagram of the building duct system along with information on smoke detection and fire suppression systems built into the HVAC ductwork at the control panel. These systems are designed to shut down the HVAC system automatically when smoke or fire is detected in the ducts. Firefighters should be familiar with the location and operation of controls that will shut down the HVAC system when necessary.

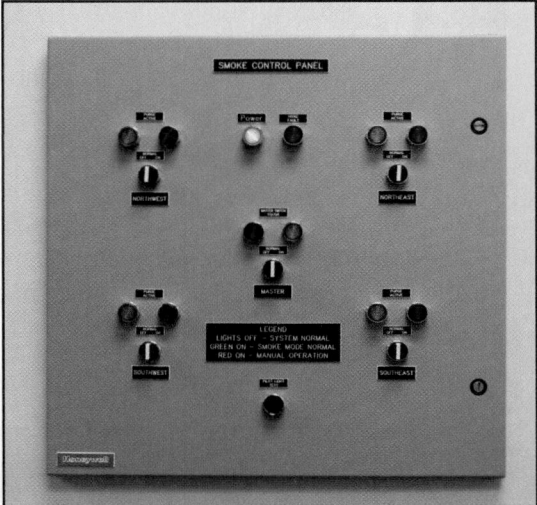

Figure 13.37 Formal smoke control systems are standard in many buildings with large open spaces.

While firefighters may need to shut down the HVAC system during a fire, clearing the system of residual smoke and restoring it to operation are the responsibilities of the building engineer or maintenance superintendent. Because an HVAC system may draw fire into the ducts along with the heat and smoke before it shuts down, the ductwork may create additional fire damage by conducting heat through the metal of the duct. Firefighters should check combustibles adjacent to the ductwork in case conduction has caused additional fires.

Many other buildings, especially high-rises, shopping malls, and buildings with open atria are equipped with built-in smoke control systems (**Figure 13.37**). These systems are designed to confine a fire to as small an area as possible by compartmentalizing the building when smoke or fire is detected. The automatic closure of doors, partitions, and windows, as well as the HVAC systems, are all methods of compartmentalizing a structure (**Figure 13.38**). Smoke control systems usually have a system diagram in the same location as the control

Figure 13.38 Automatic doors compartmentalize a building to limit exposures and damage.

panel. The panel should indicate where the alarm originated and which automatic closers were activated. Only building engineers or maintenance superintendents should operate building systems to assist in ventilation. Incorrect use of these systems can cause severe damage to them and may create a more hazardous condition.

///////////////////////////////

WARNING

Do not attempt to operate building systems that
assist in ventilation.

Chapter Summary

Tactical ventilation of a burning building allows heat, smoke, and fire gases to escape to the atmosphere. It also draws fresh air into the building. Properly applied tactical ventilation allows firefighters to see better, locate victims more easily, and find the seat of the fire sooner. Tactical ventilation also limits fire spread and channels the heat and smoke away from any trapped victims. Uncoordinated ventilation may accelerate the spread of fire. To perform horizontal and vertical ventilation safely and effectively, firefighters must understand fire behavior and know the various ventilation methods. Firefighters must have knowledge of roof construction and know how to create exhaust openings in all types of roofs that have a variety of coverings. Techniques for specialized ventilation situations involving basements, windowless buildings, and high-rise structures must be learned and practiced. Finally, the effect of building fire control systems on ventilation must be considered.

Review Questions

1. What are the reasons for tactical ventilation?

2. What considerations will affect the decision to ventilate?

3. What are the basic means used to accomplish ventilation?

4. How do smoke, air flow, heat, and flame impact fire behavior in a structure?

5. What differences are there between horizontal and vertical ventilation?

6. How do the advantages and disadvantages of natural, mechanical, and hydraulic ventilation compare to one another?

7. What are the main types of horizontal ventilation?

8. What are the types of vertical ventilation?

9. What other types of ventilation situations might firefighters encounter?

10. How can a built-in heating ventilation and air conditioning (HVAC) system affect tactical ventilation?

13-I-1
Ventilate using mechanical negative pressure in a window.

NOTE: Firefighters must note the wind direction prior to performing ventilation.

Step 1: Select the vent site (window) and ensure the opening is clear by removing all shades, drapes, blinds, screens, and other obstructions.

Step 2: Mount/hang the fan (smoke ejector or blower) in the opening either secured by hooks or ladder. Fan should be directed to exhaust building air to the outside.

Step 3: Once the fan is in position, connect to a power source.

Step 4: Ensure effectiveness of ventilation.

NOTE: Firefighters must note the wind direction prior to performing ventilation.

Step 1: Select the vent site (doorway) and ensure the opening is clear by removing all shades, drapes, blinds, screens, and other obstructions.

Step 2: Place the fan (smoke ejector or blower) in the opening; fan should be directed to exhaust building air to the outside.

Step 3: Once the fan is in position, connect to a power source.

Step 4: Ensure effectiveness of ventilation.

NOTE: Firefighters must note the wind direction prior to ventilation.

Step 2: Confirm order with officer to ventilate structure.

Step 3: Start fan(s) and temporarily direct away from opening.

Step 1: Place fan near entrance (4 feet to 6 feet [1.2 m to 1.8 m]) opening so that it will create a positive pressure within the structure when needed.

Step 4: Create exit opening approximately 2-3 times larger than the "point of entry."

Step 5: Direct fan into point of entry so that cone of air covers opening.

Step 6: Determine if smoke is moving away from point of entry and toward exit. If not: discontinue use of fan and reevaluate location of point of entry, exit, and any obstructions of the flow of air.

Step 7: Clear smoke out of building.

Step 8: Ensure effectiveness of ventilation.

NOTE: Firefighters must note the wind direction prior to ventilation.

Step 1: Ensure opening is clear of obstructions.

Step 2: Set the fog nozzle pattern wide enough to cover 85 to 90 percent of window or door opening.

Step 3: Monitor progress of ventilation.

Step 4: Ensure effectiveness of ventilation.

Step 1: Before placing any weight on the roof, be sure to check for sturdiness.

Step 2: Holding an axe vertically by the handle with the head down, let the axe bounce on the roof surface.

Step 3: Determine what the sound indicates about your location. You will hear different sounds based on your location.

    a. You should hear a hollow sound between the supports

    b. You will hear a solid sound near the supports

    c. If you hear a sound like striking rotten wood, this will indicate the roof is weakening

Step 4: Sound the roof as you proceed to your desired location to ensure integrity.

NOTE: Firefighters must note the wind direction prior to ventilation.

Step 1: While on the ground, confirm the saw is working to ensure proper operation.

CAUTION: The saw should NOT be running while ascending to roof.

Step 2: Sound roof with tool.

b. Second cut: Horizontal, top working toward firefighter

c. Third cut: Lengthwise closer to firefighter, cutting from top to bottom

d. Fourth cut: Horizontal, bottom working toward firefighter

Step 3: Locate rafters/floor joists and check roof/floor for integrity.

Step 4: Start saw and make the following cuts.

    a. First cut: Lengthwise away from firefighter, cutting from top to bottom

Step 5: Maintain secure footing while cutting.

Step 6: Lock off saw or use chain brake.

Step 7: Use tool to open ventilation hole.

Step 8: Ensure effectiveness of ventilation.

NOTE: Firefighters must note the wind direction prior to ventilation.

Step 1: Sound roof with tool.

Step 2: Locate rafters/floor joists and check roof/floor for integrity.

Step 3: Make the following cuts.

    a. First cut: Lengthwise away from firefighter, cutting from top to bottom

    b. Second cut: Horizontal, top working toward firefighter

    c. Third cut: Lengthwise closer to firefighter, cutting from top to bottom

    d. Fourth cut: Horizontal, bottom working toward firefighter

Step 4: Maintain secure footing while cutting.

Step 5: Use tool to open ventilation hole.

Step 6: Ensure effectiveness of ventilation.

NOTE: Local SOPs may require use of an inspection hole during roof ventilation.

Step 1: Size up scene for any hazards.

Step 2: Select location for ventilation.

    a. Position upwind of planned ventilation opening

    b. Sound for roof integrity before putting weight on the roof

    c. Place ventilation opening in safe working area as close to fire as feasible and away from roof-mounted equipment

Step 3: Outline ventilation opening with the pick of an axe or other similar tool.

    a. Must be at least 4 x 4 foot (1.2 m by 1.2 m) opening, expand the hole as much as needed to match fire conditions

    b. Remove gravel or other materials from outlines that may limit ability to cut opening

Step 4: Cut three-sided (triangular) inspection opening in roof to determine fire conditions.

    a. First cut parallel to farthest support

    b. Cut through decking only

    c. All cuts intersect to form a triangle

Step 5: Cut roof deck parallel to a roof truss or support on side furthest away from ladder or escape route. This is cut #1.

    a. Cut is downwind from position

    b. Cut is at least 4 feet (1.2 m) long

    c. Inspection opening cut in Step 4 incorporated into this cut

    d. Cut is completely through decking material

    e. Size up fire conditions inside roof from discharge through cut

    f. Maintain situational awareness

Step 6: Cut roof deck on one side of opening perpendicular to the first cut. Cut must intersect first cut in Step 5. This is cut #2.

    a. Begin cut away from escape route if possible

    b. Cut is at least 4 feet (1.2 m) long or three rafters wide – inside first rafter, over second rafter, and inside third rafter

    c. Cut is completely through decking material but not through structural framing

    d. Maintain situational awareness

NOTE: Local SOPs may require use of an inspection hole during roof ventilation.

Step 7: Cut roof deck on opposite side of cut made in Step 6. Cut must intersect cut made in Step 5. This is cut #3.

    a. Begin cut away from escape route if possible

    b. Cut is at least 4 feet (1.2 m) long

    c. Cut is completely through decking material

    d. Maintain situational awareness

Step 8: Complete the ventilation hole by cutting between cut #2 and cut #3.

    a. Cut is completely through decking material

    b. Maintain situational awareness

Step 9: Remove decking from the ventilation opening with axe, pike pole, or other sounding tool.

    a. Keep decking out of ventilation opening

    b. Size up fire conditions in the roof space

Step 10: Plunge through interior ceiling using pike pole working from upwind side of ventilation hole.

Step 11: Report to officer completion of assigned task.

NOTE: Local SOPs may require use of an inspection hole during roof ventilation.

Step 4: Outline ventilation opening with pick, axe, or other similar tool.

  a. Must be at least 4 x 4 feet (1.2 m by 1.2 m) opening

  b. Remove gravel, tiles, or other materials that may limit ability to cut opening

Step 5: Cut roof deck across the rafters on the high side of the roof parallel to the ridge.

  a. Cut is at least 4 feet (1.2 m) long or three rafters wide – inside first rafter, over second rafter, and inside third rafter

  b. Cut is completely through decking material but not through structural framing

  c. Size up fire conditions inside roof from discharge through cut

  d. Maintain situational awareness

Step 1: Confirm order with officer to ventilate pitched roof.

Step 2: Size up scene for any hazards.

Step 3: Select location for ventilation.

  a. Position upwind of planned ventilation opening

  b. Sound for roof integrity

  c. Observe fire and smoke conditions coming from roof openings (vents, etc.)

  d. Place ventilation opening in safe working area as close to fire as feasible and high on the roof

  e. Avoid placing opening near roof-mounted equipment

NOTE: Local SOPs may require use of an inspection hole during roof ventilation.

Step 6: Cut roof deck on furthest side of ventilation opening perpendicular to the cut made in Step 5.

    a. Begin cut at top of roof and work downward

    b. Cut is at least 4 feet (1.2 m) long inside first rafter

    c. Cut is completely through decking material

    d. Maintain situational awareness

Step 7: Cut roof deck on opposite side of cut made in Step 6.

    a. Begin cut at top of roof and work downward

    b. Cut is at least 4 feet (1.2 m) long inside third rafter

    c. Cut is completely through decking material

    d. Maintain situational awareness

Step 8: Complete the ventilation opening by cutting between the bottom of the two parallel cuts made in Steps 6 and 7.

    a. Cut is completely through decking material

    b. Maintain situational awareness

Step 9: Remove decking from the ventilation opening with axe or pike pole.

    a. Keep decking out of ventilation opening

    b. Size up fire conditions in the roof space

Step 10: Plunge down through the ceiling using pike pole working from upwind side of ventilation opening.

Step 11: Ensure effectiveness of ventilation.

NOTE: Local SOPs may require use of an inspection hole during roof ventilation.

Step 1: Confirm order with officer to make trench cut ventilation opening.

Step 2: Size up scene for any hazards.

Step 3: Ensure saw is functioning properly on ground before going to roof.

NOTE: Due to the time required to make this opening, multiple firefighters may need to operate saws simultaneously.

Step 4: Select location for ventilation.

   a. Location should be far enough from fire position to allow opening to be made before fire's arrival

   b. Cut should be made on the fire side of a firewall (if possible)

   c. Sound for roof integrity

   d. Observe fire and smoke conditions coming from roof

   e. Determine the location of ladders or other means of egress from roof

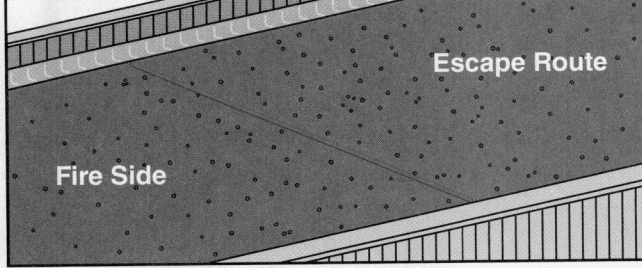

Step 5: Cut roof deck perpendicular to the length of building from exterior wall to exterior wall, rolling the saw over roof supports as encountered. This is cut #1.

   a. First perpendicular cut should be made closest to the fire

   b. Cut completely through decking material

   c. Size up fire conditions inside roof from discharge through cut

   d. Maintain situational awareness

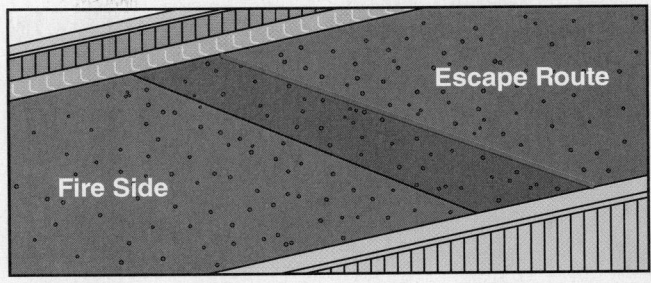

Step 6: Make a second cut perpendicular to the length of the building from exterior wall to exterior wall, rolling the saw over roof supports as encountered. This cut should be positioned at least 4 feet (1.2 m) from cut #1. This is cut #2.

   a. Second perpendicular cut should be made closest to escape route

   b. Cut completely through decking material

   c. Maintain situational awareness

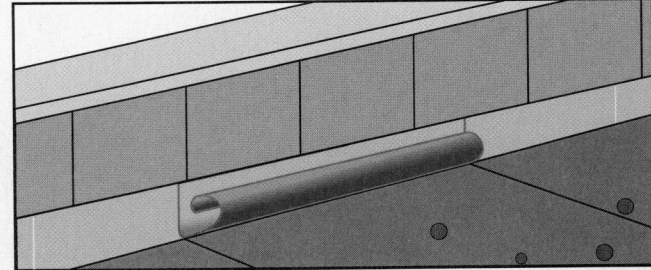

Step 7: Roll back flashing or other roofing material that may be present at exterior walls.

NOTE: Local SOPs may require use of an inspection hole during roof ventilation.

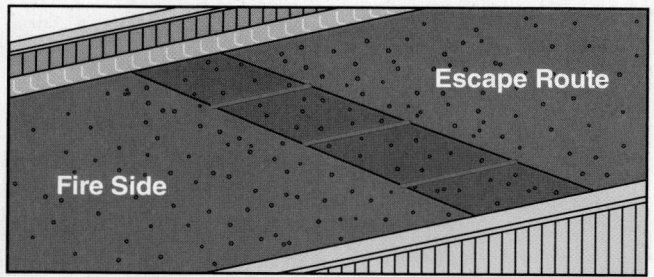

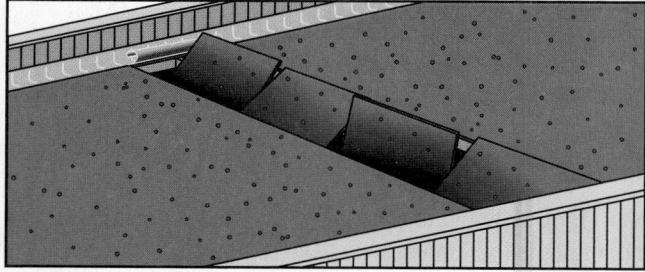

Step 8: Make cuts parallel to exterior walls that completely intersect cuts #1 and #2.

  a. Cuts should be positioned between roof supports

  b. Number of cuts needed is dictated by length of trench and spacing of roof supports

  c. Cut completely through decking material

  d. Maintain situational awareness

Step 9: Tilt roof decking into ventilation opening with axe, pike pole, or other sounding tool.

  a. Decking is still typically fastened to roof supports and will pivot into opening

  b. If possible, completely remove decking from trench

  c. Size up fire conditions in the roof space

Step 10: Plunge through interior ceiling along entire length of trench using pike pole working from escape route side of trench.

Step 11: Report to officer completion of assigned task.

Courtesy of of Bob Esposito.

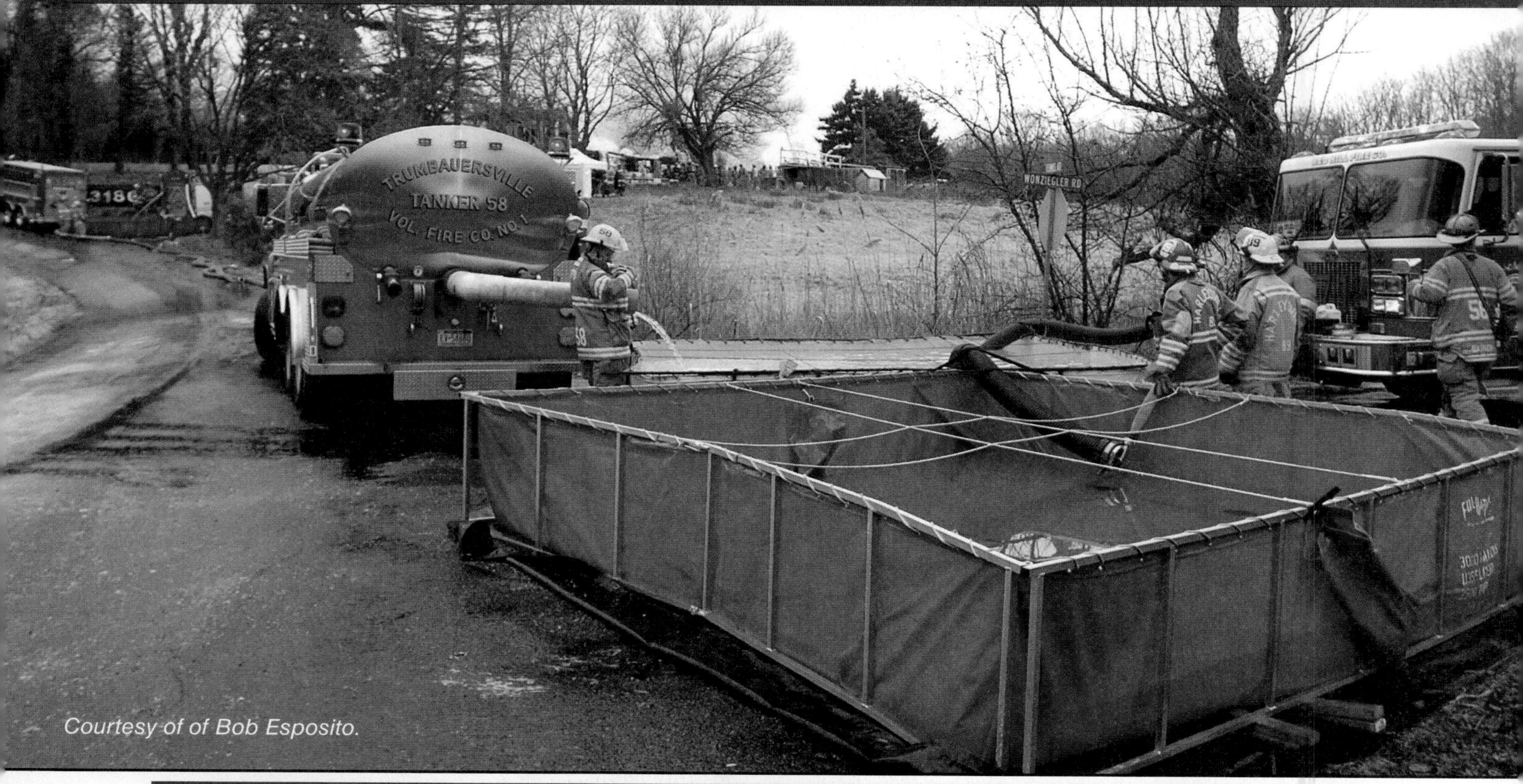

Chapter Contents

Chapter 14

Key Terms

NFPA® Job Performance Requirements

This chapter provides information that addresses the following job performance requirements of NFPA® 1001, *Standard for Fire Fighter Professional Qualifications* (2013).

Firefighter I
5.3.15

1. Explain the ways water supply system components are used by firefighters. (5.3.15)

2. Describe types of fire hydrants and hydrant markings. (5.3.15)

3. Explain fire hydrant operation and inspection considerations. (5.3.15)

4. Explain alternative water supply sources and methods of access. (5.3.15)

5. Describe methods used for rural water supply operations. (5.3.15)

6. Operate a hydrant. (Skill Sheet 14-I-1; 5.3.15)

7. Make soft-sleeve and hard-suction hydrant connections. (Skill Sheet 14-I-2; 5.3.15)

8. Connect and place a hard-suction hose for drafting from a static water source. (Skill Sheet 14-I-3; 5.3.15)

9. Deploy a portable water tank. (Skill Sheet 14-I-4; 5.3.15)

Chapter 14
Water Supply

An adequate water supply and distribution system is an essential tool firefighters need to control and extinguish fire. Water-based fire suppression systems, including automatic sprinklers, standpipes, and other types of systems, are ineffective without an adequate water supply.

Water is easily stored and can be transferred over large distances in a well-designed water distribution system. You must be familiar with the types of water supply distribution systems in your community in order to ensure that the systems are adequate to handle emergency situations.

This chapter provides the knowledge and skills about water supply systems you will need to be certified as a Firefighter I. This requires knowledge of the following:

- Loading and off-loading procedures for mobile water supply apparatus
- Fire hydrant operation
- Suitable static water supply sources

The skills required to perform these operations include the following:

- Deploying portable water tanks and associated equipment
- Connecting to various water sources
- Connecting a hard-suction hose in place for drafting operations
- Connecting a supply hose to a hydrant
- Fully opening and closing a hydrant

▌I▐ Water Supply System Components

There are two basic types of water supply systems in North America: public and private systems. Public water supply distribution systems are generally a function of local government; that is, a department of municipal government or a state-authorized water district. The local water department is usually a separate city utility whose main function is to provide sanitary water that is safe for human use. An elected board generally governs a state-authorized water district.

A private water supply system may provide water under contract to a municipality, region, or single property. Private water supply distribution systems may take a variety of forms, including systems that are specific to an industrial facility or complex such as a refinery or that provide water to a residential subdivision. In the former case, the facility owner/occupant has responsibility for the inspection, testing, and maintenance of the system. The water source may be a public supply distribution system separated from the private system by a water meter and check valve or an on-site water supply such as a well or lake. Private water supplies may also serve a particular area such as a residential subdivision. The owners' association for the area may maintain the system in these cases. The water sources may be the same as those mentioned for the industrial facility example.

The design of public and private water supply systems may vary from region to region. However, all systems are composed of the following basic components which are explained in the following sections (**Figure 14.1**):

- **Water supply source(s)** — Lakes, reservoirs, ponds, rivers, streams, wells, or springs
- **Processing or treatment facilities** — Purification or desalination plants
- **Means of moving the water** — Water pumps
- **Water distribution and storage systems** — Storage tanks, control valves, piping systems, and hydrants

Water Supply Sources

Water is supplied from natural freshwater sources such as wells, springs, rivers, lakes, and ponds (**Figure 14.2**). Streams and springs may be fed by groundwater or snowmelt at higher elevations. Rainfall replenishes lakes, ponds, and groundwater. The ocean can also be a source of water; however, seawater is 220 times saltier than freshwater. Besides the salt, ocean water contains other impurities that must be removed to make the water potable for humans and useful for agricultural use.

The amount of water that a community needs for both domestic use and fire protection can be calculated based on the history of consumption and estimates of anticipated needs. Cities and other water providers track their average and maximum daily water consumption over time. Some even track their peak hourly consumption. Engineers add to these calculations based upon anticipated fire flow requirements

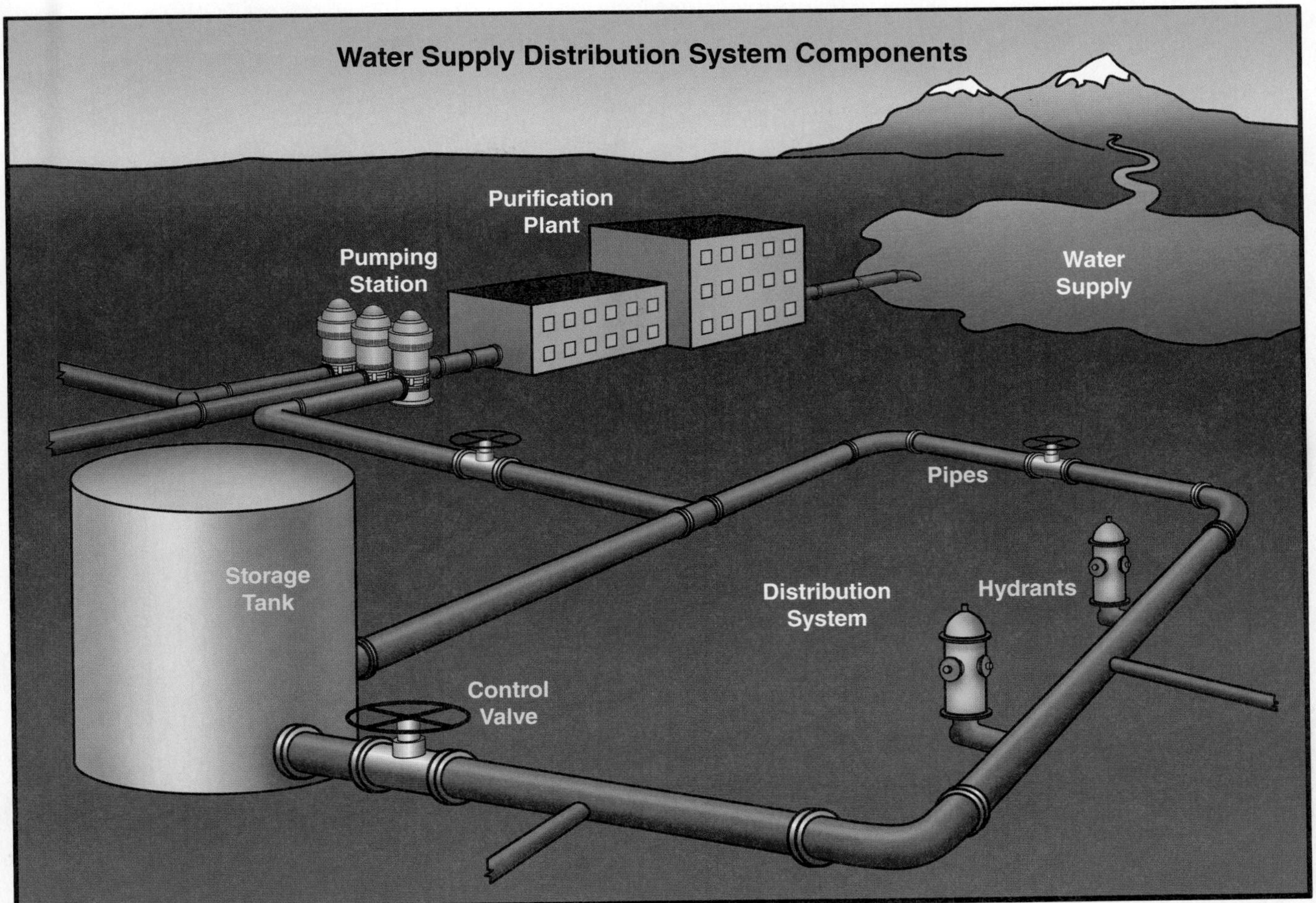

Water Supply Distribution System Components

Purification
Plant

Pumping
Station

Water
Supply

Pipes

Storage
Tank

Distribution
System

Hydrants

Control
Valve

Figure 14.1 The components of a water supply distribution system work together to connect end users with a municipality's water supply.

Figure 14.2 Water for community use is supplied from natural freshwater sources such as lakes and ponds.

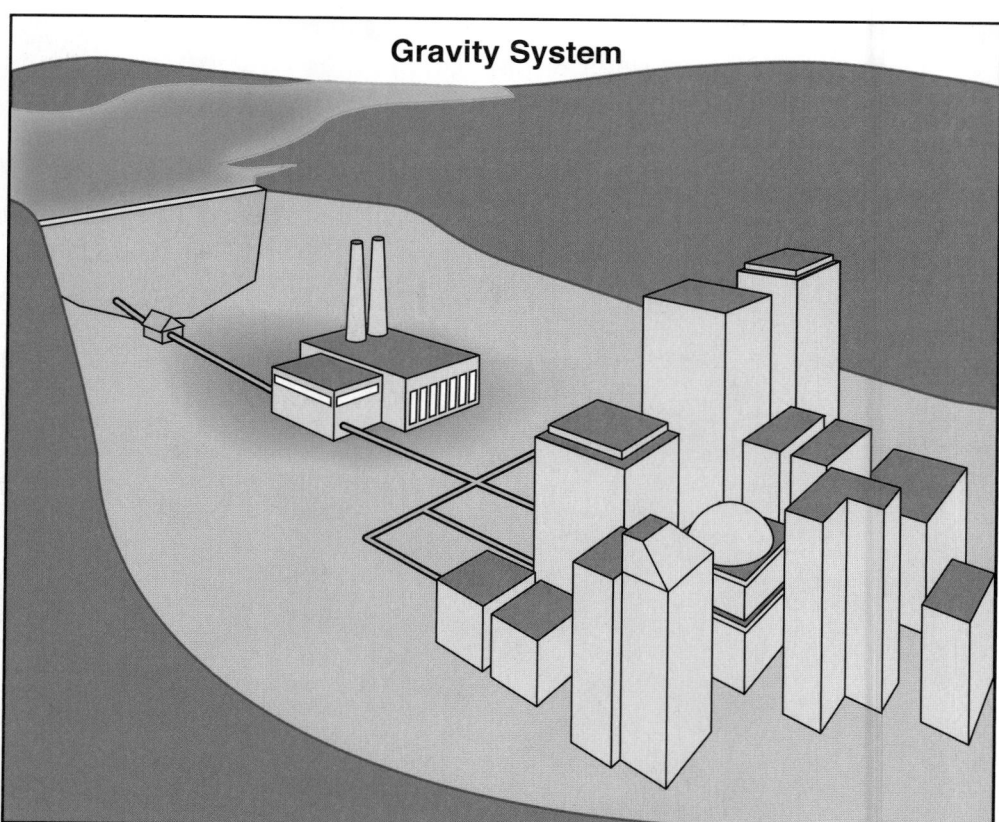

Figure 14.3 A gravity system uses the elevation of the water source to create pressure instead of relying on pumps.

needed for fire protection within the jurisdiction's boundaries. To be considered adequate, a system must be capable of supplying the water needed for fire protection in addition to the domestic requirement. In most cities, the domestic/industrial requirements exceed that required for fire protection. However, in small towns the requirements for fire protection may exceed other requirements.

Water Treatment or Processing Facilities

Before it can be used, most water must be processed to remove impurities and minerals that can be harmful to humans, animals, and plants. Most communities operate water treatment plants and/or water desalination plants. Before water can enter the distribution system, it first must pass through these facilities. A mechanical breakdown, natural disaster, loss of power supply, or fire could disable or severely limit the facility's pumping capacity and seriously reduce the volume and pressure of water available for fire fighting operations.

Means of Moving Water

Water must be moved from its original source to the treatment facilities and from there to the point it is distributed and used. Three methods for moving water through the system are gravity, direct pumping, and a combination system using both direct pumping and gravity.

Gravity Systems

A **gravity system** delivers water from the source or the treatment plant to the distribution system without pumping equipment **(Figure 14.3)**. The difference in the height of the water source and the point of use creates elevation pressure (also known as *elevation head pressure*). The elevation pressure forces the water throughout the water distribution system. However, gravity pressure is adequate only when the primary water source is located more than 100 feet (30 m) higher than the highest point in the water distribution system. The most common examples of true gravity systems are those supplied from an alpine lake or a mountain reservoir that supplies water to consumers below.

Direct Pumping Systems

When elevation pressure cannot provide sufficient pressure to meet the needs of the community, a pump is placed near the water source or treatment plant to create the required pressure within the distribution system. This system for moving water is called a **direct pumping system (Figure 14.4)**.

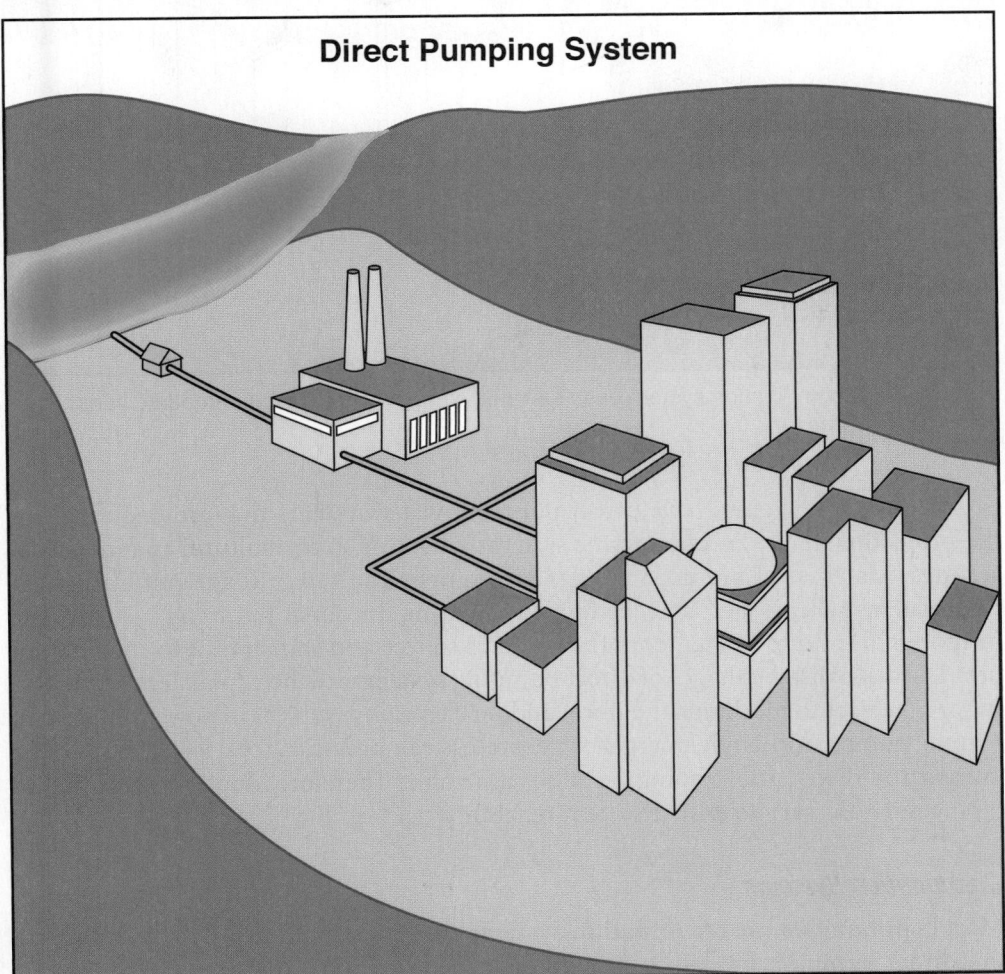

Direct Pumping System

Figure 14.4 When the surface water source lacks the required elevation to generate pressure, a series of pumps move the water.

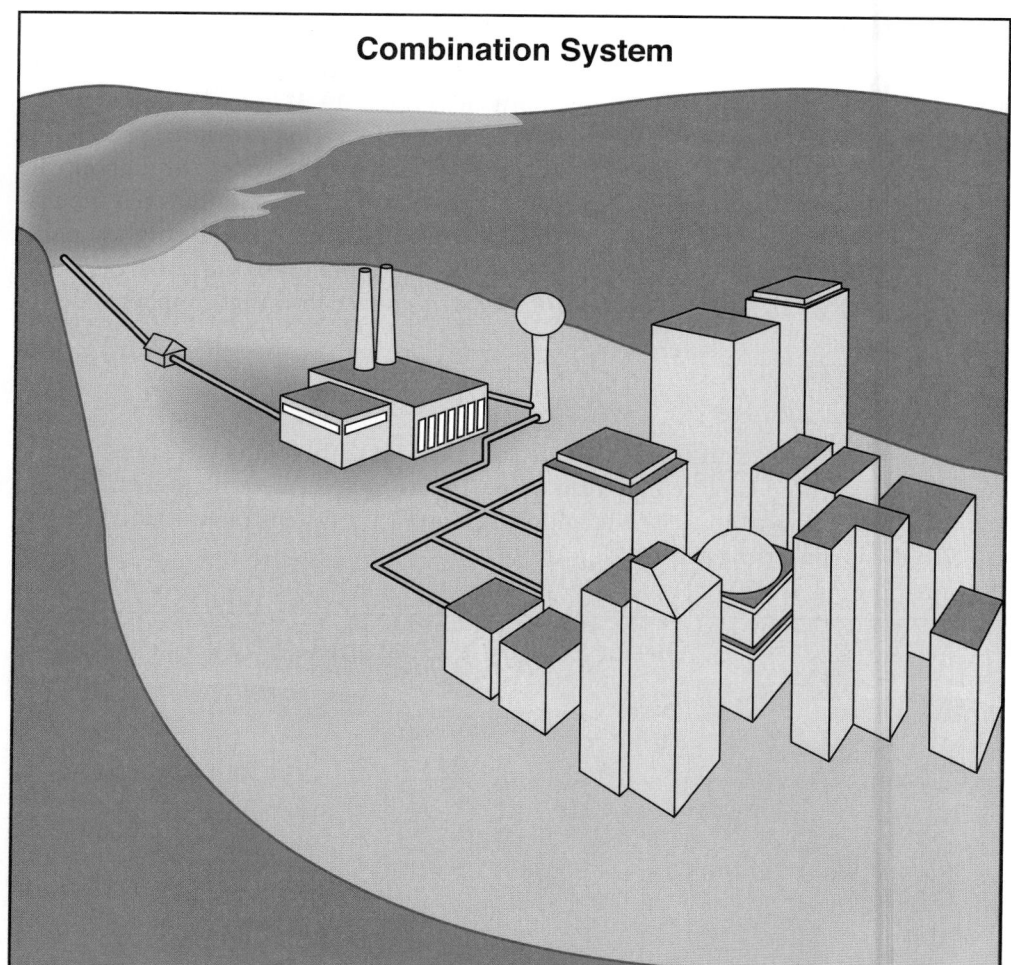

Combination System

Figure 14.5 A combination system uses elevated storage tanks to accommodate variable water usage rates.

Although a few cities have direct pumping water systems that are dedicated for fire protection, most direct pumping systems are found in agricultural and industrial settings. Many rural fire departments are equipped to tap into agricultural irrigation systems when water is needed for fire fighting. In direct pumping systems, one or more pumps draw water from the primary source and transport it to the point of use. The main disadvantages of direct pumping systems are their total dependence on pumps (subject to mechanical failure) and on electricity (subject to power outages) to run the pumps. Although emergency generators can prevent a total loss of power, the pumps and distribution piping are still vulnerable. Therefore, duplicate pumps and piping are necessary to ensure system reliability.

Combination Systems

Combination System —
Water supply system that is a combination of both gravity and direct pumping systems. It is the most common type of municipal water supply system.

Most communities use **combination systems** that consist of both gravity tanks and the direct pumping process to provide adequate pressure **(Figure 14.5)**. Water is pumped into the distribution system and elevated storage tanks (that provide gravitational pressure). When the consumption demand is greater than the rate at which the water is pumped, water flows from the storage tanks into the distribution system. Conversely, when demand is less, water is pumped into the storage tanks.

Many industrial facilities have their own combination water supply systems with elevated storage tanks. By prior agreement, these water supplies are often available to the local fire department in an emergency. Water for fire protection may be available to some communities from storage systems, such as underground cisterns (tanks used to store rainwater), that are considered a part of the distribution system. Fire department pumpers draft (draw water from a static source) from these sources and provide the pressure needed to transport the water to a fire.

Water Distribution and Storage Systems

The water distribution system consists of a network of pipes, storage tanks, isolation and control valves, and hydrants throughout the community or service area that carry the water under pressure to the points of use. Each of these system components is described in the sections that follow.

Piping

The ability of a water system to deliver a sufficient quantity of water at adequate pressure depends upon the capacity and elevation of the storage tanks and the condition and carrying capacity of the system's network of underground pipes, often called **water mains**. Underground water mains are generally made of cast iron, ductile iron, asbestos cement, steel, polyvinyl chloride (PVC) plastic, or concrete. When water flows through any pipe, its movement causes friction that reduces the water pressure. The internal surface of the pipe, regardless of the material from which it is made, offers resistance to water flow. Some materials, such as PVC plastic, have considerably less resistance to water flow than others.

The term *grid* or *gridiron* describes the interlocking network of water mains that compose a water distribution system. A water distribution system consists of three types of water mains (**Figure 14.6, p. 790**):

- **Primary feeders** — Large pipes, also known as arterial mains, with relatively widespread spacing. These mains convey large quantities of water to various points in the distribution system and supply smaller secondary feeder mains. Arterial mains can be very large, ranging from 16 inches (400 mm) to 72 inches (1 825 mm) in diameter or greater. Fire hydrants are rarely attached directly to these mains.

- **Secondary feeders** — Intermediate pipes that interconnect with the primary feeder lines to create a grid. They are 12 to 14 inches (300 mm to 350 mm) in diameter. Control valves can be used to isolate each secondary feeder.

- **Distributors** — Small water mains, 6 to 8 inches (150 to 200 mm) in diameter, that serve individual fire hydrants and commercial and residential consumers. Distributors may form an intermediate grid between secondary feeders or may be dead-end lines with the hydrant or supplied property at the end of the line.

To ensure a sufficient water supply, two or more primary feeders should run from the source of supply to the high-risk and industrial districts of the community along separate routes. Similarly, secondary feeders should provide water from the primary feeders along two directions to any end point. This arrangement increases the capacity of the supply at any given point and ensures that a break in a feeder main will not completely cut off the supply.

Water distribution systems are generally designed using computer programs and hydraulic calculations that ensure constant pressure and quantity throughout the system. The grid or loop system is designed to provide constant pressure or flow when

> **Water Main** — A principal pipe in a system of pipes for conveying water, especially one installed underground.

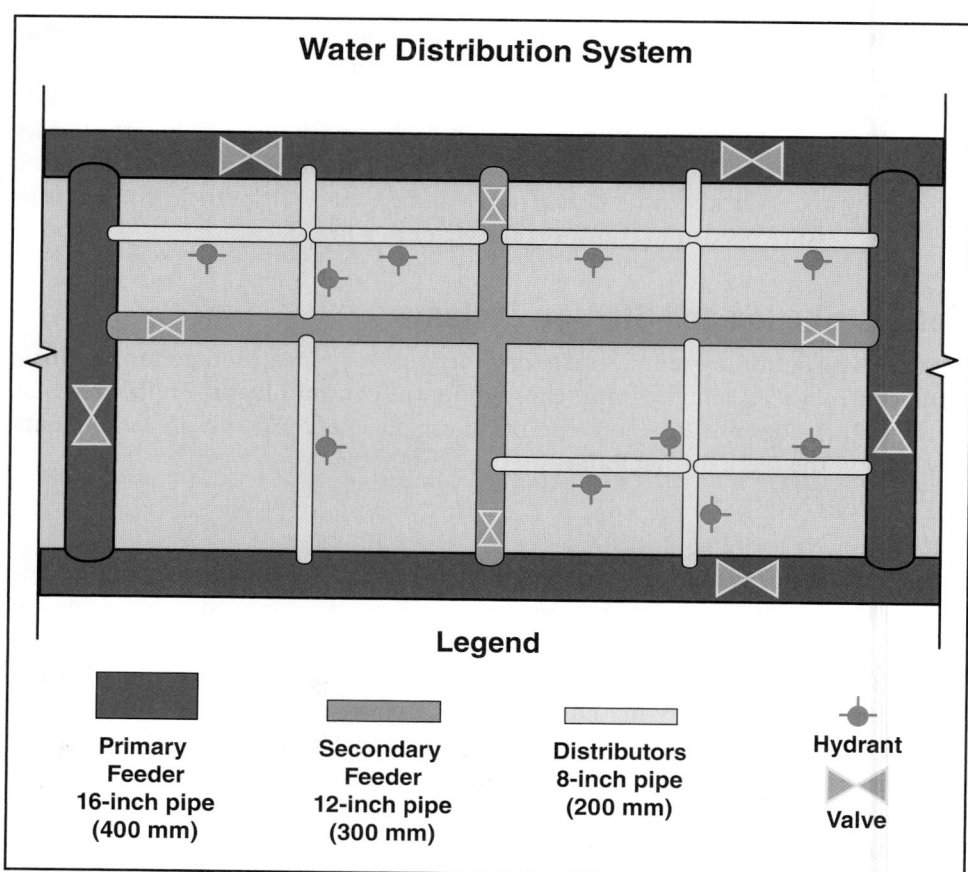

Water Distribution System

Legend

Primary Feeder 16-inch pipe (400 mm)	Secondary Feeder 12-inch pipe (300 mm)	Distributors 8-inch pipe (200 mm)	Hydrant / Valve

Figure 14.6 Three types of water mains work together to distribute water and provide shut-off points for maintenance.

pipes or the grid must be repaired. Another advantage of the grid system is that high demand in one area does not reduce water flow in other areas. Dead-end lines may exist but have disadvantages such as allowing water to stagnate in the pipes, requiring constant flushing, and causing services to be turned off when pipes are repaired.

The ability to deliver adequate quantities of water under pressure depends on the capacity of the system's network of pipes. Today, 8-inch (200 mm) pipe is often the minimum size used, although some communities are allowing 6-inch (150 mm) pipes in residential subdivisions.

Access to the water supply system is made through connections to the piping system. These connections may be through waterflow control valves and flow meters at the point that customers gain water from the system or through fire hydrants that are used for fire protection.

Storage Tanks

To ensure constant pressure, water distribution systems may have elevated storage tanks located throughout the system to create pressure through gravity. Elevated gravity storage tanks are usually constructed of steel or concrete. These tanks may be located on high towers or at ground level on hilltops. The higher the tank, the more elevation head pressure that is generated. Gravity tank capacities range from 5,000 gallons (20 000 L) to over a million gallons (greater than 4 000 000 L) (**Figure 14.7**).

Figure 14.7 Elevated water storage tanks use gravity to create water pressure throughout the system. *Courtesy of Sand Springs (OK) Fire Department.*

Isolation and Control Valves

Water supply systems contain valves to interrupt water flow to:

- Individual hydrants or properties
- Distribution lines
- Secondary feeders
- Primary feeders
- Entire water systems

Most valves are constructed of brass, steel, or cast iron. There are two types of valves generally used in public water supply systems: isolation valves and control valves.

Isolation valves. These valves may also be known as stop or shutoff valves and are either **gate valves** or **butterfly valves**. Isolation valves are used to isolate sections for maintenance and repair, to replace hydrants, or to make new connections to the system. The location of isolation valves is intended to disrupt a minimum number of customers while the system is down. The maximum lengths for valve spacing should be 500 feet (150 m) in high-value districts and 800 feet (240 m) in other areas as recommended by Commercial Risk Services, Inc.

NOTE: Commercial Risk Services, Inc. is a subsidiary of the Insurance Services Office (ISO) that conducts property rating surveys to help insurance companies develop accurate premiums.

Isolation valves should be tested (opened and closed) at least once a year to ensure that they are in good working condition. The municipal water department usually performs these tests.

Isolation valves are generally located on municipal easement (on or adjacent to streets, sidewalks, and alleys) and below ground. They are usually marked with the word *Water* or the name of the municipality or jurisdiction (**Figures 14.8a and b, p. 792**). This cover should be removed to access the nonindicating type valve. Either a residential or commercial water shutoff key can then be inserted into the opening to turn the valve stem 90 degrees to the direction of flow to shut off the water. All valves and hydrants are opened by rotating the stem or operating nut to the left or counterclockwise. Rotating the stem or nut to the right or clockwise closes the valve or hydrant.

Isolation valves for private water supply systems are usually indicating-type valves. An indicating valve shows whether the gate valve seat is open, closed, or partially closed. Two common indicating valves are the **post indicator valve (PIV)** and the **outside stem and yoke (OS&Y) valve.** The post indicator valve is a hollow metal post that houses the valve stem. A plate attached to the valve stem inside this post has the words *OPEN* and *SHUT* printed on it so that the position of the valve is shown. The OS&Y valve has a yoke on the outside with a threaded stem that opens or closes the gate inside the valve. The threaded portion of the stem is visible when the valve is open and not visible when the valve is closed.

Control valves. Control valves are also located between public water supply distribution systems and private water supply distribution systems. Typical types of control valves include:

- Pressure-reducing
- Pressure-sustaining
- Pressure-relief valves

Gate Valve — Control valve with a solid plate operated by a handle and screw mechanism; rotating the handle moves the plate into or out of the waterway.

Butterfly Valve — Control valve that uses a flat circular plate in a pipe that rotates 90 degrees across the cross section of the pipe to control the flow of water.

Post Indicator Valve (PIV) — A type of valve used to control underground water mains that provides a visual means for indicating "open" or "shut" position; found on the supply main of installed fire protection systems.

Outside Stem and Yoke (OS&Y) Valve — Outside stem and yoke valve; a type of control valve for a sprinkler system in which the position of the center screw indicates whether the valve is open or closed.

Figure 14.8a Nonindicating valves in public water systems are usually located underground and are operated through a valve box activated with special valve key.

Figure 14.8b Some water control valves are located below ground.

- Flow-control valves
- Throttling valves
- Float valves
- Check valves

In addition to these water-flow control valves, a water **flowmeter** and a **backflow preventer** will be installed on the water supply line. The water flowmeter determines the quantity of water that the facility is using for billing purposes. The backflow preventer prohibits any water from flowing back into the public water system.

Fire Hydrant Locations

Fire hydrants are found in urban, suburban, and some rural areas served by both public and private water supply systems **(Figure 14.9)**. Fire hydrants are located along all portions of the water distribution system and are generally connected at specified intervals by 6-inch (150 mm) connecting pipes. Water department personnel usually determine the location, spacing, and distribution of fire hydrants. In general, fire hydrants should not be spaced more than 300 feet (100 m) apart in high-value districts. Hydrants should be located at every other intersection so that every building on a given street is within one block of a hydrant. Additional intermediate hydrants may be required where distances between intersections exceed 350 to 400 feet (105 m to 120 m). Other factors that affect hydrant location and spacing include:

- Types of building construction
- Types of occupancies
- Building densities
- Sizes of water mains
- Required fire flows for occupancies within a given area

Flowmeter — Mechanical device installed in a discharge line that senses the amount of water flowing and provides a readout in units of gallons per minute (liters per minute).

Backflow Preventer — A check valve that prevents water from flowing back into a system and contaminating it.

Friction loss and location of hydrants can reduce pressure in the distribution system. Encrustations of minerals and sediment that accumulate inside the mains over a period of years can cause friction loss in water mains that serve fire hydrants. Friction loss reduces the volume and pressure of water available from fire hydrants. The location of hydrants can also affect the water volume and pressure. For instance, a fire hydrant that receives water from only one direction, known as a **dead-end hydrant**, has a limited water supply. To overcome this and ensure greater pressure, fire hydrants can be located so they receive water from two or more directions (**Figure 14.10**). **Circulating hydrants** that receive water from more than one direction are said to have **circulating feed** or a **looped system**. A distribution system that provides circulating feed from all directions is called a *grid system*.

The locally adopted building or fire code determines the type and location of fire hydrants on the system. Fire hydrants are discussed in greater detail in the following section.

Figure 14.9 Fire hydrants are a fire service connection point in public and private water systems.

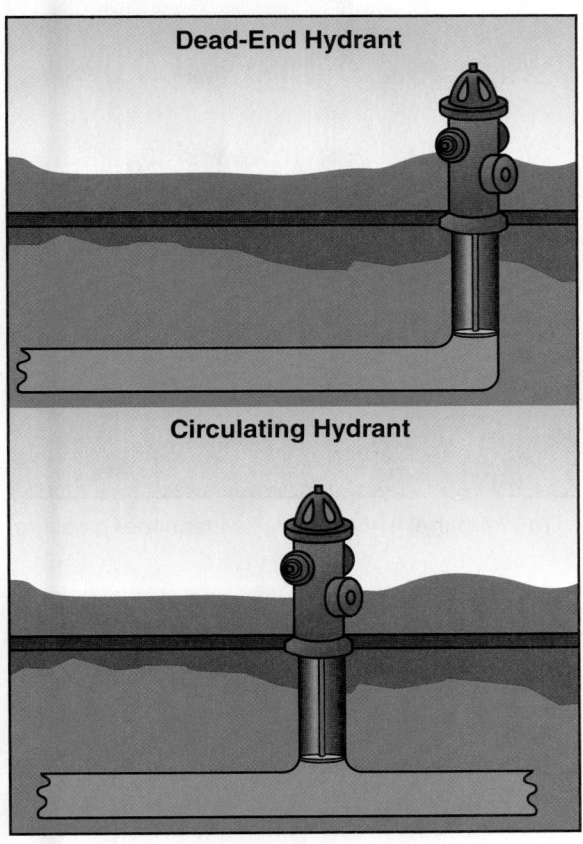

Figure 14.10 Dead-end hydrants receive water from only one direction, whereas circulating-feed hydrants receive water from two directions.

Dead-End Hydrant — Fire hydrant located on a dead-end main that receives water from only one direction.

Circulating Hydrant — Fire hydrant that is located on a secondary feeder or distributor main that receives water from two directions.

Circulating Feed — Fire hydrant that receives water from two or more directions.

Loop System — Water main arranged in a complete circuit so that water will be supplied to a given point from more than one direction. *Also known as* circle system, circulating system, or belt system.

Fire Hydrants

Although the initial source of water that firefighters use is the water tank on their pumper, the most dependable source comes from fire hydrants near the incident. Generally, fire hydrants can provide a consistent volume of water under constant pressure. However, you must realize that fire hydrants and water supply systems can fail. Failures or reduction in water supply or pressure can result from:

- Damaged hydrant valves and connections
- Broken water mains
- Greater demand than the system can provide

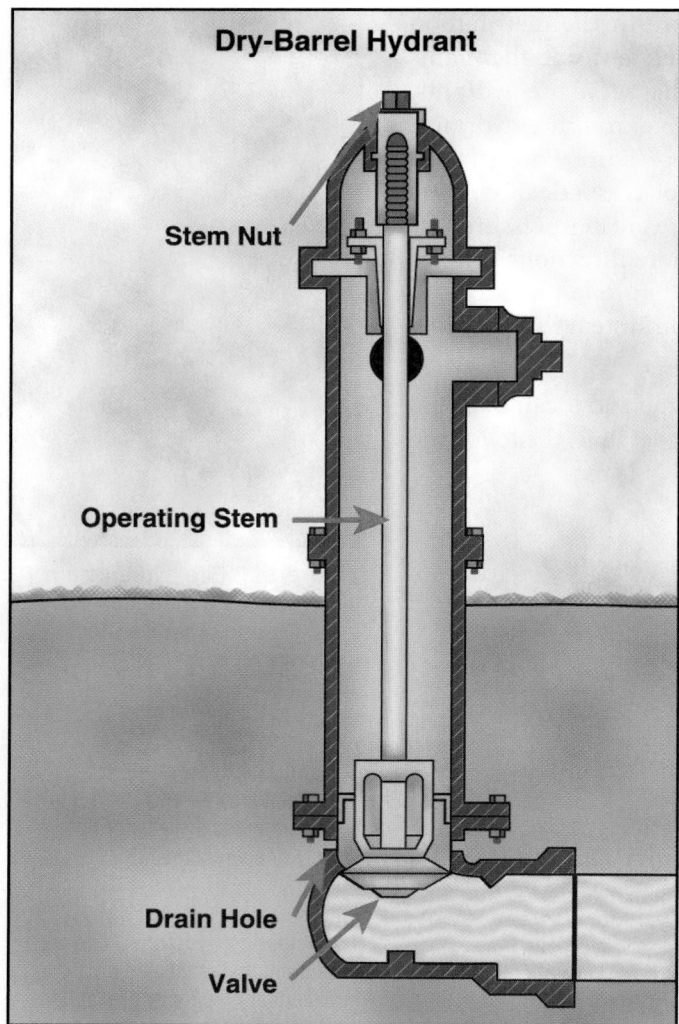

Dry-Barrel Hydrant

Stem Nut

Operating Stem

Drain Hole

Valve

Figure 14.11a To prevent water from freezing inside the hydrant, dry-barrel hydrants do not store water when they are not in use.

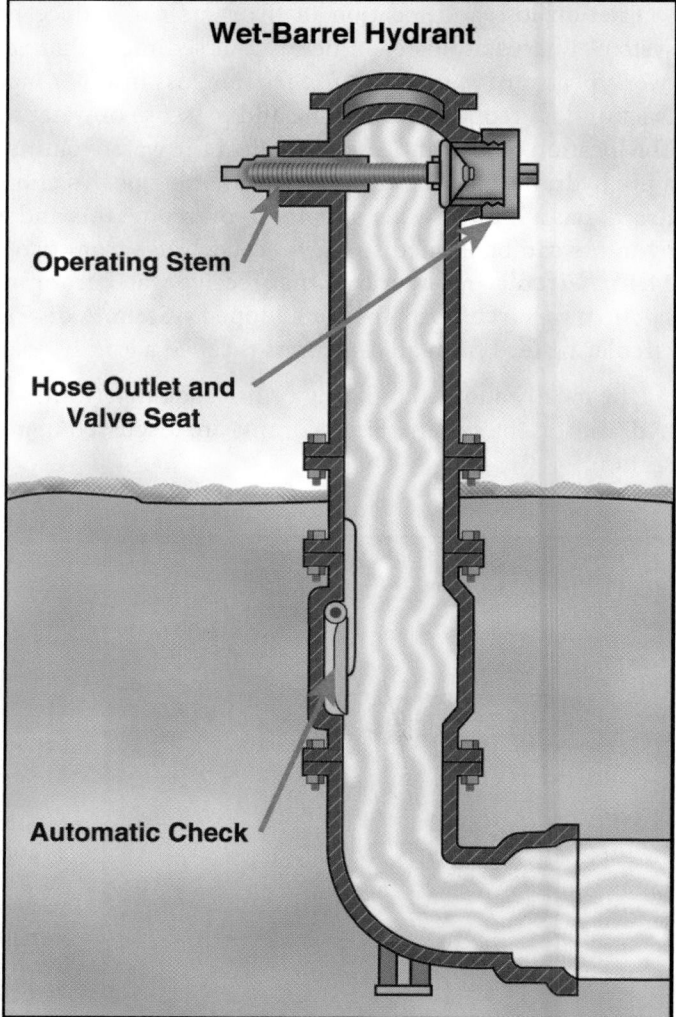

Wet-Barrel Hydrant

Operating Stem

Hose Outlet and Valve Seat

Automatic Check

Figure 14.11b Wet-barrel hydrants store water under pressure at all times.

- Hydrants located on dead-end water mains
- Closed isolation valves
- Restricted mains caused by sediment and mineral deposits
- Pipes or hydrants that are frozen

When hydrants fail to provide sufficient volume or pressure, it is necessary to select an alternate water supply. Alternate water supplies are discussed later in this chapter.

Because fire hydrants are a basic tool used in fire fighting, you must know fire hydrant types, markings, use, inspections, and maintenance. Firefighters, inspectors, water department personnel, or private contractors perform fire hydrant testing.

Types

In general, all fire hydrant bonnets, barrels, and foot pieces are made of cast iron. The internal working parts are usually made of bronze, but valve facings may be made of rubber, leather, or composite materials. The two main types of fire hydrants used in North America are **dry-barrel** and **wet-barrel** (**Figures 14.11a and b**). Even though they serve the same purpose, their designs and operating principles are quite different.

Regardless of the location, design, or type, hydrant discharge outlets are considered standard if they contain the following two components:

- At least one large (4 or 4½ inches [100 mm or 115 mm]) outlet often referred to as the **pumper outlet nozzle** or **steamer connection**
- Two hose outlet nozzles for 2½-inch (65 mm) couplings

Hydrant specifications require a 5-inch (125 mm) valve opening for standard three-way hydrants and a 6-inch (150 mm) connection to the water main. The male threads on all hydrant discharge outlets must conform to those used by the local fire department. NFPA® 1963, *Standard for Fire Hose Connections* sets regulations for the number of threads per inch and the outside diameter of the male thread.

Dry-Barrel Hydrants

Designed for use in climates that have freezing temperatures, the main control valve of the dry-barrel hydrant is located at the base or foot of the hydrant below the frost line with an isolation valve located on the distribution line. The stem nut used to open and close the control valve is located on top of the hydrant. Water is only allowed into the hydrant when the stem nut is operated. Any water remaining in a closed dry-barrel hydrant drains through a small drain valve that opens at the bottom of the hydrant when the main valve approaches a closed position. Use a hydrant wrench to open the valve by turning the stem in a counterclockwise direction or to close it by turning the stem in a clockwise direction.

Wet-Barrel Hydrants

Wet-barrel hydrants are designed to have water in the hydrant at all times. These hydrants are usually installed in warmer climates where prolonged periods of subfreezing weather are uncommon. Horizontal compression valves are usually at each outlet, but there may be another control valve in the top of the hydrant to control the water flow to all outlets.

Fire Hydrant Markings

Fire hydrant markings can be used to designate flow capacity. Hydrant colors vary in different jurisdictions. Generally, hydrants will be painted so that they are readily visible. Violet has begun to be used to designate hydrants with nonpotable water (water that is unfit for human or animal consumption).

The rate of flow from individual fire hydrants varies for several reasons. First, the size of the water main to which a hydrant is connected has a major effect on the rate of flow from that hydrant. Also, sedimentation and deposits within the water mains may decrease water flow. These problems develop over time so older water systems are more likely to experience a decline in the available flow than newer systems.

Markings used to designate flow rates are described in **Table 14.1, p. 796**. The colors of the bonnets (tops) and discharge caps on public hydrants should be painted as required in NFPA® 291, *Recommended Practice for Fire Flow Testing and Marking of Hydrants*.

Dry-Barrel Hydrant — Fire hydrant that has its operating valve located at the base or foot of the hydrant rather than in the barrel of the hydrant. When operating properly, there is no water in the barrel of the hydrant when it is not in use. These hydrants are used in areas where freezing may occur.

Wet-Barrel Hydrant — Fire hydrant that has water all the way up to the discharge outlets; may have separate valves for each discharge or one valve for all the discharges. This type of hydrant is only used in areas where there is no danger of freezing weather conditions.

Pumper Outlet Nozzle — Fire hydrant outlet that is 4 inches (100 mm) in diameter or larger.

Steamer Connection — Large-diameter outlet, usually 4½ inches (115 mm), at a hydrant or at the base of an elevated water storage container.

Table 14.1
Classifications and Markings of Municipal Fire Hydrants

Classification	Fire Flow	Barrel Color	Top and Nozzle Cap Colors	Pressure
Class AA	1,500 gpm (5 680 L/min) or greater	Chrome Yellow	Light Blue	20 psi (140 kPa)
Class A	1,000–1,499 gpm (3 785–5 675 L/min)	Chrome Yellow	Green	20 psi (140 kPa)
Class B	500–999 gpm (1 900–3 780 L/min)	Chrome Yellow	Orange	20 psi (140 kPa)
Class C	500 gpm (1 900 L/min) or less	Chrome Yellow	Red	20 psi (140 kPa)

Based on information given in NFPA® 291, *Recommended Practice for Fire Flow Testing and Marking of Hydrants,* 2007.

Hydrant Flow Rates: SLIP

An acronym that can be used to remember factors that affect hydrant flow rate is SLIP or:

S — Size of the main (diameter)

L — Length of the main (distance between hydrants)

I — Internal condition of the main (Roughness caused by age or sediment)

P — Pressure within the main (actual flow pressure from the hydrant)

Fire Hydrant Operation

To ensure that dry- and wet-barrel hydrants in your response area are ready for use, they must be inspected and operated at least twice a year. You must know how to operate fire hydrants in order to:

• Provide water through hoses for fire suppression operations

• Flow water from hydrant discharge openings to flush sediment

• Perform periodic inspections

• Ensure proper operation of valves and caps

• Assist in flow tests

Use the following steps when operating a wet-barrel hydrant:

1. Ensure that the valve is shut by placing the hydrant wrench on the operating stem and turning clockwise.

2. Tighten the discharge caps that will not be used and then remove the cap from the discharge outlet onto which you are going to attach the hose.

Figure 14.12 Attach the hose to the discharge outlet and tighten the coupling.

3. Check that the discharge is free of debris or obstructions and flush the hydrant by flowing a small amount of water.

4. Attach the hose to the discharge outlet and tighten the coupling (**Figure 14.12**).

5. Stand on the opposite side of the hydrant from the open discharge outlet.

6. Turn the operating stem slowly counterclockwise until the hydrant valve is completely open.

7. Remove any kinks or bends from the hose.

To prevent freezing, the dry-barrel hydrant main valve is located underground below the frost line. Normally, the hydrant barrel from the top of the stem down to the main valve is empty. This prevents water in the barrel freezing during extended periods of subfreezing temperatures.

When the stem nut is turned counterclockwise, the main valve moves downward allowing water to flow into the hydrant. As the main valve moves downward, a drain valve plate attached to the stem closes a drain hole located near the bottom of the hydrant, but allows water to flow past it into the hydrant barrel.

When the hydrant is shut down by slowly turning the stem nut clockwise, the main valve rises and shuts off the flow of water into the hydrant barrel. At the same time, the drain valve plate rises, opening the drain hole. The water that remains in the hydrant barrel empties through the drain hole into the surrounding soil.

Additional precautions should be taken when operating dry-barrel hydrants that are installed in areas where prolonged periods of subfreezing weather are common. If a dry-barrel hydrant is not opened fully, the drain may be left partially open. The resulting flow through the drain hole can cause erosion of the soil around the base of the hydrant, sometimes called "undermining" the hydrant. Over time, this erosion can destroy the hydrant's support and cause it to leak badly. This can put the hydrant out of service and necessitate it being reinstalled. Therefore, it is important that dry-barrel hydrants be either completely open or completely closed.

When a dry-barrel hydrant is shut down, it is also important to verify that the water left in the hydrant barrel is draining out. This can be tested by taking the following steps:

1. Close the main valve by turning the stem nut clockwise until resistance is felt; then turn it a quarter-turn counterclockwise.

2. Cap all discharges except one.

3. Place the palm of one hand over the open discharge.

If the hydrant is draining, a slight vacuum should be felt pulling the palm toward the discharge. If this vacuum is not felt, repeat the entire process and try again. If the hydrant still is not draining, the drain hole is probably plugged. Notify the water authority and have them inspect the hydrant. If this occurs in winter, the hydrant must be pumped until empty to prevent the water from freezing in the barrel before the hydrant is repaired or replaced.

In some areas, the Environmental Protection Agency (EPA) has required drain holes to be closed to prevent contamination of the water supply. In these areas, dry-barrel hydrants must be pumped until empty after each use.

If water is seen bubbling up out of the ground at the base of a dry-barrel hydrant when the hydrant is fully open, a broken component in the hydrant barrel is allowing water to get past the drain opening. This hydrant should be reported to the water authority which will mark it out-of-service until it is repaired.

Regardless of the type, all hydrants must be opened and closed slowly to prevent damage to fire hose, hydrants, and other equipment, or possible injury to firefighters. Opening a hydrant too fast may cause the fire hose connected to it to flail violently as the water pressure straightens the hose. Closing a hydrant too fast may cause a sudden increase in pressure (water hammer) within the water supply system, which can result in damage to the system piping or appliances attached to the system such as water heaters in adjacent residences.

Skill Sheet 14-I-1 provides the steps for operating a fire hydrant and **Skill Sheet 14-I-2** provides the steps for connecting a soft-sleeve suction hose and a hard-suction hose.

Fire Hydrant Inspection

In most cities, the repair and maintenance of fire hydrants are responsibilities of the water department. However, fire department personnel may be assigned the task of inspecting and flow testing fire hydrants and performing limited maintenance. Report any repair needs to your supervisor or the appropriate authority. When inspecting fire hydrants, you should look for the following potential problems (**Figure 14.13**):

• Obstructions, such as sign posts, utility poles, weeds, bushes, or fences that might interfere with pumper-to-hydrant connections or with opening the hydrant valve

• Outlets that face the wrong direction for pumper-to-hydrant connections

Figure 14.13 Fire hydrants must be maintained to preserve functionality. *Courtesy of Trent Hawkins, Stillwater (OK) Fire Department.*

- Insufficient clearance between outlets and the ground
- Damage to the hydrant
- Rusting or corrosion
- Outlet caps missing or stuck in place with paint
- Stem or operating nut that cannot be turned or turns feely with no visible result
- Obstructions (bottles, cans, rocks) inside the hydrant outlets
- Damp ground surrounding the hydrant or erosion indicating a drain valve leak
- Hydrants painted by property owners (miscolored)

Alternative Water Supplies

Even in areas with good water supply systems, the location and types of alternative water supplies should be determined during preincident surveys. The following can all be used as alternative water supplies:

- Adjacent private water systems
- Available lakes, ponds, rivers, and the ocean
- Swimming pools
- Farm stock tanks
- Rainwater collection cisterns and detention ponds

To access alternative water supplies it will be necessary to establish a **drafting** operation **(Figure 14.14, p. 800)**. Almost any static source of water can be used if it is sufficient in quantity and not contaminated to the point of creating a health hazard or damaging the fire pump. The depth of the water source is an important consideration. When drafting from any natural source, all hard-suction hoses should have strainers attached to the end. The hard-suction hose should be positioned and supported so the strainer does not rest on or near the bottom of the source. A minimum of 24 inches (600 mm) of water above and below the hard intake strainer is usually needed for it to function properly and avoid picking up silt and debris. Silt and debris can clog strainers, seize or damage pumps, and allow small sand and rocks to enter attack lines and clog fog nozzles. However, floating strainers can draft water from sources as shallow as 24 inches (600 mm) deep **(Figure 14.15, p. 800)**.

Drafting — Process of acquiring water from a static source and transferring it into a pump that is above the source's level; atmospheric pressure on the water surface forces the water into the pump where a partial vacuum was created.

Figure 14.14 Drafting is accomplished directly from a body of water into the apparatus pumps.

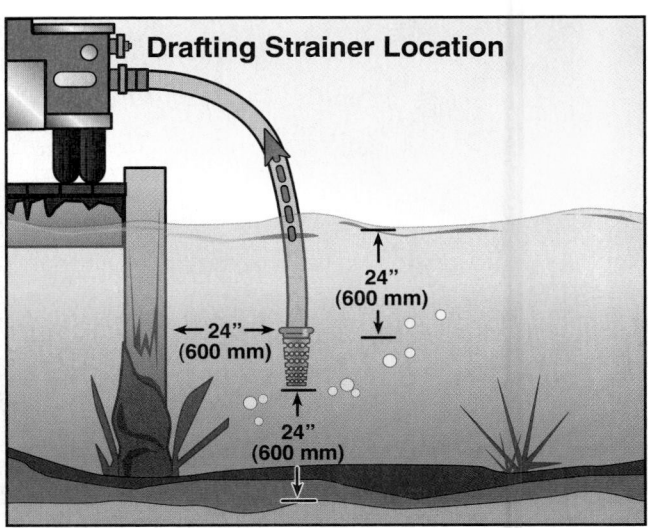

Drafting Strainer Location

24" (600 mm)

24" (600 mm)

24" (600 mm)

Figure 14.15 A hard intake strainer requires a minimum of 24 inches (600 mm) of water around the strainer.

In some areas, drafting hydrants are installed at static water sources to increase the water supply available for fire fighting. Drafting hydrants are usually constructed of steel or PVC pipe with strainers at the water source and steamer ports to connect to the pumper. They are designed to supply at least 1,000 gpm [4 000 L/min].

Fire departments should make every effort to identify, mark, and record alternative water supply sources during preincident planning. Consideration should be given to the effect that weather has on the amount of water available and the ease of access to water sources. Firefighters working within 10 feet (3 m) of the water's edge may be required to wear a life vest as discussed in Chapter 10. **Skill Sheet 14-I-3** details the steps for drafting from a static water source.

Rural Water Supply Operations

Many rural areas lack public water distribution systems or have systems with inadequate volume and pressure. In these areas, rural water supply operations must frequently be performed. Two common operations are **water shuttle operations** and relay pumping. For these operations to succeed, preincident planning and frequent practice are required. The following sections briefly explain each of these operations. For additional information on rural water supply operations, see the IFSTA **Pumping Apparatus Driver/Operator Handbook** and NFPA® 1142, *Standard on Water Supplies for Suburban and Rural Fire Fighting*.

Water Shuttle Operation — Method of water supply by which mobile water supply apparatus continuously transport water between a fill site and the dump site located near the emergency scene.

Figure 14.16 A water shuttle serves the purpose of collecting water and dumping it in portable tanks on the fire scene. *Courtesy of Bob Esposito.*

Water Shuttles

Water shuttle operations involve hauling water from a supply source (fill site) to the incident scene. The water is then transferred to the attack pumper's tank or to **portable tanks** (dump sites) from which water may be drawn to fight a fire **(Figure 14.16)**. In most cases, water shuttle operations are recommended for distances greater than ½ mile (0.8 km) from the nearest fire hydrant or water source or greater than the fire department's capability of laying supply hoselines. It is critical to have a sufficient number of mobile water supply apparatus to maintain the needed fire flow (water flow).

Fast-fill and fast-dump capabilities are critical to efficient water shuttle operations. Radio equipped water supply officers are assigned at both the fill and dump sites to coordinate these operations. Also important are traffic control, hydrant operations, hookups, and tank venting. Mobile water supply apparatus driver/operators should remain in their vehicles during filling/dumping operations.

There are three key components to water shuttle operations:

- Dump site at the fire
- Fill site at the water source
- Use mobile water supply apparatus to haul water from the fill site to the dump site

The dump site is generally located near the actual fire or incident. The dump site usually consists of one or more portable water tanks into which mobile water supply apparatus deposit water before returning to the fill site. Apparatus attacking the fire may draft directly from the portable tanks, or other apparatus may draft from the tanks and supply the attack apparatus. Low-level intake devices permit use of most of the water in the portable reservoir.

Portable tanks have capacities beginning at 1,000 gallons (4 000 L). In general, there are two types of portable water tanks **(Figure 14.17, p. 802)**. One is the collapsible or folding style that uses a square metal frame and a synthetic or canvas duck liner. Another style is a round, self-supporting synthetic tank with a floating collar that rises as the tank is filled. These frameless portable tanks are widely used in wildland fire fighting operations.

> **Portable Tank** — Storage tank used during a relay or shuttle operation to hold water from water tanks or hydrants. This water can then be used to supply attack apparatus. Also called Catch Basin, Fold-a-Tank, Portable Basin, or Porta-Tank.

Figure 14.17 Portable tanks can be set up quickly and hold large capacities of water.

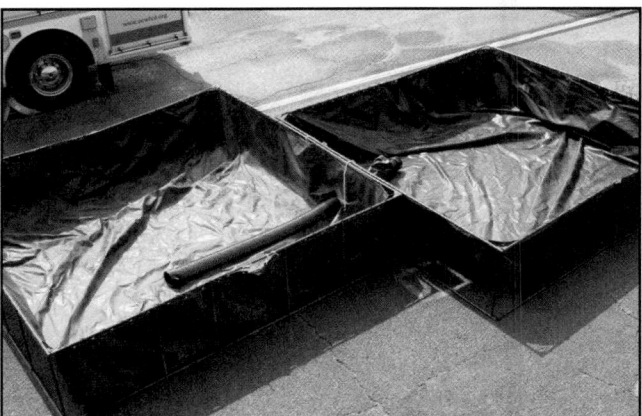

Figure 14.18 Portable tanks may serve separate but related purposes and may be connected at their drain openings.

Before opening a portable tank, spread a salvage cover or heavy tarp on the ground to help protect the liner once water is dumped into it. When the situation permits, portable tanks should be as level as possible to ensure maximum capacity. The tank should be positioned in a location that allows easy access from multiple directions but does not inhibit access of other apparatus to the fire scene. Ideally, portable tanks will be set up so that more than one mobile water supply apparatus can off-load at the same time. Ensure that the drain is located on the downhill side of the tank and away from the drafting tank. Consult **Skill Sheet 14-I-4** for steps to take in deploying a portable water tank.

When large quantities of water must be maintained, multiple portable tanks are set up. One portable tank is used for the attack pumper while the mobile water supply apparatus dump into the other tanks. When two portable tanks are used, they can be interconnected through their drain fittings **(Figure 14.18)**. If multiple portable tanks are needed, jet siphon devices can be used to transfer water from one tank to another. A jet siphon uses a 1½-inch (38 mm) discharge line connected to the siphon. The siphon is then attached to a hard sleeve placed between two tanks.

Firefighters can calculate the volume of water supplied in water shuttle operations based upon a complete round trip including fill time, travel time, and dump time. The total gallons (Liters) carried in each apparatus divided by the total time will provide the gallons per minute (gpm) (L/min) being supplied.

There are four basic methods by which mobile water supply apparatus unload:

- Gravity dumping through large (10- or 12-inch [250 mm or 300 mm]) dump valves
- Jet-assist dumps that increase the flow rate
- Apparatus-mounted pumps that off-load the water
- Combination of these methods

NFPA® 1901, *Standard for Automotive Fire Apparatus,* requires that mobile water supply apparatus on level ground be capable of dumping or filling at rates of at least 1,000 gpm (4 000 L/min). This rate requires adequate tank venting and openings in tank baffles. To avoid damage to the tank, only a trained apparatus driver/operator should pump the water from the tank of a mobile water supply apparatus. Any firefighter may activate a gravity dump which saves time in the process because the driver/operator does not have to leave the cab. Some gravity dumps can be activated from the cab, but others must be activated by a lever near the outlet.

Figure 14.19 A vacuum mobile water supply apparatus drafts water from a water supply or holding tank and is able to immediately send that water out as a fire stream.

To fill mobile water supply apparatus quickly, use the best fill site or hydrant available, large hoselines, multiple hoselines, and if necessary, a pumper for adequate flow. In some situations multiple portable pumps may be necessary. Both fill sites and dump sites should be arranged so that a minimum of backing or maneuvering of apparatus is required.

The use of self-filling vacuum mobile water supply apparatus has increased in recent years **(Figure 14.19)**. The vacuum pump on the apparatus permits rapid water filling and discharge and reduces the number of personnel required for the operation. The system also eliminates the requirement to have a pumper at the fill site for drafting.

Relay Pumping

In some situations the water source is close enough to the fire scene that **relay pumping** can be used instead of water shuttles. A combination water shuttle and relay pumping to minimize congestion of apparatus at the fire scene can also be used. Two important factors must be considered regarding the establishment of a relay operation:

- The water supply must be capable of maintaining the desired volume of water required for the duration of the incident.
- The relay must be established quickly enough to be worthwhile.

The number of pumpers needed and the distance between pumpers is determined by several factors such as:

- Required volume of water
- Distance between the water source and the fire scene
- Size of supply hose available
- Amount of hose available
- Pumper capacities

The apparatus with the greatest pumping capacity should be located at the water source. Large-diameter supply hose or multiple hoselines increase the distance and volume that a relay can supply because of reduced friction loss. A water supply officer must be appointed to determine the distance between pumpers and to coordinate water supply operations.

Relay Pumping — Use of two or more pumpers to move water over a long distance by operating them in series. Water discharged from one pumper flows through hoses to the inlet of the next pumper, and is then pumped to the next pumper in line.

After considering these factors, the water supply officer should make a calculation to determine the distance between pumpers taking into account the friction loss at particular flows for the size hose being used. The results of the calculations can be made into a chart and placed on the pumper for quick reference. The best way to prepare for relay operations is to plan them in advance and to practice them often.

Chapter Summary

Because water is the primary fire extinguishing agent used by firefighters, and because fires often occur considerable distances from major water sources, fire departments must develop ways to transport the available water from its source to the place it is needed. Firefighters must be familiar with the water supply systems in their response areas. They must know about water sources, pumping systems, gravity systems, and the system of underground water mains used to distribute the water. They must also know how to inspect, maintain, and operate the fire hydrants in their jurisdictions. Finally, they must know how to use water shuttle and relay pumping operations in areas that do not have water supply systems.

Review Questions

Firefighter I

1. What are the three main means of moving water used by firefighters?

2. How do the main components of water distribution and storage systems operate?

3. What are the main types of fire hydrants?

4. How does the operation of a dry-barrel hydrant compare to that of a wet-barrel hydrant?

5. When should alternative water supplies be identified?

6. What are the common operations for accessing rural water supplies?

Step 1: As a safety precaution, tighten hydrant outlet caps that will not be used.

  a. Turn caps clockwise

  b. Use appropriate tools

Step 2: Turn outlet nut counterclockwise and removethe cap from one outlet.

  a. Stand clear of closed caps

Step 3: Open the hydrant.

  a. Use tools to slowly turn hydrant nut appropriate direction, typically counterclockwise

  b. Continue until fully open

  c. Stand clear of closed caps

  d. Do not lean over top of hydrant

Step 4: Close the hydrant.

  a. Use tools to slowly turn hydrant nut appropriate direction, typically clockwise

  b. Continue until fully closed

  c. Check to see that the hydrant is draining if it is a dry-barrel style

Step 5: Replace cap on outlet.

  a. Turn outlet nut appropriate direction, typically clockwise, until firmly closed

  b. Stand clear of closed caps

Soft-Sleeve Connection

Step 1: Confirm order with officer to make hydrant connection.

Step 2: Remove necessary equipment from the pumper.

    a. Hydrant or spanner wrench

    b. Reducer (if necessary)

    c. Rubber mallet, if needed

Step 3: Remove the hydrant cap by turning it counterclockwise and using a spanner wrench if the cap is tight.

Step 4: Inspect the hydrant for exterior damage and check for debris or damage in inside outlet.

Step 5: Place the hydrant wrench on hydrant nut with handle pointing away from outlet.

Step 6: Determine if adapter is preconnected.

    a. If adapter is preconnected to the hose: proceed to step 7

    b. If adapter is not preconnected: Place the reducer adapter on the hydrant, turning clockwise and making hand tight

Step 7: Remove the intake hose from the pumper.

14-I-2
Make soft-sleeve and hard-suction hydrant connections.

Soft-Sleeve Connection

Step 8: Connect the intake hose to the pump intake, turning clockwise and making hand tight.

Step 9: Stretch the intake hose to the hydrant, placing two full twists in the hose to prevent kinking.

Step 10: Make the hydrant connection to steamer outlet or outlet with adapter, turning clockwise and making hand tight.

Step 11: Open the hydrant slowly until hose is full.

Step 12: Tighten any leaking connections using rubber mallet or spanner wrench.

SKILL SHEETS

14-1-2
Make soft-sleeve and hard-suction hydrant connections.

Hard-Suction Connection

Step 1: Confirm order with officer to make hydrant connection.

Step 2: Remove the intake hose from the pumper.

Step 3: Connect the intake hose to the hydrant or apparatus (depending on local preference), turning connection clockwise and making hand tight.

Step 4: Connect opposite end to the hydrant or apparatus, turning connection clockwise and making hand tight.

14-I-3
Connect and place a hard-suction hose
for drafting from a static water source.

SKILL SHEETS

Step 1: Confirm order with officer to connect hose for drafting.

Step 2: Check the hard-suction couplings.

    a. Remove any dirt or debris

    b. Replace worn or frayed coupling gaskets

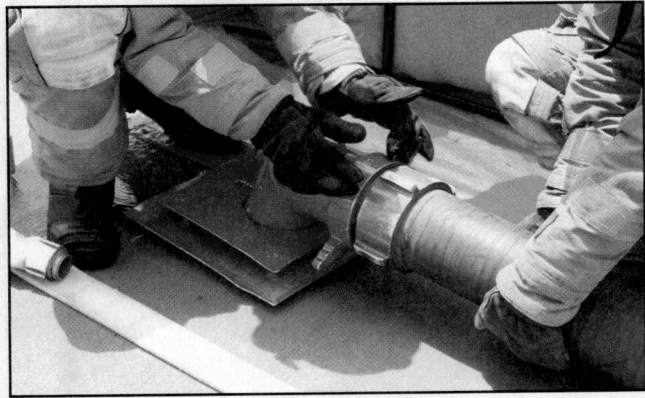

Step 3: Connect the sections of hard-suction hose.

    a. Align sections

    b. Hand tighten in clockwise direction

    c. Use rubber mallet to make airtight connection, if necessary

    d. Keep off of ground

Step 4: Connect the strainer to one end of the hard-suction hose.

    a. Hand tighten in clockwise direction

    b. Using rubber mallet to make airtight connection, if necessary

    c. Fasten rope to strainer

Step 5: Put the strainer into the water; if a barrel strainer, use the rope to maneuver the hose and to keep the strainer off the bottom.

Step 6: Prepare pump intake for coupling by removing pump intake cap and keystone intake valve from intake, if applicable.

Step 7: Connect the hard-suction hose to the pump intake, aligning the sections and hand tightening in a clockwise direction.

Step 8: Tie up strainer rope (if used) to pumper or stationary object.

Step 9: Dismantle drafting equipment and return to proper storage on pumper per departmental SOPs.

Step 1: Remove the tarps from the apparatus.

Step 2: Carry the tarps to the planned location for the water reservoirs.

    a. Location provides easy access from multiple directions

    b. Location allows other apparatus access to the fire scene

Step 3: Open the tarps and spread them flat on the ground.

Step 4: Remove the portable tank, jet siphon, and manufacturer's setup instructions from the apparatus.

    a. Use proper lifting techniques

    b. Carry to the setup location

Step 5: Set up two portable tanks within departmental time limits (if specified) per manufacturer's instructions.

Step 6: Connect the intake and discharge hoses to the jet siphon per manufacturer's instructions.

Step 7: Position the jet siphon properly to draw and discharge water per manufacturer's instructions.

Step 8: Dismantle the portable tanks following manufacturer's instructions.

Step 9: Shake and fold the tarps.

Step 10: Return equipment to the proper storage locations on the apparatus.

Key Terms

NFPA® Job Performance Requirements

This chapter provides information that addresses the following job performance requirements of NFPA® 1001, *Standard for Fire Fighter Professional Qualifications* (2013).

Firefighter I
5.3.8
5.3.10
5.3.15
5.5.2

Firefighter II
6.5.5

1. Explain basic fire hose characteristics. (5.3.8, 5.3.10)
2. Describe different causes of and prevention methods for hose damage. (5.5.2)
3. Identify basic inspection, care, and maintenance methods for fire hose. (5.5.2)
4. Compare various uses for hose appliances and tools. (5.3.8, 5.3.10)
5. Describe basic hose rolls. (5.5.2)
6. Explain basic hose loads and finishes. (5.5.2)
7. Compare various methods to make preconnected hose loads for attack lines. (5.5.2)
8. Explain the methods used for supply hose lays. (5.3.8, 5.3.15)
9. Recognize different methods for handling hoselines. (5.3.8, 5.3.10)
10. Describe methods for advancing hoselines in various ways (5.3.8, 5.3.10)
11. List the considerations that can impact operating attack hoselines. (5.3.8, 5.3.10)
12. Couple and uncouple a hose. (Skill Sheet 15-I-1, 5.3.10)
13. Inspect and maintain a fire hose. (Skill Sheet 15-I-2, 5.5.2)
14. Make a straight hose roll. (Skill Sheet 15-I-3, 5.5.2)
15. Make a donut hose roll. (Skill Sheet 15-I-4, 5.5.2)
16. Make the flat hose load. (Skill Sheet 15-I-5, 5.5.2)
17. Make the accordion hose load. (Skill Sheet 15-I-6, 5.5.2)
18. Make the horseshoe hose load. (Skill Sheet 15-I-7, 5.5.2)
19. Make a finish. (Skill Sheet 15-I-8, 5.5.2)
20. Make the preconnected flat hose load. (Skill Sheet 15-I-9, 5.5.2)
21. Make the triple layer hose load. (Skill Sheet 15-I-10, 5.5.2)
22. Make the minuteman hose load. (Skill Sheet 15-I-11, 5.5.2)
23. Make a hydrant connection from a forward lay. (Skill Sheet 15-I-12, 5.5.2)
24. Make the reverse hose lay. (Skill Sheet 15-I-13, 5.5.2)
25. Advance a hose load. (Skill Sheet 15-I-14, 5.3.10)
26. Deploy a wye-equipped hose during a reverse hose lay. (Skill Sheet 15-I-15, 5.3.15)
27. Advance a charged hoseline using the working line drag method. (Skill Sheet 15-I-16, 5.3.10)
28. Advance a line into a structure. (Skill Sheet 15-I-17, 5.3.10)
29. Advance a line up and down an interior stairway. (Skill Sheet 15-I-18, 5.3.10)
30. Connect to a stairway standpipe connection and advance an attack hoseline onto a floor. (Skill Sheet 15-I-19, 5.3.10)
31. Advance an uncharged line up a ladder into a window. (Skill Sheet 15-I-20, 5.3.10)
32. Advance a charged line up a ladder into a window. (Skill Sheet 15-I-21, 5.3.10)
33. Operate a charged attack line from a ladder. (Skill Sheet 15-I-22, 5.3.10)
34. Operate a small hoseline – One-firefighter method. (Skill Sheet 15-I-23, 5.3.10)
35. Operate a large hoseline for exposure protection – One-firefighter method. (Skill Sheet 15-I-24, 5.3.10)
36. Operate a large hoseline – Two-firefighter method. (Skill Sheet 15-I-25, 5.3.15)
37. Extend a hoseline. (Skill Sheet 15-I-26, 5.3.10)
38. Replace a burst hoseline. (Skill Sheet 15-I-27, 5.3.10)

Firefighter II Chapter Objectives

1. Describe the safety considerations taken when service testing a fire hose. (6.5.5)
2. Service test a fire hose. (Skill Sheet 15-II-1, 6.5.5)

Chapter 15
Fire Hose

During a mutual aid response to a 6,000 square foot (557.42 sq. m) residential structure, an engine company was ordered to provide supply lines to a ladder (quint) company that was starting the interior fire attack with a 2½-inch (65 mm) attack line. A supply line 500 feet (150 m) long was deployed from the hydrant to the quint. When the hydrant was opened and the supply line was charged, the ladder company moved deeper into the structure. When they reached the fire, the ladder company opened the nozzle to put water on the fire. Once the nozzle was open, a vacuum developed in the supply line reducing the flow of water in the line below what was needed for an effective attack. The attack team communicated their situation to Command and was ordered to withdraw. Upon investigation, the lack of water was attributed to the firefighter who connected the supply line to the hydrant. When he made the connection, he failed to remove the hydrant strap. The strap created a kink in the hose which limited the quantity of water flowing to the quint and the attack line.

Source: FirefighterNearMiss.com

Fire hose is a flexible, portable extension of the water distribution system that carries water from the fire hydrant or water source to the pumper/quint. From there a pump controls the pressure and forwards the water through attack hoselines to the point where water is needed. As a firefighter, you must know the following information described in this chapter:

- Fire hose characteristics including construction and sizes
- Causes of hose damage and methods for preventing that damage
- General care and maintenance of fire hose
- Types of fire hose couplings
- Types of hose appliances
- Tools used in hose operations

In addition, you need to be able to perform the following skills which are described step-by-step in this chapter:

- Rolling hose
- Loading supply hose on apparatus
- Preparing finishes

- Loading preconnected attack lines
- Deploying hose
- Handling various sizes of hose
- Advancing and operating hoselines
- Service testing fire hose

I Fire Hose Characteristics

The characteristics that are used to describe **fire hose** include the type of construction and the materials used, the internal diameter, and the type of couplings used to make connections. In general, fire hose is used as either supply hose or attack hose (**Figure 15.1**). **Supply hose** transports water from a fire hydrant or other water supply source to an apparatus equipped with a pump located at or near the fire scene. **Attack hose** transports water or other agents, at increased pressure, from the following sources:

- From the pump-equipped apparatus to a nozzle or nozzles
- From a pump-equipped apparatus to a **fire department connection** (FDC) mounted on a structure
- From a building standpipe to the point the water is applied to the fire

Hose Construction

To be reliable, fire hose must be constructed of the best materials available, used in an appropriate manner, and maintained according to the manufacturer's recommendations. Most fire hose is flexible, watertight, and has a smooth rubber or neoprene lining covered by a durable jacket. Depending on its intended use, fire hose is manufactured in a variety of configurations, the most common being single-jacket, double-jacket, rubber single-jacket, and hard-rubber or plastic noncollapsible types (**Figure 15.2**).

Figure 15.1 Supply hose and attack hose serve different purposes, but they are often needed during the same incident.

Fire Hose Construction Features

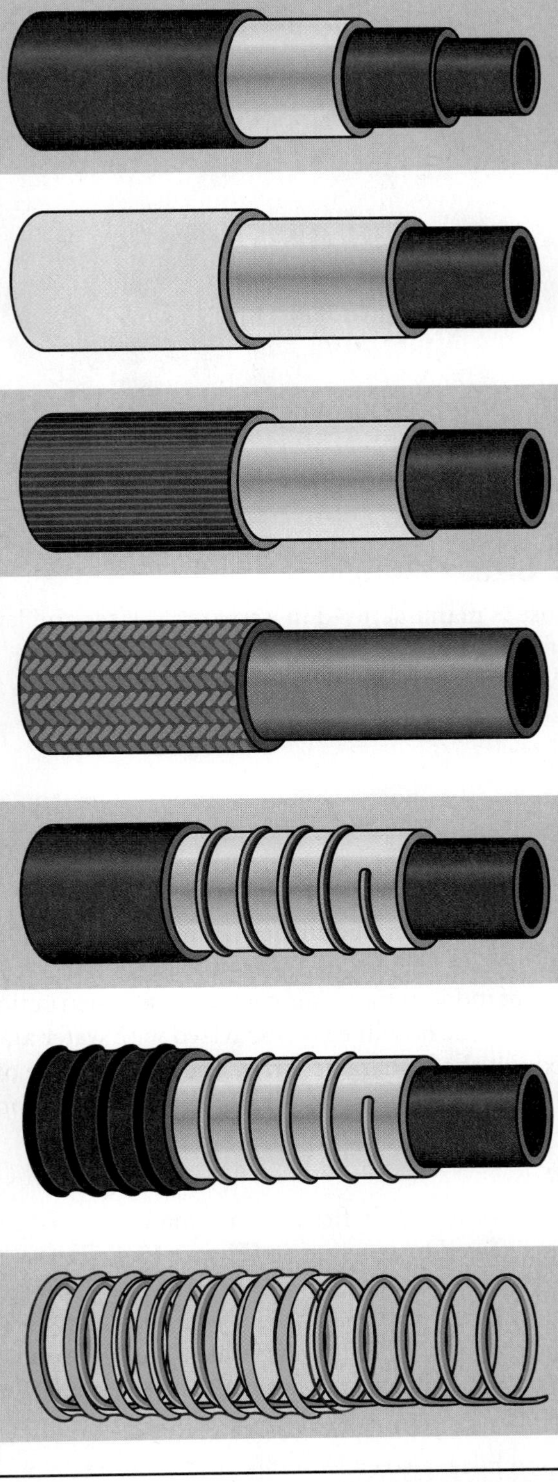

Hose Types and Diameters	Cutaway Views	Linings, Coverings, and Reinforcements
Booster Hose ¾ or 1 inch (20 mm or 25 mm)		• **Rubber Covered** • **Rubber Lined** • **Fabric Reinforcement**
Woven-Jacket Hose 1 to 6 inches (25 mm to 150 mm)		• **One or More (Two Showing) Woven-Fabric Jackets** • **Rubber Lined**
Rubber-Covered Hose 1 to 6 inches (25 mm to 150 mm)		• **Woven Polyester Tube** • **Nitrile Rubber** • **Synthetic Rubber Outer Cover**
Impregnated Single-Jacket Hose 1½ to 5 inches (38 mm to 125 mm)		• **Woven Polyester Nylon or Combination of Synthetic Fibers Form Tube** • **Polymer Covered** • **Polymer Lined**
Noncollapsible Intake Hose 2½ to 6 inches (65 mm to 150 mm)		• **Rubber Covered** • **Fabric and Wire (Helix) Reinforcement** • **Rubber Lined**
Flexible Noncollapsible Intake Hose 2½ to 6 inches (65 mm to 150 mm)		• **Rubber Covered** • **Fabric and Plastic (Helix) Reinforcement** • **Rubber Lined**
Flexible Noncollapsible Clear Intake Hose 2½ to 6 inches (65 mm to 150 mm)		• **Polyvinyl Tube** • **Polyvinyl Reinforcement (Helix)**

Figure 15.2 Hose construction features include the diameter of the hose and the number and composition of layers.

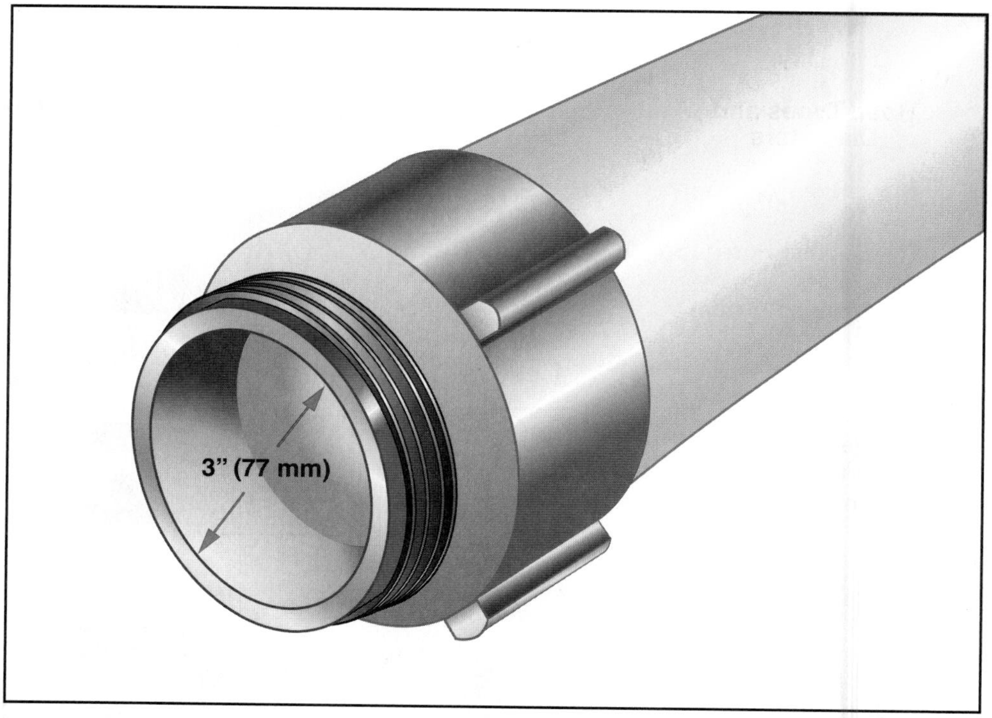

Figure 15.3 A hose's measured diameter refers to the internal measurement, not the external diameter which may vary depending on the number and composition of the external jacketing layers.

Hose Sizes

Fire hose is manufactured in a variety of sizes and lengths based upon the specifications for fire hose in NFPA® 1961, *Standard on Fire Hose*. Diameter and length are common ways to describe hose size.

Diameter

The size of a fire hose refers to its inside diameter. According to NFPA®, the internal diameter of a fire hose should not be less than the advertised or labeled size of the hose, which means that the diameter of the hose is no less than its actual internal diameter. For example, hose labeled 3 inches (77 mm) in diameter must have an internal diameter of 3 inches (77 mm) **(Figure 15.3)**. Some types of fire hose can expand beyond their actual manufactured internal diameter because of the elastic qualities of modern materials used in hose construction. Expansion results in a larger orifice when the hoseline is pressurized with water and lower friction loss. Not all fire hose exhibit these characteristics. The performance of a particular hoseline depends on the materials and methods used in its construction.

Length

Both attack and supply fire hose are manufactured in 50 or 100 feet (15 m or 30 m) lengths, referred to as a section **(Figure 15.4)**. These lengths were traditionally determined for convenience and ease of handling. However, the traditional length of fire hose in North America is 50 feet (15 m) per section. Modern hose may be carried or maneuvered in longer sections because it is often constructed of high-strength, lightweight synthetic materials that have the same relative weight of traditional sections of hose.

Suction hose (also called *intake hose*), used to connect the pumper to a hydrant or other water source, are manufactured in minimum lengths specified in NFPA® 1901, *Standard for Automotive Fire Apparatus*. Large **soft sleeve hose** is a minimum 15 feet (4.6 m) in length, has two female or nonthreaded couplings, and is used to connect

Soft Sleeve Hose — Large-diameter, collapsible piece of hose used to connect a fire pump to a pressurized water supply source; sometimes incorrectly referred to as *soft suction hose*.

Suction Hose — Intake hose that connects pumping apparatus or portable pump to a water source.

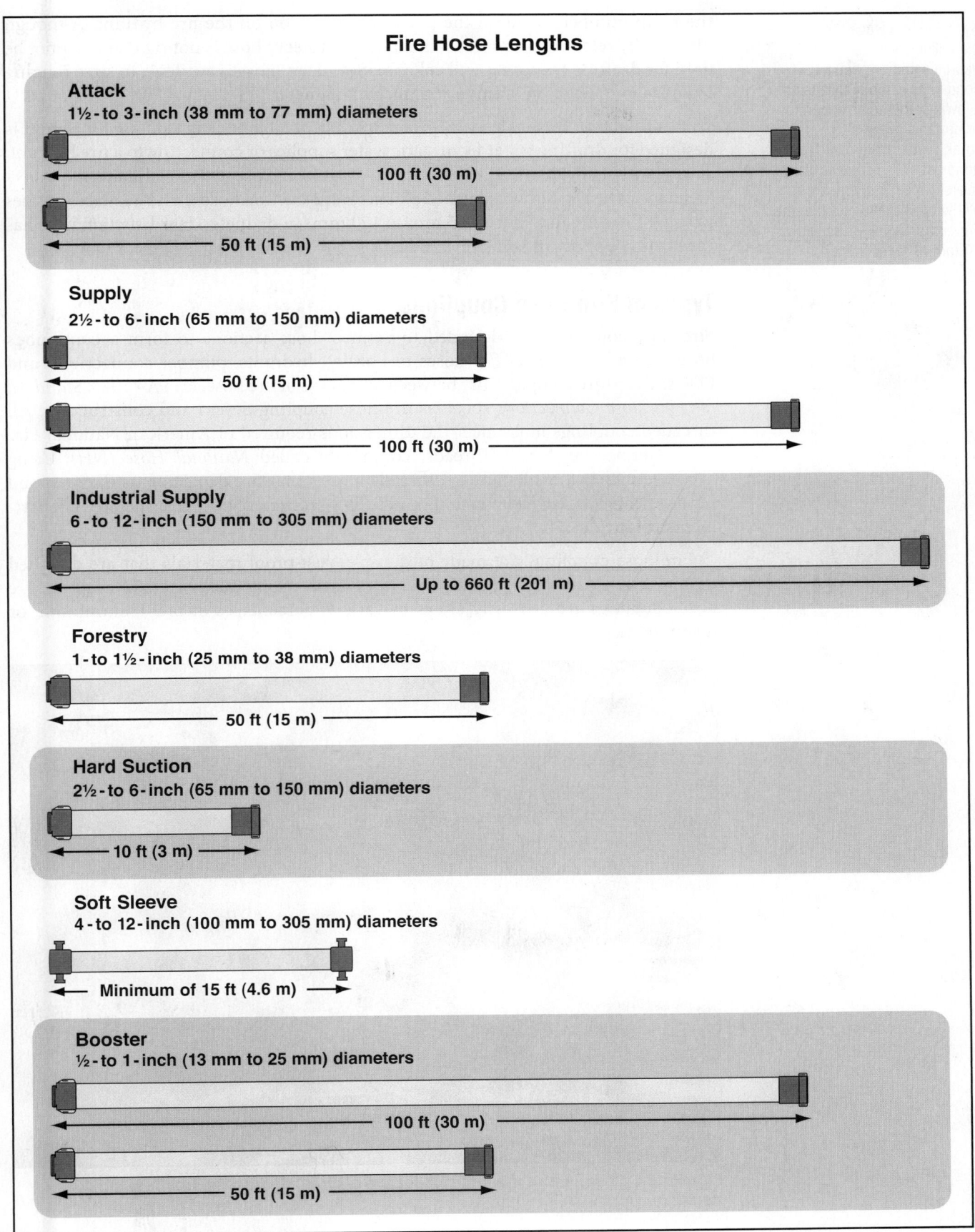

Figure 15.4 Fire hose lengths are standardized across the industry for specific purposes, depending on pressure requirements and water supply features.

the main pumper intake to the pumper connection on the fire hydrant. Although historically referred to as soft suction hose, soft sleeve hose is not rigid and cannot be used for drafting because it will collapse. Soft sleeve hose is available in sizes ranging from 2½ to 6 inches (65 mm to 150 mm) in diameter.

Hard-suction hose is generally constructed in 10 foot (3 m) long sections and is designed for drafting water from static water supplies or connecting to a fire hydrant. Some hard-suction hose is constructed of a rubberized, reinforced material; others are made of heavy-duty corrugated plastic. Hard-suction hose is also available in sizes ranging from 2½ to 6 inches (65 mm to 150 mm) in diameter. Hard-suction hose has the same couplings as soft sleeve hose.

Types of Fire Hose Couplings

Fire hose couplings are designed to connect hose sections to form a continuous hoseline and to connect fire hoses to nozzles, hydrants, pumper connections, and FDCs. To ensure compatibility between all brands of fire hose, NFPA® 1963, *Standard for Fire Hose Connections* specifies fire hose coupling design and construction. All threaded couplings must meet the dimensions required of American National Fire Hose Connection Screw Threads, commonly called *National Hose (NH)*. Using fire hose with national standard threads means that fire departments that respond together can connect hose sections and supply sections of hose from adjacent departments (**Figure 15.5**).

Fire hose couplings are made of durable, rust-proof materials that are designed to couple and uncouple quickly and with little effort. The materials used for fire hose couplings are generally alloys in various percentages of brass, aluminum, or magnesium.

Figure 15.5 Hoseline use is a common feature of an incident scene. *Courtesy of Tom Aurnhammer.*

Couplings are categorized by the way they are manufactured:

- **Cast** — Cast couplings are very weak and only found on occupant-use fire hose. They often crack if reattachment to the hose is attempted.

- **Extruded** — Extruded couplings are usually made of aluminum or aluminum alloy, allowing for their lightweight and high strength. They are somewhat stronger than cast couplings.

- **Drop forged** — Drop forged couplings are made of brass or other malleable metal and are the strongest and most expensive of the three coupling types.

Both attack hose and supply hose may be equipped with either **threaded** or **nonthreaded** couplings. Coupling and uncoupling are simple procedures for connecting and disconnecting sections of hose with either threaded or nonthreaded couplings. The need for speed and efficiency under emergency conditions requires that techniques for coupling and uncoupling hose be learned and practiced. Nozzles are connected or removed from the fire hose using the same methods as those used for coupling and uncoupling sections of hose. **Skill Sheet 15-I-1** describes methods of coupling and uncoupling threaded and nonthreaded fire hose. The sections that follow describe the characteristics of threaded and nonthreaded couplings.

Threaded Couplings

One of the oldest coupling designs involves the casting or machining of a spiral thread into the face of two distinctly different couplings — male and female. A male coupling thread is cut on the exterior surface, while a female coupling thread is on the interior surface of a free-turning ring called a swivel. The swivel permits connecting two sections of hose without twisting the entire hose. Each section of fire hose with threaded couplings has a male coupling at one end and a female coupling at the other. Together, the two couplings are referred to as a *set* (also referred to as a *three-piece coupling*). The male coupling is considered one piece, and the female coupling is a two-piece assembly allowing it to swivel during coupling with the male end (**Figure 15.6, p. 822**).

CAUTION

Connect couplings hand tight to avoid damage to the coupling and gasket.

A threaded coupling has several other parts. The portion of the coupling that serves as a point of attachment to the hose is called the **shank**. A flattened angle at the end of the threads on the male and female couplings called the **Higbee cut** (also called *blunt start*) prevents cross-threading when couplings are connected. The **Higbee indicator** (indentation) on the exterior of the coupling marks where the Higbee cut begins (**Figure 15.7, p. 822**).

Unlike common pipe threads that are relatively fine, fire hose coupling threads are coarse (with wide tolerances). Coarse threads allow the couplings to be connected quickly. Some manufacturers make the large coupling sizes (3½ inches [90 mm] and above) with either ball bearings or roller bearings under the swivel to ensure their smooth operation. A removable rubber gasket located inside the base of the female coupling ensures a tight fit and reduces the chance of water leaks.

Threaded couplings are manufactured with either lugs or handles to aid in tightening and loosening the connection. Lugs are located on the shank of a male coupling and on the swivel of a female coupling. Lugs are grasping points where firefighters

Threaded Coupling — Male or female coupling with a spiral thread.

Nonthreaded Coupling — Coupling with no distinct male or female components. *Also known as* Storz Coupling or sexless coupling.

Shank — Portion of a coupling that serves as a point of attachment to the hose.

Higbee Cut — Special cut at the beginning of the thread on a hose coupling that provides positive identification of the first thread to eliminate cross-threading.

Higbee Indicators — Notches or grooves cut into coupling lugs to identify by touch or sight the exact location of the Higbee Cut.

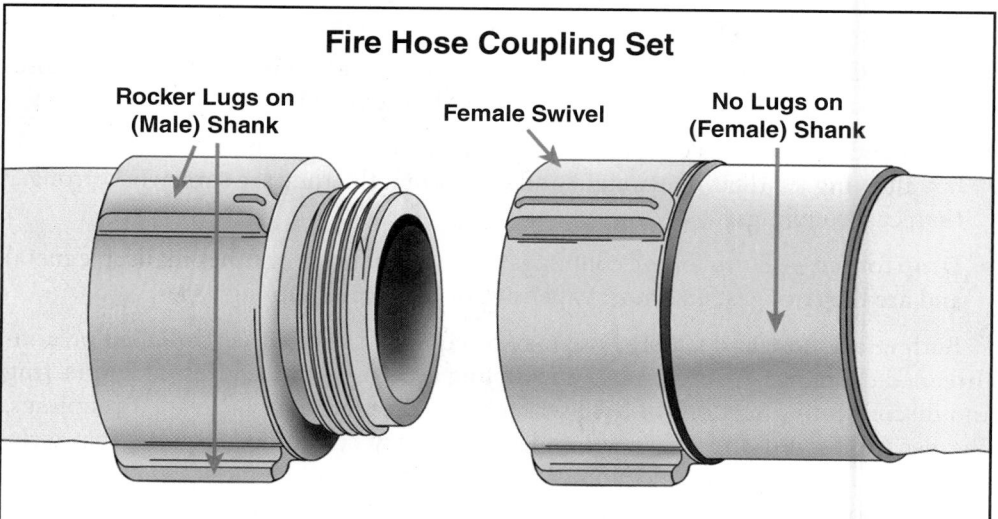

Fire Hose Coupling Set

Rocker Lugs on (Male) Shank

Female Swivel

No Lugs on (Female) Shank

Figure 15.6 Coupling sets have distinctly different features on each connecting end that swivel during attachment.

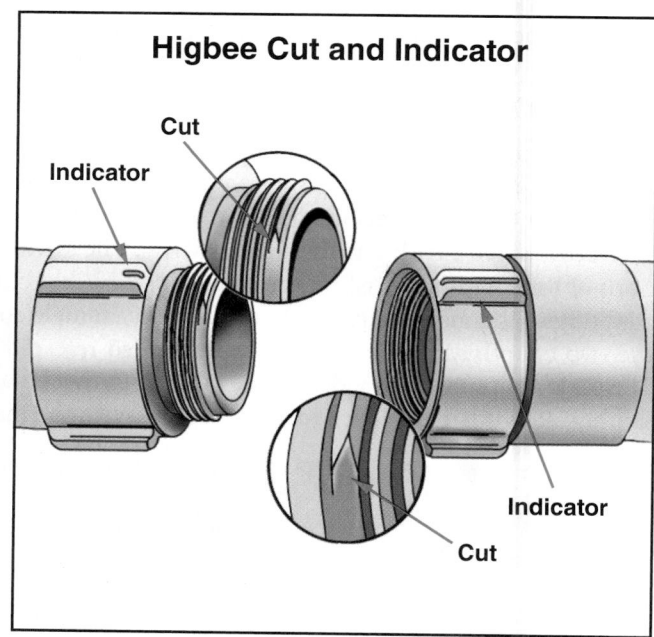

Higbee Cut and Indicator

Cut

Indicator

Indicator

Cut

Figure 15.7 The Higbee indicator provides a marker of the location of the Higbee cut to aid a team connecting lengths of fire hose to get a quick and correct connection.

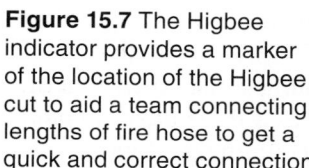

Spanner Wrench — Small tool primarily used to tighten or loosen hose couplings; can also be used as a prying tool or a gas key.

can easily hold the coupling when making and breaking coupling connections. Connections may be made manually or with the assistance of **spanner wrenches** (special wrenches that fit against the lugs; often referred to as spanners). Three types of lugs are found today: pin, recessed, and rocker (**Figure 15.8**).

Pin lugs, usually found on couplings of old fire hose, resemble small pegs. Although still available, pin-lug couplings are not commonly ordered with new fire hose because of their tendency to catch when hose is dragged over objects or deployed from the hose bed of a pumping apparatus.

Booster fire hose normally has couplings with recessed lugs that are simply shallow holes drilled into the coupling. This lug design prevents abrasion that would occur if the hose had protruding lugs and was wound onto reels. Firefighters will need to use a pin-lug spanner wrench inserted into the recessed lug holes to tighten or loosen the coupling.

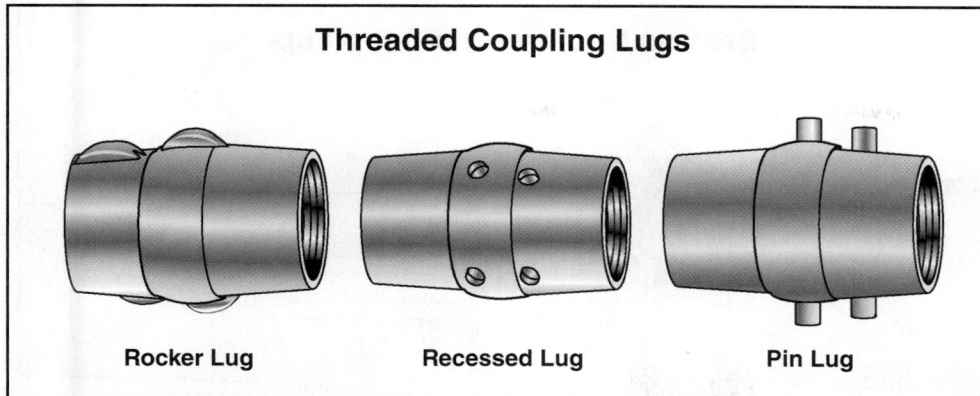

Threaded Coupling Lugs

Rocker Lug　　　Recessed Lug　　　Pin Lug

Figure 15.8 Different styles of lugs serve the same purpose of providing grip for responders connecting hose lengths.

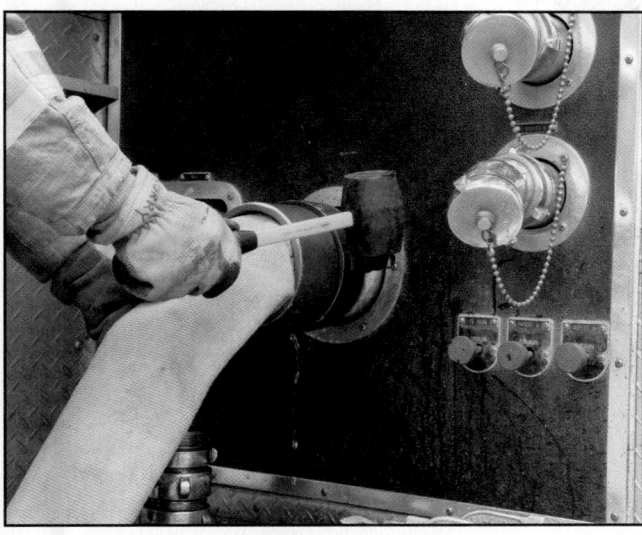

Figure 15.9 A rubber mallet may be used to loosen or tighten couplings during drafting operations.

Modern threaded couplings have rounded rocker lugs. The rounded shape of rocker lugs (unlike pin lugs) helps prevent the hose from catching on objects. On the couplings, one of the rocker lugs on the swivel is scalloped with a shallow indentation (the Higbee indicator) to mark where the Higbee cut begins. This indicator aids in matching the male coupling thread to the female coupling thread during low-light situations or when the threaded end of the coupling is not readily visible.

Handles or extended lugs are located on the swivels of large intake supply or suction hoses. Firefighters can grasp these handles when manually tightening the large coupling that connects the hose to a pump valve intake. If necessary, striking the handles with a rubber mallet can help loosen the coupling or tighten leaking couplings on a charged line. You may need to use the mallet for tightening during the setup for a drafting operation, for example (**Figure 15.9**).

Nonthreaded Couplings

Nonthreaded couplings are connected with locks or cams rather than screw threads. Although there are some nonthreaded couplings that have male and female ends, *sexless couplings* are more prevalent in North America. A sexless coupling set has no distinct male and female components so both couplings are identical. There are two kinds of sexless couplings: **quarter-turn** and **Storz** (**Figure 15.10, p. 824**).

Quarter-Turn Coupling — Nonthreaded (sexless) coupling with two hook-like lugs that slip over a ring on the opposite coupling and then rotate 90 degrees clockwise to lock.

Storz Coupling — Nonthreaded (sexless) coupling commonly found on large-diameter hose. Nonthreaded fire hose couplings have been used in the North American fire and emergency services since the early 1900s. With this type of coupling, the mating of two couplings is achieved with locks or cams without the use of screw threads.

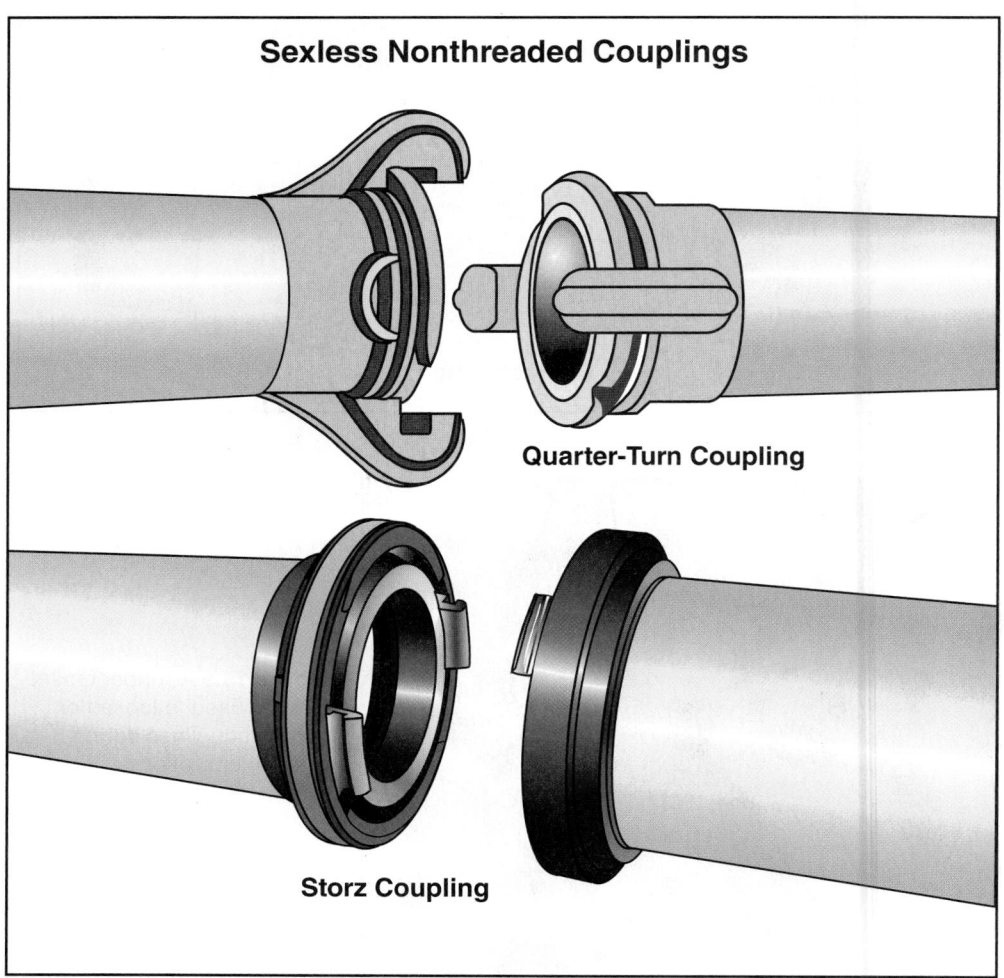

Figure 15.10 Sexless, nonthreaded couplings use identical couplings to create a seal between lengths of hose.

The quarter-turn coupling has two hook-like lugs on each coupling. The lugs, which are grooved on the underside, extend past a raised lip or ring on the open end of the coupling. When the couplings are mated, the lug of one coupling slips over the ring of the opposite coupling and then rotates 90 degrees clockwise to lock. A gasket on the face of each coupling seals the connection to prevent leakage.

Storz couplings are most commonly found on large-diameter hose. Like quarter-turn couplings, they are joined and then rotated until locked in place to form a connection. Unlike quarter-turn couplings, the locking components consist of grooved lugs and inset rings built into the face of each coupling swivel. When mated, the lugs of each coupling fit into recesses in the opposing coupling ring and then slide into locking position behind the ring with a one-third-turn rotation. External lugs at the rear of the swivel provide leverage for connecting and disconnecting couplings. On most manufacturers' couplings, the lugs align to give a visual indicator of a connected coupling (couplings properly aligned and locked in place).

Nonthreaded couplings have the following advantages:

- Fire hose can be quickly connected. However, the use of spanner wrenches to ensure a complete connection, slows the connecting operation somewhat.

- There is no risk of cross-threading a connection and damaging a coupling because there are no threads.

- Double-male or double-female adapters (adapters connecting two threaded couplings of the same thread type, size, and sex) are not needed, thus hose can be deployed from the hose bed regardless of hose load type.

Nonthreaded couplings also have certain disadvantages as follows:

- Hose can become uncoupled, often suddenly and violently, if a complete connection has not been made.

- Hydrants require an adapter to make connections with nonthreaded couplings. The time needed to attach the adapter increases the time required to connect to the hydrant and begin deploying hose.

- Dirt and other large debris can become lodged inside the coupling's grooves, giving the impression that a tight seal has been made when in fact, the hose is not connected.

NOTE: Although not widely seen, some fire and emergency service organizations that have adopted sexless couplings for their supply hose operations use permanent adapters installed on hydrants for fire hydrant connections.

Causes and Prevention of Fire Hose Damage

The types of damage that can occur to the exterior covering and the inner lining of fire hose include:

- **Mechanical damage** — Abrasions, cuts, and tears
- **Thermal damage** — Exposure to fire, high heat, or freezing temperatures
- **Organic damage** — Attack by mold and mildew
- **Chemical damage** — Deterioration due to solvent action on synthetic materials and natural fibers
- **Corrosion** — Rusting of metal couplings
- **Age deterioration** — Cracking at points where hose is folded and separation of inner liner from exterior covering

Mechanical Damage

Mechanical damage occurs when contact with an object or surface causes slices, rips, and abrasions on the exterior covering, crushed or damaged couplings, or cracked inner linings. The following practices are recommended to help prevent mechanical damage:

- Avoid laying or pulling a hose over rough, sharp edges or objects such as corners, cornices, parapets, and windowsills.

- Protect hoses by using a hose roller or placing a folded salvage cover over sharp edges.

- Clear windowsills of broken glass fragments.

- Provide traffic control to prevent vehicles from driving over hose.

- Use hose ramps or bridges to protect hose from vehicles driving over it when traffic cannot be rerouted **(Figure 15.11, p. 826)**.

- Open and close nozzles, valves, and hydrants slowly to limit excessive stress and prevent water hammer.

Figure 15.11 Hoselines that must cross areas of vehicle activity must be protected against compression (which may inhibit stream flow) and damage. *Courtesy of Sam Goldwater.*

Figure 15.12 Hose bed covers reduce thermal damage to large amounts of hose stored in an apparatus. *Courtesy of Sam Goldwater.*

- Provide chafing blocks to prevent abrasion to hose when it vibrates near the pumper.
- Avoid excessive pump pressure on hoselines.
- Deploy hoselines away from debris, or clear debris from the path of hose during overhaul operations.
- Change position of folds in hose when reloading it on apparatus.
- Clean hose before reloading it to prevent abrasions from dirt or grit.

Thermal Damage

Thermal damage to fire hose can result from exposure to excess heat or cold temperatures. Excessive heat exposure or direct flame contact can char, melt, or weaken the outer jacket and dehydrate the rubber lining. Inner linings can also be dehydrated when hose is hung to dry in a drying tower longer than necessary or when it is dried in direct sunlight. Mechanical hose dryers eliminate this concern but extend the time required to dry large amounts of hose because of their limited capacity. To prevent thermal damage, follow these recommended practices:

- Protect hose from exposure to excessive heat or fire when possible.
- Remove hose from any heated area as soon as it is dry.
- Use moderate temperature for mechanical drying. A current of warm air is much better than hot air.
- Keep the outside of woven-jacket fire hose dry when not in use.
- Run water through hose that has not been used for some time to keep the liner soft.
- Avoid laying fire hose on hot pavement to dry.
- Roll dry hose in a straight roll for storage. This keeps the liner from drying out.
- Prevent hose from coming in contact with, or being placed close to, vehicle exhaust systems.
- Use hose bed covers on apparatus to shield the hose from the sun (**Figure 15.12**).

Cold damage occurs when water on the inside and/or the outside of fire hose freezes. Fire departments located in regions that suffer severely cold temperatures should use special cold-resistant hose designed for use at temperatures down to -65°F (-54°C). Cold-resistant hose should perform with the same reliability as regular fire hose while withstanding the rigors of freezing and thawing. Regardless of the type of fire hose used in extreme cold temperatures, allow some water to flow through the nozzle to prevent water from freezing inside the hose during intermittent use at fires.

Use the following guidelines to help prevent hose from freezing:

- Maintain water flow in intake hose by circulating water from a hydrant through the fire pump, discharging it through a drain-off hose that routes water down a gutter, or to a place away from the pumping apparatus.

- Immediately drain and roll hose that is no longer needed for fire fighting.

- Tighten all hose connections to prevent couplings from leaking and freezing.

- Apply a manufacturer approved, cold-weather lubricant that contains an anti-freeze agent on the swivel and gasket portions of the couplings.

When fire hose becomes frozen to an ice-covered surface, there are three ways to remove it: (1) melt the ice with a steam-generating device, (2) chop the hose loose with axes, or (3) leave the hose until the weather warms enough to melt the ice. Guidelines for removing frozen hose are as follows:

- Make all cuts well away from the hose when chopping hose out of ice in order to minimize the chance of an axe blade striking and damaging the hose fabric.

- Avoid using exhaust manifold heat from the pumping apparatus because it can be very hot and poses a carbon monoxide hazard to firefighters working with the hose.

- Wait until hose is thawed before folding it. Folding frozen hose can damage the hose's lining and outer jacket often in ways that are not apparent during visual inspections.

- If fire hose sections can be uncoupled, carefully load them onto a flatbed vehicle and transport them to a location where they can be thawed and protected from damage.

- Perform a service test before placing thawed hose back in service to ensure that no damage has occurred.

Organic Damage

Mildew and mold are living organisms (fungus) that can rot natural fibers. When hose with a woven-jacket of cotton or other natural fiber is stored wet, rot from mold and mildew may weaken the jacket which can lead to ruptures under pressure **(Figure 15.13)**. The outer jacket of some woven-jacket fire hose is made of synthetic fibers such as Dacron™ polyester that resist organic damage. The outer jacket of some natural-fiber hose has been chemically treated to resist mildew and mold but such treatment is not always 100 percent effective. Rubber-jacket hose is not subject to organic damage.

Some methods of preventing mildew and mold on natural-fiber woven-jacket hose are as follows:

- Remove all wet hose from the apparatus after a fire and replace with dry hose or dry the wet hose thoroughly before reloading it on the apparatus.

- Inspect, wash, and dry hose that has been contaminated in any way.

Figure 15.13 Mildew and mold can destroy woven-jacket fire hose through the process of rotting.

- Remove, inspect, sweep, and reload hose if it has not been unloaded from the apparatus during a period of six months. Make sure that the hose is folded at different points than when previously loaded.
- Inspect and test hose annually and after possible damage or freezing.
- Ensure that cotton or cotton-blend fire hose is completely dry before storing or loading.
- Cover hose beds with water-repellent covers to keep hose loads dry during inclement weather.
- Inspect fire hose in storage racks and hose beds periodically
- Remove and rotate hose periodically. Even if mildew is not visible, a musty smell is often an indicator that it is hidden somewhere within the hose.
- Ventilate all areas where fire hose is kept, including pumping apparatus hose beds and compartments.
- Wash hose immediately whenever mildew is discovered. Steps:
  1. Scrub the cover jacket with a very mild soap or bleach solution (5 percent in water).
  2. Rinse well.
  3. Dry completely or to the point recommended in the manufacturer's instructions.
  4. Inspect the hose section within the next few days for the reappearance of mildew.

Chemical Damage

Certain chemicals and chemical vapors can damage the outer jacket on fire hose or cause the rubber lining to separate from the inner jacket. Common examples of chemical damage firefighters are likely to encounter include:

- Exposure to petroleum products, paints, acids, or alkalis may weaken hose to the point of bursting under pressure:
  — Motor oil, found in some quantity on most streets and highways where hose is laid, will penetrate the woven outer cover and cause a reaction that separates the inner lining of the hose.
  — Gasoline contact will also react to separate the inner lining of the hose, but has a much quicker and more severe reaction.
- Battery acid can destroy hose jacket fibers.
- Runoff water from a fire may carry foreign materials that can cause chemical damage to fire hose.
- Water that is not drained completely from the hose can form sulfuric acid, which weakens or destroys the liner.

After exposure to chemicals or chemical vapors, hose should be cleaned as soon as is practical. Some recommended practices are as follows:

- Avoid laying fire hose directly against curbs where oil, gasoline, and battery acid may accumulate or pool from parked automobiles. Hose laid in gutters may also come in contact with fire fighting runoff water that could contain harmful chemicals. Even though some types of fire hose are constructed from materials that are not affected by chemical exposure, other hose types are still used. Procedures:
  — Place the hose 2 to 4 feet (0.6 m to 1.2 m) away from the curb or gutter but not in vehicle travel lanes.
  — Move the hose, if possible, onto a sidewalk or into a median to avoid vehicle and contamination damage.

- Avoid exposing fire hose to hazardous material spills.

- Avoid exposing fire hose to spills of foam concentrate, which is mildly corrosive and can deteriorate the hose lining or cover material if it remains on the hose.

- Scrub fire hose suspected of having contacted acid or other caustic chemical thoroughly with a solution of bicarbonate of soda and water **(Figure 15.14)**. Remove the hose from service and contact the manufacturer for further maintenance procedures.

- Remove hose periodically from the apparatus, wash it with plain water, and dry it thoroughly.

- Test hose properly if there is any suspicion of damage (see Service Testing Fire Hose section).

- Dispose of hose according to departmental SOPs if it has been exposed to hazardous materials and cannot be decontaminated.

Figure 15.14 A deck brush plus a solution of bicarbonate of soda and water should be used on hose that has been exposed to acids or other chemicals.

Corrosion

Corrosion is a chemical process in which a metal is attacked by some substance in its environment and converted to an unwanted compound that gradually weakens or destroys the metal. The most common fire hose coupling metals are brass and aluminum. Each of these metals possesses a high resistance to corrosion but each will suffer some deterioration when exposed to certain conditions.

Brass is highly resistant to corrosion. Over time, however, a brass coupling will corrode when it is in contact with moist organic material or earth. The metal will darken and turn green as copper oxides are formed. Although these copper oxides are usually found only on exposed surfaces, they can form on the interior of female swivels or inside surfaces of nozzles, reducing the ease of the device's operation. Normal cleaning removes most of the surface corrosion; however, the only way to free the swivels or operating mechanisms is to lubricate moving parts according to the manufacturer's recommendations.

Figure 15.15 A hose packing strategy includes reloading the hose with folds in previously unfolded places.

Aluminum couplings develop a layer of corrosion (aluminum oxide) that in effect "seals" the metal against further oxidation. This protective layer can be scratched or abraded during normal use, resulting in a new layer being formed.

Age Deterioration

If fire hose is left in an apparatus bed for a long time, the hose can deteriorate and crack because of the sharp folds in the tightly-packed hose load. To prevent this, hose loads should be removed and repacked every six months if they are not used. Because the hose near the top of the hose bed is deployed most often, lower layers may seldom be removed. When reloading the hose, pack the hose loosely and fold the hose in places that were not previously folded (**Figure 15.15**). Because fire hose that is loaded on edge wears more quickly, manufacturers recommend using a flat load.

Fire hose will also deteriorate if it is left hanging in a hose tower for excessive periods of time. The inner lining of hose can become weakened at the point where it hangs over the support peg. Reinforced jacketed fabric hose may suffer a separation of the rubber or plastic lining from the inner reinforcement, reducing the strength of the hose at the point of the separation. To prevent this damage, remove the hose from the tower as soon as it is dry. If fire hose must remain in a tower for prolonged periods, change the hose/peg contact point periodically.

Care for Fire Hose

Always follow the manufacturer's instructions and department SOPs for hose care and maintenance.

Inspection, Care, and Maintenance of Fire Hose

Thorough inspections, care, and maintenance can significantly extend the working life of fire hose. The techniques of washing and drying and the provisions for storage are very important functions in the care of fire hose.

Inspection

According to NFPA® 1962, hose should be inspected and service-tested within 90 days before being placed in service for the first time and at least annually thereafter. Each time a section of hose is used, whether for emergency incidents or training, it needs to be inspected to ensure that it is free of visible soil or damage. Check couplings for ease of operation, any deformations, or other visible damage. While gathering equipment and rolling fire hose immediately following an incident, conduct a postincident inspection. This quick inspection allows firefighters to identify and mark damage on hose and couplings. Refer to **Skill Sheet 15-I-2** for general procedures for inspecting and maintaining fire hose.

Before fire hose and couplings are stored or placed back in service after use, correct or report any of the following deficiencies that may be present (**Figure 15.16**):

- Evidence of dirt or debris on the hose jacket or couplings
- Damage to the hose jacket
- Coupling loosened from the hose
- Damage to male and female threads
- Obstructed operation of the swivel
- Absence of a well-fitting gasket in the swivel

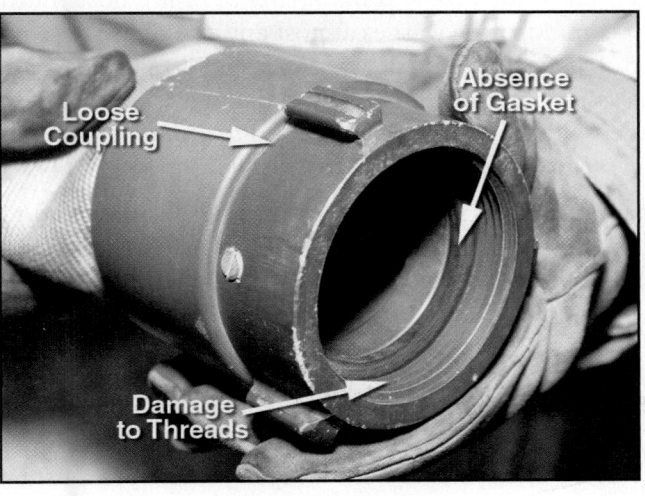

Figure 15.16 Before storing or placing fire hose and couplings into service, examine all components and correct or report any damage or deficiencies.

Figure 15.17 Following use, lightly-soiled hose may be cleaned with a low-pressure hose.

Figure 15.18 High-pressure hose washing devices may be used to scrub the surface of a fire hose.

Washing Hose

The method used to wash fire hose depends on the type of hose. Hard-rubber booster hose, hard intake hose, and rubber-jacket collapsible hose only require rinsing with clear water, although a mild soap may be used if necessary.

Most woven-jacket fire hose requires a little more care. After woven-jacket hose is used, any dust and dirt should be thoroughly brushed or swept off of the hose. If brushing or sweeping does not remove the dirt, wash with clear water while scrubbing with a stiff brush.

When fire hose has been exposed to oil, it should be washed with a mild soap or detergent using common scrub brushes or straw brooms. Make sure that the oil is completely removed. The hose should then be rinsed thoroughly with clear water **(Figure 15.17)**.

A hose-washing machine can make the care and maintenance of fire hose much easier **(Figure 15.18)**. The most common type washes almost any size of fire hose up to 3 inches (77 mm). The flow of water into this device can be adjusted as desired, and the movement of the water assists in propelling the hose through the device. The hoseline that supplies the washer with water can be connected to a pumper or used directly from a hydrant. Higher water pressure gives better results.

A cabinet-type machine that washes, rinses, and drains fire hose is designed to be used in the station. This type of machine can be operated by one person, is self-propelled, and can be used with or without detergents.

Drying Hose

Fire hose should be dried before being stored **(Figure 15.19)**. The methods used to dry hose depend on the type of hose. Hose should be dried in accordance with departmental SOPs and manufacturer's recommendations. Woven-jacket hose must be thoroughly dried before being reloaded on an apparatus. Hard-rubber booster hose, hard intake hose, and synthetic-jacket collapsible hose may be placed back on the apparatus while wet with no ill effects.

Figure 15.19 Hose drying cabinets speed the process of returning a hose to ready condition.

Hose towers and drying racks must have adequate ventilation and protection so that fire hose is not exposed to excessive temperatures or direct sunlight. Take the following actions when drying hose in hose towers or on racks:

- Remove hose from exterior hose towers as soon as it is dry to protect it from sun damage.
- Lash or tie the coupling ends of hose hung on outside drying towers together to prevent them from swinging in the wind. Swinging couplings can collide with each other or with tower supports, resulting in mechanical damage.
- Cover male threads with precut sections of tubing to provide additional protection.
- Incline drying racks enough to allow water to drain from the hose during drying.
- Avoid placing hose sections too close together or allowing them to touch, which can slow the drying process.

Storing Hose

After fire hose has been washed and dried, roll and store it in racks in a manner that protects them from damage. Mount racks permanently on the wall or stand them free on the floor. Mobile hose racks can be used to both store and move hose from storage rooms to apparatus for loading. Hose that is stored in the fire apparatus room/bay may be exposed to cleaning solvents, lubricants, oils, diesel fumes, and other airborne contaminants. If hose must be stored in the fire apparatus room/bay, inspect and clean it more frequently than fire hose that is stored in a separate space.

CAUTION
Never store solvents, petroleum products, or other chemicals close to fire hose and couplings.

Figure 15.20 Fire hose may be stored as rolls with the male coupling at the center.

Take the following precautions to prevent damage to hose stored in racks:

- Locate hose racks in a clean, well-ventilated room that is easily accessible to the apparatus room/bay.

- Store hose where it is not exposed to direct sunlight. The sun's ultraviolet rays break down the natural or synthetic fabric of the hose, reducing its expected service life.

- Pack cotton fabric hose loosely so that air circulates around it. Synthetic and rubber-jacketed hose can be stored in tight rolls after normal cleaning procedures.

- Store hose in a rack in such a way that couplings are not in walkways and will not come into contact with equipment or passing personnel.

- Roll the hose with the male end inside the roll to protect the male coupling threads **(Figure 15.20)**.

- When it is necessary to store fire hose with the male coupling on the outside of the roll, protect the exposed threads with a cap or other protective device.

- Place sexless couplings in a storage rack in a way that prevents dirt or other foreign objects from collecting in their ramp grooves. These contaminates can interfere with their ability to securely lock into place.

Care of Fire Hose Couplings

Although fire hose couplings are designed to be durable, they can be damaged. On threaded couplings, the male threads are exposed when not connected and subject to being dented. The female threads are not exposed, but the swivel can be bent into an oval shape. When either threaded or nonthreaded couplings are connected, there is less danger of damage to their parts during ordinary usage; however, they can be bent or crushed if vehicles drive over them. Some simple rules for the care of fire hose couplings are as follows:

- Avoid dropping and/or dragging couplings.

- Do not permit vehicles to drive over fire hose or couplings.

- Inspect couplings when hose is washed and dried.

- Remove the gasket and twist the swivel in warm, soapy water.
- Clean threads to remove tar, dirt, gravel, and oil.
- Inspect gasket and replace if cracked or creased.

Hose-washing machines will not clean hose couplings sufficiently when the coupling swivel becomes difficult to spin because of dirt or other foreign matter. The swivel should be submerged in a container of warm, soapy water and moved forward and backward to thoroughly clean the swivel. The male threads should be cleaned with a stiff brush **(Figure 15.21)**. It may be necessary to use a wire brush if threads are clogged by tar, asphalt, or other foreign matter. Lubricants such as graphite or silicone are usually all that is needed to maintain swivels so that they spin freely. If a gasket is cracked, scored or has become hardened and inflexible, replace it.

Figure 15.21 A stiff brush will help dislodge dirt or other foreign matter from between coupling threads.

Hose Appliances and Tools

Hose appliances and hose tools are used in conjunction with hose and nozzles to complete hose layouts. Hose appliances are devices that route water in a variety of ways and make different types of hose connections. Hose tools are devices that assist with the movement, handling, protection, and connecting of hose. A simple way to remember the difference between hose appliances and hose tools is that water flows through appliances but not through tools.

Hose appliances include valves and valve-controlled devices such as **wyes**, **Siameses**, **water thieves**, large-diameter hose appliances, and hydrant valves, as well as **fittings** such as **adapters** and intake strainers. Examples of hose tools include hose rollers, spanner wrenches, hose strap and hose rope tools, hose chain tools, hose ramps, hose jackets, blocks, and hose clamps.

Hose Appliances

A hose appliance is any piece of hardware used in conjunction with fire hose for the purpose of controlling the flow of water and creating a variety of pathways for water through hose layouts. Common hose appliances include valves and valve devices, fittings, and intake strainers.

Wye — Hose appliance with one female inlet and multiple male outlets, usually smaller than the inlet. Outlets are also usually gated.

Siamese — Hose appliance used to combine two or more hoselines into one. The Siamese has multiple female inlets and a single male outlet. An example of a Siamese is a fire department connection.

Water Thief — Any of a variety of hose appliances with one female inlet for 2½ -inch (65 mm) or larger hose and with three gated outlets, usually two 1½-inch (38 mm) outlets and one 2½-inch (65 mm) outlet.

Adapter — Device for connecting hose couplings with dissimilar threads but with the same inside diameter.

Fitting — Device that facilitates the connection of hoselines to provide an uninterrupted flow of extinguishing agent.

Figure 15.22 The internal ball shutoff component has a hole through its center, which provides a waterway designed to minimize friction loss when it is opened fully.

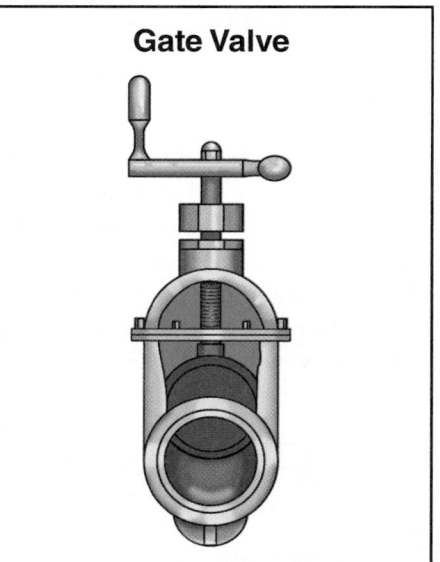

Gate Valve

Figure 15.23 A gate valve slides a disk into and out of the water flow to control the pressure allowed through the nozzle.

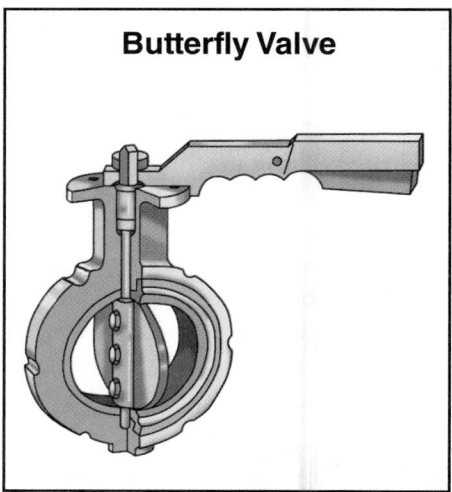

Butterfly Valve

Figure 15.24 A butterfly valve rotates a disk to control water flow.

Valves

The following valves in hoselines, at hydrants, and at pumpers are used to control the flow of water:

● **Ball valves** — Used in pumper discharges and gated wyes. Ball valves are open when the handle is in line with the hose and closed when it is at a right angle to the hose. Ball valves are also used in fire pump piping systems (**Figure 15.22**).

● **Gate valves** — Used to control the flow from a hydrant. Gate valves have a baffle that is lowered into the path of the water by turning a screw-type handle (**Figure 15.23**).

● **Butterfly valves** — Used on large pump intakes and incorporates a flat baffle that turns 90 degrees. Most are operated manually using a quarter-turn handle, but some are operated using an electric motor and can be controlled remotely. The baffle is in the center of the waterway and aligned with the flow when the valve is open (**Figure 15.24**).

● **Clapper valves** — Used in Siamese appliances and fire department connections (FDC) to allow water to flow in one direction only. Clapper valves prevent water from flowing out of unused ports when one intake hose is connected and charged before the addition of more hose. The clapper is a flat disk hinged at the top or one side which swings open and closed like a door (**Figure 15.25**).

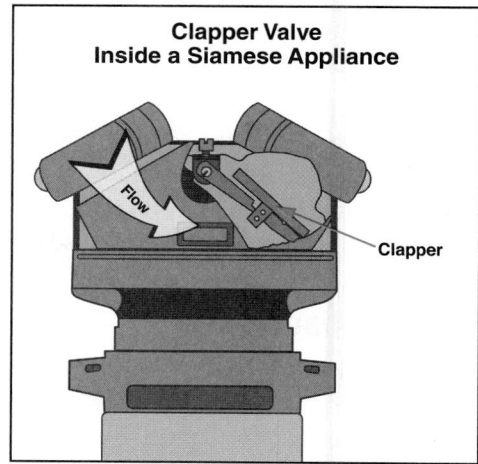

Clapper Valve
Inside a Siamese Appliance

Flow

Clapper

Figure 15.25 A clapper valve swings along a fixed point to move a seal into the opening of a Siamese appliance.

Valve Devices

Valve devices allow the number of hoselines operating on the fireground to be increased or decreased. These devices include wye appliances, Siamese appliances, water thief appliances, large-diameter hose appliances, and hydrant valves.

Wye appliances. Wye appliances are used to divide a single hoseline into two or more lines. All wyes have a single female inlet and multiple male outlet connections. Wyes that have valve-controlled outlets are called *gated wyes*. Ball valves are generally used in gated wyes. One of the most common wyes has a 2½-inch (65 mm) inlet that divides into two 1½-inch (38 mm) outlets, although many other combinations are available **(Figure 15.26)**. For high water volume operations, wyes with a large-diameter hose (LDH) inlet and two 2½-inch (65 mm) outlets are used.

Siamese appliances. Firefighters sometimes confuse Siamese and wye appliances because of their similar appearance. While wyes divide a single hoseline into multiple lines, a Siamese combines multiple lines into one line. These appliances permit multiple supply hoselines to be laid parallel to supply a pumper or high-output device.

Siamese appliances usually consist of two female inlets, with either a center clapper valve or two clapper valves (one on each side) and a single male outlet **(Figure 15.27)**. Some Siamese appliances are equipped with three clappered inlets; they are commonly called triamese appliances or manifolds. The clapper valves are used to control the flow of the inlet streams into the single outlet stream. Siamese and triamese appliances are commonly used when LDH is not available to overcome friction loss in exceptionally long hose lays or those that carry a large flow. They are also used when supplying ladder pipes that are not equipped with a permanent waterway.

Water thief appliances. In operation, the water thief is similar to the wye appliance; however, there is an inlet and outlet of matching size combined with smaller outlets that "steal" water from the main line. Larger volume water thief appliances consist of an LDH inlet and outlet and two or more 2½-inch (65 mm) valve-controlled male outlets **(Figure 15.28)**.

Figure 15.26 A gated wye is used to supply two hoses using the same water source.

Figure 15.27 A Siamese clapper valve allows the connection of one or two supply hoses while preventing water from entering the uncharged line until the second valve is opened. *Courtesy of Task Force Tips.*

Figure 15.28 A large-diameter Storz manifold permits multiple hoselines to be fed by a single large-diameter supply line.

Large-diameter hose appliances. Some fire fighting operations require water to be distributed at various points along the main supply line. In these cases, an LDH water thief can be used. In other cases, when a large volume of water is needed near the end of the main supply line, an LDH manifold appliance can be used. A typical LDH manifold consists of one LDH inlet and three 2½-inch (65 mm) valve-controlled male outlets. Depending on the locale and the configuration of the appliance, these devices are sometimes called *portable hydrants, phantom pumpers,* or *large-diameter distributors.*

Hydrant valves. A variety of hydrant valves are available for use in supply-line operations. Known by a variety of regional names, these valves are used when a forward lay is made from a low-pressure hydrant to the fire scene. A hydrant valve has four main functions **(Figure 15.29)**:

- Allow additional hoselines to be laid to the hydrant
- Connect a supply pumper to the hydrant
- Boost the pressure in the original supply line without interrupting the flow of water in the line
- Allow the original supply line to be connected to the hydrant and charged before the arrival of another pumper at the hydrant

Fittings

Fittings are used to connect hose of different diameters and thread types or to protect the couplings on standpipes and on apparatus intakes and outlets. There are two main types of fittings: adapters and reducers.

An adapter is a fitting for connecting hose couplings with similar threads and the same inside diameter. The double-male and double-female adapters are among the most often-used hose fittings **(Figure 15.30, p. 840)**. These adapters allow two male couplings or two female couplings of the same diameter and thread type to be connected. An increasingly common fitting is that used to connect a sexless coupling to a threaded outlet on a hydrant.

Reducers are another common type of hose fitting. They are used to connect a smaller-diameter hoseline to the end of a larger one. However, using a reducer limits the larger hose to supplying one smaller line only **(Figure 15.31, p. 840)**. Using a wye appliance allows the larger hose to supply two smaller ones.

Other common fittings include elbows that provide support for intake or discharge hose at the pumping apparatus. The threads on pump male discharge outlets are protected with hose caps. Female inlets on some FDCs are capped with hose plugs. They are also used on the standpipe outlets in stairway risers to prevent kinks in the attack line.

Intake Strainers

Intake strainers are devices attached to the drafting end of a hard-suction hose when pumping from a static water source. They are designed to keep debris from entering the apparatus or portable pump. Such debris can either damage the pump or pass through it to clog the nozzle.

Intake strainers must not rest on the bottom of a static water source except when the bottom is clean and hard, such as the bottom of a swimming pool. To prevent a strainer from resting on the bottom of a lake or pond, tie one end of a length of rope to the eyelet on the strainer and the other to an apparatus or another anchor point **(Figure 15.32, p. 840)**. Floating intake strainers are also available to keep the intake strainer off the bottom of a static water source.

Reducer — Fitting used to attach a smaller hose to a larger hose. The female end has the larger threads, while the male end has the smaller threads.

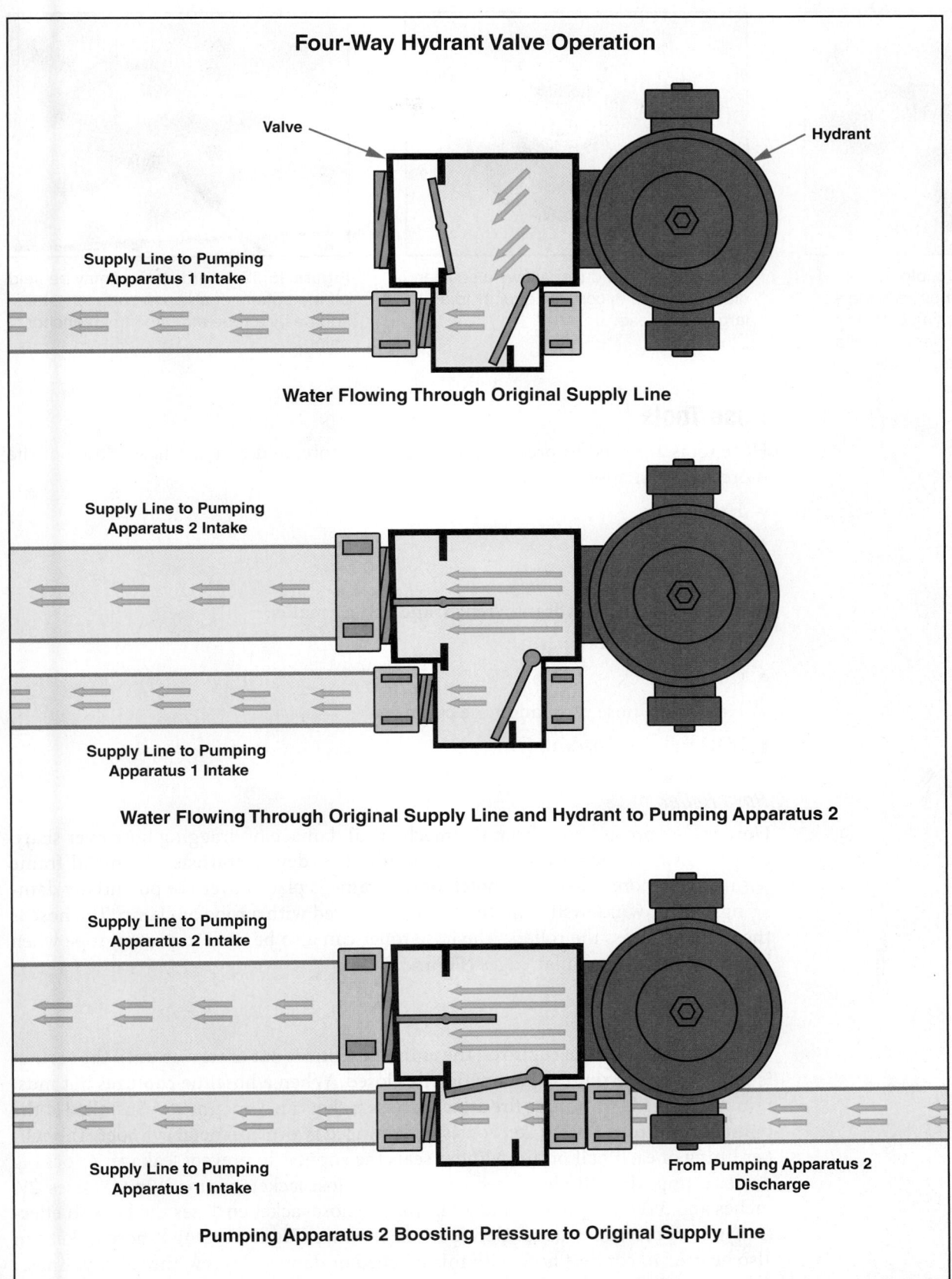

Four-Way Hydrant Valve Operation

Valve

Hydrant

Supply Line to Pumping
Apparatus 1 Intake

Water Flowing Through Original Supply Line

Supply Line to Pumping
Apparatus 2 Intake

Supply Line to Pumping
Apparatus 1 Intake

Water Flowing Through Original Supply Line and Hydrant to Pumping Apparatus 2

Supply Line to Pumping
Apparatus 2 Intake

Supply Line to Pumping
Apparatus 1 Intake

From Pumping Apparatus 2
Discharge

Pumping Apparatus 2 Boosting Pressure to Original Supply Line

Figure 15.29 Hydrant valves may be adjusted to serve several purposes, both independently and simultaneously.

Figure 15.30 Double-male and double-female hose adapters enable the coupling of hose lengths with similar threads and the same inside diameter.

Figure 15.31 A reducer allows a smaller-diameter hose to connect directly to a larger size hose.

Figure 15.32 A hard strainer may be held off the bottom of an open surface water source by a rope attached to an anchor point.

Hose Tools

Hose tools are used to protect, move, handle, store, and connect hose. Some of the more common hose tools include:

- Hose roller
- Hose jacket
- Hose clamp
- Spanner wrench, hydrant wrench, and rubber mallet
- Hose bridge or ramp
- Chafing block
- Hose strap, hose rope, and hose chain
- LDH roller for loading

Hose Roller

Hose rollers protect hose from the mechanical damage of dragging hose over sharp corners such as roof edges and windowsills. This device consists of a metal frame with two or more rollers. The notch of the frame is placed over the potentially damaging edge or windowsill, and the frame is secured with a rope or clamp. The hose is then pulled across the rollers. The hose roller can also be used to protect rope when hoisting tools over similar edges (**Figure 15.33**).

Hose Jacket

When a section of hose ruptures, the entire hoseline is out of service until the section is replaced or the rupture is temporarily closed. When a hoseline ruptures but must remain charged to continue fire attack, a hose jacket can sometimes be installed at the point of rupture. A hose jacket consists of a hinged two-piece metal cylinder. The rubber lining of each half of the cylinder seals the rupture to prevent leakage. A locking device clamps the cylinder closed when in use. Hose jackets are made in two sizes: 2½ inches and 3 inches (65 mm and 77 mm). The hose jacket encloses the hose so effectively that it can continue to operate at full pressure (**Figure 15.34**). A hose jacket can also be used to connect hose with mismatched or damaged screw-thread couplings.

Figure 15.33 Edge rollers protect hoselines from mechanical damage when crossing sharp drops such as windowsills and wall parapets.

Figure 15.34 Hose jackets may be used to allow a damaged hoseline to continue operating at full pressure despite a rupture or leak.

Hose Clamp

A hose clamp can be used to stop the flow of water in a hoseline for the following reasons:

- To prevent charging the hose bed during a forward lay from a hydrant
- To allow replacement of a burst section of hose without stopping the water supply
- To allow extension of a hoseline without stopping the water supply
- To allow extension of a charged hoseline

There are three types of hose clamps: screw-down, press-down, and hydraulic press **(Figure 15.35)**. Unless applied correctly, a hose clamp can injure firefighters or damage the hose. Some general rules that apply to hose clamps are as follows:

- Apply the hose clamp at least 20 feet (6 m) behind the apparatus.
- Apply the hose clamp approximately 5 feet (1.5 m) from the coupling on the supply side.
- Center the hose evenly in the jaws to avoid pinching the hose.
- Close and open the hose clamp slowly to avoid water hammer.
- Stand to one side when applying or releasing any type of hose clamp (the operating handle or frame can snap open suddenly).

Figure 15.35 Hose clamps stop the flow of water in a hoseline.

⚠ CAUTION

Never stand over the handle of a hose clamp when applying or releasing it. The handle or frame may pop open and swing upward violently.

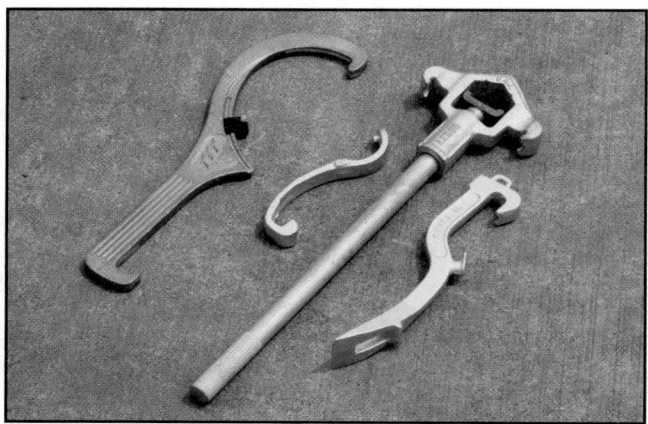

Figure 15.36 Types of tools used in conjunction with hoselines should be kept on hand when they might be needed. Specialized tools may have more than one function built into them.

Spanner, Hydrant Wrench, and Rubber Mallet

Some of the most common tools used to tighten or loosen hose couplings are the spanner wrench, **hydrant wrench**, and the rubber mallet (**Figure 15.36**). Although the primary purpose of the spanner wrench is to tighten or loosen couplings, a number of other features have been built into some spanner wrenches:

- Wedge for prying
- Opening that fits gas utility valves
- Slot for pulling nails
- Flat surface for hammering

Hydrant wrenches are primarily used to remove discharge caps from fire hydrant outlets and to open fire hydrant valves. The hydrant wrench is usually equipped with a pentagonal opening in its head that fits most standard fire hydrant operating nuts. The lever handle may be threaded into the operating head to make it adjustable, or the head and handle may be of the ratchet type. The head may also be equipped with a spanner to help make or break coupling connections.

A rubber mallet is sometimes used to strike the lugs to tighten or loosen intake hose couplings. Even though intake hose couplings have long operating lugs, using a rubber mallet to further tighten the intake connection makes it easier for firefighters to achieve a completely airtight connection with a good seal when setting up a drafting operation.

Hose Bridge or Ramp

Hose bridges or ramps help prevent damage to fire hose when vehicles must drive over it. They should be used wherever a hoseline is laid across a street or other area where it may be driven over. Hose ramps can also be positioned over small spills to keep hoselines from being contaminated, and they can be used as chafing blocks (**Figure 15.37**).

Chafing Block

Charged hoselines vibrate and rub against other surfaces which can cause abrasions. Chafing blocks are devices used to protect fire hose from these abrasions. Chafing blocks are particularly useful near pumpers where intake hose comes in contact with pavement or curbs because vibrations from the pumper may keep the intake hose in constant motion. Chafing blocks may be made of wood, leather, or sections of old truck tires (**Figure 15.38**).

Hose Bridges

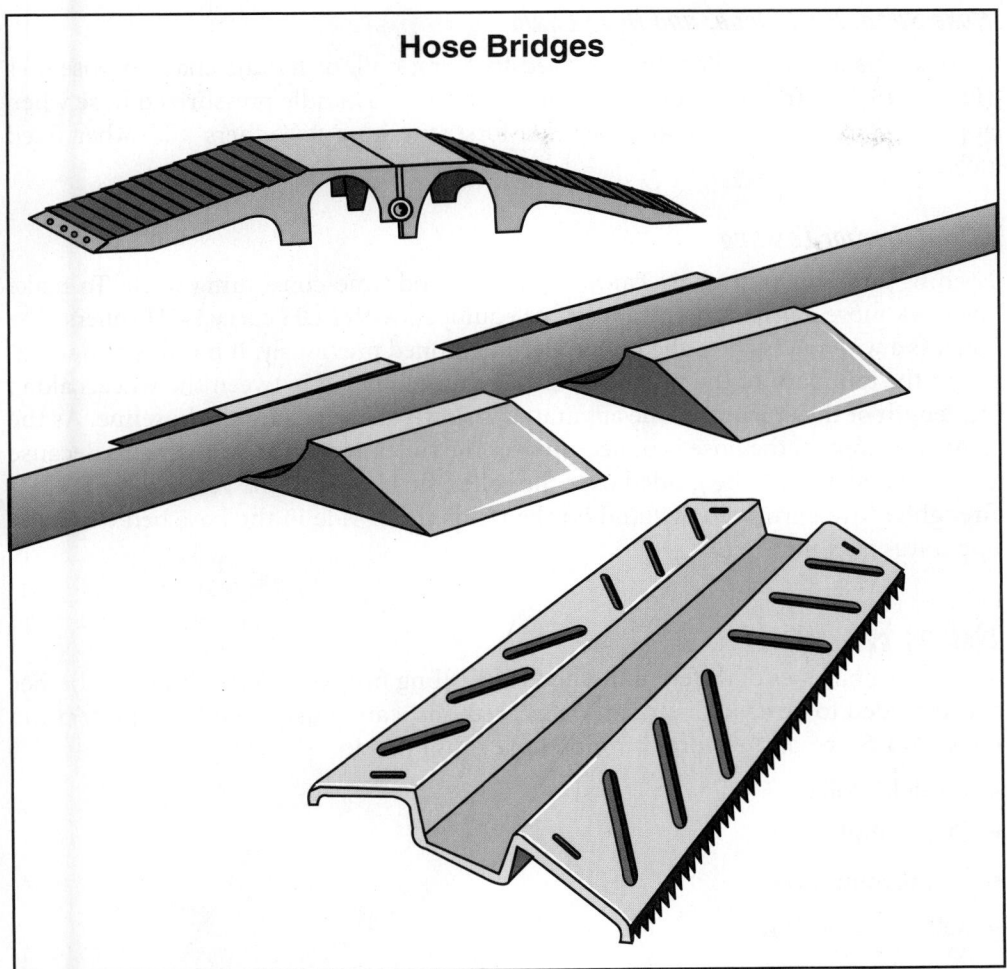

Figure 15.37 Hose bridges provide a channel for the hose to pass through to prevent the hose from becoming contaminated or damaged.

Chafing Blocks

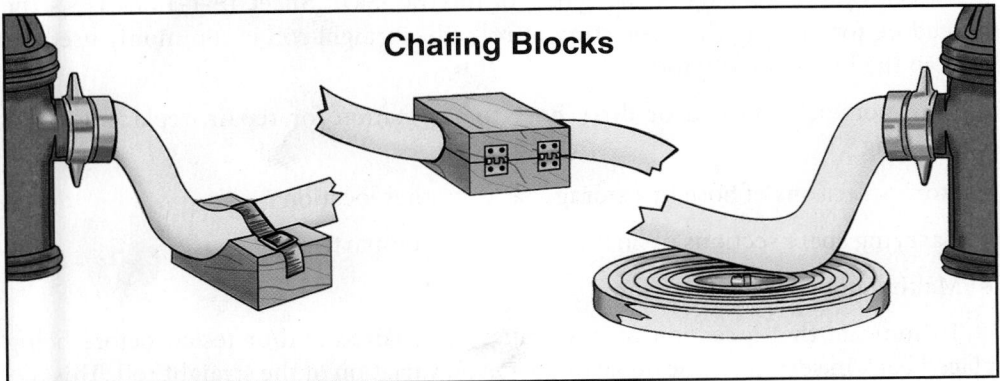

Figure 15.38 Chafing blocks protect the hoseline from abrasion during routine use.

Figure 15.39 Hose straps can be used to reduce stress on the nozzle operator and improve the mobility of a charged hoseline.

Hose Strap, Hose Rope, and Hose Chain

Hose straps, ropes, and chains are used to carry, pull, or handle charged hoselines (**Figure 15.39**). They provide a more secure means to handle pressurized hose when applying water. They may also be used to secure hose to ladders and other fixed objects.

LDH Roller for Loading

Loading large-diameter hose can be laborious and time-consuming work. To make the work more efficient, many apparatus equipped with LDH carry LDH rollers. The roller is a wider version of the hose roller mentioned previously. It mounts temporarily on the tailboard of the pumper. With the hoseline laid between the wheels along the length of the apparatus, the apparatus is slowly driven along the hoseline. As the apparatus moves, the hose is pulled up over the roller and into the hose bed. Because the hose may need to be guided over the roller, this is one of the very few times that firefighters are permitted to stand on the tailboard or ride in the hose bed while the apparatus is in motion.

Hose Rolls

There are a number of different methods for rolling fire hose, depending on whether it is intended to be used or stored. In all methods, care must be taken to protect the couplings. Some of the more common hose rolls include:

- Straight Roll
- Donut Roll
- Twin Donut Roll
- Self-Locking Twin Donut Roll

Straight Roll

The single-section straight roll is the simplest of all hose rolls. The roll is usually made by starting at the male coupling end and rolling toward the female coupling end of the hose (**Figure 15.40**). When the roll is completed, the female end is exposed and the male end is protected in the center of the roll. **Skill Sheet 15-I-3** describes the procedure for making the basic straight roll. The straight roll is commonly used for hose in the following situations:

- Transporting damaged or dirty hose to the station for repair, replacement, or cleaning
- Storing sections of hose in a storage rack or other location
- Carrying spare sections of hose in apparatus compartments
- Making hose loading easier

To indicate that a section of hose must be repaired and/or tested before being placed back in service, some departments use a variation of the straight roll. This roll is begun at the female coupling so that when the roll is completed, the male coupling is exposed. Another method is to tie a knot in the exposed end or attach a tag indicating the type and location of damage. This variation of the straight roll is also used when the hose is going to be reloaded on the apparatus for a forward (straight) lay.

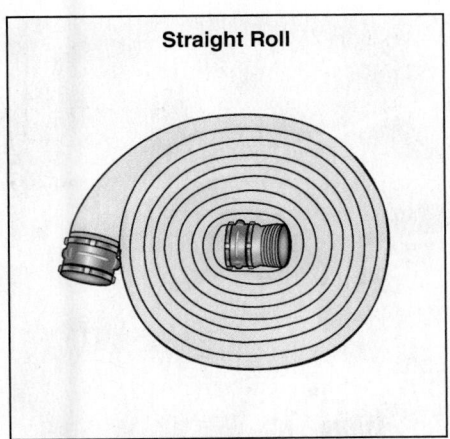

Figure 15.40 A straight roll features a simple coil of hose beginning with the male end on the center and the female at the outside.

Figure 15.41 A donut roll begins with the center of the hose in the middle and the male and female couplings at the outside at opposite ends of the roll.

Figure 15.42 A twin donut roll is primarily used on small-diameter hose for the purpose of easy transport.

Donut Roll

The *donut roll* is commonly used in situations where hose is likely to be deployed for use directly from a roll **(Figure 15.41)**. One or two firefighters can perform a donut roll as described in **Skill Sheet 15-I-4.**

The donut roll has certain advantages over the straight roll:

- The firefighter has control of both couplings, which protects them from damage.
- The hose rolls out easier with fewer twists or kinks.
- Holding both couplings enables a quicker connection to other couplings.

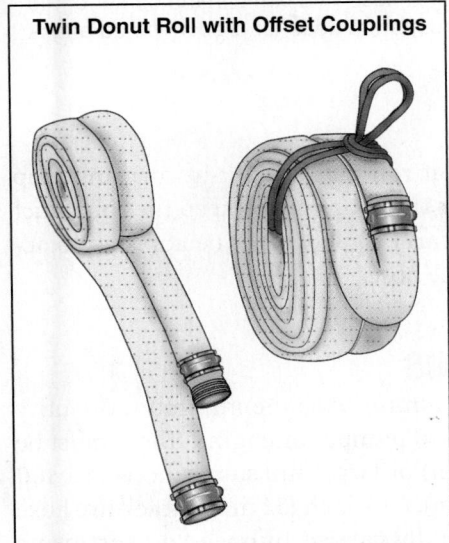

Figure 15.43 The twin donut roll with offset couplings allows a firefighter to carry the roll hands-free using a hose strap.

Twin Donut Roll

The twin donut roll usually works best on 1½-inch (38 mm) and 1¾-inch (45 mm) hose, although 2-, 2½-, or 3-inch (50 mm, 65 mm, or 77 mm) hose can also be rolled in this manner **(Figure 15.42)**. The purpose of this hose roll is to create a compact roll that can be easily transported and carried for special applications such as high-rise or standpipe operations. **Skill Sheet 15-I-4** also describes a method used to make the twin donut roll.

If the couplings are offset by about 1 foot (0.3 m) at the beginning, they can be coupled together after the roll is tied or strapped. A hose strap, inserted into the center of the roll, is used to carry the hose roll **(Figure 15.43)**.

Self-Locking Twin Donut Roll

Figure 15.44 A self-locking twin donut roll uses a length of hoseline to secure the shape of the roll and to allow the roll to be carried using the loop.

Figure 15.45 Some apparatus hose beds feature a split hose load divided by a hose bed baffle in the center of the apparatus and two separate beds for preconnected attack lines on either side.

Self-Locking Twin Donut Roll

The self-locking twin donut roll is a twin donut roll with a built-in carrying loop formed from the hose itself. This loop locks over the couplings to keep the roll intact for carrying (**Figure 15.44**). The length of the carrying loop may be adjusted to accommodate the height of the person carrying the hose.

Basic Hose Loads and Finishes

NFPA® 1901, *Standard For Automotive Fire Apparatus,* lists the minimum quantity of hose in various sizes to be carried on a standard pumper or engine. There must be a minimum 800 feet (240 m) of 2½-inch (65 mm) or larger fire supply hose and 400 feet (120 m) of 1½-inch (38 mm), 1¾-inch (45 mm), or 2-inch (52 mm) attack fire hose (**Figure 15.45**). Supply and attack hose is generally carried in open compartments called **hose beds**. Hose beds vary in location, size, and shape, and are sometimes built for specific needs. The front of the hose bed is that part of the compartment closest to the front of the apparatus, and the rear of the hose bed is that part of the compartment closest to the rear of the apparatus. Most hose beds have open aluminum slats in the bottom that allow air to circulate throughout the hose load to prevent mildew damage to woven-jacketed hose.

Hose Bed — Main hose-carrying area of a pumper or other apparatus designed for carrying hose.

Apparatus hose beds may be a single compartment, or they may be divided or separated by a vertical panel that runs from the front to the rear of the hose compartment. This division creates a split hose bed allowing the apparatus to have hose loaded that can be deployed as a single or double supply line or for both forward and reverse hose lays at the same time. Hose in a split bed should be stored so that both beds may be connected when a long hose lay is required.

The three most common loads for supply hoselines are the *flat*, *accordion*, and *horseshoe*. Hose loads may also have a **finish**: an additional section connected to the hose load and arranged on the top of the load, which can be rapidly deployed for forward or reverse supply hose lays or as an attack line.

Additional supply and attack hose may be carried in compartments on the apparatus. A common type of hose bundle is the high-rise pack, described later in this section.

Flat Load

Of the three supply hose loads, the **flat load** is the easiest to load. It is suitable for any size of supply hose and is the best way to load large-diameter hose. As the name implies, the hose is laid so that its folds lie flat rather than on edge (**Figure 15.46**). Hose loaded in this manner is less subject to wear from apparatus vibration during travel. A disadvantage of this load is that the hose folds contain sharp bends at both ends of the bed, which requires that the hose be reloaded periodically to change the location of the bends within each length to prevent damage to the lining.

In a single hose bed, the flat load may be started on either side. In a split hose bed, lay the first length against the partition with the coupling hanging far enough below the hose bed so that the coupling can be connected to the last coupling of the load

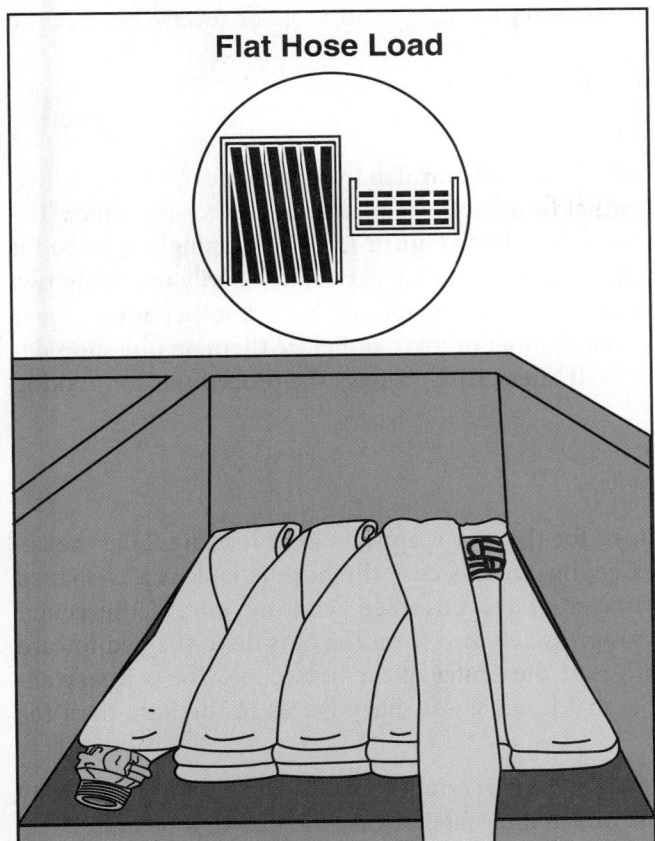

Flat Hose Load

Figure 15.46 A flat hose load is the easiest to load because the hose is laid flat instead of on its edge.

Finish — Arrangement of hose usually placed on top of a hose load and connected to the end of the load.

Flat Load — Arrangement of fire hose in a hose bed or compartment in which the hose lies flat with successive layers one upon the other.

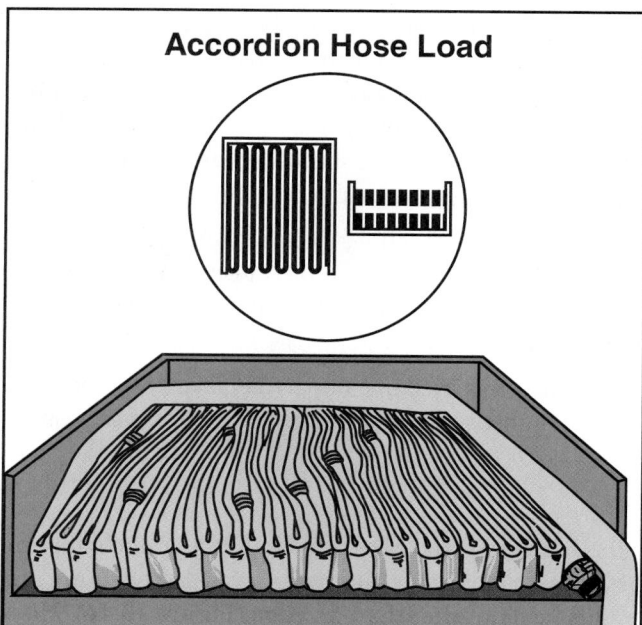

Figure 15.47 An accordion hose load is laid on edge with the first coupling in the rear of the bed.

on the opposite side and laid on top of the load. This placement allows the couplings to be easily disconnected when the load must be divided to lay dual lines. How far the coupling needs to hang is based on your estimate of the anticipated height of the hose bed. **Skill Sheet 15-I-5** demonstrates the proper method for making a flat load.

The hose load for large-diameter hose should be started 12 to 18 inches (300 mm to 450 mm) from the front of the hose bed. This extra space should be reserved for couplings, and all couplings should be laid in a manner that allows them to deploy without flipping over. It may be necessary to make a short fold or reverse bend in the hose to do this.

Accordion Load

Accordion Load — Arrangement of fire hose in a hose bed or compartment in which the hose lies on edge with the folds adjacent to each other.

The **accordion load** is named for the manner in which the hose appears after loading. The hose is laid on edge in folds that lie adjacent to each other (accordion-like). The first coupling is placed in the rear of the bed (**Figure 15.47**). In a single hose bed it can be placed in either corner. An accordion load is easy to load, only requiring two or three people to load the hose, although four people are best. Another advantage is that firefighters can easily pick up a number of folds and place them on one shoulder to carry the hose from the bed. **Skill Sheet 15-I-6** shows the procedures for making an accordion load.

Horseshoe Load

Horseshoe Load — Arrangement of fire hose in a hose bed or compartment in which the hose lies on edge in the form of a horseshoe.

The **horseshoe load** is also named for the way it appears after loading. Like the accordion load, it is loaded on edge, but in this case the hose is laid in a U-shaped configuration around the perimeter of the hose bed working toward the center (**Figure 15.48**). Each length is progressively laid from the outside of the bed toward the inside so that the last length is at the center of the horseshoe. The primary advantage of the horseshoe load is that it has fewer sharp bends in the hose than the accordion or flat loads.

Horseshoe loads in single hose beds have certain disadvantages. Excess hose may be deployed because the hose is pulled alternately from one side of a bed and then the other creating a wavy or snakelike lay. Another disadvantage is that folds for a

Horseshoe Hose Load

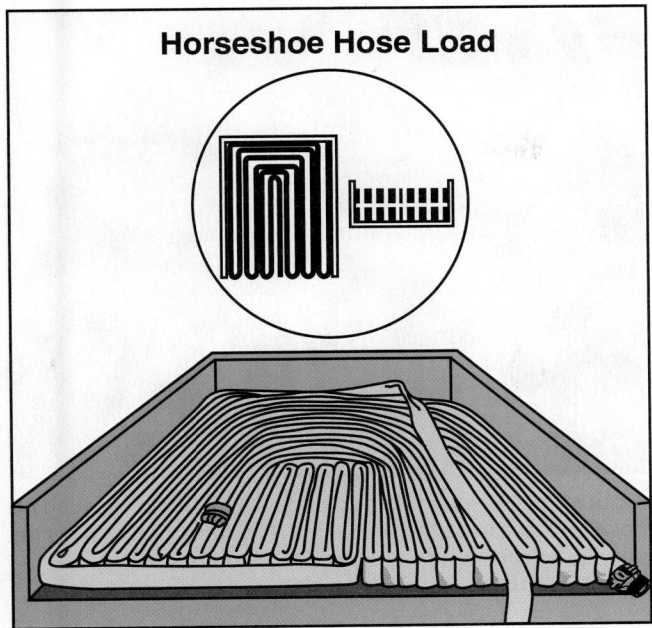

Figure 15.48 A horseshow hose load features a series of steadily-smaller passes of the hose, starting by lining the perimeter of the hose bed and then working toward the center.

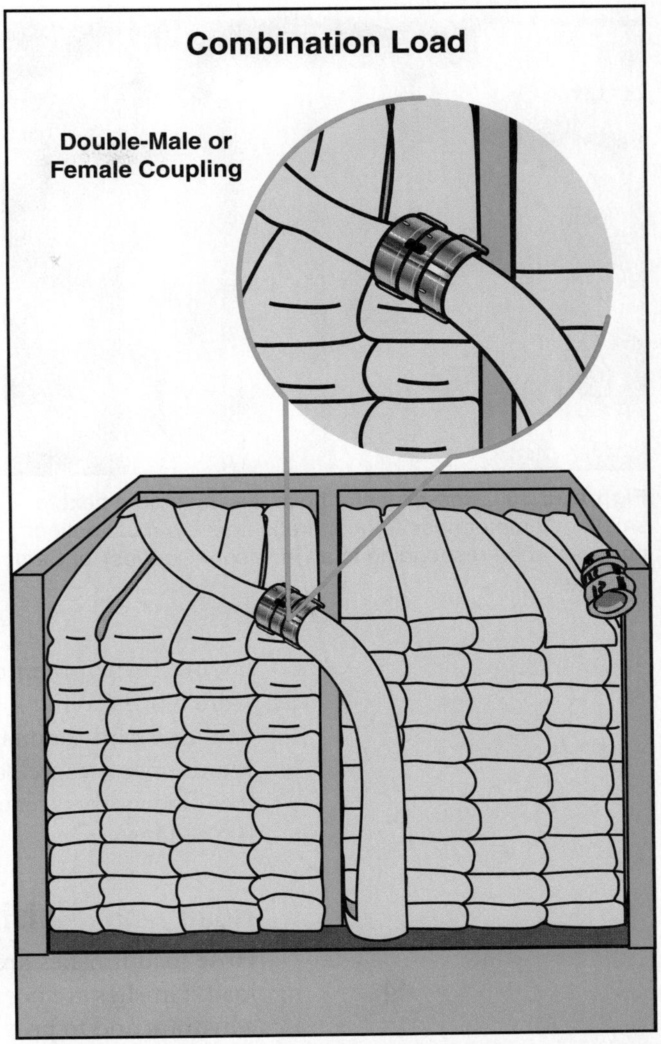

Combination Load

Double-Male or Female Coupling

Figure 15.49 A combination load allows an apparatus to perform a forward lay from the water source by exposing the female coupling on one half of the bed and the male coupling on the other half.

shoulder carry cannot be pulled as easily as they can with an accordion load. With the horseshoe load, two people are required to make the shoulder folds for the carry. As is the case with the accordion load, the hose is loaded on edge, which can result in wear on hose edges. The horseshoe load does not work for large-diameter hose because the hose remaining in the bed tends to fall flat as the hose is deployed, which can cause the hose to become entangled.

In a single hose bed, the horseshoe load may be started on either side. In a split hose bed, lay the first length against the partition with the coupling hanging far enough below the hose bed so that the coupling can be connected to the last coupling of the load on the opposite side and laid on top of the load. This placement allows the couplings to be easily disconnected when the load must be divided to lay dual lines. How far the coupling needs to hang is based on your estimate of the anticipated height of the hose bed. **Skill Sheet 15-1-7** describes the procedures for making the horseshoe hose load.

Combination Load

Combination loads are used with split hose beds that are loaded with threaded-coupling hose. This load permits the apparatus to make a forward lay from the water source to the fire followed by a reverse lay back to the water source. One half of the bed is loaded with the female coupling exposed and other half has the male coupling exposed (**Figure 15.49**). Where the two beds are connected, a double-female adapter fitting is used. This load can be a flat, accordion, or horseshoe load.

Loading large-diameter supply hose on one side of the bed and smaller-diameter hose that can be used for either supply or attack on the other side is another version of a combination load (**Figure 15.50, p. 850**). A pumper loaded in this manner can

Figure 15.50 Large-diameter hose may be stored next to smaller-diameter hose to offer firefighters the most options possible when responding to an incident. *Courtesy of Sam Goldwater.*

Figure 15.51 A straight finish consists of the last section of hose folded across the hose bay. A hydrant wrench may be used to create neat folds by holding the hose in place.

lay LDH when the fire situation requires the pumper to lay its own supply line and work alone (laying it forward so the pumper stays at the incident scene). Firefighters can use small-diameter hose as a supply line at fires with less demanding water flow requirements as well as for attack lines on large fires. Therefore, a split hose bed gives fire officers the greatest number of choices when determining the best way to use limited resources.

Hose Load Finishes

Hose load finishes are added to the basic hose load to increase the versatility of the load. Finishes are normally loaded to provide enough hose to connect the hoseline to a hydrant and to provide an attack hoseline at the fire scene.

Finishes fall into two categories: finishes used for forward hose lays (straight finish) and finishes used for reverse hose lays (reverse horseshoe and skid load finishes). Finishes for forward lays facilitate making a hydrant connection and are not as elaborate as finishes for reverse lays. Finishes for reverse lays provide an adequate amount of hose at the scene for initial fire attack. **Skill Sheet 15-I-8** describes the steps for making each of these types of finishes.

Straight Finish

A straight finish consists of the last section of hose arranged loosely back and forth across the top of the hose load **(Figure 15.51)**. A hydrant wrench, gate valve, and any necessary adapters are usually strapped to the hose at or near the female coupling.

Reverse Horseshoe Finish

This finish is similar to the horseshoe load except that the bottom of the U portion of the horseshoe is at the rear of the hose bed. It is made of one or two 100-foot (30 m) sections of hose, each connected to one side of a gated wye **(Figure 15.52)**. Any size attack hose can be used: 1½, 1¾, or 2½ inches (38 mm, 45 mm, or 65 mm). The smaller sizes require a 2½- × 1½-inch (65 mm by 38 mm) gated wye. The 2½-inch (65 mm) hose requires a 2½- × 2½-inch (65 mm by 65 mm) gated wye. A nozzle of the appropriate size is also needed for each attack line.

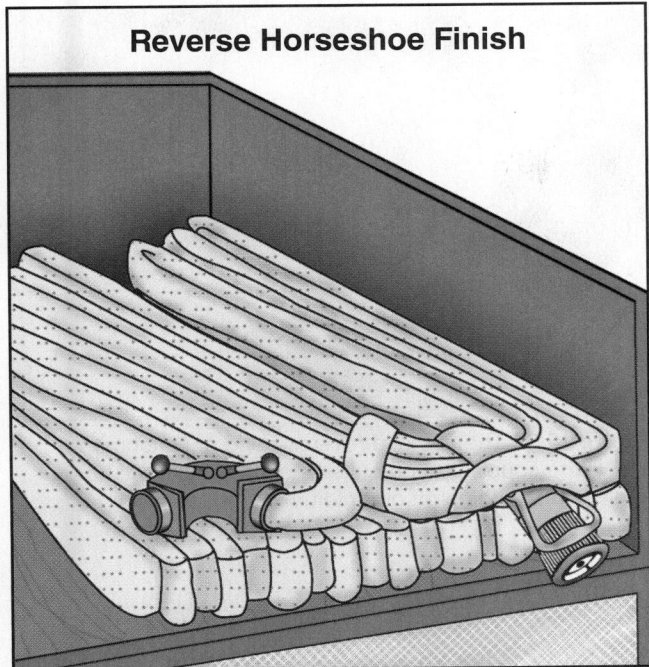

Reverse Horseshoe Finish

Figure 15.52 A reverse horseshoe finish has the closed portion of the horseshoe shape in the front of the hose bed.

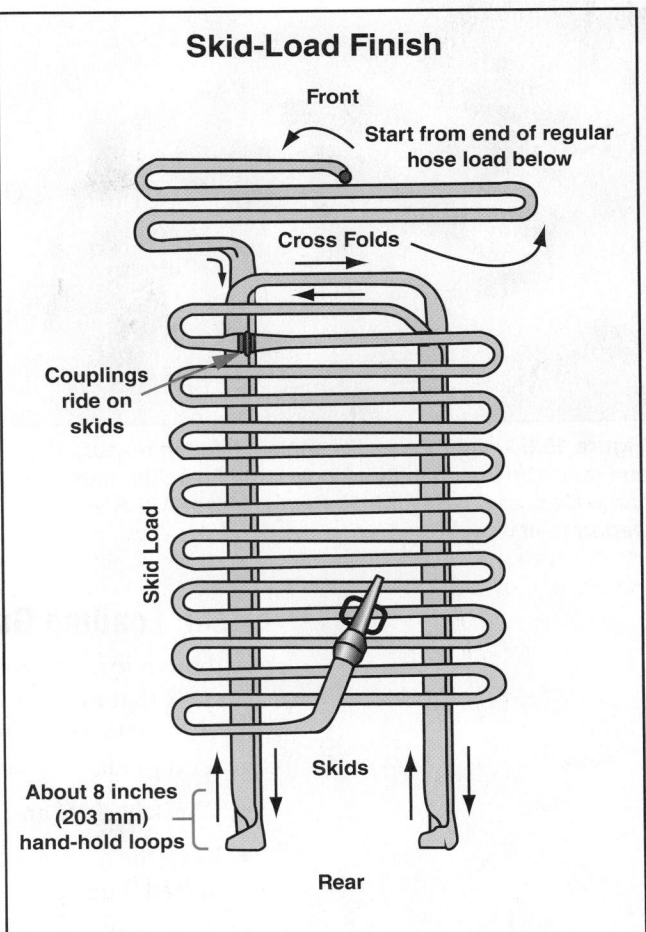

Skid-Load Finish

Front

Start from end of regular hose load below

Cross Folds

Couplings ride on skids

Skid Load

About 8 inches (203 mm) hand-hold loops

Skids

Rear

Figure 15.53 A skid-load finish features a buffer zone between the regular hose load below and the couplings of the upper load.

The reverse horseshoe finish can also be used for a preconnected line and can be loaded in two or three layers. With the nozzle extending to the rear, firefighters can place the finish over one shoulder and extend the opposite arm through the loops of the layers to pull the hose from the bed for an arm carry. A second preconnected line can be located in the hose bed below when there is sufficient depth.

Skid Load Finish

A skid load finish consists of folding the last three sections (150 feet [45 m] of 2½-inch [65 mm] hose) into a compact bundle on top of the rest of the hose load. The load begins by forming three or more pull loops that extend beyond the end of the hose load. The rest of the hose, with nozzle attached, is accordion-folded across the hose used to form the pull loops in the hose bed (**Figure 15.53**).

High-Rise Pack

High-rise packs are assembled to provide enough attack hose for firefighters to operate from a standpipe connection and still be light enough for one person to carry. The packs may be located in a compartment or secured to the apparatus exterior. The packs may be carried in a roll or strapped together in an accordion-like fashion (**Figure 15.54, p. 852**). A nozzle may be attached to the male coupling and adapters and tools may be carried separately in a bag.

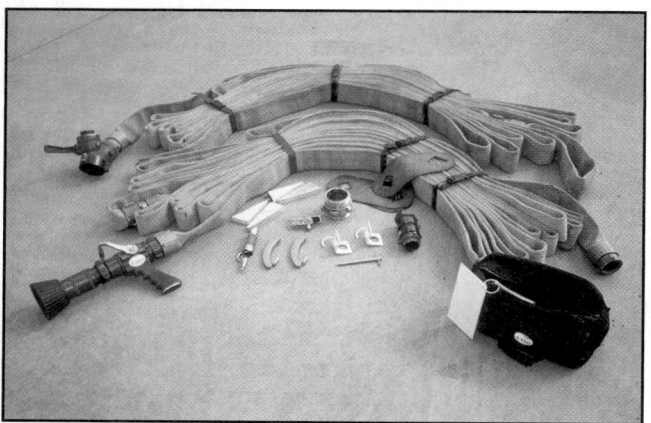

Figure 15.54 High-rise packs contain needed resources and minimize the quantity of tools each firefighter must carry. *Courtesy of Rick Montemorra, Mesa (AZ) Fire Department.*

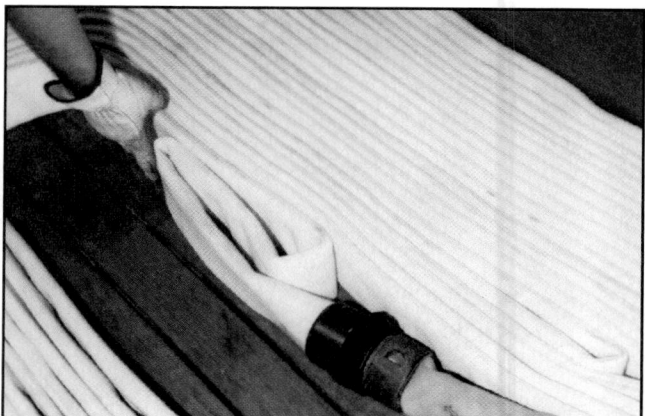

Figure 15.55 The use of a reverse bend or Dutchman allows the coupling to be pulled smoothly from the hose bed without binding on other folds in the hose.

Hose Loading Guidelines

Although loading hose on fire apparatus is not an emergency, operation, it is a critical task that must be done correctly. During an emergency, properly loaded hose can be efficiently and effectively deployed for supply or attack operations. The following general guidelines should be followed, regardless of the type of hose load used:

- Check gaskets and swivel before connecting any coupling.

- Keep the flat sides of the hose in the same plane when two sections of hose are connected. Lugs on the couplings do not need to be aligned.

- Tighten the couplings hand tight. Never use wrenches or excessive force.

- Remove kinks and twists from fire hose when it is bent to form a loop in the hose bed.

- Make a short fold or reverse bend, called a **Dutchman**, in the hose during the loading process so that couplings are not too close to the front or rear of the hose bed and will not flip over when pulled out of the bed. The Dutchman serves two purposes: (1) it changes the direction of a coupling and (2) it changes the location of a coupling. The reverse bends should not be over used in the same layer because it can result in couplings becoming wedged in the bed (**Figure 15.55**).

- Load large-diameter hose (3½-inch [90 mm] or larger) with all couplings near the front of the bed. This saves space and allows the hose to lie flat.

- Do not pack hose too tightly. This puts excess pressure on the folds of the hose and may cause couplings to snag when the hose pays out of the bed. A general rule is that the hose should be loose enough to allow a gloved hand to be easily inserted between the folds.

Dutchman — Extra fold placed along the length of a section of hose as it is loaded so that its coupling rests in proper position.

Preconnected Hose Loads for Attack Lines

Preconnected hoselines, called simply **preconnects**, are the primary lines most fire departments use for fire attack. These hoselines are connected to a discharge valve and placed in an area other than the main hose bed. Preconnected hoselines generally range from 50 to 250 feet (15 m to 75 m) in length. Preconnected attack lines can be carried in the following places (**Figures 15.56a-d**):

Preconnect — Attack hose connected to a discharge when the hose is loaded; this shortens the time it takes to deploy the hose for fire fighting.

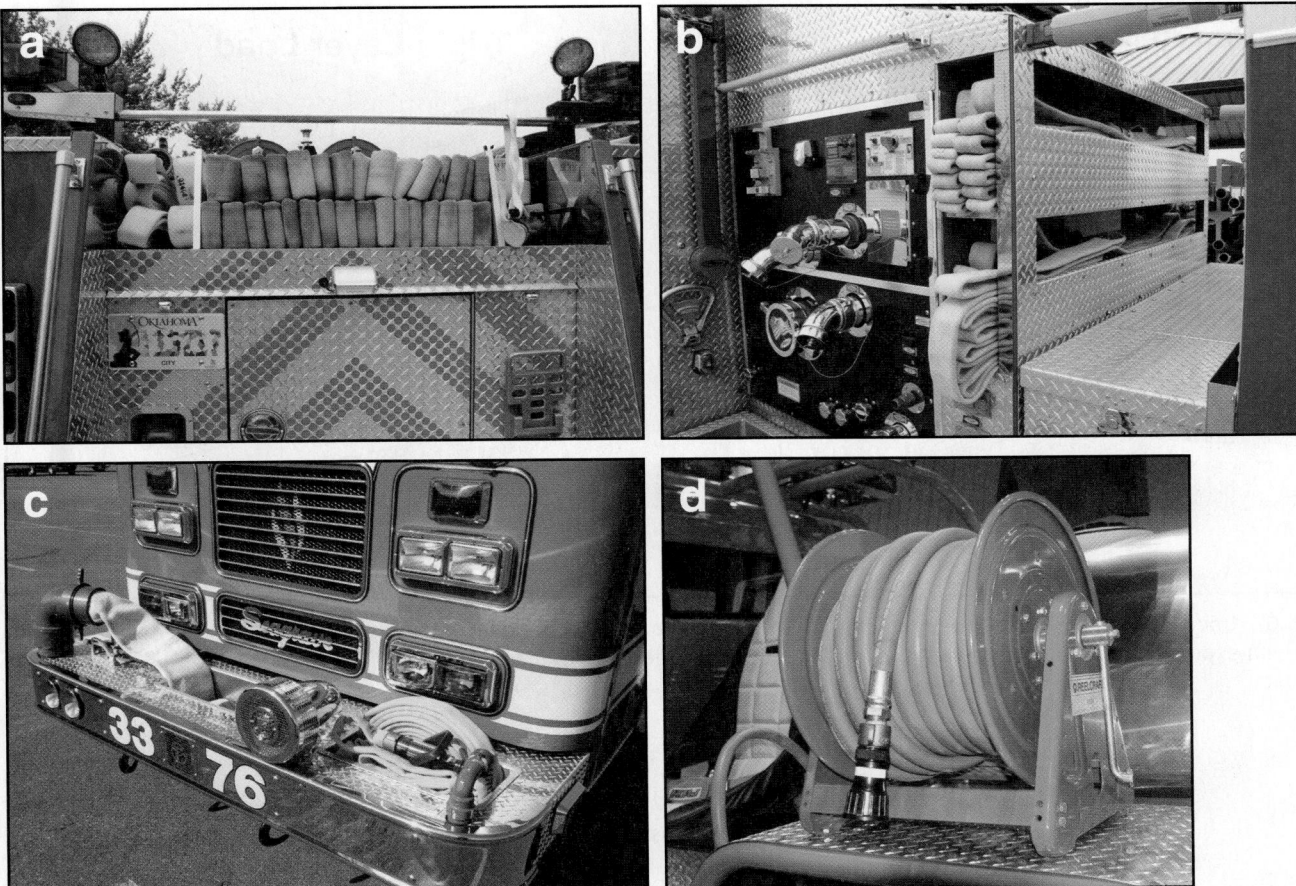

Figure 15.56a-d Preconnected hoselines may be loaded in a) longitudinal beds similar to main hose beds, b) transverse compartments, c) front bumper wells, and d) reels.

- Longitudinal beds
- Raised trays
- Transverse beds (sometimes called cross lays or Mattydale hose beds)
- Tailboard compartments
- Side compartments or bins
- Front bumper wells (sometimes called jump lines)
- Reels

The sections that follow describe some of the more common loads for preconnects. Special loads to meet local requirements may be developed based on individual experiences and apparatus configurations. Regardless of the type of load used, the preconnected attack hose must be fully deployed from the hose bed before the line is charged. In some departments this is referred to as "clearing the bed" and is the responsibility of the driver/operator or firefighter pulling the hose.

Preconnected Flat Load

The preconnected flat load is adaptable for varying sizes of hose beds and is often used in transverse beds. This load is similar to a flat load for larger supply hose except that exposed loops are provided for pulling the load from the bed (**Figure 15.57, p. 854**). Place the loops at regular intervals within the load so that equal portions of the

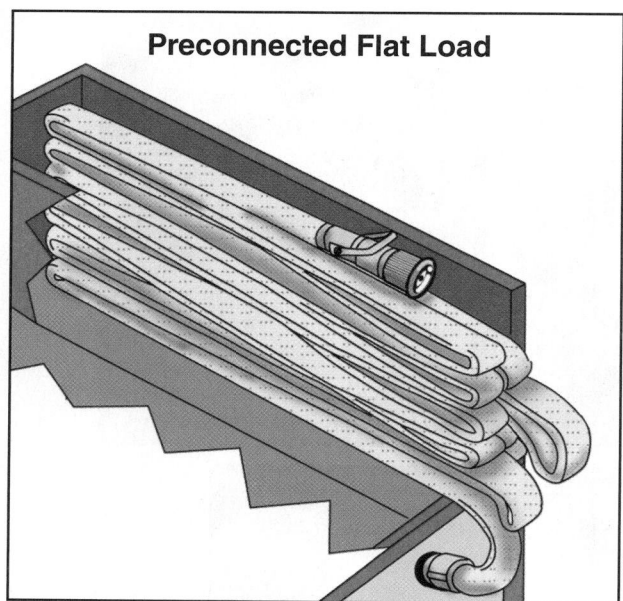

Preconnected Flat Load

Figure 15.57 Preconnected flat loads are commonly arranged with exposed loops to aid in pulling the load from the bed.

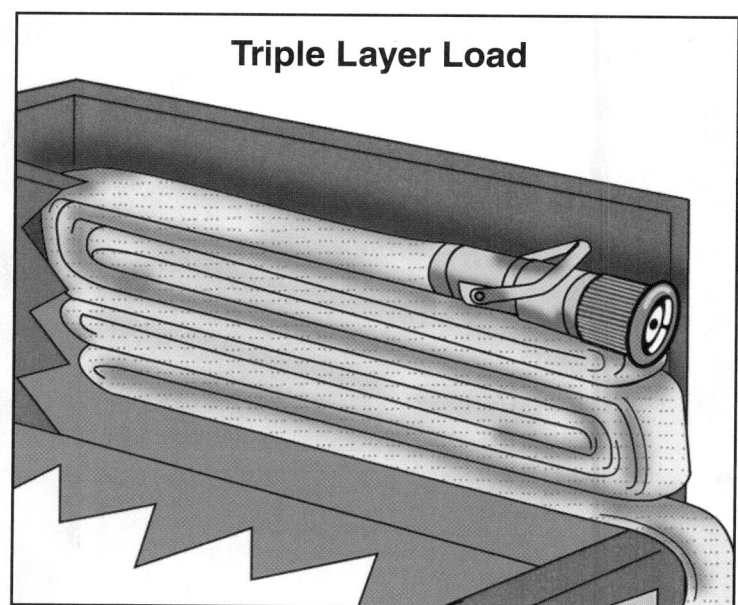

Triple Layer Load

Figure 15.58 A triple layer load is an S-shaped load designed to be pulled by one person.

load are pulled from the bed. The number of loops and the intervals at which they are placed depend on the size and total length of the hose. The procedures for creating the preconnected flat load are described in **Skill Sheet 15-I-9.**

Triple Layer Load

The triple layer load gets its name because the load begins with hose folded in three layers. The three folds are then laid into the bed in an S-shaped fashion (**Figure 15.58**). The load is designed to be pulled by one person.

The layers in a triple layer load may be as long as 50 feet (15 m) each. All of this hose must be completely removed from the bed before deploying the nozzle end of the hose which can be difficult or impossible if the space directly behind the hose bed is restricted. While this hose load can be used for all sizes of attack lines, it is often preferred for larger (2- and 2½-inch [50 mm and 65 mm]) attack lines that may be too cumbersome for shoulder carries. The procedures for making the triple layer load are given in **Skill Sheet 15-I-10.**

Minuteman Load

The minuteman load is designed to be pulled and advanced by one person. This load can be carried on the shoulder, completely clear of the ground, which makes it less likely that the hose will catch on obstacles. The load deploys from the shoulder as the firefighter advances toward the fire. The load is also particularly well-suited for a narrow hose bed (**Figure 15.59**). However, this load can be awkward to carry when wearing an SCBA. If the load is in a single stack, it may also collapse on the shoulder if not held tightly in place. The procedures for making the minuteman load are described in **Skill Sheet 15-I-11.**

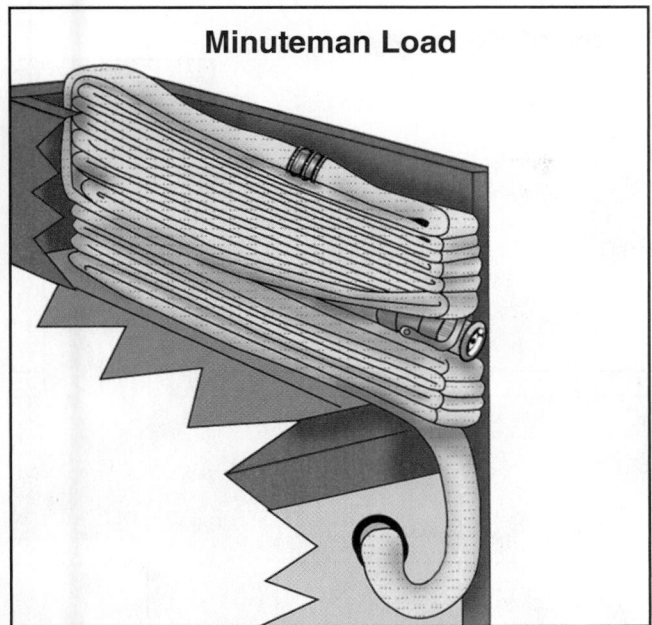

Figure 15.59 The minuteman load is designed to maximize the efficiency of narrow hose beds.

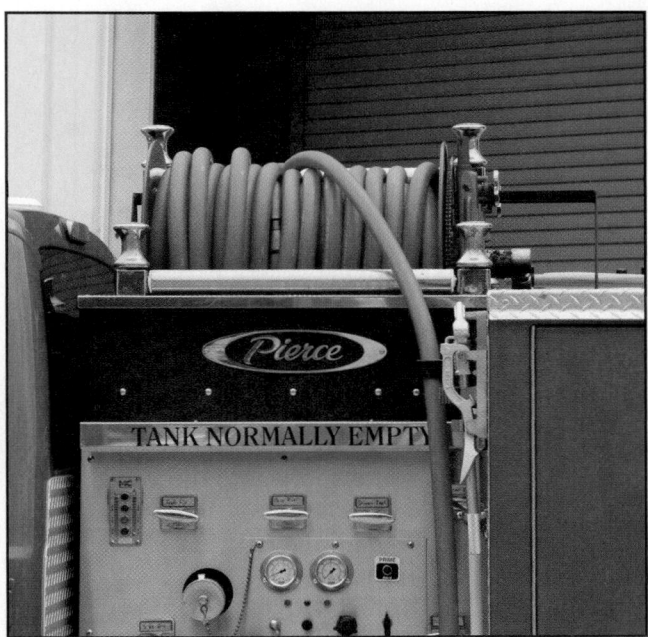

Figure 15.60 Booster hoselines are usually preconnected and may be mounted above the pump panel.

Booster Hose Reels

Booster hoselines are rubber-covered hose that are usually carried preconnected and coiled on reels. These booster hose reels may be mounted in any of several places on the apparatus according to specified needs and the design of the apparatus. Some booster hose reels are mounted above the pump panel and behind the apparatus cab (**Figure 15.60**). This arrangement provides booster hose that can be unrolled from either side of the apparatus. Other booster hose reels are mounted on the front bumper of the apparatus or in rear compartments. Manual- and power-operated reels are available. In order to load the maximum amount of hose with the easiest removal from the reel, booster hose should be wound onto the reel one layer at a time in an even manner.

> **Booster Hoseline —**
> Noncollapsible rubber-covered, rubber-lined hose usually wound on a reel and mounted somewhere on the apparatus and used for extinguishment of incipient and smoldering fires. This hose is most commonly found in ½-, ¾-, and 1-inch (13 mm, 19 mm, and 25 mm) diameters and is used for extinguishing low-intensity fires and overhaul operations.

> **WARNING**
> Booster lines are not appropriate for interior fire fighting operations or for vehicle fires because they do not deliver a sufficient volume of water to protect firefighters if conditions suddenly deteriorate.

Supply Hose Lays

The three basic hose lays for supply hose are forward lay, reverse lay, and combination lay. Hose lay procedures vary from department to department, but the basic methods of laying hose remain the same. Hose is either laid forward from a water source to the incident scene, reverse from the incident scene to a water source, or in combination with one hoseline laid from the water source to the incident scene and one back.

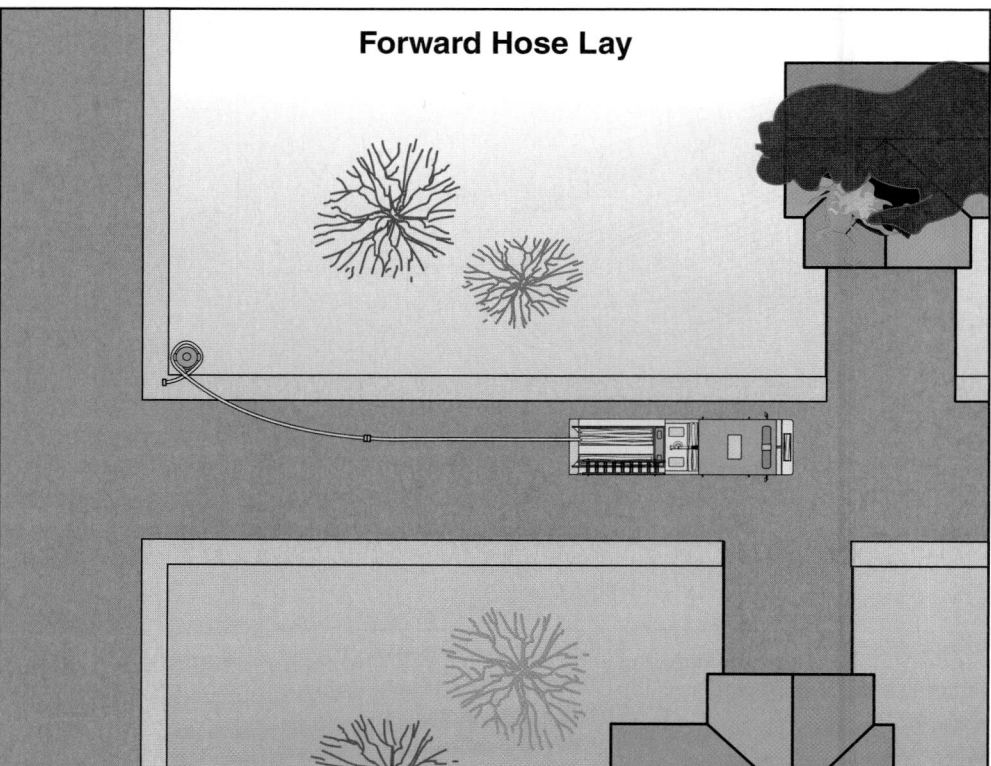

Figure 15.61 A forward hose lay allows the driver/operator to place a hose in operation using the forward movement of the apparatus.

Regardless of the method chosen, use the following guidelines when laying hose:

- Do not ride in a standing position when the apparatus is moving.
- Drive no faster than 10 mph (16 km/h) – the slower speed allows couplings to clear the tailboard as the hose leaves the bed.
- Deploy the hose to one side of the roadway (but not in the gutter) so that other apparatus are not forced to drive over it.

Forward Lay

In a forward hose lay, hose is deployed from the water source to the incident such as when the water source is a hydrant and the pumper must be positioned near the fire **(Figure 15.61)**. The first coupling to come off the hose bed in a forward lay should be female. Deploying from a forward lay consists of stopping the apparatus at the hydrant and allowing a firefighter to safely leave the apparatus and secure the hose. The apparatus then proceeds to the fire deploying either a single hoseline or parallel hoselines.

The primary advantage of a forward lay is that a pumper can remain at the incident scene so its hose, equipment, and tools are readily available if needed. The pump operator also has visual contact with the fire suppression operation and can better react to changes at the fire scene than if the pumper were at the hydrant.

If a long length of 2½- or 3-inch (65 mm or 77 mm) hose is laid, or if the hydrant has inadequate flow pressure, it may be necessary for a second pumper to be positioned at the hydrant to increase the pressure in the line. The first pumper in this scenario must have used a **four-way hydrant valve** if the transition from hydrant pressure to pump pressure is to be made without interrupting the flow of water in the supply hose. Another disadvantage is that one member of the crew is temporarily unavailable for a fire fighting assignment because that person must stay at the hydrant long enough to make the hose connection and open the hydrant.

Four-Way Hydrant Valve — Hose appliance that is attached to the hydrant to permit additional supply hoses to be attached without interrupting the flow of water.

The firefighter who is going to make the hydrant connection (also known as "catching the plug" or "making the hydrant") must know the following: (1) proper procedures for securing and connecting to the hydrant and (2) correct operation of the hydrant valve if one is used.

Making the Hydrant Connection

Local SOPs and resources dictate the method used for connecting the fire hose to the hydrant in a forward hose lay operation. In the simplest version, the firefighter takes a hydrant wrench, the finish section of hose, and preferably a portable radio with him or her when connecting to the hydrant. Departments that use the four-way hydrant valve may have the valve preconnected to the hose or in a bag stored near the finish section.

To ensure that water arrives quickly at the pump intake, some form of communication between the driver/operator and the firefighter at the hydrant is essential. The portable radio provides this link. If radios are not assigned to all personnel, some means of visual or audible signal must be established and practiced. The use of horns or sirens can be a problem when other apparatus are responding to the scene. At a minimum, the driver/operator and the firefighter making the connection should establish a time for opening the hydrant when the connecting firefighter can be sure that water will arrive after a hose clamp is placed on the hose but before the apparatus tank is empty.

The first task to be accomplished when starting a forward lay is for the hydrant catcher to remove enough hose to reach the hydrant and wrap around it. The finish section of hose is usually long enough to accomplish this task **(Figure 15.62)**. If not, place the finish section, along with the hydrant wrench, on the ground near the tailboard and pull a second section from the hose bed.

Figure 15.62 The hydrant catcher's task during a forward lay is to pull enough hose to reach and wrap around the hydrant.

Figure 15.63 Connecting a four-way hydrant valve allows one hoseline to charge immediately and leaves open the possibility for a later-arriving pumper to connect to the hydrant.

Next take the end of the finish section and wrap it around the hydrant base. The hydrant catcher can place a foot on the hoseline and against the hydrant to further anchor the hose. Then the firefighter signals the driver/operator that it is safe to proceed to the fire. The procedures for making a hydrant connection from a forward lay are given in **Skill Sheet 15-I-12**.

Using Four-Way Hydrant Valves

A four-way hydrant valve allows a forward-laid supply line to be immediately charged and allows a later-arriving pumper to connect to the hydrant (**Figure 15.63**). The second pumper can supply additional supply lines and/or increase the pressure to the original line. Typically, the four-way hydrant valve is preconnected to the end of the supply line. This allows the firefighter who is making the connection to secure the valve and the hose to the hydrant in one action.

Reverse Lay

When a pumper must first go to the fire location before laying a supply line, a reverse hose lay should be deployed from the incident scene to the water source (**Figure 15.64**). This deployment is also the most expedient way to lay hose if the apparatus that lays the hose must stay at the water source such as when drafting or boosting hydrant pressure to the supply line. Hose beds set up for reverse lays should be loaded so that the first coupling to come off the hose bed is male.

Laying hose from the incident scene to the water source has become a standard method for establishing a relay pumping operation when using 2½- or 3-inch (65 mm or 77 mm) hose as a supply line. With long lays of this size hose, it is necessary to place a pumper at the hydrant to increase the pressure in the supply hose. The reverse lay is the most direct way to supplement hydrant pressure and establish drafting operations.

Deploying a reverse hose lay can cause a delay in the initial fire attack because tools and equipment, including attack hose, must be removed and placed at the fire scene before the pumper proceeds to the water source. The reverse lay also causes the pump operator to stay with the pumper at the water source, preventing the operator from performing other essential fireground activities.

A common operation involving two pumpers — an attack pumper and a water-supply pumper — calls for the first-arriving pumper to go directly to the scene to start an initial attack on the fire using water from its tank, while the second-arriving

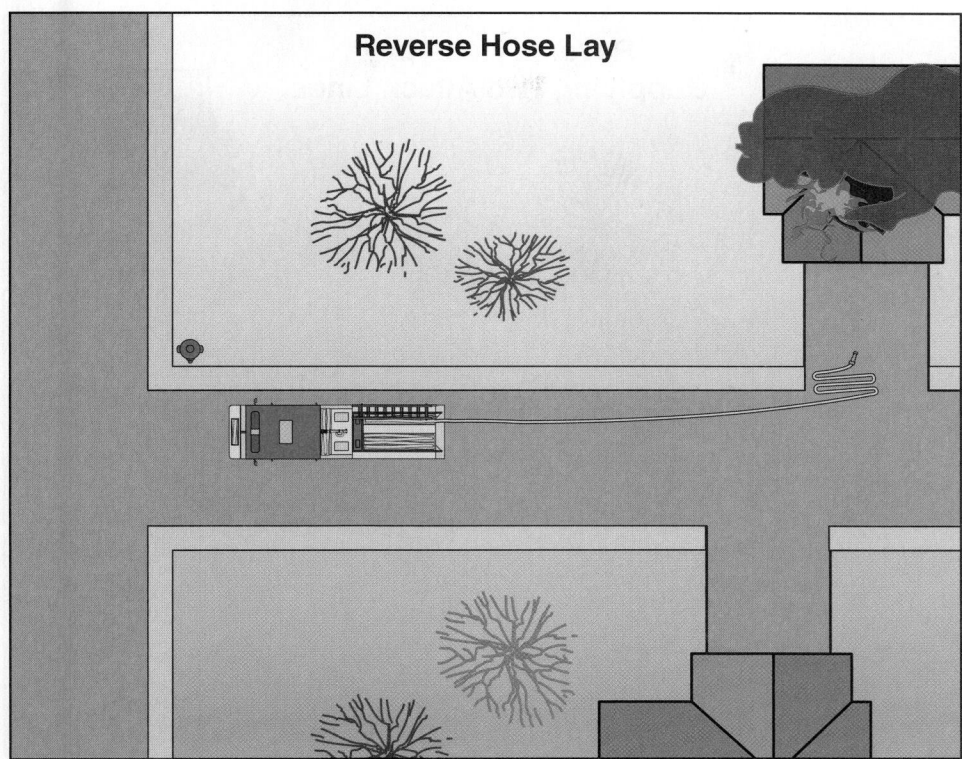

Reverse Hose Lay

Figure 15.64 A reverse hose lay is deployed from the incident scene to the water source.

pumper lays the supply line from the attack pumper back to the water source. The second pumper only needs to connect its just-laid hose to their discharge outlet, connect an intake hose, and begin pumping.

When reverse-laying a supply hose, connecting a four-way hydrant valve is optional. One can be used, however, if the pumper might have to disconnect from the supply hose later in the incident and leave the hose connected to the hydrant. Disconnecting may be desirable when the demand for water diminishes to the point that the second pumper can be made available for response to other incidents. As with a forward lay, using the four-way valve in a reverse lay provides the means to switch from pump pressure to hydrant pressure without interrupting the flow.

The reverse lay is also used when the first pumper arrives at a fire and must work alone for an extended period of time. In this case, the hose laid in reverse becomes an attack line. It is often connected to a reducing wye so that two smaller hoses can be used to make a two-directional attack on the fire **(Figure 15.65, p. 860)**. The reverse-lay procedures outlined in **Skill Sheet 15-I-13** describe how the second pumper lays a line from an attack pumper to a hydrant. They can be modified to accommodate most types of apparatus, hose, and equipment.

Frequently, firefighters will assist pumper driver/operators in making hydrant connections following a reverse lay. Either soft or hard intake hose designed for hydrant operations may be used to connect to hydrants. Hard intake hose must be used when drafting from a static water supply source.

Soft Intake Connections

Connecting the pumper to the hydrant with a soft suction hose was described in Chapter 14. However, not all hydrants have large steamer outlets capable of accepting direct connections from soft intake hose. Operations on hydrants equipped with two 2½-inch (65 mm) outlets require the use of two 2½- or 3-inch (65 mm or 77 mm)

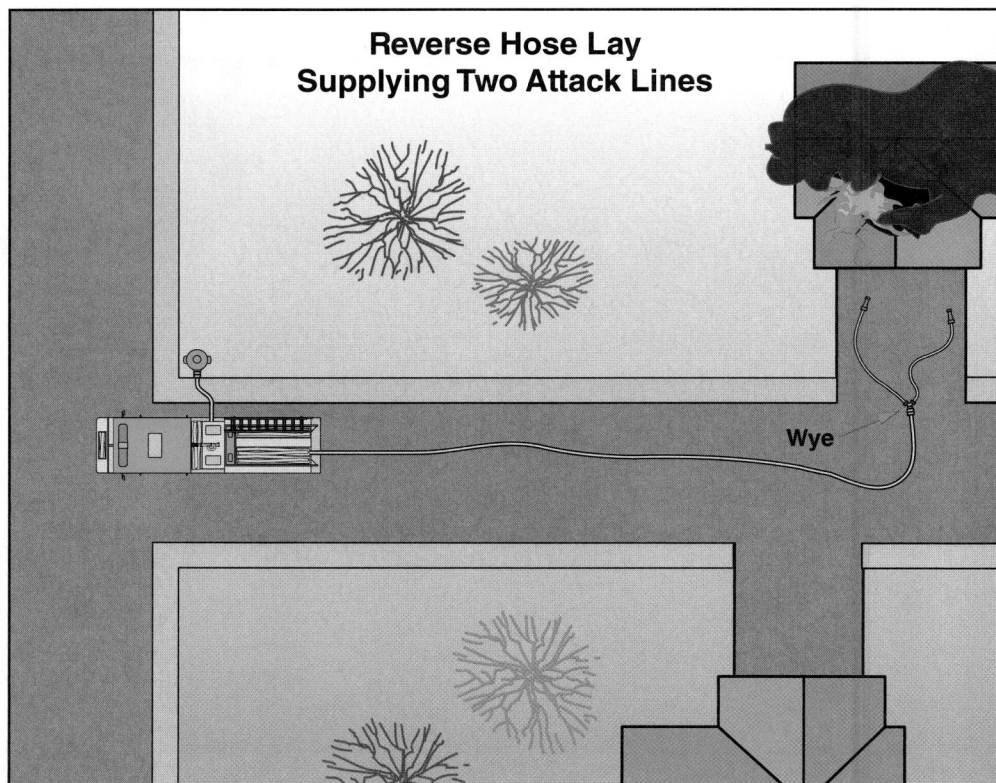

Figure 15.65 A reverse hose lay featuring a wye allows one hoseline to be used in a two-directional attack on the fire.

hoselines. These smaller intake hoses can be connected to a Siamese at the pump. It is more efficient to connect a 4½-inch (115 mm) or larger intake hose to a hydrant with only 2½-inch (65 mm) outlets. Such a connection is made by using a 4½-inch (115 mm) hose, or whatever size intake hose coupling is used, and connecting it to a 2½-inch (65 mm) reducer coupling.

Hard Intake Connections

Connecting a pumper to a fire hydrant using hard intake hose requires coordination and teamwork because more people are needed to connect hard intake hose than are needed to connect soft intake hose. Making hydrant connections with some types of hard intake hose is also considerably more difficult than making connections with a soft intake hose. The first aspect that is important is the positioning of the pumper in relation to the hydrant. No definite rule can be given to determine this distance because not all hydrants are the same distance from the curb or road edge, and the hydrant outlet may not directly face the street or road.

Another determining factor is that while most apparatus have pump intakes on both sides, others may also have one at the front or rear. It is considered good policy to stop the apparatus with the intake of choice just in front of the hydrant outlet. Depending on local protocols, the hard intake hose may be connected to either the apparatus or the hydrant first when making hydrant connections.

NOTE: If the hard intake is marked FOR VACUUM USE ONLY, do not use it for hydrant connections. This type of hard intake is for drafting operations only.

Combination Lay

The term *combination lay* refers to any of a number of ways to lay multiple supply hose with a single engine. To make a combination lay, the hose must be loaded into the hose bed in two separate hose bed compartments. Depending upon whether the beds are set up for forward or reverse lays, hoselines of the same diameter can be laid in the following ways:

- Two lines laid forward
- Two lines laid reverse
- Forward lay followed by a reverse lay
- Reverse lay followed by a forward lay
- Two lines laid forward followed by one or two lines laid reverse
- Two lines laid reverse followed by one or two lines laid forward

The basic tasks for catching the hydrant or deploying the supply hoseline are the same in each version of the combination lay. When two lines are laid at the same time, the hydrant catcher disconnects the hose at the crossover between the beds and pulls hose from both sides of the bed. Hose adapter fittings will be needed when threaded hose couplings are used in any reverse hose lay.

When using hose equipped with sexless couplings, the hose may be laid in either direction. However, firefighters and pumper/operators must make sure that the proper adapters are present at each end of the lay to make the appropriate connections.

Handling Hoselines

To effectively attack and extinguish a fire, hoselines must be removed from the apparatus and advanced to the location of the fire. The techniques used to advance hoselines depend on how the hose is loaded. Hoselines may be loaded preconnected to a discharge outlet or simply placed in the hose bed unconnected.

Deploying Preconnected Hoselines

The method used to deploy preconnected hoselines varies with the type of hose load. Both speed and efficiency will increase with practice. Your local SOPs may vary from the steps listed in this section and the referenced skill sheets.

Preconnected flat loads may deploy to either side or from the rear of the apparatus. To advance the preconnected flat load, grasp the hose loop in one hand and the nozzle in the other, pull the hose from the compartment, and walk toward the fire. Spread the hose and straighten it to remove any kinks before the line is charged with water. This procedure is described in **Skill Sheet 15-I-14.**

The minuteman load is intended to be deployed without dragging the hose on the ground. The hose deploys by unfolding from the top of the stack carried on the shoulder as the firefighter advances toward the fire. This procedure should result in the hoseline deploying with fewer kinks and bends in it.

Advancing the triple layer load involves placing the nozzle and the fold of the first tier on the firefighter's shoulder and walking away from the apparatus toward the fire. This procedure is described in **Skill Sheet 15-I-14.**

Deploying Other Hoselines

While some 2½-inch (65 mm) or larger attack hoselines may be preconnected, others may be deployed using supply hose as attack line. The hose may be deployed from either side of the hose bed and may require the addition of an adapter to mate the coupling with a nozzle or connect the hose to an FDC.

The hoselines equipped with wye appliances are normally used in connection with a reverse layout because the wye connection is fastened to the 2½- or 3-inch (65 mm or 77 mm) supply hose. One person performing two consecutive operations can unload these hoselines. Remove the attack lines in hose bundles or disconnect the preconnected hoselines and place them on the ground behind the apparatus with any necessary nozzles and adapters. Then remove the wye and enough hose to supply the smaller attack lines connected or to be connected to the wye. Kneel on the supply hose to anchor it as the driver/operator drives the apparatus slowly toward the water source. **Skill Sheet 15-I-15** describes the steps used for deploying a wye-equipped hose during a reverse hose lay.

To deploy individual sections from flat, accordion, or horseshoe loads, load one section of hose on another firefighter's shoulder one at a time. Multiple firefighters carry the hose to the desired location once it is disconnected from the remainder of the hose in the bed. Because all of the folds in an accordion load and a flat load are nearly the same length, they can be loaded on the shoulder by taking several folds at a time directly from the hose bed (**Figure 15.66**).

Figure 15.66 Accordion and flat loads feature folds that are nearly the same length, so several folds may be carried on the shoulder.

Advancing Hoselines

Once hoselines have been laid out from the attack pumper, they must be advanced into position for applying water onto the fire. Deploying hose over flat surfaces with no obstacles is very simple using most deployment methods. Advancing hoselines is considerably more difficult when hoses must be deployed up or down stairways, from standpipes, up ladders, and/or deep into buildings. Hoselines can be deployed more easily before they are charged because water adds weight and pressure that makes the hose difficult to maneuver. However, it is often unsafe to enter burning buildings with uncharged hoselines; therefore, you must know how to handle both uncharged and charged lines. You must also know how to add more hose to extend a hoseline as well as how to secure and replace a ruptured section of hose if necessary.

Advancing a Charged Hoseline

The working line drag is one of the quickest and easiest ways to advance a charged hoseline at ground level. Its use is limited by available personnel, but when adapted to certain situations, it is an acceptable method (**Figure 15.67**). **Skill Sheet 15-I-16** contains the steps for advancing a charged hoseline using the working line drag.

Advancing Hose Into a Structure

Before advancing hose into a structure, you must be alert for potential dangers such as backdraft, flashover, and structural collapse. The uncharged attack hoseline is advanced to the designated point of entry. **Skill Sheet 15-I-17** provides steps for advancing an attack hoseline into a structure. Observe the following general safety guidelines when advancing a hoseline into a burning structure:

- Check for and remove kinks and bends from the hoseline as it is advanced.

- Bleed air from the hoseline as it is being charged and before entering the building or fire area.

Figure 15.67 The working line drag may be used to advance a charged hoseline at ground level if enough personnel are available and certain situations are considered.

Figure 15.68 The act of advancing a hoseline requires a handful of safety checks before entering the structure.

- Position the nozzle operator and all members of the hose team on the same side of the hoseline **(Figure 15.68)**.
- Check for heat using the back of your gloved hand before opening the door.
- Stay low and avoid blocking ventilation openings such as doorways or windows.
- Chock self-closing doors to prevent the door from closing and pinching the hoseline.

Advancing Hose Up and Down a Stairway

Advancing hose up and down stairways can be very difficult. Therefore, when conditions allow, the hoseline should be advanced uncharged. The shoulder carry works well for stairway advancement because the hose is carried instead of dragged and is deployed as needed. The minuteman load and carry is also excellent for use on stairways. When advancing hose up a stairway, lay the uncharged hose against the outside wall to keep the stairs as clear as possible and avoid sharp bends and kinks in the hose. Advancing an uncharged hoseline down a flight of stairs is considerably easier than advancing a charged hose down the stairs. However, advancing an uncharged line down stairs is recommended only when the fire is very minor or not present.

When it is necessary to advance a charged hoseline up a stairway, excess hose should be deployed on the stairs toward the floor above the fire floor **(Figure 15.69, p. 864)**. The weight of the water and gravity will make extending the excess hoseline onto the fire floor easier. If possible, position a firefighter at every turn or point of resistance to aid in deployment of the charged hoseline.

Advancing a charged hoseline down a stairway can be almost as difficult as advancing one up stairs. Because deploying excess hose down the stairway would obstruct the stairs, excess hose should be stretched outside the stairway, such as in a hallway or room adjacent to the stairway, and firefighters positioned on the stairs to feed the hose down to the nozzle team. Firefighters must also be positioned at corners and pinch points. You must also have enough hose to reach the fire including the distance in the stairs, around corners, and onto the fire floor. **Skill Sheet 15-I-18** gives steps for advancing a hoseline up and down an interior stairway.

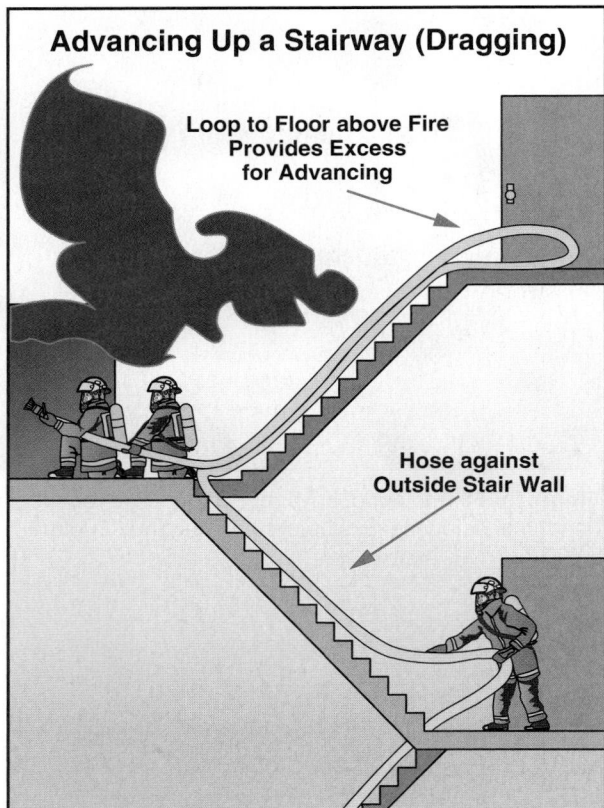

Advancing Up a Stairway (Dragging)

Loop to Floor above Fire Provides Excess for Advancing

Hose against Outside Stair Wall

Figure 15.69 Advancing up a stairway requires teamwork and a strategy which includes laying the hoseline against the outside wall.

Figure 15.70 Fire crews connect their attack lines to a standpipe outlet near the point of attack – usually on the floor below the fire floor. *Courtesy of Rick Montemorra, Mesa (AZ) FD.*

Advancing Hose from a Standpipe

Getting hose to the upper floors of high-rise buildings can be a challenge for firefighters. While preconnected hoselines may be able to access fires on the lower floors, fires beyond the reach of these lines require that hose be carried to the standpipe outlet closest to the fire. One approach is to have preassembled hose rolls, bundles, or packs on the apparatus ready to carry aloft and connect to the building's standpipe system. How these high-rise packs are constructed is a matter of local preference, but the most common are hose bundles that are easily carried on the shoulder or in specially designed hose packs complete with nozzles, fittings, and tools.

Except in rare instances, firefighters are not allowed to use elevators in burning buildings, so hose must be carried to the fire floor over an aerial ladder or up an interior stairway. Regardless of how the hose is brought up, fire crews normally stop one floor below the fire floor and connect the attack hoselines to the standpipe (**Figure 15.70**). If the standpipe connection is in an enclosed stairway, it is acceptable to connect on the fire floor. The standpipe connection is usually in or near the stairway. By observing the floor below, you can get a general idea of the layout of the fire floor.

Be alert for pressure-relief devices and follow your SOPs for removal or connection. If 1½-, 1¾-, or 2-inch (38 mm, 45 mm, or 50 mm) hose is used, placing a gated wye on the standpipe outlet will permit the attachment of a second attack hose if needed. A 2½-inch (65 mm) attack line may also be used depending on the size and nature of the fire. While the standpipe connection is being completed, any extra hose should be deployed up the stairs toward the floor above the fire. **Skill Sheet 15-I-19**

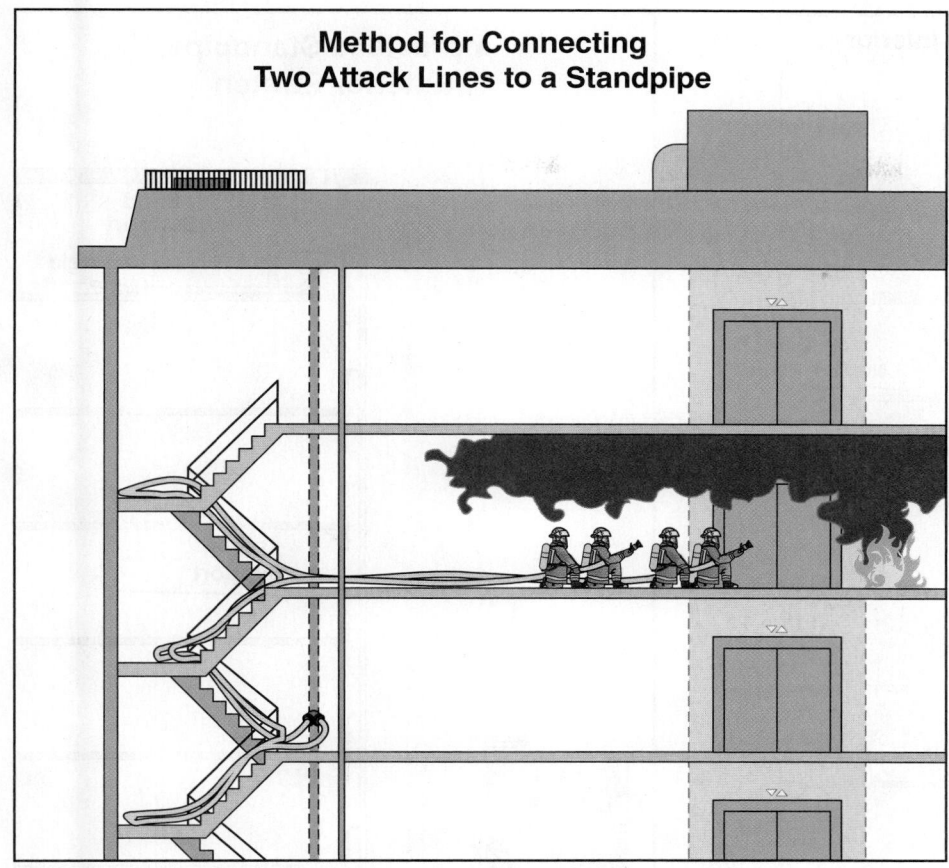

**Method for Connecting
Two Attack Lines to a Standpipe**

Figure 15.71 When advancing two lines from the same standpipe connection, one hoseline should place its slack on a higher staircase and the other hoseline should place its slack on a lower staircase to minimize the chances of entanglement.

provides procedures for connecting to a standpipe connection and advancing an attack hoseline onto a floor. When two lines are being advanced from the same standpipe connection, deploy one hoseline down the lower set of stairs and the other hoseline up the stairway in order to lessen the chances of the two hoselines becoming entangled **(Figure 15.71)**. When fire extinguishment is complete, carefully drain the water contained in the hoselines down a floor drain, out a window, or down a stairway to prevent unnecessary water damage.

Improvising a Standpipe

Most building codes mandate the installation of standpipes in structures three stories and higher. However, older buildings and those less than three stories may not have standpipes or existing connections may be obstructed or out of service as a result of construction, demolition, sabotage, or natural disasters like earthquakes. One way to supply water to a building without a standpipe system is to create an improvised standpipe. There are two methods for improvising standpipes: the interior stairway stretch and the outside stretch.

> **CAUTION**
> When firefighters must improvise a standpipe system, there will be a delay in applying water to the fire. This delay must be considered in planning the overall fire fighting strategy.

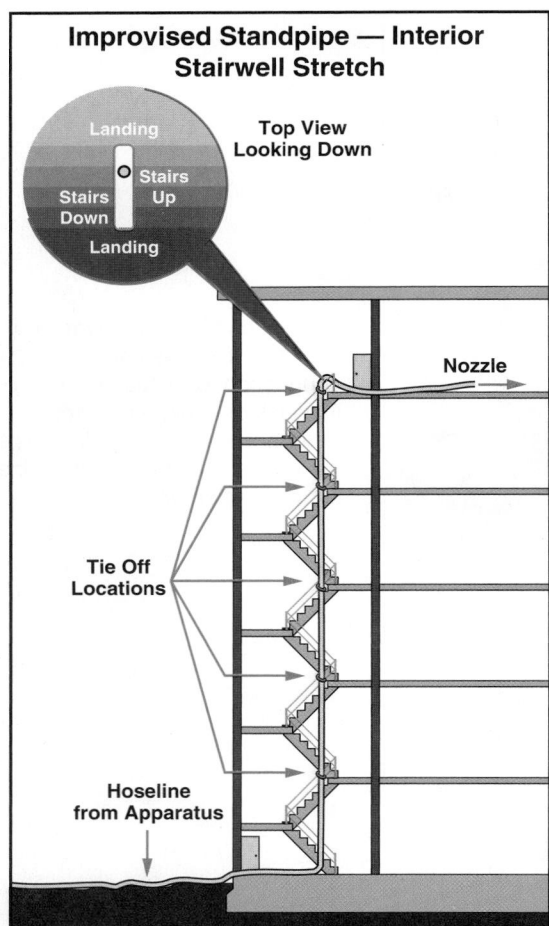

Figure 15.72 An improvised standpipe follows the rising stairs nearest the main building to follow the shortest distance between the ground and the fire floor.

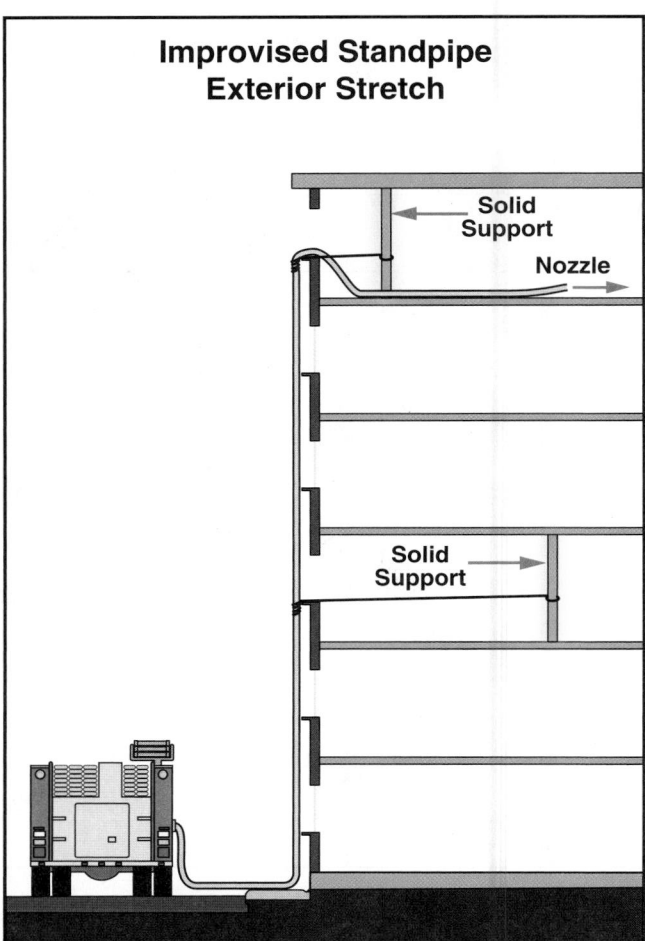

Figure 15.73 An outside stretch of hoseline should be augmented with periodic ties to an interior support to help hold the weight of a charged line.

Interior Stairway Stretch

The interior stairway stretch is a labor-intensive task used in stairways that have an open shaft or stairwell in the center. To improvise the standpipe, an uncharged hoseline is suspended in the middle of the stairs rather than laying it on the stairs and around each corner. Hose rolls or bundles can be carried up the stairs, secured to a hand rail and the end lowered to the point where another section is attached to it. Secure the hose to the hand rails for support at appropriate intervals to reduce the tendency of water weight to pull the hose back down once the hose is charged (**Figure 15.72**). Advancing a dry hoseline during interior stairway stretch should take into consideration the diameter of the pressurized hose relative to the space between the handrail openings.

Outside Stretch

An outside stretch can be used for lower floors of high-rise buildings. Supply hose can be hoisted up the exterior of the building to the desired floor using a rope. Because the weight of the water in the charged line can cause the hose to fall back down the building, some of the hoseline can be extended into windows and secured to available anchor points inside the building at an interval of about every three stories (**Figure 15.73**).

Figure 15.74 Advancing a hoseline up a ladder is easiest accomplished using a dry hoseline.

Figure 15.75 A charged hoseline may be advanced up a ladder when necessary.

Advancing Hose Up a Ladder

When standpipes are not available but stairways are accessible, one of the safest ways to get hose to an upper story is to carry it up the stairs in a bundle and lower the female end over a balcony railing or out a window to connect to a water source. Another method is to hoist the hose and attached nozzle up to a window or landing using a rope.

If standpipes are not available, stairways are not accessible, or there is no other viable option, it may be necessary to advance the hose up a ground ladder or aerial device. Advancing fire hose up a ladder is easier and safer with an uncharged line **(Figure 15.74)**. In most cases, the firefighter heeling the ground ladder can also help feed the hose to those on the ladder. If the hose is already charged with water, it may be advisable to drain the hose before advancing it up the ladder.

The best way to advance an uncharged hoseline up a ground ladder or aerial ladder is described in **Skill Sheet 15-I-20**. To avoid overloading the ladder, only one person is allowed on each section of the ladder. Rope hose tools or utility straps can also be used for this advancement. The hose can be charged once it has reached the point from which the fire attack will be made **(Figure 15.75)**. When it is absolutely necessary to advance a charged line up a ladder, firefighters should follow the steps described in **Skill Sheet 15-I-21**.

> **WARNING**
> Do not exceed the rated weight capacity of the ladder. If the hose cannot be passed up the ladder without exceeding the load limit, it should be hoisted up.

Figure 15.76 When the hose is secured to the ladder, open the nozzle slowly.

It is sometimes necessary for firefighters to operate a hoseline from a ground ladder or supported aerial ladder (the tip of the aerial ladder must be in contact with the windowsill) **(Figure 15.76)**. In this case, firefighters should follow the steps described in **Skill Sheet 15-I-22**.

Aerial platforms can be used as portable standpipes for advancing a hoseline onto a floor. A high-rise pack is placed on the platform along with the attack line crew. When the platform arrives at the desired floor and the window is opened, the hose is attached to the discharge outlet on the platform. The crew advances the hoseline onto the floor and the hose is charged.

Operating Attack Hoselines

As a firefighter assigned to an attack hoseline, it is your job to extinguish the fire quickly and effectively with a minimum amount of water while still protecting yourself and your teammates. During an interior structural fire attack, there should always be a minimum of two firefighters on an attack hoseline: the nozzle operator and the person directly behind the nozzle. The second person may be the fire officer or a senior firefighter directing the fire stream. As a team, maneuver the hoseline and direct the fire stream where it will be most effective.

In some instances, you may be operating an attack hoseline alone. This may occur during overhaul, when fighting an outdoor fire, or during trash or Dumpster™ fires, grass fires, and vehicle fires. You may also have to handle the nozzle alone when washing down flammable liquid spills. In any case, you must be able to maneuver the hoseline alone and direct the fire stream correctly. The sections that follow provide the skills for operating an attack hoseline effectively.

Operating Small Hoselines

One or two firefighters can operate small hoselines, such as booster lines and 1½-, 1¾-, and 2-inch (38 mm, 45 mm, and 50 mm) hoselines. However, even small hoselines can require additional firefighters when the hose is charged and there are obstructions that must be negotiated.

One-Firefighter Method

Assigning one firefighter to operate an attack hoseline only occurs when combating a small ground cover fire, rubbish or trash fire, vehicle fire, small structure fire, or during overhaul operations **(Figure 15.77)**. **Skill Sheet 15-I-23** illustrates the one-firefighter method for operating a small hoseline.

Figure 15.77 One firefighter may be assigned to an attack hoseline under specific conditions including small fires and overhaul.

Figure 15.78 The backup firefighter maintains the correct direction of the hoseline behind the nozzle operator. *Courtesy of Shad Cooper, Wyoming State Fire Marshal's Office.*

Two-Firefighter Method

Two firefighters are the minimum number required for handling any attack line during interior structural operations. The nozzle operator holds the nozzle with one hand and holds the hose just behind the nozzle with the other hand. The hoseline is then rested against the waist and across the hip. Holding nozzles equipped with a pistol grip is slightly different: hold the pistol grip in one hand while holding the operating bale in the other.

The backup firefighter takes a position on the same side of the hose about 3 feet (1 m) behind the nozzle operator. The second firefighter holds the hose with both hands and rests it against the waist and across the hip or braces it with the leg. The backup firefighter is responsible for keeping the hose straight behind the nozzle operator **(Figure 15.78)**. During extended operations, either one or both firefighters may apply a hose strap or rope hose tool to reduce the effects of nozzle reaction. Skills for the two-firefighter method are included in Chapter 16, Fire Streams.

Operating Large Hoselines

Once a large attack hoseline has been deployed and advanced to the fire, it must be placed in operation. The methods described in the following sections can be used with large attack hoselines of 2½- and 3-inch (65 mm and 77 mm) or larger hose.

One-Firefighter Method

During exposure protection or overhaul operations, one firefighter may be assigned to operate a large hoseline if a master stream device is not available **(Figure 15.79, p. 870)**. To reduce fatigue during extended operations, the nozzle operator can either use a hose strap or rope hose tool looped over the shoulder, or reduce the nozzle flow

Figure 15.79 One firefighter can control a large handline for an extended period of time during exposure protection or overhaul operations.

Figure 15.80 A two-firefighter method of hose control uses hose straps and has the backup firefighter work as an anchor to absorb some of the nozzle reaction.

if conditions allow. Except for limited lateral (side-to-side) motion, this method does not permit very much maneuvering of the nozzle. **Skill Sheet 15-I-24** describes the method for one firefighter operating a large hoseline for exposure protection.

Two-Firefighter Method

When two firefighters are assigned to handle a large hoseline, they may need a means of anchoring the hoseline to offset nozzle reaction. **Skill Sheet 15-I-25** illustrates the steps for handling a large hoseline with two firefighters.

Another two-firefighter method uses hose straps or rope hose tools to assist in anchoring the hose. The nozzle operator loops a hose strap or rope hose tool around the hose a short distance from the nozzle placing the large loop across the back and over the outside shoulder. The operator then holds the nozzle with one hand and the hose just behind the nozzle with the other hand. The hoseline rests against the body. Leaning slightly forward helps control the nozzle reaction. The backup firefighter again serves as an anchor about 3 feet (1 m) back. The backup firefighter also has a hose strap or rope hose tool around the hose and leans his or her shoulder forward to absorb some of the nozzle reaction (**Figure 15.80**).

Three-Firefighter Method

Several methods are used for three firefighters to control large hoselines. In all cases, the positioning of the nozzle operator is the same as previously described for the two-firefighter method. The only differences are in the position of the second and third firefighters on the hoseline. Some departments prefer the first backup firefighter to stand directly behind the nozzle operator, with the third firefighter kneeling on the hose behind the second firefighter. Another technique that allows for the most mobility is for all firefighters to use hose straps or rope hose tools and remain in a standing position.

Regardless of the techniques used, firefighters operating fire hose nozzles and those backing them up must maintain their situational awareness and not limit their focus to what is directly in front of the nozzle. Changes in fire behavior, indicators of imminent structural collapse, and other warning signs can be missed if the nozzle team fails to monitor what is going on around them.

Figure 15.81 If necessary, water flow may be stopped by kinking the hose.

Extending a Section of Hose

It may become necessary to extend an attack hoseline during interior or exterior operations. Extending the hoseline will require a hose clamp, spanner wrench, and the necessary number of hose rolls or bundles for the distance required. Firefighters should follow the procedures described in **Skill Sheet 15-I-26** to accomplish this task.

Controlling a Loose Hoseline

A loose hoseline is one in which water under pressure is flowing through a nozzle, an **open butt**, or a rupture, and is out of control. Water pressure will cause the loose hoseline to flail about or whip back and forth. The uncontrolled and whipping hoseline could strike firefighters and bystanders with great force causing serious injury or death.

Closing a valve at the pump or hydrant to stop the flow of water is the safest way to control a loose hoseline. Another method is to apply a hose clamp at a stationary point in the hoseline. It may also be possible to put a kink in the hose at a point away from the break until the appropriate valve is closed. To put a kink in a hose, form a loop in the line, press down on the top of the loop, and apply body weight to the bends in the hose (**Figure 15.81**). In most cases, this action will not completely stop the flow of water through the hose, but it will reduce the flow sufficiently for firefighters to safely gain control of the end of the hose.

NOTE: The procedure for placing a kink in hose does not apply to LDH due to its size and weight when charged.

Open Butt — The end of a charged hoseline that is flowing water without a nozzle or valve to control the flow.

Replacing Burst Sections

Mechanical damage or overpressure can cause a hoseline to burst. When a hoseline bursts, the nozzle operator should request that the pump operator close the discharge controlling the hoseline. The water flow may also be stopped at any gated wye in the line, by applying a hose clamp, or by creating a kink in the hose. Two additional sections of hose should be used to replace any one bad section because hoselines stretch to longer lengths when under pressure; the couplings in the line will always be farther apart than the length of a single replacement section. Steps to take when replacing burst sections of hose are given in **Skill Sheet 15-I-27**.

II Service Testing Fire Hose

NFPA® 1962, *Standard for the Inspection, Care, and Use of Fire Hose, Couplings, and Nozzles and the Service Testing of Fire Hose*, provides guidelines for **service testing** of fire hose. Service tests are performed annually, after repairs have been made, and after a vehicle has run over the hose.

Before being service tested, examine the hose for excessive wear or damage to the jacket, coupling damage, and defective or missing gaskets. If any defects are found, tag the hose for repair. If damage is not repairable, remove the hose from service.

Test Site Preparation

Hose should be tested in a paved area with enough room to lay out the hose in straight lines, free of kinks, bends, or twists. The site should be protected from vehicular traffic. If testing is done at night, the area should be well lighted. The test area should be smooth and free from rocks and debris. A slight grade to aid water drainage is also helpful. A water source sufficient for charging the hose is also necessary.

The following equipment is needed to service test hose:

- Hose-testing machine, portable pump, or fire department pumper equipped with gauges certified as accurate within one year before testing
- Hose-test gate valve
- Means of recording the hose numbers and test results
- Tags or other means to identify sections that fail
- Nozzles with shutoff valves
- Means of marking each length with the year of the test to easily identify which lengths have been tested and which have not without looking in the hose records

Service Test Procedure

Exercise care when working with hose, especially when it is under pressure. Pressurized hose is potentially dangerous because of its tendency to whip back and forth if a rupture occurs or a coupling pulls loose. To prevent this from happening, use a specially designed hose test gate valve **(Figure 15.82)**. These valves have a ¼-inch (6 mm) hole in the gate that permits pressurizing the hose but does not allow water to surge through the hose if it fails. Even when using the test gate valve, stand or walk near the pressurized hose only as necessary. See **Skill Sheet 15-II-1** for steps in service testing hose.

Service Test — Series of tests performed on apparatus and equipment in order to ensure operational readiness of the unit. These tests should be performed at least yearly or whenever a piece of apparatus or equipment has undergone extensive repair.

Figure 15.82 A hose-test gate valve prevents water from surging if a pressurized hose fails.

When using a fire department pumper, connect the hose to discharges on the side of the apparatus opposite the pump panel. Close all valves slowly to prevent water hammer in the hose and pump. Test lengths of hose should not exceed 300 feet (100 m) in length because it is difficult to purge air from longer lengths of hose.

Laying large-diameter hose flat on the ground before charging it helps to prevent unnecessary wear on the edges. Stand away from the discharge valve connection when charging because the hose has a tendency to twist when it is filled with water and pressurized; this twisting could cause the connection to loosen.

Keep the hose testing area as dry as possible when filling and discharging air from the hose. During testing, this air aids in detecting minor leaks around couplings.

Chapter Summary

Fire hose is a basic tool used to carry water from its source to the point it is needed to extinguish a fire. You must know the types of hose your department uses, how it is constructed, the ways hose can be damaged, and how to care for it. You must know the differences between supply hose and attack hose and how to deploy, advance, and operate both supply and attack hose. It is critical that you know the types of fire hose loads and finishes and how they relate to various hose deployments. As you advance to FirefighterI II Firefighter, you must know the procedure for testing hose.

Firefighter I

1. What are the three basic fire hose characteristics a firefighter must understand?

2. How are thermal damage and corrosion in a hose similar or different?

3. What are the steps taken to perform basic inspection and maintenance for fire hose?

4. What are the types of hose appliances and tools a firefighter may need to use?

5. When should firefighters use basic hose rolls?

6. What hose loads can a firefighter choose from when storing hose?

7. How do the various methods to make preconnected hose loads for attack lines compare with one another?

8. What methods can be used for supply hose lays?

9. How do firefighters decide what technique to use when handling hoselines?

10. What are the safety guidelines for advancing a hose into a structure?

11. What considerations can impact operating attack hoselines?

Firefighter II

1. What are the safety considerations that should be taken when service testing a fire hose?

Couple - Foot-Tilt Method

Step 1: Stand facing the two couplings so that one foot is near the male end.

Step 2: Place a foot on the hose directly behind the male coupling.

Step 3: Apply pressure to tilt it upward.

Step 4: Grasp the female end by placing one hand behind the coupling and the other hand on the coupling swivel.

Step 5: Bring the two couplings together, align the Higbee cut, and turn the swivel clockwise with thumb to make the connection.

Couple - Two-Firefighter Method

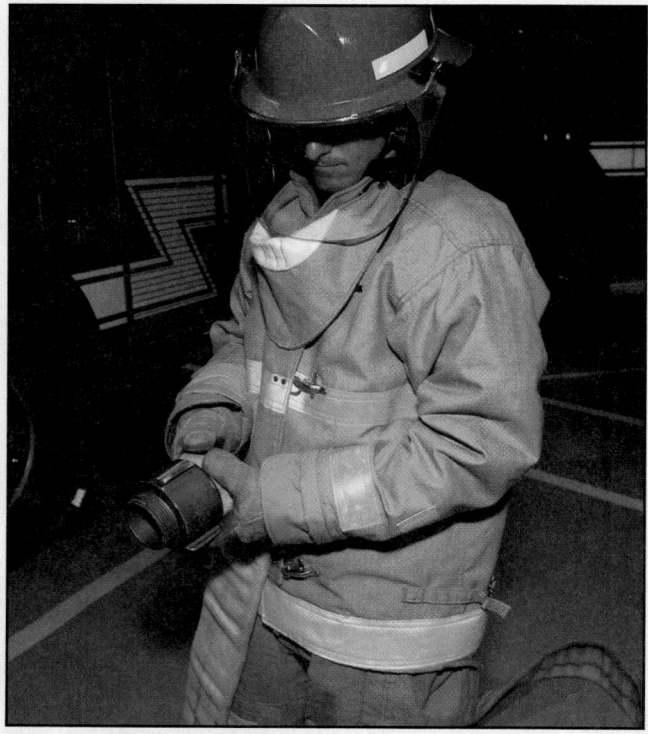

Step 1: Firefighter #1: Grasp the male coupling with both hands.

Step 2: Firefighter #1: Bend the hose directly behind the coupling.

Step 3: Firefighter #1: Hold the coupling and hose tightly against the upper thigh or midsection with the male threads pointed outward.

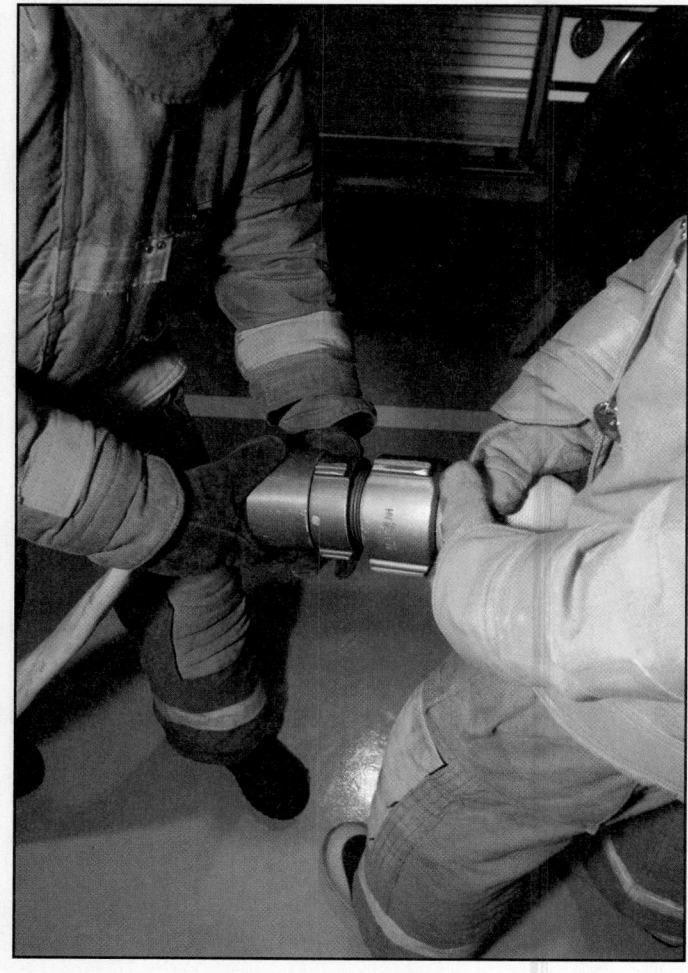

Step 4: Firefighter #2: Grasp the female coupling with both hands.

Step 5: Firefighter #2: Bring the two couplings together, and align their positions.

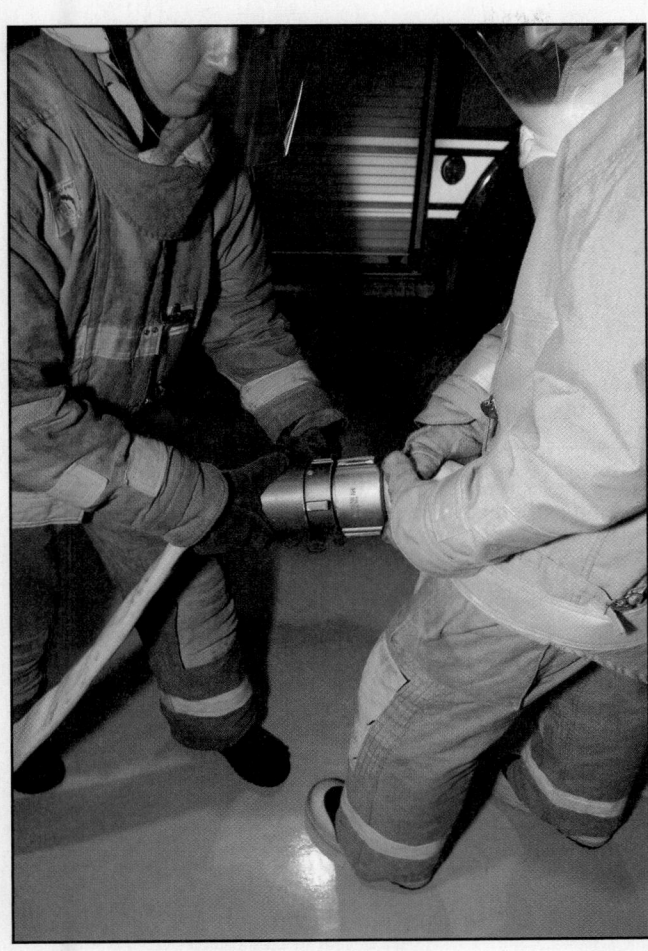

Step 6: Firefighter #2: Turn the female coupling counterclockwise until a click is heard. This indicates that the threads are aligned.

Step 7: Firefighter #2: Turn the female swivel clockwise to complete the connection.

Uncouple - Knee-Press Method

Step 5: Snap the swivel quickly in a counterclockwise direction as body weight is applied to loosen the connection.

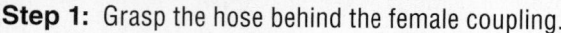

Step 1: Grasp the hose behind the female coupling.

Step 2: Stand the male coupling on end.

Step 3: Set feet well apart for balance.

Step 4: Place one knee upon the hose and shank of the female coupling.

Uncouple - Two-Firefighter Method

Step 2: Both Firefighters: Keep arms stiff, and use the weight of both bodies to turn each hose coupling counter-clockwise, thus loosening the connection.

Step 1: Both Firefighters: Take a firm two-handed grip on respective coupling and press the coupling toward the other firefighter, thereby compressing the gasket in the coupling.

Courtesy of Iowa Fire Service Training Bureau and Ames (IA) FD. Photos by Dawn Beisner.

Inspect

Step 1: Stretch hose to full length on flat, clean, dry surface.

   a. Ensure grease or other chemicals cannot come into contact with hose

   b. Attempt to unroll hose rather than drag it

Step 2: Begin at male coupling and check condition of threads, attachment to hose, and lugs.

Step 3: Carefully place male coupling back on surface, protecting threads from impact with ground.

Step 4: Slowly walk along the section of hose, visually inspecting surface for abrasion, burns, or other damage.

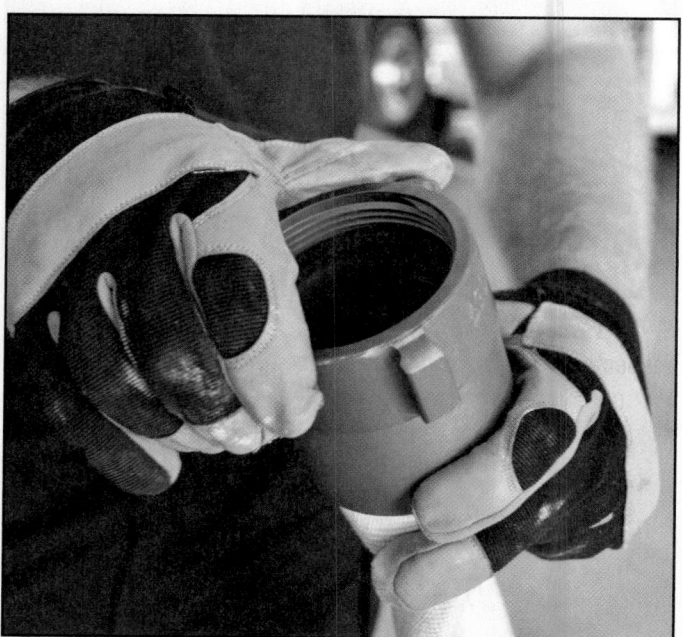

Step 5: Circle any damaged spots (on both sides of hose) with piece of chalk or nonpermanent marker.

Step 6: Inspect threads, swivel, hose coupling attachment upon reaching female hose coupling.

   a. Inspect fire hose coupling swivel gasket – Should be pliable with no cracks or deep depressions caused by overtightening

Courtesy of Iowa Fire Service Training Bureau and Ames (IA) FD. Photos by Dawn Beisner.

Step 7: Turn hose over to inspect bottom side.

    a. Follow same procedure and inspect hose back to male coupling

    b. Pay particular attention to marked locations on other side of hose

Step 8: Roll the hose.

    a. Roll from male coupling (keeping it on the inside) back to female coupling

    b. Roll with opposite side out as previously rolled

Step 9: Note general inspection results, update fire hose service log as required by local SOPs.

    a. If hose is damaged or has other defects, tag with out-of-service tag, remove from service until repaired and tested

    b. If free of damage, return to appropriate location

Courtesy of Iowa Fire Service Training Bureau and Ames (IA) FD. Photos by Dawn Beisner.

Maintain - Replace a Hose Gasket

Step 1: Remove and discard old or damaged gasket in proper receptacle.

Step 2: Pick up new gasket with middle finger and thumb, and fold loop upward with index finger.

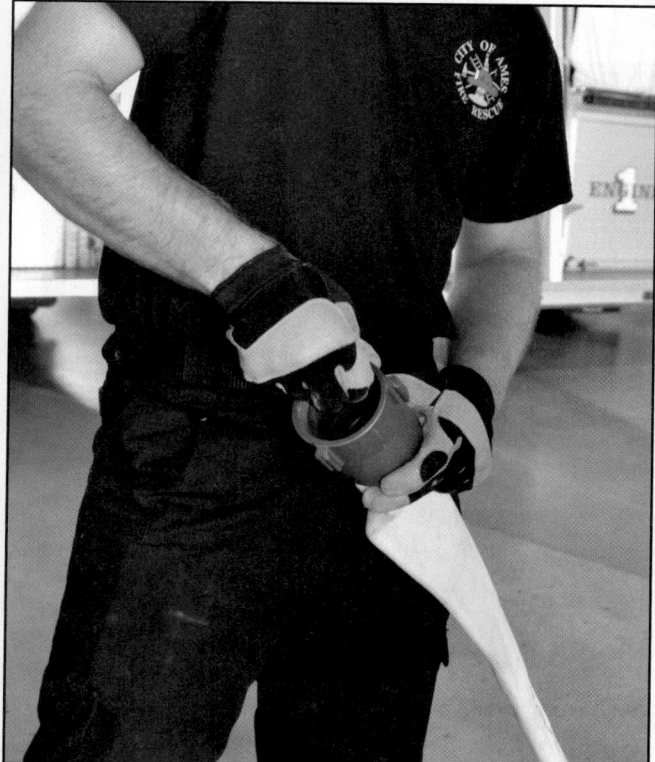

Step 3: Place gasket into swivel, large loop first, smoothing as necessary to seat.

Step 1: Lay out the hose straight and flat on a clean surface.

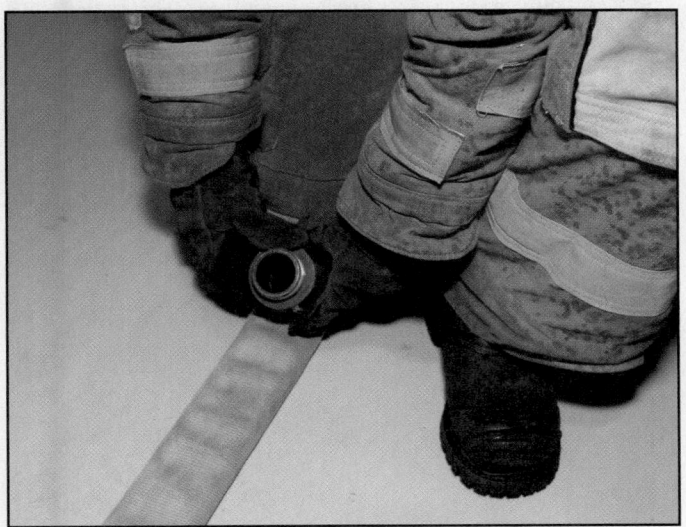

Step 2: Roll the male coupling over onto the hose to start the roll.

    a. Form a coil that is open enough to allow the fingers to be inserted

Step 3: Continue rolling the coupling over onto the hose, forming an even roll.

    a. Keep the edges of the roll aligned on the remaining hose to make a uniform roll as the roll increases in size

Step 4: Lay the completed roll on the ground.

Step 5: Tamp any protruding coils down into the roll with a foot.

Method One

Step 1: Lay the section of hose flat and in a straight line.

Step 2: Start the roll from a point 5 or 6 feet (1.5 m or 1.8 m) off center toward the male coupling.

Step 3: Roll the hose toward the female end.

    a. Leave sufficient space at the center loop to insert a hand for carrying

Step 4: Extend the short length of hose at the female end over the male threads to protect them.

Method Two

Step 1: Grasp either coupling end, and carry it to the opposite end. The looped section should lie flat, straight, and without twists.

Step 2: Face the coupling ends.

Step 3: Start the roll on the male coupling side about 2½ feet (0.8 m) from the bend (1½ feet [0.5 m] for 1½-inch [38 mm] hose).

Step 4: Roll the hose toward the male coupling.

Step 5: Pull the female side back a short distance to relieve the tension if the hose behind the roll becomes tight during the roll.

Step 6: Lay the roll flat on the ground as the roll approaches the male coupling.

Step 7: Draw the female coupling end around the male coupling to complete the roll.

Courtesy of Iowa Fire Service Training Bureau and Ames (IA) FD. Photos by Dawn Beisner.

Twin Donut Roll Method

Step 1: Lay a section of hose flat, without twisting, to form two parallel lines from the loop end to the couplings (couplings should now be next to each other).

Step 2: Start the roll by folding the loop end over and upon the two hose lengths.

Step 3: Roll both lengths simultaneously toward the coupling ends to form a twin roll with a decreased diameter.

Courtesy of Iowa Fire Service Training Bureau and Ames (IA) FD. Photos by Dawn Beisner.

Step 4: Insert a strap through the center of the roll for carrying purposes.

Courtesy of Iowa Fire Service Training Bureau and Ames (IA) FD. Photos by Dawn Beisner.

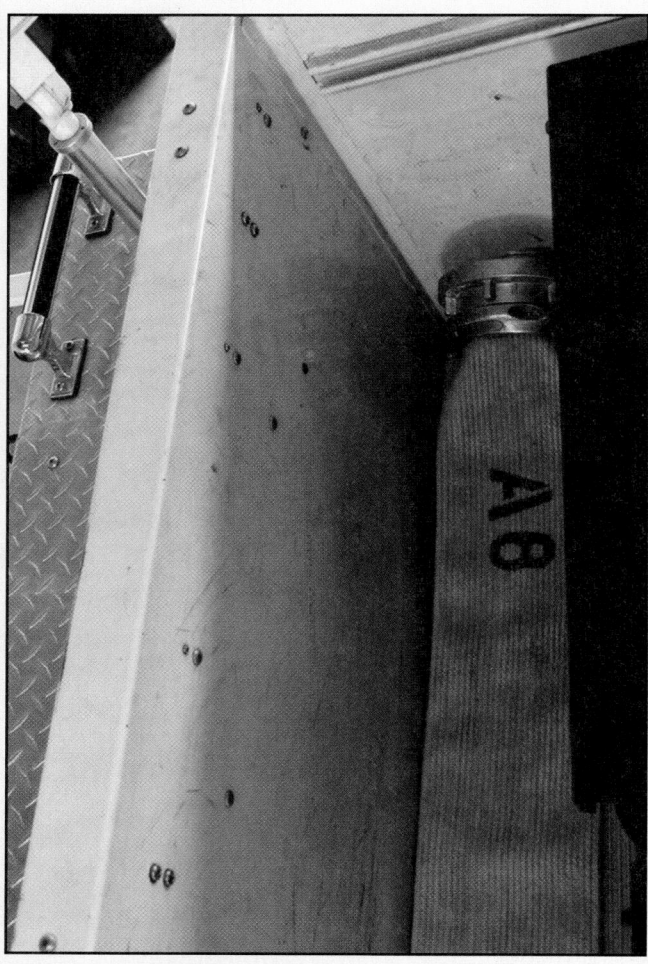

Step 1: Inspect the hose and hose couplings for damage.

Step 2: Place first coupling at a front corner of the hose bed.

Step 3: Lay the hose flat in the hose bed in a front-to-back fashion.

Step 4: Fold the hose back on itself (make a loop) and lay the hose in the opposite direction.

    a. Repeat until hose covers the bottom of the hose bed

Courtesy of Iowa Fire Service Training Bureau and Ames (IA) FD. Photos by Dawn Beisner.

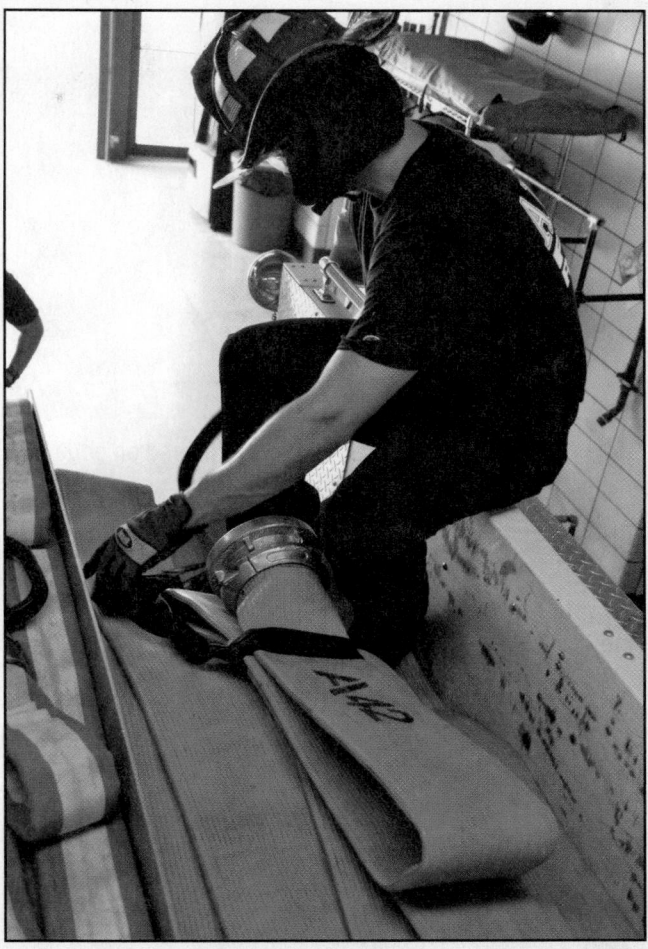

Step 5: Start second layer repeating Steps 3 and 4.

    a. Repeat until all hose is loaded

Step 6: Finish hose load with donut roll or other finish as required by local protocol.

Courtesy of Iowa Fire Service Training Bureau and Ames (IA) FD. Photos by Dawn Beisner.

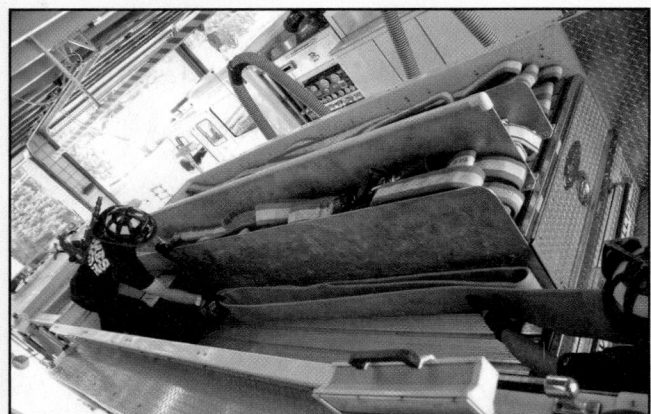

Step 1: Lay the first length of hose in the bed on edge against the partition.

Step 2: Fold the hose at the front of the hose bed back on itself.

Step 3: Lay the hose back to the rear next to the first length.

Step 4: Fold the hose at the rear of the hose bed so that the bend is even with the rear edge of the bed.

Step 5: Lay the hose back to the front.

Step 6: Continue laying the hose in folds across the hose bed.

    a. Stagger the folds at the rear edge of the bed so that every other bend is approximately 2 inches (50 mm) shorter than the edge of the bed

    b. This stagger may also be done at the front of the bed if desired

Step 7: Angle the hose upward to start the next tier.

Step 8: Make the first fold of the second tier directly over the last fold of the first tier at the rear of the bed.

Step 9: Continue with the second tier in the same manner as the first, progressively laying the hose in folds across the hose bed.

Step 10: Make the third and succeeding tiers in the same manner as the first two tiers.

Step 11: Move to the opposite hose bed.

Step 12: Load the hose in the same manner as the first side.

Step 13: Connect the last coupling on top with the female coupling from the first side when the load is completed.

Step 14: Lay the connected couplings on top of the hose load.

Step 15: Pull out the slack so that the crossover loop lies tightly against the hose load.

Courtesy of Iowa Fire Service Training Bureau and Ames (IA) FD. Photos by Dawn Beisner.

Step 1: Place the coupling in a front corner of the hose bed.

Step 2: Lay the first length of hose on edge against the wall.

Step 3: Make the first fold at the rear even with the edge of the hose bed.

Step 4: Lay the hose to the front and then around the perimeter of the bed so that it comes back to the rear along the opposite side.

Step 5: Make a fold at the rear in the same manner as done before.

Step 7: Lay succeeding lengths progressively inward toward the center until the entire space is filled.

Step 6: Lay the hose back around the perimeter of the hose bed inside the first length of hose.

Courtesy of Iowa Fire Service Training Bureau and Ames (IA) FD. Photos by Dawn Beisner.

Step 8: Start the second tier by extending the hose from the last fold directly over to a front corner of the bed, laying it flat on the hose of the first tier.

Step 9: Make the second and succeeding tiers in the same manner as the first.

Courtesy of Iowa Fire Service Training Bureau and Ames (IA) FD. Photos by Dawn Beisner.

Reverse Horseshoe Method

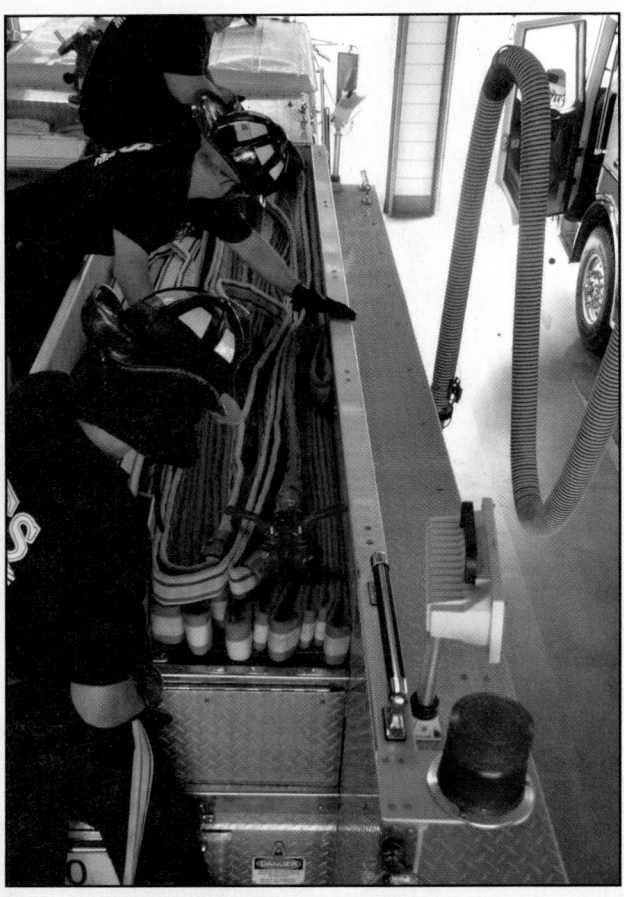

Step 5: Lay the hose back inside the previously laid length in the same manner, and continue until the entire hose length has been loaded.

Step 1: Connect the gated wye to the end (male) coupling of the hose load at the rear of the hose bed.

Step 2: Place the wye in the center of the hose load with the two male openings toward the rear of the hose bed.

Step 3: Connect one of the 1½ inch (38 mm) hoses to the gated wye. Place the hose on its edge to the front of the hose bed, and make a fold in the hose.

Step 4: Lay the hose back to the rear alongside the first length. Form a U at the rear edge of the bed, return the hose to the front, and make a fold.

Step 6: Wrap the male end of the hose once around the horseshoe-shaped loops.

Step 7: Form a small loop by bringing the end back under the center of the loops, and then over the top.

Step 8: Attach the nozzle, and place it inside the small loop.

Step 9: Pull the remaining slack hose back into the center of the horseshoe to tighten the loop against the nozzle.

Step 10: Load the second length of hose on the opposite side of the hose bed in the same manner.

15-I-8
Make a finish.

Courtesy of Iowa Fire Service Training Bureau and Ames (IA) FD. Photos by Dawn Beisner.

Skid Load Method

Step 1: Connect the first section of attack hose directly to the supply hose load.

Step 2: Place several folds (accordion like) on edge across the front of the bed to provide some slack hose for pulling the attack hose finish from the bed.

Step 3: Turn the hose flat, and lay it to the rear of the hose bed approximately one-third of the way from one side.

Step 4: Form a loop that extends approximately 6 inches (152 mm) beyond the hose bed.

Step 5: Lay the hose back to the front of the hose bed approximately one-third of the way from the opposite side to form a supporting skid.

Step 6: Make the second skid in the same manner.

Courtesy of Iowa Fire Service Training Bureau and Ames (IA) FD. Photos by Dawn Beisner.

Step 7: Connect additional attack hose, and load it (accordion like) back and forth across the two skids until all of the hose is loaded.

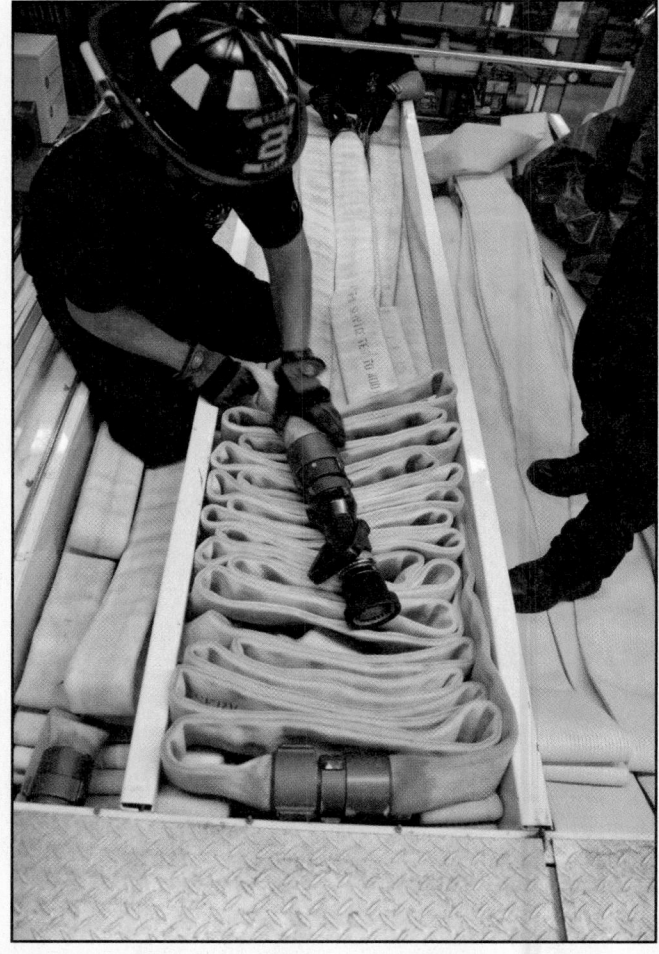

Step 8: Attach the nozzle, place it on top of the load.

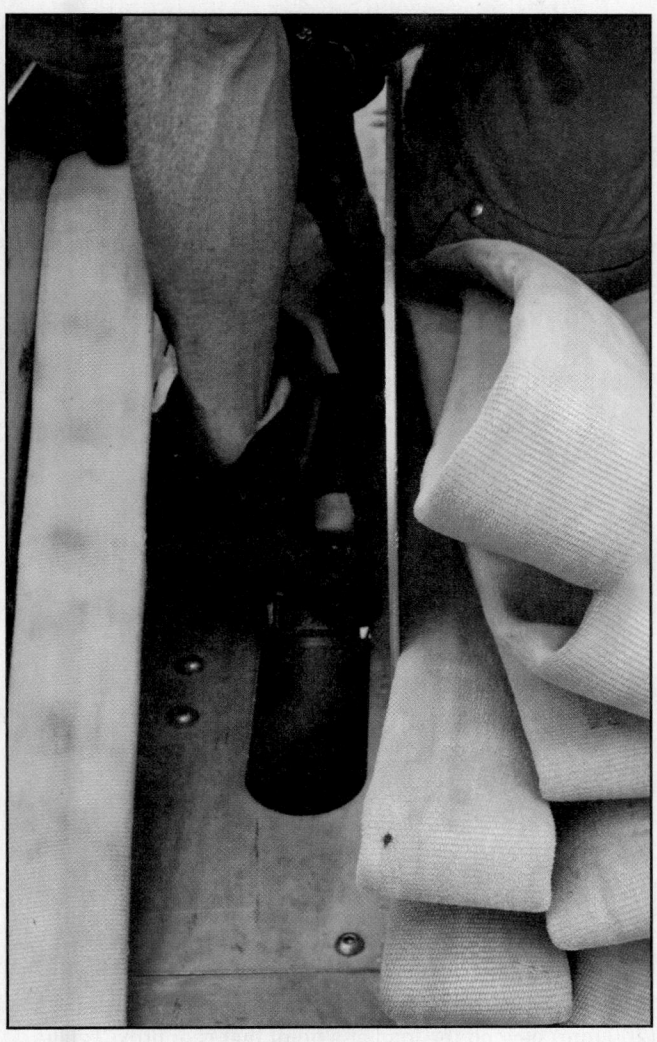

Step 1: Attach the female coupling to the discharge outlet.

Step 2: Lay the first length of hose flat in the bed against the side wall.

Step 3: Angle the hose to lay the next fold adjacent to the first fold and continue building the first tier.

Step 4: Make a fold that extends approximately 8 inches (200 mm) beyond the load at a point that is approximately one-third the total length of the load. This loop will later serve as a pull handle.

Step 5: Continue laying the hose in the same manner, building each tier with folds laid progressively across the bed.

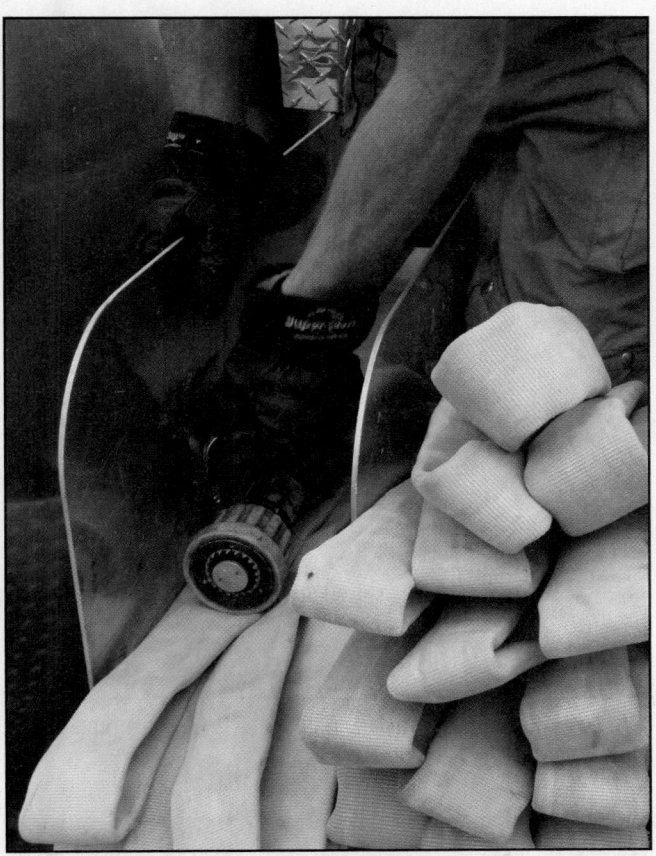

Step 6: Make a fold that extends approximately 14 inches (350 mm) beyond the load at a point that is approximately two-thirds the total length of the load. This loop will also serve as a pull handle.

Step 7: Complete the load.

Step 8: Attach the nozzle and lay it on top of the load.

Courtesy of Iowa Fire Service Training Bureau and Ames (IA) FD. Photos by Dawn Beisner.

NOTE: Start the load with the sections of hose connected and the nozzle attached.

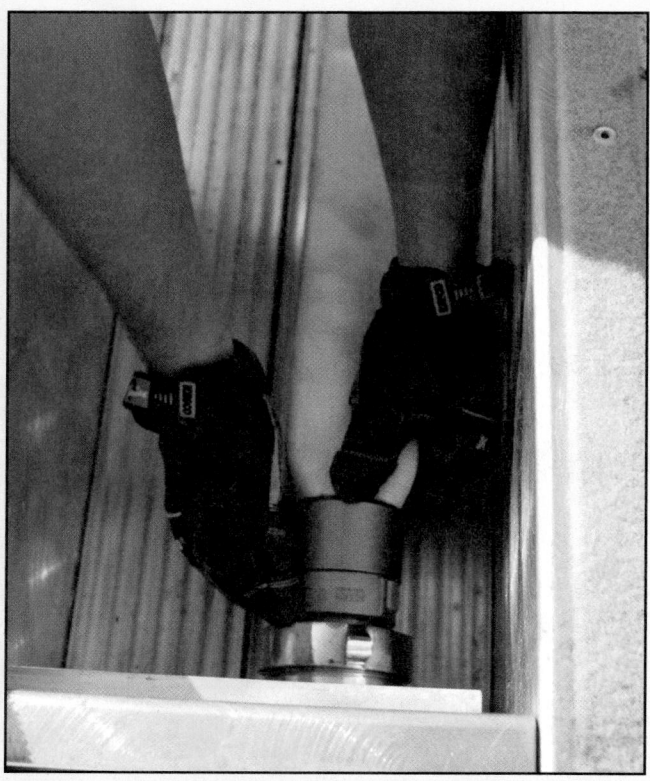

Step 1: Connect the female coupling to the discharge outlet.

Step 2: Extend the hose in a straight line to the rear.

Step 3: Pick up the hose at a point two-thirds the distance from the tailboard to the nozzle.

Courtesy of Iowa Fire Service Training Bureau and Ames (IA) FD. Photos by Dawn Beisner.

Step 4: Carry this hose to the tailboard.

Step 5: Using several firefighters, pick up the entire length of the three layers.

Courtesy of Iowa Fire Service Training Bureau and Ames (IA) FD. Photos by Dawn Beisner.

Step 6: Begin laying the hose into the bed by folding over the three layers into the hose bed.

Step 7: Fold the layers over at the front of the bed.

Step 8: Lay them back to the rear on top of the previously laid hose.

    a. If the hose compartment is wider than one hose width, alternate folds on each side of the bed

    b. Make all folds at the rear even with the edge of the hose bed

Step 9: Continue to lay the hose into the bed in an S-shaped configuration until the entire length is loaded.

Step 10: Optional: Secure the nozzle to the first set of loops using a rope or strap if desired.

Courtesy of Iowa Fire Service Training Bureau and Ames (IA) FD. Photos by Dawn Beisner.

Step 1: Connect the first section of hose to the discharge outlet.

    a. Do not connect it to the other lengths of hose

Step 2: Lay the hose flat in the bed to the front.

Step 3: Lay the remaining hose out the front of the bed to be loaded later.

    a. If the discharge outlet is at the front of the bed, lay the hose to the rear of the bed and then back to the front before it is set aside. This provides slack hose for pulling the load clear of the bed

Courtesy of Iowa Fire Service Training Bureau and Ames (IA) FD. Photos by Dawn Beisner.

Step 4: Couple the remaining hose sections together.

Step 5: Attach a nozzle to the male end.

Step 6: Place the nozzle on top of the first length at the rear.

Step 7: Angle the hose to the opposite side of the bed and make a fold.

Step 8: Lay the hose back to the rear.

Courtesy of Iowa Fire Service Training Bureau and Ames (IA) FD. Photos by Dawn Beisner.

Step 9: Make a fold at the rear of the bed.

Step 10: Angle the hose back to the other side and make a fold at the front.

NOTE: The first fold or two may be longer than the others to facilitate the pulling of the hose from the bed.

Step 11: Continue loading the hose to alternating sides of the bed in the same manner until the complete length is loaded.

Step 12: Connect the male coupling of the first section to the female coupling of the last section.

Step 13: Lay the remainder of the first section in the bed in the same manner.

Courtesy of Iowa Fire Service Training Bureau and Ames (IA) FD. Photos by Dawn Beisner.

Step 2: Firefighter #1: Signal the driver/operator to proceed and deploy the hose to the incident.

Step 1: Firefighter #1(Hydrant Firefighter): Take tools, devices, and other equipment needed to complete connection. Pull enough supply hose to reach and wrap around the hydrant. Place a loop of hose around the hydrant or other effective anchor to secure it.

NOTE: Wrap the hose around the hydrant in a manner that restrains the hose when the pumping apparatus moves away from the hydrant.

Courtesy of Iowa Fire Service Training Bureau and Ames (IA) FD. Photos by Dawn Beisner.

Step 3: Firefighter #1: Connect supply hose to hydrant.

  a. Remove the cap from the hydrant

  b. Place the hydrant wrench on the valve stem operating nut

  c. Remove the hose loop from the hydrant

  d. Connect the hose to the outlet nearest the fire

Step 4: Firefighter #2: Complete the hose lay to the scene. Apply a hose clamp and signal hydrant firefighter to charge the line.

Step 5: Firefighter #1: After slowly and fully opening the hydrant, proceed along the hose to the pumper removing kinks and checking for leaks.

Firefighter #2: Disconnect the supply hoseline from the hose bed. Connect the hose to the fire pump intake valve. Release the hose clamp.

Step 1: Firefighter #1: Pull sufficient hose to reach the intake valve on the attack pumper.

Step 2: Firefighter #1: Anchor the hose.

Step 3: Firefighter #1: Apply a hose clamp to the hose at the attack pumper, move back toward fire scene removing kinks and checking couplings.

Step 4: Firefighter #2: After the pump stops at the water source, make an intake hose connection.

Step 5: Firefighter #2: Pull the remaining length of the last section of hose from the hose bed.

Step 6: Firefighter #2: Disconnect the couplings and return the unused coupling to the hose bed.

Step 7: Firefighter #2: Connect the supply hose to a discharge valve.

Preconnected Flat Hose Method

Step 1: Put one arm through the longer loop.

Step 2: Grasp the shorter pull loop with the same hand.

Step 3: Grasp the nozzle with the opposite hand.

Step 4: Pull the load from the bed using the pull loops.

Step 5: Walk toward the fire.

Step 6: Proceed until the hose is fully extended.

Step 7: Conduct visual size-up of scene to identify hazards.

Minuteman Hose Method

Step 1: Grasp the nozzle and bottom loops, if provided.

Step 2: Pull the load approximately one-third to one-half of the way out of the hose bed.

Step 3: Face away from the apparatus.

Step 4: Place the hose load on the shoulder with the nozzle against the stomach.

Step 5: Walk away from the apparatus, pulling the hose out of the bed by the bottom loop.

Step 6: Advance toward the fire, allowing the load to pay off from the top of the pile.

Step 7: Conduct visual size-up of scene to identify hazards.

Triple Layer Method

Step 3: Walk away from the apparatus.

Step 1: Place the nozzle and fold of the first tier over the shoulder.

Step 2: Face the direction of travel.

Step 4: Pull the hose completely out of the bed.

Step 5: Drop the folded end from the shoulder when the hose bed has been cleared.

Step 6: Advance the nozzle.

Step 7: Conduct visual size-up of scene to identify hazards.

Shoulder Load Method

Step 1: Firefighter #1: Attach the nozzle to the end of the hose if desired.

Step 2: Firefighter #2: Position at the tailboard facing the direction of travel.

Step 3: Firefighter #2: Place the initial fold of hose over the shoulder so that the nozzle can be held at chest height.

Step 7: Firefighter #3: Position at the tailboard facing the direction of travel.

Step 8: Firefighter #3: Load hose onto the shoulder in the same manner as Firefighter #2, making knee-high folds until an appropriate amount of hose is loaded.

Step 9: Firefighter #1: Uncouple the hose from the hose bed, and hand the coupling to the last firefighter.

Step 4: Firefighter #2: Bring the hose from behind back over the shoulder so that the rear fold ends at the back of the knee.

Step 5: Firefighter #2: Make a fold in front that ends at knee height and bring the hose back over the shoulder.

Step 6: Firefighter #2: Move forward approximately 15 feet (5 m).

15-1-15
Deploy a wye-equipped hose during a reverse hose lay.

SKILL
SHEETS

Step 1: Firefighter #1: Grasp each side of one bundle of hose by the tie strap or rope, and pull it from the hose bed.

Step 2: Firefighter #1: Advance the bundle approximately 10 feet (3 m) from the tailboard of the pumping apparatus, and place it on the ground.

Step 3: Firefighter #2: Pull the opposite bundle from the hose bed in the same manner, and place it next to the first bundle, leaving room for the gated wye.

Step 4: Firefighter #1: Bring the gated wye and attached hose from the rear of the hose bed, and put the wye between the bundles near the rope or strap ties.

Step 5: Firefighter #1: Pick up the gated wye when the attack hose has been deployed to the ground along with enough hose to reach a location where the hose can be anchored.

Step 6: Firefighter #1: Anchor the hose, and signal the driver/operator.

Step 7: Driver/Operator: Proceed to the water source, deploying hose from the bed.

Step 8: Firefighter #1: Continue to anchor the hose to ensure that it deploys from the bed as the pumping apparatus moves away.

Step 9: Firefighters #1 and #2: Untie the bundles after the pumping apparatus makes the lay, and begin the first hose advance.

NOTE: The second attack hose bundle remains for the second attack team.

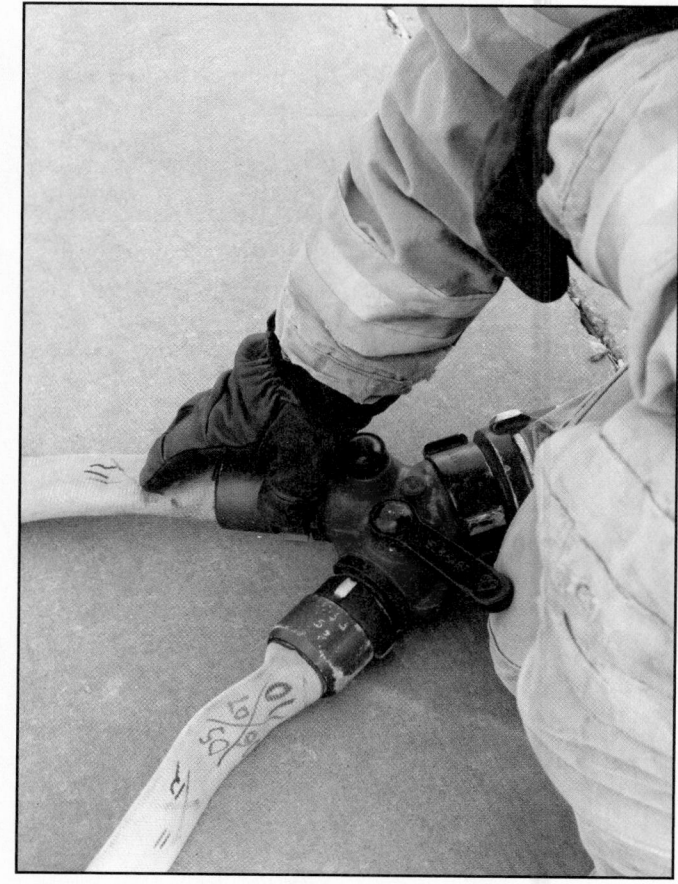

Step 10: Firefighter #1: Open the gated wye when the attack teams are in position and ready for water.

15-I-16
Advance a charged hoseline using the working line drag method.

SKILL SHEETS

Step 1: Stand alongside a single hoseline at a coupling or nozzle.

Step 2: Face the direction of travel.

Step 3: Place the hose over the shoulder with a coupling in front, resting on the chest.

Step 4: Hold the coupling in place and pull with the shoulder.

Step 5: Position additional firefighters at each coupling to assist in advancing the hose.

NOTE: Ensure appropriate SCBA and PPE are in place and don SCBA mask prior to entering IDLH atmosphere.

Step 1: Confirm order with officer to advance a line into the structure.

Step 2: Unload the hose using accordion unload.

Step 3: Horseshoe shoulder the hose, all placing the hose on the same shoulder.

    a. Firefighters spaced about 12 feet (4 m) apart on same side of hose facing the nozzle with about 15 feet (5 m) to 20 feet (6 m) of hose between each firefighter

Step 6: Direct driver/operator to charge hoseline.

Step 7: Open nozzle fully to ensure adequate water flow and to allow pump operator to set pressure.

Step 4: Fully open SCBA before approaching structure entrance or entering smoke environment.

Step 5: Advance the hose to building entrance but do not enter the building.

    a. Size up environment to identify hazards

    b. Approach door from side opposite hinges

Step 8: Set the desired nozzle pattern and bleed air from hoseline.

Step 9: Confirm readiness to enter structure with officer.

Step 10: Enter the structure while staying low and maintaining spacing.

    a. Whenever possible, position firefighters at critical points (obstructions and corners) to help feed the hose

Up Interior Stairs (Uncharged Hoseline)

Step 1: Confirm order with officer to advance a line.

Step 2: Position for shouldering the hoseline by facing the nozzle with about 15 feet (5 m) to 20 feet (6 m) of hose between each firefighter.

Step 3: Place hose bundles on same shoulders per appropriate shoulder carry.

Step 4: Position stationary firefighters along the route and on the stairs at critical points (obstructions and corners) to help feed the hose and to keep the hose on the outside of the staircase.

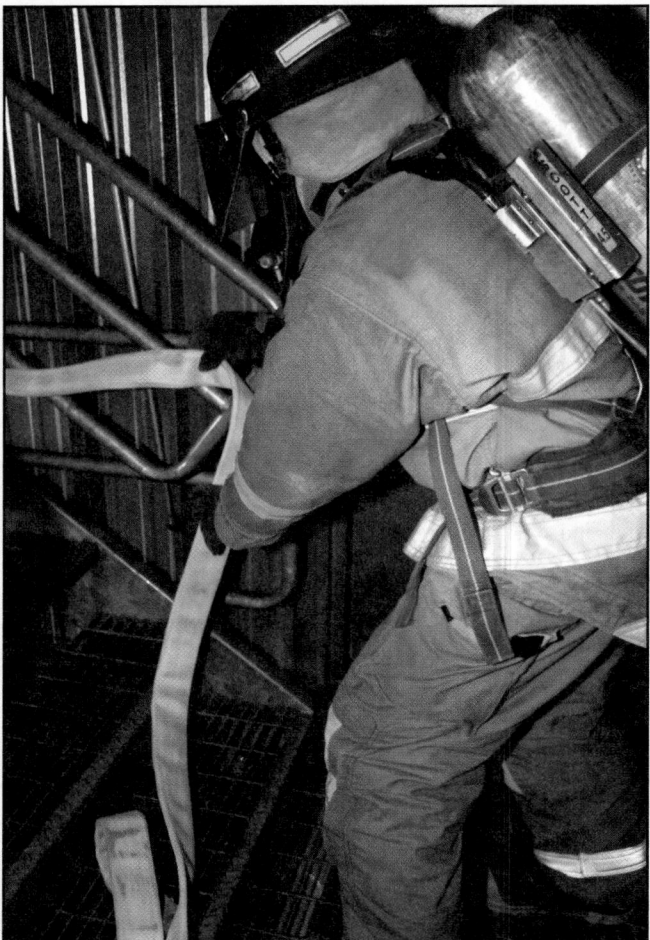

Step 5: Advance the hoseline up a flight of stairs against outside wall avoiding sharp bends and kinks and maintaining spacing between firefighters.

Step 6: Flake excess hose up the stairway leading to floor above fire to make fire floor advance easier and quicker.

Step 7: Lay the hose down the stairway along outside wall to fire floor.

Step 8: Last Firefighter: After hose supply is depleted, advance and assist nozzle operator in removing kinks, pushing hose to outside wall of stairway as necessary.

15-I-18
Advance a line up and down an interior stairway.

SKILL SHEETS

Down Interior Stairs (Uncharged Hoseline)

Step 1: Confirm order with officer to advance a line.

Step 2: Position for shouldering the hoseline by facing the nozzle with about 25 feet (7.5 m) to 30 feet (9 m) of hose between each firefighter.

Step 3: Place hose bundles on same shoulders per appropriate shoulder carry.

Step 5: Advance the hoseline down a flight of stairs against outside wall, avoiding sharp bends and kinks and maintaining spacing between firefighters.

Step 4: Position stationary firefighters along the route and at top of the stairs at critical points (obstructions and corners) to help feed the hose and to keep the hose on outside of the staircase.

Step 6: Last Firefighter: After hose supply is depleted, advance and assist nozzle operator in removing kinks, pushing hose to outside wall of stairway as necessary.

Up Interior Stairs (Charged Hoseline)

Step 1: Confirm order with officer to advance a line.

Step 2: Advance the line using the working line drag.

Step 3: Position stationary firefighters along the route and at top of the stairs at critical points (obstructions and corners) to help feed the hose and to keep the hose on outside of the staircase.

Step 4: Advance up the stairs against outside wall, avoiding sharp bends and kinks, maintaining spacing between firefighters, and using working drag to one floor above fire floor.

Step 5: Make a large loop on floor above fire floor to provide excess line for fire floor advancement.

Step 6: Advance the hose down the stairway to the fire floor, using working drag.

Step 7: Last Firefighter: After hose supply is depleted, advance and assist nozzle operator in removing kinks, pushing hose to outside wall of stairway as necessary.

Step 8: Secure the other end to the discharge pipe or other nearby anchor.

Step 9: Attach a shutoff nozzle (or any device that permits water and air to drain from the hose) to the open end of each test length.

Step 10: Fill each hoseline with water with a pump pressure of 50 psi (350 kPa) or to hydrant pressure.

Step 11: Open the nozzles as the hoselines are filling.

Step 12: Hold nozzles above the level of the pump discharge to permit all the air in the hose to discharge.

Step 13: Discharge the water away from the test area.

Step 14: Close the nozzles after all air has been purged from each test length.

Step 15: Make a chalk or pencil mark on the hose jackets against each coupling.

Step 16: Check that all hose is free of kinks and twists and that no couplings are leaking. Any length found to be leaking from BEHIND the coupling should be taken out of service and repaired before being tested.

Step 17: Retighten any couplings that are leaking at the connections. If the leak cannot be stopped by tightening the couplings, depressurize, disconnect the couplings, replace the gasket, and start over at Step 10.

Step 18: Close each hose test gate valve.

Step 19: Increase the pump pressure to the required test pressure given in NFPA® 1962.

Step 20: Closely monitor the connections for leakage as the pressure increases.

Step 21: Maintain the test pressure for the time specified in your departmental SOP.

Step 22: Inspect all couplings to check for leakage (weeping) at the point of attachment.

Step 23: Slowly reduce the pump pressure after 3 minutes.

Step 24: Close each discharge valve.

Step 25: Disengage the pump.

Step 26: Open each nozzle slowly to bleed off pressure in the test lengths.

Step 27: Break all hose connections and drain water from the test area.

Step 28: Observe marks placed on the hose at the couplings. If a coupling has moved during the test, tag the hose section for recoupling. Tag all hose that has leaked or failed in any other way.

Step 29: Record the test results for each section of hose.

Chapter Contents

Key Terms

NFPA® Job Performance Requirements

This chapter provides information that addresses the following job performance requirements of NFPA® 1001, *Standard for Fire Fighter Professional Qualifications* (2013).

Firefighter I
5.3.10

Firefighter II
6.3.1
6.3.2

Firefighter I Chapter Objectives

1. Explain the way vaporization and steam relate to the extinguishing properties of water. (5.3.10)

2. Identify the factors that create pressure loss or gain. (5.3.10)

3. Describe the impact water hammer has on fire streams. (5.3.10)

4. Explain fire stream patterns and their possible limiting factors. (5.3.10)

5. Describe the three types of fire stream nozzles. (5.3.10)

6. Compare the different types of nozzle control valves. (5.3.10)

7. Describe the factors in operating and maintaining handline nozzles. (5.3.10)

8. Operate a fog-stream nozzle. (Skill Sheet 16-I-1, 5.3.10)

9. Operate a broken-stream nozzle. (Skill Sheet 16-I-2, 5.3.10)

10. Operate a solid stream nozzle. (Skill Sheet 16-I-3, 5.3.10)

Firefighter II Chapter Objectives

1. Describe the methods by which fire fighting foam prevents or controls a hazard. (6.3.1)

2. Identify foam concentrates. (6.3.1)

3. Explain the factors that impact foam expansion and selection. (6.3.1)

4. Describe methods by which foam may be proportioned. (6.3.1)

5. Explain the advantages and disadvantages of various foam proportioners, delivery devices, and generating systems. (6.3.1, 6.3.2)

6. Identify causes of poor foam production. (6.3.1, 6.3.2)

7. Distinguish among various foam application techniques. (6.3.1, 6.3.2)

8. Identify foam hazards and ways to control them. (6.3.1, 6.3.2)

9. Place a foam line in service using an in-line eductor. (Skill Sheet 16-II-1, 6.3.1, 6.3.2)

10. Extinguish an ignitable liquid fire. (Skill Sheet 16-II-2, 6.3.1)

Chapter 16
Fire Streams

Water is the primary extinguishing agent you will use to extinguish fires. Water is readily available, inexpensive to use, and extinguishes by cooling and smothering the fire. The majority of this chapter explains the following information about water and its application:

- Extinguishing properties of water
- Principles of fire streams
- Application methods
- Types, design, operation, effect, and flow capacities of nozzles
- Correct methods for opening, closing, and adjusting nozzles
- Correct methods for achieving desired flow rates and discharge patterns

In some fire departments, foam additives are used to extinguish Class A as well as Class B fires. The last portion of the chapter provides information for Firefighter II trainees in the basic principles of fire fighting foams including:

- How and why foam works
- Types of foam concentrates
- General characteristics of foam
- How foam is mixed with water
- Foam application, equipment, and techniques

▮ Extinguishing Properties of Water

Several characteristics of water are extremely valuable for fire extinguishment including:

- Water is readily available.
- Water is relatively inexpensive.
- Water has a greater heat-absorbing capacity (high specific heat) than most other common extinguishing agents.
- Water changing into steam requires a relatively large amount of heat (high latent heat of vaporization).
- Water can be applied in a variety of ways.

To understand how water is used to extinguish fire, you must know how it is converted to vapor when heat is applied and the properties of steam. These topics are covered in the following sections.

Vaporization

The primary way that water extinguishes fire is by absorbing heat which creates a cooling effect. When heated to its boiling point, water absorbs heat and converts into water vapor or steam in a process called *vaporization*.

Specific Heat — Amount of energy required to raise the temperature of a specified unit mass of a material 1 degree in temperature.

Energy is required both to raise the temperature of water and to change its state from a liquid to a gas (steam). **Specific heat** is the amount of energy required to increase the temperature of a substance by one degree. In the customary system used in the United States, this would be expressed in terms of the number of British thermal units (Btu) required to raise the temperature of 1 pound (lb) of water by 1°F. In the Standard International System of Units (SI), specific heat is expressed in joules (J) per kilogram (kg) per Kelvin. In the SI system, water has a specific heat of 4 200 J/kg K. **(Figure 16.1)**.

Latent Heat of Vaporization — Quantity of heat absorbed by a substance at the point at which it changes from a liquid to a vapor.

Latent heat of vaporization is the amount of heat required to convert unit mass of a liquid into a vapor without a temperature change. Water has a latent heat of vaporization of 970 Btu/lb (2 257 kJ/kg) at its boiling point. For example, when water is heated to its boiling point of 212°F (100°C) additional energy is required to change it from liquid phase to gas phase, but the temperature of the water remains 212°F (100°C) as illustrated in **Figure 16.2**.

Properties of Steam

When water is vaporized into steam it expands. At 212° F (100°C) water expands approximately 1,700 times its original volume when it turns to steam **(Figure 16.3)**. As the temperature increases, steam (like any gas) continues to expand. The volume

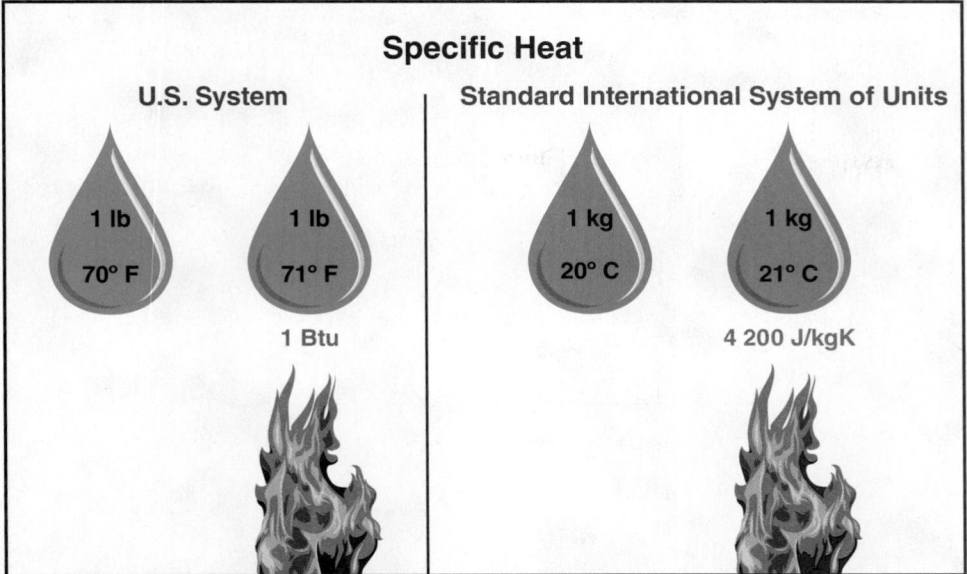

Figure 16.1 Specific heat is the amount of energy required to raise the temperature of a substance by one degree.

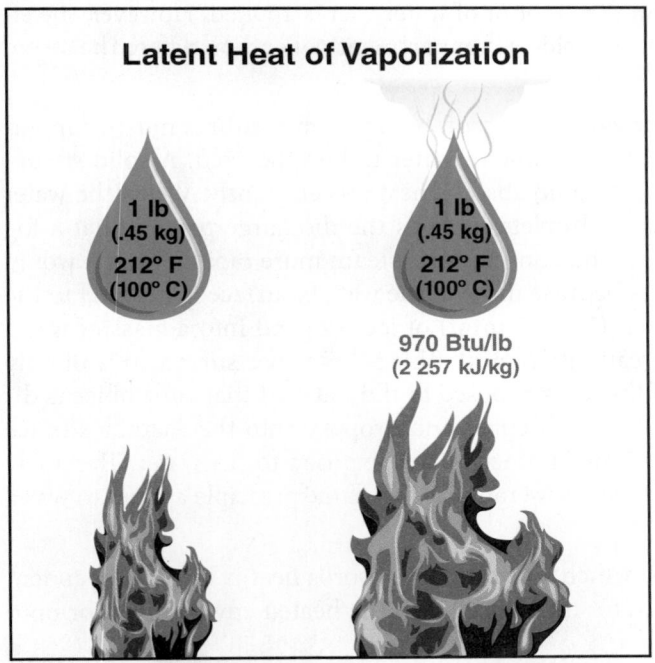

Figure 16.2 Although additional energy is required to change water from a liquid to a gas, the temperature of the water remains the same through the process.

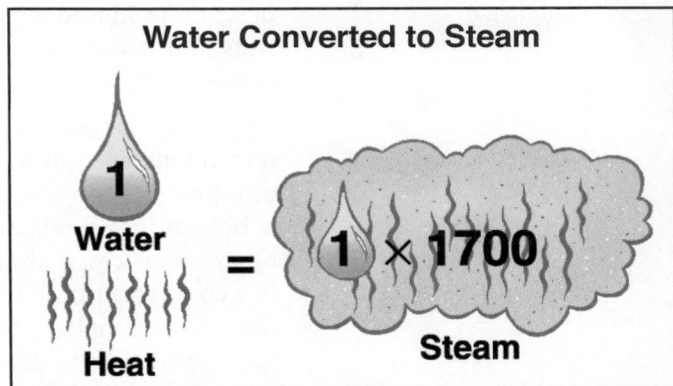

Figure 16.3 Water expands 1,700 times its original volume when converted to steam.

Figure 16.4 The total volume of smoke competes for space in the upper level of the room and expands downward as steam from a fog spray is added.

of steam produced depends on the amount of water that is applied. However, the effects of this steam on conditions inside a compartment depend on where the steam is produced.

In order for complete vaporization to occur, boiling temperatures must be maintained long enough for the entire volume of water to be vaporized. A solid stream of water has a smaller surface area and absorbs heat less efficiently. When the water is broken into small particles or droplets, such as the discharge pattern that a fog nozzle produces, it absorbs heat and converts into steam more rapidly than it would in a compact form. This occurs because more of the water's surface is exposed to the heat. For example, 1 cubic inch (1 638.7 mm³) of ice dropped into a glass of water takes some time to absorb its capacity of heat. This is because a surface area of only 6 square inches (387 mm²) of the ice is exposed to the water. If that cube of ice is divided into 1/8-cubic inch (204.8 mm³) cubes and dropped into the water, a surface area of 48 square inches (3 096 mm²) of ice is now exposed to the water. The finely divided particles of ice absorb heat more rapidly. This same principle applies to water in the liquid state.

NOTE: The efficiency with which a fire stream absorbs heat is largely dependent upon the surface area of the water introduced into the heated environment or onto the fuel surface.

Steam production during fire fighting is necessary for effective and efficient use of water as an extinguishing agent. However, care must be taken to apply the appropriate amount of water in the right place to achieve the desired effect. When steam is produced on contact with hot surfaces such as burning fuel or walls and ceiling materials that are hotter than 212°F (100°C), water is vaporized into steam which adds to the total volume of the upper layer of hot smoke and fire gases. As this total volume increases and the room fills with the mixture of hot smoke and steam, the upper layer of smoke expands downward, potentially making conditions uncomfortable or even dangerous for firefighters inside the room **(Figure 16.4)**.

When water turns to steam in the upper layer of hot smoke and fire gases, the upper layer tends to shrink rather than expand. Cooling the upper layer requires vaporizing the water while it passes through the hot gases. As the water vaporizes, the hot gases are cooled and the upper layer contracts. Because the energy required to heat and vaporize water is much greater than that required to cool the hot gases, the temperature of the hot gases drops faster and farther than the temperature of the steam rises. Because of this effect, water turned to steam in the hot gases of the upper layer actually causes the volume of the upper layer to decrease and contract toward the ceiling.

NOTE: Good nozzle control and coordination with tactical ventilation make for effective, efficient, and safer fire stream operation and fire control.

Effects of Stream Patterns on Vaporization at Ceiling Level

Consider the following two examples. Both cases involve a fire in a 10 x 20 foot (3 m by 6 m) compartment with a 10-foot (3 m) ceiling, providing a total volume of 2,000 ft³ (57 m³). The hot layer of smoke and fire gases is 5 feet (1.5 m) deep and has an average temperature of 932°F (500°C).

Example 1: In this example, the nozzle operator uses a straight stream or narrow fog pattern to discharge sufficient water upward towards the ceiling to lower the temperature in the compartment to 392°F (200°C). The straight stream or narrow fog pattern causes most of the water to reach the ceiling. If only 10 percent of the water is vaporized in the hot upper layer and 90 percent is vaporized on contact with the hot ceiling surface, the volume of the upper layer will almost double, filling the room with hot smoke, fire gases, and steam.

Example 2: In this example, the nozzle operator uses a fog nozzle set on a wider pattern and applies the same volume of water as in Example 1 using short pulses of water fog, lowering the temperature in the compartment to 392°F (200°C). However, in this case 40 percent of the water is vaporized in the upper layer of hot gases and 60 percent is vaporized on contact with the hot ceiling surface. This results in a slight decrease in the volume of the upper layer comprised of hot smoke, fire gases, and steam.

Pressure Loss or Gain

To effectively extinguish a fire, there must be enough water exiting the nozzle at a sufficient pressure to reach the fire. The water, which originates at the pumper and travels through a varying length of hose, will lose or gain pressure between the pumper and the nozzle. Pressure loss is caused by friction that is created as the water travels through the hose and by elevation if the nozzle is higher than the pumper. Pressure gain will occur when the nozzle is lower than the elevation of the pumper.

Principles of Friction Loss

Friction loss is that part of total water pressure that is lost while forcing water through pipes, fittings, fire hose, and adapters. When water flows through these items, the water molecules rub against the insides, producing friction. The friction slows the water flow and reduces its pressure at the nozzle. The loss of pressure in a hoseline between a pumper and the nozzle (excluding pressure changes due to elevation) is the most common example of friction loss (**Figure 16.5, p. 948**). Friction loss in a hoseline

Friction Loss — Loss of pressure created by the turbulence of water moving against the interior walls of fire hose, pipes, fittings, and adapters.

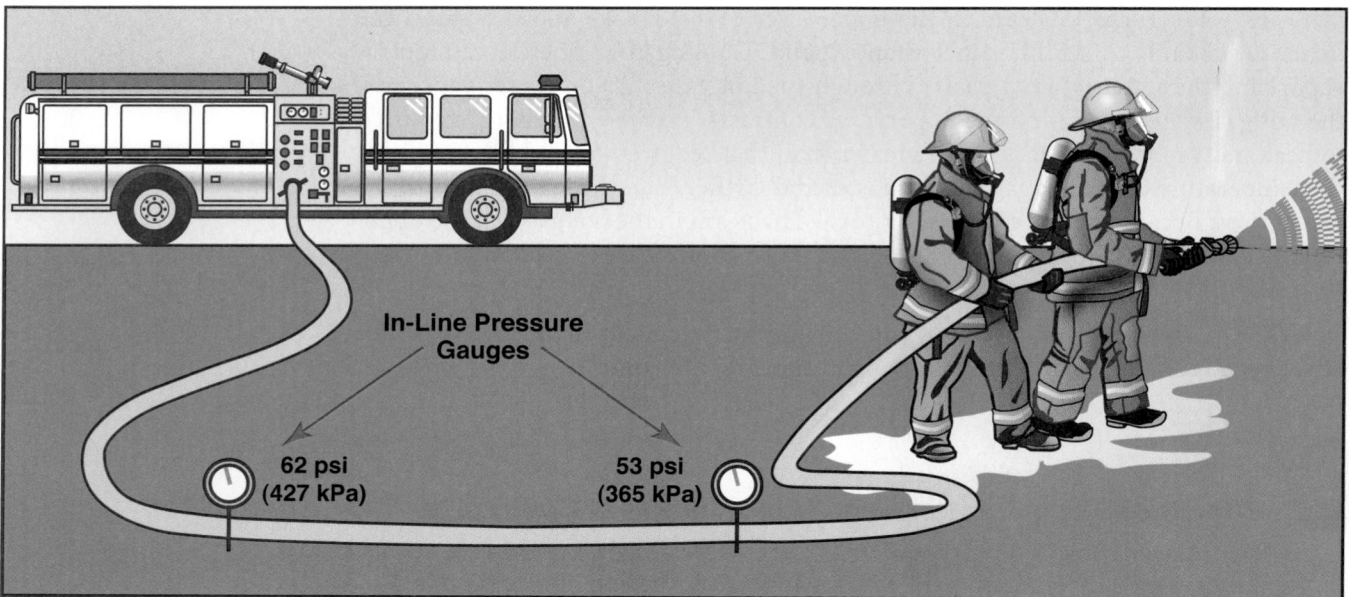

Figure 16.5 To measure friction loss between the pumper and the nozzle, insert in-line pressure gauges at intervals in the hoseline.

In-Line Pressure Gauges

62 psi
(427 kPa)

53 psi
(365 kPa)

can be measured by inserting in-line gauges at different points in a hose layout. The difference in the pressures between gauges when water is flowing through the hose is the friction loss for the length of hose between those gauges for a given rate of flow.

There is a practical limit to the velocity or speed at which water can travel through a hoseline. If the velocity is increased beyond this limit, the friction becomes so great that the water in the hoseline is agitated by resistance. Certain characteristics of hose layouts, such as hose size and length of the hose lay, also affect friction loss. In general, the smaller the hose diameter and the longer the hose lay, the higher the friction loss at a given pressure and flow rate.

Friction loss in fire hose is increased by the following conditions:

- Rough linings in fire hose
- Damaged hose couplings
- Sharp bends in hose
- Number of adapters
- Length of hose lay
- Hose diameter

Friction loss is overcome by increasing the hose size, adding additional parallel hoselines, or increasing the pump pressure. It can also be reduced by taking any kinks or sharp bends out of the hoseline.

Elevation Loss or Gain

The difference in elevation between the nozzle and the pumping apparatus causes elevation pressure. When a nozzle is *above* the fire pump, there is a *pressure loss* (**Figure 16.6**). When the nozzle is *below* the pump, there is a *pressure gain*. These changes in pressure caused by gravity must be compensated for by adjusting the pressure at the pump.

Figure 16.6 A hose nozzle produces higher pressure when it is in a position lower than the pump, and a nozzle produces lower pressure when it is higher than the pump. *Courtesy of Bob Esposito.*

Water Hammer

When the nozzle is closed quickly and suddenly, a shock wave is produced when the moving water reaches the closed nozzle and bounces back. The resulting pressure surge is referred to as **water hammer**. This sudden change in the direction creates excessive pressures that can cause considerable damage to water mains, plumbing, fire hose, hydrants, and fire pumps. At low flow rates, water hammer is minimal. At higher flow rates, the effects of water hammer increase significantly. To prevent water hammer, slowly close nozzles, hydrants, control valves, and hose clamps.

Fire Stream Patterns, Nozzles, and Control Valves

A **fire stream** is a stream of water or other extinguishing agent after it leaves a fire hose nozzle until it reaches the desired target. The following factors have an effect on a fire stream:

- Velocity of the water
- Gravity
- Wind direction and velocity
- Air friction
- Operating pressure
- Nozzle design and adjustment
- Condition of the nozzle opening

  Fire streams are used for the following:

- Apply water or foam directly onto burning material to reduce its temperature
- Apply water or foam into open flames to reduce the temperature so that firefighters can advance handlines

Water Hammer — Force created by the rapid deceleration of water causing a violent increase in pressure that can be powerful enough to rupture piping or damage fixtures. Generally results from closing a valve or nozzle too quickly.

Fire Stream — Stream of water or other water-based extinguishing agent after it leaves the fire hose and nozzle until it reaches the desired point.

- Reduce the temperature of the upper gas layers
- Disperse hot smoke and fire gases from a heated area
- Create a water curtain to protect firefighters and property from heat
- Create a barrier between a fuel and a fire by covering the fuel with a foam blanket

  Fire streams can be best described in terms of the following information:
- The patterns they form
- The nozzles that create those patterns
- The types of control valves that permit the flow of water through the nozzle
- The factors that limit a fire stream

The size of the nozzle opening or orifice and nozzle pressure determines the quantity of water flowing from the nozzle. At the same time, the size of the opening also influences the reach or distance of the fire stream. Finally, the type of nozzle determines the shape of the fire stream. A smooth bore nozzle produces a tightly-packed solid stream of water. A fog nozzle produces a fog or straight stream. The sections that follow provide more detailed information about fire streams.

Fire Stream Patterns

Fire stream patterns are defined by their size and type. The size refers to the volume or quantity of water flowing from the nozzle per minute; the type indicates the specific pattern or shape of the water after it leaves the nozzle.

Size

Fire streams are classified into one of three sizes: low-volume streams, handline streams, and master streams (**Figure 16.7**). The rate of discharge of a fire stream is measured in gallons per minute (gpm) or liters per minute (L/min) as follows:

- **Low-volume stream** — Discharges less than 40 gpm (160 L/min). Typically supplied by ¾-inch (20 mm), 1-inch (25 mm), or 1½-inch (38 mm) hoselines.
- **Handline stream** — Supplied by 1½- to 3-inch (38 mm to 77 mm) hose, with flows from 40 to 350 gpm (160 L/min to 1 400 L/min). Nozzles with flows in excess of 350 gpm (1 400 L/min) are not recommended for handlines.

Figure 16.7 A low-volume stream, a handline, and a master stream appliance each serve important functions at a fire scene.

- **Master stream** — Discharges more than 350 gpm (1 400 L/min) and is fed by one or more 2½- or 3-inch (65 mm or 77 mm) hoselines or large-diameter hoselines connected to a master stream nozzle. Nozzle pressures of 80 to 100 psi (560 to 700 kPa) are common with master stream devices. Master streams are large-volume fire streams created by master stream appliances such as apparatus-mounted deck pipes and ladder pipes.

The volume of water discharged is determined by the design of the nozzle and the water pressure at the nozzle. To be effective, a fire stream must deliver a volume of water sufficient to absorb heat faster than the fire generates it. If the heat-absorbing capability of a fire stream does not exceed the heat output from the fire, extinguishing the fire by cooling is impossible **(Figure 16.8)**.

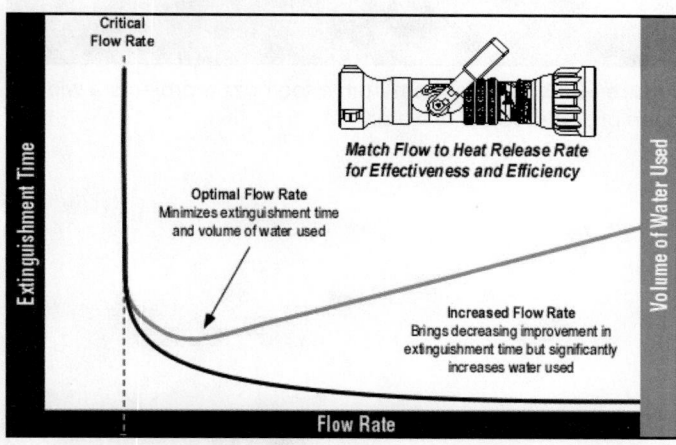

Figure 16.8 Critical and optimal water flow rates influence the extinguishment time. *Courtesy of Ed Hartin.*

Type

The type of fire stream indicates a specific pattern or shape of the stream as it leaves the nozzle. In general, the pattern must be compact enough for the majority of the water to reach the burning material. Effective fire streams must meet or exceed the **critical flow rate**. They must also have sufficient reach to put water where it is needed. The major types of fire stream patterns are solid, fog, straight, and broken **(Figures 16.9 a-d), p. 952**. The fire stream pattern may be any one of these in any size classification.

To produce an effective fire stream, regardless of type and size, several things are needed. All fire streams must have an agent (water), a pressuring device (pump), a means for the agent to reach the discharge device (hoseline), and a discharge device (nozzle) **(Figure 16.10, p. 953)**. The following sections more closely examine the characteristics of different types of fire streams.

Solid stream. A **solid stream** is a fire stream produced from a fixed orifice, **smooth bore nozzle**. Smooth bore nozzles are designed to produce a stream as compact as possible with little shower or spray. A solid stream has the ability to reach areas that other streams might not reach. It can also penetrate and saturate burning materials or debris. The reach of a solid stream can be affected by gravity, friction of the air, and wind **(Figure 16.11, p. 953)**.

Characteristics of solid streams include:
- Good reach and stream penetration
- Stream produced at low nozzle pressure
- Produces less steam conversion
- Provides less heat absorption per gallon (liter)
- More likely to conduct electricity

Critical Flow Rate — The minimum flow rate at which extinguishment can be achieved.

Solid Stream — Hose stream that stays together as a solid mass, as opposed to a fog or spray stream; a solid stream is produced by a smooth bore nozzle and should not be confused with a straight stream.

Smooth Bore Nozzle — A nozzle with a straight, smooth tip, designed to produce a solid fire stream.

CAUTION
Do not use solid streams on energized electrical equipment.

Figure 16.9a A smooth bore nozzle projects a high velocity solid stream.

Figure 16.9b A fog stream from a fog nozzle projects a wide cone of soft droplets of water.

Figure 16.9c A straight stream from a fog nozzle may serve as an equivalent to a solid stream from a smooth bore nozzle.

Figure 16.9d Spray from a broken-stream nozzle is used for special applications.

Figure 16.10 The four critical elements of a fire stream are an agent, a pressuring device (pump), a means of delivery (hose), and a discharge device (nozzle).

Figure 16.11 The breakover point of a stream can be observed as the distance where it begins to lose its forward velocity. *Courtesy of Major Danny Atchley, Oklahoma City Fire Department.*

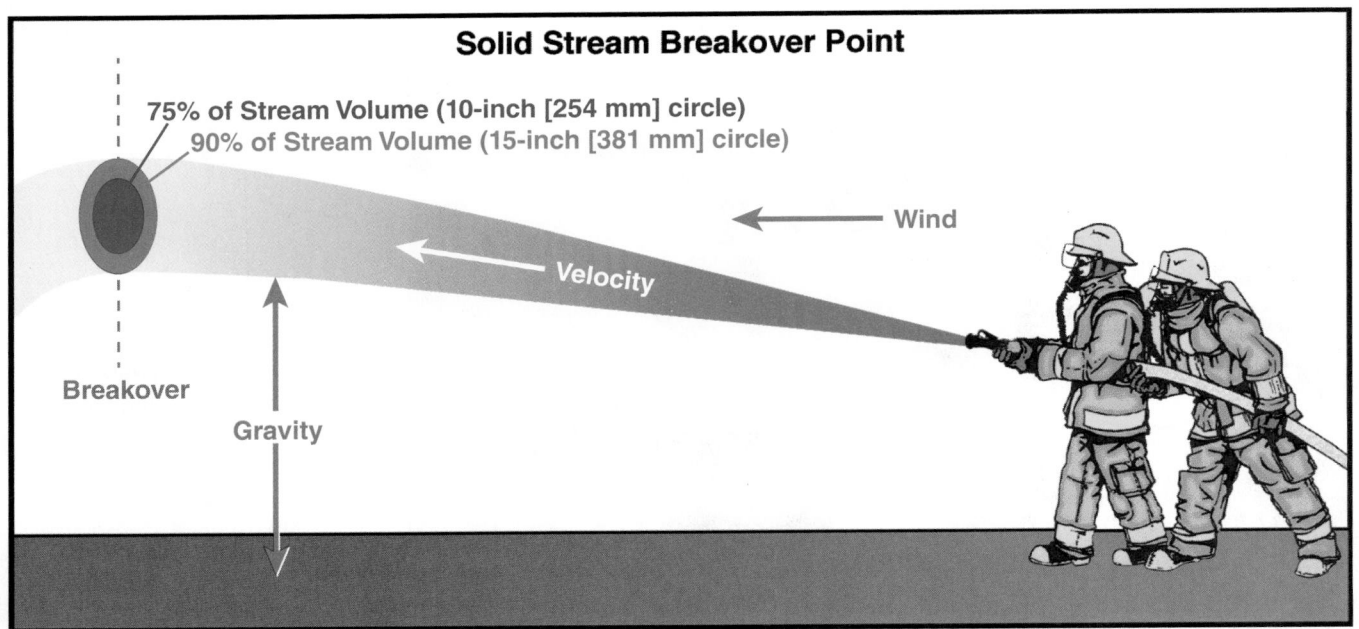

Solid Stream Breakover Point

75% of Stream Volume (10-inch [254 mm] circle)
90% of Stream Volume (15-inch [381 mm] circle)

Wind

Velocity

Breakover

Gravity

Figure 16.12 The point at which a stream loses its effectiveness may be classified under a few different parameters including the breakover point.

The extreme limit at which a solid stream of water can be classified as an effective stream is a matter of judgment. It is difficult to say just exactly where the stream ceases to be effective. Observations and tests covering the effective range of fire streams classify effective streams as:

- A stream that does not lose its continuity until it reaches the point where it loses its forward velocity (breakover) and falls into showers of spray that are easily blown away **(Figure 16.12)**.

- A stream that is cohesive enough to maintain its original shape and attain the required height even in a light, gentle wind (breeze).

The performance of a solid stream depends on the velocity of the stream resulting from the pump pressure and the size of the nozzle orifice. A **nozzle pressure** (NP) of 50 psi (350 kPa) will produce fire streams from smooth bore nozzles with good reach and volume. If greater reach and volume are needed, the nozzle pressure may be increased to 65 psi (450 kPa). Above this pressure, the nozzle and hoseline will require more personnel to handle safely.

Fog stream. A **fog stream** is a fine spray composed of tiny water droplets. **Fog nozzles** are used to produce fog streams and designed to permit adjustment of the tip to produce different fog-stream patterns. Water droplets, in either a shower or spray, are formed to expose the maximum water surface for heat absorption. The desired performance of a fog stream is characterized by the amount of heat that it absorbs and the rate by which the water is converted into steam or vapor.

Fog streams have the following characteristics:

- Fog-stream patterns can be adjusted to suit the situation.
- Fog streams can be used for hydraulic ventilation.
- Fog streams can be used for vapor dispersion.
- Fog streams can be used for crew protection.
- Fog streams reduce heat by exposing the maximum water surface for heat absorption.

Figure 16.13 A solid stream nozzle projects force behind the line of deployment, as opposed to a fog nozzle which lets the water feather into mist at the point where the straight stream reaches the breakover point. *Courtesy of Shad Cooper, Wyoming State Fire Marshal's Office.*

- Fog streams may be used to cool the hot fire gas layer as well as hot surfaces.

- Fog streams have shorter reach or penetration than solid or straight fire streams **(Figure 16.13)**.

- Fog streams are more affected by wind than are solid or straight fire streams.

- Fog streams may disturb thermal layering in a room or compartment if applied incorrectly.

- Fog streams may intensify the fire by pushing fresh air into the fire area if used incorrectly

The angle of fog streams range from narrow to wide. A narrow-angle fog pattern has the highest forward velocity and its reach varies in proportion to the pressure applied. A wide-angle fog pattern has less forward velocity and a shorter reach than the other fog settings. Like all fire streams, any fog pattern will have a maximum reach. A nozzle pressure of 100 psi (700 kPa) is the standard for fog nozzles. Once the nozzle pressure has produced a fire stream with maximum reach, further increases in nozzle pressure have little effect on the stream except to increase the volume. See **Skill Sheet 16-I-1** for procedures in using a fog nozzle to produce a variety of fog-stream patterns.

Straight stream. A **straight stream** pattern is a semi-solid steam that is produced by a fog nozzle. This is done by rotating the stream shaper until a straight stream is produced. Characteristics of straight stream patterns are similar to those of the solid stream.

Broken stream. A **broken stream** is a fire stream that has been broken into coarsely divided water droplets by specialized nozzles such as cellar nozzles, piercing (penetrating) nozzles, and chimney nozzles **(Figure 16.14, p. 956)**. While a solid stream may become a broken stream past the breakover point, a true broken stream takes on that form as it leaves the discharge device. In some cases, the effects of a broken stream can be produced by deflecting solid or straight streams off a wall or ceiling so they break up over the fire. This type of stream is used to extinguish fires in attics, cocklofts, basements, and other confined spaces. Steps for using a broken stream delivery device are given in **Skill Sheet 16-I-2**.

Straight Stream — Semi-solid stream that is produced by a fog nozzle.

Broken Stream — Stream of water that has been broken into coarsely divided drops.

Figure 16.14
A broken-stream nozzle can create a swirling pattern used to extinguish fires in confined spaces above or below ground.

Broken streams have the following characteristics:

- Coarse droplets absorb more heat per gallon (liter) than a solid stream.
- They have greater reach and penetration than a fog stream.
- They can be effective on fires in confined spaces.
- May have sufficient continuity to conduct electricity.
- Stream may not reach some fires.

Fire Stream Limiting Factors

There are five limiting factors that affect the reach of a fire stream:

- **Gravity** — Gravity not only limits the verticle and horizontal distance the fire stream will travel, it also causes solid streams to separate and lose their compact shape.
- **Water velocity** — Effective forward velocity of the fire stream ranges from 60 to 120 feet per second (18.3 to 36.6 meters per second). This velocity is generated by nozzle pressures of 25 to 100 psi (175 kPa to 700 kPa).
- **Fire stream pattern** — Solid stream patterns have greater reach than straight, fog, or broken patterns.
- **Water droplet friction with air** — Air friction has greater effect on the multiple finely-formed water droplets in a fog stream than it does on the outer surfaces of a compact solid stream.
- **Wind** — Wind direction and speed can shorten the reach and deteriorate the shape of the fire stream considerably. The negative effect is increased on fog streams.

Under ideal circumstances, the greatest horizontal reach for a fire stream is attained at 45 degrees from the horizontal plane. In actual operation, fire stream angles between 30 degrees and 34 degrees provide the maximum effective horizontal reach (**Figure 16.15**).

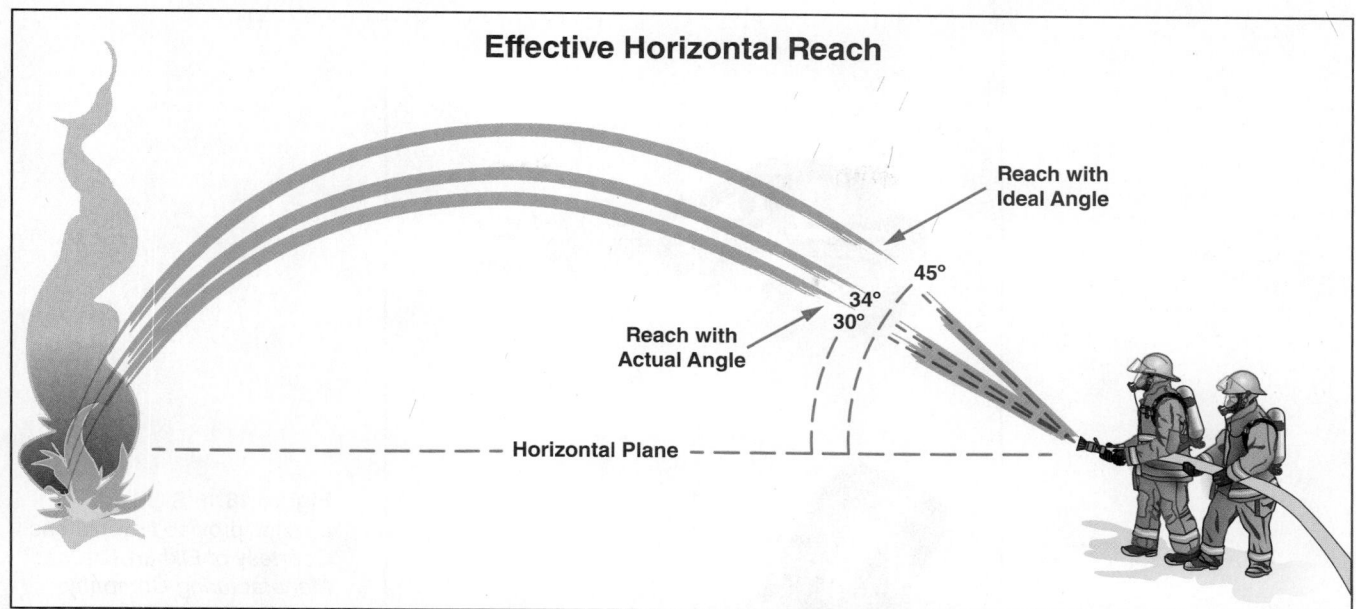

Effective Horizontal Reach

Reach with Ideal Angle

Reach with Actual Angle

45°
34°
30°

Horizontal Plane

Figure 16.15 The actual angle of greatest effectiveness is between 30 and 34 degrees, as opposed to results found in a laboratory which indicate that a 45 degree angle is more useful.

Fire Stream Nozzles

NFPA® 1963, *Standard for Fire Hose Connections* (2009), establishes two general categories of fire stream nozzles: straight tip nozzles and spray nozzles. For this text, straight tip nozzles will be referred to as *smooth bore nozzles* and spray nozzles will be referred to as *fog nozzles*. Smooth bore and fog nozzles are used on handlines and on **master stream** appliances such as fixed apparatus-mounted monitors, portable monitors, and elevated monitors mounted on aerial devices. In addition, delivery devices for broken fire streams, not covered by the standard, can be used to apply water in confined spaces that attack hoselines cannot reach. Both categories of nozzles as well as the broken-stream delivery devices perform three main functions: controlling water flow, creating reach, and shaping the fire stream.

Master Stream — Large-caliber water stream usually supplied by combining two or more hoselines into a manifold device or by fixed piping that delivers 350 gpm (1 325 L/min) or more.

Nozzle Terminology

During the history of the fire service, many terms have been used to describe nozzle types. The terms *straight tip* and *spray* are based on the current NFPA® definitions. However, other terms may be in use by different departments or agencies. They include:

NFPA® Terms	Alternate Terms
Straight Tip Nozzle	Smooth Bore Nozzle
	Solid Bore Nozzle
	Solid Stream Nozzle
Spray Nozzle	Combination Nozzle
	Fog Nozzle
	Adjustable Fog Nozzle

Figure 16.16 Stacked nozzles provide flow options. *Courtesy of Elkhart Brass Manufacturing Company, Inc., Task Force Tips, and Akron Brass Company.*

Smooth Bore Nozzles

Smooth bore nozzles are designed so that the shape of the water in the nozzle is gradually reduced until it reaches a point a short distance from the orifice. At this point, the nozzle becomes a smooth cylinder with a length between 1 to 1½ times its inside diameter. The purpose of this short, cylindrical section is to give the water its round shape before discharge.

NOTE: The smooth bore nozzle tip size should not be larger than one-half the diameter of the hose.

Characteristics of smooth bore nozzles:

- Operate at low nozzle pressures
- Are less prone to clogging with debris
- Can be used to apply compressed-air foam
- May allow hoselines to kink due to less pressure
- Do not allow for selection of different stream patterns

Table 16.1 (Customary)
Flow in gpm from Various Sized Solid Stream Nozzles

Nozzle Pressure in psi	Nozzle Diameter in Inches									
	1	1⅛	1¼	1⅜	1½	1⅝	1¾	1⅞	2	2¼
50	209	265	326	396	472	554	643	740	841	1065
55	219	277	342	415	495	581	674	765	881	1118
60	229	290	357	434	517	607	704	810	920	1168
65	239	301	372	451	537	631	732	843	958	1215
70	246	313	386	469	558	655	761	875	994	1260
75	256	324	399	485	578	678	787	905	1030	1305
80	264	335	413	500	596	700	813	935	1063	1347

The velocity of the stream is a result of the nozzle pressure. This pressure and the size of the discharge opening determine the flow from a smooth bore nozzle. When smooth bore nozzles are used on handlines, they are usually operated at 50 psi (350 kPa) nozzle pressure. Most smooth bore master stream appliances are operated at 80 psi (560 kPa).

The flow rate from smooth bore nozzles depends on the velocity of the stream resulting from the pump pressure and the size of the nozzle orifice. Some smooth bore nozzles are equipped with a single-size tip for a single flow rate and others have stacked tips to provide varied flows **(Figure 16.16)**. When using nozzles equipped with a stacked tip, remove low-flow tips before placing the nozzle in operation if higher flows are required. Changing the flow rate requires that the nozzle be shut off and the tip changed. **Table 16.1** shows the flows available through various size tips at a constant pressure. Steps for using a smooth bore nozzle to create a solid stream are given in **Skill Sheet 16-I-3**.

Fog Nozzles

Fog nozzles may be manually or automatically adjusted resulting in straight stream, narrow-angle fog, and wide-angle fog patterns. Fog nozzles should be operated at their designed operating pressures. Characteristics of fog nozzles are:

- The discharge pattern can be adjusted.
- Fog nozzles can provide protection to firefighters with a wide fog pattern.
- Fog nozzles can be used for a variety of applications.
- They offer a variety of nozzle choices.
- Fog nozzles can be used to apply certain types of foam.

There are four types of fog nozzles available to the fire service **(Figures 16.17a-d, p. 960)**:

- **Basic fog nozzle** — An adjustable-pattern fog nozzle in which the rated discharge is delivered at a designated nozzle pressure and nozzle setting
- **Constant gallonage fog nozzle** — An adjustable pattern fog nozzle that discharges a constant discharge rate throughout the range of patterns from a straight stream to a wide fog at a designed nozzle pressure

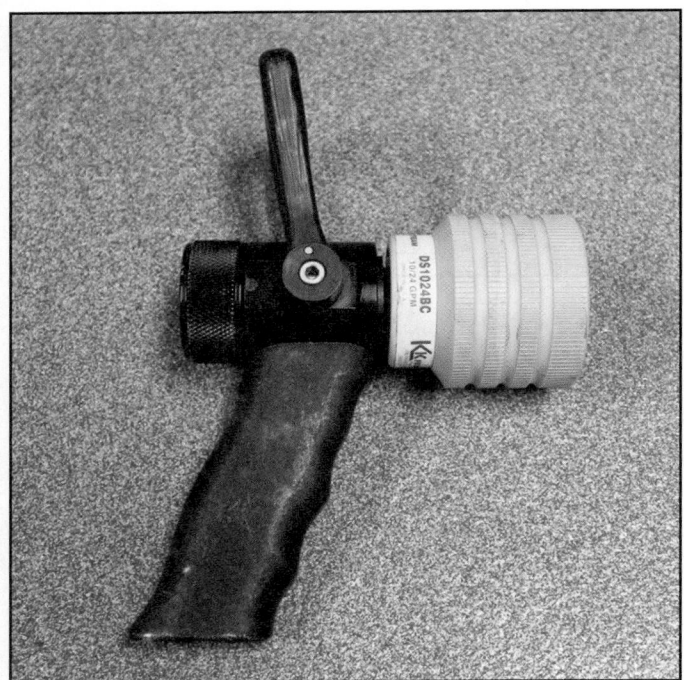

Figure 16.17a A basic fog nozzle has an adjustable ring to control output. *Courtesy of Shad Cooper, Wyoming State Fire Marshal's Office.*

Figure 16.17b A constant gallonage nozzle controls the discharge rate to maintain a steady flow rate. *Courtesy of Shad Cooper, Wyoming State Fire Marshal's Office.*

Figure 16.17c Constant pressure nozzles may be adjusted for more than one discharge rate in order to maintain a constant pressure at the nozzle.

Figure 16.17d A constant/select nozzle allows an operator to manually adjust the flow rate while the nozzle is flowing.

- **Constant pressure (automatic) fog nozzle** — An adjustable-pattern fog nozzle in which the pressure remains relatively constant through a range of discharge rates

- **Constant/select gallonage fog nozzle** — A constant discharge rate fog nozzle with a feature that allows manual adjustment of the orifice to effect a predetermined discharge rate while the nozzle is flowing

The rate of discharge from a manually adjustable fog nozzle can be changed by rotating the selector ring — usually located directly behind the nozzle tip — to a specific gpm (L/min) setting. Each setting provides a constant rate of flow as long as there is adequate nozzle pressure. The nozzle operator has the choice of making flow rate adjustments either before opening the nozzle or while water is flowing. Depending upon the size of the nozzle, the operator may adjust flow rates from 10 gpm to 250 gpm (40 L/min to 1 000 L/min) for handlines and from 350 gpm to 2,500 gpm (1 200 L/min to 10 000 L/min) for master streams. Most of these nozzles also have a "flush" setting to rinse debris from the nozzle.

Adjustments to the rate of flow should be made in small increments. Major adjustments can cause an abrupt change in the reaction force of the hoseline that may throw firefighters off balance.

CAUTION
Abrupt changes in the reaction force of the hoseline may throw firefighters off balance.

Constant-pressure fog nozzles automatically vary the rate of flow to maintain a reasonably constant nozzle pressure through a specific flow range. A specified minimum nozzle pressure is needed to maintain a good fog pattern. With this type of nozzle, the nozzle operator can change the rate of flow by opening or closing the shutoff valve. Automatic fog nozzles allow the nozzle operator to vary the flow rate while maintaining a consistent nozzle pressure.

Automatic fog nozzles for handlines are designed for the following flow rates:

- Low flows such as 10 gpm (40 L/min) to 125 gpm (500 L/min)

- Mid-range flows such as 70 gpm (280 L/min) to 200 gpm (800 L/min)

- High flows such as 70 gpm (280 L/min) to 350 gpm (1 400 L/min)

Automatic master stream fog nozzles are typically designed to flow between 350 gpm (1 400 L/min) and 1,250 gpm (5 000 L/min). These nozzles are supplied by large-diameter or multiple hoselines or are directly connected to a fire pump by piping.

NOTE: Water flow adjustments in manual and automatic fog nozzles require close coordination between the nozzle operator, the company officer, and the pump operator.

Fog nozzles are designed to operate at a variety of nozzle pressures. The designed operating pressure for most fog nozzles is 100 psi (700 kPa); nozzles with a designed operating pressure of 75, 50, or even 45 psi (525, 350, or 315 kPa) are also available. Although these nozzles have less nozzle reaction compared to nozzles designed to operate at 100 psi (700 kPa), droplet size is much greater, fog pattern density is lower, and the fire stream has less velocity.

Figure 16.18 Piercing nozzles limit the ventilation impact of access to a confined space while allowing water to access to the fire in that space.

Broken-Stream Delivery Devices

A broken stream can be used effectively to extinguish concealed space fires such as those found in basements, chimneys, attics, or other types of concealed spaces. The special nozzles that can be used to produce a broken stream include: piercing nozzle, Bresnan distributors, and Rockwood cellar pipes.

Piercing nozzles. Used to access fires in concealed spaces. This nozzle can be used to pierce material such as stucco, block, wood and lightweight steel. The nozzle consists of a piercing tip, shaft, hose connection, and striking plate at the end. A nozzle control valve can be attached between the hose connection and the supply hose. The nozzle is usually driven into place with a mallet, sledgehammer, or flathead axe **(Figure 16.18)**.

Cellar nozzles. The cellar nozzle consists of a rotating head with multiple outlets that distribute water in a circular pattern. The nozzle may be supplied by a 1½-inch (38 mm) or 2½-inch (65 mm) supply hose with a control valve located one section of hose from the nozzle. The nozzle and hose are lowered into the cellar, attic, cockloft, or confined space through a hole cut in the overhead surface. Two commonly used cellar nozzles are the Bresnan distributor and the Rockwood cellar pipe **(Figures 16.19a and b)**.

Nozzle Control Valves

Nozzle control or shutoff valves enable the nozzle operator to start, stop, increase, or decrease the flow of water while maintaining effective control of the nozzle. These valves allow the operator to open the nozzle slowly, to control the nozzle reaction increases, and to close it slowly to prevent water hammer. There are three main types of nozzle control valves found on smooth bore, fog nozzles, and broken-stream delivery devices: ball, slide, and rotary control.

Figure 16.19a Bresnan distributors are lowered into a cellar or other confined space to create an irregular discharge pattern. *Courtesy of Shad Cooper, Wyoming State Fire Marshal's Office.*

Figure 16.19b The Rockwood cellar pipe creates an irregular discharge pattern during activation.

Ball Valve

Ball valves are the most common nozzle control valves. They provide effective nozzle control with a minimum of effort. The ball, perforated by a smooth waterway, is suspended from both sides of the nozzle body and sealed against a seat **(Figure 16.20, p. 964)**. The ball can be rotated up to 90 degrees by moving the valve handle or bale backward to open it and forward to close it. With the valve in the closed position, the waterway is perpendicular to the nozzle body, blocking the flow of water through the nozzle. With the valve in the open position, the waterway is in line with the axis of the nozzle, allowing water to flow through it.

Although the nozzle will operate in any position between fully closed and fully open, operating it with the valve in the fully open position gives maximum flow and performance. When a ball valve is used with a smooth bore nozzle, the turbulence caused by a partially open valve may affect the quality of the solid stream.

Ball Valve — Valve having a ball-shaped internal component with a hole through its center that permits water to flow through when aligned with the waterway.

Slide Valve

The cylindrical slide valve control seats a movable cylinder against a shaped cone to turn off the flow of water **(Figure 16.21, p. 964)**. When the shutoff handle is in the forward position, the cylinder is closed preventing water flow past the shaped cone. As the handle is pulled back, the cylinder slides open permitting water to flow through the nozzle without creating turbulence.

Rotary Control Valve

Rotary control valves are found only on rotary control fog nozzles. They consist of an exterior barrel guided by a screw that moves the exterior barrel forward or backward, rotating around an interior barrel **(Figure 16.22, p. 965)**. Rotary control valves also control the discharge pattern of the stream. This type of nozzle is commonly found in standpipe cabinets attached to occupant-use hoselines **(Figure 16.23, p. 965)**.

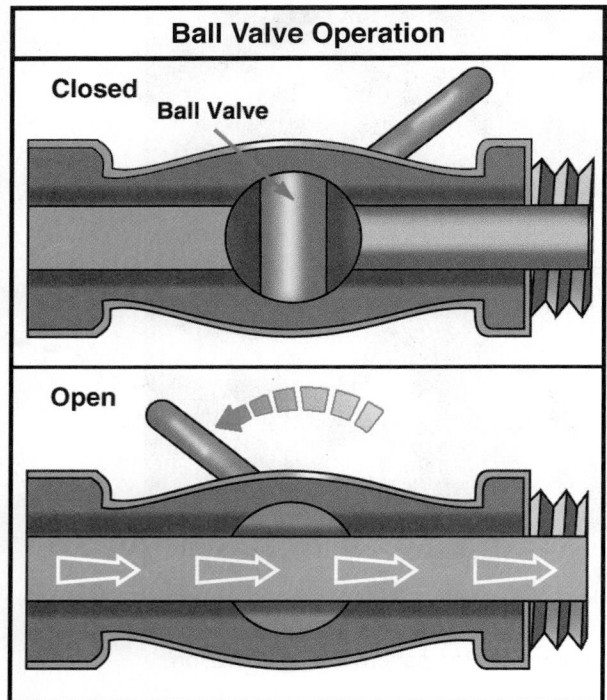

Figure 16.20 A ball valve operates by moving the handle to rotate a ball with a channel cut through one axis.

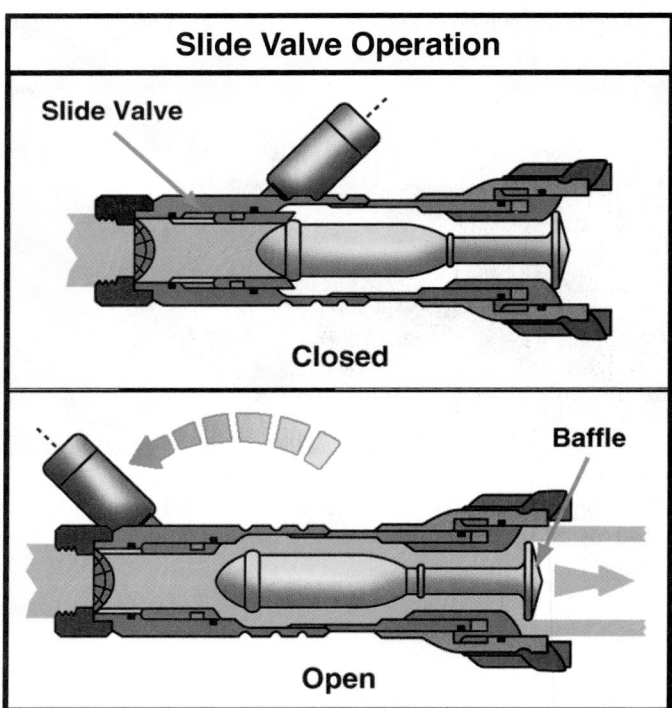

Figure 16.21 A slide valve nozzle operates by moving the handle to slide a cylinder toward and away from a cone-shaped opening.

Operating and Maintaining Handline Nozzles

Because of the differing designs of handline nozzles, each one handles somewhat differently when operated at the recommended pressure. Variable pattern nozzles may also handle differently at different settings. The water pattern that the nozzle setting produces may affect the ease with which a particular nozzle is operated. At or above standard operating pressures, handline nozzles are not always easy to control.

Operating Smooth Bore Nozzles

When water flows from a smooth bore nozzle it creates force in the direction of the stream and equal force in the opposite direction. The force in the opposite direction pushes back on the nozzle operator. This **nozzle reaction** is caused by the velocity, flow rate, and discharge pattern of the stream. The reaction acts against both the nozzle and the curves in the hoseline, sometimes making the nozzle difficult to handle. Increasing the nozzle discharge pressure and flow rate increases nozzle reaction.

When the nozzle is in place, the nozzle operator should control the nozzle as follows:

- Cradle the hoseline under one arm while holding the nozzle or nozzle pistol grip in one hand.
- Pull back slowly on the bale with the other hand to open the nozzle.
- As the action increases, lean forward with both legs apart, one foot forward, weight evenly distributed on both feet.

Nozzle Reaction — Counter force directed against a person holding a nozzle or a device holding a nozzle by the velocity of water being discharged.

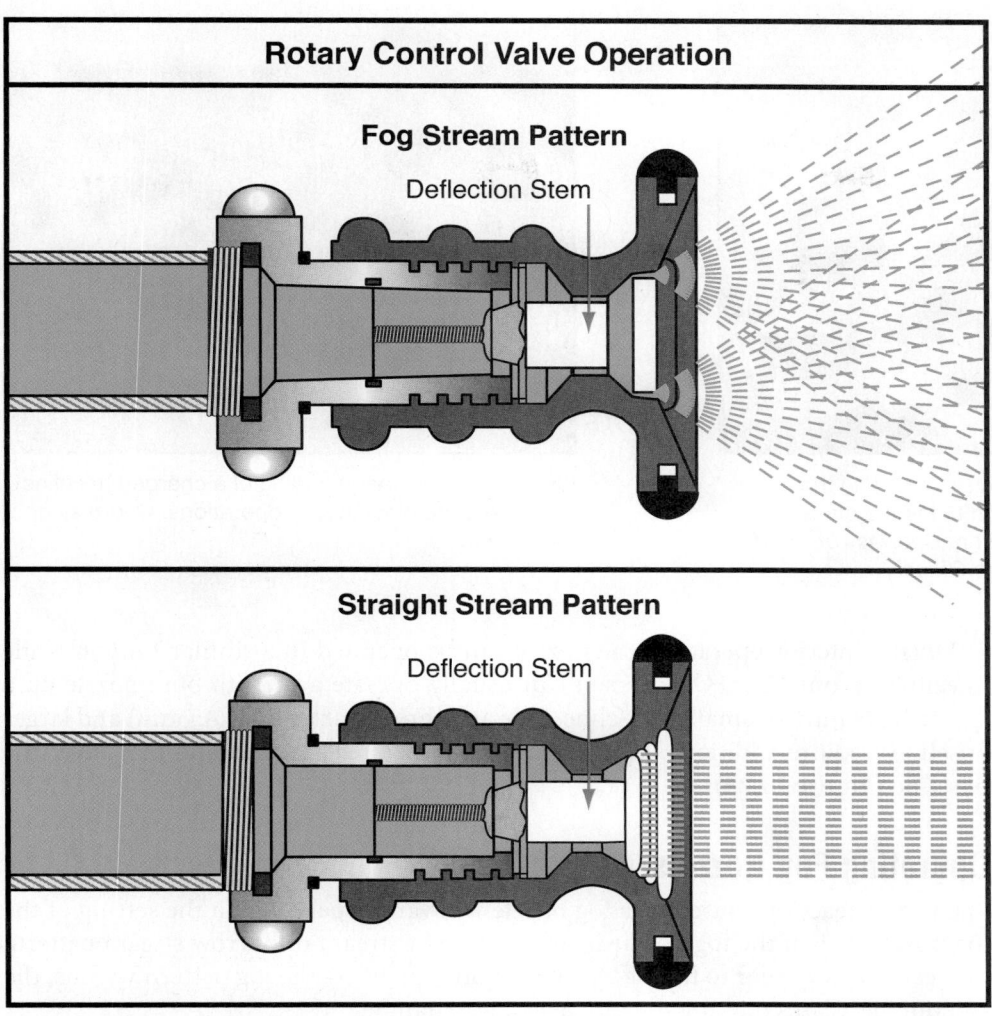

Figure 16.22 A rotary control valve on a fog nozzle can be rotated to flow a straight stream pattern or a fog stream pattern.

Figure 16.23 A rotary control valve may be brass or plastic and controls the output of water from a standpipe.

Figure 16.24 The nozzle reaction is felt as a kick to the operator when pressure in the direction of the fire stream equals the force in the opposite direction. *Courtesy of Shad Cooper, Wyoming State Fire Marshal's Office.*

Figure 16.25 Proper handling of a charged hoseline maximizes the efficiency of operations. *Courtesy of Dick Giles.*

During interior operation, the nozzle can be operated in a similar fashion while kneeling on one knee. One person can usually operate a smooth bore nozzle on a 1½-inch (38 mm) or smaller hoseline. One and three quarter inch (45 mm) and larger hoselines require additional personnel to overcome the reaction and maneuver the hoseline (**Figure 16.24**).

Operating Fog Nozzles

The nozzle reaction caused by a fog nozzle will vary depending on the setting of the fog nozzle. When the fog nozzle is set on straight stream or narrow stream pattern, the reaction is similar to that of a smooth bore nozzle. As the fog pattern widens, the reaction decreases making the nozzle easier to handle.

Handling the fog nozzle is the same as that of the smooth bore nozzle (**Figure 16.25**). Information on operating hoselines during fire attack is included in Chapter 17, Fire Control.

Maintaining Nozzles

Nozzles should be inspected after each use and at least annually to make sure they are in proper working condition. Basic maintenance, care, and cleaning should always be performed in accordance with the manufacturer's recommendations. Technical nozzle maintenance should only be performed by qualified maintenance technicians. Inspections include the following actions:

- Inspect the swivel gasket for damage or wear. Replace worn or missing gaskets.
- Look for external damage to the nozzle body, coupling, and tip (**Figure 16.26**).
- Look for internal damage and debris.
- Check for ease of operation of the nozzle parts.
- Ensure that the pistol grip (if applicable) is secured to the nozzle.
- Ensure that all parts are in place and in good condition.

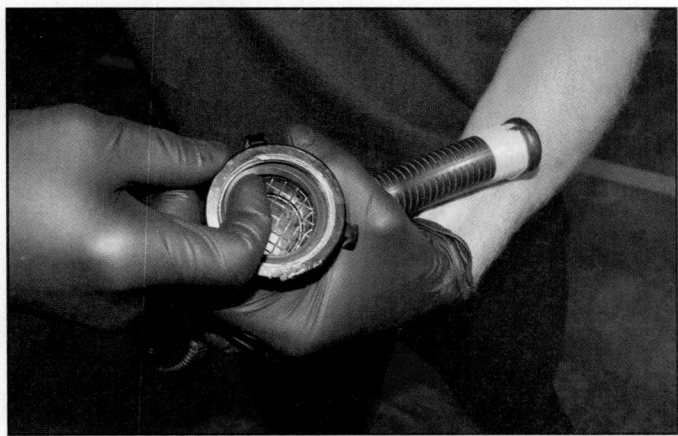

Figure 16.26 A swivel gasket should be in place and in good repair.

Figure 16.27 Nozzles should be cleaned after each use to prevent buildup of contaminants from the fire scene.

General nozzle care includes:

- Thoroughly cleaning nozzles after each use with soap and water using a soft bristle brush (**Figure 16.27**).
- Following manufacturer's recommendations for cleaning and lubricating any moving parts that are sticking.
- Storing nozzle with the control valve bale in the closed position.
- Never dropping or dragging nozzles.
- Using the flush setting on fog nozzles to remove any internal debris. If debris remains, shut off the water supply, remove the nozzle and physically remove the debris.

🔢 Fire Fighting Foam

In general, fire fighting **foam** works by forming a blanket of foam on the surface of burning fuels — both liquid and solid. Fire fighting foam extinguishes and/or prevents ignition in several ways (**Figure 16.28, p. 968**):

- **Separating** — Creates a barrier between the fuel and the fire
- **Cooling** — Lowers the temperature of the fuel and adjacent surfaces
- **Smothering** — Prevents air from reaching the fuel and mixing with the vapors and prevents the release of flammable vapors reducing the possibility of ignition or reignition
- **Penetrating** — Lowers the surface tension of water and allows it to penetrate fires in Class A materials

The majority of fire fighting foams are either Class A foams intended for use on ordinary combustibles (Class A fuels), or Class B foams intended for use on flammable liquids.

On solid fuels, Class A foam blankets and cools the fuel and stops the burning process. After controlling the flames, the water in the foam is slowly released into the fuel as the foam breaks down. This action provides a cooling effect on the fuel.

Foam — Extinguishing agent formed by mixing a foam concentrate with water and aerating the solution for expansion; for use on Class A and Class B fires. Foam may be protein, fluoroprotein, film forming fluoroprotein, synthetic, aqueous film forming, or alcohol-resistant type.

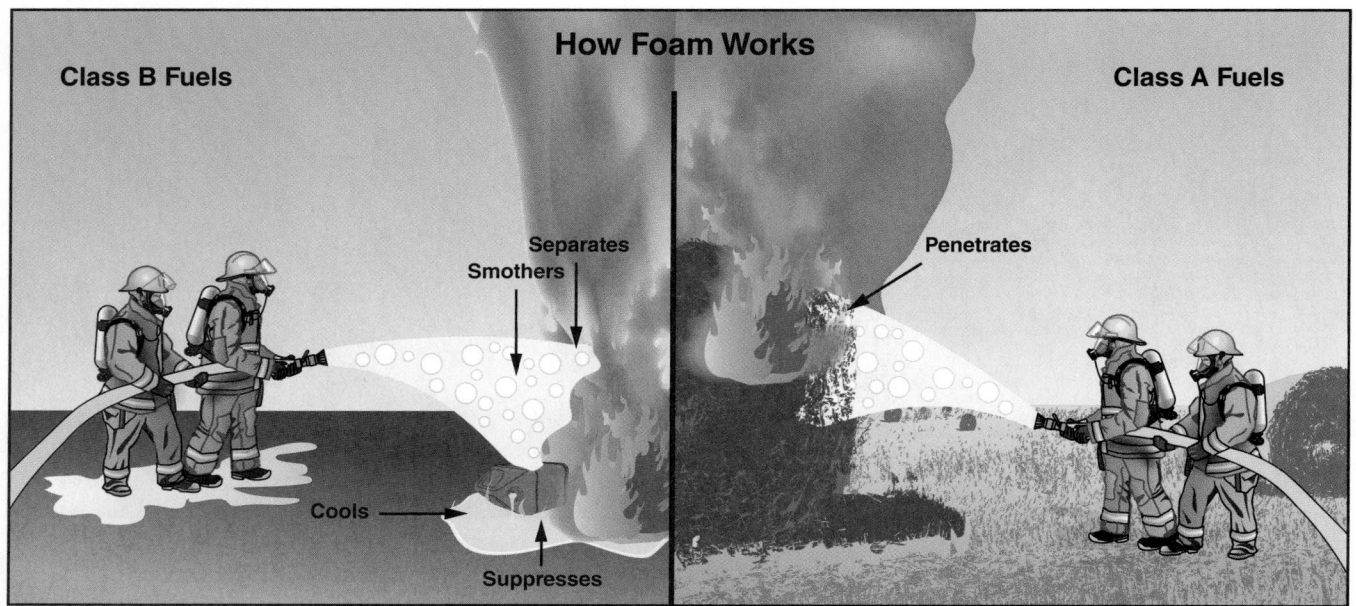

Figure 16.28 Foam agents extinguish Class B fuels by suppressing vapors, separating the fuel from the fire, cooling the fuel and other surfaces, and smothering flaming combustion. Foam agents extinguish Class A fuels by penetrating the fire.

Foam Concentrate — Chemical compound solution that is mixed with water and air to produce finished foam; may be protein, synthetic, aqueous film forming, high expansion, or alcohol types.

Foam Proportioner — Device that introduces foam concentrate into the water stream to make the foam solution.

Foam Solution — Mixture of foam concentrate and water before the introduction of air.

Finished Foam — Completed product after air is introduced into the foam solution.

Foam Expansion — Result of adding air to a foam solution consisting of water and foam concentrate. Expansion creates the foam bubbles that result in finished foam or foam blanket.

With liquid fuels, the Class B foam blanket also prevents or reduces the release of flammable vapors from the surface of the fuel. Class B foam is especially effective on the two basic categories of flammable liquids: hydrocarbon fuels and polar solvents.

How Foam Is Generated

Foam concentrate and water are mixed in the correct proportion or ratio with a **foam proportioner** to produce a **foam solution**. Air is then added to the solution through mechanical agitation, or aeration, to produce the **finished foam**. The foam concentrate, water, and air must be present and blended in the correct ratios; removing any element results in either no foam production or poor-quality foam (**Figure 16.29**).

Aeration is needed to produce an adequate amount of foam bubbles to form an effective foam blanket. Proper aeration produces uniform-sized bubbles that provide a longer-lasting blanket. A good foam blanket is required to maintain an effective cover for the period of time required for extinguishment. Even though the foam bubbles dissipate, a residual foam layer is still present.

Foam Expansion

Foam expansion refers to the increase in volume of a foam solution when it is aerated. This is a key characteristic that must be considered when a foam concentrate for a specific application is chosen. Degrees of expansion depend on the following factors:

- Type of foam concentrate used
- Accurate proportioning (mixing) of the foam concentrate in the solution
- Quality of the foam concentrate
- Method of aeration

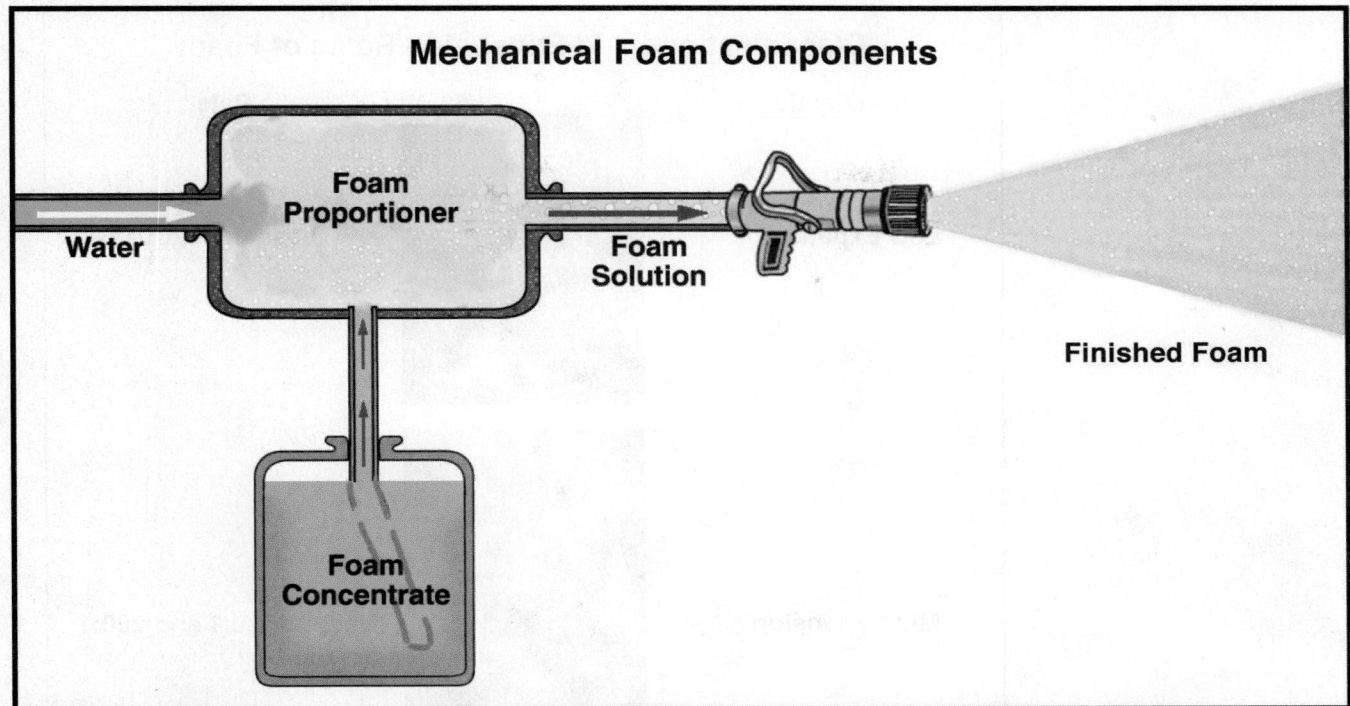

Figure 16.29 The foam-making process combines water, foam concentrate, and air to generate finished foam.

There are three classifications of foam based on their foam expansion ratio: low-expansion, medium-expansion, and high-expansion **(Figure 16.30, p. 970)**. Low-expansion foam has an air/solution ratio up to 20 parts finished foam for every part of foam solution (20-to-1 ratio). Low-expansion foams are effective for controlling and extinguishing most Class B fires. They are also effective for cooling and penetrating Class A fires.

Medium-expansion foam is most commonly used at a ratio of between 20-to-1 to 200-to-1 through hydraulically operated nozzle-style delivery devices. These foams are used to suppress vapors from hazardous materials spills when applied at expansion ratios of 30-to-1 and 55-to-1.

High-expansion foams are synthetic foaming agents created by high-expansion foam generators at ratios from 200-to-1 to 1,000-to-1. They are typically used in confined spaces such as shipboard compartments, basements, mines, and enclosed aircraft hangars.

Foam Concentrates

To be effective, foam concentrates must match the fuel to which they are applied. For example, Class A foams are not designed to extinguish Class B fires. Class B foams designed solely for hydrocarbon fires will not extinguish polar solvent fires regardless of the concentration at which they are used. Many types of foam that are intended for polar solvents may be used on hydrocarbon fires, but this should not be attempted unless the manufacturer of the particular concentrate specifically says this can be done. This incompatibility factor is why it is extremely important to identify the type of fuel involved before applying foam. **Appendix D** highlights each of the common types of foam concentrates.

Classifications and Expansion Rates of Foam

Classification		Rate
Low Expansion		Less than 20:1
Mid-Expansion		Between 20:1 and 200:1
High Expansion	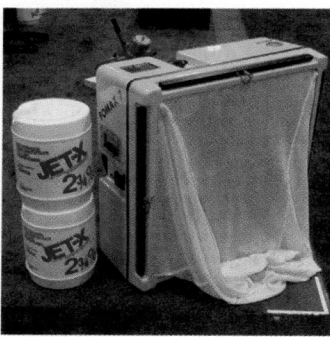	Greater than 200:1

Figure 16.30 Foam expansion may be categorized into three classes that are used under specific circumstances.

CAUTION

Failing to match the proper foam concentrate to the burning fuel will result in an unsuccessful extinguishing attempt and could endanger firefighters. Likewise, mixing different types of foam together can result in substandard quality foam.

Figure 16.31 Class A foams are increasingly used in structural fire fighting because of their improved ability to penetrate the fuel source. *Courtesy of Bob Esposito.*

Class A Foam

Foams specifically designed for use on Class A fuels (ordinary combustibles) are increasingly used in both wildland and structural fire fighting (**Figure 16.31**). **Class A foam** is a special formulation of hydrocarbon-based **surfactants**. These surfactants reduce the **surface tension** of water in the foam solution, allowing better water penetration into the fuel, thereby increasing its effectiveness. Aerated Class A foam coats and insulates fuels, preventing pyrolysis and ignition from an adjacent fire.

Class A foam may be used with fog nozzles, air-aspirating foam nozzles, medium- and high-expansion devices, and compressed air foam systems (CAFS). Class A foam concentrate has solvent characteristics and is mildly corrosive. It is important to thoroughly flush equipment after use.

Class B Foam

Class B foam is used to prevent the ignition of or to extinguish fires involving flammable and combustible liquids (**Figure 16.32**). It is also used to suppress vapors from unignited spills of these liquids. The types of liquid fuels that Class B foam is effective on are as follows:

- **Hydrocarbon fuels** — Petroleum-based combustible or flammable liquids that float on water including:
  - Crude oil
  - Fuel oil
  - Gasoline

Class A Foam — Foam specially designed for use on Class A combustibles. Class A foams, hydrocarbon-based surfactants, are essentially wetting agents that reduce the surface tension of water and allow it to soak into combustible materials more easily than plain water.

Surfactant — Chemical that lowers the surface tension of a liquid; allows water to spread more rapidly over the surface of Class A fuels and penetrate organic fuels.

Surface Tension — (1) Force minimizing a liquid surface's area. (2) The effect of a surfactant on the water/concentrate solution; allows the water to spread more rapidly over the surface of Class A fuels and to penetrate organic fuels.

Figure 16.32 Class B foam is effective against liquid fuels. *Courtesy of James Mack/Richmond International Airport.*

— Benzene
— Naptha
— Jet fuel
— Kerosene

- **Polar solvents** — Flammable liquids that mix readily with water including:
    — Alcohols
    — Acetone
    — Lacquer thinner
    — Ketones
    — Esters
    — Acids

There are several types of Class B foam concentrates; each type has its advantages and disadvantages. Class B foam concentrates are manufactured from either a synthetic or protein base. Protein-based foams are derived from animal protein. Synthetic foam is made from a mixture of fluorosurfactants. Some foam is made from a combination of synthetic and protein bases. Protein and fluoroprotein foams are effective as extinguishing agents and vapor suppressants on hydrocarbon fuels because they can float on the surface of these fuels. Alcohol-resistant foams are specially developed for polar solvents. Ethanol or ethanol-based fuels (E-10, E-85, or E-95) also require alcohol-resistant foams for extinguishment.

Like Class A foam, Class B foam may be proportioned into the fire stream through a fixed system, an apparatus-mounted system, or by portable foam proportioning equipment. Foams such as **aqueous film forming foam (AFFF)** and **film forming fluoroprotein foam (FFFP)** may be applied either with fog nozzles or with air-aspirating foam nozzles. The minimum amount of foam solution that must be applied, referred to as the *rate of application*, for Class B foam varies depending on any one of several variables:

- Type of foam concentrate used
- Whether the fuel is on fire
- Type of fuel (hydrocarbon/polar solvent) involved
- Whether the fuel is spilled or contained in a tank
- Whether the foam is applied via either a fixed system or portable equipment

    NOTE: If the fuel is in a tank, the type of tank will have a bearing on the application rate.

Figure 16.33 Fuel spills may be prevented from igniting by covering them with a film of foam to suppress vapors. *Courtesy of U.S. Air Force.*

Unignited spills create vapor hazards that may ignite. A foam blanket can be applied to suppress the vapors, separating the fuel from the oxygen (**Figure 16.33**). The depth of the foam blanket and application techniques will depend on the type of foam and manufacturer's recommendations.

Foam concentrate supplies should be on the fireground at the point of proportioning before application is started. Once application has started, it should continue uninterrupted until extinguishment is complete. Stopping and restarting may allow the fire to consume whatever foam blanket has been established.

Because polar solvent fuels have differing affinities for water, it is important to know application rates for each type of solvent. These rates also vary with the type and manufacturer of the foam concentrate selected. Foam concentrate manufacturers provide information on the proper application rates as listed by Underwriters' Laboratories (UL). For more complete information on application rates, consult NFPA® 11, the foam manufacturer's recommendations, and the IFSTA **Principles of Foam Fire Fighting** manual.

Specific Application Foams

Numerous types of foams are available for specific applications according to their properties and performance. Some are thick and **viscous** and form a tough, heat-resistant blanket over burning liquid surfaces; others are thinner and spread more rapidly. Some foams produce a vapor-sealing film of surface-active water solution on a liquid surface. Others, such as medium- and high-expansion foams, are used in large volumes to flood surfaces and fill cavities.

Specialized foams are also used for acid spills, pesticide fires, confined- or enclosed-space fires, and deep-seated Class A fires. In addition to regular fire fighting foams, there are foams designed solely for use on unignited spills of hazardous liquids. These special foams are necessary because unignited chemicals have a tendency to either change the **pH** of water or remove the water from fire fighting foams, thereby rendering them ineffective.

Viscous — Having a thick, sticky, adhesive consistency.

pH — A measure of the acidity or alkalinity of a solution.

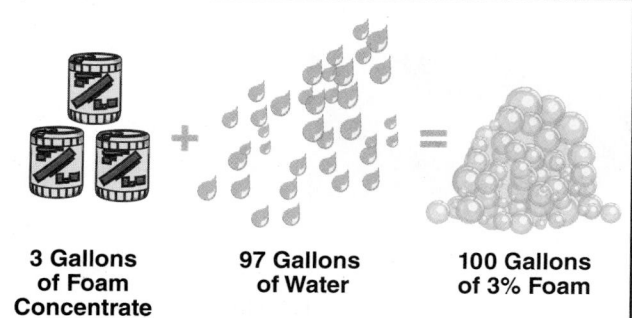

Figure 16.34 A 3 percent foam solution consists of 3 gallons (or parts) of foam concentrate and 97 gallons (or parts) of water.

3 Gallons of Foam Concentrate

97 Gallons of Water

100 Gallons of 3% Foam

Proportioning — Mixing of water with an appropriate amount of foam concentrate to form a foam solution.

Foam Proportioning

The term **proportioning** is used to describe the mixing of water with foam concentrate to form a foam solution. For maximum effectiveness, foam concentrates must be proportioned at the specific percentage for which they are designed. This percentage rate varies with the intended fuel and is written on the outside of every foam container. Failure to follow this procedure, such as trying to use 6 percent foam at a 3 percent concentration, will result in poor-quality foam that may not perform as desired.

Most fire fighting foam concentrates are intended to be mixed with 94 to 99.9 percent water. For example, when using 3 percent foam concentrate, 97 parts water mixed with 3 parts foam concentrate equals 100 parts foam solution (**Figure 16.34**). For 6 percent foam concentrate, 94 parts water mixed with 6 parts foam concentrate equals 100 percent foam solution.

The proportioning percentage for Class A foams can be adjusted (within limits recommended by the manufacturer) to achieve specific objectives. To produce a dry (thick) foam suitable for exposure protection and creating fire breaks in wildland fires, the foam concentrate can be adjusted to a higher percentage. To produce wet (thin) foam that rapidly penetrates a fuel's surface, the foam concentrate can be adjusted to a lower percentage. Most Class A foams are mixed in proportions of 1 percent or less.

Class B foams are mixed in proportions from 1 percent to 6 percent. Some multipurpose Class B foams designed for use on both hydrocarbon and polar solvent fuels can be used at different concentrations, depending on which of the two fuels is burning. These concentrates are normally used at a 3 percent rate for hydrocarbons and 6 percent for polar solvents. Newer multipurpose foams may be used at 3 percent concentrations regardless of the type of fuel. Always follow the manufacturer's recommendations for proportioning.

Some types of proportioning equipment are designed for mobile apparatus and others are designed for fixed fire protection systems. The selection of a proportioner depends on the following:

- Foam solution flow requirements
- Available water pressure
- Cost of the foam
- Intended use for the foam (truck, fixed, or portable)

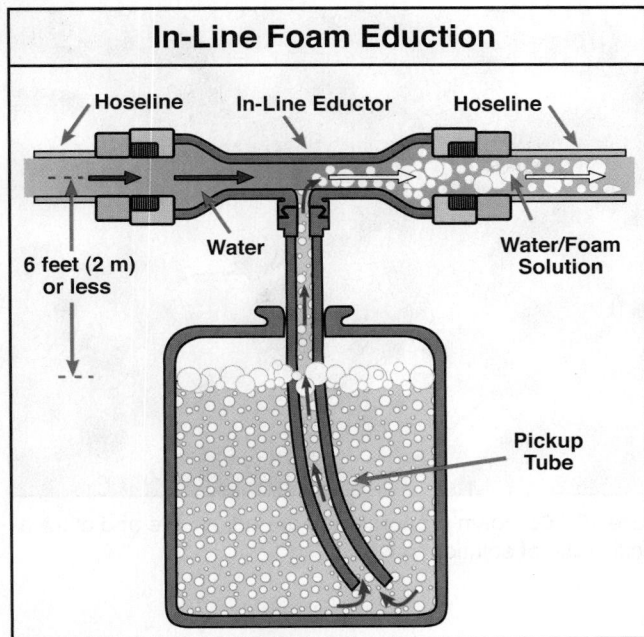

In-Line Foam Eduction

Hoseline

In-Line Eductor

Hoseline

6 feet (2 m) or less

Water

Water/Foam Solution

Pickup Tube

Figure 16.35 As water flows past the pickup tube orifice, foam concentrate is drawn into the water stream to create the foam solution.

- Foam agent to be used

Proportioners and delivery devices (foam nozzle, foam maker, etc.) are engineered to work together. Using a foam proportioner that is not compatible with the delivery device (even if the two are made by the same manufacturer) can result in unsatisfactory foam or no foam at all.

There are four basic methods by which foam may be proportioned:

- Eduction
- Injection
- Batch-mixing
- Premixing

Eduction

The **eduction** (induction) method of proportioning foam uses the pressure energy in the stream of water to induct (draft) foam concentrate into the fire stream. This is achieved by passing the stream of water through an *eductor,* a device that depends on the **Venturi Principle** to draw the foam through a hose connected to the foam concentrate container and into the water stream **(Figure 16.35)**. In-line eductors and foam nozzle eductors are examples of foam proportioners that use this method **(Figures 16.36 a and b, p. 976)**.

Injection

The **injection** method of proportioning foam uses an external pump or head pressure to force foam concentrate into the fire stream at the correct ratio for the water flow. These systems are commonly employed in apparatus-mounted or fixed fire protection system applications **(Figure 16.37, p. 976)**.

Batch-Mixing

Batch-mixing is the simplest method of mixing foam concentrate and water. It is commonly used to mix foam within a fire apparatus water tank or a portable water

Eduction — Process used to mix foam concentrate with water in a nozzle or proportioner; concentrate is drawn into the water stream by the Venturi method; also called induction.

Venturi Principle — Physical law stating that when a fluid, such as water or air, is forced under pressure through a restricted orifice, there is an increase in the velocity of the fluid passing through the orifice and a corresponding decrease in the pressure exerted against the sides of the constriction. Because the surrounding fluid is under greater pressure (atmospheric), it is forced into the area of lower pressure. *Also known as* Venturi Effect.

Injection — Method of proportioning foam that uses an external pump or head pressure to force foam concentrate into the fire stream at the correct ratio for the flow desired.

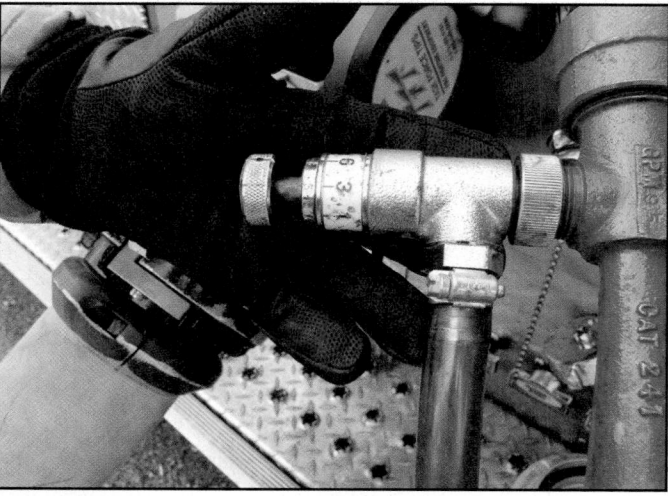

Figure 16.36b Foam proportioners are adjustable and allow a specific ratio of solution to water.

Figure 16.36a In-line eduction is the process of drawing one liquid with another; in this case, the fire stream draws a foam solution into a hoseline.

Figure 16.37 Foam may be injected into an apparatus fire stream via a mechanical system.

Figure 16.38 Foams are commonly mixed into the fire apparatus water tank.

Figure 16.39 Premixing may be accomplished using a portable system.

tank (**Figure 16.38, p. 978**). Batch-mixing is common with Class A foams but is less common with Class B foams.

Batch-mixing may not be effective on large incidents because when the tank becomes empty, the foam attack lines must be shut down until the tank is completely filled with water and more foam concentrate is added. Another drawback of batch-mixing is that Class B concentrates and tank water must be circulated for a period of time to ensure thorough mixing before the solution is discharged. The time required for mixing depends on the viscosity and solubility of the foam concentrate. Another disadvantage to the method is that it can be difficult to refill an apparatus water tank due to excessive bubbling from residual solution.

After the incident, all components in which foam was batch-mixed must be thoroughly flushed with plain water. Because the foam solution goes through the pump during batch-mixing, the pump may require additional maintenance due to the foam's degreasing capabilities.

Premixing

Premixing is one of the more commonly used methods of proportioning. Premeasured portions of water and foam concentrate are mixed in a container. Typically, the premix method is used with portable extinguishers, wheeled extinguishers, skid-mounted twin-agent units, and vehicle-mounted tank systems (**Figure 16.39**).

In most cases, premixed solutions are discharged from a pressure-rated tank using either a compressed inert gas or air. An alternative method of discharge uses a pump

> **Premixing** — Mixing premeasured portions of water and foam concentrate in a container. Typically, the premix method is used with portable extinguishers, wheeled extinguishers, skid-mounted twin-agent units, and vehicle-mounted tank systems.

Figure 16.40 An in-line eductor is commonly used to proportion foam solution and water.

and a non-pressure-rated atmospheric storage tank. The pump discharges the foam solution through piping or hose to the delivery devices. Premix systems are limited to a one-time application. When used, the tank must be completely emptied and then refilled before it can be used again. Since most Class A foam solutions are biodegradable, premixing the solution and storing it for long periods can result in decreased foaming ability.

Proportioners, Delivery Devices, and Generating Systems

In addition to a pump to supply water and fire hose to transport it, there are two other pieces of equipment needed to produce a foam fire stream: a foam proportioner and a foam delivery device (nozzle or generating system). The proportioner and delivery device/system must be compatible to produce usable foam. Foam proportioning simply introduces the appropriate amount of foam concentrate into the water to form a foam solution. A foam-generating system/nozzle adds the air into foam solutions to produce finished foam.

Foam Proportioners

There are three types of foam proportioning systems currently in use. Proportioning systems may be portable, apparatus-mounted, or **compressed air foam systems (CAFS)**.

Compressed Air Foam System (CAFS) — Generic term used to describe a high-energy foam-generation system consisting of a water pump, a foam proportioning system, and an air compressor (or other air source) that injects air into the foam solution before it enters a hoseline.

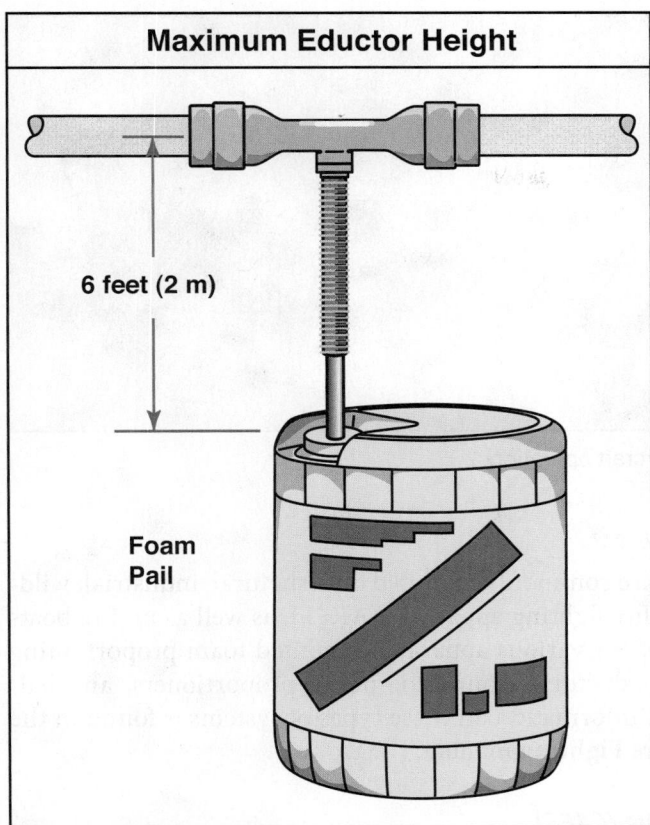

Maximum Eductor Height

6 feet (2 m)

Foam Pail

Figure 16.41 If the eductor is held more than 6 feet (2 m) above the surface of the foam concentrate, the concentration may be too lean or induction does not occur.

Portable Foam Proportioners

Portable foam proportioners are the simplest and most common foam proportioning devices in use today. Two types of portable foam proportioners are in-line foam eductors and foam nozzle eductors.

In-line foam eductors. The **in-line eductor** is the most common type of foam proportioner used in the fire service (**Figure 16.40**). It is designed to be directly attached to the pump panel discharge outlet or connected at some point in the hose lay. When using an in-line eductor, it is very important to follow the manufacturer's instructions about inlet pressure and the maximum hose lay between the eductor and the appropriate discharge nozzle.

In-line eductors use the Venturi Principle to draft foam concentrate into the water stream. The eductor pickup tube is connected to the eductor at this low-pressure point. A pickup tube submerged in the foam concentrate draws the concentrate into the water stream, creating a foam/water solution. The foam concentrate inlet to the eductor should not be more than 6 feet (2 m) above the liquid surface of the foam concentrate (**Figure 16.41**). If the inlet is too high, the foam concentration will be very lean or foam may not be inducted at all.

Foam nozzle eductors. A foam nozzle eductor operates on the same basic principle as an in-line eductor; however, this eductor is built into the nozzle rather than into the hoseline. As a result, its use requires the foam concentrate to be available where the nozzle is operated. If the foam nozzle is moved, the foam concentrate container must also be moved. The size and number of concentrate containers required magnify the logistical problems of relocation. Use of a foam nozzle eductor can also compromise firefighter safety. Firefighters cannot always move quickly, and they may have to leave their concentrate supplies behind if they are required to retreat for any reason.

In-Line Eductor — Eductor that is placed along the length of a hoseline.

Figure 16.42 An aircraft rescue apparatus is specially designed for aircraft operations.

Apparatus-Mounted Proportioners

Foam proportioning systems are commonly mounted on structural, industrial, wild-land, and aircraft rescue and fire fighting apparatus (ARFF), as well as on fire boats **(Figure 16.42)**. Three types of the various apparatus-mounted foam proportioning systems are installed in-line eductors, around-the-pump proportioners, and balanced pressure proportioners. Information on these types of systems is found in the IFSTA **Principles of Foam Fire Fighting** manual.

Compressed Air Foam Systems (CAFS)

Compressed air foam systems (CAFS) are mounted on many types of fire fighting apparatus **(Figure 16.43)**. A CAFS functions as follows:

- A standard centrifugal pump supplies the water

- A direct-injection foam-proportioning system mixes foam solution with the water on the discharge side of the pump

- An onboard air compressor adds air to the mix before it is discharged from the apparatus

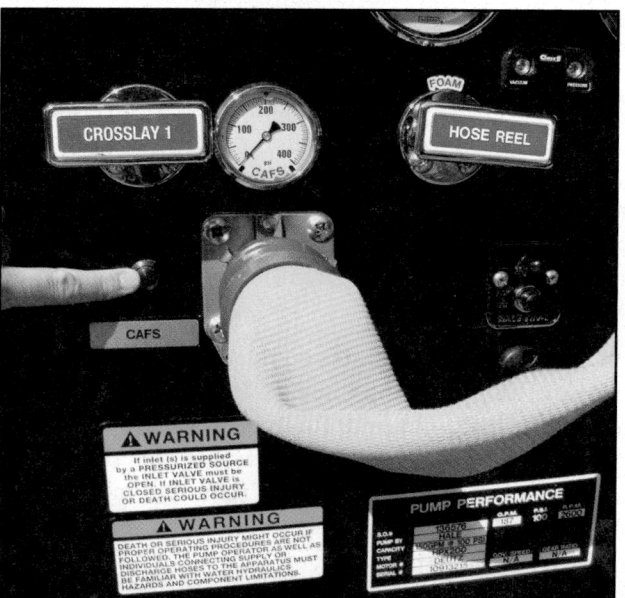

Figure 16.43 A compressed air foam system serves functions that include adding air to the foam solution before it is discharged.

Unlike other foam systems, with CAFS the hoseline contains the finished foam. Among the advantages of using compressed air foam are the following:

- Stream reach is considerably longer than with other foam systems.

- Hoselines are lighter than those full of water or foam solution.

- Foam produced is very durable.

- Foam produced adheres well to vertical surfaces.

   Some of the disadvantages include:

- CAFS add expense to the purchase and maintenance of the apparatus.

- The stored energy created by the compressed air pressure in the hose can create a high nozzle reaction when the nozzle is opened that may throw the nozzle operator off balance.

- Additional training is required for firefighters and driver/operators.

Foam Delivery Devices (Nozzles/Generating Systems)

Foam delivery devices, consisting of nozzles and generating systems, designed to discharge foam are sometimes called *foam makers*. There are many types of devices that can be used, including standard fire stream nozzles. The following sections highlight some of the more common foam application devices.

NOTE: Foam nozzle eductors are considered portable foam nozzles but they have been omitted from this section because they were covered earlier in the Portable Foam Proportioners section.

Handline Nozzles

Smooth bore and fog **handline nozzles** generally flow less than 350 gpm (1 400 L/min). Most handline foam nozzles flow considerably less than that amount.

> **Handline Nozzle** — Any nozzle that can be safely handled by one to three firefighters and flows less than 350 gpm (1 400 L/min).

   Smooth bore nozzles. The use of smooth bore nozzles is limited to certain types of Class A foam applications. In these applications, the smooth bore nozzle provides an effective fire stream with maximum reach capabilities. Smooth bore nozzles are most often used with compressed air foam systems (CAFS) **(Figure 16.44)**.

   Fog nozzles. Fog nozzles can be used with foam solutions to produce a low-expansion, short-lasting foam **(Figure 16.45, p. 982)**. These nozzles break the foam solution into tiny droplets and use the agitation of water droplets moving through air to achieve foaming action. Their best application is when used with regular AFFF

Figure 16.44 CAFS are used most often with Class A foam and smooth bore nozzles.

Figure 16.45 Fog nozzles aerate the foam solution as the hose is being discharged.

Figure 16.46 Air-aspirating foam nozzles use the Venturi Principle to induct air into the solution.

and Class A foams. These nozzles cannot be used with protein and fluoroprotein foams. Fog nozzles may be used with alcohol-resistant AFFF foams on hydrocarbon fires but should not be used on polar solvent fires because insufficient aeration occurs to handle the polar solvent fires. Some nozzle manufacturers have foam aeration attachments that can be added to the end of the nozzle to increase the aeration of the foam solution.

Air-aspirating foam nozzles. An air-aspirating foam nozzle is the most effective appliance for the generation of low-expansion foam. The air-aspirating foam nozzle inducts air into the foam solution using the Venturi Principle (**Figure 16.46**). This nozzle is especially designed to provide the aeration required to make the highest quality foam possible. These nozzles must be used with protein and fluoroprotein concentrates. They may also be used with Class A foams. These nozzles provide maximum expansion of the agent. The reach of the stream is less than that of a standard fog nozzle.

Specialized application system. Advances in foam technology have created an application system that uses a solid foam/wetting agent. The solid agent container is inserted into the specially perforated foam sleeve between the hoseline and an adjustable fog nozzle (**Figure 16.47**). The concentrate is designed for use on Class A and Class B fires. Each 1½-pound (0.68 kg) cartridge of solid agent is equal to 5 gallons (18.93 L) liquid agent and will treat approximately 660 gallons (2 498.37 L) of water.

Figure 16.47 A solid foam agent rests between the hoseline and the fog nozzle to generate foam. *Courtesy of Shad Cooper, Wyoming State Fire Marshal's Office.*

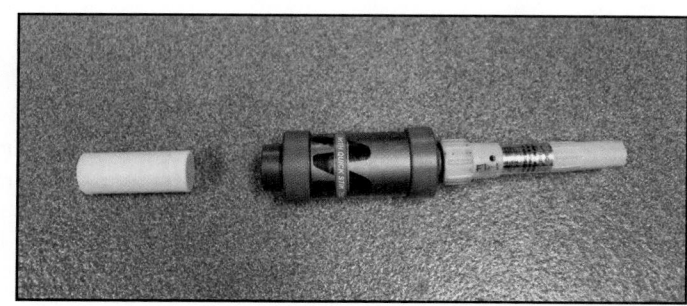

Figure 16.48 The mechanical blower generator forces large quantities of air into foam in total-flooding applications.

Medium- and High-Expansion Foam Generating Devices

Medium- and high-expansion foam generators produce a foam that is semi-stable with high air content. There are two basic types of medium- and high-expansion foam generators: the water-aspirating type nozzle and the mechanical blower.

Water-aspirating type nozzle. The water-aspirating type nozzle is very similar to the other foam-producing nozzles except it is much larger and longer. The back of the nozzle is open to allow air flow. The foam solution is pumped through the nozzle in a fine spray that mixes with air to form moderate-expansion foam. The end of the nozzle has a screen or series of screens that break up the foam even more while mixing it with air. These nozzles typically produce foam with lower air volume than do mechanical blower generators.

Mechanical blower generator. A mechanical blower generator is similar in appearance to a smoke ejector. It operates on the same principle as the water-aspirating nozzle except that a powered fan is used to force air through the foam spray instead of water movement. This device produces foam with high air content and is typically associated with total-flooding applications (**Figure 16.48**). Its use is limited to high-expansion foam.

Assembling a Foam Fire Stream System

To provide a foam fire stream, a firefighter or apparatus driver/operator must be able to correctly assemble the components of the system in addition to locating problem areas and making adjustments. There are a number of reasons for failure to generate foam or for generating poor-quality foam. The most common reasons for failure include:

- Eductor and nozzle flow ratings do not match, preventing foam concentrate from inducting into the fire stream.

- Air leaks at fittings cause a loss of suction.

- Improper cleaning of proportioning equipment causes clogged foam passages.

- Nozzle is not fully open, restricting water flow.

- Hose lay on the discharge side of the eductor is too long, creating excess back pressure and causing reduced foam pickup at the eductor.

- Hose is kinked and restricts or stops flow.
- Nozzle is too far above the eductor, which causes excessive elevation pressure.
- Mixing different types of foam concentrate in the same tank results in a mixture too viscous to pass through the eductor.

Skill Sheet 16-II-1 describes the steps for placing a foam line in service using an in-line eductor.

Foam Application Techniques

It is important to use the correct techniques when applying foam from handline or master stream nozzles. If incorrect techniques are used, such as plunging the foam into a liquid fuel, the effectiveness of the foam is reduced. The techniques for applying Class B foam to a liquid fuel fire or spill include the *roll-on method, bank-down method,* and *rain-down method.* **Skill Sheet 16-II-2** describes the steps for extinguishing an ignitable liquid fire.

Roll-On Method

The roll-on method directs the foam stream on the ground near the front edge of a burning liquid spill **(Figure 16.49)**. The foam then rolls across the surface of the fuel. Firefighters continue to apply foam until it spreads across the entire surface of the fuel and the fire is extinguished. It may be necessary to move the stream to different positions along the edge of a liquid spill to cover the entire pool. This method is used only on a pool of ignited or unignited liquid fuel on the open ground.

Figure 16.49 The roll-on method of foam application is used on ignited or unignited flammable liquids spills.

Figure 16.50 In the bank-down method, foam is applied to a vertical surface and allowed to run down and onto the surface of the fuel.

Bank-Down Method

The bank-down method may be employed when an elevated object is near or within the area of a burning pool of liquid or an unignited liquid spill. The object may be a wall, tank shell, or similar vertical structure. The foam stream is directed onto the object, allowing the foam to run down and onto the surface of the fuel. As with the roll-on method, it may be necessary to direct the stream onto various points around the fuel area to achieve total coverage and extinguishment of the fuel (**Figure 16.50**). This method is used primarily on fires contained in diked pools around storage tanks and fires involving spills around damaged or overturned transport vehicles.

Rain-Down Method

The rain-down method is used when the other two methods are not feasible because of the size of the ignited or unignited spill area or the lack of an object from which to bank the foam. The rain-down method is also the primary manual application technique used on aboveground storage tank fires. This method directs the stream into the air above the fire or spill and allows the foam to float gently down onto the surface of the fuel (**Figure 16.51, p. 986**). On small fires, the nozzle operator sweeps the stream back and forth over the entire surface of the fuel until the fuel is completely covered and the fire is extinguished. On large fires, it may be more effective for the firefighter to direct the stream at one location to allow the foam to collect there and then float out from that point.

Foam Hazards

Foam concentrates, either at full strengths or in diluted forms, pose minimal health risks to firefighters. In both forms, foam concentrates may be mildly irritating to the skin and eyes. Affected areas should be flushed with water. Some concentrates and their vapors may be harmful if ingested or inhaled. Consult the manufacturer's **safety data sheets (SDS)** for information on any specific foam concentrate.

Safety Data Sheet (SDS) — Form provided by the manufacturer and blender of chemicals that contains information about chemical composition, physical and chemical properties, health and safety hazards, emergency response procedures, and waste disposal procedures of the specified material. Formerly known as the material safety data sheet (MSDS).

Rain-Down Method

Figure 16.51 The rain-down method is effective against large flammable liquids fires or spills.

Most Class A and Class B foam concentrates are mildly corrosive. Although foam concentrate is used in small percentages and in diluted solutions, follow proper flushing procedures to prevent damage to equipment. Pumps, eductors, hoselines, and nozzles must be thoroughly flushed and washed to remove concentrate residue.

The effect of the finished foam after it has been applied to a liquid fuel fire or spill is a primary environmental concern. The biodegradability of foam is determined by the rate at which environmental bacteria cause it to decompose. This decomposition process results in the consumption of oxygen. In a river, stream, pond, or lake, the subsequent reduction in oxygen can kill fish and other aquatic creatures. Therefore, firefighters should take care to prevent foam from directly entering bodies of water. The less oxygen required to degrade a particular foam, the better or the more environmentally friendly the foam is when it enters a body of water.

The environmental effect of foam concentrates varies. Each foam concentrate manufacturer can provide information on its specific products. In the United States, Class A foams should be approved by the U.S. Department of Agriculture Forest Service for environmental suitability. The chemical properties of Class B foams and their environmental effect vary depending on the type of concentrate and the manufacturer. Generally, protein-based foams are safer for the environment. Consult the various manufacturers' data sheets for environmental impact information.

Chapter Summary

To fight fires safely and effectively, firefighters must know the capabilities and limitations of the various nozzles and extinguishing agents available in their departments. They must understand the effects that wind, gravity, velocity, and friction with the air have on a fire stream once it leaves the nozzle. Firefighters must know what operating pressure their nozzles require and how the nozzles can be adjusted during operation. Finally, firefighters must know the differences between the types of foam, how to generate foam, and how to apply foam most effectively.

Review Questions

Firefighter I

1. What are the extinguishing properties of water?

2. How can friction loss and elevation loss/gain impact fire stream pressure?

3. What impact does water hammer have on fire streams?

4. How do the four types of fire stream patterns compare with one another?

5. What are the benefits of each of the types of fire stream nozzles?

6. How do the different types of nozzle control valves compare with one another?

7. What are the main factors to consider when operating and maintaining a handline nozzle?

Firefighter II

1. How does fire fighting foam prevent or control a hazard?

2. What are the types of foam concentrates used in the fire service?

3. Which methods can be used to proportion foam?

4. What are the advantages of each type of foam delivery device?

5. What are some possible causes of poor foam production?

6. What are the three main types of foam application techniques and how do they work?

7. How can firefighters work to mitigate foam hazards?

SKILL
SHEETS

Straight Stream

Step 1: Position team members on same side of hose with one firefighter on nozzle and one as backup.

Step 2: Prior to opening nozzle, wait for backup firefighter to communicate that readiness.

Step 3: Twist the stream adjustment ring to adjust the stream pattern to a straight stream.

Step 4: Aim the nozzle at the target indicated by officer.

Straight Stream

Step 5: Open the nozzle fully.

Step 6: Hold the stream on target.

Step 7: Shut off the nozzle slowly so that water hammer is avoided.

Narrow Fog Stream

Step 1: Position team members on same side of hose with one firefighter on nozzle and one as backup.

Step 2: Prior to opening nozzle, wait for backup firefighter to communicate readiness.

Step 3: Adjust the stream pattern by twisting stream adjustment ring to a narrow fog stream (15 to 45 degrees).

Step 4: Aim the nozzle at the target indicated by officer.

Narrow Fog Stream

Step 5: Open the nozzle fully.

Step 6: Hold the stream on target.

Step 7: Shut off the nozzle slowly so that water hammer is avoided.

Wide Fog Stream

Step 1: Position yourselves on same side of hose with one firefighter on nozzle and one as backup.

Step 2: Prior to opening nozzle, wait for backup firefighter to communicate that they are ready.

Step 3: Adjust the stream pattern by twisting stream adjustment ring to a wide fog stream (45 to 80 degrees).

Step 4: Aim the nozzle at the target indicated by officer.

Step 5: Open the nozzle fully.

Step 6: Hold the stream on target.

Step 7: Shut off the nozzle slowly so that water hammer is avoided.

Step 1: Attach nozzle according to manufacturer's guidelines and/or departmental SOPs.

Step 2: Place nozzle in proper location (i.e. through a hole in floor, driven through wall, or lowered into chimney).

Step 3: Position firefighters so that control can be maintained on the hoseline.

Step 4: Call for water and/or open valve fully to flow water.

Step 5: Maintain flow until fire is extinguished.

Step 6: Shut off nozzle slowly so that water hammer is avoided.

Step 7: Remove nozzle from area of extinguishment.

Step 1: Position team members on same side of hose with one firefighter on nozzle and one as backup.

Step 2: Prior to opening nozzle wait for backup firefighter to communicate that they are ready.

Step 3: Aim the nozzle at the target indicated by officer.

Step 4: Open the nozzle fully.

Step 5: Hold the stream on target.

Step 6: Shut off the nozzle slowly so that water hammer is avoided.

Step 1: Confirm order with officer to place line in service.

Step 2: Select the proper foam concentrate for the burning fuel involved.

Step 3: Place the foam concentrate at the eductor.

Step 4: Do not begin until you are sure you have enough foam.

Step 5: Check the eductor and nozzle for hydraulic compatibility (rated for the same flow).

Step 7: Attach the eductor to a hose capable of efficiently flowing the rated capacity of the eductor and the nozzle. This should be at least 50 ft (15 m) and no more than 200 ft (60 m) from the nozzle.

Step 6: Adjust the eductor metering valve to the same percentage rating as that listed on the foam concentrate container.

SKILL SHEETS

16-II-1
Place a foam line in service – In-line eductor.

Step 8: Attach the attack hoseline and desired nozzle to the discharge end of the eductor. Avoid kinks in the hose.

Step 9: Place the eductor suction hose into the foam concentrate.

Step 10: Open nozzle fully.

Step 11: Increase the water-supply pressure to that required for the eductor. Be sure to consult the manufacturer's recommendations for the specific eductor.

Step 12: Report to officer completion of assigned task.

Ground Level Fire Attack – Rain-Down Method

Step 1: Confirm order with Incident Commander (IC) to extinguish fire.

Step 2: Size up incident scene for hazards.

    a. Fire conditions

    b. Type of fuel

    c. Wind conditions

    d. Escape route

Step 3: Verify foam type and concentration are appropriate for fuel and fire conditions.

Step 4: Verify attack line is functioning and ready for attack.

Step 5: Extend hoseline to point of fire attack.

    a. Upwind and uphill

    b. Able to apply stream as needed

Step 6: Extinguish fire by applying foam solution as directed.

    a. Direct foam stream into air above fire or spill so that foam floats gently down onto surface of fuel

    b. Maintain stream as foam spreads across surface of fuel

        i. For small fires – Sweep stream gently back and forth

        ii. For large fires – Direct stream to one location and allow foam to float out from that point

    c. Apply foam until it spreads across entire surface of fuel and extinguishes fire

Step 7: Maintain situational awareness.

Step 8: Report to officer completion of assigned task.

Gound Level Fire Attack – Bank-Down Method

Step 3: Verify foam type and concentration are appropriate for fuel and fire conditions.

Step 4: Verify attack line is functioning and ready for attack.

Step 1: Confirm order with Incident Commander (IC) to extinguish fire.

Step 2: Size up incident scene for hazards.

    a. Fire conditions

    b. Type of fuel

    c. Wind conditions

    d. Escape route

Ground Level Fire Attack — Bank-Down Method

Step 5: Extend hoseline to point of fire attack.

    a. Upwind and uphill

    b. Able to apply stream as needed

Step 6: Extinguish fire by applying foam solution as directed.

    a. Direct foam stream onto nearby elevated object; allow foam to run down onto surface of fuel

    b. Maintain stream as foam spreads across sur-face of fuel

    c. Apply foam until it spreads across entire surface of fuel and extinguishes fire

Step 7: Maintain situational awareness.

Step 8: Report to officer completion of assigned task.

Ground Level Fire Attack – Roll-On Method

Step 1: Confirm order with Incident Commander (IC) to extinguish fire.

Step 2: Size up incident scene for hazards.

    a. Fire conditions

    b. Type of fuel

    c. Wind conditions

    d. Escape route

Step 3: Verify foam type and concentration are appropriate for fuel and fire conditions.

Step 4: Verify attack line is functioning and ready for attack.

Step 5: Extend hoseline to point of fire attack.

    a. Upwind and uphill

    b. Able to apply stream as needed

Step 6: Extinguish fire by applying foam solution as directed.

    a. Direct foam stream on the ground near front edge of fire so that foam rolls across surface of fuel

    b. Maintain stream as foam rolls across surface of fuel

    c. Apply foam until it spreads across entire surface of fuel and extinguishes fire

Step 7: Maintain situational awareness.

Step 8: Report to officer completion of assigned task.

Courtesy of Mike Wieder.

Chapter Contents

Key Terms

NFPA® Job Performance Requirements

This chapter provides information that addresses the following job performance requirements of NFPA® 1001, *Standard for Fire Fighter Professional Qualifications* (2013).

Firefighter I

5.3.7
5.3.8
5.3.10
5.3.13
5.3.14
5.3.15
5.3.18
5.3.19

Firefighter II

6.3.1
6.3.2
6.3.3

1. Describe initial factors to consider when suppressing structure fires. (5.3.8, 5.3.10)

2. Summarize considerations taken when making entry. (5.3.8, 5.3.10)

3. Describe direct attack, indirect attack, combination attack, and gas cooling techniques. (5.3.8, 5.3.10)

4. Describe safety considerations that must be identified for upper level structure fires. (5.3.8, 5.3.10)

5. Explain actions taken when attacking belowground structure fires. (5.3.8, 5.3.10)

6. Discuss methods of fire control through exposure protection and controlling building utilities. (5.3.18)

7. Describe steps taken when supporting fire protection systems at protected structures. (5.3.8, 5.3.10, 5.3.14)

8. Explain considerations taken when deploying, supplying, and staffing master stream devices. (5.3.8)

9. Describe situations that may require suppression of Class C fires. (5.3.8, 5.3.10)

10. Identify hazards associated with suppressing Class C fires. (5.3.8, 5.3.10)

11. Describe actions associated with suppressing Class D fires. (5.3.8, 5.3.10)

12. Explain actions taken when suppressing a vehicle fire. (5.3.7)

13. Compare methods used to suppress fires in stacked and piled materials, small unattached structures, and trash containers. (5.3.8)

14. Summarize the main influences on ground cover fire behavior. (5.3.19)

15. Compare types of ground cover fires. (5.3.19)

16. Describe elements that influence ground cover fire behavior. (5.3.19)

17. Identify the parts of a ground cover fire. (5.3.19)

18. Describe protective clothing and equipment used in fighting ground cover fires. (5.3.19)

19. Describe methods used to attack ground cover fires. (5.3.19)

20. Summarize safety principles and practices when fighting ground cover fires. (5.3.19)

21. Attack a structure fire using a direct, indirect, or combination attack. (Skill Sheet 17-I-1; 5.3.8, 5.3.10, 5.3.13)

22. Attack a structure fire above, below, and at ground level – Interior attack. (Skill Sheet 17-I-2; 5.3.8, 5.3.10, 5.3.13)

23. Turn off building utilities. (Skill Sheet 17-I-3; 5.3.18)

24. Connect supply fire hose to a fire department connection. (Skill Sheet 17-I-4; 5.3.8, 5.3.10, 5.3.14, 5.3.15)

25. Operate a sprinkler system control valve. (Skill Sheet 17-I-5; 5.3.8, 5.3.10, 5.3.14)

26. Stop the flow of water of an activated sprinkler. (Skill Sheet 17-I-6; 5.3.8, 5.3.10, 5.3.14)

27. Deploy and operate a portable master stream device. (Skill Sheet 17-I-7; 5.3.8)

28. Attack a passenger vehicle fire. (Skill Sheet 17-I-8; 5.3.7)

29. Attack a fire in stacked or piled materials. (Skill Sheet 17-I-9; 5.3.8)

30. Attack a fire in a small unattached structure. (Skill Sheet 17-I-10; 5.3.8)

31. Extinguish a fire in a trash container. (Skill Sheet 17-I-11; 5.3.8)

32. Attack a ground cover fire. (Skill Sheet 17-I-12; 5.3.19)

1. Describe considerations taken when coordinating fireground operations. (6.1.1, 6.1.2, 6.3.2)

2. Explain fireground roles and responsibilities a FireFighter II may need to coordinate. (6.1.1, 6.1.2, 6.3.2)

3. Discuss the process of establishing and transferring Command. (6.1.1, 6.1.2, 6.3.2)

4. Describe hazards that may be present at fires in underground spaces. (6.3.2)

5. List safety precautions that should be taken at flammable/combustible liquid fire incidents. (6.3.1, 6.3.3)

6. Recognize methods used when coordinating operations at a property protected by a fire suppression system. (6.3.2)

7. Explain ways to use water to control Class B fires. (6.3.1)

8. Compare methods used to suppress bulk transport vehicle fires and flammable gas incidents. (6.3.3)

9. Establish Incident Command and coordinate interior attack of a structure fire. (Skill Sheet 17-II-1; 6.1.1, 6.1.2, 6.3.2)

10. Control a pressurized flammable gas container fire. (Skill Sheet 17-II-2; 6.3.3)

Chapter 17
Fire Control

Courtesy of Dick Giles.

Case History

On the evening of February 24, 2012, seven firefighters were seriously injured in a fire in a vacant house in Prince George's County, Maryland. Personnel arrived to find fire coming from both the basement and first floor of the one-story, single-family residence. It was a vacant structure, although firefighters believed the house may have been occupied because a car was parked in the driveway. Firefighters had initiated an interior attack on the fire when high winds introduced a sudden rush of air from the rear of the house, either from a door or window being opened or broken. According to a fire department press release, the sudden addition of a large amount of fresh air into the fire created blowtorch or furnace-like conditions inside, engulfing the firefighters. Wind gusts were measured at more than 40 mph (64 Km/h) around that time. Firefighters tried to escape but the flames overtook them. The Incident Commander immediately called for additional resources. Another team of firefighters entered the basement of the house and the fire was extinguished. Authorities subsequently determined that the fire cause was arson.

Up to this point in your training you have gained the basic knowledge and skills necessary to effectively control and extinguish fire. Now it is time to learn how to apply those skills and knowledge to control fires involving:

- Structures
- Stored Class A materials located outside of structures (outside fires)
- Ground cover (vegetation/wildland fires)
- Vehicles
- Class C materials (electrical equipment)
- Class D materials (flammable metals)

As you advance to the Firefighter II level you will learn how to extinguish flammable and combustible liquid fires. In addition, you will learn how to supervise a fire fighting team in all types of fire suppression situations.

■ Suppressing Structure Fires

Extinguishing a structure fire may take from only a few minutes to several hours. Your duties and assigned tasks will depend on your departmental standard operating procedures (SOPs), your assigned job on your company, the amount of fire involvement, and the type of structure involved. This section will focus on the strategy and tactics used to suppress structure fires including:

- Strategy and coordination of resources
- Hoseline selection
- Nozzle selection
- Making entry
- Fire attack
- Gas cooling
- Fires in upper levels of structures
- Belowground structure fires
- Exposure protection
- Controlling building utilities

Strategy and Coordination of Resources

Depending on the conditions at the fire scene, the Incident Commander (IC) determines the strategy and tactics for controlling the fire. These are based in order of importance, on three priorities: life safety, incident stabilization, and property conservation **(Figure 17.1)**. Because life safety includes the safety of firefighters, the IC makes a risk/benefit analysis to determine if lives can be saved at the risk of firefighters. Essentially, the IC must decide if risking the lives of firefighters will have a positive outcome such as saving the lives of victims inside the burning structure. This decision will often dictate whether the fire is attacked offensively or defensively.

For example, if it is believed that there may be victims near the seat of the fire or in the most involved room, it is both highly likely that those victims cannot be saved and that an attempt to do so could result in the severe injury or death of firefighters. The

Figure 17.1 The Incident Commander makes strategic and tactical decisions based on the incident priorities. *Courtesy of Bob Esposito.*

IC must make the difficult decision to attack the fire defensively and make an attempt at rescue when the fire is better controlled. However, if it is known that there may be victims in uninvolved areas of the structure or there is a clear path for firefighters to get close to the seat of the fire safely, the IC can decide to enter the structure for an offensive attack.

Offensive

The offensive strategy used at a structure fire usually means deploying resources for interior tactical operations to accomplish incident priorities. Various factors divided into specific categories will help determine the tactics used during an offensive strategy as follows:

- **Value** — Life and safety hazards at the scene, savable lives, and/or salvageable property
- **Time** — Time to accomplish selected tactics, potential for collapse and deterioration of structural stability, and potential changes in fire conditions
- **Size** — Tactical flow rates needed to control the fire, available resources, and fire conditions

Rescue and/or fire extinguishment may be the objective for an offensive strategy **(Figure 17.2)**. In some fire incidents, rescue and extinguishment will occur simultaneously. Engine crews will attack the seat of the fire while other personnel search for victims. In extreme cases where a victim is known to be trapped, rescue will become

Figure 17.2 An offensive strategy may include advancing an attack hoseline into a structure to extinguish the fire, rescuing occupants, or both.

the *primary* activity and fire attack will be performed only to protect the rescuers and the victim. If rescuers cannot be adequately protected, other tactics should be used and rescue should be attempted when it can be completed with lowered risk to firefighters.

Defensive

A defensive strategy is typically selected given one or more of the following factors:

- No threat to occupant life exists.
- Occupants are not savable.
- The property is not salvageable.
- Sufficient resources are not available for an offensive strategy.
- There is a danger of structural collapse.
- An offensive strategy would endanger the lives of firefighters because of hazardous conditions at the scene.

The defensive strategy is intended to isolate or stabilize an incident and keep it from expanding. In the case of a structure fire, a defensive strategy may mean sacrificing a building that is on fire to save adjacent buildings that are not burning. A defensive strategy is generally an exterior operation that is chosen because an interior attack is unsafe or resources are insufficient **(Figure 17.3)**. Defensive strategic operations are employed when the following conditions are present at a structure fire:

- **Excessive volume of fire** — Amount of fire exceeds the ability of available resources to confine or extinguish it. Lack of resources includes:
  - Lack of personnel or lack of trained personnel.
  - Inability to provide adequate fire flow in gpm (L/min) because of insufficient pumping capacity or availability of water supply.
  - Lack of appropriate apparatus or equipment to implement the required tactics.

Figure 17.3 Defensive strategic operations can occur when the volume of fire exceeds the resources required for an offensive attack. *Courtesy of Chris Mickal.*

- **Structural deterioration** — Structure is unsafe for interior entry.

- **Risk outweighs benefit** — The amount of risk to emergency responders is greater than the benefit.

- **Unfavorable wind conditions** — The wind conditions prohibit safe entry due to the potential development of high temperature flow paths within the structure.

Strategic Transitions

Transitions between the two strategies may occur at any time. The more common and less dangerous transition is from defensive to offensive. A transition from offensive to defensive occurs when conditions in the structure deteriorate rapidly and unexpectedly.

Defensive to offensive. In some situations, the first-arriving unit may need to deploy hoselines and begin with a defensive strategy until additional resources arrive or the amount of fire has been reduced or extinguished to safe levels. This strategy may be necessary especially when there are not enough firefighters to meet the Occupational Safety and Health Administration (OSHA) two-in, two-out regulation. While waiting for additional personnel, firefighters at the scene can use fire streams to protect exposures or limit fire development. An offensive interior strategy can be employed once additional personnel, apparatus, and adequate fire flow are available.

Offensive to defensive. When a situation rapidly changes, the IC must make the decision to transition from an offensive to a defensive strategy. How the transition occurs depends on the speed at which the situation changes. The IC will communicate the change to all personnel and units operating at the incident. Next, the IC will order a personnel accountability report (PAR) from all personnel to ensure that they have been advised and have withdrawn from offensive positions. All personnel must be made aware of the transition. Some units may need to remain in place to protect the egress/withdrawal of other units. Supervisors/company officers must always know the location of personnel assigned under their command and must conduct personnel accountability checks when withdrawal is complete.

During an orderly tactical withdrawal, hoselines should *not* be abandoned unless absolutely necessary. In an emergency evacuation, all personnel evacuate as quickly as possible. Abandoned hoselines cannot provide any protection during an evacuation.

Rapid intervention crew or team (RIC/RIT) personnel must be ready to assist any units during the transition. Companies should continue to perform their assignments during the transition.

When a transition from offensive to defensive strategy occurs, you must:

- Use or maintain situational awareness to recognize changes in fire behavior and structural stability.

- Know your department's evacuation signal (often an air horn or particular radio tone).

- Continue to monitor your radio for further orders.

- Remain calm and follow orders.

- Stay with your team.

- Use the hoseline to guide you to the exit.

- Use your hoseline and have the nozzle set on the appropriate pattern to protect yourself during a tactical withdrawal.

- Evacuate as quickly and as safely as possible.

- Respond to requests for personnel accountability reports (PAR).

- Know your department's SOPs on offensive and defensive strategies.

Resource Coordination

Fire attack during interior operations must be coordinated with forcible entry, search and rescue operations, ventilation, and control of utilities. As the operation progresses, fire attack must also be coordinated with loss control, cause determination, and victim recovery efforts. When fighting any fire, large or small, exterior or interior, firefighters should always work as a team under the direction of a supervisor. Coordination between crews performing different tactics and tasks is crucial. For example if attack hoselines are not yet in place or forcible entry personnel are not yet ready, ventilation could create air flow paths that will result in the unwanted spread of the fire. If ventilation is properly coordinated with attack and entry teams, then ventilation is an aid to their efforts rather a newly introduced hazard.

Situational Awareness Increases Your Safety

One of the most critical aspects of coordination between crews — and of your personal safety and survival — is maintaining your *situational awareness*. As mentioned in Chapter 2, Firefighter Safety and Health, your safety and that of your fellow firefighters can depend on being aware of everything going on around you. The opposite of situational awareness is *tunnel vision* where firefighters become so focused on fire fighting or other operational assignments that they fail to sense changes in their environment.

To maintain your situational awareness, act as follows:

- Look up, look down, look all around.

- Listen for new or unusual sounds and feel vibrations or other movements.

- Communicate to your supervisor, the other members of your crew, or Command any changes in the situation of which you become aware.

NOTE: Entire crews have died or suffered injuries because one member failed to communicate critical information about the fire environment.

Hoseline Selection

While there are other fire control technologies that involve the application of dry chemicals, foam, and other extinguishing agents, water is still the agent most often used to extinguish structure fires. Fire control will be successful only if the amount of water applied is sufficient to cool the burning fuels. Hoseline selection should be dependent upon fire conditions and other factors such as:

- Fire load and material involved

- Flow rate needed for extinguishment

- Stream reach needed

- Number of firefighters available to advance hoselines

- Need for speed and mobility

- Tactical requirements

- Ease of hoseline deployment

- Potential fire spread

- Size of building

- Size of fire area

- Location of fire

Selecting the correct size of hoseline is critical for efficiency and safety. For example, using a **small-diameter hoseline (SDH)** may allow water to be applied sooner than deploying a larger attack line. However, the size of the hoseline may not provide a sufficient volume of water for the size of the fire which may, in turn, delay extinguishment, exposing firefighters to danger from rapid fire development. In some heavily involved structure fires, even 1½-inch (38 mm) or 1¾-inch (45 mm) hoselines may not provide a sufficient flow rate of water to safely and effectively attack the fire. For interior fire fighting, the area involved and the fire load should dictate the size of the hoseline(s) that should be used. For an interior fire, a hoseline no smaller than 1½-inch (38 mm) should be used to provide a sufficient volume of water to cool the fire and protect the firefighters.

Table 17.1, p. 1010 gives a sample analysis of hose stream characteristics and is intended to help firefighters make decisions about which hose streams to select. It is not meant to replace the judgment of fire personnel in selecting hoselines.

Besides the primary attack hoseline, a backup hoseline must be placed in service at the same time. The backup hoseline performs three critical functions by:

1. Protecting the attack hoseline team from extreme fire behavior.

2. Protecting the means of egress for the attack hoseline team.

3. Providing additional fire suppression capability in the event that the fire increases in volume.

The backup hoseline should be at least the same size and provide the same fire flow as the attack hoseline. A fog nozzle is also preferred because it will provide the greatest protection for both teams.

CAUTION
A hoseline no smaller than 1½-inch (38 mm)
should be used on an interior fire.

Nozzle Selection

Your supervisor will tell you which nozzle to select based on the desired fire stream. The following factors will dictate the nozzle and stream selection:

- Fire conditions

- Available water supply

- Number of firefighters available to safely operate the hoseline

- Capabilities of the nozzle being used

For an interior fire attack, a fog nozzle is generally the most useful. A wide fog pattern can be used to protect firefighters from radiant heat as well as cool the hot fire gases **(Figure 17.4, p. 1011)**. When it becomes necessary to penetrate the hot gas layer and cool the compartment linings (ceiling or wall) or reach the burning fuel, the nozzle can be adjusted to a straight stream.

Small-Diameter Hose (SDH) — Hose of ¾-inch to 2 inches (20 mm to 50 mm) in diameter; used for fire fighting purposes. *Also called* Small Line.

Table 17.1

Hose Stream Characteristics

Size in (mm)	GPM (L/min)	Reach (Maximum) ft (m)	No. of Persons on Nozzle	Mobility	Control of Damage	Control of Direction	When Used	Estimated Effective Area
1½ inches (38 mm)	40–125 gpm (160 L/min to 500 L/min)	25–50 feet (8 m to 15 m)	1 or 2	Good	Good	Excellent	• Developing fire — still small enough or sufficiently confined to be stopped with relatively limited quantity of water	One to Three Rooms
1¾ inches (45 mm)	40–175 gpm (160 L/min to 700 L/min)	25–50 feet (8 m to 15 m)	2	Good to Fair	Good	Good	• For quick attack • For rapid relocation of streams • When personnel are limited • When ratio of fuel load to area is relatively light • For exposure protection	
2 inches (50 mm)	100–250 gpm (400 L/min to 1 000 L/min)	40–70 feet (12 m to 21 m)	2 or 3	Fair	Fair	Good	• When size and intensity of fire are beyond reach, flow, or penetration of 1½-inch (38 mm) line • When both water and personnel are ample	One Floor or More; Fully Involved
2½ inches (65 mm)	125–350 gpm (500 L/min to 1 400 L/min)	50–100 feet (15 m to 30 m)	2 to 4	Fair to Poor	Fair	Good	• When safety of crew dictates • When larger volumes or greater reach are required for exposure protection	
Master Stream	350–2,000 gpm (1 400 L/min to 8 000 L/min)	100–200 feet (30 m to 60 m)	1	Poor to None (Aerial Master Streams can be Good)	Poor	Good	• When size and intensity of fire are beyond reach, flow, or penetration of handlines • When water is ample, but personnel are limited • When safety of personnel dictates • When larger volumes or greater reach are required for exposure protection • When sufficient pumping capability is available • When massive runoff water can be tolerated • When interior attack can no longer be maintained	Large Structures; Fully Involved

Adapted from Joe Batchler, Maryland Fire & Rescue Institute.

Figure 17.4 Fog patterns provide firefighters with protection from radiant heat while cooling hot gases.

Figure 17.5 Straight or solid streams can be used for applying water from a distance. *Courtesy of Ron Jeffers.*

If an attack is going to be made from the structure's exterior, a solid stream nozzle may be the best choice. The solid stream will deliver the greatest amount of water over the farthest distance (**Figure 17.5**). It can also be directed through an opening at the compartment lining causing the stream to disperse into small drops and absorb more heat.

The water pressure and quantity available will also determine the type of nozzle selected. A stream that lacks the necessary pressure to reach the target will result from using a solid stream nozzle tip that is too large. If the nozzle tip is too small, the stream will not deliver the volume of water required to extinguish the fire. Generally, an automatic fog-stream nozzle will adjust to the volume and pressure of the water supplied by the pumper.

Nozzle reaction will also dictate the number of personnel required to advance the hoseline and operate the nozzle within the confines of a structure. The greater the reaction, the more firefighters who will be needed. If the number of personnel is limited, it may be necessary to reduce the size of the nozzle or use a master stream device from the exterior.

Making Entry

The IC or supervisor will decide where and from what direction to make entry for an interior fire attack. Generally, attack hoselines are placed to protect firefighters, occupants, and property. However, you should be aware of some of the factors that are used to make the decision, which include:

- Wind direction and velocity
- Building conditions
- Initial fire location
- Location of occupants
- Exposures

Before entering a burning building, every member of the crew should conduct a quick size-up and maintain a high level of situational awareness. Entry size-up begins well before reaching the entry point and may include visual observations and the use of a thermal imager (TI). The following are additional pre-entry considerations critical to firefighter safety and effectiveness:

- Reading fire behavior indicators
- Understanding the crew's tactical assignment
- Identifying potential emergency escape routes (other doors, windows, etc.)
- Evaluating forcible entry requirements
- Identifying hazards (overhead wires, structural instability, etc.)
- Verifying that radios are receiving and transmitting on the correct channel
- Ensuring that self-contained breathing apparatus (SCBA) is on, cylinder is full, and operating properly
- Ensuring that all Personal Alert Safety System (PASS) devices are on and operating properly
- Ensuring other team members are prepared to enter the structure by doing a buddy check

Interior fire attack crews advancing hoselines must carry tools and equipment needed to open interior doors, check concealed spaces for fire extension, or to make an emergency exit. This equipment could include such tools as a portable radio, hand light, pike pole, thermal imager, and forcible entry tools. Before entering the building or the fire area, the firefighter assigned to the nozzle should open the nozzle fully to ensure adequate flow, check the pattern setting, and bleed the air from the hoseline. Opening the bale slightly while waiting for water to arrive hastens this process. Firefighters should also either test the range of stream patterns of the nozzle or set the proper nozzle pattern for the attack to verify that the nozzle is working properly.

CAUTION

DO NOT open the door until you have a charged hoseline and are ready to control the conditions encountered.

When an interior attack is going to be made on a structure fire, firefighters should wait in a safe area near the building entrance. From this location extinguish visible fires in any fascia or soffit, boxed cornices, other exterior overhangs, open windows and doors, or around entry or exit points. When the attack crew moves to the building entrance, they should stay low and out of the doorway while the door is forced open. Check the door for heat before opening it by using the back of a gloved hand, a TI, or applying a small amount of water spray to the surface of the door **(Figure 17.6)**. If the door is very hot, the water will evaporate and convert to steam. Excessive heat may be obvious from the smoke, air flow, and other fire behavior indicators. The door should be kept closed until the hoseline is charged and the crew is ready to enter.

If the fire is ventilation controlled and the door is opened, a significant increase in heat release rate can quickly occur. Unburned fuel in the form of smoke will escape at the top of the doorway while fresh air will enter at the bottom, providing oxygen for fire development. In this situation, cooling the hot gases overhead can reduce the risk of ignition potentially leading to flashover and provide a safer operating environment.

Figure 17.6 A thermal imager can be used to detect heat radiating from a closed door prior to opening it.

Figure 17.7 A strap or rope can be used to control a door under fire conditions.

You must observe the smoke movement and air flow when the door is opened: fast air movement in at the bottom and smoke moving out at the top indicates an active fire in the structure. With the attack hoseline in place, open the door slightly, apply water to the hot gas layers, and wait 5-10 seconds to observe any reactions before entering the structure.

Firefighters must maintain control of the door as it is opened. Place a rope hose tool or utility strap over the doorknob so that it can be quickly pulled closed if necessary **(Figure 17.7)**. Once the door is open and entry is made, chock the door to prevent it from closing on the hoseline.

A conventional guideline in the fire service for many years has been to "attack the fire from the unburned side." Major fire departments and the National Institute of Science and Technology (NIST) have performed fire analysis and laboratory tests that greatly disprove this traditional guideline. The following two factors have changed the fire service's understanding of fire behavior in structure fires:

- Greater heat release rates of modern building construction materials and modern furnishings
- Increased effect of wind on fire expansion and development

Wind creates air flow patterns within the structure that directly increase fire expansion and cause firefighter casualties. Because wind can cause unpredictable changes to the fire, you should attack with the wind to your back. To determine the best entry point based upon wind direction, the IC should complete a thorough size-up and do a 360-degree survey, whenever possible, before deploying attack hoselines.

Fire Attack

Depending on the nature and size of the fire, firefighters may use a direct, indirect, or combination method of attacking the fire. Hoseline selection and stream selection are also made when the fire attack is conducted. **Skill Sheet 17-I-1** describes the methods for making interior direct, indirect, and combination attacks on a structure fire.

Direct Attack

A **direct attack** on the fire using a solid or straight stream uses water most efficiently on free-burning fires **(Figure 17.8, p. 1014)**. The water is applied directly onto the burning fuels until the fire is extinguished. Another effective technique is to direct the stream onto the ceiling and walls which can slow or stop the pyrolysis process on

Direct Attack (Structural) — Attack method that involves the discharge of water or a foam stream directly onto the burning fuel.

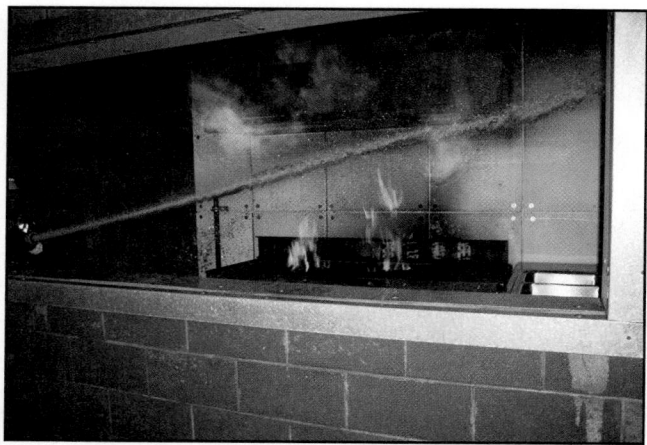

Figure 17.8 A direct attack is used to quickly place water on the fire.

Figure 17.9 An indirect attack involves directing a stream toward the ceiling of the room. *Courtesy of Dick Giles.*

these hot surfaces. Water should not be applied long enough to upset the thermal layering (sometimes called *thermal balance*) in the compartment; the steam produced will begin to condense, causing the smoke and heat to drop rapidly to the floor and move sluggishly thereafter.

Indirect Attack

Indirect Attack (Structural) — Form of fire attack that involves directing fire streams toward the ceiling of a compartment in order to generate a large amount of steam in order to cool the compartment. Converting the water to steam displaces oxygen, absorbs the heat of the fire, and cools the hot gas layer sufficiently for firefighters to safely enter and make a direct attack on the fire.

When firefighters are unable to enter a burning building or compartment because of the intense heat inside, an **indirect attack** can be made from outside the structure or involved area. The attack is made through a window or other opening, directing the stream toward the ceiling to cool the room **(Figure 17.9)**. This method of attack produces large quantities of steam and must be coordinated with ventilation. While an indirect attack cools the fire environment, it results in a fairly uniform temperature from floor to ceiling and fills the compartment with the combined mixture of smoke and steam. However, this may be the only method of attack possible until temperatures are reduced.

To make an indirect attack on the fire, a fog stream is introduced through an opening and directed at the ceiling where the temperature is the highest. The heat converts the water spray to steam, which fills the compartment and absorbs the majority of the heat. Once the majority of the fire has been reduced in quantity and the space has been ventilated, hoselines can be advanced inside and firefighters can make a direct attack on the body of the fire.

Combination Attack

Combination Attack — Extinguishing a fire by using both a direct and an indirect attack. This method combines the steam-generating technique of a ceiling level attack with an attack on the burning materials near floor level.

A **combination attack** combines cooling the hot gas layer at the ceiling level using an indirect attack with a direct attack on the fuels burning near the floor level. To combine both attacks, move the nozzle from the area overhead to the floor in a Z, inverted T, or rotational manner **(Figure 17.10)**. Excessive application of water to smoke does not extinguish the fire and may cause unnecessary water damage and disturbance of the thermal layering. Applying water to smoke that is not heated may disrupt the thermal layering which has a tendency to decrease visibility if ventilation is not accomplished.

Figure 17.10 A combination attack adds the cooling effect of an indirect attack to the suppression effect of a direct attack.

Gas Cooling

Gas cooling is not a fire extinguishment method but is a way of reducing heat release from the hot gas layer. This technique is effective when faced with a **shielded fire**; that is, a fire you cannot see from the doorway because it is located in a remote part of the structure or objects are shielding the fire. In these situations, you cannot apply water directly onto the burning material without entering the room and working under the hot gas layer.

The hot gas layer accumulating in the upper levels of the compartment can present a number of problems for you and the other members of the hoseline crew. Remember that *smoke is fuel*, and it may transition to rollover, flashover, or a smoke explosion at any time. In addition, hot smoke radiates heat to furniture and other combustibles in the compartment. This increases pyrolysis which adds more flammable fuel to the gas layer. Cooling the hot gas layer slows the transfer of heat to other combustibles and reduces the chances of the overhead gases igniting.

To cool the hot gas layer, direct short bursts or pulses of water fog into it. With the nozzle set on a fog pattern, direct it upward toward the gas layer and quickly and smoothly open and close it (**Figure 17.11**). The length of the pulse will depend on the size of the space, varying from less than a second to much longer. The nozzle pattern may need to be adjusted based on the fire conditions in the compartment and its configuration and size. In narrow hallways, the fog pattern may need to be restricted. In large-volume compartments or when the upper layer temperature is extremely high, the duration of the pulses may need to be increased.

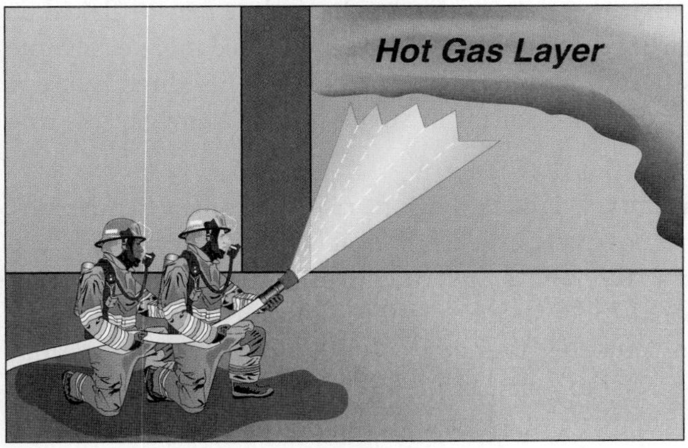

Shielded Fire – A fire that is located in a remote part of the structure or hidden from view by objects in the compartment.

Hot Gas Layer

Figure 17.11 The hot gas layer can be cooled by directing a fog pattern into it and opening and closing the nozzle to provide short bursts.

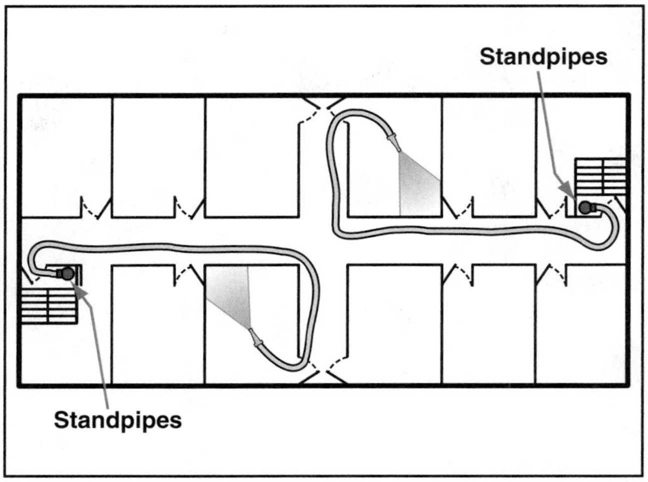

Figure 17.12 In modern buildings, standpipe connections may be found in protected stairways allowing hoselines to reach all areas of the floor.

The reach of the stream is also important for cooling the gas layer. Your intent is to cool the gases, *not* to cool the ceiling. Cooling the ceiling will produce a large volume of steam. If water droplets fall out of the overhead smoke layer, it means that the gases have been cooled and you can stop spraying water into the smoke. If the fire continues to burn unchecked, the gas layer will regain its heat and the gas-cooling technique may have to be repeated. The gas-cooling technique should be repeated as necessary while the hose team advances under the gas layer toward the fire.

Fires in Upper Levels of Structures

Multistory structures may or may not contain standpipe systems. Depending on the local building code and age of the structure, a standpipe may be required in structures three stories or higher. In structures that lack standpipes, fire attack proceeds through the main entrance and up the closest stairway to the fire location. Always check for fire extension below the fire floor before advancing up the stairs. Additional personnel will be required to advance the fire hose and ensure that kinks or bends do not prevent full fire flow at the nozzle.

In structures equipped with standpipe systems, the location of the standpipe connection will determine the method of fire attack. Although standpipe connections in older structures may be located in corridors or near open stairwells, most modern structures have the standpipes located in protected stairways. If standpipe connections are located in unprotected locations, the attack hoseline is connected on the floor below the fire floor and advanced up the nearest stairwell.

If the standpipe connection is in a protected stairway, hoselines may be connected on the fire floor (**Figure 17.12**). Extra sections of the attack hoseline may be flaked up the stairway to the first landing above the fire floor so that it will feed more easily into the fire floor as the line is advanced. In addition to attacking the fire directly, crews should be checking floors above the fire floor for fire extension and any victims who may have been unable to escape.

Fires in upper levels of structures, especially high-rise buildings, can require large numbers of personnel to conduct large-scale evacuations, carry tools and equipment to upper levels, and maintain a sustained fire attack. In many cases, firefighters must carry additional tools and equipment up many flights of stairs. Elevators must not be used to transport fire crews to the fire floor. The fire may damage the elevator or its controls and strand the firefighters in the elevator car or deposit them directly onto the fire floor. Some departments allow elevators to be used to transport personnel, fire fighting tools, and equipment to a staging area normally located two floors below the fire floor. Always follow your department SOPs when using elevators in fire buildings.

Personnel must exercise caution in the streets around the outside perimeter of a high-rise building on fire. Glass and other debris falling from many stories above the street can severely damage equipment, cut hoselines, and injure or kill firefighters. To minimize the danger from falling objects or debris, cordon off safe paths of entry into the building. Conditions dictate how large an area needs to be cordoned off.

Belowground Structure Fires

Fires originating in basements or subfloors are some of the most difficult and dangerous structure fires that you may encounter. Because of the high level of risk, the IC must make a thorough and accurate risk/benefit analysis before committing personnel into a structure experiencing a basement fire.

Residential Basements

Residential basements exist in many communities. They may be used as:

- Self-contained living quarters
- Entertainment rooms
- Utility spaces for:
  - Storage
  - Heating, ventilation, and air-conditioning equipment (HVAC)
  - Water heaters
  - Coal or fuel oil storage
- Garages
- Storm shelters

The spaces may be totally unfinished, partially finished, or completely finished (**Figure 17.13**). In an unfinished basement, the first floor joists are exposed to the fire and will fail sooner than a ceiling protected with drywall. Because unfinished basements are generally unoccupied, the fire may have the chance to spread before it is

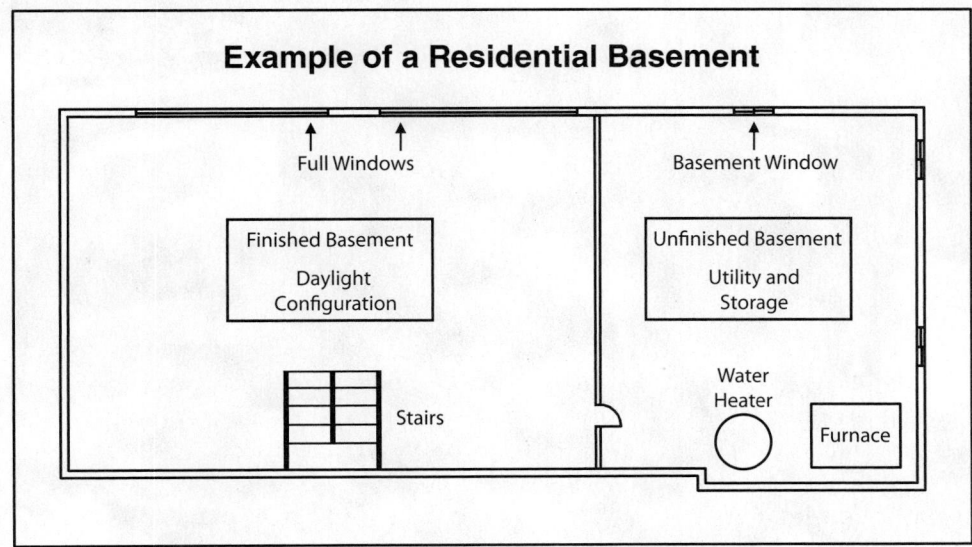

Figure 17.13 Residential basements may include finished rooms used as living quarters and/or unfinished space used for storage and mechanical equipment.

found. In some partially or completely finished basements, the ceiling may be composed of a metal grid system to support drop-in tiles that offer minimal fire resistance and may add to the fuel load.

Other factors that contribute to basement fires include:

- Fuel loading, especially the fuel load on the floor above the basement
- Age of exposed joists
- Hidden paths for fire in walls and ducts that could be exposed in basements
- Use of lightweight construction materials that are susceptible to rapid collapse

Basements that have been converted to living spaces, including sleeping areas, create life safety hazards for the occupants as well as firefighters. Although some jurisdictions require windows that can be used for egress, owner/occupants may not comply with these regulations.

Initial size-up of any basement fire is especially important because basement fires are extremely dangerous, especially with regards to structural collapse. The floor assemblies over basements can reach a point of collapse *before* firefighters arrive on the scene. In tests that Underwriters Laboratories Inc. (UL) performed, supporting joists collapsed before the floor decking burned through, resulting in failure of the floor **(Figure 17.14)**. This test shows that both firefighters working on the floor above a basement fire and those fighting the fire are at a heightened risk of floor collapse both during operations and after the fire has been extinguished.

WARNING
Basement fires weaken the main floor of a structure creating a constant danger of structural collapse.

Figure 17.14 Basement fires can weaken floor joists and burn through floor decking allowing the fire to extend into other levels of the house. *Courtesy of NIST.*

During initial fire fighting operations, sounding the floor and using a thermal imager (TI) have been used to determine if the floor is safe to walk on. The UL tests indicated that these two practices are not sufficient to ensure the integrity of the floor system. When the fire is extinguished, a visual inspection of the floor joists should be made before personnel are permitted to work on the first floor.

///////////////////////////
CAUTION
Thermal imagers (TI) will not always provide an accurate assessment of the structural integrity of the floor system.

Accessing the basement is also very dangerous. Depending on the design, basements may be accessed through:

- Interior enclosed stairwells
- Exterior enclosed stairwells
- Exterior open stairwells
- Window wells
- Ground-level walk-in doors or windows

The interior and exterior enclosed stairwells act as a flow path for smoke, flames, and heated gases, much like a fireplace chimney **(Figure 17.15)**. Attempting to advance an attack hoseline down an enclosed stairwell may be the only avenue avail-

Figure 17.15 Smoke from a basement fire will flow up a stairwell in much the same way as it flows up a chimney.

able but it exposes firefighters to tremendous hazards. If the first floor is determined to be unsafe, an exterior attack can be made through basement windows. If any opening into the basement exists, use it to apply water to the area before ventilating or entering the first floor. A penetrating or cellar nozzle can also be used if it can be installed without placing personnel on the weakened floor. Reduce the amount of fire and then make another assessment of the basement conditions.

Skill Sheet 17-I-2 describes how to attack a structure fire above, below, and at ground level with an interior stairwell attack. It is critical that you have enough hoseline to reach the base of the stairs and an additional 6 to 8 feet (1.8 to 2.4 m) at the bottom to get through the door and out of the narrow area created by the doorway and stairwell.

Depending on the fire's heat release rate, the water from a 1½- to 1¾-inch (38 mm to 45 mm) hoseline may not provide enough cooling to overcome the gases venting up the stairway. If the fire is ventilation limited, the added ventilation may result in a flashover of the basement and be fatal to any firefighters on the stairs or in the basement. Underwriters' Laboratories (UL) data shows that the conditions in the basement may be untenable even if a flashover has not occurred. Basement fires are hot and opening a vent point to allow oxygen to enter is going to create more problems.

Commercial Basements and Subfloors

Basements and subfloors in commercial structures may have similar construction as those in residential structures, though they may be more robust if the fuel load on the main floor is significant. In older Type III construction, the first floor joists may be exposed wood joists or even heavy timbers. In more modern structures, the floors may be exposed or unexposed concrete panels or metal C-joists. In many cases, these floors are designed to support the weight of machinery, products, and upper stories. Some structures may have multiple subfloors used for mechanical spaces or even parking garages **(Figure 17.16)**.

Exposure to fire may weaken metal floor supports. Heavy objects on the floor above the fire can increase the chances of floor collapse because the added weight accelerates the failure of supporting members. Unprotected steel girders and other supports elongate when exposed to temperatures of 1,000°F (538°C) or more. These supports have been known to topple walls during a fire. The longer steel supports are subjected to fire, the more likely they are to fail, regardless of their configuration.

In some commercial structures, standpipe connections will exist in the stairwells leading to the subfloors. Attacking a fire in these lower levels follows the same process used for upper floors.

A risk/benefit analysis of a commercial basement fire is performed in the same way as one for a residential basement fire. Preincident surveys and inspections will help to determine the type of basement ceiling construction and the amount of fire it can withstand before it collapses. The location of standpipe connections, potential ventilation air flow paths, and the amount of breathing air needed to access a basement fire will influence the tactics used.

Exposure Protection

Preventing a fire from spreading to unaffected areas is an important tactic. Unaffected areas are referred to as **exposures** and may exist inside or outside of a structure. **Exposure protection** can take a number of forms depending on the location and type of exposure and the resources that are available to the fire department.

Exposures — Structure or separate part of the fireground to which a fire could spread.

Exposure Protection — Covering any object in the immediate vicinity of the fire with water or foam.

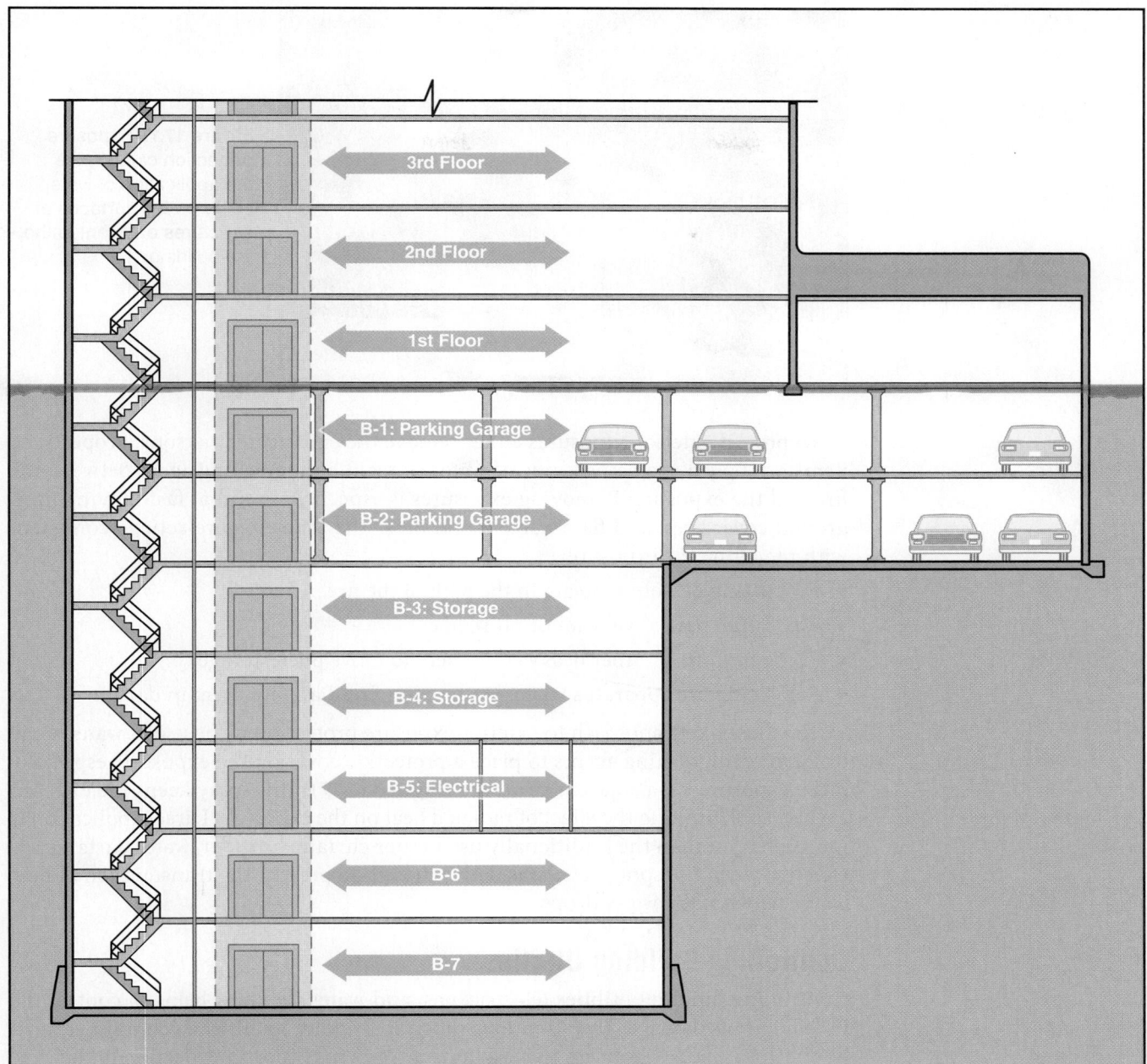

Figure 17.16 Basements and subfloors in commercial structures can have many uses, including parking, electrical/mechanical, and storage.

Interior exposure protection generally involves closing doors or other openings between the fire area and the unaffected area and the proper use of tactical ventilation to ensure that smoke movement is limited. Passive forms of exposure protection, such as fire-rated walls and doors, are also used to prohibit fire and smoke movement. You must be aware of the arrangement of the building you are working in, the location of the fire, and how tactical ventilation is being used. This awareness can help you to determine which doors and openings must be kept closed and where to locate ventilation fans.

Figure 17.17 Exposure protection can involve the application of water spray on the surfaces of structures adjacent to the fire building.

To protect exterior exposures either remove the endangered persons, property, or items or apply a protective spray of water or foam extinguishing agent between the fire and the exposure. Removing exposures is especially useful at fast-moving fires, ground cover fires, and flammable liquid fires. The following are actions consistent with removing the exposure:

- Evacuating persons who are in the path of the fire
- Relocating parked vehicles or railroad cars
- Using forklifts or other heavy equipment to move piled storage
- Relocating fire apparatus when fast-moving fires have put them in danger

The most likely approach to exterior exposure protection is the use of water spray or foam extinguishing agents to place a protective cover on the exposure, especially if the exposure is an adjacent structure (**Figure 17.17**). This spray keeps the exposed surface cool, limiting the effect of radiated heat on the exposure. Direct application is more effective than the traditionally used water curtain; however, water curtains are effective when the spray pattern is dense enough to prevent the transmission of heat waves between the water drops.

Controlling Building Utilities

Controlling building utilities (electric, gas, and water supplies) helps to control the fire and limits damage. The IC or your supervisor will determine which utilities must be controlled and will assign the task to you. You must know the location of the control switches and valves and how to shut them off. The locations and means of control will vary among types of structures, although some general concepts can be applied and will be introduced in the appropriate section below. For more information on turning off building utilities, see **Skill Sheet 17-I-3**.

Fire department personnel are not responsible for turning utilities back on and should not attempt to do so. The utility provider will have to determine if the building and utility distribution system is safe before the service can be reestablished. Turning the electricity or gas back on could cause a reignition of the fire due to damaged wiring, pipes, or appliances.

Electricity

The sources of electricity are usually a commercial power company or an alternative power source in the form of solar panels, wind generators, or fuel-powered generators. In any case, electric service must be disconnected when there is a fire in the structure. In some cases, structures are total electric, meaning that light, heat, and cooking are solely powered by electricity. In other cases, electricity only provides lighting while a form of gas provides hot water, heating, and cooking. Firefighters must be able to control the flow of electricity into structures where emergency operations are being performed.

Commercial power supply. Commercial power companies provide electricity in urban, suburban, and rural areas. Electric lines that may be aboveground or buried connect structures to the main power grid. If the lines are aboveground, they will run from the power company pole to a weather head and service mast (a metal pipe extending above the roof of the house) which will extend down the side of the house to the electric meter box and shutoff. The electric meter is the primary location for shutting off the power. If the lines are buried, the meter box and shutoff may be on the side of the pole where the line extends underground. In multifamily dwellings, the meter boxes may be clustered in one location on the side of the building. In commercial, industrial, and institutional occupancies, preincident surveys should identify the location of the power connection.

When shutting off the power, pull the handle on the side of the meter box down **(Figure 17.18)**. In a residential structure, you can shut off the main circuit breakers and cut power to the structure. However, turning off individual circuit breakers will not cut all the power off and may not isolate the area where the fire occurred. The meter box is the only location that will shut off power to the entire structure. Shutting off the meter is appropriate, but only the power company should ever attempt to remove a meter. The IC should notify the power company when power to the structure is shut off so that a representative can be dispatched. In some residential and commercial occupancies, removing the electric meter does not completely stop the flow of electricity to the structure.

Figure 17.18 Electric service control panels have a master power shut off on the outside of the panel, and the handle may be locked in place.

Avoid Service Masts

Never touch a steel service mast. Some older brands of fuse panels were widely used and are commonly known for their main breaker shorting and literally exploding, causing the wires to come in contact with the steel panel and electrifying the mast pipe.

Electrical power to the entire building should not be shut off until ordered because electrical power is necessary to operate elevators, air-handling equipment, and other essential systems in all types of occupancies. In some instances, securing the electrical power in commercial/industrial occupancies can stop essential chemical processes or safety systems and create an unsafe situation. The decision to shut off the power is based on knowledge gained through preincident surveys.

Both **high-voltage** and **low-voltage** systems may be found in many buildings, each supplying a portion of the power needed to operate the building services. If power is shut off to the entire building or any device in it, the main power switch should be locked out and tagged out to prevent it from being turned back on before it is safe to do so. If lockout/tagout devices are not available, assign a firefighter equipped with a portable radio to stand by at the switch until power can be safely restored.

In many structure fires it is advantageous for electrical power to remain on to provide power for lighting, ventilation equipment, fire pumps, and other essential systems. The IC will make this decision in consultation with the Incident Safety Officer (ISO). When a fire involves only one area of a structure, it would be counterproductive

High-Voltage — Any voltage in excess of 600 volts.

Low-Voltage — Any voltage that is less than 600 volts and safe enough for domestic use, typically 120 volts or less.

Figure 17.19 Solar panel arrays retain power even if the shutoff switches are closed. *Courtesy of Ron Moore and McKinney (TX) FD.*

Figure 17.20 Fuel-powered generators supplement the main source of power for many types of buildings. *Courtesy of Ron Moore and McKinney (TX) FD.*

to shut off power to the entire building. When the building becomes damaged to the point that service is interrupted or an electrical hazard is created, however, power should be turned off at the main panel — preferably by a power utility employee. As always, firefighters must follow their departmental SOPs.

Alternative sources. Firefighters should be alert for installations with alternative emergency power capabilities such as solar panel arrays, usually located on the roof, wind turbines, or liquid/gas-fueled generators. Where these sources are present, removing the meter or turning off the master switch may not turn off the power entirely.

One solar panel can generate enough power to kill a person; therefore, solar panels and photovoltaic panels (PV panels) are marked with red warning labels. Most solar panel arrays have two shutoff switches, one on each side of the power inverter **(Figure 17.19)**. There is also a shutoff switch on the electric meter. Closing any of these switches will shut off power inside the structure. But, it will not turn off the solar panel array. As long as there is any amount of sunlight, the panel is always on and generating power. When performing tactical vertical ventilation on a roof with solar panels, completely avoid these panels.

> **WARNING**
> Solar panels generate current whenever there is a light source (sunlight, moonlight, artificial lighting) and are always energized.

You may encounter a rural or urban structure with a wind turbine providing alternative power in addition to or in place of the power company. In any case, the power can be turned off at the meter box with the main power shutoff. The power line from the wind turbine to the meter will remain energized, however.

Fuel-powered generators are primarily used to replace the power company's service when service is interrupted **(Figure 17.20)**. In some locations, the generator may be the sole source of power for a structure. These systems can be controlled by simply turning off the generator. If the system supplements the power company's service, shut off the supply at the meter box.

NOTE: When shutting off building utilities do not assume there is no backup generator or alternative source of energy present. Always use CAUTION when performing this skill and refer to local policy for further guidance.

Figure 17.21 To shut off natural gas, turn the tang 90 degrees to the pipe.

Gas Utilities

To control gas utilities, you must have a working knowledge of the hazards and correct procedures for handling incidents involving natural gas and liquefied petroleum gas (LPG). Many houses, manufactured homes, businesses, and industrial properties use natural gas or LPG for cooking, heating, or industrial processes.

Natural gas. Natural gas in its pure form is methane, which has a flammability range of 5 to 15 percent but is nontoxic. Natural gas is lighter than air so it tends to rise and diffuse in the open. While natural gas is nontoxic, it is classified as an asphyxiant because it may displace normal breathing air in a confined space and lead to suffocation. Although natural gas has no odor of its own, mercaptan is added by utility companies, which causes a very distinctive sulphur-like odor, much like rotten eggs.

When ordered, the natural gas supply to a structure must be shut off at the meter. Generally, the meter is located outside the structure near the foundation or on the easement near the property line. However, it may be found inside the structure in a basement or mechanical space. In some industrial and institutional occupancies, critical equipment and processes depend upon an uninterrupted supply of natural gas. For example, the emergency generators in some hospitals are fueled by natural gas.

The shutoff is an inline valve located on the owner supply side of the meter; that is, between the distribution system and the meter. When the valve is open, the tang (a rectangular bar) is in line with the pipe. To close the valve, use a spanner wrench, pipe wrench, or similar tool to turn the tang until it is 90 degrees to the pipe (**Figure 17.21**).

Contact the local utility company when the gas has been shut off or when any emergency involving natural gas occurs in its service area. The local utility will provide an emergency response crew equipped with special (nonsparking) tools, maps of the distribution system, and the training and experience needed to help control the flow of gas. The response of these crews is usually prompt, but their response may take somewhat longer in rural areas or in times of great demand such as following an earthquake, hurricanes, or other large-scale events. It is the responsibility of the utility company, not the fire department, to turn gas utilities back on after they have been shut off.

CAUTION
Natural gas that leaks underground in wet soil can lose its odorant and become difficult to detect without instruments.

Liquefied Petroleum Gas (LPG). Also known as bottled gas, **Liquefied Petroleum Gas (LPG)** refers to fuel gases stored in a liquid state under pressure. While there are two main gases in this category — butane and propane — propane is the most widely used. Propane is used primarily as a fuel gas in campers, manufactured homes, agricultural applications, rural homes, and businesses. It is also used as a fuel for motor vehicles. Propane gas has no natural odor of its own, but mercaptan is added to give it a very distinctive odor. The gas is nontoxic, but it is classified as an asphyxiant because it may displace normal breathing air in a confined space and lead to suffocation.

LPG is about one and one-half times as heavy as air so it will sink to the lowest point possible. The gas is explosive in concentrations between 1.5 and 10 percent. LPG is shipped from its distribution point to its point of usage in cylinders and in tanks on cargo trucks. It is stored in cylinders and tanks near its point of use. The tank or cylinder is then connected by steel piping and copper tubing to the appliances the gas serves. The supply of gas into a structure may be stopped by shutting the tank valve. An LPG leak will produce a visible cloud of vapor that hugs the ground. A fog stream of at least 100 gpm (400 L/min) can be used to dissipate this cloud of unburned gas.

Depending on the local building code, a shutoff valve should be located at the point where the supply line from the LPG tank enters the structure. This valve may be similar to the type used for natural gas or water supply lines. There will also be a control valve located on the LPG tank (**Figure 17.22**). Do not attempt to open the valve after the fire or emergency has been terminated. Turning on the gas is the responsibility of the owner and the LPG supplier.

Figure 17.22 Liquid petroleum gas tanks generally have a control valve located on the top of the tank.

Water

It will be necessary to shut off the water supply to prevent water damage from broken pipes. Water shutoff valves are located underground with the water meter. Their location will depend on the location of water distribution lines in the jurisdiction. Residential water shutoff keys or pipe wrenches are used to turn the tang 90 degrees to the pipe. In some jurisdictions, a valve may be located inside the structure. Use caution when touching water pipes because the electrical ground wire may be connected to the water pipe in residential structures.

Commercial structures and large institutional and industrial facilities have larger diameter supply lines than residential single-family structures. Therefore, the shutoff valve has a larger, usually square-shaped, tang. This will require a special water shutoff key that the fire department may assign to certain units or the water department may have to provide. Restoring the water supply will be the responsibility of the water department or the owner.

Supporting Fire Protection Systems at Protected Structures

Protected structures feature some sort of fire protection system such as an automatic sprinkler system and/or standpipe system. Fire protection systems can be found in both residential and commercial structures. When these systems are present, you should work with the system to control the fire. Automatic sprinkler systems are designed to increase the survivability of occupants and firefighters in a structure. Therefore, activated systems should not be discontinued before the fire is under control and the shutdown order has been given by the IC. The sections that follow describe ways firefighters can support protection systems to control a fire at protected structures.

Connecting to a FDC

One of the first priorities at a fire in a protected structure is to connect the fire department pumper to the fire department connection (FDC). FDCs allow a pumper to supplement the water supply and pressure in a structure's sprinkler or standpipe systems. Each FDC is labeled for the system and/or building zone that it serves (**Figure 17.23**). Once the order is given to connect to the FDC, firefighters should locate the appropriate FDC, select the appropriate supply hose, and make the connection. Preincident plans should identify the location and size of FDCs. Steps for connecting a supply hoseline to a FDC can be found in **Skill Sheet 17-I-4**.

Shutting a Control Valve

One means of stopping the water flow from activated sprinkler heads is to close the system's control valve. The control valve is located between the sprinkler system and the main water supply and is used to shut down water supply to the entire system. The control valve is usually located immediately under the sprinkler alarm valve, the dry-pipe or deluge valve, or outside the building near the sprinkler system it controls. Some systems are designed with sectional valves that are used to shut down small areas or individual sprinklers. Sprinkler control valves are either secured in the open position with a chain and padlock or are electronically supervised to make sure they are not inadvertently closed (**Figure 17.24**).

Figure 17.23 Fire department connections have labels indicating the type of system or zone of the building that they serve.

Figure 17.24 Electronic switches located on sprinkler control valves are activated when the valve is shut, sending a signal to the fire control panel.

Shutting the sprinkler system's control valve should only be done once the fire has been brought under control and the IC has given the order to do so. Closing the valve before the fire is under control can lead to a rapid increase in fire conditions. Steps for operating a sprinkler control valve can be found in **Skill Sheet 17-I-5**.

A firefighter with a portable radio should be stationed at any control valve that has been closed so that the valve can be reopened if necessary. The firefighter stationed at the valve should be familiar with the various types of control valves that are common on fire detection systems.

NOTE: Some departments prefer to plug active sprinklers individually rather than close the fire suppression system's control valve.

Several different types of control valves are used in fire suppression systems. It is important that you recognize each type and understand its operation. Valves used in sprinkler systems are typically of the indicating type because they indicate at a glance whether they are open or closed. Types of indicating control valves include the following:

- **Outside stem and yoke (OS&Y) valve** — Has a yoke on the outside with a threaded stem that opens and closes the gate inside the valve housing. The threaded portion of the stem is visible beyond the yoke when the valve is open and not visible when the valve is closed.

- **Post indicator valve (PIV)** — Hollow metal post that houses the valve stem. Attached to the valve stem is a movable plate with the words *OPEN* or *SHUT* visible through a small glass window on the side of the housing. When not in use, the operating handle is locked to the valve housing.

- **Wall post indicator valve (WPIV)** — Similar to a PIV except that it extends horizontally through the wall with the target and valve operating nut on the outside of the building.

- **Post indicator valve assembly (PIVA)** — The PIVA does not use a plate with words *OPEN* and *SHUT* as does a PIV. Instead, a PIVA uses a circular disk inside a flat plate on top of the valve housing. When the valve is open, the disk is perpendicular to the surrounding plate. When the valve is closed, the disk is in line with the plate that surrounds it. Unlike the PIV or WPIV, the PIVA is operated with a built-in crank.

Stopping the Flow from a Sprinkler

Once the fire has been brought under control in a protected structure, firefighters should stop the flow from any activated sprinklers in order to minimize the amount of water damage to the structure and its contents. Wooden wedges, sprinkler tongs, and other devices can be used to stop the flow of water (**Figure 17.25**). **Skill Sheet 17-I-6** illustrates the steps for stopping the flow of an activated sprinkler.

Deploying Master Stream Devices

Master streams are usually deployed in situations where the fire is beyond the effectiveness of handlines or there is a need for fire streams in areas that are unsafe for firefighters. The four main uses for a master stream are as follows:

- Direct fire attack
- Indirect fire attack
- Supplement handlines that are already attacking the fire from the exterior
- Provide exposure protection

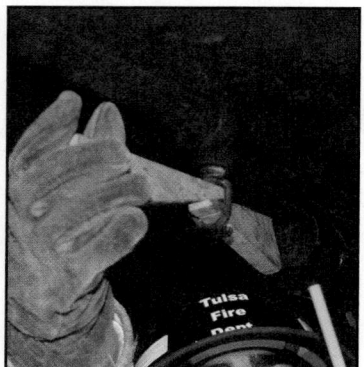

Figure 17.25 Wooden wedges are inserted into sprinkler heads to stop the flow of water.

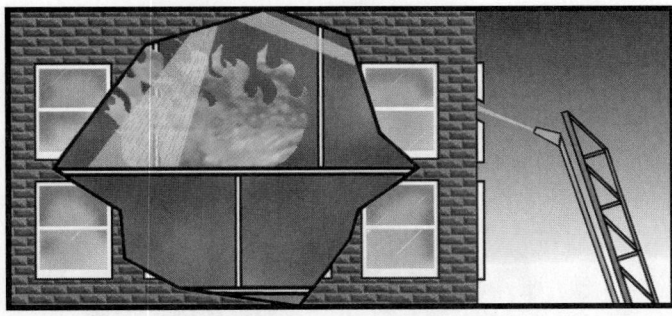

Figure 17.26 Elevated master streams can be directed against the ceiling to break up the streams and spread over the fire.

Proper placement of the master stream device is extremely important. First, master stream devices must be properly positioned to apply an effective stream on a fire, particularly when using a fog nozzle because fog streams do not have the stream reach and penetration of solid streams. Even though the master stream nozzle can be adjusted up and down and right and left, if it is necessary to move the device it must be shut down. Moving a master stream device can be a time-consuming and labor-intensive process, causing the device to be out of operation for a period of time.

The second consideration for master stream placement is the angle at which the stream enters the structure. Firefighters should aim the stream so that it enters the structure at an upward angle causing it to deflect off the ceiling or other overhead objects. This angle makes the stream diffuse into smaller droplets that rain down on the fire, providing maximum extinguishing effectiveness **(Figure 17.26)**. Streams that enter the opening at an angle that is horizontal or less are not as effective.

Finally, place the master stream device in a location that allows the stream to cover the most surface area of the building, especially where there is a large volume of fire and a limited number of master stream devices. Doing so allows firefighters to change the direction of the stream and to direct it into more than one opening if necessary. See **Skill Sheet 17-I-7** for information on deploying and operating a portable master stream device.

A master stream device can be very effective for providing exposure protection to other structures. There are two approaches for providing exposure protection. The first approach, and the most effective, is to direct the stream at the surface of the structure that faces the fire. The stream should strike the surface and run down it. If the surface is wide, multiple devices can be used or one unit can sweep the face constantly keeping it wet. The second approach is to create a water curtain between the fire and the exposure. This can be effective if the exposure is not a single surface, such as densely placed trees. The water curtain will be able to stop the radiated heat as long as the water drops are compact.

Supplying Master Streams

Master stream devices flow a minimum of 350 gpm (1 400 L/min) which can mean high friction loss in supply hose. Therefore, except for small quick-attack devices that are designed to operate from a single 2½-inch (65 mm) line, it is not practical to supply master stream appliances with anything less than two 2½-inch (65 mm) hoselines **(Figure 17.27, p. 1030)**. Conventional master stream devices may be temporarily supplied by one 2½-inch (65 mm) line while adding additional hoselines. When greater quantities of water are required, a third 2½-inch (65 mm) or large-diameter supply line will be required. Some master stream devices are equipped to handle one large-diameter (4-inch [100 mm] or larger) supply line.

Figure 17.27 Master stream devices are generally supplied by 2½ inch (65 mm) hoselines.

The operation of master stream devices consumes large volumes of water that accumulate inside structures when master streams are directed into buildings. This water accumulates on floors, and the building contents may absorb some of this water as well. Both accumulation and absorption of water add weight that affects structural integrity and increases the potential for structural collapse during overhaul and fire investigation activities.

> **CAUTION**
> Added water weight from master stream operations increases the potential for structural collapse.

Because master streams are used primarily in defensive operations, it may be necessary to shut down other handlines in order to maintain effective fire streams from the master stream devices. The IC will make the decision on joint handline and master stream operations based on your local SOPs.

Staffing Master Stream Devices

Except for apparatus-mounted deck guns, deploying a master stream device and the necessary water supply hoselines will usually require a minimum of two firefighters, although more firefighters can accomplish it faster. Once a portable master stream device is in place, one firefighter can operate it. When water is flowing, at least one firefighter should be stationed at the master stream device at all times unless the device is being used in a hazardous position (close to a fire-weakened wall or near an LPG tank, for example). The firefighter tending the device can change the direction of the stream when required and prevent pressure in hoselines from moving the device.

If the situation is too dangerous to have firefighters stationed at the device, it can be securely anchored in position. Once the device is deployed, hoselines attached and charged and the desired stream developed, personnel can be withdrawn to a safe distance. If the device starts to move, the pump operator can decrease the pressure at the apparatus to stop any movement.

Figure 17.28 Elevated master stream devices are used to apply large quantities of water during defensive operations. *Courtesy of Chris Mickal.*

Elevated Master Streams

Elevated master stream devices are used to apply water to the upper stories of multistory buildings, either as a direct fire attack or as a water supply for handlines. They can also provide exposure protection to endangered structures. A number of different types of aerial apparatus can deliver elevated master streams, most commonly quints, aerial ladders, aerial platforms, and water towers **(Figure 17.28)**. Under a variety of circumstances, you may be assigned to operate an elevated master stream device or to support such an operation.

Suppressing Class C Fires

Class C fires involve energized electrical equipment. Firefighters often fail to recognize the danger from electrocution Class C fires present or take appropriate steps for self-protection. However, when the electrical equipment is de-energized, these fires can be handled with relative ease. Once the electrical power is turned off, these fires may self-extinguish. If they do continue to burn, they will become either Class A or Class B fires. As mentioned earlier in this chapter, lockout/tagout devices should be used whenever possible to prevent main power from being turned back on before it is safe to do so **(Figure 17.29)**.

Figure 17.29 Lockout/tagout devices prevent power from being turned on before it is safe to do so.

Class C fires may occur in the following locations:

- Electric power stations or substations **(Figure 17.30, p. 1032)**
- Commercial high-voltage installations
- Telephone relay switch stations
- Photovoltaic arrays
- Electrical substations
- Railroad lines and yards with electric engines
- Streetcar and subway tracks or stations
- Vehicle incidents that involve hybrid or electric vehicles
- Computer server rooms or data centers

Figure 17.30 Electrical stations and substations are subject to Class C fires.

///////////////////////////////
WARNING
Before initiating fire suppression activities, stop the flow of electricity to the device involved.

When handling fires in delicate electronic or computer equipment, clean extinguishing agents, such as Halotron®, should be used to prevent further damage to the equipment. Multipurpose dry chemical agents are very effective at extinguishing Class C fires, but some are chemically reactive with electrical components. All of these agents require considerable cleanup. Using water on energized equipment is inappropriate because of the inherent shock hazard and resulting damage to the electrical equipment. Any voltage greater than 40 volts is potentially dangerous. However, if water *must* be used, apply it from a distance in the form of a fog or spray stream.

Class C fire suppression techniques are also needed for fires involving transmission lines and equipment, underground lines, and alternative energy sources such as solar panels, and wind- or engine-powered generators. In addition, departmental operating procedures should clearly state the responsibilities for controlling electrical power, the dangers of electric shock, and the guidelines for electrical emergencies.

Transmission Lines and Equipment

Electrical transmission lines can be damaged during earthquakes, snow and ice storms, high winds, tornados, hurricanes, or traffic accidents. Broken electrical power lines can start fires in grass and other vegetation, on the exterior of structures, or in vehicles.

To reduce the risk of shock from electric current in the ground, a circle with a radius equal to the distance between the power poles should be cordoned off around the point where the power line contacts the earth. If a ground cover fire does start, firefighters should wait for the fire to burn away from the point of contact before attempting to extinguish it. If there is not a risk to life and property, many fire departments notify the responsible electrical utility company, control the scene, and deny entry until utility personnel arrive. For maximum safety, only utility personnel should cut electrical power lines.

///////////////////////////////
WARNING
Assume that all power lines are energized until the power company informs you otherwise.

Figure 17.31 Electrical transformers may be located in the open, concealed behind fences, or in vaults.

Fires in electrical transformers are relatively common **(Figure 17.31)**. Some older transformers can present a serious health and environmental hazard because of coolant liquids that contain **polychlorinated biphenyls (PCBs)**. These liquids are both flammable and carcinogenic (cancer-causing). Even though transformers containing PCBs have been outlawed for many years and most have been replaced, firefighters should assume that any transformer contains PCBs until proven otherwise. Even transformers that are marked as containing no PCBs can legally contain up to 49 parts per million (ppm).

Use a dry chemical or carbon dioxide extinguisher to extinguish fires in transformers at ground level. Allow pole-top transformer fires to burn until utility personnel can extinguish the fire with a dry chemical extinguisher from an aerial device. If the fire is burning through a wooden cross-arm that would cause the power line to fall, it is SOPs in some fire departments to extinguish the fire with a fog stream. Firefighters must always follow their departmental SOPs in these situations.

Underground Transmission Lines

Underground transmission systems consist of conduits and vaults below ground. Explosions caused when a fuse opens (called *tripping* or *blowing*) or a short circuit ignites accumulated gases are the most serious hazards underground transmission systems present. Such explosions may throw utility access covers a considerable distance endangering both the public and firefighters. Firefighters and the public should stay at least 300 feet (91 m) away from the site and make sure that apparatus is not positioned over a utility access cover **(Figure 17.32, p. 1034)**.

Because fire suppression activities can be undertaken from outside an electrical utility vault, firefighters should not enter one until a qualified person has shut off the power. The configuration of underground vaults creates a higher-than-normal risk of backdraft conditions developing in these spaces.

> **WARNING**
> *Only* personnel who are properly trained and equipped for confined-space entry should enter a utility vault.

Polychlorinated Biphenyl (PCB) — Toxic compound found in some older oil-filled electric transformers.

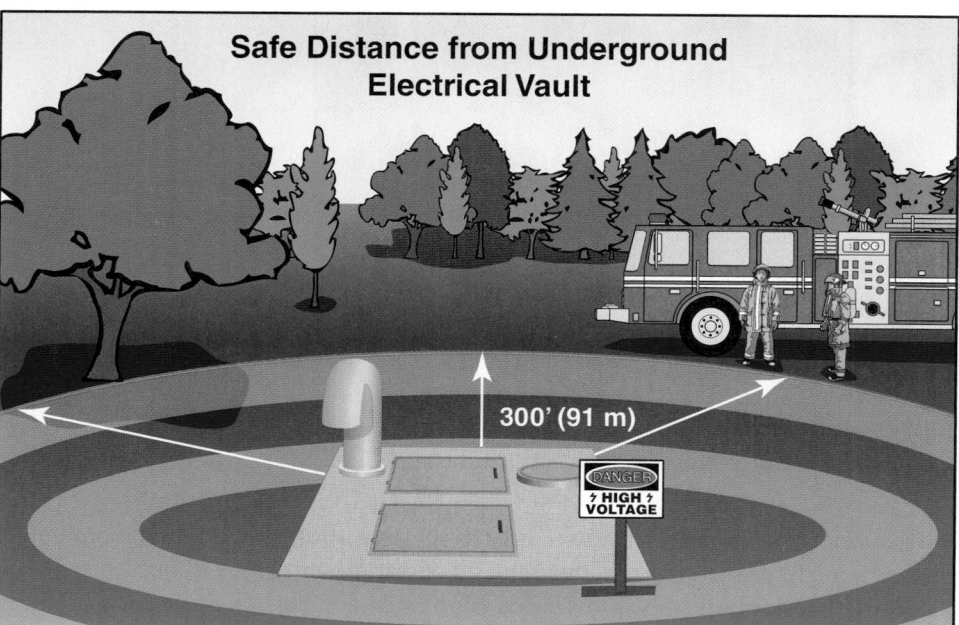

Figure 17.32 Firefighters should remain at least 300 feet (91 m) away from underground vaults until the power to the vault has been shut off by a qualified person.

Commercial High-Voltage Installations

Many commercial and industrial complexes have electrical equipment that requires current in excess of 600 volts. High-voltage equipment such as transformers or large electric motors are usually housed in vaults or fire-resistive rooms with *High-Voltage* warning signs on the entry doors (**Figure 17.33**). Some transformers use flammable coolants that are themselves hazardous. Water (even in the form of fog) should not be used in these situations because of the damage it may cause to electrical equipment not involved in the fire.

Smoke from these fires may contain toxic chemicals emitted from the plastic installations and coolants used in high-voltage installations. Firefighters properly trained in confined-space rescues should enter these installations only when rescue operations require it and rescue is possible without jeopardizing the life of the firefighter. Entry personnel must wear full PPE including SCBA, and wear a tag line monitored by an attendant outside the enclosure. A rapid intervention crew or team (RIC/RIT) is also required in these situations. Entrants should search with a clenched fist or the back of the hand to prevent the reflex action of grabbing energized equipment if it is touched accidentally. Because the smoke may be highly toxic, use the appropriate decontamination procedures for all PPE and equipment that has been exposed to the contaminants.

Figure 17.33 High-voltage electrical equipment is common in many commercial and industrial complexes.

Electrical Hazards

To avoid injuries to themselves and to protect electrical equipment, firefighters should be familiar with electrical transmission systems and their hazards. While electrocution is usually associated with high-voltage equipment, conventional residential current is sufficient to deliver a fatal shock. In addition to reducing the risk of injury or electrocution, controlling electrical power reduces the danger of igniting adjacent combustibles or accidentally energizing electrical equipment.

The consequences of electrical shock can include the following:

- Cardiac arrest
- Ventricular fibrillation
- Respiratory arrest
- Involuntary muscle contractions
- Paralysis
- Surface or internal burns
- Damage to joints
- Ultraviolet arc burns to the eyes

Factors most affecting the seriousness of electrical shock include the following:

- Path of electricity through the body
- Degree of skin resistance — Wet (low) or dry (high)
- Length of exposure
- Available current — Amperage flow
- Available voltage — Electromotive force
- Frequency — Alternating current (AC) or direct current (DC)

Guidelines for Electrical Emergencies

The first rule in dealing with electrical hazards is to assume that all electrical wires, equipment, and devices are energized until proven otherwise. The following list contains additional guidelines for dealing with electrical emergencies. The list is not all-inclusive but gives principles that should be considered to maintain a safe working environment for personnel:

- Establish an exclusion zone equal to the distance between power poles (one span) in all directions from downed power lines **(Figure 17.34, p. 1036)**.
- Be aware that a short circuit may have weakened other wires, and they could fall at any time.
- Wear full protective clothing and use only tested and approved tools with insulated handles.

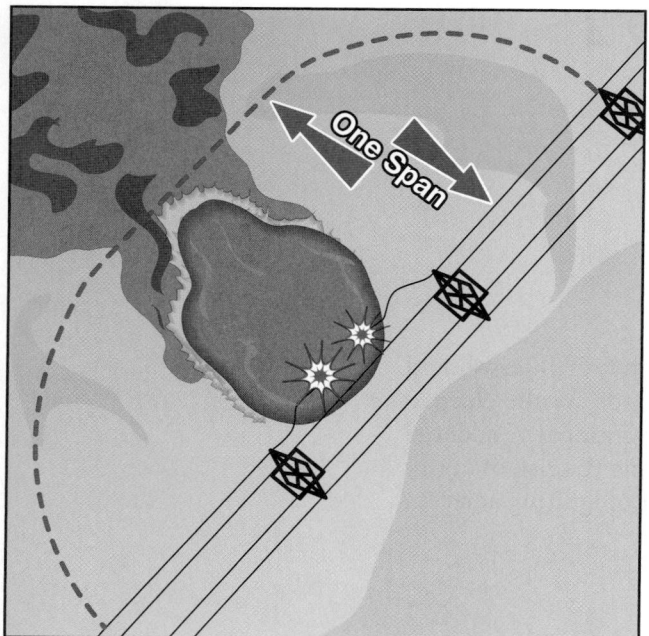

Figure 17.34 An exclusion zone equal to the distance between power poles should be established around downed power lines.

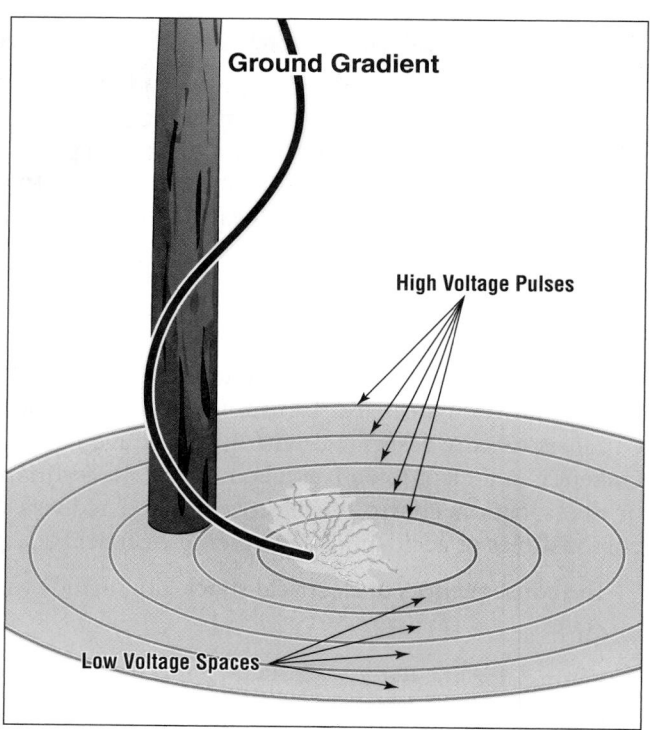

Figure 17.35 Electrical current from a downed power line radiates in ripples from the point of contact dissipating the farther it travels.

- Guard against electrical shocks, burns, and eye injuries from electrical arcs.
- Wait for utility company workers to cut any power lines.
- Use lockout/tagout devices when working on electrical equipment.
- Stay at least 10 feet (3 m) away from power lines when raising or lowering ground ladders or aerial devices.
- Do not touch any vehicle or apparatus that is in contact with electrical wires.
- Do not use solid and straight streams on fires in energized electrical equipment.
- Use fog streams with at least 100 psi (700 kPa) nozzle pressure on energized electrical equipment.
- Be aware that live wires outside your field of view may be in contact with wire mesh or steel rail fences, energizing them and posing an electrocution hazard.
- Where wires are down, heed any tingling sensation felt in the feet and back away.
- Maintain a large safety zone around downed electrical wires to avoid ground gradient hazards.
- Remain inside a vehicle or apparatus that is in contact with power lines. If you must leave the vehicle or apparatus, jump clear of the apparatus, landing with both feet together.

Ground gradient is an electrical behavior that produces electrical pulses in the ground starting at the point where the power line contacts the earth and expanding out in concentric circles. Because the gradients are invisible, you must imagine that they are similar to the ripples that result when you throw a stone into a pond. Each ripple is a pulse of electric current alternating from high to low voltage (**Figure 17.35**). Stepping from one ripple to another creates an electrical differential that will result in shock and physical injury or death. Likewise, dragging a hoseline, ladder, pike pole, or other object in the area of a downed wire also risks injury from a ground gradient condition.

If you find yourself in a gradient field or feel a tingling in your legs, put or place your feet close together and hop or shuffle until you are out of the danger area. Do not attempt to walk or crawl as both may place you astraddle of the ripples. Depending on the voltage of the downed power line, the distance you may need to travel can be upwards of 150 feet (45.7 m).

Your fire fighting boots are designed to meet NFPA® 1971, *Standard on Protective Ensembles for Structural Fire Fighting and Proximity Fire Fighting*, which requires footwear to provide a certain level of protection from electrical shock. This level of protection will diminish over time with wear and may not meet the hazard to which you are exposed.

CAUTION

To exit a ground gradient area, keep both feet in contact with each other and hop or shuffle out of the affected area.

Suppressing Class D Fires

Class D combustible metal fires present the dual problem of burning at extremely high temperatures and being reactive to water **(Figure 17.36)**. Directing hose streams at burning metal can result in the violent decomposition of the water and subsequent release of flammable hydrogen gas. Small metal chips or metal dust are more reactive to water than are large castings and other finished products. When a combustible metal is burning, water is only effective for keeping nearby exposures below their ignition temperatures.

In some situations, if a Class D extinguishing agent is not available, firefighters can simply protect the exposures and allow the metal to burn itself out. Class D extinguishing agents can be manually shoveled or scooped onto the burning metal or applied using Class D fire extinguishers in sufficient quantity to completely cover the burning metal.

Figure 17.36 Class D fires react violently when they come in contact with water. *Courtesy of NIST.*

Figure 17.37 Full PPE must be worn when extinguishing vehicle fires to protect against toxic and nontoxic smoke and vapors. *Courtesy of Bob Esposito.*

Combustible metal fires emit a characteristic brilliant white light that only diminishes when an ash layer covers the burning material. Once this layer has formed, it may appear that the fire is out. Do not assume that these fires are extinguished just because flames are not visible. It may be an extended period of time before the area or substance cools to safe levels. Combustible metal fires are very hot — greater than 2,000°F (1 093°C) — even after they appear to be extinguished.

Suppressing Vehicle Fires

Fires in small passenger vehicles such as automobiles, minivans, and sport utility vehicles are among the most common types of fires to which you will be called. These fires may be the result of a collision, a malfunction of the vehicle propulsion system, or an intentional act. In all situations, you must take the time to size up the incident, provide crew protection, and wear full PPE including SCBA (**Figure 17.37**). Because of the materials used to construct contemporary vehicles, vehicle fires generate a wide variety of toxic and nontoxic smoke and vapors. Until the atmosphere is tested, you must treat the incident with the same care you treat a structure fire.

Most vehicles have some variant of an internal combustion engine; however, completely electric engines are becoming more common. Modern vehicles use a number of different fuel sources:

- Gasoline
- Diesel
- Electricity
- Hybrids (fuel and electricity)
- Compressed or liquefied natural gas (CNG or LNG)
- Biofuels
- Hydrogen

Vehicle Incident Size-Up

The first action upon arrival at a vehicle accident is to size up the incident. Decide if the incident scene will necessitate traffic being diverted and request assistance. Follow the U.S. Department of Transportation (DOT) guidelines for protecting the scene from vehicular traffic in order to establish a safe working zone around the incident. These guidelines were presented in Chapter 2, Firefighter Safety and Health.

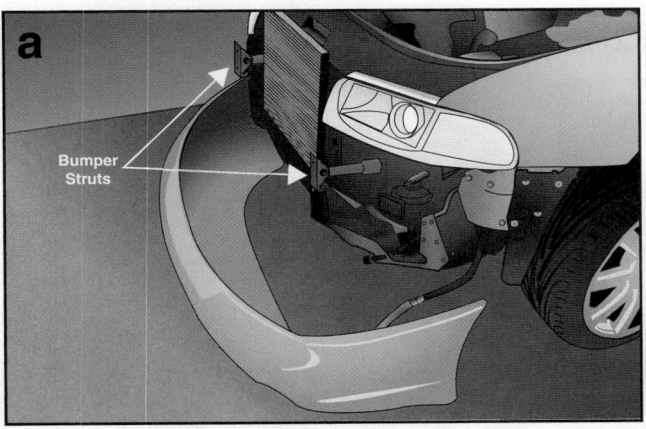

Figure 17.38 a and b Hydraulic a) bumper and b) trunk struts can explode and launch deadly missiles when heated by a fire.

Once scene safety has been established, you can focus on saving the vehicle occupants and extinguishing the fire. Determine if there are victims in the vehicle and if they will require extrication. At the same time, determine if the vehicle is on fire or leaking fuel. Confirm the type of fuel and select the appropriate extinguishing agent.

When approaching the vehicle, avoid components in modern vehicles that are under constant pressure such as bumpers and sometimes hoods and trunk lids. These components incorporate hydraulic or pneumatic struts intended to absorb the shock of minor collisions. If the fire heats these struts, they can explode and catapult the bumper to which they are attached considerable distances. Likewise, the struts used to support the engine hood and trunk lid can also be launched from the vehicle with tremendous force (**Figures 17.38a and b**). Anyone standing in the path of travel could be injured or killed.

Finally, before attacking the fire or commencing with extrication, follow your department's SOPs for establishing scene protection and then isolate the vehicle from any ignition sources or eliminate the ignition source. Next, stabilize the vehicle (if safely possible), control any downed power lines, and address any additional hazards. In some situations, it may be necessary to use defensive fire fighting techniques such as deploying unstaffed master steam devices and isolating and denying entry to the area.

Vehicle Fire Attack

The basic procedures for attacking a fire in a vehicle are as follows:

- Position a hoseline between the burning vehicle and any exposures.
- Attack the fire from a 45-degree angle to avoid the potential for injuries from exploding hydraulic or pneumatic struts (**Figure 17.39, p. 1040**).
- Extinguish any fire near the vehicle occupants first.
- Issue an "all clear" when all occupants are out of the vehicle.
- Extinguish any ground fire around or under the vehicle.
- Extinguish any fire remaining in or around the vehicle.

Deploy an attack hoseline that will provide a minimum 95 gpm (360 L/min) flow rate, such as 1½- or 1¾ inch- (38 mm to 45 mm) attack hose. Booster hoselines do not provide enough protection or rapid cooling to effectively and safely fight a vehicle fire

Danger Zones Around Vehicle Fires

SAFE APPROACH

DANGER ZONE

SAFE APPROACH

DANGER ZONE

DANGER ZONE

DANGER ZONE

SAFE APPROACH

SAFE APPROACH

Figure 17.39 To avoid potential injuries caused by exploding hydraulic or pneumatic struts, always approach a vehicle fire from a 45-degree angle.

and should not be used. Approach the fire at a 45-degree angle to the vehicle's side and from upwind and uphill when possible. A backup hoseline should be deployed as soon as possible. Portable fire extinguishers can extinguish some fires in the vehicle's engine compartment or electrical system and some alternative fuel types. Extinguishment will be complete when flaming and smoldering combustion have ceased. For more information on how to attack a passenger vehicle fire, see **Skill Sheet 17-I-8**.

Apply water to cool combustible metal components that are not burning but are exposed to fire. If combustible metal components do become involved, apply large amounts of water to protect adjacent combustibles while applying Class D extinguishing agent to the burning metal.

Be aware that vehicles may contain extraordinary hazards. In many of these situations, it may be necessary to isolate the area and request the assistance of a hazardous materials team. Extraordinary hazards include the following:

- Large-capacity saddle fuel tanks
- Pressurized natural gas tanks
- Alternative fuel tanks
- Hazardous contents such as explosives or other hazardous materials

- Radioactive materials such as medical isotopes being delivered to medical facilities

- Munitions, especially when military vehicles are involved

- Vehicles disguised to hide mobile, illegal drug labs

Once the fire has been controlled, conduct overhaul as soon as possible to check for extension and hidden fires. Other overhaul considerations include disconnecting the battery, securing air bags (Supplemental Restraint System [SRS] or Side-Impact Protection System [SIPS]), and cooling fuel tanks and any intact, sealed components. Be aware that air bags can deploy from the steering wheel, dashboard, or door of the vehicle.

Fires that originate in certain compartments of a vehicle require special tactics or skills. The sections that follow highlight some of these special fire attack situations.

Engine or Trunk Compartment Fires

If a fire is isolated to the trunk or engine compartment of a vehicle, you will need to gain access in order to extinguish the fire. To gain access to the engine compartment or trunk, first use normal methods such as the release lever or button near the driver's seat. If these methods do not work, then use forcible entry methods similar to those listed in Chapter 11. Cool the front and rear bumper struts to prevent accidental activation from heat exposure. Forcing entry into the engine compartment or trunk can be accomplished with manual forcible entry tools or power tools. Pry the hood or trunk free with a Halligan or crow bar or use an air chisel or reciprocating saw to cut the metal around the key or latch (**Figure 17.40**). You can also use the spike from a Halligan tool or a pick-head axe to knock the lock barrel out of place. Insert a screwdriver or similar object into the remaining locking mechanism and turn to release the lock. Once the trunk is open, direct the hose stream into the space until the fire is extinguished.

In many engine compartment fires, the fire must be controlled before the hood can be opened using one of the following methods:

- Direct a hose stream through the grill or air scoop.

- Drive a piercing nozzle through the hood, fenders, or wheel wells.

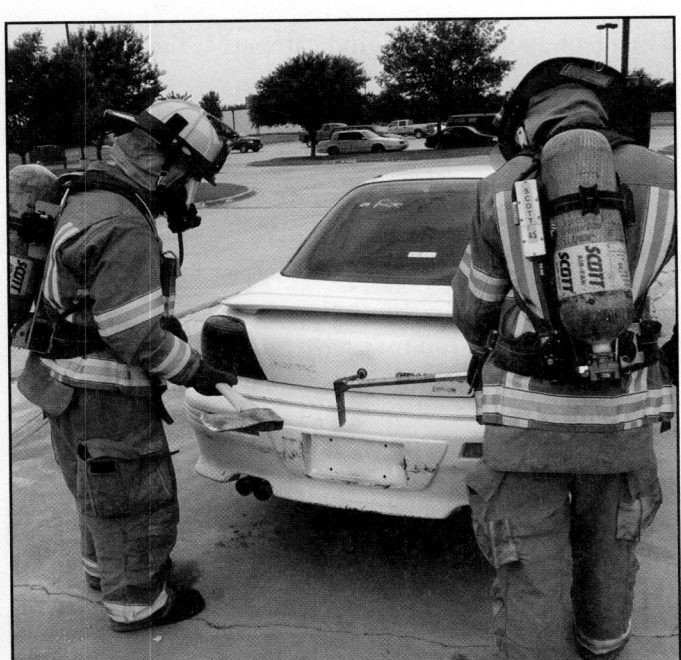

Figure 17.40 Forcible entry tools can be used to open the lock on trunk lids.

- Make or cut an opening large enough for a hose stream to be introduced.

- Use a pry tool to create an opening between the hood and the fender, and then direct a straight stream or narrow fog nozzle in the opening.

Passenger Compartment Fires

When attacking a fire in the passenger compartment of an unoccupied vehicle, use the most appropriate nozzle and pattern for the situation. Attempt to open the door, but if it is locked, the driver may have the key. If normal entry is not possible, break a window and attack the fire with a medium fog pattern.

Undercarriage Fires

The following three methods can be used for fires in the undercarriage:

1. If there is a hazard in getting close to the vehicle, use a straight stream from a distance to reach under the vehicle.

2. If the vehicle is on a hard surface such a concrete or asphalt, direct the stream downward and allow the water to deflect up toward the underside of the vehicle.

3. Open the hood and direct the stream through the engine compartment.

Alternative Fuel Vehicles

Alternative fuels create different risks to emergency responders, making it even more important to accurately size up every motor vehicle incident. Alternative fuels currently include:

- Natural gas (CNG and LNG)

- Liquefied Petroleum Gas (LPG)

- Electric or hybrid electric

- Ethanol/Methanol

- Biodiesel

- Hydrogen

During the size-up of the incident, look for visual indicators that the vehicle uses an alternative fuel source as follows:

- Vehicle logos

- Fuel-specific logos

- Special fuel ports

- Distinctive vehicle profiles

There is no national requirement for markings to indicate the type of fuel used to power cars, trucks, and buses. As a result, you may have no idea about a vehicle's fuel source despite your best efforts. Some states/provinces may regulate markings, however.

CAUTION
There may be no visual indicators that a vehicle
uses an alternative fuel source.

Additional tactics that should be considered when dealing with alternative fuel vehicles include:

- Park apparatus a minimum of 100 feet (30 m) from the incident.
- Approach from uphill and upwind, if possible.
- Approach from a 45-degree angle to the vehicle.
- Wear full PPE including SCBA.
- Use nonsparking extrication tools.
- Do not use flares.
- Deploy a backup hoseline.
- Select extinguishing agent specific to the type of fuel or battery pack.

Natural Gas (CNG and LNG)

Natural gas is used in the form of compressed natural gas (CNG) and liquefied natural gas (LNG). Natural gas properties include:

- Clean burning
- High ignition temperature
- Narrow explosive range
- Nontoxic
- Noncorrosive
- Colorless and odorless in its natural state (mercaptan is added as a safety measure)
- Lighter than air
- Stored under pressure
- Visible flame

Currently, government agencies and taxi cab, utility, refuse, and mass transit bus companies are most likely to have CNG or LNG vehicles. A CNG or LNG diamond logo may be affixed to the front and rear of the vehicle **(Figure 17.41)**. Fuel tanks are usually located in the trunk area, under side panels, or in the open bed of pickup trucks. Both CNG and LNG tanks can rupture if exposed to fire resulting in an explosion. A pressure-relief device and vent and a fuel shutoff valve may be located in the wheel well with a placard nearby.

Figure 17.41 Look for compressed natural gas logos on the rear of alternative fuel vehicles.

CNG is stored under high pressure in a gaseous state. Tactics for fires or leaks involving CNG vehicles include the following:

- If no fire is visible:
  - Use a gas detector to locate leaks, shutoff valves, and eliminate any ignition sources.
  - Stay clear of any detected vapor clouds.
- If fire is visible:
  - Allow fuel to burn itself out.
  - Use water or foam to extinguish if necessary.
  - Use fog stream to disperse vapor clouds.
  - Avoid contact with high velocity jet of escaping gas.

LNG is stored in a liquid state by cooling to –260°F (–162°C) in double-walled, vacuum-insulated pressure tanks. It is lighter than water and has a vapor cloud that is heavier than air. Frost on the fuel tank exterior indicates tank failure. If there is no fire or leak, stabilize the vehicle, set the emergency brake or chock the tires, turn off the ignition, and shut off the gas cylinder valve handle.

Tactics and guidelines for fires or leaks involving LNG include the following:

- Avoid any contact with LNG.
- Stay clear of vapor clouds identified.
- Shut off the ignition to top the fuel flow to a leak or fire.
- Use Purple K dry-chemical agent or high-expansion foam on the surface of LNG fire.
- Use sand or dirt to prevent LNG from entering storm drains.

Liquefied Petroleum Gas (LPG)

Liquefied petroleum gas (LPG), also known as propane, is the third most common vehicle fuel type after gasoline and diesel and has the following characteristics:

- Clean burning
- Safer than gasoline
- Colorless and odorless in its natural state (mercaptan is added as a safety measure)
- Stored under pressure

LPG expands rapidly when heated, 1.5 times for every 10 degrees of increase in temperature. Because it is stored in pressurized tanks, this rapid expansion creates the conditions for Boiling Liquid Expanding Vapor Explosion (BLEVE) when an LPG tank is exposed to heat.

An LPG vehicle may be marked with a logo. The following tactics should be used at incidents involving LPG vehicles:

- Approach only from the sides at a 45-degree angle, never from the ends.
- Use gas detectors to determine leaks and isolate leaks from ignition sources.
- Allow the fire (if present) to self-extinguish.
- Use foam or water when necessary for extinguishment.
- Direct fire streams at the top of the LPG tank to provide adequate cooling.
- Stay clear of any identified vapor clouds.

Figure 17.42 Electric and hybrid electric vehicles have charging ports located on the exterior of the body.

Electric or Hybrid Electric

Electric or hybrid electric vehicles should have certain visible indicators such as the vehicle name, logo, charging port on a side or the front of vehicle, and a distinctive profile (**Figure 17.42**). Batteries may be located in the engine compartment, trunk area, or under the vehicle. When the engine is running, there may not be any noise. Remember that most electric and hybrid electric vehicles also contain a 12-volt battery system with separate battery and wiring harness.

If no fire is visible, secure the vehicle, chock wheels, turn off the ignition, and remove the key. If smoke is visible, you must wear full PPE and SCBA because the fumes are toxic until air monitoring indicates that the atmosphere is clear. Do not approach if the vehicle is on fire or there is arcing under the hood. Instead, establish scene security and protect exposures. Avoid contact with all fluids because they may include battery acid that can cause injuries.

Electric vehicles run solely on electricity stored in batteries that must be recharged periodically. There are many types, designs, and locations of vehicle battery packs. Use inertia switches and pilot circuits to shut off a high-voltage system. It will take approximately five minutes for energy in the system to dissipate. Do not cut orange high-voltage cables. These are high-voltage systems and electrocution is possible. Blue and yellow color-coded cables also present an electrocution hazard although they do not carry high voltage. During fire extinguishment, wear full PPE, insulated tools, water or foam to extinguish, and specific recommended extinguishing agent for battery pack fires.

Hybrids combine battery powered electrical systems with gasoline, diesel, bio-diesel, and natural gas to run the engine. Some hybrid vehicles use a photovoltaic solar panel mounted in the roof as a power source. In all cases, shut off power with the ignition or power switch and remove the ignition key. Water is the recommended extinguishing agent although specific agents or tactics may be required for specific fuels or battery types.

> **WARNING**
>
> Do not cut or contact any orange, blue, or yellow color-coded electrical cables or components in electric or hybrid electric vehicles.

Ethanol/Methanol

Ethanol and methanol are gasoline blends. They are water-soluble, electrically-conductive, clear liquids that have a slight gasoline odor. Ethanol and methanol fires burn bright blue and may be hard to see during the day. As a result, you will need to use a TI to see the flames and locate the fire. Currently, over 50 percent of the gasoline sold in the U. S. is an ethanol blend. Ethanol and methanol use the same fuel tanks as conventional gasoline engines. The vehicles probably will not have any visible logos.

If no fire or leak is visible, secure the vehicle, chock the tires, and turn off the ignition. If a fuel leak is suspected, use caution and approach in full PPE and SCBA, with hoselines deployed and charged. If the vehicle is on fire, establish a control zone and use only Alcohol Resistant (AR) Class B foam, such as AR-AFFF, to extinguish the fire. In both fire and leak situations, request a hazardous materials team response.

Biodiesel

Biodiesel is a blend of liquids made from natural plants and diesel. It is a yellow liquid with the odor of cooking oil that is nontoxic, biodegradable, and sulfur free. It is slightly lighter than water and has a flash point 266°F (130°C). Biodiesel is used in any vehicle designed for diesel fuel and no logo is required. If there is a fuel leak, control the leak per local SOPs and request a hazardous materials team. If the vehicle is on fire, use dry chemical, CO_2, water fog spray, or foam to extinguish.

Hydrogen

Hydrogen-fueled vehicles are in use in some areas of North America though most are still in the concept stage of development. Hydrogen is colorless, odorless, nontoxic, and energy efficient. It has a self-ignition temperature of 550°F (287.7°C) with a flammability range between 4 to 75 percent. Because the flame is invisible during the day, you should use a TI to see the flame. Vehicles are marked with a manufacturer's logo and the vented fuel cell is in the trunk. Tactics for both a leak and a fire include shutting off the ignition, isolating the fuel from ignition sources, and chocking the wheels. Do not extinguish the fire. Instead, protect the exposures and allow the fuel to burn off. If extrication is required, do not cut C-posts which contain the vents.

Suppressing Fires in Other Class A Materials

Exterior fires may occur in stacked and piled materials, small unattached structures, and trash containers. These fires can create a hazard to nearby structures, flammable/combustible storage tanks, parked vehicles, and vegetation. How you extinguish these fires will depend on the type of material involved, the weather, and the type and quantity of extinguishing agent you use.

Stacked and Piled Materials

Stacked and piled combustible materials can be found around all types of occupancies and in all types of jurisdictions including:

- Raw materials such as those found at sawmills, lumberyards, and manufacturing factilities
- Bales of used cardboard or pallets near large retail outlets (**Figure 17.43**)
- Miscellaneous and varied materials stored outdoors at residencies
- Bales or large rolls of hay on farms
- Loose flammable materials, such as mulch or fertilizer, at nurseries or garden centers

Figure 17.43 Large piles of cardboard and waste materials may be found near retail outlets.

Figure 17.44 Fires in small structures such as tool and garden sheds should be prevented from spreading to other exposures.

The value of these materials will vary widely but the possibility that any portion can be salvaged once a fire starts is small. The greatest danger is to the exposures, primarily nearby structures and ground cover. Therefore, the goal is to confine the fire to the pile or building of origin. Depending on the material and how involved the fire is, fire streams should be directed at the extreme edge of the fire, controlling the spread. Use a straight stream from a distance and then shift to a fog pattern.

As the quantity of fire is reduced, move the nozzle closer to the stacked material using the fog pattern to protect yourself. Other personnel can use pike poles and hay hooks to pull the material apart so that the stream can reach all the burning material. Flying embers may also spread the fire causing the need for additional units to perform spark patrol downwind of the fire. For information on attacking a fire in stacked or piled materials, see **Skill Sheet 17-I-9**.

Class A foam, applied with an eductor or through a compressed-air foam (CAF) system, is very effective for these types of fires. The foam will soak into the materials and also coat the ground cover and exposures preventing fire spread.

Small Unattached Structures

Small unattached structures can be found in all jurisdictions, in all shapes, sizes, and uses. They range from storage sheds to outhouses, pump houses to bath houses, and fishing houses to ice houses. Their age, construction type and material, and value will also cover the spectrum for structures.

NFPA® does not define or specify what constitutes a small structure, only that the fire should be attacked from the exterior **(Figure 17.44)**. Like stacked and piled materials, unless there is some overwhelming reason to try to save the structure, the primary mission is to prevent fire spread to exposures and then extinguish the fire. Class A foam and fog streams can be very effective for exposure protection, advancing close to the fire, and extinguishment.

Because small structures are generally used for storage of miscellaneous materials, you should assume that chemicals, flammable/combustible liquids, explosives, or illegal materials may be inside the building. The volume of smoke and fire as well as the color of the smoke can provide an indication of the primary materials that are on fire. If there is any question of the hazard, water should be applied through a straight stream, the exposures protected, and the structure allowed to self-extinguish. For information on attacking a fire in a small unattached structure, see **Skill Sheet 17-I-10**.

Clandestine Drug Labs

The presence of clandestine drug labs, indoor marijuana-growing operations, and explosives labs are increasing in the U.S. and Canada. These operations are often wired illegally to the main power supplies of adjacent buildings. This is done both to steal electricity and to prevent the power company from noticing a huge increase in the use of electrical power at a given location. Firefighters should be aware of the following possible hazards in these locations:

- Electrical hazards due to makeshift wiring
- Volatile chemicals that may be toxic and/or flammable
- Booby traps set by the occupants

CAUTION

Chemicals used in the production of some illegal drugs are extremely toxic and volatile. Incidents involving them may require the assistance of trained hazardous materials personnel.

Trash Container Fires

Trash containers may be as small as a garbage can or a large-capacity Dumpster™. Toxic products of combustion will be present in trash container fires of all types, so full PPE and SCBA should be worn when attacking any trash container fire (**Figure 17.45**). The refuse may include:

- Hazardous materials or plastics that emit highly toxic smoke and gases
- Aerosol cans and batteries that may explode when exposed to heat
- Biological waste in marked or unmarked containers

Once the fire has been controlled, it may be possible to use standard overhaul techniques to complete the extinguishment. However, it may be advantageous to attack the fire using Class A foam. In some departments, it is SOPs to use a master stream to flood the container with water to drown any fire that might be hidden in the contents.

Figure 17.45 Always wear full PPE including respiratory protection when extinguishing fires in waste containers.

However, this technique can present containment problems if the water used to fill the container becomes contaminated with a hazardous substance. Extinguishment of fire in a trash container is described in more detail in **Skill Sheet 17-I-11**.

Combating Ground Cover Fires

Ground cover or wildland fires can be as small as a backyard grass fire and as large and devastating as the fires in the western U.S. in June 2012, which destroyed millions of acres of grass and forest land (**Figure 17.46**). They may occur in vacant urban lots, parks, cemeteries, farm lands, or forests. Ground cover fires include those in weeds, grass, field crops, brush, forests, and similar vegetation. Not only can they destroy hundreds of acres of natural resources, destroy houses and farm structures, and kill livestock and wild game, they also place firefighters at risk.

Ground cover fires have natural and human causes as follows:

Figure 17.46 Ground cover, wildland, and forest fires have devastated large areas of North America. *Courtesy of the Los Angeles Fire Department – ISTS.*

- Lightning strikes
- Autoignition
- Volcanic activity
- Sparks from rockslides
- Arson
- Discarded smoking materials
- Campfires
- Sparks or other ignition sources from machinery
- Electrical shorts in power lines and fences

Because ground cover fires are unconfined, they have characteristics that are different from fires in burning buildings. The three main influences on ground cover fire behavior are fuel, weather, and topography. Of these, weather is the most significant. Because local topography, fuel types, water availability, and predominant weather patterns vary from one region to another, the tools and techniques used to control ground cover fires also vary.

Once a ground cover fire starts, burning can be rapid and continuous. For your own survival, you must learn how ground cover fires behave in a variety of conditions and how temperature and humidity, winds, fuel types, and terrain features influence flame lengths and rate of fire spread.

WARNING
Ground cover fires can be deadly to firefighters even if they are working in very light fuels or working during the overhaul phase of an operation.

Types of Ground Cover Fires

There are three basic types of ground cover fires based on the type and location of the fuel: ground fires, surface fires, and crown fires.

Figure 17.47 Forest areas contain layers of dead organic material that can provide fuel for ground cover fires.

Figure 17.48 Grass, shrubs, and other vegetation provide fuel for surface or crawling fires.

Ground Fire

Ground fires burn in the layer of dead organic matter (called humus) that generally covers the soil in forested areas **(Figure 17.47)**. These are slow-moving, smoldering fires that can go undetected for months before they enter a flaming stage. Due to the composition of the fuel, these fires are generally limited to forests and are very difficult to extinguish.

Surface Fire

The surface or crawling fire is the most common type of ground cover fire, burning on the soil surface consuming low-lying grass, shrubs, and other vegetation **(Figure 17.48)**. Surface fires can occur anywhere and can be natural or human caused.

Crown Fire

Crown or canopy fires are wind-driven, high-intensity fires that move through the tree tops of heavily forested areas **(Figure 17.49)**. The typical causes of crown fires are lightning strikes or extensions from ground or surface fires. These extensions are called *ladder fires* because the fire spreads upward through small trees, fallen timber, and vines to reach the forest canopy.

NOTE: Some documents include the ladder fire as a fourth type of ground cover fire.

Figure 17.49 Crown fires move through the tops of trees in forested areas.

Ground Cover Fire Behavior

Typically, two of the elements of the fire triangle — oxygen and fuel — are always present where ground cover is found. It is the addition of an ignition source, either naturally or by human action, that will result in a ground cover fire. In addition to fuel, oxygen, and heat, weather and topography also contribute to the intensity and spread of the fire.

Fuel

Ground cover fuels are typically categorized based upon the location of the fuels as follows:

- **Subsurface fuels** — Roots, peat, and other partially decomposed organic matter that lie under the surface of the ground

- **Surface fuels** — Needles, duff, twigs, grass, field crops, brush up to 6 feet (2 m) in height, downed limbs, logging slash, and small trees on or immediately adjacent to the surface of the ground

- **Aerial fuels** — Suspended and upright fuels (brush over 6 feet [2 m], leaves and needles on tree limbs, branches, hanging moss, etc.) physically separated from the ground's surface (and sometimes from each other) to the extent that air can circulate freely between them and the ground

The following factors affect the burning characteristics of ground cover fuels:

- **Fuel size** — Small or light fuels burn faster than heavier ones.

- **Compactness** — Tightly compacted fuels burn slower than those that are loosely piled.

- **Continuity** — When fuels are close together, the fire spreads faster because of heat transfer. In patchy fuels (those growing in clumps), the rate of spread is less predictable than in continuous fuels.

- **Volume** — The amount of fuel present in a given area (its volume) influences the fire's intensity and the amount of water needed to achieve extinguishment.

- **Fuel moisture content** — Fuels that contain less moisture ignite more easily and burn with greater intensity (amount of heat produced) than those with a higher moisture content.

Weather

Some weather factors that influence ground cover fire behavior are the following:

- **Wind** — Fans the flames into greater intensity and supplies fresh air that speeds combustion; very large-sized fires create their own winds **(Figure 17.50, p. 1052)**.

- **Temperature** — Effects on wind and is closely related to relative humidity; primarily affects the fuels as a result of long-term drying.

- **Relative humidity** — Significantly affects dead fuels that only gain moisture from surrounding air rather than their root system.

- **Precipitation** — Largely determines the moisture content of live fuels. Dead fuels (those easily ignited) may dry quickly; large, dead fuels retain this moisture longer and burn slower.

Figure 17.50 Wind can drive fires through all types of fuels, leaping obstacles like roads and streams.

Figure 17.51 Fires will spread rapidly up slopes. *Courtesy of California Office of Emergency Services.*

Topography

Topography — Physical configuration of the land or terrain.

Topography refers to the features of the earth's surface, and it has a decided effect upon fire behavior. The steepness of a slope affects both the rate and direction of a ground cover fire's spread. Fires will usually spread faster uphill than downhill, and the steeper the slope, the faster the fire spreads **(Figure 17.51)**. Other topographical factors influencing ground cover fire behavior include the following:

- **Aspect** — The compass direction a slope faces (aspect) determines the effects of solar heating. In North America, full southern exposures receive more of the sun's direct rays and therefore receive more heat. Ground cover fires typically burn faster on southern exposures.

- **Local terrain features** — Features such as canyons, ridges, ravines, and even large rock outcroppings may alter air flow and cause turbulence or eddies, resulting in erratic fire behavior.

- **Drainages (or other areas with wind-flow restrictions)** — These steep ravines are terrain features that create turbulent updrafts causing a chimney effect. Wind movement can be critical in *chutes* (narrow V-shaped ravines) and saddles (depression between two adjacent hilltops). Fires in these areas can spread at an extremely fast rate, even in the absence of winds, and are potentially very dangerous.

Parts of a Ground Cover Fire

The names used to identify the parts of a typical ground cover fire are shown in **Figure 17.52**. Every ground cover fire contains at least two or more of the following parts:

- **Origin** — The *origin* is the area where the fire started and the point from which it spreads.

- **Head** — The *head* is the part of a ground cover fire that spreads most rapidly. The head is usually found on the opposite side of the fire from the area of origin and in the direction toward which the wind is blowing. The head burns intensely and usually does the most damage. Usually, the key to controlling the fire is to control the head and prevent the formation of a new head.

- **Finger** — *Fingers* are long narrow strips of fire extending from the main fire. They usually occur when the fire burns into an area that has both light fuel and patches of heavy fuel. Light fuel burns faster than the heavy fuel, which gives the finger effect. When not controlled, these fingers can form new heads.

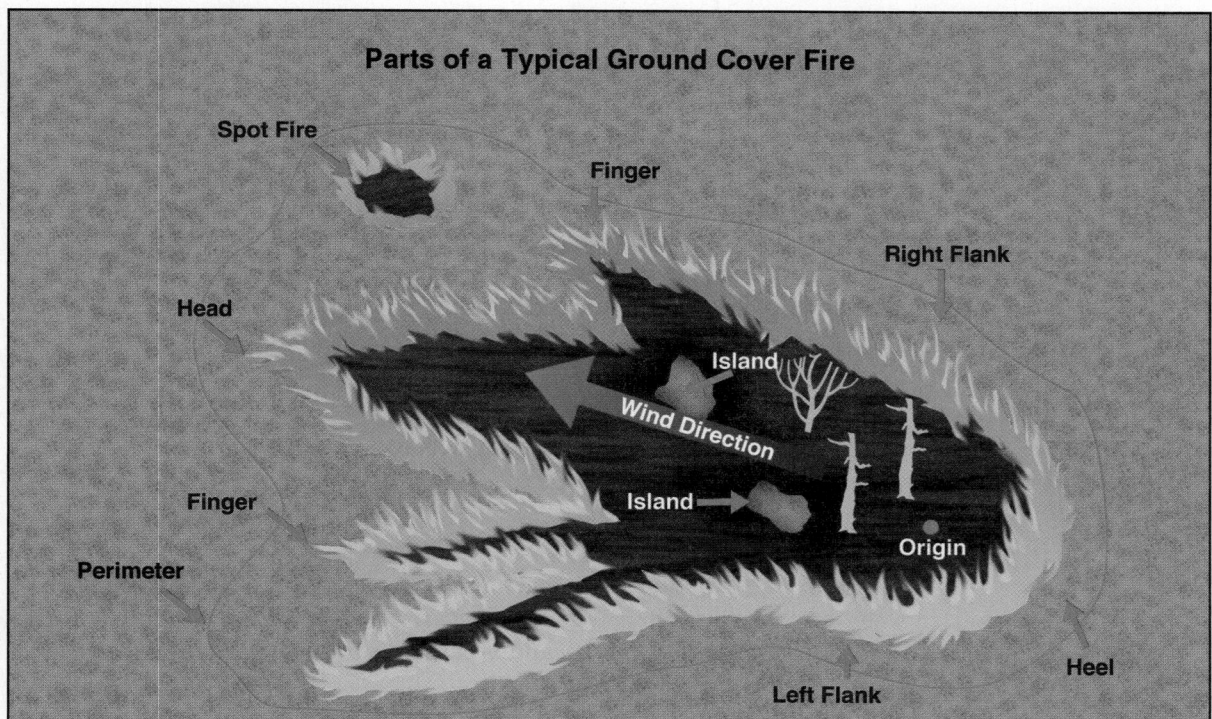

Figure 17.52 All ground cover fires share two or more of a set of common parts.

- **Perimeter** — The *perimeter* is the outer boundary, or the distance around the outside edge, of the burning or burned area. It will continue to grow until the fire is suppressed. Also commonly called the *fire edge.*

- **Heel** — The *heel,* or *rear,* of a ground cover fire is the side opposite the head. Because the heel usually burns downhill or against the wind, it burns slowly and quietly and is easier to control than the head.

- **Flanks** — The *flanks* are the sides of a ground cover fire, roughly parallel to the main direction of fire spread. The right and left flanks separate the head from the heel. It is from these flanks that fingers can form. A shift in wind direction can change a flank into a head.

- **Spot fire** — *Spot fires* are caused by flying sparks or embers landing outside the main fire. Spot fires present a hazard to personnel (and equipment) working on the main fire because they could become trapped between the two fires. Spot fires must be extinguished quickly or they will form a new head and continue to grow in size.

- **Islands** — Patches of unburned fuel inside the fire perimeter are called *islands.* Because they are unburned potential fuels for more fire, they must be patrolled frequently and checked for spot fires.

- **Green** — The area of unburned fuels next to the involved area is called the *green.* The green area of a ground cover fire should not be confused with the *green zone* often indicated at hazardous materials incidents or fire scenes. Green at ground cover fires is simply the opposite of the burned area (the black) and does not indicate that the area is safe.

- **Black** — The opposite of the green, the *black* is the area in which the fire has consumed or "blackened" the fuels. The black can sometimes be a relatively safe area during a fire but can be a very hot and smoky environment.

Protective Clothing and Equipment

Before attacking a ground cover fire, firefighters need to be wearing wildland (ground cover) fire protective clothing because standard structural turnout clothing is inappropriate and can even be dangerous. When fighting a ground cover fire, firefighters should wear PPE that meets the requirements of NFPA® 1977, *Standard on Protective Clothing and Equipment for Wildland Fire Fighting*. NFPA® 1500, *Standard on Fire Department Occupational Safety and Health Program*, specifies the minimum PPE for firefighters to participate in ground cover fire fighting (**Figure 17.53**). This standard requires firefighters to be equipped with the following:

Figure 17.53 Specialized PPE including respiratory protection is required when fighting ground cover fires.

- Helmet with eye protection and neck shroud
- Flame retardant shirt and pants (or one-piece jumpsuit)
- Protective footwear (sturdy boots without steel toes)
- Gloves
- Fire shelter (in crush-resistive case)

In addition, most wildland fire agencies also provide firefighters with a canteen or bottled water and a backpack or web belt for carrying extra gear. Firefighters carry **fusees**, extra food, water, clean socks, and other items in these packs.

Fusee — A friction match with a large head capable of burning in a wind.

Direct Attack (Ground Cover) — Operation where action is taken directly on burning fuels by applying an extinguishing agent to the edge of the fire or close to it.

Indirect Attack (Ground Cover) — A method of controlling a ground cover fire in which a control line is constructed or located some distance from the edge of the main fire, and the fuel between the two points is burned.

Attacking the Fire

The methods used to attack ground cover fires revolve around perimeter control. The control line may be established at the burning edge of the fire, next to it, or at a considerable distance away. The objective is to establish a control line that completely encircles the fire with all the fuel inside rendered harmless.

The *direct* and *indirect* approaches are the two most basic methods for attacking ground cover fires. A **direct attack** is action taken directly against the flames at its edge or closely parallel to it (**Figure 17.54**). The **indirect attack** is used at varying distances from the advancing fire. Starting from an anchor point (road, highway, body of water, previous burn), a line is constructed some distance from the fire's edge and the unburned intervening fuel is allowed to self-extinguish. This method is generally used against fires that are too hot, too fast, or too big for a direct attack.

Because a ground cover fire is constantly changing, it is quite possible to begin with one attack method and switch to another. As with any other type fire, size-up must continue throughout a ground cover fire so that these adjustments can be made when required. The steps for attacking a ground cover fire can be located in **Skill Sheet 17-I-12**.

Figure 17.54 A direct attack is conducted at the edge of the flames. *Courtesy of National Interagency Fire Center.*

Safety Principles and Practices

The first and most important principle of ground cover fire safety is life safety, both for the civilian population and firefighters. Firefighter safety and your safety begins with an accurate size-up of the incident. Size-up must be conducted on all ground cover fires. Size-up information includes:

- Fire location
- Fire type
- Incident access
- Exposures
- Weather conditions
- Wind direction
- Wind velocity
- Topography
- Visibility
- Resources
  — Water supply
  — Personnel
  — Apparatus/equipment

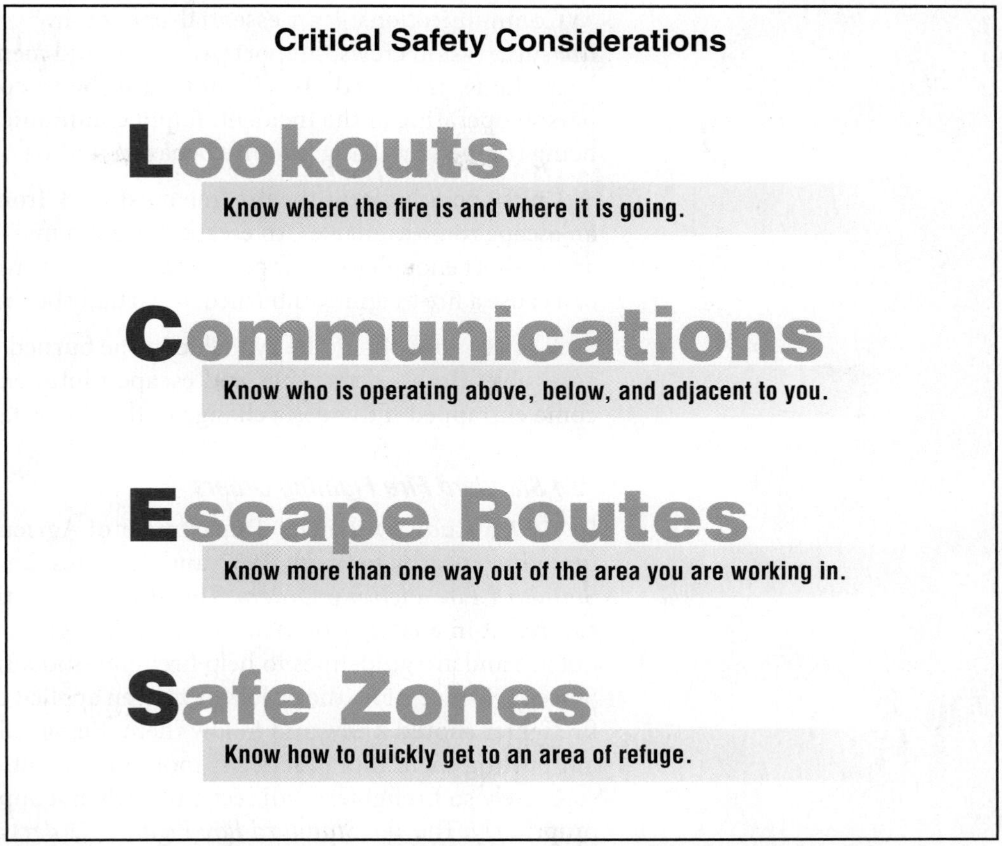

Critical Safety Considerations

Lookouts
Know where the fire is and where it is going.

Communications
Know who is operating above, below, and adjacent to you.

Escape Routes
Know more than one way out of the area you are working in.

Safe Zones
Know how to quickly get to an area of refuge.

Figure 17.55 The components of LCES should be applied to any ground cover fire.

Once the size-up has been conducted, you must continue to practice situational awareness. Situational awareness must continue throughout the incident to counter any changes in fire behavior, weather, or topography.

Lookouts, Communications, Escape Routes, and Safety Zones (LCES)

Applying situational awareness can be more effective if the concept of LCES is applied. LCES is a situational awareness technique that stands for the following (**Figure 17.55**):

- **L**ookout
- **C**ommunications
- **E**scape Routes
- **S**afety Zones

Developed for large-scale wildland fires, LCES can be used on any size ground cover fire.

Lookouts are used to monitor fire development and spread. Watch for rekindles within the burned area and keep the IC informed of these changes. Lookouts are placed at locations that can observe the fire without being in front of the fire. In some jurisdictions, airborne lookouts in helicopters or airplanes are used to monitor the fire.

Communications is an essential part of any ground cover operation. Lookouts, fire suppression crews, support personnel, and members of the Command Staff must keep the IC informed. The IC must also be in communication with every unit or person operating at the incident. Rapid communication can prevent personnel from being trapped by changes in fire behavior and weather patterns.

Units operating near the fire line must work from the burned side and always have an escape route available. An escape route is a marked path that leads to a safety zone and is short enough to allow personnel to safely travel to it. A change in wind direction can cause a fire to ignite unburned or partially burned materials behind fire line crews.

A safety zone should be available in the burned area if it is sufficiently cooled and accessible. Both safety zones and escape routes ensure that firefighters will not become entrapped if there is a change in the fire behavior or direction.

Ten Standard Fire Fighting Orders

In 1957, a study by the U.S. Department of Agriculture's Forest Service of wildland firefighter deaths between 1937 and 1956 resulted in the development of the *Ten Standard Fire Fighting Orders*. The study found that violating any of these orders can result in a fatality or serious injury. The orders are organized in a deliberate sequence and are guidelines to help firefighters identify and avoid high-risk situations. Since their inception, the orders have been applied to all fire fighting situations. Every firefighter should know and follow them. Being able to recite them is commendable, but putting them into practice is more important. Each order should be considered separately so firefighters will recognize when it applies during a fire and respond appropriately. The *Ten Standard Fire Fighting Orders* are as follows:

1. Keep informed on fire weather conditions and forecasts.
2. Know what the fire is doing at all times.
3. Base all actions on current and expected behavior of the fire.
4. Identify escape routes and safety zones, and make them known.
5. Post lookouts when there is possible danger.
6. Be alert, keep calm, think clearly, and act decisively.
7. Maintain prompt communications with your forces, your supervisor, and adjoining forces.
8. Give clear instructions and ensure that they are understood.
9. Maintain control of your forces at all times.
10. Fight fire aggressively, providing for safety first.

In addition to these orders, there are 18 Watch Out Situations that should be reviewed and practiced during ground cover fires. This list is provided in **Appendix E**.

Non Fire Hazards

Other hazards may be encountered during ground cover fire operations as follows:

- **Unstable hazard trees** — Trees that have been weakened by age or fire and may collapse.
- **Animals** — Animals that have escaped the fire as well as reptiles may be found in caves and confined spaces.

- **Insects** — Usually more of a nuisance than a hazard; some insect stings can be fatal to persons with allergies. Repellant sprays and first aid creams should be carried when needed.

- **Electrified fences** — Electrified fences have caused numerous firefighter deaths. All wire strand fences should be considered to be electrified until proven otherwise.

- **Electrical power lines** — Ground cover fires can cause power poles to fall and power lines to break. See the section on Class C fires for safety tips.

- **Explosives** — Explosives and unexploded ordinance may be found around military training areas, near construction sites, and in areas open to hunting. Do not touch or move any explosives that have been exposed to fire and establish a perimeter around them.

- **Hazardous materials** — Broken pipelines, storage tanks, oil and gas wells, and storage buildings can create hazardous materials hazards when exposed to fire. Treat these situations like hazardous materials incidents, establishing a perimeter, and withdrawing a designated distance.

- **Rolling or falling debris** — In rough terrain, rocks, burned vegetation, and limbs can fall and strike you or create a slipping or tripping hazard.

- **Pits or shafts** — Abandoned mine shafts, pits, and natural sink holes may be covered by loose debris.

- **Animal traps** — Traps used for hunting may be hidden under brush much like booby traps. In areas where illegal activities have been reported, these devices may be used to protect them.

Lightning is a unique, nonfire hazard that requires special precautions. The lightning that started a ground cover fire can also injure or kill firefighters. Take the following precautions to protect yourself from lightning hazards:

- Be aware of the weather.

- Do not stand under tall, isolated trees.

- Stay away from open water, metal objects, equipment, or wire fences.

In a forested area, seek shelter in a low ravine. If you are in a flat field and feel your hair stand on end, it is an indication that lightning is about to strike: drop to your knees and bend forward putting your hands on your knees. Do not lie flat on the ground.

Finally, the majority of line-of-duty deaths at ground cover fires result from heart failure. See Chapter 2 for information on heart attacks and how to prevent them. At ground cover fires, monitor your own stress level, check heart and lung rates, stay hydrated, and use the rehabilitation facilities that are provided.

Ⅱ Coordinating Fireground Operations

The incident priorities for any emergency are life safety, incident stabilization, and property conservation. To accomplish these priorities, fire officers and firefighters apply tactics traditionally known by the acronym RECEO-VS as follows:

- Rescue
- Overhaul
- Exposures
- Ventilation
- Confinement
- Salvage
- Extinguishment

As a Firefighter I, you learned to perform these tactics under the supervision of a fire officer or Firefighter II. Now you must be able to assemble a team and coordinate an interior fire attack. In order to accomplish these tasks effectively, you need to apply your knowledge of fire behavior and building construction to the interior attack. You also need to know the responsibilities of other companies functioning at the scene so that you can coordinate your interior attack with their assignments.

Situational Awareness

As a Firefighter II you will be directly responsible for the safety of firefighters assigned to you. First, you must continue to apply situational awareness to your environment. Because you have additional training and experience you should be able to recognize changes and predict the effect the changes can have on your surroundings.

Second, and in some ways more important, you must be able to listen to the concerns and observations of the firefighters with you. Remember that they may see something that you overlooked, or they may have additional information they gained from their own experience. They may also have a different physical point of view than you do and are able to see something you cannot from your location.

Situational awareness depends on open communication among all members of the crew. The use of situational awareness allows the crew leader to make informed decisions. However, it is not meant as an opportunity for debate or a vote. You are the crew leader with the final authority. Gather information and make a decision based on your crew's input.

Fireground Roles and Responsibilities

Fireground tactics must be performed in a coordinated way for the operation to be both safe and successful. Depending on your company assignment and department SOPs, your role and responsibility will vary. The following information describes a typical response to a fire in a residential structure and details the typical responsibilities of each unit involved.

First-Arriving Engine Company

The first-arriving fire officer or firefighter will make decisions and take actions that will determine the outcome of a structure fire. In some instances the first-arriving unit may be an engine that is fully staffed, a firefighter driving the water tender, or a chief officer in a command vehicle. Your department SOPs will define the actions that the first-arriving person or unit is expected to take, but typically these include:

- Establish Command
- Making the initial size-up
- Deploying available resources
- Communicating the situation to the communication center and other responding units

NOTE: Establishing Command is described in more detail later in this chapter.

If smoke or fire is visible as the first-arriving engine approaches the scene, departmental SOPs may require that unit to stop and lay a supply line from the closest fire hydrant into the scene (**Figure 17.56**). The fire hydrant can be opened and the supply line charged as soon as a hose clamp is applied at the scene. In other departments it is SOPs to give the officer in charge of the first-arriving engine the option of deploying a supply line or proceeding directly to the scene and initiating a quick attack using the water supply on the apparatus.

Figure 17.56 Depending on local procedures, the first-arriving engine may be required to lay a supply line to the scene.

Even though the size-up of the fire actually began much earlier, the first-arriving fire officer or firefighter will conduct a rapid initial assessment of the situation based on incident priorities, strategies, and tactics. The assessment is a quick mental evaluation. The first-arriving unit should base their actions on the answers to certain questions such as:

- Are there occupants in need of immediate rescue?
- Does the fire threaten other exposures?
- What does the visible fire and smoke indicate?
- Are only the contents involved or is the structure burning?
- Are there sufficient resources on scene or en route to handle this situation?

If there are too few resources assigned to the incident to meet incident objectives established during size-up, then additional resources must be requested. For example, if an interior fire attack or search and rescue operation is needed, fire departments must adhere to the OSHA established, *two-in, two-out* regulation. This regulation requires that when two firefighters enter a structure fire that has developed beyond the incipient stage, there must be two fully equipped firefighters outside to rescue the attack team in an emergency. If there are not enough personnel to conform to this rule, then no one enters the structure until additional resources arrive. However, if there is a known life safety hazard to a victim that can be saved without undue risk to firefighters, this rule can be amended to allow the two-person attack team to enter the structure leaving one person outside. The fire officer or firefighter giving the order must be able to justify his or her actions based upon local SOPs or NFPA® standards.

If there are no obvious and immediate life safety concerns and the fire is threatening a nearby exposure, the person in charge may order hoselines deployed to apply water to protect the exposure. A master stream appliance or handline may be put into operation to keep the exposure cool while hoselines are deployed for a direct attack on the fire.

Figure 17.57 A 360-degree survey should be made of the structure when possible.

Once the size-up and 360-degree survey is complete and the location of the fire is known, the first engine company will deploy the initial attack hoseline to accomplish the following priorities **(Figure 17.57)**:

- Intervene between trapped occupants and the fire.
- Protect rescuers.
- Protect primary means of egress.
- Protect interior exposures (other rooms).
- Protect exterior exposures (other buildings).
- Initiate extinguishment.
- Operate master streams.

Second-Arriving Engine Company

Unless otherwise assigned, the second engine company must first make sure that adequate water supply is established to the fireground. The second-arriving engine company may be required to complete tasks that the first engine company began such as:

- Finishing a hose lay **(Figure 17.58)**
- Deploying an additional hoseline
- Connecting to a hydrant to support hoselines that are already deployed

The need to pump the hoselines from the hydrant depends on local factors including:

- Size and quantity of hoselines
- Distance from the fire hydrant to the scene
- Available water pressure in the distribution system

Figure 17.58 Additional supply hoseline may need to be hand laid between apparatus at the scene.

Once the water supply has been established, the personnel from the second company will perform tasks that the IC assigns including:

- Assist advancing first attack hoseline.
- Back up the initial attack line.
- Protect secondary means of egress.
- Prevent fire extension (confinement).
- Protect the most threatened exposure.
- Assist in extinguishment.
- Assist with fireground support company operations.
- Form the rapid intervention crew/team (RIC/RIT).

Fireground Support Company

If a support company (aerial apparatus or rescue unit) has been dispatched to the fire, it may arrive before, with, or after the first engine company. Once at the scene, the situation will dictate the tasks the support company will perform including any of the following:

- Forcible entry
- Search and rescue
- Property conservation (salvage)
- Ground or aerial ladder placement
- Ventilation
- Scene lighting
- Utility control (electric, gas, and water)
- Checking for fire extension
- Operating elevated fire streams
- Overhaul

Figure 17.59 Blitz attacks must be coordinated to avoid spreading fire into uninvolved areas.

These functions may be performed by engine company personnel when support companies are not available. Initially, the support company may be assigned to check the outside of the building for victims needing immediate rescue and raise the ladders needed for rescues or roof access for ventilation. They may be assigned to force entry into the building to facilitate simultaneous interior fire attack and search and rescue operations.

Depending upon the situation, teams may first search areas closest to the fire if doing so will not put them at risk of severe injury or death. Search patterns can also begin in areas that are most likely to be inhabited or where victims are known to be trapped. Regardless of the pattern used, searches must be conducted systematically and in accordance with departmental SOPs. In most cases, the search priorities are:

- The areas most severely threatened (assuming this area is searchable without undue risk to firefighter safety)

- The areas where the largest number are threatened (assuming this area is searchable without undue risk to firefighter safety)

- Remainder of fire area

- Exposures

In addition to search and ventilation operations, it is SOPs in some departments for support company personnel to assist engine companies in making the fire attack. In some cases, support company personnel equipped with forcible entry tools accompany the hose team as they advance toward the seat of the fire. In other cases, support company assistance can be limited to the placement of ground ladders for fire attack, setting up scene lighting, and similar exterior functions.

Blitz Attack — To aggressively attack a fire from the exterior with a large-diameter (2½-inch [65 mm] or larger) fire stream.

In other situations, a master stream can be used to attack the fire in what is called a **blitz attack (Figure 17.59)**. Blitz attacks must be coordinated with other operations to avoid spreading the fire to uninvolved parts of the building. In addition, steam from master stream attacks can injure interior attack teams. Poorly directed streams can force the interior teams to retreat. Elevating devices such as aerial ladders, platforms, and ladder towers can also be used to conduct blitz attacks on upper floors and as substitute standpipes from which engine company personnel can advance hoselines.

Figure 17.60 In some jurisdictions, RIC/RIT units are designated, trained, and equipped to perform firefighter rescues.

Rapid Intervention Crew/Team (RIC/RIT)

Rapid intervention crews or teams are tasked with locating and assisting firefighters who have become trapped, lost, or incapacitated during interior structural fire fighting. Rapid intervention crews or teams (RIC/RIT) may be any engine, ladder, or rescue company that is equipped and assigned the task once they arrive on scene. In some departments, units may be designated and permanently assigned as the RIC/RIT (**Figure 17.60**).

Either departmental SOPs or the IC will establish the exact number of RIC/RITs needed at the scene. Crews are added as necessary if the incident escalates or the number of operations increases, which allows flexibility in RIC/RIT composition based on the type of incident and number of personnel on scene.

Each RIC/RIT consists of two or more members wearing complete PPE and respiratory protection. The team should be equipped with a radio, any special rescue tools needed, a spare SCBA or air cylinder, and equipment necessary to perform a rescue of other emergency personnel. Although individual RIC/RIT members may be assigned other minor emergency scene duties, they must be prepared to stop whatever they are doing and deploy immediately if needed.

The RIC/RIT should report to the IC and perform such tasks as:

- Staging equipment
- Sizing up the building for possible paths of egress
- Completing a 360-degree survey if possible
- Removing barriers to egress
- Monitoring radio traffic for distress calls
- Clearing windows
- Placing ladders
- Opening exits
- Illuminating the building

CAUTION

Do not allow additional assigned duties to prevent you from deploying in your primary rescue capacity when working as a RIC/RIT member.

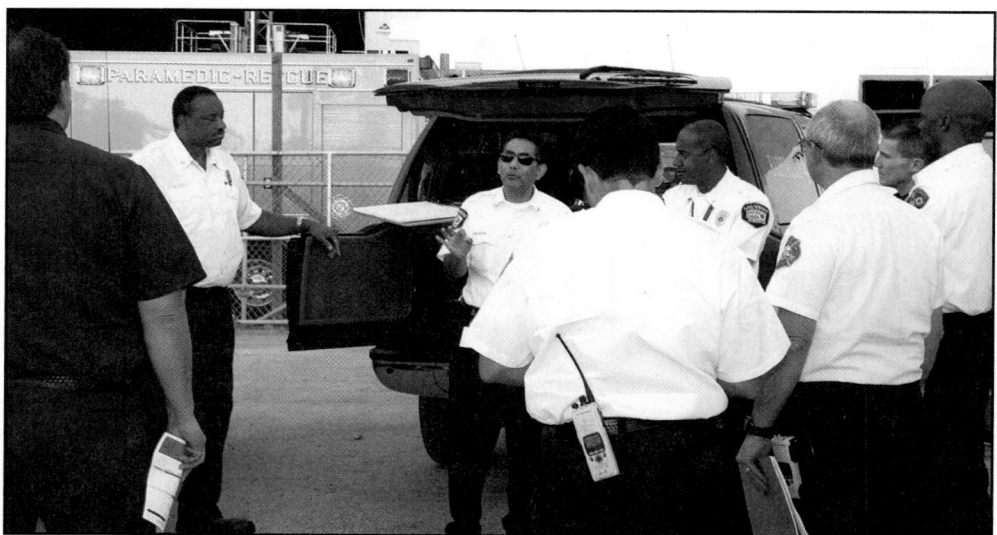

Figure 17.61 The Incident Command structure is designed to allow the chief officer to take Command or act as a liaison.

Chief Officer/Incident Commander

Upon arrival at the scene, a chief officer may choose to assume Command from the original IC and take responsibility for all on-scene operations. Alternatively, if the original IC has the incident well organized and reasonable progress toward incident stabilization is being made, the chief officer may choose to assume another role in the Incident Command structure (**Figure 17.61**). For example, the chief may choose to act as liaison with technical experts, utility crews, and members of the media.

Establishing Command

The first-arriving fire officer or firefighter must establish Command. **Skill Sheet 17-II-1** explains how to establish Incident Command and transfer Command at a structure fire.

Your local SOPs contain the specific activities you must follow when you establish Command. Generally, there are three initial Command options that should be relayed when Command is taken as described in the following sections.

Nothing Showing

When the problem generating the response is not obvious to the first-arriving unit, the officer or firefighter should assume Command of the incident and broadcast on the radio that "nothing is showing." That person should direct the other responding units to assume predetermined positions at the scene or stage at the last intersection in their route of travel. Staging resources in this way allows for a maximum of deployment flexibility and applies to all types of emergencies. The officer/firefighter then accompanies unit personnel on an investigation of the situation and maintains command using a portable radio (**Figure 17.62**). This option is also referred to as the *Investigation mode.*

Fast-Attack

When the officer or firefighter's direct involvement is necessary for the unit to take immediate action to save a life or stabilize the situation, the officer/firefighter should take Command and announce that the unit is initiating a fast attack (**Figure 17.63**).

Figure 17.62 In the Investigation mode, the officer in command enters the structure to determine the type of action that needs to be taken.

Figure 17.63 In the Fast Attack mode the officer is directly involved in fire suppression or rescue activities.

Personnel will continue the fast attack, which usually lasts only a short time, until one of the following situations occurs:

- Incident is stabilized.
- Incident is not stabilized, but the officer/firefighter must withdraw outside the hazardous area to establish a formal Incident Command Post (ICP).
- Command is transferred.

Depending upon the incident, the balance of the unit may be left inside the hazardous area when they can function safely and effectively *and* have radio communications capability. However, under no circumstances may fewer than two responders be left in a hazardous area. When only two responders (including the officer/firefighter) are in the hazardous area and one must leave, both must leave.

Name the Incident and Establish the ICP

Because of the nature and/or scope of some incidents, immediate and strong overall Command is needed. In these incidents, the first-arriving officer/firefighter should assume Command by naming the incident and designating an ICP, giving an initial report on conditions and requesting the additional resources needed.

- **Combat command** — Involves the officer/firefighter performing multiple tasks such as serving as Incident Commander (IC), developing the Incident Action Plan (IAP), and performing active tasks such as advancing a hoseline.

- **Formal command** — Involves the company officer remaining at the mobile radio in the apparatus, assigning tasks to unit personnel, communicating with other responding units, and expanding the NIMS-ICS as needed by the complexity of the incident (**Figure 17.64, p. 1066**). In addition, the officer/firefighter must decide how to deploy the remainder of the unit. Three options are normally available:

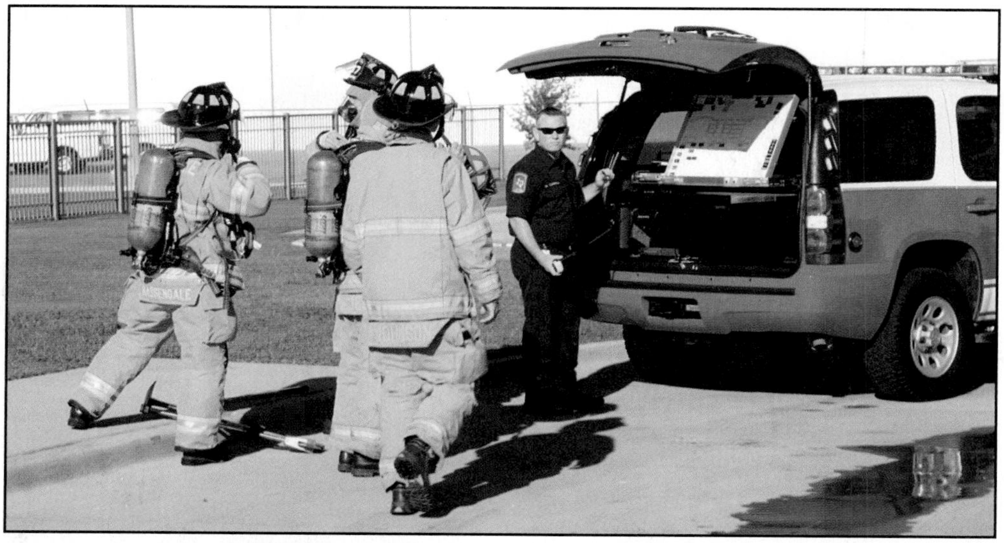

Figure 17.64 A Command Post is established in the formal Command Mode and the Incident Commander remains in that location to direct operations.

1. Appoint one unit member to supervise the rest of the unit, provide a portable radio, and make an assignment such as performing a search of the incident site.

2. Assign unit members to work under the supervision of another company officer, performing the tasks assigned to that officer by the IC.

3. Use unit members to perform staff functions in support of Command such as the Incident Safety Officer (ISO) or accountability officer.

Transferring Command

When an officer/firefighter IC needs to transfer Command of an incident to another fire officer, the transfer must be done correctly to avoid confusion (**Figure 17.65**). The fire officer assuming Command must communicate either face-to-face or over the radio with the officer/firefighter being relieved. Face-to-face communication of a transfer of Command is preferred. *Command should never be transferred to anyone who is not on the scene.* When transferring Command, the officer/firefighter being relieved should brief the relieving fire officer on the following items:

- Name of incident
- Incident status (such as fire conditions, number of victims, etc.)
- Safety considerations
- Goals and objectives listed in the Incident Action Plan (IAP)
- Progress toward completion of tactical objectives
- Deployment of assigned resources
- Assessment of the need for additional resources

Fires in Underground Spaces

Underground spaces include:

- Transformer vaults
- Mechanical spaces
- Parking garages
- Tunnels
- Caves
- Sewers
- Storage tanks
- Trenches

Figure 17.65 To avoid confusion, the transfer of Command must be done in person or by radio.

The single most important factor in safely operating at these emergencies is recognition of the inherent hazards of confined spaces **(Figure 17.66)**. Electrical equipment such as flashlights, portable fans, portable lights, and radios should be intrinsically safe for use in flammable atmospheres. The atmospheric and physical hazards that may be expected include the following:

- Oxygen deficiencies
- Flammable gases and vapors
- Toxic gases
- Extreme temperatures
- Explosive dusts
- Limited means of entry and egress
- Cave-ins or unstable support members
- Standing water or other liquids
- Utility hazards — Electricity, gas, sewage

Hazards Common to Confined Spaces

- Oxygen deficiency
- Flammable gases and vapors
- Toxic gases
- Extreme temperatures
- Explosive dusts

Figure 17.66 Underground spaces share types of hazards with confined spaces.

Figure 17.67 The Accountability Officer records all personnel entering or leaving a confined-space operation.

Plant or building supervisors or other knowledgeable people at the scene may provide valuable information on the fire, its probable location, and hazards that are present. Likewise, preincident plans of existing confined spaces in the fire department's jurisdiction reduce guesswork and should be referred to during operations in these locations. You should be ready to implement prearranged hazard mitigation plans, rescues, and extinguishment efforts without delay. These plans should include provisions for victim and rescuer protection, control of utilities and other physical hazards, communications, fire extinguishment methods, ventilation, and lighting.

Because of the hazards that may exist inside these spaces, the Command Post and Staging area must be established outside the hot zone. The Staging area should be near but not obstructing the entrance. Firefighters must not be allowed to enter these spaces until an IAP has been developed and communicated to on-scene personnel. An Accountability Officer or Incident Safety Officer must be stationed at the confined-space entrance to track personnel and equipment entering and leaving the space **(Figure 17.67)**.

Firefighters may also attack confined-space fires indirectly using penetrating nozzles, cellar nozzles, or distributor nozzles. Due to the difficulty of venting heat from some confined spaces, firefighters may tire more quickly and consume their air supply faster. In these conditions, firefighters must be relieved before they are out of air. An effective air-management system should be part of the IAP to prevent firefighters from advancing into confined spaces farther than their air supplies will safely allow.

Suppressing Class B Fires

Class B fires involve flammable and combustible liquids and gases. Flammable liquids have flash points of less than 100°F (38°C); examples are gasoline and acetone. Flammable liquids can be ignited without being preheated. Combustible liquids have flash points higher than 100°F (38°C); examples are kerosene and vegetable oil. Combustible liquids must be heated above their flashpoint before they can be ignited. Flammable and combustible liquids can be further divided into hydrocarbons (that do not mix with water) and polar solvents (that do mix with water).

Class B fires may begin as a spill or leak resulting from a vehicle accident, natural disaster, or opened valve among others. Because wind currents can spread the vapors or gases, your first action should be to determine the wind direction. After determining wind direction, take the following actions:

- Locate your apparatus upwind and uphill of the incident.

- Establish a perimeter.

- Report current conditions to all responding units and the communications center.

- Evacuate any civilians in the affected area.

- Request a hazardous materials response company and remain outside the hot zone.

- Establish a water supply and deploy attack hoselines as required.

Safety Precautions at Flammable/Combustible Liquid Fire Incidents

Firefighters must exercise caution when attacking large fires involving flammable and combustible liquids. The first precaution is to avoid standing in pools of fuel or runoff water contaminated with fuel floating on top. Protective clothing can absorb fuel in a "wicking" action, which can lead to skin irritation and even to the clothing catching on fire if an ignition source is present. Even if the wicking action does not occur, extreme danger exists if the pool of liquid ignites. In addition, benzene in petroleum product fumes is a known carcinogen. Remove from service PPE that is soaked with flammable or combustible liquids until it is thoroughly cleaned according to the manufacturer's recommendations.

> **WARNING**
> PPE soiled with flammable and combustible liquids may ignite when exposed to heat.

Flammable/combustible liquid fires burning around relief valves or piping must not be extinguished until the leak is controlled **(Figure 17.68)**. Unburned vapors are usually heavier than air and form pools or pockets of gas in low areas where they may ignite. Firefighters must attempt to control all ignition sources in a leak area. Vehicles, smoking materials, electrical fixtures, and sparks from tools can provide an ignition source sufficient to ignite flammable liquid vapors.

Figure 17.68 To extinguish a flammable gas fire, keep tanks and piping cool while control valves are locked out/tagged out.

Figure 17.69 Unattended master stream devices can be used to cool the upper portions of tanks to prevent tank failure due to a BLEVE.

An increase in the intensity of sound or fire issuing from a relief valve may indicate that the vessel is overheating and rupture is imminent. Firefighters should not assume that relief valves are sufficient to safely relieve excess pressures under severe fire conditions. The rupture of both large and small liquid vessels has caused firefighter fatalities.

If a closed container, such as a pressure vessel containing liquids or liquefied gases (LPG) is heated, the liquid inside begins expanding. When the liquid reaches its boiling point, it begins to return to its gaseous state. The change from a liquid to a gas in the confined space begins to increase the internal pressure on the vessel. When too much pressure builds up, the vessel loses its structural integrity and ruptures, releasing massive amounts of pressure and the flammable contents of the vessel. The release and subsequent vaporization of these flammable liquids can result in a BLEVE. For a BLEVE to occur, the liquid or liquefied gas must be above its boiling point (at standard temperature and pressure) when the container failure occurs.

A BLEVE produces a violent explosion that sends large pieces of the tank flying in all directions and a huge fireball with radiant heat sufficient to incinerate anything near the site. Tank failure may occur as a result of mechanical damage to the tank or from direct flame impingement on the vapor space in the tank. The most common cause of a BLEVE is when flames contact the tank shell above the liquid level and when insufficient water is applied to keep a tank shell cool. When attacking these fires, apply water to the upper portions of the tank, preferably from unattended master stream devices **(Figure 17.69)**.

Applying foam is the method most often used to control flammable liquid fires (see Chapter 16, Fire Streams). Class B fire fighting techniques are also needed for fires in gas utility facilities and accidents involving fuel transport trucks and rail tank cars. Water can be used in several forms (cooling agent, mechanical tool, and crew protection) to control Class B fires.

Fires in Properties Protected by Fixed Systems

You should familiarize yourself with the fixed fire suppression systems at locations containing Class B materials in your jurisdiction. When coordinating operations at a property protected by a fire suppression system, you should support the system rather than working against it. You should also recognize hazards particular to these systems that may harm the firefighters under your Command:

- Oxygen depletion following activation of carbon dioxide flooding systems
- Poor visibility

- Energized electrical equipment
- Toxic environments

Standard operating procedures used at these occupancies are often contained in a preincident plan. Some preincident plans specify the procedures for each emergency response unit to follow according to the conditions that are found. A building site plan showing water supplies, protection system connections, and unit placement is an integral part of the plan and must be updated regularly to reflect changes affecting fire department operations. The sections that follow offer guidelines for making Command decisions when working with water-based and non-water-based systems.

Figure 17.70 To ensure adequate operating pressure, fire department pumpers supply sprinkler, standpipe, and foam extinguishing systems through the FDC.

Water-Based Systems

Water-based systems include:

- Automatic sprinkler systems
- Standpipe systems
- Foam systems

All these systems depend on water from underground pipes or water storage tanks to operate. To supplement the water or to maintain a constant operating pressure, fire department pumpers deploy supply hoselines from nearby fire hydrants and connect to the fire department connection (FDC) **(Figure 17.70)**.

Automatic sprinkler system. When a fire is burning in a building equipped with an automatic sprinkler system, support company personnel are often used to manage the system's operation. While these personnel must always follow their departmental SOPs regarding what they can and cannot do to the sprinkler system, some of the possible actions they may take are as follows:

- Connect pumper to FDC to supplement the water supply and maintain constant pressure on the system.
- Assign a radio-equipped firefighter to the sprinkler control valve to close or re-open it as ordered and to prevent it from being closed prematurely.
- Install wooden wedges or sprinkler stops to halt the flow of water from open sprinklers.
- Replace open sprinklers to allow the system to be restored to normal (if allowed by SOPs).
- Restore the sprinkler system to normal (if allowed by SOPs).
- Monitor the building after the fire has been extinguished and while waiting for the owner or designee to restore the sprinkler system.

Standpipe systems. Standpipes permit firefighters to deploy attack hoselines on upper stories of structures and in large area structures and industrial sites. Fire department pumpers connect supply hoses to the standpipe connection and increase the system pressure based on the amount of friction loss that the height of the structure creates. High-rise structures may also have pumps installed on upper stories to help maintain the required pressure. To coordinate a fire attack using a standpipe, you must perform the following tasks:

- Connect pumper to FDC to supplement the water supply and maintain constant pressure on the system.
- Determine the location of the fire.
- Don full PPE and respiratory protection.

Figure 17.71 Foam discharge nozzles mounted in the ceiling of aircraft hangars can rapidly flood the structure when activated.

- Take the necessary tools, hose, nozzle, and equipment to the standpipe outlet located in the stairwell below the fire floor.

- Connect the attack hoseline to the outlet and advance the line up the stairs to the door into the fire floor.

- Charge the hoseline, bleed air off the line, and adjust the nozzle for the desired pattern.

- Advance the hoseline onto the fire floor and used the appropriate attack method to extinguish the fire.

Foam systems. Foam fire suppression systems are installed in locations where large quantities of Class B flammable/combustible liquids are stored or used. These locations include refineries, aircraft hangars, and automotive paint booths, to name a few **(Figure 17.71)**. These systems may include connections to FDCs or they may be self-contained. When foam systems have activated, it is critical that the foam blanket not be disturbed. Additional foam of the same type can be added to the layer. In some locations, such as aircraft hangars and transformer vaults, a **total flooding system** will be installed. In these sites the entire compartment or structure is filled with foam. Until it is determined that the fire is extinguished, the foam should remain in place and not be disturbed.

> **Total Flooding System** — Fire suppression system designed to protect hazards within enclosed structures; foam is released into a compartment or area and fills it completely, extinguishing the fire.

Nonwater-Based Systems

Non-water-based systems are often found in industrial occupancies such as coal-fired power plants. Aircraft hangars and maintenance facilities are protected by wet chemical systems. These systems are also commonly found on large cargo vessels.

Non-water-based systems include carbon dioxide, clean agent, and dry and wet chemical systems. When they activate they either fill the compartment with extinguishing agent, as in the case of carbon dioxide and clean agent, or blanket the fire, as in the case of dry and wet chemical systems. In both cases the extinguishing agents should not be disturbed or removed until it is determined that the fire is completely extinguished. Attack hoselines should be deployed and in position in the event that fires in adjacent Class A materials need to be extinguished. Self-contained breathing apparatus must be worn when entering or working in areas where these agents have been used.

Cooling a Vessel

Figure 17.72 A protective coat of water can be used to protect exposures as well as cool the tank shell to prevent failure of the storage tank itself.

Using Water to Control Class B Fires

Even though water alone is an ineffective extinguishing agent for Class B fires, it can be used in various ways to safely control them. These techniques require a basic understanding of the properties of Class B fuels and the effects water has on them. The important thing for you to remember is that hydrocarbons (gasoline, kerosene, and other petroleum products) do not mix with water, and polar solvents (alcohols, lacquers, etc.) do mix with water. These differences affect how water can be used to control and even extinguish Class B fires.

Cooling Agent

Water can be used as a cooling agent to control Class B fires and to protect exposures. Water without foam additives is not particularly effective on lighter petroleum distillates (such as gasoline or kerosene) or alcohols. However, water applied as droplets in sufficient quantities can absorb the heat from fires in heavier oils such as raw crude and extinguish the fires.

Water is most useful as a cooling agent for protecting exposures. To be effective, water streams need to be applied so they form a protective water film on materials that might weaken or collapse such as metal tanks or support beams **(Figure 17.72)**. Water applied to burning storage tanks should be directed above the level of the contained liquid to achieve the most efficient use of the water.

Mechanical Tool

Water from hoselines can be used to move Class B fuels (burning or not) to areas where they can safely burn or where ignition sources are more easily controlled. Class B fuels must *never* be flushed down storm drains or into sewers. Use appropriate fog patterns for protection from radiant heat and to prevent "plunging" the stream into the liquid. Plunging a stream into burning flammable liquids causes increased production of flammable vapors and greatly increases fire intensity. Slowly move the stream from side to side and "sweep" the fuel or fire to the desired location. Keep the leading edge of the fog pattern in contact with the fuel surface or the fire may flash under the stream back toward the attack crew. Through the use of fog streams, water may also be used to dissipate flammable vapors. Because fog streams aid in dilution and dispersion, these streams can be used to influence the movement of the vapors to a target location.

Figure 17.73 Multiple hoselines are used when extinguishing a flammable liquids fire or shutting off a leaking control valve.

Crew Protection

Fog-stream patterns can be used as crew protection when advancing to shut off liquid or gas control valves. Coordination and slow, deliberate movements provide relative safety from flames and heat. Although one hoseline can be used for crew protection, two lines with a backup line are preferred for fire control and safety **(Figure 17.73)**.

> **WARNING**
> Only firefighters who have practiced using hoselines for crew protection should do so during an emergency.

When pressure vessels containing flammable/combustible liquids or compressed gases (including nonflammable gases) are exposed to flame impingement, apply solid streams from their maximum effective reach until relief valves close. According to an NFPA® study, a minimum of 500 gpm (2 000 L/min) must be applied at each point of flame impingement. To meet this requirement, arch a stream along the top of the vessel so that water runs down on both sides. This film of water cools the vapor space inside the tank, the tank shell, and the steel supports under the tanks. Tank supports must be cooled to prevent their collapse.

There is no angle or direction from which to approach a fire-involved pressurized storage tank that is safer than any other angle. Current research from the NFPA®, Transport Canada, and the Canadian Gas Association indicates that when a tank ruptures it will cause metal fragments to fly in all directions. If it is absolutely necessary to make temporary repairs or shut off the fuel source, make a hoseline advance toward a tank that is on fire using a wide protective fog pattern. Use a separate pump and water source to supply a backup hoseline to protect firefighters in case other lines fail or additional tank cooling is needed.

Bulk Transport Vehicle Fires

Always follow preincident plans for transportation emergencies to reduce life loss, property damage, and environmental pollution. The techniques of extinguishment for fires in vehicles transporting flammable fuels are similar in many ways to fires in flammable fuel storage facilities. The amount of fuel available to burn, the possibility of vessel failure, and danger to exposures presents similar difficulties at both types of fires. The major differences in fires in vehicles transporting flammable fuels and fires in flammable fuel storage facilities include the following:

- Increased life safety risks to firefighters from traffic
- Increased life safety risks to passing motorists
- Reduced water supply
- Difficulty in identifying the products involved
- Difficulty in containing spills and runoff
- Force of collisions can weaken or damage tanks and piping
- Instability of vehicles
- The location of the incident raises additional concerns for civilians and civilian structures

Although a serious accident may bring traffic to a halt, many incidents are handled while traffic passes the scene at near-normal speeds. At least one lane of traffic in addition to the incident lane should be closed during initial emergency operations. Avoid using road flares because of the possibility of igniting leaking fuels. When law enforcement personnel are unavailable, one or more firefighters trained in traffic control should be assigned to direct traffic and control access to the scene.

The techniques of approaching and controlling leaks or fires involving transport vehicles and rail tank cars are the same as for storage vessels; however, there are notable, additional considerations at vehicle fires **(Figure 17.74)**. Vehicle tires could fail which, in turn, causes the flammable load to shift. The status and limitations of the water supply vary much more than at fixed facilities. It may also be necessary to protect trapped victims with hose streams until they can be rescued.

Figure 17.74 Storage vessel fires share features in common with vehicle fires, but vehicle fires include additional considerations.

Firefighters must determine the exact nature of cargos as soon as possible from bills of lading, manifests, placards, or the drivers of the transport vehicles. Unfortunately, cases exist where these items could not be found, placards were either wrong or obscured, and drivers were unable to identify their cargos. In these instances, contact should be made with the shippers or manufacturers responsible for the vehicles.

During fire control procedures, the fire department is obligated to protect the environment as well. To prevent runoff of contaminated water and spilled liquids, storm water drains should be blocked.

Flammable Gas Incidents

The distribution system for natural gas consists of a vast network of surface and subsurface pipes. Gas pressure in these pipes ranges from 1,000 psi (7 000 kPa) in the distribution network to 0.25 psi (2 kPa) at the point of use. However, the pressure is usually below 50 psi (350 kPa) in local distribution piping. Natural gas may also be compressed, stored, and shipped in cylinders marked as compressed natural gas (CNG). Natural gas is also shipped and stored as a liquid (LNG) and is subject to BLEVE in this form. Information on controlling a pressurized flammable gas container fire can be located in **Skill Sheet 17-II-2**.

Excavation equipment breaking through underground pipes is generally the cause of most natural gas (CNG) and liquefied petroleum gas (LPG) incidents. When these breaks occur, contact the utility company immediately. Even if the gas has not yet ignited, apparatus should approach from and stage on the upwind side, on the side from which the wind is blowing. Firefighters must wear full PPE and be prepared in the event of an explosion and accompanying fire.

The first concerns are the evacuation of the area immediately around the break, evacuation of the area downwind, and elimination of ignition sources. Service connections near the break may have been damaged; therefore, check surrounding buildings for the odor of gas inside. Firefighters should follow their departmental SOPs regarding crimping a gas line to stop a leak. If gas is burning, the flame *should not be extinguished*. If necessary, use hose streams to protect exposures. The utility company should be contacted and an attempt to shut off the pressurized gas supply should be made. A hazardous materials response team should also be requested if available.

> **WARNING**
> If gas is burning from a broken gas pipe,
> do not extinguish the fire.
> Provide protection for exposures.

Chapter Summary

Attacking fires early in their development is an important aspect of a successful fire fighting operation. Likewise, selecting and applying the most effective fire attack strategy and tactics are also important. Failing to do any of these things can result in a fire growing out of control, an increase in fire damage and loss, and possibly in firefighter injuries. They need to know how to safely and effectively attack and extinguish fires involving structures, vehicles, stacked and piled materials, and ground cover. Firefighter II firefighters must be able to supervise teams involved in structure fires and Class B liquids and gases.

Firefighter I

1. What initial factors must be considered when suppressing structure fires?

2. What are the factors that must be considered when making entry?

3. How do direct attack and combination attack techniques compare with one another?

4. What are the main differences between indirect attack and gas cooling techniques?

5. How does the presence or absence of a standpipe system impact upper level structure fires?

6. What are the main actions that should be taken when attacking a belowground structure fire?

7. How quickly can floor assemblies over basements reach a point of collapse?

8. How can using exposure protection or controlling building utilities help in fire control?

9. What are the steps that must be taken when supporting a fire protection system at a protected structure?

10. How should a master stream device be properly deployed?

11. What situations may require suppression of a Class C fire?

12. What are some safety guidelines that can be used when suppressing Class C fires?

13. How can a Class D fire be suppressed?

14. What steps should be taken when suppressing a vehicle fire?

15. What are the factors that influence suppression methods in stacked and piled materials, small unattached structures, and trash containers?

16. What are a few of the main causes of ground cover fires?

17. How do surface fires and crown fires compare with ground fires?

18. What three elements influence ground cover fire behavior?

19. What are the parts of a typical ground cover fire?

20. What types of protective clothing and equipment can be used when fighting ground cover fires?

21. How do direct attack and indirect attack methods for ground fires compare with one another?

22. What safety principles and practices should firefighters use when fighting ground cover fires?

Firefighter II

1. What are the priorities that must be considered when beginning fireground operations?

2. What are the fireground roles a Firefighter II may need to coordinate at an incident?

3. How should Command be established at an incident?

4. What hazards may be present at fires in underground spaces?

5. What safety precautions should be taken at flammable/combustible liquid fire incidents?

6. How do suppression methods for water-based and non-water-based suppression systems differ?

7. What are the ways water can be used to control Class B fires?

8. How do suppression methods for bulk transport vehicle fires and flammable gas incidents compare with one another?

SKILL SHEETS

17-I-1

Attack a structure fire using a direct, indirect, or combination attack.

Direct Attack Method

Step 10: Cool hot gases overhead as needed when accessing a shielded fire using short applications of water fog.

Step 11: Using a straight or solid stream or a narrow fog pattern, direct the water onto the base of the fire.

Step 12: Observe fire conditions.

Step 13: Shut off nozzle when fire is extinguished.

Step 14: Report to officer completion of task.

Step 1: Confirm order with officer to attack fire.

Step 2: Don appropriate PPE, including SCBA.

Step 3: Ensure PPE and SCBA have been checked by safety officer.

Step 4: Select the proper attack hoseline and nozzle based on the location and size of the fire.

Step 5: Deploy and advance uncharged attack hoseline as directed by supervisor.

Step 6: Don SCBA facepiece, activate air supply, and activate PASS device when attack hoseline is in place.

Step 7: Signal the pump operator when ready for water.

Step 8: Open nozzle to purge air, ensure that water has reached the nozzle, and then close the nozzle.

Step 9: When ordered, enter the structure and advance to the seat of the fire while crouching or crawling, extinguishing any fires that are encountered.

17-I-1
Attack a structure fire using a direct, indirect,
or combination attack.

SKILL SHEETS

Indirect Attack Method

Step 8: Open nozzle to purge air and ensure that water has reached the nozzle.

Step 9: Select correct fog pattern, and close the nozzle.

Step 1: Confirm order with officer to attack fire.

Step 2: Don appropriate PPE, including SCBA.

Step 3. Ensure PPE and SCBA have been checked by safety officer.

Step 4: Select the proper attack hoseline and nozzle based on the location and size of the fire.

Step 5: Deploy and advance uncharged attack hoseline as directed by supervisor.

Step 6: Don SCBA facepiece, activate air supply, and activate PASS device when attack hoseline is in place.

Step 7: Signal the pump operator when ready for water.

SKILL SHEETS

17-I-1
Attack a structure fire using a direct, indirect, or combination attack.

Indirect Attack Method

Step 10: When ordered, enter the structure and advance toward the seat of the fire while crouching or crawling, extinguishing any fires that are encountered.

Step 11: Cool hot gases overhead as needed when accessing a shielded fire using short applications of water fog.

Step 12: When in place, open nozzle and direct fog pattern toward the ceiling and upper area of the walls.

Step 13: Close the interior door to the room allowing the steam to develop. Crack the door to observe the conditions.

Step 14: Continue to apply water to the compartment linings (walls and ceiling) until fire is reduced.

Step 15: Shut off nozzle when fire is extinguished.

Step 16: Report to officer completion of task.

17-I-1
Attack a structure fire using a direct, indirect, or combination attack.

 SKILL SHEETS

Combination Attack Method

Step 1: Confirm order with officer to attack fire.

Step 2: Don appropriate PPE, including SCBA.

Step 3: Ensure PPE and SCBA have been checked by safety officer.

Step 4: Select the proper attack hoseline and nozzle based on the location and size of the fire.

Step 5: Deploy and advance uncharged attack hoseline as directed by supervisor.

Step 6: Don SCBA facepiece, activate air supply, and activate PASS device when attack hoseline is in place.

Step 7: Signal the pump operator when ready for water.

Step 8: Open nozzle to purge air and ensure that water has reached the nozzle.

Step 9: Select correct fog pattern and close nozzle.

Step 10: When ordered, enter the structure and advance toward the seat of the fire while crouching or crawling, extinguishing any fires that are encountered.

Step 11: Cool hot gases overhead as needed when accessing a shielded fire using short applications of water fog.

17-I-1

Attack a structure fire using a direct, indirect,
or combination attack.

Combination Attack Method

Step 12: When in place, open nozzle and direct fog pattern toward the upper edge of the fire at the ceiling level.

Step 15: Apply water using the direct attack method as needed.

Step 16: Shut off nozzle when fire is extinguished.

Step 17: Report to officer completion of task.

Step 13: Use a T, Z, or O pattern, moving the fire stream from high to low.

Step 14: Shut off the nozzle when the room begins to darken.

17-I-2
Attack a structure fire above, below, and at ground level - Interior attack.

SKILL SHEETS

Ground Level Fire Attack

Step 1: Confirm order with officer to attack fire.

Step 2: Don appropriate PPE, including SCBA.

    a. Connect SCBA facepiece to regulator

    b. Activate PASS device

Step 3: Ensure PPE and SCBA have been checked by safety officer.

Step 4: Prior to entry, check nozzle pattern and bleed air from hoseline.

Step 5: Size up building/fire conditions, fire location, wind direction, and environment for hazards.

Step 6: Extinguish burning fascia, boxed cornices, or other doorway overhangs as necessary before entering.

    a. Ensure that fire is not overhead or behind attack hoseline crew

Step 7: Advance hoseline into the structure.

    a. At signal from officer

    b. All firefighters on same side of hose

c. Whenever possible and safe to do so, leave one firefighter at each 90-degree turn to assist in advancing hose

d. Approach fire from windward side

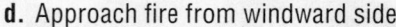

Step 8: Maintain situational awareness and monitor for changing conditions.

Step 9: Open nozzle fully and extinguish fire with a direct, indirect, or combination attack as directed by officer.

Step 10: Communicate effectiveness/ineffectiveness of attack to Incident Commander.

SKILL SHEETS

17-1-2

Attack a structure fire above, below, and at ground level - Interior attack.

Above Ground Fire Attack

Step 1: Confirm order with officer to attack fire.

Step 2: Don appropriate PPE, including SCBA.

    a. Connect SCBA facepiece to regulator

    b. Activate PASS device

Step 3: Ensure PPE and SCBA have been checked by safety officer.

Step 4: Prior to entry, check nozzle pattern and bleed air from hoseline.

Step 5: Size-up building/fire conditions, fire location, wind direction, and environment for hazards.

Step 6: Extinguish burning fascia, boxed cornices, or other doorway overhangs as necessary before entering.

    a. Ensure that fire is not overhead or behind attack hoseline crew

Step 7: Advance hoseline into the structure.

    a. At signal from officer

    b. Adequate firefighters to advance line to upper fire floor

    c. Whenever possible and safe to do so, leave one firefighter at each 90-degree turn to assist in advancing hose

    d. Approach fire from windward side

Step 8: Advance hoseline up stairwell to fire floor.

    a. If possible, lay extra hoseline in stairwell above fire floor

Step 9: Maintain situational awareness and monitor for changing conditions.

Step 10: Extinguish fire with a direct, indirect, or combination attack as directed by officer.

    a. All firefighters on same side of hose

    b. Approach fire from unburned side

Step 11: Communicate effectiveness/ineffectiveness of attack to Incident Commander.

17-I-2
Attack a structure fire above, below, and at ground level - Interior attack.

Below Ground Fire Attack

Step 1: Confirm order with officer to attack fire.

Step 2: Don appropriate PPE, including SCBA.

    a. Connect SCBA facepiece to regulator

    b. Activate PASS device

Step 3: Ensure PPE and SCBA have been checked by safety officer.

Step 4: Prior to entry, check nozzle pattern and bleed air from hoseline.

Step 5: Size up building/fire conditions, fire location, wind direction, and environment for hazards.

Step 6: Extinguish burning fascia, boxed cornices, or other doorway overhangs as necessary before entering.

    a. Ensure that fire is not overhead or behind attack hoseline crew

Step 7: Ventilate the basement before entry if possible.

Step 8: Advance hoseline into the structure.

    a. At signal from officer

    b. All firefighters on same side of hose

    c. Whenever possible and safe to do so, leave one firefighter at each 90-degree turn to assist in advancing hose

    d. Approach fire from windward side

Step 9: Advance hoseline down stairwell into the basement.

    a. Deploy enough hose at top of stairs to reach and extend into basement

    b. Evaluate fire conditions including stability of stair-well prior to advancing

    c. Coordinate attack with ventilation

    d. Maintain contact with other firefighters

    e. Advance down stairwell quickly to limit exposure to heated gases and smoke

Step 10: Maintain situational awareness and monitor for changing conditions.

Step 11: Open nozzle fully and extinguish fire with a direct, indirect, or combination attack as directed by officer.

Step 12: Communicate effectiveness/ineffectiveness of attack to Incident Commander.

NOTE: These steps are to be performed assuming there is no back-up generator or alternative source of energy present. Always use CAUTION when performing this skill and refer to local policy for further guidance.

Step 1: Confirm order with officer to turn off utilities.

Step 2: Locate and shut off electricity by closing the main breaker switch at main service panel.

  a. Individual breakers may need to be used if there is not main breaker switch

  b. Note any tripped breakers

  c. Always use caution; backup or alternative energy sources may be present

Step 3: Locate natural gas meter and/or LPG/CNG storage tank/cylinder and shut off.

Step 4: Locate water meter box and shut off water meter.

Step 5: Report to officer completion of assigned task.

17-I-4
Connect supply fire hose to a fire department connection.

SKILL SHEETS

Step 1: Wearing appropriate PPE, confirm order with officer to connect line.

Step 2: Extend hoselines to fire department connection.
   a. Male thread toward fire department connection
   b. Obtain double-male appliances if required
   c. Carry spanner wrenches

Step 3: Lay down hose couplings near fire department connection, gently folding back hose so that male threads do not hit pavement.

Step 4: Remove caps from fire department connection.
   a. Unscrew screw-in caps
   b. Break breakaway caps
   c. Strike center with end of wrench handle

Step 5: Inspect the fire department connection for debris.
   a. Check threads and clapper
   b. Check and replace gasket, if necessary

Step 6: Connect hoselines to left inlet of wye (if present) or inlet with lowest fitting first.

Step 7: Tighten connections with spanner wrench.

Step 8: Report to officer completion of assigned task.

OS&Y

Step 1: Confirm order with officer to operate valve.

Step 2: Unlock and remove chain if required.

Step 3: Close the OS&Y valve by turning it clockwise until the valve is fully closed and the stem is flush with the wheel.

Step 4: Open the OS&Y valve by turning it counterclockwise until fully opened.

Step 5: Back off the OS&Y valve one-quarter turn clockwise.

PIV

Step 1: Confirm order with officer to operate valve.

Step 2: Unlock the PIV wrench from the PIV body.

Step 3: Position the PIV wrench on stem nut.

Step 4: Close the PIV valve, turning it clockwise slowly until the target window indicates CLOSED or SHUT.

Step 5: Open the PIV valve, turning it counterclockwise until fully open and target window indicates OPEN

Step 6: Back off the PIV valve, turning it clockwise one-quarter turn ensuring that the target window remains OPEN.

Step 7: Replace and lock the wrench onto the PIV body.

17-I-6
Stop the flow of water of an activated sprinkler.

SKILL SHEETS

Wedge

Step 2: Insert the wedges between the sprinkler arms, flat sides against sprinkler.

Step 3: Drive the wedges into the sprinkler with the heel of hand until water flow stops.

Step 1: Wearing appropriate PPE including eye protection, place and climb ladder safely to within reach of the sprinkler.

Clamp-Type Sprinkler Tongs

Step 1: Wearing appropriate PPE including eye protection, place and climb ladder safely to within reach of sprinkler.

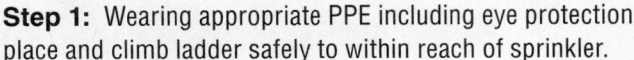

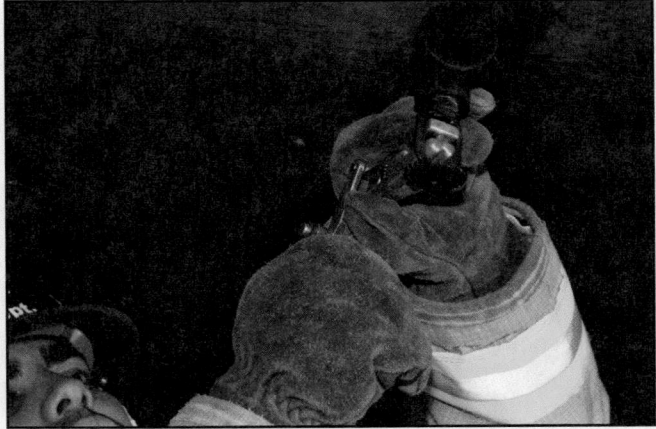

Step 3: Open the tongs (by clamping the handles together) until water flow stops.

Step 4: Lock the tongs in open position, with the keeper pulled as far as it will go toward end of handles.

Step 2: Insert the tongs into the sprinkler between the arms.

Swivel-Type Sprinkler Tongs

Step 1: Wearing appropriate PPE including eye protection, place and climb the ladder safely to within reach of the sprinkler.

Step 2: Insert the tongs into the sprinkler between the arms

Step 3: Open the tongs with handle swiveled upward

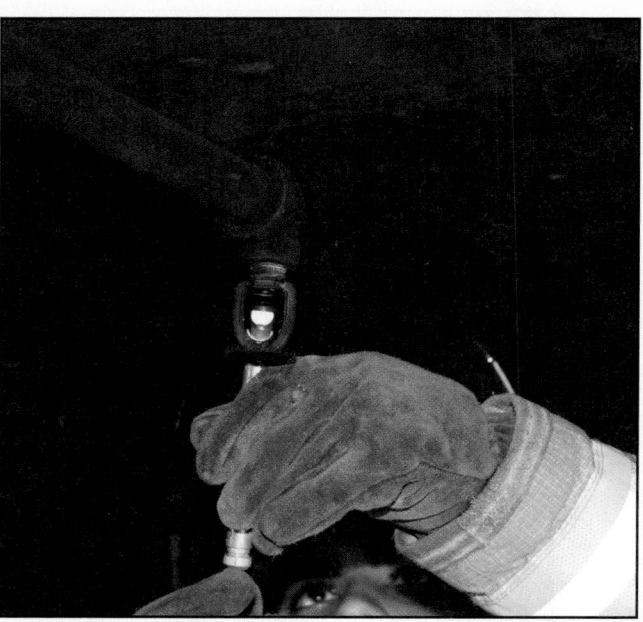

Step 4: Turn the locking knob clockwise to lock the tongs in the open position.

17-I-7
Deploy and operate a portable master stream device.

SKILL SHEETS

Step 1: Confirm order with officer to deploy master stream device.

Step 2: Remove needed tools and appliances from the apparatus.

Step 3: With assistance, remove the monitor from the apparatus, using proper lifting techniques.

Step 4: Carry the monitor unit to the set-up area.

Step 5: Position the monitor on a solid, level surface.

Step 6: Secure monitor according to manufacturer's guidelines.

Step 7: Adjust the nozzle to the proper elevation.

Step 8: Secure the anchor lock, if applicable.

Step 9: Extend hoseline to the monitor.

Step 13: Signal the pumper driver/operator to charge the line.

Step 14: Steady the monitor.

Step 15: Adjust the direction of water flow as necessary.

Step 10: Connect the hoselines to the monitor unit.

Step 11: Tighten the swivel couplings using spanner wrenches.

Step 12: Check the tip size, ensuring proper tip for situation, or select desired fog pattern stream.

Step 16: Operate master stream device by aiming stream in correct direction and hitting designated target.

Step 1: Confirm order with officer to attack passenger vehicle fire.

Step 2: Ensure vehicle is secure, chock wheels if necessary.

Step 3: Lay out attack line for fire attack.

    a. Use appropriate PPE including SCBA

    b. Select appropriate hoseline and nozzle

    c. Select appropriate hand tool(s)

Step 4: Charge attack line.

    a. Bleed air from hoseline

    b. Select appropriate fog pattern

Step 5: Advance attack line to vehicle.

    a. Approach at a 45-degree angle from upwind and uphill if possible

    b. Size up scene for hazards

    c. Use fog pattern for personnel protection

Step 6: Extinguish any fire under vehicle or in line of approach.

    a. Use a narrow fog pattern or straight stream for attack

Step 7: Extinguish fire in passenger compartment.

    a. Break window to gain entry and ventilate

    b. Use a narrow fog pattern or straight stream for attack

    c. Check for victims

    d. Maintain situational awareness

Step 8: Extinguish fire in engine compartment.

    a. Approach from side of vehicle

    b. Open hood at corner using tool such as Halligan

    c. Use a narrow fog pattern or straight stream for attack

    d. When possible, open hood using latch and prop open

    e. Maintain situational awareness

Step 10: Overhaul hidden and smoldering fires.

    a. Preserve fire cause evidence

    b. Extinguishment is complete – no hidden or smoldering fires remain

    c. All other hazards such as leaking fuel addressed

    d. Maintain situational awareness

Step 11: Report to officer completion of task.

Step 9: Extinguish fire in trunk.

    a. Approach from side of vehicle

    b. Knock out locking mechanism and open latch

    c. Open trunk and prop open

    d. Maintain situational awareness

Step 1: Wearing appropriate PPE, confirm order with officer to attack fire.

Step 2: Size up environment for hazards.

    a. Identify and verbalize collapse zone

    b. Work outside of collapse zone

Step 3: Check nozzle pattern and bleed air from hoseline.

Step 4: Check for threat to exposures and cool as necessary.

Step 7: Overhaul debris using pike pole or trash hook.

Step 8: Report to officer completion of assigned task.

Step 5: Advance to position to make fire attack.

Step 6: Extinguish fire with straight stream.

    a. At directions of officer

Step 1: Wearing appropriate PPE, deploy attack hoseline as directed by officer.

Step 2: Straighten hoseline.

Step 3: Advance toward structure, open nozzle slightly to allow air to escape, and wait for water to reach nozzle.

Step 4: Using a straight (solid) stream nozzle setting, direct the stream at the structure.

Step 5: As you advance close to the fire, if needed, adjust the nozzle pattern to fog setting to protect from heat.

Step 6: Work the nozzle around to extinguish fire.

Step 7: Use straight (solid) stream setting to extinguish fire in debris.

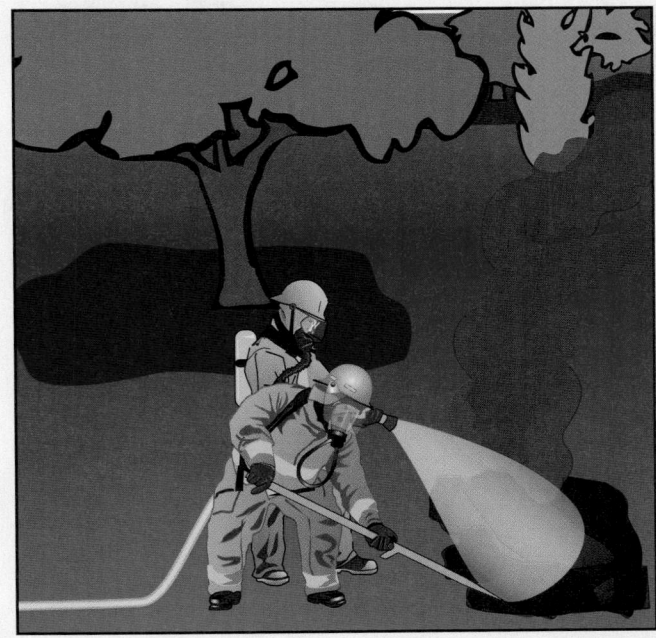

Step 8: Work with partner, Firefighter #1 uses rake or pike pole to move debris while Firefighter #2 directs the stream into the debris.

Step 1: Wearing appropriate PPE, confirm order with officer to extinguish fire.

Step 2: Set the nozzle flow to a straight stream.

Step 3. Open the nozzle fully, briefly, and aim stream to side to test fog pattern and expel air.

Step 4: Size up environment for hazards.

Step 5: Advance to the trash container from uphill and upwind.

    a. Keep the stream between the container and any exposures

    b. Maintain situational awareness

Step 6: Cool outside of container and any exposures.

Step 7: Attack the fire with a medium fog pattern until it is knocked down.

Step 8: Perform overhaul.

    a. Firefighter #2: Break up material and probe with pike pole for hot spots

    b. Firefighter #1: Extinguish hot spots

    c. Do not enter container and avoid placing head into or over any openings

Step 1: While wearing PPE, confirm order with officer to attack fire.

Step 4: Approach flame edge and apply water with hand-line or extinguisher or use hand tools.

Step 5: Maintain situational awareness.

Step 6: Extinguish fire.

    a. Maintain communication with officer

    b. Monitor weather

    c. Monitor fire and smoke conditions

Step 2: Size up environment for hazards.

    a. Identify and verbalize safe zones and escape routes

Step 3: Position at perimeter of hot zone and approach from the burned area (black).

Step 7: Mop up hot spots.

Step 8: Exit hazard area to safe area.

Step 9: Report to officer completion of assigned task.

SKILL SHEETS

17-II-1
Establish Incident Command and coordinate interior attack of a structure fire.

Step 1: Size up incident scene on arrival.

  a. Fire conditions

  b. Type of occupancy

  c. Hazards

  d. Rescue potential

Step 2: Transmit initial report over radio.

  a. Situation found

  b. Actions to be taken/assignments made

  c. Command status

  d. Resources needed

Step 3: Establish Incident Command.

  a. Place and name of Command

  b. Announce assigned tactical radio channel

Step 4: Identify incident objectives and strategies.

Step 5: Assign available resources to tasks.

  a. RIC/RIT assigned if required

  b. Verify understanding of assignments

Step 6: Request additional resources as required.

Step 7: Monitor progress of assignments.

  a. Firefighter accountability

  b. Monitor communications

  c. Verify completion of incident objectives and strategies as defined in Incident Action Plan

  d. Evaluate Incident Action Plan

Step 8: Maintain situational awareness of incident.

  a. Evaluate fire and structural conditions

  b. Evaluate physical condition of personnel

Step 9: Provide briefing to senior officer that is assuming Command.

  a. Current incident situation

  b. Incident action plan

  c. Accountability status

  d. Potential hazards

  e. Additional resources needed, requested, or en route

17-II-2
Control a pressurized flammable gas container fire.

SKILL SHEETS

Step 1: While wearing appropriate PPE, confirm order with officer to extinguish fire.

Step 2: Size up incident scene for hazards.

    a. Fire conditions

    b. Type of fuel

    c. Integrity of container

    d. Wind conditions

    e. Escape route and safe haven

Step 3: Deploy handlines.

    a. Bleed air from hoselines

    b. Ensure adequate hoseline to reach container

Step 7: Close control valve.

    a. Shut valve completely

    b. Report to officer that control valve is closed

Step 8: Cool container from safe distance.

    a. Withdraw hoselines

    b. Apply straight stream to container

Step 9: Report to officer completion of assigned task.

Step 4: Cool cylinder or storage tank.

    a. Apply straight stream to container

Step 5: Extend hoselines to isolate control valve.

    a. Approach upwind and uphill

    b. Approach container from the side

    c. Push flame away from valve with fog stream (30-degree pattern)

NOTE: If team is unable to push flame away from the valve, immediately withdraw to a safe location and continue to cool the container.

Step 6: Maintain situational awareness.

Loss Control

Chapter Contents

Key Terms

NFPA® Job Performance Requirements

This chapter provides information that addresses the following job performance requirements of NFPA® 1001, *Standard for Fire Fighter Professional Qualifications* (2013).

Firefighter I

5.3.10

5.3.13

5.3.14

5.5.1

1. Explain the philosophy of loss control. (5.3.14)

2. Describe the ways preincident planning impacts loss control. (5.3.14)

3. Determine appropriate salvage procedures. (5.3.14)

4. Compare and contrast different types of salvage covers. (5.3.14)

5. Explain ways to fold, roll, spread, and improvise with salvage covers. (5.3.14)

6. Describe ways to cover openings during salvage operations. (5.3.14)

7. Explain methods used to maintain fire safety during overhaul. (5.3.13)

8. Describe factors that influence locating hidden fires. (5.3.10, 5.3.13)

9. Identify different overhaul procedures. (5.3.13)

10. Indicate the ways a thermal imager can be used during overhaul. (5.3.13)

11. Clean, inspect, and repair a salvage cover. (Skill Sheet 18-I-1; 5.3.14)

12. Roll a salvage cover for a one-firefighter spread. (Skill Sheet 18-I-2; 5.3.14)

13. Spread a rolled salvage cover — One-firefighter method. (Skill Sheet 18-I-3; 5.3.14)

14. Fold a salvage cover for a one-firefighter spread. (Skill Sheet 18-I-4; 5.3.14)

15. Spread a folded salvage cover — One-firefighter method. (Skill Sheet 18-I-5; 5.3.14)

16. Fold a salvage cover for a two-firefighter spread. (Skill Sheet 18-I-6; 5.3.14)

17. Spread a folded salvage cover — Two-firefighter balloon throw. (Skill Sheet 18-I-7; 5.3.14)

18. Construct a water chute without pike poles. (Skill Sheet 18-I-8; 5.3.14)

19. Construct a water chute with pike poles. (Skill Sheet 18-I-9; 5.3.14)

20. Construct a catchall. (Skill Sheet 18-I-10; 5.3.14)

21. Make a chute and attach it to a catchall. (Skill Sheet 18-I-11; 5.3.14)

22. Locate and extinguish hidden fires. (Skill Sheet 18-I-12; 5.3.10, 5.3.13)

Chapter 18
Loss Control

Courtesy of Iowa Fire Service Training Bureau.

Case History

A fire department added a fire scene air-monitoring policy to its standard operating procedures, requiring air-monitoring of the environment after any working fire. After implementing the policy, the department responded to two structure fires.

The first fire involved an eight-unit apartment building with four apartments on the first floor and four on the second connected by a central stairwell. The fully-involved fire was confined to a second-story apartment. Air-monitoring was conducted after extinguishment and ventilation. All units in the building were monitored prior to allowing occupants to return to their apartments.

The results of the air-monitoring surprised everyone. In a first-floor apartment on the opposite side of the building from the fire, with no visible smoke or strong odor, dangerous levels of hydrogen cyanide were detected. Despite the lack of visible smoke or odor, extensive ventilation had to be performed before it was deemed safe to allow the occupants to return.

The second incident involved a kitchen fire. A pot on the stove ignited, catching the cabinet above it on fire. The fire was out upon arrival, but crews reported light to moderate smoke inside the apartment. Air-monitoring was conducted and extremely hazardous levels of Carbon Monoxide (CO) and Hydrogen Cyanide (HCN) were detected. If fire fighting crews had not been wearing their SCBAs, they would have been exposed to the toxic gases very quickly.

These incidents illustrate that conducting air-monitoring and wearing full PPE, including SCBA, at all fires is extremely important. Only air-monitoring can ensure that it is safe to work in an environment without wearing an SCBA. The lack of visible smoke and smoke odor is not a guarantee of safety. Concentrations of CO, HCN, and other toxic gases generated by even the smallest fire will be present and can travel into apparently unaffected areas of a structure. In some cases, concentrations could be lethal. Even a small fire confined to a stove and cabinet in a kitchen can be deadly.

Property conservation is second only to life safety in the fire service. Firefighters must do everything they can to protect property without compromising their first priority, protecting life and health. Unfortunately, fire and smoke are not the only factors that contribute to property damage. If not carefully performed, fire suppression activities can cause more damage than the initial fire.

Protecting property requires firefighters to practice good loss control techniques during all phases of an incident. This chapter describes the philosophy behind loss control and describes two traditional tactics that firefighters use: salvage and overhaul.

▐ I ▐ Philosophy of Loss Control

Loss control is a term used in the fire service to describe the activities performed before, during, and after a fire has been extinguished to minimize losses to property. Quality loss control practices are a sign of professionalism and exhibit the good customer service for which the fire service is renowned. Properly applied loss control activities include:

- Minimizing damage to the structure, exposures, and contents
- Eliminating the chance that a fire will reignite in the structure
- Reducing the amount of time needed to repair and reopen the business
- Reducing the stress on the owner/occupants of the structure
- Creating goodwill for the fire department within the community
- Minimizing financial loss for the owner, occupant, insurance company, and the community

There are two types of damage that result from a structure fire (**Figures 18.1a and b**). **Primary damage** is caused by the fire and smoke. **Secondary damage** results from fire suppression activities such as forcible entry, ventilation, and fire extinguishment operations. The vulnerability of a structure to weather and vandalism following fire suppression is also considered a form of secondary damage.

Loss control consists of two tactics intended to reduce property damage: **salvage** and **overhaul**. When well-coordinated, salvage and overhaul can be very effective at reducing the overall damage to a structure and its contents due to fire (**Figures 18.2a and b**). Salvage consists of those operations associated with fire fighting that aid in reducing primary and secondary damage during fire fighting operations. Overhaul consists of operations involved in searching for and extinguishing hidden or remaining fires after the main body of the fire is extinguished.

Preincident Planning for Loss Control

Effective loss control depends on good planning. When conducting a preincident survey, special loss control-related concerns should be identified and addressed. While it may not be your responsibility to develop loss control preincident plans, you will be required to implement the plan as directed by the Incident Commander (IC). Examples of loss control concerns that should be identified during a preincident plan include:

- The most effective and least destructive means of gaining access to the structure (lockboxes, etc.).
- The most effective means of evacuating or protecting building occupants during a fire.

Loss Control — Practice of minimizing damage and providing customer service through effective mitigation and recovery efforts before, during, and after an incident.

Primary Damage — Damage caused by a fire itself and not by actions taken to fight the fire.

Secondary Damage — Damage caused by or resulting from actions taken to fight a fire and leaving the property unprotected.

Salvage — Methods and operating procedures by which firefighters attempt to save property and reduce further damage from water, smoke, heat, and exposure during or immediately after a fire; may be accomplished by removing property from a fire area, by covering it, or by other means.

Overhaul — Operations conducted once the main body of fire has been extinguished; consists of searching for and extinguishing hidden or remaining fire, placing the building and its contents in a safe condition, determining the cause of the fire, and recognizing and preserving evidence of arson.

Figure 18.1a Primary damage results directly from fire and smoke. *Courtesy of Donny Howard.*

Figure 18.1b Secondary damage results from the activities taken to access and suppress fire. Courtesy of Bob Esposito.

Figure 18.2a Salvage operations serve to protect property and minimize the fuel load available to a fire.

Figure 18.2b Overhaul operations search for and extinguish fires in hidden areas.

- The location of vital business records (computer servers, filing cabinets, vaults) in the structure and how to best protect them during a fire.

- When and how built-in fire suppression systems, such as automatic sprinklers and standpipes, are to be supported and used.

- How building contents are to be protected from smoke and water damage (salvage covers, plastic sheeting, removal from the structure, etc.).

Figure 18.3 High-value contents that may be permanently damaged by water and smoke should be protected by special preincident plans.

Special preincident plans should be developed for buildings with high-value contents that are especially susceptible to water and smoke damage. High-value contents include electronic equipment, computer systems, art work, and documents, among others **(Figure 18.3)**.

In residential occupancies, preincident plans should include covering upholstered furniture, bedding, and other water-absorbent objects. Plans should also be made to protect items such as photographs, important documents, computer equipment, and artwork that could be of particular monetary or sentimental value to the residents. Protecting these items may involve covering them, moving them to a dry unaffected area in the house, or removing them from the structure altogether.

In commercial occupancies, preincident plans should reflect an awareness of the value of contents vital to business survival. The business owner or representative is an excellent resource when determining items of vital importance in a structure. For example, the owner may accept losing the structure itself as long as proprietary items such as client information and financial records are preserved. Some businesses never recover from a fire due to the loss of such critical items, even if they are fully insured.

Interacting with the business owner or representative is also a good opportunity for fire department personnel to recommend continual loss control practices in the facility. For example, suggesting loss-control measures, such as inventory protection or storing duplicate copies of records off site, can benefit both the business and the fire department.

Salvage

Salvage is perhaps the most important aspect in fire department loss control. In addition to preincident planning, proper salvage operations require knowledge of the procedures needed as well as the tools and equipment necessary to perform the job. Improvisation is often necessary when presented with unique situations and limited equipment. In addition, the protection of damaged property from weather and trespassers is also critical.

Salvage Procedures

Salvage operations begin upon arrival and continue until the last unit leaves the scene. In situations where the on-scene resources are sufficient and the situation permits, the IC may order salvage operations to be conducted at the same time suppression activities are underway. In some instances, the contents of the room(s) immediately below the fire floor can be protected with salvage covers while fire suppression operations are being conducted on the floor above. In other instances, it may even be necessary to delay suppression activities for a short time in order to remove vital contents. In these cases, the financial loss caused by greater primary damage is often small compared to the cost of secondary damage due to suppression efforts. The IC should make any decisions about delaying fire suppression after taking into consideration the preincident plan, fire development, and any unsafe structural conditions.

The choice of salvage procedures will depend on:

- Number of personnel available
- Extent and location of the fire
- Type, size, and quantity of the contents
- Current weather conditions

Salvage procedures include:

- Moving contents to a safe location in the structure
- Removing contents from the structure
- Protecting the contents in place with salvage covers (**Figure 18.4**)

One useful salvage technique is to move contents inside the structure to areas that are away from concentrations of smoke, not in danger of fire extension, and where water will not spread. This method is best used when the fire is limited and not likely to spread, and weather conditions would damage contents if they were moved outside. It may still be necessary to cover these contents with salvage covers or raise them off the floor as a precaution against secondary damage.

Removing contents from the structure will help protect them from further primary and potential secondary damage. However, this method may interfere with fire suppression and ventilation crews that are using the same doors to enter the structure. Contents should be stacked on surfaces that are dry, such as a parking lot or driveway, and not near areas where firefighters may be collecting debris for disposal. Contents that are stored outside must also be protected from theft or vandalism once

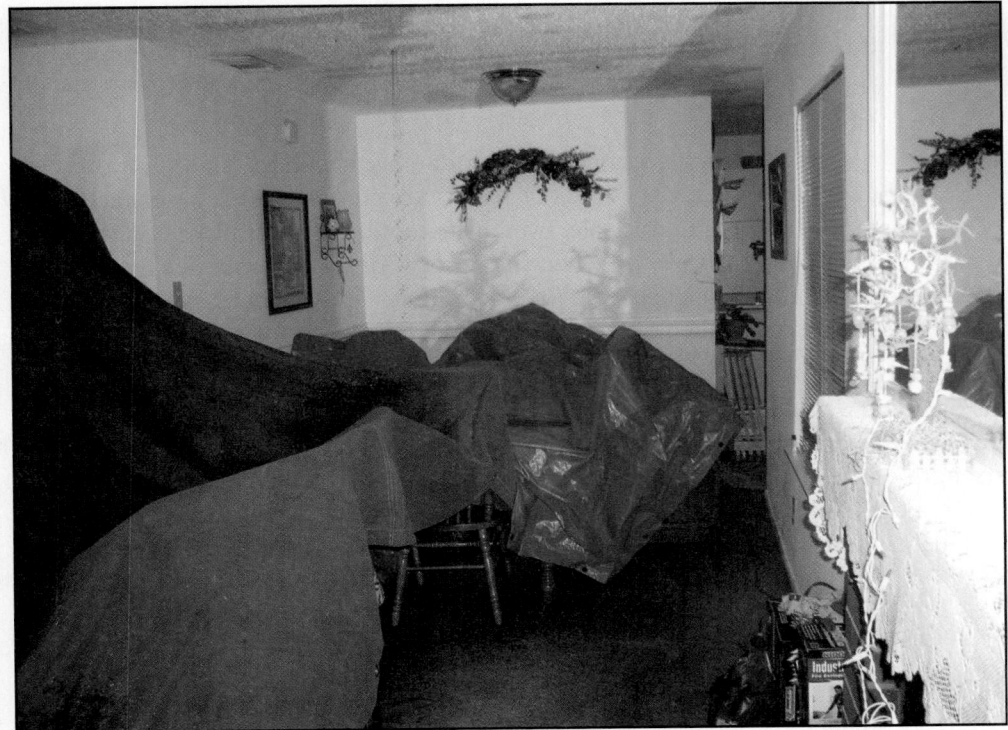

Figure 18.4 Salvage operations include protecting a building's contents where they are instead of having to move them to another location. *Courtesy of Ron Moore and McKinney (TX) FD.*

the fire is extinguished. The owner/occupant must be made aware that the contents have been stored outside or the contents should be secured in some fashion.

The method most often used to protect contents is to leave them in the room in which they are found. Contents are gathered into compact piles that can be covered with a minimum of salvage covers. Grouping contents in this manner allows more items to be protected than if they were covered in their original position. If possible, group household furnishings in the center of the room when arranging for salvage. In many cases, one salvage cover can protect the contents of one residential room. If the floor covering is a removable rug, slip the rug out from under furniture as each piece is moved and roll the rug to make it easier to move.

Figure 18.5 Salvage operations should prevent water damage caused to porous items resting on a floor that may become saturated during fire fighting operations.

When arranging grouped items in a residence, a dresser, chest, or high object should be placed at the end of a bed with other furniture grouped close by. Creating one high point in the furniture group allows water to run off without it collecting in depressions. Pictures, curtains, lamps, clothing, and other fragile items can be placed on the bed. It may sometimes be necessary to place a salvage cover into position before some articles are placed on the bed. The first cover protects the bed while the second cover protects the contents placed upon the bed.

Furniture sitting on wet carpet can absorb water and become damaged even though it is well covered. To prevent this damage, furniture should be raised off the wet floor with water-resistant materials such as precut plastic or foam blocks **(Figure 18.5)**. If blocks are not available, you can improvise with items such as canned goods from the kitchen.

Commercial occupancies present challenges for firefighters who are trying to perform salvage functions. It may be difficult to cover contents in these occupancies when large stocks and display features are involved. In addition, display shelves are frequently built to the ceiling and directly against the wall. This construction feature makes contents difficult to cover because when water flows down a wall, it naturally comes into contact with shelving and wets the contents. Contents stacked too close to the ceiling also present a salvage problem. Ideally, there should be enough space between the inventory and ceiling to allow firefighters to easily apply salvage covers.

Stock that is susceptible to water damage should be placed off the floor to prevent saturation. Skids or pallets are commonly used for stacking in these instances if available. Even if susceptible stock is placed off the floor, it still must be covered. However, when the number of salvage covers is limited, it is good practice to use available covers for water chutes and catchalls even though the water must be routed to the floor and removed later.

Firefighters must be extremely cautious of high-piled stock such as boxed materials or rolled paper that has become wet at the bottom. The wetness often causes the material to expand and push out interior or exterior walls. Wetness also reduces the strength of the material and may cause the piles to collapse. Some rolls of paper can weigh a ton (900 kg) or more. If one of these rolls were to fall on firefighters it could seriously injure or kill them.

You can remove large quantities of water in the following ways:

- Locate and clean clogged drains.
- Remove toilet fixtures.

- Create **scuppers**.
- Make use of existing sanitary piping systems.
- Create chutes made of salvage covers, plastic, or other available materials to route water into other areas.

Water left on cabinets and other horizontal surfaces may ruin finishes over a period of hours. Wiping cabinets and tabletops with disposable paper towels is a quick and easy way to save the building owner/occupant a great deal of potential loss.

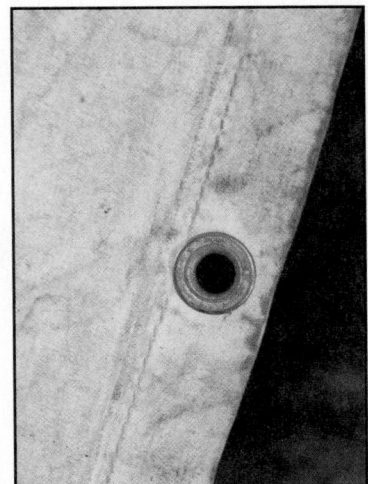

Scupper — Form of drain opening provided in outer walls at floor or roof level to remove water to the exterior of a building in order to reduce water damage.

Salvage Covers and Equipment

Depending on the size and organization of the fire department, salvage operations are generally assigned to ladder companies and specially designated salvage and overhaul companies. Engine companies will carry salvage covers, some hand tools, and buckets used for salvage operations. All firefighters are trained in the use of salvage covers and equipment and should be familiar with the type and locations of all salvage equipment used by the department.

Salvage Covers

Salvage covers are made of waterproof canvas or vinyl and are manufactured in various sizes **(Figure 18.6)**. These covers have reinforced corners and edge hems into which grommets are placed for hanging or draping. Vinyl synthetic covers are lightweight, easy to handle, economical, and practical for both indoor and outdoor use.

Figure 18.6 Salvage covers are made of canvas or vinyl and feature grommets and reinforced edges.

Many departments use disposable heavy-duty plastic covers. The plastic is available on rolls and can be extended as needed to cover large areas. Covers can also be cut from the rolls in different shapes and sizes as needed.

Salvage Cover Maintenance

Properly cleaning, drying, and repairing of salvage covers increases their service life **(Figure 18.7)**. Typically, the only cleaning required for canvas salvage covers is wetting or rinsing with a hose stream and scrubbing with a broom. Extremely dirty or stained covers may be scrubbed with a detergent solution and then thoroughly rinsed. Some foreign materials, such as roofing tar, are difficult to remove (even with a detergent) when they have been allowed to dry on the cover.

Canvas salvage covers should be clean and completely dry before they are folded and stored on the apparatus. This practice is essential to prevent mildew and rot. Permitting canvas salvage covers to dry when they are dirty is not a good practice because carbon and ash stains can rot the canvas. It is typically acceptable to dry salvage covers outdoors, but avoid doing so when it is windy. **Skill Sheet 18-I-1** describes how to clean, inspect, and repair a salvage cover.

NOTE: Long-term exposure of canvas to sunlight will result in damage from ultraviolet rays. Drying in direct sunlight may degrade the material over time.

Figure 18.7 Maintenance of salvage covers reduces the spread of contamination and increases service life.

Synthetic salvage covers do not require as much maintenance as those made of canvas. Synthetic covers may be folded when wet, but it is usually better to let them dry first so they will not mildew.

After salvage covers are dry, they should be examined for damage. To inspect for holes, three or four firefighters stand side-by-side along one end of the cover. The firefighters pick up the end and pass it back over their heads while walking toward the other end, looking up at the underside of the cover. Light will show through even the smallest holes. Mark any holes with chalk or a marking pen (**Figure 18.8**). Use chalk for canvas salvage covers and a marking pen for vinyl covers. Depending upon the cover's material, firefighters can place duct tape or mastic tape over holes to repair them or patch the cover with iron-on or sew-on patches.

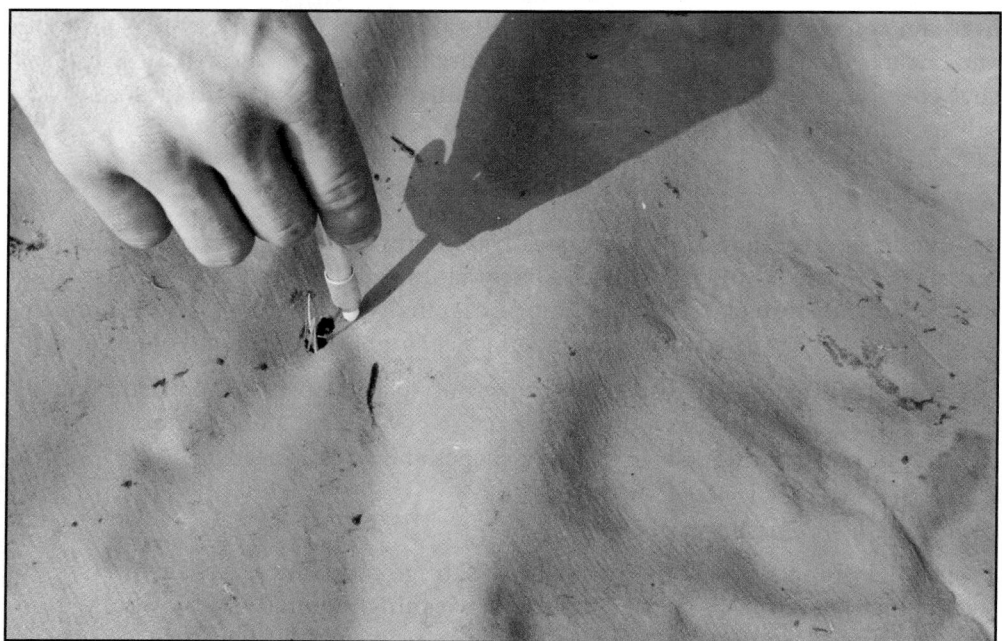

Figure 18.8 Mark any holes in a salvage cover with an approved marking method.

Salvage Equipment

Performing salvage operations requires a specific collection of tools and equipment. Small tools and equipment should be stored in a specially designated salvage toolbox or other containers to make them easier to carry. Salvage materials and supplies may be kept in a plastic tub and brought into the structure as soon as possible. The tub itself provides a useful water-resistant container to protect items such as computers, pictures, and other water-sensitive materials.

Typical tools and equipment used in salvage operations include electrical, mechanical, plumbing, and general carpentry tools. Mops, squeegees, and buckets are also useful for removing water.

Automatic sprinkler kit. The tools in a sprinkler kit are used to stop the flow of water from an open sprinkler. A flow of water from an open sprinkler can do considerable damage to merchandise on lower floors after a fire has been controlled in a commercial building. Sprinkler tongs or stoppers and wooden sprinkler wedges are suggested tools for a sprinkler kit (**Figures 18.9a-c**).

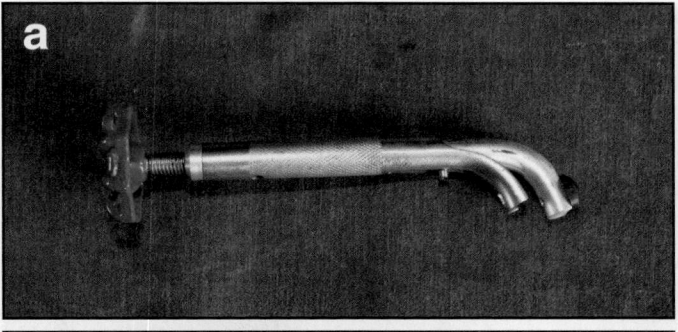

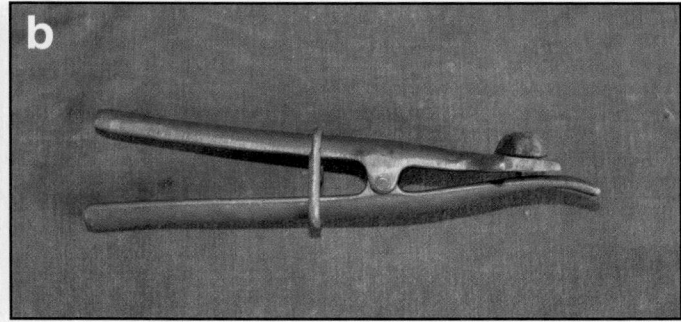

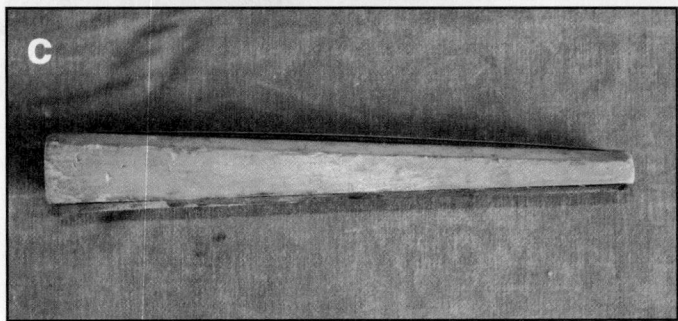

Figures 18.9a-c Activated sprinklers can be stopped using several tools including a) sprinkler stoppers, b) sprinkler tongs, and c) wooden sprinkler wedges.

Routing Water Out of a Sprinklered Building

After the sprinkler system main control valve is shut, water will continue to flow from open sprinklers. Catchalls and chutes can be used to capture and route the water out of the building. In addition, where only one sprinkler is flowing, a 2½- or 3-inch (65 mm or 77 mm) hoseline can be fitted over the sprinkler and directed out a door or window.

NOTE: Responsibility for restoring automatic sprinkler systems to service is determined by the authority having jurisdiction. You may only restore systems if you are authorized and trained to do so.

Carryalls. **Carryalls** are used to carry debris, catch falling debris, and provide a water basin for immersing small burning objects. Carryalls should be constructed of nonflammable material (**Figure 18.10, p 1112**).

Floor runners. Firefighters often unintentionally damage flooring with their boots and equipment during fire suppression operations. These floor coverings may be protected by using floor runners. Floor runners can be unrolled from an entrance to almost any part of a building (**Figure 18.11, p. 1112**). Commercially prepared vinyl-laminated nylon floor runners are lightweight, flexible, tough, heat and water resistant, and easy to maintain.

Dewatering devices. Dewatering devices are pumps used to remove water from basements, elevator shafts, and sumps. Portable pumps capable of passing grey water filled with debris, jet-siphons, and submersible pumps are best suited for salvage operations (**Figure 18.12, p. 1113**). These devices can be moved to any point where a line of hose can be placed and an outlet for water can be provided.

Water vacuum. One of the easiest and fastest ways to remove water is with the use of a water-vacuum device. These devices can be used to dewater floors, carpets, and other areas where the water is not deep enough to be picked up by a submersible pump or siphon ejector. The water vacuum appliance consists of a tank (worn on the

Carryall — Waterproof carrier used to carry and catch debris or used as a water sump basin for immersing small burning objects.

Figure 18.10 Carryalls catch and carry debris and may also be used as a water basin during salvage operations.

Figure 18.11 Floor runners protect delicate flooring from water, heat, and equipment used during fire suppression activities.

back or placed on wheels) and a nozzle. Backpack-type tanks normally have a capacity of 4 to 5 gallons (15 L to 20 L) and can be emptied by pulling a lanyard that empties the water through the nozzle or through a separate drain hose. Floor models on rollers may have capacities up to 20 gallons (80 L) (**Figure 18.13**).

J-hooks. J-hooks are designed to be driven into walls or wooden framing to provide a strong point from which to hang objects. They are most often used to hang salvage covers on walls to protect wall-mounted book cases and other shelving units.

S-hooks. S-hooks are most often used for the same purpose as J-hooks, but in a slightly different way. S-hooks cannot be driven into walls or framing but must have a horizontal ledge (such as the top of a secure shelving unit or structural member) from which to hang.

Folding/Rolling and Spreading Salvage Covers

A key factor in successful salvage operations is the proper folding and deploying of salvage covers. The following sections describe the basic methods firefighters can use to fold and deploy salvage covers.

Figure 18.13 Water-safe vacuums remove water that is too shallow for a submersible pump.

Figure 18.12 Dewatering devices include pumps and siphons that move water from one location to another.

One-Firefighter Spread with a Rolled Salvage Cover

The main advantage of the one-firefighter salvage cover roll is that one person can quickly unroll a cover across the top of an object **(Figure 18.14, p. 1114)**. **Skill Sheet 18-I-2** describes the procedure for two firefighters to roll a salvage cover so that it can be deployed by one firefighter. A salvage cover rolled for a one-firefighter spread may be carried on the shoulder or under the arm. Use the steps described in **Skill Sheet 18-I-3** for a one-firefighter spread with a rolled salvage cover.

One-Firefighter Spread with a Folded Salvage Cover

Some departments prefer to carry salvage covers that have been folded as opposed to rolled **(Figure 18.15, p. 1114)**. The procedures in **Skill Sheet 18-I-4** describe the steps for folding a salvage cover for one-firefighter deployment. Two firefighters are needed to make this fold, performing the same functions simultaneously. Carrying a folded salvage cover on the shoulder is typically most convenient but any safe carrying method is acceptable. **Skill Sheet 18-I-5** illustrates the procedure for deployment of a folded salvage cover by one firefighter.

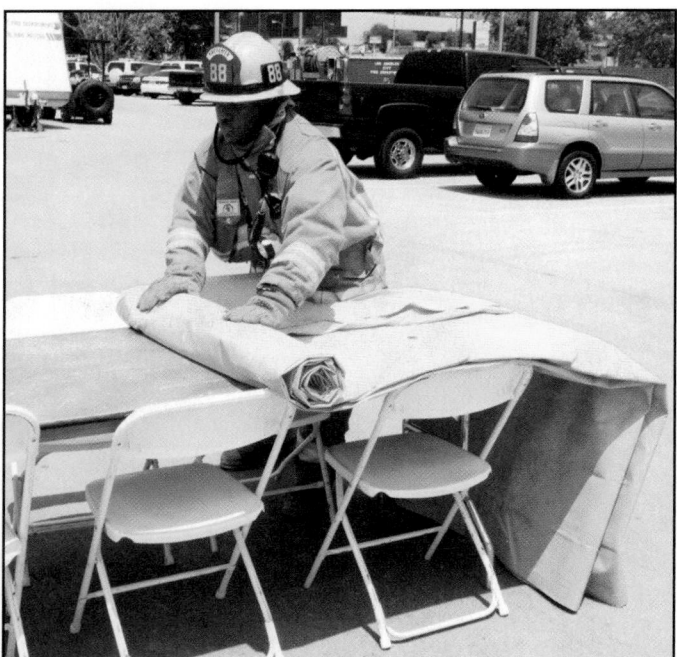

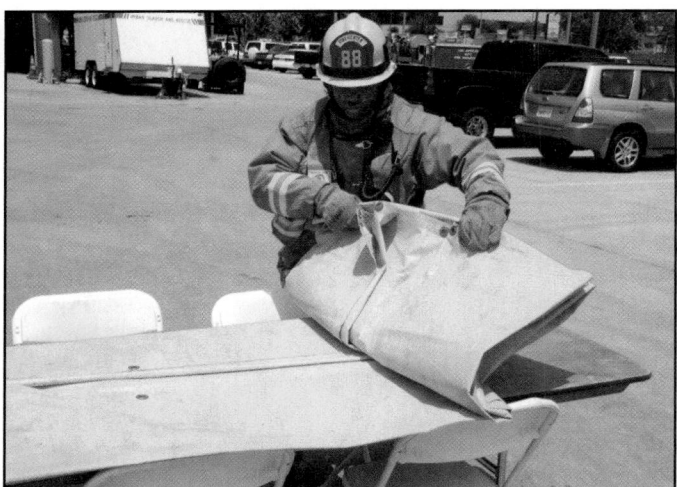

Figure 18.15 Salvage covers are often carried folded instead of rolled.

Figure 18.14 The one-firefighter salvage cover roll enables one person to protect an object without assistance.

Two-Firefighter Spread with a Folded Salvage Cover

A single firefighter cannot easily handle a large salvage cover. Therefore, these covers should be folded for two-firefighter deployment. The procedure in **Skill Sheet 18-I-6** can be used to make the two-firefighter salvage cover fold. The most convenient way to carry this fold is on the shoulder with the open edges next to the neck. It makes little difference which end of the folded cover is placed in front of the carrier because two open-end folds will be exposed. Position the cover so that the firefighter carrying it holds the lower pair of corners and the second firefighter holds the uppermost pair.

A method called the *balloon throw* is most commonly used for two firefighter deployment of a large salvage cover (**Figure 18.16**). The balloon throw works best when sufficient air is pocketed under the cover. This pocketed air gives the cover a parachute effect that allows it to float into place over the article to be covered. **Skill Sheet 18-I-7** highlights the procedure for making the balloon throw.

Improvising with Salvage Covers

While salvage covers are typically used to cover building contents, they may also be used to catch and route water from fire fighting operations or other structural flooding situations. The following sections describe improvised uses of salvage covers.

Removing Water with Chutes

A water chute is one of the most practical methods of removing water that comes through the ceiling from upper floors. Water chutes may be constructed on the floor below fire fighting operations to drain runoff out of the structure through windows or doors (**Figure 18.17**). Some fire departments carry prepared chutes, approximately 10 feet (3 m) long, as regular equipment, but others find it more practical to construct

Figure 18.16 The balloon throw is commonly used as a deployment of a salvage cover by two firefighters.

Figure 18.17 Water chutes may be deployed on the floor below the fire to drain runoff.

chutes when and where needed using floor runners or one or more covers. Plastic sheeting, a heavy-duty stapler, and duct tape can be used to construct water diversion chutes. **Skill Sheets 18-I-8** and **18-I-9** describe procedures for constructing water chutes.

Constructing a Catchall

Catchalls are constructed from a salvage cover that has been placed on the floor to hold small amounts of water. The catchall may also be used as a temporary means to control large amounts of water until chutes can be constructed to route the water outside the structure. Properly constructed catchalls can hold several hundred gallons (liters) of water and often save considerable time during salvage operations. The steps required to make a catchall are shown in **Skill Sheet 18-I-10**. In order to catch as much water as possible, place the cover into position as soon as possible, even if the sides of the cover have not been uniformly rolled. Two firefighters are usually needed to construct a catchall **(Figure 18.18)**.

Splicing Covers

When objects or groupings are too large to be covered using a single cover or when long chutes or catchalls need to be made, it will be necessary to splice covers with watertight joints. There are many different methods for splicing salvage covers, and your department will train you on the specific procedure to be used. Many departments use disposable rolled plastic sheeting that can be cut to size as needed. The use of plastic rolls saves time and property because it eliminates the need for splicing and reduces the risk of leakage.

Splicing a Chute to a Catchall

A plan should be developed to remove water from a catchall as soon as it is constructed, especially if it appears that the volume of water will be greater than the catchall's capacity. Submersible pumps may be used if available and there is a significant and constant flow of water into the catchall. A commonly used method of water removal is to splice a water chute to the catchall **(Figure 18.19)**. An advantage to this system is that as soon as water accumulates in the catchall it is drained to the outside. The steps to splice a chute to a catchall are shown in **Skill Sheet 18-I-11.**

Covering Openings

Another critical aspect of salvage operations is covering openings to prevent further damage to the property by weather and trespassers. Doors or windows that have been broken or removed during suppression activities should be covered with plywood, heavy plastic, or some similar materials to keep out rain **(Figure 18.20)**. Plywood, hinges, a hasp, and a padlock can be used to fashion a temporary door. Openings in roofs should be covered with plywood, roofing paper, heavy plastic sheeting, or tar paper. Use appropriate roofing nails if roofing, tar paper, or plastic is used. Place strips of lath along the edges of the material and nail them in place. Covering openings cut in floors of upper stories or over basements and crawl spaces is very important. These openings must be covered with lumber or thick plywood that will support a person's weight.

Figure 18.18 A catchall may hold a significant amount of water and usually requires two firefighters to be constructed.

Figure 18.19 A water chute can function as a drain when spliced onto a catchall.

Figure 18.20 After salvage operations conclude, openings should be protected in a way to prevent damage from weather and trespassers.

Overhaul

Overhaul includes the following activities that are conducted once the main body of fire has been extinguished:

- Searching for and extinguishing hidden or remaining fire
- Placing the building and its contents in a safe condition
- Determining the cause of the fire
- Recognizing and preserving evidence of arson

The IC and the individual responsible for fire investigation should authorize when overhaul should begin. Once the order is given, firefighters should attempt to put the building, its contents, and the fire area in as safe and habitable condition as possible.

Salvage operations performed during fire fighting will directly affect any overhaul work that may be needed later. Many of the tools and equipment used for overhaul are the same as those used for forcible entry, ventilation, and salvage operations. Some of the tools and equipment used specifically for overhaul, along with their uses, may include the following:

- **Pike poles and plaster hooks** — Open ceilings to inspect for fire extension.
- **Axes** — Open walls and floors.
- **Prying tools** — Remove door frames, window frames, and baseboards (**Figure 18.21**).
- **Power saws, drills, and screwdrivers** — Install temporary doors and window coverings.
- **Carryalls, buckets, and tubs** — Carry debris or provide a basin for immersing smoldering material.
- **Shovels, bale hooks, and pitchforks** — Move baled or loose materials.
- **Thermal imager (TI)** — Check void spaces and look for hot spots.

A supervisor or officer not directly engaged in overhaul should direct overhaul operations. If a fire investigator is on the scene, he or she should be involved in planning and supervising the overhaul activities to avoid disturbing potential evidence needed to determine fire cause.

Thermal Imager — Electronic device that forms images using infrared radiation. *Also known as as a* thermal imaging camera.

Figure 18.21
Prying tools serve many purposes including the removal of hazards and constrictions in building openings.

Fire Safety During Overhaul

The first consideration before beginning overhaul operations is safety. After a fire has been brought under control, there is time to plan and organize overhaul activities. The overhaul plan should provide the highest possible degree of safety to firefighters and others who might be allowed on the scene. The steps required to establish safe conditions include the following:

- Inspecting the premises

- Developing an operational plan

- Providing needed tools and equipment

- Eliminating or mitigating hazards (including securing any remaining utilities)

- Monitoring the atmosphere for carbon monoxide (CO) and hydrogen cyanide (HCN) levels before removing SCBA

Toxic gases that continue to be produced from a smoldering fire are a significant threat to firefighters during overhaul operations. Even if the air in a structure appears to be without smoke, toxic products of combustion can still exist in dangerous concentrations. Carbon monoxide (CO) and hydrogen cyanide (HCN) are commonly encountered toxic gases and countless others can also be present depending on the building contents involved in the fire. Air monitoring devices should be deployed based on your department's SOPs, and all fire department personnel should continue to use their self-contained breathing apparatus (SCBA) until the atmosphere in the structure has been determined acceptable **(Figure 18.22)**. Once air monitoring confirms that SCBAs can be safely removed, firefighters should wear particulate masks to protect them from nontoxic, airborne particles. In addition, property owners/occupants should not be allowed to enter the structure until the atmosphere has been deemed acceptable.

Figure 18.22 Air-monitoring activities should be continued through all phases of response to a fire incident, and the use of SCBA should continue until the atmosphere is safe.

///////////////////////////////////////

CAUTION
Wear proper protective clothing including self-contained breathing apparatus (SCBA) until the atmosphere has been proven safe.

Many other hazards exist for firefighters performing overhaul. For instance, personnel may fall and become injured when fire-weakened floors collapse or fail. When any potentially hazardous areas are identified, they should be marked or barricaded immediately. Firefighters can also be injured by stepping on broken glass, nails, or other sharp objects. Because fire debris must often be handled during overhaul, firefighters are susceptible to cuts, punctures, and thermal burns if they are not wearing gloves. Likewise, eye protection is critical to avoid injuries to the eyes. Injuries such as strains and sprains can be prevented through physical conditioning and practicing safe lifting techniques.

Another preventable cause of injury is fatigue. Exhausted firefighters are more susceptible to injury than those who are rested. Firefighters who were not directly involved in the rescue and fire control efforts should conduct overhaul operations if resources allow.

Figure 18.23 Charged hoselines continue to play an important role during overhaul operations.

Due to the threat of reignition, charged hoselines should still be present during overhaul operations. Typically, 1½-inch (38 mm) or 1¾-inch (45 mm) attack lines are used for overhaul (**Figure 18.23**).

At least one attack line should still be available in the event of a rekindle. Regardless of the type of hose being used, place the nozzle so that it will not cause additional water damage. In addition, hoselines should be constantly monitored for leakage, especially at the couplings. Using a 100-foot (30 m) section of hose as the first section on attack lines greatly reduces the chances that any couplings other than those at the nozzle would even be inside a building.

To protect yourself during overhaul operations, you must continue to maintain your situational awareness and focus on safety. The following are some additional safety considerations during overhaul operations:

- Continue to work in teams of two or more.
- Maintain awareness of available exit routes.
- Maintain a rapid intervention crew or team (RIC/RIT) throughout the operation.
- Monitor personnel for the need for rehabilitation.
- Beware of hidden gas or electrical utilities.
- Continue using the accountability system until the incident is terminated.

Locating Hidden Fires

Before starting a search for hidden fires, evaluate the structural condition of the area to be searched. The intensity of the fire and the amount of water used for its control are two important factors that affect the condition of the building. The intensity of the fire determines the extent to which structural members have been weakened. The amount of water used determines the additional weight placed on floors and walls due to the absorbent properties of the building contents. These factors should be considered along with appropriate measures to ensure firefighter safety during overhaul.

When evaluating the stability of a fire-damaged structure, look for the following indicators of possible loss of structural integrity:

- Weakened floors due to floor joists being burned away
- Concrete that has spalled due to heat
- Weakened steel roof members
- Walls offset because of elongation of steel roof supports
- Weakened roof trusses due to burn-through of key members
- Mortar in wall joints opened due to excessive heat
- Wall ties holding veneer/curtain walls melted from heat
- Heavy storage on mezzanines or upper floors
- Water pooled on upper floors
- Large quantities of wet insulation

Firefighters can often detect hidden fires by sight, touch, sound, or electronic sensors **(Figure 18.24)**. The following are some of the indicators for each:

- Sight:
  - — Discoloration of materials
  - — Peeling paint
  - — Smoke emissions from cracks
  - — Cracked plaster
  - — Rippled wallpaper
  - — Burned areas
- Touch:
  - — Heat felt through walls and floors
- Sound:
  - — Popping or cracking of fire burning
  - — Hissing of steam
- Electronic sensors:
  - — Thermal (heat) signature detection with thermal imager
  - — Infrared heat detection

Overhaul Procedures

Overhaul typically begins in the area of most severe fire involvement. The process of looking for fire extension should begin as soon as possible after the IC gives the order. If the fire has extended to other areas in the structure, firefighters must determine the path through which it traveled (concealed wall spaces, unsealed pipe chases, etc.). If floor beams have burned at their ends where they enter a party wall, overhaul the ends by flushing the voids in the wall with water. Also inspect the other side of the wall to determine whether the fire or water has come through. Thoroughly check insulation materials because they can retain hidden fires for a prolonged period. It is usually necessary to remove insulation material in order to extinguish fire in it. Do not start making random openings in walls or ceilings without cause. Your actions must be justifiable.

Understanding the basic concepts of building construction will help you in searching for hidden fires. If the fire has burned around windows or doors, pull open these areas to expose the inner parts of the frame or casing and visually verify full extinguishment. When fire has burned around a combustible roof or cornice, it is advisable to open the cornice and inspect for hidden fires. In structures using balloon construction, check the attic and basement for fire extension.

NOTE: Review information in Chapter 4 concerning building construction.

It is often necessary to search for hidden fires in concealed spaces below floors, above ceilings, or within walls and partitions. In these instances, first move the furnishings of the room to locations where they will not be damaged. If it is not possible to move the contents, protect them with salvage covers. Remove only enough wall, ceiling, or floor covering to verify complete extinguishment **(Figure 18.25, p. 1122)**. Weight-bearing members should not be disturbed. Inspect wall openings such as electrical receptacles and switches, return air ducts, heating vents, and telephone and

Figure 18.24 Sensors extend a firefighter's ability to detect hidden fires.

Figure 18.25 Removal of ceiling, wall, or floor coverings should be conducted carefully with the aim of minimizing how much of the material is disturbed in the process.

cable connections for possible fire extension into the wall cavity. The walls and ceilings in kitchens, bathrooms, and utility rooms contain ventilation fans and pipes, ducts, and other passages that will permit fire to extend. If these rooms show evidence of fire spread, the walls and ceiling should be inspected.

When opening concealed spaces, consider whether the space contains electrical wiring, gas piping, or plumbing. Electrical outlets, gas connections, and water faucets are all indicators of the presence of utilities. Consideration should also be given to the future repair of the structure. While openings must often be made in construction to check for extension and allow extinguishment, they should be made in a neat and planned manner. This reduces the amount of work that will be necessary for future restoration and is a sign of professionalism on the part of the firefighter.

Ceilings may be opened from below using a pike pole or other appropriate overhaul tool. To open lath and plaster ceilings, first break the plaster and then pull off the lath. Some plaster ceilings have wire mesh imbedded in the plaster. When these ceilings start to come down, they may fall in one very large piece. Some newer plaster ceilings are backed with gypsum wallboard instead of wooden lath. Metal or composition ceilings may be pulled from the joists in a similar manner.

When pulling any ceiling, do not stand directly under the area to be opened. Always position yourself between the area being pulled and the doorway to keep the exit route from being blocked with falling debris. Always wear full PPE including respiratory protection when pulling ceilings.

CAUTION
When pulling any ceiling, stand clear of any falling debris.

Small burning objects are frequently uncovered during overhaul. Because of their size and condition, it is often more effective to submerge an entire object in containers of water than to drench it with hose streams. Bathtubs, sinks, lavatories, and wash tubs are all useful for this purpose. Large smoldering items such as mattresses, stuffed furniture, and bed linens should be taken outside the structure where they can be thoroughly extinguished **(Figure 18.26)**. Scorched or partially burned articles may prove helpful to an investigator in preparing an inventory or determining the cause of the fire. Firefighters need to work in close coordination with the fire investigator to ensure potential evidence is not disturbed.

The use of wetting agents such as Class A foam is of considerable value when extinguishing hidden fires. The penetrating qualities of wetting agents facilitate complete extinguishment in cotton, upholstery, and baled goods. The only way to ensure that fires in bales of items such as rags, cotton, and hay are extinguished is to break them apart. See **Skill Sheet 18-I-12** for more information on how to locate and extinguish hidden fires.

Figure 18.26 Large removable items that contribute to the fire load may be extinguished away from other combustibles.

Overhaul with Thermal Imagers (TIs)

Thermal imagers (TIs) identify the heat signature of items and project the resulting image onto a screen for the firefighter to see **(Figure 18.27)**. Firefighters can use TIs to find hidden fires in concealed spaces such as floors, ceilings, and walls without having to open up the areas and visually inspect them. This reduces the time needed to perform a search and limits the secondary damage to the structure.

Changing Terminology

Since thermal imaging technology first entered fire service use, the term *thermal imaging camera (TIC)* has been in common usage. Many of these devices looked and operated much like cameras, although some did not. However, the 2010 edition of NFPA® 1801, *Standard on Thermal Imagers for the Fire Service*, has standardized the term as *thermal imager (TI)*. Therefore, this is the term that will be used throughout this IFSTA manual.

Because of the way in which they operate, TIs sometimes do not provide quality images of items behind reflective materials such as metal, mirrors, and glass. In these instances, traditional methods should be used to reveal hidden fire. While TIs are extremely useful, they are only tools. If there are discrepancies between the image shown on the TI and signs of fire in a concealed space, the space should be opened up and inspected visually.

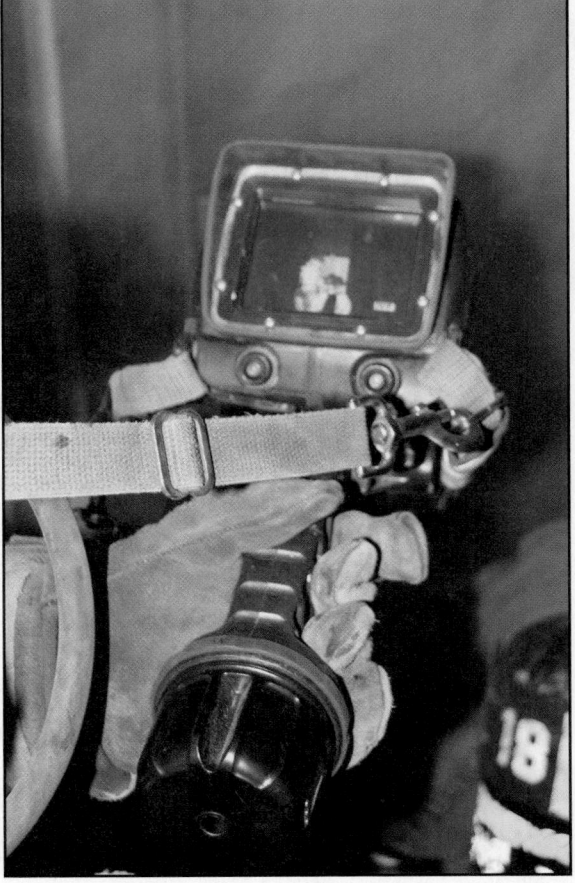

Figure 18.27 Thermal imagers use heat sensors to identify items and display an image on a screen to help firefighters identify the source of the heat.

Thermal Imagers Must Not Replace Your Senses

The TI is a tool and does not replace your senses. When in doubt, open up the structural member. The TI cannot see heat through a cooler object. Insulation may cover a heat signature and cause you to miss a small smoldering fire.

Chapter Summary

Loss control is an important component of fire department service delivery. The philosophy of loss control is to minimize secondary damage to structures and their contents during and after fire control operations. Salvage and overhaul operations are two of the most effective means of loss control. It is imperative that fire personnel identify and protect valuable contents in structures affected by fire. It is also important to search for areas of hidden fire to ensure that a rekindle of the fire does not occur. Taking a customer-service-oriented approach to loss control ensures that citizens' property is adequately protected and that the reputation of the fire service is held in the highest regard.

Review Questions

Firefighter I

1. How does the philosophy of loss control impact fire suppression?

2. In what ways can preincident planning influence loss control?

3. What is the best way to determine appropriate salvage procedures?

4. What are the different types of salvage covers commonly used in the fire service?

5. Why is it necessary to know several ways to fold, roll, spread, and improvise with salvage covers?

6. What ways can firefighters cover openings during salvage operations?

7. What methods can be used to maintain fire safety during overhaul operations?

8. How can a firefighter describe the factors that influence locating hidden fires?

9. What are some of the overhaul procedures used in the fire service?

10. How can using a thermal imager be useful during overhaul?

Step 1: Wash salvage cover with clean water and detergent by using a scrub brush.

Step 2: Rinse thoroughly with clean water.

Step 3: Hang to dry.

Step 4: Inspect salvage cover.

    a. Firefighters: Raise salvage cover at each corner

    b. Firefighter: Inspect underneath of cover for light coming through holes or tears

Step 5: Mark holes with chalk or marker.

Step 6: Patch according to manufacturer or departmental guidelines.

Step 7: Put away cleaning supplies and salvage cover according to departmental procedures.

NOTE: Two firefighters must make initial folds to reduce the width of the cover to form this roll. Steps 1 through 8 are performed simultaneously by both firefighters on opposite sides of the cover. Steps 9 through 12 may be performed by both firefighters who are stationed at the same end of the roll.

Step 1: Grasp the cover with the outside hand midway between the center and the edge to be folded.

Step 2: Place the other hand on the cover as a pivot midway between the outside hand and the center.

Step 4: Grasp the cover corner with the outside hand.

Step 5: Place the other hand as a pivot on the cover over the outside fold.

Step 6: Bring this outside edge over to the center, and place it on top of and in line with the previously placed first fold.

Step 3: Bring the fold over to the center of the cover; this creates an inside fold (center) and an outside fold.

Step 7: Fold the other half of the cover in the same manner by using Steps 1 through 6.

Step 8: Straighten the folds if they are not straight.

Step 9: Fold over about 12 inches (300 mm) at each end of the cover to make clean, even ends for the completed roll.

Step 10: Start the roll by rolling and compressing one end into a tight compact roll; roll toward the opposite end.

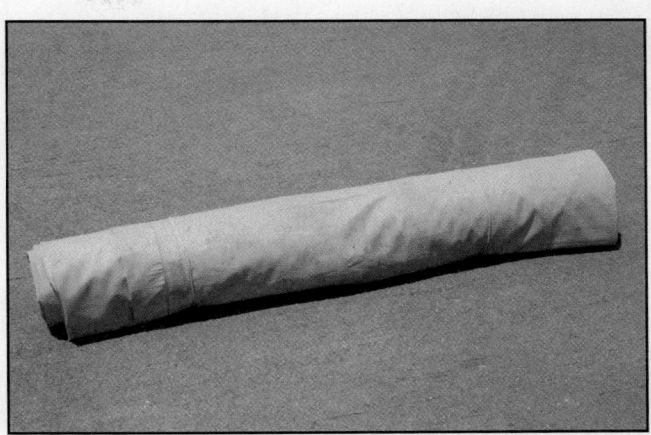

Step 11: Tuck in any wrinkles that form ahead of the roll as the roll progresses.

Step 12: Secure the completed roll with inner tube bands or Velcro® straps or tie with cords.

Step 13: Store salvage cover according to departmental procedures.

Step 1: Start at one end of the object to be covered.

Step 2: Unroll a sufficient amount to cover the end.

Step 3: Unroll toward the opposite end and let the rest of the roll fall into place at the other end.

Step 4: Stand at one end.

Step 5: Grasp the open edges where convenient, one edge in each hand.

Step 6: Open the sides of the cover over the object by snapping both hands up and out.

Step 7: Open the other end of the cover over the object in the same manner.

Step 8: Tuck in all loose edges at the bottom.

Step 9: Store according to departmental procedures.

18-I-4
Fold a salvage cover for a one-firefighter spread.

SKILL SHEETS

NOTE: Two firefighters must make initial folds to reduce the width of the cover. Steps 1 through 7 are performed simultaneously by both firefighters on opposite sides of the cover. Steps 8 through 13 may be performed by both firefighters who are stationed at the same end of the fold.

Step 1: Grasp the cover with the outside hand midway between the center and the edge to be folded.

Step 2: Place the other hand on the cover as a pivot midway between the outside hand and the center.

Step 4: Grasp the cover corner with the outside hand.

Step 5: Place the other hand as a pivot on the cover over the outside fold.

Step 6: Bring this outside edge over to the center, and place it on top of and in line with the previously placed first fold.

Step 3: Bring the fold over to the center of the cover. This will create an inside fold (center) and an outside fold.

Step 7: Fold the other half of the cover in the same manner by using Steps 1 through 6.

Step 8: Straighten the folds if they are not straight.

SKILL SHEETS

18-I-4
Fold a salvage cover for a one-firefighter spread.

Step 9: Grasp the same end of the cover, with the cover folded to reduce width.

Step 10: Bring this end to a point just short of the center.

Step 11: Use one hand as a pivot and bring the folded end over and place on top of the first fold.

Step 12: Fold the other end of the cover toward the center, leaving about 4 inches (100 mm) between the two folds.

Step 13: Place one fold on top of the other for the completed fold; the space between the folds now serves as a hinge.

Step 14: Store according to departmental procedures.

18-1-5
Spread a folded salvage cover — One-firefighter method.

🔥 **SKILL SHEETS**

Step 1: Lay the folded cover on top of and near the center of the object to be covered.

Step 2: Separate the cover at the first fold.

Step 3: Select either end and continue to unfold the salvage cover by separating the next fold.

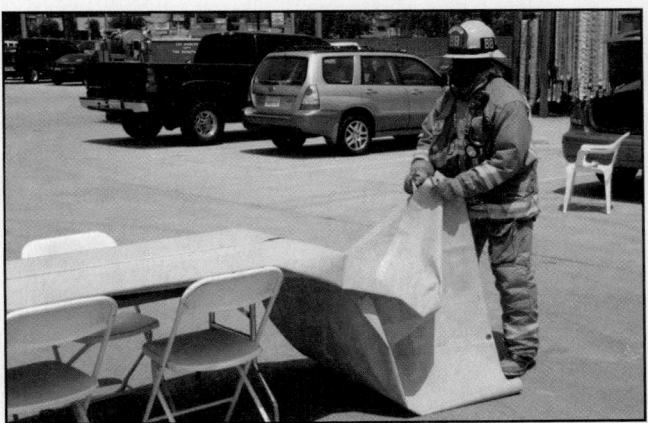

Step 7: Unfold the other end of the cover in the same manner over the object.

Step 8: Stand at one end.

Step 9: Grasp the open edges where convenient, one edge in each hand.

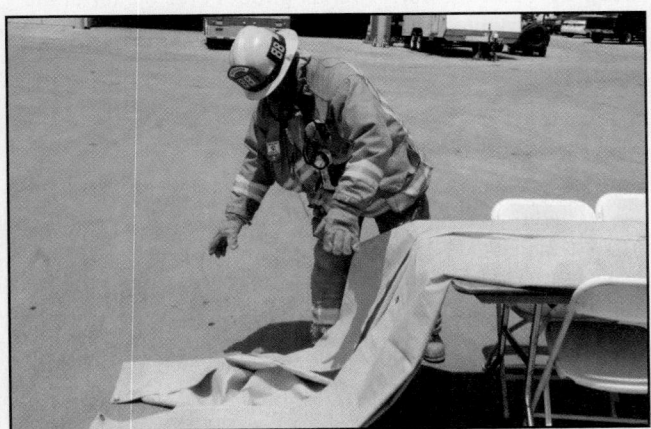

Step 4: Unfold this same end toward the end of the object to be covered.

Step 5: Grasp the end of the cover near the center with both hands to prevent the corners from falling outward.

Step 6: Bring the end of the cover into position over the end of the object being covered.

Step 10: Open the sides of the cover over the object by snapping both hands up and out.

Step 11: Open the other end of the cover over the object in the same manner.

Step 12: Tuck in all loose edges at the bottom.

Step 13: Store according to local procedures.

SKILL SHEETS

18-I-6
Fold a salvage cover for a two-firefighter spread.

NOTE: Two firefighters must make initial folds to reduce the width of the cover. Steps 1 through 11 are performed simultaneously by both firefighters. Steps 12 through 19 are performed by the respective firefighters. Steps 20 through 24 are performed simultaneously by both firefighters.

Step 1: Grasp opposite ends of the cover at the center grommet with the cover stretched lengthwise.

Step 2: Pull the cover tightly between each firefighter.

Step 3: Raise this center fold high above the ground.

Step 4: Shake out the wrinkles to form the first half-fold.

Step 5: Spread the half-fold on the ground.

Step 6: Smooth the half-fold flat to remove the wrinkles.

Step 7: Stand at each end of the half-fold and face the cover.

Step 8: Grasp the open-edge corners with the hand nearest to these corners.

Step 9: Place the corresponding foot at the center of the half-fold, making a pivot for the next fold.

Step 10: Stretch that part of the cover being folded tightly between each firefighter.

Step 11: Make the quarter-fold by folding the open edges over the folded edge.

Step 12: Firefighter #1: Stand on one end of the quarter-fold.

Step 13: Firefighter #2: Grasp the opposite end and shake out all the wrinkles.

Step 14: Firefighter #2: Carry this end to the opposite end, maintaining alignment of outside edges.

18-I-6
Fold a salvage cover for a two-firefighter spread.

SKILL SHEETS

Step 15: Both Firefighters: Place the carried end on the opposite end, aligning all edges.

Step 16: Both Firefighters: Position at opposite ends.

Step 17: Firefighter #2: Stand on the folded end of the cover.

Step 18: Firefighter #1: Shake out all wrinkles.

Step 19: Firefighter #1: Align all edges.

Step 20: Grasp the open ends and use the inside foot as a pivot for the next fold.

Step 21: Bring these open ends over and place them just short of the folded center fold.

Step 22: Continue this folding process by bringing the open ends over and just short of the folded end.

Step 23: Complete the operation by making one more fold in the same manner.

Step 24: Bring the open ends over and to the folded end using the free hand as a pivot during the fold.

Step 25: Store according to local procedures.

NOTE: These steps are done with both firefighters performing the steps simultaneously.

Step 1: Stretch the cover along one side of the object to be covered.

Step 2: Separate the last half-fold by grasping each side of the cover near the ends.

Step 3: Lay the side of the cover closest to the furniture on the ground.

Step 5: Place the outside hand about midway down the end hem.

Step 6: Place the inside foot on the corner of the cover to hold it in place.

Step 7: Pull the cover tightly between each firefighter.

Step 8: Swing the folded part down, up, and out in one sweeping movement in order to pocket as much air as possible.

Step 4: Make several accordion folds in the inside hand.

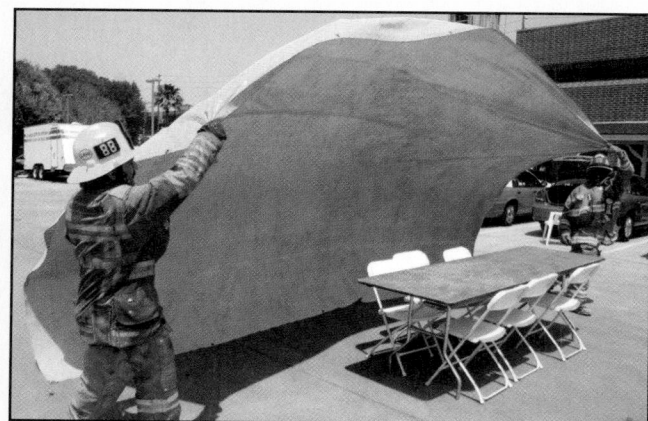

Step 9: Pitch or carry the accordion folds across the object when the cover is as high as each firefighter can reach. This action causes the cover to float over the object.

Step 10: Guide the cover into position as it floats over the object.

Step 11: Straighten the sides for better water runoff.

Step 1: Open the salvage cover.

Step 2: Lay the cover flat at the desired location.

Step 3: Roll the opposite edges of the salvage cover toward the middle until there is a 3 foot (1 m) width between the rolls.

Step 4: Turn the cover over.

Step 5: Adjust the chute to collect and channel water by elevating one end.

Step 6: Extend the other end out a door or window.

Step 1: Open the salvage cover.

Step 2: Lay the cover flat at the desired location.

Step 4: Roll the edges over the pike poles toward the middle until there is a 3-foot (1 m) width between the rolls.

Step 3: Place pike poles at opposite edges of the salvage cover with the pike extending off the end of the cover.

Step 5: Turn the cover over, keeping the folds in place.

Step 6: Place the chute to collect and channel water.

Step 7: Extend the other end out a door or window.

Step 1: Open the salvage cover.

Step 2: Lay the cover flat at the desired location.

Step 3: Roll the sides inward approximately 3 feet (1 m).

Step 4: Lay the ends of the side rolls over at a 90-degree angle to form the corners of the basin.

Step 5: Roll one end into a tight roll on top of the side roll and form a projected flap.

Step 6: Lift the edge roll.

Step 7: Tuck the end roll to lock the corners.

Step 8: Roll the other end in a like manner.

Step 9: Lock the corners.

Step 1: Open the salvage cover.

Step 2: Lay the cover flat at the desired location.

Step 3: Roll one edge (along the long axis) tightly inward toward the center of the cover.

Step 4: Roll the opposite edge tightly inward toward the center leaving a 1 to 3 foot (.34 m to 1 m) space between them.

Step 6: Slide the end of the chute under one corner of the catchall, about 1 to 2 feet (.34 m to .69 m).

Step 5: Turn the chute upside down flattening the center to the floor.

Step 7: Unfold the corner of the catchall.

Step 8: Flatten the corner of the catchall to form a seamless path for the water.

Step 1: Confirm order with officer to overhaul.

Step 2: Locate area(s) with potential hidden or smoldering fire.

  a. Use heat detector or thermal imaging device

  b. Observe fire area to detect smoking or smoldering materials – watch, listen, feel

  c. Observe burn and smoke patterns

  d. Wear appropriate personal protective equipment, including respiratory protection

Step 3: Remove ceiling and wall covering and insulation.

  a. Begin with area closest to hidden or smoldering fire

  b. Overhaul area until unburned structural materials are visible

  c. Preserve potential evidence for fire cause investigation

  d. Minimize damage when possible

Step 4: Extinguish hidden and smoldering fires with small handline.

    a. Use minimal water for extinguishment

    b. Complete extinguishment – no hidden or smoldering fires remain

    c. Remove stuffed materials, such as mattresses, from structure and overhaul outside

Step 5: Report to officer completion of assigned task.

Courtesy of Donny Howard

Chapter Contents

Key Terms

NFPA® Job Performance Requirements

This chapter provides information that addresses the following job performance requirements of NFPA® 1001, *Standard for Fire Fighter Professional Qualifications* (2013).

Firefighter I
5.3.8
5.3.13
5.3.14

Firefighter II
6.3.4

Firefighter I Chapter Objectives

1. Explain ways to recognize obvious signs of the area of origin. (5.3.8, 5.3.14)

2. Describe the relationship between fire cause classifications and cause determination. (5.3.8, 5.3.13)

3. Recognize signs of arson. (5.3.13)

4. Describe the importance of preserving evidence. (5.3.8, 5.3.14)

5. Explain techniques for preserving evidence. (5.3.8, 5.3.14)

Firefighter II Chapter Objectives

1. Describe types of evidence used to indicate the area of origin or fire cause. (6.3.4)

2. Recognize fire cause evidence. (6.3.4)

3. Explain the roles and responsibilities of responders and investigators involved in fire investigations. (6.3.4)

4. Tell what legal issues impact location and collection of evidence during a fire investigation. (6.3.4)

5. Protect evidence of fire cause and origin. (Skill Sheet 19-II-1, 6.3.4)

Chapter 19
Fire Origin and
Cause Determination

*Courtesy of Iowa Fire
Service Training Bureau.*

Case History

In August 2011, Marion County (FL) firefighters responded to an early morning fire in a home in Ocala, Florida. Upon arrival, the firefighters reported that approximately 25 percent of the 1,500 square foot (140 m²) structure was involved in fire. During the interior fire attack, firefighters discovered the bodies of three victims. A fourth body was uncovered during overhaul operations.

Fire suppression lasted for approximately 3 hours. Following suppression activities, fire investigators from the Marion County Sheriff's office and the Florida State Fire Marshal's Office began removing debris and placing it on tarps in the front yard. Autopsies later determined that the victims, two women and two children, had all been shot.

The cause of death of the victims and physical evidence uncovered at the scene indicated that the fire had been intentionally set. This incident is an example of how one type of crime, arson, can be used to destroy evidence of another crime and in this case, murder.

Source: Marion County (FL) Fire Department news release and Ocala newspaper articles, link provided by Shad Cooper.

It is every firefighter's responsibility to recognize the area where the fire began, note indicators of the cause of the fire, and to protect any physical evidence relating to the cause. These responsibilities apply to all types of fires including fires involving ground cover, exterior piles of stored Class A materials, vehicles, and structure interiors and contents. This chapter will provide you with the knowledge needed to accomplish these tasks.

Area of Origin

The **area of origin** is the general location where the fire began. It will contain the precise **point of origin** where ignition took place **(Figure 19.1, p. 1146)**. The area of origin will often contain the greatest amount of damage and debris. Until air monitoring indicates that it is safe, you must always assume that potentially lethal concentrations

Area of Origin — The general location (room or area) where the ignition source and the material first ignited actually came together for the first time.

Point of Origin — Exact physical location where the heat source and fuel come in contact with each other and a fire begins.

Figure 19.1 The area of origin is a general location that contains the point of origin. *Courtesy of Donny Howard.*

Area of Origin

of unburned gases such as carbon monoxide (CO) and hydrogen cyanide (HCN) are present. You must always wear full personal protective equipment (PPE) including the appropriate level of respiratory protection when working in this area.

The debris in the area of origin will contain an indication of the **fire cause**. Fire cause indicators include an ignition source, such as matches, and a fuel source, such as a waste paper container. It may also contain material that may be used as evidence in a court case. While not all intentional fires are illegal, those that are can result in a criminal charge of **arson**.

Until there are orders to move it, all debris should remain in place. If necessary, it may be removed and placed in a specific area where it can be controlled and sorted by a fire investigator, as in the Case History at the first of this chapter. Control of all evidence must be maintained as part of the **chain of custody** required in legal cases. As part of the fire suppression team, you may become part of the chain of custody. It is important to remember what you see, smell, and hear during the incident.

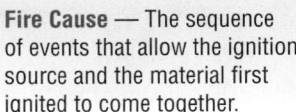

CAUTION
You should minimize fire suppression and overhaul activities that could destroy important evidence regarding the origin and cause of the fire.

In structure and vehicle fires, the area of origin is often readily apparent. If it is not, firefighters can follow physical indicators from the area of least damage to the area of most damage to locate the area of origin. They can also rely upon witness accounts. Identifying witnesses, securing the scene, and noting initial scene observations are critical to the overall success of the process. Once the fire is under control, the Incident Commander (IC), and sometimes a fire or criminal investigator, determines the point the fire started, the fire cause, and protects or collects evidence that may be used in a legal proceeding. The area of origin in wildland or ground cover fires may be less apparent and require an experienced investigator to locate **(Figure 19.2)**.

Structure Fires

The IC or your supervisor may search for the area of origin and request your help in finding it. The search for the area of origin starts outside the structure or from the unburned portion of the interior. First, examine the exterior and then the interior of the structure **(Figure 19.3)**. Indicators, such as **fire patterns**, provide important information regarding the location of the fire origin.

In situations where the area of origin cannot be accurately determined, firefighters should delay overhaul operations beyond locating and extinguishing fires, protect the scene, and establish scene security. A fire investigator should be requested to perform a thorough investigation of the scene.

Fire Pattern — The apparent and obvious design of burned material and the burning path of travel from a point of fire origin. Previously known as a *burn pattern*.

Figure 19.2 In wildland fires, an experienced investigator may have to use all of his or her expertise to discover the origin and cause.

Figure 19.3 Fire patterns provide indicators of the area of origin and the types of materials that burned. *Courtesy of Donny Howard.*

Preliminary Scene Assessment and Exterior Examination

The examination of a structure fire starts with a preliminary scene assessment. This assessment consists of an examination of the entire incident scene in order to determine its size and scope and to determine whether it is safe to continue working in the structure. The assessment should begin with the exterior of the structure and continue, if possible, completely around the entire incident scene, including the roof. During this examination, potential physical evidence and its location should be documented or preserved. Observations should be made regarding the following:

- Building damage (including openings such as broken windows, forced doors, and other damage to the building) as well as structural stability
- Fire and ventilation patterns around windows and doors and under the roof eaves
- Means of ingress and egress (for safety purposes)
- Utility services including natural gas meters, electrical meters and connections, and telephone and cable television connections
- Tire tracks or footprints, especially near windows or other openings
- Discarded containers, such as gas cans
- Indications of forcible entry around doors and windows
- Presence or location of surveillance cameras

Interior Examination

After the exterior examination has been completed, begin an interior examination, working from the area of least damage to the area of greatest damage. Using this method you can use fire indicators to determine the path of fire spread and the area of fire origin. These indicators include:

- Fire patterns
- Melted metal and glass
- Degree of damage to structure and contents

Due to the normal upward movement of heat, floors are normally less damaged than ceilings during a fire (**Figure 19.4**). The ceiling or roof structure may have the most severe damage above the area where the fire burned the longest or most intensely. Areas of heavily charred flooring only indicate that a fuel burned at that location. Floor charring may be the result of a number of factors, including flashover, fuel at a greater height falling downward in a flaming state, ventilation, or ignitable liquids (whether introduced intentionally or normally present) (**Figure 19.5**).

Figure 19.4 A common pattern shows a clear delineation between the smoke damage and the floor where cooler air remained. *Courtesy of Donny Howard.*

Figure 19.5 Charring from flammable liquids may take the shape of a circle with irregular edges. *Courtesy of Donny Howard.*

Charring on vertical surfaces of walls, closed doors, and objects will face toward the area or point of origin. Charring is usually deeper at or near the area of origin **(Figure 19.6)**. Doors that were open during the fire and objects within the area of origin may have equal charring on all sides.

The discovery of what appears to be multiple areas of origin may indicate an intentionally set fire. Unintentional fires, however, may also ignite items that burn intensely and give the same appearance. The following factors affect fire spread and should be evaluated during the scene examination:

- The nature and composition of the combustible materials in the fire's path
- Building features and layout that assist in or resist fire spread
- Ventilation openings and their size and location
- Combustible materials present (fire load)
- Fire suppression tactics
- Activation of fire suppression systems

Recent laboratory tests on fires started with accelerants indicate that fire and burn patterns traditionally used as fire indicators may not be as reliable as once thought. You should be aware of general patterns that will exist and realize that a fire investigator should be called if there is any question concerning the area of origin or fire cause.

Figure 19.6 Char depth is often deepest facing the area of origin.

Figure 19.7 A total loss structure may offer indicators of the origin of the fire and its ventilation factors. *Courtesy of Donny Howard.*

Total Structural Fire Loss

In fires that result in total destruction of the structure and its contents, traditional fire indicators may not exist. In other cases, indicators may be present but be more difficult to locate and identify (**Figure 19.7**). Determining the location of greatest damage is even more difficult due to the complete consumption of combustible materials and the unknown ventilation factors that existed during the fire.

Exterior Fires

Exterior fires may involve stacks of stored materials, debris, trash, rubbish, or ground cover or vegetation. The fire may originate in the ground cover and spread to stored materials and structures or travel the other direction. When dealing with ground cover fires, you must know basic fire behavior and the effects of winds, topography, and natural fuels on fire spread.

On flat ground with a consistent fuel bed and no wind, a fire would burn equally in all directions, and the point of origin would be in the center of the circular fire pattern (**Figure 19.8**). However, ground cover fires rarely occur in these conditions. In reality, the rate and direction of spread of ground cover fires are affected by:

- Wind direction
- Wind velocity
- Terrain/topography
- Types of fuel
- Ambient temperature
- Relative humidity
- Moisture content of fuel

The area of origin may display evidence of slower and less intense fire growth than at the head or leading edge of the fire. More unburned materials may remain in the area of origin, and the effects of flame on the fuels may be considerably less than at the head.

Radiant heat from burning materials affects other adjacent fuels. As the fire burns past any given area, the flames scorch or char surfaces exposed to the fire. But the backsides of grass stems or tree trunks are protected. Grass will fall toward the area of origin as it burns (**Figure 19.9**). Even if the grass completely burns, the remaining ash will still point toward the area of origin. White ash is a product of more complete combustion, and it appears on the exposed sides of the remaining debris.

Vehicle Fires

Determining the area of origin and cause of a fire in a vehicle, truck, motor home, or boat is similar to searching for the origin and cause of a structure fire. Along with interviewing witnesses and firefighters, it is necessary to examine both the exterior and the interior of the movable property. By surveying the damage to the vehicle, you will gain valuable clues as to where the fire started.

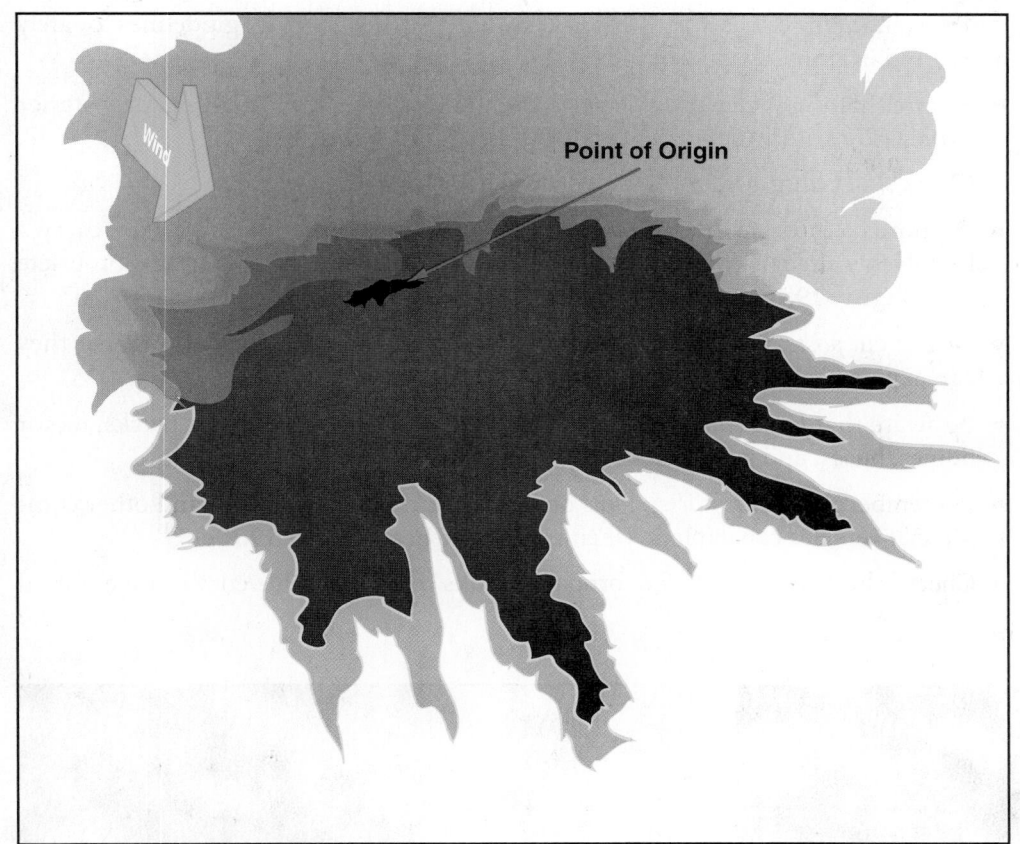

Figure 19.8 From its point of origin, a fire spreads outward in a pattern influenced by any wind and ground cover.

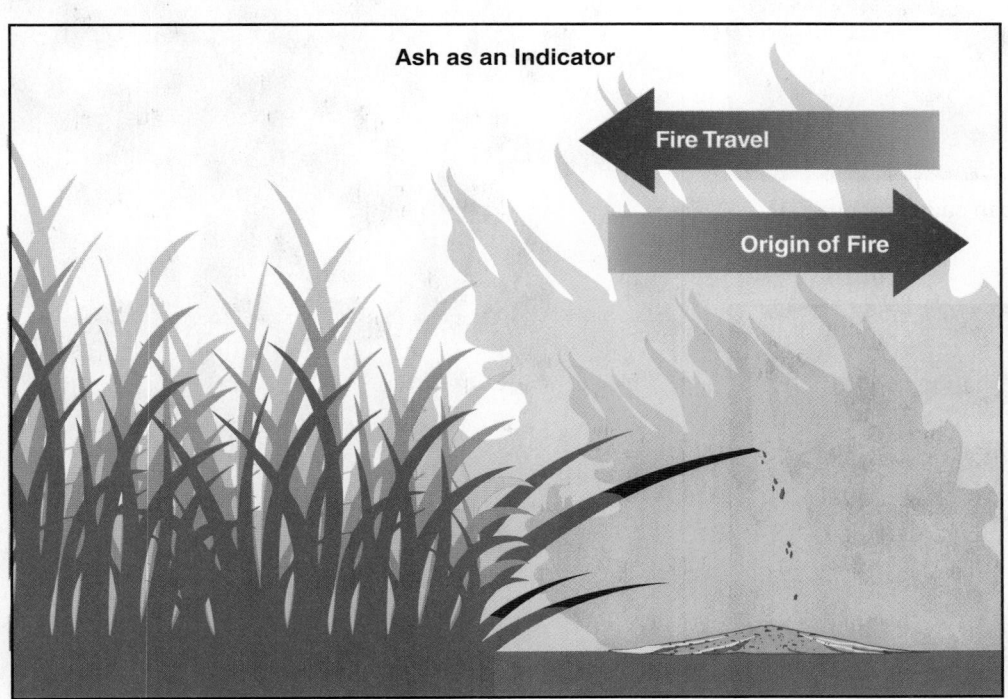

Figure 19.9 Grass shows the patterns of fire travel by falling toward the area of origin as it burns.

To avoid injury or death, it is crucial that you use the following guidelines to safely access and examine a vehicle:

- Ensure that the *undeployed air bags* are inoperable before beginning any interior survey (**Figure 19.10a**).

- Check *shock absorber bumpers* to ensure that they are inoperable.

- Do not cut through door posts on *hybrid and electric cars* because they carry a high eletrical charge. Check for voltage that is present. Always wear appropriate PPE (**Figure 19.10b**).

- Do not cut *posts* on some cars that contain seat belt restraint systems because they can explode if they are cut (**Figure 19.10c**).

- Be aware that *large capacity or multiple fuel cells* may exist on light trucks, motor homes, buses, and large trucks (**Figure 19.10d**).

- Remember that *hydraulic systems* on refuse trucks, dump trucks, and other commercial vehicles can explode when exposed to heat (**Figure 19.10e**).

- Check the *trunk area* for propane tanks or flammable/combustible liquid containers.

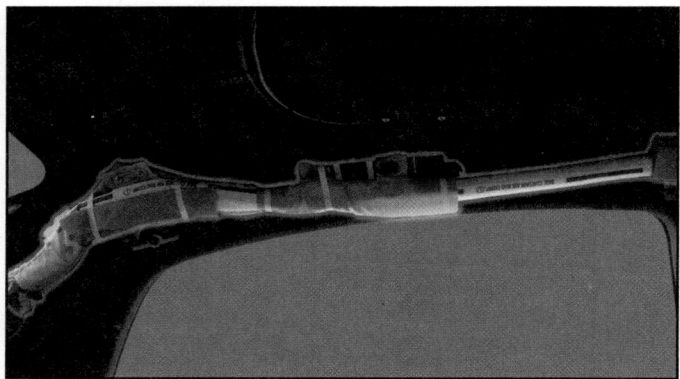

Figure 19.10a Undeployed air bags can cause serious damage if they deploy suddenly.

Figure 19.10b Hybrid and electric cars must be assumed to carry live, high voltage electrical charges.

Figure 19.10c Some newer cars are configured with pyrotechnic seat belt restraints in the support posts.

Figure 19.10d High capacity fuel tanks may be plainly visible on large vehicles. *Courtesy of Alan Braun, Missouri Fire Rescue Training.*

Figure 19.10e Hydraulic cylinders can react explosively under specific circumstances. *Courtesy of James Nilo, Richmond International Airport.*

WARNING

Do not attempt to access or examine a vehicle unless you know how to do so safely.

Fire Cause Determination

To determine the fire cause, you must look for the **competent ignition source**, the material that first ignited, and the action that brought the two together, referred to as the **ignition sequence**. Determining the fire cause can accomplish the following:

- Document the causes of fires which help to prevent similar fires from occurring in the future

- Indicate trends in unsafe behavior that can be corrected through educational programs

- Indicate the existence of defective equipment or design flaws that need to be corrected

- Indicate malicious or illegal behavior to be used in arson cases

According to data collected through National Fire Incident Reporting System (NFIRS) reports, compiled by the U.S. Fire Administration (USFA), the apparent causes of fires in structures include:

- Incendiary (human-caused, intentionally set)

- Open flame, ember, or torch

- Other heat, flame, or spark

- Smoking materials

- Lightning, natural

- Heating equipment

- Cooking equipment

Competent Ignition Source — A competent ignition source will have sufficient temperature and energy and be in contact with the fuel long enough to raise it to its ignition temperature.

Ignition Sequence — History of the fire, beginning when the ignition source and the first fuel ignited meet at the area of origin, and proceeding through the entire duration of fire spread through the scene.

- Electrical or lighting equipment
- Heating, ventilating, and air-conditioning (HVAC) appliances
- Other equipment
- Exposure (anything in the immediate range of a fire that is not burning but could start burning if the fire is not contained)
- Children playing with matches, lighters, or other ignition sources

Types of Fire Cause Classifications

Types of fire cause can be divided into four generally accepted classifications:

- Accidental
- Natural
- Incendiary
- Undetermined

Accidental

Accidental fires do not involve a deliberate human act to ignite or spread the fire into an area where the fire should not be. Accidental fires can result from unsafe human behavior such as poor housekeeping, careless use of flammable liquids, falling asleep while smoking in bed, or placing too much wrapping paper into a fireplace. They can also result from an unsafe condition such as overloaded electrical circuits, worn electrical wiring, or overheated machinery **(Figure 19.11)**. Finally, they can result from a hazardous process such as flammable vapors from a paint booth or a concentration of dust in a grain silo contacting an ignition source.

Natural

Natural fires occur when humans are not involved in the ignition process. Lightning strikes, earthquakes, tornados, or floods can all cause an ignition source to contact a fuel **(Figure 19.12)**. Materials reaching autoignition temperature is another type of natural cause. For example, bales of hay stored in a barn can ignite when moisture in the center of the bale causes the hay to decompose and generate heat. When the heat is sufficient for autoignition, the hay may pyrolize and begin to burn.

Incendiary

Incendiary fires are deliberately set under circumstances in which the responsible party knows that the fire should not be ignited **(Figure 19.13)**. If there is sufficient evidence to prove that the fire was incendiary, a criminal charge of arson can be filed against the suspect. Multiple reasons and motives cause both adults and juveniles to set fires, such as revenge, vandalism, or profit, among others.

While proving that a fire was intentionally set is relatively easy, proving who set it is extremely difficult. Much of the physical evidence is usually destroyed in the fire making it difficult to connect the suspect to the act of setting the fire. This is why it is extremely important that you secure the area of origin and protect all evidence.

Undetermined

The undetermined classification is used when the specific cause has not been determined to a reasonable degree of probability. This classification may refer to situations in which each of the specific components of the ignition sequence is not specifically

Figure 19.11 Accidental fires may result from misuse of electrical equipment. *Courtesy of Donny Howard.*

Figure 19.12 Natural fires result from phenomena outside of human causality. *Source: NOAA — Public Domain.*

identified. A classification of undetermined may result from total destruction of the structure and contents or removal of evidence during the overhaul process. However, a fire that is undetermined should not be referred to as suspicious. Suspicious tends to be misinterpreted as meaning the fire resulted from arson.

Obvious Signs of Cause

Of the elements in the fire tetrahedron that are necessary for a fire to start, oxygen and fuel are almost always present. Therefore, the most obvious sign of cause will be the presence of a competent ignition source. A competent ignition source will have sufficient temperature and energy and be in contact with the fuel long enough to raise it to its ignition temperature. For example, if the area of origin is determined to be the kitchen of a structure, ignition sources might include the following (**Figure 19.14, p. 1156**):

- Oven
- Cook top
- Electrical appliances
- Exposed wiring
- Electrical receptacles
- Open fires

In a ground cover fire, the area of origin might include visual signs of a lightning strike, a camp fire, or the remains of smoking materials or matches. If the fire was incendiary in nature, a flammable liquids container might be present.

Figure 19.13 An incendiary fire is set by someone who knows that a fire should not be set in those particular circumstances. *Courtesy of the Iowa Fire Service Training Bureau.*

Figure 19.14 Ignition sources may be found at any of several locations that have sufficient temperature and energy to raise the temperature of fuel to its ignition temperature. *Courtesy of Bob Esposito.*

In some cases, the absence of an obvious ignition source will help to eliminate some causes and focus on others. For instance, an electrical short or power surge will not be the cause of a fire in an abandoned house that does not have an electrical service or power meter.

Signs of Arson

When the initial cause determination indicates that the fire was incendiary or undetermined, it will be necessary to gather evidence based on your observations and those of other firefighters. Depending on local protocol, local or state fire investigators or law enforcement officials may be assigned to perform a formal investigation. In addition to gathering physical evidence, these investigators will need all the information you can provide to them. Some of the things you should notice include:

- **Time of day** — Are people and circumstances at the scene as they normally would be at this time of day? For example, if a fire is in a dwelling at 0300 hrs. the building occupants would probably be wearing night clothes, not street clothes. If a fire is in an office building well after working hours, the owner or employees may need to explain why they are present at that hour.

- **Weather and natural hazards** — Is it hot, cold, or stormy? Is there heavy snow, ice, high water, or fog? If the outside temperature is high, the furnace in the structure would not be operating. If the outside temperature is low, the windows normally would not be wide open. Arsonists sometimes set fires during inclement weather because the fire department's response may be delayed.

- **Man-made barriers** — Are there any barriers such as barricades, fallen trees, cables, trash bins, or vehicles blocking access to hydrants, sprinkler and standpipe connections, streets, or driveways? These situations could indicate an arsonist's attempt to impede firefighters' access and delay fire suppression efforts (**Figure 19.15**).

- **People leaving the scene** — Are people leaving the scene in haste? Most people are intrigued by a fire and will remain in the area to watch. People leaving the scene by vehicle or on foot may be an important observation to an investigation. If a vehicle is seen leaving the scene (especially at high speed), note its color and as many details about it as possible, especially the license plate number. If possible, note how many occupants are in the vehicle. If someone is seen leaving the scene on foot, try to remember that person's attire, general physical appearance, and any peculiarities such as trying to leave undetected, walking briskly, or looking over his or her shoulder.

Additional information that firefighters may notice while carrying out their duties include the following:

- **Time of arrival and extent of fire** — Note the extent of fire involvement at the time of arrival. Observe the color and movement of smoke and flames.

- **Wind direction and velocity** — Note wind direction and velocity. These factors may have a great effect on the natural path of fire spread.

- **Doors or windows locked or unlocked** — Note the position and condition of doors and windows upon arrival. Before opening doors and windows, determine whether they are locked, unlocked, or show any signs of forced entry such as broken glass or damaged frames. In some cases the insides of windows may be covered with blankets, paint, or paper to delay discovery of the fire.

- **Location of the fire** — Observe the location of the fire. This information may help to identify the area of origin. Also note whether there were separate, seemingly unconnected fires. If so, the fire might have been set in several locations or spread by **trailers** (combustible material used to spread fire from one area to another) **(Figure 19.16)**.

- **Containers or cans** — Note metal or plastic containers found inside or outside the structure. They may have been used to transport accelerants **(Figure 19.17)**.

Figure 19.15 Whether intentional or not, blocked egress from a structure impedes first responders from responding to a fire. *Courtesy of Donny Howard.*

Trailer — Combustible material, such as rolled rags, blankets, newspapers, or flammable liquid, often used in intentionally set fires in order to spread fire from one area to other points or areas.

Figure 19.16 Trailers burn a clear path, often in a straight line, characterized by minor feathering around the edges. *Courtesy of Donny Howard.*

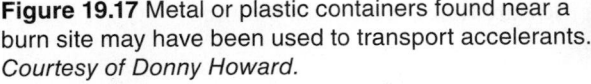

Figure 19.17 Metal or plastic containers found near a burn site may have been used to transport accelerants. *Courtesy of Donny Howard.*

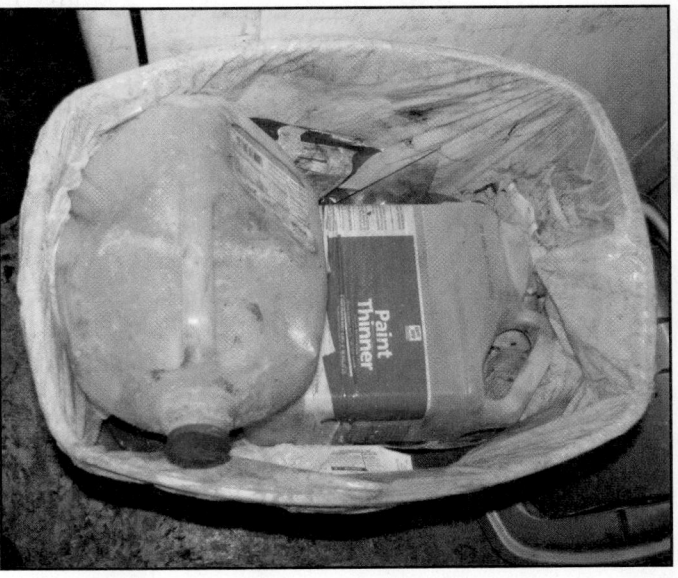

- **Burglary tools** — Note any tools such as pry bars or screwdrivers found in areas away from workshops. They may have been used to break into the facility.

- **Familiar faces** — Notice familiar faces in the crowd of bystanders — people who are seen at numerous fires in the area. They may be individuals who like to watch fire incidents, or they may be habitual firesetters.

As the operation continues, you should continue to observe conditions that may lead to the determination of the fire cause:

- **Unusual odors** — Note any unusual odors. Even though firefighters wear SCBA during interior fire fighting and overhaul operations, they may smell unusual odors outside the structure. In addition, odors may cling to personal protective equipment and be detectable after firefighters come out of the smoke.

- **Abnormal behavior of fire when water is applied** — Observe how the fire behaves when water is applied to it. Flashbacks, reignition, several rekindles in the same area, and an increase in the intensity of the fire may indicate possible accelerant use. Water applied to a burning liquid accelerant may cause it to splatter, allowing flame intensity to increase and the fire to spread in several directions. Water applied to fires involving ordinary combustibles usually reduces flame spread.

- **Obstacles hindering fire fighting** — Note whether any doors are nailed shut or furniture is placed in doorways and hallways to hinder fire fighting efforts. Holes may be cut in the floors that not only hinder fire suppression activities but also spread the fire.

- **Incendiary devices** — Note any pieces of glass, fragments of bottles or containers, and metal parts of electrical or mechanical devices. Most **incendiary devices** (any device designed and used to start a fire) leave evidence of their existence (**Figure 19.18**). More than one device may be found, and sometimes a malfunctioning device can be found during a thorough search.

- **Trailers** — Note combustible materials such as rolled rags, blankets, newspapers, or ignitable liquid that could be used to spread fire from one area to another. Trailers usually leave char or burn patterns and may be used with incendiary ignition devices.

- **Structural alterations** — Observe any alterations to the structure: removal of plaster or drywall to expose wood; holes made in ceilings, walls, and floors; and fire doors secured in an open position. All of these methods can be used to allow a fire to spread quickly through the structure.

- **Fire patterns** — Note the fire's movement and intensity patterns. These can trace how the fire spread, identify the original ignition source, and determine the fuel(s) involved. Carefully note any areas of irregular burning or locally heavy charring in areas of little fuel (**Figure 19.19**).

- **Heat intensity** — Look for evidence of high heat intensity, especially in relation to other areas of the same room. This may indicate the use of accelerants or intentionally disconnected gas lines. However, other factors may contribute to variations in heat intensity. One of these factors is synthetic materials, such as polyurethane, that may produce abnormally high heat intensity and may be confused with the use of accelerants.

- **Availability of documents** — Note anyone who conveniently produces insurance policies, inventory lists, deeds, or other legal documents that would normally be locked away. This may indicate that the fire was planned.

- **Fire detection and protection systems** — Check for evidence of tampering or intentional damage if fire detection and suppression systems and devices are inoperable.

- **Intrusion alarms** — Check intrusion alarms to see whether they have been tampered with or intentionally disabled.

- **Location of fire** — Note any possible ignition sources in the area of the fire. Fires burning in areas remote from normal ignition sources demand an explanation. Some examples are fires in closets, bathtubs, file drawers, or in the center of the floor.

- **Personal possessions** — Note anything that might suggest that preparations were made for a fire: absence or shortage of clothing, furnishings, appliances, food, and dishes; absence of personal possessions such as diplomas, financial papers, and toys; absence of items of sentimental value such as photo albums, special collections, wedding pictures, and heirlooms; absence of pets that would ordinarily be in the structure (**Figure 19.20**).

  NOTE: Do not read too much into a shortage of material possessions. A person's economic status may dictate his or her lifestyle, and some people just do not have as much as others.

- **Household items** — Note whether any major household items appear to be removed or replaced with those of lesser value or of inferior quality. Check to see whether major appliances were disconnected or unplugged and determine why they were in this condition.

- **Equipment or inventory** — Look for obsolete equipment or inventory, fixtures, display cases, and raw materials.

- **Business records** — Determine whether important business records are out of their normal places and left where they would be endangered by fire. Check safes, fire-resistant files, etc., to determine whether they are open and exposing the contents.

Preserving Evidence

Preserving evidence is the responsibility of all fire officers and firefighters at the scene. Evidence of the location of the area of origin and fire cause is necessary to determine the fire cause classification. Remember, unless you are qualified/certified to do so, you are not in a position to determine what is evidence, nor are you in a position to determine what is admissible in court. You should protect everything that looks odd or suspicious until those

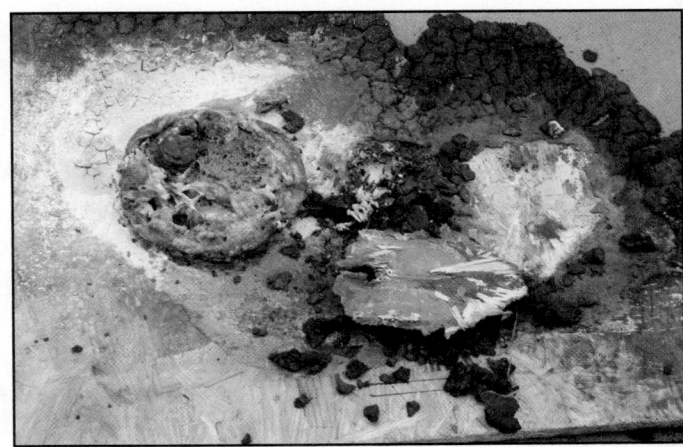

Figure 19.18 Incendiary devices designed to start a fire often leave traces of their existence.

Figure 19.19 Irregular burn patterns may contain indicators of fuel content and the ignition point. *Courtesy of Donny Howard.*

Figure 19.20 A cleared out closet may indicate a hasty collection of clothing before evacuating the site. *Courtesy of Donny Howard.*

conducting the investigation have examined it. You must be familiar with the importance of preserving evidence, the methods for preserving it, and the methods for securing the fire scene.

Reasons for Protecting the Fire Scene

Fire scene security is an important method for protecting evidence and ensuring that it is not damaged, altered, or removed. In a sense, scene security is the first step in establishing the chain of custody of the evidence.

A secure fire scene is one with a recognizable perimeter and someone to maintain that perimeter. Fire scene security is initially the responsibility of the fire suppression personnel who respond to extinguish the fire. Early security measures should include the following:

- Restricting access to the scene

- Protecting any potential evidence located in the area

- Minimizing fire suppression and overhaul activities that could destroy important information regarding the origin and cause of the fire

The first-arriving fire investigator may have to adjust the security measures already in place or implement additional measures to protect the scene. This decision will be based on the investigator's assessment of the scene, the circumstances surrounding the fire, and the measures in place. On incidents requiring the assistance of law enforcement agencies such as those involving injuries or fatalities, an investigator should request an officer to monitor each entrance and exit to the scene to document every individual who has entered the secured area.

There are several guidelines for establishing a perimeter of the proper size in the following situations:

1. **Explosions** — The perimeter for explosions should be established at 1.5 times the distance from the farthest piece of debris found (**Figure 19.21**). As the investigation continues, this perimeter may expand as additional debris is located.

2. **Structure fires** — Firefighters may establish a fire scene perimeter to limit access to the fire and keep bystanders at a safe distance. The perimeter may be expanded to encompass the area surrounding the building and any potential evidence that is located outside the building (**Figure 19.22**). The perimeter should extend beyond the farthest piece of evidence located during the exterior examination of the structure. If no evidence is found outside the building, it may be sufficient to restrict access into the building or an area of the building. The restricted area should provide fire investigators with room to work and protect all known or suspected evidence without being hampered by unnecessary personnel or bystanders.

To be effective, fire scene perimeters must be both recognizable and enforceable. Common ways to establish these perimeters are as follows:

- Ensure that the initial perimeter is larger than is necessary for investigations. The perimeter can always be made smaller if needed, but increasing the size of a perimeter is much more difficult.

- Ensure that the perimeter is visible and recognizable to everyone on the scene. Methods used to mark the perimeter should be easy to use. To accommodate the rapid establishment of a perimeter that is visible and recognizable, many public safety organizations use rope, traffic cones, or marked barrier tape.

- Use uniformed law enforcement officers or firefighters to control access into the established perimeter. For long-term operations, private security guards and/or the installation of construction barriers or fences may be necessary.

If the investigation is considered to be criminal in nature, the investigator may institute the following procedures:

- Keep a log of all persons who enter and leave the incident perimeter whenever it is necessary to limit the number of personnel entering a scene.

- Permit access only to those individuals who are authorized to be in the area.

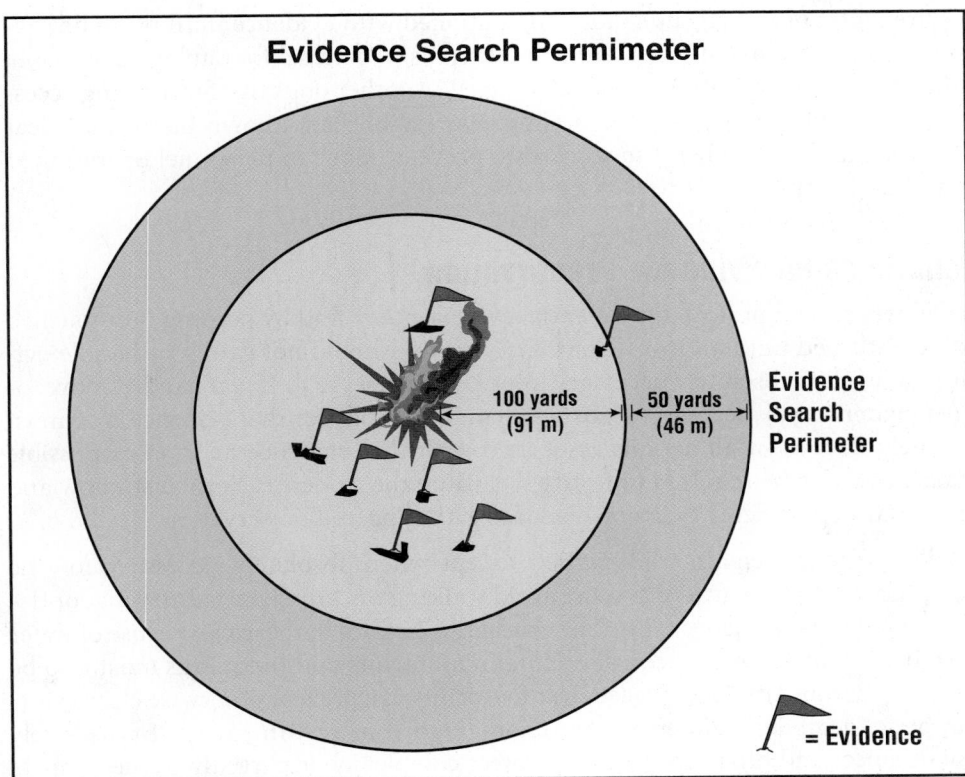

Figure 19.21 Explosion scenes require an extended perimeter to accommodate debris scattered over large distances.

Figure 19.22 A scene perimeter separates a burned structure and its immediate surroundings from public access. *Courtesy of Ron Moore, McKinney (TX) Fire Department.*

- Ensure that firefighters and other emergency personnel move to a staging area outside the perimeter when they have completed their tasks in the area to wait for additional assignments or release from the incident.

- Ensure that others brought into the area are always escorted.

- Mark potential evidence located within the perimeter so that it will not be disturbed before detailed examination, documentation, and collection. Use available materials, such as rope, traffic cones, or barrier tape, to provide this protection.

A firefighter or investigator should be stationed with evidence until it can be processed if personnel are still operating and there is a potential for damage due to foot traffic or from the operations being conducted. Another objective for limiting access to the fire scene is safety. As the perimeter is established, known hazardous areas should be marked or otherwise secured to prevent injury to personnel operating in the area and bystanders.

Techniques for Evidence Preservation

Firefighters should protect any potential evidence they find by keeping it untouched and undisturbed until an investigator arrives. You should not gather or handle evidence unless it is absolutely necessary in order to preserve it. If you handle, move, or gather evidence, you become part of the chain of custody for that evidence. You must accurately document all actions associated with that evidence as soon as possible because you may be required to testify regarding the evidence. You must know and follow your department's SOPs on evidence gathering and preservation.

Evidence must remain undisturbed except when absolutely necessary for the extinguishment of the fire. You must avoid walking on, cross-contaminating, or destroying potential evidence. The same precaution applies to the excessive use of water to avoid washing the evidence away. Human footprints and tire marks must also be protected. Cardboard boxes placed over footprints can prevent otherwise clear prints from being degraded before they are photographed or cast in plaster. Immediately close dampers and other openings to protect completely or partially burned papers found in a furnace, stove, or fireplace. Leave charred documents found in containers such as wastebaskets, small file cabinets, and binders that can be moved easily. These documents may have been used to start the fire (**Figure 19.23**). Also, though the paper may be burned or appear destroyed, important data may still be recovered during laboratory analysis. Protect these items from air flow that can disturb or scatter them. Plastic tarps or salvage covers may be placed over evidence to indicate its location and protect it.

Overhaul operations can be detrimental to the fire cause investigation. Some departments take great pride in their overhaul work and boast that they leave a building neater, cleaner, and more orderly than it was before the fire. This thoroughness in overhaul is admirable, but in some cases it destroys evidence of how a fire started. To avoid disturbing or destroying evidence, delay overhaul operations until the origin and cause of the fire have been determined. Once critical evidence has been identified and protected, overhaul operations can begin (**Figure 19.24**). Limit or postpone nonessential overhaul operations until the IC or whoever is in charge of the investigation authorizes it.

Even if a scene is secured and access is restricted to only emergency responders or other authorized personnel, contamination or spoliation can occur that would threaten the value of the evidence. Sometimes these contaminations arise as unforeseen consequences of routine activities.

Contamination is a broad concept, encompassing anything that can taint physical evidence **(Figure 19.25)**. Examples of potential sources of contamination include the following:

- Smoking materials such as cigarette butts or matchsticks that have been dropped by firefighters, spectators, or investigators
- Ignitable liquid traces introduced into the scene by items or equipment that have been used or stored at the scene including:
  — Boots/gloves
  — Power cords
  — Tools
  — Power equipment

Figure 19.23 Charred documents may be salvageable if handled by skilled laboratory analysts. *Courtesy of Nicole Fuge, Portland Arson Investigation, Portland, Oregon.*

Figure 19.25 Contamination of a scene includes negligent disposal of common items. *Courtesy of Donny Howard.*

Figure 19.24 Overhaul operations include acts that may significantly change the scene of a fire. *Courtesy of Bob Esposito.*

Spoliation refers to evidence that is destroyed, damaged, altered, or otherwise not preserved by someone who has the responsibility to preserve it. Spoliation occurs when the movement, change, or alteration of the evidence prevents another investigator or interested party from obtaining the same evidentiary, interpretive, or analytical value from the evidence as the initial investigator **(Figure 19.26)**. The legal definition and the consequences of spoliation vary among jurisdictions.

Depending on local policies and procedures, once the fire is out and the investigation complete, the structure may be secured and turned over to the owner/occupant or to law enforcement officials. Leaving the fire scene intact can assist insurance and private investigators who perform investigations on behalf of the owner or an insurance company.

If local protocol is to remove debris from the structure, this is the final activity before securing the structure. Remove charred materials to prevent the possibility of rekindle and to help reduce smoke damage. Unburned materials can be separated from the debris and cleaned. Debris may be shoveled into large containers to reduce the number of trips between the fire area and the debris collection location. Dumping debris onto streets, sidewalks, or shrubbery can cause poor public relations. If a backyard or alley not visible from the street is not available, it may be necessary to dump the debris in a driveway until it can be removed permanently. Dumping debris on inexpensive plastic tarps makes it easier to clean up, protects the drive or yard, and is good for public relations.

Figure 19.26 A scene may be considered spoiled if an investigator did not approve overhaul actions before they took place. *Courtesy of Donny Howard.*

II Advanced Origin, Cause, and Arson Topics

To this point we have discussed the information that a Firefighter I needs to know to recognize the area of origin, fire cause, and evidence indicating an incendiary or arson fire. As a Firefighter II, you must expand your knowledge to include the:

- Types of evidence
- Recognizing important fire cause evidence
- Roles and responsibilities of responders and investigators
- Legal issues of evidence preservation

Types of Evidence

While primarily associated with court cases or legal actions, evidence indicating the area of origin or fire cause can be important in other ways. Generally, evidence is any means of proof that may be presented to prove or disprove a certain hypothesis. Evidence is usually used to support testimony but can sometimes speak for itself. Three primary categories of evidence are direct, circumstantial, and physical.

Direct Evidence

Direct evidence is composed of facts to which a person can attest without further support. Direct evidence is found through the five physical senses. Examples of direct evidence are a person *seeing* (witnessing) another individual pour and ignite gasoline on a floor or observing a coffeemaker as it erupts into flame. Another example is someone smelling propane odors in a structure.

Circumstantial Evidence

Circumstantial evidence supports an inference formed from direct evidence. For example, one could infer that a person set a fire in a building if there was direct evidence that the person was seen carrying a container of ignitable liquid into the building and was seen running from the building as the fire started. One could also infer that a smoldering cigarette started a fire noticed shortly after an ashtray had been emptied into a trash receptacle.

A common misconception concerning circumstantial evidence is that it is not as valuable as direct evidence. Both criminal prosecutions and civil litigation traditionally rely heavily on circumstantial evidence. On occasion, cases may be proven using only circumstantial evidence because direct evidence may not be available.

Physical Evidence

Another form of evidence is known as **physical evidence** and includes physical objects available for inspection. Following are some examples of physical (real) evidence that may relate to fire cases:

- Electrical conductors (**Figure 19.27a, p. 1166**)
- Photographs, film, tape recordings, or videotapes
- Gas can (**Figure 19.27b, p. 1166**)
- Closed sprinkler valve or other disabled device
- Fire patterns (**Figure 19.27c p. 1166**)

Direct Evidence — Type of evidence provided by a witness who obtained it through his or her senses.

Circumstantial Evidence — Evidence presented in a trial that tends to prove a factual matter through inference by proving other events or circumstances.

Physical Evidence — Tangible or real objects that are related to the incident.

- Footwear impressions
- Lack of personal property normally found in an occupied structure
- Warning labels and owner's manuals
- Damaged gas piping
- Appliances (**Figure 19.27d**)

Figure 19.27a The condition of electrical conductors may be submitted as physical evidence. *Courtesy of Donny Howard.*

Figure 19.27b A fuel container in an unexpected location should be considered physical evidence. *Courtesy of Donny Howard.*

Figure 19.27c The proximity of a plausible fire source — in this case, the remains of a grill — near a pattern indicating that it is the point of origin must be considered physical evidence. *Courtesy of Donny Howard.*

Figure 19.27d Charred appliances may show fire patterns that indicate the point of origin and may be submitted as physical evidence. *Courtesy of Donny Howard.*

Recognizing Important Fire Cause Evidence

The particular evidence and information that is collected at the scene will be up to the judgment of the fire investigator. However, you should be able to recognize important evidence and information that will interest the investigator. The following is a partial list of items of evidence and information the investigator may find:

- **Separate fires** — Signs of separate, unrelated fires in different locations and rooms.

- **Timing devices** — Cigarette-match combinations and candles are frequently used timing devices. Wax from a candle often soaks into the floor and can be detected on the scene or from a laboratory test. The spot beneath the candle often is not burned as badly as the surrounding floor. Alarm clocks are not often used as timing devices but should not be discounted. The metal parts of alarm clocks and similar devices are seldom destroyed by fire (**Figures 19.28a and b**).

- **Trailers** — Trailers are used to spread the fire from one area of a structure to another or from one floor level to another. Trailers are usually ordinary combustibles that may be soaked with ignitable liquids. Items used to make trailers may include:
  - Toilet paper
  - Newspapers
  - Black gunpowder
  - Flammable or combustible liquids
  - Wax paper
  - Excelsior
  - Blasting or hobby fuse
  - String
  - Cotton, paper, and similar materials

Figure 19.28a Improvised ignition devices may include the use of common household items.

Figure 19.28b The metal pieces of timing devices are seldom completely destroyed by fire.

- **Chemicals** — Some examples are oxidizers such as swimming pool chlorine products, wood stain, and chemicals used in clandestine drug labs (**Figure 19.29**).

- **Matches** — Matches are not always consumed by the fire. Unburned matchbooks can be compared to a match found at the scene or may carry fingerprints; therefore, handle them carefully.

- **Ignitable liquids** — Flammable and combustible liquids include gasoline, kerosene, solvent, alcohol, paint thinner, acetone, ether, and others.

- **Bottles** — Bottles may be used to hold ignitable liquids such as those used to make Molotov cocktails (**Figure 19.30**). Unburned cloth might be found in the bottle's neck. Some remains of the bottle may be found.

- **Rubber or latex items** — Examination gloves, toy balloons, hot-water bottles, and similar rubber items used to hold ignitable liquids or other flammable products (**Figure 19.31**).

Figure 19.29 Clandestine drug labs have unique markers including specific combinations of chemicals and ignition sources. *Courtesy of MSA.*

Figure 19.31 Rubber or latex items may be used to hold ignitable liquids.

Figure 19.30 Glass bottles that held Molotov cocktails may leave remains in the form of glass shards and unburned cloth.

- **Containers** — Other containers that could have held ignitable liquids can often be found in the structure or on the grounds around the structure.

- **Glass** — Glass that focused the sun's rays on a combustible substance may be recovered at the scene. This type of evidence may be more common in wildland fires than in structural fires.

- **Lighters** — The plastic container from a lighter may be consumed, but the metal portions of the lighter may still be located.

- **Electrical sources** — There can be evidence of electrical heating appliances in contact with common combustibles; overloads on circuits can also be observed.

- **Modified equipment** — Appliances, safety devices, fuel supplies, and other controls that have been tampered with or improperly repaired may malfunction.

- **Items that should not be present and items that should be present but are missing** — Examples of items present would include all of the physical evidence previously mentioned. Items missing may include clothes, firearms, baseball card collection, sentimental items, etc.

- **Oily rags** — The ash of an oily rag, tablecloth, or clothing may retain its shape and may be readily identifiable.

- **Fire patterns** — Fire patterns are physical evidence and should be documented.

Roles and Responsibilities of Responders and Investigators

Determining the cause of a fire may be a relatively simple procedure that requires only the expertise of the firefighters on the scene. More complex fire incidents or possible crime scenes often require the additional assistance of law enforcement personnel and qualified fire investigators at the local, state, provincial, or federal level. The guidelines for requesting the assistance of each of these personnel vary among jurisdictions and according to the nature of the incident. You should be aware of your role and responsibilities as well as those of fire, criminal, and insurance investigators.

The Firefighter

In most jurisdictions, the fire chief has the legal responsibility for determining the origin and cause of a fire. The fire chief delegates this authority to the fire officers and firefighters at the scene. Properly trained firefighters should be able to recognize and collect important information about the fire and its behavior during the response, upon arrival, when entering the structure or fire scene, and while locating and extinguishing the fire. The first-arriving firefighters are in the best position to observe unusual conditions that may indicate an incendiary fire.

First-arriving firefighters should make mental notes of the following:

- Vehicles and people present in the area

- Status of doors and windows (locked or open)

- Evidence of forced entry by anyone other than firefighters (**Figure 19.32**)

- Condition of contents of the rooms, whether in usual order, ransacked, or unusually bare

- Indications of unusual fire behavior or more than one point of origin

Figure 19.32 Evidence of forced entry may indicate illegal forced entry. *Courtesy of Donny Howard.*

During fireground operations, firefighters must be aware that what they do and how they do it can affect the determination of the origin and cause of the fire. Having an alert and open mind combined with performing judicious and careful overhaul might also uncover or preserve important evidence that would otherwise be lost. **Skill Sheet 19-II-1** describes how a Firefighter II can protect evidence of fire cause and origin.

You should refrain from expressing personal opinions to anyone (even other firefighters) about the probable cause. The property owner, news media, or other bystanders could overhear these opinions and report them as fact. Unauthorized remarks that are published or broadcast can be very embarrassing to the fire department. Such remarks can also impede the efforts of an investigator to prove malicious intent as the fire cause. A sufficient reply to any question concerning cause is, "The fire is under investigation."

After the fire investigator arrives, you should make your statements only to him or her. Any public statement regarding the fire cause should be made only after the investigator and fire officer in charge have agreed to its accuracy and validity and have given permission for it to be released.

Figure 19.33 A fire investigation unit carries specialty tools and equipment. *Courtesy of Ron Jeffers.*

The Fire or Criminal Investigator

Fire marshals, fire inspectors, fire investigators, or other members of the fire department are responsible for conducting detailed investigations and analysis beyond the initial determination of fire origin and cause. In many departments, fire investigators are also sworn peace officers who are authorized to carry weapons and make arrests. Firefighters may be interviewed by an investigator or asked to assist in some aspect of an investigation.

Some fire departments have special fire investigation or arson units (**Figure 19.33**). In other departments, fire department and law enforcement personnel work together. In some localities the police department or sheriff's office has sole responsibility for handling an arson investigation. In other areas, the responsibility for cause determination and investigation lies with the state fire marshal, state police, or some other state agency rather than with local agencies.

Assisting these investigators is often necessary when the fire cause cannot be determined, there is a fatality resulting from the fire, or when an incendiary fire is suspected. You must provide professional assistance, accurate information, and be courteous in your relationship with them. Remember that both firefighters and fire investigators have the same mission: to protect the public.

Insurance Investigators

Insurance companies employ investigators who determine the cause, amount of loss, and in some cases liability for fires involving property they insure. They generally begin their investigations after the fire is over and the fire department has left the scene. Private investigators may also conduct separate investigations on behalf of owners, occupants, or other interested parties. You may be interviewed by any of these officials following the incident. You may even be required to provide testimony in court based on what you saw or experienced during or after the incident. You must work with these officials in a professional and courteous manner, providing unbiased and factual testimony.

Insurance companies may employ or retain insurance adjusters. The insurance adjuster's responsibility is usually to visit the incident site, collect police and fire reports, create diagrams, interview witnesses, and take photos of the scene on the insurer's behalf. Insurance adjusters may, upon notification of a loss, contract the services of specialized fire and arson investigators, bringing their expertise into the investigation. This investigation, even though it is often made simultaneously with law enforcement/fire investigations, remains independent. It is not uncommon for each group to compare their findings and hypotheses of origin and cause. Care must always be exercised to ensure the impartiality of each report. Contradictory conclusions usually result from differing investigative methodologies or in some instances, due to an oversight by one team of investigators.

Legal Issues of Evidence Preservation

The location and collection of evidence used in fire cause determination and arson prosecution is based on a variety of laws. These laws change constantly and vary among jurisdictions. You must be familiar with the most common legal issues that govern the collection of evidence and entry into private property. In each of these cases, the use or assistance by law enforcement agencies may be necessary. Important legal issues include:

- **Right of entry** — When responding to an emergency, fire department personnel have the right to enter and remain on the subject premises. They have the continued right to remain on the scene while dealing with the emergency. Once control of the scene has been relinquished, however, fire department personnel can only reenter the scene after they have secured permission from the owner/occupant, have obtained a court-authorized warrant, or **exigent circumstances** exist, such as the appearance of smoke or the rekindle of the fire. Therefore, it is essential that control of the property be retained until all evidence has been collected or photographed.

- **Search and seizure** — A fire scene investigation by its nature is a search in legal terms. A person must have the right of entry to search property, either as part of mitigating an emergency, with the consent of the owner or occupant, or with a warrant. In order to remove property that may be relevant to a fire origin and cause investigation, fire department personnel will need either consent of the owner or a court-authorized warrant. In Canada, after an emergency has ended, either a warrant or legislative authority is required to search or seize property.

- **Statements and the Miranda warning** — If a fire investigation turns into a criminal investigation and an individual becomes the focus of the investigation, the individual must be advised of his or her constitutional rights before being placed in custody or interviewed. You must recognize if and when an investigation may involve a crime. You should then obtain direction from a law enforcement official before proceeding.

- **Chain of custody and continuity of evidence** — Evidence taken from a fire scene must be documented and a chain of custody developed and maintained to ensure the integrity of the evidence.

- **Spoliation** — Evidence that is damaged, altered, lost, or destroyed during the cause determination and collection process is considered to have been affected by spoliation. Spoliation will jeopardize any legal proceedings.

Exigent Circumstances — Right of entry stating that the fire department does not require a warrant to enter a property to suppress a fire, or to remain on the property for a reasonable amount of time afterward in order to determine the origin and cause of the fire.

U.S. Constitutional Protection

The Fourth Amendment to the Constitution of the United States guarantees people protection from unreasonable search of their persons or property or seizure of their property. Therefore, a court-issued warrant is required for fire department personnel to reenter a structure once the incident has been terminated.

Chapter Summary

Before an investigation into the origin and cause of a fire can be conducted, there must be evidence to evaluate. As a firefighter, one of your most important responsibilities is to avoid damaging, altering, or moving evidence while fighting the fire. In the area of origin, you must use appropriate caution when spraying water, moving debris, and even walking around. Once the area of origin is known, a more thorough investigation can be conducted to determine the exact cause of the fire. As a firefighter, you might be assigned to determine the cause of the fire; more likely, you may be assigned to assist your supervisor or a fire investigator in making that determination. If the fire origin and cause investigation reveals evidence of an incendiary fire, the property becomes a crime scene and must be treated as one. You and your fellow firefighters must cooperate fully with whoever is assigned to investigate the crime.

Review Questions

Firefighter I

1. What ways can a Firefighter I recognize obvious signs of a fire's area of origin?

2. What is the relationship between fire cause classifications and cause determination?

3. How can a Firefighter I recognize signs of arson?

4. Why is preserving evidence an important task?

5. What are some techniques firefighters can use to preserve evidence?

Firefighter II

1. What types of evidence are used to indicate an area of origin or a fire's cause?

2. How can a Firefighter II recognize fire cause evidence?

3. Who are the types of responders and investigators that may be involved in a fire investigation?

4. What legal issues impact the location and collection of evidence during a fire investigation?

Step 1: Protect potential evidence.

  a. Avoid touching, disturbing, or tramping on evidence

  b. Avoid using excessive water during extinguishment once fire is under control and evidence has been identified

  c. Leave evidence in place unless it must be moved to preserve it

Step 2: Preserve evidence as necessary.

  a. Move evidence only as necessary to preserve it

  b. Provide security for the evidence until an investigator is available

Step 3: Move evidence as necessary.

  a. Avoid damage to evidence

  b. Provide security for the evidence until an investigator is available

Step 4: Record information about evidence.

  a. Document information about location and appearance of evidence if it must be moved or cannot be preserved

  b. Document location evidence was moved to and the appearance of the location it was moved from

  c. Initiate chain of custody record

Step 5: Provide evidence and records to investigator before leaving incident site.

Courtesy of Ron Moore, McKinney (TX) FD.

Chapter Contents

Key Terms

NFPA® Job Performance Requirements

This chapter provides information that addresses the following job performance requirements of NFPA® 1001, *Standard for Fire Fighter Professional Qualifications* (2013).

Firefighter II

6.5.3

1. Describe fire alarm systems. (6.5.3)

2. Identify alarm initiating devices. (6.5.3)

3. Explain the ways automatic sprinkler systems work. (6.5.3)

4. Describe standpipe and hose systems. (6.5.3)

5. Explain the ways smoke management systems work. (6.5.3)

Chapter 20
Fire Protection Systems

Case History

Over a period of 10 days in January 2012, the Pearland (TX) Fire Department responded to three structure fires that were controlled by automatic sprinkler systems. In each case, the activation of the sprinklers prevented the fire from extending into other portions of the structures.

The first fire occurred at approximately 0830 hrs. on Monday, January 2. Fire department units responded to the report of a fire at a recycling center and found fire and smoke coming from under an exterior awning at the rear of the building. The automatic sprinkler system had activated preventing the fire from spreading into the building.

The second fire occurred in an apartment kitchen at approximately 1600 hrs. on Friday, January 6. A cooking fire in the apartment kitchen was contained when the residential sprinkler system activated. There were no injuries reported and damage was limited to the area of origin.

The final fire occurred in a commercial strip shopping center. It was reported at approximately 0500 hrs. on Tuesday, January 10. An unattended candle caused the accidental fire that was controlled by the sprinkler system.

Each incident is an example of the importance of sprinkler systems in protecting property and reducing fire damage. Combined with early notification and rapid response by fire units, secondary water damage can also be limited. According to the NFPA®, the operation of one sprinkler can control 90 percent of fires.

As a Firefighter II, you must be able to identify fire protection systems during preincident surveys. You must also understand how they operate and be able to shut down the systems to limit water damage to the structure.

Fire protection systems provide some of the following functions:

- Notifying building occupants to take necessary action to escape in the event of a fire
- Detecting the presence of fire or products of combustion
- Summoning organized assistance to initiate or assist in fire control activities
- Initiating automatic fire control and suppression systems and sounding an alarm

- Supervising fire control and suppression systems to ensure that operational status is maintained
- Operating ventilation systems to remove smoke and other products of combustion
- Initiating a wide variety of auxiliary functions involving environmental, utility, and process controls including control of elevators

  Fire protection systems include:
- Alarm systems
- Alarm-initiating devices
- Automatic sprinkler systems
- Standpipe and hose systems
- Smoke management systems
- Building controls used in a coordinated fire protection system

Ⅱ Fire Alarm Systems

An important life safety component of **fire protection systems** is the fire alarm system. Fire alarm systems provide notification of an emergency condition to building occupants and in some cases the local emergency response organization. These systems can range from simple to complex. The authority having jurisdiction (AHJ) typically establishes requirements for system design and installation and often requires systems to be tested regularly.

Alarm System Components

There are several components that make up a typical alarm system. These include the fire alarm control panel (FACP), power supplies, initiating devices, and notification appliances. The components of alarm systems, including the fire command center, are described in more detail in the following sections.

Fire Alarm Control Panel (FACP)

The **fire alarm control panel (FACP)** contains the electronics that control and monitor the fire alarm system (**Figure 20.1**). In general terms, the FACP basically serves as the "brain" for the alarm system. It receives signals from alarm initiating devices, processes the signals, and produces output signals. Power and fire alarm circuits are connected directly into this panel. In addition, the power boosters and power supplies for the notification system are considered to be part of the FACP.

Controls for the alarm system are located in the FACP. The FACP can also perform other functions such as the control of a remote **annunciator panel**, the operation of relays that capture and recall elevators, or public address and mass notifications (**Figure 20.2**).

Power Supplies

Alarm systems are served by both a primary and secondary power supply (**Figure 20.3**). The primary power supply is usually obtained from the building's main connection to the local utility provider. If the primary power supply is interrupted due to a loss of power, an alarm signal must be activated.

A secondary power supply must also be provided for the system. This secondary supply ensures that the system will still operate even if the primary system fails. Secondary power sources can include batteries with chargers and auxiliary generators.

Fire Protection System — System designed to protect structure and minimize loss due to fire.

Fire Alarm Control Panel (FACP) — System component that receives input from automatic and manual fire alarm devices and may provide power to detection devices or communication devices.

Annunciator Panel — Electrical device used to indicate the source or location of an activated fire alarm initiating device or the status of the system. The panel may include individual lights located on a schematic map and an audible alarm signal.

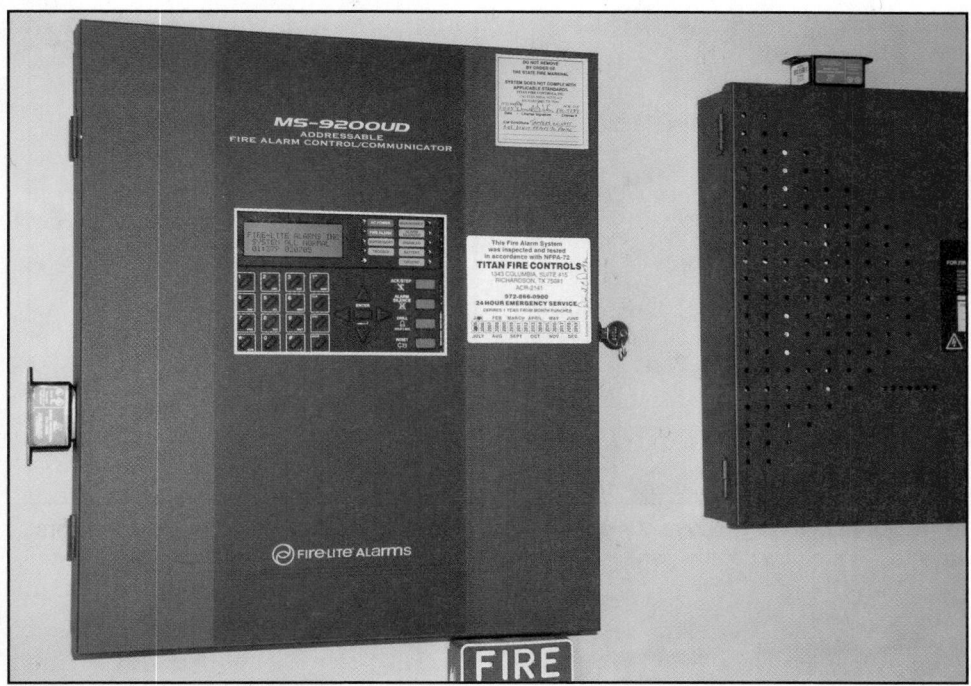

Figure 20.2 An annunciator panel indicates the location of an activated alarm.

Figure 20.1 A fire alarm control panel controls and monitors the fire alarm system for a building or series of buildings. *Courtesy of Ron Moore, McKinney (TX) Fire Department.*

Figure 20.3 Primary and secondary power supplies provide alarm systems with electricity during normal and interrupted conditions. *Courtesy of Ron Moore, McKinney (TX) Fire Department.*

Figure 20.4a Manual pull stations send a signal to a fire alarm control panel.

Figure 20.4b Automatic smoke detectors trigger an alarm when specific conditions are present.

Initiating Devices

Initiating devices may be manually operated (pull stations) or automatic (smoke detectors). Automatic devices sense the presence of products of combustion or other hazardous conditions **(Figures 20.4a and b)**. These devices send a signal to the FACP and may either be hard-wired or connected over a special radio frequency. Initiating devices will be described in more detail later in this chapter.

Notification Appliances

Once an initiating device sends a signal to the FACP, the control unit processes the signal and activates local and remote notification appliances. The system may include the ability to send a signal to a central alarm monitoring center or directly to the fire department. Local notification devices that alert the building occupants include **(Figures 20.5a-d)**:

- Bells
- Horns
- Recorded voice messages
- Strobe lights
- Speakers
- Buzzers

Audible notification signaling appliances are the most common types of alarm-signaling systems used. Depending on the design of the system, the local alarm may either sound only in the area of the activated detector or sound in the entire facility. Notification appliances fall under the following categories which can be used in any combination:

- **Audible** — Approved sounding devices such as horns, bells, or speakers that indicate a fire or emergency condition
- **Visual** — Approved lighting devices such as strobes or flashing lights that indicate a fire or emergency condition
- **Textual** — Visual text or symbols indicating a fire or emergency condition
- **Tactile** — Indication of a fire or emergency condition through sense of touch or vibration

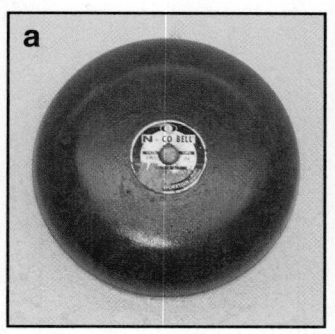

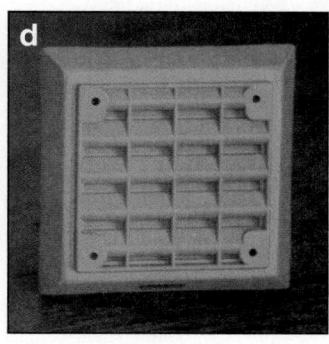

Figure 20.5a-d Local notification devices sound an alarm via sound and/or light by the following devices: a) A bell; b) A horn; c) A strobe/speaker combination unit; d) A buzzer.

Figure 20.6 A fire command center contains equipment used to monitor and control fire protection systems.

Fire Command Center

Many large buildings now incorporate a **fire command center (FCS)** (fire control room or station) that consolidates all of the fire protection system controls for the structure in one location (**Figure 20.6**). A central location allows the various fire protection systems in the structure to be conveniently monitored and controlled as needed. Items contained in the FCS can include the following:

● Fire alarm control panel

● Smoke-control station

● Fire pump status indicators

● Emergency elevator controls

● Emergency communication systems

● Spare sprinklers and fuses

● Building plans and system diagrams

Fire Command Center — Designated room or area in a structure where the status of the fire detection, alarm, and protection systems is displayed and the systems can be manually controlled; may be staffed or unstaffed and can be accessed by the fire department.

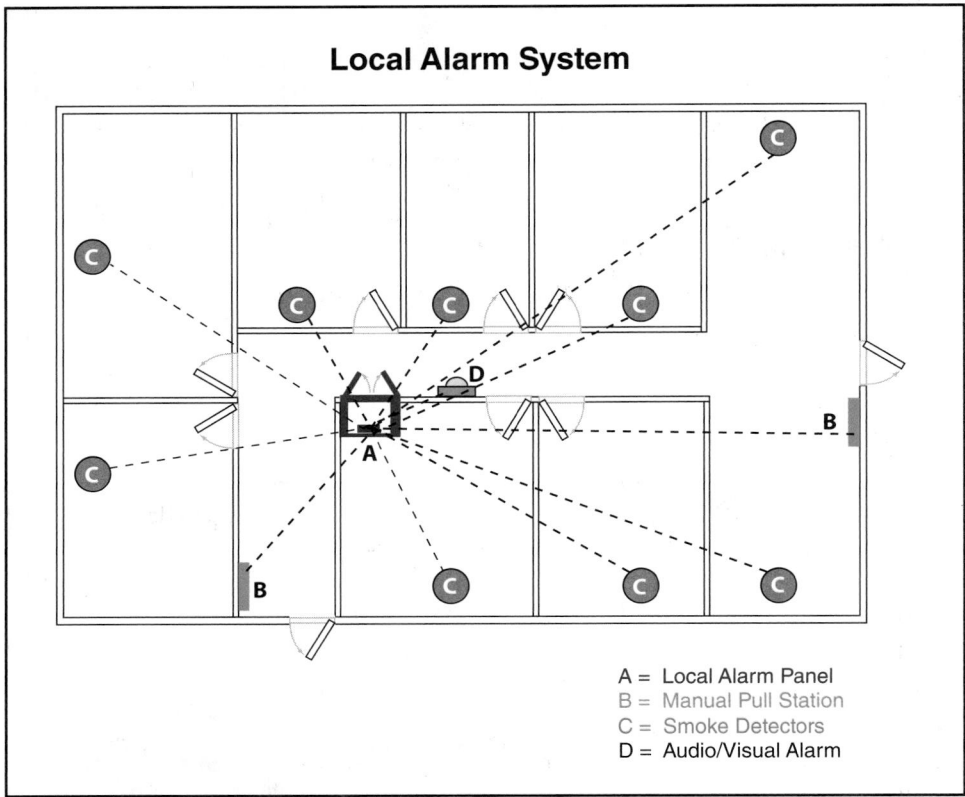

Local Alarm System

A = Local Alarm Panel
B = Manual Pull Station
C = Smoke Detectors
D = Audio/Visual Alarm

Figure 20.7 A local alarm system alerts occupants of one building to an incident occurring on that property.

Types of Alarm Signaling Systems

Once the alarm system receives an indication of an emergency condition, it must signal the appropriate alarm. Simple systems may only sound a local evacuation alarm while complex systems may sound a local alarm, activate building services, and notify appropriate fire and emergency services agencies to respond. Types of alarm signaling systems are introduced in the following sections.

Protected Premises (Local)

A **protected premises system** or *local alarm system* is designed to provide notification only to building occupants on the immediate premises **(Figure 20.7)**. Where these systems are allowed, there are no provisions for automatic off-site reporting. Either manual initiating devices or automatic initiating devices may be used to activate the system. A local system may also be capable of annunciating a supervisory or trouble condition to ensure that service interruptions do not go unnoticed. Local systems can be designed to activate the auxiliary services described later in this chapter. There are three basic types of local alarm systems:

- Noncoded
- Zoned/annunciated
- Addressable

Noncoded alarm. A noncoded system is the simplest type of local alarm. When an alarm-initiating device, such as a smoke detector, sends a signal to the FACP, all of the alarm-signaling devices operate simultaneously. The signaling devices usually operate continuously until the FACP is reset. The FACP is not capable of identifying which initiating device triggered the alarm; therefore, building and fire department

Protected Premises System —
Alarm system that alerts and notifies only occupants on the premises of the existence of a fire so that they can safely exit the building and call the fire department. If a response by a public safety agency (police or fire department) is required, an occupant hearing the alarm must notify the agency.

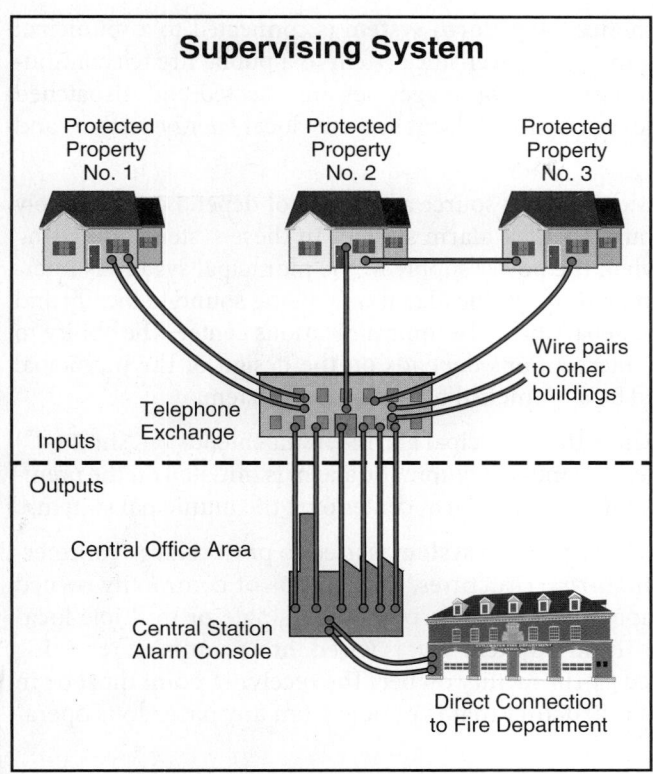

Supervising System

Protected Property No. 1

Protected Property No. 2

Protected Property No. 3

Wire pairs to other buildings

Telephone Exchange

Inputs

Outputs

Central Office Area

Central Station Alarm Console

Direct Connection to Fire Department

Figure 20.8 A supervising system monitors several properties for the purpose of relaying alarms to the appropriate responders.

personnel must walk around the entire facility and visually check to see which device was activated. These systems are only practical in small occupancies with a limited number of rooms and initiating devices.

Zoned/annunciated alarm. Fire-alarm system annunciation enables emergency responders to identify the general location (zone) of alarm device activation. In this type of system, an annunciator panel screen, FACP, or a computer printout visibly indicates the building, floor, fire zone, or other area that coincides with the location of an operating alarm-initiating device. Alarm-initiating devices in common areas are arranged in circuits or zones. Each zone has its own indicator light or display on the FACP. When an initiating device in a particular zone is triggered, the notification devices are activated, and the corresponding indicator is illuminated on the FACP. This signal gives responders a better idea of where the problem is located.

Addressable alarm systems. Addressable alarm systems display the location of each initiating device on the FACP or annunciator panel. Firefighters or building personnel responding to the alarm can use the system to pinpoint the specific device that has been activated. Addressable systems reduce the amount of time it takes to respond to emergency situations. These systems also allow for quick location and correction of malfunctions in the system.

Supervising Station Alarm Systems

A supervising station alarm system continuously monitors a remote location for the purpose of reporting a supervisory, trouble, or alarm signal to the appropriate authorities (**Figure 20.8**). Supervising station alarm systems are the predominant types of signal-monitoring systems used in the United States. Types of supervising station systems include:

- Auxiliary alarm systems
- Central station systems
- Proprietary systems
- Remote receiving systems

Auxiliary alarm systems. An **auxiliary alarm system** is connected to a municipal fire alarm system. Alarms are transmitted over this system to a public fire telecommunications center where the appropriate response agencies are selected and dispatched to the alarm. There are two types of auxiliary alarm systems: local energy systems and shunt systems.

A **local energy system** has its own power source and does not depend on the supply source that powers the entire municipal fire alarm system. In these systems, initiating devices can be activated even when the power supply to the municipal system is interrupted. However, interruption may result in the alarm only being sounded locally and not being transmitted to the fire department telecommunications center. The ability to transmit alarms during power interruptions depends on the design of the municipal system. These systems are served by a municipal fire alarm box system.

Shunt systems are those in which the municipal alarm circuit extends (is "shunted") into the protected property. When a manual or automatic alarm is initiated on the premises, the alarm is instantly transmitted to the alarm center over the municipal system.

Proprietary systems. A **proprietary alarm system** is used to protect large commercial and industrial buildings, high-rise structures, and groups of commonly owned facilities such as a college campus or industrial complex in single or multiple locations. Each building or area has its own system that is wired into a common receiving point that is owned and operated by the facility owner. The receiving point must be in a separate structure or a part of a structure that is remote from any hazardous operations **(Figure 20.9)**.

Personnel who are trained in the system's operation and can take appropriate actions when alarms activate should continuously staff the receiving station. The operator should be able to automatically summon a fire department response through using system controls or the telephone.

Central station systems. A **central station system** is monitored by contracted services at a receiving point called the *central station*. When an alarm is activated at a particular client's location, central station employees receive that information and contact local emergency services and representatives of the occupancy. A central station is required to have the ability to quickly remedy issues with the system.

Remote receiving systems. **Remote receiving systems** are common in jurisdictions that do not require central station systems. These systems are not connected to the emergency services telecommunications center through a municipal alarm box system. Instead, the remote system is connected by another means, usually a telephone line. Where permitted, a radio signal over a dedicated radio frequency may also be used. Remote receiving systems do not incorporate the use of a runner service.

Depending on local requirements, the fire department may approve other organizations to monitor the remote system. In some small communities, particularly those with volunteer fire departments, the local emergency services telecommunications center will monitor the remote system. In these cases, it is important that emergency services telecommunications personnel are aware of the importance of these alarm signals and trained in the actions that must be taken upon alarm receipt.

Alarm-Initiating Devices

Alarm-initiating devices identify the presence of fire or products of combustion and send a signal to the alarm system. Manual devices allow occupants to activate the alarm system. Automatic initiating devices are designed to detect heat, smoke, fire gases, flame, or a combination of these. The type of device used depends greatly on the type of occupancy and its contents.

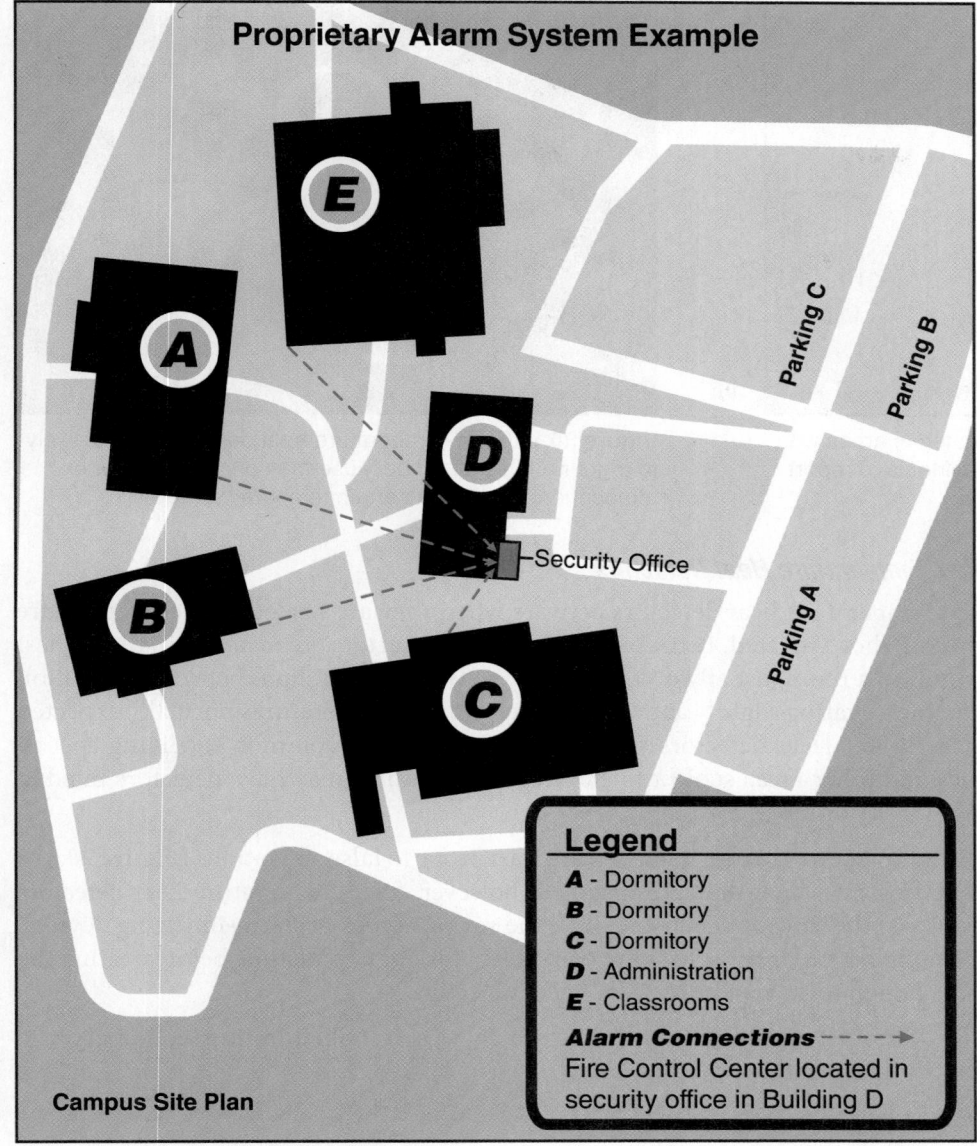

Proprietary Alarm System Example

Parking C

Parking B

Parking A

Security Office

Legend

A - Dormitory

B - Dormitory

C - Dormitory

D - Administration

E - Classrooms

Alarm Connections - - - - →

Fire Control Center located in security office in Building D

Campus Site Plan

Figure 20.9 A proprietary alarm system protects structures and commonly owned facilities for the purpose of maintaining a single relay point.

Pull Stations

Pull stations are alarm-initiating devices that allow occupants to manually initiate the fire signaling system. **Manual pull stations** may be connected to systems that sound local alarms, off-premise alarm signals, or both. Although they come in a variety of shapes and sizes, pull stations are required to be red with white lettering that specifies what the device is and how it is to be used **(Figure 20.10, p. 1186)**. Some jurisdictions allow firefighters to reset pull stations after activation while others require the building owner or their representative to reset the system.

Heat Detectors

Heat detectors activate when the temperature in a monitored area reaches a predetermined threshold. While there are many different types of heat detectors, they all operate on two different principles: fixed-temperature and rate-of-rise.

Alarm-Initiating Device — Alarm system component that transmits a signal when a change occurs; change may be the result of an action such as the activation of a manual fire alarm box, the presence of products of combustion in the atmosphere, or the automatic activation of a supervisory switch.

Manual Pull Station — Manual fire alarm activator.

Heat Detector — Alarm-initiating device that is designed to be responsive to a predetermined rate of temperature increase or to a predetermined temperature level.

Figure 20.10 Manual pull stations are located in areas with a high population density to allow local residents to report a fire.

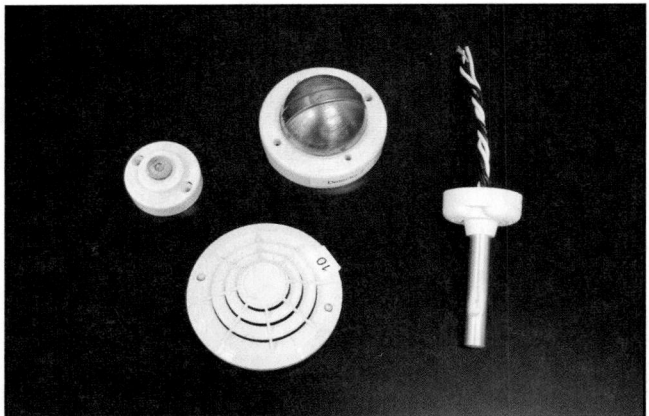

Figure 20.11 Fixed-temperature heat detectors are unlikely to trigger a false alarm because they only activate when the atmosphere reaches a certain temperature.

Fixed-Temperature Heat Detectors

Fixed-temperature heat detectors activate when they are heated to the temperature for which they are rated. Because heat rises, heat detectors are installed in the highest portions of a room, usually on the ceiling. A heat detector should have an activation temperature rating slightly above the highest ceiling temperatures normally expected in that space. Heat detectors rated at 165°F (74°C) are common for living spaces. Attics and other areas subject to elevated temperatures may have detectors rated at 200°F (94°C) or more.

Fixed-temperature heat detectors are least prone to false activations **(Figure 20.11)**. Depending on where they are installed, however, fixed-temperature heat detectors can also be the slowest to activate of all the various types of alarm-initiating devices. For example, in a large room, a fire could burn for quite some time before heating the room enough to activate this type of heat detector.

The various types of fixed-temperature devices described in this section activate by one or more of three mechanisms:

- Expansion of heated material
- Melting of heated material
- Changes in resistance of heated material

Fusible devices/frangible bulbs. While these two devices are more commonly associated with automatic sprinklers, they are also used in fire detection and signaling systems. As will be explained in the section on automatic sprinkler systems, the operating principles of these devices are identical to the **fusible links** and **frangible bulbs** used with automatic sprinklers.

A fusible link is a metal strip normally held in place by a solder with a known melting (fusing) temperature **(Figure 20.12)**. Under normal conditions, the device holds a spring-operated contact inside the detector in the open position. When the ambient temperature is raised to the fusing temperature of the device, the solder melts which allows the spring to close the contact points. This action completes the circuit, which sends a signal to the FACP. Once activated, some of these detectors may be restored by replacing the fusible device. Others require the entire heat detector to be replaced.

A frangible bulb in a heat detector holds the electrical contacts apart in a similar manner to a fusible device. A small glass vial (frangible bulb) contains a liquid with a small air bubble **(Figure 20.13)**. The bulb is designed to break when the liquid is

Fusible Link — Connecting link device that fuses or melts when exposed to heat; used in sprinklers, fire doors, dampers, heat detectors, and ventilators.

Frangible Bulb — Small glass vial fitted into a detector or sprinkler. The glass vial is partly filled with a liquid that expands as heat increases. At a predetermined temperature, vapor pressure causes the glass bulb to break, which activates the device.

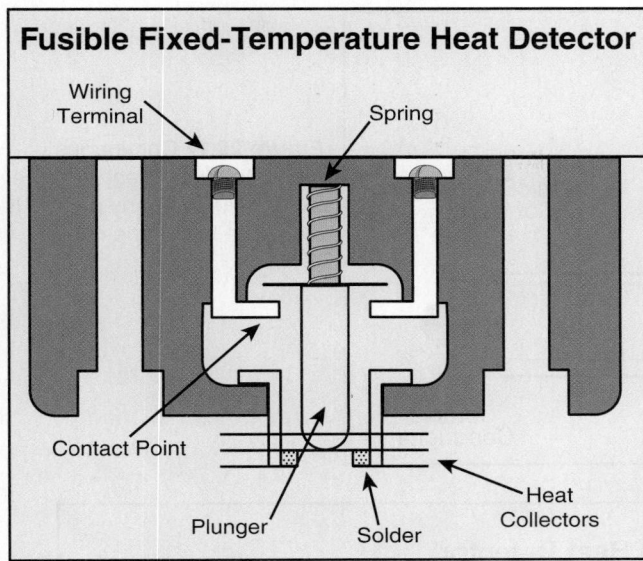

Fusible Fixed-Temperature Heat Detector

Wiring Terminal

Spring

Contact Point

Plunger

Solder

Heat Collectors

Figure 20.12 A fusible link is a fixed-temperature detector that uses solder with a known melting point to separate a spring from the contact points.

Figure 20.13 Frangible bulb detectors are still in service, but their manufacture has been discontinued.

heated to a predetermined temperature. As the liquid in the bulb is heated, it expands and absorbs the air bubble. When the rated temperature is reached, the bulb fractures and falls out, allowing the contacts to complete the circuit, and initiate an alarm. In order to restore the system, the entire detector must be replaced. While detectors of this type are still in service, their manufacture has been discontinued.

Continuous line detector. Most of the detectors described in this chapter are of the spot type; that is, they are designed to detect heat only in a relatively small area surrounding the specific spot where they are installed. However, continuous line detection devices can detect heat over a linear area parallel to the detector.

One such device consists of a cable with a conductive metal inner core sheathed with stainless steel tubing (**Figure 20.14, p. 1188**). The inner core and the sheath are separated by an electrically insulating semiconductor material that keeps them from touching but allows a small amount of current to flow between the two. This insulation loses some of its electrical resistance capabilities when the ambient temperature reaches a predetermined level. When this happens, the current flow between the two components increases and a signal is transmitted to the FACP. This type of detection device restores itself when the ambient temperature is reduced.

Another type of continuous line detector uses two insulated wires with an outer covering (**Figure 20.15, p. 1188**). When the rated temperature is reached, the insulation melts and allows the two wires to touch which completes the circuit and sends an alarm signal to the FACP. To restore this type of line detector, the fused portion of the wires must be cut out and replaced with new wire.

Bimetallic detector. Bimetallic heat detectors use two metals that have different thermal expansion characteristics. Thin strips of the metals are bonded together, and one or both ends of the strips are attached to the alarm circuit. When heated, one metal expands faster than the other, causing the strip to arch or bend. This bending of the strip either makes or breaks contact in the alarm circuit, sending a signal to the alarm panel. Most bimetallic detectors will reset automatically when cooled but must be inspected postfire to ensure that they were not damaged.

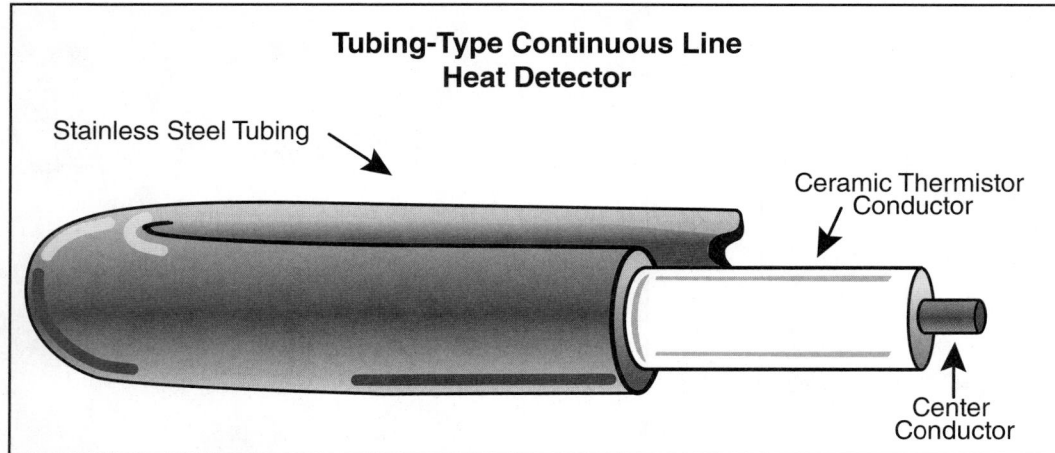

Tubing-Type Continuous Line Heat Detector

Stainless Steel Tubing

Ceramic Thermistor Conductor

Center Conductor

Figure 20.14 Continuous line detectors detect extreme temperatures at any point along the line of the cable.

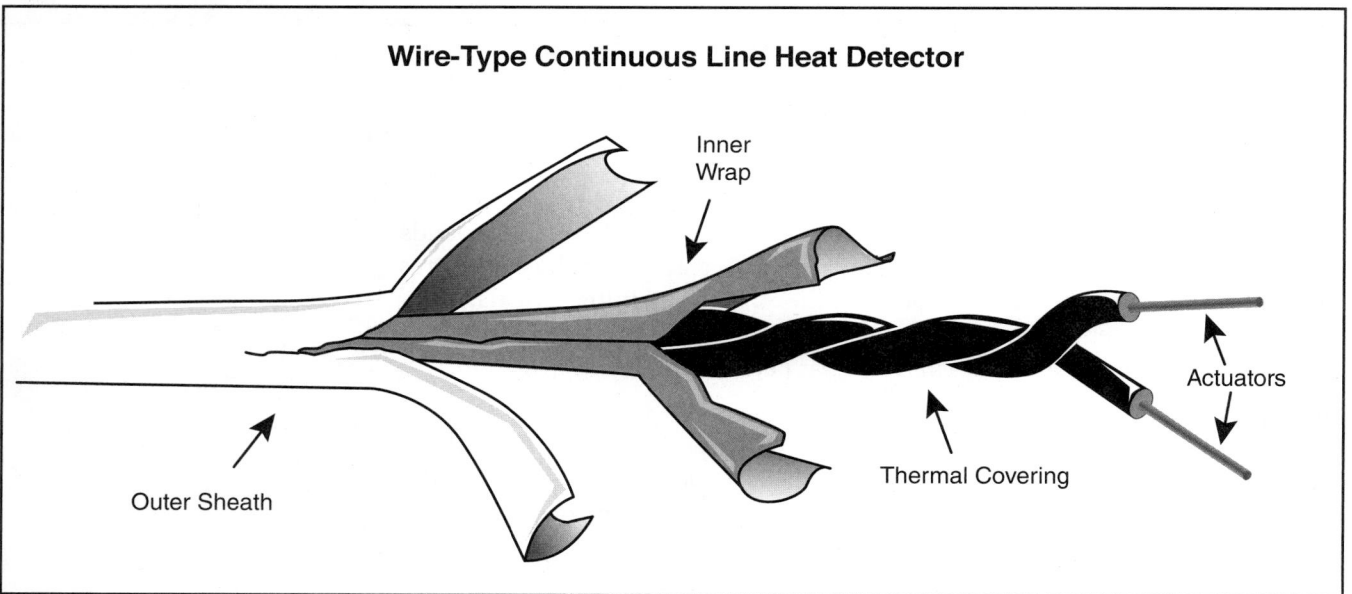

Wire-Type Continuous Line Heat Detector

Inner Wrap

Actuators

Thermal Covering

Outer Sheath

Figure 20.15 Some continuous line detectors are constructed of a pair of wires individually wrapped in material that loses its insulation properties at a certain temperature.

Rate-of-Rise Heat Detectors

Rate-of-Rise Heat Detector — Temperature-sensitive device that sounds an alarm when the temperature changes at a preset value, such as 12°F to 15°F (7°C to 8°C) per minute.

Rate-of-rise heat detectors operate on the assumption that the temperature in a room will increase faster from a fire than from normal atmospheric heating. Typically, rate-of-rise heat detectors are designed to initiate a signal when the rise in temperature exceeds 12°F to 15°F (7°C to 8°C) in one minute. Because the alarm is initiated by a sudden rise in temperature regardless of the initial temperature, an alarm can be initiated at a room temperature far below that required for initiating a fixed-temperature device.

Most rate-of-rise heat detectors are reliable and not subject to false activations. If they are not properly installed, however, they can be activated under nonfire conditions. For example, if a rate-of-rise heat detector is installed just inside an exterior door in an air-conditioned building, opening the door on a hot day can initiate an alarm because of the influx of heated air. Relocating the detector farther from the doorway or choosing a different type of device should alleviate the problem. Rate-of-rise heat detectors automatically reset if they are undamaged.

The following rate-of-rise heat detectors are described in the sections that follow:

- Pneumatic rate-of-rise line heat detector
- Pneumatic rate-of-rise spot detector
- Rate-compensated detector
- Electronic spot-type heat detector

Pneumatic rate-of-rise line heat detector. Pneumatic rate-of-rise line heat detectors are used to monitor large areas of a building. These detectors consist of a system of pneumatic tubing arranged over a wide area of coverage. The space inside the tubing acts as a pressurized air chamber that allows the contained air to expand as it heats. The heat detector contains a flexible diaphragm that responds to an increase in pressure from the tubing when the air expands at a rate that exceeds the relief capacity of the vent. When the tubing is subjected to a rapid increase in temperature, the pressure increases and the alarm is activated. The tubing in this system must be limited to about 1,000 feet (300 m) in length and should be arranged in rows that are not more than 30 feet (10 m) apart and 15 feet (5 m) from walls.

Pneumatic rate-of-rise spot detector. Pneumatic rate-of-rise spot detectors operate on the same principle as the pneumatic rate-of-rise line heat detector. The major difference between the two is that the spot heat detector is self-contained in one unit that monitors a specific location **(Figure 20.16)**. Alarm wiring extends from the detector back to the alarm panel.

Rate-compensated detector. This detector is designed for use in areas normally subject to regular temperature changes that are slower than those under fire conditions **(Figure 20.17)**. The detector consists of an outer metallic sleeve that encases two bowed struts with electrical contacts that have a slower expansion rate than the sleeve. In the normal position, these contacts do not touch. When the detector is heated rapidly, the outer sleeve expands in length. This expansion reduces the tension on the inner strips and allows the contacts to come together, thus initiating a signal to the FACP.

If the rate of temperature rise is fairly slow, such as 5°F to 6°F (2°C to 3°C) per minute, the sleeve expands slowly enough to maintain tension on the inner strips. This tension prevents unnecessary system activations. However, regardless of the rate of temperature increase, when the surrounding temperature reaches a predetermined point, an alarm signal will be initiated.

Figure 20.16 A spot detector monitors a specific location and sends a signal to the alarm panel.

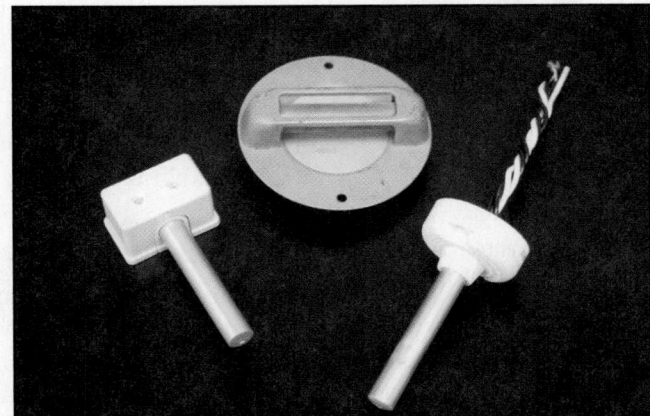

Figure 20.17 A rate-compensated detector triggers an alarm during a rapid rise in temperature. It can accommodate a slower temperature increase.

Figure 20.18 Smoke detectors only monitor an environment for smoke and are incapable of sounding their own alarm; they send a signal to another device to sound the alarm.

Figure 20.19 Smoke alarms monitor an environment and sound a local alarm.

Electronic spot-type heat detector. This type of heat detector consists of one or more temperature-sensitive wires called *thermistors* that produce a marked change in electrical resistance when exposed to heat. The rate at which thermistors are heated determines the amount of current that is generated. Greater changes in temperature result in larger amounts of current flowing and activation of the alarm system. These heat detectors can be calibrated to operate as rate-of-rise detectors and also function at a fixed temperature. They are designed to "bleed" or dissipate small amounts of current, which reduces the chance of a small temperature change activating an alarm.

Smoke Detectors/Alarms

Both **smoke detectors** and **smoke alarms** detect the presence of smoke or other products of combustion. Smoke detectors are only capable of detection and must transmit a signal to another device that sounds an alarm (**Figure 20.18**). In most cases, these devices are installed in nonresidential and large multifamily residential occupancies. Smoke alarms are self-contained units capable of both detecting the presence of smoke and sounding an alarm (**Figure 20.19**). These devices are typically installed in single-family residences and smaller multifamily residential occupancies.

Because a smoke detector can respond to smoke or other products of combustion generated very early in the growth stage — well before sufficient heat is produced to initiate an alarm — it can initiate an alarm much more quickly than a heat detector. For this reason, the smoke detector is the preferred type of detector in many types of occupancies especially residences. The two basic types of smoke detectors (photoelectric and ionization) are described in the following sections, along with an explanation of power sources for smoke alarms.

Photoelectric Smoke Detectors

A **photoelectric smoke detector**, sometimes called a *visible products-of-combustion detector*, works well on all types of fires and usually responds more quickly to smoldering fires than ionization-type detectors. Photoelectric smoke detectors automatically reset when conditions return to normal.

A photoelectric smoke detector consists of a photoelectric cell coupled with a specific light source. The photoelectric cell functions in one of two ways to detect smoke — projected beam application (obscuration) or refractory application (scattered).

Smoke Detector — Alarm-initiating device designed to actuate when visible or invisible products of combustion (other than fire gases) are present in the room or space where the unit is installed.

Smoke Alarm — Device designed to sound an alarm when the products of combustion are present in the room where the device is installed. The alarm is built into the device rather than being a separate system.

Photoelectric Smoke Detector — Type of smoke detector that uses a small light source, either an incandescent bulb or a light-emitting diode (LED), to detect smoke by shining light through the detector's chamber: smoke particles reflect the light into a light-sensitive device called a photocell.

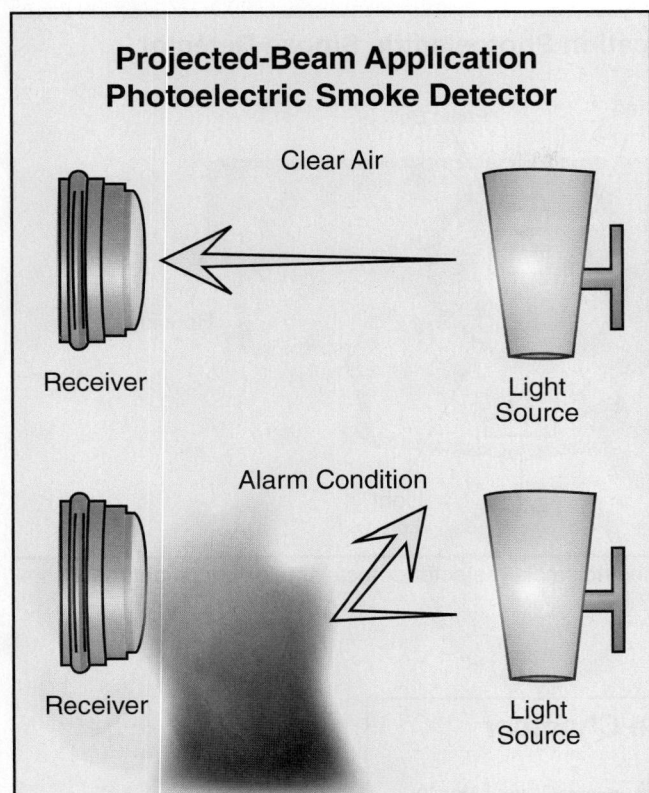

Projected-Beam Application Photoelectric Smoke Detector

Clear Air

Receiver

Light Source

Alarm Condition

Receiver

Light Source

Figure 20.20 A projected-beam photoelectric detector senses smoke when light is blocked by smoke from reaching a sensor.

The projected-beam, or light obscuration style of photoelectric detector, uses a beam of light that is focused across the area being monitored onto a photoelectric cell (**Figure 20.20**). The cell constantly converts the beam into electrical current, which keeps a switch open. When smoke interferes with or obscures the light beam, the amount of current produced is lessened. The detector's circuitry senses the change in current and initiates an alarm when a current change threshold is reached. It is important that projected-beam detectors be mounted on a stable stationary surface. Any movement due to temperature variations, structural movement, and vibrations can cause the light beam to misalign with the receiving device.

A refractory application or light-scattering smoke detector uses a beam of light from a light-emitting diode (LED) that passes through a small chamber at a point away from the light source (**Figure 20.21, p. 1192**). Normally, the light does not strike the photocell. When smoke particles enter the light beam, light strikes the particles and reflects in random directions onto the photosensitive device, causing the detector to generate an alarm signal.

Ionization Smoke Detectors

During combustion, minute particles and aerosols too small to be seen by the naked eye are produced. These invisible products of combustion can be detected by devices that use a tiny amount of radioactive material to ionize air molecules as they enter a chamber within the detector. These ionized particles allow an electrical current to flow between negative and positive plates within the chamber. When the particulate products of combustion (smoke) enter the chamber, they attach themselves to electrically charged molecules of air (ions), making the air within the chamber less conductive. The decrease in current flowing between the plates transmits an alarm-initiating signal (**Figure 20.22, p. 1192**).

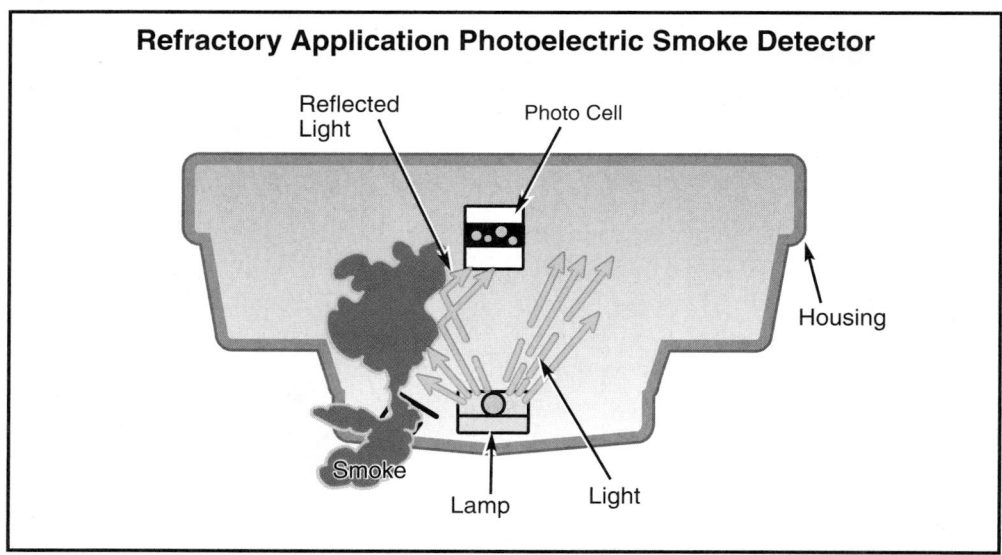

Refractory Application Photoelectric Smoke Detector

Reflected Light

Photo Cell

Housing

Smoke

Lamp

Light

Figure 20.21 A refractory photoelectric smoke detector detects smoke when light reaches a sensor after reflecting off smoke.

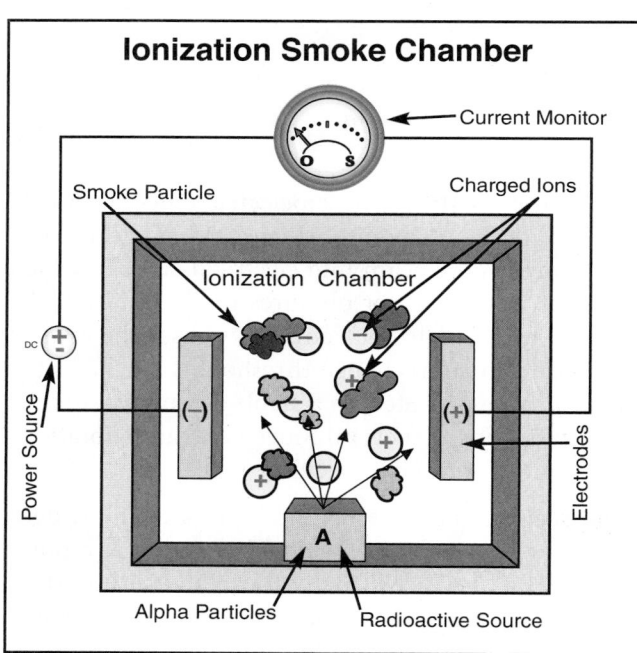

Ionization Smoke Chamber

Current Monitor

Smoke Particle

Charged Ions

Ionization Chamber

Power Source

DC

(−)

(+)

Electrodes

A

Alpha Particles

Radioactive Source

Figure 20.22 An ionization smoke detector monitors charged particles inside the detector.

Ionization Smoke Detector — Type of smoke detector that uses a small amount of radioactive material to make the air within a sensing chamber conduct electricity.

Ionization smoke detectors respond satisfactorily to most fires but these types of detectors generally respond faster to flaming fires than to smoldering ones. They automatically reset when the atmosphere has cleared.

Power Sources

Batteries or household current can be used to power residential smoke alarms. Battery-operated alarms offer the advantage of easy installation and reliability because they can operate during power failures. This feature is especially important when the cause of the fire is a malfunction in the house wiring. However, these alarms are only as reliable as the batteries they use. Traditional alarms use batteries that should be tested monthly and replaced twice a year. Firefighters may suggest that citizens change their smoke alarm batteries each spring and fall when they reset their clocks for daylight saving time. Newer smoke alarms may come equipped with lithium batteries rated for a 10-year service life.

Firefighters should be aware of any state/province or local laws that deal with smoke alarms. Such legislation, in addition to specifying minimum installation requirements for given occupancies (including homes), may designate the power source to be used. Laws requiring hard-wired systems have been enacted because statistics show a growing lack of maintenance in battery-operated alarms (depleted batteries not being replaced). Consequently, many codes requiring alarms in newly constructed homes specify 110-volt, hard-wired units. Alarms powered by household current are usually more reliable. In some rural areas and areas with high thunderstorm occurrence, power failures may be more frequent and battery-operated units may be more reliable. Some hard-wired units also incorporate a battery backup should power failure occur.

Flame Detectors

Three basic types of **flame detectors** (sometimes called *light detectors*) are used to detect the following **(Figure 20.23)**:

- Light in the ultraviolet wave spectrum (UV detectors)
- Light in the infrared wave spectrum (IR detectors)
- Light in both ultraviolet and infrared wave spectrums

While these types of detectors are among the fastest to respond to fires, nonfire conditions such as welding, sunlight, and other sources of bright light may initiate false activations in these detectors. To combat this problem, flame detectors are usually located in areas where these other light sources are unlikely. They are positioned so they have an unobstructed view of the protected area. If their line of sight is blocked by an opaque object, they will not activate.

Because some single-band infrared (IR) detectors are sensitive to sunlight, they are usually installed in fully enclosed areas. To reduce the likelihood of false alarms, most IR detectors are designed to require the flickering motion of a flame to initiate an alarm.

Ultraviolet detectors are virtually insensitive to sunlight, so they can be used in areas not suitable for IR detectors. They are not suitable for areas where arc welding is done or where intense mercury-vapor lamps are used.

Fire-Gas Detectors

When a fire burns in a confined space, it changes the composition of the atmosphere within the space. Depending on the fuel, some of the gases released by a fire may include the following:

- Water vapor (H_2O)
- Carbon dioxide (CO_2)
- Carbon monoxide (CO)
- Hydrogen chloride (HCl)
- Hydrogen cyanide (HCN)
- Hydrogen fluoride (HF)
- Hydrogen sulfide (H_2S)

Flame Detector — Detection and alarm device used in some fire detection systems (generally in high-hazard areas) that detect light/flames in the ultraviolet wave spectrum (UV detectors) or detect light in the infrared wave spectrum (IR detectors).

Figure 20.23 Flame detectors detect ultraviolet and/or infrared wave spectrums.

Figure 20.24 Fire-gas detectors monitor the levels of carbon monoxide and carbon dioxide in the environment.

Figure 20.25 A system of pipes, sprinklers, and control valves are designed to automatically discharge water or extinguishing agent to prevent the spread of fire while first responders travel to the scene. *Courtesy of the U.S. Navy.*

Only water vapor, carbon dioxide, and carbon monoxide are released from all fires. Other gases released vary with the specific chemical makeup of the fuel. Therefore, it is only practical to monitor the levels of carbon dioxide and carbon monoxide for general fire detection purposes. **Fire-gas detectors** will initiate an alarm signal somewhat faster than a heat detector but not as quickly as a smoke detector **(Figure 20.24)**.

Of more importance than the speed of response is the fact that a fire-gas detector can be more discriminating than other types of detectors. A fire-gas detector can be designed to be sensitive only to the gases produced by specific types of hostile fires and to ignore those produced by friendly fires. This feature is needed in industrial occupancies such as refineries, chemical plants, and electronics factories. Compared to the number of other types of detectors, few fire-gas detectors are in use.

Combination Detectors

Depending on the design of the system, various combinations of the previously described means of detection may be used in a single device. These combinations include fixed temperature/rate-of-rise heat detectors, heat/smoke detectors, and smoke/fire-gas detectors. The different combinations make these detectors more versatile and more responsive to fire conditions.

Automatic Sprinkler Systems

An **automatic sprinkler system** is an integrated system of pipes, **sprinklers** (sometimes called *sprinkler heads*), and control valves designed to activate during fires by automatically discharging enough water or extinguishing agent to extinguish the fire or prevent its spread until firefighters arrive **(Figure 20.25)**. The system consists of a series of sprinklers that are systematically arranged so that the system will adequately distribute enough water or extinguishing agent in the protected area. Sprinklers can either extend from exposed pipes or protrude through the ceiling or walls from hidden pipes.

There are two general types of sprinkler coverage: complete sprinkler coverage and partial sprinkler coverage. A complete sprinkler system protects the entire building. A partial sprinkler system protects only certain areas such as high-hazard areas, exit routes, or places designated by code or by the AHJ.

The following NFPA® standards are the primary design and installation criteria for sprinkler systems in most occupancies:

- NFPA® 13, *Standard for the Installation of Sprinkler Systems*
- NFPA® 13D, *Standard for the Installation of Sprinkler Systems in One- and Two-Family Dwellings and Manufactured Homes*
- NFPA® 13R, *Standard for the Installation of Sprinkler Systems in Residential Occupancies Up To and Including Four Stories in Height*

These standards have requirements on the spacing of sprinklers in a building, the size of pipe to be used, the proper method of hanging the pipe, and all other details concerning the installation of a sprinkler system. They also specify the minimum design area that should be used to calculate the system based upon the maximum number of sprinklers that might be expected to activate at one time. A minimum design area is necessary because installing the piping and other components required to adequately supply 500 or 1,000 operating sprinklers would be prohibitively expensive. Thus, the design of the system is based on the assumption that only a portion of the sprinklers will operate during a fire.

In general, reports reveal that only in rare instances do automatic sprinkler systems fail to operate. When failures are reported, the reason is rarely because of failure of the actual sprinklers. A sprinkler system may not perform properly because of:

- Partially or completely closed main water control valve
- Interruption to the municipal water supply
- Damaged or painted-over sprinklers
- Frozen or broken pipes
- Excess debris or sediment in the pipes
- Failure of a secondary water supply
- Tampering and vandalism
- Sprinklers obstructed by objects stacked too close

All components of an automatic sprinkler system should be tested and certified by a nationally recognized testing laboratory such as Underwriters Laboratories Inc. or FM Global. Automatic sprinkler systems are recognized as the most reliable of all fire protection devices and it is essential for firefighters to understand the basic system and the operation of pipes and valves **(Figure 20.26, p. 1196)**. Firefighters should also know the various applications of sprinkler systems along with their effects on life safety.

Effects of Sprinkler Systems on Life Safety

The safety of building occupants is greatly enhanced by the presence of a sprinkler system because it discharges water directly onto a fire while it is still relatively small. Because the fire is extinguished or controlled in the early growth stage, products of combustion are limited. Sprinklers are also effective in preventing the spread of fire upwards in multistory buildings and protecting the lives of occupants in other parts of the building.

There are also times when sprinklers alone are not as effective, such as in the following situations:

- Fires are too small to activate the sprinkler system.
- Smoke generation reaches occupants before the sprinkler system activates.
- Building occupants are sleeping, handicapped, impaired, or incarcerated.

Figure 20.26 All components of a sprinkler system serve specific purposes, and first responders should know how to use the equipment.

Automatic sprinkler systems are also important to the life safety of firefighters. Controlling a fire in its early stages increases the safety of the environment for firefighters to conduct search and rescue and suppression activities. Therefore, it is important that firefighters understand how automatic systems operate so that operations can be conducted more efficiently in sprinkler protected buildings.

Sprinkler System Components

The principal parts of an automatic wet-pipe sprinkler system are illustrated in **Figure 20.27**. The system includes many more components starting with a water main and continuing into the control valve. The **riser** is the vertical piping to which the sprinkler valve, one-way check valve, fire department connection (FDC), alarm valve, main drain, and other components are attached. The **feed main** is the pipe connecting the riser to the **cross mains**. The cross mains directly service a number of branch lines on which the sprinklers are installed. Cross mains extend past the last branch lines and are capped to facilitate flushing. System piping decreases in size from the riser outward. The entire system is supported by hangers and clamps.

The following sections describe many of the components of a sprinkler system, including:

- Sprinklers
- Sprinkler deflectors
- Sprinkler storage
- Control valves
- Operating valves
- Waterflow alarms

Sprinklers

Sprinklers are fixed-spray nozzles that are opened individually. When a heat-responsive element such as a fusible link activates, the cap or plug in the sprinkler opens allowing water to discharge. There are numerous types and designs of sprinklers.

Riser — Vertical water pipe used to carry water for fire protection systems aboveground such as a standpipe riser or sprinkler riser.

Feed Main — Pipe connecting the sprinkler system riser to the cross mains.

Cross Main — Pipe connecting the feed main to the branch lines on which the sprinklers are located.

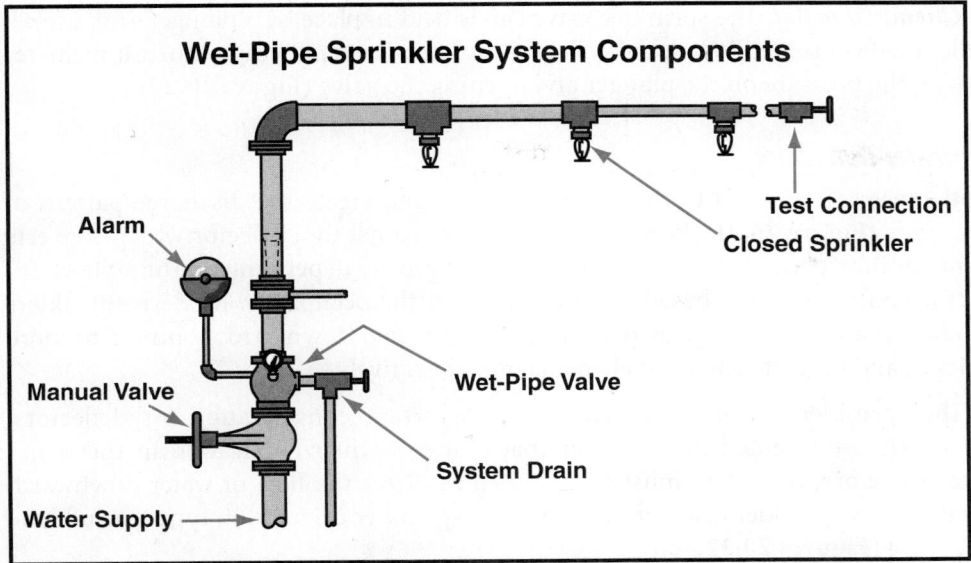

Wet-Pipe Sprinkler System Components

Alarm

Manual Valve

Water Supply

Wet-Pipe Valve

System Drain

Test Connection

Closed Sprinkler

Figure 20.27 A wet-pipe sprinkler system includes several essential components.

Open-type sprinklers with no heat-responsive element can also be encountered, especially in deluge systems. Deluge systems will be described in greater detail in a later section.

To speed the activation of sprinklers, engineers have designed devices known as *early-suppression fast-response* (ESFR) sprinklers. These sprinklers typically react five to ten times faster than traditional sprinklers. ESFR sprinklers can usually be quickly identified because they are larger than traditional sprinklers.

Sprinklers are commonly rated according to the temperature at which they are designed to operate. This temperature is usually identified in one of the following ways:

• Color-coding the sprinkler frame arms
• Color-coding the liquid in frangible bulb-type sprinklers
• Stamping the temperature into the sprinkler itself

Three of the most commonly used release mechanisms to activate sprinklers are fusible links, frangible bulbs, and **chemical pellets**. Each of these sprinkler mechanisms open in response to heat.

Fusible link. The design of a sprinkler using a fusible link involves a frame that is screwed into the sprinkler piping (**Figure 20.28**). Two levers press against the frame and a cap over the orifice from which the water flows. The fusible link holds the levers together until heat from a fire melts the link, after which the water or air pressure in the pipe pushes the levers and cap out of the way. Water then flows from the orifice and strikes the deflector attached to the frame. The deflector converts the standard ½-inch (13 mm) stream into water spray for more efficient extinguishment.

Frangible bulb. Some sprinklers generally incorporate a small bulb filled with heat sensitive alcohol or glycerol liquid and an air bubble to hold the orifice shut (**Figure 20.29**). The air bubble prevents false activations due to normal rise and fall of atmospheric temperature. In a fire, heat expands the liquid until the bubble is absorbed into the liquid, which increases the internal pressure until the bulb shatters at the proper temperature. The type of liquid and the size of the bubble in the bulb regulate the breaking temperature. The liquid is color-coded to designate the designed breaking temperature. When the bulb shatters, the valve cap is released and water is allowed to flow.

Chemical Pellet — A pellet of solder, under compression, within a small cylinder, that melts at a predetermined temperature, allowing a plunger to move down and release the valve cap parts.

Figure 20.28 A fusible link sprinkler activates when the solder fusing the link melts at a specific temperature.

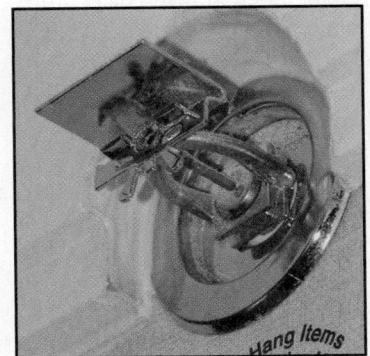

Figure 20.29 A frangible bulb breaks out of the sprinkler at a specific temperature, at which time the extinguishing agent will be able to flow from the affected sprinklers.

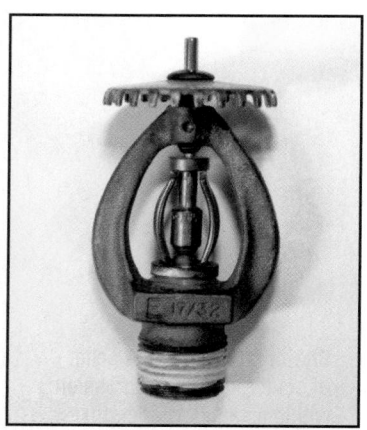

Figure 20.30 A chemical pellet sprinkler activates when a pellet of solder melts at a specific temperature and allows an extinguishing agent to flow.

Deflector — Part of the sprinkler assembly that creates the discharge pattern of the water.

Figure 20.31 A sprinkler deflector creates the discharge pattern of the water.

Chemical pellet. The sprinkler valve cap is held in place by a plunger and a small pellet made of solder. When the pellet reaches its operating temperature, it melts releasing the pressure on the plunger and opening the valve (**Figure 20.30**).

Sprinkler Deflectors

Deflectors are attached to the sprinkler frame and create the discharge pattern of the water (**Figure 20.31**). Pressure forces water against the deflector which converts it into a spray pattern. Discharge patterns vary greatly depending on the style of deflector and are selected based on risk factors in the occupancy. Modern sprinklers produce a uniform discharge pattern that is directed downward, control fire more quickly, and protect structural elements more effectively.

The sprinkler's orientation is also an important consideration for deflectors. Sprinklers are oriented in a manner that best suits their application in the structure. Therefore, deflectors must be designed to direct the flow of water downward, regardless of sprinkler orientation. Common sprinkler orientation types include the following (**Figures 20.32a-e**):

- **Upright** — Designed to deflect the spray of water downward in a hemispherical pattern. Upright sprinklers cannot be inverted for use in the hanging or pendant position because the spray would be deflected toward the ceiling. These are typically used in dry systems.

- **Pendant** — Used where it is impractical or unsightly to use sprinklers in an upright position, such as below a suspended ceiling. The deflector on this type of sprinkler breaks the pattern of water into a circular pattern of small water droplets and directs the water downward.

- **Sidewall** — Used in instances where it may be desirable or required to install sprinklers on the wall at the side of a room or space. This may be for cost-savings or appearance. By modifying the deflector, a sprinkler can be made to discharge most of its water to one side. Sidewall sprinklers are useful in areas such as corridors, offices, hotel rooms, and residential occupancies.

- **Concealed** — Hidden by a removable decorative cover that releases when exposed to a specific level of heat.

- **Flush** — Mounted in a ceiling with the body of the sprinkler, including the threaded shank, above the plane of the ceiling.

- **Recessed** — Installed in recessed housing within the ceiling of a compartment or space; all or part of the sprinkler other than the threaded shank is mounted in the housing.

- **In-Rack** — Typically used in storage facilities. In-rack sprinklers incorporate a protective disk that shields the heat-sensing element from water that is discharged from the sprinklers above.

Sprinkler Storage

A storage cabinet to house spare sprinklers and a sprinkler wrench is usually installed near the sprinkler riser and water control valve (**Figure 20.33**). Sprinkler storage cabinets can also be found in the fire command center (fire control center) if applicable. Normally, these cabinets hold a minimum of six sprinklers and a sprinkler wrench in accordance with NFPA® 13 and 13D. In many jurisdictions, the job of changing sprinklers is the responsibility of the building representatives or individuals who are

Figure 20.32a-e Sprinkler orientations include the following: **a** Upright sprinkler; **b** Pendant sprinkler; **c** Sidewall sprinkler; **d** Hidden/concealed sprinkler; **e** Flush sprinkler

Figure 20.33 A sprinkler storage cabinet is often located near a riser and valve. *Courtesy of Ron Moore, McKinney (TX) Fire Department.*

Figure 20.34a An outside stem and yoke (OS&Y) valve shows the threaded stem when the valve is open.

Figure 20.34b A post indicator valve (PIV) uses a movable plate to indicate in words if the valve is open or shut.

Figure 20.34c A wall post indicator valve (WPIV) extends horizontally through a structure and uses a movable plate to indicate in words if the valve is open or shut.

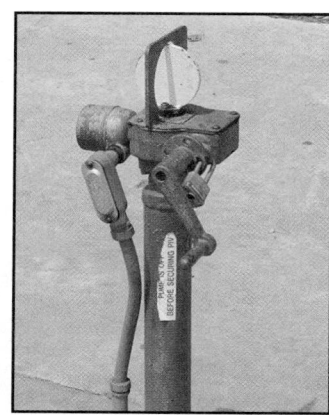

Figure 20.34d A post indicator valve assembly (PIVA) uses a circular disk that rotates on an axis on top of the valve housing to indicate the position of the valve.

qualified to perform work on sprinkler systems. In other jurisdictions, firefighters are allowed to replace fused or damaged sprinklers to restore the system to service sooner. Firefighters must follow their departmental SOPs in this regard.

Control Valves

Every sprinkler system is equipped with a main water control valve. Control valves are used to stop supplying water to the system in order to replace sprinklers, perform maintenance, or interrupt operations.

Most main water control valves are of the indicating type and are manually operated. An **indicating valve** shows at a glance whether it is open or closed. Types of indicating control valves include the following (**Figures 20.34a-d**):

Indicating Valve — Water main valve that visually shows the open or closed status of the valve.

- **Outside stem and yoke (OS&Y) valve** — Has a yoke on the outside with a threaded stem that opens and closes the gate inside the valve housing. The threaded portion of the stem is visible beyond the yoke when the valve is open and not visible when the valve is closed.

- **Post indicator valve (PIV)** — Hollow metal post that houses the valve stem. Attached to the valve stem is a movable plate with the words *OPEN* or *SHUT* visible through a small glass window on the side of the housing. When not in use, the operating handle is locked to the valve housing.

- **Wall post indicator valve (WPIV)** — Similar to a PIV except that it extends horizontally through the wall with the target and valve operating nut on the outside of the building.

- **Post indicator valve assembly (PIVA)** — The PIVA does not use a plate with words *OPEN* and *SHUT* as does a PIV. Instead, a PIVA uses a circular disk inside a flat plate on top of the valve housing. When the valve is open, the disk is perpendicular to the surrounding plate. When the valve is closed, the disk is in line with the plate that surrounds it. Unlike the PIV or WPIV, the PIVA is operated with a built-in crank.

Operating Valves

Sprinkler systems employ various valves such as the alarm test valve, inspector's test valve, and main **drain valve (Figures 20.35a-c)**. The alarm test valve is located on a pipe that connects the supply side of the **alarm check valve** to the **retard chamber** (void that contains excess water from momentary water pressure surges). The alarm test valve can be used to simulate the actuation of the system. It allows water to flow into the retard chamber and operate the water flow alarm devices.

An inspector's test valve is located in a remote part of the sprinkler system. This valve is equipped with the same size orifice as one sprinkler and is used to simulate the activation of one sprinkler. The water from the inspector's test valve normally discharges outside the building.

Every sprinkler system riser has a main drain valve. The primary purpose of the main drain is to allow sprinkler service personnel to drain water from the system for maintenance purposes. Because a large volume of water will flow when the main drain valve is opened, it can also be used to check the system water supply.

NOTE: The water contained in sprinkler systems may be extremely dirty and contaminated. You should wear full PPE including eye protection when operating test valves or drains. Care should also be taken to direct the flow of water away from areas that can be damaged.

Drain Valve — Valve that allows piping to drain when pressure is relieved in the pipe.

Alarm Check Valve — Type of check valve installed in the riser of an automatic sprinkler system that transmits a water-flow alarm when the water flow in the system lifts the valve clapper.

Retard Chamber — Chamber that catches and slows the excess water that may be sent through the alarm valve of an automatic sprinkler system during momentary water pressure surges. This reduces the chance of false-alarm activation. The retard chamber is installed between the alarm check valve and alarm-signaling equipment.

Figure 20.35a An alarm test valve allows water to flow enough to trigger the water flow alarm.

Figure 20.35b An inspector's test valve is located remotely from the main standpipe and is used to simulate the activation of one sprinkler.

Figure 20.35c The main drain valve is the primary means to drain a sprinkler system.

Waterflow Alarms

Automatic sprinkler systems are typically equipped with a **waterflow device** that initiates an alarm when water begins to flow in the system **(Figure 20.36)**. A waterflow device consists of a vane or paddle that protrudes through the riser into the waterway. The vane is connected to an alarm switch on the outside of the riser. When flowing water moves the vane, the alarm switch operates initiating an alarm. The vane must be thin and pliable so that if many sprinklers operate, the water flow will flatten the vane against the wall of the riser, resulting in a clear waterway.

Water Supply

A minimum water supply has to deliver the required volume of water to the highest sprinkler in a building at a residual pressure of 15 psi (105 kPa). The minimum flow depends on the hazard to be protected, the occupancy, and the building contents. A connection to a public water system that has adequate volume, pressure, and reliability is a good source of water for automatic sprinklers. This type of connection is often the only water supply available.

To ensure that adequate water volume and pressure are maintained during periods of demand due to fire, a fire pump is typically incorporated into the sprinkler system **(Figure 20.37)**. Fire pumps are powered by an electric, diesel, or steam pump driver. The pump controller activates the driver and pump when the system pressure falls below a predetermined level.

In many cases, the water supply for sprinkler systems is designed to supply only a portion of the sprinklers actually installed on the system. If a large fire occurs or a pipe breaks, the sprinkler system will need an outside source of water and pressure to do its job effectively. A fire department pumper that is connected to the sprinkler fire department connection (FDC) can provide additional water and pressure. FDCs for sprinklers usually consist of a Siamese inlet with at least two 2½-inch (65 mm) female connections with a clapper valve in each connection, or one large-diameter connection that is attached to a clappered inlet **(Figure 20.38)**.

Figure 20.36 A waterflow alarm indicates the flow of water through the system, and may be accompanied by a system drain valve.

Figure 20.37 A fire pump may be incorporated into a sprinkler system to ensure that adequate pressure is maintained during an incident. *Courtesy of Ron Moore, McKinney (TX) Fire Department.*

Figure 20.38 A clappered inlet allows water to flow in only one direction.

Figure 20.39 A check valve ensures the proper direction of water flow during operations that use a fire department connection.

Sprinkler FDCs should be supplied with water from pumpers that have a capacity of at least 1,000 gpm (4 000 L/min) or greater. A minimum of two 2½-inch (65 mm) or larger hoses should be attached to the FDC. Whenever possible, fire department pumpers supplying attack lines should operate from hydrants connected to mains other than the main supplying the sprinkler system.

After water flows through the fire department connection into the system, it passes through a **check valve (Figure 20.39)**. This valve prevents water flowing from the sprinkler system back into the FDC; however, it does allow water from the FDC to flow into the sprinkler system. The proper direction of water flow through a check valve is usually indicated by an arrow on the valve or by observing the appearance of the valve casing. A ball drip valve may also be installed at the check valve and FDC. This valve is designed to keep both the valve and connection dry and operating properly during freezing conditions.

> **Check Valve** — Automatic valve that permits liquid flow in only one direction.

Departmental preincident plans may identify the pressure at which a sprinkler system should be supported as well as any special circumstances. Such a plan cannot be established until fire department personnel have become familiar with properties protected by automatic sprinkler systems in their jurisdiction. A standard plan of operation should cover the buildings in the department's jurisdiction, including type of occupancy, type of system, and extent of the system. Therefore, a preincident survey is a prerequisite for a plan of operation. A thorough knowledge of the public water system is also important, including knowing the volume and pressure available.

Types of Sprinkler Systems

The following sections highlight the major applications of sprinkler systems. Firefighters should have a basic understanding of the operation of each of the following:

- Wet-pipe
- Dry-pipe
- Deluge
- Preaction
- Special extinguishing systems
- Residential

Figure 20.40 Wet-pipe sprinkler systems are low-maintenance and are found in areas that will not freeze.

Wet-Pipe Systems

Wet-pipe sprinkler systems, sometimes referred to as *straight stick systems*, are used in locations where temperatures remain above 40°F (4°C). A wet-pipe sprinkler system is the simplest type of automatic fire sprinkler system and generally requires little maintenance. This system contains water under pressure at all times. It is connected to a public or private water supply so that an open sprinkler will immediately discharge a water spray in the area and activate an alarm. This type of system includes, at a minimum, an alarm valve, OS&Y valve, and FDC. The system may also include an electronic waterflow alarm, also known as a paddle-type waterflow device **(Figure 20.40)**.

A wet-pipe sprinkler system may be equipped with a retard chamber as part of the alarm check valve. This chamber catches excess water that may be sent through the alarm valve during momentary water pressure surges. This reduces the chances of false alarm activations. Wet-pipe systems are extremely efficient and reliable. Research by the NFPA® indicates that 99 percent of fires in structures protected by these systems are controlled with less than ten sprinklers activated.

Dry-Pipe Systems

Dry-pipe sprinkler systems are used in locations where the piping may be subjected to temperatures below 40°F (4°C). All pipes in dry-pipe systems are pitched (sloped) to help drain the water in the system back toward the main drain. In these systems, air or nitrogen under pressure replaces water in the sprinkler piping above the dry-pipe valve (device that keeps water out of the sprinkler piping until a fire actuates a sprinkler) **(Figure 20.41)**. When a sprinkler activates, the pressurized air escapes first. Then the dry-pipe valve automatically opens to permit water into the piping system.

A dry-pipe valve has larger surface area on the air side of the clapper valve than on the water side. This design allows a small amount of air-pressure above the dry-pipe valve to hold back a much greater water pressure on the water supply side. The valve is equipped with an air-pressure gauge above the clapper and a water-pressure gauge below the clapper. The required air pressure for dry-pipe systems is usually about

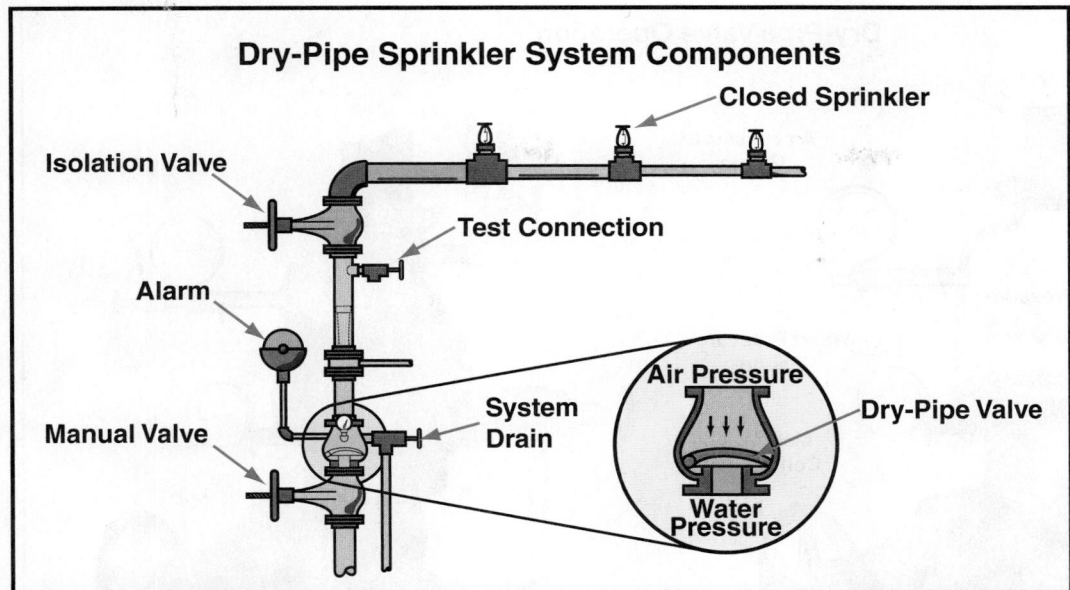

Dry-Pipe Sprinkler System Components

Closed Sprinkler

Isolation Valve

Test Connection

Alarm

Air Pressure

Dry-Pipe Valve

Manual Valve

System Drain

Water Pressure

Figure 20.41 A dry-pipe sprinkler system uses air pressure to seal the water from flowing into the pipes before activation.

20 psi (140 kPa) above the trip pressure. Under normal circumstances, the air pressure gauge will read a pressure that is substantially lower than the water-pressure gauge. If the gauges read the same, the system has been tripped and water has been allowed to enter the pipes.

Figure 20.42, p. 1206 illustrates the dry-pipe valve in both the standby and fire positions. The dry-pipe valve must be located in a heated area of the structure that is maintained at a temperature of 40°F (4°C) or greater. Dry-pipe systems are equipped with either electric or hydraulic alarm-signaling equipment.

In a large dry-pipe system, several minutes could be lost while the air is being expelled from the system. Standards require that a quick-opening device be installed in systems that have a water capacity of over 500 gallons (2 000 L). An accelerator is one type of quick-opening device. The basic purpose of this device is to redirect system air to accelerate the opening of the dry-pipe valve which allows water into the sprinkler system more quickly. An exhauster is another type of quick-opening device. Exhausters quickly expel air from the system to allow water to flow in its place.

Deluge Systems

Deluge sprinkler systems are similar to dry-pipe systems in that there is no water in the distribution piping before system activation. However, in a deluge system, all sprinklers are open all the time (these are known as open-head sprinklers). This means that when the system is activated and water enters the piping, the water will discharge from all of the sprinklers simultaneously. A **deluge valve** controls the flow of water into the system. Fire detection devices (heat, smoke, or flame detectors) installed in the area protected by the system control the operation of the deluge valve **(Figure 20.43, p. 1206)**.

Normally installed in high-hazard occupancies such as aircraft hangars, deluge systems are designed to quickly supply a large volume of water or extinguishing agent to the protected area. A variation of this system is the partial deluge system in which some sprinklers are open and some are not. Another form of deluge system designed to provide exposure protection is called a *water curtain system*.

Deluge Sprinkler System — Fire suppression system that consists of piping and open sprinklers. A fire detection system is used to activate the water or foam control valve. When the system activates, the extinguishing agent expels from all sprinkler heads in the designated area.

Deluge Valve — Automatic valve used to control water to a deluge sprinkler system.

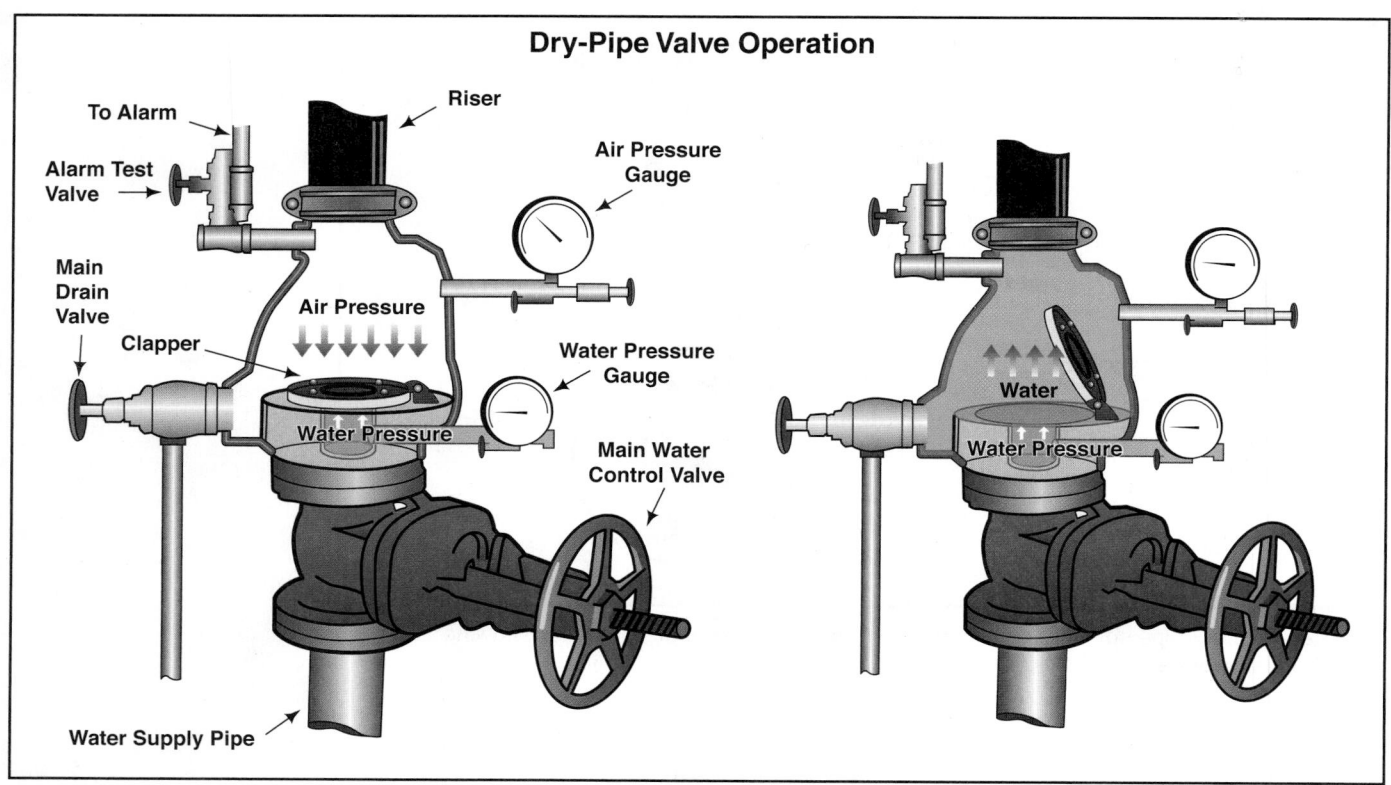

Figure 20.42 A dry-pipe shows different pressures on the air and water gauges before activation.

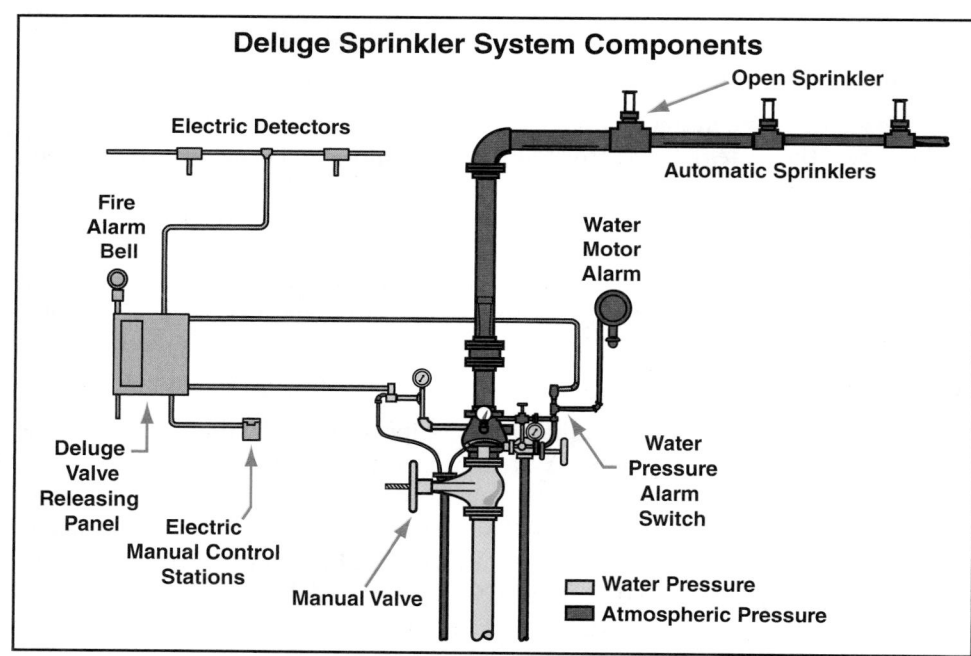

Figure 20.43 A deluge system maintains all of the sprinklers in an open position at all times, so that when the system activates, all sprinklers discharge simultaneously.

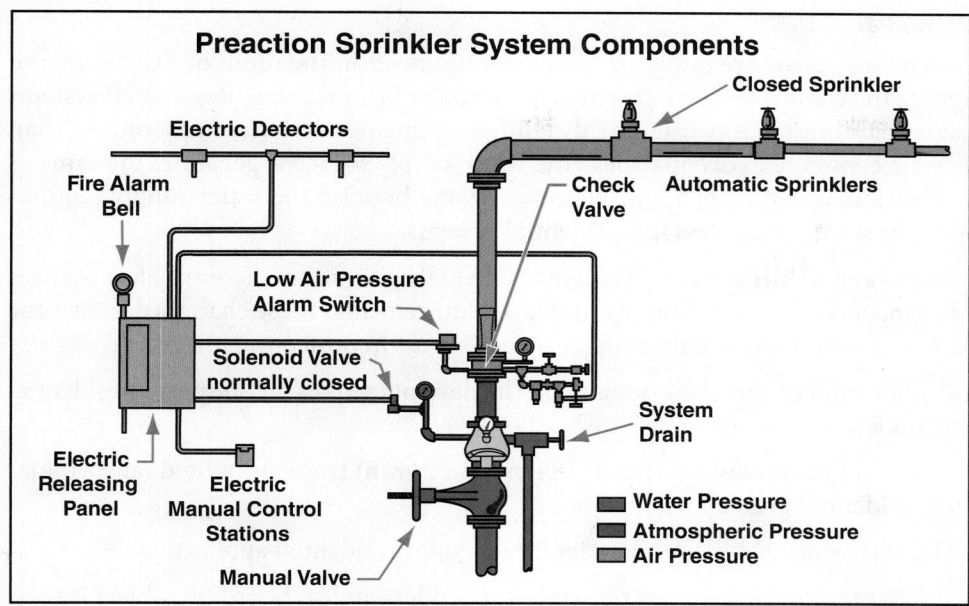

Preaction Sprinkler System Components

Closed Sprinkler

Electric Detectors

Fire Alarm Bell

Check Valve

Automatic Sprinklers

Low Air Pressure Alarm Switch

Solenoid Valve normally closed

System Drain

Electric Releasing Panel

Electric Manual Control Stations

Water Pressure
Atmospheric Pressure
Air Pressure

Manual Valve

Figure 20.44 Preaction systems require sprinklers to be activated by smoke- or heat-detection systems before water will flow to the sprinklers.

Preaction Systems

Preaction sprinkler systems are dry systems that employ a deluge-type valve, fire detection devices, and closed sprinklers (**Figure 20.44**). This type of system is used when it is especially important to prevent water damage, even if pipes are broken. The system will not discharge water into the sprinkler piping except in response to either smoke- or heat-detection system actuation. Fire detection devices operate a release switch located in the system actuation unit. This release switch opens the deluge valve and permits water to enter the distribution system so that water is ready when the sprinklers activate. Once water is in the system it will only discharge through sprinklers that have been activated.

Special Extinguishing Systems

Special extinguishing systems typically use an extinguishing agent other than water or in addition to water. These extinguishing agents are unique to their application in specific occupancies. Some facilities such as industrial manufacturing plants may incorporate several special extinguishing systems depending on the materials and processes that are used.

Types of special extinguishing systems include the following:

- Wet chemical
- Dry chemical
- Clean-agent
- Carbon dioxide
- Water-mist and hybrid systems
- Foam

NOTE: For more information on special extinguishing systems, refer to IFSTA's **Fire Detection and Suppression Systems** manual.

> **Preaction Sprinkler System** — Fire suppression system that consists of closed sprinkler heads attached to a piping system that contains air under pressure and a secondary detection system; both must operate before the extinguishing agent is released into the system.

Residential Systems

Residential systems are designed to prevent flashover in the room of fire origin and improve the chance for the occupants to escape or be evacuated. Residential systems designed for one- and two-family dwellings are smaller and more economical than those for commercial occupancies. The water supply source is generally the same as the domestic water supply. This approach works because the water supply requirements are substantially less for residential systems.

To make sprinkler systems useful in residential application, there are a few changes in design, operation, water supply, and flow requirements. These changes decrease the cost of the system while enhancing its effectiveness in protecting life and property:

- Modification of sprinkler design and the development of fast-response residential sprinklers

- Minimum flow requirements of 18 gpm (68 L/min) from an individual sprinkler for residential protection

- Alarms that are simpler and better designed for residential applications

A major difference between residential sprinklers and standard sprinklers is their sensitivity or speed of operation. Residential sprinklers operate more quickly than standard sprinklers. By designing the fusible link to activate when the ceiling reaches a temperature of 165°F (74°C), the sprinkler will operate before conditions in the room become so untenable that occupants cannot survive.

Sprinkler coverage in residential systems is not as extensive as in standard commercial systems. Sprinklers can be omitted from areas such as garages, carports, closets, and small bathrooms, for example. However, local codes may be amended to include some of these traditionally exempted areas. Residential sprinklers are designed to discharge water higher on the walls of a room to prevent a fire from traveling above the spray, which might occur with burning drapes or in preflashover conditions.

These systems are installed with a variety of piping materials not generally found in commercial installations **(Figure 20.45)**. In addition, piping methods such as multipurpose or combination domestic and sprinkler water lines are used to supply both the domestic needs of a residence and the fire sprinklers.

To be of value, a residential sprinkler system must continually be in service. As with a standard system, inadvertent or deliberate closing of valves renders the system useless. Therefore, using one valve to control both the sprinklers and the domestic water service for the residence eliminates the possibility of the supply valve being closed. The sprinklers cannot be turned off without the household water supply being turned off as well.

The water supply for residential sprinklers may be taken from several sources as follows:

- **Connection to the public water system** — Very reliable and usually provides adequate volume

- **On-site pressure tank** — Often found at rural homes that do not have public water service

- **Storage tank with an automatic pump** — May also be found at rural homes or large homes in urban areas where the public system must be augmented for the size of the residence; may also need a fire pump; should be based upon hydraulic calculations

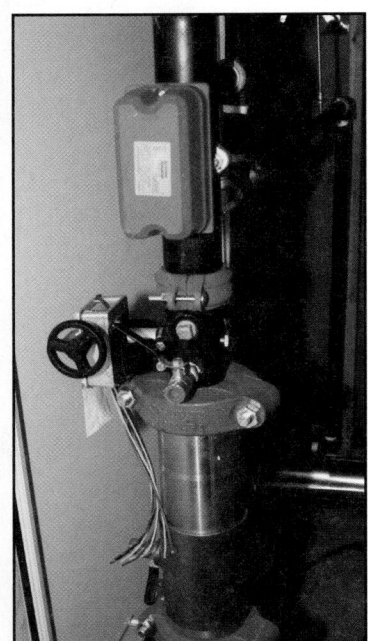

Figure 20.45 Residential sprinkler systems may employ features not found in industrial applications because residential fires tend to include hazards higher along the walls.

Figure 20.46 Standpipe systems allow firefighters to rapidly deploy fire hose from locations remote from the fire apparatus.

Standpipe and Hose Systems

Standpipe and hose systems are designed to provide a means for rapidly deploying fire hoses and operating fire streams at locations that are remote from the fire apparatus **(Figure 20.46)**. **Standpipe systems** can be found in a variety of buildings and structures where required by the AHJ.

The value of standpipes in large area one-story structures is primarily based on expediency. Horizontal standpipes reduce the time and effort needed to manually advance a hoseline several hundred feet (meters) to reach the seat of the fire. During overhaul, horizontal standpipes can also reduce the amount of hose needed to reach areas already controlled by sprinklers. In many high-rise buildings, a standpipe is the primary means for manual extinguishment and overhaul of a fire and is an essential aspect of the building's design.

Depending on the type of system installed, either firefighters or properly trained occupants may use standpipe systems. A reliable water supply may be in place to service the standpipe system. Fire departments may provide the water supply or augment the existing supply using a FDC. The standpipe system may also be part of or separate from an automatic sprinkler, water spray, water mist, or foam-water system.

Although standpipe systems are required in many buildings, they do not take the place of automatic sprinkler systems, nor do they lessen the need for sprinklers. Automatic sprinklers continue to be the most effective method of fire control.

Components and Classifications of Standpipe Systems

Components found in standpipe systems commonly include the following:

- Hose stations
- Water supply
- Waterflow control valves
- Risers
- Pressure-regulating devices
- Fire department connection (FDC)

Standpipe System — Wet or dry system of pipes in a large single-story or multistory building with fire hose outlets installed in different areas or on different levels of a building to be used by firefighters and/or building occupants. The system is used to provide for quick deployment of hoselines during fire fighting operations.

Figure 20.47 A Class I standpipe connection may supply up to a 2½-inch (65 mm) hose.

Figure 20.48 A Class II system may be used by building occupants or fire department personnel, and may support a 1½-inch (38 mm) hose.

NFPA® 14, *Standard for the Installation of Standpipes and Hose Systems*, describes the design and installation of standpipes. This standard establishes three classes of standpipe systems. These classifications are based on the intended use of the hose stations or discharge outlets. NFPA® 13, *Standard for the Installation of Sprinkler Systems*, also contains information on hose stations.

Class I

Class I standpipe systems are primarily for use by fire suppression personnel trained in handling large hoselines **(Figure 20.47)**. Class I systems must be capable of supplying effective fire streams during the more advanced stages of fire within a building. A Class I system provides 2½-inch (65 mm) hose connections or hose stations attached to the standpipe riser. The 2½-inch (65 mm) hose connections may be equipped with a reducer on the cap that allows for the connection of a 1½-inch (38 mm) hose coupling as well.

Class II

A Class II system is primarily designed for use either by building occupants who are trained in its use or by fire department personnel **(Figure 20.48)**. These systems are equipped with 1½-inch (38 mm) hose and nozzle and stored on a hose rack system. The hose used in these systems is typically a single-jacket type and equipped with a lightweight, twist-type shut-off nozzle. These systems are sometimes referred to as **house lines**.

House Line — Permanently fixed, private standpipe hoseline.

Figure 20.49 A Class III system combines the features of Class I and Class II systems and allows those features to be used simultaneously.

There is some disagreement over the value of a Class II system. The presence of the small hose may give a false sense of security to building occupants and create the impression that they should attempt to fight a fire even though the safer course would be to escape. Fire department personnel should follow their department's standard operating procedures (SOPs) concerning the use of these systems.

Class III

A Class III system combines the features of Class I and Class II systems (**Figure 20.49**). Class III systems provide 1½-inch (38 mm) hose stations to supply water for use by building occupants who have been trained and 2½-inch (65 mm) hose connections to supply a larger volume of water for use by fire departments and those trained in handling heavy fire streams. The design of the system must allow both Class I and Class II services to be used simultaneously. It is possible that the local jurisdiction may request the removal of the hose, nozzles, and rack leaving only the 2½-inch (65 mm) and 1½-inch (38 mm) discharge connections.

Types of Standpipe Systems

Within the three classes of standpipe systems, there are different types. These types of standpipe systems include the following:

- **Automatic wet** — This system contains water at all times. The water supply is capable of meeting the system demand automatically (**Figure 20.50, p. 1212**). The water-supply control valve is open and pressure is maintained in the system at all times. A wet standpipe with an automatic water supply is most desirable because water is constantly available at the hose station. Wet standpipe systems cannot be used in cold environments.

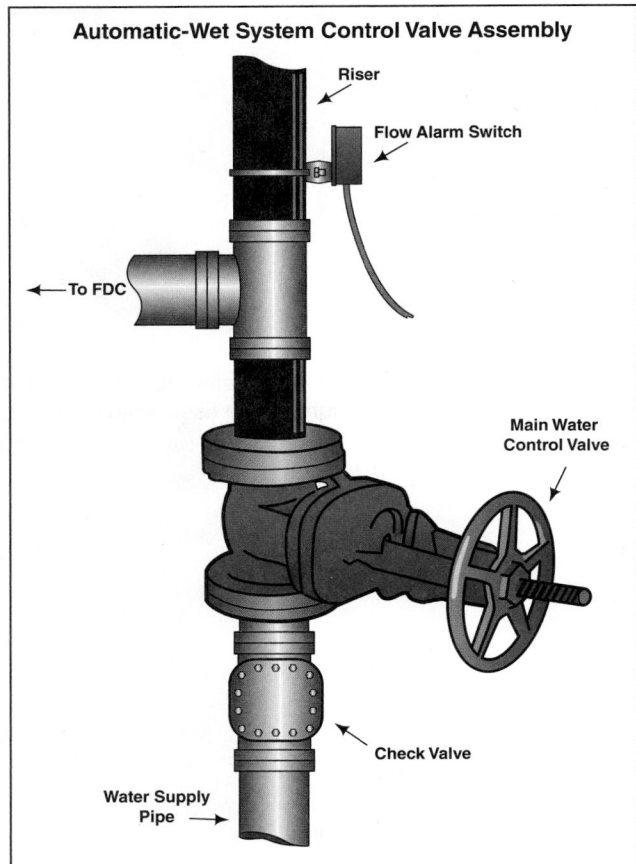

Figure 20.50 An automatic-wet system contains water at all times and is capable of activating immediately.

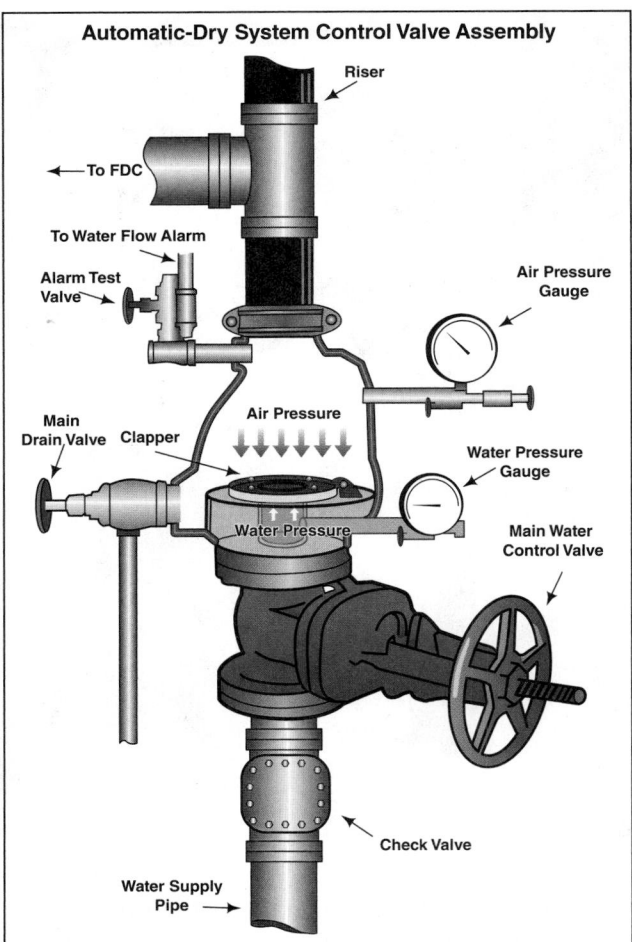

Figure 20.51 An automatic-dry system contains air under pressure to prevent water from freezing in the pipes.

- **Automatic dry** — This system contains air under pressure to maintain the integrity of the piping **(Figure 20.51)**. Water is admitted to the system through a dry-pipe valve upon the opening of a hose valve. Automatic dry systems have a permanently attached water supply. This type of system has the disadvantage of greater cost and maintenance requirements.

- **Semiautomatic dry** — Standpipe system attached to a water supply that is capable of supplying the system demand at all times; it requires activation of a control device to provide water at hose connections. The system is designed to admit water into the system when a dry-pipe valve is activated at the hose station.

- **Manual dry** — This system does not have a permanent water supply. It is designed to have water only when the system is being supplied through the FDC.

- **Manual wet** — Standpipe system that is maintained full of water but has no water supply; the water in the system is maintained to identify leaks. The fire department must provide water to the system.

Smoke Management Systems

The purpose of a **smoke management system** is to remove smoke and/or control its spread. These systems may reduce injuries and fatalities due to smoke inhalation and may also reduce property losses due to smoke damage. Most systems are designed for life safety purposes, though some may be more for property protection, especially where high-value contents are at risk. Smoke management systems are designed to route smoke away from safe escape routes or safe refuge areas.

Smoke management systems have the following functions:

- Maintaining a tenable environment in the area of egress during the time required for evacuation.

- Controlling and reducing the migration of smoke from the fire area.

- Providing conditions outside the fire zone that will assist emergency response personnel in conducting search-and-rescue operations and in locating and controlling the fire.

- Contributing to the protection of life and reduction of property loss.

Contemporary fire protection includes smoke management or smoke-control as a part of the overall fire protection system of the building especially in occupancies such as high-rise buildings, covered malls, buildings with atriums, and warehouses with high-piled storage. Smoke management is an all-inclusive term that can include compartmentation, pressurization, exhaustion, ejection, dilution, and buoyancy elements (**Figure 20.52**). Smoke management also includes smoke barriers and exhaust fans and vents. Smoke-control refers to any effort to change the pressure in spaces adjacent to the fire area to compartmentalize or exhaust smoke from the area of the fire's origin.

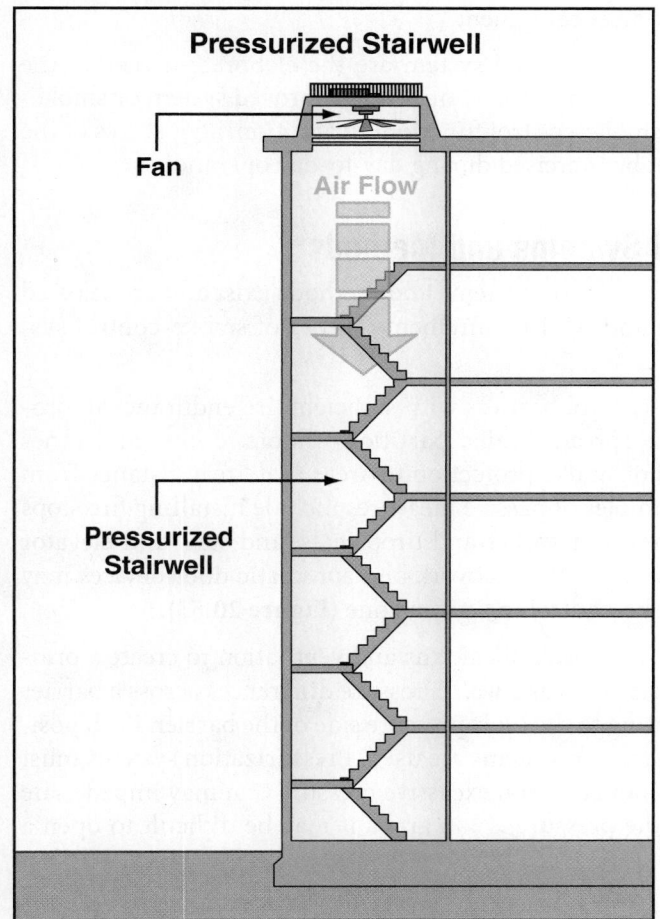

Figure 20.52 Smoke control in a tall building may be accomplished with a pressurized stairway.

Smoke Control Strategies

Different strategies for smoke-control include passive (including compartmentation), pressurization, zoned, dilution, exhaust, and opposed air flow. Some building systems are dedicated specifically to smoke-control while others can be repurposed to help control smoke.

Dedicated smoke-control systems are those that are intended and specifically listed for smoke-control purposes. These systems allow for separate air movement and distribution and do not function under normal operating conditions in the building. Upon activation, these systems operate specifically for smoke-control. Advantages of a dedicated smoke-control system include:

- Operation and control are generally simpler than other systems.

- Modification of the controls during system maintenance is less likely to occur than with other systems.

- It is less likely to be affected by the modification or failures of other building systems.

Non dedicated smoke-control systems are those that share components with other systems such as the building's heating, ventilation, and air conditioning (HVAC) system. Activation causes the system to change its mode of operation from heating and cooling to smoke-control. Potential advantages of non dedicated smoke-control systems can include the following:

- Less chance for component failure due to regular use and maintenance

- Lower cost

- Less space needed for mechanical equipment

The disadvantages of the non dedicated system are the elaborate nature of the system control and the possibility of modification of the approved system or smoke-controls that might affect the smoke-control function. In addition, all features of the smoke-control system may not be exercised during day-to-day operations.

Types of Smoke Control Systems and Methods

Several different types of smoke-control systems and methods exist and are selected based on the occupancy type and AHJ requirements. Types of smoke-control systems include the following:

- **Passive Systems** — Smoke-control barriers with sufficient fire endurance to provide protection against fire spread. Walls, partitions, floors, doors, and other barriers provide some level of smoke protection to areas that are a distance from the fire's area of origin. Examples of passive measures include installing fire stops around barrier penetrations, door gasket and drop seals, and stair and elevator vestibules. Smoke dampers in HVAC ductwork and automatic door devices may also be used to provide smoke-control in a given zone (**Figure 20.53**).

- **Pressurization Systems** — Use mechanical fans and ventilation to create a pressure difference across a barrier such as a wall. Pressure differences across a barrier prevent smoke from infiltrating to the high-pressure side of the barrier. Both positive pressure and negative pressure systems are used. Pressurization systems must be designed so that they do not create an excessive pressure that may impede safe egress from a building. If the pressure is too great, it may be difficult to open a door.

Figure 20.53 Smoke dampers may be standard equipment in passive systems.

- **Zoned Smoke-Control** — Designed to limit the movement of smoke from one compartment of a building to another. A building is divided into a number of smoke zones, separated by partitions and floors. During a fire, mechanical fans are used to contain smoke in the zone of fire origin. In most situations, each floor of a building is chosen as a separate smoke-control zone. However, a zone can be multiple floors or can be a division of a single floor. During a fire, all of the nonsmoke zones, or only those adjacent to the fire area, may be pressurized. Changes to building codes have eliminated this requirement and now require that smokeproof enclosures be used. However, many zoned systems still exist in older buildings.

- **Dilution** — In some areas, such as building atriums and highway tunnels, smoke control systems may be designed to dilute the contaminants. Fresh air dilutes contaminates to an acceptable level for humans to breathe.

- **Exhaust Method** — Uses mechanical ventilation along with the properties of smoke to collect smoke at the highest point in a large space. Because smoke rises to the upper levels of a space, this area serves as a smoke reservoir to contain the smoke. A properly designed system should allow the smoke to be maintained at a level of 6 to 10 feet (2 m to 3 m) above the highest occupied floor.

- **Opposed Air Flow Method** — Used in large spaces where smoke migration from the fire zone is limited by an opposed air flow. High velocity air aimed at the area of fire origin keeps the smoke from migrating into unaffected areas. This method must not result in air flow toward the fire, which would intensify the fire or interfere with the egress of the building's occupants. These systems have been used successfully in settings such as subway, railroad, and highway tunnels.

Firefighters' Smoke Control Station (FSCS)

A firefighters' smoke-control station (FSCS) provides full monitoring and manual control of all smoke-control systems and equipment. Firefighters can use the FSCS to control all smoke-control system equipment or zones within the building. Wherever practical, the FSCS should allow firefighters control of individual zones rather than individual pieces of equipment. Controlling entire zones helps to ensure that firefighters do not activate equipment in the wrong sequence or neglect to control a critical component.

The FSCS should be located in the Fire Command Center or other location as approved by the AHJ. Means should be provided to ensure that only authorized individuals have access to the FSCS. The FSCS should contain a building diagram that clearly indicates the type and location of all smoke-control equipment as well as the building areas affected by the equipment. The status of the systems and equipment that are activated should be clearly indicated at the FSCS. Manual override switches should be provided at the FSCS to restart or shut down the operation of any smoke-control equipment.

Firefighter's Smoke-Control Station (FSCS) — Interface between the smoke management system and the fire response forces.

Chapter Summary

Many of the buildings to which firefighters are called for emergency operations are protected partially or fully by fire protection systems. These systems can be simple or complex depending on the type of occupancy of the structure and AHJ requirements. Fire protection systems can detect the presence of fire or products of combustion, alert building occupants and emergency services, suppress or extinguish the fire, provide water supply for fire fighting operations, and remove smoke from the structure. It is important that systems are identified during preincident surveys and that firefighters understand how to operate in structures protected by these systems.

Review Questions

Firefighter II

1. What are the basic components of an alarm system?

2. How do protected premises and station alarm systems differ?

3. What are the major categories of alarm initiating devices?

4. What are the fundamental aspects of a sprinkler system?

5. How do the application methods for sprinkler systems differ?

6. How are standpipe systems classified?

7. What types of smoke control systems can be used at an incident?

Chapter Contents

Key Terms

Model Code 1238
THIRA 1222

NFPA® Job Performance Requirements

This chapter provides information that addresses the following job performance requirements of NFPA® 1001, *Standard for Fire Fighter Professional Qualifications* (2013).

Firefighter I
5.1.1

Firefighter II
6.5.1
6.5.2
6.5.3

Firefighter I Chapter Objectives

1. Explain the steps taken during fire and life safety program development. (5.1.1)

2. Describe the components involved in fire and life safety program delivery. (5.1.1)

3. Explain the impact of safety hazards, messages, and target audiences on creating fire and life safety education programs. (5.1.1)

4. Indicate ways to identify and prevent firesetter development. (5.1.1)

5. Describe the role of a Firefighter I in enforcing fire and life safety codes. (5.1.1)

Firefighter II Chapter Objectives

1. Describe the role of a Firefighter II in planning for and conducting private dwelling fire safety surveys. (6.5.1)

2. Explain the components that must be considered when developing fire and life safety presentations. (6.5.2)

3. Recognize considerations that must be addressed when giving presentations to young children and fire station tours. (6.5.1, 6.5.2)

4. Describe the role of a Firefighter II in planning for and conducting preincident planning surveys. (6.5.1, 6.5.3)

5. Conduct a fire safety survey in an occupied structure. (Skill Sheet 21-II-1; 6.5.1)

6. Make a fire and life safety presentation. (Skill Sheet 21-II-2; 6.5.1, 6.5.2)

7. Conduct a fire station tour. (Skill Sheet 21-II-3; 6.5.1, 6.5.2)

8. Prepare a preincident planning survey. (Skill Sheet 21-II-4; 6.5.1, 6.5.3)

Chapter 21
Fire and Life Safety Initiatives

Courtesy of Brett Noakes.

Case History

The value of fire and life safety initiatives was demonstrated in 2009 when a 9-year-old boy saved his own life when he remembered to "stop, drop, and roll" when his clothes caught fire. About 0930 hrs., the boy walked by a propane space heater in his living room and his clothes ignited. He dropped to the floor and rolled around before his grandmother called an ambulance. The boy suffered second-degree burns to his back and side, but recovered fully from his injuries. "He did exactly what he was supposed to do," said a fire department spokesperson.

Fire and life safety initiative programs are a comprehensive approach to protecting the public. These programs primarily work to establish and develop the resources needed to counter hazards and threats to safety through prevention, protection, mitigation, response, and recovery. Local fire departments are heavily involved in each of these areas. In addition, the programs bring the community and the fire department together in a number of efforts: mentoring, fire prevention, community education, health awareness (i.e. diabetes and stroke prevention), and large-scale disaster preparedness.

Traditionally, a fire department's prevention activities were limited to gathering information from fire code inspections and fire investigations and presenting public fire education programs. As the types of hazards that communities face have increased and the services delegated to fire departments have changed, these traditional activities have expanded to meet the community's needs.

As a Firefighter I, you will be asked to assist in the delivery of your jurisdiction or department's fire and life safety initiative program. To participate effectively, you should understand the value of the program and the program's development. You should also have basic knowledge about your jurisdiction's education programs and the fire and life safety codes that they have adopted.

As you continue in your fire service career as a Firefighter II, your role in the program will expand to include the following responsibilities:

- Conducting fire and life safety surveys of private dwellings
- Presenting fire and life safety messages to the public
- Leading fire station tours
- Conducting preincident planning surveys

I Program Value and Development

In order to meet your responsibilities as part of the fire and life safety initiatives, you need to understand the value of the program and also the development process used to create the program. The majority of fire department mission statements include phrases such as "protection of life and property from hazards" and advancing "public safety through fire prevention, investigation, and education programs." The fire and life safety initiative program helps to meet this mission statement. More specifically, the goal of the program is to encourage and empower citizens in your community to act in a safe manner to reduce the potential for fires, accidents, and injuries. When the community is better informed, they can also be prepared for the hazards that humans cannot prevent, such as severe weather and intentional hostile acts.

Similarly, the program raises awareness among your department about possible hazards firefighters will face in the community. It also emphasizes firefighter safety. Finally, the program helps firefighters prepare for threats beyond their control, make plans to reduce the potential for injury or loss, and respond efficiently during the postincident period.

In order to be effective, jurisdictions should use a systematic process to develop their risk reduction programs. In April 2012, FEMA published a comprehensive preparedness guide that can be used as a model for fire and life safety initiatives programs. The guide refers to this model as a **Threat and Hazard Identification and Risk Assessment (THIRA)**. THIRA follows a five-step process that communities can follow to develop and implement the program. The five steps are:

THIRA — Acronym for Threat and Hazard Identification and Risk Assessment.

- **Identify the threats and hazards** — Answer the question: "What could happen in my community?"

- **Give the threats and hazards context** — Describe how a threat or hazard could happen in my community, and when and where it could happen.

- **Examine the core capabilities** — What effect would each threat or hazard have on the jurisdiction's core capabilities as identified in the National Preparedness Goal?

- **Set capability targets** — Using the information above, set the level of capability a community needs to prevent, protect against, mitigate, respond to, and recover from its risks.

- **Apply the results** — Use the capability targets to decide how to use resources from the whole community.

Identify Threats and Hazards

Hazard identification requires that the community determine the specific types of hazards it will face, determine the threat the hazard will pose, and make an estimate of the portion of the population that may be affected. Identifying the hazards requires a review of historical data, trends, and statistical analysis. THIRA threats and hazards are divided into three categories (**Figures 21.1a-c**):

- **Natural hazards** — Resulting from acts of nature, such as hurricanes, earthquakes, or tornadoes and disease outbreaks or epidemics.

- **Technological hazards** — Resulting from accidents or the failures of systems and structures, such as hazardous materials spills or dam failures.

- **Threats or human-caused hazards** — Resulting from the intentional actions of an adversary, such as a threatened or actual chemical, biological, or cyber attack.

Figure 21.1a-c The three categories of the Threat and Hazard Identification Assessment (THIRA) include a) natural hazards such as flooding, b) technological hazards such as accidents and engineering failures, and c) human-caused hazards such as threatened and actual attacks. a) *Courtesy of Chris Mikal.* b) *Courtesy of Rich Mahaney.*

Give the Threats Context

In this step, a brief explanation of the conditions under which the threat or hazard might occur is developed. These explanations would include a possible time and location as well as the portion of the population that would be at greatest risk. For example, a coastal community might list a hurricane making landfall in a developed residential area during the peak tourist season on a Saturday morning in September. These types of descriptions will help in the development of possible effects and responses.

Examine Core Capabilities

Working with the list of descriptions, the jurisdiction determines the effect each threat and hazard will have on the community. Effects will vary among communities and regions but may include considerations such as:

- Displaced persons
- Mass casualties
- Devastation of property
- Long-term loss of electrical service
- Direct economic effects
- Indirect economic effects
- Disruption of supply system
- Disruption to infrastructure

These effects are then combined with the outcomes that the community wants to reach within a reasonable time. An assessment of the local response capabilities will also be needed to determine if the outcomes can be met. If the local resources are not sufficient, contingency plans will need to be developed through mutual aid agreements or reliance on other governmental agencies. For instance, with the aid of power companies located in other states, electrical service can be restored to 50 percent of the habitable structures within 15 days. With FEMA's assistance, temporary housing for displaced families can be available within 5 days.

Set Capability Targets

With the outcomes defined, the next step is to set targets or goals that will result in the greatest benefit within the desired time frame. For instance, if the greatest damage results in one thousand people losing their homes, the target may be to provide temporary shelter in an arena within 72 hours and housing within 5 days.

Apply Results

With the assessment results in hand, the jurisdiction can develop action plans, coordinate activities with other levels of government, purchase materials, and educate the public. In an area that has experience with fire deaths in a high-risk population, the fire department might purchase and install smoke detectors in residential occupancies during voluntary home safety inspections. In areas that are prone to tornadoes, the state government may offer tax incentives for the installation of safe rooms or shelters.

Program Delivery

Once the THIRA has determined the areas where resources are needed to prevent and respond to the hazards, program delivery can begin. The fire department may be responsible for some programs while other government, nongovernment, and private sector agencies may provide others. As a Firefighter I, you may be assigned to assist in program delivery. A good model for program delivery is to follow the Five Es (**Figure 21.2**):

- Education
- Enforcement

- Economic Incentives
- Engineering
- Emergency Response

Education

Fire departments have been a key element in the education component. As a part of the community, firefighters must be able to answer questions on fire and life safety topics not traditionally covered under the title of fire prevention. As a Firefighter I, you may be involved with public education programs related to fire safety and risk reduction as follows:

- Fire prevention presentations
- CPR training classes
- Fire prevention week activities annually in October
- Fire extinguisher training
- Home safety education
- Slip and fall prevention for older adults
- General health education for the community or fire department
- Practical bicycle safety for youths

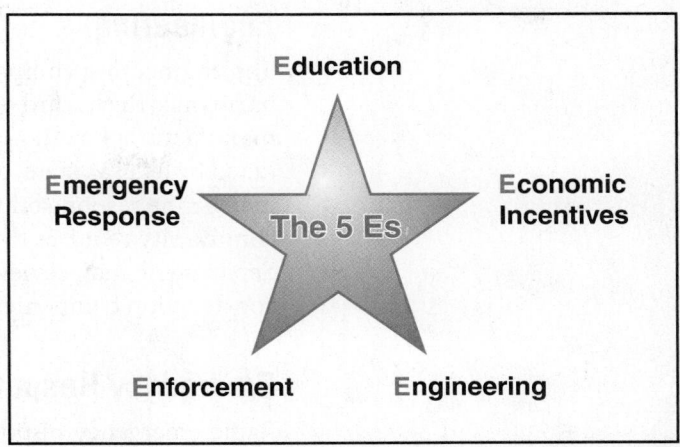

Figure 21.2 The Five Es of program delivery work together to address hazards before they cause harm.

Enforcement

Enforcement activities assigned to the fire department are generally limited to building, fire, and life safety code enforcement:

- Performing periodic inspections of occupied properties to ensure that the structure meets the locally adopted building and fire codes.
- Reviewing and approving new construction and renovation plans.
- Investigating the cause of fires and documenting the results. Providing law enforcement agencies with the findings if there is reason to believe the fire was intentional and illegal.

Fire departments also have a responsibility to enforce regulations internally, especially establishing and adhering to departmental SOPs for both emergency and nonemergency operations. Some examples of regulations that help protect firefighters and the general public are observing speed limits during emergency response and wearing seat belts whenever apparatus are in motion.

Economic Incentives

Economic incentives pay the individual for participating in the program. Incentives include tax deductions for installing sprinkler systems, manufacturer's rebates when a smoke detector is purchased, or decreased insurance rates when the community improves the water distribution system.

Engineering

The engineering component provides solutions to prevent the hazard or reduce the harm once the hazard occurs. While this component is primarily the responsibility of manufacturers, designers, and contractors, the installation and use are usually mandated by federal, state, or local laws and ordinances. Enforcing these local ordinances may be the responsibility of the fire department or another agency. For example, if a community requires that residential construction include sprinkler systems, the fire department may review building plans and construction processes to ensure that construction companies are complying with the local code.

Emergency Response

Rapid emergency response to fires continues to be the primary responsibility of the fire service. Emergency response has expanded to include responses to other types of emergencies. Therefore, local fire departments have been equipped and trained to handle all types of hazardous conditions from extrication and medical care to structural collapse and specialized rescues. Effective emergency response, while not a preventative measure, does serve the same purpose as the fire and life safety initiatives program. Well-trained firefighters who can respond to a variety of emergencies help to protect the public from hazardous conditions when prevention measures have been ineffective.

Fire and Life Safety Education

Fire and life safety education is designed to inform citizens about unsafe behaviors and provide information on how to change those behaviors. It can include formal classes in CPR, fire station tours, and voluntary home safety surveys. To be effective, you must know the hazards you are attempting to prevent, the messages that are typically provided, and various audiences for those messages. In addition, you should understand how structural surveys can be used as educational tools and how juvenile firesetters pose a particular danger to themselves and to others.

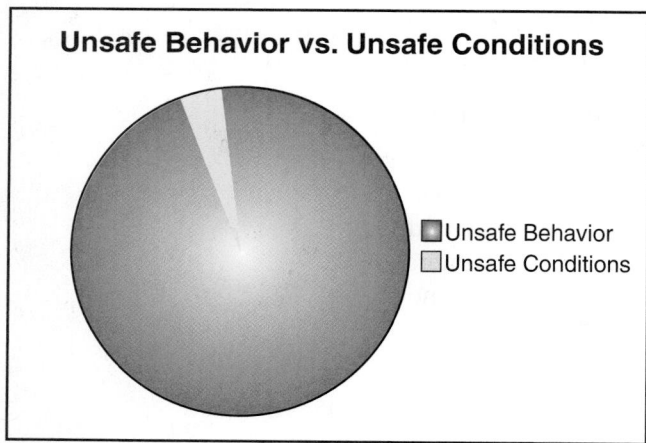

Figure 21.3 Ninety-six percent of all accidents may be attributed to unsafe behaviors.

Fire and Life Safety Hazards

Fire and life safety hazards can be broadly divided between fire hazards and safety hazards. A fire hazard is a condition that would increase the likelihood of a fire starting or would increase the extent or severity of a fire.

Safety hazards include unsafe behaviors or conditions that can result in injury, death, or property damage not associated with fire. According one study, ninety-six percent of all accidents are caused by unsafe behavior while only four percent are the result of unsafe conditions (**Figure 21.3**). Non-fire-related messages will also be a part of the fire and life safety presentations that you will give.

There are three categories of hazards that can benefit from fire and life safety education:

- Unsafe behaviors
- Unsafe conditions
- Hazardous processes

Unsafe Behaviors

Unsafe behaviors include a variety of activities that can cause fire or accidents leading to injuries or death. Unsafe behaviors are usually corrected by educating people in the consequences of their actions and the safest way to act. During preincident planning surveys, home safety surveys, and following emergency incidents, you should be aware of such unsafe behaviors as (**Figures 21.4a-c**):

- Poor housekeeping
- Hoarding combustible materials
- Ignoring ignition sources
- Open burning
- Improper use of electricity
- Careless use of flammable and combustible liquids
- Unsafe use of smoking materials
- Unattended cooking materials

Examples of unsafe behaviors not related to fire include:

- Reckless operation of a motor vehicle
- Not wearing safety belts
- Texting while operating a motor vehicle
- Speeding

Figure 21.4a Excessive trash buildup is a common fire hazard.

Figure 21.4b Overloading electrical outlets may cause a fire hazard.

Figure 21.4c Improperly storing liquid fuel is an unsafe behavior.

- Not wearing a safety helmet while riding a bicycle
- Not holding onto a handrail while descending a stairway
- Lifting or carrying an object that is too heavy or large for one person
- Improper operation of a power tool
- Removing or altering safe guards on power tools and equipment

Unsafe Conditions

Unsafe conditions that are fire hazards may arise from the design or use of a building, facility, or equipment or may be related to behaviors or processes. Unsafe conditions may be specific to a certain type of occupancy or found in all occupancies. Unsafe conditions may result from mechanical failure or damage or an environmental condition. Proper application of the local fire and life safety codes and standards will reduce sources of potential unsafe conditions including the following **(Figure 21.5)**:

- Electrical hazard conditions
- Material storage facilities
- Heating, ventilating, and air conditioning equipment/systems
- Cooking equipment
- Industrial furnaces and ovens
- Loose handrail on a stairway
- Ice on sidewalk or exterior stairs
- Unsecured swimming pool access
- Dead battery in or nonworking smoke detector **(Figure 21.6)**

Hazardous Processes

Hazardous processes may exist temporarily or as a constant function of an industrial or manufacturing site. Welding, cutting, grinding, or painting can generate ignition sources or hazardous atmospheres at construction or demolition sites **(Figure 21.7)**. Welding and the application of flammable finishes as well as dipping and quenching operations may be found in manufacturing facilities, auto repair and painting shops, fabrication businesses, and furniture repair shops. Dry cleaning operations are present in most communities and pose a hazard based on the type of chemicals used to clean clothing. Enforcement of fire and life safety codes, periodic inspections, and education work together to reduce hazards associated with these processes.

Fire and Life Safety Messages

There are many fire and life safety messages that must be communicated to educate the public. Your fire department will select the ones that are most important based on an assessment of your community's risks. Fire and life safety messages can generally be divided into four categories: Prevent (Fire and Burns), Prepare (For a Fire Emergency), Protect (Yourself in an Emergency), and Persuade (Others to be Safe) as follows:

- **Prevent (Fire and Burns) — Examples:**
  - Smoke outside - Extinguish all cigarettes and other smoking materials with water after you smoke.
  - Lock matches out of the reach of young children.

Figure 21.5 Even when properly stored, a large quantity of materials in an open room contribute substantially to the fire load of a building.

Figure 21.6 A nonworking smoke detector cannot alert residents to fire conditions.

Figure 21.7 Hazardous processes, such as welding, can generate ignition sources.

— Correct fire and burn hazards in the kitchen.

— Watch what you cook – Turn off the stove every time you leave the kitchen.

— Keep children and pets out of the "DANGER ZONE" – 3 feet (1 m) around the stove.

— Keep handles of pots and pans turned to the back of the stove.

— Keep stove top and ovens clean – No food or grease accumulation.

— Keep things that burn away from the stove top (paper towels, kitchen rags, boxes of food).

— Exercise safe candle use:

  ▪ Keep candles at least 12 inches (300 mm) from things that will burn.

  ▪ Extinguish candles when you leave the room.

  ▪ Use large, noncombustible candle holders.

  ▪ Use flameless candles.

— Use a flashlight when the power is out.

— Keep a 3-foot (about 1 m) space between all heating equipment and anything that will burn.

— Use flammable liquids safely outdoors.

— Set hot water heaters at 120°F (48°C).

— Keep medical oxygen separated from smokers and open flames.

— Use heat-limiting electric stove modification system to prevent stove top fires.

- **Prepare (For a Fire Emergency) — Examples:**

  — Install a home fire sprinkler system.

  — Install and test smoke alarms.

  — Install and test carbon monoxide alarms.

  — Know the sound of your smoke alarm or other alert devices.

  — Plan and practice an Exit Drill In The Home (EDITH).

  — Know who will wake and assist children.

  — Keep the way out clear of obstacles that could delay your escape.

  — Clear a defensible space around your home in wildland or urban interface areas.

  — Practice a school fire drill.

  — Recognize/know/find two ways of exiting a public place.

- **Protect (Yourself in an Emergency) — Examples:**

  — Crawl Low Under Smoke.

  — Get Out Fast, Stay Out.

  — Have a family meeting place outside and in front of your home.

  — Have adults wake and help children escape.

- Call 9-1-1 or local emergency number after escape.
- Stop, Drop, and Roll if your clothes catch on fire.
- Cool a burn with cool water.

- **Persuade (Others to be Safe) — Examples:**
  - Encourage outside smoking only.
  - Install, test, and maintain residential smoke alarms.
  - Attend public fireworks displays and avoid personal use of fireworks.

Fire and Life Safety Audiences

Published fire and life safety materials usually organize these topics based upon the age range of the intended audience:

- Preschool children
- Elementary age children
- Middle school children
- High school students
- Adults
- Older adults

Preschool Children

Preschool children, specifically three-to five-year-olds, are the youngest group to whom fire and life safety messages can effectively be taught. Preschool children learn by seeing and doing; what they see will last longer than what they hear. What children do will last longer than what they see or hear.

The following are appropriate messages for this age group:

- **Match and lighter safety** — This message should positively inform children of things to do when they see matches or lighters rather than tell them what not to do. The emphasis should be on telling an adult when the child finds matches or lighters within their reach. Children must be told to never touch them.

- **When clothes are on fire, Stop, Drop, and Roll** — This activity is often remembered and practiced; however, many children confuse this behavior with activities or behaviors to perform when their house is on fire. It is important to make the distinction between the two behaviors and to present the behaviors at different times.

- **Home escape** — Young children do not understand cause and effect, nor do they perform well when having to make choices. Teaching young children about "crawling low under smoke," "feeling the door," "if it's hot, don't open it — use your second way out" presents too many options, steps, and decisions for the young child. However, it is possible to teach this process of home escape in small segments, with children mastering the first behavior, mastering the second, and putting them together. This instructional process will take time.

- **Firefighters as helpers** — Children commonly fear a firefighter wearing protective clothing and breathing apparatus. It is important to eliminate this fear at a young age. To eliminate fear, the firefighter can have the children help put the equipment on a piece at a time.

- **Bicycle helmets** — Because they are just learning to ride tricycles, scooters, or bicycles, teaching helmet use to young children helps to create a lifetime habit of wearing a helmet when on a bicycle **(Figure 21.8)**. These presentations may also be good opportunities to talk about other safety rules of riding bicycles, including rules of the road.

- **Poison prevention** — Most poisonings occur among children five years of age and younger. The dangers of taking medications, vitamins, or other drugs without a parent present should be emphasized, as well as look-alike products, drinks, and foods.

Figure 21.8 When young children are taught to wear bicycle helmets, they learn safe behaviors. *Courtesy of Brett Noakes.*

Elementary Age Children

Fire and life safety messages for elementary age children, while similar, are more complex than those for preschool children. Many fire and life safety programs and curricula have been published to educate elementary age children. These programs can serve as excellent resources. Sample age-appropriate messages include:

- **Smoke alarm response** — Children can learn what they should do when a smoke alarm is activated.

- **Survival skills** — Children are taught the correct actions to take at home in the event of a fire, tornado, earthquake, or other natural disaster.

- **Use and misuse of fire** — Children are taught responsible use of fire under the supervision of an adult and the result of misuse of fire.

- **Home escape plan** — Children create a home fire escape plan.

Middle School Children

Middle school students are neither adults nor children and are hard to reach as a result. They may be just beginning to take on more adult responsibilities such as caring for younger siblings but still have some of the motivations of elementary children. To this group, new safety messages can be presented, such as:

- Burn care and prevention

- Fire science, making the connection between fire behavior and classroom science

- Health and safety education, such as CPR training

- Cooking safety

- Babysitting classes
- Firesetting and other risky activities
- The importance of home exit drills

High School Students

High school students will respond to fire and life safety messages if they believe the information is relevant to them. High school students are looking forward to a future of independence after they leave home. As a result, there are some messages that will seem more relevant to them as they look to an independent future:

- **Safe driving habits** — The results that come from not wearing seat belts, texting when driving, inattentive driving, and using alcohol before or while driving are all important messages for teens.

- **Smoke alarm safety** — Young adults need to learn how to maintain smoke alarms, replace batteries, install alarms, and respond to an activated alarm.

- **Fire service careers** — Young adults should be educated on the advantages of following a career in the fire service whether volunteer or full-time.

- **Preparation for living independently** — Young adults can be receptive to information on cooking safety, home security, proper use of medication, and proper use of portable fire extinguishers.

- **College dormitory fire safety** — College dormitories, fraternity/sorority houses, and apartments all contain fire hazards of which young adults must be aware.

Adults

Most adults want practical programs that they can apply in their professional or personal lives. These programs should involve solving problems, completing tasks, or handling lifestyle choices (**Table 21.1, p. 1234**).

All fire and life safety topics and messages are appropriate for adults although the presentation of these messages will need to change for specific audiences. A sample of the most common topics for adults includes:

- **Smoke alarms** — Information should include installation, maintenance, testing, and replacement.

- **Residential sprinklers for new and existing construction** — Firefighters must be prepared to refute myths about residential sprinklers especially those concerning cost and performance.

- **Home escape plans and fire drills, particularly for parents of younger children** — Many families do not have a home escape plan and may never have thought about creating a fire drill. These are lifesaving measures that families and other members of the household need to implement.

- **Disaster planning and preparedness for all ages** — Developing a disaster plan, creating a disaster supply kit, and establishing meeting places outside the neighborhood are important features that need to be prepared.

- **Overall unintentional injury prevention** — Information includes fall prevention for older adults as well as accident prevention for children.

Table 21.1
Eleven Keys to Engaging the Adult Learner

Make it meaningful.	Answer the questions: How does this apply to me, and how will I benefit?
Ask them what they want.	Be prepared to modify the presentation to meet their expectations.
See, hear, do.	Let them hear and see through audiovisuals. Let them experience it by doing something with new information such as application, problem solving, etc.
Discuss more, lecture less.	Discussion allows the participant to connect the new information with the information and experiences already stored in their memory.
Present, apply, and repeat as necessary.	Present one chunk of information and then apply it. Present the next chunk and then apply. This allows for a transference of information to long-term memory.
Connect with their experiences.	Connect the new information with their past experiences. In addition, make sure there is a relationship between the educator's past experiences and those of the participants.
Be a real person.	The only person to be is oneself.
Chunks, please.	Present six or seven pieces of information at a time and then ensure understanding.
More dos, fewer don'ts.	Express the message in a positive format. Tell the person what to do instead of what *not* to do.
Expect success.	Expect participants to be successful and to incorporate the training into their lives. Avoid negative attitudes or comments.
Say "thank you."	When done, thank the learners for their time and attention and for the opportunity to improve their skills and help them be safe.

Source: Ed Kirtley, Oklahoma FST

Older Adults

Older adults are one of the three high-risk groups in a community. Older adults are twice as likely to die from fire, leading all other age groups in fire-related deaths. They are at greater risk for injuries due to fire and burn injuries as well as injuries related to trips and falls.

The use of smoking materials is the leading cause of fire fatalities among older adults, while cooking fires are the leading cause of injuries. Messages for this audience should focus on the following issues (**Figure 21.9**):

- **Safe cooking** — Many older adults experience fires or burns due to unattended cooking. Due to a diminishing sense of sight, smell, and touch, they may be less likely to detect dangers in the kitchen.

- **Careless smoking** — Drowsiness, overmedication, and the combination of prescription medications and alcohol all contribute to a greater risk for fires due to careless smoking.

- **Electrical appliances and equipment** — Older adults may not understand the dangers in using equipment or appliances that are old or in need of repair. They may be reluctant to replace old items even though they may no longer be safe to use.

- **Falls** — For many older adults, falls are a common occurrence. Most falls take place inside the home. Falls may be caused by weakness, dizziness, diminishing vision, and medications.

Figure 21.9 A presentation to older adults should focus on hazards common to the group.

Structure Surveys as Part of Fire and Life Safety Education

Structure surveys are performed by fire companies for two reasons. The first is to become familiar with public-access structures and workplaces. The second is to provide a public service to homeowners and renters in their place of residence. Surveys of private residential structures are not required by law. However, the fire department may conduct voluntary home fire and life safety surveys.

Company-level preincident surveys are generally conducted on most occupied structures that are not private residences. These surveys include the public (nonoccupant rooms) areas of hotels, motels, and apartment complexes. Commercial, industrial, institutional, and educational structures are also subject to surveys **(Figure 21.10)**.

You may be involved in surveys of both public and private occupied structures during your career. As a Firefighter I, these surveys are an opportunity for you to learn about your community. They are also opportunities for you to help distribute fire and life safety literature and information to community members and businesses. In this way, structure surveys are both an educational opportunity for you and an opportunity for homeowners and business owners to learn about improving the safety of their structures.

NOTE: Information on conducting these surveys is provided in the Firefighter II section of this chapter.

Figure 21.10 A preincident survey takes account of structures that are occupied but are not residences, including commercial structures.

Juvenile Firesetters

A juvenile firesetters program is intended to reduce fires set by children by identifying at-risk children, assessing the cause of their obsession, and intervening in their behavioral development. Statistics provided by the U. S. Fire Administration (USFA) indicate that:

- Children under 18 years of age account for 54 percent of all arson arrests.

- Juvenile arson arrests are proportionally higher than any other type of crime.

- The average property loss in fires set by juveniles in approximately 250,000 incidents exceeds $20,000.

- Juvenile firesetting is the second leading cause of fatalities in residential fires.

Juvenile firesetters pose an immediate risk in a community. To save lives and eliminate this risk, parents, educators, caregivers, law enforcement, and firefighters must work together.

NOTE: For more information about juvenile firesetters, see the IFSTA **Fire and Life Safety Educator** manual.

Identifying Firesetters

Specialists in the field have identified four categories of juvenile firesetters as follows (**Table 21.2**):

- Curiosity/experimental

- Troubled/crisis

- Delinquent/criminal

- Pathological/emotionally disturbed

By being aware of these categories and their indicators you can help to locate children who are at risk. First, include juvenile firesetter information in fire and life safety presentations and station tours. Second, look for indicators during fire cause determination at emergency responses. Third, be aware of the actions of family members and friends that may be an indication of a problem. Finally, demonstrate fire-safe behavior to children. You should also be familiar with the juvenile firesetter program in your community and the process for implementing it.

While parents, caregivers, and teachers may become aware of a developing problem by observing the actions of children, firefighters may not know it until fires occur. Firefighters should watch for trends in fires. Trends include locations, types of structures, and time of day the fire occurs. Discuss patterns with personnel on other work shifts and stations as well as fire investigators. Watch for familiar faces (of juveniles) at fires.

Occasionally, parents, teachers, or neighbors may provide information indicating that a child is showing signs of inappropriate behavior with matches or lighters. Gather information and follow your department's juvenile firesetter policy.

Preventing Firesetter Development

Education is the best tool to use to prevent the development of firesetters in your community. Using your department's fire and life safety information, teach children:

- That matches and lighters are tools and not toys.

- How to safely use fire under adult supervision.

- About the destructiveness of fire when it is not properly used.

Table 21.2
Four Categories of Juvenile Firesetters

- **Curiosity/experimental**
  - Boys and girls between 2 and 10 years of age.
  - They lack understanding of the destructive potential of fire.
  - They have ready access to lighters, matches, or open flames.
  - They are unsupervised.
  - They hide when lighting fires.
  - They imitate the actions of adults they see lighting grills, smoking materials, candles, or fireplaces.

- **Troubled/crisis**
  - Mostly boys of all ages.
  - They have set two or more relatively simple fires using matches or lighters.
  - Use fires to express emotions such as anger, sadness, or frustration.
  - They may not understand the consequences of their actions.
  - They will most likely continue setting fires if they are not stopped and professionally treated.
  - Sometimes referred to as "cry for help" firesetters.

- **Delinquent/criminal**
  - Usually teens with a history of firesetting, truancy, antisocial behavior, or drug/alcohol abuse.
  - Juvenile knows the danger of fires and the consequences of his/her actions.
  - Fires are acts of vandalism with the intention to destroy property.
  - Firesetting may involve peers.
  - Targets are typically schools, abandoned buildings, open fields, or Dumpsters™.
  - Punishment usually involves criminal prosecution.

- **Pathological/emotionally disturbed**
  - May be boys or girls of all ages who have a life-long fascination with fire.
  - They set multiple fires that are very destructive and sophisticated.
  - Fires may be random, ritualized, or intended to destroy property.
  - Chronic history of emotional problems.
  - Requires psychiatric diagnosis.

Provide adults with information from your department's juvenile firesetter program which describes:

- The categories of juvenile firesetters and the indicators of each
- The importance of keeping matches and lighters out of the reach of small children
- The supervision of young children and the importance of never leaving children unattended with open flames, such as candles
- Making children aware of the dangers of fire when improperly used
- The reasons that only trained babysitters should be hired

When a trend begins to develop, report it to the fire investigations personnel or a juvenile firesetter intervention professional. Always follow your department's SOPs while handling suspected juvenile firesetters.

Fire and Life Safety Codes

Codes and standards are developed to establish minimum requirements for the design, construction, and use of buildings, structures, and facilities and the installation of equipment. They are part of both the engineering and enforcement components of the fire and life safety initiative program. The authority having jurisdiction (AHJ) adopts codes and standards and the building department or fire department enforces them. Fire and life safety codes establish the minimum level of safety that should be present in a structure. Fire and life safety codes regulate both construction materials and designs and occupant behavior and processes. These codes must be current and consistent to provide the desired level of protection established by the local jurisdiction.

NOTE: Refer to Chapter 1 for more information on codes and standards.

You should be familiar with the fire and life safety codes adopted by your jurisdiction and the adoption process. You will refer to these codes when making preincident surveys or assisting fire prevention inspectors with their duties. You should also know the standard operating procedures (SOPs) that regulate building inspections and fire investigations.

Code Adoption

Most jurisdictions use **model codes** as the starting point for their code adoption process. The term model code is used to describe a set of requirements similar to a standard. Model codes came into existence as a way for communities to standardize the adoption process rather than rewriting codes themselves. A consensus organization such as NFPA® or the International Code Council® (ICC®) develops model codes that contain agreed-upon requirements for areas such as fire and life safety or electrical equipment designs and installations.

Currently, there are two model code organizations in the United States and one in Canada. Each code organization has a series of codes that, when adopted, can be used to regulate building components — structural, mechanical, electrical, and plumbing — as well as fire and life safety:

- Canadian Commission on Building and Fire Codes (CCBFC)
  - *National Fire Code of Canada (NFC)*
  - *National Building Code of Canada (NBC)*
- International Code Council® (ICC®)
  - *International Fire Code® (IFC®)*
  - *International Building Code® (IBC®)*
- National Fire Protection Association® (NFPA®)
  - NFPA® 1, *Fire Code Handbook*
  - NFPA® 101, *Life Safety Code®* (occupancy classifications)
  - NFPA® 5000, *Building Construction and Safety Code®*

> **Model Code** — Consensus-based standards or codes established to provide uniformity in regulations in regards to construction, design, and use. When adopted by the local jurisdiction, these codes become enforceable laws.

Department of Defense (DoD)

Facilities that are under the jurisdiction of the United States Department of Defense (DoD) enforce a different set of model codes. Those codes are known as the *Unified Facility Criteria* (*UFC*). Elements of other model codes and standards may be part of the *UFC*.

Codes are only enforceable when the AHJ adopts them. In some communities, a model code may be adopted intact with no changes. In other instances, the code may be amended to meet local requirements. Some codes may be created locally, such as a code that regulates the size of advertising signs along streets or highways. The jurisdiction may also adopt model codes, such as building and fire codes, either as written or with amendments.

Fire department personnel must know whether the codes and standards in their jurisdiction are current. Most of the model codes that jurisdictions adopt are revised on a regular basis, typically every three to five years. A revised edition of the model code in effect in a community does not automatically take effect when the model code organization releases it. The previous (existing) edition remains in effect until the AHJ formally and legally adopts the new edition.

Once the fire and life safety code official has received the final draft of the proposed model fire and life safety code, the process of preparing a legislative resolution for its formal adoption begins. Each jurisdiction has a legislative procedure for presenting resolutions for approval of the code. Generally, the process includes preparation of the proposed legislation, study by the legislative body, formal presentation, discussion, and legal action. Once the code is adopted, the date is set for implementation and a public announcement is made.

Code Enforcement

There are two main aspects of code enforcement: inspection and investigation. Periodic inspections identify occupancies that are not in compliance with building and fire codes that the AHJ has adopted. Fire investigations provide data that can inform the code adoption process and also identify intentionally set fires to assist law enforcement with apprehending arsonists.

Inspections

To ensure that building codes are enforced in the jurisdiction, inspectors who are trained in code interpretation monitor the construction or renovation process. Fire department inspectors, qualified to NFPA® 1031, *Standard for Professional Qualifications for Fire Inspector and Plan Examiner*, may also be involved in plan review and construction inspections **(Figure 21.11)**. Fire department inspectors also make periodic inspections of occupancies. These inspections are used to ensure that the structures are in compliance with all locally adopted fire and life safety codes. In some jurisdictions, inspections of single-family dwellings and the private living areas of multifamily dwellings are determined by the use of the structure and may occur periodically.

While you will not be assigned the task of fire and life safety inspections, you will need to be aware of fire and life safety code requirements. During preincident surveys and emergency responses, you may become aware of potential life-threatening conditions. When that occurs, you should follow local SOPs and report the problem to the fire and life safety inspections office.

Figure 21.11 Building plan reviews should be conducted periodically to ensure that fire codes meet the life safety standards in a jurisdiction.

Figure 21.12 The information collected by fire investigators is used to develop and monitor the fire and life safety initiative program.

Investigations

Fire investigations provide data about the types and causes of fires in the community. This information can be used to develop and monitor the effectiveness of the fire and life safety initiative program (**Figure 21.12**). Investigations are also law enforcement tools that attempt to determine responsibility for intentionally set fires and bring the responsible parties to justice. Finally, they gather data that will help in making products, processes, and materials safer.

Firefighters and their supervisors will be involved in the initial portion of cause determination, as outlined in Chapter 20. In some cases, it will be necessary to request the assistance of a qualified fire investigator from the fire department, law enforcement department, or an outside agency. You will work with fire investigators in locating and protecting evidence, gathering information, and providing your own impression of emergency incidents. You will need to use your situational awareness to provide accurate information.

ⅠⅠ Private Dwelling Fire Safety Surveys

Private dwelling fire safety surveys (*home safety surveys*) can be performed on a voluntary basis to determine the extent to which community members are implementing fire and life safety behaviors where they live. Home safety surveys can provide information about the amount and kind of safety equipment in the home, such as smoke alarms, safety gates across stairs, fire extinguishers, and grab bars in bathrooms (**Figure 21.13**).

In most jurisdictions, single-family residences (houses and apartments) are not subject to the requirements of local fire codes to the same degree as are businesses, institutions, and industrial properties. In these jurisdictions, residential fire and life safety surveys are provided as a public service, and homeowner or occupant participation is voluntary. When a homeowner or occupant requests a safety survey, firefighters survey the residence indicating potential fire and life safety hazards and suggesting corrective actions. In some jurisdictions, residential fire and life safety surveys may be conducted as part of the fire and life safety initiative program. **Skill Sheet 21-II-1** outlines the steps for conducting a fire safety survey in an occupied structure.

Home safety surveys must be treated as an educational opportunity and not an enforcement tool. They are voluntary and residents should be informed about the purpose of the safety surveys. Furthermore, those conducting the safety surveys must be trained in proper survey practices.

Residential Fire and Life Safety Hazards

Fire statistics collected by the NFPA® for 2010 show that more than ninety-two percent of all fire deaths occur in residential structures. Firefighters conducting residential surveys should look for the most common causes of fires (**Table 21.3, p. 1242**). It is a good idea to complete a survey form for each residence and provide a copy to the occupant. If the form includes a checklist of common hazards, it can serve as a guide to summarize the survey results. The following are common causes of residential fires:

- Malfunctioning heating appliances and water heaters
- Combustibles too close to heating appliances or lamps
- Unsafe cooking procedures

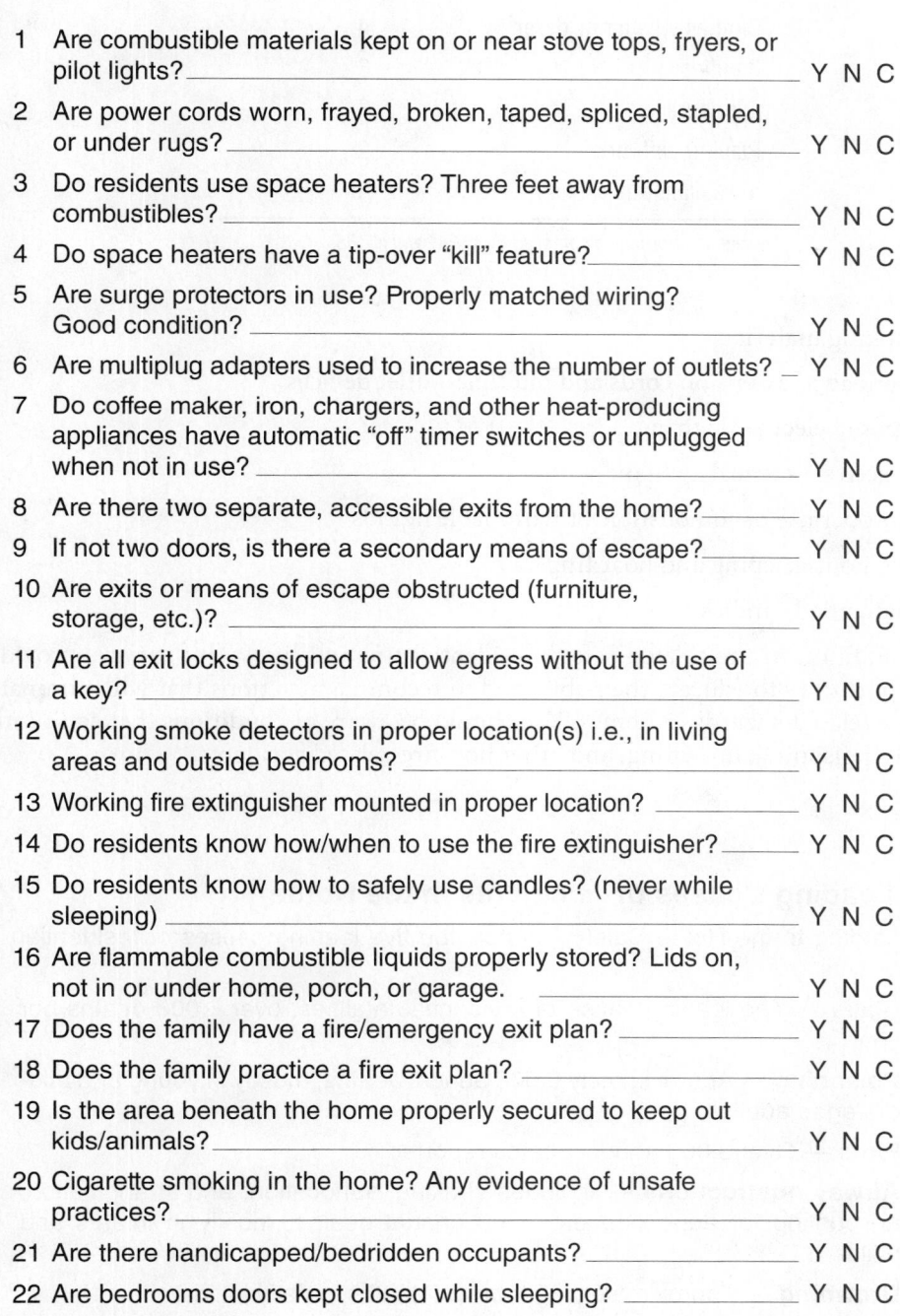

Palm Beach Gardens Fire Rescue
Residential Fire Safety Survey

1 Are combustible materials kept on or near stove tops, fryers, or pilot lights?_____ Y N C

2 Are power cords worn, frayed, broken, taped, spliced, stapled, or under rugs?_____ Y N C

3 Do residents use space heaters? Three feet away from combustibles?_____ Y N C

4 Do space heaters have a tip-over "kill" feature?_____ Y N C

5 Are surge protectors in use? Properly matched wiring? Good condition? _____ Y N C

6 Are multiplug adapters used to increase the number of outlets? _ Y N C

7 Do coffee maker, iron, chargers, and other heat-producing appliances have automatic "off" timer switches or unplugged when not in use? _____ Y N C

8 Are there two separate, accessible exits from the home?_____ Y N C

9 If not two doors, is there a secondary means of escape?_____ Y N C

10 Are exits or means of escape obstructed (furniture, storage, etc.)? _____ Y N C

11 Are all exit locks designed to allow egress without the use of a key? _____ Y N C

12 Working smoke detectors in proper location(s) i.e., in living areas and outside bedrooms? _____ Y N C

13 Working fire extinguisher mounted in proper location? _____ Y N C

14 Do residents know how/when to use the fire extinguisher?_____ Y N C

15 Do residents know how to safely use candles? (never while sleeping) _____ Y N C

16 Are flammable combustible liquids properly stored? Lids on, not in or under home, porch, or garage. _____ Y N C

17 Does the family have a fire/emergency exit plan? _____ Y N C

18 Does the family practice a fire exit plan? _____ Y N C

19 Is the area beneath the home properly secured to keep out kids/animals?_____ Y N C

20 Cigarette smoking in the home? Any evidence of unsafe practices? _____ Y N C

21 Are there handicapped/bedridden occupants? _____ Y N C

22 Are bedrooms doors kept closed while sleeping? _____ Y N C

Figure 21.13 A residential fire safety survey form follows a checklist to document the conditions found and whether they have been corrected.

Table 21.3 Causes of Residential Fires Between 2005 and 2009	
Cause	**Percent of Residential Fires**
Cooking	42
Heating	17
Electrical malfunction	6
Intentional	8
Smoking	5
Clothes washer or dryer	4
Candles	3
Exposure	3
Playing with fire	2
Miscellaneous causes	10

Based on information compiled by NFPA® and NFIRS.

- Smoking materials
- Overloaded extension cords and multiple-outlet devices
- Exposed electrical wiring
- Defective electrical appliances
- Improper use of combustible or flammable liquids
- Poor housekeeping and hoarding
- Unattended candles

In addition to fire-related safety, residential fire and life safety surveys provide an opportunity to educate the public and to recommend actions that will eliminate nonfire related hazards in homes. You should be aware of conditions that can result in falls, poisoning, drowning, and other non-fire-related accidents.

Leading Causes of Accidents in the Home

According to the Home Safety Council, the five leading causes of residential accidents are:

- **Falls** — The leading cause of residential fatalities, over 6,000 deaths per year

- **Poisoning** — Approximately 5,000 annual deaths, mostly in young and middle-aged adults

- **Fires** — Over 3,000 annual deaths reported

- **Airway obstructions** — Includes choking, suffocation, and strangulation, accounting for approximately 1,000 annual deaths, mostly in infants and children

- **Drowning** — Approximately 800 annual deaths, mostly children

Planning and Public Relations

When residential fire and life safety surveys are conducted as part of an organized public awareness and education program, a great deal of advanced planning and publicity is necessary to gain full acceptance by the community. The public must be made aware that the program is a public service, not a code enforcement activity. In other words, the firefighters are coming to make family members aware of safety hazards, not to cite them for code violations. When firefighters enter the home to conduct a residential fire and life safety survey, their main objectives include the following:

- Preventing accidental fires
- Improving life safety conditions
- Helping the owner or occupant to understand and improve existing conditions

When firefighters conduct residential fire and life safety surveys, citizens get to know and trust their firefighters. Fire and life safety surveys allow the residents to feel that their fire department is really concerned about their welfare. An increase in goodwill translates into support when annual budgets are developed or during community fund-raising events.

Educational Opportunities

After identifying hazards during a fire and life safety survey, firefighters can do the following to promote fire and life safety education:

- Provide customized safety information in a way the resident can understand and embrace.
- Distribute fire and life safety literature.
- Check for or distribute emergency telephone number stickers.
- Discuss the correct installation and maintenance of smoke detectors.
- Discuss the advantages of residential sprinkler systems.
- Determine effective escape routes from that home.
- Help develop an exit plan and determine a meeting place.
- Provide information about carbon monoxide (CO) detectors.
- Provide information on the prevention and proper treatment of burns.
- Recommend the correct portable fire extinguisher and placement of extinguishers.
- Demonstrate how to prevent fires in kitchens.
- Provide information on the safe use of candles.

Firefighters also educate themselves and their departments during home safety surveys. They can gain the following valuable information about homes in their jurisdictions:

- Home construction
- Presence of security grilles and gates
- Occupancy conditions
- Local trends in housing development
- Location of streets, fire hydrants, and water supplies
- Identification of special needs residents
- Residents who do not speak English in the community

This information can be applied in a variety of ways. For example, older residents and people with disabilities have specific needs including text telephones (TTYs) and telecommunications devices designed for people who are deaf, including visual alerting systems, and sign language interpreters. Firefighters may be able to keep these residents informed of their options for locating these special devices. If residents are encountered who do not speak English as a first language, translators can be requested and brochures provided in the appropriate language. Firefighters can also provide residents with information about other community services that are available.

Additional safety messages appropriate to the specific home conditions can also be provided. Families with small children can benefit from information on child safety seats, bicycle helmets, and child monitoring devices. Older adults may need to know about slip and fall prevention, personal alert devices that can notify caregivers if the wearer needs help, or methods to prevent smoking-related fires. Residents who have physical, mental, or sensory limitations will need to know how to notify emergency responders in an emergency.

Some fire departments also give special cards or slips to compliment homeowners whose dwellings are found in a safe condition. Other cards saying, *"We're sorry we missed you"* are used to notify owners or occupants that firefighters were in the neighborhood conducting safety surveys. The cards list the appropriate telephone number that residents can call if they would like to schedule a fire and life safety survey. In addition, some departments use the "time change/battery change" as an opportunity to promote surveys for older adults and persons with disabilities by also offering to replace batteries in their smoke alarms or install new ones.

Firefighter Responsibilities

Firefighters should take home safety surveys seriously and behave professionally and conscientiously during the survey. The public expects firefighters to be subject matter experts on home fire and life safety issues. You should be prepared to answer a wide variety of questions. When conducting residential fire and safety surveys, firefighters should use the following guidelines:

- Conduct surveys in teams of two or more (**Figure 21.14**).
- Dress and act professionally.
- Introduce yourself and your partner and provide proper identification.
- Explain the survey procedure.
- Maintain a courteous and businesslike attitude at all times.
- Focus on preventing fires and eliminating life safety hazards.
- Compliment the occupants when favorable conditions are found.
- Offer constructive suggestions for correcting or eliminating hazardous conditions.
- Ask to survey all rooms including the garage.
- If accessible, survey the basement.
- Ask to survey the attic if it is used for storage or contains a heating or cooling unit and is accessible.
- Ask the occupant to open any closed doors.
- Discuss the survey results with the owner/occupant and answer any questions (**Figure 21.15**).

- Thank the owners or occupants for the invitation into their home.

- Leave behind educational materials appropriate for the occupants.

- Keep the survey results confidential; do not share the results with any outside entity.

Residential fire and life safety programs usually include a checklist of hazards commonly found in homes. When the checklist is completed, give the form, or a copy, to the resident with recommendations for corrections of any hazards. You may also need to recommend that a professional, such as an electrician or heating and cooling technician, be called to repair some items. Follow your local SOPs and do not recommend specific professionals.

If no one is at home, leave appropriate materials and a notice that you tried to contact the residents. Remember that the U.S. Postal Service prohibits placing un-stamped materials in mailboxes. Finally, document the survey per your local SOPs.

Figure 21.14 Residential surveys are conducted in teams of two or more.

Figure 21.15 Share the residential fire safety survey with home owners to improve their awareness of safe and unsafe behaviors and conditions in the home.

Fire and Life Safety Presentations

Fire and life safety education fulfills the educational component of the community risk reduction. Firefighters deliver this information through presentations and fire station tours. You need to be familiar with basic presentation skills, how to make presentations to young children, and how to conduct fire station tours. **Skill Sheet 21-II-2** provides a list of steps for making a fire and life safety presentation.

Basic Presentation Skills

Presentation is the art of clearly and concisely explaining information in ways that your audience can understand. Good presentation skills include:

- **Audience-centered** — You must know the audience and adapt the topic and presentation style to this audience.

- **Good development of ideas** — Use interesting, appealing, and memorable ways to present your information. This may include the following:

  — Using relevant examples

  — Telling stories to which the audience can relate

- **Good organization of ideas** — Organize your material so that the audience is never lost during the presentation. For instance, if you are talking to adults about the importance of smoke detectors, start with an attention grabber, then provide necessary background information, followed by an illustration of how smoke detectors save lives, and finally how they can benefit by installing smoke detectors.

- **Best choice of words** — Do not to speak above the intellectual level of the audience by using words they might not understand. It is just as important not to talk below the level of the audience members, insulting their intelligence by being too basic. Also, do not use jargon that only firefighters would easily understand.

- **Good delivery skills** — Use the following communication techniques to enhance your presentation:

  — Keep appropriate eye contact with the audience members.

  — Speak to the entire audience, not just one section or one side of the group.

  — Use appropriate gestures to illustrate mental pictures or emphasize key points **(Figure 21.16)**.

  — Refrain from adding too many "war stories" to presentations.

- **Good vocal characteristics** — Major elements are as follows:

  — Pronunciation: Pronouncing each word correctly, stressing the right words or syllables, and pausing where appropriate.

  — Good grammar: Correct tense, possession, pronoun agreement, etc.

  — Inflection: Varying the tone (pitch) of words, syllables, or phrases to emphasize important points.

  — Variety: Changes in loudness, tone, and rate of speech.

  — Enunciation: Clearly emphasizing each syllable, accent, and pause. The opposite of enunciating is slurring.

  — Projection: Speaking loudly and clearly enough to be heard in the back of the room or auditorium.

- **Conversational tone** — A relaxed tone helps the audience feel at ease and ready to receive information.

- **Positive attitude** — Always display a positive attitude about the subject matter you are presenting.

- **Appropriate use of humor** — Appropriate humor can create a relaxed atmosphere and get the attention of the audience. However, you should avoid inappropriate humor that may offend members of the audience.

- **Personal style** — Use a personal style, capitalizing on your own unique experiences and abilities.

Finally, you must know your topic. Use the following guidelines to ensure that you are prepared to present:

- Read and know the material your department provides.

- Do additional research to supplement the materials.

- Gather additional information that could be relevant to the presentation.

- Discuss the presentation with other members of your crew.

- Anticipate questions that the audience may ask.

Figure 21.16 Nonverbal gestures help to emphasize the message.

Audience

Your audience may be an individual or group of people. The message and how you present it will depend on the demographics or composition of the audience. Some of these categories overlap, for example older adults who are part of an ethnic minority. When the audience falls into more than one category, tailoring the message to that group becomes more challenging. Audiences for fire and life safety presentations can be divided according to the following demographic characteristics:

- Age

- Education

- Cultural diversity

- Socioeconomic level

- Physical ability

Age

Age is a primary factor in choosing the topic and presentation style. Each age group has a different learning style, background experience, and interest (**Figure 21.17, p. 1248**). Two of the three high-risk or at-risk groups are based on age and include children under age 5 and adults over age 65.

Educational Level

Like age, a person's educational level will shape his or her ability to understand fire and life safety topics. A message for a child will need to be simple while one for a college student can be more detailed and complex.

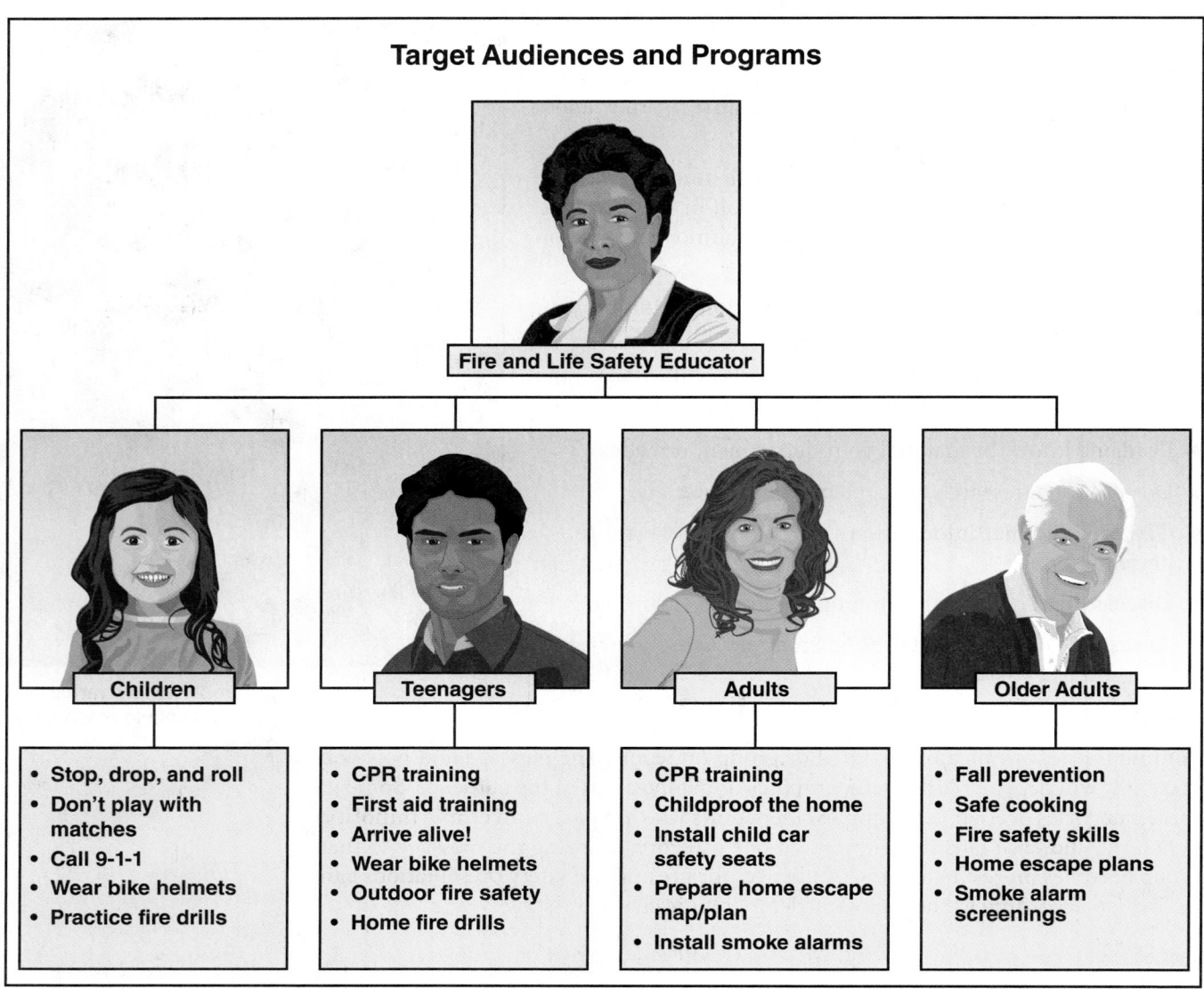

Target Audiences and Programs

Fire and Life Safety Educator

Children
- Stop, drop, and roll
- Don't play with matches
- Call 9-1-1
- Wear bike helmets
- Practice fire drills

Teenagers
- CPR training
- First aid training
- Arrive alive!
- Wear bike helmets
- Outdoor fire safety
- Home fire drills

Adults
- CPR training
- Childproof the home
- Install child car safety seats
- Prepare home escape map/plan
- Install smoke alarms

Older Adults
- Fall prevention
- Safe cooking
- Fire safety skills
- Home escape plans
- Smoke alarm screenings

Figure 21.17 Fire and life safety educators must try to engage different age groups with their presentations.

Cultural Diversity

A person's heritage or ethnicity will determine their worldview or outlook on life. It will also have an effect on their experience with fires and illness. The diversity of the North American population also means that many people may not speak English as a first language. Fire and life safety materials and presentations may require translations to be effective.

Socioeconomic Level

The socioeconomic level is a contributing factor that can determine the message to a particular audience. Poverty is a contributing risk factor that can result in higher rates of fires and fire-related casualties. For example, people who cannot afford to buy food for their families will not have money to replace batteries in smoke detectors.

Physical Ability

People with disabilities or special needs are considered the third of the three high-risk or at-risk groups within the community. This group, as well as adults over sixty-five, can have reduced physical, mental, and sensory abilities. In addition, presentations given to people who are hearing-impaired will require translators who can use American Sign Language.

Effective Fire and Life Safety Messages

The effectiveness of a fire and life safety presentation is only as effective as the message that is delivered. To be effective, safety messages must be accurate, positive, and specific to the target audience.

Accurate Messages

Presenting accurate fire and life safety messages for children and adults is the responsibility of every firefighter. There are several sources available to assist each fire department in verifying the accuracy of the content in fire and life safety programs and presentations. One source is the NFPA® Educational Messages Advisory Committee. The committee annually reviews current educational fire and life safety messages to determine if the messages are still valid and appropriate or if new research is needed to revise the message. The committee's work is a good baseline of information that firefighters and fire departments can use as a ready resource to prepare and update fire and life safety materials and presentations.

Changes in Messages over Time

Messages may change in subtle ways over time. For instance, one message that was in use for a while was, "Crawl low *in* smoke." Because the point is to stay away from the deadly smoke, the message has been modified to become more accurate as, "Crawl low *under* smoke."

Finally, when answering questions during a presentation, make sure that you provide accurate answers. If you are asked a question that you are not prepared to answer, simply say that you will find out the answer and get back to the person asking it. Never fabricate an answer or try to bluff your way through.

Positive Messages

Behavioral scientists have determined that people remember positive statements more than negative statements. Positive messages instruct what action *to* take rather than emphasizing actions *not* to take. The positive approach avoids any confusion about which actions are dangerous and which actions are not. In an emergency such as a home fire escape, people must act quickly and remember the correct behavior. If the presentation included what NOT to do, people may confuse correct and incorrect actions to take during an emergency. It is best to say, "crawl low under smoke" instead of "do not stand up in smoke." It is best to say "Get out! Get out!" instead of "Do not hide in the closet." It is best to teach "Call 9-1-1" instead of "Do not call the operator."

Targeted Messages

A targeted message has particular importance to the audience listening to the presentation. Fire and life safety messages should be targeted to be most effective based upon a number of factors:

- Age of the audience

- Seasonal fire and life safety hazards

- Recent fire incidents that have occurred in the community

- Emerging trends or issues such as increases in fires started from neglected candles, security bars on windows leading to increased fatalities, or a growing occurrence of Dumpster™ fires at a local school

- Topics in the public eye such as hoarding behavior or unsafe sleeping environments for infants

Big, life-changing events in an adult's life such as moving to a new home, having a baby, or moving an aging parent into the home are good opportunities for fire departments to partner with other community organizations to find and deliver fire and life safety information. Community welcome packets should include safety pamphlets and information about the fire department and its services.

Some fire departments give fire and life safety presentations in a neighborhood immediately after a fire. This provides an opportunity for the neighbors to ask questions and for the fire department to discuss important fire and life safety concepts such as the importance of installing and testing smoke alarms.

Seasonal messages are an effective approach, linking topics to specific holidays, events, or celebrations. Depending on the geographic region, some of these messages may be year round rather than season specific. Seasonal target messages include the following:

- Winter:
  - Space heaters
  - Wood burning stoves and fireplaces
  - Holiday decorations, especially Christmas trees
  - Candles
  - Carbon monoxide detectors

- Spring:
  - Bicycle and pedestrian safety

- Summer:
  - Grilling/cooking outside
  - Flammable/combustible liquids
  - Camping
  - Boating safety
  - Pool safety
  - Fireworks

- Fall:
  - Halloween safety
  - Turkey fryers

Organizing the Message

Depending on your topic and audience, you should organize the information in a pattern or sequence that will make it easy to understand. There are four generally accepted organizational sequences that you can use:

- **Known-to-unknown** — Begin with information that the audience is familiar with before moving into unfamiliar or unknown material.

- **Simple-to-complex** — Begin with basic knowledge, then move to more difficult or complex knowledge as the presentation progresses. Basic knowledge is a necessary foundation for learning more complex ideas.

- **Whole-part-whole** — Begin this sequence with an overview of the entire topic. Next, divide the topic into subsections and describe each of them. Close by providing a summary of the entire topic.

- **Step-by-step** — Teach each step in the correct order and then have members of the audience practice them in the same order.

Presentations for Young Children

Young children are especially at risk for fire deaths and burn injuries. The National Association for the Education of Young Children (NAEYC) defines "young children" as from birth to eight years old. They may be found in child care centers and pre-kindergarten through 2nd grade classrooms. Because of the high risk of this population, fire and life safety presentations for this group are very important.

Classroom Considerations

Children have limited attention spans so a presentation that is 15 minutes or less is ideal for them. Keep your presentation short and interesting to the children (**Figure 21.18**). Decide with the teacher how you will manage the children's questions. Not responding to the children who have their hands raised to ask questions is disrespectful and not appropriate. You may want to say that you will answer questions at the end of your time together or you may provide a few minutes at the halfway point and then again at the end. One approach is to have the children and the teacher prepare a list of questions before you arrive.

When you arrive in the classroom the children may already be assembled on the floor in a space used for group time. Get down on their eye level. You should sit on the floor, kneel, squat, or sit on a low chair. Begin with a sincere greeting and ask if everyone can see you. You can then say, "Let me see everyone's eyes so I know you are ready to see and listen."

Figure 21.18 Firefighters must present information in a way that is interesting to their audience.

Working with a small group of children is best, one classroom at a time. Large group presentations in a gym or a cafeteria are not educationally appropriate for children younger than eight years old. When your presentation is less than 15 minutes, you can teach in three or four classrooms in a little more than an hour.

Children immediately know when you are not being sincere. If you gesture or roll your eyes they will see it! You are a role model. Your language and attitude are important. Be relaxed and feel comfortable with the children. They will know when you are glad to be with them or if you are not. A bad attitude is always apparent to children.

Teachers have high expectations for firefighters, and firefighters need to have expectations for classroom teachers. It is not appropriate for a teacher to leave any guest in the classroom alone with the children. Teachers can share age-appropriate questions and bits of information as the firefighter is presenting. Take cues from the teacher.

You can expect that sometimes a child will misbehave. Do not focus on or challenge the poor behavior. Most teachers will handle a misbehaving child. A simple comment like, "I would appreciate your help," may be all that you need to say.

Children's Common Fears

The most common complaint voiced by early childhood classroom teachers is that the firefighter scared the children. You should take special precautions not to frighten children, whether intentionally or unintentionally. A scared child cannot listen or learn. Do not frighten children with explicit stories about home fires or show images of burned clothing, toys, or household items. Do not talk about burned pets.

Loud noises are frightening to children. This includes the sound of a smoke alarm, the bell ringing on an SCBA when you turn on the bottle, and sirens. If you are going to test a smoke alarm or make any other loud noise, tell the children first. Let them know to expect a loud noise and encourage them to cover their ears or stand close to the teacher to get ready for the loud sound.

One other very common fear for children is to see a firefighter wearing personal protective equipment (PPE) and breathing apparatus. Firefighters must never approach children or enter a classroom dressed out in full PPE. Always don your personal protective equipment while the children watch. Step by step, don each piece, including the backpack and air cylinder, the face piece, and the regulator. To protect children and others from contaminants, always use clean equipment for safety demonstrations and classroom presentations.

Appropriate Vocabulary

Use simple language when working with young children. Do not use baby talk, acronyms, or fire department jargon. Use short simple sentences. Use the words *fire truck* and *fire engine*, not *apparatus*. Children do not think about the differences in the type of fire department equipment. Statements like, "Fire engines carry water and get water from a fire hydrant," and "Ladder trucks carry tools including lots of ladders," are perfectly appropriate for kindergartners. A large picture of each type of vehicle from your department would be a good teaching tool.

How we use water is another topic that will interest children. Tell them how the fire hose carries the water from the truck to the place where it is needed to put out the fire. Do not use details about pressure and gallons per minute. Talk about water in terms to which small children might be able to relate: "The hose on the fire engine can spray fifty bathtubs full of water in one minute," or "The engine carries enough water to fill a small swimming pool."

Learning Style

The very nature of young children makes teaching fire and life safety especially challenging. They see and understand the world around them very differently than adults. Children interpret what you say and what you do very literally. For example, a child knows that the letter carrier brings the mail and may think that a firefighter brings fire. You should not assume that something that makes sense to you will also make sense to children.

Young children do NOT learn by repeating catchy phrases about fire safety. They can often repeat phrases, such as Stop, Drop, and Roll or Crawl Low Under Smoke, but they don't really know what they mean. Have children show you what they know, instead of just telling you **(Figure 21.19)**. Children are active learners, and more involvement means more learning and more fun, such as performing Crawl Low Under Smoke with a sheet billowing in the air as the pretend smoke, or a pantomime of putting on firefighter equipment. Children need to practice, practice, practice — one time is not enough for any fire or life safety message.

Figure 21.19 Young children learn by practicing skills.

Children learn best through experience, using all five senses. For example, you can pass around a clean firefighter's helmet and a pair of gloves. Children need to touch and feel to know that the helmet is heavy and the gloves are big and thick. Tell the children it is your helmet and they can pass it around, but only you can wear it. Make sure that every child is an active participant, not just an observer.

Repeat and reinforce fire and life safety messages at every opportunity. Young children love repetition, such as hearing the same stories over and over. Children gain confidence and a sense of security when they hear and see the same material repeated. Firefighters are encouraged to repeat their presentations and demonstrations. A pre-K student will learn additional information in kindergarten from the very same presentation and then again in the first grade.

Children also learn through watching adults and modeling the adult's behavior. Showing wrong behavior such as a clown running when their clothes are on fire might seem like a joke because it is illogical and ridiculous to a firefighter, but young children do not know the difference between real and pretend. Showing the wrong behavior is very risky for all children. Children are mimics and they learn though watching YOU and everything you do.

To best meet the learning style of children, fire and life safety messages should be presented in this order:

1. Recognize the firefighter as a helper and friend.

2. Stay away from hot things that hurt.

3. Cool a burn with cool water.

4. Tell a grownup when you find matches or a lighter — do not touch.

5. Stop, Drop, and Roll if your clothes catch on fire.

6. Know the sound and purpose of a smoke alarm! Get out, Get out!

7. Practice a home fire drill using your escape map.

8. Crawl low under smoke.

9. Know where to meet when you escape.

10. Call 9-1-1 for help.

Fire Station Tours

Fire station tours are opportunities to enhance the department's public image, provide fire and life safety messages, and distribute safety awareness literature. These tours may be spur-of-the-moment visits from people who walk in off the street or scheduled visits from organized groups. Station tours for groups of children during Fire Prevention Week (the week including October 9th) are quite common (**Figure 21.20**). **Skill Sheet 21-II-3** outlines the steps for conducting a fire station tour with the assistance of an adult civilian.

In preparation for the tour, you or your supervisor should:

- Confirm the age of the children, the number of children, the number of adults, the time available, and any key fire and life safety messages that the adults expect you to reinforce.

- Decide on and receive approval for areas of the station to show and highlight.

- Check to see that these selected areas are clean and orderly.

- Provide a designated place like a long table or empty wall where children can leave their coats and other personal effects.

- Designate a bathroom that may be used by the children.

- Inform station personnel when a tour is scheduled, when the group arrives, and when the group leaves.

- Ensure that the apparatus you will showcase is clean and easy to access.

- Follow departmental SOPs for civilians accessing an apparatus. For example, know the limit on the number of people who can be in or on the apparatus at any one time.

- Ensure that the protective clothing you show is clean and safe for children to touch.

- Set aside handouts to provide to the children and adults. Place these items out-of-sight of the children and in an easily accessible location near the end of the tour route.

Of particular concern is where a tour group should go to remain safe and out of the way if an alarm sounds and firefighters must respond to an emergency. Decide where the children should go and clear this decision with the officer in charge so all personnel are aware. Before the tour, explain to the adult that if a fire alarm sounds, you will escort the group to a safe area. They can watch the firefighters and the equipment leave and be safe and out of the way. Depending on your department's SOPs, instruct the group leaders on what action to take once the units have cleared the station.

Whenever civilians are in the fire station, firefighters should be dressed appropriately and should conduct themselves with courtesy and professionalism. The impressions firefighters will make on any visitors will be strong ones, so all television sets should be turned off and other activities should be as positive as possible. The goal is to present a professional workplace atmosphere to the visitor.

Figure 21.20 Fire station tours provide firefighters a chance to interact with the public in a very positive way.

⚠️ **CAUTION**
Provide safety instructions at the beginning of the tour about what to do and where to go if an alarm sounds during the tour.

Firefighters should answer all questions courteously and to the best of their ability. Fire and life safety information can be distributed to visitors at this time. While some departments allow visitors to climb on apparatus or don equipment items, many others do not. Many parents love to photograph their children wearing fire helmets and other gear; however, this activity exposes children to unhealthy contaminants such as smoke, soot, and sweat. Also, some fire helmets are too heavy for some children's neck muscles to adequately support which can lead to injuries. An alternative approach is to maintain a set of clean, used protective clothing specifically for children to wear.

Visitors, especially children, should never be allowed to roam around the fire station unescorted. A firefighter or officer should be assigned to meet visiting groups as soon as they arrive at the station. This individual should keep the group together until the tour begins. During the tour, closely supervise curious children or other individuals around shop areas or slide poles. All groups should be kept together and, if necessary, larger groups should be divided into smaller groups with a firefighter assigned to each.

Equipment and apparatus should be demonstrated with appropriate caution to ensure that no one is endangered. Positioning firefighters at strategic locations will help to prevent visitors from straying into hazardous areas during demonstrations. Many fire departments prohibit taking visitors up on elevating platforms or aerial ladders. Appropriate caution should be exercised when activating apparatus sirens in the presence of visitors because the noise level produced can be harmful to their hearing.

Remember that station mascots (dogs, cats, etc.) can be potential safety and liability hazards. Excited animals have been known to bite or scratch visitors; therefore, many fire departments restrict the presence of animals during station tours.

Preincident Planning Surveys

Preincident planning surveys allow fire department personnel to gather information about occupied structures before an emergency occurs which could assist with any of the following at that structure:

- Locating and controlling a fire
- Locating occupants
- Determining potential hazards
- Improving emergency operations
- Improving both firefighter and occupant safety

During these surveys, firefighters document the following details about a structure using maps, drawings, photographs, and written notes:

- Construction type
- Floor plan or layout
- Contents
- Occupancy type
- Hazardous materials storage
- Special processes
- Fire detection and suppression systems
- Fuel load

The ultimate purpose of survey documentation is to help firefighters accomplish the following goals:

- Become familiar with structures in their district, their uses, and their associated hazards.
- Recognize existing hazards.
- Visualize how standard tactics may or may not apply in various occupancies.
- Develop new tactics if necessary.
- Determine if occupants have disabilities or medical conditions that may prohibit self-evacuation.
- Determine if there are occupants with language barriers that may require translators.

Preincident surveys and code enforcement inspections of structures that are open to the public are sometimes conducted by the same personnel because some departments require their companies to perform both functions in one visit. However, preincident surveys and code enforcement inspections are conducted for entirely different purposes and the information should not be combined. Company-level personnel conduct preincident surveys to become familiar with a structure or facility. Inspections are conducted to see that owners/occupants have complied with the applicable codes require to prevent fires from occurring and allow occupants to escape during an emergency. If code violations are discovered during a survey, the fire officer may request that the owner/occupant correct the violation or simply report the problem to the authority's fire inspection division.

Preincident planning surveys are scheduled in advance and should not be surprise visits. Generally the fire department will assign fire companies based on a predetermined cycle. Surveys that occur when a structure is remodeled or when the occupancy type changes can be very effective. The fire inspection and code enforcement division may also notify fire companies when a structure should be visited.

Conducting Preincident Planning Surveys

The firefighter is part of a team conducting preincident surveys under the supervision of the company officer. Local SOPs will establish the procedures for performing preincident planning surveys. **Skill Sheet 21-II-4** outlines the steps for preparing a preincident planning survey. Effective preincident surveys generally follow these steps:

- Contact owner/occupant before arriving.

- Meet the owner/occupant or designee upon arrival.

- Begin the survey on the outside of the building to make general observations
  — Make preliminary notes and sketches.
  — Take photographs if permission has been granted by the owner/occupant.

- Use detailed site plans of the exterior from previous surveys.

- Locate the primary and secondary entrances to the building.

- Note windows and doors that are covered by security grilles and gates.

- Locate utility shutoffs.

- Note the location of fire hydrants, fire department connections, water control valves, and water supply sources **(Figure 21.21)**.

- Note the location of fire control room and alarm panels.

- Determine fire hydrant water flow.

- Record building construction type, height, occupancy, and proximity of adjacent exposures.

- Note conditions for planning fire apparatus response and positioning:
  — Property accessibility on all sides
  — Street conditions
  — Distance from the street to the building (setback)
  — Barriers to aerial device operations
  — Landscaping, including plants, water features, and topography that could make access difficult
  — Overhead obstructions and power lines that would restrict the use of ladders

Figure 21.21 The location and condition of fire fighting connections and resources must be documented during a preincident planning survey.

When the survey of the exterior is completed, the survey team should go directly to the roof or basement and proceed with a systematic survey. It does not matter if the team starts on the roof and works downward or starts in the basement and works up. From a practical standpoint, however, many firefighters find it helpful and less confusing to start on the roof. Regardless of which procedure is used, progression should be planned so that the firefighters can systematically look at each floor in succession.

If floor plan drawings are not available from the building owner/occupant, firefighters may be able to obtain them from the fire inspection or plans review office. To conduct a thorough survey, firefighters must make notes and take photographs of observed hazards and unsafe conditions. Drawings of the interior layout, high-hazard areas, egress routes, and important features should be made (or upgraded on existing drawings) or photographed. Make sure to record any changes that have been made and update the floor plan drawings accordingly. The drawings are particularly important when survey information is used to develop written preincident operational plans or is entered into computer-aided dispatch (CAD) systems. A complete set of notes, photographs, and well-prepared drawings of the building provides dependable information from which a complete report can be written.

In large or complex buildings, firefighters may need to make more than one visit to complete the survey. If the property includes several buildings, each should be surveyed separately. It is a good idea to start on the roof of the highest building so you can get a good overall view.

Taking the time to discuss the survey results with the owner or occupant is also beneficial. If fire or life safety hazards are uncovered, the owner/occupant should be informed of them and asked to correct them. The information should also be provided to the fire inspection and code enforcement division of the department.

Maps, Drawings, and Photographs

Maps, drawings, and photographs can show a great deal of information as part of the documentation in a preincident survey. Including as much of this information as possible during a survey enhances the content of the survey and is very beneficial for the fire department.

Maps or site plans that contain information regarding building construction, fire-protection systems, occupancy, fuel loads, special hazards, and other details are very helpful to firefighters. The location of building entrances and exits are essential to firefighter safety. Large occupancies or complexes may already have maps that were prepared by their insurance carriers. These maps normally use some form of common map symbols (Figure 21.22).

For buildings where existing maps are unavailable or outdated, firefighters should include a simple plot plan drawing that shows the general arrangement of the property with respect to streets and other buildings. They should also note any other important features and information that might affect fire fighting tactics in that occupancy. Drawings must be accurate because they are often the most important documents created during surveys. A clipboard and a ruler can be helpful aids when making drawings. Before the preincident surveys, a Google™ map or other aerial photo of the area is helpful and may provide the plot plan needed.

Using common symbols on a floor plan allows firefighters to show the type of construction, thickness of walls, partitions, openings, roof types, parapets, and other important features. In addition, fire protection systems, water mains, automatic sprinkler control valves, and other potentially important features of fire protection

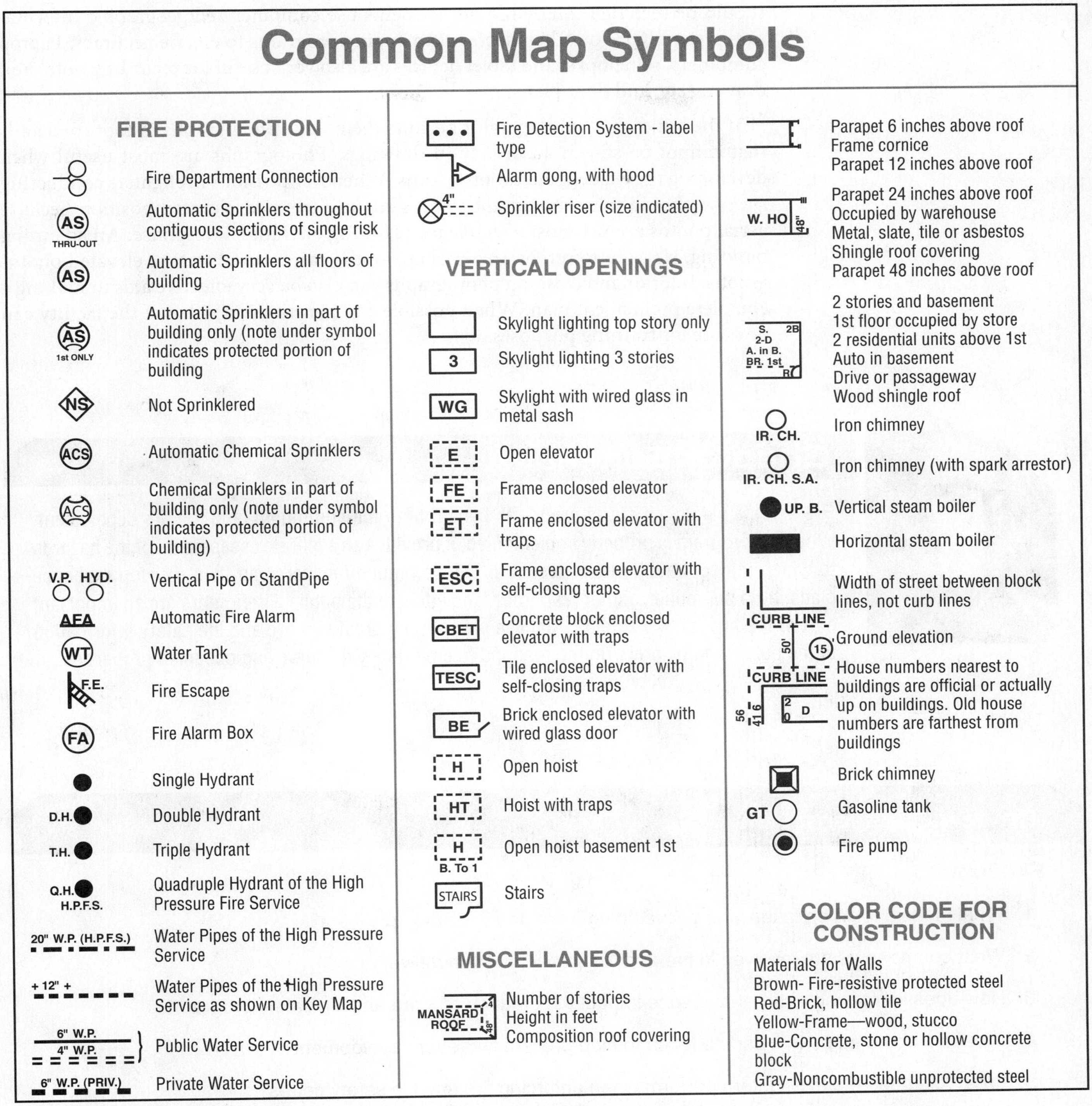

Common Map Symbols

FIRE PROTECTION

Fire Department Connection

(AS) THRU-OUT — Automatic Sprinklers throughout contiguous sections of single risk

(AS) — Automatic Sprinklers all floors of building

(AS) 1st ONLY — Automatic Sprinklers in part of building only (note under symbol indicates protected portion of building

(NS) — Not Sprinklered

(ACS) — Automatic Chemical Sprinklers

(ACS) — Chemical Sprinklers in part of building only (note under symbol indicates protected portion of building)

V.P. HYD. — Vertical Pipe or StandPipe

AFA — Automatic Fire Alarm

(WT) — Water Tank

F.E. — Fire Escape

(FA) — Fire Alarm Box

● — Single Hydrant

D.H. ● — Double Hydrant

T.H. ● — Triple Hydrant

Q.H. ● H.P.F.S. — Quadruple Hydrant of the High Pressure Fire Service

20" W.P. (H.P.F.S.) — Water Pipes of the High Pressure Service

+ 12" + — Water Pipes of the High Pressure Service as shown on Key Map

6" W.P. / 4" W.P. — Public Water Service

6" W.P. (PRIV.) — Private Water Service

⋯ Fire Detection System - label type

⚑ Alarm gong, with hood

⊗ 4" Sprinkler riser (size indicated)

VERTICAL OPENINGS

▭ Skylight lighting top story only

3 Skylight lighting 3 stories

WG Skylight with wired glass in metal sash

E Open elevator

FE Frame enclosed elevator

ET Frame enclosed elevator with traps

ESC Frame enclosed elevator with self-closing traps

CBET Concrete block enclosed elevator with traps

TESC Tile enclosed elevator with self-closing traps

BE Brick enclosed elevator with wired glass door

H Open hoist

HT Hoist with traps

H B. To 1 Open hoist basement 1st

STAIRS Stairs

MISCELLANEOUS

MANSARD ROOF — Number of stories / Height in feet / Composition roof covering

---

Parapet 6 inches above roof
Frame cornice
Parapet 12 inches above roof

W. HO — Parapet 24 inches above roof / Occupied by warehouse / Metal, slate, tile or asbestos / Shingle roof covering / Parapet 48 inches above roof

S. 2B 2-D A. in B. BR. 1st R — 2 stories and basement / 1st floor occupied by store / 2 residential units above 1st / Auto in basement / Drive or passageway / Wood shingle roof

IR. CH. — Iron chimney

IR. CH. S.A. — Iron chimney (with spark arrestor)

● UP. B. — Vertical steam boiler

▬ Horizontal steam boiler

CURB LINE — Width of street between block lines, not curb lines

CURB LINE 50' (15) — Ground elevation

56 41'6 2 D 0 — House numbers nearest to buildings are official or actually up on buildings. Old house numbers are farthest from buildings

▣ Brick chimney

GT ○ Gasoline tank

◉ Fire pump

COLOR CODE FOR CONSTRUCTION

Materials for Walls
Brown- Fire-resistive protected steel
Red-Brick, hollow tile
Yellow-Frame—wood, stucco
Blue-Concrete, stone or hollow concrete block
Gray-Noncombustible unprotected steel

Figure 21.22 The use of common map symbols adds clarity to the map. *Courtesy of the Sanborn Map Company.*

should be included. Many fire departments use computerized geographic information system (GIS) or other electronic mapping programs to ensure accuracy. Laptop computers, notebooks, and tablet devices are also very useful for recording notes and drawing site and floor plans.

If the building owner/operator permits them, take photographs to capture details that cannot be shown accurately in drawings. Photographs are most useful when developing preincident operational plans. Whenever possible, firefighters conducting surveys should take photographs of a structure from an elevated position because aerial photos are the most effective for planning operational response. An adjoining building, elevated tower, or an aerial apparatus can be used to take elevated photographs. Interior and close-up photographs can also be very effective aids in making a complete preincident plan. When possible and permitted, a video of the facility can be made for training purposes.

Chapter Summary

Fire and life safety initiatives programs benefit both the community and the fire department. When the program is properly implemented, it provides an accurate assessment of the hazards to which the community must respond. The community leaders can then develop response plans, train personnel, gather resources, and educate the public. Firefighters are an important part of the fire and life safety initiatives program by providing fire and life safety information to the public and by preparing to respond to emergency by surveying occupied properties.

Review Questions

Firefighter I

1. What steps should be taken when developing a fire and life safety program?

2. What components are involved in fire and life safety program delivery?

3. How does identifying a target audience impact the creation of a fire and life safety program?

4. What are some steps that can be taken to help prevent firesetter development?

5. What role does the Firefighter I perform when enforcing fire and life safety codes?

Firefighter II

1. How can a Firefighter II plan for conducting a private dwelling fire safety survey?

2. What components must be considered when developing fire and life safety presentations?

3. How can firefighters help address the fears of small children during fire and life safety presentations?

4. What can a Firefighter II do to prepare for conducting a preincident planning survey?

21-11-1
Conduct a fire safety survey in an occupied structure.

SKILL
SHEETS

Step 3: Explain the purpose and benefits of the survey to the resident.

    a. Emphasis on voluntary nature of survey

    b. Explain reason for survey

Step 1: Gather equipment and informational materials required to conduct the survey.

Step 2: Contact the resident.

    a. Approach residence on sidewalk or entryway

    b. Respect all notices and signs such as 'No Soliciting'

    c. Avoid dangerous situations such as possible dog bites

SKILL SHEETS

21-II-1
Conduct a fire safety survey in an occupied structure.

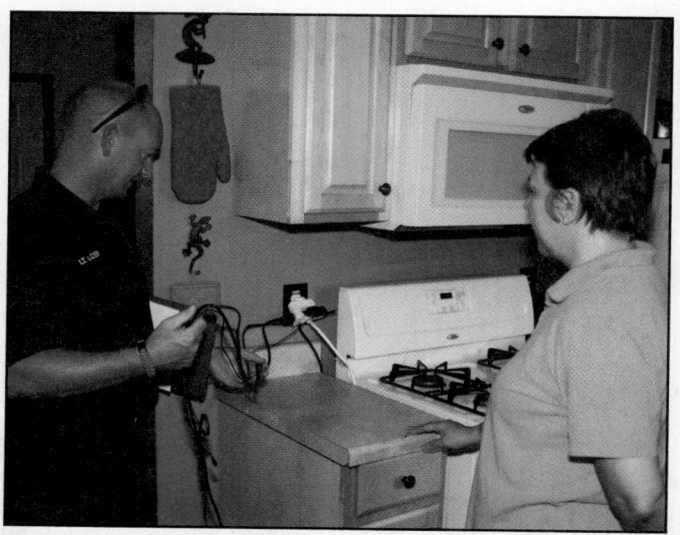

Step 4: Conduct survey of the residence.

    a. Survey attic, utility rooms, storage areas, kitchen, living-room, garage, and basement

    b. Take notes of hazards

Step 5: Identify fire hazards and recommend appropriate solutions to the resident.

    a. Explain the nature of the hazard

    b. Explain solution(s) to the hazard

    c. Correct the hazard immediately, if possible

    d. Mount smoke alarms, if needed

Step 6: Discuss general fire safety information with the resident.

    a. Address home escape planning, maintenance of smoke alarms, storage of flammable and toxic liquids, gate/control mechanisms around outdoor pools, fire-safe cooking procedures, residential sprinkler systems (if present), and other security devices

    b. Provide printed fire safety information

Step 7: Conclude survey.

    a. Thank resident for cooperation

    b. Review any issues that require follow-up by the department

Step 8: Record information on the survey in appropriate department database.

Step 1: Determine an appropriate fire or life safety topic for the audience.

Step 2: Select location, date, and time for the presentation.

Step 3: Review prepared lesson outline and double check that all necessary equipment and materials are available.

Step 4: Notify the group or audience of the presentation details prior to the date of the presentation.

Step 5: Conduct the presentation according to the lesson outline.

    a. Educational methods used are developmentally appropriate

    b. All steps in outline are followed

    c. Questions are answered

    d. Participants are engaged by the presentation

Step 6: Return equipment and materials according to department policy.

Step 7: Record information about presentation in appropriate department database.

Step 1: Notify the group point of contact of the date and time of tour.

Step 2: Determine characteristics of the group touring the station.

   a. Age of group

   b. Developmental characteristics

   c. Number of visitors

   d. Purpose of visit

Step 3: Select appropriate fire safety message(s) to be presented during the tour.

Step 4: Select written materials, handouts, etc. that supports the message(s) from Step 2 to be distributed during the tour.

Step 5: Reconfirm the date and time of the tour with the group point of contact.

   a. Contact at least one shift prior to visit

   b. Inform officer and crew members about tour

Step 6: Inspect station in preparation for the tour.

   a. Remove any safety hazards

   b. Clean station and apparatus

Step 7: Welcome the group to the station.

   a. Introduce yourself

   b. Give basic department background and introduce on-duty station personnel

   c. Inform group of tour rules

Step 8: Provide time at the end of the tour for questions.

Step 1: Contact the business owner or manager to gain permission to conduct the survey.

    a. Emergency contact information

    b. Correct address

Step 2: Record initial observations of the outside of the building.

    a. Number and location of fire hydrants, fire department connections, fire alarm boxes, rapid entry key systems, etc.

    b. Type of building construction and materials

    c. Types of exposures

    d. Access and egress from the site

    e. Occupancy of building

    f. Any construction or environmental features which could negatively impact fire suppression

Step 3: Prepare a sketch of the building, streets, hydrants, etc.

Step 4: Calculate and record hydrant fire flow.

Step 5: Survey the interior of the building beginning on the lowest floor or roof.

Step 6: Record any features or conditions related to life safety and fire suppression.

    a. Location of fire protection systems, alarm panel, control valves, standpipes, etc.

    b. Location of exit stairwells, corridors, doors, etc.

    c. Hazardous operations, equipment, or materials

    d. Electrical control panels

    e. Life safety risks

    f. Roof access

    g. Potential ventilation openings

    h. Elevators

    i. High value content or merchandise

    j. Potential fuel loads

    k. Any other potential hazards present

Step 7: Draw floor plan of building to include all pertinent information from Step 6.

Step 8: Discuss results of survey with owner/manager.

    a. Thank manager for allowing fire department to conduct survey

    b. Offer to provide a copy of the preincident plan for the building's underwriter

    c. Comment on conditions found

    d. Answer any questions

Step 9: Disseminate completed preincident plan to other companies and stations according to local protocols.

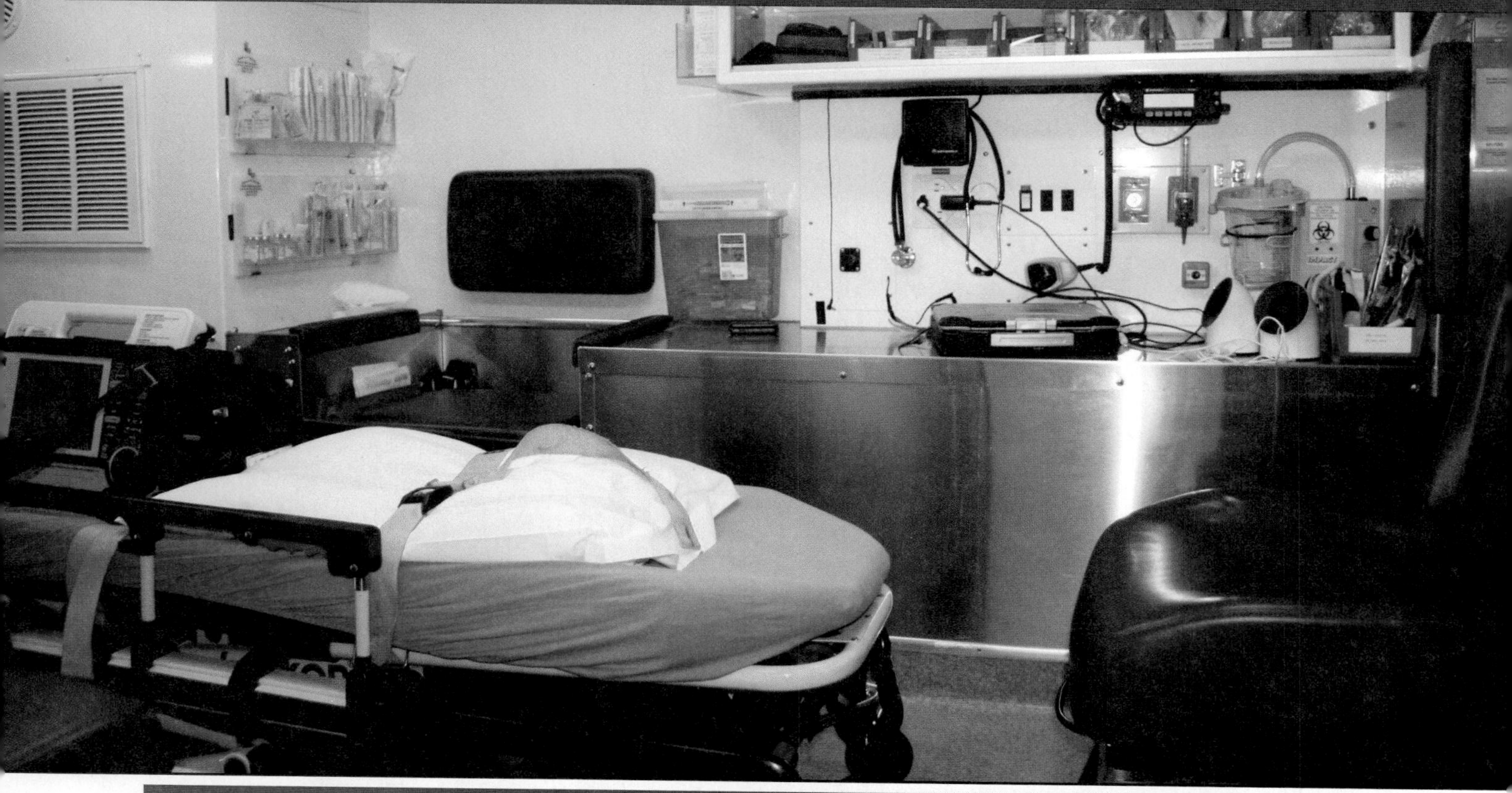

Chapter Contents

Key Terms

NFPA® Entrance Requirements

This chapter provides information that addresses the following entrance requirements of NFPA® 1001, *Standard for Fire Fighter Professional Qualifications* (2013).

4.3

1. Describe the roles the fire service may take in providing emergency medical care. (4.3)

2. Summarize patient confidentiality requirements. (4.3)

3. Distinguish among commonly encountered communicable diseases. (4.3)

4. Summarize immunization considerations for first responders (4.3)

5. Explain the importance of body substance isolation (BSI). (4.3)

6. Explain actions taken for basic patient assessment. (4.3)

7. Compare and contrast CPR techniques for adults, children, and infants. (4.3)

8. Explain when to administer and when to discontinue CPR. (4.3)

9. Describe basic types of external bleeding. (4.3)

10. Explain the use of direct pressure and elevation to control external bleeding. (4.3)

11. Describe the signs and symptoms of internal bleeding. (4.3)

12. Describe the role that recognizing the types, signs, and symptoms of shock plays in shock management. (4.3)

Chapter 22
Emergency Medical Care
for Fire Department
First Responders

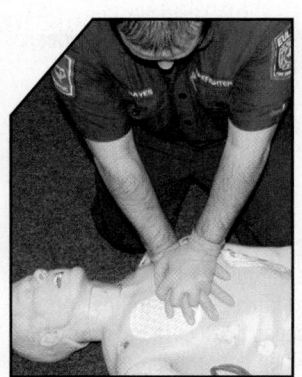

Case History

On December 31, 2011, Tulsa (OK) firefighters were dispatched to a residential structure fire. While conducting interior fire suppression operations, a member of the attack team became unresponsive. Members of the Rapid Intervention Team removed the firefighter from the structure and found him to be in cardiac arrest. Fire department paramedics immediately began treating the firefighter and were able to resuscitate him prior to the arrival of an ambulance. The firefighter was conscious upon arrival at the hospital and returned to work in February, 2012. Because of the medical training and actions of fire department personnel on scene, a tragedy was averted.

To accomplish the mission of the fire service, it is your responsibility as a firefighter to save lives. Nowhere in the fire service is this responsibility more relevant than in the delivery of medical services. The fire service provides emergency medical care in a variety of ways. You must be familiar with the way your department provides medical care. You must also be proficient in the medical skills and duties assigned to you to function effectively in your role as a firefighter.

NFPA® 1001 establishes entrance requirements for firefighters and provides a list of minimum capabilities for emergency medical care that a jurisdiction should set for its personnel. These capabilities include the following:

- Infection control
- Cardiopulmonary resuscitation (CPR)
- Bleeding control
- Shock management

The information and skills contained in this chapter are intended to satisfy the entrance requirements for firefighter training required by NFPA® 1001. However, it is not meant to substitute for formal training as an emergency medical care provider. The chapter begins with an introduction to fire service based emergency medical care and important laws regarding patient confidentiality. It continues with information about infection control practices, patient assessment, and CPR and concludes with content on bleeding control and shock management.

Fire Service-Based Emergency Medical Care

Emergency medical service (EMS) is commonly known as the treatment on scene and the transport of victims who are ill or injured to a definitive healthcare facility. While the fire service has a long tradition in the United States, EMS has only had formal recognition since the 1960s. However, the fire service has been a key participant in providing emergency medical care since its inception and continues to be involved today.

Fire departments provide EMS in a number of different ways. Even if the organization does not operate ambulance units, its personnel often respond as medical first responders in order to begin treatment until an ambulance arrives. Ambulance services are typically provided in one of the following ways:

- **Fire-based EMS** — Ambulance service is provided as a function of the fire department. Staffing is typically provided by firefighters who have been **cross-trained** as **emergency medical technicians (EMTs)** or **paramedics**. In some instances, the ambulance service is provided by the fire department but staffing is provided by EMTs and paramedics who do not have fire fighting responsibilities.

- **Third-service EMS** — Ambulance service is provided by an organization that is separate from the fire and police services and has its own administration and personnel. In most instances, this organization is a function of a municipal, county, provincial, or regional government. The service may also be provided by a for-profit or not-for-profit organization that is under contract.

- **Hospital-based EMS** — Ambulance service is contracted to a hospital by the local government. Personnel are employees of the hospital and patients are typically transported to the contracted hospital for treatment.

Regardless of the type of ambulance service that is provided in your jurisdiction, you must be familiar with the equipment, protocols, and standards of care that are used (**Figure 22.1**). In areas where ambulance service is not provided by the fire department, it is especially important that firefighters train regularly and have a good working relationship with EMS personnel.

In many jurisdictions, fire department personnel are often able to respond to an emergency medical call faster than EMS units. In most areas, there are multiple fire stations and apparatus but only a few ambulances. In addition, some areas routinely have more medical calls than there are ambulances available. For these and other reasons, fire departments are often requested to respond to the scene and begin care until EMS personnel arrive.

Some departments provide this service with personnel who have limited EMS training while others provide fully-functioning paramedics who carry most of the same equipment and medications on the fire apparatus that are carried on an ambulance (**Figure 22.2**). Regardless of the level of service provided by your department, you will likely be called upon to provide some type of emergency medical care. With the aging population in North America, calls for EMS assistance continue to increase, following a trend established over the last several years.

Patient Confidentiality

As an emergency medical care provider, your actions are governed by applicable laws. In particular, healthcare providers must safeguard the **protected medical information (PMI)** of a patient. In the U.S., the *Health Insurance Portability and Accountability Act* (HIPAA) of 1996 has instituted greater regulations on the use, distribution, and storage of PMI. The law also establishes civil and criminal penalties for

Emergency Medical Services (EMS) — Initial medical evaluation/treatment provided to employees and others who become ill or are injured.

Cross-Training — Training emergency services personnel to function in more than one capacity. Occurs most often when personnel are trained as firefighters and EMTs or paramedics.

Emergency Medical Technician (EMT) — Professional-level provider of basic life support emergency medical care. Requires certification by some authority.

Paramedic — Professional level of certification for emergency medical personnel who are trained in advanced life support procedures.

Protected Medical Information (PMI) — Information of a patient that includes personal data (name, birth date, social security number, address), medical history, and condition.

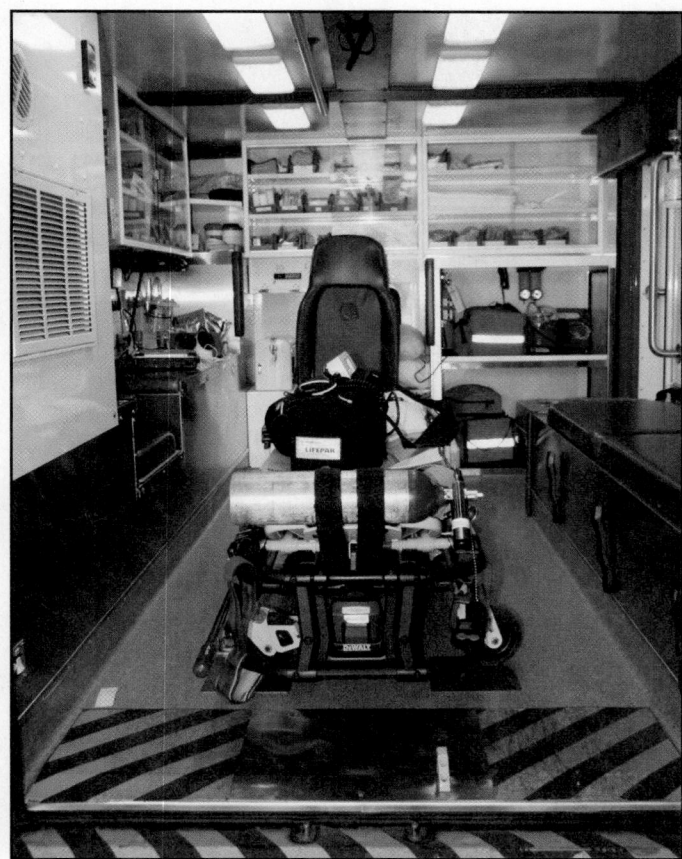

Figure 22.1 Responders must be familiar with the equipment used in their jurisdiction.

Figure 22.2 Some fire departments operate fire apparatus staffed with paramedics and equipped with advanced life support equipment.

noncompliance. Fire departments are typically required to follow these regulations if they provide medical care and bill electronically for their services. HIPAA regulations dictate how and to whom PMI can be shared and how the information is to be protected from unauthorized disclosure.

If your department is bound by HIPAA regulations, it is required to have a designated privacy officer and to provide you with compliance training. Regardless of your department's obligations under HIPAA, it is good professional practice to be as discreet as possible with regard to PMI. This means that PMI should only be disclosed to those who have a legitimate need for the information. Individuals who often have need for this information include law enforcement personnel, EMS personnel, and hospital staff **(Figure 22.3, p. 1274)**. Release of PMI to other individuals such as friends or family members can only be done with the consent of the patient. As someone who will be expected to provide medical care, it is important that you become familiar with and adhere to privacy guidelines established by your department and AHJ.

Infection Control

As a firefighter, you may be called on to provide medical care to patients with communicable diseases. **Communicable diseases** are not always obvious; however, the ability to identify signs and symptoms of common communicable diseases along with the use of proper **body substance isolation (BSI)** procedures will greatly reduce your risk of infection. Immunizations are also an important safeguard against infection.

Communicable Disease — Disease that is transmissible from one person to another.

Body Substance Isolation (BSI) — The practice of taking proactive, protective measures to isolate body substances in order to prevent the spread of infectious disease.

Figure 22.3 Law enforcement personnel often need to know a patient's protected medical information.

Commonly Encountered Communicable Diseases

While firefighters can potentially be exposed to numerous communicable diseases, a few are regularly encountered. Commonly encountered communicable diseases include hepatitis, tuberculosis, HIV/AIDS, and multidrug-resistant organisms (MDRO) such as drug-resistant staph infections. Several different types of influenza can also be encountered.

Hepatitis

Hepatitis is an inflammation of the liver. Viral hepatitis is most common, but drugs, alcohol, hazardous chemicals, or heredity can also cause the disease. Symptoms include a viral-like illness with yellowish discoloration of the eyes and skin. The different types of viral hepatitis are:

- **Hepatitis A** — Typically caused by consuming food or water that has been contaminated, particularly by fecal matter. Hepatitis A can be transmitted by close contact with infected individuals. Signs and symptoms of Hepatitis A include: fatigue, abdominal pain, fever, dark urine, and a marked yellowing of the skin and/or eyes. This is generally a short-term disease and typically has no long-term consequences to the individual. Hepatitis A is the least serious form of viral hepatitis.

- **Hepatitis B** — Typically transmitted through blood and other body fluids. Hepatitis B infections can either be short-term or long-term in nature and can potentially cause serious scarring and injury to the liver. This can eventually progress to liver failure. Signs and symptoms of Hepatitis B are similar to those of Hepatitis A. However, unlike Hepatitis A, Hepatitis B is a serious and potentially life-long infection.

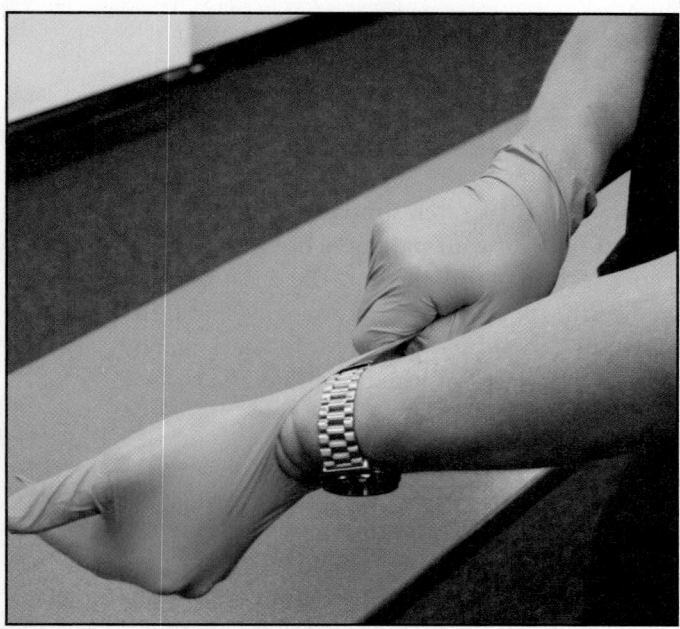

Figure 22.4 The use of disposable gloves serves as one facet of BSI procedures that combat infection and protect first responders.

Figure 22.5 Tuberculosis is spread via airborne droplets and is prevalent in nursing homes and other places with dense populations. *Courtesy of Ron Moore, McKinney (TX) FD.*

- **Hepatitis C** — Typically transmitted through blood and other bodily fluids. Hepatitis C differs from other types of Hepatitis because infected individuals can often go for many years without exhibiting symptoms. While the individual feels fine and is symptom free, the virus has more than likely done serious damage to the liver. In some cases this damage can be long-term or permanent. Liver failure can occur as with Hepatitis B.

- **Hepatitis D (Delta)** — Uncommon rare strain of hepatitis and only occurs in individuals who are also infected by Hepatitis B. This virus makes the effects of Hepatitis B much worse.

Vaccines are available for Hepatitis A and B, but those for Hepatitis C are still in the development stage. It is important that firefighters be immunized against hepatitis and other infectious diseases. Perhaps the best way to combat infection by hepatitis is by thoroughly washing your hands and maintaining proper BSI procedures that will be detailed later in this chapter **(Figure 22.4)**.

Tuberculosis

Tuberculosis (TB) is a bacterial infection that primarily affects the respiratory system. TB is contagious and is spread through droplets in the air produced by the breathing and coughing of an infected person. Bystanders have a chance of becoming infected when these droplets are inhaled. Because the disease is transmitted by airborne droplets, incidences of TB are more prevalent in high-density living areas such as nursing homes and prisons **(Figure 22.5)**.

TB is considered to be active when the infected individual exhibits signs and symptoms of the illness. Signs and symptoms of active TB include:

- Fever
- Fatigue
- Chills

- Weight loss
- Painful breathing
- Productive cough (often with traces of blood)
- Coughing that lasts several weeks

Because the signs and symptoms of TB are often similar to other illnesses, take BSI precautions and use an N-95 mask, even if your patient is not known to be infected. N-95 masks are detailed in a later section of this chapter. There is no widely used vaccine to protect against TB, but healthcare providers are typically skin tested to check for occupational exposure to the disease annually. More information about TB testing is provided in the section of this chapter on immunizations.

HIV/AIDS

Acquired immune deficiency syndrome (AIDS) is a result of prolonged exposure to the human immunodeficiency virus (HIV). This condition weakens the immune system to the point where the body is unable to fight off diseases that a person with a normal immune system could usually overcome. Early symptoms are similar to other viral illnesses but more advanced stages of the disease lead to severe muscle wasting and increased likelihood of contracting other infections. Because of this weakened immune defense, many persons who are infected by HIV are also likely to be infected by diseases such as hepatitis and TB.

HIV is spread through contact with infected blood and body fluids. While HIV/ AIDS are serious diseases, they pose far less risk to emergency responders than hepatitis and TB due largely to the short amount of time that HIV can survive outside of the body. However, it is still important to practice proper BSI procedures and to adequately decontaminate any medical equipment before it is used on another patient. Information on the cleaning and disposal of contaminated items is provided later in this chapter.

CAUTION
Medical equipment and surfaces should be decontaminated immediately following soiling to prevent the possible spread of infectious diseases to responders and other patients.

Multi Drug-Resistant Organisms (MDRO)

Multi drug-resistant organisms (MDROs) are an increasing concern in healthcare settings because they are difficult to control and do not typically respond to normal antibiotic treatment. Drug-resistant staph infections, better known as *methicillin-resistant staphylococcus aureus (MRSA)* are a commonly encountered type of MDRO. MRSA infections typically occur in healthcare settings, but there have been more recent outbreaks in communities. These infections can develop in numerous ways but are usually initially seen as abscesses in the skin that are commonly mistaken for spider bites. Because MRSA infections are easily spread, it is important that proper BSI precautions be taken. In addition, all reusable medical equipment such as stethoscopes, blood pressure cuffs, and backboards must be sanitized after each use to prevent these types of infections from spreading.

Other Diseases

From time to time there are outbreaks of infectious diseases that occur in our communities. Recent outbreaks in North America include H1N1 influenza ("swine flu"), avian influenza ("bird flu") and others. Chances are that different infectious disease outbreaks will occur during your time as a firefighter. Regardless of the disease, emergency responders should be prepared to treat affected patients. Proper sanitation and BSI procedures will go a long way toward keeping you and your patients safe.

Immunizations

As an emergency responder, your jurisdiction may require you to be immunized against certain infectious diseases if you have not already done so (**Figure 22.6**). **Immunizations** that may be required or recommended include:

- Hepatitis B
- Measles, mumps, and rubella (MMR)
- Varicella (chickenpox)
- Tetanus/diphtheria
- Influenza

In addition, an annual purified protein derivative (PPD) test may be administered to determine if you have been exposed to tuberculosis. This test is performed by injecting an inactive bacterium just under your skin. You will be required to return to have the results determined within 48-72 hours. A reaction to this test may indicate that you have been exposed to tuberculosis. In these instances, a chest x-ray and follow-up with a physician may be necessary to determine if there is an active infection. However, there are occasionally individuals whose result is a false-positive.

Body Substance Isolation (BSI)

Diseases are spread by **pathogens**, which are organisms that carry the disease and can cause infection. Pathogens exist in the body fluids of infected persons and can be transmitted from one person to another through contact with these fluids. While any type of body fluid can carry pathogens, bloodborne pathogens are most commonly encountered. Airborne pathogens are also a threat and are spread by tiny droplets when an infected individual coughs or breathes.

Exposure occurs when the body fluid of an infected individual comes in contact with an exposed area of another person. There are numerous routes where emergency responders are susceptible to exposure. These include open wounds, cuts, and sores on the body. Other routes of exposure include contact with the eyes, nose, or mouth. Any break in the skin is a potential route of exposure (**Figure 22.7, p. 1278**).

With the threat of exposure to communicable diseases, it is critical that firefighters take all precautions necessary to protect themselves from exposure. The best way to do this is by maintaining proper BSI procedures. BSI is an all-encompassing group of procedures that include hand washing, proper use of personal protective equipment, and proper disposal and/or cleaning of soiled items. The United States Centers for Disease Control and Prevention (CDC) and Canadian provincial health authorities are good sources of information for establishing BSI procedures.

Immunization — Process or procedure by which a subject (person, animal, or plant) is rendered immune or resistant to a specific disease. This term is often used interchangeably with *vaccination* or *inoculation*, although the act of inoculation does not always result in immunity.

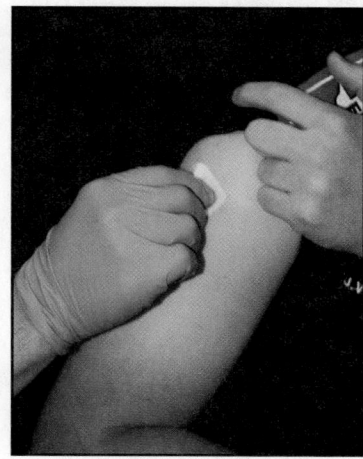

Figure 22.6 Many jurisdictions require responders to be immunized against commonly encountered infectious diseases.

Pathogens — Organisms that cause infection such as viruses and bacteria.

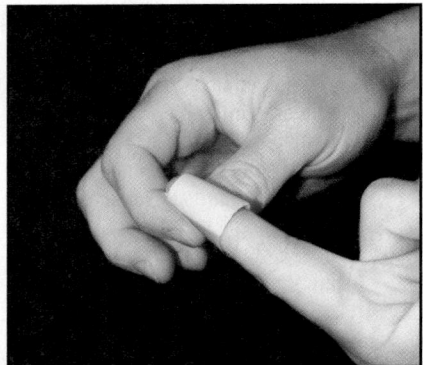

Figure 22.7 Any break in skin, no matter how minor, may serve as a route of exposure.

Figure 22.8 Hands should be thoroughly washed with warm water and soap.

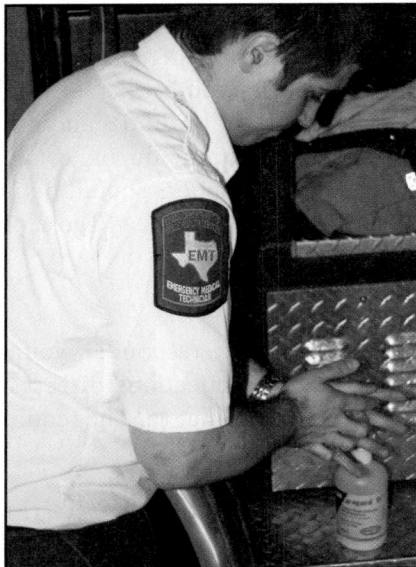

Figure 22.9 Using an alcohol-based hand sanitizer is a good alternative when hand-washing facilities are unavailable.

Hand Washing

Proper hand washing has been shown to greatly reduce the transmission of disease and should be done frequently. This is especially true before and after coming in contact with a patient.

Hands should be washed in a methodical manner with warm water and soap (**Figure 22.8**). Special attention should be paid to areas on the hands that are soiled, the areas between fingers, and to the fingernails. In addition, it is good practice to wash the wrists and forearms as well. A specially designed scrub pad can also be used if available. Hands should be washed for no less than 30 seconds in order to ensure that they are cleaned thoroughly.

Hand washing is often not an option when responding to emergency calls. In these cases, an alcohol-based hand cleaning solution should be used (**Figure 22.9**). This is especially true after providing patient care. It is a good practice to clean your hands with alcohol solution before entering the passenger compartment of the apparatus.

Personal Protective Equipment (PPE)

In order to prevent transmission of infectious diseases, it is imperative that emergency responders use adequate personal protective equipment (PPE). The numerous types of PPE allow emergency responders to adjust their level of protection based on the potential threat of infection (**Figure 22.10**). In some instances, this may mean simply wearing gloves. In other cases, a gown, mask, eye protection, and gloves may be needed. Your jurisdiction will dictate the level of PPE that should be worn on emergency medical calls.

Figure 22.10 Various types of PPE can be selected and used based on the potential threat of infection.

When selecting PPE, it is always better to wear too much than not enough. Refer to your department's SOPs regarding PPE use on emergency medical responses. The following sections will discuss types of PPE in more detail.

Gloves. Gloves are an essential component of PPE and should always be worn during patient contact. In the past, medical gloves were typically made out of latex. Unfortunately, some patients and healthcare providers have an allergic reaction to

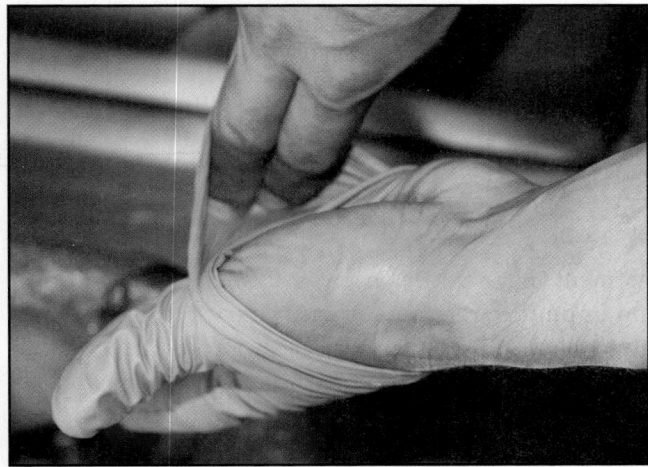

Figure 22.11 Removing soiled gloves should be accomplished using two clean fingers on the inside of the glove to avoid exposure.

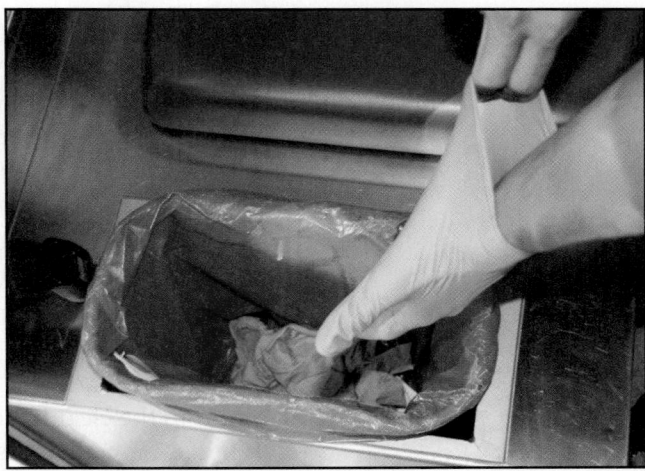

Figure 22.12 Gloves soiled with bodily fluids including blood should be disposed of in a waste container specifically designed for the disposal of biohazards.

this material. Therefore, many jurisdictions supply emergency responders with gloves made out of vinyl or other materials. If your jurisdiction uses latex gloves, you should ask the patient if he or she has a latex allergy. Non latex gloves should be available in these instances.

The same gloves should never be used when treating multiple patients. When moving from one patient to another, you should remove the old gloves and don a clean pair before providing treatment. Many responders carry multiple pairs of gloves for these instances.

To remove soiled gloves, place two clean fingers on the inside of the glove and peel off (**Figure 22.11**). Gloves that are soiled with blood or other body fluids should be disposed of in a sealed container that is specifically used for the disposal of biohazards. These containers are typically red in color and have a red liner (**Figure 22.12**).

Eye protection. Because the eyes are a route for disease entry, eye protection is necessary when there is the chance for blood or other body fluids to be splashed, sprayed, or spattered. Safety glasses are a common type of eye protection used during medical responses (**Figure 22.13**). These are designed to keep fluids away from the eyes and usually provide some type of impact resistance as well. Accessories are available for those who wear prescription glasses so that they can be modified for use. In addition, a combination mask and eye shield is available and widely used by emergency response agencies (**Figure 22.14**). Helmet shields and Bourke Eye Shields do not provide suitable eye protection for emergency medical use.

Masks. Masks protect against respiratory hazards, airborne pathogens, and body fluids. In situations where contact with blood or other body fluids are a concern, surgical-style masks typically suffice. However, if you suspect that a patient has a communicable respiratory disease such as TB, it is best to use a mask that provides a greater level of respiratory protection. In these instances, placing a mask on the patient can also prevent the transmission of the disease to others.

N-95 respirators may be provided by your jurisdiction for use when treating patients where TB is suspected. These respirators are termed N-95 because they are tested and shown to block at least 95 percent of airborne particles (**Figure 22.15, p. 1280**). Because a proper seal is necessary in order to provide protection, potential

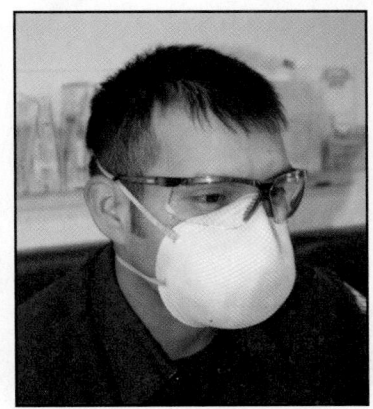

Figure 22.13 Eyes are a direct route for disease entry and must be protected.

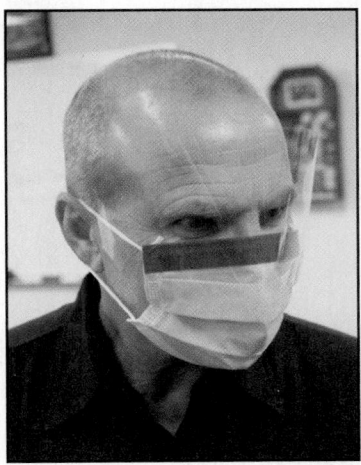

Figure 22.14 Combination mask and eye shields are widely used by emergency response agencies.

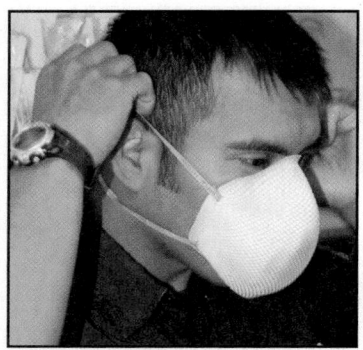

Figure 22.15 N-95 masks can block at least 95 percent of airborne particles and greatly reduce the risk of airborne disease exposure for first responders.

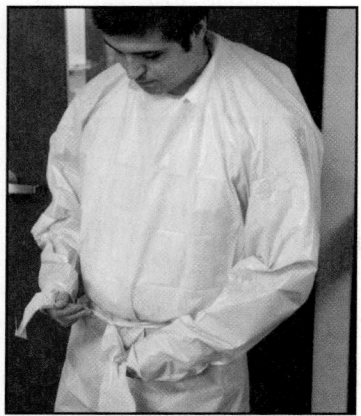

Figure 22.16 Gowns protect a responder's uniform and skin from gross exposure to body fluids and are commonly used during trauma care and childbirth.

Figure 22.17 A fresh uniform in an easily-accessible location helps maintain the image of emergency responders as professionals and limits the potential spread of disease.

wearers should be fit tested to ensure that they are able to obtain a complete seal with the respirator and that they are able to perform normal activities while wearing it.

Gowns. Gowns are available to protect exposed skin and your uniform from the spray and spatter of body fluids. They are especially useful in situations such as trauma and emergency childbirth. As with all PPE, you should know where gowns are stored on the apparatus or ambulance and should be able to don a gown quickly **(Figure 22.16)**.

Spare uniforms. Your department may require that you have an extra uniform ensemble available at the station or on the apparatus in case your clothing becomes soiled **(Figure 22.17)**. Not only does a soiled uniform look unprofessional, it can potentially allow for the spread of disease. Follow departmental guidelines with regard to the cleaning of soiled uniforms. In instances of extreme contamination, you should immediately shower and notify your department's infection control officer.

Cleaning and Disposing of Contaminated Items

Medical equipment and nondisposable PPE that come into contact with any patient must be considered to be contaminated. Decontamination must follow the requirements of NFPA® 1581, *Standard on Fire Department Infection Control Program*, and your department's infection control protocol. Follow these general steps for decontaminating equipment and PPE:

1. Wear disposable gloves, mask, and eye protection to place contaminated items in a biohazard bag or container.

2. Transport the contaminated items to the designated disinfecting area of the fire station or other facility.

3. Don the appropriate splash-resistive eyewear, cleaning gloves, and fluid-resistant clothing designated for decontaminating equipment.

4. Use a solution of bleach and water or disinfectant in accordance with the equipment or PPE manufacturer's instructions **(Figure 22.18)**.

5. Do not decontaminate equipment or PPE in the kitchen, bathroom, or other living areas of the fire station.

6. Contaminated station uniforms or structural PPE must not be taken home or cleaned in normal laundry facilities **(Figure 22.19)**.

7. Decontaminate the sink and cleaning area, remove PPE worn during cleaning, and dispose of in accordance with fire department protocol.

Contaminated disposable equipment and PPE must be placed in medical waste containers and disposed of in accordance with local protocol. Needles and blades, must be placed in sharps containers designated for such items. Sharps containers must be closable, puncture-proof, leakproof, and labeled according to federal, state, or local regulations.

Patient Assessment

Before medical treatment can begin, the patient's condition must be assessed. The most basic assessment is to determine if the critical functions of the body are working properly. If the patient is responsive, the assessment involves an evaluation of the patient's airway, breathing, and circulation, which has traditionally been known as the ABCs. The vast majority of patients you will care for as a firefighter will be responsive and assessed in this manner.

Figure 22.18 Some equipment may be cleaned with a disinfectant solution.

Figure 22.19 The use of approved laundry facilities limits contamination from uniforms and other PPE.

In patients that are unresponsive and not breathing, or breathing with irregular gasping breaths (also known as agonal respirations), the assessment sequence should be rearranged to <u>C</u>irculation, <u>A</u>irway, then <u>B</u>reathing (CAB). This change of the usual assessment and intervention sequence is based on research that shows that *any* delay in the delivery of high-quality chest compressions reduces the likelihood of a successful resuscitation from cardiac arrest.

If a pulse is present (and this will be most of the time), then assess the airway and follow the normal ABC sequence. If no pulse is found, immediately begin a cycle of chest compressions and then open the airway and ventilate the patient: the cardiac arrest CAB sequence. The following sections identify the ABCs of patient assessment in greater detail. Remember that formal CPR training will provide more definitive information on the performance of assessment techniques.

Airway

A patient's **airway** is the passage between the lungs and the nose and mouth where air travels during breathing (**Figure 22.20**). If the airway is obstructed by the tongue, a foreign object, or fluid, air cannot travel freely. If the patient is able to talk and appears to be breathing without difficulty, it can be assumed that the airway is clear. However, if the patient is unresponsive, you may need to perform a technique to open the airway. Techniques for opening the airway are introduced during formal CPR training.

Airway — A passage for carrying air from the nose or mouth to the lungs.

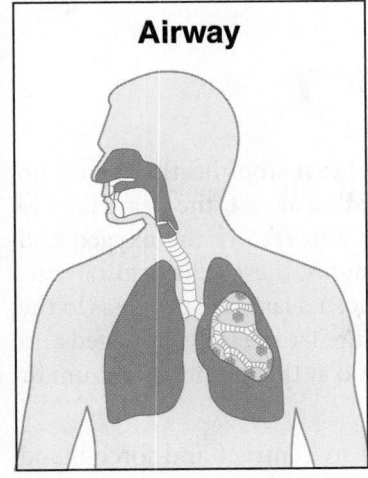

Figure 22.20 The airway is the passage between the lungs and the nose and mouth where air travels during breathing.

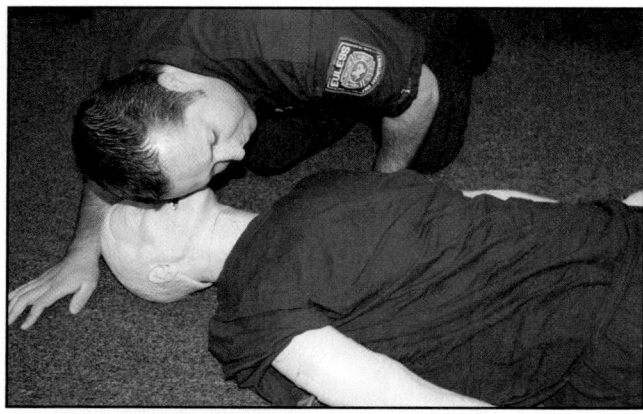

Figure 22.21 Responders should look, listen, and feel for breathing.

Pulse — Rhythmic throbbing caused by expansion and contraction of arterial walls as blood passes through them.

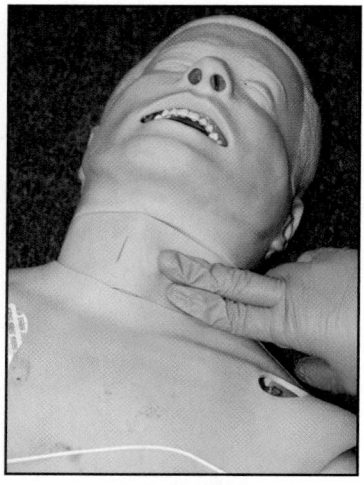

Figure 22.22 Taking a patient's pulse is the easiest way to determine the presence of circulation.

Cardiac Arrest — Sudden cessation of heartbeat.

Clinical Death — Term that refers to the lack of signs of life, where there is no pulse and no blood pressure; occurs immediately after the onset of cardiac arrest.

Biological Death — Condition present when irreversible brain damage has occurred, usually 4 to 10 minutes after cardiac arrest.

Breathing

After determining that the airway is open, it is necessary to determine if the patient is breathing. The best way to determine this is to look, listen, and feel for breathing by placing your ear near the patient's nose and mouth and looking at the chest (**Figure 22.21**). You should see the patient's chest rise and fall with each breath. When listening, you should be able to hear the patient breathing through the mouth or nose. When feeling, you should be able to feel exhaled air. If the patient is not breathing or no air is being moved during breathing, you will need to provide assistance. Techniques for rescue breathing are introduced during formal CPR training.

Circulation/Compressions

Circulation/compressions is the flow of blood through the body. The easiest way to determine the presence of circulation is by feeling for a **pulse** (**Figure 22.22**). Common locations to find a pulse are shown in **Figure 22.23**. Most pulses are assessed at the radial and carotid arteries. Because of the distance of the radial artery from the heart, it typically takes stronger blood flow for a pulse to be felt in this area than at the carotid artery. Therefore, just because a pulse cannot be found at the radial artery does not mean that the patient is pulseless. In these instances, you should attempt to find a pulse at the carotid artery as well. If a pulse cannot be found, chest compressions should be initiated immediately.

Cardiac Arrest and Cardiopulmonary Resuscitation (CPR)

A patient is considered **clinically dead** when his or her heart stops beating and is no longer breathing. During this condition, known as **cardiac arrest**, the heart fails to circulate blood through the body, and cells are not able to receive the oxygen and nutrients they need to survive. In the span of a few minutes, these cells begin to die. If this condition is allowed to continue for much longer, cell death will progress to the point where organs such as the brain, heart, and lungs are irreversibly damaged and cannot be revived. This is known as **biological death** and at this point no amount or type of medical intervention can revive the patient.

The heart produces electrical impulses that cause it to contract and force blood throughout the body. Heart failure that causes clinical death is in many instances the result of a problem with the heart's ability to produce and/or transmit these impulses. This condition often persists unless an outside electrical stimulus is applied. This outside stimulus is commonly known as **defibrillation** (**Figure 22.24**).

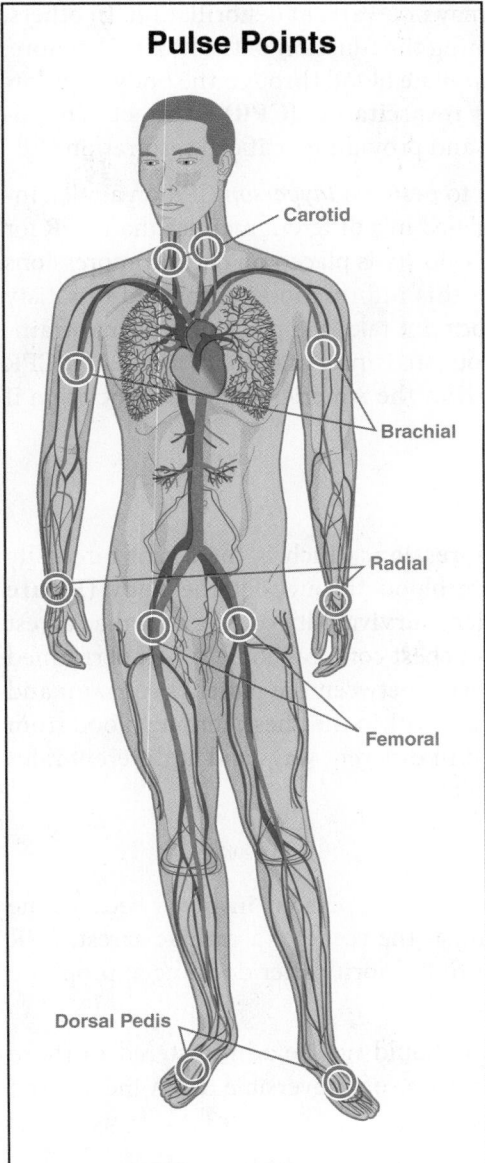

Pulse Points

Carotid

Brachial

Radial

Femoral

Dorsal Pedis

Figure 22.23 Common locations to find a pulse.

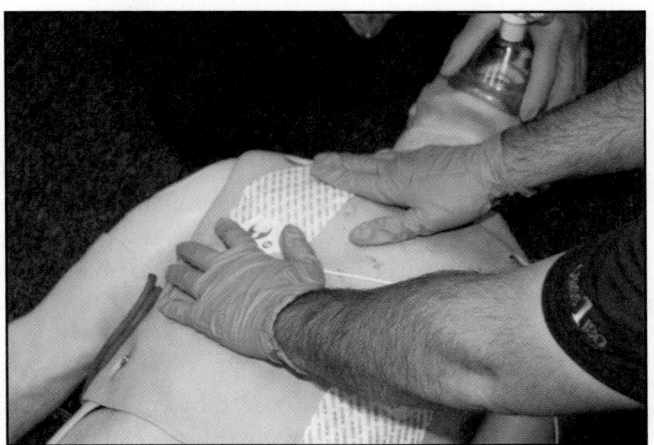

Figure 22.24 Defibrillation is an application of electrical stimulus to resume normal function of the heart.

Figure 22.25 AEDs are often provided in public places for use in case of an emergency.

Because the heart will often not resume operation without defibrillation, it is critical that it be done as soon as possible. Advanced-level EMS providers are trained to recognize electrical issues with the heart and manually administer defibrillation and medications that treat these issues. Unfortunately, it may take several minutes for these personnel to arrive on scene. During this time, the patient may have already transitioned from clinical death to biological death.

Realizing the importance of early defibrillation, many public facilities such as shopping malls, schools, and airports now have **automated external defibrillators (AEDs)** available for layperson use **(Figure 22.25)**. These devices instruct the layperson on how to use the AED, automatically determine if defibrillation is needed, and notify the user to push a button that delivers the shock. It is also common to find AEDs on fire apparatus that do not carry paramedics and their equipment. The use of AEDs is beyond the scope of this chapter; however, your department will provide further training if you are expected to use AEDs during medical responses.

Defibrillation — The delivery of a measured dose of electrical current by a special machine in order to regain normal function of the heart.

Automated External Defibrillator (AED) — Cardiac defibrillator designed for layperson use that analyzes the cardiac rhythm and determines if defibrillation is warranted.

In some instances, the heart's condition may not warrant defibrillation. In others, defibrillation may not be successful in resuming the function of the heart. Therefore, outside intervention is needed in order to circulate blood through the body and slow the rate of cellular death. **Cardiopulmonary resuscitation (CPR)** is the act of physically forcing blood through a patient's body and providing artificial respiration.

The following sections will highlight how to perform *layperson CPR* on adults, infants, and children. Layperson CPR is performed in a different manner than CPR for healthcare providers. In particular, a higher priority is placed on chest compressions than patient assessment. Keep in mind that this information is provided to satisfy entry requirements for NFPA® 1001 and does not take the place of a formal training program on CPR. It is imperative that you participate and are certified in a CPR course administered by your department and/or the American Heart Association if you are to provide this type of care.

Chest Compressions

The main component of CPR is **chest compression**, which is the act of forcefully compressing the heart in order to circulate blood throughout the body **(Figure 22.26)**. Research has clearly shown that patient survivability during a cardiac arrest greatly improves with early administration of chest compressions. When performed properly, chest compressions compress the heart between the patient's sternum and spine in rapid succession and alter the pressure within the chest, forcing blood from the heart. Chest compressions are performed in different ways and at different rates and depths, depending on the age of the patient.

Administering Chest Compressions

Events of cardiac arrest are extremely stressful for all persons involved because the event is typically unexpected. When arriving at the scene of a cardiac arrest, CPR may or may not already be in progress. Your first priority after donning appropriate PPE is to determine the condition of the patient.

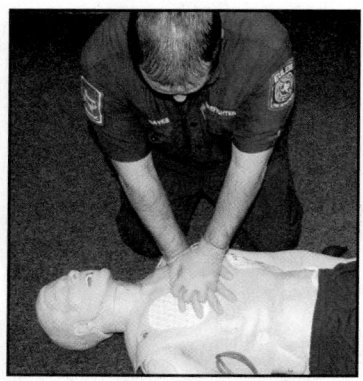

Figure 22.26 Chest compressions circulate blood throughout the body and improve patient survivability.

While CPR is a lifesaving intervention, it should not be administered to those who have obvious signs of irreversible death. Signs of irreversible death include the following:

- **Rigor mortis**
- Obvious wounds that are not compatible with life (such as decapitation)
- Decomposition
- **A line of lividity**

Remember that the techniques identified in this chapter are for *layperson* CPR. As such, there is a greater emphasis on providing chest compressions than providing ventilation. Therefore, rescue breathing will not be addressed in this chapter. Research reviewed and endorsed by the American Heart Association shows that compression-only CPR is extremely effective and still provides some air movement into and out of the lungs. Full CPR incorporates both chest compressions and rescue breathing. Most emergency responders are certified in CPR to the Basic Life Support (BLS) for Healthcare Providers-Level. It is critical that you attend and obtain certification from a formal CPR training course provided through your jurisdiction or the American Heart Association. The following sections will detail chest compression administration for adults, children, and infants.

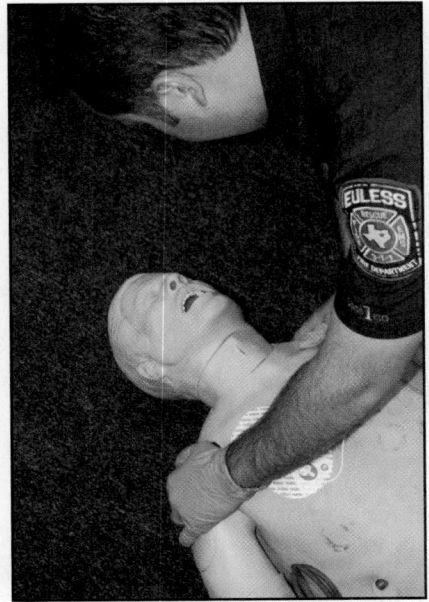

Figure 22.27 Upon arrival for a suspected cardiac arrest, the patient's level of consciousness should be determined.

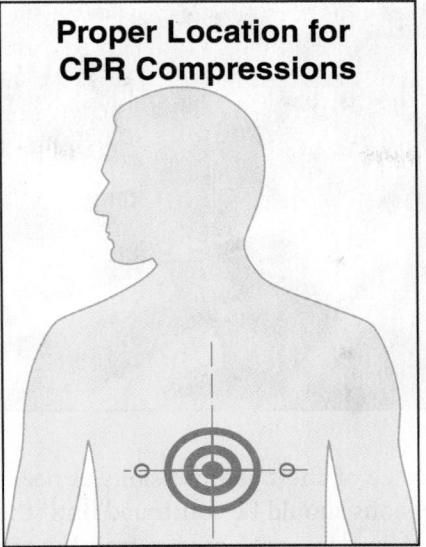

Proper Location for CPR Compressions

Figure 22.28 The correct location for the administration of chest compressions is in the center of the patient's chest between the nipples.

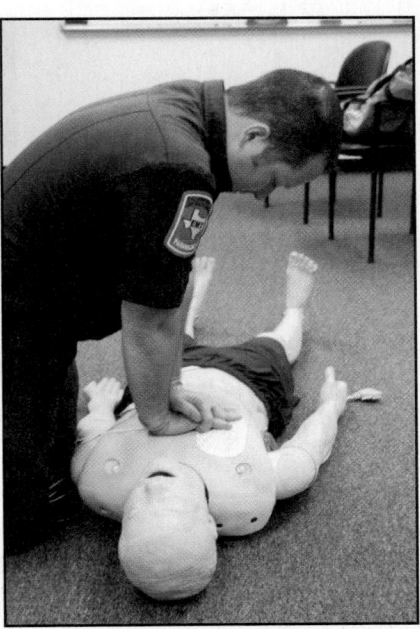

Figure 22.29 Proper alignment of shoulders and hands ensures that the compression is properly delivered.

Chest Compressions for Adults

Upon arrival for a suspected cardiac arrest, it is critical that you identify the patient and determine the patient's status. Your first determination should be if the patient is conscious or unconscious. This can be performed by gently shaking or tapping the patient and asking them if they are OK **(Figure 22.27)**. If there is no physical or verbal response, you should determine if the patient has a pulse. If no pulse can be found at the radial or carotid arteries, chest compressions should be started. The communications center should also be notified of the condition and any additional resources that are needed.

When performing chest compressions on an adult, the patient must first be lying on their back on a hard surface such as the floor or a backboard. A patient who is lying on a bed should be moved to the floor because performing chest compressions on someone lying on a soft surface is typically ineffective. Next, find the correct location to administer compressions. This location is in the center of the patient's chest between the nipples **(Figure 22.28)**. In order to perform compressions, place one hand on top of the other and complete the following steps:

Step 1 — Straighten arms and lock elbows. Elbows should remain locked while performing compressions.

Step 2 — Align your shoulders directly over your hands. This ensures that compressions are delivered straight down **(Figure 22.29)**.

Step 3 — Press straight down on the chest hard enough to depress the sternum at least 2 inches (50 mm) **(Figure 22.30)**.

Step 4 — Release the compression by lifting up. However, keep the elbows locked at all times and do not remove the hands from the compression location.

Compressions should be administered by quickly and firmly depressing the chest at a rate of at least 100 compressions per minute. Responders should rotate the

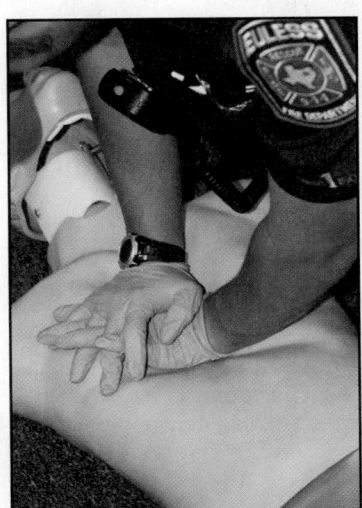

Figure 22.30 The sternum should be depressed at least 2 inches (50 mm).

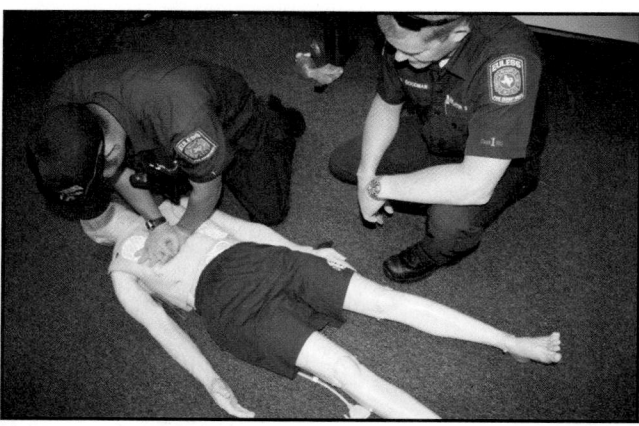

Figure 22.31 Responders should rotate the performance of chest compressions as needed to prevent exhaustion.

performance of chest compressions as needed to prevent exhaustion (**Figure 22.31**). Compressions should be continued until the patient moves or regains consciousness or until responders with greater training arrive.

Chest compressions should be performed in a rhythmic manner and it should never feel as though you are jabbing at the patient. It is possible that some type of cracking of the sternum or ribs may be heard or felt during compressions, but this is normal.

NOTE: While performing chest compressions, it is often common to hear or feel a patient's ribs break due to the pressure. While this sensation is not pleasant, it is a sign of good quality compressions and should not discourage you from continuing.

Chest Compressions for Children

When arriving on scene for a cardiac arrest involving a child, first determine the patient's status. Gently shake or tap the child to determine if conscious. Speak to the child and listen for a response. If the patient is unresponsive, check for a pulse. If no pulse can be found, begin chest compressions. You should move the patient to the floor or onto a backboard if the patient is not lying on a solid surface. This would be a good time for you or another responder to update dispatch with the condition and request any additional resources that may be needed.

Chest compressions are performed for children ages 1-8 in a slightly different manner than those for adults. Instead of performing compressions with 2 hands, rescuers should only use one hand (**Figure 22.32**). The hand should be placed in the center of the chest in line with the patient's nipples. When performing compressions, the patient's chest should be compressed about ⅓ the depth between the chest and the back or about 2 inches (50 mm). This ensures that adequate blood flow is achieved. Compressions should be administered firmly and quickly at a rate of at least 100 compressions per minute. Continue administering compressions until the child begins to move, regains consciousness, or rescuers with greater training arrive.

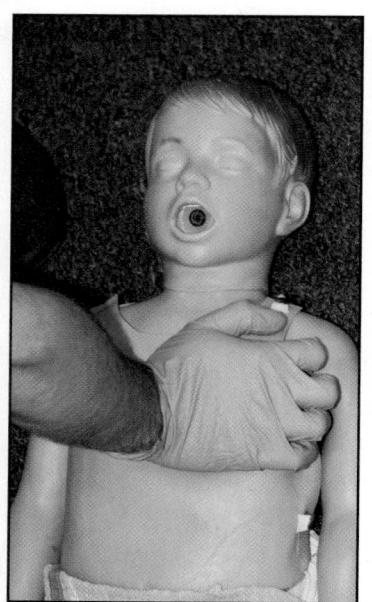

Figure 22.32 When performing chest compressions for a child ages 1-8, use only one hand.

Chest Compressions for Infants

When responding to a cardiac arrest in an infant patient (less than one year old), it is important to first determine the patient's status. Most incidences of cardiac arrest in infants are due to airway obstructions. If an obstruction can be seen in the mouth, remove it (**Figure 22.33**). If not, continue in determining the patient's status by speaking to and stroking the infant and looking for any type of movement or other response. If there is no response, check for a pulse. If no pulse can be found, begin chest compressions. While beginning compressions, have another responder update dispatch on the patient's status and request any additional resources that may be necessary.

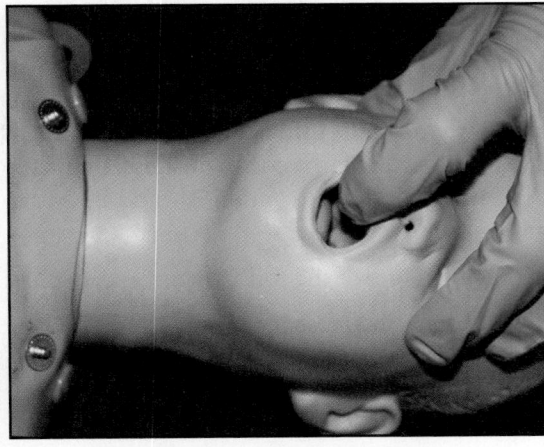

Figure 22.33 Remove any obstructions that can be seen in the mouth.

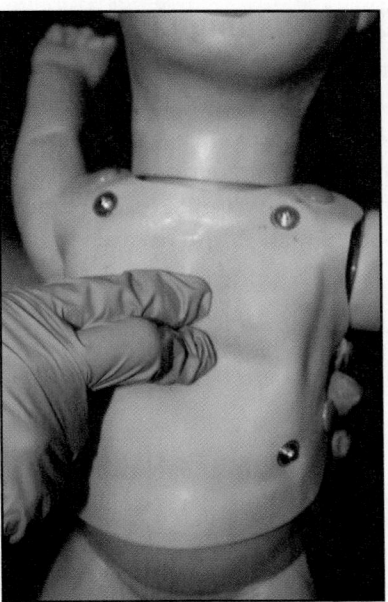

Figure 22.34 Chest compressions for an infant use only the index and middle fingers of one hand.

Chest compressions for infants are performed with the index and middle fingers of one hand (**Figure 22.34**). This allows the chest to be better compressed than with the heel of the hand. As with adults and children, the compressions should be delivered in the center of the chest in line with the patient's nipples. Compressions should be delivered at a rate of at least 100 compressions per minute and the infant's chest should be compressed to ⅓ the depth of the torso, about 1.5 inches (38 mm). Periodically check the patient's mouth to see if any foreign bodies are present. If so, remove them from the infant's mouth and reassess. If not, continue performing compressions until the infant begins to move, regains consciousness, or until responders with greater training arrive.

Chest compressions for infants can also be performed in an "around the chest" manner where the thumbs are used to perform compressions (**Figure 22.35**). The same compression rate and depth as the traditional method still applies.

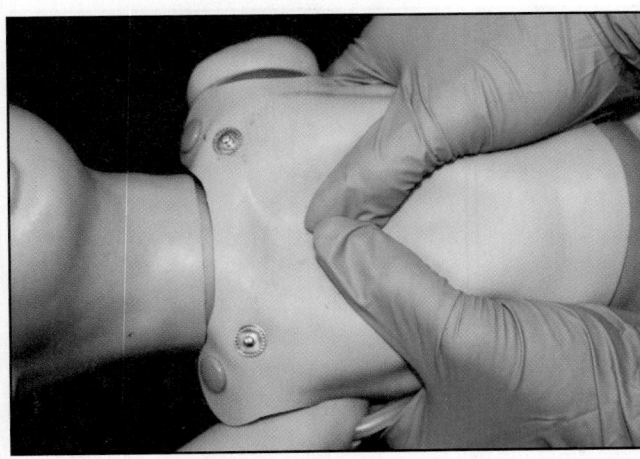

Figure 22.35 An "around the chest" method for infants may also be performed using the thumbs to deliver compressions.

Discontinuing CPR

It is important to periodically reassess the patient during CPR to determine if it is working. The main way to determine if CPR is successful is if the patient regains consciousness or begins to move. Another way to determine the success of CPR is by reassessing periodically for a pulse. If the patient regains consciousness, begins to move, or has a pulse, CPR should be discontinued and the patient should be transported to a definitive care facility.

Once CPR is initiated, it should continue until one of the following events occurs:

- The patient begins to move and/or regains consciousness
- The patient has a pulse
- You are unable to continue due to exhaustion
- You hand over care to a rescuer with higher training
- You are instructed to stop CPR by a medical control physician

As a rescuer, you must understand the emotional nature of cardiac arrest events. This is especially true if the patient's family or friends are present. When deciding not to begin CPR or when discontinuing CPR without patient response, you may be subject to anger or other emotional responses from bystanders. It is important to speak calmly to family and friends of the deceased and to keep in mind their feelings when discussing the patient with other emergency responders.

Bleeding Control

Blood in the body is critical to transport oxygen and nutrients to cells. Maintaining blood flow and volume is extremely important; therefore, emergency responders should be prepared to control bleeding in victims. There are two types of bleeding: external and internal. This section will describe the types of bleeding as well as methods for bleeding control.

Types of External Bleeding

External bleeding is that which occurs outside of the body. It is typically caused by lacerations, punctures, and other openings in the skin (**Figure 22.36**). There are three types of external bleeding: arterial, venous, and capillary.

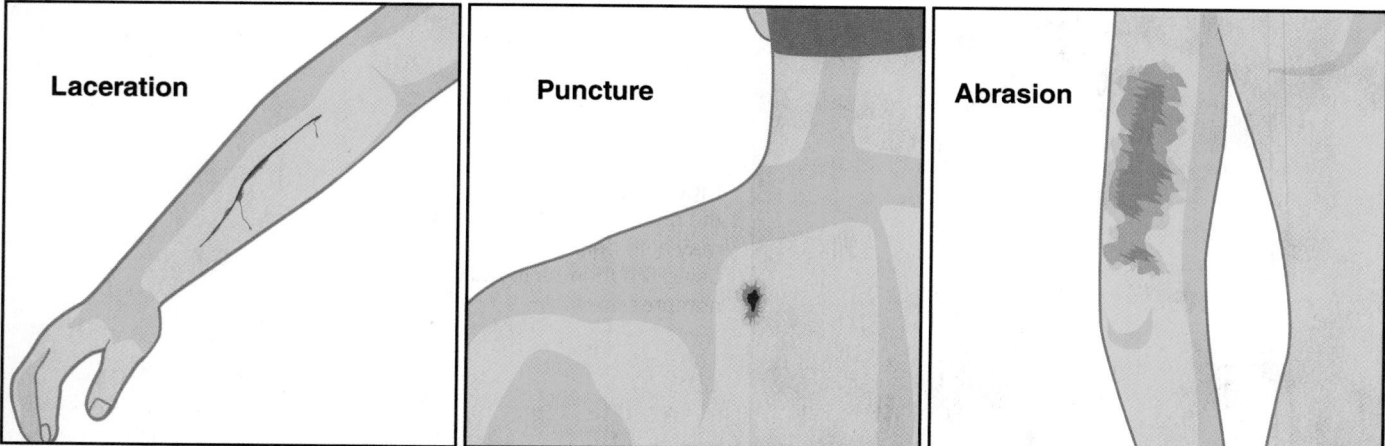

Figure 22.36 Lacerations, punctures, and abrasions are commonly encountered sources of external bleeding.

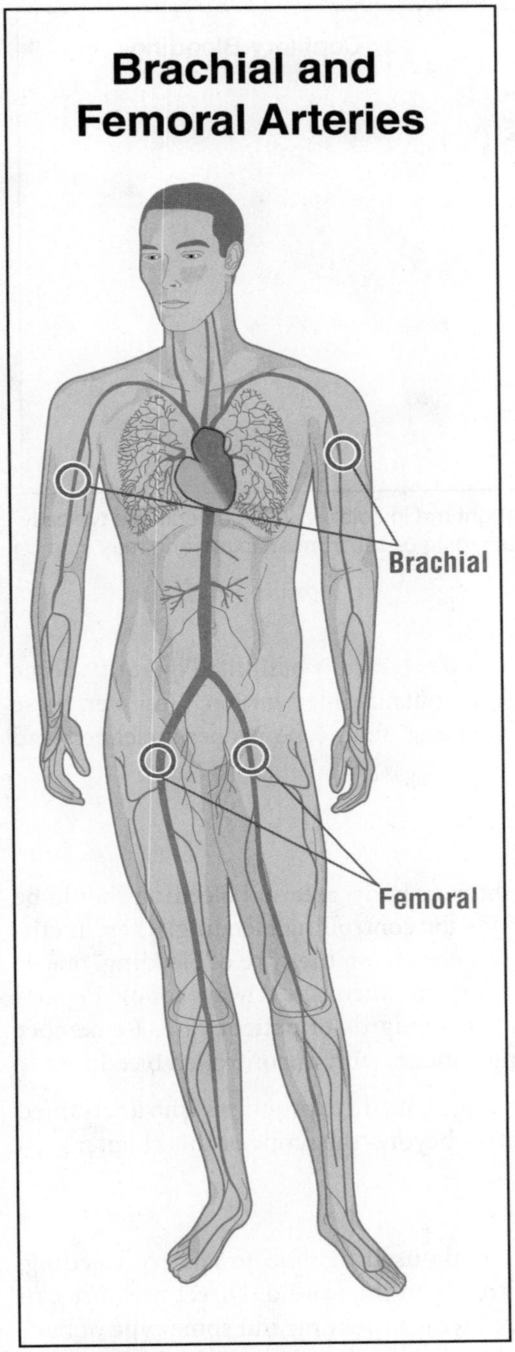

Brachial and Femoral Arteries

Brachial

Femoral

Figure 22.37 Major arteries in the body.

Arterial Bleeding

Blood is transported away from the heart through vessels called **arteries** (**Figure 22.37**). The blood transported through arteries is under high pressure and is bright red in color due to the large amount of oxygen it contains. Arterial bleeding occurs when the wall of the artery has been ruptured.

Arterial bleeding can be identified when the blood is bright red and spurting or pulsing (**Figure 22.38, p. 1290**). This spurting bleeding coincides with each contraction of the heart. As significant quantities of blood are lost, the force of the blood decreases. Arterial bleeding is often difficult to control due to the substantial force of the blood and is a true medical emergency. Patients with arterial bleeding have the potential to lose a substantial amount of blood in a short period of time. Methods for controlling bleeding are discussed later in this chapter. Keep in mind, however, that it is sometimes not possible to stop arterial bleeding without surgical intervention. Therefore, there should be no delay in transporting patients with arterial bleeding to a hospital by ambulance.

Venous Bleeding

Veins are responsible for returning blood that has been used by cells back to the heart. While arteries carry blood with substantial force, blood movement through veins occurs at a slow and steady flow. Therefore, venous bleeding is typically easier to control than arterial bleeding.

Patients with venous bleeding will have a steady flow of blood. The venous blood will also be much darker than arterial blood because oxygen has been removed by the cells and carbon dioxide and waste have been added.

Capillary Bleeding

Capillaries are the areas between arteries and veins where the oxygen and nutrients in blood are delivered to cells. In addition, carbon dioxide and waste are removed by blood through the capillaries. Because of this exchange, blood moves relatively slowly in capillaries. The majority of injuries to capillaries come in the form of scrapes or superficial lacerations.

Artery — Blood vessel that carries oxygen-rich blood away from the heart.

Vein — Any blood vessel that carries blood from the tissues to the heart.

Capillaries — Tiny blood vessels in the body's tissues in which the exchange of oxygen and carbon dioxide take place.

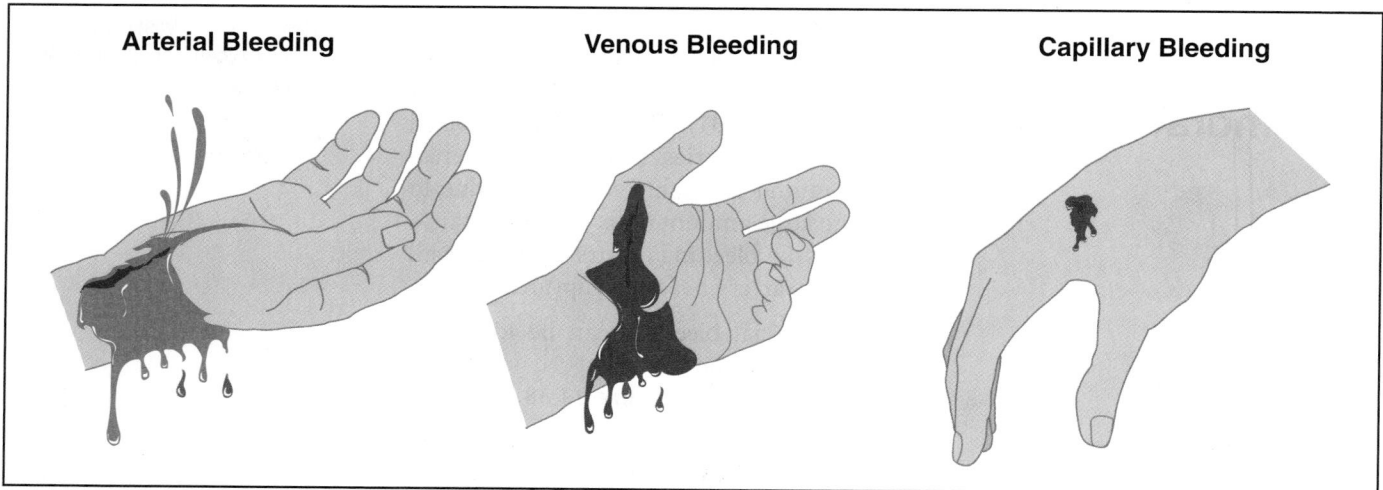

Figure 22.38 Arterial bleeding spurts or pulses with each heart contraction and is bright red in color. Venous bleeding is typically a slow and steady flow of blood and is darker in color than arterial blood. Capillary bleeding oozes from an injury and may stop on its own.

Blood "oozes" from injuries to capillaries and is typically of limited quantity. Some capillary bleeding can stop on its own without outside intervention. However, these injuries provide an opportunity for infection and should be properly cleaned and bandaged.

Controlling External Bleeding

Because of the importance of blood in the human body, external bleeding should be controlled quickly. There are several methods for controlling bleeding. These methods include direct pressure and elevation. Depending on the type of bleeding, one or more of these methods may need to be used simultaneously. Always follow departmental medical protocols and recommended standards of patient care. Remember that proper PPE must be used when treating patients with uncontrolled bleeding.

Tourniquets are also used to control bleeding by medical providers who are trained in their use. However, the use of tourniquets is beyond the scope of this chapter.

Direct Pressure

Direct pressure is the first and most commonly used method to control bleeding. As the name implies, pressure is applied directly to the wound. Direct pressure can be applied with a gloved hand, a **dressing**, or with a dressing and some type of bandage (**Figures 22.39a and b**). With minor bleeding, the application of a dressing to the wound may be all that is necessary to stop the bleeding. If blood soaks through the dressing, more dressings can be applied on top instead of removing the soiled dressings. This procedure should continue until blood ceases to soak through. At this point, a bandage can be applied to hold the dressings in place and to keep pressure on the wound.

In more severe instances of bleeding, especially arterial bleeding, direct pressure should be applied with a gloved hand immediately. Blood can be lost so quickly in these instances that no time should be wasted in applying pressure. Once dressings and bandages are located and opened they can be applied to the wound.

Dressing — Clean or sterile covering applied directly to a wound; used to stop bleeding and to prevent contamination of the wound.

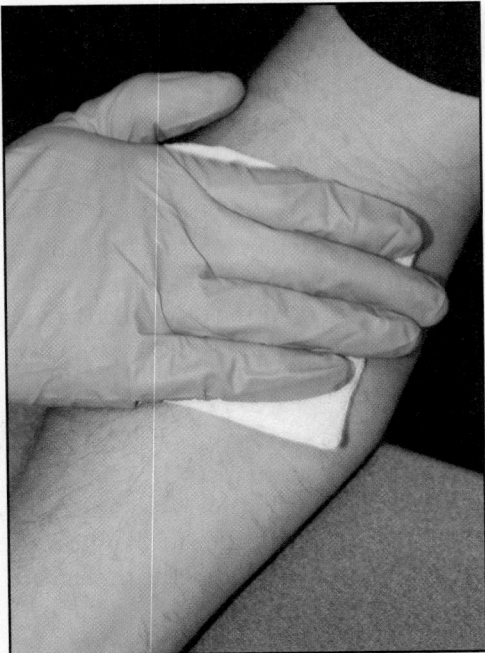

Figure 22.39a Direct pressure should be initially applied with a dressing if possible.

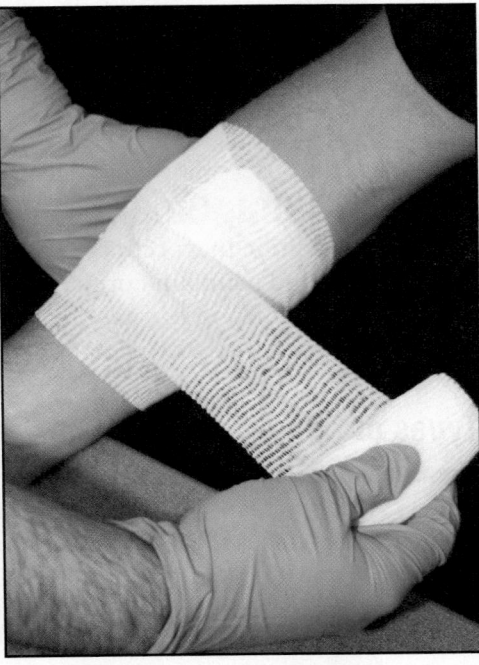

Figure 22.39b Dressings applied to a wound should be secured with a bandage.

Elevation

When injuries occur to extremities, elevation can be used in conjunction with direct pressure. Elevation is simply the act of raising the extremity above the level of the patient's heart **(Figure 22.40)**. The effect of gravity helps to slow the amount of blood traveling through the extremity, in turn slowing the bleeding. Elevating the extremity may not be possible in some instances, such as with fractures and injuries where spinal immobilization is needed.

Internal Bleeding

When blood vessels rupture inside the body, internal bleeding may occur. Internal bleeding is dangerous because spaces within the body can hold a considerable amount of blood before there is evidence of a problem. While this bleeding cannot be seen externally, there are several signs and symptoms that internal bleeding may be present. These include:

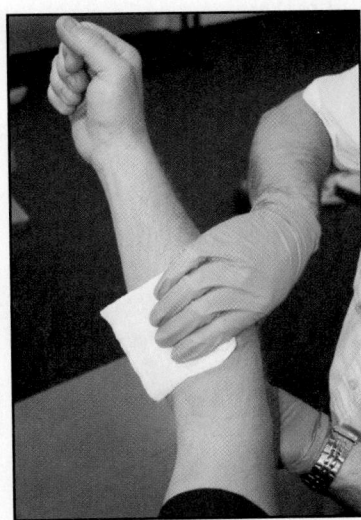

Figure 22.40 Elevation of an injury above the patient's heart can work to slow the amount of bleeding.

- Significant bruising or other indications of injury, especially involving the patient's torso

- Bleeding from openings in the body such as the mouth, nose, or rectum

- Bloody stool or urine

- A painful, swollen, or rigid abdomen

- Vomiting of a substance that looks like coffee grounds

- Cool, pale, and clammy skin

Internal bleeding can be caused by both trauma and medical issues. Traumatic causes of internal bleeding can include a penetrating gunshot or stab wound, motor vehicle accident, or fall. Medical causes of internal bleeding include ruptured blood vessels or organs.

Uncontrolled internal bleeding is a true medical emergency. Patients with internal bleeding should be treated for shock. Patients who show signs and symptoms of internal bleeding should be immediately transported by ambulance to a hospital for treatment.

Shock Management

Shock is a condition that occurs when the body for one reason or another is unable to regulate itself and maintain normal function. In particular, a body in shock is unable to supply enough blood to vital organs to keep them functioning. This section will address the types of shock, the signs and symptoms of shock, and shock treatment.

Types of Shock

There are numerous types of shock. However, the most common types are hypovolemic, cardiogenic, and neurogenic. Shock can be caused by both trauma and other medical emergencies. Other causes of shock can be anaphylactic (due to allergic reaction) and septic (due to infection). The following sections will address the three main types of shock in more detail.

Hypovolemic Shock

When a patient loses a substantial amount of blood, he or she is at risk of suffering from **hypovolemic shock**. Shock occurs when the body is unable to supply blood to tissue. If there is significant blood loss as a result of internal and/or external bleeding, the body is unable to adequately supply the tissue. Therefore, hypovolemic shock is a "volume" issue because the amount of blood in the body is substantially reduced. The majority of instances of shock are due to hypovolemia.

Cardiogenic Shock

Even if the body has an adequate amount of blood, it still must be forced through the body by the heart. **Cardiogenic shock** occurs when the heart is unable to force enough blood to tissue for it to continue functioning. This lack of supply is said to be a "pump" issue because the heart is unable to pump enough blood. A myocardial infarction (MI), also known as a *heart attack*, is a main cause of heart impairment and can cause the body to quickly go into cardiogenic shock.

Neurogenic Shock

Blood vessels such as arteries expand and contract due to signals from the brain. Blood flow is carefully regulated by the body in this manner. When blood vessels are expanded, significant amounts of blood can quickly travel through the body. Likewise, when blood vessels are constricted, very little blood is made available to organs and tissue.

Neurogenic shock occurs when there is damage to the brain, spinal cord, or other nerves that control or regulate the blood vessels. This damage prevents the body from controlling the expansion and contraction of blood vessels as it normally would. Neurogenic shock is typically the result of the overexpansion of blood vessels. As the vessels expand, the pressure of the blood decreases due to the increased capacity. Therefore, neurogenic shock is said to be a "container" issue. Common causes for neurogenic shock are spinal trauma and head injuries.

Shock — Failure of the circulatory system to produce sufficient blood to all parts of the body; results in depression of bodily functions, and eventually death if not controlled.

Hypovolemic Shock — Shock caused by loss of blood.

Cardiogenic Shock — Shock caused by poor cardiac output.

Neurogenic Shock — Shock caused by the overexpansion of blood vessels due to damage to the brain, spinal cord, or other nerves.

Other Types of Shock

There are several other types of shock that can be encountered. One type is **anaphylactic shock**. Anaphylactic shock occurs due to a severe allergic reaction. Allergic reactions can be the result of issues such as food allergies, environmental allergies, and insect bites or stings. Another type of shock is **septic shock**. Sepsis is the result of a severe infection in the body. While these types of shock are not as commonly encountered as the types listed earlier, they are still dangerous and should be treated as true medical emergencies.

Signs and Symptoms of Shock

Internal functions of the body such as pulse, blood pressure, and respiration are carefully adjusted in order to achieve optimal results. When the body is affected by a negative condition, it will adjust functions accordingly in order to maintain ideal results. However, if the condition continues, the body may not be able to maintain normal operations as it once did. Gradually, the body will begin to exhibit negative effects of the condition.

There are two stages of shock: compensated and decompensated. When the body is affected by a negative condition and is adjusting functions in order to maintain normal levels, the body is considered to be in compensated shock. When the body is no longer able to adjust to maintain these levels, it is said to be in decompensated shock. Compensated shock can often be difficult to identify because the patient may exhibit only slight changes (such as an increased pulse rate) or may exhibit no visible changes at all. However, when a patient is in decompensated shock, the body will produce signs and symptoms that are readily identifiable. Signs and symptoms of decompensated shock may include one or more of the following:

- Pale, cool, and/or clammy skin
- Confusion (also known as altered mental status)
- Nausea and/or vomiting
- Changes to vital signs (typically increased pulse, increased breathing, and decreased blood pressure)
- Others such as dizziness, dilated pupils, and loss of consciousness

Whenever shock is suspected, the patient should be quickly treated and prepared for transport by ambulance to a hospital.

Shock Treatment

Because shock is a life-threatening condition, patients exhibiting signs and symptoms should be treated immediately and prepared for transport. While advanced life support treatments by paramedics are extremely important, they should not delay transport to the hospital. Transport to a hospital can be viewed as a treatment and in most cases is the best treatment that can be provided to a shock victim. For this reason, transport of any patient exhibiting symptoms of shock should never be delayed.

Any bleeding that is noticed should be controlled immediately by using the methods detailed in the bleeding control section. A blanket should be used to cover the patient in order to maintain body temperature and prevent hypothermia **(Figure 22.41)**. EMT intermediates and paramedics are able to provide other advanced treatments for shock in these instances.

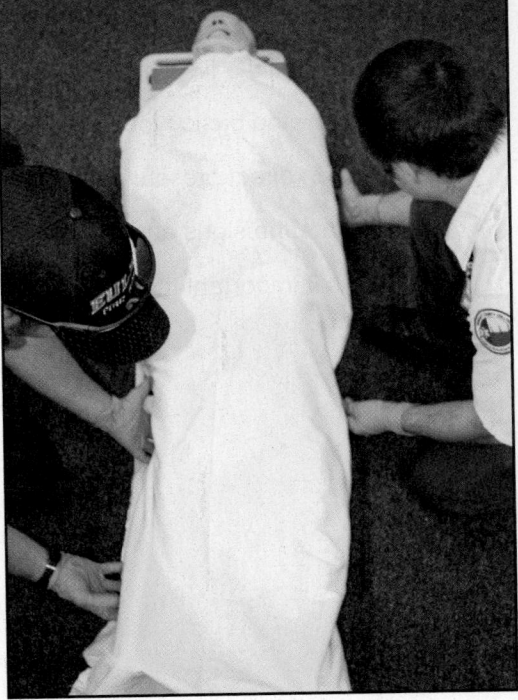

Figure 22.41 Patients exhibiting signs of shock should be kept warm to maintain body temperature and prevent hypothermia.

Chapter Summary

The provision of emergency medical services is a critical function of the fire service. With the aging population in North America, demand for emergency medical care from fire department resources will only increase. It is important that fire department personnel be properly trained to respond to these types of emergencies. Properly trained firefighters are also able to assist other firefighters who may become sick or injured performing their duties. In many of these situations, having trained personnel on scene can make the difference between life and death.

Review Questions

1. What are the roles the fire service may take in providing emergency medical care?

2. What does the *Health Information Portability and Accountability Act (HIPAA)* of 1996 require regarding patient confidentiality?

3. What are the communicable diseases commonly encountered by firefighters?

4. What types of immunizations may first responders be required to have?

5. What are some basic ways first responders can practice body substance isolation (BSI)?

6. What actions are taken to assess a patient's condition?

7. How do CPR techniques for adults, children, and infants differ from one another?

8. When should a first responder administer and discontinue CPR?

9. What are the basic types of external bleeding?

10. How can direct pressure and elevation control external bleeding?

11. What are the signs and symptoms of internal bleeding?

12. Why is it important that a first responder recognize the types, signs, and symptoms of shock?

Chapter Contents

Key Terms

NFPA® Core Competencies

This chapter provides information that addresses the following requirements of NFPA® 472: *Standard for Competence of Responders to Hazardous Materials/ Weapons of Mass Destruction Incidents.*

4.2.1	5.2.1	5.2.1.1.2	5.2.1.1.5	5.2.1.2.1	5.2.1.3.1	5.2.1.6	5.2.4
4.2.2	5.2.1.1	5.2.1.1.3	5.2.1.1.6	5.2.1.2.2	5.2.1.3.2	5.2.2	5.3.1
4.4.1	5.2.1.1.1	5.2.1.1.4	5.2.1.2	5.2.1.3	5.2.1.3.3	5.2.3	

1. Recognize introductory information regarding hazardous materials. [NFPA® 472, 4.2.1]

2. Explain the six types of hazardous materials hazards. [NFPA® 472, 4.4.1, 5.2.2, 5.2.3]

3. Describe routes of entry for hazardous materials. [NFPA® 472, 4.4.1]

4. Describe the physical properties of hazardous materials. [NFPA® 472, 5.2.3]

5. Explain the six stages of the General Emergency Behavior Model (GEBMO) used to describe typical hazardous materials events. [NFPA® 472, 5.2.3]

6. Identify the seven categories of clues to the presence of hazardous materials/weapons of mass destruction. [NFPA® 472, 4.2.1, 4.2.2, 5.2.1, 5.2.1.1, 5.2.1.2, 5.2.1.3, 5.2.1.1.1, 5.2.1.1.2, 5.2.1.1.3, 5.2.1.1.4, 5.2.1.1.5, 5.2.1.1.6, 5.2.1.3.3, 5.2.2, 5.2.1.2.1, 5.2.1.2.2, 5.2.1.3.1, 5.2.1.3.2]

7. Describe the written resources used to identify hazardous materials. [NFPA® 472, 4.2.2, 5.2.2]

8. Explain the ways to safely use the five senses, along with monitoring and detection equipment, to detect the presence of hazardous materials. [NFPA® 472, 4.2.1, 5.2.4]

9. Identify common indicators of terrorist attacks. [NFPA® 472, 4.2.1, 5.2.1.6, 5.2.3]

10. Describe the common indicators and types of illicit laboratories. [NFPA® 472, 4.2.1]

11. Explain ways to protect against secondary attacks and booby traps. [NFPA® 472, 4.2.1, 5.3.1]

Chapter 23
Hazards, Behavior, and Identification of Haz Mat/WMD

Courtesy of the Mohave Museum of History and Arts.

Case History

On January 19, 2003, firefighters near Houston, Texas, responded to a structure fire at an auto repair shop. A four-member crew initiated an interior attack, but within minutes the fire intensified and rolled over their heads. As one member exited the building with burning hands, an air horn sounded, warning the remaining three to exit the building. Two quickly made it outside, but they were engulfed in flames that had to be extinguished with a second handline. Seconds after they exited, a nitrous oxide cylinder attached to a race car inside the shop exploded **(Figure 23.1)**. A Rapid Intervention Crew/Team (RIC/RIT) made two attempts to rescue the crew's fourth member, but the blaze had intensified even further and they had to withdraw. After 40 minutes of master stream application, more teams were finally able to enter the structure. They found their missing colleague lying dead near an office door.

Figure 23.1 Compressed gas cylinders may create an explosion with shrapnel when overheated. *Courtesy of NIOSH.*

The NIOSH investigation revealed that the nitrous oxide cylinder had exploded with a force close to 4 pounds (2 kg) of TNT, causing significant primary blast injuries. The shock wave had ruptured the victim's eardrums and caused severe concussive damage to the lungs. These injuries had not been immediately fatal, but investigators believe that in this injured state, the firefighter became disoriented while trying to escape. The intensity of the blaze prevented rescuers from helping him to safety.

Most fatalities at haz mat incidents are caused by explosions. The effect of explosive shock waves intensifies in closed spaces, such as in a building or vehicle. Blast waves are also reflected by solid surfaces, so anyone standing next to a wall or vehicle may suffer increased primary blast injury. In this case, the force of the explosion had propelled a fire extinguisher through the side of a car before it wrapped around a metal support post.

The intense heat that prevented an earlier rescue may also have been related to hazardous materials. The fire had begun when a flammable liquid (lacquer thinner) had ignited inside the shop.

Source: NIOSH

Several million tons of chemicals are manufactured every year for use by consumers, industry, the military, and the government. Although they have beneficial uses, many of these materials can be hazardous to both people and the environment. This chapter discusses the following aspects of hazardous materials:

- Health effects of exposure
- Routes of entry
- Properties and behavior
- Identification

Introduction to Hazardous Materials

Harmful substances are known as **hazardous materials** (or *haz mat*) in the United States and **dangerous goods** in Canada and other countries. Hazardous materials that are particularly dangerous, such as chemical, biological, radiological, nuclear, or explosive (CBRNE) materials, can be used as weapons by terrorists. They are sometimes referred to as **weapons of mass destruction (WMD)** because they can cause mass casualties and damage.

NOTE: While specific WMDs are detailed in the Terrorist Attacks and Illicit Laboratories section of this chapter, it should be understood that all WMDs are hazardous materials. Any general discussion of hazardous materials in this chapter therefore includes WMDs.

Haz mat incidents may require specialized equipment, procedures, training, and personal protective equipment (PPE) **(Figure 23.2)**. Hazardous materials can be extremely difficult to control and may be dangerous even in very small quantities. Sometimes they are also difficult to detect, even with sophisticated equipment.

Haz mat incidents can be caused by:

- Human error
- Mechanical malfunction
- Container failure
- Transportation accidents
- Vandalism or **terrorism**

To safely mitigate hazardous materials incidents, you must understand the variety of hazardous materials you may encounter, the potential health effects of the materials, and the physical **hazards** associated with them. You must also know how to identify different types of hazardous materials and how to recognize their presence in a terrorist attack or illicit laboratory.

Hazardous Material — Substance that can be dangerous to human health or the environment if not properly controlled.

Dangerous Goods — (1) Alternate term for hazardous materials, used in Canada and other countries. (2) U.S. or Canadian term for hazardous materials aboard an aircraft.

Weapon of Mass Destruction (WMD) — Weapon or device that can cause death or serious injury to a large number of people; may include chemical, biological, radiological, nuclear, or explosive (CBRNE) type weapons.

Terrorism — Unlawful force or violence against people or property to coerce or intimidate a government or its citizens, for social or political purposes (as defined by the U.S. *Code of Federal Regulations*).

Hazard — Condition, substance, or device that can directly cause injury or loss; the source of a risk.

CAUTION
The presence of hazardous materials may significantly change mitigation strategies and tactics.

Figure 23.2 Haz mat incidents present unique challenges that require the use of specialized equipment and responses.

Hazard Types

Exposure to hazardous materials can have a severely negative effect on human health. The resulting health effects can be either acute or chronic. **Acute health effects** are short-term conditions that appear within hours or days, such as vomiting or diarrhea. **Chronic health effects** are long-term conditions that may take years to appear, such as cancer.

The chance of developing health effects, and the severity of those effects, depends on the health of the person who has been exposed. The elderly and persons with chronic diseases are more likely to suffer effects, and more likely to have serious effects.

Use the acronym TRACEM to remember the six haz mat hazard types that can damage your health:

- Thermal hazards
- Radiological hazards
- Asphyxiation hazards
- Chemical hazards
- Etiological/biological hazards
- Mechanical hazards

Acute Health Effects — Health effects that develop rapidly, often after exposure to a hazardous substance.

Chronic Health Effects — Long-term health effects, often resulting from exposure to a hazardous substance.

Figures 23.3a and b Caution must be exercised while shipping materials that are either (a) extremely hot (elevated-temperature) or (b) extremely cold (cryogenic). *(a) Courtesy of Rich Mahaney. (b) Courtesy of Steve Irby, Owasso (OK) Fire Department.*

Polymerization — Chemical reaction in which two or more molecules combine to form larger molecules; this reaction can often be violent.

Thermal Hazards

Thermal hazards involve extreme heat or cold. They may be produced by the materials themselves, as in the case of elevated-temperature materials and cryogenic liquids. The materials may also cause secondary fires, explosions, or exothermic reactions, which are the sudden release of heat energy that typically occurs during oxidation or **polymerization**.

Elevated Temperatures

Elevated-Temperature Material — Material that meets any of the following criteria during transport: (a) in a liquid phase, at or above 212°F (100°C), (b) intentionally heated at or above its liquid phase flash point of 100°F (38°C), or (c) in a solid phase, at or above 464°F (240°C).

Elevated-temperature materials include molten sulfur and molten aluminum, which are typically shipped at 1,300°F (704°C) **(Figure 23.3a)**. These materials can cause lethal burns and create fire, steam, and explosions. Working around these materials also increases the heat stress caused by wearing personal protective equipment.

Low Temperatures

Cryogens — Gases that convert to liquids when cooled at or below -130°C (-90°C). *Also known as* Refrigerated Liquids and Cryogenic Liquids.

Low temperature hazards are caused by liquefied gases and cryogenic liquids, which absorb heat so quickly they can freeze nearby objects **(Figure 23.3b)**. Liquefied gases become partially liquid at 70°F (21°C) under charging pressure. Examples include propane and carbon dioxide. **Cryogens** are gases that turn into liquid at or below -130°F (-90°C) at 14.7 psi (101 kPa) {1.01 bar}. Examples include nitrogen, helium, hydrogen, argon, liquid oxygen (LOX), and liquefied natural gas (LNG). These substances are typically stored and transported in their liquid states, at extremely cold temperatures. They can freeze other materials on contact, including human tissue. Some cryogens pose additional hazards by being oxidizers or **poisons**.

Poison — Any material that is injurious to health when taken into the body.

Radiological Hazards

Ionizing Radiation — Radiation that causes a chemical change in atoms by removing their electrons.

Radiological hazards are most common at medical centers, nuclear power plants, research facilities, and transportation incidents. They may also be present in terrorist attacks and some industrial operations. The least energetic type of radiation is **nonionizing radiation**, such as visible light or radio waves **(Figure 23.4)**. The most energetic type is **ionizing radiation**. This type is also the most hazardous.

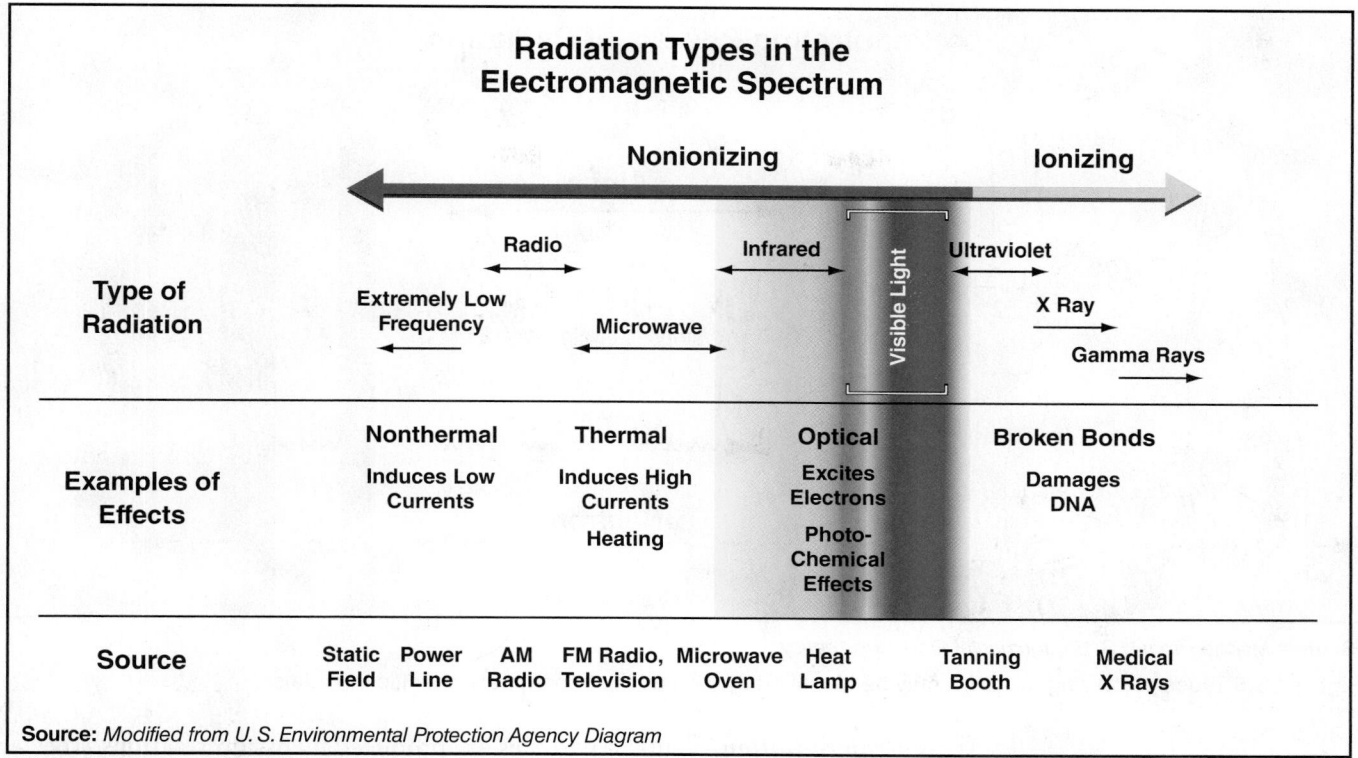

Radiation Types in the Electromagnetic Spectrum

	Nonionizing					Ionizing	
Type of Radiation	Extremely Low Frequency	Radio	Microwave	Infrared	Visible Light	Ultraviolet / X Ray	Gamma Rays
Examples of Effects	Nonthermal Induces Low Currents		Thermal Induces High Currents Heating		Optical Excites Electrons Photo-Chemical Effects	Broken Bonds Damages DNA	
Source	Static Field · Power Line	AM Radio	FM Radio, Television · Microwave Oven	Heat Lamp		Tanning Booth	Medical X Rays

Source: *Modified from U. S. Environmental Protection Agency Diagram*

Figure 23.4 Generally speaking, ionizing radiation is of more concern to first responders because of the potential health effects associated with exposure.

Types of Ionizing Radiation

The four types of ionizing radiation are alpha, beta, gamma, and neutron radiation (**Figure 23.5, p.1304**). All four types can be extremely hazardous.

Alpha radiation. Positively charged alpha particles are emitted from the nucleus of heavy radioactive elements, such as uranium and radium. Alpha particles lose energy rapidly when traveling through matter; for example, they can be stopped by a sheet of paper. They can also be completely blocked by the outer layer of human skin, so they are not a hazard outside the body. However, they can be very harmful if ingested or inhaled. Materials that emit alpha radiation are typically sealed, encased, and shielded so that the presence of the radiation cannot be detected. Sources of this radiation include some types of smoke detectors.

Beta radiation. Positively or negatively charged beta particles (**protons** and **electrons**) are emitted from the nucleus of radioactive elements such as tritium, carbon-14, and strontium-90. These fast-moving particles are more penetrating than alpha particles. They can penetrate the outer layer of skin and cause radiation damage to the body, and they are even more harmful if ingested or inhaled. Beta particles can travel up to 20 feet (6 m) through the air, but lose energy quickly as they do so. After traveling that distance, they can be stopped by a layer of clothing or .08 inches (2 mm) of aluminum.

Gamma radiation. Gamma rays consist of high-energy **photons** that often accompany the emission of alpha or beta particles from a nucleus. Naturally occurring sources include potassium-40, and industrial sources include cobalt-60, iridium-192, and cesium-137. Gamma rays have no mass and electric charge, but they are highly penetrating. They can easily pass through the human body, inflicting severe radiation damage. They can be stopped by approximately 2 inches (50 mm) of lead, 2 feet (0.6 m) of concrete, or several feet (meters) of earth. However, standard fire fighting protective clothing provides *no* protection.

Nonionizing Radiation — Energy waves composed of oscillating electric and magnetic fields traveling at the speed of light; examples include visible light, radio waves, microwaves, infrared radiation, and ultraviolet radiation.

Protons — Subatomic particle that possesses a positive electric charge.

Electron — Subatomic particle that possesses a negative electric charge.

Photon — Weightless packet of electromagnetic energy, such as X-rays or visible light.

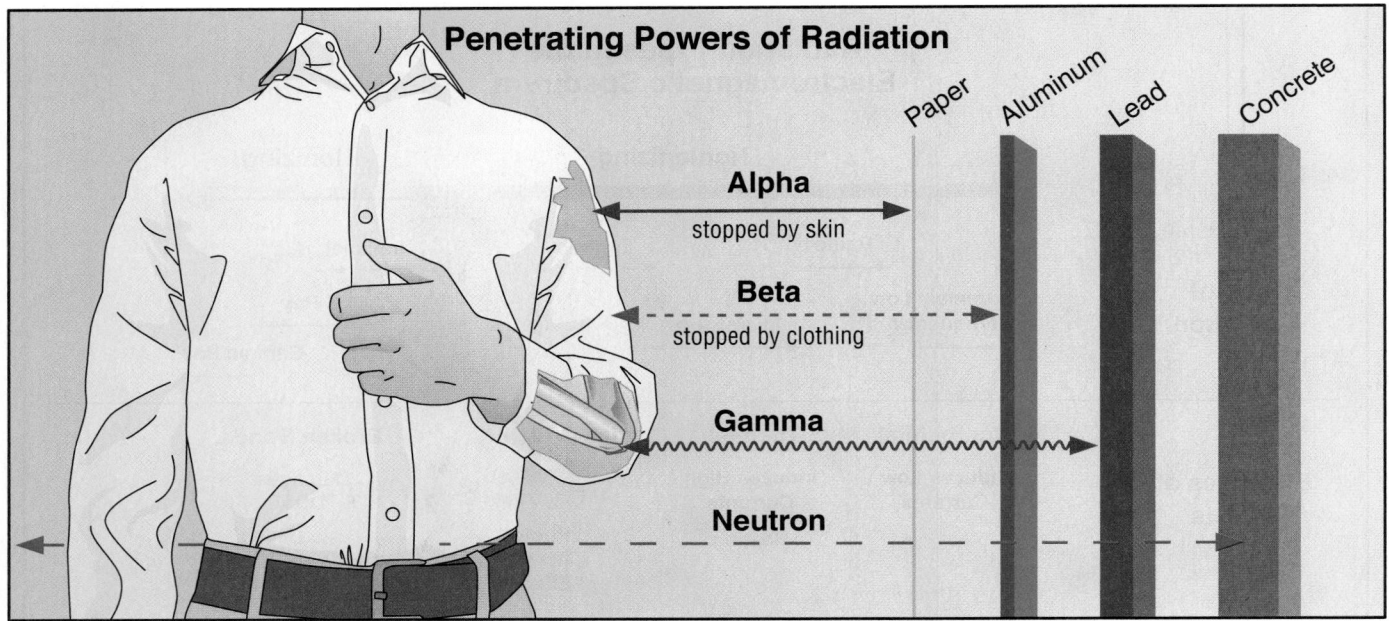

Source: *Modified from U.S. Environmental Protection Agency*

Figure 23.5 Types of ionizing radiation may be identified by their ability to penetrate specific materials.

Neutron radiation. Neutron particles are produced by fission reactions. They are typically found in research laboratories and nuclear power plants. Other sources include oil moisture density gauges found at construction sites. These ultra-high energy particles have mass but no electrical charge, and they are even more penetrating than gamma rays. Neutron radiation is difficult to measure in the field. It is usually estimated based on gamma measurements because gamma rays are also produced by fission reactions. Neutron radiation is hazardous to health because it creates secondary radiation as it passes through human tissue.

Exposure — Contact with a hazardous material, causing biological damage; typically by swallowing, breathing, or touching. Exposure may be short-term (acute exposure), of intermediate duration, or long-term (chronic exposure).

Routes of Entry — Means by which hazardous materials enter the body; inhalation, ingestion, skin contact, injection, absorption, and penetration (for radiation).

What's the Difference Between Exposure to Radioactive Material and Radioactive Contamination?

Radioactive contamination occurs when radioactive material is deposited on surfaces, skin, or clothing. It is important to remember that radiation does not spread; rather, it is radioactive material/contamination that spreads. **Exposure** to radiation occurs when a person is exposed to radiatioactive materials, causing biological damage. Without biological damage, there is no exposure. For example, it is not recommended that you touch a source of alpha radiation, but there is no exposure because your skin protects you from biological damage. If you touch a source of beta or gamma radiation, however, exposure does occur because the radiation causes biological damage.

But exposure does not cause contamination. Contamination only occurs if radioactive material remains after contact. A person can become contaminated externally and/or internally, with radioactive material entering the body via one or more **routes of entry** (see Routes of Entry section). If you are contaminated with radioactive material that your PPE or outer skin does not protect you from, you will receive radiation exposure until the material is removed. There are three main types of contamination:

(continued)

- External contamination occurs when radioactive material is on the skin or clothing; if these do not protect against radiation damage, the contamination results in external exposure.

- Internal contamination occurs when radioactive material is inhaled, swallowed, or absorbed through wounds, causing internal exposure.

- Environmental contamination occurs when radioactive material spreads or is unshielded, creating another potential source of external exposure.

Exposure to radiation is like exposure to sunlight. Once you remove yourself from the source, you are no longer exposed.

Radiation Health Hazards

Ionizing radiation damages the atoms of the body's cells. It changes their atomic structure and chemical properties, making them unable to properly function. A sufficiently high **radiation dose** can cause cancer or genetic mutation. Radiation doses can be acute or chronic, depending on the duration of exposure.

Radiation Dose — Quantity of radiation energy absorbed into the body.

Acute doses. Acute doses are received over a short period of time. Some acute doses of radiation have no long-term health effects, but others can produce hair loss, nausea, vomiting, diarrhea, fatigue, or reduced blood count. Acute doses can also be lethal if the quantity of radiation is great enough. The radiation from a nuclear explosion can kill within a few hours.

Chronic doses. Chronic doses are small amounts received over a long period of time. Health effects of these doses typically take years to develop because the body is able to repair some of the damage done by low levels of radiation. But chronic doses of even low levels of radiation can eventually be lethal. Everyday background exposure, such as the type experienced by workers in nuclear facilities, has been shown to cause cancer.

Responders rarely develop health effects from exposure at a haz mat incident, especially if they take proper precautions. Even terrorist attacks are unlikely to produce dangerous levels of radiation. However, it is important to monitor for radiation at any incident involving explosions or suspected terrorism.

Asphyxiation Hazards

Asphyxiants prevent the body from absorbing oxygen, which can result in death by suffocation. *Simple asphyxiants* are gases, such as nitrogen, that displace oxygen **(Figure 23.6, p. 1306)**. Large quantities are required to lower oxygen concentration beneath the level required to sustain human life. *Chemical asphyxiants* prohibit the body from processing available oxygen. Even small amounts of common chemical asphyxiants such as carbon monoxide and cyanide can be lethal. Chemical asphyxiants are sometimes used in terrorist attacks.

Chemical Hazards

The sections that follow discuss types of hazardous chemicals or substances. These include:

- Poisons/toxins
- Corrosives
- Irritants
- Convulsants
- Carcinogens
- Sensitizers/allergens

Figure 23.6 Asphyxiants, like nitrogen, are dangerous at high concentrations because they displace oxygen.

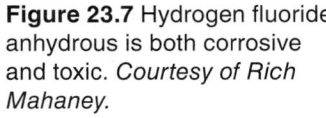

Corrosive Material — Chemical that severely corrodes steel and damages human tissue. *Also known as Corrosive.*

Systemic Effect — Damage spread through an entire system; opposite of a local effect, which is limited to a single location.

Poisons/Toxins

Toxic chemicals can cause injury at the site where they contact the body, typically the skin and the mucous membranes of the eyes, nose, mouth, and respiratory tract. These injuries are known as *local toxic effects.* **Corrosive materials** cause local damage to the skin when touched, while irritant gases such as chlorine produce localized toxic effects in the respiratory tract when inhaled **(Figure 23.7)**. But when toxins are absorbed into the bloodstream they are distributed throughout the body, producing **systemic effects**. Many pesticides are absorbed and distributed this way, causing seizures and cardiopulmonary problems.

Figure 23.7 Hydrogen fluoride anhydrous is both corrosive and toxic. *Courtesy of Rich Mahaney.*

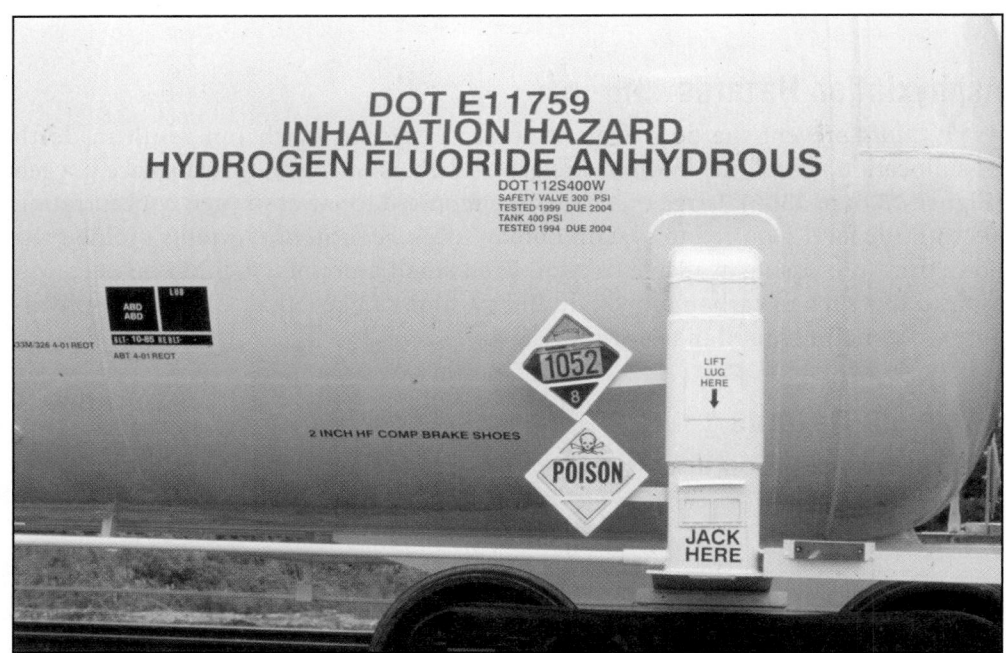

Many toxins have immediate, acute effects, but others have delayed, chronic effects. Some toxins also produce a combination of local and systemic effects, some of which do not manifest for many years. Neurotoxins produce systemic effects in the nervous system, by disrupting nerve impulses. Types of toxins, their target organs, and chemical examples are given in **Table 23.1**.

Even small amounts of some toxins can be lethal. Concentrations that can cause death or serious illness/injury are described as *immediately dangerous to life or health (IDLH)*.

CAUTION
All personnel working at hazardous materials incidents must use appropriate personal protective equipment, including respiratory protection.

Table 23.1
Types of Toxins and Their Target Organs

Toxin	Target Organ	Chemical Examples
Nephrotoxins	Kidney	Halogenated Hydrocarbons, Mercury, Carbon Tetrachloride
Hemotoxins	Blood	Carbon Monoxide, Cyanides, Benzene, Nitrates, Arsine, Naphthalene, Cocaine
Neurotoxins	Nervous System	Organophosphates, Mercury, Carbon Disulphide, Carbon Monoxide, Sarin
Hepatoxins	Liver	Alcohol, Carbon Tetrachloride, Trichloroethylene, Vinyl Chloride, Chlorinated HC
Immunotoxins	Immune System	Benzene, Polybrominated Biphenyls (PBBs), Polychlorinated Biphenyls (PCBs), Dioxins, Dieldrin
Endocrine Toxins	Endocrine System (including the pituitary, hypothalamus, thyroid adrenals, pancreas, thymus, ovaries, and testes)	Benzene, Cadmium, Chlordane, Chloroform, Ethanol, Kerosene, Iodine, Parathion
Musculoskeletal Toxins	Muscles/Bones	Fluorides, Sulfuric Acid, Phosphine
Respiratory Toxins	Lungs	Hydrogen Sulfide, Xylene, Ammonia, Boric Acid, Chlorine
Cutaneous Hazards	Skin	Gasoline, Xylene, Ketones, Chlorinated Compounds
Eye Hazards	Eyes	Organic Solvents, Corrosives, Acids
Mutagens	DNA	Aluminum Chloride, Beryllium, Dioxins
Teratogens	Embryo/Fetus	Lead, Lead Compounds, Benzene
Carcinogens	All	Tobacco Smoke, Benzene, Arsenic, Radon, Vinyl Chloride

pH Scale

Concentration of Hydrogen Ions Compared to Distilled Water	pH Scale	Examples of Solutions at this pH
Acids	0	
	1	Battery Acid
	2	Vinegar
	3	Orange Juice
	4	Acid rain, Wine
	5	Black Coffee
	6	Milk
Neutral	7	Distilled Water
Bases	8	Seawater
	9	Baking Soda
	10	Milk of Magnesia
	11	Ammonia
	12	Lime
	13	Lye
	14	Sodium Hydroxide

Figure 23.8 The pH scale is used to express the corrosivity of acids and bases.

Corrosives

Corrosives are chemicals that cause permanent damage to anything they touch. They can corrode metal and burn human tissue. These common industrial chemicals are divided into two broad categories: acids and bases. However some corrosives, such as hydrogen peroxide, are neither acids nor bases. Corrosivity levels are expressed in terms of pH (**Figure 23.8**).

Acids. Acids are chemicals that break down (ionize) in water to yield hydrogen **ions**. They have pH values of 0 to 6.9. Common examples include nitric, sulfuric, and hydrochloric acid. Contact with an acid typically causes immediate pain, and possible severe chemical burns or permanent eye damage.

Bases. Bases, also known as *alkalis* or *caustics*, are water-soluble compounds that break apart in water, forming negatively charged hydroxide ions. They have pH values of 8 to 14. Common examples include caustic soda and potassium hydroxide, both of which are used in common household drain cleaners. Contact with a base typically does not cause immediate pain. However, bases can cause severe eye damage because they adhere to the tissues in the eye, making them difficult to remove. They also break down fatty skin tissues and can penetrate deeply into the body. A common sign of exposure to a base is a greasy or slick feeling of the skin, which is caused by the breakdown of fatty tissues.

Ion — Atom that has lost or gained an electron, giving it a positive or negative charge.

The corrosive action of strong acids generates tremendous heat, sometimes enough to start a fire or cause an explosion. Hydrochloric and other acids can also react with metal to form explosive hydrogen gas. Acids and bases can react violently when mixed together or with water. This reaction is an important consideration during spill cleanup and decontamination.

Irritants

Irritants cause temporary but sometimes severe inflammation to the eyes, skin, or respiratory system. Common irritants may be found in fertilizers, disinfectants, and solvents.

Convulsants

Convulsants cause involuntary muscle contraction (convulsions). Examples include strychnine, organophosphates, carbamates, and uncommon drugs such as picrotoxin. They can cause death through asphyxiation or exhaustion.

Carcinogens

Carcinogens are cancer-causing agents. Examples of known or suspected carcinogens include:

- Polyvinyl chloride (PVC)
- Benzene
- Asbestos
- Some chlorinated hydrocarbons
- Arsenic
- Nickel
- Some pesticides
- Many plastics

Even small amounts of some carcinogens can cause fatal cancers, sometimes as long as 10 to 40 years after exposure. However, exact data is unavailable regarding the duration or dose of exposure needed.

Sensitizers/Allergens

Sensitizers and *allergens* cause allergic reactions. Common examples include latex, bleach, and urushiol, which is a natural chemical found in poison ivy, poison oak, and poison sumac (**Figure 23.9, p. 1310**). Some people experience no symptoms at the first exposure, but develop severe reactions after subsequent exposures.

Etiological or Biological Hazards

Etiological hazards, also known as *biological hazards*, are microorganisms that cause severe illness or disease. Examples include:

- **Viruses** — Viruses reproduce in the living cells of their hosts and do not respond to antibiotics (**Figure 23.10, p. 1310**).
- **Bacteria** — Bacteria are single-celled organisms that do not typically cause disease (**Figure 23.11, p. 1310**). However, harmful bacteria can invade tissues or produce toxins.

Etiological Hazards — Harmful viruses and bacteria; when used deliberately, they are *also known as* Biological Weapons.

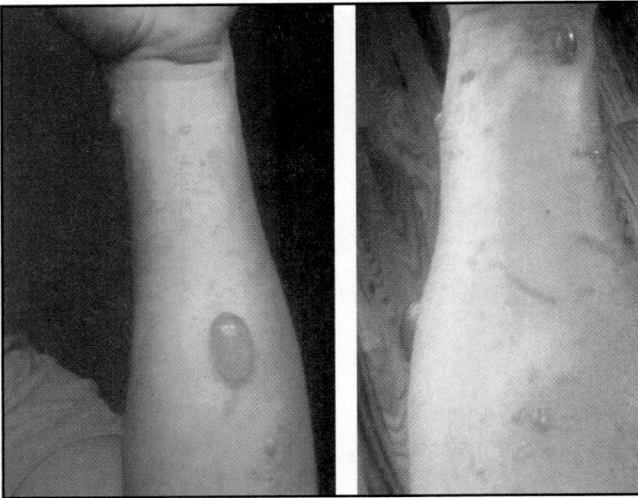

Figure 23.9 Urushiol, a chemical found in the sap of poison ivy, is an allergen that causes skin reactions in many people. *Source: CDC – Public Domain.*

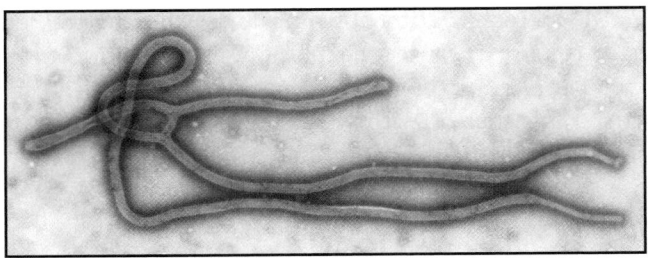

Figure 23.10 Ebola, a deadly virus, is unaffected by antibiotics. *Courtesy of the CDC Public Health Image Library.*

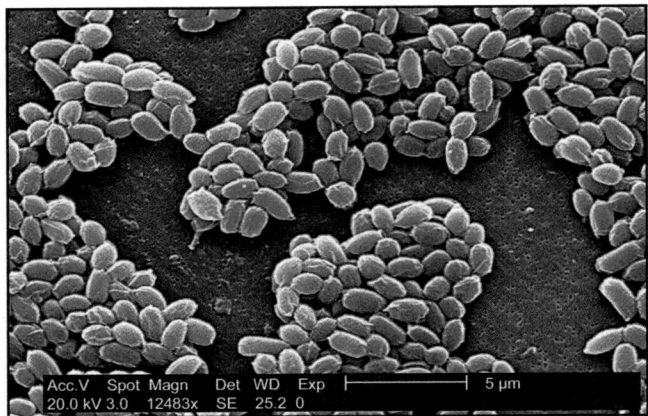

Figure 23.11 Anthrax, bacteria that causes infections, can be treated by antibiotics. *Courtesy of the CDC Public Health Image Library.*

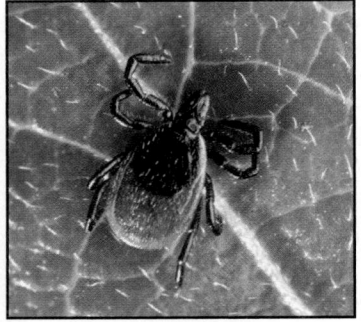

Figure 23.12 Ticks, fleas, and other arthropods may carry and transmit rickettsias, which include diseases such as Rocky Mountain Spotted Fever or Typhus. *Couresy of the U.S. Department of Agriculture.*

Infectious — Transmittable; able to infect people.

Contagious — Easily transmitted from one person to another, either through contact or close proximity.

- **Rickettsias** — Rickettsias are specialized bacteria spread by infected fleas, ticks, and lice (**Figure 23.12**). Like other bacteria, they are single-celled organisms that respond to antibiotics. But like viruses, they reproduce only in living cells.

- **Biological toxins** — Biological toxins are produced by living organisms, even though the organism itself is usually not harmful. For example, ricin is a lethal toxin derived from castor beans. Other biological toxins are produced in laboratories to be used as biological weapons (**Figure 23.13**).

Viruses, bacteria, and some biological toxins cause **infectious** diseases, some of which are **contagious**. Other biological toxins do not cause infectious disease, but their toxicity can cause severe illness.

Diseases associated with etiological hazards include malaria, AIDS, and typhoid. Exposure to these hazards is most likely in encounters with infected people, who transmit the disease through their bodily fluids. Accidental exposure can also occur at incidents involving medical or biological laboratories. Deliberate exposure can result from biological weapons used during criminal activity or terrorist attack.

The 2001 anthrax attacks in the United States were an example of a biological attack. Most biological attacks target people, although some target plants and animals. Potential biological weapons include:

- Smallpox (virus) (**Figure 23.14**)

- Anthrax (bacteria)

- Botulism (toxin from the bacteria *Clostridium botulinum*)

Figure 23.13 Ricin, a lethal biological toxin, is made from castor beans.

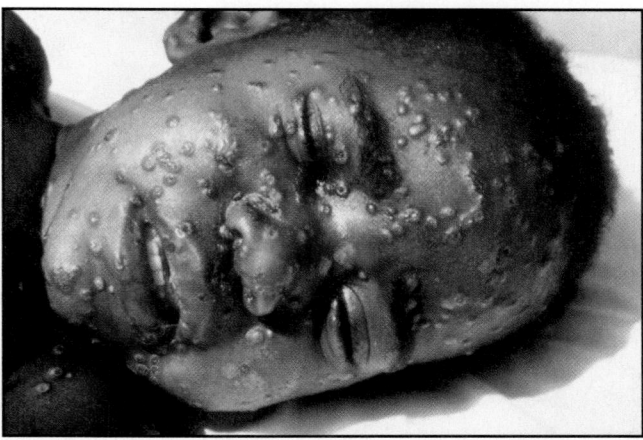

Figure 23.14 Small children are affected strongly by the smallpox virus. *Courtesy of the CDC Public Health Image Library.*

Mechanical Hazards

Mechanical hazards cause injury through blunt physical force. At haz mat incidents this typically results from an explosion caused by a bomb or **improvised explosive device (IED)**, the failure of a pressurized container, or the **reactivity** of the hazardous material. Explosions create three types of mechanical hazard (**Figure 23.15**):

- **Blast-pressure wave (shock wave)** — Rapidly released gases create a shock wave that travels outward from the center of the blast, causing most of the injuries and damage. The farther the wave travels, the more its force diminishes.

> **Improvised Explosive Device (IED)** — Homemade bomb that is not deployed in a conventional military fashion.

> **Reactivity** — Ability of a substance to chemically react with other materials, and the speed with which that reaction takes place.

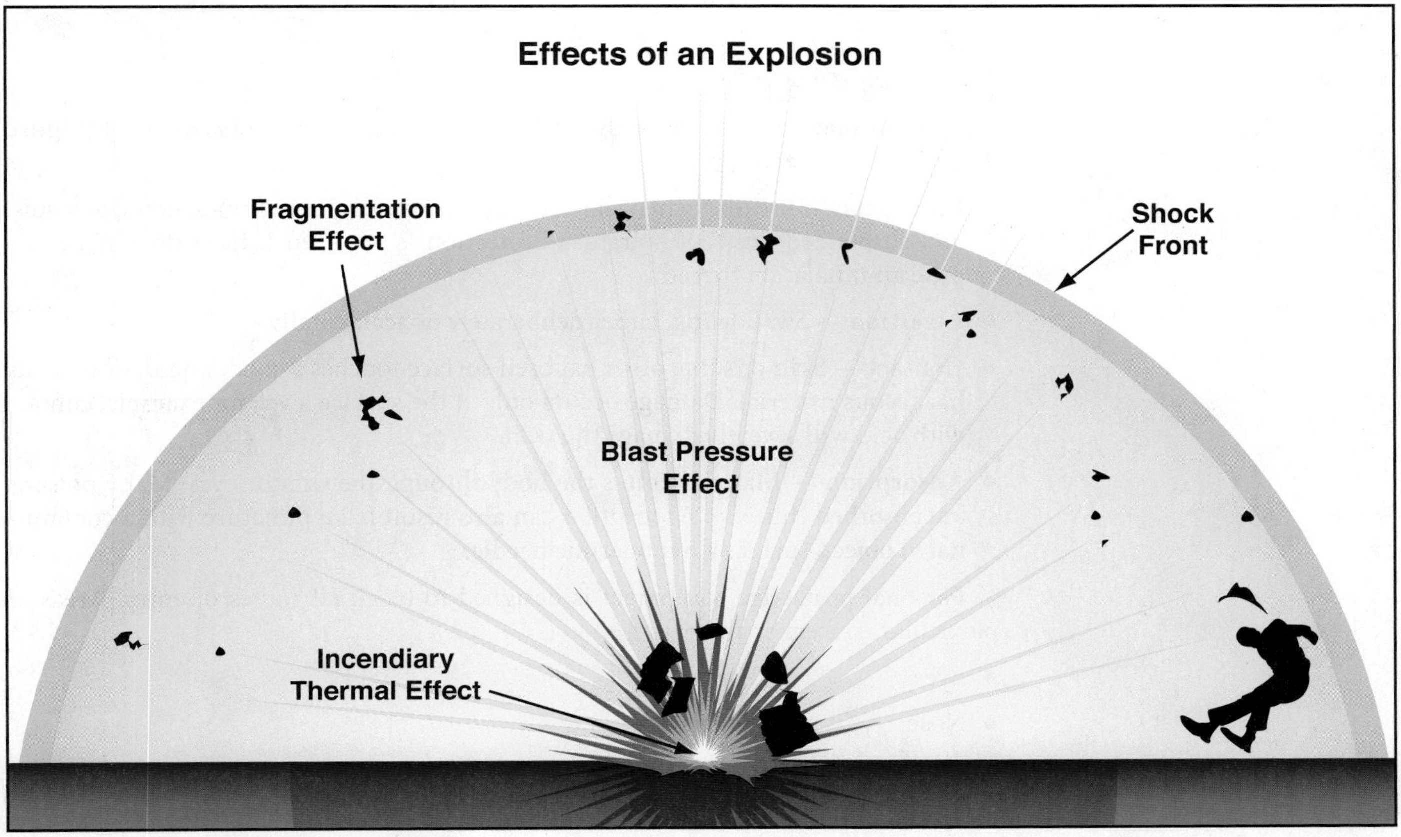

Effects of an Explosion

Fragmentation Effect

Shock Front

Blast Pressure Effect

Incendiary Thermal Effect

Figure 23.15 Mechanical effects of an explosion include the blast pressure wave, fragmentation effect, and the shock, or seismic, effect.

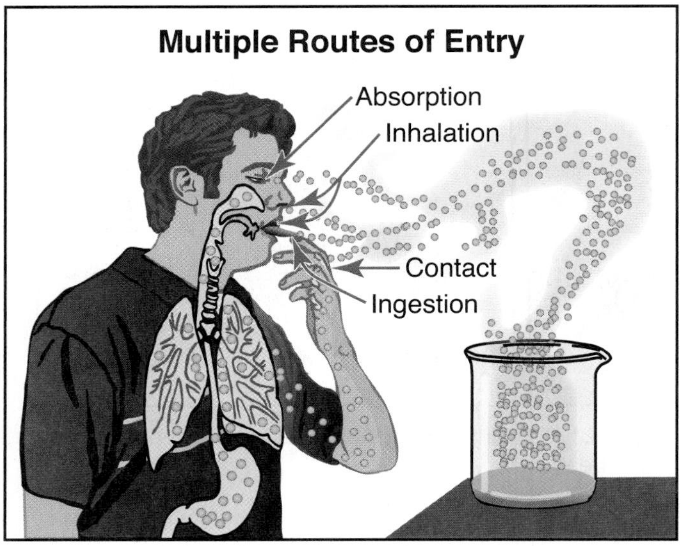

Figure 23.16 Hazardous materials have four common routes of entry into the body: inhalation, ingestion, contact, and absorption.

- **Shrapnel fragmentation** — Debris thrown outwards in all directions, typically small pieces of the ruptured container or structure. Damages nearby objects and causes potentially fatal injuries.

- **Seismic effect** — Blasts near ground level create shock waves, causing the ground to shake as it would during an earthquake.

Explosions can also create a temporary thermal hazard, an *incendiary thermal effect*. Combustible gases combine with superheated ambient air to form a fireball, which quickly dissipates.

Routes of Entry

Hazardous materials can enter the body through four main routes of entry (**Figure 23.16**):

- **Inhalation** — Breathing in hazardous vapors, smoke, gases, fumes, liquid aerosols, and suspended dusts. Respiratory protection is required if hazardous materials pose an inhalation threat.

- **Ingestion** — Swallowing, either deliberately or accidentally.

- **Contact** — Skin or some other exposed surface touches a solid, liquid, or gaseous hazardous material. Damage occurs only at the surface level; for example, contact with acid will severely damage the skin.

- **Absorption** — Material enters the body through the skin or eyes. Many poisons are absorbed this way. Absorption can also result from puncture with a contaminated object, such as a hypodermic needle.

Personal protective equipment is designed to block all routes of entry. Areas of particular concern include:

- Eyes
- Nose
- Mouth
- Wrists
- Neck
- Ears
- Hands
- Groin
- Underarms

Properties and Behavior

A material's behavior is determined by its physical properties. Understanding a material's properties and behavior helps responders understand the nature of the hazards they will face at an incident. The most critical properties of hazardous materials are:

- State of matter
- Flammability
- Vapor pressure
- Boiling point
- Vapor density
- Solubility
- Specific gravity
- Persistence
- Reactivity

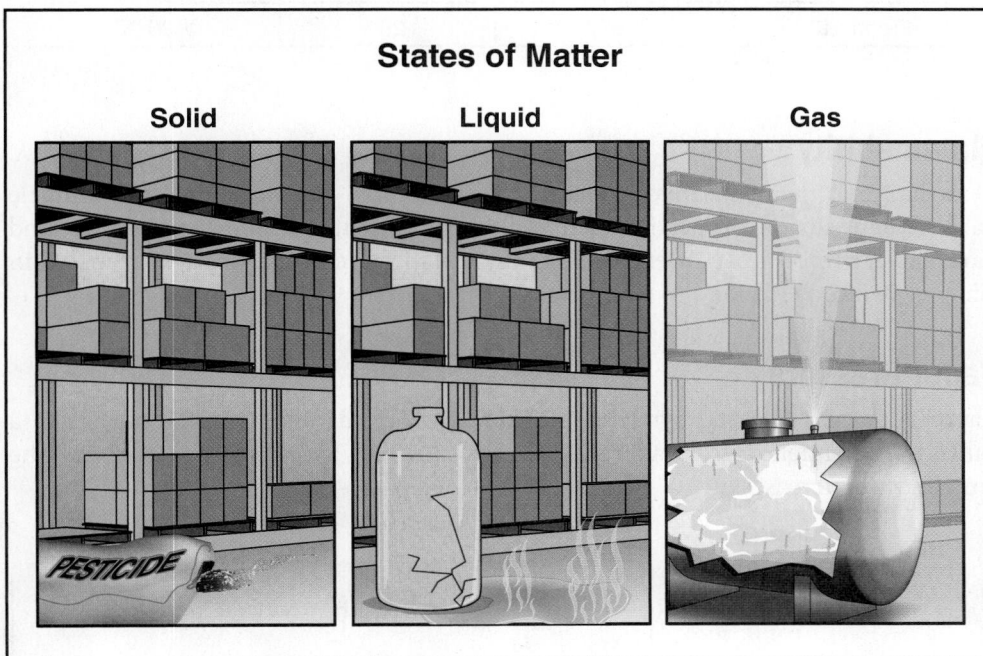

States of Matter

Solid	Liquid	Gas

Figure 23.17 The physical state of matter (solid, liquid, or gas) influences the way it behaves.

States of Matter

Hazardous materials can exist in any of the three states of matter: gas, liquid, and solid **(Figure 23.17)**. Different states of matter pose different hazard types.

Gaseous hazardous materials are present in the air, creating a potential inhalation hazard. Some gases also pose a contact hazard. Gases move according to wind and air movement, making them extremely difficult to contain. Compressed gases expand rapidly when released, potentially threatening large areas.

Liquid hazardous materials are primarily a contact hazard (through splashing). They may also give off vapors that pose inhalation hazards. Liquids flow or pool along contours in the floor or ground, making them easier to contain.

Solids with a small particle size remain suspended in the air for much longer, whereas large particles settle more quickly. Solids typically remain in place, but they can be moved by exterior forces such as wind and water.

Table 23.2
Vapor Pressure of Common Gases

Gas	Vapor Pressure (bar)	Vapor Pressure (mmHg)	Temperature
Helium	1	750	@ -269.15 °C
Propane	22	16500	@ 55 °C
Butane	2.2	1650	@ 20 °C
Carbonyl sulfide	12.55	9412	@ 25 °C
Acetaldehyde	0.987	740	@ 20 °C
Freon 113	0.379	284	@ 20 °C

Flammability

Most haz mat incidents involve flammable materials. The behavior of flammable hazards depends on properties such as flash point, autoignition temperature, and flammable, combustible, or explosive range. All of these properties are discussed in Chapter 5, Fire Behavior.

Vapor Pressure

Vapor is the gas released by a liquid (**Table 23.2**), and vapor pressure measures a substance's tendency to evaporate (**Figure 23.18**). Vapor pressure rises as the temperature increases.

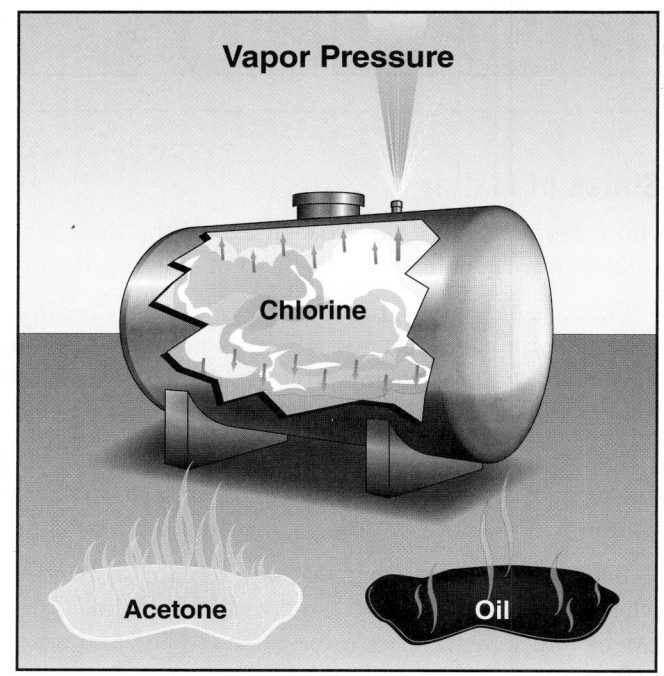

Figure 23.18 Chlorine has an extremely high vapor pressure and will escape as a gas if it is released from its container.

Boiling Point

The **boiling point** is the temperature at which vapor pressure is equal to or greater than atmospheric pressure. In other words, it is the temperature at which a liquid changes to a gas. The boiling point is usually expressed in degrees Fahrenheit (Celsius) at sea-level air pressure. For mixtures, the initial boiling point or boiling-point range may be given; this is because mixtures often contain materials with different boiling points, making the exact boiling point of the mixture difficult to estimate. Flammable materials with low boiling points generally present special fire hazards, because they give off more vapors into the air.

When a liquid in a container begins to boil, the increase in vapor pressure may exceed the vessel's ability to relieve excess pressure. The container may fail catastrophically, creating a boiling liquid expanding vapor explosion (BLEVE) **(Figure 23.19)**. As the vapor is released, it expands rapidly and if flammable, ignites, sending flames and container pieces in all directions. BLEVEs typically occur when flames contact the tank shell above the liquid level, or there is not enough water to cool the tank shell.

Vapor Density

Vapor density is the weight of pure vapor or gas compared to the weight of an equal volume of dry air, at the same temperature and pressure **(Figure 23.20)**. Ambient air has a vapor density of 1. A vapor density less than 1 indicates a vapor lighter than air, while a vapor density greater than 1 indicates a vapor heavier than air. Most gases have a vapor density greater than 1.

Figure 23.19 Moments after this photograph was taken, the railway tank car in Kingman, Arizona, exploded in a BLEVE killing 12 people and injuring more than 90 others. *Courtesy of the Mohave Museum of History and Arts.*

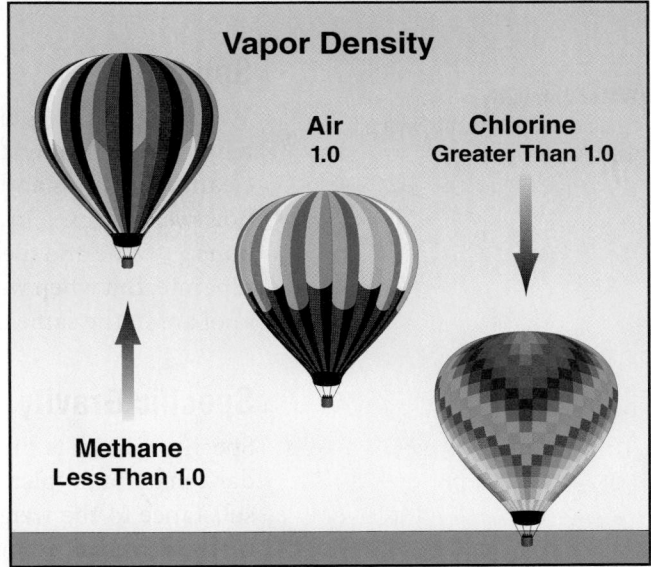

Figure 23.20 Methane will rise under normal atmospheric conditions while chlorine will sink.

All vapors and gases mix with air, but lighter materials tend to rise and dissipate. Heavier vapors and gases tend to concentrate along or under floors. This concentration creates health and/or fire hazards in sumps, sewers, manholes, trenches, and ditches.

Solubility

Water Solubility — Ability of a liquid or solid to mix with or dissolve in water.

Water solubility is the percentage of a material that dissolves in water at ambient temperature. Knowing a material's solubility can help to determine appropriate spill cleanup methods and extinguishing agents. The solubility of a substance determines how well it mixes in water. When non-water-soluble liquids, such as the hydrocarbons gasoline and diesel are in the same container with water, the two liquids remain separate. But when water-soluble liquids, such as the polar solvents alcohol and methanol are in the same container with water, the two liquids mix easily.

Specific Gravity

Specific gravity is the density of a substance compared to the density of some standard material, typically water. Most commonly, it is a comparison of the weight of a substance to the weight of an equal volume of water. For example, if a volume of a material weighs 8 pounds (3.6 kg), and an equal volume of water weighs 10 pounds (4.5 kg), the material has a specific gravity of 0.8. Water has a specific gravity of 1. Material with a specific gravity of less than 1 will float on water, whereas material with a specific gravity greater than 1 will sink. Most flammable liquids have specific gravities less than 1 (**Figure 23.21**).

Persistence

Persistence — Length of time a chemical agent remains effective without dissipating.

Dispersion — Process of being spread widely.

The **persistence** of a chemical is its ability to remain in the environment. Persistent chemicals remain effective at their point of **dispersion** for a long time, whereas less persistent chemicals quickly dissipate or break down (**Figure 23.22**).

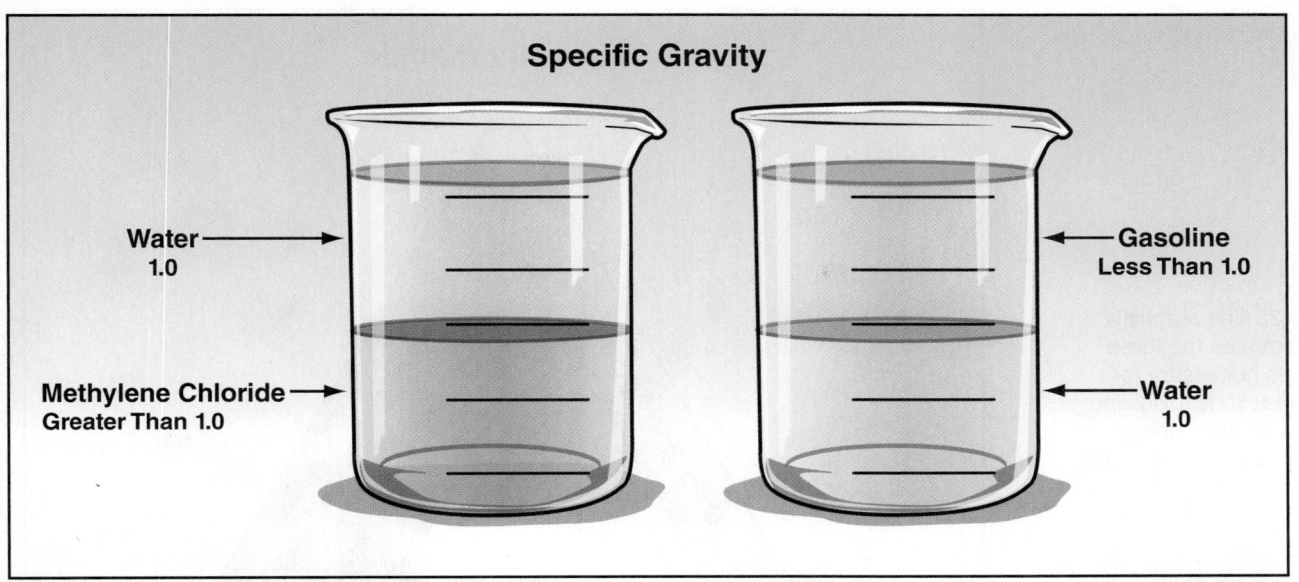

Specific Gravity

Water
1.0

Gasoline
Less Than 1.0

Methylene Chloride
Greater Than 1.0

Water
1.0

Figure 23.21 In general, hydrocarbons, such as gasoline, float on water, and chlorinated solvents, such as methylene chloride, sink.

Persistence

Figure 23.22 Persistent materials do not dissipate readily. Contact with those materials should be avoided when possible.

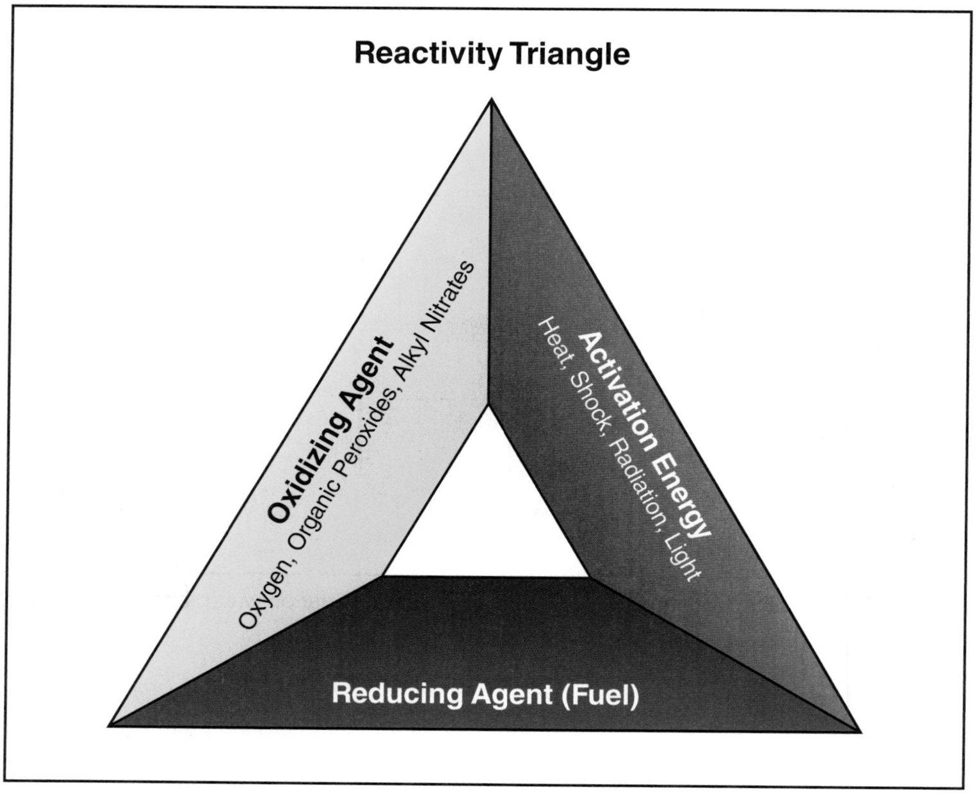

Figure 23.23 The reactivity triangle illustrates the three components necessary for chemical reactions: oxidizing agent, an activation energy source, and a reducing agent.

Reactivity

Reactivity is the ability of a substance to chemically react with other materials. In industry, *reactive materials* are unstable chemicals that react violently when combined with air, water, heat, light, or other materials. With any type of material, chemical reactions can create undesirable pressure buildup, temperature increase, or harmful byproducts.

By now, you are familiar with the fire tetrahedron or the four components necessary to produce combustion: oxygen, fuel, heat, and a chemical chain reaction. Fire is just one type of chemical reaction, and a *reactivity triangle* can be used to explain the basic components of many chemical reactions: an oxidizing agent (oxygen), a reducing agent (fuel), and an activation energy source (typically heat) **(Figure 23.23)**.

All chemical reactions require **activation energy** to get started. This is usually supplied by heat, but it can also come from radio waves, radiation, shock waves, or a pressure change. Sometimes very little energy is necessary. For example, when liquid oxygen is spilled on a roadway, a firefighter's boot striking the asphalt may initiate an explosive reaction. Also, materials that react when mixed with water (water-reactive materials) will do so at room temperature because the heat of the ambient air is sufficient to start the reaction. See **Table 23.3** for a summary of the different ways in which chemicals can be reactive. This table supplies the definition and chemical examples of nine reactive hazard classes.

The oxidizing agent in the reactivity triangle provides the oxygen necessary for the chemical reaction. **Strong oxidizers** are materials that readily give off large quantities of oxygen; they produce a strong reaction by readily accepting electrons from a reducing agent (fuel). The additional oxygen can make a fire burn hotter, faster, and brighter. Petroleum products and other hydrocarbons will ignite spontaneously when they come into contact with a strong oxidizer, such as nitrates, perchlorates, chlorine, and fluorine.

Activation Energy — Energy that starts a chemical reaction when added to an atomic or molecular system.

Strong Oxidizer — Substance that readily gives off large quantities of oxygen, thereby stimulating combustion; produces a strong reaction by readily accepting electrons from a reducing agent (fuel).

Table 23.3
Nine Reactive Hazard Classes

Reactive Hazard Class	Definition	Chemical Examples
Highly Flammable	Substances having flash points less than 100°F (38°C) and mixtures that include substances with flash points less than 100°F (38°C).	Gasoline, Acetone, Pentane, Ethyl Ether, Toluene, Methyl Ethyl Ketone (MEK), Turpentine
Explosive	A material synthesized or mixed deliberately to allow the very rapid release of chemical energy; also, a chemical substance that is intrinsically unstable and liable to detonate under conditions that might reasonably be encountered.	Dynamite, Nitroglycerin, Perchloric Acid, Picric Acid, Fulminates, Azide
Polymerizable	Capable of undergoing self-reactions that release energy; some polymerization reactions generate a great deal of heat. (The products of polymerization reactions are generally less reactive than the starting materials.)	Acrylic Acid, Butadiene, Ethylene, Styrene, Vinyl Chloride, Epoxies
Strong Oxidizing Agent	Oxidizing agents gain electrons from other substances and are themselves thereby chemically reduced, but strong oxidizing agents accept electrons particularly well from a large range of other substances. The ensuing oxidation-reduction reactions may be vigorous or violent and may release new substances that may take part in further additional reactions. Keep strong oxidizing agents well separated from strong reducing agents. In some cases, the presence of a strong oxidizing agent can greatly enhance the progress of a fire.	Hydrogen Peroxide, Fluorine, Bromine, Calcium Chlorate, Chromic Acid, Ammonium Perchlorate
Strong Reducing Agent	Reducing agents give up electrons to other substances and are thereby oxidized, but strong reducing agents donate electrons particularly well to a large range of other substances. The ensuing oxidation-reduction reactions may be vigorous or violent and may generate new substances that take part in further additional reactions.	Alkali metals (Sodium, Magnesium, Lithium, Potassium), Beryllium, Calcium, Barium, Phosphorus, Radium, Lithium Aluminum Hydride
Water-Reactive	Substances that may react rapidly or violently with liquid water and steam, producing heat (or fire) and often toxic reaction products.	Alkali metals (Sodium, Magnesium, Lithium, Potassium), Sodium Peroxide, Anhydrides, Carbides
Air-Reactive	Likely to react rapidly or violently with dry air or moist air; may generate toxic and corrosive fumes upon exposure to air or catch fire.	Finely divided metal dusts (Nickel, Zinc, Titanium), Alkali metals (Sodium, Magnesium, Lithium, Potassium), Hydrides (Diborane, Barium Hydrides, Diisobutyl Aluminum Hydride)
Peroxidizable Compound	Apt to undergo spontaneous reaction with oxygen at room temperature, to form peroxides and other products. Most such auto-oxidations are accelerated by light or trace impurities. Many peroxides are explosive, which makes peroxidizable compounds a particular hazard. Ethers and aldehydes are particularly subject to peroxide formation (the peroxides generally form slowly after evaporation of the solvent in which a peroxidizable material had been stored).	Isopropyl Ether, Furan, Acrylic Acid, Styrene, Vinyl Chloride, Methyl Isobutyl Ketone, Ethers, Aldehydes
Radioactive Material	Spontaneously and continuously emitting ions or ionizing radiation. Radioactivity is not a chemical property, but hazard that exists in addition to the chemical properties of a material.	Radon, Uranium

Source: U.S. Environmental Protection Agency's CEPPO (Chemical Emergency Preparedness and Prevention Office) Computer-Aided Management of Emergency Operations (CAMEO) software was used to identify this information.

Figure 23.24 Chemicals prone to polymerization, like styrene, may react violently in the presence of a catalyst. *Courtesy of Rich Mahaney.*

Catalyst — Substance that modifies (usually increases) the rate of a chemical reaction, without being consumed in the process.

Inhibitor — Substance that slows down or prevents a chemical reaction; typically added to materials that are prone to polymerization. *Also known as Stabilizer.*

The reducing agent in the reactivity triangle is the fuel source for the reaction. As it combines with the oxygen (or loses electrons to the oxidizer), energy is released. Some reducing agents are more volatile than others. Oxidation-reduction reactions can be extremely violent because they release a tremendous amount of energy.

Another type of chemical reaction is polymerization, in which a **catalyst** causes molecules to combine into a chain. If this reaction is uncontrolled, a tremendous amount of energy can be released. Materials that are prone to polymerization include styrene and propylene oxide. Many substances can be catalysts, including heat, light, water, acids, and other chemicals (**Figure 23.24**).

Inhibitors are materials that slow down or prevent unwanted chemical reactions. They are typically added to polymerizing materials, especially during shipping, when such materials can become extremely unstable. Many inhibitors are *time-sensitive*, meaning that they become *exhausted* (lose effectiveness) gradually over time; they may also be *consumed* more rapidly when *overwhelmed* by exposure to heat or other catalysts. The most common inhibitor is phenol, also known as carbolic acid.

General Emergency Behavior Model (GEBMO)

To successfully mitigate a haz mat incident, firefighters must be able to predict the hazardous material's behavior. The General Emergency Behavior Model (GEBMO) describes the typical sequence of events at an incident, which is broken into six stages (**Figure 23.25**):

1. **Stress** — A container that is stressed beyond its design strength may breach (fail). The three types of container stress are:

   • **Thermal** — Excessive heating or cooling of the container, causing it to expand or contract. Thermal stress may also increase internal pressure and reduce container integrity, resulting in sudden failure. Possible signs of thermal stress include flame impingement on the container, operation of a relief valve, or ambient temperature changes.

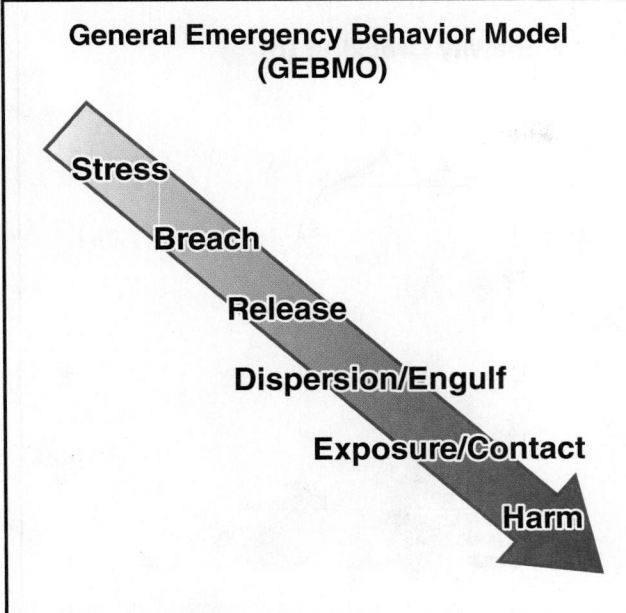

General Emergency Behavior Model (GEBMO)

Stress

Breach

Release

Dispersion/Engulf

Exposure/Contact

Harm

Figure 23.25 Haz mat events can be categorized into six typical stages of development, in which the situation worsens to the point of harm to the environment and humans as the container condition deteriorates.

- **Chemical** — Chemical reactions produce heat and pressure, causing the container to suddenly or gradually deteriorate. Chemical stress can result from external exposure to corrosives, or an unsuitable mix of chemicals being stored within the container. Possible signs of chemical stress include visible corrosion. However, if the stress is produced internally, the container may not show any signs of impending failure.

- **Mechanical** — Physical force that crushes, cracks, penetrates, or weakens the container wall; also, physical force that affects valves and piping. Common causes include collision, impact, and excessive internal pressure. Possible signs of mechanical stress include physical damage or operation of relief devices.

2. **Breach** — The way in which a container breaches is based on its construction material, the type of stress it is exposed to, and pressure inside the container at the time it fails. A breach or failure of the container may be partial or total resulting in the release of the contents into the environment. There are five types of container breach (**Figures 23.26a–e, p. 1322**):

- **Disintegration** — General loss of integrity, such as a glass bottle shattering or a grenade exploding.

- **Runaway cracking** — Single crack that grows rapidly, breaking the container into large pieces.

- **Attachments (closures) open or break** — Stress applied to attachments, such as pressure-relief devices or discharge valves; this causes them to open or break off, leading to total container failure.

- **Puncture** — Mechanical force that punctures the container; for example, a forklift puncturing a drum.

- **Split or tear** — Failure of the welded seam on a tank (split) or the stitched seam on a bag of fertilizer (tear); typically caused by mechanical or thermal stress.

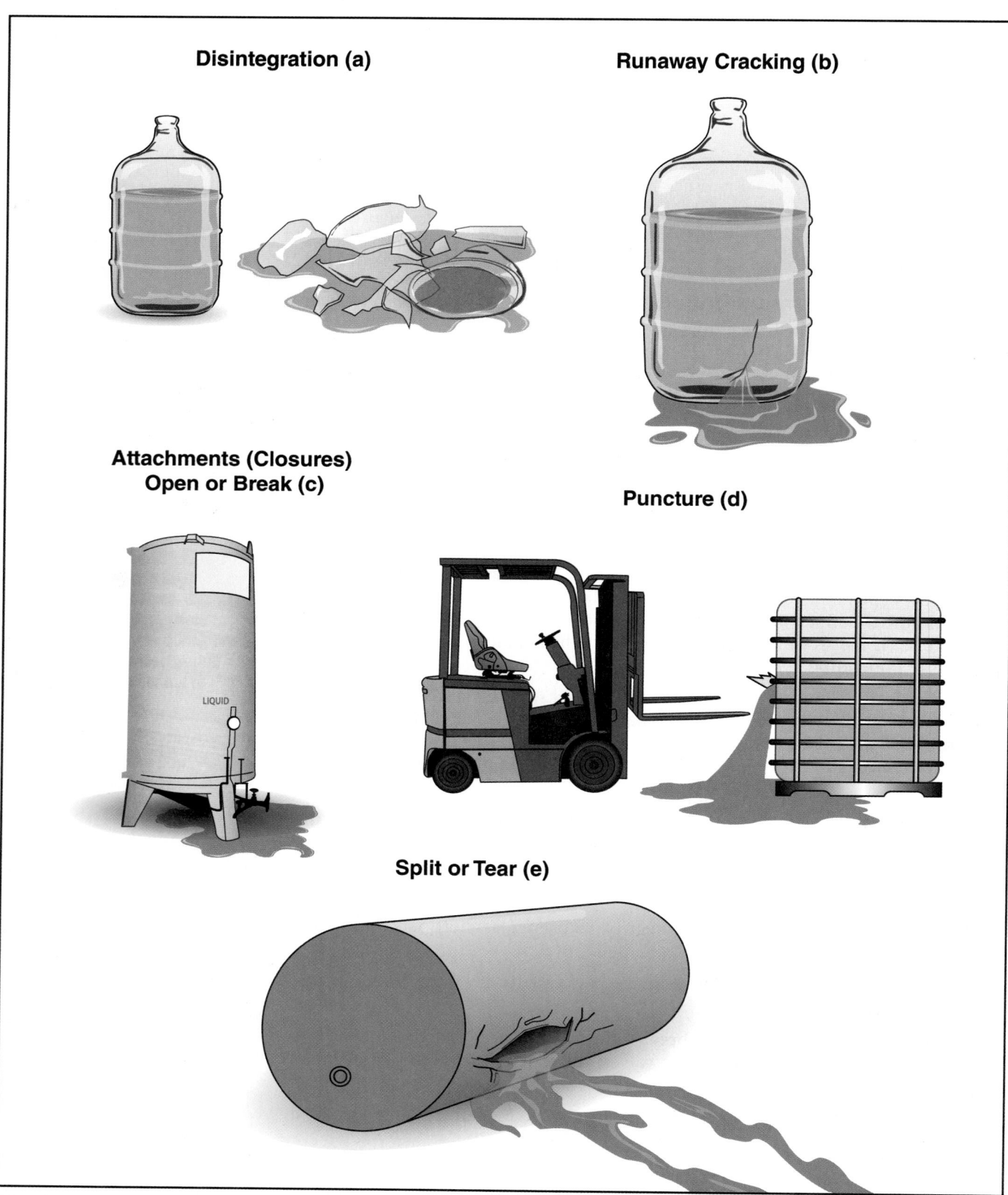

Disintegration (a)

Runaway Cracking (b)

**Attachments (Closures)
Open or Break (c)**

LIQUID

Puncture (d)

Split or Tear (e)

Figure 23.26a–e (a) Disintegration is characterized by general loss of integrity. (b) Runaway cracking begins as a single point of structural weakness and develops into large shards of broken material. (c) Containers with broken closures may be otherwise structurally sound but they cannot fulfill their intended function. (d) Puncture damage creates an uncontrolled opening. (e) Mechanical or thermal stress may cause a split or torn seam.

3. **Release** — After a container breaches, its contents and stored energy are released. Pieces of the container may also be projected outward. The four types of container release are:

- **Detonation** — Explosive release of stored chemical energy of a hazardous material.

- **Violent rupture** — Sudden release of chemical or mechanical energy caused by runaway cracks. Pieces of the container are expelled outward and so are its contents, as in a BLEVE.

- **Rapid relief** — Sudden release of a pressurized material through holes in the container or through damaged valves, piping, or attachments.

- **Spill/leak** — Slow release of pressurized material through holes, rips, tears, attachments or usual openings.

4. **Dispersion/Engulf** — After being released, the container's contents disperse and engulf the area. Use the acronym CCHIPPS to remember the seven common dispersion patterns (**Figures 23.27a – g**):

- **Cloud** — Ball-shaped pattern in which materials collectively rise above the surface (**Figure 23.27a**).

- **Cone** — Cone-shaped pattern with a point source at the breach and a wide base downwind (**Figure 23.27b, p. 1324**).

- **Hemispheric** — Dome-shaped pattern in which some material stays on the surface, but the rest becomes airborne (**Figure 23.27c, p. 1324**).

- **Plume** — Irregular airborne pattern influenced by wind and/or terrain (**Figure 23.27d, p. 1324**).

- **Pool** — Three-dimensional (including depth), slow-flowing liquid dispersion. Liquids assume the shape of the area they flow into and pool in low areas (**Figure 23.27e, p. 1324**).

- **Stream** — Pattern in which liquids flow along the surface, affected by gravity and terrain (**Figure 23.27f, p. 1325**).

- **Irregular** — A nongeometric pattern, often caused when material is spread by contaminated responders (**Figure 23.27g, p. 1325**).

Figure 23.27a A vapor cloud is a ball-shaped pattern of an airborne hazardous material where the material has collectively risen above the ground or water.

Cone

Figure 23.27b A cone is a triangular-shaped pattern of a hazardous material with a point source at the breach and a wide base downrange.

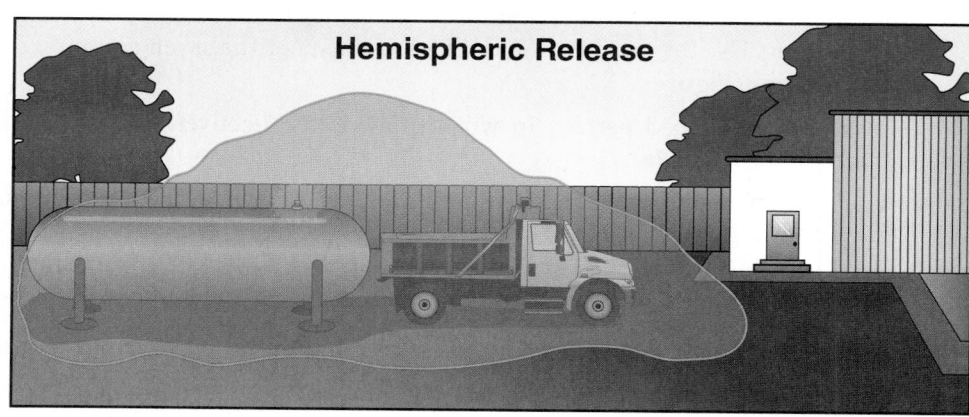

Hemispheric Release

Figure 23.27c A hemispheric release is a semicircular or dome-shaped pattern of an airborne hazardous material that is still partially in contact with the ground or water.

Plume

WIND

Figure 23.27d A plume is an irregularly-shaped pattern of an airborne hazardous material influenced by wind and/or topography in its downrange course.

Pool

Figure 23.27e In a pool, liquids assume the shape of their container, typically accumulating in low areas.

Stream

Figure 23.27f A stream is pulled by gravity, following the topographical contours of the surface.

Irregular Dispersion

Figure 23.27g Irregular dispersion results from indiscriminate deposit of a hazardous material such as that caused by contaminated vehicles or responders.

5. **Exposure/Contact** — Dispersed materials may come into contact with people, property, and the surrounding environment. The time frame of exposure can be:

- **Short-term** — Seconds, minutes, and hours
- **Medium-term** — Days, weeks, and months
- **Long-term** — Years and generations, as seen in the high cancer rates among children whose parents were exposed to high doses of radiation

6. **Harm** — Some exposures may be harmful to the environment and/or human health. Estimations of potential harm should always include a worst-case scenario.

Releases in urban environments are affected by wind speed and direction, as well as structures in the path of the plume. **Figures 23.27h and i, p. 1326** provide guidelines for plume modeling in urban environments.

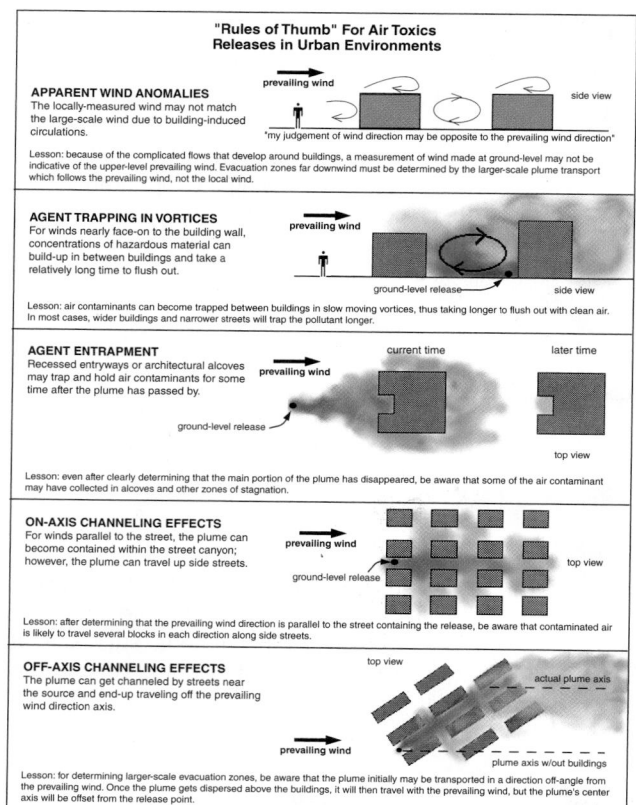

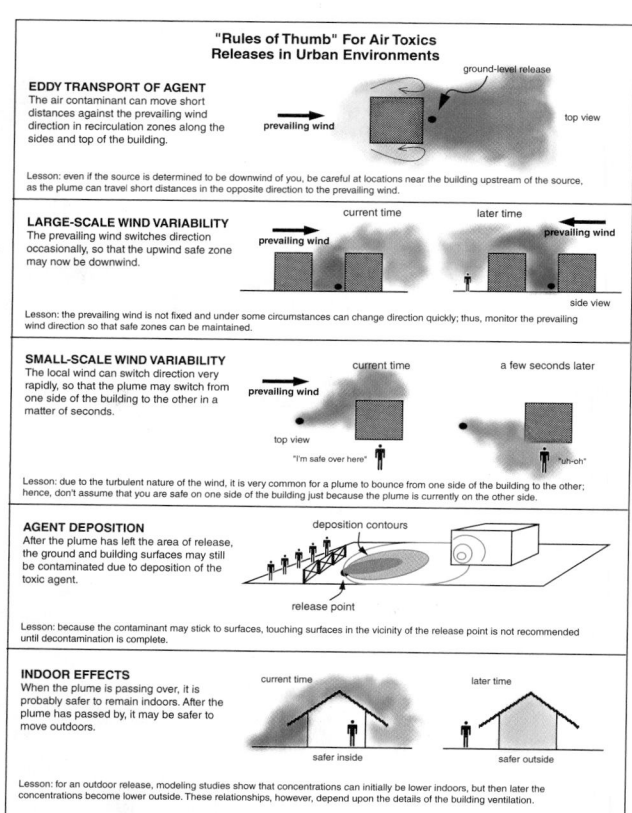

Figure 23.27h and i Responders should be aware of the rules of thumb regarding plume modeling behavior in urban environments. *Courtesy of Los Alamos National Laboratory.*

Identifying Hazardous Materials

There will often be clues at the scene to help you determine whether hazardous materials are involved in an incident, and if so, which type. These clues can be grouped into seven categories:

1. Locations and occupancies

2. Container types and shapes

3. Transportation placards, labels, and markings

4. Other markings and colors (nontransportation)

5. Written resources

6. Senses

7. Monitoring and detection devices

The clues are numbered in this order for two reasons. Higher numbers indicate that the clue will be more difficult to detect; they also indicate increasing risk to firefighters (**Figure 23.28**). For example, even from a distance, you can tell that you are likely to encounter toxic chemicals because you know that you are en route to a chemical plant. But to use monitoring devices you must be much closer to the hazard, where you are at greater risk.

Although these clues can be extremely helpful, at some incidents there will be nothing to indicate the presence or type of hazardous materials. Placards, labels, and signs may not be visible from a safe distance. Identification markings may be destroyed in the incident. Vehicles carrying mixed loads on a roadway may have

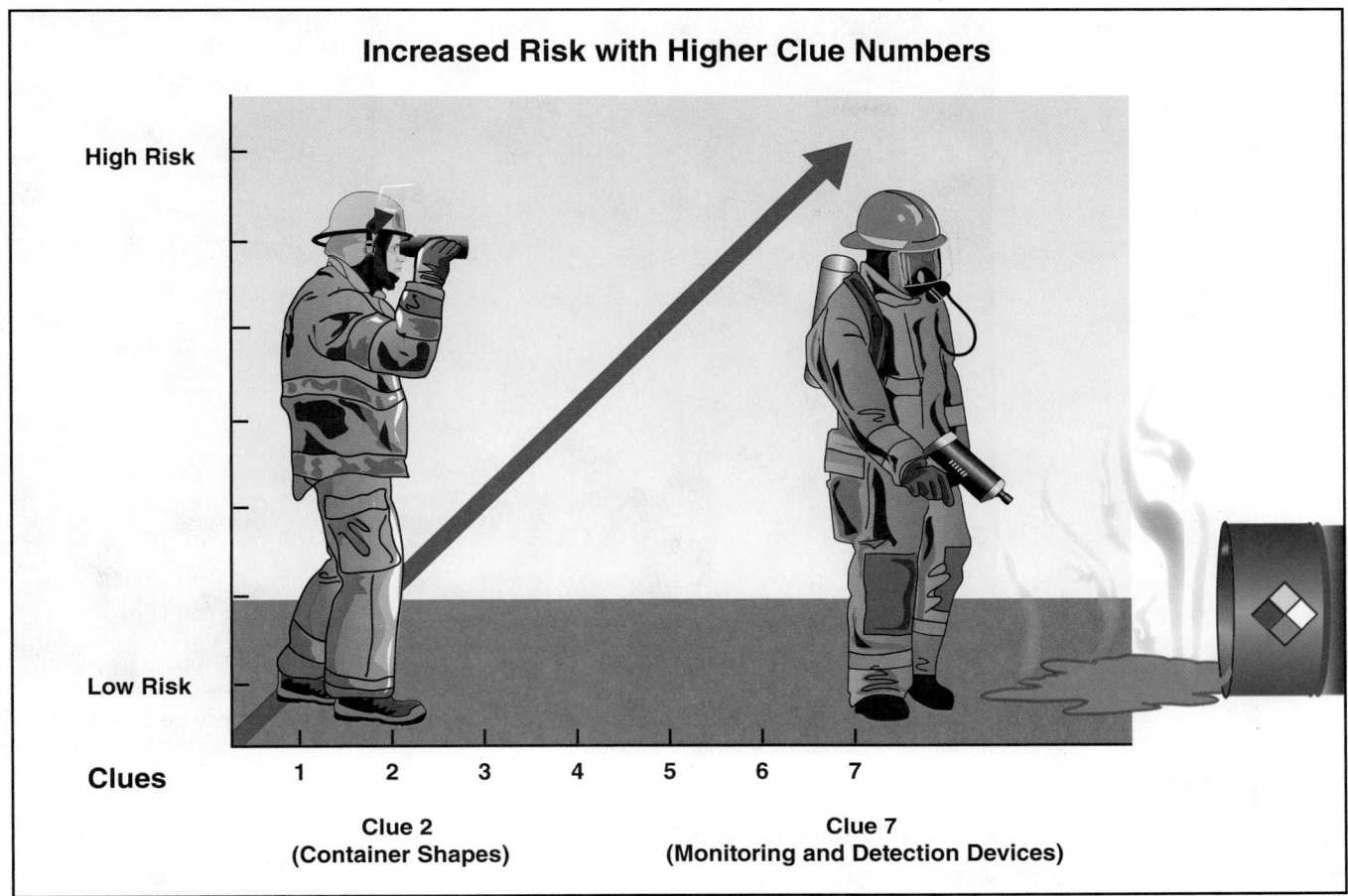

Increased Risk with Higher Clue Numbers

High Risk

Low Risk

Clues   1   2   3   4   5   6   7

Clue 2
(Container Shapes)

Clue 7
(Monitoring and Detection Devices)

Figure 23.28 The risk to responders increases as they move closer to the hazardous material.

no markings at all (**Figure 23.29, p. 1328**). Shipping papers may be inaccessible. At warehouses or industrial plants, inventories may have changed since the preincident survey and containers may be improperly labeled.

Locations and Occupancies

Hazardous materials can be found anywhere. Preincident surveys and emergency response plans are the best and safest ways to determine the location of hazardous materials. However, you are particularly likely to encounter these materials in the following types of occupancies:

- Refineries and fuel storage facilities
- Gas/service stations and convenience stores
- Paint supply stores
- Plant nurseries, garden centers, and agricultural facilities
- Pest control and lawn care companies
- Swimming pool supply companies
- Medical facilities
- Photo processing laboratories
- Dry cleaners

Figure 23.29 Small quantities and mixed loads of hazardous materials may be transported without any placards or other markings.

- Plastics and high-technology factories
- Metal-plating businesses
- Mercantile occupancies such as hardware stores, groceries stores, and department stores
- Chemistry laboratories in educational facilities
- Lumberyards
- Feed/farm stores
- Veterinary clinics
- Print shops
- Warehouses
- Industrial and utility plants
- Transportation centers including railroad yards, airports, seaports, and truck transport lines
- Aircraft maintenance facilities
- Treatment storage disposal (TSD) facilities

Residential occupancies may also have small quantities of hazardous materials, such as drain cleaners, paint products, and flammable liquids. Rural residences and farms may have propane tanks, and large quantities of pesticides and fertilizers.

Other occupancy types are more likely to be the target of terrorist attacks, which may include hazardous materials. Terrorism intends to do harm by:

- Causing the greatest number of casualties
- Causing panic and/or disruption

- Damaging the economy
- Destroying property
- Demoralizing the community

Places with large public gatherings are potential targets for attack, as are sites with historic, economic, or symbolic significance **(Figure 23.30)**. Specific targets include:

- **Mass transportation** — Airports, ferry terminals, ports, bus and train stations, planes, subways, buses, commuter trains

- **Critical infrastructure** — Dams, water treatment facilities, power plants, electrical substations, nuclear power plants, trans-oceanic cable landings, telecommunication switch centers (telecom hotels), financial institutions, rail and road bridges, tunnels, levees, liquefied natural gas (LNG) terminals, natural gas (NG) compressor stations, petroleum pumping stations, petroleum storage tank farms

- **Areas of public assembly and recreation** — Convention centers, hotels, casinos, shopping malls, theaters, stadiums, theme parks

- **High profile buildings and locations** — Monuments, high-rise buildings, buildings/structures of historic or national significance

- **Industrial sites** — Chemical manufacturing facilities, shipping facilities, warehouses

- **Educational sites** — Colleges, universities, community colleges, vocational/training facilities, primary and secondary schools

- **Medical and science facilities** — Hospitals, clinics, nuclear research labs, other research facilities, nonpower nuclear reactors, national health stockpile sites

Incidents reported at these occupancies should be scrutinized closely for potential terrorist involvement. If terrorism is suspected, law enforcement authorities must be

Figure 23.30 Target locations include public areas that may gather large numbers of people. *Courtesy of U.S. Customs and Border Protection, photo by Gerald L. Nino.*

notified immediately. Your department must be able to identify potential targets and predict the consequences of an attack. This information can be found in emergency response plans and preincident surveys.

Container Types and Shapes

Familiarize yourself with the types of containers, vessels, tanks, packages, and vehicles in which hazardous materials are stored and transported. Containers can provide useful information about the materials they contain; for example, cylinder-shaped containers typically hold pressurized gases or liquids.

This section discusses the five main container types:

- Bulk-capacity containment systems at fixed facilities
- Bulk transportation packaging
- Intermediate bulk containers
- Ton containers
- Nonbulk containers

Carboy — Cylindrical container of about 5 to 15 gallon (19 L to 57 L) capacity used for pure or corrosive liquids. Made of glass, plastic, or metal, with a neck and sometimes a pouring tip; cushioned in a wooden box, wicker, basket, or special drum.

Bulk and Nonbulk Packaging

In bulk packaging, materials are loaded with no intermediate form of containment onto a transport vehicle or freight container, such as a portable tank, cargo tank, railcar, or intermodal (IM) container **(Figure 23.31)**. Bulk packaging must meet at least one of the following criteria:

- Maximum capacity is greater than 119 gallons (450 L) as a receptacle for a liquid
- Maximum net mass is greater than 882 pounds (400 kg) or maximum capacity is greater than 119 gallons (450 L) as a receptacle for a solid
- Water capacity is 1,001 pounds (454 kg) or greater as a receptacle for a gas

Nonbulk packaging is packaging that is smaller than the minimum criteria established for bulk packaging **(Figure 23.32)**. Drums, boxes, **carboys**, and bags are examples. Composite packages (packages with an outer packaging and an inner receptacle) and combination packages (multiple packages grouped together in a single outer container such as bottles of acid packed inside a cardboard box) may also be classified as nonbulk packaging.

Bulk-Capacity Fixed-Facility Containers

Containers at fixed facilities include:

- Buildings
- Above and underground storage tanks
- Machinery
- Pipelines
- Reactors
- Bins
- Vats
- Storage cabinets
- Other fixed, on-site containers

Figure 23.31 Bulk packaging is characterized by a lack of intermediate packaging in a large capacity container. *Courtesy of Rich Mahaney.*

Figure 23.32 Nonbulk packaging consists of packaging with a smaller capacity than the bulk packaging minimum criteria.

Aboveground storage tanks are divided into two major categories:

● **Nonpressure or atmospheric tanks** — If these tanks are storing any product, they will normally have a small amount of pressure (up to 0.5 psi [3.45 kPa] {0.03 bar}) inside.

● **Pressure tanks** — These tanks are divided into:

— Low-pressure storage tanks that have pressures between 0.5 psi to 15 psi (3.45 kPa to 103 kPa) {0.03 bar to 1.03 bar}.

— Pressure vessels that have pressures above 15 psi (103 kPa) {1.03 bar}.

NOTE: Underground storage tanks may be atmospheric or pressurized.

Table 23.4, p. 1332 provides pictures and examples of various atmospheric/nonpressure storage tanks and also describes underground storage caverns. **Table 23.5, p. 1334** provides pictures and examples of various pressure tanks.

Bulk-capacity fixed-facility storage tanks used for cryogenic liquids can have varying pressures, but some can be very high. They are usually heavily insulated with a vacuum in the space between the outer and inner shells.

Bulk Transportation Containers

Bulk transportation containers can be divided into three main categories determined by the mode of transportation:

● Tank and other rail cars (railroad) **(Tables 23.6, p. 1335, 23.7, p. 1336, 23.8, p. 1336, and 23.9, p. 1337)**

— Nonpressure tank cars (also known as general service or low-pressure tank cars) with vapor pressures below 25 psi (172 kPa) {1.7 bar} at 105° to 115°F (41°C to 46°C)

— Pressure tank cars with pressures greater than 25 psi (172 kPa) {1.7 bar} at 68°F (20°C)

— Cryogenic liquid tank cars

— Other cars, including hopper cars and box cars

— Special service cars

Table 23.4
Atmospheric/Nonpressure Storage Tanks

Tank Type	Descriptions
	**Horizontal Tank** Cylindrical tanks sitting on legs, blocks, cement pads, or something similar; typically constructed of steel with flat ends. Horizontal tanks are commonly used for bulk storage in conjunction with fuel-dispensing operations. Old tanks (pre-1950s) have bolted seams, whereas new tanks are generally welded. A horizontal tank supported by unprotected steel supports or stilts (prohibited by most current fire codes) may fail quickly during fire conditions. **Contents:** Flammable and combustible liquids, corrosives, poisons, etc.
	**Cone Roof Tank** Have cone-shaped, pointed roofs with weak roof-to-shell seams that break when or if the container becomes overpressurized. When it is partially full, the remaining portion of the tank contains a potentially dangerous vapor space. **Contents:** Flammable, combustible, and corrosive liquids
 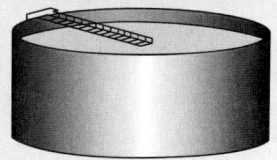	**Open Top Floating Roof Tank** Large-capacity, aboveground holding tanks. They are usually much wider than they are tall. As with all floating roof tanks, the roof actually floats on the surface of the liquid and moves up and down depending on the liquid's level. This roof eliminates the potentially dangerous vapor space found in cone roof tanks. A fabric or rubber seal around the circumference of the roof provides a weather-tight seal. **Contents:** Flammable and combustible liquids
	**Covered Top Floating Roof Tank** Have fixed cone roofs with either a pan or deck-type float inside that rides directly on the product surface. This tank is a combination of the open top floating roof tank and the ordinary cone roof tank. **Contents:** Flammable and combustible liquids

Vents around rim provide differentation
from Open Top Floating Roof Tanks

Continued

Table 23.4 (concluded)

Tank Type	Descriptions
	**Covered Top Floating Roof Tank with Geodesic Dome** Floating roof tanks covered by geodesic domes are used to store flammable liquids.
	**Lifter Roof Tank** Have roofs that float within a series of vertical guides that allow only a few feet (meters) of travel. The roof is designed so that when the vapor pressure exceeds a designated limit, the roof lifts slightly and relieves the excess pressure. **Contents:** Flammable and combustible liquids
	**Vapordome Roof Tank** Vertical storage tanks that have lightweight aluminum geodesic domes on their tops. Attached to the underside of the dome is a flexible diaphragm that moves in conjunction with changes in vapor pressure. **Contents:** Combustible liquids of medium volatility and other nonhazardous materials
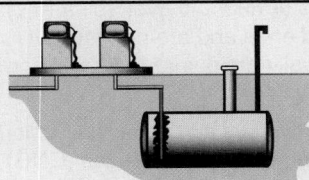 **Fill Connections Cover** 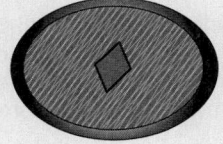	**Atmospheric Underground Storage Tank** Constructed of steel, fiberglass, or steel with a fiberglass coating. Underground tanks will have more than 10 percent of their surface areas underground. They can be buried under a building or driveway or adjacent to the occupancy. This tank has fill and vent connections located near the tank. Vents, fill points, and occupancy type (gas/service stations, private garages, and fleet maintenance stations) provide visual clues. Many commercial and private tanks have been abandoned, some with product still in them. These tanks are presenting major problems to many communities. **Contents:** Petroleum products Rare and technically are not "tanks." First responders should be aware that some natural and manmade caverns are used to store natural gas. The locations of such caverns should be noted in local emergency response plans.

Table 23.5
Low-Pressure Storage Tanks and Pressure Vessels

Tank/Vessel Type	Descriptions
	**Dome Roof Tank** Generally classified as low-pressure tanks with operating pressures as high as 15 psi (103 kPa). They have domes on their tops. **Contents:** Flammable liquids, combustible liquids, fertilizers, solvents, etc.
	**Spheroid Tank** Low-pressure storage tanks. They can store 3,000,000 gallons (11 356 200 L) or more of liquid. **Contents:** Liquefied petroleum gas (LPG), methane, and some flammable liquids such as gasoline and crude oil
	**Noded Spheroid Tank** Low-pressure storage tanks. They are similar in use to spheroid tanks, but they can be substantially larger and flatter in shape. These tanks are held together by a series of internal ties and supports that reduce stresses on the external shells. **Contents:** LPG, methane, and some flammable liquids such as gasoline and crude oil
	**Horizontal Pressure Vessel*** Have high pressures and capacities from 500 to over 40,000 gallons (1 893 L to over 151 416 L). They have rounded ends and are not usually insulated. They usually are painted white or some other highly reflective color. **Contents:** LPG, anhydrous ammonia, vinyl chloride, butane, ethane, compressed natural gas (CNG), chlorine, hydrogen chloride, and other similar products
	**Spherical Pressure Vessel** Have high pressures and capacities up to 600,000 gallons (2 271 240 L). They are often supported off the ground by a series of concrete or steel legs. They usually are painted white or some other highly reflective color. **Contents:** Liquefied petroleum gases and vinyl chloride

Continued

Table 23.5 (concluded)

Tank/Vessel Type	Descriptions
	**Cryogenic-Liquid Storage Tank** Insulated, vacuum-jacketed tanks with safety-relief valves and rupture disks. Capacities can range from 300 to 400,000 gallons (1 136 L to 1 514 160 L). Pressures vary according to the materials stored and their uses. **Contents:** Cryogenic carbon dioxide, liquid oxygen, liquid nitrogen, etc.

* It is becoming more common for horizontal propane tanks to be buried underground. Underground residential tanks usually have capacities of 500 or 1,000 gallons (1 893 L or 3 785 L). Once buried, the tank may be noticeable only because of a small access dome protruding a few inches (millimeters) above the ground.

Table 23.6
Nonpressure Tank Cars

Photograph	Descriptions
Nonpressure Tank Car 	**Nonpressure tank car without an expansion dome:** Fittings are visible. **Carries:** Flammable liquids, flammable solids, reactive liquids, reactive solids, oxidizers, organic peroxides, poisons, irritants, corrosive materials, and similar products
Nonpressure Tank Car With an Expansion Dome 	**Older Model** **Nonpressure tank car with an expansion dome** **Carries:** Flammable liquids, flammable solids, reactive liquids, reactive solids, oxidizers, organic peroxides, poisons, irritants, corrosive materials, and similar products

Source: Information courtesy of Union Pacific Railroad.

Table 23.7
Pressure Tank Cars

Photograph	Descriptions
Typical Pressure Tank Car 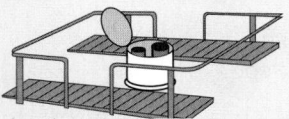 	**Typical pressure tank car:** Fittings are inside the protective housing. **Carries:** Flammable, nonflammable, and poison gases as well as flammable liquids
Hydrogen Cyanide (Hydrocyanic Acid), HCN *Candy Stripe* Car 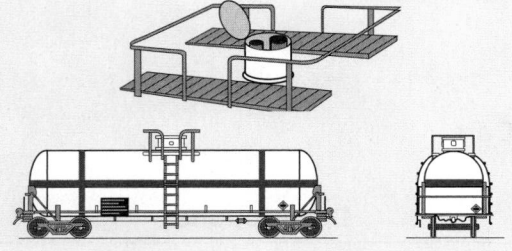	**Hydrogen cyanide (hydrocyanic acid), HCN *candy stripe car*:** In the past, some HCN cars were painted white with red bands (called *candy stripes*). This practice is becoming obsolete due to concerns about terrorism.

Table 23.8
Cryogenic Liquid Tank Car

Photograph	Descriptions
Typical Cryogenic Liquid Tank Car 	**Typical cryogenic liquid tank car:** Fittings are in ground-level cabinets at diagonal corners of the car or in the center of one end of the car. **Carries:** Argon, hydrogen, nitrogen, oxygen, LNG, and ethylene **NOTE:** Cryogenic liquid tanks can also be enclosed inside a boxcar.

Table 23.9
Other Railroad Cars

Illustration	Photograph
Covered Hopper Car 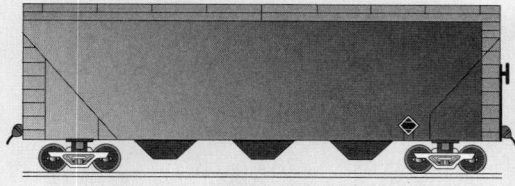	 **Carries:** Calcium carbide, cement, and grain
Open Top Hopper 	 **Carries:** Coal, rock, gravel, and sand
Miscellaneous: Boxcar 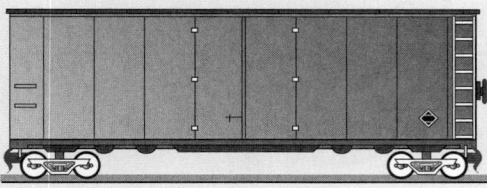	 **Carries:** All types of materials and finished goods
Miscellaneous: Gondola 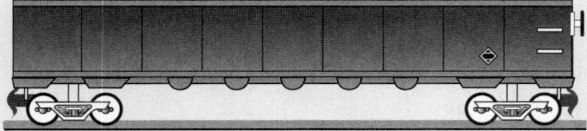	 **Carries:** Sand, rolled steel, and other materials that do not require protection from the weather

Continued

Table 23.9 (concluded)
Other Railroad Cars

Illustration	Photograph
Miscellaneous: Flat Bed Car with Intermodal Containers 	 **Carries:** 1 ton containers, intermodal containers (shown), large vehicles, and other commodities that do not require protection from the weather
Pneumatically Unloaded Hopper Car 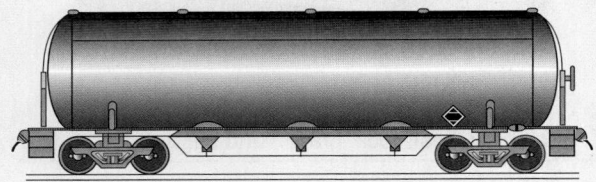	 **Carries:** Dry caustic soda, ammonium nitrate fertilizer, other fine-powdered materials, plastic pellets, and flour

- Cargo tanks (highway) — Motor Carrier (MC) and Department of Transportation (DOT)/Transport Canada (TC) designations indicate the tank's manufacturing specifications (**Table 23.10, p. 1340**):
  - Nonpressure liquid tanks (DOT/TC 406, MC 306)
    Low-pressure liquid tanks (DOT/TC 407, MC 307)
  - Corrosive liquid tanks (DOT/TC 412, MC 312)
  - High-pressure tanks (MC/TC 331)
  - Cryogenic liquid tanks (MC/TC 338)
  - Vacuum Loaded Tank (DOT/TC 407, DOT/TC 412)
  - Compressed-gas/tube trailers (no designation)
  - Dry bulk cargo tanks (no designation)

- Intermodal containers (highway, railroad, or marine vessel) **(Table 23.11, p. 1344)**
  - — Nonpressure intermodal tanks
  - — Pressure intermodal tanks
  - — Specialized intermodal tanks
    - Cryogenic intermodal tanks
    - Tube modules
  - — Freight containers **(Figure 23.33)**

Intermediate Bulk Containers

An *intermediate bulk container (IBC)* is a rigid or flexible portable packaging (other than a cylinder or portable tank) designed for mechanical handling. The maximum capacity of an IBC is not more than 3 cubic meters (3 000 L, 793 gal, or 106 ft³). The minimum capacity is not less than 0.45 cubic meters (450 L, 119 gal, or 15.9 ft³) or a maximum net mass of not less than 400 kilograms (882 lbs).

IBCs are divided into two types: flexible intermediate bulk containers (FIBCs) **(Figure 23.34, p.1345)** and rigid intermediate bulk containers (RIBCs) **(Figure 23.35, p. 1345)**. Both types are often called *totes,* although only FIBCs are truly totes. RIBCs may be used to carry liquids, fertilizers, solvents, and other chemicals. They may have capacities up to 400 gallons (1 514 L) and pressures up to 100 psi (689 kPa) {6.9 bar} **(Figure 23.36, p. 1345)**.

Ton Containers

Ton containers are tanks that have capacities of 1 short ton or approximately 2,000 pounds (907 kg or 0.9 tonne). They are typically stored on their sides, and the ends (heads) of the containers are convex or concave with two valves in the center of one end **(Figure 23.37, p. 1345)**. Ton containers typically contain chlorine and are often found at water treatment plants and commercial swimming pools. They may also contain other products such as sulfur dioxide, anhydrous ammonia, or Freon® refrigerant.

> ## CAUTION
> Structural fire fighting PPE does not provide adequate protection against the hazardous materials commonly stored in ton containers.

Intermodal Freight Containers

Dry Van *(box containers)*

Refrigerated *(reefers)*

Open Top

Flat

Figure 23.33 Freight containers come in a variety of different types including refrigerated containers, flat containers, dry van containers, and open top containers.

Table 23.10
Cargo Tank Trucks

Truck	Descriptions

Nonpressure Liquid Tank
DOT406, TC406, SCT-306
(MC306, TC306)

- Pressure less than 4 psig (28 kPa)
- Typical maximum capacity: 9,000 gallons (34 069 L)
- New tanks made of aluminum
- Old tanks made of steel
- Oval shape
- Multiple compartments
- Recessed manways
- Rollover protection
- Bottom valves
- Longitudinal rollover protection
- Valve assembly and unloading control box under tank
- Vapor-recovery system on curb side and rear, if present
- Manway assemblies, and vapor-recovery valves on top for each compartment
- Possible permanent markings for ownership that are locally identifiable

Carries: Gasoline, fuel oil, alcohol, other flammable/combustible liquids, other liquids, and liquid fuel products

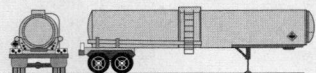

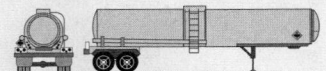

Low-Pressure Chemical Tank
DOT407, TC407, SCT-307
(MC307, TC307)

- Pressure under 40 psi (172 kPa to 276 kPa)
- Typical maximum capacity: 7,000 gallons (26 498 L) [per NFPA®]
- Rubber lined or steel
- Typically double shell
- Stiffening rings may be visible or covered
- Circumferential rollover protection
- Single or multiple compartments
- Single- or double-top manway assembly protected by a flash box that also provides rollover protection
- Single-outlet discharge piping at midship or rear
- Fusible plugs, frangible disks, or vents outside the flash box on top of the tank
- Drain hose from the flash box down the side of the tank
- Rounded or horse-shaped ends

Carries: Flammable liquids, combustible liquids, acids, caustics, and poisons

Continued

Table 23.10 (continued)

Truck	Descriptions

Corrosive Liquid Tank
DOT412, TC412, SCT-412
(MC312, TC312)

- Pressure less than 75 psi (517 kPa)
- Typical maximum capacity: 7,000 gallons (26 498 L) [per NFPA®]
- Rubber lined or steel
- Typically single compartment
- Small-diameter round shape
- Exterior stiffening rings may be visible on uninsulated tanks
- Typical rear station with exterior piping extending to the bottom of the tank
- Splashguard serving as rollover protection around valve assembly
- Flange-type rupture disk vent either inside or outside the splashguard
- May have discoloration around loading/unloading area or area painted or coated with corrosive-resistant material
- Permanent ownership markings that are locally identifiable

Carries: Corrosive liquids (usually acids)

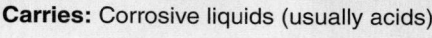
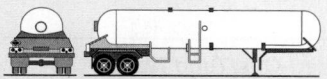

High-Pressure Tank
MC-331, TC331, SCT-331

- Pressure above 100 psi (689 kPa)
- Typical maximum capacity: 11,500 gallons (43 532 L)
- Single steel compartment
- Noninsulated
- Bolted manway at front or rear
- Internal and rear outlet valves
- Typically painted white or other reflective color
- Large hemispherical heads on both ends
- Guard cage around the bottom loading/unloading piping
- Uninsulated tanks, single-shell vessels
- Permanent markings such as *FLAMMABLE GAS, COMPRESSED GAS,* or identifiable manufacturer or distributor names

Carries: Pressurized gases and liquids, anhydrous ammonia, propane, butane, and other gases that have been liquefied under pressure

High-Pressure Bobtail Tank

Used for local delivery of liquefied petroleum gas and anhydrous ammonia

Continued

Table 23.10 (continued)

Truck	Descriptions

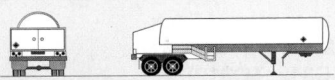

Cryogenic Liquid Tank
MC338, TC338, SCT-338
(TC341, CGA341)

- Pressure less than 22 psi (152 kPa)
- Well-insulated steel tank
- Possibly discharging vapor from relief valves
- Round tank with flat ends
- Large and bulky double shelling and heavy insulation
- Loading/unloading station attached either at the rear or in front of the rear dual wheels, typically called the doghouse in the field
- Permanent markings such as *REFRIGERATED LIQUID* or an identifiable manufacturer name

Carries: Liquid oxygen, liquid nitrogen, liquid carbon dioxide, liquid hydrogen, and other gases that have been liquefied by lowering their temperatures

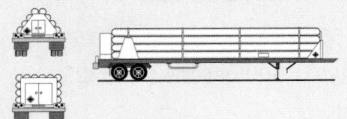

Compressed-Gas/Tube Trailer

- Pressure at 3,000 to 5,000 psi (20 684 kPa to 34 474 kPa) (gas only)
- Individual steel cylinders stacked and banded together
- Typically has over-pressure device for each cylinder
- Bolted manway at front or rear
- Valves at rear (protected)
- Manifold enclosed at the rear
- Permanent markings for the material or ownership that is locally identifiable

Carries: Helium, hydrogen, methane, oxygen, and other gases

Continued

Table 23.10 (concluded)

Truck	Descriptions
 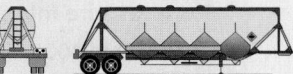   	**Dry Bulk Cargo Trailer** • Pressure less than 22 psi (152 kPa) • Typically not under pressure • Bottom valves • Shapes vary, but has V-shaped bottom-unloading compartments • Rear-mounted, auxiliary-engine-powered compressor or tractor-mounted power-take-off air compressor • Air-assisted, exterior loading and bottom unloading pipes • Top manway assemblies **Carries:** Calcium carbide, oxidizers, corrosive solids, cement, plastic pellets, and fertilizers

Table 23.11
Intermodal Tanks

Tank Type	Descriptions
	**Nonpressure Intermodal Tank** • IM-101: 25.4 to 100 psi (175 kPa to 689 kPa) • IM-102: 14.5 to 25.4 psi (100 kPa to 175 kPa) **Contents:** Liquids or solids (both hazardous and nonhazardous)
	**Pressure Intermodal Tank** 100 to 500 psi (689 kPa to 3 447 kPa) **Contents:** Liquefied gases, liquefied petroleum gas, anhydrous ammonia, and other liquids
	**Cryogenic Intermodal Tank** **Contents:** Refrigerated liquid gases, argon, oxygen, helium
	**Tube Module Intermodal Container** **Contents:** Gases in high-pressure cylinders (3,000 or 5,000 psi [20 684 kPa or 34 474 kPa]) mounted in the frame

Figure 23.34 Intermediate bulk containers come in a variety of styles and may carry both solid materials and fluids. This is a flexible intermediate bulk container. *Courtesy of Rich Mahaney.*

Figure 23.35 This tank is a rigid intermediate bulk container (RIBC). *Courtesy of Rich Mahaney.*

Figure 23.36 RIBCs may be large, square or rectangular boxes or bins carrying liquids, fertilizers, solvents, or other chemicals, with capacities up to 400 gallons (1 514 L). *Courtesy of Rich Mahaney.*

Figure 23.37 Ton containers may be identified by their size, concave or convex ends, and two valves in the center of one end. *Courtesy of Rich Mahaney.*

Nonbulk Packaging

Nonbulk packaging is used to transport smaller quantities of hazardous materials than bulk or IBCs. The most common types of nonbulk packaging include the following types of containers (see **Table 23.12, p. 1346**):

- Bags
- Carboys and jerrycans
- Cylinders
- Drums
- Dewar flasks (cryogenic liquids)

Table 23.12
Nonbulk Packaging

Package Type	Descriptions
	**Bags** • Made of paper, plastic, film, textiles, woven material, or others • Sizes vary **Contents:** Explosives, flammable solids, oxidizers, organic peroxides, fertilizers, pesticides, and other regulated materials.
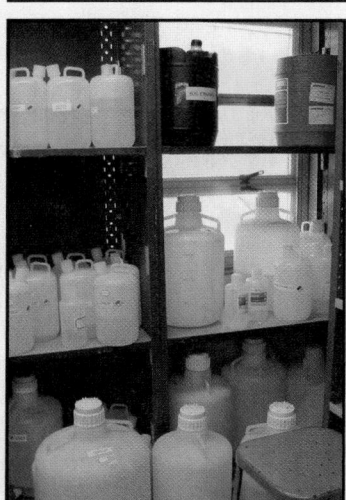	**Carboys and Jerrycans** • Made of glass or plastic • Often encased in a basket or box • Sizes vary **Contents:** Flammable and combustible liquids, corrosives.
	**Cylinders** • Pressures higher than 40 psi (276 kPa) {2.76 bar} but vary • Sizes range from lecture bottle size to very large **Contents:** Compressed gases.

Continued

Table 23.12 (concluded)

Package Type	Descriptions
	**Drums** • Made of metal, fiberboard, plastic, plywood, or other materials • May have open heads (removable tops) or tight (closed) heads with small openings • Sizes range from 55 gallons (208 L) to 100 gallons (379 L) **Contents:** Hazardous and nonhazardous liquids and solids.
	**Dewar Flasks** • Vacuum insulated • Made of glass, metal, or plastic with hollow walls from which the air has been removed • Sizes vary **Contents:** Cryogenic liquids; thermoses may contain nonhazardous liquids.

Containers for Radioactive Materials

All shipments of radioactive materials (sometimes called *RAM*) must be packaged and transported according to strict regulations. The type of packaging used to transport radioactive materials is determined by its type, form, and activity (the number of nuclear decays or disintegrations the radioactive material undergoes in a given amount of time). Depending on these factors, most radioactive material is shipped in one of five basic types of containers/packaging: Type A, Type B, Type C, industrial, or excepted **(Table 23.13, p. 1348)**.

Type C packages are very rare. They are used for high-activity materials, such as plutonium, that are transported by aircraft. They are designed to withstand plane crashes without loss of containment.

Transportation Placards, Labels, and Markings

In North America, the transportation of hazardous materials is regulated by the U.S. Department of Transportation (DOT), Transport Canada (TC), and Mexico's Ministry of Communications and Transport. These identification systems are very similar, since they are all based on United Nations (UN) recommendations.

Table 23.14, p. 1349 provides the U.S. DOT placard hazard classes and divisions. It also lists the hazards associated with each hazard class and examples of products for each division. Placards are required on specific bulk quantities of hazardous

Table 23.13
Radioactive Container Descriptions

Excepted Packaging	Industrial Packaging (IP)	Type A Packaging	Type B Packaging
Designed to survive normal conditions of transport	Designed to survive normal conditions of transport (IP-1) and at least the DROP test and stacking test for Type A packaging (IP-2 and IP-3)	• Designed to survive normal transportation, handling, and minor accidents • Certified as Type A on the basis of performance requirements, which means it must survive certain tests	Must be able to survive severe accidents
Used for transportation of materials that are either Low Specific Activity (LSA) or Surface Contaminated Objects (SCO) and that are limited quantity shipments, instruments or articles, articles manufactured from natural or depleted uranium, or natural thorium; empty packagings are also excepted (49 *CFR* 173.421–428)	Used for transportation of materials with very small amounts of radioactivity (Low Specific Activity [LSA] or Surface Contaminated Objects [SCO])	Used for the transportation of limited quantities of radioactive material (RAM) that would not result in significant health effects if they were released	Used for the transportation of large quantities of radioactive material
Can be almost any packaging that meets the basic requirements, with any of the above contents; they are excepted from several labeling and documentation requirements	Usually metal boxes or drums	• May be cardboard boxes, wooden crates, or drums • Shipper and carrier must have documentation of the certification of the packages being transported	• May be a metal drum or a huge, massive shielded transport container • Must meet severe accident performance standards that are considerably more rigorous than those required for Type A packages • Either has a Certificate of Compliance (COC) by the Nuclear Regulatory Commission (NRC) or Certificate of Competent Authority (COCA) by the Department of Transportation (DOT)

Source: Courtesy of U.S. Department of Energy (DOE)/Sandia National Laboratories.

Table 23.14
U.S. DOT Placard Hazard Classes and Divisions

Class 1: Explosives (49 *CFR* 173.50)

**Compatibility Group
Letter will vary**

An **explosive** is any substance or article (including a device) that is designed to function by explosion (that is, an extremely rapid release of gas and heat) or (by chemical reaction within itself) is able to function in a similar manner even if not designed to function by explosion.

Explosive placards will have a *compatibility group* letter on them, which is a designated alphabetical letter used to categorize different types of explosive substances and articles for purposes of stowage and segregation. However, it is the division number that is of primary concern to first responders.

The primary hazards of explosives are thermal (heat) and mechanical, but may include the following:

- Blast pressure wave
- Shrapnel fragmentation
- Incendiary thermal effect
- Seismic effect
- Chemical hazards from the production of toxic gases and vapors
- Ability to self-contaminate with age, which increases their sensitivity and instability
- Sensitivity to shock and friction

	**Division 1.1** — Explosives that have a mass explosion hazard. A mass explosion is one that affects almost the entire load instantaneously. *Examples:* dynamite, mines, wetted mercury fulminate
	**Division 1.2** — Explosives that have a projection hazard but not a mass explosion hazard. *Examples:* detonation cord, rockets (with bursting charge), flares, fireworks
	**Division 1.3** — Explosives that have a fire hazard and either a minor blast hazard or a minor projection hazard or both, but not a mass explosion hazard. *Examples:* liquid-fueled rocket motors, smokeless powder, practice grenades, aerial flares
	**Division 1.4** — Explosives that present a minor explosion hazard. The explosive effects are largely confined to the package and no projection of fragments of appreciable size or range is expected. An external fire must not cause virtually instantaneous explosion of almost the entire contents of the package. *Examples:* signal cartridges, cap type primers, igniter fuses, fireworks
	**Division 1.5** — Substances that have a mass explosion hazard but are so insensitive that there is very little probability of initiation or of transition from burning to detonation under normal conditions of transport. **Examples:** prilled ammonium nitrate fertilizer/fuel oil (ANFO) mixtures and blasting agents
	**Division 1.6** — Extremely insensitive articles that do not have a mass explosive hazard. This division is comprised of articles that contain only extremely insensitive detonating substances and that demonstrate a negligible probability of accidental initiation or propagation.

Continued

Table 23.14 (continued)

Class 2: Gases (49 *CFR* 173.115)

DOT defines *gas* as a material that has a vapor pressure greater than 43.5 psi (300 kPa) at 122°F (50°C) or is completely gaseous at 68°F (20°C) at a standard pressure of 14.7 psi (101.3 kPa).

NOTE: The DOT definition for gas is much more specific than the definition provided in Chapter 2, Hazardous Materials Properties and Hazards.

The potential hazards of gases may include thermal, asphyxiation, chemical, and mechanical hazards:

- Thermal hazards (heat) from fires, particularly associated with Division 2.1 and oxygen
- Thermal hazards (cold) associated with exposure to cryogens in Division 2.2
- Asphyxiation caused by leaking/released gases displacing oxygen in a confined space
- Chemical hazards from toxic and/or corrosive gases and vapors, particularly associated with Division 2.3
- Mechanical hazards from a boiling liquid expanding vapor explosion (BLEVE) for containers exposed to heat or flame
- Mechanical hazards from a ruptured cylinder rocketing after exposure to heat or flame

	**Division 2.1: Flammable Gas** — Consists of any material that is a gas at 68°F (20°C) or less at normal atmospheric pressure or a material that has a boiling point of 68°F (20°C) or less at normal atmospheric pressure and that (1) Is ignitable at normal atmospheric pressure when in a mixture of 13 percent or less by volume with air, or (2) Has a flammable range at normal atmospheric pressure with air of at least 12 percent, regardless of the lower limit. *Examples:* compressed hydrogen, isobutene, methane, and propane
	**Division 2.2: Nonflammable, Nonpoisonous Gas** — Nonflammable, nonpoisonous compressed gas, including compressed gas, liquefied gas, pressurized cryogenic gas, and compressed gas in solution, asphyxiant gas and oxidizing gas; means any material (or mixture) which exerts in the packaging an absolute pressure of 40.6 psi (280 kPa) or greater at 68°F (20°C) and does not meet the definition of Divisions 2.1 or 2.3. *Examples:* carbon dioxide, helium, compressed neon, refrigerated liquid nitrogen, cryogenic argon, anhydrous ammonia, pepper spray
	**Division 2.3: Gas Poisonous by Inhalation** — Material that is a gas at 68°F (20°C) or less and a pressure of 14.7 psi (101.3 kPa) (a material that has a boiling point of 68°F [20°C] or less at 14.7 psi [101.3 kPa]), and that is known to be so toxic to humans as to pose a hazard to health during transportation; or (in the absence of adequate data on human toxicity) is presumed to be toxic to humans because of specific test criteria on laboratory animals. Division 2.3 has *ERG*-designated hazard zones associated with it, determined by the concentration of gas in the air: • Hazard Zone A — LC_{50} less than or equal to 200 ppm • Hazard Zone B — LC_{50} greater than 200 ppm and less than or equal to 1,000 ppm • Hazard Zone C — LC_{50} greater than 1,000 ppm and less than or equal to 3,000 ppm • Hazard Zone D — LC_{50} greater than 3,000 ppm and less than or equal to 5,000 ppm *Examples:* cyanide, diphosgene, germane, phosphine, selenium hexafluoride, and hydrocyanic acid, arsine, cyanogen chloride, phosgene, chlorine
	**Oxygen Placard** — Oxygen is not a separate division under Class 2, but first responders may see this oxygen placard on containers with 1,001 lbs (454 kg) or more gross weight of either compressed gas or refrigerated liquid.

Continued

Table 23.14 (continued)

Class 3: Flammable and Combustible Liquids (49 *CFR* 173.120)

A *flammable liquid* is generally a liquid having a flash point of not more than 140°F (60°C), or any material in a liquid state with a flash point at or above 100°F (37.8°C) that is intentionally heated and offered for transportation or transported at or above its flash point in a bulk packaging.

A *combustible liquid* is any liquid that does not meet the definition of any other hazard class and has a flash point above 141°F (60.5°C) and below 200°F (93°C). A flammable liquid with a flash point at or above 100°F (32.8°C) that does not meet the definition of any other hazard class may be reclassified as a combustible liquid. This provision does not apply to transportation by vessel or aircraft, except where other means of transportation is impracticable. An elevated-temperature material that meets the definition of a Class 3 material because it is intentionally heated and offered for transportation or transported at or above its flash point may not be reclassified as a combustible liquid.

The primary hazards of flammable and combustible liquids are thermal, asphyxiation, chemical, and mechanical, and may include the following:

- Thermal hazards (heat) from fires and vapor explosions
- Asphyxiation from heavier than air vapors displacing oxygen in low-lying, and/or confined spaces
- Chemical hazards from toxic and/or corrosive gases and vapors
- Chemical hazards from the production of toxic and/or corrosive gases and vapors during fires
- Mechanical hazards from a BLEVE, for containers exposed to heat or flame
- Mechanical hazards caused by a vapor explosion
- Vapors that can mix with air and travel great distances to an ignition source
- Environmental hazards (pollution) caused by runoff from fire control

When responding to a transportation incident, first responders must keep in mind that a flammable liquid placard can indicate a product with a flash point as high as 140°F (60°C).

FLAMMABLE 3	**Flammable Placard** *Examples:* gasoline, methyl ethyl ketone
GASOLINE 3	**Gasoline Placard** — May be used in the place of a flammable placard on a cargo tank or a portable tank being used to transport gasoline by highway
COMBUSTIBLE 3	**Combustible Placard** *Examples:* diesel, fuel oils, pine oil
FUEL OIL 3	**Fuel Oil Placard** — May be used in place of a combustible placard on a cargo tank or portable tank being used to transport fuel oil by highway.

Class 4: Flammable Solids, Spontaneously Combustible Materials, and Dangerous-When-Wet Materials (49 *CFR* 173.124)

This class is divided into three divisions: 4.1 Flammable Solids, 4.2 Spontaneously Combustible Materials, and 4.3 Dangerous-When-Wet (see definitions below).

First responders must be aware that fires involving Class 4 materials may be extremely difficult to extinguish. The primary hazards of Class 4 materials are thermal, chemical, and mechanical and may also include the following hazards:

Continued

Table 23.14 (continued)

Class 4: (continued)

- Thermal hazards (heat) from fires that may start spontaneously or upon contact with air or water
- Thermal hazards (heat) from fires and vapor explosions
- Thermal hazards (heat) from molten substances
- Chemical hazards from irritating, corrosive, and/or highly toxic gases and vapors produced by fire or decomposition
- Severe chemical burns
- Mechanical effects from unexpected, violent chemical reactions and explosions
- Mechanical hazards from a BLEVE, for containers exposed to heat or flame (or if contaminated with water, particularly for Division 4.3)
- Production of hydrogen gas from contact with metal
- Production of corrosive solutions on contact with water, for Division 4.3
- May spontaneously reignite after fire is extinguished
- Environmental hazards (pollution) caused by runoff from fire control

	**Division 4.1: Flammable Solid Material** — Includes (1) wetted explosives, (2) self-reactive materials that can undergo a strongly exothermal decomposition, and (3) readily combustible solids that may cause a fire through friction, certain metal powders that can be ignited and react over the whole length of a sample in 10 minutes or less, or readily combustible solids that burn faster than 2.2 mm/second: • **Wetted explosives:** Explosives with their explosive properties suppressed by wetting with sufficient alcohol, plasticizers, or water • **Self-reactive materials:** Materials liable to undergo a strong exothermic decomposition at normal or elevated temperatures due to excessively high transport temperatures or to contamination • **Readily combustible solids:** Solids that may ignite through friction or any metal powders that can be ignited *Examples:* phosphorus heptasulfide, paraformaldehyde, magnesium
	**Division 4.2: Spontaneous Combustible Material** — Includes (1) a pyrophoric material (liquid or solid) that, without an external ignition source, can ignite within 5 minutes after coming in contact with air and (2) a self-heating material that, when in contact with air and without an energy supply, is liable to self-heat *Examples:* sodium sulfide, potassium sulfide, phosphorus (white or yellow, dry), aluminum and magnesium alkyls, charcoal briquettes
	**Division 4.3: Dangerous-When-Wet Material** — Material that, by contact with water, is liable to become spontaneously flammable or to release flammable or toxic gas at a rate greater than 1 liter per kilogram of the material per hour *Examples:* magnesium powder, lithium, ethyldichlorosilane, calcium carbide, potassium

Class 5: Oxidizers and Organic Peroxides (49 *CFR* 173.127 and 128)

This class is divided into two divisions: 5.1 Oxidizers and 5.2 Organic Peroxides (see definitions below). Oxygen supports combustion, so the primary hazards of Class 5 materials are fires and explosions with their associated thermal and mechanical hazards:

- Thermal hazards (heat) from fires that may explode or burn extremely hot and fast
- Explosive reactions to contact with hydrocarbons (fuels)
- Chemical hazards from toxic gases, vapors, and dust
- Chemicals hazards from toxic products of combustion
- Chemical burns
- Ignition of combustibles (including paper, cloth, wood, etc.)

Continued

Table 23.14 (continued)

Class 5: (continued)

- Mechanical hazards from violent reactions and explosions
- Accumulation of toxic fumes and dusts in confined spaces
- Sensitivity to heat, friction, shock, and/or contamination with other materials

	**Division 5.1: Oxidizer** — Material that may, generally by yielding oxygen, cause or enhance the combustion of other material *Examples:* chromium nitrate, copper chlorate, calcium permanganate, ammonium nitrate fertilizer
 	**Division 5.2: Organic Peroxide** — Any organic compound containing oxygen (O) in the bivalent -O-O- structure and which may be considered a derivative of hydrogen peroxide, where one or more of the hydrogen atoms has been replaced by organic radicals *Examples*: liquid organic peroxide type B

Class 6: Poison (Toxic) and Poison Inhalation Hazard (49 *CFR* 173.132 and 134)

A poisonous material is a material, other than a gas, that is known to be toxic to humans. The primary hazards of Class 6 materials are chemical and thermal and may include the following:

- Toxic effects due to exposure via all routes of entry
- Chemicals hazards from toxic and/or corrosive products of combustion
- Thermal effects (heat) from substances transported in molten form
- Flammability and its associated thermal hazards (heat) from fires

	**Division 6.1: Poisonous Material** — Material, other than a gas, that is known to be so toxic to humans as to afford a hazard to health during transportation or that is presumed to be toxic to humans based on toxicity tests on laboratory animals *Examples:* aniline, arsenic, mustard agents, nerve agents, hydrogen cyanide, most riot control agents
No Placard for Division 6.2, see labels	**Division 6.2: Infectious Substance** — Material known to contain or suspected of containing a pathogen. A pathogen is a virus or microorganism (including its viruses, plasmids, or other genetic elements, if any) or a proteinaceous infectious particle (prion) that has the potential to cause disease in humans or animals.
	**Inhalation Hazard Placard** — Used for any quantity of Division 6.1, Zones A or B inhalation hazard only (see Division 2.3 for hazard zones)
	**PG III** — For Division 6.1, packing group III* (PG III) materials, a POISON placard may be modified to display the text "PG III" below the mid line of the placard rather than the word "POISON" **A packing group is a DOT packaging category based on the degree of danger presented by the hazardous material. Packing Group I indicates great danger; Packing Group II, medium danger; and Packing Group III, minor danger. The PG III placard, then, might be used for materials that are not as dangerous as those that would be placarded with the "POISON" placard*

Continued

Table 23.14 (concluded)

Class 7: Radioactive Materials (49 *CFR* 173.403)

A radioactive material means any material having a specific activity greater than 70 becquerels per gram (0.002 microcurie per gram). The primary hazard of Class 7 materials is radiological, including burns and biological effects.

Radioactive Placard — Is required on certain shipments of radioactive materials; vehicles with this placard are carrying "highway route controlled quantities" of radioactive materials and must follow prescribed, predetermined transportation routes

Examples: solid thorium nitrate, uranium hexafluoride

Class 8: Corrosive Materials (49 *CFR* 173.136)

A corrosive material means a liquid or solid that causes full thickness destruction of human skin at the site of contact within a specific period of time or a liquid that has a severe corrosion rate on steel or aluminum. The primary hazards of Class 8 materials are chemical and thermal, and may include the following hazards:

- Chemical burns
- Toxic effects due to exposure via all routes of entry
- Thermal effects, including fire, caused by chemical reactions generating heat
- Reactivity to water
- Mechanical effects caused by BLEVEs and violent chemical reactions

Corrosive Placard

Examples: battery fluid, chromic acid solution, soda lime, sulfuric acid, hydrochloric acid (muriatic acid), sodium hydroxide, potassium hydroxide

Class 9: Miscellaneous Dangerous Goods (49 *CFR* 173.140)

A miscellaneous dangerous good is a material that (1) has an anesthetic, noxious, or other similar property that could cause extreme annoyance or discomfort to flight crew members and would prevent their correct performance of assigned duties; (2) is a hazardous substance or a hazardous waste; or (3) is an elevated- temperature material; or (4) is a marine pollutant.

Miscellaneous dangerous goods will primarily have thermal and chemical hazards. For example, polychlorinated biphenyls (PCBs) are carcinogenic, while elevated-temperature materials may present some thermal hazards. However, hazardous wastes may present any of the hazards associated with the materials in normal use.

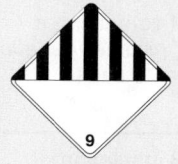

Miscellaneous Placard

Examples: blue asbestos, polychlorinated biphenyls (PCBs), solid carbon dioxide (dry ice)

Dangerous Placard — A freight container, unit load device, transport vehicle, or railcar that contains nonbulk packaging with two or more DOT Chart 12, Table 2 categories of hazardous materials may be placarded *DANGEROUS*. However, when 2,205 lbs (1,000 kg) or more of one category of material is loaded at one loading facility, the placard specified in DOT Chart 12, Table 2 must be applied.

materials that are transported by tank cars, cargo tanks, and other bulk packages and portable tanks. Some nonbulk containers may also require placarding. **Table 23.15, p. 1356** provides Canadian transportation placards, labels, and markings.

Anhydrous Ammonia

Placarded materials may have many hazards not reflected by the placard classification. For example, anhydrous ammonia is placarded in the U.S. as a nonflammable gas, but under certain conditions (particularly inside where fumes can become concentrated), it will burn.

In 1984, in Shreveport, Louisiana, two hazardous materials response team members entered a cold-storage facility to stop a leak of anhydrous ammonia. With a lower explosive limit (LEL) of 16 percent and an explosive range of 16 to 25 percent, the fumes inside the facility reached a flammable concentration. A spark ignited the vapors, killing one team member and seriously burning the other.

Concentrated anhydrous ammonia fumes can also be fatal if inhaled, even though anhydrous ammonia is not classified as an inhalation hazard by the DOT. In liquid form, it also causes severe corrosion. In many other countries, anhydrous ammonia is classified as a poisonous gas and a corrosive (caustic) liquid.

Table 23.16, p. 1361 provides examples of the unique U.S. DOT labels. Other labels for the nine hazard classes and subdivisions are essentially the same as the placards shown earlier in Table 23.14. Labels are 3.9-inch (100 mm), square-on-point diamonds, which may or may not have written text that identifies the hazardous material within the packaging (Class 7 Radioactive labels must always contain text).

Table 23.17, p. 1363 provides examples of U.S. DOT markings. A marking is a descriptive name, an identification number, a weight, or a specification. It includes instructions, cautions, UN marks, or combinations thereof on the outer packaging of hazardous materials. Additional markings found on tank cars, intermodal containers, and other packaging will be discussed in later sections.

Four-Digit UN Identification Numbers

The UN has a system of four-digit identification numbers that is used in conjunction with illustrated placards in North America. Each individual hazardous material is assigned a unique four-digit number. This number will often be displayed on placards, orange panels, and markings on containers or packaging.

The four-digit ID number must be displayed in one of three ways, as illustrated in **Figure 23.38, p. 1365**. In North America, the numbers must be displayed on the following containers/packages:

- Rail tank cars
- Cargo tank trucks
- Portable tanks
- Bulk packages
- Vehicle containers containing large quantities (at least 8,820 lbs or 4 000 kg) of the same hazardous material in nonbulk packages
- Certain nonbulk packages (such as poisonous gases)

Table 23.15
Canadian Transportation Placards, Labels, and Markings

Class 1: Explosives

Placard and Label	**Class 1.1** — Mass explosion hazard
Placard and Label	**Class 1.2** — Projection hazard but not a mass explosion hazard
Placard and Label	**Class 1.3** — Fire hazard and either a minor blast hazard or a minor projection hazard or both but not a mass explosion hazard
Placard and Label	**Class 1.4** — No significant hazard beyond the package in the event of ignition or initiation during transport * = Compatibility group letter
Placard and Label	**Class 1.5** — Very insensitive substances with a mass explosion hazard * = Compatibility group letter
Placard and Label	**Class 1.6** — Extremely insensitive articles with no mass explosion hazard * = Compatibility group letter

Class 2: Gases

Placard and Label	**Class 2.1 — Flammable Gases**
Placard and Label	**Class 2.2 — Nonflammable and nontoxic Gases**

Continued

Table 23.15 (continued)

Class 2: Gases (continued)

	Class 2.3 — Toxic Gases
1005 Placard and Label	**Anhydrous Ammonia**
Placard and Label	**Oxidizing Gases**

Class 3: Flammable Liquids

Placard and Label	**Class 3 — Flammable Liquids**

Class 4: Flammable Solids, Substances Liable to Spontaneous Combustion, and Substances that on Contact with Water Emit Flammable Gases (Water-Reactive Substances)

Placard and Label	**Class 4.1 — Flammable Solids**
Placard and Label	**Class 4.2 — Substances Liable to Spontaneous Combustion**
Placard and Label	**Class 4.3 — Water-Reactive Substances**

Class 5: Oxidizing Substances and Organic Peroxides

Placard and Label	**Class 5.1 — Oxidizing Substances**

Continued

Table 23.15 (continued)

Class 5: Oxidizing Substances and Organic Peroxides (continued)

Placard and Label

Class 5.2 — Organic Peroxides

Class 6: Toxic and Infectious Substances

Placard and Label

Class 6.1 — Toxic Substances

Label Only

Class 6.2 — Infectious Substances
Text:
INFECTIOUS
In case of damage or leakage, Immediately notify local authorities AND
INFECTIEUX
En cas de Dommage ou de fuite communiquer Immédiatement avec les autorités locales ET
CANUTEC
613-996-6666

Placard Only

Class 6.2 — Infectious Substances

Class 7: Radioactive Materials

Label and Optional Placard

Class 7 — Radioactive Materials
Category I — White
RADIOACTIVE
CONTENTS......................CONTENU
ACTIVITY.......................... ACTIVITÉ

Label and Optional Placard

Class 7 — Radioactive Materials
Category II — Yellow
RADIOACTIVE
CONTENTS......................CONTENU
ACTIVITY..........................ACTIVITÉ
INDICE DE TRANSPORT INDEX

Continued

Table 23.15 (concluded)

Class 7: Radioactive Materials (continued)

 **Label and Optional Placard**	**Class 7 — Radioactive Materials** **Category III** — Yellow RADIOACTIVE CONTENTS.....................CONTENU ACTIVITY.........................ACTIVITÉ INDICE DE TRANSPORT INDEX
 **Placard**	**Class 7 — Radioactive Materials** The word RADIOACTIVE is optional.

Class 8: Corrosives

 **Placard and Label**	**Class 8 — Corrosives**

Class 9: Miscellaneous Products, Substances, or Organisms

 **Placard and Label**	**Class 9 — Miscellaneous Products, Substances, or Organisms**

Other Placards, Labels, and Markings

	**Danger Placard**
	**Elevated-Temperature Sign**
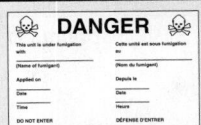	**Fumigation Sign** Text is in both English and French
	**Marine Pollutant Mark** The text is MARINE POLLUTANT or POLLUANT MARIN.

Table 23.16
Unique U.S. DOT Labels

Subsidiary Risk Labels

Subsidiary risk labels may be used for the following classes: Explosives, Flammable Gases, Flammable Liquids, Flammable Solids, Corrosives, Oxidizers, Poisons, Spontaneously Combustible Materials, and Dangerous-When-Wet Materials.

Class 1: Explosives

Explosive Subsidiary Risk Label

Class 3: Flammable Liquid

Flammable Liquid Label — Marks packages containing flammable liquids.

Examples: gasoline, methyl ethyl ketone

Class 6: Poison (Toxic), Poison Inhalation Hazard, Infectious Substance

Infectious Substances Label — Marks packages with infectious substances (viable micro-organism, or its toxin, which causes or may cause disease in humans or animals).

This label may be used to mark packages of Class 6.2 materials as defined in 49 *CFR* 172.432.

Examples: anthrax, hepatitis B virus, *escherichia* coli (E. coli)

Biohazard Label — Marks bulk packaging containing a regulated medical waste as defined in 49 *CFR* 173.134(a)(5).

Examples: used needles/syringes, human blood or blood products, human tissue or anatomical waste, carcasses of animals intentionally infected with human pathogens for medical research

Etiological Agents Label — Marks packages containing etiologic agents transported in interstate traffic per 42 *CFR* 72.3 and 72.6

Examples: rabies virus, rickettsia, Ebola virus, salmonella bacteria

Continued

Table 23.16 (continued)
Unique U.S. DOT Labels

Class 7: Radioactive Materials

Packages of radioactive materials must be labeled on two opposite sides, with a distinctive warning label. Each of the three label categories — RADIOACTIVE WHITE-I, RADIOACTIVE YELLOW-II, or RADIOACTIVE YELLOW-III — bears the unique trefoil symbol for radiation.

Class 7 Radioactive I, II, and III labels must always contain the following additional information:

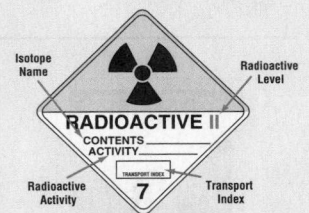

- Isotope name
- Radioactive activity

Radioactive II and III labels will also provide the Transport Index (TI) indicating the degree of control to be excercised by the carrier during transportation. The number in the Transport Index box indicates the maximum radiation level measured (in mrem/hr) at one meter from the surface of the package. Packages with the Radioactive I label have a Transport Index of 0.

	Radioactive I Label — Label with an all-white background color that indicates that the external radiation level is low and no special stowage controls or handling are required.
	Radioactive II Label — Upper half of the label is yellow, which indicates that the package has an external radiation level or fissile (nuclear safety criticality) characteristic that requires consideration during stowage in transportation.
	Radioactive III Label — Yellow label with three red stripes indicates the transport vehicle must be placarded RADIOACTIVE.
	Fissile Label — Used on containers of fissile materials (materials capable of undergoing fission such as uranium-233, uranium-235, and plutonium-239). The Criticality Safety Index (CSI) must be listed on this label. The CSI is used to provide control over the accumulation of packages, overpacks, or freight containers containing fissile material.
	Empty Label — Used on containers that have been emptied of their radioactive materials, but still contain residual radioactivity.

Continued

Table 23.16 (concluded)
Unique U.S. DOT Labels

Aircraft Labels

Old  **New**  *Mandatory January 1, 2013*	**Danger - Cargo Aircraft Only** — Used to indicate materials that cannot be transported on passenger aircraft.

Table 23.17
Unique U.S. DOT Markings

Marking	Description
HOT	**Hot Marking** — Has the same dimensions as a placard and is used on elevated- temperature materials. *Note:* Bulk containers of molten aluminum or molten sulfur must be marked MOLTEN ALUMINUM or MOLTEN SULFUR, respectively.
(marine pollutant symbol)	**Marine Pollutant Marking** — Must be displayed on packages of substances designated as marine pollutants. *Examples:* cadmium compounds, copper cyanide, mercury based pesticides
INHALATION HAZARD	**Inhalation Hazard Marking** — Used to mark materials that are poisonous by inhalation. *Examples:* anhydrous ammonia, methyl bromide, hydrogen cyanide, hydrogen sulfide
DANGER THIS UNIT IS UNDER FUMIGATION WITH " _____ APPLIED ON Date _____ Time _____ Ventilated on _____ DO NOT ENTER	**Fumigant Marking** — Warning affixed on or near each door of a transport vehicle, freight container, or railcar in which the lading has been fumigated or is undergoing fumigation with any material. The vehicle, container, or railcar is considered a package containing a hazardous material unless it has been sufficiently aerated so that it does not pose a risk to health and safety.
(orientation arrows)	**Orientation Markings** — Markings used to designate the orientation of the package. Sometimes these markings will be accompanied by words such as "this side up."
CONSUMER COMMODITY ORM-D	**ORM-D** — Used on packages of ORM-D materials. *Examples:* consumer commodities, small arms cartridges
CONSUMER COMMODITY ORM-D-AIR	**ORM-D-AIR** — Used on packages of ORM-D materials shipped via air.

Continued

Table 23.17 (concluded)
Unique U.S. DOT Markings

Marking	Description
OVERPACK	**Inner Packaging** — Used on authorized packages containing hazardous materials being transported in an overpack as defined in 49 *CFR* 171.8 and 49 *CFR* 173.25 (a) (4).
	Excepted Quantity — Excepted quantities of hazardous materials. The "*" must be replaced by the primary hazard class, or when assigned, the division of each of the hazardous materials contained in the package. The "**" must be replaced by the name of the shipper or consignee if not shown elsewhere on the package.
UN3373	**Category B Biological Substances** — Diagnostic and clinical specimens that do not cause permanent disability or life-threatening or fatal disease to humans or animals when exposure occurs.
keep away from heat	**Keep Away From Heat Marking** — Used for aircraft transportation of packages containing self-reactive substances of Division 4.1 or organic peroxides of Division 5.2.
... kg max	**Marking of IBCs** — For IBCs not designed for stacking, the figure "0" and the symbol for IBCs not capable of being stacked must be displayed. For IBCs designed for stacking, the maximum permitted stacking load applicable (in kilograms) when the IBC is in use must be included with the symbol for IBCs capable of being stacked.

Sample Displays of Four-Digit UN Identification Numbers

FLAMMABLE 3

1090 3

1993 3

1090

Figure 23.38 UN numbers may be displayed in a variety of ways on bulk containers and certain nonbulk packages.

The *Emergency Response Guidebook (ERG)* provides a key to the four-digit identification numbers in the yellow-bordered section (see Chapter 24). It includes information about how to respond at incidents involving different types of hazardous materials. The four-digit identification number also appears on shipping papers, and it should match the numbers displayed on the exteriors of tanks or shipping containers.

It should be noted that common reference materials such as the *ERG* do not list all four-digit UN Identification numbers. For example, the *ERG* does not list any numbers below 1000. In the U.S., the entire list of UN Identification numbers is included in 49 *CFR* 172.101.

Other North American Highway Vehicle Identification Markings

Highway vehicles may have other identification markings, such as company names, logos, specific tank colors for certain tanks, stenciled commodity names (such as *Liquefied Petroleum Gas [LPG]*), and manufacturers' specification plates (**Figure 23.39, p. 1366**). Specification plates provide information about the standards to which the container/tank was built.

North American Railroad Tank Car Markings

Railroad tank car markings provide valuable information about the tank and its contents. These markings include:

● Initials (reporting marks) and number

● Capacity stencil

● Specification marking

Reporting marks, also called *railcar initials and numbers*, may be used to get information about the car's contents and construction from the railroad or the shipper. Tank cars are marked with their own unique set of reporting marks, which should

Emergency Response Guidebook (ERG) — Manual that identifies hazardous materials labels/placards; also outlines initial actions to be taken at haz mat incidents. Developed jointly by U.S., Mexican, and Canadian transportation agencies. Formerly known as *North American Emergency Response Guidebook (NAERG)*.

Reporting Marks — Combination of letters and numbers stenciled on rail tank cars that may be used to get information about the car's contents from the railroad or the shipper. *Also known as* Railcar Initials and Numbers.

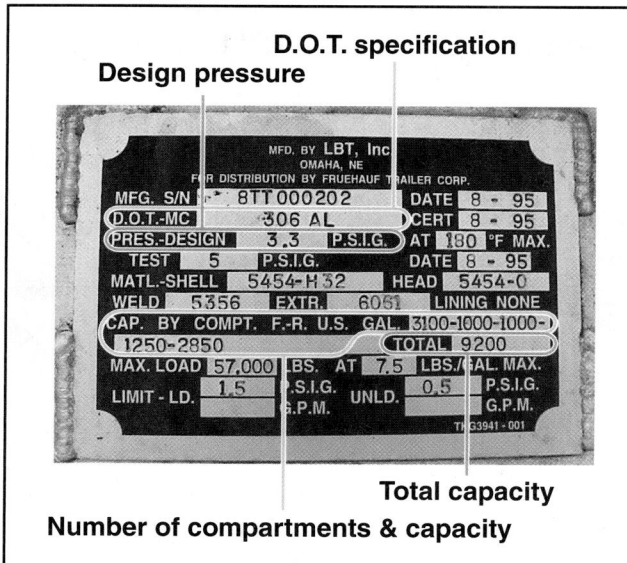

Figure 23.39
Manufacturers' specification plates provide a wealth of information about the standards to which the container/tank was built including pressure, capacity, number of compartments, and DOT specification. *Courtesy of Rich Mahaney.*

match the initials and numbers provided on the car's shipping papers. They are stenciled on all four sides of the tank: to the left on the longer sides, and in the upper center of the front and back (**Figures 23.40a and b**). Reporting marks are sometimes stenciled on the top of the car to help identify its contents in case an accident turns it on its side.

The *ERG* provides a key to these markings in the railcar identification chart, and more information is provided in sections that follow. Cars transporting hazardous materials sometimes have the name of the material stenciled on the sides of the car, in 4-inch (102 mm) letters; this is especially true for high-pressure tank cars. Dedicated railcars transporting a single material should have the name of that material painted on the car. Manufacturers' names on cars may also provide some contact information.

Figures 23.40a and b
Reporting marks can be used to identify the specific car in order to get information about the car's contents from the shipper or shipping papers. *Courtesy of Rich Mahaney.*

Figure 23.41 The capacity stencil can be seen on the end of the tank beneath the reporting marks. *Courtesy of Rich Mahaney.*

The **capacity stencil** shows the volume of the tank. The volume in gallons/liters is stenciled on both ends of the car under the car's initials and number. The volume in pounds/kilograms is stenciled on the sides of the car under the car's initials and number. The term *load limit* may be used to mean the same thing as *capacity*. A tank's water capacity (in pounds/kilograms) may also be stenciled on the sides of the tank, near the center of the car **(Figure 23.41)**.

The **specification marking** indicates the standards to which a tank car was built. It is stenciled on both sides of the tank, to the right (opposite from the initials and numbers) **(Figure 23.42)**. It is also stamped into the tank heads, where it is not readily visible. **Figure 23.43, p. 1368** provides a brief explanation of tank car specification markings.

> **Capacity Stencil** — Number stenciled on the exterior of a tank car to indicate the volume of the tank.

> **Specification Marking** — Stencil on the exterior of a tank car indicating the standards to which the tank car was built; may also be found on intermodal containers and cargo tank trucks.

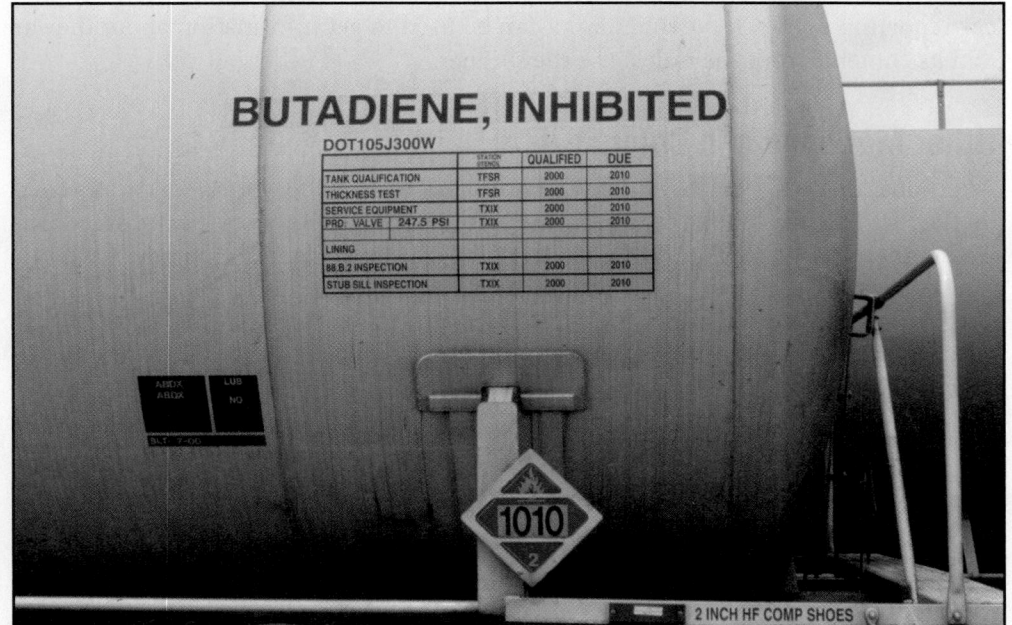

Figure 23.42 Specification markings are clearly visible on compliant tank cars. *Courtesy of Rich Mahaney.*

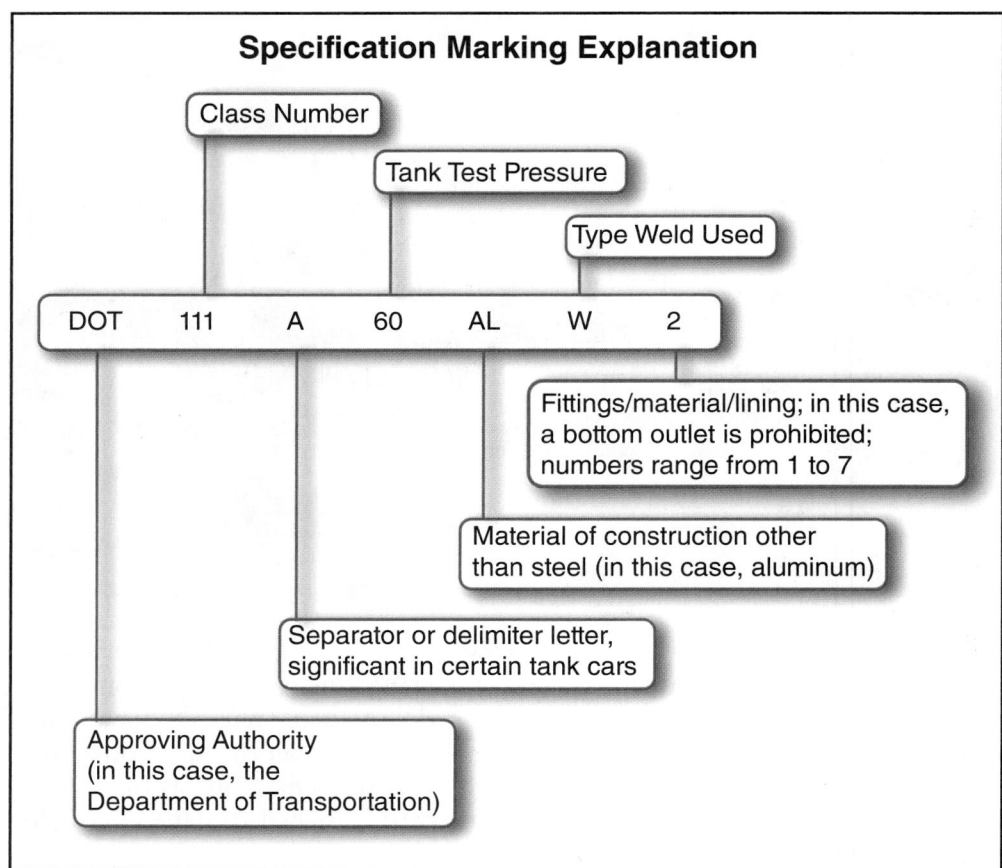

Specification Marking Explanation

Class Number

Tank Test Pressure

Type Weld Used

DOT   111   A   60   AL   W   2

Fittings/material/lining; in this case, a bottom outlet is prohibited; numbers range from 1 to 7

Material of construction other than steel (in this case, aluminum)

Separator or delimiter letter, significant in certain tank cars

Approving Authority (in this case, the Department of Transportation)

Figure 23.43 First responders are most likely to need the DOT class number.

International Intermodal Container/Tank Markings

Intermodal tanks and containers must also have reporting marks (initials and numbers) and DOT placards (**Figure 23.44**). They are generally found on all four sides of the car, to the right. They may be on the tank/container or the frame. As with tank car reporting marks, reporting marks can be used to get information about the car and its contents from the railway or the shipper.

Other Markings and Colors

Additional markings can be simple and informal, such as the word *chlorine* stenciled on the side of a fixed-facility tank. However, they can also be complex and standardized. Other widely used marking systems for hazardous materials include:

- NFPA® 704
- Hazardous communication labels
- Canadian Workplace Hazardous Materials Information System
- Manufacturers' labels and signal words
- Military markings
- Pipeline identifications
- Pesticide labels

Figure 23.44 Similar to tank car reporting marks, intermodal reporting marks identify the specific container. *Courtesy of Rich Mahaney.*

NFPA® 704 System

Local ordinances often require that hazardous materials be marked using the system described in NFPA® 704, *Standard System for the Identification of the Hazards of Materials for Emergency Response.* It is typically used in fixed-storage facilities. However, it is *not* designed for the following situations or hazards:

- Transportation
- General public use
- Nonemergency occupational exposures
- Explosives and blasting agents, including commercial explosive materials
- Chronic health hazards
- Biological agents and other similar hazards

NFPA® 704 Limitations

NFPA® markings indicate the type of hazard, but they do not specify the type of hazardous material or its quantity. Also, if the markings are used for a large area or structure as opposed to a specific container, they do not say where the hazardous material is located. Responders must identify the materials and their location through other means, such as container markings, company records, preincident surveys, and information from employees.

The NFPA® 704 system rates the severity of hazards on a scale from 0 to 4; the higher the number, the greater the hazard. Three numbers are assigned to represent flammability, instability, and health hazards. The numbers are displayed on a diamond-shaped marker or sign, with the health rating on the left, flammability rating in the center,

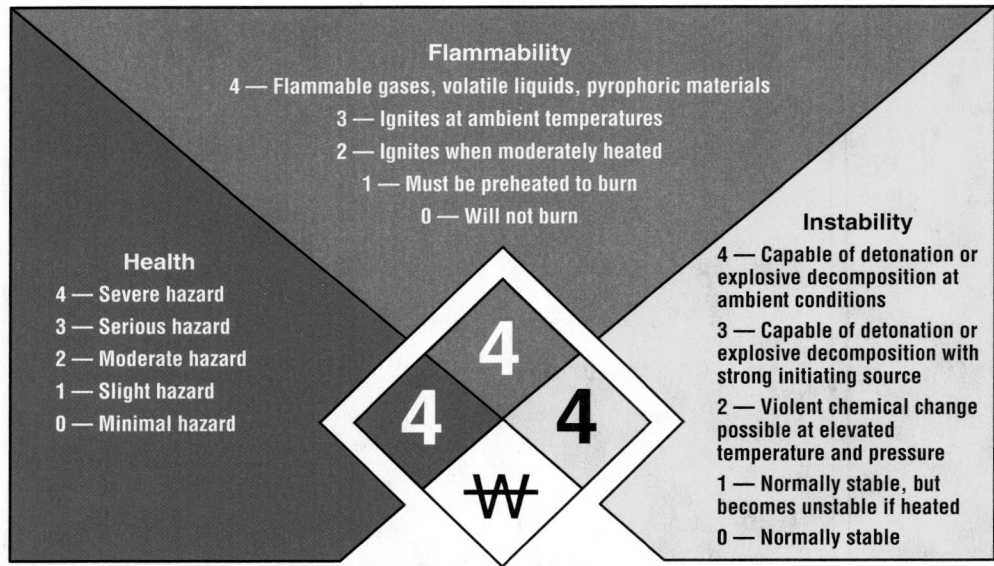

Figure 23.45 The NFPA® 704 hazard identification system is color and number coded.

Flammability
4 — Flammable gases, volatile liquids, pyrophoric materials
3 — Ignites at ambient temperatures
2 — Ignites when moderately heated
1 — Must be preheated to burn
0 — Will not burn

Health
4 — Severe hazard
3 — Serious hazard
2 — Moderate hazard
1 — Slight hazard
0 — Minimal hazard

Instability
4 — Capable of detonation or explosive decomposition at ambient conditions
3 — Capable of detonation or explosive decomposition with strong initiating source
2 — Violent chemical change possible at elevated temperature and pressure
1 — Normally stable, but becomes unstable if heated
0 — Normally stable

and instability rating on the right (**Figure 23.45**). The backgrounds are typically blue, red, and yellow respectively, but in some cases, other contrasting colors may be used.

Two special hazards may also be identified in the six o'clock position, on a white background: W̶ for water-reactive materials and *OX* for oxidizers. On older placards, other symbols may also be located in this position, including the trefoil radiation symbol. If more than one special hazard is present, multiple symbols may be present.

Hazard Communications Labels and Markings

According to OSHA's Hazard Communication Standard (HCS), all hazardous material containers in the workplace must be appropriately labeled. The type of hazardous material must be identified, along with appropriate hazard warnings. However, no particular marking system is specified, so you may encounter a variety of labeling and marking systems in your jurisdiction (**Figure 23.46**). Conducting preincident surveys should help you to locate, identify, and understand these systems.

Figure 23.46 First responders may encounter a variety of different identification systems used by employers in their response district.

Canadian Workplace Hazardous Materials Information System

In Canada, the Workplace Hazardous Materials Information System (WHMIS) requires **safety data sheets (SDSs)** for all hazardous chemicals and markings for all hazardous products. Markings are typically provided through supplier labels and workplace labels, which include information such as the product name, hazard class, or information about the supplier (**Figure 23.47**). **Table 23.18, p. 1372** shows the WHMIS symbols and hazard classes.

Safety Data Sheet (SDS) — Form provided by chemical manufacturers and blenders; contains information about chemical composition, physical and chemical properties, health and safety hazards, emergency response procedures, and waste disposal procedures. *Also known as* Material Safety Data Sheet (MSDS).

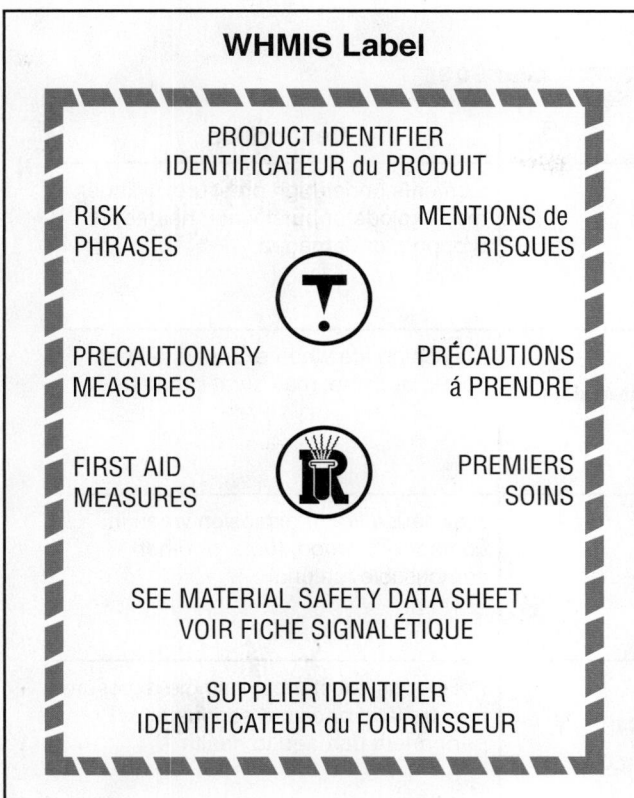

WHMIS Label

PRODUCT IDENTIFIER
IDENTIFICATEUR du PRODUIT

RISK PHRASES

MENTIONS de RISQUES

PRECAUTIONARY MEASURES

PRÉCAUTIONS á PRENDRE

FIRST AID MEASURES

PREMIERS SOINS

SEE MATERIAL SAFETY DATA SHEET
VOIR FICHE SIGNALÉTIQUE

SUPPLIER IDENTIFIER
IDENTIFICATEUR du FOURNISSEUR

Figure 23.47 Workplace Hazardous Materials Information System (WHMIS) labels must be completed on all controlled products in Canadian workplaces.

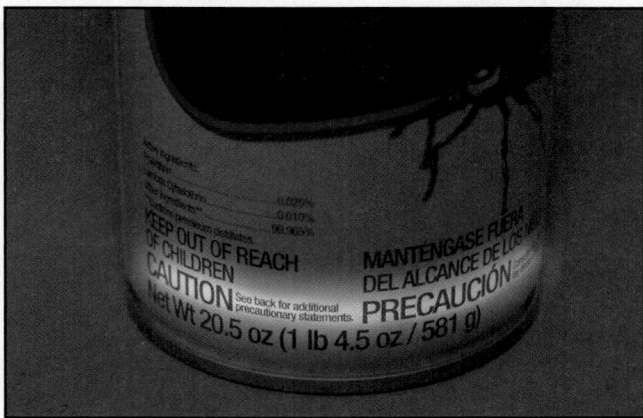

Figure 23.48 Signal words are used on the packaging of consumer products that are potentially hazardous.

Manufacturers' Labels and Signal Words

U.S. chemical manufacturers and importers must include labels on all product containers. Labels must include the name of the product, the manufacturer's contact information, and warnings of any associated hazards. These labels may also provide directions for use and handling, names of active ingredients, first aid instructions, and other pertinent information.

If a consumer product is potentially hazardous, its label must incorporate a **signal word** that indicates the degree of hazard (**Figure 23.48**):

- *CAUTION* — Minor health effects, such as eye or skin irritation
- *WARNING* — Moderate hazards, such as significant health effects or flammability
- *DANGER* — Highest degree of hazard, used for products with severe or deadly health effects, or products that explode when exposed to heat
- *POISON* — Required in addition to *DANGER* on the labels of highly toxic materials

Signal Word — Government-mandated warning (CAUTION, WARNING, DANGER, or POISON) on the labels of hazardous products.

Military Markings

The U.S. and Canadian militaries have their own marking systems for hazardous materials. These markings are required in fixed facilities, but optional on military vehicles. Exercise caution when dealing with the military placard system, as it is not

Table 23.18
WHMIS Symbols and Hazard Classes

Symbol	Hazard Class	Description
	Class A: **Compressed Gas**	Contents under high pressure; cylinder may explode or burst when heated, dropped, or damaged
	Class B: **Flammable and Combustible Material**	May catch fire when exposed to heat, spark, or flame; may burst into flames
	Class C: **Oxidizing Material**	May cause fire or explosion when in contact with wood, fuels, or other combustible material
	Class D, Division 1: **Poisonous and Infectious Material:** **Immediate and serious toxic effects**	Poisonous substance; a single exposure may be fatal or cause serious or permanent damage to health
	Class D, Division 2: **Poisonous and Infectious Material:** **Other toxic effects**	Poisonous substance; may cause irritation; repeated exposure may cause cancer, birth defects, or other permanent damage
	Class D, Division 3: **Poisonous and Infectious Material:** **Biohazardous infectious materials**	May cause disease or serious illness; drastic exposures may result in death
	Class E: **Corrosive Material**	Can cause burns to eyes, skin, or respiratory system
	Class F: **Dangerously Reactive Material**	May react violently, causing explosion, fire, or release of toxic gases when exposed to light, heat, vibration, or extreme temperatures

Source: WHMIS = Canadian Workplace Hazardous Materials Information System. Table adapted from Canadian Centre for Occupational Health and Safety (CCOHS) with pictograms from Health Canada.

entirely uniform. Some hazardous areas may also be unmarked for security reasons. **Table 23.19, p. 1374** provides the U.S. and Canadian military markings for explosive ordnance, fire hazards, chemical hazards, and PPE requirements.

CAUTION

When the military ships hazardous materials by common carrier, it is not required to use DOT and TC transportation markings.

Pipeline Identification

Petroleum products and other materials are transported across North America through pipelines, most of which are buried in the ground. When pipelines cross under (or over) roads, railroads, and waterways, pipeline companies must put up a marker to indicate that a pipeline runs through the area. However, the markers do not always show the pipeline's exact location. Pipeline markers in the U.S. and Canada include the signal words *CAUTION*, *WARNING*, or *DANGER* and contain information describing the pipeline's contents, along with contact information for the carrier **(Figure 23.49)**.

Pesticide Labels

The U.S. Environmental Protection Agency (EPA) regulates the manufacture and labeling of pesticides. Each EPA label must contain the manufacturer's name for the pesticide and one of the following signal words: *DANGER/POISON*, *WARNING*, or *CAUTION*. The words *DANGER/POISON* are used for highly toxic materials, *WARNING* means moderate toxicity, and *CAUTION* is used for chemicals with relatively low toxicity. The words *EXTREMELY FLAMMABLE* are also displayed on the label if the contents have a flash point below 80°F (27°C).

Figure 23.49 Pipeline markers include signal words, a description of the pipeline's contents, and contact information for the carrier. *Courtesy of Rich Mahaney.*

Table 23.19
U.S. and Canadian Military Symbols

Symbol	Fire (Ordnance) Divisions
1 (octagon)	**Division 1: Mass Explosion** Fire Division 1 indicates the greatest hazard. This division is equivalent to DOT/UN Class 1.1 Explosives Division **Also, this exact symbol may be used for:** **Division 5: Mass Explosion — very insensitive explosives (blasting agents)** This division is equivalent to DOT/UN Class 1.5 Explosives Division
2 (X shape)	**Division 2: Explosion with Fragment Hazard** This division is equivalent to DOT/UN Class 1.2 Explosives Division **Also, this exact symbol may be used for:** **Division 6: Nonmass Explosion — extremely insensitive ammunition** This division is equivalent to DOT/UN Class 1.6 Explosives Division
3 (inverted triangle)	**Division 3: Mass Fire** This division is equivalent to DOT/UN Class 1.3 Explosives Division
4 (diamond)	**Division 4: Moderate Fire — no blast** This division is equivalent to DOT/UN Class 1.4 Explosives Division

Symbol	Chemical Hazards
"Red You're Dead"	**Wear Full Protective Clothing (Set One)** Indicates the presence of highly toxic chemical agents that may cause death or serious damage to body functions.
"Yellow You're Mellow"	**Wear Full Protective Clothing (Set Two)** Indicates the presence of harassing agents (riot control agents and smokes).
"White is Bright"	**Wear Full Protective Clothing (Set Three)** Indicates the presence of white phosphorus and other spontaneously combustible material.

Continued

Table 23.19 (concluded)

Symbol	Chemical Hazards
	Wear Breathing Apparatus Indicates the presence of incendiary and readily flammable chemical agents that present an intense heat hazard. This hazard and sign may be present with any of the other fire or chemical hazards/symbols.
	Apply No Water Indicates a dangerous reaction will occur if water is used in an attempt to extinguish the fire. This symbol may be posted together with any of the other hazard symbols.

Symbol	Supplemental Chemical Hazards
G	**G-Type Nerve Agents** — persistent and nonpersistent nerve agents *Examples: sarin (GB), tabun (GA), soman (GD)*
VX	**VX Nerve Agents** — persistent and nonpersistent V-nerve agents *Example: V-agents (VE, VG, VS)*
BZ	**Incapacitating Nerve Agent** *Examples: lacrymatory agent (BBC), vomiting agent (DM)*
H	**H-Type Mustard Agent/Blister Agent** *Example: persistent mustard/lewisite mixture (HL)*
L	**Lewisite Blister Agent** *Examples: nonpersistent choking agent (PFIB), nonpersistent blood agent (SA)*

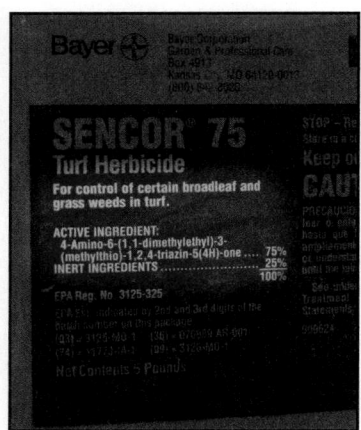

Figure 23.50 The active ingredients listed on a label indicate the chemical content of the product.

The label also lists an EPA registration number, which can be used to obtain information about the product from the manufacturer's 24-hour emergency contact. Another requirement is an establishment number that identifies the manufacturing facility. Other information that may be found on these labels includes:

- Routes of entry into the body
- Precautionary statements (such as *Keep out of the reach of children*)
- Active ingredients, which are listed by percentage (**Figure 23.50**).
- Requirements for storage and disposal
- First aid information
- Antidotes for poisoning (if known)
- Hazard statements indicating product poses an environmental hazard

Materials originating in Canada carry a Pest Control Products (PCP) Act number, which can be used to obtain information about the materials from the Canadian Transport Emergency Centre (CANUTEC), operated by TC. Canadian product labels also have the same signal words and required information as in the U.S.

Written Resources

A variety of written resources can help you identify hazardous materials at an incident. Fixed facilities should have safety data sheets, inventory records, and other facility documents, in addition to signs, markings, and other labels. At transportation incidents, you should be able to use the current *ERG* as well as shipping papers. If you need response information directly from the manufacturer or shipper, their contact information should be on shipping papers and safety data sheets. If not, you can get the response information by contacting an emergency response center such as CHEMTREC® (see Chapter 24, Emergency Response Centers).

Shipping Papers

Shipments of hazardous materials must be accompanied by shipping papers that describe them. The information can be provided on a **bill of lading**, waybill, or similar document. The general location and type of paperwork change according to the mode of transport (**Table 23.20**). Hazardous waste shipments must be accompanied by a Uniform Hazardous Waste Manifest document.

Bill of Lading — Shipping paper that describes the cargo, origin, destination, route; used in trucking and other industries, and typically placed in the cab of every truck tractor. Establishes the terms of a contract between a shipper and a carrier.

Table 23.20 Shipping Paper Identification			
Transportation Mode	**Shipping Paper Name**	**Location of Papers**	**Party Responsible**
Air	Air Bill	Cockpit	Pilot
Highway	Bill of Lading	Vehicle Cab	Driver
Rail	Trainlist/Consist	Engine or Caboose	Conductor
Water	Dangerous Cargo Manifest	Bridge or Pilot House	Captain or Master

If a close approach to an incident is safe, you can examine the cargo shipping papers. You may need to check with the person responsible for the shipment in order to locate these documents. If that person is not available, you will need to check some general locations. In trucks and airplanes, these papers are placed near the driver or pilot. In aircraft they can also be attached to the package itself. On ships and barges, the papers are called Dangerous Cargo Manifests, and they are placed on the ship's bridge or in the pilothouse of a controlling tugboat. **Figure 23.51, p. 1378** provides an example of shipping paper requirements.

Railroad train crews should have *train consists* that list the entire cargo; these documents are also known as *train lists* or *wheel reports*. If you cannot locate the crew, call the railroad's emergency phone number for a copy of the train consist. The engine may also have a copy. Most train consists will list the cars from the front of the train to the back. For more information about train shipping documents, contact the railroads that pass through your jurisdiction.

Railroad papers also use Standard Transportation Commodity Code numbers (STCC numbers) to identify chemicals. These are seven digit numbers; hazardous waste numbers start with 48, and hazardous material numbers start with 49. A guide to these numbers can be found in many hazardous materials reference sources.

Safety Data Sheets

A safety data sheet (SDS) contains detailed information about a product or chemical. It is prepared by the manufacturer or importer. U.S. and Canadian regulations previously required a material safety data sheet (MSDS), but both are switching to the Globally Harmonized System (GHS) format, which uses the SDS. Until the change is complete, responders may encounter MSDSs based on American National Standards Institute (ANSI) standards; MSDSs based on OSHA standards; MSDSs based on Canadian standards; and SDSs based on the new GHS standard.

Safety data sheets are often attached to shipping papers and containers. You can also acquire them from chemical manufacturers, suppliers, and shippers, facility hazard communication plans **(Figure 23.52, p. 1379)**, and emergency response centers such as CHEMTREC® and CANUTEC (see the Emergency Response Centers section of Chapter 24).

Under GHS guidelines, SDSs must include the following numerically identified sections:

1. Identification
2. Hazard(s) identification
3. Composition/information on ingredients
4. First aid measures
5. Fire fighting measures
6. Accidental release measures
7. Handling and storage
8. Exposure controls/personal protection
9. Physical and chemical properties
10. Stability and reactivity
11. Toxicological information
12. Ecological information

STRAIGHT BILL OF LADING
ORIGINAL NOT NEGOTIABLE

Shipper No. _____

Carrier No. _____

Date _____

Page _____ of _____

_____ (Name of carrier) _____ (SCAC)

TO:
Consignee _____

Street _____

City _____ State _____ Zip Code _____

FROM:
Shipper _____

Street _____

City _____ State _____ Zip Code _____

24 hr. Emergency Contact Tel. No. **800-555-1234**

Route _____

Vehicle Number _____

No. of Units & Container Type	HM	BASIC DESCRIPTION Proper Shipping Name, Hazard Class, Identification Number (UN or NA), Packing Group, per 172.101, 172.202, 172.203	TOTAL QUANTITY (Weight, Volume, Gallons, etc.)	WEIGHT (Subject to Correction)	RATE	CHARGES (For Carrier Use Only)
		Ⓘ Ⓢ Ⓗ Ⓟ				
4 Drums	X	UN1805, Phosphoric acid solution, 8, PGIII	4 gal			
		OR				
		Ⓢ Ⓗ Ⓘ Ⓟ				
4 Drums	X	Phosphoric acid solution, 8, UN1805, PGIII	4 gal			

Basic Description sequence and UN Harmonization.

The examples shown above are allowed in § 172.202(b). For international shipments, the **ISHP** sequence is mandatory January 1, 2007. Voluntary compliance for Domestic shipments begins January 1, 2007. Mandatory compliance for the **ISHP** sequence is January 1, 2013. This guide provides examples using the **ISHP** sequence.

PLACARDS TENDERED: YES ☐ NO ☐

Note – (1) Where the rate is dependent on value, shippers are required to state specifically in writing the agreed or declared value of the property, as follows: "The agreed or declared value of the property is hereby specifically stated by the shipper to be not exceeding _____ per _____.
(2) Where the applicable tariff provisions specify a limitation of the carrier's liability absent a release or a value declaration by the shipper and the shipper does not release the carrier's liability or declares a value, the carrier's liability shall be limited to the extent provided by such provisions. See section NMFC Item 172.
(3)Commodities requiring special or additional care or attention in handling or stowing must be so marked and packaged as to ensure safe transportation. See section 2(e) or sect. 360. Bills of Lading, Freight Bills and Statements of Charges and Section 1(a) of the Contract Terms and Conditions for a list of such articles.

This is to certify that the above named materials are properly classified, described, packaged, marked and labeled/placarded, and are in all respects in proper condition for transport according to applicable international and national governmental regulations.

_____ Signature

REMIT C.O.D. TO: ADDRESS

COD Amt: $ _____

C.O.D. FEE PREPAID ☐ COLLECT ☐ $

Subject to Section 7 of the conditions, if this shipment is to be delivered to the consignee without recourse on the consignor, the consignor shall sign the following statement:
The carrier shall not make delivery of this shipment without payment of freight and all other lawful charges.

_____ (Signature of Consignor)

TOTAL CHARGES: $ _____

FREIGHT CHARGES
FREIGHT PREPAID Check box if charges are to be collect. ☐
COLLECT

RECEIVED, subject to classifications and tariffs in effect on the date of the issue of this Bill of Lading, the property described above in apparent good order, except as noted (contents and condition of contents of packages unknown), marked consigned, and destined as indicated above which said carrier (the word carrier being understood throughout this contract as meaning any person or corporation in possession of the property under contract) agrees to carry to its usual place of delivery at said destination, if on its route, otherwise to deliver to another carrier on the route to said destination. It is mutually agreed as to each carrier of all or any of, said property over all or any portion of said route to des

tination and as to each party at any time interested in all or any said property, that every service to be performed hereunder shall be subject to all the bill of lading terms and conditions in the governing classification on the date of shipment.
Shipper hereby certifies that he is familiar with all the bill of lading terms and conditions in the governing classification and the said terms and conditions are hereby agreed to by the shipper and accepted for himself and his assigns.

SHIPPER _____

PER _____

CARRIER _____

PER _____

DATE _____

Permanent post-office address of shipper _____

PRINTED ON RECYCLED PAPER USING SOYBEAN INK   PRINTED WITH SOY INK

Figure 23.51 Cargo shipping paper requirements are illustrated on this straight Bill of Lading. Note the X placed in the column captioned HM to indicate the presence of a hazardous material.

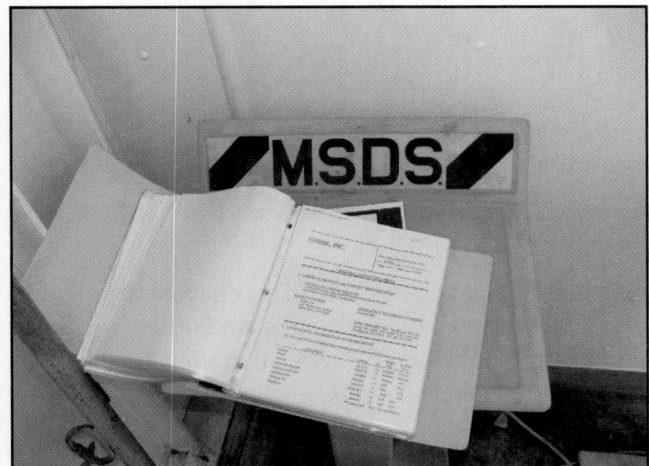

Figure 23.52 New safety data sheets (SDS) are based on a Globally Harmonized System (GHS), but the older style material safety data sheets (MSDS) may still be encountered in some workplaces.

Figure 23.53 When provided with the four-digit UN identification number, first responders can quickly identify a material using the *Emergency Response Guidebook* (*ERG*). *Courtesy of Rich Mahaney.*

13. Disposal considerations
14. Transport information
15. Regulatory information
16. Other information

Emergency Response Guidebook

The current *ERG* was developed jointly by TC, DOT, and the Secretaría de Comunicaciones y Transportes (SCT) of Mexico with the collaboration of Centro de Información Química para Emergencias [CIQUIME] of Argentina; for use by firefighters, law enforcement, and other emergency services personnel who may be the first to arrive at the scene of a transportation incident involving dangerous goods/hazardous materials **(Figure 23.53)**. The *ERG* is primarily a guide to aid emergency responders in quickly identifying the specific or generic hazards of materials involved in an emergency incident and protecting themselves and the general public during the initial response phase of the incident. For information on how to use the *ERG*, see Chapter 24.

The *ERG* does not address all possible circumstances that may be associated with a dangerous goods/hazardous materials incident. It is primarily designed for use at a dangerous goods/hazardous materials incident occurring on a highway or railroad. There may be limited value in its application at fixed-facility locations.

Firefighters at the scene of a haz mat incident should seek additional, specific information about any material in question as soon as possible. The information received by contacting the appropriate emergency response agency, calling the emergency response number on the shipping document, or consulting the information on or accompanying the shipping document may be more specific and accurate than the guidebook in providing guidance for the materials involved.

Senses

Do not rely on your sight, hearing, smell, taste, or touch to detect the presence of hazardous materials. This method is unreliable and extremely dangerous. Physical contact with a hazardous material or its mists, vapors, dusts, or fumes can be lethal.

> ⚠️ **WARNING!**
>
> Never use your senses to detect hazardous materials. This method is unreliable and extremely dangerous.

Vision is the safest of the senses, especially at a distance. This chapter has already discussed visual clues such as container shapes, placards, signs, and other markings. Other visual indicators are related to the physical or chemical actions/reactions of hazardous materials. These indicators include:

- Spreading vapor cloud or smoke (**Figure 23.54**)
- Unusual colored smoke
- Flames
- Changes in vegetation
- Container deterioration
- Containers bulging
- Victims who are physically ill
- Dead or dying birds, animals, insects, or fish
- Discolored valves or piping

Figure 23.54 Spreading smoke or a vapor cloud is a visual indicator that a chemical reaction or physical change is taking place.

Figure 23.55 A rainbow sheen on water surfaces is a physical indication that a hazardous material is present. *Courtesy of FEMA News Photos, photo by Liz Roll.*

Physical actions do not change the materials' elemental composition. For example, a liquefied material changes into a gas as it escapes from a vessel; the visual indicator of this process would be a white vapor cloud. Other indicators of physical actions include:

- Rainbow sheen on water surfaces (**Figure 23.55**)
- Wavy vapors over a volatile liquid
- Frost or ice buildup near a leak
- Containers deformed by the force of an accident
- Activated pressure-relief devices

Chemical reactions convert one substance to another. Visual evidence of chemical reactions include:

- Extraordinary fire conditions
- Peeling or discoloration of a container's finish
- Spattering or boiling of unheated materials
- Distinctively colored vapor clouds
- Smoking or self-igniting materials
- Unexpected deterioration of equipment
- Unexplained changes in ordinary materials
- Symptoms of chemical exposure exhibited by victims or responders

Your hearing can also provide clues, such as the hiss of a gas escaping a valve at high pressure. Heat-exposed vessels may also ping and pop, which indicate that a physical action is occurring.

Some hazards produce a noticeable smell, especially chemical reactions. However, to detect this you must be very close — so close that you are not safe. Most hazardous materials cannot be detected by smell, so odorants are often added to make them easier to detect. For example, natural gas is odorless, but the distinct odor we associate with natural gas is actually caused by an additive called mercaptan. Some toxins can also cause **olfactory fatigue**, which makes you unable to detect an odor soon after initial exposure.

Other hazards can be tasted. Some gases produce a pungent taste, and carbon dioxide (CO_2) tastes acidic. As with your sense of smell, if you are close enough to taste a hazard, you are too close to be safe. Always withdraw if you can taste (or smell) the hazard.

The most common clues from your sense of touch involve temperature changes. You may feel the heat of an exothermic chemical reaction, or the temperature drop caused by an endothermic reaction. You may also feel your gloves melting if they come in contact with a hazardous substance.

Other people's senses may also provide important clues. Take note of any smells, tastes, or symptoms reported by victims, witnesses, and other responders. Symptoms can occur separately or in clusters, depending on the chemical involved. Symptoms of chemical exposure include:

- **Changes in respiration**
  - Difficulty breathing
  - Increase or decrease in respiration rate
  - Tightness in the chest
  - Irritation of the nose and throat
  - Respiratory arrest

- **Changes in level of consciousness**
  - Dizziness
  - Lightheadedness
  - Drowsiness
  - Confusion
  - Fainting
  - Unconsciousness

- **Abdominal distress**
  - Nausea
  - Vomiting
  - Cramping

- **Change in activity level**
  - Fatigue
  - Weakness
  - Stupor
  - Hyperactivity
  - Restlessness
  - Anxiety
  - Giddiness
  - Faulty judgment

- **Visual disturbances**
  - Double vision
  - Blurred vision
  - Cloudy vision
  - Burning of the eyes
  - Dilated or constricted pupils
- **Skin changes**
  - Burning sensations
  - Reddening
  - Paleness
  - Fever
  - Chills
- **Changes in excretion or thirst**
  - Uncontrolled tears
  - Profuse sweating
  - Mucus flowing from the nose
  - Diarrhea
  - Frequent urination
  - Bloody stool
  - Intense thirst
- **Pain**
  - Headache
  - Muscle ache
  - Stomachache
  - Chest pain
  - Pain in areas that came in contact with the hazard

Monitoring and Detection Devices

Monitoring and detection devices can be effective, but only if they are in actual contact with the material (or its dusts, vapors, mists, or fumes). Devices such as combustible gas indicators and multigas meters can be used to determine the presence and concentration of hazardous materials. Specialized training is required to use these devices.

Terrorist Attacks and Illicit Laboratories

Terrorist attacks and **illicit clandestine laboratories** pose unique challenges for first responders. Although these incidents have much in common with other haz mat incidents, there are important differences. For example, the sites of both incidents are crime scenes, which means that law enforcement officers must be included in the initial response, and firefighters must preserve the crime scene after operations have concluded. Terrorist attacks may also involve far more hazardous materials than typical haz mat incidents, such as nerve agents and **toxic industrial materials (TIMs)**. This section describes the different types of terrorist attacks and illicit labs and provides guidance on how to recognize them.

Illicit Clandestine Laboratory — Laboratory that produces illegal or controlled substances, such as drugs, explosives, biological agents, or chemical warfare agents. *Also known as* Illegal Clandestine Laboratory.

Toxic Industrial Material (TIM) — Toxic chemical that is produced in large quantities (at least 30 million tons per year at a single production facility). They are intended for industrial use, but can be used by terrorists to cause deliberate harm. *Also known as* Toxic Industrial Chemical (TIC).

Terrorist Attacks

Most terrorist attacks deliberately target people, so compared to other haz mat incidents, the potential for civilian casualties is much higher. There is also greater risk from contaminated victims and structural collapse (**Figure 23.56**), and responders may face booby traps, armed resistance, and secondary devices. Therefore it is important to identify terrorist incidents as quickly as possible.

Circumstances that may lead you to consider the possibility of terrorism include:

- An unusually large number of people with similar symptoms reporting to emergency rooms or medical offices

- An explosion at a movie theater, government building, or location with historical or symbolic significance

- Two or more medical emergencies in a transportation hub, office building, or other public place (see the *Locations and Occupancies* section earlier in this chapter)

Some terrorist incidents are preceded by the threat of attack from terrorist groups, or a warning of attack from an intelligence agency. But this is not always the case, so responders must be aware of clues at the scene that may help to determine whether hazardous materials are present. Chemical, biological, radiological, nuclear, and explosive (CBRNE) attacks each have unique indicators, which are summarized in **Table 23.21** and explained in greater detail in the sections that follow.

Table 23.21 Terrorist Attacks at a Glance	
Chemical Attack	**Biological Attack**
• Victims in a concentrated area • Symptoms immediate (seconds to hours after exposure) • Symptoms very similar (SLUDGEM) • May have observable features such as chemical residue, dead foliage, dead animals/insects, and pungent odor	• Victims dispersed over a wide area • Symptons delayed (days to weeks after exposure) • Symptoms most likely vague and flu-like • No observable features
Explosive Attack	**Radiological Attack**
• Explosion self evident (debris field, fire, etc.) • Victims in a concentrated area • Mechanical and thermal injuries • Potential radiation and chemical agent risk — monitoring for both is necessary	• Explosion self evident (debris field, fire, etc.) • Victims in a concentrated area • Mechanical and thermal injuries initially, radiological symptoms (if any) will likely be delayed • Radiation detected through monitoring

Figure 23.56 Incidents created by terrorist attacks may present a higher degree and wider variety of risk to responders than more common types of emergencies. *Courtesy of FEMA News Photos, photo by Mike Rieger.*

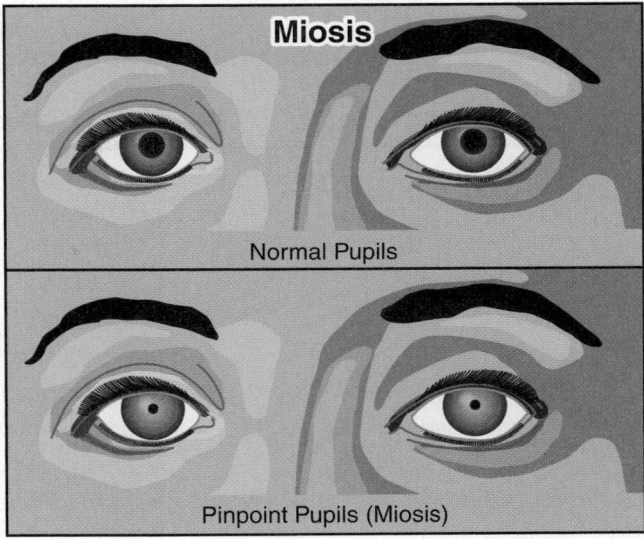

Figure 23.57 Exposure to chemical agents may cause miosis (pinpoint pupils).

Chemical Attack Indicators

Chemical attacks involve either toxic industrial materials or **chemical warfare agents (Table 23.22, p. 1386)**. People exposed to chemical attacks quickly develop symptoms, such as:

- Skin, eye, or airway irritation
- Nausea
- Vomiting
- Twitching
- Tightness in chest
- Sweating
- Pinpoint pupils (miosis) **(Figure 23.57)**
- Runny nose
- Disorientation
- Difficulty breathing
- Convulsions
- Blisters and/or rashes

If multiple victims report unexplained onset of these symptoms, they may have been exposed to a chemical attack. Other clues of a possible chemical attack include:

- Hazardous materials or lab equipment that seem inappropriate for the location
- Unexplained odors or tastes that seem inappropriate for the location
- Unexplained bomb or munitions-like material, especially if it contains a liquid
- Unexplained vapor clouds, mists, or plumes
- Unexplained casualties
- Casualties distributed downwind (outdoors) or near ventilation systems (indoors).
- Oily film on water surfaces, or oily droplets/films on other surfaces

Chemical Warfare Agent — Chemical substance intended for use in warfare or terrorist activity; designed to kill, injure, or incapacitate.

Table 23.22
Toxic Industrial Materials Listed by Hazard Index Ranking

High Hazard	Medium Hazard	Low Hazard
Ammonia	Acetonic cyanohydrin	Allyl isothiocyanate
Arsine	Acrolein	Arsenic trichloride
Boron trichloride	Acrylonitrile	Bromine
Boron trifluoride	Allyl alcohol	Bromine chloride
Carbon disulfide	Allylamine	Bromine pentafluoride
Chlorine	Allyl chlorocarbonate	Bromine trifluoride
Diborane	Boron tribromide	Carbonyl fluoride
Ethylene oxide	Carbon monoxide	Chlorine pentafluoride
Fluorine	Carbonyl sulfide	Chlorine trifluoride
Formaldehyde	Chloroacetone	Chloroacetaldehyde
Hydrogen bromide	Chloroacelonitrile	Chloroacetyl chloride
Hydrogen chloride	Chlorosulfonic acid	Crotonaldehyde
Hydrogen cyanide	Diketene	Cyanogen chloride
Hydrogen fluoride	1,2-Dimethylhydrazine	Dimethyl sulfate
Hydrogen sulfide	Ethylene dibromide	Diphenylmethane-4,4'-diisocyanate
Nitric acid, fuming	Hydrogen selenide	Ethyl chloroformate
Phosgene	Methanesulfonyl chloride	Ethyl chlorothioformate
Phosphorus trichloride	Methyl bromide	Ethyl phosphonothioic dichloride
Sulfur dioxide	Methyl chloroformate	Ethyl phosphonic dichloride
Sulfuric acid	Methyl chlorosilane	Ethyleneimine
Tungsten hexafluoride	Methyl hydrazine	Hexachlorocyclopentadiene
	Methyl isocyanate	Hydrogen iodide
	Methyl mercaptan	Iron pentacarbonyl
	Nitrogen dioxide	Isobutyl chloroformate
	Phosphine	Isopropyl chloroformate
	Phosphorus oxychloride	Isopropyl isocyanate
	Phosphorus pentafluoride	n-Butyl chloroformate
	Selenium hexafluoride	n-Butyl isocyanate
	Silicon tetrafluoride	Nitric oxide
	Stibine	n-Propyl chloroformate
	Sulfur trioxide	Parathion
	Sulfuryl chloride	Perchloromethyl mercaptan
	Sulfuryl fluoride	sec-Butyl chloroformate
	Tellurium hexafluoride	tert-Butyl isocyanate
	n-Octyl mercaptan	Tetraethyl lead
	Titanium tetrachloride	Tetraethyl pyrophosphate
	Trichloroacetyl chloride	Tetramethyl lead
	Trifluoroacetyl chloride	Toluene 2,4-diisocyanate
		Toluene 2,6-diisocyanate

Source: "Summary of the Final Report of the International Task Force 25: Hazard from Industrial Chemicals," April 15, 1999.

Figure 23.58 Chemical attacks kill people and animals indiscriminately. This photo was taken after a chemical attack in Halabja, Iraq. *Courtesy of Sayeed Janbozorgi, published under the Creative Commons Attribution ShareAlike 3.0 License, http://creativecommons.org/licenses/by-sa/3.0/.*

- Unusual security, locks, bars on windows, covered windows, and barbed wire enclosures

- Damaged or dead trees, crops, lawns, and other plant life

- Sick or dead birds, fish, or other animals **(Figure 23.58)**

- Unexplained patterns of sudden, similar, nontraumatic illnesses or deaths; patterns can relate to geography, employer, methods of dissemination

SLUDGEM or DUMBELS

These acronyms, *SLUDGEM* or *DUMBELS,* can be used to remember the symptoms of exposure to chemical warfare agents:

Salivation (drooling)

Lacrimation (tearing)

Urination

Defecation

Gastrointestinal upset/aggravation (cramping)

Emesis (vomiting)

Miosis (pinpointed pupils) and **M**uscular twitching/spasms

or

Defecation

Urination

Miosis and **M**uscular twitching/spasms

Bronchospasm (wheezing)

Emesis

Lacrimation

Salivation

Toxic industrial materials and chemicals used as chemical weapons may be identified through traditional methods, such as:

- Identification of occupancy types and locations
- Container shapes
- Hazardous materials placards, labels, and markings
- Written resources
- Sensory indicators
- Use of monitoring and detection devices

For example, if terrorists blow up a rail car, you will find the same placards you would find at a regular incident. If terrorists blow up a chlorine tanker, there will still be written resources, such as shipping papers.

Biological Attack Indicators

Biological attacks utilize viruses, bacteria, and/or biological toxins. The effects of biological attacks may not be readily noticeable. Signs and symptoms may take many days to develop (**Figure 23.59**). Biological attack indicators include:

- Diseases that do not normally occur in the geographic area, or unusual diseases (such as smallpox)
- Unusual number of sick or dying people or animals (often of multiple species)
- Multiple casualties with similar signs or symptoms
- Unusual spraying activity (**Figure 23.60**)
- Abandoned spray devices

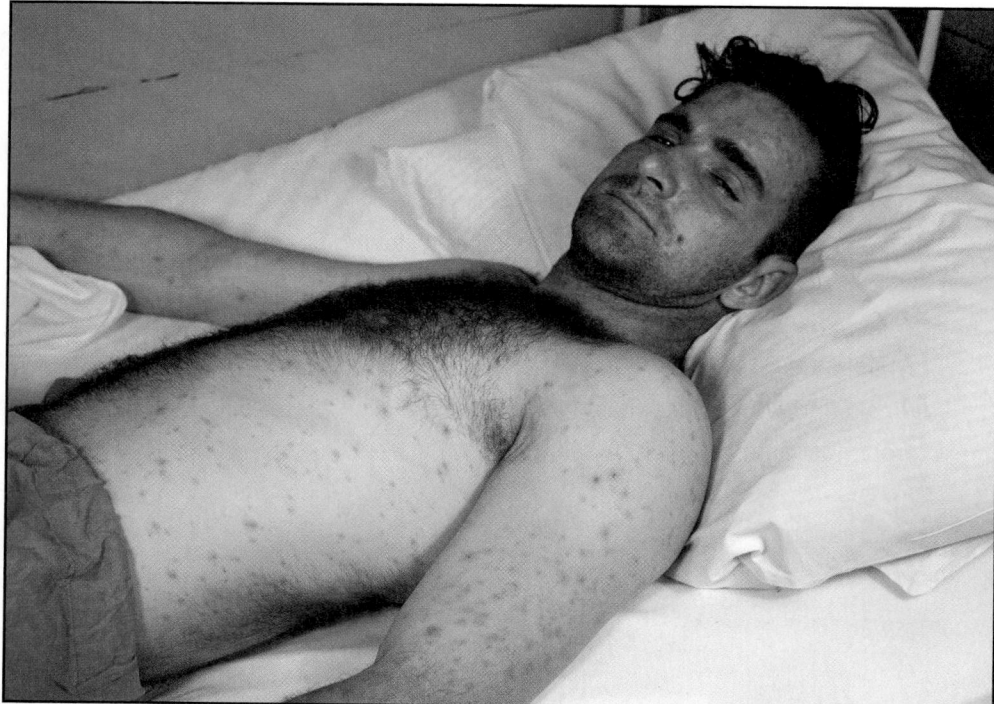

Figure 23.59 Symptoms of biological attack (such as smallpox) may take days to develop. *Courtesy of the CDC Public Health Image Library.*

- Casualty distribution that matches wind direction
- Illnesses associated with a common source of food, water, or location
- Large numbers of people with flu-like symptoms outside of flu season

Healthcare and emergency medical personnel may be the first to realize that there has been a biological attack. In some cases there may be reliable evidence, such as a witness to an attack or the discovery of a delivery system. If you suspect a biological attack, notify your local public health agency immediately.

NOTE: Most biological agents and toxins are classified as UN/DOT Class 6.2.

Radiological Attack Indicators

Radiological attacks use weapons that release radioactive materials, most likely in the form of dust or powder. Dispersal may be accomplished by including the material in an explosive device, referred to as a **radiological dispersal device (RDD)** or "dirty bomb." Radiological attack indicators include:

- People with symptoms of radiation exposure
- Radioactive materials packaging left unattended in a public location **(Figure 23.61)**
- Suspicious packages that weigh more than they should (may contain lead to shield a radiation source)
- Activation of radiation detection devices, with or without an explosion
- Material that is hot without any sign of an external heat source
- Glowing material

Nuclear Attack Indicators

A nuclear attack is the intentional detonation of a nuclear weapon. Its indicators include:

- Mushroom cloud **(Figure 23.62)**
- Exceptionally powerful explosion
- **Electromagnetic pulse (EMP)**

Figure 23.60 Unscheduled or unusual spray activity (for example, over a crowded metropolitan area) could be a sign of a biological attack. *Courtesy of the U.S. Department of Agriculture.*

Radiological Dispersal Device (RDD) — Conventional high explosives wrapped with radioactive materials; designed to spread radioactive contamination over a wide area. *Also known as* Dirty Bomb.

Electromagnetic Pulse (EMP) — Burst of electromagnetic energy produced by a nuclear explosion; EMPs damage electronic systems by causing voltage and current surges.

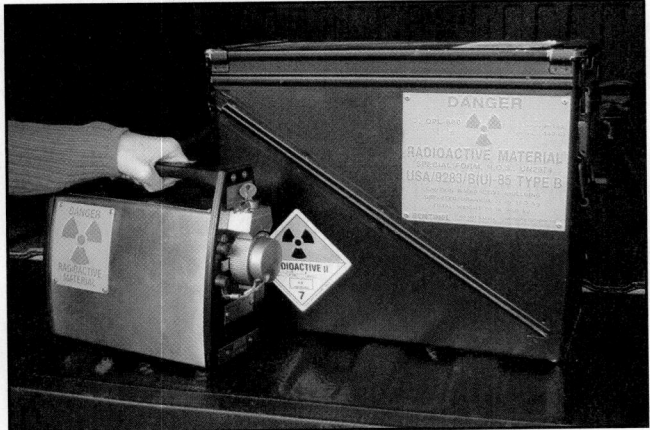

Figure 23.61 Unattended radiological packaging left in public places may be evidence of radiological attack. *Courtesy of Tom Clawson.*

Figure 23.62 A mushroom cloud is a highly visible indicator of nuclear attack. *Courtesy of the U.S. Department of Energy.*

Figure 23.63 The majority of terrorist attacks utilize weapons such as explosives and incendiary devices like the improvised explosive devices (IEDs) pictured here. *Courtesy of the U.S. Department of Defense.*

Explosive/Incendiary Attack Indicators

Most terrorist attacks involve explosive or incendiary devices (**Figure 23.63**). These are typically considered conventional attacks, but when used to inflict high casualties and large-scale damage, explosives may be classified as weapons of mass destruction (WMD). Explosives may also be used to disseminate chemical, biological, and radiological materials.

Explosive/incendiary attack indicators include:

- Accelerant odors
- Multiple fires or explosions
- Incendiary/explosive device components, such as wreckage from a car bomb or broken glass from a Molotov cocktail
- Fires that are unusually intense, hot, or fast-burning
- Unusually colored smoke or flames
- Propane or other flammable gas cylinders in unusual locations
- Unattended packages/backpacks/objects left in public areas
- Fragmentation damage or injury
- Unusually heavy structural damage, such as shattered reinforced concrete or bent structural steel (**Figure 23.64**)
- Crater (**Figure 23.65**)
- Scattering of small metal objects such as nuts, bolts, and/or nails used as shrapnel

Illicit Laboratories

Illicit clandestine laboratories produce illegal or controlled substances, such as drugs, explosives, biological agents, and chemical warfare agents. They can be found almost anywhere, including abandoned buildings, hotel rooms, rural farms, urban apartments, rental storage units, or upscale residential neighborhoods. Illicit **methamphetamine (meth) labs** can be so portable that they have even been found in campgrounds, highway rest stops, and vehicles (**Figure 23.66**).

Many of the products used in clandestine labs are toxic, explosive, or highly flammable. Responders may also face booby traps or armed resistance when responding to emergencies at illicit labs. If an illicit lab is discovered or suspected, contact law enforcement authorities immediately. Exercise extreme caution and withdraw if it is appropriate.

Meth Lab — Illicit clandestine laboratory that produces methamphetamine (meth).

Figure 23.64 Car and truck bombs can cause greater damage than accidental gas explosions. Indicators include shattered reinforced concrete and bent structural steel. *Courtesy of U.S. Air Force, photo by Senior Airman Sean Worrell.*

Figure 23.65 Craters often indicate the use of explosives. *Courtesy of the U.S. Department of Defense.*

Figure 23.66 Meth labs may be as portable as a small box. *Courtesy of MSA.*

Drug Labs

Approximately 80 to 90 percent of illicit drug labs produce methamphetamine (METH). However, they may also make other illegal drugs, such as:

- Ecstasy
- Cocaine
- Phenyl-2-propanone (P2P)
- Phencyclidine (PCP)
- Heroin
- Lysergic acid diethylamide (LSD or LSD-25)
- Amphetamines

Meth is easy to make and uses commercially available ingredients. However, many U.S. states restrict the purchase of these ingredients. Making meth is also known as *cooking*, and many different methods exist, such as the *Red P, Nazi/Birch, Shake and Bake,* and *One Pot* methods **(Figure 23.67, p. 1392)**.

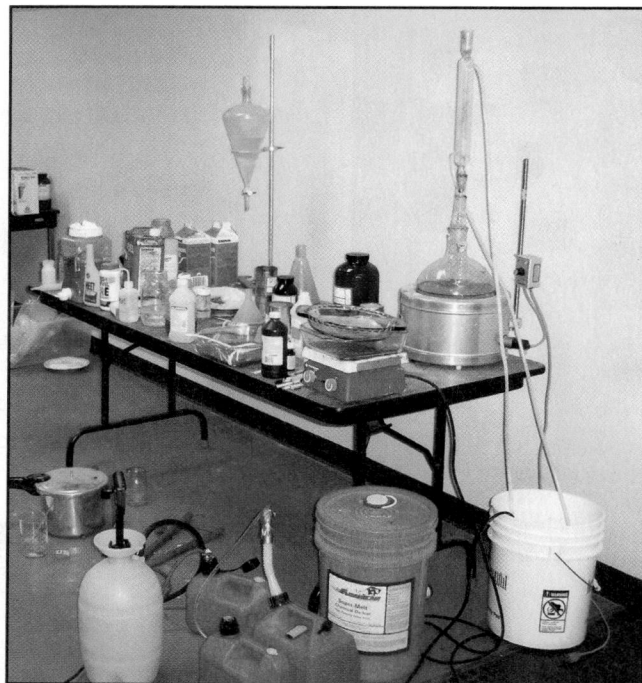

Figure 23.67 A lab set up to make *Red P* meth has some unique components compared to other meth labs. *Courtesy of MSA.*

Phosphine — Colorless, flammable, and toxic gas with an odor of garlic or decaying fish; ignites spontaneously on contact with air. Phosphine is a respiratory tract irritant that attacks the cardiovascular and respiratory systems, causing pulmonary edema, peripheral vascular collapse, and cardiac arrest and failure.

Choking Agent — Chemical warfare agent that attacks the lungs, causing tissue damage.

Meth labs are hazardous to the meth cook, the community surrounding the lab, and emergency responders who discover it. Flammability is the most common hazard, and many labs are only discovered after a fire or explosion. Many ingredients and byproducts are oxidizers, and others are corrosive or toxic. For example, **phosphine** gas is sometimes classified as a chemical warfare **choking agent**. Meth lab locations pose serious health and environmental hazards for many years, unless they are properly decontaminated. A summary of the products commonly used in making meth and the hazards associated with them is provided in **Table 23.23**.

///////////////////////////////////

CAUTION
Fire suppression at illicit labs can be difficult because chemicals such as sodium and lithium are highly water-reactive.

Equipment used in meth labs includes (**Figures 23.68a and b, p. 1396**):

- **Condenser tubes** — Used to cool vapors produced during cooking
- **Filters** — Coffee filters, cloth, and cheesecloth
- **Funnels/turkey basters** — Used to separate layers of liquids
- **Gas containers** — Propane bottles, fire extinguishers, self-contained underwater breathing apparatus (SCUBA) tanks, plastic drink bottles (often attached to some sort of tubing) (**Figure 23.69, p. 1396**)
- **Glassware** — Particularly Pyrex® or Visions® cookware, Mason jars, and other laboratory glassware that can tolerate heating and violent chemical reactions

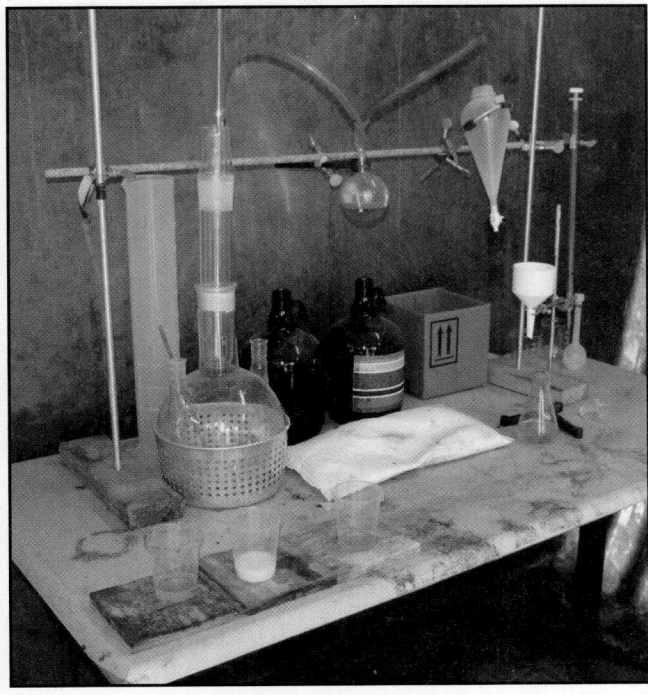

Figure 23.70 Sophisticated lab equipment in an informal environment may be a clue to a chemical agent lab.

- Discoloration of structures, pavement, and soil
- Strong odor of solvents
- Smell of ammonia, starting fluid, or ether
- Iodine- or chemical-stained fixtures in the kitchen or bathroom

For every pound (0.5 kg) of meth produced, approximately 6 pounds (2.7 kg) of hazardous waste is generated. Meth producers typically dump waste with residential trash, flush it into the sewage system, or leave it beside the roadway, on a vacant property, or in nearby bodies of water. First responders and other public agencies typically dispose of the waste, a process that is both expensive and very dangerous.

Chemical Labs

Some chemical warfare agents can be made in illicit laboratories. Recipes for many chemical agents are easy to find, but access to some materials is restricted. Indicators of a chemical lab include:

- Sophisticated lab equipment (**Figure 23.70**)
- Military manuals
- Underground "cookbooks"
- Cyanides or acids
- Chemicals that are not used to make illegal drugs, such as **organophosphate pesticides**
- Chemicals such as methyl iodide and phosphorus trichloride, which are used to make **sarin**

Organophosphate Pesticides — Chemicals that kill insects by disrupting their central nervous systems; because they have the same effect on humans, they are sometimes used in terrorist attacks.

Sarin (GB) — Fluorinated phosphinate that attacks the central nervous system; classified as a chemical warfare agent.

Figure 23.71 Recipes for making explosives are readily available to the public. *Courtesy of August Vernon.*

Explosives Labs

Explosive labs are the second most common type of lab discovered, after drug labs. They are sometimes mistaken for drug labs because both lab types use household chemicals and some explosive materials can be mistaken for narcotics. Explosive labs can be discovered anywhere because they require little equipment. Recipes are easy to find on the Internet and in anarchist literature (**Figure 23.71**). Common explosive materials such as black powder and smokeless powder can be easily incorporated into an improvised explosive device (IED).

Labs that make chemical explosives may look like industrial or university chemistry labs, and labs that make peroxide-based explosives often resemble meth or drug labs (see information box). However, some labs do not need to heat or cook their materials, so they do not have traditional lab equipment. For example, a garage workshop that is used to manufacture pyrotechnics will not have glassware, tubing, Bunsen burners, and chemical bottles.

Homemade/Improvised Explosive Materials

Most of the explosives firefighters encounter are homemade or improvised. Improvised explosives are made by combining fuel with an oxidizer (**Figure 23.72**). Most do not require technical expertise or specialized equipment, but the materials are often highly unstable.

Peroxide-based explosives such as acetone peroxide (triacetonetriperoxide or TATP) and hexamethylene triperoxide diamine (HMTD) have been used in several recent terrorist attacks, including the 2005 mass transit bombings in London. They are made by mixing concentrated hydrogen peroxide, acetone, and either hydrochloric or sulfuric acid. Recipes are easily accessible, and materials are cheap and easy to purchase commercially. TATP and HMTD are very unstable, both during manufacturing and when handling the finished product. TATP is typically a crystalline powder with a distinct acrid smell. Its color can range from white to yellowish (**Figures 23.73a - c, p. 1400**).

Potassium chlorate is another white crystalline powder with approximately 83 percent of the power of TNT. It is a common ingredient in some fireworks and can be purchased in bulk from fireworks/chemical supply houses. It is also used in printing, dying, steel, weed killers, matches and the explosive industry.

Urea nitrate is a fertilizer-based explosive composed of nitric acid and urea. Sulfuric acid is sometimes added as a catalyst. Urea is used in sidewalk de-icers and can also be derived from urine. Urea nitrate has a destructive power similar to ammonium nitrate.

Components of Improvised Explosives

🔥 Potential Fuels   +   🔥 Potential Oxidizers   =   💥 Explosive Blends (Oxidizer + Fuel)

Hydrocarbons:
Alcohol
Carbon Black
Charcoal
Dextrin
Diesel
Ethylene Glycol
Gas
Kerosene
Naphtha
Rosin
Sawdust
Shellac
Sugar
Vaseline
Wax/Parfin

Energetic Hydrocarbons:
Nitrobenzene
Nitromethane
Nitrocellulose

Elemental "Hot" Fuels:
Powdered Metals
- Aluminum
- Magnesium
- Zirconium
- Copper
Phosphorus
Sulfur
Antimony Trisulfide

Oxidizers:
Perchlorate
Chlorate
Hypochlorite
Nitrate
Peroxide
Iodate
Chromate
Dichromate
Permaganate
Sodium Chlorate
Potassium Chlorate
Ammonium Nitrate
Potassium Nitrate
Hydrogen Peroxide
Barium Peroxide
Ammonium Perchlorate
Calcium Hypochlorite
Nitric Acid
Lead Iodate
Sodium Chlorate
Potassium Permanganate
Lithium Chromate
Potassium Dichromate

Nitrate Blends:
ANFO (Ammonium Nitrate + Diesel Fuel)

ANAl (Ammonium Nitrate + Aluminum Powder)

ANS (Ammonium Nitrate + Sulfur Powder

ANIS (Ammonium Nitrate + Icing Sugar)

Black Powder (Potassium Nitrate + Charcoal + Sulfur)

Chlorate/Perchlorate Blends:
Flash Powder (Potassium Chlorate/Perchlorate + Aluminum Powder + Magnesium Powder + Sulfur)

Poor Man's C-4 (Potassium Chlorate + Vaseline)

Armstrong's Mixture (Potassium Chlorate + Red Phosphorus)

Liquid Blend:
Hellhoffite (Nitric Acid + Nitrobenzene)

Common Precursors Used To Make Explosives

💥 Precursors:
Hydrogen Peroxide
Sulfuric Acid (battery acid)
Nitric Acid
Hydrochloric Acid (muriatic acid)
Urea
Acetone
Methyl Ethyl Ketone
Alcohol (Ethyl or Methyl)
Ethylene Glycol (antifreeze)
Glycerin(e)
Hexamine (camp stove tablets)
Citric Acid (sour salt)

💥 Nitrated Explosives:
Nitroglycerine (Glycerine + Mixed Acid [Nitric Acid + Sulfuric Acid])

Ethylene Glycol Dinitrate (EGDN) (Ethylene Glycol + Mixed Acid [Nitric Acid + Sulfuric Acid])

Methyl Nitrate (Methyl Alcohol [methanol] + Mixed Acid [Nitric Acid + Sulfuric Acid])

Urea Nitrate (Urea + Nitric Acid)

Nitrocotton (Gun Cotton) (Cotton + Mixed Acid [Nitric Acid + Sulfuric Acid])

Peroxide Explosives:
Triacetone Triperoxide (TATP) (Acetone + Hydrogen Peroxide + Strong Acid [Sulfuric, Nitric, or Hydrochloric])

Hexamethylene Triperoxide Diamine (HMDT) (Hexamine + Hydrogen Peroxide + Citric Acid)

Methyl Ethyl Ketone Peroxide (MEKP) (Methyl Ethyl Ketone + Hydrogen Peroxide + Strong Acid [Sulfuric, Nitric, or Hydrochloric])

Figure 23.72 Most homemade explosives are made by combining an oxidizer with a fuel.

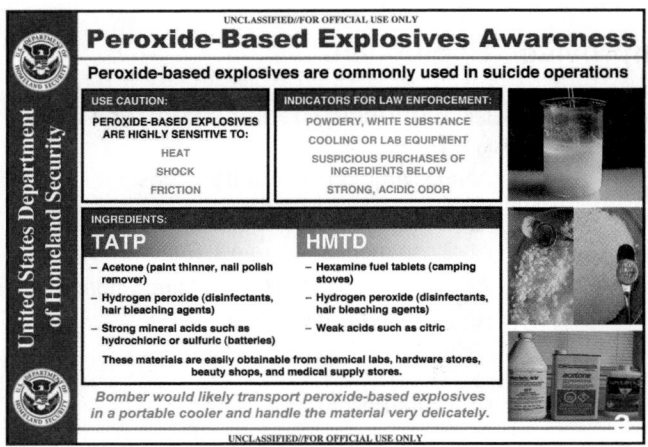

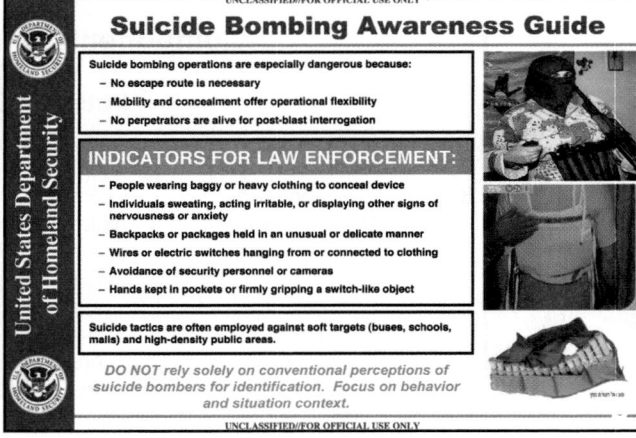

Figures 23.73a - c
Peroxide-based explosives and the labs that make them are a valid concern for emergency responders. *(a) Courtesy of the U.S. Department of Homeland Security. (b and c) Courtesy of TATP.*

Some improvised explosive materials do require a production lab, but the lab equipment can be purchased in drug and hardware stores **(Figure 23.74)**. Peroxide-based explosives are particularly likely to require these facilities because they are extremely sensitive to heat, shock, and friction.

Explosives lab indicators include:

- Refrigerators/coolers/ice baths
- Glassware and laboratory equipment
- Blenders
- Blasting caps/batteries/fuses/switches
- Pipes/end caps/storage containers
- Shrapnel-type materials
- Strong acidic odors
- Explosives, military ordnance
- Bomb-making literature
- Large quantities of matches, flares, or fireworks
- Ammunition, such as shotgun shells

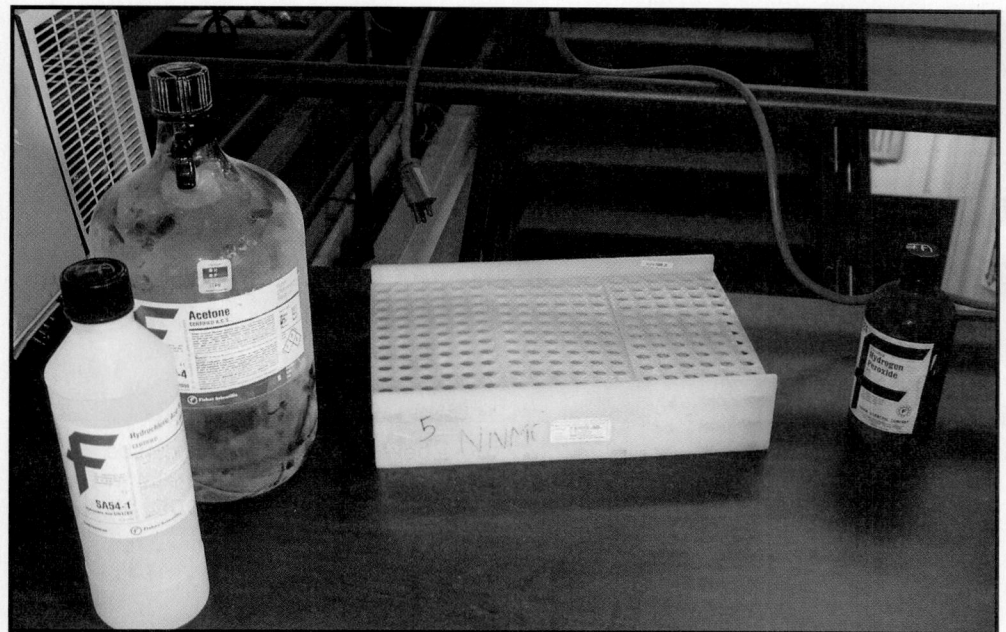

Figure 23.74 Peroxide explosives require a controlled lab environment as they are sensitive to heat, shock, and friction.

Figure 23.75 The presence of commercial explosives, fireworks, large quantities of black powder, and/or other incendiary materials should raise suspicion of an explosives lab. *Courtesy of the U.S. Bureau of Alcohol, Tobacco, Firearms and Explosives* and the *Oklahoma Highway Patrol Bomb Squad.*

- Black powder

- Smokeless powder

- Commercial explosives **(Figure 23.75)**

- Incendiary materials

- Electronic components that could be used in an IED, such as wires, circuit boards, and cell phones

- Containers that could be used to house an IED, such as empty fire extinguishers and propane containers

    A peroxide-based explosives lab is also likely to contain:

- Acetone

- Ethanol

- Hexamine (solid fuel for camp stoves)

- Hydrogen peroxide

- Strong or weak acids such as sulfuric or citric acids

Exercise extreme caution inside any clandestine lab. In explosives labs, the primary hazard you will face is unstable explosive materials. Be especially careful with materials that are being cooled. Some materials must be kept cold to remain stable, and raw materials are typically transported on ice, in a cooler. Mishandled materials can be deadly.

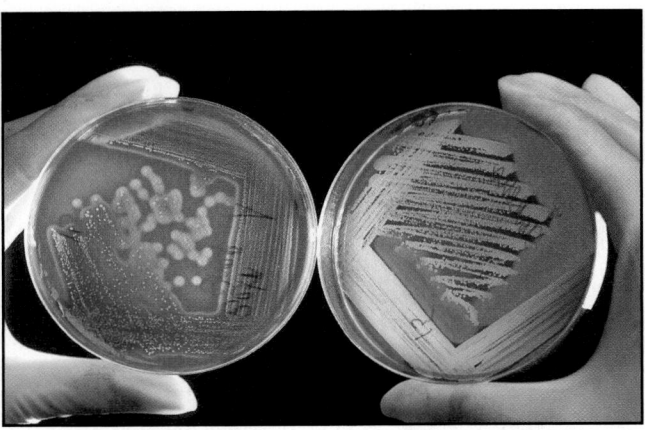

Figure 23.76 Petri dishes and agar plates are used to grow biological cultures. *Courtesy of the National Cancer Institute, photo by Bill Branson.*

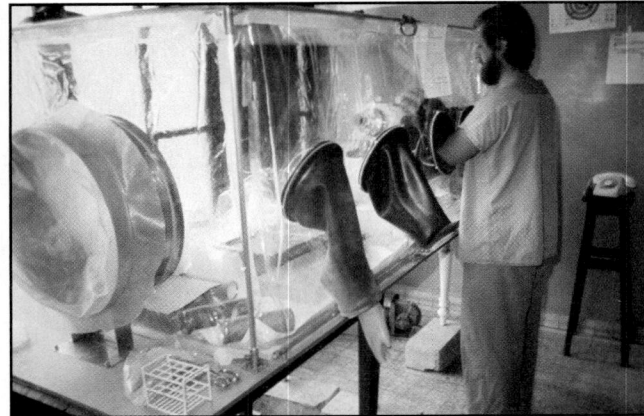

Figure 23.77 Glove boxes such as this one can be improvised with Plexiglas, plastic sheeting, duct tape, and other common materials. *Courtesy of the CDC, photo by Joel G. Breman, M.D., D.T.P.H.*

Biological Labs

Biological labs use specialized materials and equipment that are very different from chemical labs. Indicators of a biological lab include:

- Microscopes
- Antibiotics and vaccines
- Personal protective equipment such as masks, rubber gloves, and respirators (particularly with HEPA filters)
- Laboratory animals or related materials, such as cages and food
- Viral/bacterial cultures, growth containers (Petri dishes, agar plates), and growth mediums (agar, meat broth, gelatin) **(Figure 23.76)**
- Biological materials known to be the source of toxins, such as castor beans
- **Glove boxes**, biological safety cabinets, or improvised setups using plastic sheeting, Plexiglas®, duct tape, and fans **(Figure 23.77)**
- Incubators, refrigerators, and bench top fermenters **(Figure 23.78)**
- Bleach and other sterilization supplies, such as antiseptics, pressure cookers, or **autoclaves**
- Alterations to building ventilation systems
- Instruction manuals and other printed materials relating to biological agents
- Sprayers, **nebulizers**, or other delivery devices

Secondary Devices and Booby Traps

Secondary attacks target responders and bystanders who are already at the scene of an incident. They may be used to create chaos, or to divert responders away from the primary attack area. **Secondary devices** may be used during terrorist attacks or criminal events, or at illicit labs. They are typically improvised explosives.

Glove Box — Sealed container that allows scientists to safely manipulate hazardous microorganisms. Two glove ports are built into the side of the box, with gloves extending inside; the user is therefore able to perform tasks inside the box without breaking the seal.

Autoclave — Device that uses high-pressure steam to sterilize objects.

Nebulizer — Electrically powered machine that turns liquid medication into a mist.

Secondary Device — Bomb or other weapon that targets responders and bystanders who are already at the scene of an incident.

Booby traps are often set to protect illicit laboratories (**Figure 23.79, p. 1404**). Traps utilizing other weapons are also possible, including the use of chemical, biological, or radiological materials, and even animals such as snakes or guard dogs.

Secondary devices and booby traps are typically detonated using a timer or a radio-controlled or cell phone-activated device. In some cases, an obvious IED may be used to lure personnel to a specific area where a less obvious IED is hidden.

Secondary devices and booby traps will be hidden or camouflaged. They can be disguised as almost anything, but responders should look for things that may seem out of place. Be very cautious of any item at an incident that arouses your curiosity, including the following:

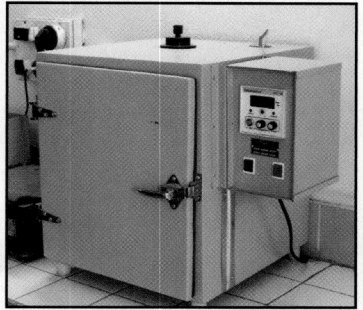

Figure 23.78 Bacteriological incubators indicate that biological lab work may be occurring nearby.

- Containers with unknown liquids or materials

- Unusual devices or containers with electronic components such as wires, circuit boards, cellular phones, antennas, and other items attached or exposed

- Devices containing:
  — Fuses
  — Fireworks
  — Match heads
  — Black powder
  — Smokeless powder
  — Incendiary materials
  — Other unusual materials

- Materials attached to or surrounding an item that could be used for shrapnel; for example, nails, bolts, drill bits, and marbles (**Figure 23.80, p. 1405**)

- Ordnance such as:
  — Blasting caps
  — Detonating cord
  — Military explosives
  — Commercial explosives
  — Grenades

Guidelines for protecting against possible secondary devices include the following:

- Anticipate the presence of a secondary device at any suspicious incident.

- Perform a visual search for suspicious items before moving into the incident area. Limit the number of nonessential personnel allowed on scene until the search is complete.

- Never touch or move any item that may conceal an explosive device, such as a backpack or purse.

- Establish scene security and scene control zones.

- Evacuate victims and nonessential personnel as quickly as possible.

Booby traps can be inside or outside illicit labs, and may include any of the following:

- Explosives, such as grenades and dynamite

- Wires attached to explosives or alerting devices

- Weapons tied to doors

Booby Traps

Figure 23.79 Illicit labs may be guarded by a wide variety of dangerous booby traps.

- Bottles that will break, thereby mixing chemicals to produce toxic fumes
- On/off switches that have been reversed
- Holes in floors (trap doors to snake pits)
- Electrified door handles
- Exposed wiring
- Spikes
- Hooks
- Acid

Always maintain your situational awareness and avoid complacency. Other guidelines for avoiding booby traps include:

- Use explosion-proof equipment
- Take aerial reconnaissance photographs before entering
- Do not touch or move loose items
- Check doors and openings for wires and/or traps

Turning lab equipment on or off may also be a trigger for booby traps. Also, leave electrical pumps (such as those used in cooling baths) turned on to prevent overheating, which can ignite unstable or combustible materials.

If you find anything suspicious at an incident scene or illicit lab, treat the item with caution, evacuate the area immediately, and notify law enforcement personnel. Only bomb squad/Explosive Ordnance Disposal (EOD) personnel should attempt to neutralize secondary devices and booby traps.

NOTE: Preserve the scene for evidence collection and crime investigation.

Figure 23.80 This replica of an actual IED that detonated in Israel shows the materials that were added to the device to increase shrapnel injuries.

Chapter Summary

Hazardous materials can be present at any incident scene, and you must be aware of the unique challenges they pose. You must be knowledgeable about the types of hazards they present and how they can enter the body. You must be able to recognize identifying placards, labels, and signs, and distinguish between different types of containers. You must be able to predict their behavior based on an understanding of their physical properties. You must also know how to recognize and respond to incidents involving terrorist attacks and illicit labs.

1. What are the main causes of hazardous materials incidents?

2. What do the four types of ionizing radiation have in common; how do they differ?

3. What are the common categories for chemical hazards?

4. What types of mechanical hazards are created by explosions?

5. How do hazardous materials enter the body?

6. How do the nine physical properties of hazardous materials help predict the behavior of the material?

7. What are the six stages of the General Emergency Behavior Model (GEBMO)?

8. What types of occupancies and locations are most likely to have hazardous materials?

9. How can container types and shapes give clues about the presence of hazardous materials/ weapons of mass destruction?

10. What types of transportation, placards, labels, and markings are used when transporting hazardous materials?

11. What are three common written resources used to identify hazardous materials?

12. How are the senses, as well as monitoring and detection equipment, used to safely detect the presence of hazardous materials?

13. What are the five categories of terrorist attack indicators?

14. What are the indicators a responder should look for when trying to identify the difference between a drug lab and an explosives lab?

15. What types of devices may be commonly used as booby traps or for a secondary attack?

Chapter Contents

Key Terms

NFPA® Core Competencies

This chapter provides information that addresses the following requirements of NFPA® 472: *Standard for Competence of Responders to Hazardous Materials/Weapons of Mass Destruction Incidents.*

4.2.1	5.2.1	5.2.3	5.3.3	5.4.3	6.2.3.1	6.6.3.2
4.2.3	5.2.1.4	5.2.4	5.3.4	5.4.4	6.2.4.1	6.6.4.1
4.4.1	5.2.1.5	5.3.1	5.4.1	5.5.1	6.2.5.1	
4.4.2	5.2.2	5.3.2	5.4.2	5.5.2	6.6.3.1	

1. Summarize first responder roles at haz mat/WMD incidents. [NFPA® 472, 4.4.1, 5.4.3]

2. Summarize incident priorities for haz mat/WMD incidents.

3. Explain the management structure used for haz mat/WMD incidents. [NFPA® 472, 4.2.1, 4.4.1, 5.2.2, 5.4.3]

4. Explain the considerations that must be taken into account during the analysis stage of haz mat/WMD incidents. [NFPA® 472, 5.2.1, 5.2.1.4, 5.2.4, 5.3.1, 5.4.3]

5. Describe the steps used for planning the appropriate response at haz mat/WMD incidents. [NFPA® 472, 5.3.1, 5.3.2]

6. Describe the process for evaluating and communicating the progress at haz mat/WMD events. [NFPA® 472, 5.5.1, 5.5.2]

7. Explain how the *Emergency Response Guidebook (ERG)* is used at haz mat/WMD incidents. [NFPA® 472, 4.4.1, 4.2.3, 5.2.1.5]

8. Summarize the role of emergency response centers during haz mat/WMD incidents. [NFPA® 472, 5.2.2]

9. Explain the considerations that must be taken when choosing personal protective equipment at haz mat/WMD incidents. [NFPA® 472, 5.3.3, 6.2.3.1]

10. Distinguish among the four levels of EPA defined protection. [NFPA® 472, 6.2.3.1]

11. Describe Mission-Oriented Protective Posture (MOPP) ensembles. [NFPA® 472, 6.2.3.1]

12. Describe the selection factors that must be considered when selecting personal protective equipment at haz mat/WMD incidents. [NFPA® 472, 6.2.3.1, 6.6.3.2]

13. Explain safety and emergency procedures used for personnel wearing protective clothing. [NFPA® 472, 6.2.4.1]

14. Explain proper procedures for PPE inspection, storage, testing, and maintenance. [NFPA® 472, 5.4.4, 6.2.4.1, 6.2.5.1]

15. Describe the techniques used for isolation and scene control. [NFPA® 472, 4.4.1]

16. Identify basic notification considerations at haz mat/WMD incidents. [NFPA® 472, 4.4.2, 5.2.2, 5.4.3]

17. Describe methods that help ensure the protection of responders during haz mat/WMD incidents. [NFPA® 472, 6.2.4.1, 5.4.4, 5.2.4, 5.5.1, 5.5.2, 5.4.3]

18. Describe methods that help ensure the protection of the public during haz mat/WMD incidents. [NFPA® 472, 5.4.1, 4.4.1]

19. Describe the considerations and limitations of emergency and technical decontamination. [NFPA® 472, 5.3.4, 5.2.3, 5.4.1, 5.3.2, 6.2.3.1, 6.2.4.1]

20. Tell what rescue actions can be taken at haz mat/WMD incidents by personnel without specialized training. [NFPA® 472, 5.3.1]

21. Explain the strategic goal of spill control and confinement. [NFPA® 472, 6.6.3.1]

22. Describe methods used to complete the strategic goal of leak control and containment. [NFPA® 472, 6.6.3.1, 6.6.4.1]

23. Summarize the actions necessary when an incident is suspected to involve terrorist activity. [NFPA® 472, 4.4.1]

24. Explain how to preserve crime scene evidence. [NFPA® 472, 5.4.2]

25. Explain the goals for the recovery and termination phases of haz mat/WMD incidents.

26. Obtain information about a hazardous material using the Emergency Response Guidebook (ERG). [Skill Sheet 24-I-1, NFPA® 472, 4.4.1, 4.2.3]

27. Perform emergency decontamination. [Skill Sheet 24-I-2, NFPA® 472, 5.3.4]

28. Perform absorption. [Skill Sheet 24-I-3, NFPA® 472, 6.6.3.1]

29. Perform absorption. [Skill Sheet 24-I-4, NFPA® 472, 6.6.3.1]

30. Perform diking operations. [Skill Sheet 24-I-5 , NFPA® 472, 6.6.3.1]

31. Perform damming operations. [Skill Sheet 24-I-6 , NFPA® 472, 6.6.3.1]

32. Perform diversion operations. [Skill Sheet 24-I-7, NFPA® 472, 6.6.3.1]

33. Perform retention operations. [Skill Sheet 24-I-8, NFPA® 472, 6.6.3.1;]

34. Perform dilution operations. [Skill Sheet 24-I-9 , NFPA® 472, 6.6.3.1]

35. Perform vapor dispersion. [Skill Sheet 24-I-10, NFPA® 472, 6.6.3.1]

36. Perform a remote valve shutoff. [Skill Sheet 24-I-11, NFPA® 472, 6.6.3.1, 6.6.4.1]

Chapter 24
Mitigating Haz Mat/ WMD Incidents

Courtesy of the U.S. Coast Guard, photo by PA3 Brent Erb.

Case History

On January 6, 2005, two trains collided in Graniteville, South Carolina. Sixty tons of chlorine gas were released after a tank ruptured during the crash. Nine people were killed, 250 were treated for chlorine exposure, and over 5,000 residents nearby were evacuated. Rescue and decontamination took 14 days and involved collaboration between local, state, and federal emergency response crews **(Figures 24.1a and b)**.

Firefighters arrived on the scene three minutes after receiving the first 9-1-1 emergency call. Within minutes, the senior officer on scene had difficulty breathing, and other response teams were asked to stand by until a better size-up could be conducted. Haz mat teams were requested 3 minutes after the first responders arrived on scene. One minute later, a Reverse 9-1-1 Emergency Notification was initiated, advising residents to shelter indoors. A Command center was quickly established near the incident, with a 1-mile (1.6 km) perimeter. Due to light winds in the area, the escaping gas extended at least 2,500 feet (762 m) to the north, 1,000 feet (305 m) to the east, 900 feet (274 m) to the south, and 1,000 feet (305 m) to the west.

For the next several hours, crews evacuated nearby residents. The crews were drawn from nearby jurisdictions because the closest fire station was only 100 yards (91 m) from the accident site, so all its personnel and equipment had been contaminated. After assembling a small fleet of privately-owned pickup trucks, entry teams wearing PPE rode through nearby neighborhoods for several hours, transporting anyone they found to four decontamination sites.

Figures 24.1a and b A chlorine leak caused by a train derailment in Graniteville, South Carolina, required collaboration from all levels of emergency response to stabilize and decontaminate the scene. a) and b) *Courtesy of the U.S. EPA.*

Meanwhile, haz mat response teams began to stabilize and decontaminate both the wreckage and the surrounding area. This process took nine days to complete.

Post incident reports emphasized two lessons to be learned from this episode. The first is that response crews should not rush in without proper PPE. Doing so creates extra risk to fire personnel, and the response teams who are there to help can become victims themselves without appropriate protection. The second lesson is to establish a Unified Command early. In a large incident with multiple crews working together, it is vital that rescue and decontamination efforts be coordinated in order to achieve the greatest level of safety for rescue crews and those in the affected area.

Figure 24.2 Successful incident mitigation requires the skillful use of incident management elements. *Courtesy of Dennis Walus, www.detroitfiregroundimages.com.*

In many ways, haz mat incidents are similar to other emergencies to which firefighters respond. Successful incident mitigation is achieved through incident management elements such as setting priorities, implementing an incident management structure, and following a problem-solving process that includes emphasis on firefighter safety from beginning to end **(Figure 24.2)**. However, if these incident management elements are not in place or are used improperly, a haz mat incident can rapidly spiral out of control. ***Mistakes made in the initial response to the incident can mean the difference between solving the problem and becoming part of it.*** For this reason, you must understand your role and mission at haz mat incidents.

In order to effectively mitigate a hazardous materials/WMD incident, Incident Commanders use a problem-solving process that emphasizes life safety, incident stabilization, and protection of the environment and property. This chapter addresses some of the strategic goals and tactical objectives commonly used at haz mat incidents as part of the problem-solving process. Strategic goals are broad statements of what must be done to resolve an incident. Tactical objectives are specific operations that must be done in order to accomplish those goals.

This chapter will discuss:

- First responder roles
- Priorities
- Management structure
- Haz mat incident mitigation
- Using the *Emergency Response Guidebook (ERG)*
- Personal protective clothing
- Isolation and scene control
- Notification
- Protection
- Decontamination
- Rescue
- Spill control and confinement
- Crime scene preservation
- Recovery and termination

First Responder Roles

The United States (U.S.) **Occupational Safety and Health Administration (OSHA)** and the U.S. **Environmental Protection Agency (EPA)** require that responders to hazardous materials incidents meet specific training standards. The OSHA versions of these legislative mandates are outlined in paragraph (q) of Title 29 (Labor) *Code of Federal Regulations (CFR)* 1910.120, **Hazardous Waste Operations and Emergency Response (HAZWOPER)**. The training requirements found in 29 *CFR* 1910.120 are included by reference in the EPA regulations in Title 40 (Protection of Environment) *CFR* 311, *Worker Protection*. This EPA regulation provides protection to those responders not covered by an OSHA-approved State Occupational Health and Safety Plan.

In addition to U.S. Government regulations, the National Fire Protection Association® (NFPA®) has several standards that apply to personnel who respond to hazardous materials emergencies. The requirements in these standards are recommendations, not laws or regulations, unless they are adopted as such by the authority having jurisdiction (AHJ). However, because they are a national standard, they can be used as a basis for *accepted practice*. The NFPA®'s hazardous materials requirements are detailed in the following standards:

- NFPA® 472, *Standard for Competence of Responders to Hazardous Materials/ Weapons of Mass Destruction Incidents* (2013)

- NFPA® 473, *Standard for Competencies for EMS Personnel Responding to Hazardous Materials/Weapons of Mass Destruction Incidents* (2013)

In Canada, the Ministry of Labour (in most provinces) or the Workers Compensation Board (WCB) in British Columbia are the regulatory bodies governing response to haz mat incidents and the training requirements for first responders. These provincial bodies also require employers to provide standard operating procedures (SOPs) or standard operating guidelines (SOGs) to protect their employees. Canadian firefighters and most emergency responders are trained to the same NFPA® standards as their U.S. counterparts. While Canada does not have the definitive equivalent of OSHA 29 *CFR* 1910.120, the minimum acceptable level of training for first responders is NFPA® 472.

Occupational Safety and Health Administration (OSHA) — Division of the U.S. Department of Labor (DOL) that enforces occupational safety regulations.

Environmental Protection Agency (EPA) — U.S. government agency that creates and enforces laws designed to protect the air, water, and soil from contamination; responsible for researching and setting national standards for a variety of environmental programs.

Hazardous Waste Operations and Emergency Response (HAZWOPER) — U.S. regulations in Title 29 (Labor) *CFR* 1910.120 for cleanup operations involving hazardous substances and emergency response operations for releases of hazardous substances.

What This Means To You

If you are a first responder to haz mat incidents in the U.S., by law your employer must meet the requirements set forth in the HAZWOPER regulation (29 *CFR* 1910.120). If your AHJ has formally adopted the applicable NFPA® standards as law, your employer is required to meet them as well. If you belong to a volunteer fire and emergency services organization, you will also have to meet these regulations. Under 40 *CFR* 311, volunteers are considered employees.

If you are a first responder to haz mat incidents in Canada, your employer must provide you with standard operating procedures (or standard operating guidelines) and the training required by your province. If you are a firefighter, you must be trained in accordance with the requirements in NFPA® 472.

NFPA® 472 and the OSHA regulations in 29 *CFR* 1910.120 identify two levels of training: **Awareness** and **Operations**. These are detailed in following sections.

Awareness-Level Personnel

Personnel trained and certified to the Awareness Level are individuals who, in the course of their normal duties, may be the first to arrive at or witness a haz mat/WMD incident. To summarize both OSHA and NFPA® requirements, individuals trained to the Awareness Level are expected to assume the following responsibilities when faced with an incident involving hazardous materials:

• Recognize the presence or potential presence of a hazardous material.

• Recognize the type of container at a site and identify the material in it if possible.

• Transmit information to an appropriate authority and call for appropriate assistance.

• Identify actions to protect themselves and others from hazards.

• Establish scene control by isolating the hazardous area and denying entry.

Operations-Level Responders

Responders who are trained and certified to the Operations Level respond to releases (or potential releases) of hazardous materials as part of their normal duties. Operations-Level responders are expected to protect individuals, the environment, and property from the effects of the release in a defensive manner (**Figure 24.3**).

Figure 24.3 Operations-Level responders perform some tasks that directly affect a hazardous materials incident. *Courtesy of Danny Atchley.*

Responsibilities of the Operations-Level responder include the Awareness-Level responsibilities. The Operations Level adds confining a release in a defensive fashion from a safe distance. To summarize both OSHA and NFPA® requirements, first responders at the Operational Level must be able to perform the following actions:

- Identify the hazardous material(s) involved in an incident if possible.

- Analyze an incident to determine the nature and extent of the problem.

- Protect themselves, nearby persons, the environment, and property from the effects of a release.

- Develop a defensive plan of action to address the problems presented by the incident (plan a response).

- Implement the planned response to mitigate or control a release from a safe distance (initiate defensive actions to lessen the harmful incident) and keep it from spreading.

- Evaluate the progress of the actions taken to ensure that response objectives are safely met.

Per NFPA® 472, Operations-Level responders may be trained to a set of core competencies (Operations Core) or beyond, incorporating mission-specific competencies (Operations-Mission-Specific) for actions personnel may be trained to perform at haz mat/WMD incidents. The mission-specific competencies identified in NFPA® 472 are:

- Personal Protective Equipment

- Mass Decontamination

- Technical Decontamination

- Evidence Preservation and Sampling

- Product Control

- Air Monitoring and Sampling

- Victim Rescue and Recovery

- Response to Illicit Laboratory Incidents

Priorities

There are three incident priorities for all haz mat and terrorist incidents, and all decisions must be made with these priorities in mind:

- Life safety

- Incident stabilization

- Protection of property and the environment

The first priority is the safety of emergency responders. If responders do not protect themselves first, they cannot protect the public. A dead, injured, or unexpectedly contaminated firefighter becomes part of the problem, not the solution. For this reason, with the exception of certain situations involving flammable and combustible liquids, firefighters are limited to defensive or nonintervention actions at haz mat/WMD incidents unless they receive additional training. They should not come in contact with the hazardous materials involved. Hazardous materials response teams and personnel with additional training are needed to conduct offensive operations.

Dam/Dike — Temporary or permanent barrier that contains or directs the flow of liquids.

Divert — Actions to control movement of a hazardous material to an area that will produce less harm.

If there is no immediate threat to either responders or civilians, the next consideration is incident stabilization. Stabilizing actions can include establishing scene perimeters to prevent the spread of contamination and constructing **dams** and **dikes** to prevent or **divert** hazardous liquids from entering streams or water supplies. Stabilizing the incident can also minimize environmental and property damage.

At haz mat incidents, property conservation includes both protection of property and the environment. Hazardous materials can seriously damage the environment, and cleanup from an incident can be very expensive. The hazards presented by these materials can continue to cause problems for a long time, so environmental protection must be a consideration from the beginning of the mitigation process.

Management Structure

At any incident, firefighters must operate within the standard Incident Command System. However, incident management at haz mat/WMD incidents may differ from other incidents in the following ways:

- A haz mat division/group led by a Haz Mat Supervisor may be required.
- A Unified Command structure is often used because of the complexity of the response.
- The National Incident Management System-Incident Command System (NIMS-ICS) must be used at any U.S. incident involving terrorism or a federal response.
- Federal responses may incorporate federal resource teams.

In addition to the Incident Command System, firefighters must also operate in accordance with predetermined procedures such as their agency's SOPs and their local emergency response plan (LERP). In the U.S., per 29 *CFR* 1910.120, *Hazardous Waste Operations and Emergency Response* (HAZWOPER), all fire and emergency services organizations that respond to haz mat incidents are required to have an emergency response plan. This plan includes predetermined guidelines or procedures for managing incidents involving terrorist attacks and hazardous materials releases.

ICS Haz Mat Positions

The ICS organization will be determined by the complexity of the incident and the resources needed for mitigation. A hazardous material branch or group may be added to the ICS structure when resources are needed to specifically address the problems created by the hazardous material(s) **(Figure 24.4)**. For example, if a hazardous materials response team responds to mitigate a spill or leak, they would be assigned to the Haz Mat Branch or Group.

Haz Mat Branch or Group positions are at the Technician Level, but Awareness- and Operations-Level responders should be familiar with them. Standard ICS positions particular to hazardous materials incidents include:

- **Hazardous Materials Branch Director/Group Supervisor** — Manages the resources assigned to the Branch or Group and directs the primary tactical functions (called the Haz Mat Group Supervisor hereafter).
- **Entry Team Leader** — Supervises all companies and personnel operating in the hazardous area, with the responsibility to direct all tactics and control the positions and functions of all personnel in the hazardous area.
- **Decontamination Team Leader** — Supervises operations in the scene control zone where decontamination is conducted and ensures that all rescued citizens, response personnel, and equipment have been decontaminated before leaving the incident.

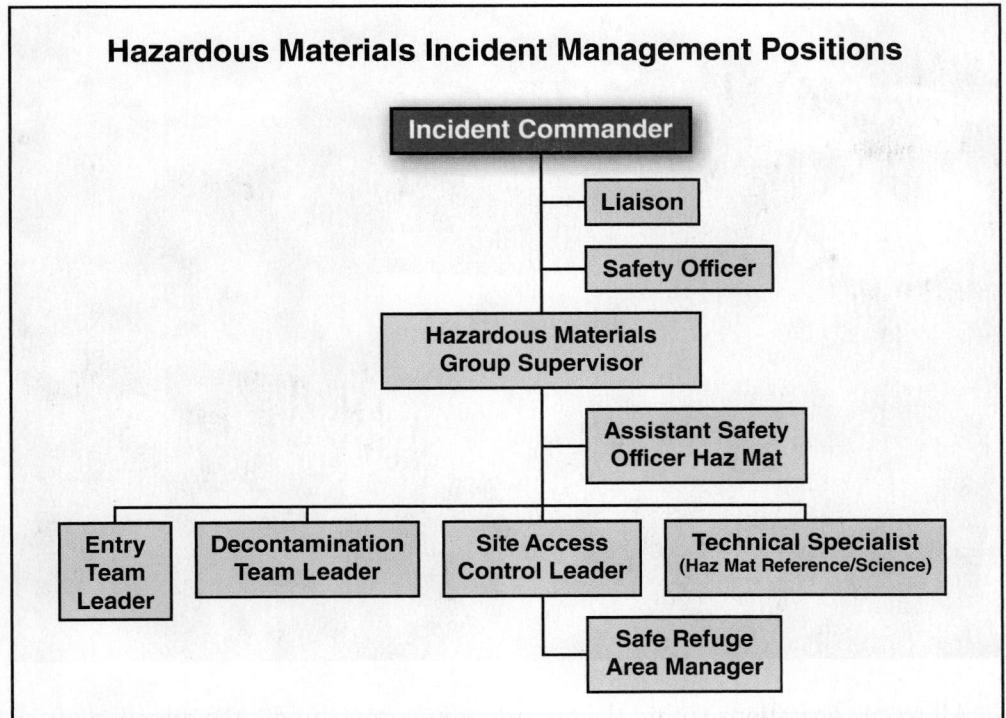

Hazardous Materials Incident Management Positions

- Incident Commander
  - Liaison
  - Safety Officer
  - Hazardous Materials Group Supervisor
    - Assistant Safety Officer Haz Mat
      - Entry Team Leader
      - Decontamination Team Leader
      - Site Access Control Leader
      - Technical Specialist (Haz Mat Reference/Science)
      - Safe Refuge Area Manager

Figure 24.4 At incidents involving hazardous materials, the Incident Management System IMS structure may include a dedicated haz mat branch or group.

- **Site Access Control Leader** — Controls all movement of personnel and equipment between the control zones and is responsible for isolating the control zones and ensuring proper routes; also has the responsibility for the control, care, and movement of people before they are decontaminated; may appoint a Safe Refuge Area Manager.

- **Assistant Safety Officer (Hazardous Materials)** — Is responsible for the overall safety of assigned personnel within the Hazardous Materials Group and reports directly to the Incident Safety Officer; must be appointed at hazardous materials incidents and have the requisite knowledge to function as the Assistant Safety Officer at a haz mat incident.

- **Technical Specialist (Hazardous Materials Reference/Science Technical Specialist)** — Is responsible for providing technical information and assistance to the Hazardous Materials Group and the Planning Section using various sources such as computer databases, technical journals, public and private technical information agencies, facility representatives, and product specialists.

- **Safe Refuge Area Manager** — Is responsible for evaluating and prioritizing victims for treatment, collecting information from the victims, and preventing the spread of contamination by these victims; also it is recommended that this person have an EMS background.

The functional positions of the Hazardous Materials Branch/Group (Entry Team Leader, Decontamination Team Leader, and Site Access Control Leader) require a high degree of control and close supervision. The Haz Mat Group Supervisor manages the functional responsibilities, which includes all tactical operations carried out in the hazardous area.

Figure 24.5 Unified Command is accomplished through the establishment of an easily identifiable Command Post. *Courtesy of Ron Jeffers.*

All rescue operations within the hazardous area come under the direction of the Haz Mat Group Supervisor. In addition to the primary functions, the Haz Mat Group Supervisor works with an Assistant Safety Officer who is trained in hazardous materials and must be present at the hazardous site. The Haz Mat Group Supervisor may also supervise one or more Technical Specialists. Evacuation and all other tactical objectives that are outside the scene control zones are not responsibilities of the Haz Mat Group Supervisor. These tactical operations as well as many other hazardous materials related functions are managed by the Operations Section.

Unified Command

Control of a large-scale incident involving multiple agencies with overlapping authority and responsibility is accomplished through the use of **Unified Command** (**Figure 24.5**). The concept of Unified Command means that all agencies that have a jurisdictional responsibility at a multijurisdictional, multiagency, and/or multidisciplinary incident contribute to the process by taking the following actions:

- Establish one set of incident objectives.
- Select strategies.
- Accomplish joint planning.
- Ensure integrated tactical operations.
- Use all assigned resources effectively.

Controlling hazardous material incidents may require the coordinated efforts of the following agencies/organizations with missions at haz mat/WMD incidents:

- Fire service provides rescue and fire, spill, and leak control.
- Law enforcement secures crime scenes and provides crowd control, force protection, evidence preservation and sampling, and bomb/explosives removal.
- EMS provides medical assistance and monitoring.

Unified Command (UC) — In the Incident Command System, a shared command role in which all agencies with geographical or functional responsibility establish a common set of incident objectives and strategies. In unified command there is a single incident command post and a single operations chief at any given time.

- Private concerns:
  - Material's manufacturer provides information about haz mat involved.
  - Material's shipper provides information about haz mat containers and shipping information.
  - Facility manager provides information about haz mat involved and incident facilities and location.
- Government agencies (local, tribal, state/provincial, federal) with mandated interests in health and environmental issues provide a variety of technical expertise and authority.
- Privately contracted cleanup and salvage companies provide remediation, cleanup, and salvage.
- Specialized emergency response groups, organizations, and technical support groups provide specialized services.
- Utilities and public works provide assistance in shutting off utilities, heavy equipment and supplies, and knowledge of public works systems (such as drainage, sewers, and water supplies).

To avoid jurisdictional and command disputes, the specific agency/organization responsible for handling and coordinating response activities should be identified in a mutual aid contract before an incident happens. It is important to know what the contracts do and do not cover. Preincident coordination should be done at the local level so that jurisdictional disputes can be avoided.

U.S. NIMS and the National Response Framework

After the terrorist attacks of September 11, 2001, the United States government decided that all U.S. emergency services organizations needed to have common terminology and command structures. Homeland Security Presidential Directive/HSPD-5 formalized this idea by requiring federal departments and agencies to make adoption of NIMS by state, tribal, and local organizations a condition for federal preparedness assistance (through grants, contracts, and other activities).

NIMS-ICS is designed to be applicable to small, single-unit incidents that may last only a few minutes as well as complex, large-scale incidents involving several agencies and many mutual aid units that could last for days or weeks. NIMS-ICS builds from the ground up and is the basic operating system for all incidents within each facility or agency. By design, the NIMS-ICS can grow from a small-scale organization to a large-scale organization depending on the needs of the incident.

Depending on incident complexity, the IC may delegate responsibilities and assign personnel to subordinate management roles. If the functions are not delegated, the IC retains these responsibilities and must ensure that all requisite functions are completed as part of the **Incident Action Plan (IAP)**.

In addition to the five general ICS functions, NIMS-ICS adds another position, Intelligence, which is responsible for gathering information relating to the incident. Depending on the type and complexity of the incident, the Intelligence function may be staffed as a staff officer within the Command Staff, as part of the Operations Section or the Planning Section, or assigned to a Section level position on its own. Because terrorist incidents are criminal acts and fall under the jurisdiction of the FBI in the U.S., it is likely the Intelligence position will be staffed.

Incident Action Plan (IAP) — Written or unwritten plan for the disposition of an incident; contains the overall strategic goals, tactical objectives, and support requirements for a given operational period during an incident. All incidents require an action plan. On relatively small incidents, the IAP is usually not in writing; on larger, more complex incidents, a written IAP is created for each operational period and disseminated to all assigned units. Written IAPs may have a number of forms as attachments.

In addition to NIMS-ICS, emergency responders in the U.S. should be familiar with the **National Response Framework (NRF)**. The NRF is a guide to managing all-hazards response at the local, state, and federal level, or within the private sector. The NRF, using NIMS-ICS, can be partially or fully implemented in the context of a threat, anticipation of a significant event, or the response to a significant event. Selective implementation through the activation of one or more of the system's components allows maximum flexibility in meeting the unique operational and information-sharing requirements of the situation at hand and enabling effective interaction between various federal and nonfederal entities.

The following teams are resources that have been established by the NRF:

- **Weapons of Mass Destruction-Civil Support Teams (WMD-CST)** — Teams that support civil authorities at a domestic chemical, biological, radiological, nuclear, or high-yield explosive incident site by identifying CBRNE agents/substances. The National Guard Bureau fosters the development of WMD-CSTs. There are plans for at least one CST in each state. *Duties:*

  — Assess current and projected consequences

  — Advise on response measures

  — Assist with appropriate requests for state support

  — Provide an extensive communications capability

- **Disaster Medical Assistance Teams (DMAT)** — Groups of professional and paraprofessional medical personnel (supported by a cadre of logistical and administrative staff) designed to provide emergency medical care during a disaster or other event **(Figure 24.6)**. The National Disaster Medical System (NDMS), through the U.S. Public Health Service (PHS), fosters the development of DMATs.

- **Disaster Mortuary Operational Response Teams (DMORT)** — Teams that work under the guidance of local authorities by providing technical assistance and personnel to recover, identify, and process deceased victims. The teams are composed of private citizens, each with a particular field of expertise, who are activated in the event of a disaster. The NDMS, through the PHS and the National Association for Search and Rescue (NASAR), fosters the development of DMORTs.

Figure 24.6 DMAT teams provide emergency medical care in the aftermath of a natural disaster. *Courtesy of FEMA News Photos, photo by Andrea Booher.*

- **National Medical Response Team-Weapons of Mass Destruction (NMRT-WMD)** — Specialized response forces designed to provide medical care following a nuclear, biological, and/or chemical incident. Four teams are geographically dispersed throughout the U.S. The NDMS, through the PHS, fosters the development of NMRT-WMD. These units are capable of providing the following services:

  — Mass casualty decontamination

  — Medical triage

  — Primary and secondary medical care to stabilize victims for transportation to tertiary-care facilities in a hazardous material environment

- **Urban Search and Rescue (USAR) Task Forces** — Highly trained teams that provide search-and-rescue operations in damaged or collapsed structures and stabilization of damaged structures. They can also provide emergency medical care to the injured. Currently there are twenty-eight federal USAR teams and numerous state teams that follow the DHS-FEMA USAR model regarding training, equipment, and personnel. The task forces are a partnership among the following entities:

  — Local fire departments

  — Law enforcement agencies

  — Federal and local governmental agencies

  — Private companies

- **Incident Management Teams (IMT)** — Teams of highly trained, experienced individuals who are organized to manage large and/or complex incidents. They provide full logistical support for receiving and distribution centers. National IMTs are hosted and managed by Geographic Area Coordination Centers. The teams are hosted by the U.S. Forest Service (USFS) during wildland fires. Both states and regions can have IMTs.

Haz Mat Incident Mitigation

At this stage in your career, you will not be responsible for determining your department's response to a haz mat incident. However, it is important that you understand how such decisions are made. Incident Commanders use a four-step problem-solving approach at any haz mat incident, which closely resembles the competency tasks required by NFPA® 472 **(Table 24.1, p. 1422)**, known as *APIE*:

- A — Analyze
- P — Plan
- I — Implement
- E — Evaluate (and repeat)

Analyzing the Situation

The first thing that should be done is to approach safely and isolate/deny entry. Then first responders must determine the type of incident and the scope (size) of the incident. Responders will then need to follow internal notification protocols depending on the type of incident to provide critical information and/or request additional resources. The following sections discuss the initial size-up of the incident including hazard and risk assessment and incident levels.

Table 24.1
Four-Step Problem-Solving Process

Problem-Solving Stages at Haz Mat and Terrorist Incidents

Analysis Stage and Information Gathering.
- Recognize the type of incident (Is it a haz mat incident or terrorist attack? If so, what materials are involved?)
- Identify all the hazards presented by the incident (chemical, mechanical, electrical, etc.)
- Predict the likely behavior of chemical, explosive, biological, and radiological materials
- Estimate potential harm

Planning Stage.
- Determine if additional help is needed
- Identify actions to protect emergency responders and others from hazards
- If appropriate, consult the *ERG* for instructions and guidelines regarding hazardous materials
- Determine strategies and tactics to stabilize the incident
- Determine appropriate personal protective equipment
- Determine appropriate decontamination methods
- Devise the Incident Action Plan

Implementation Stage.
- Implement the incident management system
- Transmit information to an appropriate authority and call for appropriate assistance
- Establish and enforce scene control perimeters
- Implement the Incident Action Plan
- Implement strategies and tactics appropriate to training such as isolating the area and denying entry, conducting rescues, conducting mass decontamination, etc.
- Identify and preserve evidence

Evaluation and Review Stage.
- Evaluate effectiveness of approach (is the incident stabilizing?)
- Process and provide feedback to IC

Incident Command Post Location

At haz mat/WMD incidents the Incident Command Post (ICP) should be located uphill, upwind, and upstream of the incident. It should be easily identified and accessible (either directly or indirectly), although access to the Command Post needs to be controlled. An ICP can be a predetermined location such as at a facility, a conveniently located building, or a radio-equipped vehicle located in a safe area.

Initial Size-Up

Size-up is the initial and ongoing assessment of an incident. It involves considering all the factors that will affect an incident and your response. First responders on the scene typically try to determine:

- Type of incident
- Type of hazardous material involved
- Exact location of the incident/hazards
- Type, size, and occupancy of structure
- Building construction
- Atmospheric conditions and risks
- Utilities
- Interior and exterior access to the scene
- Vertical/horizontal openings, shafts, voids, and tunnels
- Structural damage and collapse
- Existing hazards
- Special hazards
- Fire spread and direction of travel
- Number, type, and location of casualties/trapped victims
- Resources needed for search and rescue
- Resources needed for fire suppression
- Resources needed for haz mat operations
- Specialized equipment needed for recovery
- Potential for secondary attacks at criminal or terrorist incidents

To begin the size-up process, approach the incident carefully from upwind, uphill, and upstream. Focus on identifying the material involved (**Figure 24.7**). Use available resources, such as the *Emergency Response Guidebook* (*ERG*) or plume modeling software, to estimate the size of the endangered area. Predict potential exposures, including the number of people in the area, the number of affected structures, and any environmental concerns.

Figure 24.7 Responders may be able to predict the behavior of hazardous materials based on the material's properties and conditions at the scene. *Courtesy of Steve Irby, Owasso (OK) Fire Department.*

Figure 24.8 Using available resources, responders must be able to estimate the size of the area endangered by a haz mat incident and work to limit that area. *Courtesy of Rich Mahaney.*

Hazard and Risk Assessment — Formal review of the hazards and risks that may be encountered by firefighters or emergency responders; used to determine the appropriate level and type of personal and respiratory protection that must be worn. *Also known as Hazard Assessment.*

When the potential area of effect and number of exposures is determined, a number of strategic goals and tactical objectives (response options or objectives) can be considered for achieving incident stabilization **(Figure 24.8)**. Decisions about which options to use are based on the availability of resources and a risk/benefit analysis. For example, rescue may be an initial priority, but evacuation and sheltering-in-place may be the best way to save additional lives.

Size-up is frequently complicated by a lack of information. The IC's view of the incident may be limited by the scope of the hazard area or location of the event (that is, inside a vehicle or structure). Sometimes, the scope of the incident is so great that hazard assessment becomes difficult because it is impossible to see the entire incident scene. In such cases, more than one agency may be involved in the assessment of hazards, and additional resources such as aerial reconnaissance may be needed.

In addition, limited or conflicting information regarding the nature and scope of the incident is possible. Initial assessment may be based on anticipated conditions and updated as additional information becomes available such as data provided by monitoring and detection activities. For example, monitoring and detection can determine the hazardous concentrations present in various locations thereby establishing the scope of the hazard zone with greater certainty. Coupled with information from safety data sheets and other information sources, the safety and health hazards associated with released materials may be determined.

Hazard and risk assessment is a crucial part of the incident size-up. It is a continuous and ongoing process. It starts with preincident planning, extends through the receipt of an alarm, and continues throughout the course of an incident. Once on the scene, the IC conducts a hazard assessment and repeats this process as the incident continues. Final pieces of the hazard and risk assessment are added to the information made available before and after arrival.

Incident Levels

After the initial size-up has determined the scope of an incident, the level of the incident can be determined in accordance with the local emergency response plan. The NFPA® 472 incident level model identifies three levels of response graduating from Level I (least severe) to Level III (most severe).

NOTE: ICS levels of incident severity are reversed from this model (see Information Box on p. 1426). Predetermined levels help an IC to quickly identify necessary resources.

Level I. This type of incident is within the capabilities of the fire or emergency services organization or other first responders. A Level I incident is the least serious and the easiest to handle. It may pose a serious threat to life or property, although this is not usually the case. Evacuation, if required, is limited to the immediate area of the incident. The following are examples of Level I incidents:

- Small amount of gasoline or diesel fuel spilled from a vehicle **(Figure 24.9)**
- Leak from domestic natural gas line on the consumer side of the meter
- Broken containers of consumer commodities such as paint, thinners, bleach, swimming pool chemicals, and fertilizers (owner or proprietor is responsible for cleanup and disposal)

Level II. This type of incident is beyond the capabilities of the first responders on the scene and may be beyond the capabilities of the first response agency/organization having jurisdiction. Level II incidents require the services of a formal haz mat response team. A properly trained and equipped response team could be expected to perform the following tasks:

Figure 24.9 Level I incidents include small fuel spills. *Courtesy of Rich Mahaney.*

Figure 24.10 Unknown substances should be left for evaluation by Level II-responders with appropriate training and equipment.

- Use chemical-protective clothing (CPC).
- Dike and confine within the contaminated areas.
- Perform **plugging** and patching activities.
- Sample and test unknown substances **(Figure 24.10)**.
- Perform various levels of decontamination.

  The following are examples of Level II incidents:
- Spill or leak requiring limited-scale evacuation
- Any major accident, spillage, or overflow of flammable liquids
- Spill or leak of unfamiliar or unknown chemicals
- Accident involving extremely hazardous substances
- Rupture of an underground pipeline
- Fire that is posing a boiling liquid expanding vapor explosion (BLEVE) threat in a storage tank

Level III. This type of incident requires resources from state/provincial agencies, federal agencies, and/or private industry in addition to Unified Command. A Level III incident is the most serious of all hazardous material incidents. A large-scale evacuation may be required. Most likely, the incident will not be concluded by any one agency. Successful handling of the incident requires a collective effort from several of the following resources/procedures:

- Specialists from industry and governmental agencies
- Sophisticated sampling and monitoring equipment
- Specialized leak and spill control techniques
- Decontamination on a large scale

Plug — Patch to seal a small leak in a container.

The following are examples of Level III incidents:

- Incidents that require an evacuation extending across jurisdictional boundaries
- Incidents beyond the capabilities of the local hazardous material response team
- Incidents that activate (in part or in whole) the federal response plan

U.S. NIMS Incident Types

Per U.S. NIMS, incidents may be typed in order to make decisions about resource requirements. Incident types are based on the following five levels of complexity:

Type 5 – *Details*:

- The incident can be handled with one or two single resources with up to six personnel.
- Command and General Staff positions (other than the Incident Commander) are not activated.
- No written Incident Action Plan (IAP) is required.
- The incident is contained within the first operational period and often within an hour to a few hours after resources arrive on scene.
- Examples include a vehicle fire, an injured person, or a police traffic stop.

Type 4 – *Details:*

- Command Staff and General Staff functions are activated only if needed.
- Several resources are required to mitigate the incident.
- The incident is usually limited to one operational period in the control phase.
- The Agency Administrator may have briefings, and ensure the complexity analysis and delegation of authority is updated.
- No written IAP is required but a documented operational briefing will be completed for all incoming resources.
- The role of the Agency Administrator includes operational plans including objectives and priorities.

Type 3 – *Details*:

- When capabilities exceed initial attack, the appropriate ICS positions should be added to match the complexity of the incident.
- Some or all of the Command and General Staff positions may be activated, as well as Division/Group Supervisor and/or Unit Leader level positions.
- A Type 3 Incident Management Team (IMT) or Incident Command Organization manages initial action incidents with a significant number of resources, an extended attack incident until containment/control is achieved, or an expanding incident until transition to a Type 1 or 2 team.
- The incident may extend into multiple operational periods.
- A written IAP may be required for each operational period.

Type 2 – *Details*:

- This type of incident extends beyond the capabilities for local control and is expected to go into multiple operational periods. A Type 2 incident may require the response of resources out of area, including regional and/or national resources, to effectively manage the operations, command, and general staffing.

- Most or all of the Command and General Staff positions are filled.

- A written IAP is required for each operational period.

- Many of the functional units are needed and staffed.

- Operations personnel normally do not exceed 200 per operational period and total incident personnel do not exceed 500 (guidelines only).

- The Agency Administrator is responsible for the incident complexity analysis, Agency Administrator briefings, and the written delegation of authority.

Type 1 – *Details:*

- This type of incident is the most complex, requiring national resources to safely and effectively manage and operate.

- All Command and General Staff positions are activated.

- Operations personnel often exceed 500 per operational period and total personnel will usually exceed 1,000.

- Branches need to be established.

- The Agency Administrator will have briefings and ensure that the complexity analysis and delegation of authority are updated.

- Use of resource advisors at the Incident Base is recommended.

- There is a high impact on the local jurisdiction, requiring additional staff for office administrative and support functions.

Source: U.S. Fire Administration

Planning the Appropriate Response

As part of the problem-solving process, ICs must select the strategic goals and the tactical objectives used to mitigate a haz mat/WMD incident. Strategic goals are broad statements of what must be done to resolve the incident, while tactics are the specific operations that must be done in order to accomplish those goals.

Strategic goals are prioritized depending on available resources and the particulars of the incident. Some of these goals may not be needed if the hazard is not present at the incident. For example, if the incident involves nonflammable chemicals, fire control may not be an issue. Some goals may require the use of specialized resources (such as chemical-protective clothing or a bomb squad) that are not yet available and therefore must be postponed. Others may require the use of so many available resources that the ability to complete other goals in an expedient time frame might be compromised.

Strategies

The following are strategic goals commonly used to mitigate haz mat and terrorist incidents:

- Isolation and scene control
- Notification
- Identification
- Protection of responders and the public
- Rescue
- Spill control and leak containment
- Fire control
- Crime scene management and evidence preservation
- Recovery and termination

It should be understood that this list is not all-inclusive, nor must these strategies be undertaken in the order presented. ICs set the goals they deem appropriate by using strategies that will achieve successful mitigation of the incident. For example, rescue might be considered an important strategic goal at one incident but not at another. If conditions at an incident suddenly worsen, evacuation and accountability might become strategic goals that spring to the top of the priority list.

These strategies are discussed in further detail in later sections of this chapter. PPE and decontamination, although falling under the umbrella of *Protection*, are discussed in detail in their own sections.

Modes of Operation

Strategies (and their resulting tactical operations) are divided into three options that relate to modes of operation: **defensive**, **offensive**, and **nonintervention**. A defensive strategy provides confinement of the hazard to a given area by performing defensive actions such as containing the hazard, for example, by diking or damming a storm drain if a toxic industrial material is involved. An offensive strategy includes actions to actively control the hazard, for example, haz mat technicians plugging or patching a leaking container **(Figure 24.11)**. A nonintervention strategy isolates the area to protect the public and emergency responders, but allows the incident to run its course on its own. Nonintervention is the only safe strategy in many types of incidents and the best strategy in certain types of incidents when mitigation is failing or otherwise impossible.

Incident Action Plans

Incident Action Plans (IAPs) are critical to the rapid, effective control of emergency operations. An IAP is a well-thought-out, organized course of events developed to address all phases of incident control within a specified time. Written IAPs may not be necessary for short-term, routine operations; however, large-scale or complex incidents require the creation and maintenance of a written plan for each operational period. Regardless of the form of the IAP, its contents must be disseminated throughout the incident organization.

Action planning starts with identifying the strategy to achieve a solution to the confronted problems. Strategy is broad in nature and defines what has to be done. Once the strategy has been defined, the Command Staff needs to select the tactics (how, where, and when) to achieve the strategy. Tactics are measurable in both time

Defensive Strategy — Overall plan for incident control established by the Incident Commander that involves protection of exposures, as opposed to aggressive, offensive intervention.

Offensive Strategy — Overall plan for incident control established by the Incident Commander (IC) in which responders take aggressive, direct action on the material, container, or process equipment involved in an incident.

Nonintervention Strategy — Strategy for handling fires involving hazardous materials, in which the fire is allowed to burn until all of the fuel is consumed.

Figure 24.11 A haz mat team should be called to conduct offensive operations at a Level II incident. *Courtesy of New South Wales Fire Brigades.*

and performance. An IAP also provides for necessary support resources such as water supply, utility control, self-contained breathing apparatus (SCBA) cylinder filling, and the like.

The IAP essentially ties the entire problem-solving process together by stating what the analysis has found, what the plan is, and how it will be implemented. Once the plan is established and resources are committed, it is necessary to assess its effectiveness. Information must be gathered and analyzed so that necessary modifications may be made to improve the plan if necessary. This step is part of a continuous size-up process. Elements of an IAP include the following:

- Strategies
- Current situation summary
- Resource assignment and needs
- Accomplishments
- Hazard statement
- Risk assessment
- Safety plan and message
- Protective measures
- Current and projected weather conditions
- Status of injuries
- Communications plan
- Medical plan

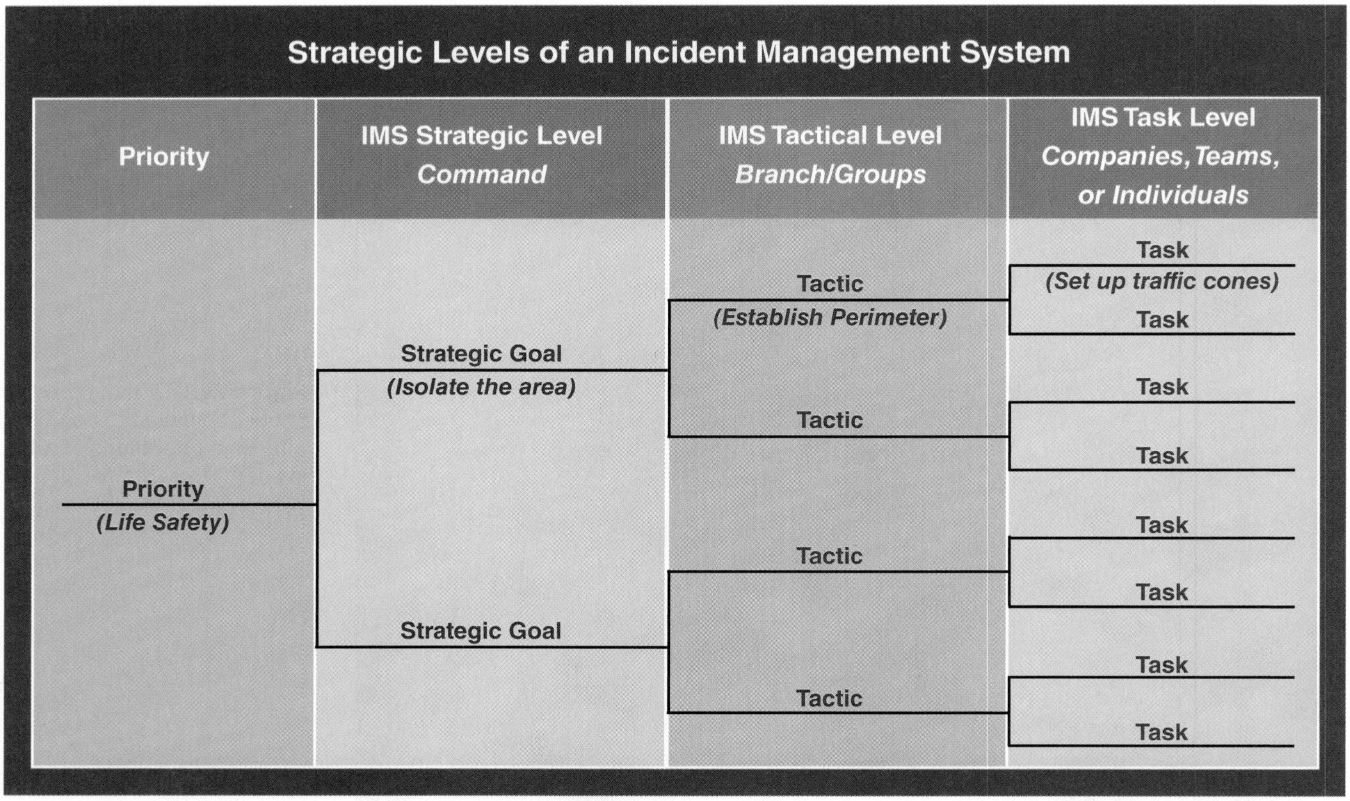

Strategic Levels of an Incident Management System

Priority	IMS Strategic Level *Command*	IMS Tactical Level *Branch/Groups*	IMS Task Level *Companies, Teams, or Individuals*
Priority *(Life Safety)*	**Strategic Goal** *(Isolate the area)*	**Tactic** *(Establish Perimeter)*	**Task** *(Set up traffic cones)* **Task**
		Tactic	**Task** **Task**
	Strategic Goal	**Tactic**	**Task** **Task**
		Tactic	**Task** **Task**

Figure 24.12 The relationship between IMS priorities, strategic goals, tactics, and tasks may be organized in a flowchart.

All incident personnel must function according to the IAP. Company officers or Group/Division Supervisors should follow predetermined procedures, and every action should be directed toward achieving the goals and objectives specified in the plan.

Implementing the Incident Action Plan

After strategic goals have been selected and the IAP formulated, the IC can begin to implement the plan. Strategic goals are met by achieving tactical objectives. Tactical objectives are accomplished or conducted by performing specific tasks (**Figure 24.12**).

Tactics, like strategies, can also be offensive, defensive, or nonintervening. Tactics related to controlling chemical and radiological releases basically fall into two categories: **confinement** and **containment**, with the majority of defensive control options being related to confinement. Other tactics may include such things as establishing scene control zones, monitoring for radiation and other hazards, calling for additional resources, wearing appropriate personal protective equipment (PPE), conducting decontamination, and extinguishing fires.

Evaluating Progress

The final part of the problem-solving process is reviewing or evaluating progress. When an IAP is effective, the IC should receive favorable progress reports from tactical and/or task supervisors, and the incident should begin to stabilize. If, on the other hand, mitigation efforts are failing or the situation is getting worse (or more intense), the plan must be reevaluated and very possibly revised by selecting new strategies or by changing the tactics used to achieve them. The plan must also be reevaluated as new information becomes available and circumstances change. If the

Confinement — The process of controlling the flow of a spill and capturing it at some specified location.

Containment — The act of stopping the further release of a material from its container.

situation deteriorates rapidly or conditions change to endanger personnel, it may be necessary to withdraw from offensive and defensive operations. In accordance with predetermined communication procedures (such as designated radio channels), it is important for firefighters to communicate the status of the planned response and the progress of their actions to the IC.

Using the *Emergency Response Guidebook*

The *ERG* is primarily a guide to aid emergency responders in quickly identifying the specific or generic hazards of materials involved in an emergency incident and protecting themselves and the general public during the initial response phase of the incident. The *ERG* does not address all possible circumstances that may be associated with a dangerous goods/hazardous materials incident. It is primarily designed for use at hazardous materials incidents occurring on a highway or railroad. Isolation and protective distances in the *ERG* are based on conditions commonly associated with transportation incidents in open areas and may be of limited value when applied to fixed-facility locations or in urban settings.

There are several ways to locate the appropriate initial action guide in the *ERG*:

- Identify the four-digit U.N. identification number on a placard or shipping papers, then look up the appropriate guide in the yellow-bordered pages of the guidebook.

- Use the name of the material involved (if known) in the blue-bordered section of the guidebook (see *ERG* Material Name Index section). Many chemical names differ only by a few letters, so **exact spelling is important** when using this method **(Figure 24.13)**.

- Identify the transportation placard of the material, then use the three-digit guide number associated with the placard in the Table of Placards and Initial Response Guide to Use On-Scene. This table is located in the front of the *ERG*.

- As a last resort, use the container profiles provided in the white pages in the front of the book. First responders can identify container shapes, then reference the guide number to the orange-bordered page provided in the nearest circle **(Figure 24.14)**.

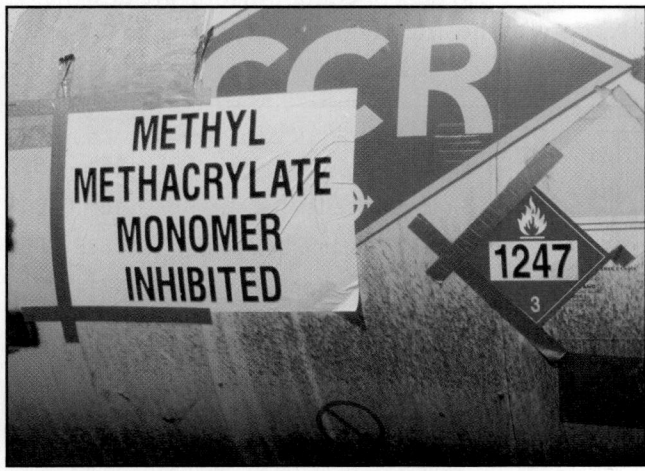

Figure 24.13 Misspelling a chemical name can lead responders toward inappropriate (and potentially dangerous) actions at an incident. *Courtesy of Rich Mahaney.*

Figure 24.14 If placards or four-digit ID numbers are not visible, first responders can use container profiles to identify the proper page in the *ERG*. *Courtesy of Rich Mahaney.*

Using the four-digit ID number or the chemical name allows you to locate the most specific initial action guide. **Skill Sheet 24-I-1** provides examples of how to use the *ERG*. The sections that follow describe the design and layout of the *ERG*.

Multiple Information Sources

Always seek multiple sources of information at any haz mat incident, and do so as quickly as possible. Do not rely on the *ERG* alone. The information received by contacting the appropriate emergency response agency, calling the emergency response number on the shipping document, or consulting the accompanying shipping document may be more specific and accurate than the guidebook in providing guidance for the materials involved.

When using chemical reference sources for information about a particular substance, more than one reference source should be consulted to ensure information is complete and accurate. Reference books may be written for a specific purpose (such as compiling information about the most dangerous workplace chemicals), and many chemicals may be left out. Absence from one reference book does *not* mean that the substance is safe. You should check multiple sources.

ERG ID Number Index (Yellow Pages)

The yellow-bordered pages of the *ERG* list hazardous materials in numerical order, based on their four-digit UN/NA ID numbers. This index displays the four-digit UN/NA ID number of the material followed by its assigned *Emergency Response Guide* and the material's name. Responders can use this section to quickly determine the appropriate action Guide (in the Orange Pages) based on the ID number.

Toxic Inhalation Hazard (TIH) — Volatile liquid or gas known to be a severe hazard to human health during transportation.

If a material in the yellow or blue index is highlighted, it means that it releases gases that are **toxic inhalation hazard (TIH)** materials. These materials require the application of additional emergency response distances (see *ERG Table of Initial Isolation and Protective Action Distances [Green Pages]* section). Materials that may undergo violent polymerization if subjected to heat or contamination are designated with a *P* in the blue and yellow sections of the *ERG* (**Figure 24.15**).

Sodium potassium alloys, solid	138	3404	Strontium phosphide	139	2013
Sodium selenite	151	2630	Strychnine	151	1692
Sodium silicofluoride	154	2674	Strychnine salts	151	1692
Sodium sulfide, anhydrous	135	1385	Styrene monomer, stabilized	128P	2055
Sodium sulfide, hydrated, with not less than 30% water	153	1849	Substituted nitrophenol pesticide, liquid, flammable, poisonous	131	2780
Sodium sulfide, with less than 30% water of crystallization	135	1385			
Sodium sulphide, anhydrous	135	1385	Substituted nitrophenol pesticide, liquid, flammable, toxic	131	2780
Sodium sulphide, hydrated, with	153	1849			

Figure 24.15 Materials designated with a *P* in the blue and yellow pages of the *ERG* may undergo rapid polymerization and become dangerously unstable.

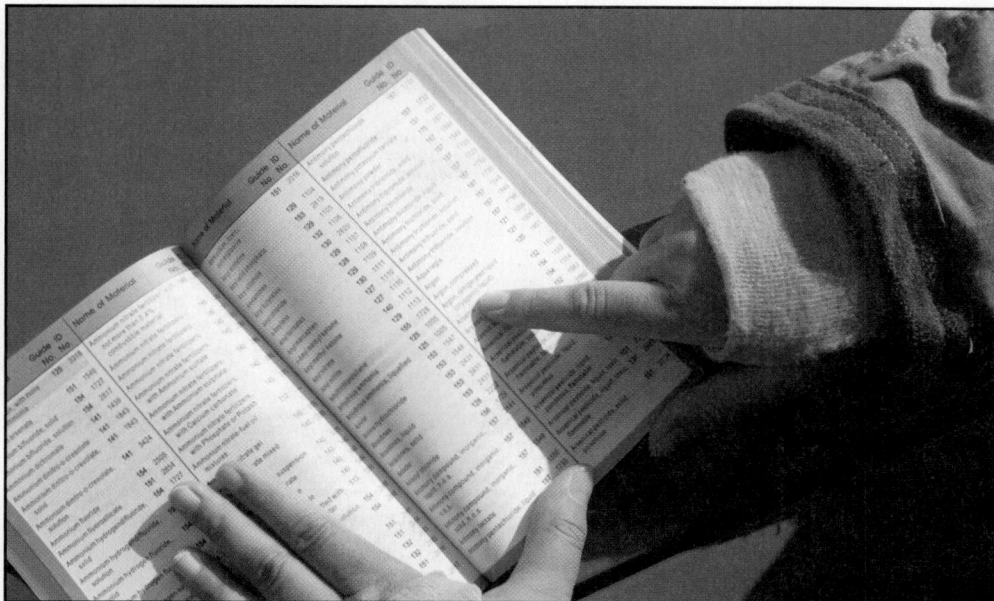

Figure 24.16 The blue pages of the *ERG* provide an index of hazardous materials in alphabetical order.

ERG Material Name Index (Blue Pages)

The blue-bordered pages of the *ERG* list hazardous materials by name. Each entry includes the material's four-digit ID and appropriate action Guide (in the Orange Pages) **(Figure 24.16)**. Responders can use this section to quickly determine the appropriate Guide based on the name of the material. As in the Yellow Pages, the designation *P* indicates risk of polymerization, and highlighted entries are toxic inhalation hazards.

ERG Initial Action Guides (Orange Pages)

The orange-bordered section of the book is the most important because it provides safety recommendations and general hazards information. It is comprised of individual guides presented in a two-page format. The left-hand page provides safety-related information, whereas the right-hand page provides emergency response guidance and activities for fire situations, spill or leak incidents, and first aid. Each Guide is designed to cover a group of materials that possess similar chemical and toxicological characteristics. The Guide title identifies the general hazards of the hazardous materials covered. Each Guide is divided into three main sections:

- *Potential Hazards*
- *Public Safety*
- *Emergency Response*

Guide 111

If a reference to a Guide cannot be found and the incident is believed to involve hazardous materials (or mixed loads/unidentified cargo), turn to Guide 111, and use it until additional information becomes available.

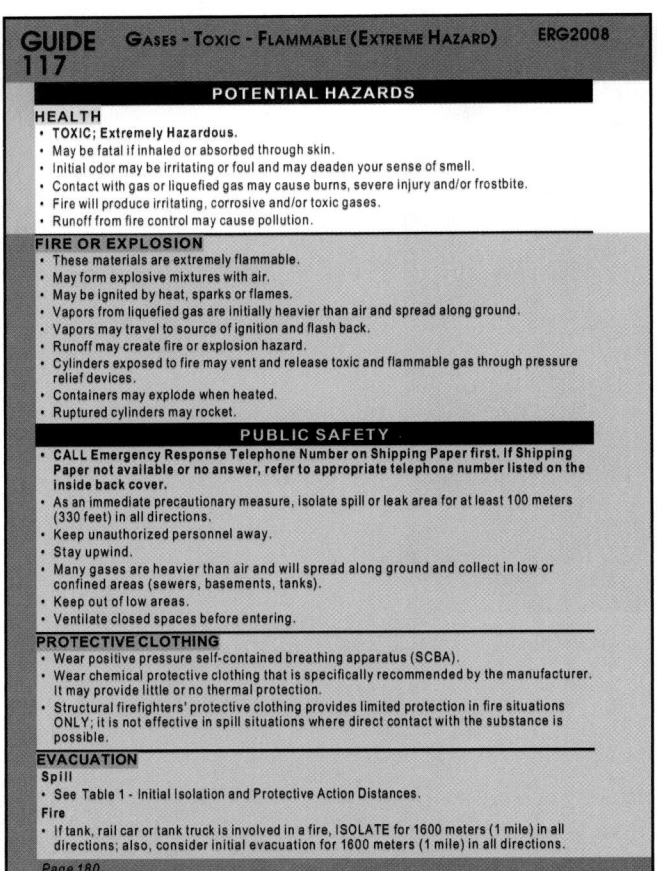

Figure 24.17 The highest potential hazard is listed first on the orange *ERG* pages.

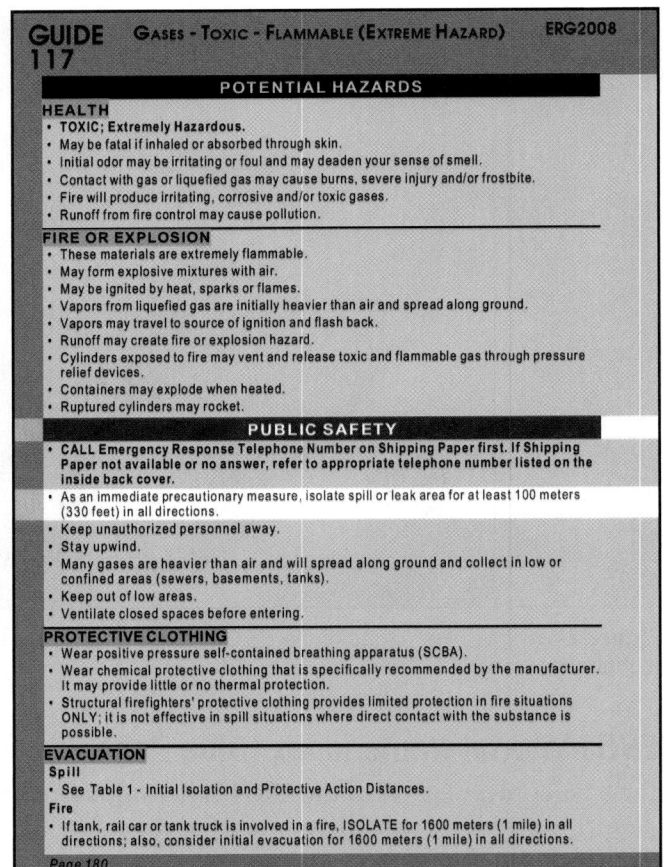

Figure 24.18 Initial isolation distances are provided as a bulleted point in the *Public Safety* section.

Potential Hazards Section

The *Potential Hazards* section describes potential health and fire/explosion hazards that the material may display (**Figure 24.17**). The highest potential hazard is listed first. This section should be consulted first because it will assist in making decisions regarding the protection of individuals at the incident. Examples of information provided include: *TOXIC; **may be fatal if inhaled or absorbed through skin***, or ***EXTREMELY FLAMMABLE***.

Public Safety Section

The *Public Safety* section provides general information regarding immediate isolation of the incident site and recommended type of protective clothing and respiratory protection. This section also lists suggested evacuation distances for small and large spills and for fire situations (which include distances for fragmentation hazards for tanks that might explode).

Isolation distances are provided in the bullet points immediately below the Public Safety section heading (**Figure 24.18**). The **initial isolation distance** is a distance within which all persons should be considered for evacuation in all directions from the haz mat spill or leak source. The **initial isolation zone** is a circular zone (with a radius equivalent to the initial isolation distance) within which persons may be exposed to dangerous concentrations upwind of the source and life-threatening concentrations downwind of the source (**Figure 24.19**). If safe to do so, firefighters should evacuate people from the hazard area to this safe distance (at a minimum) (**Figure 24.20**). They should then prevent others from entering this area by denying entry/access and securing the scene.

Initial Isolation Distance — Distance within which all persons are considered for evacuation in all directions from a hazardous materials incident.

Initial Isolation Zone — Circular zone, with a radius equivalent to the initial isolation distance, within which persons may be exposed to dangerous concentrations upwind of the source and may be exposed to life-threatening concentrations downwind of the source.

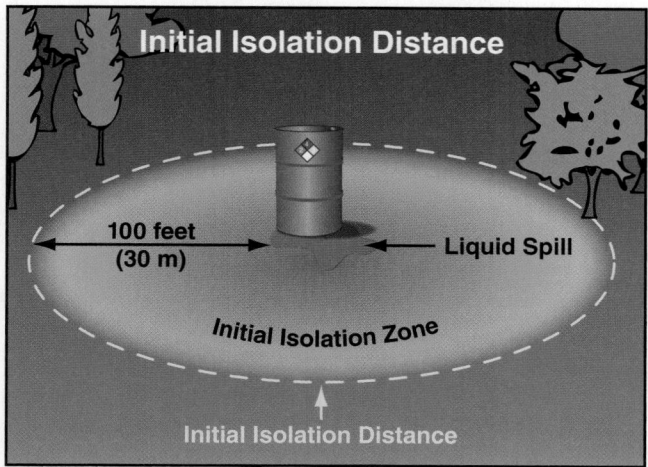

Figure 24.19 All persons within the initial isolation zone are considered at risk.

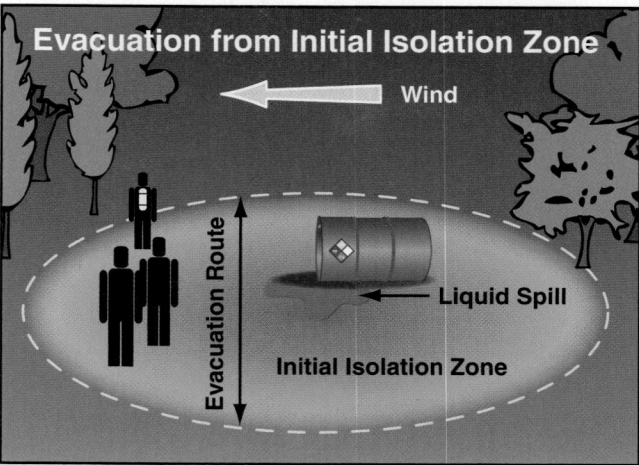

Figure 24.20 When people are downwind from the incident, evacuation from the initial isolation zone should be conducted at right angles to the prevailing wind direction, if possible.

Protective Clothing Section. This section recommends the type of personal protective clothing and equipment that should be worn at incidents involving these products (**Figure 24.21, p. 1436**). Examples include the following:

- **Street clothing** and work uniforms
- Structural firefighters' protective clothing
- Positive pressure self-contained breathing apparatus (SCBA)
- **Chemical-protective clothing (CPC)**

Evacuation Section. This section provides **evacuation** recommendations for spills/large spills and fires (**Figure 24.22, p. 1436**). When the material is a green-highlighted chemical in the yellow-bordered and blue-bordered pages, this section also directs the reader to consult the tables on the green-bordered pages listing TIH materials and water-reactive materials. Evacuation, sheltering-in-place/in-place protection, and protecting/defending in place are discussed in greater detail in the Protection of the Public sections.

Emergency Response Section

The third section, *Emergency Response*, describes emergency response topics, including precautions for incidents involving fire, spills or leaks, and first aid. Several recommendations are listed under each of these areas to further assist in the decision-making process. The information on first aid is general guidance before seeking medical care.

Fire section. This section recommends appropriate extinguishing agents for large fires, small fires, and fires involving bulk containers (**Figure 24.23, p. 1437**). Examples might include foam or water, or a specific type of fire extinguisher for small fires. If foam is recommended, it will specify the type of foam to be used. Recommendations vary by Guide, but additional information may include such things as using unmanned hose holders or cooling containers with flooding quantities of water.

Spill or leak section. This section provides actions to take in regards to spills and leaks (**Figure 24.24, p. 1437**). If a flammable liquid is involved, for example, it would recommend eliminating all ignition sources (see information box). It will also provide basic information needed to mitigate a spill, such as what materials to use to absorb the spill.

Street Clothes — Clothing that is anything other than chemical-protective clothing or structural firefighters' protective clothing, including work uniforms and ordinary civilian clothing.

Chemical-Protective Clothing (CPC) — Clothing designed to shield or isolate individuals from the chemical, physical, and biological hazards that may be encountered during operations involving hazardous materials.

Evacuation — Controlled process of leaving or being removed from a potentially hazardous location, typically involving relocating people from an area of danger or potential risk to a safer place.

CAUTION

You must be properly trained and equipped before attempting the actions recommended in the *ERG*.

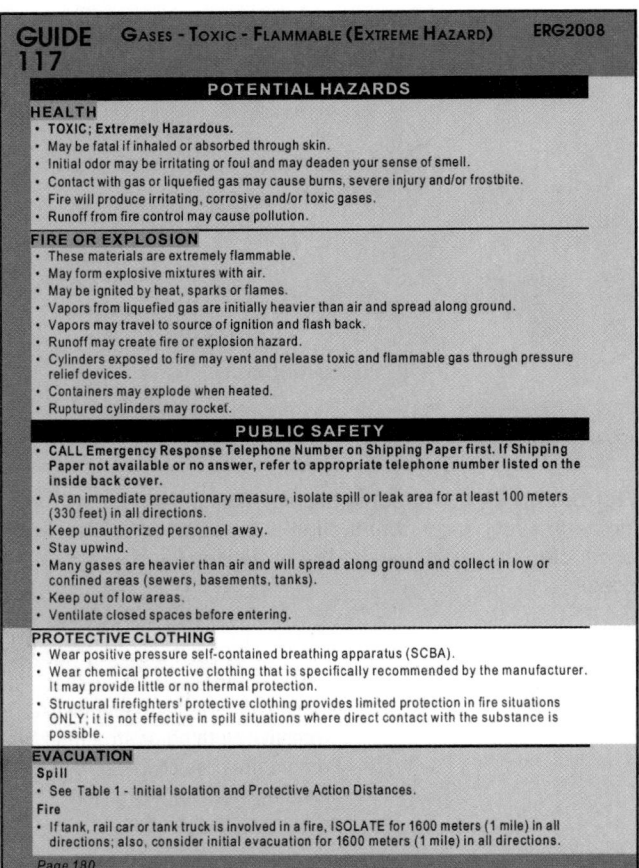

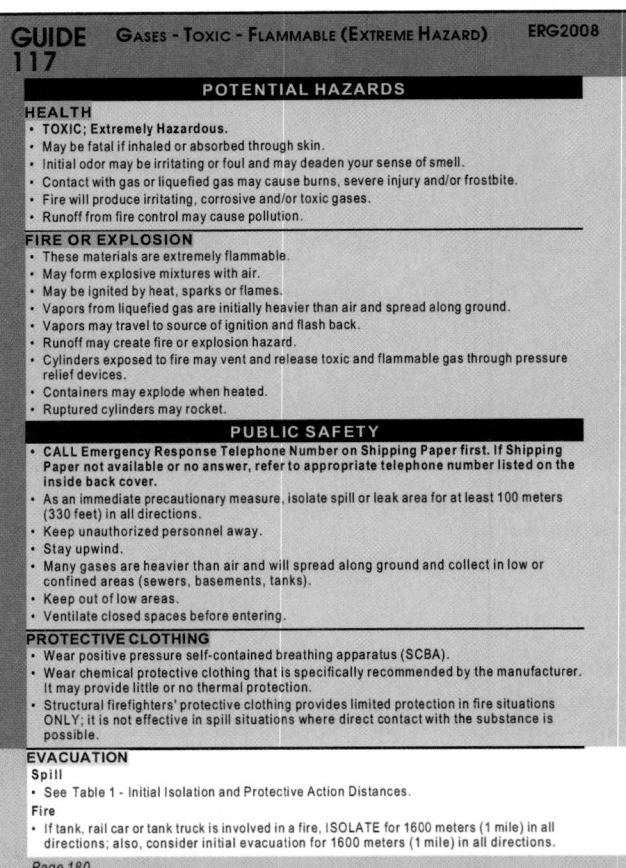

Figure 24.21 The *Protective Clothing* section of the ERG includes recommendations regarding which PPE is appropriate and which to avoid at an incident involving a specific hazardous material.

Figure 24.22 When evacuation distance differs from the initial isolation distance in the case of a spill, the user is referred to the *Table of Initial Isolation and Protective Action Distances* (green-bordered pages) for more information.

Ignition Sources at an Incident Scene

Be aware of potential ignition sources at the scene, such as:

- Open flames
- Static electricity
- Existing pilot lights
- Electrical sources, including non-explosion-proof electrical equipment
- Internal combustion engines in vehicles and generators
- Heated surfaces
- Cutting and welding operations
- Radiant and frictional heat
- Heat caused by friction or chemical reactions
- Cigarettes and other smoking materials
- Cameras
- Road flares

Avoid actions that can ignite an explosive atmosphere, such as:

- Opening or closing a switch or electrical circuit (for example, a light switch)
- Turning on a flashlight
- Operating a radio
- Activating a cell phone

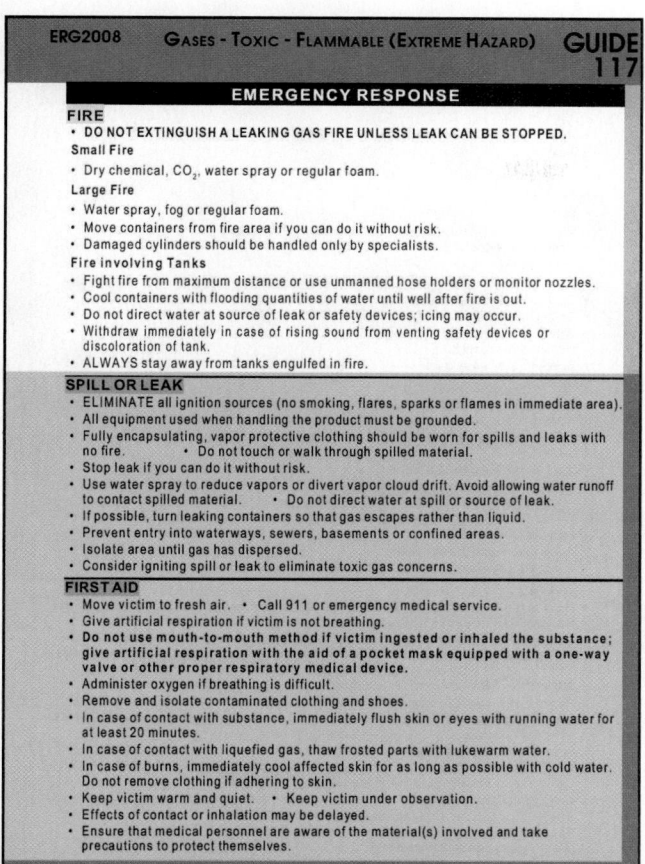

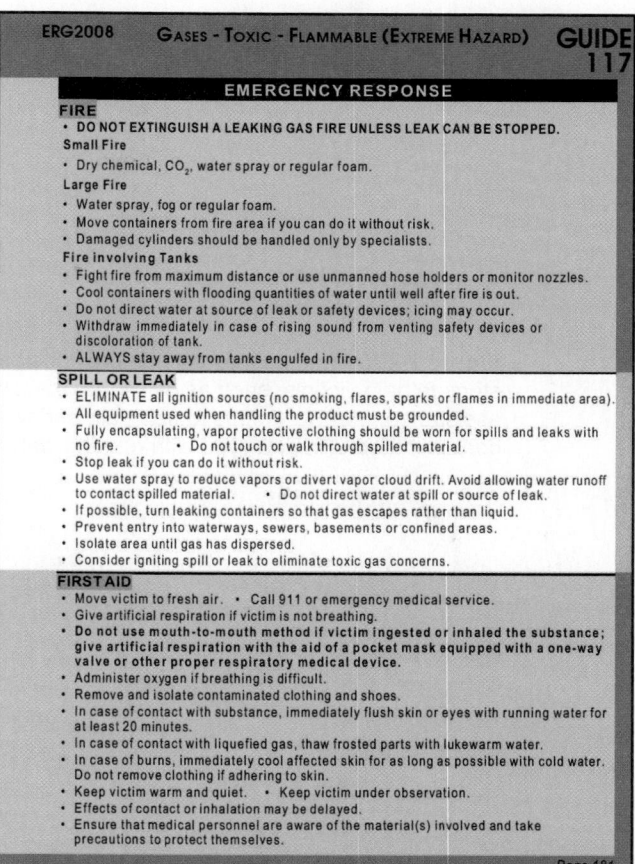

Figure 24.23 The *Fire* section provides information for firefighters, including appropriate extinguishing agents, types of foam to use, and actions to take or avoid.

Figure 24.24 The *Spill or Leak* section provides actions to take to mitigate a spill, such as what materials to use to absorb the material.

First aid section. This section provides basic steps to help victims affected by the hazardous material (**Figure 24.25, p. 1438**). Common recommendations include calling for emergency medical service assistance, moving victims to fresh air, and flushing contaminated skin and eyes with running water (**decontamination**). Avoiding direct contact with the hazardous material is also emphasized. Many of the recommendations provided in this section require specialized training due to the necessity of wearing personal protective equipment (such as chemical-protective clothing), the dangers of **cross contamination**, and the necessity of decontaminating victims before first aid is provided. For example, victims at haz mat incidents may present serious hazards to rescuers because they may be contaminated with the hazardous material. Only personnel with appropriate training, personal protective clothing, and equipment should contact these victims directly.

ERG Table of Initial Isolation and Protective Action Distances (Green Pages)

This section of the *ERG* contains a table that lists (by ID number) TIH materials, including certain chemical warfare agents and water-reactive materials that produce toxic gases upon contact with water. The table provides two different types of recommended safe distances: initial isolation distances and **protective action distances** (**Figure 24.26, p. 1438**). These materials are highlighted for easy identification in both numeric (yellow-bordered) and alphabetic (blue-bordered) *ERG* indexes.

Decontamination — Process of removing a hazardous foreign substance from a person, clothing, or area. *Also known as* Decon.

Cross Contamination — Contamination of people, equipment, or the environment outside the hot zone without contacting the primary source of contamination. *Also known as* Secondary Contamination.

Protective Action Distance — Downwind distance from a hazardous materials incident within which protective actions should be implemented.

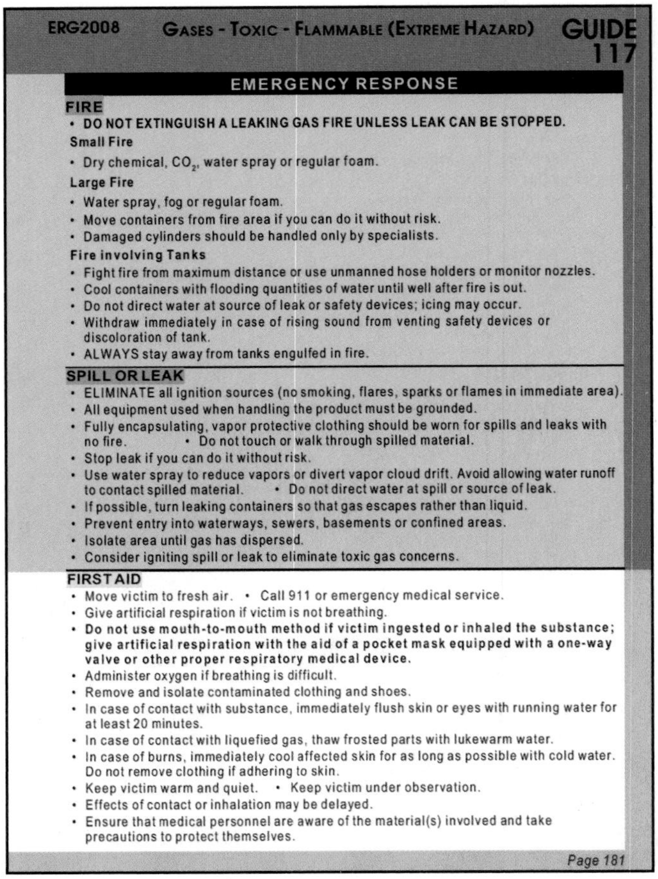

Figure 24.25 The *First Aid* section provides basic steps to help victims, such as calling for medical assistance, moving victims to fresh air, and flushing contaminated skin and eyes with running water.

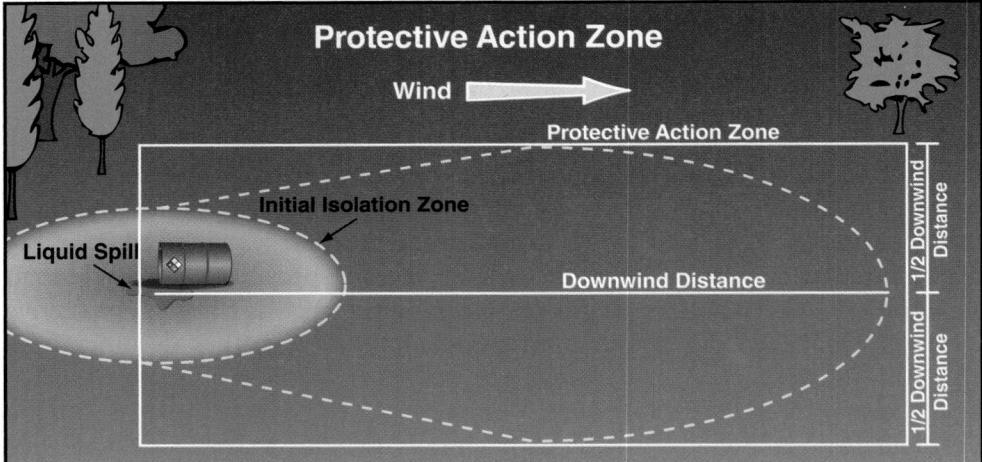

ERG2008   GASES - TOXIC - FLAMMABLE (EXTREME HAZARD)   GUIDE 117

EMERGENCY RESPONSE

FIRE
- DO NOT EXTINGUISH A LEAKING GAS FIRE UNLESS LEAK CAN BE STOPPED.

Small Fire
- Dry chemical, CO_2, water spray or regular foam.

Large Fire
- Water spray, fog or regular foam.
- Move containers from fire area if you can do it without risk.
- Damaged cylinders should be handled only by specialists.

Fire involving Tanks
- Fight fire from maximum distance or use unmanned hose holders or monitor nozzles.
- Cool containers with flooding quantities of water until well after fire is out.
- Do not direct water at source of leak or safety devices; icing may occur.
- Withdraw immediately in case of rising sound from venting safety devices or discoloration of tank.
- ALWAYS stay away from tanks engulfed in fire.

SPILL OR LEAK
- ELIMINATE all ignition sources (no smoking, flares, sparks or flames in immediate area).
- All equipment used when handling the product must be grounded.
- Fully encapsulating, vapor protective clothing should be worn for spills and leaks with no fire.   • Do not touch or walk through spilled material.
- Stop leak if you can do it without risk.
- Use water spray to reduce vapors or divert vapor cloud drift. Avoid allowing water runoff to contact spilled material.   • Do not direct water at spill or source of leak.
- If possible, turn leaking containers so that gas escapes rather than liquid.
- Prevent entry into waterways, sewers, basements or confined areas.
- Isolate area until gas has dispersed.
- Consider igniting spill or leak to eliminate toxic gas concerns.

FIRST AID
- Move victim to fresh air.   • Call 911 or emergency medical service.
- Give artificial respiration if victim is not breathing.
- Do not use mouth-to-mouth method if victim ingested or inhaled the substance; give artificial respiration with the aid of a pocket mask equipped with a one-way valve or other proper respiratory medical device.
- Administer oxygen if breathing is difficult.
- Remove and isolate contaminated clothing and shoes.
- In case of contact with substance, immediately flush skin or eyes with running water for at least 20 minutes.
- In case of contact with liquefied gas, thaw frosted parts with lukewarm water.
- In case of burns, immediately cool affected skin for as long as possible with cold water. Do not remove clothing if adhering to skin.
- Keep victim warm and quiet.   • Keep victim under observation.
- Effects of contact or inhalation may be delayed.
- Ensure that medical personnel are aware of the material(s) involved and take precautions to protect themselves.

Page 181

Figure 24.26 The protective action distance must be monitored against wind-spread toxins.

TABLE 1 - INITIAL ISOLATION AND PROTECTIVE ACTION DISTANCES

		SMALL SPILLS (From a small package or small leak from a large package)				LARGE SPILLS (From a large package or from many small packages)			
		First ISOLATE in all Directions		Then PROTECT persons Downwind during-		First ISOLATE in all Directions		Then PROTECT persons Downwind during-	
ID No.	NAME OF MATERIAL	Meters (Feet)		DAY Kilometers (Miles)	NIGHT Kilometers (Miles)	Meters (Feet)		DAY Kilometers (Miles)	NIGHT Kilometers (Miles)
1005 1005	Ammonia, anhydrous Anhydrous ammonia	30 m	(100 ft)	0.1 km (0.1 mi)	0.2 km (0.1 mi)	150 m (500 ft)		0.8 km (0.5 mi)	2.3 km (1.4 mi)
1008 1008	Boron trifluoride Boron trifluoride, compressed	30 m	(100 ft)	0.1 km (0.1 mi)	0.6 km (0.4 mi)	300 m (1000 ft)		1.9 km (1.2 mi)	4.8 km (3.0 mi)
1016 1016	Carbon monoxide Carbon monoxide, compressed	30 m	(100 ft)	0.1 km (0.1 mi)	0.1 km (0.1 mi)	150 m (500 ft)		0.7 km (0.5 mi)	2.7 km (1.7 mi)
1017	Chlorine	60 m	(200 ft)	0.4 km (0.3 mi)	1.6 km (1.0 mi)	600 m (2000 ft)		3.5 km (2.2 mi)	8.0 km (5.0 mi)
1023 1023	Coal gas Coal gas, compressed	30 m	(100 ft)	0.1 km (0.1 mi)	0.1 km (0.1 mi)	60 m (200 ft)		0.3 km (0.2 mi)	0.4 km (0.3 mi)
1026 1026	Cyanogen Cyanogen gas	30 m	(100 ft)	0.2 km (0.1 mi)	0.9 km (0.5 mi)	150 m (500 ft)		1.0 km (0.7 mi)	3.5 km (2.2 mi)
1040 1040	Ethylene oxide Ethylene oxide with Nitrogen	30 m	(100 ft)	0.1 km (0.1 mi)	0.2 km (0.1 mi)	150 m (500 ft)		0.8 km (0.5 mi)	2.5 km (1.6 mi)
1045 1045	Fluorine Fluorine, compressed	30 m	(100 ft)	0.1 km (0.1 mi)	0.3 km (0.2 mi)	150 m (500 ft)		0.8 km (0.5 mi)	3.1 km (1.9 mi)
1048	Hydrogen bromide, anhydrous	30 m	(100 ft)	0.1 km (0.1 mi)	0.4 km (0.3 mi)	300 m (1000 ft)		1.5 km (1.0 mi)	4.5 km (2.8 mi)
1050	Hydrogen chloride, anhydrous	30 m	(100 ft)	0.1 km (0.1 mi)	0.4 km (0.2 mi)	60 m (200 ft)		0.3 km (0.2 mi)	1.4 km (0.9 mi)
1051	AC (when used as a weapon)	100 m	(300 ft)	0.3 km (0.2 mi)	1.1 km (0.7 mi)	1000 m (3000 ft)		3.8 km (2.4 mi)	7.2 km (4.5 mi)
1051 1051 1051	Hydrocyanic acid, aqueous solutions, with more than 20% Hydrogen cyanide Hydrogen cyanide, anhydrous, stabilized Hydrogen cyanide, stabilized	60 m	(200 ft)	0.2 km (0.1 mi)	0.6 km (0.4 mi)	400 m (1250 ft)		1.6 km (1.0 mi)	4.1 km (2.5 mi)
1052	Hydrogen fluoride, anhydrous	30 m	(100 ft)	0.1 km (0.1 mi)	0.5 km (0.3 mi)	300 m (1000 ft)		1.7 km (1.1 mi)	3.6 km (2.2 mi)

Figure 24.27 The green border pages provide isolation and protective action distances for several variables divided into categories for small and large spills and protective distances based on day and night.

The table provides isolation and protective action distances for both small (approximately 53 gallons [200 L] or less) and large spills (more than 53 gallons [200 L]) **(Figure 24.27)**. A *small spill* is one that involves a single, small package (up to a 55-gallon [208 L] drum), small cylinder, or small leak from a large package. A *large spill* is one that involves a spill from a large package or multiple spills from many small packages.

What This Means To You

It's very important to understand what TIH materials are and how to use the *ERG* when confronted with them at an incident. A *toxic inhalation hazard (TIH)* material is a liquid or a gas presumed or known to be so toxic to humans as to pose a severe hazard to health during transportation. Small amounts of these products can kill you.

In the *ERG*, isolation or evacuation distances are shown in the Guide's (orange-bordered pages) and in the Table of Initial Isolation and Protective Action Distances (green-bordered pages). These distances may be confusing if you aren't thoroughly familiar with the *ERG*.

Some Guides refer to non-TIH materials only (40 Guides) and some refer to both TIH and non-TIH materials (22 Guides). A Guide refers to both TIH and non-TIH materials only when the following sentences appear under the title *EVACUATION* and then *Spill*: See the Table of Initial Isolation and Protective Action Distances for highlighted substances. For nonhighlighted substances, increase, in the downwind direction, as necessary, the isolation distance shown under "PUBLIC SAFETY." If these sentences do not appear, then this particular guide refers to non-TIH materials only.

(continued)

> (concluded)
>
> When dealing with a TIH material (highlighted entries in the index lists), the isolation and evacuation distances are found directly in the green-bordered pages. The orange-bordered Guide pages also remind the user to refer to the green-bordered pages for evacuation-specific information involving highlighted materials.
>
> When dealing with a non-TIH material and the Guide refers to both TIH and non-TIH materials, an immediate isolation distance is provided under the heading *PUBLIC SAFETY*. It applies to the non-TIH materials only. In addition, for evacuation purposes, the Guide informs you under the title *EVACUATION* and then *Spill* to increase (for nonhighlighted substances) in the downwind direction, if necessary, the immediate isolation distance listed under *PUBLIC SAFETY*.
>
> For example, Guide 123, Gases -Toxic and/or Corrosive, the *PUBLIC SAFETY* section instructs the user as follows: *Isolate spill or leak area immediately for at least 100 meters (330 feet) in all directions*. In case of a large spill, the isolation area could be expanded from 100 meters (330 feet) to a distance deemed safe by the on-scene Incident Commander (IC) and emergency responders.
>
> If you are dealing with a non-TIH material and the Guide refers only to non-TIH materials, the immediate isolation and evacuation distances are specified as actual distances in the Guide (orange-bordered pages) and are not referenced in the green-bordered pages.

The list is further subdivided into daytime and nighttime situations. This division is necessary because atmospheric conditions significantly affect the size of a chemically hazardous area, and differences can be generally associated with typical daytime and nighttime conditions. The warmer, more active atmosphere normal during the day disperses chemical contaminants more readily than the cooler, calmer conditions common at night. Therefore, during the day, lower toxic concentrations may be spread over a larger area than at night, when higher concentrations may exist in a smaller area. The quantity of material spilled or released and the area affected are both important, but the single most critical factor is the concentration of the contaminant in the air.

As with the isolation distances provided in the orange-bordered pages, the initial isolation distances provided in the green-bordered pages are the distance within which all persons should be considered for evacuation in all directions from an actual hazardous materials spill/leak source. This distance will always be at least 100 feet (30 m). People in this area could be evacuated and/or sheltered-in-place (see Public Protection sections).

NOTE: If hazardous materials are on fire or have been leaking for longer than 30 minutes, this *ERG* table does not apply. Seek more detailed information on the involved material on the appropriate orange-bordered page in the *ERG*. Also, the orange-bordered pages provide recommended isolation and evacuation distances for nonhighlighted chemicals with poisonous vapors and situations where the containers are exposed to fire.

Emergency Response Centers

The *ERG* provides contact information for emergency response centers that can provide valuable assistance to first responders at haz mat incidents. Canada, Mexico, and the U.S. have government-operated emergency response centers. Contact numbers are provided in the white pages in both the front and the back of the *ERG*.

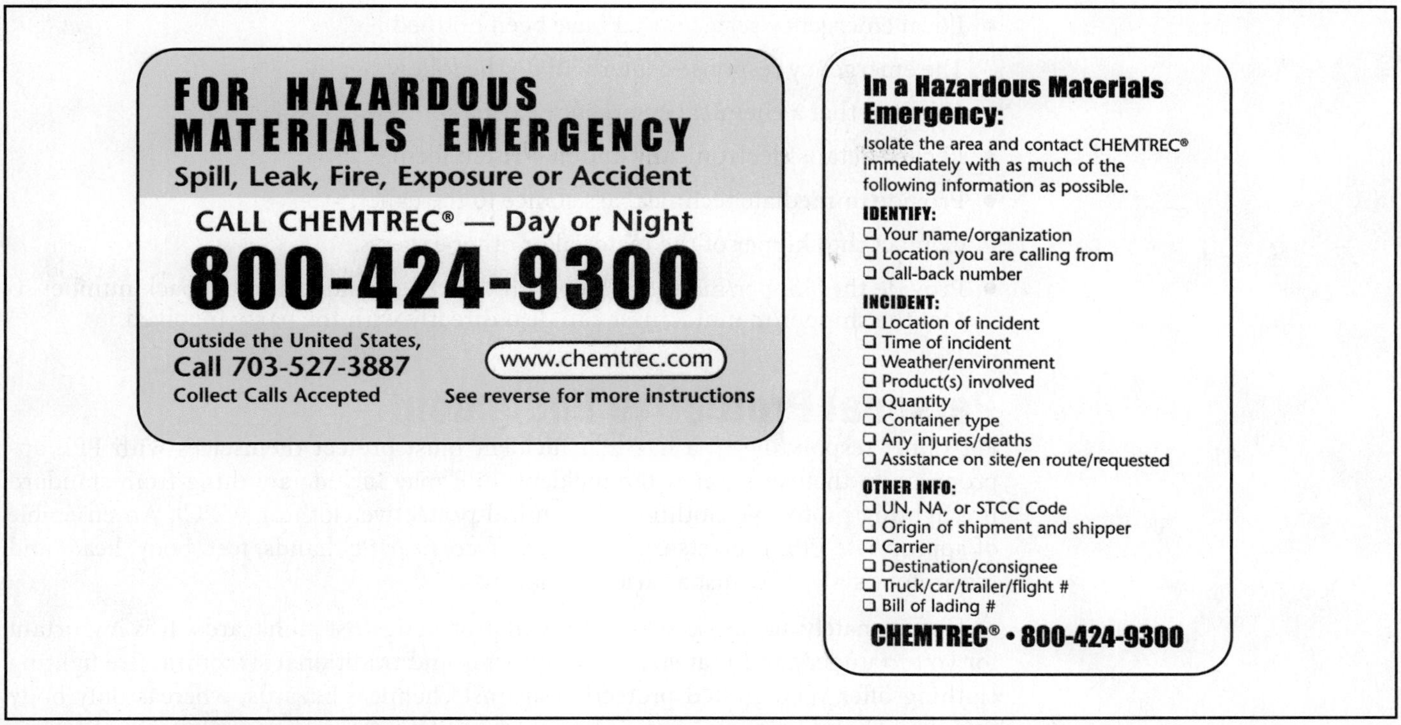

FOR HAZARDOUS MATERIALS EMERGENCY

Spill, Leak, Fire, Exposure or Accident

CALL CHEMTREC® — Day or Night

800-424-9300

Outside the United States,
Call 703-527-3887
Collect Calls Accepted

www.chemtrec.com

See reverse for more instructions

In a Hazardous Materials Emergency:

Isolate the area and contact CHEMTREC® immediately with as much of the following information as possible.

IDENTIFY:
❑ Your name/organization
❑ Location you are calling from
❑ Call-back number

INCIDENT:
❑ Location of incident
❑ Time of incident
❑ Weather/environment
❑ Product(s) involved
❑ Quantity
❑ Container type
❑ Any injuries/deaths
❑ Assistance on site/en route/requested

OTHER INFO:
❑ UN, NA, or STCC Code
❑ Origin of shipment and shipper
❑ Carrier
❑ Destination/consignee
❑ Truck/car/trailer/flight #
❑ Bill of lading #

CHEMTREC® • 800-424-9300

Figure 24.28 CHEMTREC® can provide first responders with information and assistance at haz mat incidents.

In the U.S., several emergency response centers, such as the Chemical Transportation Emergency Center (CHEMTREC®), are not government operated. CHEMTREC® was established by the chemical industry as a public service hotline for emergency responders to obtain information and assistance for emergency incidents involving chemicals and hazardous materials **(Figure 24.28)**. The Canadian Transport Emergency Centre (CANUTEC) is operated by Transport Canada. This national, bilingual (English and French) advisory center is part of the Transportation of Dangerous Goods Directorate. Mexico has two emergency response centers: (1) National Center for Communications of the Civil Protection Agency (CENACOM) and (2) Emergency Transportation System for the Chemical Industry (SETIQ), which is operated by the National Association of Chemical Industries. These centers are staffed with experts who can provide 24-hour assistance to emergency responders dealing with haz mat emergencies.

A list of emergency response centers and their telephone numbers is provided in the *ERG*. At an incident involving hazardous materials, you should collect and provide to the center as much of the following information as safely possible:

- Caller's name, callback telephone number, and FAX number
- Location and nature of problem (spill, fire, etc.)
- Name and identification number of material(s) involved
- Shipper/consignee/point of origin
- Carrier name, railcar reporting marks (letters and numbers), or truck number
- Container type and size
- Quantity of material transported/released
- Local conditions (weather, terrain, proximity to schools, hospitals, waterways, etc.)
- Injuries, exposures, current conditions involving spills, leaks, fires, explosions, and vapor clouds, etc.

- Local emergency services that have been notified

The emergency response center will do the following:

- Confirm that a chemical emergency exists.
- Record details electronically and in written form.
- Provide immediate technical assistance to the caller.
- Contact the shipper of the material or other experts.
- Provide the shipper/manufacturer with the caller's name and callback number so that the shipper/manufacturer can deal directly with the party involved.

Personal Protective Equipment

Personnel responding to a haz mat incident must protect themselves with PPE appropriate to their mission at the incident. PPE may include anything from standard fire fighting protective clothing to chemical-protective clothing (CPC). An ensemble of appropriate PPE protects the skin, eyes, face, hearing, hands, feet, body, head, and respiratory system against a variety of hazards.

Unfortunately, no single set of PPE will protect against all hazards. It is important for you to understand that standard uniforms and traditional structural fire fighting clothing offer very limited protection against chemical hazards, whereas only body armor and bomb suits can provide limited protection against projectiles and explosives that might be a concern at a terrorist attack. A combination (ensemble) of CPC and respiratory protections may offer protection against hazardous materials by protecting the routes of exposure **(Figure 24.29)**, but it will not provide adequate protection against thermal hazards (fire) or explosives. The *ERG* provides guidance on appropriate PPE for specific hazardous materials in the Protective Clothing section of the Orange-Bordered guide pages.

You must be aware of how your PPE is likely to perform in most situations. Currently, some structural and proximity PPE are certified for use at CBRN incidents based on the design requirements in NFPA® 1971. PPE specifically designed for CBRN incidents are regulated by NFPA® 1994, *Standard On Protective Ensembles For First Responders To CBRN Terrorism Incidents*. In any case, each type of PPE requires special training and instruction in its use.

Respiratory Protection

Inhalation is one of the major routes of exposure to hazardous substances. When used correctly, protective breathing equipment prevents the inhalation of chemical, biological, and radiological materials. All respiratory protective equipment has limitations; for example, self-contained breathing apparatus (SCBA) offer a limited working duration for supply of air. The basic types of protective breathing equipment used by firefighters at haz mat/WMD incidents are as follows:

- SCBA (open- and closed-circuit)
- Supplied-air respirators (SARs)
- Air-purifying respirators (APRs)
  — Particulate-removing
  — Vapor- and gas-removing
  — Combination particulate-removing and vapor- and-gas-removing
- Powered air-purifying respirators (PAPRs)

PPE Guards the Routes of Entry

**Respiratory Protection
Protects Against Inhalation
and Ingestion**

**Protective Clothing
Protects Against Skin
Contact**

Figure 24.29 An appropriate combination of protective equipment and clothing limits exposure to hazardous materials.

Responders may also need to be familiar with powered-air hoods and escape respirators. All these types of respiratory equipment are discussed in the sections that follow.

Self-Contained Breathing Apparatus (SCBA)

SCBA used for hazardous materials incidents are the same units that are used for structural fire fighting with a few exceptions. SCBA units that are used with hazardous materials chemical-protective clothing (CPC) must be compatible with that clothing. That is, quick-fill fittings on the suit must match those on the SCBA. In addition, the SCBA breathing air cylinders are higher capacity allowing for longer working time. The limitations of limited vision, communications, and mobility and added weight can affect their effectiveness.

Some SCBA can be specially certified for use against chemical, biological, radiological, and nuclear agents. Units that meet the strict certification guidelines receive a certification label from NIOSH. This label is placed in a visible location on the SCBA backplate, such as the upper corner or near the cylinder neck (**Figure 24.30, p. 1444**).

Figure 24.30 SCBA marked with this NIOSH label show compliance with CBRN criteria.

Figure 24.31 The EBSS should provide at least 5 minutes of air in case of an emergency, which should be enough to escape the hazard area into a safe atmosphere. *Courtesy of MSA.*

Supplied-Air Respirators

The supplied-air respirator (SAR) or airline respirator is an atmosphere-supplying respirator where the user does not carry the breathing air source. The apparatus usually consists of a facepiece, a belt- or facepiece-mounted regulator, a voice communications system, up to 300 feet (91 m) of air supply hose, an emergency escape pack or **emergency breathing support system (EBSS)**, and a breathing air source (either cylinders mounted on a cart or a portable breathing-air compressor) **(Figure 24.31)**. Because of the potential for damage to the air-supply hose, the EBSS provides enough air, usually 5, 10, or 15 minutes' worth, for the user to escape a hazardous atmosphere. SAR apparatus are not certified for fire fighting operations because of the potential damage to the airline from heat, fire, or debris.

NIOSH classifies SARs as Type C respirators. Type C respirators are further divided into two approved types: One type consists of a regulator and facepiece only, while the second consists of a regulator, facepiece, and EBSS. This second type may also be referred to as a SAR with escape (egress) capabilities. It is used in confined-space environments, IDLH environments, or potential IDLH environments. SARs used at haz mat or CBR incidents must provide positive pressure to the facepiece.

SAR apparatus have the advantage of reducing physical stress to the wearer by removing the weight of the SCBA. The air supply line is a limitation because of the potential for mechanical or heat damage. In addition, the length of the airline (no more than 300 feet [91 m] from the air source) restricts mobility. Problems with hose entanglement must be addressed, and it is necessary to exit the work area in exactly the same way it was entered. Other limitations are the same as those for SCBA: restricted vision and communications.

Air-Purifying Respirators

Air-purifying respirators (APRs) contain an air-purifying filter, canister, or cartridge that removes specific contaminants found in ambient air as it passes through the air-purifying element. Based on which respirator is being used, these purifying elements are generally divided into the three following types:

- Particulate-removing APRs
- Vapor- and gas-removing APRs
- Combination particulate-removing and vapor- and gas-removing APRs

APRs may be powered (PAPRs) or nonpowered. APRs do not supply oxygen or air from a separate source, and they protect only against specific contaminants at or below certain concentrations. Combination filters combine particulate-removing elements with vapor-and-gas-removing elements in the same cartridge or canister.

Respirators with air-purifying filters may have either full facepieces that provide a complete seal to the face and protect the eyes, nose, and mouth or half facepieces that provide a complete seal to the face and protect the nose and mouth **(Figure 24.32)**. It should be noted that half-face respirators will NOT protect against chemical, biological, and radiological materials that can be absorbed through the skin or eyes and therefore are not recommended for use at terrorist incidents except in very specific situations (such as explosive attacks where the primary hazard is dust).

Figure 24.32 Half-face APRs will provide respiratory protection in accordance with the filter system used, but they are not appropriate for use in hazardous atmospheres where skin contact and absorption through mucous membranes is a concern. *Courtesy of FEMA News Photos, photo by Jocelyn Augustino.*

Disposable filters, canisters, or cartridges are mounted on one or both sides of the facepiece. Canister or cartridge respirators pass the air through a filter, sorbent, catalyst, or combination of these items to remove specific contaminants from the air. The air can enter the system either from the external atmosphere through the filter or sorbent or when the user's exhalation combines with a catalyst to provide breathable air.

No single canister, filter, or cartridge protects against all chemical hazards. Therefore, you must know the hazards present in the atmosphere in order to select the appropriate canister, filter, or cartridge. You should be able to answer the following questions before deciding to use APRs for protection at an incident:

- What is the hazard?
- Is the hazard a vapor or a gas?
- Is the hazard a particle or dust?
- Is there some combination of dust and vapors present?
- What concentrations are present?

APRs do *not* protect against oxygen-deficient or oxygen-enriched atmospheres, and they must not be used in situations where the atmosphere is immediately dangerous to life and health. The three primary limitations of an APR are as follows:

- Limited life of its filters and canisters
- Need for constant monitoring of the contaminated atmosphere
- Need for a normal oxygen content of the atmosphere before use

The following precautions should be taken before using APRs:

- Know what chemicals/air contaminants are in the air.
- Know how much of the chemicals/air contaminants are in the air.
- Ensure that the oxygen level is between 19.5 and 23.5 percent.
- Ensure that atmospheric hazards are below IDLH conditions.

At haz mat incidents, APRs may be used after emergency operations are over and the hazards at the scene have been properly identified. APRs used for these situations should utilize a combination organic vapor/high-efficiency particulate air (OV/HEPA) cartridge (see following sections).

Powered Air-Purifying Respirators

Powered air-purifying respirators (PAPRs) use a blower to pass contaminated air through a canister or filter to remove the contaminants and supply the purified air to the full facepiece **(Figure 24.33)**. Because the facepiece is supplied with positive pressure, PAPRs offer a greater degree of safety than standard APRs in case of leaks or poor facial seals. Therefore they may be of use at CBR incidents for personnel conducting decontamination operations and long-term operations. Air flow also makes PAPRs more comfortable to wear for many people.

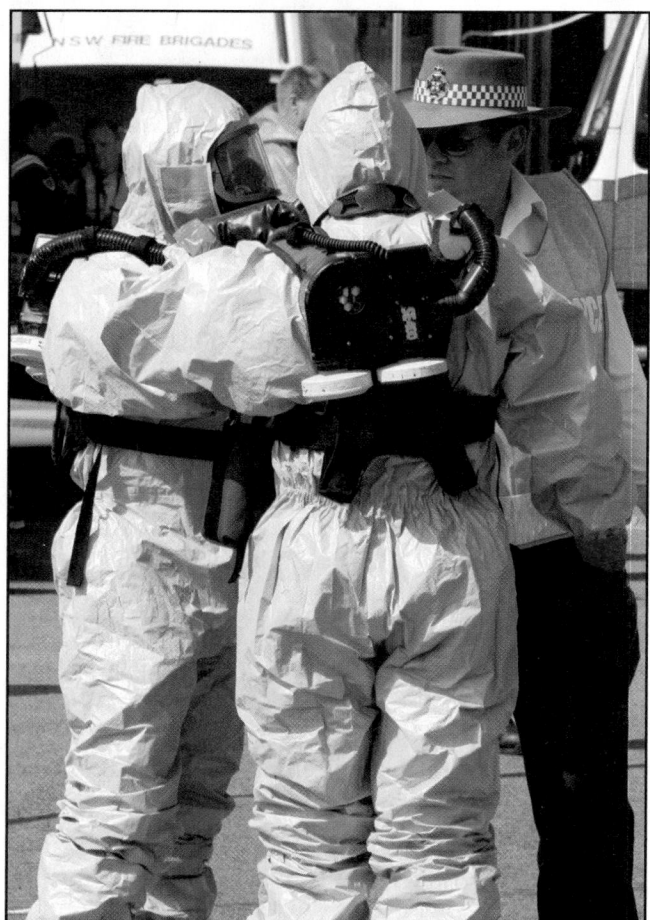

Figure 24.33 Powered air-purifying respirators may be more effective and more comfortable than other APRs for long-term incidents. *Courtesy of New South Wales Fire Brigades.*

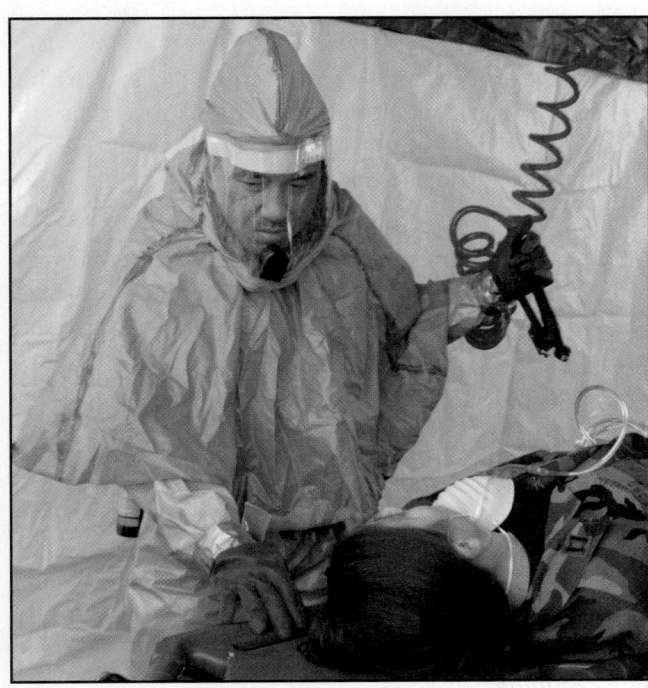

Figure 24.34 Powered- and supplied-air hoods are easy to use and may accommodate glasses and facial hair. *Courtesy of the U.S. Air Force, photo by Airman 1st class Bradley A. Lail.*

Several types of PAPR are available. Some units have a small battery-powered blower and are worn on a belt. Other units have a stationary blower (usually mounted on a vehicle) that is connected by a long, flexible tube to the respirator facepiece.

As with all APRs, PAPRs should only be used in situations where the atmospheric hazards are understood and at least 19.5 percent oxygen is present. PAPRs are not safe to wear in atmospheres where potential respiratory hazards are unidentified, nor should they be used during initial emergency operations. **Do *not* use PAPRs in explosive or potentially explosive atmospheres.** Continuous atmospheric monitoring is needed to ensure the safety of the responder.

Supplied-Air Hoods

Powered- and supplied-air hoods provide loose fitting, lightweight respiratory protection that can be worn with glasses, facial hair, and beards **(Figure 24.34)**. Hospitals, emergency rooms, and fire and emergency services organizations use these hoods as an alternative to other respirators, in part because they require no fit testing and are simple to use.

Escape Respirators

Escape respirators — sometimes called *personal escape canisters* or *escape hoods* — are designed to provide a short duration of protection, just enough for the wearer to escape the hot zone. Self-contained models typically utilize rebreathing technology, while air-purifying models are usually equipped with a HEPA filter and chemical filter, such as activated carbon. Both types are typically hooded, making it easy to put them on quickly. The hoods have a flexible seal around the neck and can accommodate glasses, facial hair, and beards.

The filter canisters of APR-style escape respirators are usually designed for single use and must be disposed of properly. Some manufacturers provide escape respirator cases that can be strapped onto the body and worn as part of an emergency PPE ensemble.

Protective Clothing

Protective clothing must be worn whenever personnel face potential hazards arising from exposure to hazardous materials. Skin contact with hazardous materials can cause a variety of problems, including chemical burns, allergic reactions and rashes, diseases, and absorption of toxic materials into the body. Protective clothing is designed to protect against these problems.

No single combination of protective equipment can protect against all hazards. For example, the fumes and vapors of many chemicals (including chemical weapons) can easily penetrate structural fire fighting protective clothing. Similarly, CPC offers limited flash protection from fires.

Several types of personal protective clothing are available:

- Structural fire fighting protective clothing
- High-temperature protective clothing
- Chemical-protective clothing (CPC)
  — Liquid-splash protective clothing
  — Vapor-protective clothing

Structural Fire Fighting Protective Clothing

Structural fire fighting clothing is not a substitute for CPC, but it may provide limited protection against some hazardous materials. The multiple layers of the coat and trousers may provide short-term exposure protection; however, to avoid harmful exposures, responders must recognize the limitations of this level of protection. For example, structural fire fighting clothing is neither corrosive-resistant nor vapor-tight. Liquids can soak through, acids and bases can dissolve or deteriorate the outer layers, and gases and vapors can penetrate the garment. Gaps in structural fire fighting clothing occur at the neck, wrists, waist, and the point where the trousers and boots overlap (**Figure 24.35**).

Inadequate Vapor Protection

Figure 24.35 While wearing structural fire fighting protective clothing, fumes and gases can penetrate through gaps in the material and ensemble.

Firefighters should also be aware that some hazardous materials can **permeate** and remain in the protective equipment. Chemicals absorbed into the equipment can subject the wearer to repeated exposure or to a later reaction with another chemical. In addition, the rubber or neoprene in boots, gloves, kneepads, and SCBA facepieces can become permeated by chemicals and rendered unsafe for use. It may be necessary to discard any equipment exposed to permeating types of chemicals.

Permeation — Process in which a chemical passes through a protective material on a molecular level.

Structural fire fighting protective clothing may be appropriate for use at haz mat incidents when the following conditions are met:

- Contact with splashes of extremely hazardous materials is unlikely.

- The material's hazards have been identified, and they will not rapidly damage or permeate structural fire fighting protective clothing.

- Total atmospheric concentrations do not contain high levels of chemicals that are toxic to the skin, and there are no adverse effects from chemical exposure to small areas of unprotected skin.

- There is a chance of fire or there is a fire (for example, a flammable liquid fire), and this type of protection is appropriate.

- When structural fire fighting protective clothing is the only PPE available and CPC is not immediately available.

At terrorism events, structural fire fighting protective clothing will provide protection against thermal damage in an explosive attack, but limited or no protection against projectiles, shrapnel, and other mechanical effects from a blast. It will provide adequate protection against some types of radiological materials, but not others. In cases where biological agents are strictly respiratory hazards, structural fire fighting protective clothing with SCBA may provide adequate protection. However, in any case where skin contact is potentially hazardous, the protection provided is not sufficient. Obviously, hazardous materials must be properly identified in order to make this determination, and any time a terrorist attack is suspected but not positively identified, it should be assumed that firefighters wearing structural fire fighting protective clothing with SCBA are still at some degree of risk.

Structural fire fighting PPE is being designed with greater chemical, biological, and radiological protection in mind. Closure and interface systems (such as magnetic seals) intended to prevent exposure to hazardous materials penetrating through gaps are currently available on CBRN-certified clothing (**Figure 24.36**). More chemical-resistant materials are being designed to mitigate problems with permeation and degradation, thereby making it safer for firefighters to respond to incidents involving hazardous materials wearing structural fire fighting PPE alone.

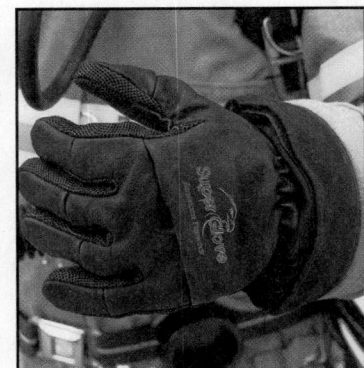

Figure 24.36 Specially designed glove/sleeve rings use strong, heat-resistant magnets to create a secure seal. *Courtesy of IAFF and Morning Pride Manufacturing.*

High-Temperature Protective Clothing

High-temperature protective clothing is designed to withstand short-term exposure to extremely high temperatures that are beyond the capabilities of standard fire fighting protective clothing (**Figure 24.37, p. 1450**). This type of clothing is usually of limited use in dealing with chemical hazards. Two basic types of high-temperature clothing that are available are as follows:

- **Proximity suits** — Permit close approach to fires for rescue, fire suppression, and property conservation activities such as in aircraft rescue and fire fighting or other fire fighting operations involving flammable liquids. Such suits provide greater heat protection than standard structural fire fighting protective clothing.

Figure 24.37 High temperature protective clothing is designed for operations involving temperatures outside the effectiveness range of standard fire fighting protective clothing. *Courtesy of William D. Stewart.*

- **Fire-entry suits** — Allow a person to work in total flame environments for short periods of time; provide short-duration and close-proximity protection at radiant heat temperatures as high as 2,000°F (1 093°C). Each suit has a specific use and is not interchangeable. Fire-entry suits are not designed to protect the wearer against chemical hazards.

High-temperature protective clothing has the following limitations:

- It contributes to heat stress on the wearer by not allowing the body to release excess heat.
- It is bulky.
- It limits wearer's vision.
- It limits wearer's mobility.
- It limits communication.
- It requires frequent and extensive training for efficient and safe use.
- It is expensive to purchase.

Chemical-Protective Clothing

The purpose of chemical-protective clothing and equipment is to shield or isolate individuals from chemical and biological hazards that may be encountered during hazardous materials operations. CPC is made from a variety of different materials, ***none of which protects against all types of chemicals*** (**Figure 24.38**). Each material provides protection against certain chemicals or products, but only limited or no protection against others. The manufacturer of a particular suit must provide a list of chemicals for which the suit is effective. Selection of appropriate CPC depends on

Figure 24.38 Each type of chemical protection material is effective against a specific range of hazards. *Courtesy of U.S. Air Force, photo by A1C Jason Epley.*

the specific chemical and on the specific tasks to be performed by the wearer. CPC provides little, if any resistance to mechanical hazards (cuts, tears, and punctures), and, with some exceptions, very little resistance to heat and flame.

WARNING
No single type of CPC protects against all chemical hazards.

CPC is designed to afford the wearer a known degree of protection from a known type, concentration, and length of exposure to a hazardous material, but only if it is fitted properly and worn correctly. Improperly worn equipment can expose and endanger the wearer.

Most protective clothing is designed to be impermeable to moisture, thus limiting the transfer of heat from the body through natural evaporation. This can contribute to heat disorders in hot environments. Other factors include the garment's degradation, permeation, and penetration abilities and its service life (see information box). A written management program regarding selection and use of CPC is recommended. Regardless of the type of CPC worn at an incident, it must be decontaminated before storage or disposal. Responders who are required to wear CPC must be familiar with their local procedures for technical decontamination.

WARNING
Responders must have sufficient training to operate in conditions requiring the use of chemical-protective clothing.

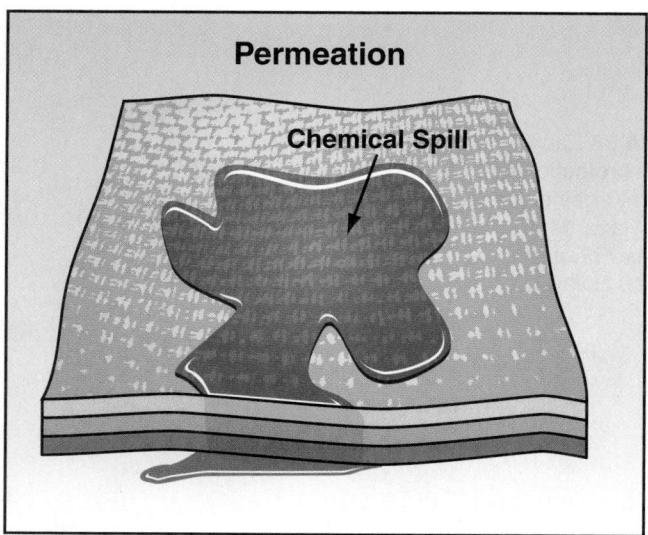

Figure 24.39 Chemical permeation of fabric may go completely unnoticed.

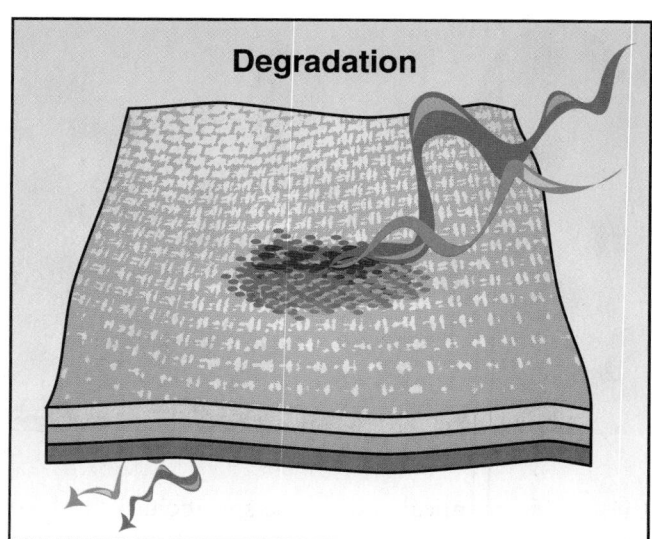

Figure 24.40 Brittleness in the outer layers of protective clothing is an indicator of chemical degradation.

Permeation, Degradation, and Penetration

The effectiveness of CPC can be reduced by three actions: permeation, degradation, and penetration. These are also characteristics that must be considered when choosing and using protective ensembles.

Permeation is a process that occurs when a hazardous material passes through a fabric on a molecular level **(Figure 24.39)**. In most cases, there is no visible evidence of chemicals permeating a material. The rate at which a chemical permeates CPC depends on factors such as the chemical properties of the hazardous material, nature of the protective barrier in the CPC, and concentration of the chemical on the surface of the CPC. Most CPC manufacturers provide charts on **breakthrough time** for a wide range of chemical compounds. Permeation data also includes information about the permeation rate or the speed at which the chemical moves through the CPC material after it breaks through.

Degradation occurs when the characteristics of a material are altered after contact with chemical substances. Examples include cracking, brittleness, and other changes in the structural characteristics of the garment **(Figure 24.40)**. The most common observations of material degradation are discoloration, swelling, loss of physical strength, or deterioration.

Penetration is a process that occurs when a hazardous material enters an opening or a puncture in a protective material. Rips, tears, and cuts in protective materials as well as unsealed seams, buttonholes, and zippers are considered penetration failures. Often such openings are the result of faulty manufacture, repair, or problems inherent to the design of the suit **(Figure 24.41)**.

Breakthrough Time — Time required for a chemical to permeate the material of a protective suit.

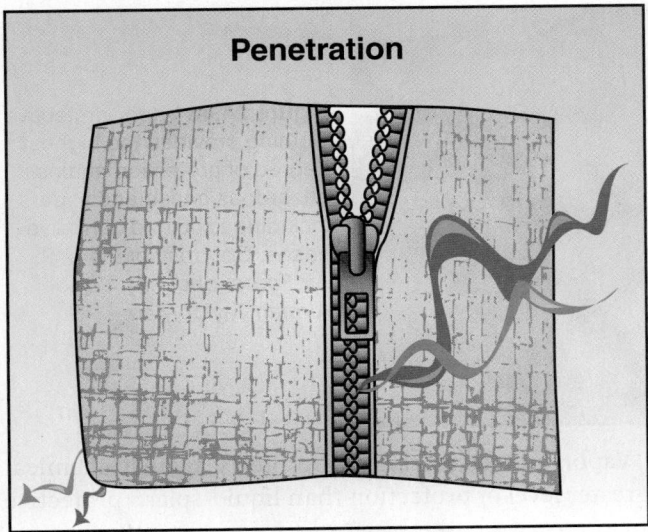

Penetration

Figure 24.41 Hazardous materials can penetrate PPE through openings in the fabric, including unsealed buttonholes and zippers.

Design and testing standards generally recognize two types of CPC: **liquid-splash protective clothing** and **vapor-protective clothing**. The sections that follow describe these two types.

Liquid-splash protective clothing. Liquid-splash protective clothing is primarily designed to protect users from chemical liquid splashes, but not against chemical vapors or gases. The material of liquid-splash protective clothing is made from the same types of material used for vapor-protective suits. When used as part of a protective ensemble, liquid-splash protective ensembles may use an SCBA, an airline (SAR), or a full-face, air-purifying, canister-equipped respirator. This type of protective clothing is a component of the U.S. Environmental Protection Agency (EPA) Level B chemical protection ensembles (see U.S. EPA Levels of Protection section), and Class 3 ensembles to be used at chemical and biological terrorist incidents as specified in NFPA® 1994. Liquid-splash protective clothing can be **encapsulating** or nonencapsulating **(Figure 24.42)**.

Fully encapsulating suits have two primary limitations:

1. They impair worker mobility, vision, and communication.

2. They trap body heat; this necessitates the use of a cooling vest, particularly when SCBA is also worn.

A nonencapsulating suit commonly consists of a one-piece coverall, but sometimes is composed of individual pieces such as a jacket, hood, pants, or bib overalls. Gaps between pant cuffs and boots and between gloves and sleeves are usually taped closed. Limitations of nonencapsulating suits include:

• They protect against splashes and dusts but not against gases and vapors.

• They do not provide full body coverage: parts of head and neck are often exposed.

• They trap body heat and contribute to heat stress.

Neither encapsulating nor nonencapsulating liquid-splash protective clothing are resistant to heat or flame exposure. They also do not protect against projectiles or shrapnel.

Figure 24.42 Nonencapsulating protective clothing may be used in some liquid-splash applications.

Figure 24.43 Vapor-protective clothing provides the highest degree of protection against hazardous gases and vapors including toxic and corrosive gases. *Courtesy of the U.S. Air Force, photo by Senior Airman Taylor Marr.*

Vapor-protective clothing. Vapor-protective clothing protects against chemical vapors and gases and offers a greater level of protection than liquid-splash protective clothing **(Figure 24.43)**. Vapor-protective ensembles must be worn with positive-pressure SCBA or combination SCBA/SAR. Vapor-protective clothing is a component of Class 1 and 2 ensembles to be used at chemical and biological terrorist incidents, as specified in NFPA® 1994. These suits are also used as part of an EPA Level A protective ensemble, providing the greatest degree of protection against respiratory, eye, or skin damage from hazardous vapors, gases, particulates, sudden splash, immersion, or contact with hazardous materials. Limitations of vapor-protective suits include the following:

- They do not protect against all chemical hazards.

- They impair mobility, vision, and communication.

- They do not allow body heat to escape; this can contribute to heat stress, which may necessitate the use of a cooling vest.

- Many suits cannot withstand high temperatures or fire conditions.

Physical Limitations of PPE

Wearing PPE with respiratory equipment causes physiological and psychological stress and may induce claustrophobia. To wear this equipment, personnel must be mentally sound, emotionally stable, and in good physical condition. Wearing this equipment has significant limitations, such as:

- **Limited visibility** — Facepieces reduce peripheral vision, and facepiece fogging can reduce overall vision.

- **Decreased ability to communicate** — Facepieces hinder voice communication.

- **Increased weight** — Depending on the model, the protective breathing equipment can add 25 to 35 pounds (11 kg to 16 kg) of weight.

- **Decreased mobility** — The increase in weight reduces the wearer's mobility.

- **Inadequate oxygen levels** — APRs cannot be worn in IDLH or oxygen-deficient atmospheres.

- **Chemical specific** — APRs can only protect against specific chemicals. The type of cartridge required depends on the chemical to be protected against.

- **Chemical limitations** — No one type of PPE will protect against all hazards.

Additionally, SCBA have maximum air-supply durations that limit the amount of time a first responder has to perform tasks. SCBAs not certified to NIOSH's CBRN standard may offer only limited protection in environments containing chemical warfare agents.

PPE Ensembles: U.S. EPA Levels of Protection

The approach in selecting PPE must encompass an *ensemble* of clothing and equipment items that are easily integrated to provide both an appropriate level of protection and still allow you to perform activities involving hazardous materials. For example, simple protective clothing such as gloves and a work uniform in combination with a faceshield (or safety goggles) may be sufficient to prevent exposure to certain etiological agents (such as bloodborne pathogens). At the other end of the spectrum, the use of vapor-protective, totally-encapsulating suits combined with positive-pressure SCBA is considered the minimum level of protection necessary when dealing with vapors, gases, or particulates of material that are harmful to skin or capable of being absorbed through the skin.

The U.S. EPA has established the following levels of protective equipment to be used at incidents involving CBR materials: Level A, Level B, Level C, and Level D (**Table 24.2, p. 1456**). NIOSH, OSHA, and the U.S. Coast Guard (USCG) also recognize these levels. They can be used as the starting point for ensemble creation; however, each ensemble must be tailored to the specific situation in order to provide the most appropriate level of protection. **Table 24.3, p. 1458** matches EPA levels of protection with the NFPA®'s applicable performance standards.

Level A

Level A ensembles must thoroughly protect the skin, eyes, and respiratory tract; they provide the highest level of protection against vapors, gases, mists, and particulates. Components include positive-pressure SCBA, totally encapsulating chemical-protective suit, inner and outer gloves, and chemical-resistant boots. Operations-Level responders are generally not allowed to operate in situations requiring **Level A protection**. If they are required to wear it, they must be appropriately trained to wear Level A PPE.

Level A ensembles are used in the following situations:

- Chemical hazards are unknown or unidentified.

- Identified chemicals are severely hazardous to the skin, eyes, and respiratory system.

- Site operations and work functions involve a high potential for splash, immersion, or exposure to unexpected vapors, gases, or particulates of material that are harmful to skin or capable of being absorbed through intact skin.

- Substances are present with known or suspected skin toxicity or carcinogenicity.

- Operations are conducted in confined or poorly ventilated areas.

Level B

Level B protection requires a garment that includes an SCBA or a SAR and provides protection against splashes from a hazardous chemical. The components of this level consist of positive-pressure SCBA, hooded chemical-resistant suit, inner and outer gloves, and chemical-resistant boots. This ensemble is worn when the highest level of respiratory protection is necessary but a lesser level of skin protection is needed. A Level B ensemble provides liquid-splash protection, but little or no protection against chemical vapors or gases to the skin. Level B CPC may be encapsulating or nonencapsulating.

Level A Protection — Highest level of skin, respiratory, and eye protection that can be provided by personal protective equipment (PPE), as specified by the U.S. Environmental Protection Agency (EPA). Consists of positive-pressure self-contained breathing apparatus, totally encapsulating chemical-protective suit, inner and outer gloves, and chemical-resistant boots.

Level B Protection — Personal protective equipment that provides the highest level of respiratory protection, but a lesser level of skin protection. Consists of positive-pressure self-contained breathing apparatus, hooded chemical-resistant suit, inner and outer gloves, and chemical-resistant boots.

Table 24.2
Personal Protective Clothing

Hazardous Materials Protective Clothing
(Based upon U.S. Environmental Protection Agency [EPA] Levels of Protection)

Type	Uses
Level A (also called *vapor-protective*)	To protect emergency responders at hazardous materials incidents when: - Unknown or unidentified chemical hazards. - Identified chemical(s) are highly hazardous to respiratory system, skin, and eyes. - A high potential for splash, immersion, or exposure to unexpected vapors, gases, particulates of material that are harmful to skin or capable of being absorbed through the intact skin. - Substances are present with known or suspected skin toxicity or carcinogenicity. - Operations are conducted in confined or poorly ventilated areas.

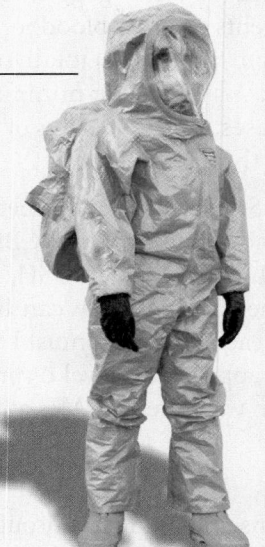

Type	Uses
Level B (also called *liquid-splash protective*)	To protect emergency responders at hazardous materials incidents when: - Substances identified and require a high level of respiratory protection but less skin protection. - Atmosphere contains less than 19.5 percent oxygen or more than 23.5 percent oxygen. - Presence of incompletely identified vapors or gases is indicated by a direct-reading organic vapor detection instrument, but the vapors and gases are known not to contain high levels of chemicals harmful to skin or capable of being absorbed through intact skin. - Presence of liquids or particulates is indicated, but they are known not to contain high levels of chemicals harmful to skin or capable of being absorbed through intact skin.

Table 24.2
Personal Protective Clothing (continued)

Hazardous Materials Protective Clothing
(Based upon U.S. Environmental Protection Agency [EPA] Levels of Protection)

Type	Uses
Level C	To protect emergency responders at hazardous materials incidents when: - Atmospheric contaminants, liquid splashes, or other direct contact will not adversely affect exposed skin or be absorbed through any exposed skin. - Types of air contaminants have been identified, concentrations have been measured, and an APR is available that can remove the contaminants. - All criteria for the use of APRs are met. - Atmospheric concentration of chemicals does not exceed IDLH levels. The atmosphere must contain between 19.5 and 23.5 percent oxygen.

Type	Uses
Level D (work uniforms, street clothing, coveralls, or firefighter structural protective clothing)	To protect emergency responders at hazardous materials incidents when: - Atmosphere contains no hazard. - Work functions preclude splashes, immersion, or the potential for unexpected inhalation of or contact with hazardous levels of any chemicals. **NOTE:** May not be worn in the hot zone and are not acceptable for haz mat emergency response above the Awareness Level.

Table 24.3
NFPA®/EPA Level Match

Performance-Based Standard	EPA Level
NFPA® 1991 (2005 Edition) worn with NIOSH CBRN SCBA	A
NFPA® 1994 (2007 Edition) Class 2 worn with NIOSH CBRN SCBA	B
NFPA® 1971 (2007 Edition) with CBRN option worn with NIOSH CBRN SCBA	B
NFPA® 1994 (2007 Edition) CLASS 3 worn with NIOSH CBRN APR/PAPR	C
NFPA® 1994 (2007 Edition) Class 4 worn with NIOSH CBRN APR/PAPR	C
NFPA® 1951 (2007 Edition) CBRN technical rescue ensemble worn with NIOSH CBRN APR/PAPR	C

Level B ensembles are used in the following situations:

- Type and atmospheric concentration of substances have been identified and require a high level of respiratory protection but less skin protection.

- Atmosphere contains less than 19.5 percent oxygen or more than 23.5 percent oxygen.

- Presence of incompletely identified vapors or gases is indicated by a direct-reading organic vapor detection instrument, but the vapors and gases are known not to contain high levels of chemicals harmful to skin or capable of being absorbed through intact skin.

- Presence of liquids or particulates is indicated, but they are known not to contain high levels of chemicals harmful to skin or capable of being absorbed through intact skin.

Level C

Level C Protection — Personal protective equipment that provides a lesser level of respiratory and skin protection than levels A or B. Consists of full-face or half-mask APR, hooded chemical-resistant suit, inner and outer gloves, and chemical-resistant boots.

Level C protection differs from Level B in the area of equipment needed for respiratory protection. Level C is composed of a full-face or half-mask APR (or PAPR), hooded splash-protecting garment, inner and outer gloves, and chemical-resistant boots. Emergency response personnel would not use this level of protection unless the specific material is known, it has been measured, and this protection level is approved by the Incident Commander (IC). All qualifying conditions for APRs and PAPRs would need to be met (that is, the product is known, an appropriate filter is available, the atmospheric oxygen concentration is between 19.5 to 23.5 percent, and the atmosphere is not IDLH). Periodic air monitoring is required when using this level of PPE.

Level C ensembles are used in the following situations:

- Atmospheric contaminants, liquid splashes, or other direct contact will not adversely affect exposed skin or be absorbed through any exposed skin.

- Types of air contaminants have been identified, concentrations have been measured, and an APR is available that can remove the contaminants.

- All criteria for the use of APRs are met.

- Atmospheric concentration of chemicals does not exceed IDLH levels. The atmosphere must contain between 19.5 and 23.5 percent oxygen.

Figure 24.44 MOPP ensembles are composed of five component pieces that may be individually selected for overall effectiveness against a specific CBR hazard.

Level D

Level D protection consist of coveralls (or work uniforms), gloves, and chemical-resistant boots or shoes. This PPE level is used only for nuisance contamination with no atmospheric hazards. There must be no potential for splashes, immersion, and contact or inhalation of hazardous materials. Level D ensembles may not be worn in the hot zone and are not acceptable for haz mat response above the Awareness Level.

NOTE: Structural fire fighting protective clothing is sometimes classified as Level D.

Mission-Oriented Protective Posture (MOPP)

The U.S. military uses Mission-Oriented Protective Posture (MOPP) ensembles to protect against chemical, biological, and radiological hazards. MOPP ensembles consist of an overgarment, mask, hood, overboots, and protective gloves (**Figure 24.44**). MOPPs provide six flexible levels of protection (0-4, plus Alpha) based on threat level, work rate for the mission, temperature, and humidity (**Table 24.4, p. 1460**). The higher the MOPP level, the greater the protection (and the lower the rate of work productivity and efficiency). MOPP Level Alpha is designed for use in situations upwind from any threat with little danger of exposure to hazardous vapors.

Permeable garments such as the joint service lightweight integrated suit technology (JSLIST) protective overgarment provide protection against liquid, solid, and/or vapor CB agents and radioactive alpha and beta particles. JSLIST overgarments are manufactured of lightweight 50 percent nylon and 50 percent cotton ripstop water repellant permeable materials. They are equipped with a charcoal/carbon lining designed to absorb harmful materials, much like the carbon filter/canister on an APR. The JSLIST protective garment will be degraded when in contact with certain solvents such as sweat and petroleum products. JSLIST can be laundered up to six times for personal hygiene. The JSLIST ensemble may be worn 45 consecutive days with a total out-of-the-bag available usage of 120 days. The protective mask is a hooded APR and the boots and gloves are made of butyl rubber.

Table 24.4
MOPP Levels

Non-firefighter MOPP Levels					
	MOPP 0	MOPP 1	MOPP 2	MOPP 3	MOPP 4
JSLIST	Carried	Worn	Worn	Worn	Worn
Protective Mask	Carried	Carried	Carried	Worn	Worn
Cotton Insert Gloves	Carried	Carried	Carried	Carried	Worn
Butyl Rubber Gloves	Carried	Carried	Carried	Carried	Worn
Protective Overgarment Boots	Carried	Carried	Worn	Worn	Worn

Firefighter MOPP Levels				
	MOPP 0	MOPP 1	MOPP 2	MOPP 4 Fire Fighting Mode
JSLIST	Carried	Worn	Worn	Worn
Nomex Hood	Carried	Worn	Worn	Worn
Fire Fighting Protective Trousers	Carried	Carried	Worn	Worn
Fire Fighting Protective Jacket	Carried	Carried	Carried	Worn
Fire Fighting Protective Footwear	Carried	Carried	Worn	Worn
CW Mask	Carried	Carried	Worn	Worn
Fire- & Chemical-Protective Gloves	Carried	Carried	Worn	Worn
Cotton Insert Gloves *	Carried	Carried	Worn	Worn
Butyl Rubber Gloves *	Carried	Carried	Worn	Worn
Fire Fighting Gloves *	Carried	Carried	Carried	Worn
Structural ARFF Helmet	Carried	Carried	Carried	Worn
SCBA	Carried	Carried	Carried	Worn

* Per current military Technical Orders

Firefighters who wear the JSLIST overgarment will find that the MOPP levels are modified to suit their capabilities. With the additional wear and tear of fire fighting protective equipment that is commonly found by firefighters wearing the JSLIST, firefighters will use only three MOPP levels (0, 1, 4). When the typical wearer is in MOPP 2, firefighters will remain in MOPP 1 until directed to proceed to a higher level. When MOPP 3 is directed, firefighters will immediately proceed from MOPP 1 and don the appropriate MOPP 4 attire. When a need to conduct fire fighting operations arises, firefighters will proceed to MOPP 4 firefighter mode where they will don their fire fighting protective jacket and SCBA.

PPE Selection Factors

The risks and potential hazards present at an incident will determine the PPE needed. In accordance with local procedures, many available sources can be consulted to determine which type and what level of PPE to use at haz mat/WMD incidents depending on the circumstances and hazards at the scene. First-arriving responders often rely upon information in the *ERG* to determine the minimum type of protection required for defensive operations. SOPs also provide guidance for situations involving defensive operations, rescue, and initial responses. Knowing the name and exposure hazard of the material aids in selecting proper PPE as well.

NOTE: The *ERG* will only reference EPA ensemble Levels A, B, and D, since SCBA is always recommended.

In general, the greater the associated risks, the higher the required level of PPE. For any given situation, select equipment and clothing that provide an adequate level of protection. **Overprotection as well as underprotection can be hazardous and should be avoided.**

Health Issues

Most types of PPE inhibit the body's ability to disperse heat. Because you are usually performing strenuous work while wearing PPE, your risk of developing heat-related disorders increases. However, when working in cold climates, considerations must be taken to protect yourself from cold-related disorders, as well. For example, CPC is not designed to provide insulation against the cold. It is important to take preventive measures to reduce the effects of any temperature extreme. Medical monitoring of responders is required when they are at risk because of environmental hazards.

Heat Disorders

Wearing PPE or other special full-body protective clothing puts the wearer at considerable risk of developing **heat stress**. It is important for you to understand the effects of heat stress. This stress can result in health effects ranging from transient heat fatigue to serious illness or death. Keep in mind that humidity has an effect on how you perceive temperature; the higher the humidity, the hotter it feels to you **(Table 24.5)**.

You need to be aware of several heat disorders, including **heat stroke** (the most serious), **heat exhaustion**, **heat cramps**, and **heat rashes**. In addition, you should know how to prevent the effects of heat exposure.

Preventing Heat Exposure

Methods of preventing or reducing the effects of heat exposure include:

- **Fluid consumption** — Drink generous amounts of water or electrolyte-replacement drinks to avoid dehydration. Drinking 7 ounces (200 ml) of fluid every 15 to 20 minutes is better than drinking large quantities once an hour. Before working, chilled fluids are good. But during and after work in an extremely hot environment, room temperature fluids are better, because they are less of a shock to the body. Balanced diets normally provide enough salts to avoid cramping problems.

- **Air cooling** — Wear long cotton undergarments or similar types of clothing to provide natural body ventilation. Once PPE has been removed, blowing air can help to evaporate sweat, thereby cooling the skin. Wind, fans, blowers, and misters can provide air movement. However, when ambient air temperatures and humidity are high, air movement provides limited benefits.

- **Ice cooling** — Use ice to cool the body. Be careful with this method, as direct contact can damage skin, and lowering body temperature too quickly can be harmful. Keep in mind that ice melts quickly. Ice cooling vests are also available.

- **Water cooling** — Use water to cool the body. Evaporating water (even sweat) cools the skin. Provide mobile showers and misting facilities or evaporative cooling vests. Water cooling becomes less effective as air humidity increases and water temperatures rise.

Heat Stress — Combination of environmental and physical work factors that compose the heat load imposed on the body; environmental factors include air, temperature, radiant heat exchange, air movement, and water vapor pressure. Physical work contributes because of the metabolic heat in the body; clothing also has an effect.

Heat Stroke — Heat illness in which the body's heat regulating mechanism fails; symptoms include (a) high fever of 105° to 106°F (40.5° to 41.1°C), (b) dry, red, hot skin, (c) rapid, strong pulse, and (d) deep breaths or convulsions. May result in coma or even death. *Also known as Sunstroke.*

Heat Exhaustion — Heat illness caused by exposure to excessive heat; symptoms include weakness, cold and clammy skin, heavy perspiration, rapid and shallow breathing, weak pulse, dizziness, and sometimes unconsciousness.

Heat Cramps — Heat illness resulting from prolonged exposure to high temperatures; characterized by excessive sweating, muscle cramps in the abdomen and legs, faintness, dizziness, and exhaustion.

Heat Rash — Condition that develops from continuous exposure to heat and humid air; aggravated by clothing that rubs the skin. Reduces the individual's tolerance to heat.

Table 24.5
NOAA's National Weather Service Heat Index

Temperature (°F)

Relative Humidity (%)	80	82	84	86	88	90	92	94	96	98	100	102	104	106	108	110
40	80	81	83	85	88	91	94	97	101	105	109	114	119	124	130	136
45	80	82	84	87	89	93	96	100	104	109	114	119	124	130	137	
50	81	83	85	88	91	95	99	103	108	113	118	124	131	137		
55	81	84	86	89	93	97	101	106	112	117	124	130	137			
60	82	84	88	91	95	100	105	110	116	123	129	137				
65	82	85	89	93	98	103	108	114	121	126	130					
70	83	86	90	95	100	105	112	119	126	134						
75	84	88	92	97	103	109	116	124	132							
80	84	89	94	100	106	113	121	129								
85	85	90	96	102	110	117	126	135								
90	86	91	98	105	113	122	131									
95	86	93	100	108	117	127										
100	87	95	103	112	121	132										

Likelihood of Heat Disorders with Prolonged Exposure or Strenuous Activity

☐ **Caution** ☐ **Extreme Caution** ☐ **Danger** ☐ **Extreme Danger**

Courtesy of NOAA

- **Cooling vests** — Wear cooling vests beneath PPE. These vests may use ice, fluids, evaporation, gels, or phase change cooling technology. Unlike the lower temperatures provided by ice or gel vests, phase change cooling technology vests interact with body heat to maintain the garment at a consistent temperature of 59°F (15°C). These vests may be bulky, cumbersome, and they may impair movement. Forced air cooling vest systems blow air through tubes close to the skin to cool the body.

  NOTE: Use of cooling vests is being reviewed in Canada and the U.S. due to various health concerns, and several haz mat teams have disallowed them.

- **Rest/rehab areas** — Provide shade, misters, and air-conditioned areas for resting.

- **Work rotation** — Rotate responders frequently if they are exposed to extreme temperatures or performing difficult tasks.

- **Proper liquids** — Before work, avoid alcohol, coffee, and other caffeinated drinks. These beverages can contribute to dehydration and heat stress.

- **Physical fitness** — Maintain good physical fitness.

Table 24.6
Wind Chill Chart

Temperature (°F)

Calm	40	35	30	25	20	15	10	5	0	-5	-10	-15	-20	-25	-30	-35	-40	-45
5	36	31	25	19	13	7	1	-5	-11	-16	-22	-28	-34	-40	-46	-52	-57	-63
10	34	27	21	15	9	3	-4	-10	-16	-22	-28	-35	-41	-47	-53	-59	-66	-72
15	32	25	19	13	6	0	-7	-13	-19	-26	-32	-39	-45	-51	-58	-64	-71	-77
20	30	24	17	11	4	-2	-9	-15	-22	-29	-35	-42	-48	-55	-61	-68	-74	-81
25	29	23	16	9	3	-4	-11	-17	-24	-31	-37	-44	-51	-58	-64	-71	-78	-84
30	28	22	15	8	1	-5	-12	-19	-26	-33	-39	-46	-53	-60	-67	-73	-80	-87
35	28	21	14	7	0	-7	-14	-21	-27	-34	-41	-48	-55	-62	-69	-76	-82	-89
40	27	20	13	6	-1	-8	-15	-22	-29	-36	-43	-50	-57	-64	-71	-78	-84	-91
45	26	19	12	5	-2	-9	-16	-23	-30	-37	-44	-51	-58	-65	-72	-79	-86	-93
50	26	19	12	4	-3	-10	-17	-24	-31	-38	-45	-52	-60	-67	-74	-81	-88	-95
55	25	18	11	4	-3	-11	-18	-25	-32	-39	-46	-54	-61	-68	-75	-82	-89	-97
60	25	17	10	3	-4	-11	-19	-26	-33	-40	-48	-55	-62	-69	-76	-84	-91	-98

WIND SPEED (mph)

Frostbite occurs in 15 minutes or less

Courtesy of NOAA

Cold Disorders

Cold temperatures caused by weather and/or other conditions such as exposure to cryogenic liquids must be considered when selecting PPE. Prolonged exposure to freezing temperatures can result in health problems as serious as **trench foot, frostbite,** and hypothermia. Protection from the cold must be a priority when conditions warrant.

The four primary environmental conditions that cause cold-related stress are low temperatures, high/cool winds, dampness, and cold water. When working outdoors, evaluate the possibility of wind chill, which is is a combination of temperature and velocity. When the air temperature of the wind is 40°F (4.4°C) and its velocity is 35 mph (56 km/h), exposed skin experiences conditions equivalent to the still-air temperature of 28°F (-2°C) (**Table 24.6**). This can cause dangerously rapid heat loss.

Medical Monitoring

Responders exposed to environmental hazards or hazardous materials should receive ongoing medical monitoring. Pre-entry monitoring should be provided before responders wearing chemical liquid-splash or vapor-protective clothing enter the hot zone; postentry monitoring should occur after they leave these zones, in a manner to be determined by the AHJ (**Figure 24.45, p. 1467**). The evaluation will check such things as vital signs, hydration, skin, mental status, and medical history. A postmedical monitoring follow-up is also recommended.

Trench Foot — Foot condition resulting from prolonged exposure to damp conditions or immersion in water; symptoms include tingling and/or itching, pain, swelling, cold and blotchy skin, numbness, and a prickly or heavy feeling in the foot. In severe cases, blisters can form, after which skin and tissue die and fall off.

Frostbite — Local tissue damage caused by prolonged exposure to extreme cold.

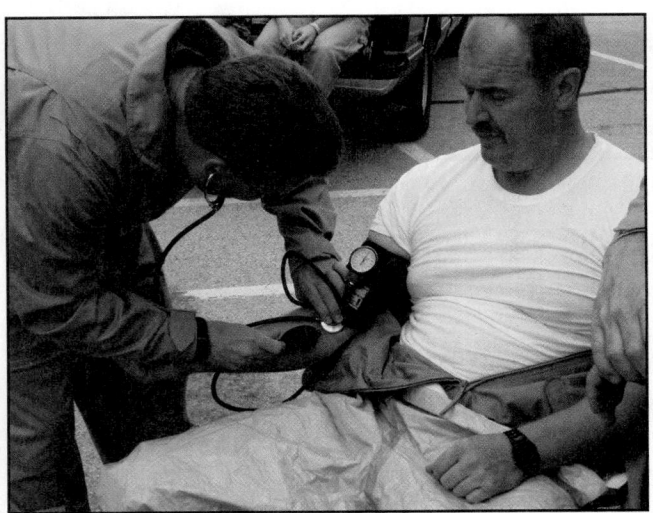

Figure 24.45 Medical monitoring is an essential function of a hazardous incident scene response plan.

Figure 24.46 The management plan for limited air supplies must include the time required for travel to and from the work zone. *Courtesy of the U.S. Navy, photo by JO2 Mark Schultz.*

Safety and Emergency Procedures

In addition to issues such as cooling, preventing dehydration, and medical monitoring, there are other safety and emergency issues involved with wearing PPE. The safety briefing will cover relevant information such as the status of the incident (based on the preliminary evaluation and subsequent updates), the hazards identified, a description of the site, the tasks to be performed, and the expected duration of the tasks. It must also cover PPE requirements, monitoring requirements, notification of identified risks, and any other pertinent information.

After using PPE at an incident, it is important to fill out any associated reports or documentation as required by the AHJ. Additional safety measures such as using buddy systems, accountability systems, and having backup personnel standing by at all times are discussed in the Protection of Responders section later in this chapter.

Anytime a limited air supply such as SCBA is worn, air management is an important consideration. Time taken to walk to the incident, time taken to return from the incident, decon time, safety time (extra time allocated for emergency use), and work time must be calculated **(Figure 24.46)**. Air must be allocated for these estimated times. Many organizations have SOPs that explain calculations for doing this and/or designate maximum entry times (such as 20 minutes) based on the air supply available.

Communications are also a concern. Have predesignated hand signals, motions, and gestures to communicate in case of problems, for example, a way to ask if everything is okay followed by a thumbs up or thumbs down. Signals for emergencies should also be designated (such as loss of air supply, medical emergency, or suit failure). If possible, entry teams, backup personnel, and appropriate safety personnel at the scene should have their own designated radio channel. Responders should always operate with buddy systems and with backups dressed in appropriate PPE (same level as the entry team) standing by. Responders should also be familiar with established evacuation signals.

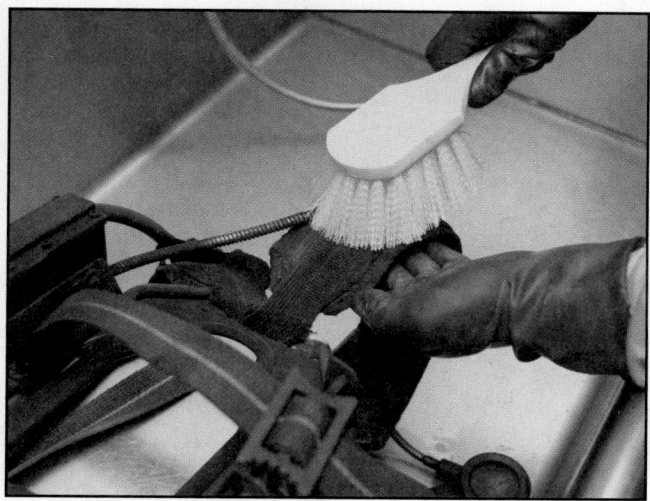

Figure 24.47 Personnel should wear protective gloves and other equipment while handling respiratory protection equipment (such as this SCBA backpack harness) that has been exposed to contaminants.

In case of an emergency, such as loss of air supply, suit integrity, or an injury and illness, all responders should follow local protocols for such situations. Typically, these protocols will involve notifying the appropriate personnel (such as the Entry Team Leader and/or Haz Mat Safety Officer) and exiting the hot zone as quickly as possible. If air supply is lost while wearing a vapor-protective suit, there is a limited amount of air in the suit itself that can be breathed if the SCBA facepiece or regulator is removed. Responders using PPE at haz mat incidents must be familiar with their local procedures for going through the technical decontamination process.

PPE Inspection, Storage, Testing, and Maintenance

A responder wearing PPE must ensure beforehand that the ensemble will perform as expected. Follow a standard program for inspection, testing, storage, maintenance, and cleaning. All inspections, testing, and maintenance of PPE must be conducted in accordance with manufacturer's recommendations. Know the limitations of your PPE, as this is the best way to avoid exposure to dangerous materials during an emergency response.

Keep detailed records of all inspections, and review them periodically; this will allow you to identify PPE that requires excessive maintenance or that is susceptible to failure. Records for each inspection should include:

- Identification number of the clothing or equipment
- Date of inspection
- Person making the inspection
- Results of the inspection
- Any unusual conditions noted

Respiratory equipment is initially inspected when it is purchased. Once the equipment is placed into service, the organization's personnel perform periodic inspections. Operational inspections of respiratory protection equipment occur after each use, daily or weekly, monthly, and annually. The organization must define the frequency and type of inspection in the respiratory protection policy, and should follow manufacturer's recommendations.

The care, cleaning, and maintenance schedules of respiratory protection equipment should be based on the manufacturer's recommendation, NFPA® standards, or OSHA requirements (**Figure 24.47**).

Figure 24.48 An isolation perimeter may be enforced using response vehicles to divert traffic away from specific areas.

Isolation and Scene Control

Isolation and scene control is one of the primary strategic goals at haz mat incidents and one of the most important means by which you can ensure the safety of yourself and others. Separating people from the potential source of harm is necessary to protect the life safety of all involved. It is also necessary to prevent the spread of hazardous materials through cross contamination (sometimes called *secondary contamination,* see Decontamination section). Isolation involves physically securing and maintaining the emergency scene by establishing isolation perimeters and denying entry to unauthorized persons. It also includes preventing contaminated or potentially contaminated individuals from leaving the scene in order to stop the spread of hazardous materials. The process may continue with either evacuation, defending in place, or shelter-in-place of people located within protective-action zones (see Protection section). Controlling the scene of a haz mat incident also include the establishment of hazard-control zones.

Isolation Perimeter

The *isolation perimeter* (sometimes called the *outer perimeter* or *outer cordon*) is the boundary established to prevent access by the public and unauthorized persons. It may be established even before the type of incident or attack has been positively identified. If an incident is inside a building, the isolation perimeter might be set at the exterior entrances, with personnel posted to deny unauthorized entry. If the incident is outside, the perimeter might be set at the surrounding street intersections, with response vehicles or law enforcement officers diverting traffic and pedestrians (**Figure 24.48**). In some cases, a traffic cordon may be set up beyond the isolation perimeter to prevent unauthorized vehicle access only, whereas pedestrian traffic is still allowed.

The isolation perimeter can be expanded or reduced as needed. For example, in some cases, the initial isolation perimeter established by first responders is expanded outward as additional help arrives. Law enforcement officers are often used to establish and maintain isolation perimeters. Once hazard-control zones are established, the isolation perimeter is generally considered to be the boundary between the public and the cold (safe) zone (see Hazard-Control Zones section).

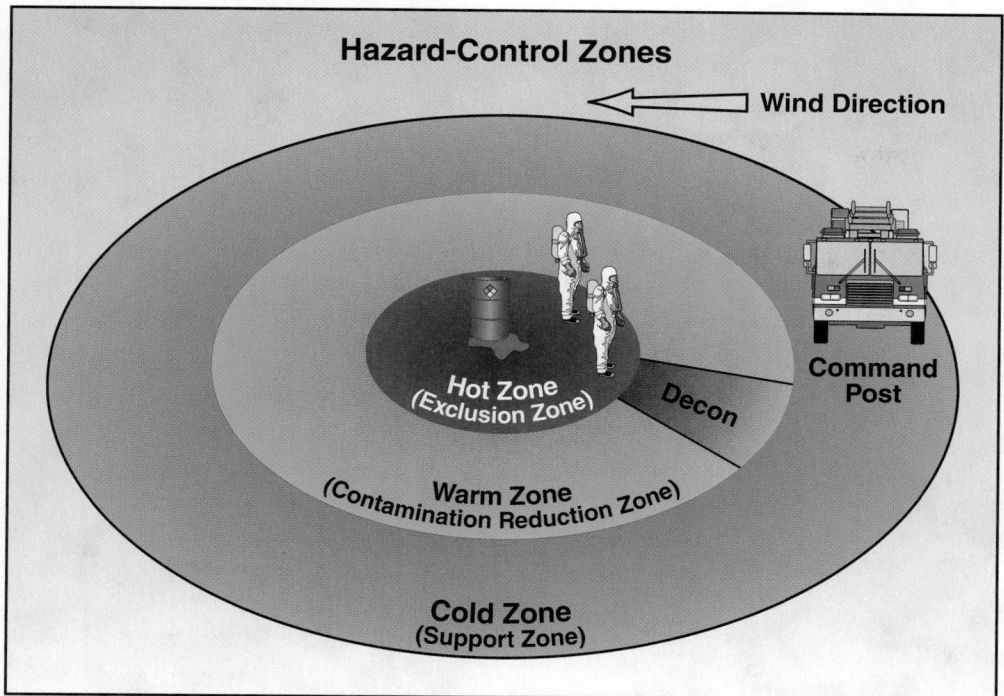

Figure 24.49 Hazard-control zones divide the levels of hazard at an incident into hot, warm, and cold zones with the hot zone indicating the highest concentration of hazard.

Hazard-Control Zones

Hazard-control zones provide the scene control required at haz mat/WMD incidents to protect responders from interference by unauthorized persons, help regulate movement of first responders within the zones, and minimize cross contamination. At large, multiagency response incidents, establishing hard perimeters for hazard-control zones helps to ensure accountability for all involved personnel.

These control zones are not necessarily static and can be adjusted as the incident changes. It should be noted that different agencies may have different needs in terms of establishing control zones. For example, law enforcement may designate a zone to incorporate an entire crime scene at a terrorist attack that would not correspond to traditional fire service activities. These law enforcement zones might change as evidence is processed and the crime scene is released. These dynamics create an extreme need for flexibility on the part of all agencies in the Unified Command when establishing these zones.

Zones divide the levels of hazard of an incident, and the designation of a zone generally depicts this level. These zones are often referred to as hot, warm, and cold **(Figure 24.49)**. It should be noted that there may be multiple control zones within the perimeter of the incident. For example, at a massive explosion, there may be multiple building collapses that might require their own set of control zones within the overall incident area.

U.S. OSHA and the U.S. EPA refer to these zones collectively as *site work zones*. They are sometimes called *scene-control zones* as well. Other countries may use different terminology for these zones with which responders must be familiar.

Figure 24.50 In most incidents, PPE specialized for the response environment will be required for entry into the hot zone. *Courtesy of the U.S. Coast Guard, photo by PA3 Brent Erb.*

Hot Zone

Hot Zone — Potentially hazardous area immediately surrounding the incident site; requires appropriate protective clothing and equipment and other safety precautions for entry. Typically limited to technician-level personnel. *Also known as* Exclusion Zone.

Traditionally, the **hot zone** (also called *exclusion zone*) is an area surrounding an incident that is potentially very dangerous either because it presents a threat in the form of a hazardous material or the effects thereof **(Figure 24.50)**. The area may be contaminated or have the potential to become contaminated by hazardous materials or chemical warfare agents. Responders must have proper training and appropriate personal protective equipment (PPE) to work in the hot zone or to support work being done inside the hot zone. Access and egress points will be established to ensure both accountability and designated PPE prior to entry.

The hot zone extends far enough to prevent people outside the zone from suffering ill effects from the released material, explosion, or other threat. Work performed inside the hot zone is often limited to highly trained personnel such as SWAT teams, US&R teams, hazardous materials technicians, Joint Hazard Assessment Teams (JHAT), and bomb technicians.

Warm Zone

Warm Zone — Area between the hot and cold zones that usually contains the decontamination corridor; typically requires a lesser degree of personal protective equipment than the Hot Zone. *Also known as* Contamination Reduction Zone or Contamination Reduction Corridor.

The **warm zone** (also called *contamination reduction zone* or *corridor*) is an area adjoining the hot zone and extending to the cold zone. The warm zone is used as a buffer between the hot and cold zones and is the place to decontaminate personnel and equipment exiting the hot zone. Decontamination usually takes place within a decon corridor located in the warm zone **(Figure 24.51)**. At incidents involving crimes, parts of the warm zone may also be considered to be part of the crime scene, with emphasis on minimal disturbance of possible evidence. PPE will normally be required in

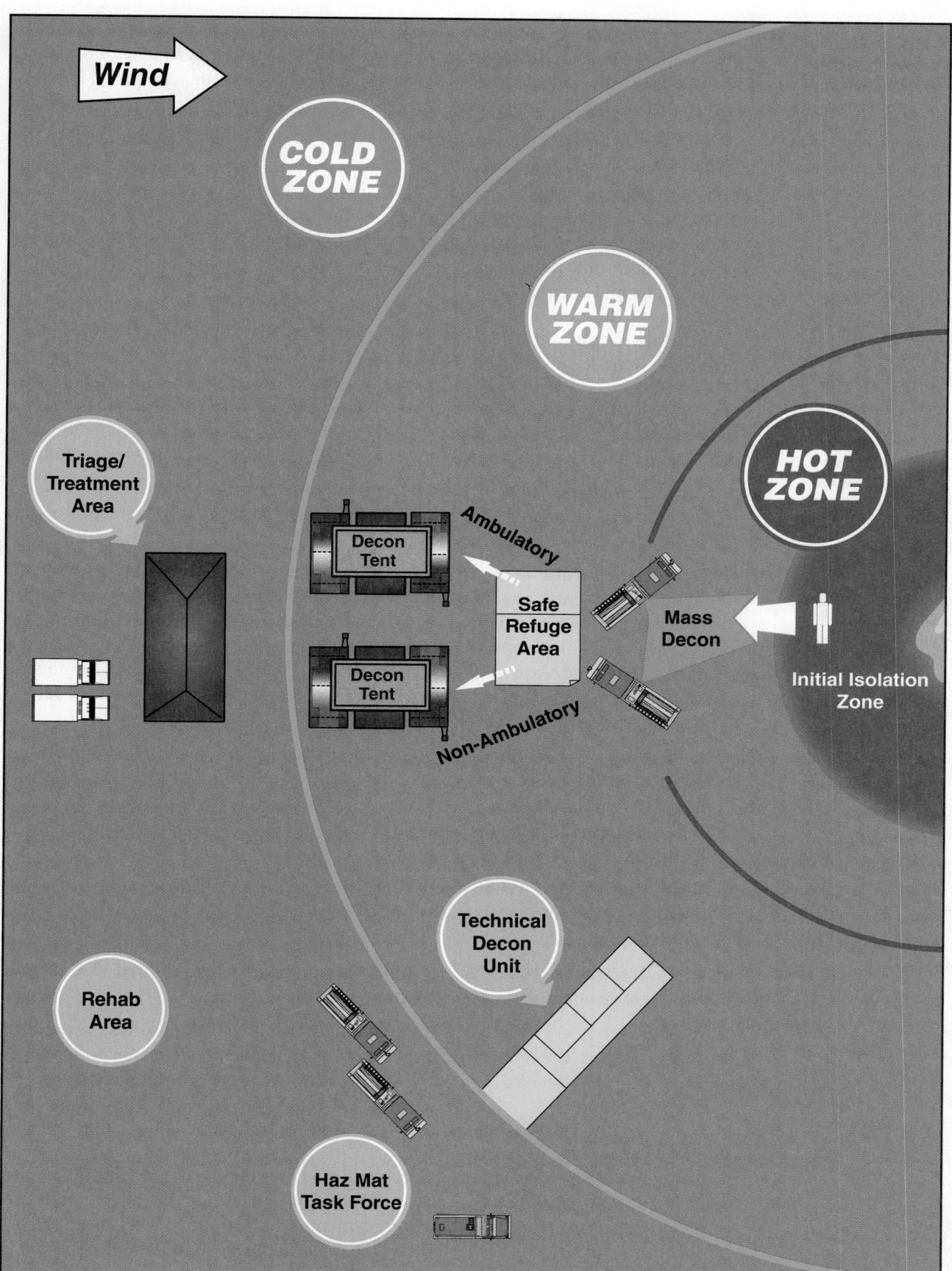

Figure 24.51 The warm zone of an incident contains decontamination stations as part of the transition between the hot zone and the cold zone.

Cold Zone — Safe area outside of the warm zone where equipment and personnel are not expected to become contaminated and special protective clothing is not required; the Incident Command Post and other support functions are typically located in this zone. *Also known as* Support Zone.

Staging Area — Prearranged, temporary strategic location, away from the emergency scene, where units assemble and wait until they are assigned a position on the emergency scene; these resources (personnel, apparatus, tools, and equipment) must then be able to respond within 3 minutes of being assigned. Staging area managers report to the Incident Commander or Operations Section Chief, if one has been established.

Transportation Area — Location where accident casualties are held after receiving medical care or triage before being transported to medical facilities.

this zone, although in some circumstances it may be at a reduced level from the hot zone. Monitoring and detection may be conducted around the perimeter of the warm zone to determine the extent of the hazards. The level of PPE to be required for work within this zone will be approved by the IC.

Cold Zone

The **cold zone** (also called *support zone*) surrounds the warm zone and is used to carry out all logistical support functions of the incident. Workers in the cold zone are not required to wear personal protective clothing because the zone is considered safe, although some personnel may still be wearing PPE to ensure safe evacuation in the case of rapid expansion of the hot zone. The multiagency Command Post (CP), **staging area**, donning/doffing area, backup teams, research teams, logistical support, criminal investigation teams, triage/treatment/rehabilitation (rehab), and **transportation areas** are located within the cold zone.

Additional Zones

Additional zones or areas may be required within the isolation perimeter, including:

- **Decontamination zone** — Area located in the warm zone where contaminated clothing, people, and equipment can be cleaned or secured; see Decontamination section.

- **Area of safe refuge** — Traditionally in the warm zone (with greater flexibility at explosive incidents), this is primarily an area serving as a safe place to wait for evacuation and/or decon; used at haz mat incident as a safe location (or locations) where evacuated persons are directed to gather while potential emergencies are assessed, decisions are made, and mitigating activities are begun. In bombings, hostage situations, or sniper attacks, law enforcement will create areas of refuge that are rendered safe.

- **Staging area** — Area(s) where personnel and equipment awaiting assignment to the incident are held to keep responders and equipment out of the way and safe until assigned, minimizing confusion and freelancing at the scene; located at an isolated spot in the cold zone where evacuated persons cannot interfere with ongoing operations.

- **Rehabilitation area (rehab area)** — Safe location located in the cold zone where emergency personnel can rest, sit or lie down, have food and drink, and have medical conditions evaluated.

- **Triage/treatment area** — Area within the cold zone where injured victims and personnel can be medically assessed and treated.

The size, nature, and scope of an incident determine which of these zones and areas are needed. Predetermined procedures help responders to determine which zones and areas to establish.

Notification

In North America, notification may be as simple as dialing 9-1-1 to report an incident. But it may also include notifying multiple agencies, identifying the incident level, and informing the public. Law enforcement must be notified whenever a terrorist or criminal incident is suspected, or if their assistance is required when handling suspected explosives or suspicious objects. In other circumstances, responders may also need to notify the public works department or the local emergency operations center (EOC). Haz mat incidents and terrorist attacks may overwhelm local responders, so it is important to know how to request additional resources. Different agencies have different procedures, so follow the guidelines contained in emergency response plans, local SOPs, and **mutual/automatic aid** agreements. Methods for notifying the public are provided in the Evacuation section of this chapter.

In the U.S. the notification process is spelled out in the National Response Plan (NRP), and all local, state, and federal emergency response plans must comply with these provisions. According to the NRP, all incidents are handled at the lowest geographic, organizational, and jurisdictional level, placing an emphasis on local response. However, when local agencies need additional assistance, they must contact their state authorities to request help. State governors have the authority to mobilize state resources and may include activation of mutual aid agreements with other states. If state resources are overwhelmed, state governors have the authority to request federal aid.

The LERP should be the first resource to turn to if you need to request outside assistance for a significant incident. Per the NRP, the local response agency should be closely tied with the community's EOC. If local assets are insufficient to manage the emergency, requests for additional assistance (such as activation of National Guard units) will be made to the state EOC. States may then request federal assistance through the Department of Homeland Security (DHS). Even if additional assistance is not required for an incident, the proper authorities at all levels must be informed that an incident has occurred.

Protection

Protection is the overall goal of ensuring safety of responders, the public, property, and the environment. Protection goals are accomplished through such tactics as:

- Identifying and controlling materials and hazards
- Using and wearing appropriate PPE
- Conducting rescues
- Implementing shoring and stabilization at incidents involving structural collapses
- Providing emergency decontamination
- Providing emergency medical care and first aid
- Taking any other measures to protect responders and the public, including conducting evacuations and sheltering-in-place

Protection of Responders

The protection and safety of emergency responders is the first priority at any incident. Injured or incapacitated responders are unable to assist in mitigation efforts or protection of the public.

Measures to protect responders include the following:

- Staying uphill, upstream, and upwind of hazardous materials
- Wearing appropriate PPE
- Using time, distance, and shielding for protection
- Decontaminating responders when necessary (**Figure 24.52, p. 1472**)

Mutual Aid — Reciprocal assistance from one fire and emergency services agency to another during an emergency, based upon a prearranged agreement; generally made upon the request of the receiving agency.

Automatic Aid — Written agreement between two or more agencies to automatically dispatch predetermined resources to any fire or other emergency reported in the geographic area covered by the agreement. These areas are generally located near jurisdictional boundaries or in jurisdictional "islands."

Figure 24.52 Decontamination practices apply to first responders in some incidents. *Courtesy of the U.S. Air Force.*

- Ensuring accountability of all personnel
- Tracking and identifying all personnel working at an incident
- Working as part of a team or buddy system
- Assigning safety officers
- Putting evacuation and escape procedures in place

Accountability Systems

One of the most important functions of NIMS-ICS is to provide a means of tracking all personnel and equipment assigned to the incident. Most units responding to an incident arrive fully staffed and ready to be assigned an operational objective; other personnel may have to be formed into units at the scene. To handle these and other differences in the resources available, the IAP must contain a tracking and accountability system that has the following elements:

- Procedure for checking in at the scene
- Way of identifying and tracking the location of each unit and all personnel on scene, including those entering and exiting hazard-control zones
- Procedure for releasing people, equipment, and apparatus that are no longer needed

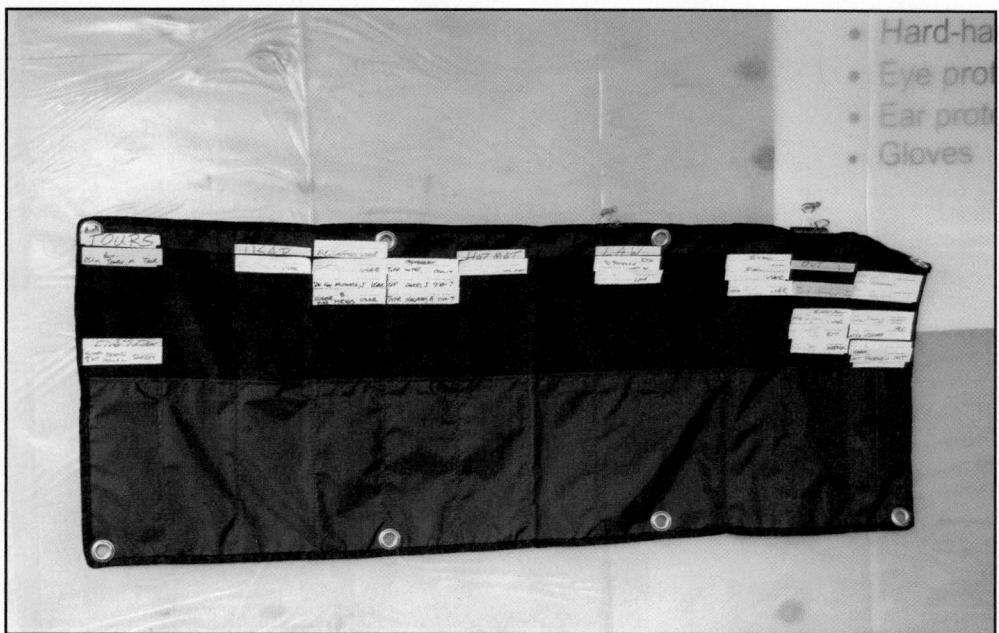

Figure 24.53 All personnel assigned to the incident must check in and out via the established accountability system.

Accountability systems are especially important when multiple agencies/organizations have responded to the incident. The agency/organization in command is responsible for tracking all responders, not just its own. Methods for tracking accountability should be determined in preplans and implemented as soon as possible at the incident scene. Types of accountability systems include traditional systems such as the fire service passport system, T-card systems, and systems that utilize newer technologies such as global positioning systems (GPSs). Requirements for accountability systems are addressed in NFPA® 1500 and 1561 (**Figure 24.53**).

Buddy Systems and Backup Personnel

Use of buddy systems and backup personnel are mandated by NFPA® and OSHA at haz mat incidents. A *buddy system* is a system of organizing personnel into workgroups in such a manner that each member has a buddy or partner, so that nobody is working alone (**Figure 24.54, p. 1474**). The purpose of the buddy system is to provide rapid help in the event of an emergency.

In addition to using the buddy system, backup personnel shall be standing by with equipment ready to provide assistance or rescue if needed. Any haz mat team working within the hazardous area must have at least two members. The minimum number of personnel necessary for performing tasks in the hazardous area is four — two working in the area itself and two standing by as backup. Backup personnel must be dressed in the same level of personal protective clothing as entry personnel. At a minimum, qualified basic life support personnel shall also be standing by with medical equipment and transportation capability.

Time, Distance, and Shielding

Using time, distance, and shielding is an effective protection strategy for first responders at hazardous materials incidents (**Figure 24.55, p. 1474**). Responders can protect themselves by utilizing the following:

Figure 24.54 Responders working in the hot zone must always work with a partner, never alone.

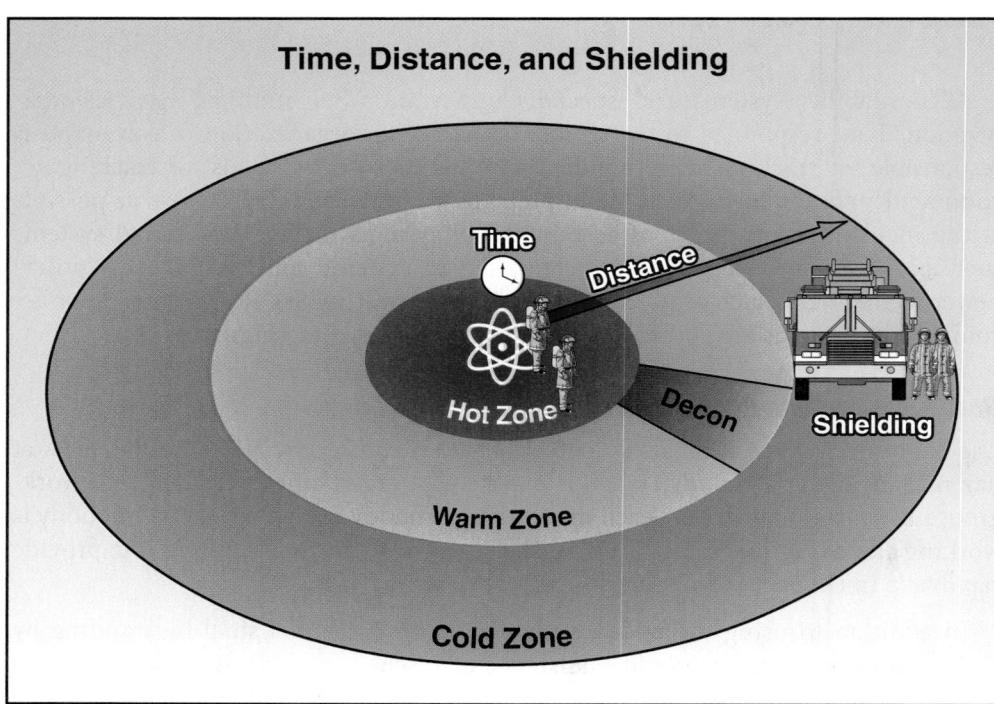

Figure 24.55 Time, distance, and shielding should be used to protect responders from exposure to hazards.

- **Time** — Limiting the time they are exposed (or potentially exposed) to hazards and hazardous materials reduces the likelihood of serious harm. This can be accomplished by restricting work times in the hot zone and frequently rotating personnel on work groups.

- **Distance** — Maximizing distance from potential hazards will often prevent or reduce harm. For example, the closer a responder is to the source of an explosion, the greater the harmful effects. Staying well away from hazardous areas will also prevent harmful exposures. Distance may be controlled by implementing hazard-control zones.

- **Shielding** — Shielding places a physical barrier between a responder and the hazard. Shielding may consist of wearing PPE or positioning personnel so that another object such as a wall, building, or apparatus is between them and the hazard, thereby minimizing the chance of contact or effect.

These tactics are particularly effective at incidents involving radiation. Using time, distance, and shielding to limit radiation exposure is sometimes known as the *ALARA* (*As Low As Reasonably Achievable*) *method* or *principle*. The following rules apply:

- For nuclear fallout following a nuclear detonation, the "rule of seven" applies. This rule states that for every sevenfold increase in time, the radioactivity level due to fallout decreases by a factor of 10. For example, if the level at 1 hour after detonation was 100 rem/hour, it will decline to 10 rem/hour in 7 hours and to 1 rem/hour in 49 hours (7x7).

- Doubling the distance from a point source divides the dose by a factor of four. This calculation is sometimes referred to as the **inverse square law**. When the radius doubles, the radiation spreads over four times as much area, so the dose is only one-fourth as much **(Figure 24.56)**.

- Exposure from fallout is reduced by about 50 percent inside a one-story building and by about 90 percent at a level belowground.

Inverse Square Law — Physical law that states that the amount of radiation present is inversely proportional to the square of the distance from the source of radiation.

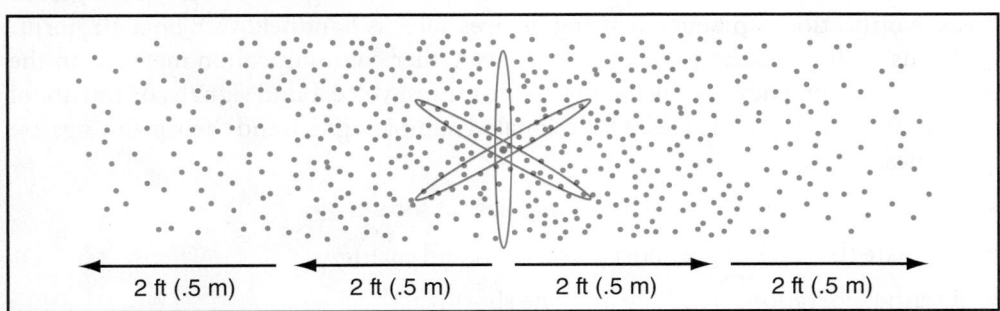

2 ft (.5 m)   2 ft (.5 m)   2 ft (.5 m)   2 ft (.5 m)

Figure 24.56 The distance from the point source is a valuable piece of information, because every time the distance doubles, the exposure dose decreases by a factor of four.

Figure 24.57 An officer dedicated to the safety of operations may be assigned to incidents and training exercises of any size.

Withdrawal/Escape Procedures

Because haz mat and terrorist incidents have a high potential for rapidly changing conditions, secondary devices, secondary collapse, or discovery of other hazardous situations, it is important to adopt and use a signaling system that will advise personnel inside the danger area when to evacuate and/or withdraw. The FEMA USAR Task Force program has developed a system for evacuating rescuers from dangerous areas. Notification can be made using devices such as handheld CO^2 boat air horns, air horns on fire apparatus, or vehicle horns. Other communication methods in the event of an emergency can include portable radios, voice, hand signals, or the use of predetermined signals. The USAR designated audible signals and their meanings are as follows:

- Cease Operations/All Quiet: one long blast (3 seconds)

- Evacuate the Area: three short blasts (1 second each)

- Resume Operations: one long and one short blast

Responders must also plan escape procedures. If the primary means of egress becomes blocked, rescuers should determine the feasibility of using the alternate route.

Safety Officers

While IC's may take responsibility for overseeing all aspects of safety at small incidents, at larger, more complex incidents an Incident Safety Officer (ISO) may be appointed to ensure the safety of operations **(Figure 24.57)**. The ISO is responsible for monitoring and identifying hazardous and unsafe situations and developing measures for ensuring personnel safety. The (ISO) must be trained to the level and scope of operations conducted at the incident and is required to perform the following duties:

- Obtain a briefing from the IC.

- Review IAPs for safety issues.

- Identify and assess hazards at the incident scene and report them to the IC.

- Provide the IC with risk assessments.

- Participate in the preparation and monitoring of incident safety considerations, including medical monitoring of entry team personnel before and after entry, and rehabilitation.

- Communicate information about injuries, illnesses, or exposure to the IC.

- Maintain communications with the IC, and advise the IC of deviations from the incident safety considerations and of any dangerous situations.

- Monitor incident communications for events that might pose safety concerns.

- Alter, suspend, or terminate any activity that is judged to be unsafe.

The ISO also ensures that safety briefings are conducted for personnel entering hazard zones. Safety briefings include:

- Information about the status of the incident, including the preliminary evaluation and subsequent updates

- The hazards identified

- A description of the site

- The tasks to be performed

- The expected duration of the tasks

- PPE requirements

- Monitoring requirements

- Notification of identified risks

- Any other pertinent information

At incidents involving potential criminal or terrorist activities, the safety briefing should also cover the following items:

- Being alert for secondary devices

- Not touching or moving any suspicious-looking articles (bags, boxes, briefcases, soda cans, etc.)

- Not touching or entering any damp, wet, or oily areas

- Wearing full PPE, including SCBA

- Limiting the number of personnel entering the crime scene

- Documenting all actions

- Not picking up or taking *any* souvenirs

- Photographing or videotaping anything suspicious

- Not destroying any possible evidence

- Seeking professional crime-scene assistance

Protection of the Public

Responders have numerous means of protecting the public at the incident scene. These include:

- Evacuation
- Sheltering-in-place
- Protecting/defending in place
- Isolating the area and denying entry
- Rescues
- Mass decontamination
- Emergency medical care and first aid

Evacuation

To *evacuate* means to move all people from a threatened area to a safer place. Evacuation is usually the best protective action, but there must be enough time to warn the evacuees, have them get ready, and then get them to safety. Firefighters should begin evacuating people who are most threatened by the incident in accordance with distances recommended by the *ERG*, preincident surveys, or other sources. Even after people move the recommended distances, they are not necessarily completely safe from harm. Do not permit evacuees to congregate at these distances. Send them by a specific route to a designated area of safe refuge.

The number of responders needed to perform an evacuation varies with the size of the area and number of people to evacuate. Evacuation can be an expensive, labor-intensive operation, so it is important to assign enough personnel resources to an incident to conduct it.

Evacuation and traffic-control activities on the downwind side of the incident could cause responders and evacuees to become contaminated and, consequently, need decontamination. Responders may also need to wear PPE to safely conduct the evacuation. Evacuation plans (including casualties) for likely terrorist targets such as stadiums and other public gathering places should be made in advance as part of the LERP.

Large-scale evacuations present many factors that must be addressed by the IC including:

- **Notification** — The public must be alerted of the need to evacuate, plus they must understand where to go. Methods to be used should be spelled out in the LERP. When notifications are made, clear and concise information must be relayed to avoid confusion or additional panic. Notification methods include:
  - Knocking on doors
  - Public address systems
  - Broadcast media, such as radio and TV
  - Sirens
  - Building alarm systems
  - Cell phone short message service (SMS)
  - Community Emergency Notification System (CENS) or Reverse 9-1-1®
  - Emergency Alerting System (EAS)
  - Loudspeakers mounted on helicopters or emergency vehicles
  - Electronic billboards

- **Transportation** — Mass transportation must be planned in advance, including school buses, public transit, boats, trains, and ferries **(Figure 24.58)**. As responders discovered during the evacuation of New Orleans before Hurricane Katrina, not everybody owns a car.

- **Relocation Facilities and Temporary Shelters** — Evacuees must have some place to go. These places must be able to provide and care for the people relocated to them, including providing food, water, medicine, bathroom and shower facilities, and places to sleep (for evacuations of long duration) **(Figure 24.59)**. Many evacuees may bring pets. Appropriate evacuation shelters must be designated in the LERP, and staffing arrangements should be determined in advance. An information/registration system should also be established to track the whereabouts of evacuees so their friends and relatives can find them.

- **Prevention of Looting** — Protection of property, while not as high a priority as life safety, must be taken into consideration. When home and business owners are no longer able to protect their property, the IC must ensure that steps are taken to prevent looting by unscrupulous individuals who are willing to risk staying in the evacuation area in order to take advantage of the situation.

Figure 24.58 Plans must be made in advance to provide transportation for individuals who are not otherwise able to leave the hazard area. *Courtesy of FEMA News Photos, photo by Win Henderson.*

Figure 24.59 In large-scale incidents, evacuees may need to have a safe place and basic amenities provided. *Courtesy of FEMA News Photos, photo by Andrea Booher.*

Evacuating Contaminated Victims

Individuals who have been exposed or potentially exposed to chemical, biological, or radiological agents must be decontaminated. While it may be impossible to keep these individuals at the scene, efforts to keep them in place should be made in order to prevent the spread of harmful or potentially deadly materials to other locations. Evacuate contaminated or potentially contaminated individuals to an area of safe refuge (or a triage and treatment area as appropriate) within the isolation perimeter to await decontamination. Because victims may leave the scene before emergency responders arrive (or ignore requests to stay in order to undergo decon), shelters, hospitals, and other public healthcare facilities must be prepared to conduct decon of **self-presenters** at their facilities.

Self-Presenters — People seeking medical attention who were not treated or decontaminated at the incident scene.

Sheltering-in-Place

Sheltering-in-place means directing people to go quickly inside a building and remain there until danger passes. Shelter-in-place may be more effective than evacuation if:

- The hazardous material is spreading too rapidly to allow time for evacuation.
- The hazardous material release will be short in duration.
- The hazardous material is too toxic to risk any exposure.
- Vapors are heavier than air, and people are in a high-rise structure.
- The population is unable to initiate evacuation because of healthcare, detention, or educational occupancies.

When protecting people inside a structure, close all doors, windows, and heating, ventilation and air-conditioning systems. Shelter-in-place may not be the best option if vapors or gases are explosive because it may take a long time for them to dissipate. Additionally, vapor or gases may permeate into any building that cannot be sealed from the outside atmosphere. Vehicles are not as effective as buildings for sheltering-in-place, but they can offer temporary protection if windows are closed and the ventilation system is turned off.

Whether using evacuation or shelter-in-place, the public needs to be informed as early as possible and receive additional instructions and information throughout the course of an emergency. Shelter-in-place may be more effective if public education has been done ahead of time through emergency planning.

If there is reason to suspect that an indoor environment is insufficiently sealed, adhesive tape and polyethylene sheets can be used to improve the seal. The public can be instructed of this technique in advance and told to store these materials along with drinking water, medication, and food.

Firefighters should also pay attention to the condition of surrounding buildings before ordering sheltering-in-place. For example, some areas may have dilapidated structures without air-conditioning or with openings between floorboards. Sheltering-in-place might not provide sufficient protection in such cases, making evacuation the better option.

Figure 24.60 Decontamination practices remove or reduce hazardous materials from victims, responders, and the environment. *Courtesy of Boca Raton (FL) Fire Rescue.*

Protecting/Defending-in-Place

Protecting/defending-in-place is an active (offensive) role or aggressive posture to physically protect those in harm's way. For example, using hose streams to diffuse a plume, or sending in law enforcement to secure a neighborhood or area. When appropriate and safe to do so, defending-in-place eliminates the need for unnecessary evacuations which, if initiated, will require additional logistical support to ensure the health and safety of evacuees.

Decontamination

Decontamination is performed at haz mat incidents to remove hazardous materials from victims, PPE, tools, equipment, and anything else that has been contaminated. Realistically, since removing *all* contaminants may be impossible in many cases, decontamination is done simply to reduce contamination to a level that is no longer harmful (**Figure 24.60**). Decon operations prevent harmful exposures and reduce or eliminate the spread of contaminants outside the hot zone.

Decon also provides victims with psychological reassurance. Some individuals who have been potentially exposed to hazardous materials may develop psychologically-based symptoms (i.e., shortness of breath) even if they have not actually been exposed to harmful levels of contamination. Conducting decon can reduce or prevent these types of problems.

The type of decon operations conducted at an incident will be determined by the size of the incident, the type of hazardous materials involved, weather, personnel available, and a variety of other factors. However, regardless of the many variables that may be encountered at the incident, the three basic principles of any decontamination operation are:

(1) Get it off.

(2) Keep it off.

(3) Contain it (prevent cross-contamination).

Because decon procedures, terminology, and other details may differ greatly from organization to organization, responders must be familiar with their organization's decon policies and procedures. The following sections will discuss **emergency decontamination** operations, including general decon guidelines and victim triage.

Decontamination Terminology

Decon terminology varies greatly from country to country and region to region. However, for purposes of this book, the following definitions will be used:

- *Contamination* — The direct transfer of a hazardous material to persons, equipment, and the environment in greater than acceptable quantities.

- *Cross (or secondary) contamination* — Indirect contamination of people, equipment, or the environment outside the hot zone. In secondary contamination, the contaminant is transferred to previously uncontaminated surfaces, persons, or equipment, etc. For example, contaminated victims who drive themselves to a hospital prior to decontamination could cause cross (or secondary) contamination of medical personnel at the hospital.

- *Decontamination (or contamination reduction)* — The process of removing hazardous materials to prevent the spread of contaminants beyond a specific area and reduce the level of contamination to levels that are no longer harmful.

- *Exposure* — The process by which people, property, animals, or the environment are put at risk of harm from contact (or potential contact) with hazardous materials.

CPC is designed to protect a person from exposure to chemicals, but in the process of doing so, it may become contaminated. Just because the CPC has become contaminated does not mean that the person has been exposed to the material. When a person *is* exposed to a chemical or other product, it is the *hazard* of the material (or the harm it can do) based on the nature of the exposure (amount inhaled, duration of exposure, and the like) that determines how it may ultimately affect a person's health.

Emergency Decon

The purpose of emergency decontamination is to remove the threatening contaminant from the victim or rescuers as quickly as possible without regard for the environment or property (**Figure 24.61**). Contaminated individuals are stripped of their clothing and washed quickly. Victims may need immediate medical treatment, so they cannot wait for a formal decontamination corridor to be established. The following situations are examples of instances where emergency decontamination is needed:

- Failure of protective clothing.
- Accidental contamination of first responders.
- Heat illness or other injury suffered by emergency workers in the hot zone.
- Immediate medical attention is required.

Emergency decontamination could be considered a *quick fix,* which is a definite limitation. Removal of all contaminants may not occur, and a more thorough decontamination must follow. Emergency decontamination can definitely harm the environment. However, the advantage of eliminating a life-threatening situation far outweighs any negative effects that may result.

Figure 24.61 Emergency decon serves a limited function and should only be used in extreme cases where the clear benefit to the victim outweighs potential negative effects on the environment.

There are times when what appears to be a normal incident really involves hazardous materials. Firefighters may become contaminated before they realize what the situation really is. When this situation occurs, firefighters need to withdraw immediately. They need to remove their turnout clothing and get emergency decon even if there is no apparent contamination evidence. These responders should remain isolated until someone with the proper expertise can ensure that they have been adequately decontaminated.

Emergency Decontamination: Advantages and Limitations

Advantages:

- Requires minimal equipment (usually just a water source such as a hoseline)
- Reduces contamination quickly
- Does not require a formal decontamination corridor

Limitations:

- Does not always totally decontaminate the victim
- Creates contaminated runoff that can harm the environment and other exposures

Emergency decontamination procedures may differ depending on the circumstances and hazards present at the scene. However, a basic set of emergency decontamination procedures is provided in **Skill Sheet 24-I-2**.

General Guidelines for Decon Operations

General guidelines for decontamination operations include the following:

- Start decon operations as quickly as possible — the longer the exposure to hazardous materials, generally the worse the victim outcome.
- **Always** wear appropriate PPE for rescue and decon operations.
- Avoid contact with hazardous materials, including contaminated victims.
- Decontaminate anyone who moves from the hot zone to the cold zone, regardless of symptoms or exposure.
- Decon emergency responders separately from victims (establish separate decon lines when possible and practical).
- Remove as much clothing as possible (disrobing is effective decon by itself) (**Figure 24.62**).
- Communicate with victims by using hand signals, signs with pictures, apparatus public address systems, megaphones or other methods to direct them to decon gathering areas as well as through the decon process itself (**Figure 24.63**). It is very important to provide clear and easily understood directions because people may be traumatized and/or suffering from exposures.

Figure 24.63 Verbal communication may be limited due to PPE, but hand signals, signs with pictures, and other forms of communication may be used to provide clear directions.

Figure 24.62 Victims should remove as much clothing as possible during Emergency Decontamination.

- Provide privacy whenever possible (including from overhead vantage points, for example, from circling news helicopters or upper stories of nearby buildings).

- If possible, provide warm water for washing. If water is cold, allow victims to gradually get wet in order to acclimate to the temperature and avoid cold shock.

NOTE: Washing off contaminants with a hose stream may not be a viable option in subzero temperatures.

Triage

Procedures for conducting **triage** of victims needing emergency decon should be predetermined by the local emergency response plan. In most instances, triage will be conducted in the cold zone after decontamination has been performed. However, in some cases (such as at explosive incidents), it may be conducted in the hot zone prior to decon.

Several factors may influence the priority for patients, including:

- Victims with serious medical symptoms (such as shortness of breath or chest tightness)

- Victims closest to the point of release

- Victims reporting exposure to the hazardous material

- Victims with evidence of contamination on their clothing or skin

- Victims with conventional injuries (broken bones and open wounds)

Triage — System used for sorting and classifying accident casualties to determine the priority for medical treatment and transportation.

Technical Decon

Technical decontamination uses chemical or physical methods to thoroughly remove or neutralize contaminants from responders' PPE and equipment (**Figure 24.64, p. 1486**). It may also be used on victims in non-life-threatening situations. Responders must be familiar with the AHJ's procedures for implementing technical decon within the Incident Command System.

Technical decon is usually conducted within a formal decon line or corridor. The contaminants involved at the incident determine the type and scope of technical decon. Firefighters must know what to do when they proceed through a decontamination corridor or line. Technical decon corridors vary in the number of stations, depending on the needs of the situation (**Figure 24.65, p. 1486**).

Rescue

Due to the defensive nature of most Operations-Level actions, rescue can be a difficult strategy to implement for first responders, particularly in the initial stages of a response. Search and rescue attempts should be made within the framework of the IAP with appropriate training, PPE, backup personnel, and other safety considerations in place.

When a rescue is too dangerous or responders do not have appropriate training, PPE, or equipment to conduct a rescue, the proper decision may be to protect the victims in place and/or wait for more resources. This rule may directly conflict with the fire fighting incident priority of *rescue first* as well as with many responders' natural desire to help victims as quickly as possible. However, because of the dangers presented by hazardous materials, responders who rush to the rescue often require rescue themselves.

Figure 24.64 Technical decon uses chemical or physical methods to thoroughly remove or neutralize contaminants carried by entry team members. *Courtesy of the U.S. Air Force, photo by Chiaki Iramina.*

Figure 24.65 Specialized stations may be added to technical decon corridors as the situation dictates.

Chapter 23 provided information about the hazards associated with each DOT hazard class, so that you can assess potential risks at such incidents involving these materials. For example, at an incident involving corrosives, you can determine that chemical burns are probably one of the major hazards. But unless you have additional training in haz mat rescue, avoid all contact with hazardous materials, and do not physically touch or move a victim who is contaminated, potentially contaminated, or located within the warm, hot, or initial isolation zone. The only rescue actions you should take are telling people what to do and/or where to go, or performing supportive action from a distance. For example, you can:

- Direct people to an area of safe refuge or evacuation point located in a safe place within the hot zone that is upwind and uphill of the hazard area.

- Instruct victims to move to an area that is less dangerous before moving them to an area that offers complete safety.

- Direct contaminated or potentially contaminated victims to an isolation point, safe refuge area, safety shower, eyewash facility, or decontamination area (**Figure 24.66**).

- Give directions to a large number of people during mass decontamination (**Figure 24.67**).

- Conduct searches during reconnaissance or defensive activities.

- Conduct searches on the edge of the hot zone.

Spill Control and Confinement

To understand the principles behind spill control/confinement, it is useful to return to the information in the haz mat behavior section in the previous chapter. Container stress can result in a breach (rupture) and loss of the container's contents. The released material then disperses according to its chemical and physical properties, topography, prevailing weather conditions, and the amount and duration of the release. Depending on its hazardous properties, the material can then harm whatever it contacts.

Spill control involves limiting the dispersion of material that has already been released from its container, so that it does not come in contact with people, property, or the environment. It may also be used to reduce the amount of harm caused by contact

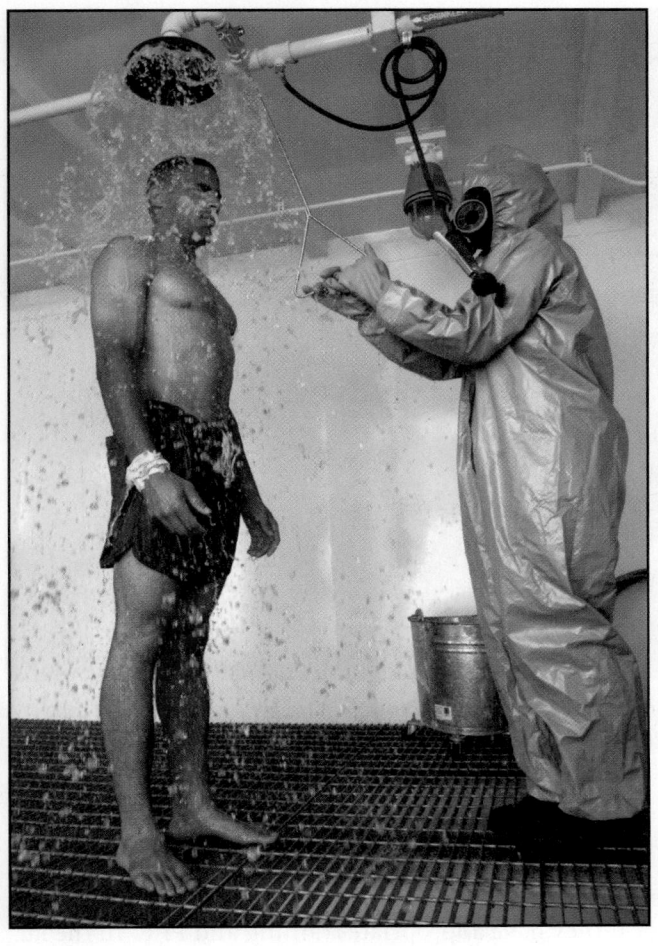

Figure 24.66 First responders can direct contaminated or potentially contaminated victims to safety showers or areas of refuge. *Courtesy of the U.S. Marine Corps photo by Sgt J. A. Lee II.*

Figure 24.67 Operations-Level responders can provide necessary directions and instructions for mass decon operations.

Figure 24.68 Defensive spill-control tactics are used to contain hazardous materials that have already escaped their containers.

(Figure 24.68). Spill control is typically defensive, and the highest priority during these operations is firefighter life safety.

Operations-Level firefighters may perform spill-control activities if there is no risk of contact with the material, or if they have appropriate training and PPE. In the latter case, they must limit their contact with the material as much as possible.

Spills may involve gases, liquids, or solids that have been released into the air, into water, or onto a surface. Spill control tactics are determined by the material involved and the type of dispersion. For example, at a flammable liquid spill, you must address the liquid spreading on the ground and the vapors being released into the air. If the spill pours into a stream or sewer system, you will also need to address the contamination of the water.

Spill Control and Confinement Tactics

Hazardous materials may be confined by building dams or dikes near the source, catching the material in another container, or directing (diverting) the flow to a remote location for collection. Generally, the fire apparatus carries the following necessary tools:

- Shovels for building earthen dams

- Salvage covers for making catch basins

- Charged hoselines for creating diversion channels

Before using the equipment to confine spilled materials, the IC needs to seek advice from technical sources to determine if the spilled materials will adversely affect the equipment. Large or rapidly spreading spills may require the use of heavy construction-type equipment, floating confinement booms, or special sewer and storm drain plugs.

Confinement is not restricted to controlling liquids. Dusts, vapors, and gases can also be confined. A protective covering consisting of a fine spray of water, a layer of earth, plastic sheets, or a salvage cover can keep dusts from blowing about at incidents. Foam blankets can be used on liquids to reduce the release of vapors. Strategically placed fire streams can direct gases or allow the water to absorb them. Check reference sources for the proper procedures for confining gases. The material type, rate of release, speed of spread, number of personnel available, tools and equipment needed, weather, and topography dictate confinement efforts.

Defensive confinement and spill control actions are discussed in the following sections:

- Absorption
- Adsorption
- Blanketing/covering
- Dam, dike, diversion, and retention
- Dilution
- Dissolution
- Vapor dispersion
- Vapor suppression
- Ventilation

CAUTION

Do not attempt confinement actions unless you are reasonably certain that you will not contact or be exposed to the hazardous material.

Absorption

Absorption occurs when one material enters the cell structure of another and is retained within. Materials typically used as absorbents include sawdust, clays, charcoal, and polyolefin-type fibers **(Figure 24.69, p. 1490)**. These materials are spread directly onto the hazardous material or in a location where the material is expected to flow. Absorbents retain the properties of the materials they absorb, so after use they must be treated and disposed of as hazardous materials. For more information about performing absorption, see **Skill Sheet 24-I-3**.

Adsorption

Adsorption is different from absorption in that the molecules of the hazardous material physically adhere to the *ad*sorbent material rather than being absorbed into the inner spaces of an *ab*sorbent material. Adsorbents do not swell like absorbents, and they are often organic-based materials such as activated charcoal or carbon. Adsorbents are primarily used to control shallow liquid spills. It is important to make sure that the adsorbent used is compatible with the spilled material in order to avoid potentially dangerous reactions (such as producing heat and causing spontaneous combustion). For more information about performing adsorption, see **Skill Sheet 24-I-4**.

Figure 24.69 Absorbent materials are used to soak up hazardous materials to facilitate containment and disposal.

Blanketing/Covering

As the title implies, this spill-control measure involves blanketing or covering the surface of the spill to prevent dispersion of materials such as powders or dusts. Blanketing or covering of solids can be done with tarps, plastic sheeting, salvage covers, or other materials (including foam), but consideration must be given to compatibility between the material being covered and the material covering it. Covering may also be done as a form of temporary mitigation for radioactive and biological substances.

Blanketing of liquids is essentially the same as vapor suppression (see Vapor Suppression section), because it uses an appropriate aqueous (water) foam agent to cover the surface of the spill. Operations-Level responders may or may not be allowed to perform blanketing/covering actions, depending on the hazards of the material, the nature of the incident, and the distance from which they must operate to ensure their safety.

Dike, Dam, Diversion, and Retention

Diking, damming, diverting, and retaining are ways to control the flow of liquid hazardous materials away from the point of discharge. Responders can use available earthen materials or materials carried on their response vehicles to construct curbs that direct or divert the flow away from gutters, drains, storm sewers, and flood-control channels (**Figure 24.70**). In some cases, it may be desirable to direct the flow into certain locations in order to capture and retain the material for later pickup and disposal. Dams may be built that permit surface water or runoff to pass over (or under) the dam while holding back the hazardous material (**Figure 24.71**). Any construction materials that contact the spilled material must be properly discarded. See **Skill Sheets 24-I-5**, **24-I-6**, **24-I-7**, and **24-I-8** for instructions on how to perform diking, damming, diverting, and retaining.

Figure 24.70 Hazardous materials may be diverted from storm drains to prevent widespread contamination.

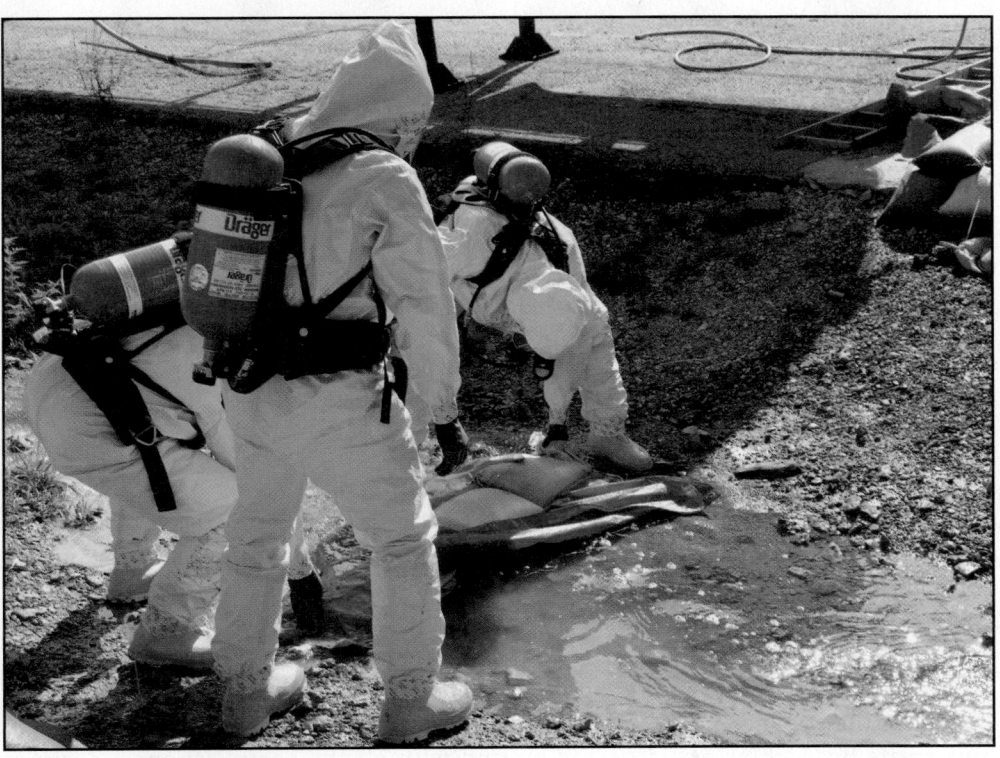

Figure 24.71 Dams may be constructed to trap materials that are heavier or lighter than water.

Dilution

Dilution involves applying water to a water-soluble material in order to reduce the hazard. It is a useful way of performing decontamination, but it is not a practical spill control method at most haz mat incidents. Effective dilution typically requires so much water that it creates a runoff problem, especially if used with slightly water-soluble liquids. Dilution may be useful with very small amounts of corrosive material, such as minor accidents in a laboratory or cases of irregular dispersion, but even then, it is generally considered for use only after other methods have been rejected. **Skill Sheet 24-I-9** describes a procedure to perform dilution.

Figure 24.72 The process of vapor dispersion uses pressurized streams of water to blend air into the vapors and therefore reduce the concentration of hazardous materials.

Dissolution

The process of dissolving a gas in water is called *dissolution*. This tactic can only be used on water-soluble gases such as anhydrous ammonia or chlorine and is generally conducted by applying a fog stream to a breach in a container or directly onto the spill. Ideally, the escaping gas passes through the water and dissolves. This process may create additional problems with contaminated runoff water and other issues. For example, using water spray at a chlorine incident to bring the vapors to the ground may have the beneficial effect of reducing or eliminating a toxic plume, but it may also create hydrochloric acid on the ground with all the complications associated with that chemical.

Vapor Dispersion

Vapor dispersion is the action taken to direct or influence the course of airborne hazardous materials. Pressurized streams of water from hoselines or unattended master streams may be used to help disperse vapors **(Figure 24.72)**. These streams create **turbulence**, which increases the rate of mixing with air and reduces the concentration of the hazardous material. After using hoselines for vapor dispersion, it is necessary for first responders to confine and analyze runoff water for possible contamination. The steps listed in **Skill Sheet 24-I-10** may be followed to perform vapor dispersion.

Turbulence — Irregular motion of the atmosphere usually produced when air flows over a comparatively uneven surface, such as the surface of the earth; when two currents of air flow past or over each other in different directions or at different speeds.

Figure 24.73 Fire fighting foam agents may be used for vapor suppression activities provided the agent is appropriate for the hazardous material.

Vapor Suppression

Vapor suppression is the action taken to reduce the emission of vapors at a haz mat spill. Fire fighting foams are effective on spills of flammable and combustible liquids if the foam concentrate is compatible with the material. Water-miscible materials such as alcohols, esters, and ketones destroy regular fire fighting foams and require an alcohol-resistant foam agent **(Figure 24.73)**. In general, the required application rate for applying foam to control an unignited liquid spill is substantially less than that required to extinguish a spill fire.

NOTE: More information on foam application is contained in the Fire Fighting Foam section of Chapter 16.

Ventilation

Ventilation is any natural or mechanical means of controlling air movement. It is used inside structures to remove and/or disperse harmful airborne particles, vapors, or gases. The same ventilation techniques used for smoke removal can also be used for haz mat incidents. When conducting negative-pressure ventilation, fans and other ventilators must be compatible with the surrounding atmosphere. For example, in a flammable atmosphere, all such equipment must be explosion-proof. Positive-pressure ventilation typically removes atmospheric contaminants more effectively than negative-pressure ventilation.

Leak Control and Containment

Leak control is intended to prevent a material from escaping its container, or to contain the release, either within the original container or by transferring it to a new one. Specific tactics are determined by the type of breach, material, and container. Leak control is also known as containment and is considered an offensive action. Operations-Level personnel do not typically attempt offensive actions, but may do so to deal with situations involving gasoline, diesel, liquefied petroleum gas (LPG), and natural gas fuels **(Figure 24.74, p. 1494)**. But even in these circumstances, they must be appropriately trained and equipped before attempting offensive action.

Figure 24.74 Appropriately trained Operations-Level responders may provide containment support by shutting off valves to natural gas lines.

Figure 24.75 Operations-Level responders may be assigned to operate emergency remote shutoff valves on cargo tank trucks.

In some situations it may be safe and acceptable for Operations-Level responders to operate emergency remote shutoff valves on cargo tank trucks (specifically, low-pressure liquid tanks and high-pressure tanks) **(Figure 24.75)**. The locations and types of remote shutoff valves vary depending on the truck **(Figures 24.76a-d)**.

Nonpressure liquid tanks (MC/DOT-306/406), low-pressure chemical tanks (MC/DOT-307/407), and high pressure tanks (MC-331) have an emergency shutoff switch on the left front corner of the tank (behind the driver). Some will also have one on either the right or the left rear corner. For example, MC-331s of 3,500 gallon (13 249 L) capacity or larger should have two emergency shutoff valves located remotely from each other, typically in this configuration: one on the tank behind the driver, the other on the rear of the tank, often on the passenger side. MC-331s of this capacity may also have an electronically operated shut-down device that can be activated 150 feet (46 m) from the vehicle. This device may also stop the engine and perform other functions.

Some cargo tanks may have emergency shutoffs in the center of the tank near valves and piping, or built into the valve box. Activation of these shutoff valves varies by device but is usually as simple as flipping a switch (or handle) or breaking off a fusible device.

NOTE: Corrosive liquid tanks (MC/DOT 312) do not normally have emergency shutoff valves.

Pipelines and piping systems carrying hazardous materials may have remote shutoff or control valves that can be activated without entering the hot zone. Typically, an IC will direct onsite maintenance personnel or local utility workers to close these valves. These personnel usually know where the valves are located and how to operate them. Trained and authorized Operations-Level responders may also do so, as long as they follow SOPs. **Skill Sheet 24-I-11** covers steps for shutting off a remote valve.

Figure 24.76a Nonpressure liquid tanks will have emergency shutoff valves on the left front corner, and they may also have them located on the rear of the tank, in the center of the tank near the valves, or in the valve box. *Courtesy of Rich Mahaney.*

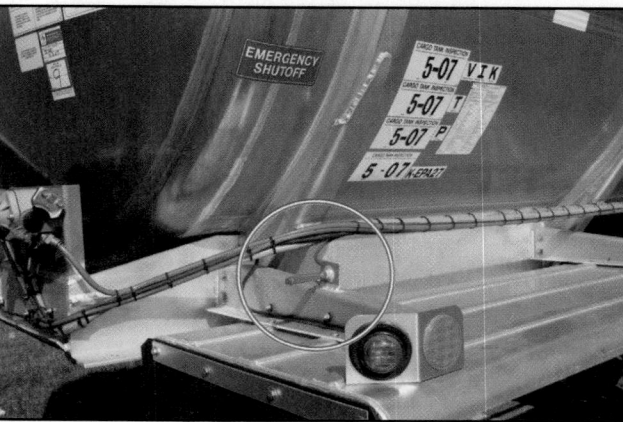

Figure 24.76b Low-pressure chemical tanks will have emergency shutoff valves on the left front corner (behind the driver). *Courtesy of Rich Mahaney.*

Figure 24.76c High-pressure chemical tanks will have two emergency shutoff valves located remotely from each other, one on the left front corner of the tank, the other typically on the right rear.

Figure 24.76d Shutoff valves can be switches, handles, pull levers, or fusible devices. Often, they will be labeled or painted red.

Terrorist Incidents

At terrorist and criminal incidents, responders have additional responsibilities. For example, they must immediately notify law enforcement and be alert for secondary devices and booby traps that target first responders or crowds **(Figure 24.77, p. 1496)**. Guidelines for terrorist and criminal incidents include:

- Do not contact contaminants or contaminated surfaces.
- Protect evidence at the crime scene.
- Document your observations.
- Take pictures, if possible.
- Make note of other witnesses and observers at the scene.
- Protect yourself and others by isolating the incident and denying entry.
- Prevent contaminated people and animals from leaving the scene, and direct them to a safe area to wait for help.
- Remember that WMD agents may be deadly in very small amounts, and biological agents may not cause symptoms for several days.

Figure 24.77 Response to terrorist and criminal attacks requires a higher level of awareness and documentation than most natural emergencies. *Courtesy of the U.S. Department of Defense.*

Figure 24.78 First responders conducting operations at crime scenes must be aware of the presence and delicate nature of evidence. Note the aircraft engine parts which are located in the debris on the left side of this photo, taken at the Pentagon after the 9/11 terrorist attack. *Courtesy of FEMA News Photos, photo by Jocelyn Augustino.*

Crime Scene Preservation

A response to a terrorist or criminal incident is essentially the same as a response to any other hazardous materials incident, but responders must protect and preserve evidence at the crime scene. Also, because a crime is involved, law enforcement organizations must be notified and included in the response. Notifying law enforcement ensures that the proper state/province and federal/national agencies respond to the incident (**Figure 24.78**). See Chapter 19, *Protecting Fire Scene Evidence*, for more information on crime scene management and evidence preservation.

Recovery and Termination

The final strategic goals at a haz mat incident are typically recovery and termination. Recovery involves ensuring that the scene is safe and returning responders to a preincident level of readiness. Termination involves documenting the incident and evaluating the response in order to improve performance.

Recovery

The major goals of the recovery phase are to:

- Return the scene to a safe condition.
- Debrief personnel before they leave the scene.
- Return all equipment and personnel to their preincident condition.

On-Scene Recovery

On-scene recovery efforts are directed toward returning the scene to a safe condition. These activities may require the coordinated effort of numerous agencies, technical experts, and contractors. Generally, fire and emergency services organizations do not conduct remedial cleanup actions unless those actions are absolutely necessary to eliminate conditions that present an imminent threat to public health and safety. If such imminent threats do not exist, contracted remediation firms under the oversight of local, state/provincial and federal environmental regulators generally provide for these cleanup activities. In these situations, the fire and emergency services organization may also provide control and safety oversight according to local SOPs.

On-Scene Debriefing

On-scene debriefing is performed to gather information from all operating personnel, including law enforcement, public works, and EMS responders. A group discussion takes place in which these personnel describe:

- Important observations
- Actions taken
- The timeline of those actions

OSHA also requires that all personnel take part in the *hazardous communication briefing*, in which responders are informed about the effects of exposure to the involved hazardous material. This briefing must be thoroughly documented, and all personnel must sign a document stating that they have received and understood the provided information, including:

- Identity of material involved
- Potential adverse effects of exposure to the material
- Actions to be taken for further decontamination
- Signs and symptoms of an exposure
- Mechanism by which a responder can obtain medical evaluation and treatment
- Exposure documentation procedures

Operational Recovery

Operational recovery involves those actions necessary to return the emergency responders to a level of preincident readiness. These actions involve the release of units, resupply of materials and equipment, decontamination of equipment and PPE, and preliminary actions necessary for obtaining financial restitution.

The financial effect of hazardous materials emergencies can be far greater than any other activity conducted by the fire and emergency services. Normally, a fire and emergency services organization's revenues obtained from taxes or subscriber fees are calculated based upon the equipment and personnel needs necessary to conduct fire suppression and other emergency activities. It is recommended that communities have in place the necessary ordinances to allow for the recovery of costs incurred from such emergencies. In addition, the proper documentation of costs through the use of forms such as the *IAP* and other tracking mechanisms is a vital part of this process.

Termination

The IC cannot terminate the incident until all strategic goals and legal requirements have been met. This involves thorough documentation, analysis, and evaluation. The termination phase involves two procedural actions: critiques and after action analysis or *After Action Report (AAR)*.

Critiques

All incidents must be critiqued so that departments can learn from their mistakes and identify operational deficiencies. All critiques must occur as soon as possible after the incident, and they must involve all responders, including law enforcement, public works, and EMS personnel. Documentation of the critique must include a list of people present and any identified operational deficiencies. Critiques are mandated by OSHA Title 29 *CFR* 1910.120.

After Action Analysis

The after action analysis process compiles the information obtained from debriefings, postincident reports, and critiques to identify trends in operational strengths and weaknesses. Once trends have been identified, recommendations for improvements are made. These recommendations may be made in the following categories:

- Operational weaknesses
- Training needs
- Necessary procedural changes
- Required additional resources
- Plan updates and/or required changes

Also included in the after-action analysis is the completion of necessary reporting procedures required to document personal exposures, equipment exposures, incident reports, and staff analysis reports. After action analysis forms the basis for improved response. Therefore, any recommendations for change or improvement are benchmarked for further consideration. Follow-up activities are scheduled to ensure successful implementation.

Chapter Summary

Hazardous materials incidents are similar in many ways to other emergencies. An incident management system is required, and the same incident priorities still apply: life safety, incident stabilization, and property conservation.

But there are also important differences. Size-up may have to be performed from a considerable distance, personnel can be at risk even far from the point of release, and there is an increased need for environmental protection.

Hazardous materials pose extreme health risks, so many personnel may not be properly trained or equipped to mitigate the incident. Instead they must establish a safe perimeter, notify haz mat specialists, and support these specialists through activities such as fire protection, damming and diking, and decontamination.

1. What are the basic responsibilities of both Awareness-Level and Operations-Level personnel at haz mat/WMD incidents?

2. What are the three incident priorities for haz mat/WMD incidents?

3. What kind of management structure is used for haz mat/WMD incidents?

4. What are the four problem-solving stages of haz mat/WMD incident mitigation?

5. What role does size-up and identifying incident levels play in the analysis stage of haz mat/WMD incidents?

6. What are the different modes of operation that can be used at a haz mat/WMD incident?

7. When should a plan be reevaluated or revised during haz mat/WMD incidents?

8. What are the three ways a firefighter can identify a hazardous material or WMD using the *Emergency Response Guidebook (ERG)*?

9. What can an emergency response center offer to responders during haz mat/WMD incidents?

10. What considerations must be taken when choosing personal protective equipment at haz mat/WMD incidents?

11. In what situations are each level of EPA defined protection used?

12. What are Mission-Oriented Protective Posture (MOPP) ensembles?

13. What factors must be considered when selecting personal protective equipment for use at haz mat/WMD incidents?

14. What safety and emergency procedures can be implemented when personnel are wearing protective clothing?

15. What type of information should be included in inspection records for PPE?

16. How do isolation and scene control help mitigate the impact of haz mat/WMD incidents?

17. What agencies may need to be notified in the event of haz mat/WMD incidents?

18. What methods are used to help ensure the protection of responders during haz mat/WMD incidents?

19. What means of protecting the public during haz mat/WMD incidents do responders have to choose from?

20. When is it best to use emergency decontamination instead of technical decontamination?

21. What rescue actions can a firefighter take without additional training in hazardous materials rescue?

22. What tactics can be used to implement the strategic goal of spill control and confinement?

23. What type of strategy is leak control and containment?

24. What are the guidelines to use when an incident is suspected to involve terrorist activity?

25. What other agency will need to be involved during a response to a terrorist or criminal incident?

26. What are the goals during recovery and termination of haz mat/WMD incidents?

24-I-1
Obtain information about a hazardous material
using the *Emergency Response Guidebook (ERG).*

Using the U.N. Identification Number

Step 1: Identify the four-digit U.N. identification number.

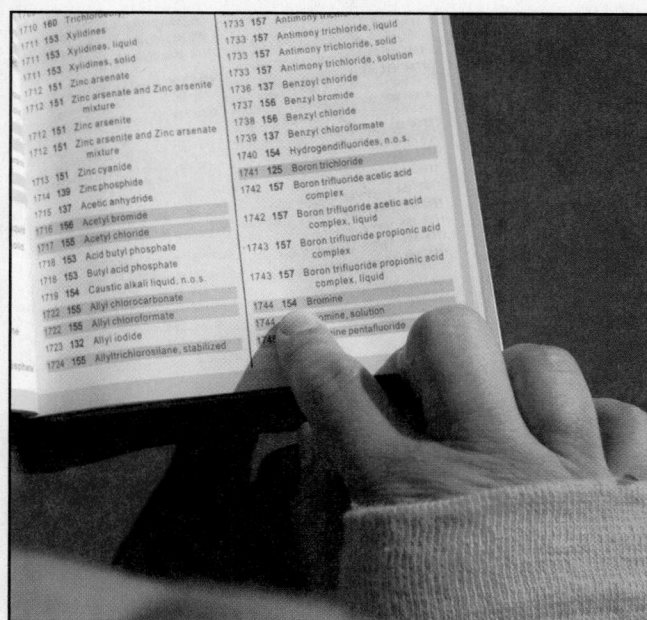

Step 2: Refer to the appropriate yellow-bordered pages to find the correct reference guide number.

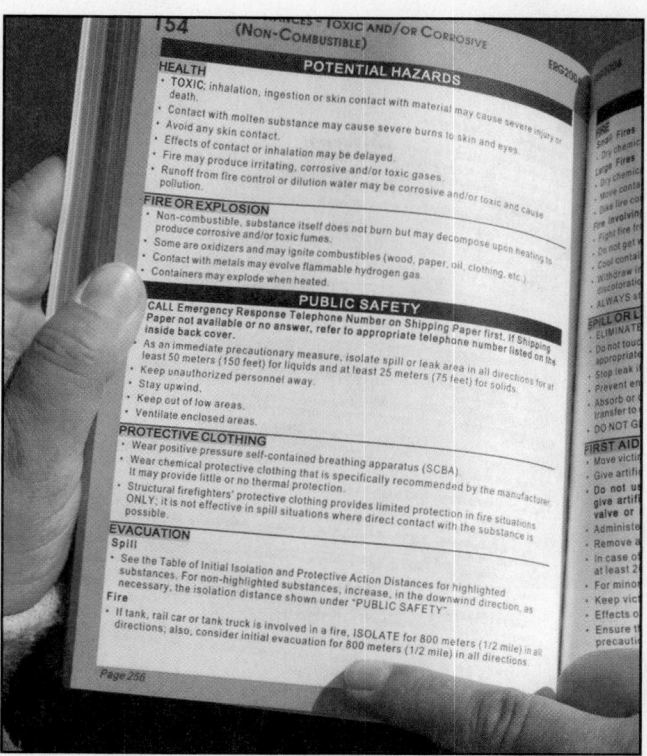

Step 3: Refer to the orange-bordered page with the appropriate guide number for information on managing the incident.

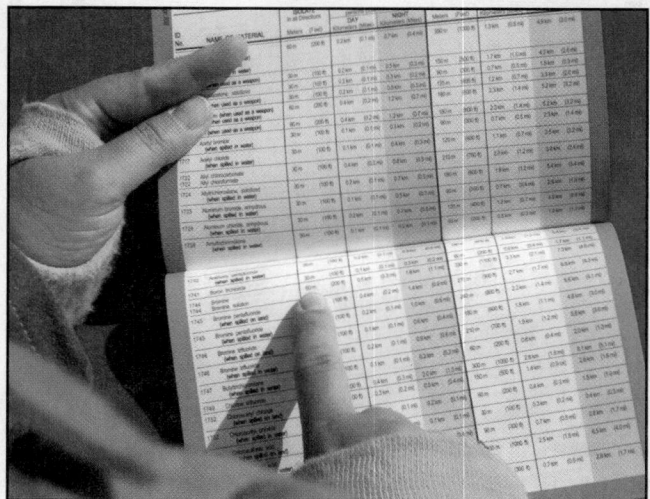

Step 4: For highlighted chemicals refer to the green-bordered pages for initial isolation by looking up the identification number.

SKILL SHEETS

24-I-1

Obtain information about a hazardous material
using the *Emergency Response Guidebook (ERG)*.

Using the Material Name

Step 1: Identify the name of the material.

Step 2: Refer to the name of the material in the blue-bordered pages to locate the correct guide number.

Step 3: Refer to the orange-bordered page with the appropriate guide number for information on managing the incident.

Step 4: For highlighted chemicals, refer to the green-bordered pages for initial isolation by looking up the identification number.

24-I-1

Obtain information about a hazardous material
using the *Emergency Response Guidebook (ERG)*.

SKILL SHEETS

Using the Container Profile

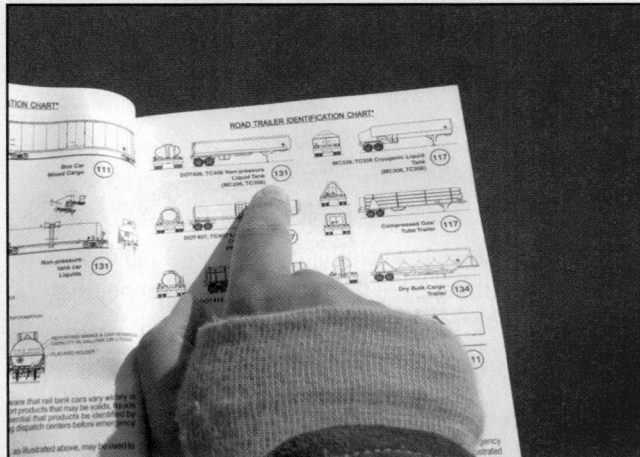

Step 1: Identify the profile of the container and locate the profile in the white pages of the ERG.

Step 2: Refer to the appropriate guide number in the circle and go to the appropriate orange-bordered page.

Using the Placard

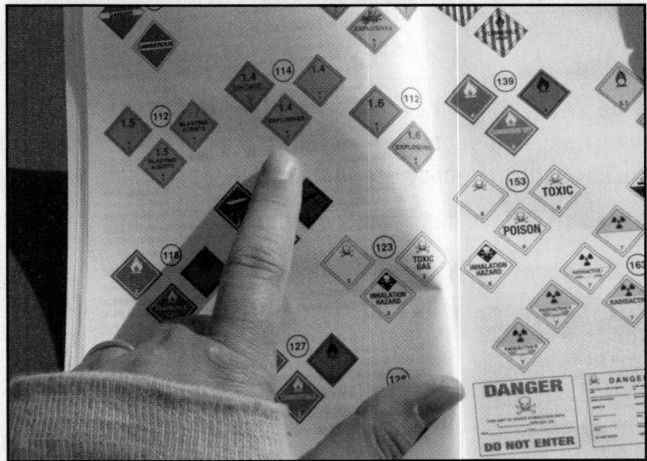

Step 1: Identify the placard and locate it in the white pages of the ERG.

Step 2: Refer to the appropriate guide number in the circle and go to the appropriate orange-bordered page.

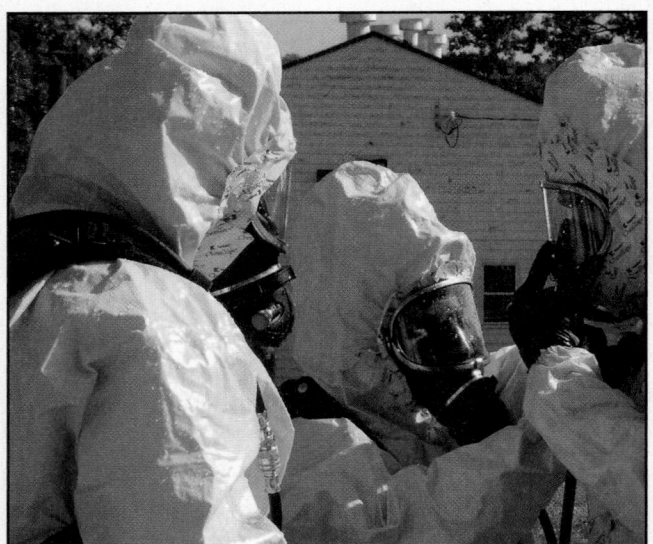

Step 1: Ensure that all responders involved in decontamination operations are wearing appropriate PPE for performing emergency decontamination operations.

Step 2: Remove the victim from the contaminated area.

Step 4: Remove victims clothing and/or PPE rapidly – if necessary, cutting from the top down in a manner that minimizes the spread of contaminants.

Step 5: Perform a quick cycle of head-to-toe rinse, wash, and rinse.

Step 6: Transfer the victim to treatment personnel for assessment, first aid, and medical treatment.

Step 7: Ensure that ambulance and hospital personnel are told about the contaminant involved.

Step 8: Decontaminate tools.

Step 9: Proceed to decontamination line for decontamination.

Step 3: Wash immediately any contaminated clothing or exposed body parts with flooding quantities of water.

Step 1: Verify that all responders involved in the control function are wearing appropriate PPE for performing absorption operations and that appropriate hand tools have been selected.

Step 2: Select a location to efficiently and safely perform the absorption operation.

Step 3: Select the most appropriate sorbent.

Step 4: Deploy the sorbent in a manner that most efficiently controls the spill.

Step 5: Upon mitigation of the incident, place any contaminated material, such as clothing, in an approved container for transportation to a disposal location. Seal and label the container and document appropriate information for department records.

Step 6: Decontaminate tools.

Step 7: Advance to decontamination line for decontamination.

Step 1: Verify that all responders involved in the control function are wearing appropriate PPE for performing adsorption operations and that appropriate hand tools have been selected.

Step 2: Select a location to efficiently and safely perform the adsorption operation.

Step 3: Select the most appropriate adsorbent.

Step 4: Deploy the adsorbent in a manner that most efficiently controls the spill.

Step 5: Upon mitigation of the incident, place any contaminated material, such as clothing, in an approved container for transportation to a disposal location. Seal and label the container and document appropriate information for department records.

Step 6: Decontaminate tools.

Step 7: Advance to decontamination line for decontamination.

Step 1: Verify that all responders involved in the control function are wearing appropriate PPE for performing diking operations and that appropriate hand tools have been selected.

Step 2: Select a location to efficiently and safely perform the diking operation.

Step 3: Construct the dike in a location and manner that most efficiently controls and directs the spill to a desired location.

Step 4: Upon mitigation of the incident, place any contaminated material, such as clothing, in an approved container for transportation to a disposal location. Seal and label the container and document appropriate information for department records.

Step 5: Decontaminate tools.

Step 6: Advance to decontamination line for decontamination.

Step 1: Verify that all responders involved in the control function are wearing appropriate PPE for performing damming operations and that appropriate hand tools have been selected.

Step 2: Select a location to efficiently and safely perform the damming operation.

Step 3: Construct the dam in a location and manner that most efficiently controls the spill.

Step 4: Upon mitigation of the incident, place any contaminated material, such as clothing, in an approved container for transportation to a disposal location. Seal and label the container and document appropriate information for department records.

Step 5: Decontaminate tools.

Step 6: Advance to decontamination line for decontamination.

Step 1: Verify that all responders involved in the control function are wearing appropriate PPE for performing diversion operations and that appropriate hand tools have been selected.

Step 2: Select a location to efficiently and safely perform the diversion operation.

Step 3: Construct the diversion in a location and manner that most efficiently controls and directs the spill to a desired location. Working as a team, use hand tools to break the soil, remove the soil, pile the soil, and pack the soil tightly.

Step 4: Upon mitigation of the incident, place any contaminated material, such as clothing, in an approved container for transportation to a disposal location. Seal and label the container and document appropriate information for department records.

Step 5: Decontaminate tools.

Step 6: Advance to decontamination line for decontamination.

Step 1: Verify that all responders involved in the control function are wearing appropriate PPE for performing retention operations and that appropriate hand tools have been selected.

Step 2: Select a location to efficiently and safely perform the retention operation.

Step 3: Evaluate the rate of flow of the leak to determine the required capacity of the retention vessel.

Step 5: Upon mitigation of the incident, place any contaminated material, such as clothing, in an approved container for transport to a disposal location. Seal and label the container and document appropriate information for department records.

Step 6: Decontaminate tools.

Step 7: Advance to decontamination line for decontamination.

Step 4: Working as a team, retain the hazardous liquid so that it can no longer flow.

Step 1: Verify that all responders involved in the control function are wearing appropriate PPE for performing dilution operations.

Step 2: Select a location to efficiently and safely perform dilution operations.

Step 3: Evaluate the rate of flow of the leak to determine the required capacity of the retention area and the quantity of water required to dilute the material.

Step 4: Working as a team, monitor and assess the leak, and advance hoselines and tools to retention area.

Step 5: Flow water to dilute spilled material.

Step 6: Monitor any diking or dams to ensure integrity of retention area.

Step 7: Upon mitigation of the incident, place any contaminated material, such as clothing, in an approved container for transportation to a disposal location. Seal and label the container and document appropriate information for department records.

Step 8: Decontaminate tools.

Step 9: Advance to decontamination line for decontamination.

Step 1: Verify that all responders involved in the control function are wearing appropriate PPE for performing vapor dispersion operations.

Step 2: Select a location to efficiently and safely perform the vapor dispersion operation.

Step 3: Working as a team, advance the hoseline to a position to apply agent through vapor cloud to disperse vapors.

Step 4: Constantly monitor the leak concentration, wind direction, exposed personnel, environmental impact, and water stream effectiveness.

Step 5: Upon mitigation of the incident, place any contaminated material, such as clothing, in an approved container for transportation to a disposal location. Seal and label the container and document appropriate information for department records.

Step 6: Decontaminate tools.

Step 7: Advance to decontamination line for decontamination.

Step 1: Ensure that all responders involved in control functions are wearing appropriate PPE for performing remote valve shutoff operations.

Step 2: Identify and locate the emergency remote shutoff device.

Step 3: Operate the emergency remote shutoff device properly.

Step 4: Notify the Incident Commander of the completed objective.

Step 5: Document the activity log.

Courtesy of District Chief Chris Mickal, New Orleans (LA) Photo Unit.

Contents

Appendix A

NFPA® Job Performance Requirements (JPRs) with Chapter and Page References

Courtesy of District Chief Chris Mickal, New Orleans (LA) Photo Unit.

NFPA® 1001 JPR Numbers	Chapter References	Page References
4.3	22	1272-1294
5.1.1	1, 2, 6, 21	20-21, 22-24, 25-26, 26-33, 33-35, 35-36, 37-40, 46-54, 55-59, 60-62, 62-63, 63-66, 73-74, 76-78, 78-80, 80-85, 259-276, 277-280, 303-307, 1222-1224, 1224-1226, 1226-1237, 1237-1240
5.1.2	1, 6, 8	35-36, 317, 324-325, 384-385, 386-389, 389-396, 410-411, 412, 413, 414, 415, 416
5.2.1	3	96-103, 103-109, 112-115, 124-125, 126-129
5.2.2	3	96-103, 103-109, 112-115, 124-125
5.2.3	3	96-103, 109-112, 112-115, 126-127
5.2.4	2, 6 , 9	466
5.3.1	6, 9	280, 281-287, 287-295, 295-296, 297-302, 307-311, 312-315, 317, 317-319, 320-321, 322-323, 328-329, 330-331, 332, 334, 443-457, 449-459, 467-469, 471, 472
5.3.2	2, 6	66-71, 72, 88-89, 259-276, 297-302, 317, 317-319, 320-321, 322-323, 324-325, 326, 327, 370-371
5.3.3	2, 6	66-71, 78-80, 80-85, 88-89, 90-91, 259-276, 277-280, 317, 317-319, 320-321, 322-323, 324-325
5.3.4	2, 4, 11	75-76, 134, 142, 157, 573-575, 575-581, 581-593, 596-608, 609-616, 616-619, 619-621, 621-623, 627, 628, 629, 630, 631, 632, 633, 634, 635, 636, 637, 638, 639, 640-641, 642-643, 644, 645, 646, 647, 648-649
5.3.5	2, 9	85-86, 439-443, 443-459, 466, 467-469, 470, 471, 472, 473, 474-475
5.3.6	12	654-659, 660-663, 666-670, 670-675, 675-677, 678-682, 682-683, 684-685, 685-686, 692, 693, 694, 695, 696, 698-706, 707-708, 709-710, 711-712
5.3.7	17	1038-1046, 1092-1093
5.3.8	15, 17, 18, 19	816-825, 835-849, 855-861, 861-862, 862-868, 868-871, 1004-1011, 1011-1013, 1013-1016, 1016-1017, 1017-1020, 1028-1031, 1031-1035, 1035-1037, 1037-1038, 1046-1049, 1078-1082, 1083-1085, 1089-1090, 1091, 1094, 1095, 1096, , 1145-1153, 1153-1156, 1159-1164
5.3.9	9, 12	422-425, 425-426, 426-428, 428-431, 431-439, 439-443, 443-457, 449-457, 459, 460, 461, 462, 463, 464, 465, 466, 467-469, 470, 471, 472, 473, 474-475, 686-687, 687-688, 724-725, 726

NFPA® 1001 JPR Numbers	Chapter References	Page References
5.3.10	2, 4, 5, 6, 11, 15, 16, 17, 18	134, 142, 157, 816-825, 835-844, 861-862, 862-863, 868-871, 875-879, 909-914, 917-918, 919, 920-923, 924-925, 926-927, 928-929, 930-931, 932-933, 934, 936, 937, 944-947, 947-949, 949-957, 957-962, 962-964, 964-967, 988-990, 991, 992, 1004-1011, 1011-1013, 1013-1016, 1016-1017, 1017-1020, 1031-1035, 1078-1082, 1083-1085, 1089-1090, 1118-1124, 1139-1140
5.3.11	5, 12, 13	208-216, 223-231, 231-233, 233-234, 234-241, 241-247, 247-250, 250-253, 692, 693, 694, 695, 696, 697, 698-706, 707-708, 709-710, 711-712, 713-714, 715, 716-717, 718-719, 720-721, 722-723, 732-734, 734-742, 742-744, 744-751, 751-759, 760-763, 763-764, 765, 766, 767, 768
5.3.12	4, 5, 8, 12, 13	134, 142, 157, 216-223, 250-253, 692, 693, 694, 695, 696, 697, 698-706, 707-708, 709-710, 711-712, 713-714, 715, 716-717, 718-719, 720-721, 722-723, 734-742, 742-744, 744-751, 751-759, 763-764, 765, 766, 767, 768, 769, 770, 771, 772-774, 775-776, 777-778
5.3.13	8, 11, 16, 17, 18, 19	1078-1082, 1083-1085, 1118-1124, 1139-1140, 1153-1156, 1156-1159
5.3.14	17, 18, 19	627, 628, 629, 630, 631, 632, 633, 634, 635, 636, 637, 638, 639, 640-641, 642-643, 644, 645, 646, 647, 648-649, 1027-1028, 1087, 1088, 1089-1090, 1104, 1104-1106, 1106-1112, 1112-1117, 1125, 1126-1127, 1128, 1129-1130, 1131, 1132-1133, 1134, 1135, 1136, 1137, 1138, 1145-1153, 1159-1164
5.3.15	14, 15, 17	784-795, 793-796, 796-799, 799-800, 800-804, 805, 806-808, 809, 810, 855-861, 906-907, 915-916, 935, 1087
5.3.16	7, 17	340-342, 342-351, 351-353, 354-356, 357-358, 360-361, 362-363, 364-365
5.3.17	10	479-484
5.3.18	17	1020-1027, 1086
5.3.19	17	1049, 1049-1050, 1050-1052, 1052-1053, 1054, 1055-1057, 1097
5.3.20	8	376-381, 381-389, 389-396, 399, 400, 401, 402, 403, 404, 405, 406, 407, 408, 409, 410-411, 412, 413, 414, 415, 416
5.5.1	6, 7, 8, 11, 12, 15, 16, 18	277-280, 295-296, 303-307, 307-311, 326, 327, 357-358, 376-381, 397-398, 593-596, 624-625, 626, 663-666, 690-691
5.5.2	15	825-831, 831-835, 844-866, 879-882, 883, 884-887, 888-889, 890, 891-893, 894-896, 897-899, 900-902, 903-905, 906-907, 908
6.1.1	1, 17	26-33, 1057-1058, 1058-1064, 1064-1068, 1098
6.1.2	17	1057-1058, 1058-1064, 1064-1068, 1098
6.2.1	3	121-123, 128
6.2.2	3	116-121
6.3.1	16, 17	967-978, 978-983, 983-984, 984-985, 985-986, 993-994, 995-998, 1068-1073, 1073-1076
6.3.2	4, 16, 17	190-195, 195-201, 978-983, 983-984, 984-985, 985-986, 993-994, 1057-1058, 1058-1064, 1064-1068, 1068-1073, 1098
6.3.3	17	1068-1073, 1073-1076, 1099

NFPA® 1001 JPR Numbers	Chapter References	Page References
6.3.4	19	1165-1166, 1167-1169, 1169-1171, 1171-1172, 1173
6.4.1	10	486-504, 504-512, 512-522, 522-531, 553, 554, 555-556, 557, 558-559, 560-561, 562, 563-565, 566, 567-568
6.4.2	10	484-486, 486-504, 531-550
6.5.1	21	1240-1245, 1251-1256, 1256-1260, 1261-1262, 1263, 1264, 1265-1266
6.5.2	21	1246-1251, 1251-1256, 1263, 1264
6.5.3	20, 21	1178-1180, 1180-1194, 1194-1209, 1209-1213, 1213-1216, 1256-1260, 1265-1266
6.5.4	10	484-486, 486-504, 551-552
6.5.5	15	872-873, 938-939

NFPA® 472 Competencies	Chapter References	Page References
4.2.1	23, 24	1300-1301, 1326-1376, 1379-1383, 1383-1390, 1390-1402, 1402-1405, 1416-1421
4.2.2	23	1326-1376, 1376-1379
4.2.3	24	1431-1440, 1501-1503
4.4.1	23, 24	1301-1312, 1312, 1413-1415, 1416-1421, 1431-1440, 1466-1470, 1478-1481, 1495, 1501-1503
4.4.2	24	1470-1471
5.2.1	23, 24	1326-1376, 1421-1427
5.2.1.1	23	1326-1376
5.2.1.1.1	23	1326-1376
5.2.1.1.2	23	1326-1376
5.2.1.1.3	23	1326-1376
5.2.1.1.4	23	1326-1376
5.2.1.1.5	23	1326-1376
5.2.1.1.6	23	1326-1376
5.2.1.2	23	1326-1376
5.2.1.2.1	23	1326-1376
5.2.1.2.2	23	1326-1376
5.2.1.3	23	1326-1376
5.2.1.3.1	23	1326-1376
5.2.1.3.2	23	1326-1376
5.2.1.3.3	23	1326-1376
5.2.1.4	24	1421-129

NFPA® 472 Competencies	Chapter References	Page References
5.2.1.5	24	1431-1440
5.2.1.6	23	1383-1390
5.2.2	23, 24	1301-1312, 1326-1376, 1376-1379, 1416-1421, 1440-1442, 1470-1471
5.2.3	23, 24	1301-1312, 1313-1320, 1320-1326, 1383-1390, 1481-1485
5.2.4	23, 24	1379-1383, 1421-1427, 1471-1479
5.3.1	23, 24	1402-1405, 1421-1427, 1427-1430, 1485-1486
5.3.2	24	1427-1430, 1481-1485
5.3.3	24	1442-1454
5.3.4	24	1481-1485, 1504
5.4.1	24	1478-1481, 1481-1485
5.4.2	24	1496
5.4.3	24	1413-1415, 1416-1421, 1421-1427, 1470-1471, 1471-1477
5.4.4	24	1465-1466, 1471-1477
5.5.1	24	1430-1431, 1471-1477
5.5.2	24	1430-1431, 1471-1477
6.2.3.1	24	1442-1454, 1455-1459, 1459-1460, 1460-1461, 1481-1485
6.2.4.1	24	1461-1465, 1465-1466, 1471-1477, 1481-1485
6.2.5.1	24	1465-1466
6.6.3.1	24	1486-1493, 1493-1495, 1505, 1506, 1507, 1508, 1509, 1510, 1511, 1512, 1513
6.6.3.2	24	1460-1461
6.6.4.1	24	1493-1495, 1513
10.6.1	24	1496-1498

1520 Appendix A • NFPA® Job Performance Requirements

Appendix B
Prehospital 9-1-1 Emergency Medical Response

PREHOSPITAL 9-1-1
EMERGENCY MEDICAL RESPONSE:

The Role of the United States Fire Service in Delivery and Coordination

Authors and Contributors

Franklin D. Pratt, M.D., FACEP

Medical Director
Los Angeles County Fire Department

Medical Director
Emergency Department
Torrance Memorial Medical Center
Torrance, CA 90505

Assistant Clinical Professor
Geffen School of Medicine
UCLA

Paul E. Pepe, MD, MPH

Professor of Medicine, Surgery, Public
Health and Chair, Emergency Medicine
University of Texas Southwestern
Medical Center and the Parkland Health
and Hospital System

Director, City of Dallas Medical
Emergency Services
for Public Health, Public Safety and
Homeland Security

Steven Katz, M.D., FACEP, EMT-P

Associate Medical Director
Palm Beach County Fire Rescue
West Palm Beach, FL

President
National Paramedic Institute
Boynton Beach, FL

Chairman
Department of Emergency Medicine
Memorial Hospital West
Pembroke Pines, FL

David Persse, MD, EMT-P, FACEP

Physician Director
Houston Fire Department
Emergency Medical Services

Public Health Authority
Houston Department of Health and
Human Services

Associate Professor of Surgery
Baylor College of Medicine

Associate Professor of Emergency
Medicine
University of Texas Medical School
Houston

2

PREHOSPITAL 9-1-1 EMERGENCY MEDICAL RESPONSE: THE ROLE OF THE UNITED STATES FIRE SERVICE IN DELIVERY AND COORDINATION

ABSTRACT

Prehospital 9-1-1 emergency response is one of the essential public safety functions provided by the United States fire service in support of community health, security and prosperity. Fire service-based emergency medical services (EMS) systems are strategically positioned to deliver time critical response and effective patient care. Fire service-based EMS provides this pivotal public safety service while also emphasizing responder safety, competent and compassionate workers, and cost-effective operations. As the federal, state, and local governments consider their strategic plans for an 'all hazards' emergency response system, EMS should be included in those considerations and decision makers should recognize that the U.S. fire service is the most ideal prehospital 9-1-1 emergency response agency.

INTRODUCTION TO FIRE SERVICE-BASED EMS

EMS is an essential component of the public services provided in the United States. The Federal EMS Act of 1973 defined an EMS system as "an entity that provides for the arrangement of personnel, facilities, and equipment for the effective and coordinated delivery of health care services under emergency conditions in an appropriate geographic area" (EMS Act 1973, (P.L. 93-154)). Much of the dialogue in the public arena today concerning prehospital 9-1-1 emergency medical care often focuses on ambulance services and, accordingly, may ignore the important distinction between prehospital 9-1-1 emergency medical response and the other key uses of the ambulance-based, out-of-hospital providers for non-emergency medical and transportation services.

3

The primary purpose of this discussion is to underscore the reality today that the fire service has become the first-line medical responder for critical illness and injury in virtually every community in America. Regardless of whatever agency provides medical transportation services, the fire service is the agency that first delivers on-scene health care services under most true emergency conditions. Therefore, prehospital 9-1-1 emergency response, in support of community health, security and prosperity, is not only a key function of each community; it has become, almost universally, a principal duty of the fire service as well. In addition, fire service-based EMS systems are strategically positioned to deliver time critical response and effective patient care rapidly. Furthermore, the fire service-based EMS accomplishes this rapid first response while emphasizing responder safety, sending competent and compassionate workers, and delivering cost-effective operations.

Although the role of the fire service is central in 9-1-1 emergency medical response, financial, political, cultural and organizational factors often can make the conversation about prehospital care providers confusing and complex for many decision makers in local communities. The goal of this discussion is to resolve and demonstrate that the use of fire service equipment and personnel to provide 9-1-1 emergency response is the best approach for a community regardless of size. This basic premise is consistent with recent Institute of Medicine publications that have placed EMS at the intersection of public safety, public health, and medical care. The U.S. Fire Service is uniquely qualified to be at that intersection and in the following pages, the history, evolution, and current medical capabilities of the fire service will be reviewed.

The Maltese Cross and Its Legacy for Fire-Service Based EMS

During the Middle Ages, the Knights of Malta, the forerunners of the fire service, took care of travelers and specifically burn victims from the Crusades and associated battles. Eventually, the Knights of Malta adopted the Maltese Cross as their emblem and it has created a revered legacy for fire departments.

4

The Knights originally began their work as the creators, administrators and care givers in a hospital in Jerusalem. As such, they were known as the Hospitallers of Jerusalem, starting their work before the year 1000 AD. For the next two hundred years, they helped the sick and poor and they set up hospitals and hospices across Europe.

Eventually, the Hospitallers became firefighters out of necessity. The conflict of the Crusades often threatened the hospitals that they had founded. So, they adapted and even engaged in battle to protect their hospitals. As a result, they also became firefighters because one of the weapons of war at that time was the glass fire bomb. The fire bomb, thrown by the enemy, created a horrendous inferno. After rescuing a fellow knight from the inferno and extinguishing the fire, a Hospitaller was awarded a medal, shaped like a Maltese Cross to honor those actions.

As conflict continued, the Hospitallers needed an identifying mark for their armor. This was necessary because without identifying markings, it was difficult to tell who was who because everyone was wearing similar armor in battle. They adopted the Maltese Cross as their identifying mark. (Maltese Cross, 2007, Foster, 2007)

In essence, more than 1200 years ago, some of the earliest ancestors of the fire service were "all-hazards responders." They initially started as caregivers for the sick and then became firefighters to protect their own. These are two of the concepts firefighters still believe in today and hold as their most sacred responsibilities—caring for the sick and caring for their own.

Longstanding History of Fire Service-Based Medical Care in the U.S.

The fire service has formally been part of the 9-1-1 emergency care delivery system since EMS began in the late 1960's. Many of the original prehospital EMS providers were firefighters, who had "special" additional training in providing medical services during emergencies that occurred outside the hospital. Today, essentially every firefighter receives emergency medical training and the fire service provides the majority of medical services during emergencies that occur out of the hospital, just as it has done for the past

5

four decades. Of the 200 largest cities in the United States, 97% have fire service-based prehospital 9-1-1 emergency medical response (*JEMS* 200-City Survey, 2006) and the fire service provides advanced life support (ALS) response and care in 90% of the 30 most populated U.S. jurisdictions (cities and counties) (IAFF/IAFC Fire Operations Survey, 2005).

Although the origin of the modern relationship between emergency medicine and fire departments is cited as the 1960's, the involvement of the fire service in patient care began much earlier. For example, in 1937 a fire department ambulance in New York transported famous song writer Cole Porter to the hospital after a horseback riding accident.

While the fire service was involved in many famous anecdotal events, other accounts demonstrate its profound effect on public safety and patient care procedures. In 1921 Claude Beck, M.D., a surgeon at Western Reserve University in Cleveland, called the fire department so he could apply a "pulmotor," an artificial breathing apparatus, to attempt resuscitation in a patient who died unexpectedly during surgery (Beck, 1941). Dr. Beck continued to be involved in resuscitation and today is recognized as one of the founders of the science of resuscitation.

The following quote from the *Journal of the American Medical Association* in 1928 summarized the evolving relationship between fire department-based out-of-hospital emergency care and subsequent resuscitation in the hospital.

"...inhalators are introduced: Cases of gas asphyxiation occur; the rescue crew of the fire department is called and resuscitates the patient. A physician sees the resuscitation and is impressed by the effectiveness of the treatment. Some time thereafter he finds himself confronted with a child which he has delivered, and which has come through a prolonged labor. It refuses to breath effectively, in spite of the application of all the ancient practices. The respiratory center has been depressed by the diminished blood supply to the brain resulting from compression of the head, and needs more than the

6

normal amount of carbon dioxide to stimulate breathing. So the physician calls in the fire department. If, as is often the case, the fire department succeeds where his medical skill and knowledge have failed, he calls for it again the next time. Now the hospitals in some cities are adopting the practice of calling for the inhalator whenever they have a baby who breathes poorly. In effect, *they add the rescue crew of the fire department to their board of consultants, and these new consultants thus contribute another service to the community over and above that for which the fire department is primarily organized.* Obviously, it is the hospitals that should be equipped to treat asphyxia -- asphyxia of every form -- and thus to help firemen overcome by smoke and gas, instead of relying on the fire department to help the hospital in such a matter as asphyxia of the newborn." (Henderson, *JAMA* 1928. note: italics added).

According to a historical account on the City's website, in 1947 in the city of Virginia, Minnesota, "the Fire Department took full possession of the ambulance along with a Pulmotor Resuscitator … This would be the first time that ambulance personnel would be properly trained in first aid, and resuscitation procedures of that time" (City of Virginia, MN, 2007).

Such widespread anecdotes not only indicate longstanding involvement of the fire service in medical care, but it demonstrates the often-quoted mission of the fire service established in the 19th Century, "To Protect and Save Lives and Property." Clearly, protecting and saving lives is the first and foremost mission for these dedicated first responders.

Growth and Specialization of Fire Service-Based EMS

As illustrated by its history, the fire service has continuously adapted and changed to meet the current needs of a community. As EMS developed, the fire service was integrally involved. In the early stages, firefighters were chosen by expert physicians to take on the role of paramedic. This era of EMS in the fire service is represented well by

7

looking at the City of Miami Fire Rescue Department nearly half a century ago. The age-old firefighter mantra to "protect and save lives and property" is well-illustrated within the history books of the City of Miami Fire Rescue and serves as an important example of Fire Rescue today in the United States. Miami was the first city to call itself a "Fire Rescue" department. Miami Fire Rescue was also revolutionary in using the advancements of technology in 2-way radios to bridge physicians in the hospital with firefighter-paramedics in the prehospital setting.

In fact, the Rescue Division of Miami Fire Rescue was established in 1939 in order to give first aid to firefighters. Rescue One, the department's first rescue truck used to treat citizens, came on-line in 1941. In these early days, "Rescue" services were limited to basic first aid with transportation usually performed by funeral homes.

In 1964, Dr. Eugene Nagel, started to teach first aid and basic cardiopulmonary resuscitation (CPR) to the firefighters of Miami Fire Rescue. Dr. Nagel's goal was to improve out-of-hospital cardiac arrest survival in the community by using lessons learned from the "quick response" system in the hospitals and apply it to the prehospital setting. Dr. Nagel still reflects, "We chose firefighters because they were there, they were available, they were willing, and they were motivated. It was really quite simple" (Nagel interview, February 2007). According to Dr. Nagel, "The fire service is dispersed throughout America and is everywhere in our country. It is an efficient method for offering emergency care rather than creating a completely separate service with separate communications, vehicles, housing, and personnel. It worked well in Miami in the 1960's and continues to work well when integrated into the fire service. It is a natural fit" (Nagel interview, February 2007). Firefighters in Miami clearly demonstrated in the pioneer days of EMS that firefighters are ideal candidates and willing dispensers of high-quality EMS. In Miami, this started with basic first aid, and progressed to CPR, intravenous therapy, electrocardiographs, telemetry, and advanced airway intervention. During this same time period, similar efforts were underway using firefighters in the cities such as Baltimore, Columbus, Seattle, and Los Angeles. Providing firefighters training in lifesaving techniques and procedures has allowed them to deliver advances in medicine to

8

the prehospital 9-1-1 emergency care patient in a cost-effective and time-sensitive manner. Just as fire departments have evolved since the 1960's to provide prehospital emergency medical care, government oversight must evolve to cohesively organize, coordinate, and supervise the integrated delivery of emergency medical care from the scene to the hospital and even the rehabilitation and recovery phase. A critical link in that chain of survival and recovery is the rapid on-scene response of the Fire service, a service that cannot be underestimated and truly emphasized in planning, funding, support, research, and quality assurance.

The protection of life and property has been the mission of the fire service for over 200 years, but the fire department of the 21st century is evolving into a multidisciplinary public safety department. It not only handles most aspects of public safety (beyond law enforcement security issues), but it also will continue to provide advances in emergency medical care and many developing public health needs such as preparations for pandemics, disasters, and weapons of mass effect.

Today, the community-based fire station, with its ready availability of personnel 24 hours a day, coupled with the unique nature of medicine outside of the hospital, creates a symbiotic blend of the traditional public concepts and duties of the fire service with the potential for the most rapid delivery of advanced prehospital 9-1-1 emergency response and care. Traditionally, fire stations are strategically placed across geographic regions, typically commensurate with population densities and workload needs. This creates an all-hazard response infrastructure meeting the routine and catastrophic emergency needs of all communities regardless of the nature of the emergency. Accordingly, the fire service helps ensure the prosperity and security of all communities and providing prehospital 9-1-1 emergency medical care is consistent with its legacy going back 1200 years.

Types of Fire Service-Based EMS Systems

The fire service can be configured many ways to deliver prehospital 9-1-1 emergency medical care such as the following general configurations:

9

- Fire service-based system using cross-trained/multi-role firefighters. Firefighters are all-hazards responders, prepared to handle any situation that may arise at a scene including patient care and transport.

- Fire service-based system using employees who are not cross-trained as fire suppression personnel. Single role EMS-trained responders accompanying firefighter first-responders on 9-1-1 emergency medical calls.

- Combined system using the fire department for emergency response and a private or "third service" (police, fire, EMS) provider for transportation support. Single role emergency medical technicians and paramedics accompany firefighter first responders to emergency scenes to provide patient transport in a private or third service ambulance.

While there are pros and cons to the various system approaches, the emergency medicine (EM) literature indicates that the most likely time to create error in medical care is when care is transferred from one provider to another in a relatively short encounter time. Such circumstances require that the fire service regularly exercise the leadership needed to ensure that integration of the parts of the prehospital emergency care system are coordinated well, with maximum benefit to the patient and minimum risk to the community. For example, in the fire service-based EMS model in which the fire department provides extrication, triage and treatment services, and a separate private provider transports the patients, appropriate quality assurance measures must be in place. This quality assurance is most effective when the fire department, as the public agency, administers and monitors the performance requirements on-scene and within the transportation agreement.

10

National Incident Management System

The U.S. Fire Service-based emergency response and medical care system is the most effective, coordinated system worldwide. The National Incident Management System (NIMS) and other nationally-defined coordination plans ensure that fire service-based 9-1-1 emergency response and medical care always provides skilled medical services to the patient regardless of the circumstances surrounding the location and condition of the patient. In addition, the fire service has the day-to-day experience and ability to work smoothly with other participants in the prehospital 9-1-1 emergency medical care arena: private ambulance companies, law enforcement agencies, health departments, public works departments, the American Red Cross and other government and non-government agencies involved in medical care, disaster response and patient services. This type of universal coordination takes leadership, work, and the willingness to subordinate fire service prerogatives to those of the greater public need. The fire service is the creator of the unified command concept that brings everyone to the table, at the same time. Using the National Incident Management System, the fire service has superior ability to coordinate incidents of any size. As a result, it provides the best return on investment of public dollars to provide the delivery of prehospital 9-1-1 emergency medical service.

Emergency 9-1-1 Response is Different from Non-emergency and In-hospital Care

For government decision makers who do not work in the public safety environment on a day-to-day basis, it may be difficult to appreciate the differences between emergency response and ambulance transport. Unless one actually has used the EMS system in a medical emergency, he or she might be likely to define a call to 9-1-1 in a medical emergency as 'needing an ambulance.' However, with the recent advances in resuscitative medical care, particularly in cardiac emergencies, we now know that what occurs in the first few minutes after onset of the medical emergency will change the long term outcome. In many of these critical circumstances, what happens on-scene determines whether the patient lives or dies. Therefore, rapid, efficient and effective delivery of emergency response and care is dependent on immediately sending nearby

11

trained personnel to the scene of an emergency regardless of the vehicle or mode of transportation.

Ambulances, of course, are necessary to transport patients to a hospital where more definitive care may be needed. However, because ambulances are often busy evacuating, transporting and turning over patients at the hospital, the most reliable vehicle to ensure a rapid response generally is the neighborhood fire truck. It should be realized that the first emergency care provider who is responsible for competent care may arrive on a fire truck separate from an ambulance. This is the case in most communities in America.

There are sub-specialties of ambulance service in the out-of-hospital arena that must not be confused with 9-1-1 emergency response. For example, ambulance services are often employed for interfacility transfers for specialty care or the need to transfer patients from one hospital to another can provide a higher level of required care. These transfers may include critical care transfers between hospitals, but more often they may also be non-emergent interfacility transports or day transport for persons with home-delivered chronic care services. Such services typically are not performed by fire departments as a fundamental public policy device to better ensure dedicated 9-1-1 emergency services and thus provide security and prosperity for the community served.

Multi-Role Firefighters: Patient Safety from Multiple Perspectives

To further emphasize that the prehospital 9-1-1 emergency care patient should be considered a separate and distinct type of patient in the continuum of health care, consider the setting and the circumstances of emergency medical care delivery. These patients not only have medical needs, but they also need simultaneous physical rescue, protection from the elements and the creation of a safe physical environment as well as management of non-medical surrounding sociologic concerns. The fire service is uniquely equipped to simultaneously address all of these needs.

The mission of the fire service is to protect and save lives and property. There are no other conflicting agendas. The fire service-based prehospital, 9-1-1 emergency response

12

medical care system is designed to be part of society's safety net. Fire and prehospital 9-1-1 emergency response medical care are intimately intertwined. Separating them from the EMS focus only serves to polarize our country's already fragmented emergency response system.

All out-of-hospital emergency care and ambulance transport professionals are taught that scene safety is the primary objective at every emergency scene. However, many of today's non-fire service-based EMS professionals do not have the additional resources and often do not have the training to effectively secure a scene. When there is a strict medical orientation in their professional training and practice, adequate preparation to appropriately and safely provide emergency medical care to an emergency patient may be compromised. Scene safety issues are often not apparent until a crew is on-scene to assess the incident.

Decision makers should consider, 'What does a non-fire based EMS crew do on the scene of a motor vehicle accident when the car is engulfed in flames and occupants are trapped inside, and fire crews were not dispatched?' In many cases, a non-fire service-based EMS provider would need to request dispatch of a fire company after the initial scene size-up, further delaying care, and further increasing risk to rescuers and victims. Streamlining this approach into the fire service-based prehospital 9-1-1 emergency medical care system is quite arguably more effective from the perspective of scene safety, short response time, integrated rescue and treatment, and then transport to a medical facility. Regardless, the firefighter response is a key element of patient safety, both medically and environmentally.

In the era of homeland security threats and the spiraling growth of the commercial transport industry, the threat of hazardous materials (Haz-Mat) is center-stage. Again, fire service Haz-Mat teams are the front-line of protection and rapid delivery of medical care can be pre-empted by such chem-bio threats, but where rapid care can be given, it can be expedited directly by cross-trained fire-service Haz-Mat care providers.

13

Fire Service-Based EMS as the Health Care System Safety Net

Prehospital 9-1-1 emergency patient medical care is a major part of the safety net for the American healthcare system. They may be the provider of last resort for the needy, yet they can be one more mechanism for overloading the health care system. Nevertheless, to its credit, the fire service-based, prehospital 9-1-1 emergency patient medical care provides unconditional service to all members of our population. Therefore, the fire service must now become an integral part of the public health system and work closely with medical and public health experts to help alleviate unnecessary burdens on already overburdened hospital, medical and public health systems. Already part of local government, the fire service may be best positioned to sit at the table and help provide important data to facilitate creating solutions to pressing health care public policy issues.

Above all, rapid response times are a pivotal advantage of fire service-based, prehospital 9-1-1 emergency EMS systems. Now equipped with automated defibrillators to reverse sudden cardiac arrest, the fire truck, coupled with bystander CPR, has become one of the greatest life-saving tools in medical history. With stroke centers to treat stroke within the golden 3 hour window, cardiac catheterization centers to treat heart attack in the 90 minute door-to-balloon time, and trauma centers to treat hemorrhaging patients, time efficiency is a key component of the best designed EMS systems. The service most capable of rapid multi-faceted response, rapid identification and triage to the appropriate facility is a fire service-based EMS system.

EMS is Not an Ambulance Ride

One of the central themes of this discussion is concern over the common misconception that EMS begins with the transport of a patient in an ambulance to a hospital. This misunderstanding resulted essentially in funding of transport service providers but not providers of emergency medical care rendered at the scene. This funding aberrancy occurred in the 1960s as Medicare provided reimbursement for transportation of trauma patients to the hospital, long before the contemporary EMS system developed. About the same time, fire service delivery of 9-1-1 emergency medical care was becoming part of the fabric of the fire service. It was managed and funded as an integral component of

14

public safety service provided by a fire department. Thus, it was funded solely as part of the fire department budget.

Payment for transportation does not fairly portray the full picture of 9-1-1 emergency response and medical care. As the need to pay for EMS was realized, federal dollars for "emergency medical services" went to the perceived greatest area of need at that time, the need for transportation. These federal dollars even provided payment of non-emergency ambulance transport for the care of chronic medical problems. Even though much of the life-saving effect of EMS in today's circumstances will play out routinely on the scene long before ambulance arrival, the focus on transport and not medical care delivery remains. This distinction has been lost and, to this date, never totally reconciled. Especially considering the resource impact, educating the public and government officials about this distinction within the EMS system in the U.S. is a critical and timely issue in the era of homeland security and Haz-Mat threats.

Funding for Prehospital EMS

The fire service supports the recent Institute of Medicine recommendations for ensuring federal payment for emergency medical care not associated with transport. Although not labeled specifically for EMS activities, grant funds are received by fire departments and emergency management agencies to enhance EMS response capabilities throughout the United States. It is deceptive to imply that only funds awarded to single function EMS delivery agencies are the only dollars benefiting those receiving prehospital 9-1-1 emergency medical care services.

For example, Assistance to Firefighter Grants (AFG) are essential to ensuring that fire departments have the baseline response capability that prepares them to respond not only to local incidents but also to effectively participate in broader, national responses. Fire department 'response' is considered 'all-hazards', inclusive of emergency, prehospital 9-1-1 medical care services. The program is extraordinarily cost-effective, with low administrative overhead and direct payments to local fire departments. As almost all fire departments provide EMS at some level, AFG dollars support equipment purchases,

15

training efforts as well as public safety education and injury prevention efforts. In fiscal year 2006 (FY 2006), 4,726 grants were awarded to fire departments throughout the United States totaling $461,092,358.

Another example of federal funding of local emergency response systems is the Staffing for Adequate Fire and Emergency Response (SAFER) Grants. The single most important obligation the federal government should fulfill to enhance local preparedness and protect Americans against all-hazards—natural and man-made—is to assure that every fire department in the nation has sufficient numbers of adequately trained and equipped fire fighter/ EMS responders. In FY 2006, there were 242 SAFER awards totaling $96,151,433 provided to fire departments throughout the United States.

Both AFG and SAFER grants present the federal government with its best opportunity to assure a strong, emergency response component in every community in America.

Federal Oversight and Administration of EMS

EMS has many voices at the federal level including the Department of Health and Human Services, Department of Transportation, Department of Justice, and Department of Homeland Security. Each voice advocates for specific entities that provide EMS as part of its services. Congress appropriately has empowered all EMS-related agencies under the Federal Interagency Committee on Emergency Medical Services (FICEMS). Recently, the FICEMS has been strengthened and provides the mechanism to accomplish this "coordination of the voices." The leadership challenge is to bring all of the voices together. The FICEMS can do this, if given a chance and a mandate.

Conclusion

In terms of the rapid delivery of emergency medical care in the out-of-hospital environment, fire departments have the advantage of having a free-standing army ready to respond anytime and anywhere. Prehospital, 9-1-1 emergency response in support of community prosperity and security is one of the essential public safety functions provided by the United States fire service. Fire service-based EMS systems are strategically

16

positioned to deliver time critical response and effective patient care and scene safety. Fire service-based EMS accomplishes this while emphasizing responder and patient safety, providing competent and compassionate workers, and delivering cost-effective operations.

References

Beck CS. Resuscitation for cardiac standstill and ventricular fibrillation occurring during operation. Am J Surg 53 (4):273-279, 1941.

City of Virginia, Minnesota, VFD History, Emergency Medical Services History, http://www.virginiamn.us/VFD%20History.htm , April 2007

Emergency Medical Services Systems Act of 1973. (P.L. 93-154). 93rd Congress S 2410.

Eugene Nagel, Personal Interview, February 2007

Foster, M. *History of the Maltese Cross, as used by the Order of St John of Jerusalem* http://www2.prestel.co.uk/church/oosj/cross.htm   April 2007.

Henderson Y. *The Prevention and Treatment of Asphyxia in the Newborn.* JAMA 90(8):383-386, 1928.

Maltese Cross, http://en.wikipedia.org/wiki/Maltese_Cross_(symbol) April 2007.

Moore-Merrell, L., IAFF/IAFC Fire Department Operations Survey, March 2007

Pepe PE, Roppolo LP, Cobb LA. Successful systems for out-of-hospital resuscitation. In: Cardiopulmonary Arrest. Ornato JP and Peberdy MA, (eds); Humana Press, Totowa, NJ 2004; pp 649-681.

Williams, D.M., *2006 JEMS 200-City Survey: EMS From All Angles.* 2007, 38-53

Appendix C
Firefighter Cancer Awareness
By Thomas Beers

We know by now that fire fighting is an inherently dangerous occupation. During the operational phases of rescue, suppression, ventilation, salvage, and overhaul, there are many dangers a firefighter faces from falls, collapse, heat, and smoke. Of course that list is incomplete, but you get the idea.

Studies have shown that there are also dangers to a firefighter long after the call, long after the overhaul process, and sometimes long after the shift and even the career is over. These dangers are silent and manifest themselves over years and quite often, but not always, are the result from poor past practices of firefighters. This danger is known to the world as *cancer*.

Because of the amount of carcinogens present at *every* fire, firefighters are exposed to these cancer causing agents more frequently than the general population. Scientific studies have shown that firefighters have a 100 percent higher risk of developing testicular cancer, a 50 percent higher risk for non-Hodgkin's Lymphoma, and a 28 percent higher risk for prostate cancer. In all related studies, firefighters were found to be more prone to ten different types of cancer than the average civilian they serve.

However, the amount of exposure to cancer causing factors can be limited through simple practices such as proper cleaning of personal protective equipment (PPE) after every fire, showering immediately after every fire to cleanse invisible oils and chemical residues found in all fires (whether Dumpster™, residential, car/auto), and of course the proper use of SCBA and respiratory protection during the crucial and sometimes long process of overhaul.

While firefighters who work in a full-time department may have the ability to clean their PPE in the station, part-time and volunteer firefighters often store their PPE in their vehicles or even their homes. Dirty PPE can "off gas" carcinogens and expose firefighters over a longer period of time. The simple rule for storing PPE in your personal vehicle or home is: If you can smell the by products of combustion, you are putting yourself at higher risk for cancer.

New firefighters may find that the more seasoned or veteran firefighters tend to have dirty PPE or try to instill the concept that dirty equipment and blackened helmets are a sign of true warriors; battle scars that show how tough they are. Blackened helmets, dirty PPE, and soiled uniforms are like giant carcinogenic sponges, exposing not just the firefighter wearing the equipment but his/her fellow firefighters as well. The NFPA® has suggestions and regulations pertaining to the proper cleaning of PPE. Furthermore, your department and the equipment manufacturer will have SOP's and recommendations for proper cleaning and care.

Your department may have a health care or insurance program that provides annual physical exams. Whether you are younger or older, physically fit or on your way to becoming physically fit, it is highly recommended that you take full advantage of these programs. Make sure you consult with your physician about your career as a firefighter and guarantee that the proper tests and blood samples are taken for cancer screening.

Finally, the sign of a professional firefighter is the one who keeps his or her body healthy by reducing the risk of disease and injury. Many scientific studies show that firefighters who manage their weight, control their diet, and exercise regularly, reduce their risk of cancer and other chronic diseases that can end a career of public service.

Remember, it is the most fit and healthy firefighter who makes the big save when the call for help needs to be answered. Proper care of your equipment, body, and overall health will ensure that firefighter is you.

IFSTA recognizes and supports the Firefighter Cancer Support Network (FCSN). This non-profit organization is a valuable resource which reaches out and supports firefighters (active & retired) and their immediate family members who are faced with cancer. Additionally the FCSN promotes and educates fire service members on the importance of many issues regarding cancer and promoting a pro-active approach regarding this disease. The FCSN can be reached at www.FirefighterCancerSupport.org.

Appendix D
Foam Concentrate
Characteristics/Application Techniques

	Appendix D				
	Foam Concentrate Characteristics/Application Techniques				
Type	**Characteristics**	**Storage Range**	**Application Rate**	**Application Techniques**	**Primary Uses**
Protein Foam (3% and 6%)	• Protein based • Low expansion • Good reignition (burnback) resistance • Excellent water retention • High heat resistance and stability • Performance can be affected by freezing and thawing • Can freeze protect with antifreeze • Not as mobile or fluid on fuel surface as other low-expansion foams	35–120°F (2°C to 49°C)	0.16 gpm/ft² (6.5 L/min/m²)	• Indirect foam stream; do not mix fuel with foam • Avoid agitating fuel during application; static spark ignition of volatile hydrocarbons can result from plunging and turbulence • Use alcohol-resistant type within seconds of proportioning • Not compatible with dry chemical extinguishing agents	• Class B fires involving hydrocarbons • Protecting flammable and combustible liquids where they are stored, transported, and processed
Fluoroprotein Foam (3% and 6%)	• Protein and synthetic based; derived from protein foam • Fuel shedding • Long-term vapor suppression • Good water retention • Excellent, long-lasting heat resistance • Performance not affected by freezing and thawing • Maintains low viscosity at low temperatures • Can freeze protect with antifreeze • Use either freshwater or saltwater • Nontoxic and biodegradable after dilution • Good mobility and fluidity on fuel surface • Premixable for short periods of time	35–120°F (2°C to 49°C)	0.16 gpm/ft² (6.5 L/min/m²)	• Direct plunge technique • Subsurface injection • Compatible with simultaneous application of dry chemical extinguishing agents • Deliver through air-aspirating equipment	• Hydrocarbon vapor suppression • Subsurface application to hydrocarbon fuel storage tanks • Extinguishing in-depth crude petroleum or other hydrocarbon fuel fires

Continued on next page.

Appendix D
Foam Concentrate Characteristics/Application Techniques

Type	Characteristics	Storage Range	Application Rate	Application Techniques	Primary Uses
Film Forming Fluoroprotein Foam (FFFP) (3% and 6%)	• Protein based; fortified with additional surfactants that reduce the burnback characteristics of other protein-based foams • Fuel shedding • Develops a fast-healing, continuous-floating film on hydrocarbon fuel surfaces • Excellent, long-lasting heat resistance • Good low-temperature viscosity • Fast fire knockdown • Affected by freezing and thawing • Use either freshwater or saltwater • Can store premixed • Can freeze protect with antifreeze • Use alcohol-resistant type on polar solvents at 6% solution and on hydrocarbon fuels at 3% solution • Nontoxic and biodegradable after dilution	35–120°F (2°C to 49°C)	**Ignited Hydrocarbon Fuel:** 0.10 gpm/ft^2 (4.1 L/min/m^2) **Polar Solvent Fuel:** 0.24 gpm/ft^2 (9.8 L/min/m^2)	• Cover entire fuel surface • May apply with dry chemical agents • May apply with spray nozzles • Subsurface injection • Can plunge into fuel during application	• Suppressing vapors in unignited spills of hazardous liquids • Extinguishing fires in hydrocarbon fuels
Aqueous Film Forming Foam (AFFF) (1%, 3%, and 6%)	• Synthetic based • Good penetrating capabilities • Spreads vapor-sealing film over and floats on hydrocarbon fuels • Can use nonaerating nozzles • Performance may be adversely affected by freezing and storing • Has good low-temperature viscosity • Can freeze protect with antifreeze • Use either freshwater or saltwater • Can premix	25–120°F (-4°F to 49°C)	0.10 gpm/ft^2 (4.1 L/min/m^2)	• May apply directly onto fuel surface • May apply indirectly by bouncing it off a wall and allowing it to float onto fuel surface • Subsurface injection • May apply with dry chemical agents	• Controlling and extinguishing Class B fires • Handling land or sea crash rescues involving spills • Extinguishing most transportation-related fires • Wetting and penetrating Class A fuels • Securing unignited hydrocarbon spills

Continued on next page.

Appendix D
Foam Concentrate Characteristics/Application Techniques

Type	Characteristics	Storage Range	Application Rate	Application Techniques	Primary Uses
Alcohol-Resistant AFFF (3% and 6%)	• Polymer has been added to AFFF concentrate • Multipurpose: Use on both polar solvents and hydrocarbon fuels (use on polar solvents at 6% solution and on hydrocarbon fuels at 3% solution) • Forms a membrane on polar solvent fuels that prevents destruction of the foam blanket • Forms same aqueous film on hydrocarbon fuels as AFFF • Fast flame knockdown • Good burnback resistance on both fuels • Not easily premixed	25–120°F (-4°C to 49°C) (May become viscous at temperatures under 50°F [10°C])	**Ignited Hydrocarbon Fuel:** 0.10 gpm/ft^2 (4.1 L/min/m^2) **Polar Solvent Fuel:** 0.24 gpm/ft^2 (9.8 L/min/m^2)	• Apply directly but gently onto fuel surface • May apply indirectly by bouncing it off a wall and allowing it to float onto fuel surface • Subsurface injection	Fires or spills of both hydrocarbon and polar solvent fuels
High-Expansion Foam	• Synthetic detergent based • Special-purpose, low water content • High air-to-solution ratios: 200:1 to 1,000:1 • Performance not affected by freezing and thawing • Poor heat resistance • Prolonged contact with galvanized or raw steel may attack these surfaces	27–110°F (-3°C to 43°C)	Sufficient to quickly cover the fuel or fill the space	• Gentle application; do not mix foam with fuel • Cover entire fuel surface • Usually fills entire space in confined-space incidents	• Extinguishing Class A and some Class B fires • Flooding confined spaces • Volumetrically displacing vapor, heat, and smoke • Reducing vaporization from liquefied natural gas spills • Extinguishing pesticide fires • Suppressing fuming acid vapors • Suppressing vapors in coal mines and other subterranean spaces and concealed spaces in basements • Extinguishing agent in fixed extinguishing systems • Not recommended for outdoor use

Continued on next page.

Type	Characteristics	Storage Range	Application Rate	Application Techniques	Primary Uses
Class A Foam	• Synthetic • Wetting agent that reduces surface tension of water and allows it to soak into combustible materials • Rapid extinguishment with less water use than other foams • Use regular water stream equipment • Can premix with water • Mildly corrosive • Requires lower percentage of concentration (0.2 to 1.0) than other foams • Outstanding insulating qualities • Good penetrating capabilities	25–120°F (-4°C to 49°C) (Concentrate is subject to freezing but can be thawed and used if freezing occurs)	Same as the minimum critical flow rate for plain water on similar Class A Fuels; flow rates are not reduced when using Class A foam	• Can propel with compressed-air systems • Can apply with conventional nozzles	Extinguishing Class A combustibles only

Appendix E
18 Watch Out Situations

Used for ground cover, wildland, and forest fire incidents.

1. The fire has not been scouted and sized up.

2. You are in country that you have not seen in daylight.

3. Safety zones and escape routes have not been identified.

4. You are unfamiliar with weather and local factors influencing fire behavior.

5. You have not been informed of strategy, tactics, and hazards.

6. Your instructions and assignments are not clear.

7. No communication link has been established with crewmembers or your supervisor.

8. You are constructing a fireline without safe anchor point.

9. You are building a fireline downhill with fire below.

10. You are attempting a frontal assault on the fire.

11. There is unburned fuel between you and the fire.

12. You cannot see the main fire and are not in contact with someone who can.

13. You are on a hillside where rolling material can ignite fuel below you.

14. The weather is getting hotter and drier.

15. The wind is increasing and/or changing direction.

16. You are getting frequent spot fires across the fireline.

17. The terrain and fuels make escape to safety zones difficult.

18. You are taking a nap near the fire line.

Glossary

A

Accordion Load — Arrangement of fire hose in a hose bed or compartment in which the hose lies on edge with the folds adjacent to each other.

Activation Energy — Energy that starts a chemical reaction when added to an atomic or molecular system.

Acute — Sharp or severe; having a rapid onset and short duration.

Acute Health Effects — Health effects that develop rapidly, often after exposure to a hazardous substance.

Adapter — Device for connecting hose couplings with dissimilar threads but with the same inside diameter.

Adz — A wedge-shaped blade attached at right angles to the handle of the tool.

Air-Aspirating Foam Nozzle — Foam nozzle especially designed to provide the aeration required to make the highest quality foam possible; most effective appliance for the generation of low-expansion foam.

Airborne Pathogens — Disease-causing microorganisms (viruses, bacteria, or fungi) that are suspended in the air.

Air Flow — The movement of air toward burning fuel and the movement of smoke out of the compartment or structure.

Air-Purifying Respirator (APR) — Respirator that removes contaminants by passing ambient air through a filter, cartridge, or canister; may have a full or partial facepiece.

Airway — A passage for carrying air from the nose or mouth to the lungs.

Alarm Assignment — Predetermined number and type of fire units assigned to respond to an emergency.

Alarm Check Valve — Type of check valve installed in the riser of an automatic sprinkler system that transmits a water-flow alarm when the water flow in the system lifts the valve clapper.

Alarm-Initiating Device — Alarm system component that transmits a signal when a change occurs; change may be the result of an action such as the activation of a manual fire alarm box, the presence of products of combustion in the atmosphere, or the automatic activation of a supervisory switch.

All-Hazard Concept — Provides a coordinated approach to a wide variety of incidents; all responders use a similar, coordinated approach with a common set of authorities, protections, and resources.

Alloy — Substance or mixture composed of two or more metals (or a metal and nonmetallic elements) fused together and dissolved into each other when molten intended to enhance the properties or usefulness of the base metal.

Ambient Conditions — Common, prevailing, and uncontrolled atmospheric weather conditions. The term may refer to the conditions inside or outside of the structure.

Anaphylactic Shock — Shock caused by a severe allergic reaction.

Annunciator Panel — Electrical device used to indicate the source or location of an activated fire alarm initiating device or the status of the system. The panel may include individual lights located on a schematic map and an audible alarm signal.

Aqueous Film Forming Foam (AFFF) — Synthetic foam concentrate that, when combined with water, can form a complete vapor barrier over fuel spills and fires and is a highly effective extinguishing and blanketing agent on hydrocarbon fuels.

Arc — High-temperature luminous electric discharge across a gap or through a medium such as charred insulation.

Area of Origin — The general location (room or area) where the ignition source and the material first ignited actually came together for the first time.

Arson — Crime of willfully, maliciously, and intentionally starting an incendiary fire or causing an explosion to destroy one's property or the property of another. Precise legal definitions vary among jurisdictions, wherein it is defined by statutes and judicial decisions.

Artery — Blood vessel that carries oxygen-rich blood away from the heart.

Asphyxiation — Fatal condition caused by severe oxygen deficiency and an excess of carbon monoxide and/or other gases in the blood.

Attack Hose — Hose that is used by trained firefighters to combat fires.

Autoclave — Device that uses high-pressure steam to sterilize objects.

Authority Having Jurisdiction (AHJ) — Term used in codes and standards to identify the legal entity, such as a building or fire official, that has the statutory authority to enforce a code and to approve or require equipment; may be a unit of a local, state, or federal government, depending on where the work occurs. In the insurance industry it may refer to an insurance rating bureau or an insurance company inspection department.

Autoignition — Initiation of combustion by heat but without a spark or flame. (NFPA® 921)

Autoignition Temperature — The lowest temperature at which a combustible material ignites in air without a spark or flame. (NFPA® 921)

Automated External Defibrillator (AED) — Cardiac defibrillator designed for layperson use that analyzes the cardiac rhythm and determines if defibrillation is warranted.

Automatic Aid — Written agreement between two or more agencies to automatically dispatch predetermined resources to any fire or other emergency reported in the geographic area covered by the agreement. These areas are generally located near jurisdictional boundaries or in jurisdictional "islands."

Automatic Location Identification (ALI) — Enhanced 9-1-1 feature that displays the address of the party calling 9-1-1 on a screen for use by the public safety telecommunicator. This feature is also used to route calls to the appropriate public safety answering point (PSAP) and can even store information in its database regarding the appropriate emergency services (police, fire, and medical) that respond to that address.

Automatic Sprinkler System — System of water pipes, discharge nozzles, and control valves designed to activate during fires by automatically discharging enough water to control or extinguish a fire.

Auxiliary Alarm System — System that connects the protected property with the fire department alarm communications center by a municipal master fire alarm box or over a dedicated telephone line.

Awareness Level — Lowest level of training established by the National Fire Protection Association® for first responders at hazardous materials incidents.

B

Backdraft — The explosive burning of heated gases that occurs when oxygen is introduced into a compartment that has a high concentration of flammable gases and a depleted supply of oxygen due to an existing fire.

Backflow Preventer — A check valve that prevents water from flowing back into a system and contaminating it.

Baffle — Intermediate partial bulkhead that reduces the surge effect in a partially loaded liquid tank.

Ball Valve — Valve having a ball-shaped internal component with a hole through its center that permits water to flow through when aligned with the waterway.

Balloon Frame Construction — A construction method using long continuous studs that run from the sill plate (located on the foundation) to the roof eave line. All intermediate floor structures are attached to the studs. Requires the use of long lumber and generally lacks any type of fire stopping within the wall cavity.

Base Station Radios — Fixed, nonmobile radio at a central location.

Battering Ram — Solid steel bar with handles and guards, a fork on one end, and a blunt end on the other, used to break down doors or create holes in walls. The tool weighs 30 to 40 pounds (13.6 to 18.1 kg) and can be operated by one or more firefighters.

Becket Bend — Knot used for joining two ropes; particularly well suited for joining ropes of unequal diameters or joining a rope and a chain. *Also known as* Sheet Bend.

Bedded Position — Extension ladder with the fly section(s) fully retracted.

Bill of Lading — Shipping paper that describes the cargo, origin, destination, route; used in trucking and other industries, and typically placed in the cab of every truck tractor. Establishes the terms of a contract between a shipper and a carrier.

Biological Death — Condition present when irreversible brain damage has occurred, usually 4 to 10 minutes after cardiac arrest.

Blitz Attack — To aggressively attack a fire from the exterior with a large diameter (2½-inch [65 mm] or larger) fire stream.

Block Creel Construction — Method of manufacturing rope without any knots or splices; a continuous strand of fiber runs the entire length of the rope's core.

Blowers — Fans that are used to push fresh air into a structure. They may be powered by electricity, gasoline engines, or hydraulics. Blowers that are not intrinsically safe may only be used to push fresh air into the structure.

Body Substance Isolation (BSI) — The practice of taking proactive, protective measures to isolate body substances in order to prevent the spread of infectious disease.

Boiling Point — Temperature at which a liquid begins to boil, and vapor pressure exceeds atmospheric pressure. The rate of evaporation exceeds the rate of condensation, so more liquid is turning into gas than gas is turning into liquid.

Booster Hoseline — Noncollapsible rubber-covered, rubber-lined hose usually wound on a reel and mounted somewhere on the apparatus and used for extinguishment of incipient and smoldering fires. This hose is most commonly found in ½-, ¾-, and 1-inch (13 mm, 19 mm, and 25 mm) diameters and is used for extinguishing low-intensity fires and overhaul operations.

Bowline Knot — Knot used to form a loop; it is easy to tie and untie, and does not constrict.

B-Post — Post between the front and rear doors on a four-door vehicle, or the door-handle-end post on a two-door car.

Braided Rope — Rope constructed by uniformly intertwining strands of rope together (similar to braiding hair).

Braid-on-Braid Rope — Rope that consists of a braided core enclosed in a braided, herringbone patterned sheath.

Breaching — The act of creating a hole in a wall or floor to gain access to a structure or portion of a structure.

Breakthrough Time — Time required for a chemical to permeate the material of a protective suit.

British Thermal Unit (Btu) — Amount of heat energy required to raise the temperature of one pound of water one degree Fahrenheit. One Btu = 1.055 kilo joules (kJ).

Broken Stream — Stream of water that has been broken into coarsely divided drops.

Buoyant — The tendency or capacity to remain afloat in a liquid or rise in air or gas.

Butterfly Valve — Control valve that uses a flat circular plate in a pipe that rotates 90 degrees across the cross section of the pipe to control the flow of water.

C

Capacity Stencil — Number stenciled on the exterior of a tank car to indicate the volume of the tank.

Capillaries — Tiny blood vessels in the body's tissues in which the exchange of oxygen and carbon dioxide take place.

Carbon Dioxide (CO_2) — Colorless, odorless, heavier than air gas that neither supports combustion nor burns; used in portable fire extinguishers as an extinguishing agent to extinguish Class B or C fires by smothering or displacing the oxygen.

Carbon Monoxide (CO) — Colorless, odorless, dangerous gas (both toxic and flammable) formed by the incomplete combustion of carbon. It combines with hemoglobin more than 200 times faster than oxygen does, thus decreasing the blood's ability to carry oxygen.

Carboy — Cylindrical container of about 5 to 15 gallon (19 L to 57 L) capacity used for pure or corrosive liquids. Made of glass, plastic, or metal, with a neck and sometimes a pouring tip; cushioned in a wooden box, wicker, basket, or special drum.

Cardiac Arrest — Sudden cessation of heartbeat.

Cardiogenic Shock — Shock caused by poor cardiac output.

Cardiopulmonary Resuscitation (CPR) — Application of rescue breathing and external cardiac compression used on patients in cardiac arrest to provide an adequate circulation and oxygen to support life.

Carryall — Waterproof carrier used to carry and catch debris or used as a water sump basin for immersing small burning objects.

Cascade System — Three or more large, interconnected air cylinders, from which smaller SCBA cylinders are recharged; the larger cylinders typically have a capacity of 300 cubic feet (8,490 L).

Case-Hardened Steel — Steel used in vehicle construction whose exterior has been heat treated, making it much harder than the interior metal.

Catalyst — Substance that modifies (usually increases) the rate of a chemical reaction, without being consumed in the process.

Ceiling Jet — A relatively thin layer of flowing hot gases that develops under a horizontal surface (e.g., ceiling) as a result of plume impingement and the flowing gas being forced to move horizontally. (NFPA® 921)

Celsius Scale — International temperature scale on which the freezing point is 0°C (32°F) and the boiling point is 100°C (212°F) at normal atmospheric pressure at sea level. *Also known as* Centigrade Scale.

Central Station System — Alarm system that functions through a constantly attended location (central station) operated by an alarm company. Alarm signals from the protected property are received in the central station and are then retransmitted by trained personnel to the fire department alarm communications center.

Chain of Custody — Continuous changes of possession of physical evidence that must be established in court to admit such material into evidence. In order for physical evidence to be admissible in court, there must be an evidence log of accountability that documents each change of possession from the evidence's discovery until it is presented in court.

Check Valve — Automatic valve that permits liquid flow in only one direction.

Chemical Flame Inhibition — Extinguishment of a fire by interruption of the chemical chain reaction.

Chemical Pellet — A pellet of solder, under compression, within a small cylinder, that melts at a predetermined temperature, allowing a plunger to move down and release the valve cap parts.

Chemical-Protective Clothing (CPC) — Clothing designed to shield or isolate individuals from the chemical, physical, and biological hazards that may be encountered during operations involving hazardous materials.

Chemical Warfare Agent — Chemical substance intended for use in warfare or terrorist activity; designed to kill, injure, or incapacitate.

Chest Compression — The act of forcefully compressing the heart in a rhythmic manner in order to circulate blood throughout the body.

Choking Agent — Chemical warfare agent that attacks the lungs, causing tissue damage.

Chronic — Long-term and reoccurring.

Chronic Health Effects — Long-term health effects, often resulting from exposure to a hazardous substance.

Circulating Feed — Fire hydrant that receives water from two or more directions.

Circulating Hydrant — Fire hydrant that is located on a secondary feeder or distributor main that receives water from two directions.

Circumstantial Evidence — Evidence presented in a trial that tends to prove a factual matter through inference by proving other events or circumstances.

Class A Foam — Foam specially designed for use on Class A combustibles. Class A foams, hydrocarbon-based surfactants, are essentially wetting agents that reduce the surface tension of water and allow it to soak into combustible materials more easily than plain water.

Class B Foam — Foam fire-suppression agent designed for use on unignited or ignited Class B flammable or combustible liquids.

Clear Text — Use of plain language, including certain standard words and phrases, in radio communications transmissions.

Clinical Death — Term that refers to the lack of signs of life, where there is no pulse and no blood pressure; occurs immediately after the onset of cardiac arrest.

Closed-Circuit Self-Contained Breathing Apparatus — SCBA that recycles exhaled air; removes carbon dioxide and restores compressed, chemical, or liquid oxygen. Not approved for fire fighting operations. *Also known as* Oxygen-Breathing Apparatus (OBA) or Oxygen-Generating Apparatus.

Clove Hitch — Knot that consists of two half-hitches; its principal use is to attach a rope to an object such as a pole, post, or hose.

Cockloft — Concealed space between the top floor and the roof of a structure.

Code — A collection of rules and regulations that has been enacted by law in a particular jurisdiction. Codes typically address a single subject area; examples include a mechanical, electrical, building, or fire code.

Code of Federal Regulations (CFR) — Rules and regulations published by executive agencies of the U.S. federal government. These administrative laws are just as enforceable as statutory laws (known collectively as federal law), which must be passed by Congress.

Cold Zone — Safe area outside of the warm zone where equipment and personnel are not expected to become contaminated and special protective clothing is not required; the incident command post and other support functions are typically located in this zone. *Also known as* Support Zone.

Collapse Zone — Area beneath a wall in which the wall is likely to land if it loses structural integrity.

Combination Attack — Extinguishing a fire by using both a direct and an indirect attack. This method combines the steam-generating technique of a ceiling level attack with an attack on the burning materials near floor level.

Combination Ladder — Ladder that can be used as a single, extension, or A-frame ladder.

Combination System — Water supply system that is a combination of both gravity and direct pumping systems. It is the most common type of municipal water supply system.

Combustible Liquid — Liquid having a flash point at or above 100°F (37.8°C) and below 200°F (93.3°C).

Combustion — A chemical process of oxidation that occurs at a rate fast enough to produce heat and usually light in the form of either a glow or flame. (NFPA® 921)

Communicable Disease — Disease that is transmissible from one person to another.

Competent Ignition Source — A competent ignition source will have sufficient temperature and energy and be in contact with the fuel long enough to raise it to its ignition temperature.

Compressed Air Foam System (CAFS) — Generic term used to describe a high-energy foam-generation system consisting of a water pump, a foam proportioning system, and an air compressor (or other air source) that injects air into the foam solution before it enters a hoseline.

Conduction — Transfer of heat through or between solids that are in direct contact.

Conductivity — The ability of a substance to conduct an electrical current.

Confinement — The process of controlling the flow of a spill and capturing it at some specified location.

Containment — The act of stopping the further release of a material from its container.

Contagious — Easily transmitted from one person to another, either through contact or close proximity.

Contamination — General term referring to anything that can taint physical evidence.

Convection — Heat transfer by circulation within a medium such as a gas or a liquid. (NFPA® 921)

Corrosive — Capable of causing corrosion by gradually eroding, rusting, or destroying a material.

Corrosive Material — Chemical that severely corrodes steel and damages human tissue. *Also known as* Corrosive.

Cribbing — Wooden or plastic blocks used to stabilize a vehicle during vehicle extrication or debris following a structural collapse; typically 4 x 4 (100 mm) inches or larger and between 16 to 26 inches long.

Critical Flow Rate — The minimum flow rate at which extinguishment can be achieved.

Cross Contamination — Contamination of people, equipment, or the environment outside the hot zone without contacting the primary source of contamination. *Also known as* Secondary Contamination.

Cross Main — Pipe connecting the feed main to the branch lines on which the sprinklers are located.

Cross-Training — Training emergency services personnel to function in more than one capacity. Occurs most often when personnel are trained as firefighters and EMTs or Paramedics.

Crowd Control — Limiting the access of nonemergency personnel to the emergency scene.

Cryogens — Gases that convert to liquids when cooled at or below -130°F (-101°C). *Also known as* Refrigerated Liquids and Cryogenic Liquids.

Culture — The shared assumptions, beliefs, and values of a group or organization.

Curtain Wall — A non-load-bearing wall, often of glass and steel, fixed to the outside of a building and serving especially as cladding.

D

Dam/Dike — Temporary or permanent barrier that contains or directs the flow of liquids.

Dangerous Goods — (1) Alternate term for hazardous materials, used in Canada and other countries. (2) U.S. or Canadian term for hazardous materials aboard an aircraft.

Dead-End Hydrant — Fire hydrant located on a dead-end main that receives water from only one direction.

Dead Load — Weight of the structure, structural members, building components, and any other features permanently attached to the building that are constant and immobile.

Decontamination — Process of removing a hazardous foreign substance from a person, clothing, or area. *Also known as* Decon.

Defensive Strategy — Overall plan for incident control established by the Incident Commander that involves protection of exposures, as opposed to aggressive, offensive intervention.

Defibrillation — The delivery of a measured dose of electrical current by a special machine in order to regain normal function of the heart.

Deflector — Part of the sprinkler assembly that creates the discharge pattern of the water.

Deionized Water — Water from which ionic salts, minerals, and impurities have been removed by ion exchange.

Deluge Sprinkler System — Fire suppression system that consists of piping and open sprinklers. A fire detection system is used to activate the water or foam control valve. When the system activates, the extinguishing agent expels from all sprinkler heads in the designated area.

Deluge Valve — Automatic valve used to control water to a deluge sprinkler system.

Direct Attack (Ground Cover) — Operation where action is taken directly on burning fuels by applying an extinguishing agent to the edge of the fire or close to it.

Direct Attack (Structural) — Attack method that involves the discharge of water or a foam stream directly onto the burning fuel.

Direct Evidence — Type of evidence provided by a witness who obtained it through his or her senses.

Direct Pumping System — Water supply system supplied directly by a system of pumps rather than elevated storage tanks.

Dispersion — Process of being spread widely.

Divert — Actions to control movement of a hazardous material to an area that will produce less harm.

Drafting — Process of acquiring water from a static source and transferring it into a pump that is above the source's level; atmospheric pressure on the water surface forces the water into the pump where a partial vacuum was created.

Drain Valve — Valve that allows piping to drain when pressure is relieved in the pipe.

Dressing — Clean or sterile covering applied directly to a wound; used to stop bleeding and to prevent contamination of the wound.

Dry-Barrel Hydrant — Fire hydrant that has its operating valve located at the base or foot of the hydrant rather than in the barrel of the hydrant. When operating properly, there is no water in the barrel of the hydrant when it is not in use. These hydrants are used in areas where freezing may occur.

Dry-Pipe Sprinkler System — Fire-suppression system that consists of closed sprinklers attached to a piping system that contains air under pressure. When a sprinkler activates, air is released that activates the water or foam control valve and fills the piping with extinguishing agent.

Dry Chemical — Extinguishing system that uses dry chemical as the primary extinguishing agent; often used to protect areas containing volatile flammable liquids.

Dry Powder — Extinguishing agent suitable for use on combustible metal fires.

Dutchman — Extra fold placed along the length of a section of hose as it is loaded so that its coupling rests in proper position.

Dynamic Rope — Rope designed to stretch under load, reducing the shock of impact after a fall.

E

Eave — The edge of a pitched roof that overhangs an outside wall. Attic vents in typical eaves provide an avenue for an exterior fire to enter the attic.

Eduction — Process used to mix foam concentrate with water in a nozzle or proportioner; concentrate is drawn into the water stream by the Venturi method; *also known as* Induction.

Electromagnetic Pulse (EMP) — Burst of electromagnetic energy produced by a nuclear explosion; EMPs damage electronic systems by causing voltage and current surges.

Electron — Subatomic particle that possesses a negative electric charge.

Elevated-Temperature Material — Material that meets any of the following criteria during transport: (a) in a liquid phase, at or above 212°F (100°C), (b) intentionally heated at or above its liquid phase flash point of 100°F (38°C), or (c) in a solid phase, at or above 464°F (240°C).

Emergency Breathing Support System (EBSS) — Escape-only respirator that provides sufficient self-contained breathing air to permit the wearer to safely exit the hazardous area; usually integrated into an airline supplied-air respirator system.

Emergency Decontamination — The physical process of immediately reducing contamination of individuals in potentially life-threatening situations, with or without the formal establishment of a decontamination corridor.

Emergency Medical Services (EMS) — Initial medical evaluation/treatment provided to employees and others who become ill or are injured.

Emergency Medical Technician (EMT) — Professional-level provider of basic life support emergency medical care. Requires certification by some authority.

Emergency Response Guidebook (ERG) — Manual that identifies hazardous materials labels/placards; also outlines initial actions to be taken at haz mat incidents. Developed jointly by U.S., Mexican, and Canadian transportation agencies. *Formerly known as North American Emergency Response Guidebook (NAERG)*.

Employee Assistance Program (EAP) — Program that an employer may provide to employees and their families to help with work or personal problems.

Encapsulating — Completely enclosed or surrounded, as in a capsule.

Endothermic Reaction — Chemical reaction that absorbs thermal energy or heat.

Energy — Capacity to perform work; occurs when a force is applied to an object over a distance, or when a chemical, biological, or physical transformation is made in a substance.

Environmental Protection Agency (EPA) — U.S. government agency that creates and enforces laws designed to protect the air, water, and soil from contamination; responsible for researching and setting national standards for a variety of environmental programs.

Etiological Hazards — Harmful viruses and bacteria; when used deliberately, they are *also known as* Biological Weapons.

Evacuation — Controlled process of leaving or being removed from a potentially hazardous location, typically involving relocating people from an area of danger or potential risk to a safer place.

Exhaust Opening — Intended and controlled exhaust locations that are created or improved at or near the fire to allow products of combustion to escape the building.

Exigent Circumstances — Right of entry stating that the fire department does not require a warrant to enter a property to suppress a fire, or to remain on the property for a reasonable amount of time afterward in order to determine the origin and cause of the fire.

Exothermic Reaction — Chemical reaction that releases thermal energy or heat.

Exposure — Contact with a hazardous material, causing biological damage; typically by swallowing, breathing, or touching. Exposure may be short-term (acute exposure), of intermediate duration, or long-term (chronic exposure).

Exposure Protection — Covering any object in the immediate vicinity of the fire with water or foam.

Exposures — Structure or separate part of the fireground to which a fire could spread.

Extension Ladder — Variable-length ladder of two or more sections that can be extended to a desired height.

External Customers — Citizens of the service area protected by the organization.

Extinguishing Agent — Any substance used for the purpose of controlling or extinguishing a fire.

Extrication — Incident in which a trapped victim must be removed from a vehicle or other type of machinery.

F

Fahrenheit Scale — Temperature scale on which the freezing point is 32°F (0°C) and the boiling point at sea level is 212°F (100°C) at normal atmospheric pressure.

Feed Main — Pipe connecting the sprinkler system riser to the cross mains.

Film Forming Fluoroprotein Foam (FFFP) — Foam concentrate that combines the qualities of fluoroprotein foam with those of aqueous film forming foam.

Finish — Arrangement of hose usually placed on top of a hose load and connected to the end of the load.

Finished Foam — Completed product after air is introduced into the foam solution.

Fire — A rapid oxidation process, which is a chemical reaction resulting in the evolution of light and heat in varying intensities. (NFPA® 921)

Fire Alarm Control Panel (FACP) — System component that receives input from automatic and manual fire alarm devices and may provide power to detection devices or communication devices.

Fire Cause — The sequence of events that allow the ignition source and the material first ignited to come together.

Fire Command Center — Designated room or area in a structure where the status of the fire detection, alarm, and protection systems is displayed and the systems can be manually controlled; may be staffed or unstaffed and can be accessed by the fire department.

Fire Department Connection (FDC) — Point at which the fire department can connect into a sprinkler or standpipe system to boost the water flow in the system.

Fire Extinguisher — Portable fire fighting device designed to combat incipient fires.

Firefighter's Smoke-Control Station (FSCS) — Interface between the smoke management system and the fire response forces.

Fire Fighting Boots — Protective footwear meeting the design requirements of NFPA®, OSHA, and CAN/CSA Z195-02 (R2008).

Fire-Gas Detector — Device used to detect gases produced by a fire within a confined space.

Fire Hose — A flexible, portable tube manufactured from water tight materials in 50 to 100 foot (15 to 30 m) lengths that is used to transport water from a source or pump to the point it is discharged to extinguish fire.

Fire Pattern — The apparent and obvious design of burned material and the burning path of travel from a point of fire origin.

Fire Point — Temperature at which a liquid fuel produces sufficient vapors to support combustion once the fuel is ignited. Fire point must exceed 5 seconds of burning duration during the test. The fire point is usually a few degrees above the flash point.

Fireproof — Obsolete term for resistance to fire; inappropriate because all materials except water will burn. Other terms such as *fire resistive* or *fire resistant* should be used.

Fire Protection System — System designed to protect structure and minimize loss due to fire.

Fire-Resistance Rating — Rating assigned to a material or assembly after standardized testing by an independent testing organization; identifies the amount of time a material or assembly will resist a typical fire, as measured on a standard time-temperature curve.

Fire Stop — Solid materials, such as wood blocks, used to prevent or limit the vertical and horizontal spread of fire and the products of combustion in hollow walls or floors, above false ceilings, in penetrations for plumbing or electrical installations, in penetrations of a fire-rated assembly, or in cocklofts and crawl spaces.

Fire Stream — Stream of water or other water-based extinguishing agent after it leaves the fire hose and nozzle until it reaches the desired point.

Fire Tetrahedron — Model of the four elements/conditions required to have a fire. The four sides of the tetrahedron represent fuel, heat, oxygen, and self-sustaining chemical chain reaction.

Fire Triangle — A model used to explain the elements/conditions necessary for combustion. The sides of the triangle represent heat, oxygen, and fuel.

Fire Wall — Fire-rated wall with a specified degree of fire resistance, built of fire-resistive materials and usually extending from the foundation up to and through the roof of a building, that is designed to limit the spread of a fire within a structure or between adjacent structures.

Fitting — Device that facilitates the connection of hoselines to provide an uninterrupted flow of extinguishing agent.

Flame — Visible, luminous body of a burning gas emitting radiant energy including light of various colors given off by burning gases or vapors during the combustion process.

Flame Detector — Detection and alarm device used in some fire detection systems (generally in high-hazard areas) that detect light/flames in the ultraviolet wave spectrum (UV detectors) or detect light in the infrared wave spectrum (IR detectors).

Flammable Liquid — Any liquid having a flash point below 100°F (37.8°C) and a vapor pressure not exceeding 40 psi absolute (276 kPa) {2.76 bar}.

Flammable (Explosive) Range — The range between the upper flammable limit and lower flammable limit in which a substance can be ignited.

Flashover — A rapid transition from the growth stage to the fully developed stage.

Flash Point — Minimum temperature at which a liquid gives off enough vapors to form an ignitable mixture with air near the liquid's surface.

Flat Load — Arrangement of fire hose in a hose bed or compartment in which the hose lies flat with successive layers one upon the other.

Flowmeter — Mechanical device installed in a discharge line that senses the amount of water flowing and provides a readout in units of gallons per minute (liters per minute).

Flow Path — Composed of at least one inlet opening, one exhaust opening, and the connecting volume between the openings. The direction of the flow is determined by difference in pressure. Heat and smoke in a high-pressure area will flow toward areas of lower pressure (NIST)

Foam — Extinguishing agent formed by mixing a foam concentrate with water and aerating the solution for expansion; for use on Class A and Class B fires. Foam may be protein, fluoroprotein, film forming fluoroprotein, synthetic, aqueous film forming, or alcohol-resistant type.

Foam Concentrate — Chemical compound solution that is mixed with water and air to produce finished foam; may be protein, synthetic, aqueous film forming, high expansion, or alcohol types.

Foam Expansion — Result of adding air to a foam solution consisting of water and foam concentrate. Expansion creates the foam bubbles that result in finished foam or foam blanket.

Foam Proportioner — Device that introduces foam concentrate into the water stream to make the foam solution.

Foam Solution — Mixture of foam concentrate and water before the introduction of air.

Fog Nozzle — An adjustable pattern nozzle equipped with a shutoff control device.

Fog Stream — Fire stream of finely divided particles used for fire control.

Folding Ladder — Single-section, collapsible ladder that is easy to maneuver in restricted places such as access openings for attics and lofts.

Forcible Entry — Techniques used by fire personnel to gain entry into buildings, vehicles, aircraft, or other areas of confinement when normal means of entry are locked or blocked.

Four-Way Hydrant Valve — Hose appliance that is attached to the hydrant to permit additional supply hoses to be attached without interrupting the flow of water.

Frangible Bulb — Small glass vial fitted into a detector or sprinkler. The glass vial is partly filled with a liquid that expands as heat increases. At a predetermined temperature, vapor pressure causes the glass bulb to break, which activates the device.

Freelance — To operate independently of the Incident Commander's command and control.

Free Radicals — Molecular fragments that are highly reactive.

Friction Loss — Loss of pressure created by the turbulence of water moving against the interior walls of fire hose, pipes, fittings, and adapters.

Frostbite — Local tissue damage caused by prolonged exposure to extreme cold.

Fuel — A material that will maintain combustion under specified environmental conditions. (NFPA® 921)

Fuel-Controlled — A fire with adequate oxygen in which the heat release rate and growth rate are determined by the characteristics of the fuel, such as quantity and geometry. (Paraphrased from NFPA 921)

Fuel Load — The total quantity of combustible contents of a building, space, or fire area, including interior finish and trim, expressed in heat units of the equivalent weight in wood.

Fulcrum — Support or point of support on which a lever turns in raising or moving something.

Fusee — A friction match with a large head capable of burning in a wind.

Fusible Link — Connecting link device that fuses or melts when exposed to heat; used in sprinklers, fire doors, dampers, heat detectors, and ventilators.

G

Gas — Compressible substance, with no specific volume, that tends to assume the shape of the container. Molecules move about most rapidly in this state.

Gate Valve — Control valve with a solid plate operated by a handle and screw mechanism; rotating the handle moves the plate into or out of the waterway.

Generator — Portable device for generating auxiliary electrical power; generators are powered by gasoline or diesel engines and typically have 110- and/or 220-volt capacity outlets.

Global Positioning System (GPS) — System for determining position on the earth's surface by calculating the difference in time for the signal from a number of satellites to reach a receiver on the ground.

Glove Box — Sealed container that allows scientists to safely manipulate hazardous microorganisms. Two glove ports are built into the side of the box, with gloves extending inside; the user is therefore able to perform tasks inside the box without breaking the seal.

Gravity System — Water supply system that relies entirely on the force of gravity to create pressure and cause water to flow through the system. The water supply, which is often an elevated tank, is at a higher level than the system.

Green Wood — Wood with high moisture content.

Ground Fault Circuit Interrupter (GFCI) — Device designed to protect against electrical shock; when grounding occurs, the device opens a circuit to shut off the flow of electricity. *Also known as* Ground Fault Indicator (GFI) Receptacle.

Ground Gradient — Electrical field that radiates outward from where the current enters the ground; its intensity dissipates rapidly as distance increases from the point of entry.

Gusset Plates — Metal or wooden plates used to connect and strengthen the joints of two or more separate components (such as metal or wooden truss components or roof or floor components) into a load-bearing unit.

H

Half-Hitch — Knot typically used to stabilize long objects that are being hoisted; always used in conjunction with another knot.

Halligan Tool — Prying tool with a claw at one end and a spike or point at a right angle to a wedge at the other end.

Halogenated Extinguishing Agents — Chemical compounds (halogenated hydrocarbons) that contain carbon plus one or more elements from the halogen series. Halon 1301 and Halon 1211 are most commonly used as extinguishing agents for Class B and Class C fires. *Also known as* Halogenated Hydrocarbons.

Handcuff (Rescue) Knot — Knot tied in a bight with two adjustable loops in opposing directions; used during rescues to secure hands or feet, so that a victim can be raised or dragged to safety.

Handline Nozzle — Any nozzle that can be safely handled by one to three firefighters and flows less than 350 gpm (1 400 L/min).

Hard-Suction Hose — Rigid, noncollapsible hose that operates under vacuum conditions without collapsing, allowing a pumping apparatus or portable pump to "draft" water from static or nonpressurized sources (lakes, rivers, wells, etc.) that are below the level of the fire pump, usually available in 10-foot (3 m) sections.

Hazard — Condition, substance, or device that can directly cause injury or loss; the source of a risk.

Hazard and Risk Assessment — Formal review of the hazards and risks that may be encountered by firefighters or emergency responders; used to determine the appropriate level and type of personal and respiratory protection that must be worn. *Also known as* Hazard Assessment.

Hazardous Material — Substance that can be dangerous to human health or the environment if not properly controlled.

Hazardous Waste Operations and Emergency Response (HAZWOPER) — U.S. regulations in Title 29 (Labor) *CFR* 1910.120 for cleanup operations involving hazardous substances and emergency response operations for releases of hazardous substances. *See Code of Federal Regulations (CFR).*

Hearing Protection — Device that limits noise-induced hearing loss when firefighters are exposed to extremely loud environments, such as apparatus engine noise, audible warning devices, and the use of power tools and equipment.

Heat — A form of energy characterized by vibration of molecules and capable of initiating and supporting chemical changes and changes of state. (NFPA® 921)

Heat Cramps — Heat illness resulting from prolonged exposure to high temperatures; characterized by excessive sweating, muscle cramps in the abdomen and legs, faintness, dizziness, and exhaustion.

Heat Detector — Alarm-initiating device that is designed to be responsive to a predetermined rate of temperature increase or to a predetermined temperature level.

Heat Exhaustion — Heat illness caused by exposure to excessive heat; symptoms include weakness, cold and clammy skin, heavy perspiration, rapid and shallow breathing, weak pulse, dizziness, and sometimes unconsciousness.

Heat Flux — The measure of the rate of heat transfer to a surface, expressed in kilowatts/m2, kilojoules/m2 · sec, or Btu/ft2 · sec. (NFPA® 921)

Heat of Combustion — Total amount of thermal energy (heat) that could be generated by the combustion (oxidation) reaction if a fuel were completely burned. The heat of combustion is measured in British Thermal Units (Btu) per pound or Megajoules per kilogram.

Heat Rash — Condition that develops from continuous exposure to heat and humid air; aggravated by clothing that rubs the skin. Reduces the individual's tolerance to heat.

Heat Release Rate (HRR) — Total amount of heat released per unit time. The HRR is measured in kilowatts (kW) and megawatts (MW) of output.

Heat Sensor Label — Label affixed to the ladder beam near the tip to provide a warning that the ladder has been subjected to excessive heat.

Heat Stress — Combination of environmental and physical work factors that compose the heat load imposed on the body; environmental factors include air, temperature, radiant heat exchange, air movement, and water vapor pressure. Physical work contributes because of the metabolic heat in the body; clothing also has an effect.

Heat Stroke — Heat illness in which the body's heat regulating mechanism fails; symptoms include (a) high fever of 105° to 106° F (40.5° to 41.1° C), (b) dry, red, hot skin, (c) rapid, strong pulse, and (d) deep breaths or convulsions. May result in coma or even death. *Also known as* Sunstroke.

Helmet — Headgear worn by firefighters that provides protection from falling objects, side blows, elevated temperatures, and heated water.

Higbee Cut — Special cut at the beginning of the thread on a hose coupling that provides positive identification of the first thread to eliminate cross-threading.

Higbee Indicators — Notches or grooves cut into coupling lugs to identify by touch or sight the exact location of the Higbee Cut.

High-Efficiency Particulate Air (HEPA) Filter — Respiratory filter that is certified to remove at least 99.97 percent of monodisperse particles of 0.3 micrometers in diameter.

High-Voltage — Any voltage in excess of 600 volts.

Hitch — (1) Temporary knot that falls apart if the object held by the rope is removed. (2) Loop that secures the rope but is not part of a standard rope knot.

Horizontal Smoke Spread — Tendency of heat, smoke, and other products of combustion to rise until they encounter a horizontal obstruction. At this point they will spread laterally (ceiling jet) until they encounter vertical obstructions and begin to bank downward (hot gas layer development).

Horizontal Ventilation — Any technique by which heat, smoke, and other products of combustion are channeled horizontally out of a structure by way of existing or created horizontal openings such as windows, doors, or other openings in walls. Typically portions of one or more of the horizontal openings will also serve as an air inlet.

Horseshoe Load — Arrangement of fire hose in a hose bed or compartment in which the hose lies on edge in the form of a horseshoe.

Hose Bed — Main hose-carrying area of a pumper or other apparatus designed for carrying hose.

Hot Zone — Potentially hazardous area immediately surrounding the incident site; requires appropriate protective clothing and equipment and other safety precautions for entry. Typically limited to technician-level personnel. *Also known as* Exclusion Zone.

House Line — Permanently fixed, private standpipe hoseline.

Hybrid Construction — Type of building construction that uses renewable, environmentally friendly or recycled materials. *Also known as* Natural or Green Construction.

Hydrant Wrench — Specially designed tool used to open or close a hydrant and to remove hydrant caps.

Hydraulic Ventilation — Ventilation accomplished by using a spray stream to draw the smoke from a compartment through an exterior opening.

Hydrocarbon Fuel — Petroleum-based organic compound that contains only hydrogen and carbon.

Hydrogen Cyanide (HCN) — Colorless, toxic, and flammable liquid until it reaches 79° F (26° C). Above that temperature, it becomes a gas with a faint odor similar to bitter almonds; produced by the combustion of nitrogen-bearing substances.

Hydrostatic Test — Testing method that uses water under pressure to check the integrity of pressure vessels.

Hypothermia — Abnormally low body temperature.

Hypovolemic Shock — Shock caused by loss of blood.

Hypoxia — Potentially fatal condition caused by lack of oxygen.

I

Ignition — The process of initiating self-sustained combustion. (NFPA® 921)

Ignition Sequence — History of the fire, beginning when the ignition source and the first fuel ignited meet at the area of origin, and proceeding through the entire duration of fire spread through the scene.

Illicit Clandestine Laboratory — Laboratory that produces illegal or controlled substances, such as drugs, explosives, biological agents, or chemical warfare agents. *Also known as* Illegal Clandestine Laboratory.

Immediately Dangerous to Life and Health (IDLH) — Description of any atmosphere that poses an immediate hazard to life or produces immediate, irreversible, debilitating effects on health.

Immunization — Process or procedure by which a subject (person, animal, or plant) is rendered immune or resistant to a specific disease. This term is often used interchangeably with *vaccination* or *inoculation*, although the act of inoculation does not always result in immunity.

Impact Load — Dynamic and sudden load placed on a rope, typically during a fall.

Improvised Explosive Device (IED) — Homemade bomb that is not deployed in a conventional military fashion.

Incendiary Device — Material or chemicals designed and used to start a fire.

Incident Action Plan (IAP) — Written or unwritten plan for the disposition of an incident; contains the overall strategic goals, tactical objectives, and support requirements for a given operational period during an incident. All incidents require an action plan. On relatively small incidents, the IAP is usually not in writing; on larger, more complex incidents, a written IAP is created for each operational period and disseminated to all assigned units. Written IAPs may have a number of forms as attachments.

Incident Command Post (ICP) — Location at which the Incident Commander and Command staff direct, order, and control resources at an incident; may be co-located with the incident base.

Incident Command System (ICS) — Standardized approach to incident management that facilitates interaction between cooperating agencies; adaptable to incidents of any size or type.

Incipient Stage — First stage of the burning process in a compartment in which the substance being oxidized is producing some heat, but the heat has not spread to other substances nearby. During this phase, the oxygen content of the air has not been significantly reduced and the temperature within the compartment is not significantly higher than ambient temperature.

Indicating Valve — Water main valve that visually shows the open or closed status of the valve.

Indirect Attack (Ground Cover) — A method of controlling a ground cover fire in which a control line is constructed or located some distance from the edge of the main fire, and the fuel between the two points is burned.

Indirect Attack (Structural) — Form of fire attack that involves directing fire streams toward the ceiling of a compartment in order to generate a large amount of steam in order to cool the compartment. Converting the water to steam displaces oxygen, absorbs the heat of the fire, and cools the hot gas layer sufficiently for firefighters to safely enter and make a direct attack on the fire.

Infectious — Transmittable; able to infect people.

Inhibitor — Substance that slows down or prevents a chemical reaction; typically added to materials that are prone to polymerization. *Also known as* Stabilizer.

Initial Isolation Distance — Distance within which all persons are considered for evacuation in all directions from a hazardous materials incident.

Initial Isolation Zone — Circular zone, with a radius equivalent to the initial isolation distance, within which persons may be exposed to dangerous concentrations upwind of the source and may be exposed to life-threatening concentrations downwind of the source.

Injection — Method of proportioning foam that uses an external pump or head pressure to force foam concentrate into the fire stream at the correct ratio for the flow desired.

In-Line Eductor — Eductor that is placed along the length of a hoseline.

Internal Customers — Employees and membership of the organization.

Interoperability — Ability of two or more systems or components to exchange information and use the information that has been exchanged.

Intrinsically Safe — Describes equipment that is approved for use in flammable atmospheres; must be incapable of releasing enough electrical energy to ignite the flammable atmosphere.

Inverse Square Law — Physical law that states that the amount of radiation present is inversely proportional to the square of the distance from the source of radiation.

Inverter — Step-up transformer that converts a vehicle's 12- or 24-volt DC current into 110- or 220-volt AC current.

Ion — Atom that has lost or gained an electron, giving it a positive or negative charge.

Ionization Smoke Detector — Type of smoke detector that uses a small amount of radioactive material to make the air within a sensing chamber conduct electricity.

Ionizing Radiation — Radiation that causes a chemical change in atoms by removing their electrons.

J

Jargon — The specialized or technical language of a trade, profession, or similar group.

Joists — Horizontal structural members used to support a ceiling or floor. Drywall materials are nailed or screwed to the ceiling joists, and the subfloor is nailed or screwed to the floor joists.

Joules (J) — Joules are defined in terms of mechanical energy. It is equal to the energy expended in applying a force of one newton through a distance of one meter. However, it is more useful for firefighters to think about the energy required to increase temperature. 4.2 Joules are required to raise the temperature of one gram of water one degree Celsius.

K

Kerf Cut — A single cut the width of the saw blade made in a roof to check for fire extension.

Kernmantle Rope — Rope that consists of a protective shield (mantle) over the load-bearing core strands (kern).

Kinetic Energy — The energy possessed by a body because of its motion.

Knot — Term used for tying a rope around itself.

L

Ladder Belt — Belt with a hook that secures the firefighter to the ladder.

Laid Rope — Rope constructed by twisting several groups of individual strands together.

Latent Heat of Vaporization — Quantity of heat absorbed by a substance at the point at which it changes from a liquid to a vapor.

Leeward Side — Protected side; the direction opposite from which the wind is blowing.

Level A Protection — Highest level of skin, respiratory, and eye protection that can be provided by personal protective equipment (PPE), as specified by the U.S. Environmental Protection Agency (EPA). Consists of positive-pressure self-contained breathing apparatus, totally encapsulating chemical-protective suit, inner and outer gloves, and chemical-resistant boots.

Level B Protection — Personal protective equipment that provides the highest level of respiratory protection, but a lesser level of skin protection. Consists of positive-pressure self-contained breathing apparatus, hooded chemical-resistant suit, inner and outer gloves, and chemical-resistant boots.

Level C Protection — Personal protective equipment that provides a lesser level of respiratory and skin protection than levels A or B. Consists of full-face or half-mask APR, hooded chemical-resistant suit, inner and outer gloves, and chemical-resistant boots.

Level D Protection — Personal protective equipment that provides the lowest level of respiratory and skin protection. Consists of coveralls, gloves, and chemical-resistant boots or shoes.

Lever — Device consisting of a bar turning about a fixed point (fulcrum), using power or force applied at a second point to lift or sustain an object at a third point.

Life Safety Rope — Rope designed exclusively for rescue and other emergency operations; used to raise, lower, and support people at an incident or during training. Must meet the requirements established in NFPA® 1983, *Standard on Life Safety Rope and Equipment for Emergency Services. Also known as* Lifeline.

Lightweight Steel Truss — Structural support made from a long steel bar that is bent at a 90-degree angle with flat or angular pieces welded to the top and bottom.

Lightweight Wood Truss — Structural supports constructed of 2- × 3-inch or 2- × 4-inch (50 mm by 75 mm or 50 mm by 100 mm) members that are connected by gusset plates.

Line Functions — Personnel who provide emergency services to external customers (the public).

Line of Lividity — A colored area of the corpse that is noticeably contrasted from the rest of the body caused by the pooling of blood.

Liquefied Petroleum Gas (LPG) — Any of several petroleum products, such as propane or butane, stored under pressure as a liquid.

Liquid-Splash Protective Clothing — Chemical-protective clothing designed to protect against liquid splashes per the requirements of NFPA® 1992, *Standard on Liquid Splash-Protective Suits for Hazardous Chemical Emergencies*; part of an EPA Level B ensemble.

Live Load — (1) Items within a building that are movable but are not included as a permanent part of the structure; merchandise, stock, furnishings, occupants, firefighters, and the water used for fire suppression are examples of live loads. (2) Force placed upon a structure by the addition of people, objects, or weather.

Load-Bearing Walls — Walls of a building that by design carry at least some part of the structural load of the building in the direction of the ground or base.

Local Energy System — Auxiliary fire alarm system that has its own power source.

Lockout/Tagout Device — Device used to secure a machine's power switches, in order to prevent accidental re-start of the machine.

Loop System — Water main arranged in a complete circuit so that water will be supplied to a given point from more than one direction. *Also known as* circle system, circulating system, or belt system.

Loss Control — Practice of minimizing damage and providing customer service through effective mitigation and recovery efforts before, during, and after an incident.

Louver Cut or Vent — Rectangular exhaust opening cut in a roof, allowing a section of roof deck (still nailed to a center rafter) to be tilted, thus creating an opening similar to a louver. *Also known as* center rafter cut.

Lower Flammable (Explosive) Limit (LFL) — Lower limit at which a flammable gas or vapor will ignite and support combustion; below this limit the gas or vapor is too *lean* or *thin* to burn (lacks the proper quantity of fuel). *Also known as* Lower Explosive Limit (LEL).

Low-Head Dam — Wall-like concrete structure across a river or stream that is designed to back up water; allows water to flow over the crest and drop into a lower level. *Also known as* Low-Water Dam.

Low-Pressure Alarm — Alarm that sounds when SCBA air supply is low, typically 25 percent.

Low-Voltage — Any voltage that is less than 600 volts and safe enough for domestic use, typically 120 volts or less.

M

Maintenance — Keeping equipment or apparatus in a state of usefulness or readiness.

Manual Pull Station — Manual fire alarm activator.

Masonry — Bricks, blocks, stones, and unreinforced and reinforced concrete products.

Master Stream — Large-caliber water stream usually supplied by combining two or more hoselines into a manifold device or by fixed piping that delivers 350 gpm (1 325 L/min) or more.

Matter — Anything that occupies space and has mass.

MAYDAY — Internationally recognized distress signal.

Means of Egress — Continuous and unobstructed way of exit travel from any point in a building or structure to a public way, consisting of three separate and distinct parts: exit access, exit, and exit discharge.

Mechanical Advantage — (1) Advantage created when levers, pulleys, and tools are used to make work easier during rope rescue or while lifting heavy objects. (2) The ratio of the force applied by a simple machine, such as a lever or pulley, to the force applied to the machine by the user.

Mechanical Ventilation — Any means other than natural ventilation. This type of ventilation may involve the use of blowers, and smoke ejectors.

Meth Lab — Illicit clandestine laboratory that produces methamphetamine (meth).

Miscible — Materials that are capable of being mixed in all proportions.

Mitigate — To make less harsh or intense; to alleviate.

Model Code — Consensus-based standards or codes established to provide uniformity in regulations in regards to construction, design, and use. When adopted by the local jurisdiction, these codes become enforceable laws.

Mutual Aid — Reciprocal assistance from one fire and emergency services agency to another during an emergency, based upon a prearranged agreement; generally made upon the request of the receiving agency.

N

Nader Pin — Bolt on a vehicle's door frame that the door latches onto in order to close.

National Response Framework (NRF) — Document that provides guidance on how communities, states, the U.S. federal government, and private-sector and non-governmental partners conduct all-hazards emergency response.

Natural Fiber Rope — Utility rope made of manila, sisal, or cotton; not accepted for life safety applications.

Natural Ventilation — Techniques that use the wind, convection currents, and other natural phenomena to ventilate a structure without the use of fans, blowers, and smoke ejectors.

Nebulizer — Electrically powered machine that turns liquid medication into a mist.

Negative-Pressure Ventilation (NPV) — Technique using smoke ejectors to develop artificial air flow and to pull smoke out of a structure. Smoke ejectors are placed in windows, doors, or roof vent holes to pull the smoke, heat, and gases from inside the building and eject them to the exterior.

Neurogenic Shock — Shock caused by the overexpansion of blood vessels due to damage to the brain, spinal cord, or other nerves.

Neutral Plane — The level at a compartment opening where the difference in pressure exerted by expansion and buoyancy of hot smoke flowing out of the opening and the inward pressure of cooler, ambient temperature air flowing in through the opening is equal.

Nonintervention Strategy — Strategy for handling fires involving hazardous materials, in which the fire is allowed to burn until all of the fuel is consumed.

Nonionizing Radiation — Energy waves composed of oscillating electric and magnetic fields traveling at the speed of light; examples include visible light, radio waves, microwaves, infrared radiation, and ultraviolet radiation.

Nonload-Bearing Wall — Wall, usually interior, that supports only its own weight.

Nonthreaded Coupling — Coupling with no distinct male or female components. *Also known as* Storz Coupling or sexless coupling.

Nozzle Pressure — Velocity pressure at which water is discharged from the nozzle.

Nozzle Reaction — Counter force directed against a person holding a nozzle or a device holding a nozzle by the velocity of water being discharged.

O

Occupational Safety And Health Administration (OSHA) — Division of the U.S. Department of Labor (DOL) that enforces occupational safety regulations.

Offensive Strategy — Overall plan for incident control established by the Incident Commander (IC) in which responders take aggressive, direct action on the material, container, or process equipment involved in an incident.

Olfactory Fatigue — Gradual inability to detect odors after initial exposure; can be extremely rapid with some toxins, such as hydrogen sulfide.

Open Butt — The end of a charged hoseline that is flowing water without a nozzle or valve to control the flow.

Open-Circuit Self-Contained Breathing Apparatus — SCBA that allows exhaled air to be discharged or vented into the atmosphere.

Operations Level — Level of training established by the National Fire Protection Association® allowing first responders to take defensive actions at hazardous materials incidents.

Organophosphate Pesticides — Chemicals that kill insects by disrupting their central nervous systems; because they have the same effect on humans, they are sometimes used in terrorist attacks.

Oriented Strand Board (OSB) — A wooden structural panel formed by gluing and compressing wood strands together under pressure. This material has replaced plywood and planking in the majority of construction applications. Roof decks, walls, and subfloors are all commonly made of OSB.

Outside Stem and Yoke (OS&Y) Valve — Outside stem and yoke valve; a type of control valve for a sprinkler system in which the position of the center screw indicates whether the valve is open or closed.

Overhand Safety Knot — Supplemental knot tied to prevent the primary knot from failing; prevents the running end of the rope from slipping back through the primary knot.

Overhaul — Operations conducted once the main body of fire has been extinguished; consists of searching for and extinguishing hidden or remaining fire, placing the building and its contents in a safe condition, determining the cause of the fire, and recognizing and preserving evidence of arson.

Oxidation — Chemical process that occurs when a substance combines with an ozidizer such as oxygen in the air; a common example is the formation of rust on metal.

Oxidizer — Any material that readily yields oxygen or other oxidizing gas, or that readily reacts to promote or initiate combustion of combustible materials. (NFPA® 430)

Oxygen-Deficient Atmosphere — Atmosphere containing less than the normal 19.5 percent oxygen. At least 16 percent oxygen is needed to produce flames or sustain human life.

Oxygen-Enriched Atmosphere — Area in which the concentration of oxygen is in excess of 21 percent by volume or 21.3 kPa; typically 23.5 percent for confined spaces, as defined by the Occupational Safety and Health Administration (OSHA).

P

Panic Hardware — Hardware mounted on exit doors in public buildings that unlock from the inside and enable doors to be opened when pressure is applied to the release mechanism.

Parallel Chord Truss — A truss constructed with the top and bottom chords parallel. These trusses are used as floor joists in multistory buildings and as ceiling joists in buildings with flat roofs.

Paramedic — Professional level of certification for emergency medical personnel who are trained in advanced life support procedures.

Parapet — Portion of the exterior walls of a building that extends above the roof. A low wall at the edge of a roof.

Particulate — Very small particle of solid material, such as dust, that is suspended in the atmosphere.

Partition Wall — Interior non-load bearing wall that separates a space into rooms.

Party Wall — A load-bearing wall shared by two adjacent structures.

Passive Agent — Materials that absorb heat but do not participate actively in the combustion process.

Pathogens — Organisms that cause infection such as viruses and bacteria.

Permeation — Process in which a chemical passes through a protective material on a molecular level.

Permissible Exposure Limit (PEL) — Legal term for the maximum amount of a chemical substance or other hazard that an employee can be exposed to; typically expressed in parts per million (ppm) or milligrams per cubic meter (mg/m3). If exposed to this concentration for an entire 40-hour work week, 95 percent of healthy adults would not suffer health consequences.

Persistence — Length of time a chemical agent remains effective without dissipating.

Personal Alert Safety System (PASS) — Electronic lack-of-motion sensor that sounds a loud alarm when a firefighter becomes motionless. It can also be manually activated.

Personal Protective Equipment (PPE) — General term for the equipment worn by fire and emergency services responders; includes helmets, coats, trousers, boots, eye protection, hearing protection, protective gloves, protective hoods, self-contained breathing apparatus (SCBA), and personal alert safety system (PASS) devices. *Also known as* Bunker Clothes, Chemical-Protective Clothing, Full Structural Protective Clothing, Protective Clothing, Turnout Clothing, or Turnout Gear.

Personnel Accountability Report (PAR) — Roll call of all units assigned to an incident. PARs may be required at specific intervals, or may be requested at any time by the IC or incident safety officer (ISO).

pH — A measure of the acidity or alkalinity of a solution.

Phosphine — Colorless, flammable, and toxic gas with an odor of garlic or decaying fish; ignites spontaneously on contact with air. Phosphine is a respiratory tract irritant that attacks the cardiovascular and respiratory systems, causing pulmonary edema, peripheral vascular collapse, and cardiac arrest and failure.

Photoelectric Smoke Detector — Type of smoke detector that uses a small light source, either an incandescent bulb or a light-emitting diode (LED), to detect smoke by shining light through the detector's chamber: smoke particles reflect the light into a light-sensitive device called a photocell.

Photon — Weightless packet of electromagnetic energy, such as X-rays or visible light.

Physical Evidence — Tangible or real objects that are related to the incident.

Pike Pole — Sharp prong and hook of steel, on a wood, metal, fiberglass, or plastic handle of varying length, used for pulling, dragging, and probing.

Piloted Ignition — Moment when a mixture of fuel and oxygen encounters an external heat (ignition) source with sufficient heat or thermal energy to start the combustion reaction.

Platform Frame Construction — A construction method in which a floor assembly creates an individual platform that rest on the foundation. Wall assemblies the height of one story are placed on this platform and a second platform rests on top of the wall unit. Each platform creates fire stops at each floor level restricting the spread of fire within the wall cavity.

Plug — Patch to seal a small leak in a container.

Plume — The column of hot gases, flames, and smoke rising above a fire; *also known as* convection column, thermal updraft, or thermal column. (NFPA® 921)

Point of No Return — Point at which air in the SCBA will last only long enough to exit a hazardous atmosphere.

Point of Origin — Exact physical location where the heat source and fuel come in contact with each other and a fire begins.

Poison — Any material that is injurious to health when taken into the body.

Polar Solvents — Flammable liquids that have an attraction for water, much like a positive magnetic pole attracts a negative pole; examples include alcohol, ketone, and lacquer.

Policy — Guide to decision making in an organization.

Polychlorinated Biphenyl (PCB) — Toxic compound found in some older oil-filled electric transformers.

Polymerization — Chemical reaction in which two or more molecules combine to form larger molecules; this reaction can often be violent.

Portable Tank — Storage tank used during a relay or shuttle operation to hold water from water tanks or hydrants. This water can then be used to supply attack apparatus. *Also known as* Catch Basin, Fold-a-Tank, Portable Basin, or Porta-Tank.

Positive-Pressure Ventilation (PPV) — Method of ventilating a room or structure by mechanically blowing fresh air through an inlet opening into the space in sufficient volume to create a slight positive pressure within and thereby forcing the contaminated atmosphere out the exit opening.

Post Indicator Valve (PIV) — A type of valve used to control underground water mains that provides a visual means for indicating "open" or "shut" position; found on the supply main of installed fire protection systems.

Potential Energy — Stored energy possessed by an object that can be released in the future to perform work once released.

Powered Air-Purifying Respirator (PAPR) — Motorized respirator that uses a filter to clean surrounding air, then delivers it to the wearer to breathe; typically includes a headpiece, breathing tube, and a blower/battery box that is worn on the belt.

Power Take-Off (PTO) System — Mechanism that allows a vehicle engine to power equipment such as a pump, winch, or portable tool; it is typically attached to the transmission.

Preaction Sprinkler System — Fire suppression system that consists of closed sprinkler heads attached to a piping system that contains air under pressure and a secondary detection system; both must operate before the extinguishing agent is released into the system.

Preconnect — Attack hose connected to a discharge when the hose is loaded; this shortens the time it takes to deploy the hose for fire fighting.

Preincident Survey — Thorough and systematic inspection of a building for the purpose of identifying significant structural and/or occupancy characteristics to assist in the development of a pre incident plan for that building.

Premixing — Mixing premeasured portions of water and foam concentrate in a container. Typically, the premix method is used with portable extinguishers, wheeled extinguishers, skid-mounted twin-agent units, and vehicle-mounted tank systems.

Pretensioner — Device that takes up slack in a seatbelt; prevents the passenger from being thrown forward in the event of a crash.

Primary Damage — Damage caused by a fire itself and not by actions taken to fight the fire.

Primary Search — Rapid but thorough search to determine the location of victims; performed either before or during fire suppression operations. May be conducted with or without a charged hoseline, depending on local policy.

Procedure — Step-by-step written plan that is closely related to a policy. Procedures help an organization to ensure that it consistently approaches a task in the correct way, in order to accomplish a specific objective.

Products of Combustion — Materials produced and released during burning.

Proportioning — Mixing of water with an appropriate amount of foam concentrate to form a foam solution.

Proprietary Alarm System — Fire protection system owned and operated by the property owner.

Protected Medical Information (PMI) — Information of a patient that includes personal data (name, birth date, social security number, address), medical history, and condition.

Protected Premises System — Alarm system that alerts and notifies only occupants on the premises of the existence of a fire so that they can safely exit the building and call the fire department. If a response by a public safety agency (police or fire department) is required, an occupant hearing the alarm must notify the agency.

Protected Steel — Steel structural members that are covered with either spray-on fire proofing (an insulating barrier) or fully encased in an Underwriters Laboratories Inc. (UL) tested and approved system.

Protective Action Distance — Downwind distance from a hazardous materials incident within which protective actions should be implemented.

Protective Coat — Coat worn during fire fighting, rescue, and extrication operations.

Protective Gloves — Protective clothing designed to protect the hands.

Protective Hood — Hood designed to protect the firefighter's ears, neck, and face from heat and debris; typically made of Nomex®, Kevlar®, or PBI®, and available in long or short styles.

Protective Trousers — Trousers worn to protect the lower torso and legs during emergency operations. *Also known as* Bunker Pants or Turnout Pants.

Protons — Subatomic particle that possesses a positive electric charge.

Proximity Fire Fighting — Activities required for rescue, fire suppression, and property conservation at fires that produce high radiant, conductive, or convective heat; includes aircraft, hazardous materials transport, and storage tank fires.

Public Safety Answering Point (PSAP) — Any location or facility at which 9-1-1 calls are answered either by direct calling, rerouting, or diversion.

Pulmonary Edema — Accumulation of fluids in the lungs.

Pulse — Rhythmic throbbing caused by expansion and contraction of arterial walls as blood passes through them.

Pumper Outlet Nozzle — Fire hydrant outlet that is 4 inches (100 mm) in diameter or larger.

Purlin — Horizontal member between trusses that support the roof.

Pyrolysis — The chemical decomposition of a solid material by heating. Pyrolysis often precedes combustion.

Q

Qualitative Fit Test (QLFT) — Respirator fit test that measures the wearer's response to a test agent, such as irritant smoke or odorous vapor. If the wearer detects the test agent, such as through smell or taste, the respirator fit is inadequate.

Quantitative Fit Test (QNFT) — Fit test in which instruments measure the amount of a test agent that has leaked into the respirator from the ambient atmosphere. If the leakage measures above a pre set amount, the respirator fit is inadequate.

Quarter-Turn Coupling — Nonthreaded (sexless) coupling with two hook-like lugs that slip over a ring on the opposite coupling and then rotate 90 degrees clockwise to lock.

R

Rabbit Tool — Hydraulic spreading tool that is specially designed to open doors that swing inward.

Radiation — Heat transfer by way of electromagnetic energy. (NFPA® 921)

Radiation Dose — Quantity of radiation energy absorbed into the body.

Radiological Dispersal Device (RDD) — Conventional high explosives wrapped with radioactive materials; designed to spread radioactive contamination over a wide area. *Also known as* Dirty Bomb.

Rafter — Inclined beam that supports a roof, runs parallel to the slope of the roof, and to which the roof decking is attached.

Rain Roof — A second roof constructed over an existing roof.

Rapid Intervention Crew or Team (RIC/RIT) — Two or more firefighters designated to perform firefighter rescue; they are stationed outside the hazard and must be standing by throughout the incident.

Rate-of-Rise Heat Detector — Temperature-sensitive device that sounds an alarm when the temperature changes at a preset value, such as 12°F to 15°F (7°C to 8°C) per minute.

Reactivity — Ability of a substance to chemically react with other materials, and the speed with which that reaction takes place.

Rebar — Short for reinforcing bar. These steel bars are placed in concrete forms before the cement is poured. When the concrete sets (hardens) the rebar within it adds considerable strength.

Recirculation — Movement of smoke being blown out of a ventilation opening only to be drawn back inside by the negative pressure created by the ejector because the open area around the ejector has not been sealed.

Reducer — Fitting used to attach a smaller hose to a larger hose. The female end has the larger threads, while the male end has the smaller threads.

Reducing Agent — The fuel that is being oxidized or burned during combustion.

Rehabilitation (Rehab) — Allowing firefighters to rest, rehydrate, and recover during an incident.

Rekindle — To reignite because of latent heat, sparks, or smoldering embers; rekindling can be prevented by proper overhaul.

Relay Pumping — Use of two or more pumpers to move water over a long distance by operating them in series. Water discharged from one pumper flows through hoses to the inlet of the next pumper, and is then pumped to the next pumper in line.

Remote Receiving System — System in which alarm signals from the protected premises are transmitted over a leased telephone line or by radio signal to a remote receiving station with a 24-hour staff; usually the municipal fire department's alarm communications center.

Repair — To restore or put together that which has become inoperable or out of place.

Reporting Marks — Combination of letters and numbers stenciled on rail tank cars that may be used to get information about the car's contents from the railroad or the shipper. *Also known as* Railcar Initials and Numbers.

Respiratory Hazards — Exposure to conditions that create a hazard to the respiratory system, including products of combustion, toxic gases, and superheated or oxygen-deficient atmospheres.

Retard Chamber — Chamber that catches and slows the excess water that may be sent through the alarm valve of an automatic sprinkler system during momentary water pressure surges. This reduces the chance of false-alarm activation. The retard chamber is installed between the alarm check valve and alarm-signaling equipment.

Ridge — The horizontal line at the junction of the top edges of two sloping roof surfaces.

Rigor Mortis — A sign of death in a deceased individual in which the muscles cause the body to be stiff and difficult to move. Rigor mortis begins within a few hours of death and recedes within a few days.

Riser — Vertical water pipe used to carry water for fire protection systems aboveground such as a standpipe riser or sprinkler riser.

Risk Management Plan — Written plan that analyzes the exposure to hazards, implements appropriate risk management techniques, and establishes criteria for monitoring their effectiveness.

Rollover — A condition where the unburned fire gases that have accumulated at the top of a compartment ignite and flames propagate through the hot-gas layer or across the ceiling.

Roof Ladder — Straight ladder with folding hooks at the top end; the hooks anchor the ladder over the roof ridge.

Rope Hose Tool — Piece of rope spliced to form a loop through the eye of a metal hook; used to secure hose to ladders or other objects.

Rope Log — Record of all use, maintenance, and inspection throughout a rope's working life; also includes the product label and manufacturer's recommendations.

Routes of Entry — Means by which hazardous materials enter the body; inhalation, ingestion, skin contact, injection, absorption, and penetration (for radiation).

Running Part — Free end of the rope used for hoisting, pulling, or belaying.

S

Safety Data Sheet (SDS) — Form provided by the manufacturer and blender of chemicals that contains information about chemical composition, physical and chemical properties, health and safety hazards, emergency response procedures, and waste disposal procedures of the specified material. *Formerly known as* Material Safety Data Sheet (MSDS).

Safety Glass — Two sheets of glass laminated to a sheet of plastic sandwiched between them; the plastic layer makes the glass stronger and more shatter resistant. Most commonly used in windshields and rear windows. *Also known as* Laminated Glass.

Salvage — Methods and operating procedures by which firefighters attempt to save property and reduce further damage from water, smoke, heat, and exposure during or immediately after a fire; may be accomplished by removing property from a fire area, by covering it, or by other means.

Saponification — A phenomenon that occurs when mixtures of alkaline based chemicals and certain cooking oils come into contact resulting in the formation of a soapy film.

Sarin (GB) — Fluorinated phosphinate that attacks the central nervous system; classified as a chemical warfare agent.

Scupper — Form of drain opening provided in outer walls at floor or roof level to remove water to the exterior of a building in order to reduce water damage.

Search Line — Nonload-bearing rope that is anchored to a safe, exterior location and attached to a firefighter during search operations to act as a safety line.

Secondary Collapse — Collapse that occurs after the initial collapse of a structure; common causes include aftershock (earthquake), weather conditions, and the movement of structural members.

Secondary Damage — Damage caused by or resulting from those actions taken to fight a fire and leaving the property unprotected.

Secondary Device — Bomb or other weapon that targets responders and bystanders who are already at the scene of an incident.

Secondary Search — Slow, thorough search to ensure that no occupants were overlooked during the primary search; conducted after the fire is under control by personnel who did not conduct the primary search.

Self-Heating — The result of exothermic reactions, occurring spontaneously in some materials under certain conditions, whereby heat is generated at a rate sufficient to raise the temperature of the material. (NFPA® 921)

Self-Presenters — People seeking medical attention who were not treated or decontaminated at the incident scene.

Septic Shock — Shock caused by a severe infection in the body.

Service Test — Series of tests performed on apparatus and equipment in order to ensure operational readiness of the unit. These tests should be performed at least yearly or whenever a piece of apparatus or equipment has undergone extensive repair.

Shank — Portion of a coupling that serves as a point of attachment to the hose.

Shielded Fire — A fire that is located in a remote part of the structure or hidden from view by objects in the compartment.

Shock — Failure of the circulatory system to produce sufficient blood to all parts of the body; results in depression of bodily functions, and eventually death if not controlled.

Shunt System — An auxiliary fire alarm system that connects a public fire alarm reporting system to initiating devices within a protected premises. When an initiating device in the protected property operates, it activates the public fire alarm sending an alarm to the public communication center.

Siamese — Hose appliance used to combine two or more hoselines into one. The Siamese has multiple female inlets and a single male outlet. An example of a Siamese is a fire department connection.

Signal Word — Government-mandated warning (CAUTION, WARNING, DANGER, or POISON) on the labels of hazardous products.

Single Ladder — One-section nonadjustable ladder. *Also known as* Wall or Straight Ladder.

Situational Awareness — Perception of the surrounding environment, and the ability to anticipate future events.

Size-Up — Ongoing evaluation of influential factors at the scene of an incident.

Small-Diameter Hose (SDH) — Hose of ¾-inch to 2 inches (20 mm to 50 mm) in diameter; used for fire fighting purposes. Also called Small Line.

Smoke Alarm — Device designed to sound an alarm when the products of combustion are present in the room where the device is installed. The alarm is built into the device rather than being a separate system.

Smoke Detector — Alarm-initiating device designed to actuate when visible or invisible products of combustion (other than fire gases) are present in the room or space where the unit is installed.

Smoke Ejectors — These are electrically powered fans that have intrinsically safe motors that are placed in the smoke-filled atmosphere to push the smoke out. They can also be used to push fresh air into the structure. They require the use of electrical power cords and generators to operate.

Smoke Explosion — Form of fire gas ignition; the ignition of accumulated flammable products of combustion and air that are within their flammable range.

Smoke Management System — System that limits the exposure of building occupants to smoke. May include a combination of compartmentation, control of smoke migration from the affected area, and a means of removing smoke to the exterior of the building.

Smooth Bore Nozzle — A nozzle with a straight, smooth tip, designed to produce a solid fire stream.

Soft Sleeve Hose — Large-diameter, collapsible piece of hose used to connect a fire pump to a pressurized water supply source; sometimes incorrectly referred to as *soft suction hose*.

Solid Stream — Hose stream that stays together as a solid mass, as opposed to a fog or spray stream; a solid stream is produced by a smooth bore nozzle and should not be confused with a straight stream.

Solubility — Degree to which a solid, liquid, or gas dissolves in a solvent (usually water).

Sounding — Striking the surface of a roof or floor to determine its structural integrity or locate underlying support members; the blunt end of a hand tool is used for this purpose.

Spalling — Expansion of excess moisture within masonry materials due to exposure to the heat of a fire, resulting in tensile forces within the material, and causing it to break apart. The expansion causes sections of the material's surface to violently disintegrate, resulting in explosive pitting or chipping of the material's surface.

Spanner Wrench — Small tool primarily used to tighten or loosen hose couplings; can also be used as a prying tool or a gas key.

Specific Gravity — Mass (weight) of a substance compared to the mass of an equal volume of water at a given temperature. A specific gravity less than 1 indicates a substance lighter than water; a specific gravity greater than 1 indicates a substance heavier than water.

Specific Heat — Amount of energy required to raise the temperature of a specified unit mass of a material 1 degree in temperature.

Specification Marking — Stencil on the exterior of a tank car indicating the standards to which the tank car was built; may also be found on intermodal containers and cargo tank trucks.

Spoliation — Term that refers to evidence that is destroyed, damaged, altered, or otherwise not preserved by someone who has responsibility for the evidence.

Spontaneous Ignition — Initiation of combustion of a material by an internal chemical or biological reaction that has produced sufficient heat to ignite the material. (NFPA® 921)

Sprinkler — Water flow discharge device in a sprinkler system; consists of a threaded intake nipple, a discharge orifice, heat-actuated plug, and a deflector to create an effective fire stream pattern that is suitable for fire control.

Stabilization — Preventing unwanted movement; accomplished by supporting key places between an object and the ground (or other solid anchor points).

Stack Effect — Phenomenon of a strong air draft moving from ground level to the roof level of a building. Affected by building height, configuration, and temperature differences between inside and outside air.

Staff/Support Functions — Personnel who provide administrative and logistical support to line units (internal customers).

Staging Area — Prearranged, temporary strategic location, away from the emergency scene, where units assemble and wait until they are assigned a position on the emergency scene; these resources (personnel, apparatus, tools, and equipment) must then be able to respond within three minutes of being assigned. Staging area managers report to the Incident Commander or operations section chief, if one has been established.

Standard — A set of principles, protocols, or procedures that explain how to do something or provide a set of minimum standards to be followed. Adhering to a standard is not required by law, although standards may be incorporated in codes, which are legally enforceable.

Standard of Care — Level of care that all persons should receive; care that does not meet this standard is considered inadequate.

Standard Operating Procedures (SOPs) — Rules for how personnel should perform routine functions or emergency operations. Procedures are typically written in a handbook, so that all firefighters can become familiar with them. *Also known as* Operating Instructions (OI), Predetermined Procedures, or Standard Operating Guidelines (SOGs).

Standing Part — Middle of the rope, between the working end and the running part.

Standpipe System — Wet or dry system of pipes in a large single-story or multistory building with fire hose outlets installed in different areas or on different levels of a building to be used by firefighters and/or building occupants. The system is used to provide for quick deployment of hoselines during fire fighting operations.

Static Rope — Rope designed not to stretch under load.

Steamer Connection — Large-diameter outlet, usually 4½ inches (115 mm), at a hydrant or at the base of an elevated water storage container.

Storz Coupling — Nonthreaded (sexless) coupling commonly found on large-diameter hose.

Straight Stream — Semi-solid stream that is produced by a fog nozzle.

Street Clothes — Clothing that is anything other than chemical-protective clothing or structural firefighters' protective clothing, including work uniforms and ordinary civilian clothing.

Stressor — Any agent, condition, or experience that causes stress.

Strong Oxidizer — Substance that readily gives off large quantities of oxygen, thereby stimulating combustion; produces a strong reaction by readily accepting electrons from a reducing agent (fuel).

Structural Collapse — Structural failure of a building or any portion of it resulting from a fire, snow, wind, water, or damage from other forces.

Structural Fire Fighting — Activities required for rescue, fire suppression, and property conservation in structures, vehicles, vessels, and similar types of properties.

Stud — An upright post in the framework of a wall for supporting sheets of lath and plaster, wallboard, or similar material.

Suction Hose — Intake hose that connects pumping apparatus or portable pump to a water source.

Supply Hose — Hose that is designed for the purpose of moving water between a water source and a pump that is supplying attack hoselines or fire suppression systems.

Surface Tension — (1) Force minimizing a liquid surface's area. (2) The effect of a surfactant on the water/concentrate solution; allows the water to spread more rapidly over the surface of Class A fuels and to penetrate organic fuels.

Surfactant — Chemical that lowers the surface tension of a liquid; allows water to spread more rapidly over the surface of Class A fuels and penetrate organic fuels.

Systemic Effect — Damage spread through an entire system; opposite of a local effect, which is limited to a single location.

Synthetic Fiber Rope — Rope made from continuous, synthetic fibers running the entire length of the rope; it is strong, easy to maintain, and resists mildew and rotting.

T

THIRA — Acronym for Threat and Hazard Identification and Risk Assessment.

Tactical Ventilation — Planned, systematic, and coordinated removal of heated air, smoke, gases or other airborne contaminants from a structure, replacing them with cooler and/or fresher air to meet the incident priorities of life safety, incident stabilization, and property conservation.

Tag Line — Non-load-bearing rope attached to a hoisted object to help steer it in a desired direction, prevent it from spinning or snagging on obstructions, or act as a safety line.

Target Hazard — Any facility in which a fire, accident, or natural disaster could cause substantial casualties or significant economic harm, through either property or infrastructure damage.

Temperature — Measure of a material's ability to transfer heat energy to other objects; the greater the energy, the higher the temperature. Measure of the average kinetic energy of the particles in a sample of matter, expressed in terms of units or degrees designated on a standard scale. *See* Celsius Scale and Fahrenheit Scale.

Tempered Plate Glass — Treated glass that is stronger than plate glass or a single sheet of laminated glass; safer than regular glass because it crumbles into chunks when broken, instead of splintering into jagged shards. Most commonly used in a vehicle's side and rear windows.

Terrorism — Unlawful force or violence against people or property to coerce or intimidate a government or its citizens, for social or political purposes (as defined by the U.S. *Code of Federal Regulations*).

Thermal Energy — The kinetic energy associated with the random motions of the molecules of a material or object; often used interchangeably with the terms heat and heat energy. Measured in joules or Btu.

Thermal Imager — Electronic device that forms images using infrared radiation.

Thermal Layering — Outcome of combustion in a confined space in which gases tend to form into layers, according to temperature, with the hottest gases found at the ceiling and the coolest gases at floor level.

Threaded Coupling — Male or female coupling with a spiral thread.

Topography — Physical configuration of the land or terrain.

Total Flooding System — Fire suppression system designed to protect hazards within enclosed structures; foam is released into a compartment or area and fills it completely, extinguishing the fire.

Toxic Industrial Material (TIM) — Toxic chemical that is produced in large quantities (at least 30 million tons per year at a single production facility). They are intended for industrial use, but can be used by terrorists to cause deliberate harm. *Also known as* Toxic Industrial Chemical (TIC).

Toxic Inhalation Hazard (TIH) — Volatile liquid or gas known to be a severe hazard to human health during transportation.

Trailer — Combustible material, such as rolled rags, blankets, newspapers, or flammable liquid, often used in intentionally set fires in order to spread fire from one area to other points or areas.

Training Evolution — Operation of fire service training or suppression covering one or several aspects of fire fighting. *Also known as* Practical Training Evolution.

Transportation Area — Location where accident casualties are held after receiving medical care or triage before being transported to medical facilities.

Trench Foot — Foot condition resulting from prolonged exposure to damp conditions or immersion in water; symptoms include tingling and/or itching, pain, swelling, cold and blotchy skin, numbness, and a prickly or heavy feeling in the foot. In severe cases, blisters can form, after which skin and tissue die and fall off.

Trench Ventilation — Defensive tactic that involves cutting an exhaust opening in the roof of a burning building, extending from one outside wall to the other, to create an opening at which a spreading fire may be cut off.

Triage — System used for sorting and classifying accident casualties to determine the priority for medical treatment and transportation.

Turbulence — Irregular motion of the atmosphere usually produced when air flows over a comparatively uneven surface, such as the surface of the earth; when two currents of air flow past or over each other in different directions or at different speeds.

U

Unibody Construction — Method of automobile construction in which the frame and body form one integral unit; used on most modern cars. *Also known as* Bird Cage Construction, Integral Frame Construction, or Unitized Construction.

Unified Command (UC) — In the Incident Command System, a shared command role in which all agencies with geographical or functional responsibility establish a common set of incident objectives and strategies. In unified command there is a single incident command post and a single operations chief at any given time.

Upper Flammable (Explosive) Limit (UFL) — Upper limit at which a flammable gas or vapor will ignite; above this limit the gas or vapor is too *rich* to burn (lacks the proper quantity of oxygen). *Also known as* Upper Explosive Limit (UEL).

Upper Layer — Buoyant layer of hot gases and smoke produced by a fire in a compartment.

Utility Rope — Rope designed for any use except rescue; can be used to hoist equipment, secure unstable objects, or cordon off an area.

V

Vapor — Gaseous form of a substance that is normally in a solid or liquid state at room temperature and pressure; formed by evaporation from a liquid or sublimation from a solid.

Vapor Density — Weight of a given volume of pure vapor or gas compared to the weight of an equal volume of dry air at the same temperature and pressure. A vapor density less than 1 indicates a vapor lighter than air; a vapor density greater than 1 indicates a vapor heavier than air.

Vaporization — Physical process that changes a liquid into a gaseous state; the rate of vaporization depends on the substance involved, heat, pressure, and exposed surface area.

Vapor Pressure — (1) Measure of the tendency of a substance to evaporate. (2) The pressure at which a vapor is in equilibrium with its liquid phase for a given temperature; liquids that have a greater tendency to evaporate have higher vapor pressures for a given temperature.

Vapor-Protective Clothing — Gas-tight chemical-protective clothing designed to meet NFPA® 1991, *Standard on Vapor-Protective Suits for Hazardous Chemical Emergencies*; part of an EPA Level A ensemble.

Vein — Any blood vessel that carries blood from the tissues to the heart.

Veneer Walls — Walls with a surface layer of attractive material laid over a base of common material.

Ventilation Controlled — A fire with limited ventilation in which the heat release rate or growth is limited by the amount of oxygen available to the fire. (Paraphrased from NFPA 921)

Venturi Principle — Physical law stating that when a fluid, such as water or air, is forced under pressure through a restricted orifice, there is an increase in the velocity of the fluid passing through the orifice and a corresponding decrease in the pressure exerted against the sides of the constriction. Because the surrounding fluid is under greater pressure (atmospheric), it is forced into the area of lower pressure. *Also known as* Venturi Effect.

Vertical Ventilation — Ventilating at a point above the fire through existing or created openings and channeling the contaminated atmosphere vertically within the structure and out the top. Done with openings in the roof, skylights, roof vents, or roof doors.

Viscous — Having a thick, sticky, adhesive consistency.

W

Warm Zone — Area between the hot and cold zones that usually contains the decontamination corridor; typically requires a lesser degree of personal protective equipment than the Hot Zone. *Also known as* Contamination Reduction Zone or Contamination Reduction Corridor.

Waterflow Device — Detector that recognizes movement of water within the sprinkler or standpipe system. Once movement is noted, the waterflow detector gives a local alarm and/or may transmit the alarm.

Water Hammer — Force created by the rapid deceleration of water causing a violent increase in pressure that can be powerful enough to rupture piping or damage fixtures. Generally results from closing a valve or nozzle too quickly.

Water Main — A principal pipe in a system of pipes for conveying water, especially one installed underground.

Water-Mist — In the fire service, water mist is associated with a fire extinguisher capable of atomizing water through a special applicator. Water-mist fire extinguishers use distilled water, while backpack pump-type water-mist extinguishers use ordinary water.

Water Shuttle Operation — Method of water supply by which mobile water supply apparatus continuously transport water between a fill site and the dump site located near the emergency scene.

Water Solubility — Ability of a liquid or solid to mix with or dissolve in water.

Water Thief — Any of a variety of hose appliances with one female inlet for 2½ -inch (65 mm) or larger hose and with three gated outlets, usually two 1½-inch (38 mm) outlets and one 2½-inch (65 mm) outlet.

Watt — A unit of measure of power or rate of work equal to one joule per second (J/s).

Weapon of Mass Destruction (WMD) — Weapon or device that can cause death or serious injury to a large number of people; may include chemical, biological, radiological, nuclear, or explosive (CBRNE) type weapons.

Webbing — Device used for creating anchors and lashings, or for packaging patients and rescuers; typically constructed from the same material as synthetic rope.

Wet-Barrel Hydrant — Fire hydrant that has water all the way up to the discharge outlets; may have separate valves for each discharge or one valve for all the discharges. This type of hydrant is only used in areas where there is no danger of freezing weather conditions.

Wet Chemical System — Extinguishing system that uses a wet chemical solution as the primary extinguishing agent; usually installed in range hoods and associated ducting where grease may accumulate.

Wet-Pipe Sprinkler System — Fire suppression system that is built into a structure or site; piping contains either water or foam solution continuously; activation of a sprinkler causes the extinguishing agent to flow from the open sprinkler.

Wheel Chock — Block placed against the outer curve of a tire to prevent the apparatus from rolling; can be wooden, plastic, or metal. *Also known as* Wheel Block.

Wildland/Urban Interface — Line, area, or zone where an undeveloped wildland area meets a human development area. Also known as Urban/Wildland Interface.

Windward Side — The side or direction from which the wind is blowing.

Working End — End of the rope used to tie a knot. *Also known as* Bitter End or Loose End.

Wye — Hose appliance with one female inlet and multiple male outlets, usually smaller than the inlet. Outlets are also usually gated.

Glossary

Index by Nancy Kopper

NOTES

NOTES

NOTES